Power Electronics

Advanced Conversion Technologies

Power Electronics

Advanced Conversion Technologies

Fang Lin Luo
Hong Ye

CRC Press is an imprint of the
Taylor & Francis Group, an **informa** business

CRC Press
Taylor & Francis Group
6000 Broken Sound Parkway NW, Suite 300
Boca Raton, FL 33487-2742

Printed in the United States of America on acid-free paper
10 9 8 7 6 5 4 3 2 1

International Standard Book Number: 978-1-4200-9429-9 (Hardback)

Library of Congress Cataloging-in-Publication Data

Luo, Fang Lin.
Power electronics : advanced conversion technologies / authors, Fang Lin Luo, Hong Ye.
p. cm.
Includes bibliographical references and index.
ISBN 978-1-4200-9429-9 (hardcover : alk. paper)
1. Power electronics. I. Ye, Hong, 1973- II. Title.

TK7881.15.L86 2010
621.31'7--dc22 2009043846

Visit the Taylor & Francis Web site at
http://www.taylorandfrancis.com

and the CRC Press Web site at
http://www.crcpress.com

Contents

Preface

This book is aimed at both engineering students and practicing professionals specializing in power electronics and provides useful and concise information with regard to advanced converters. It contains more than 200 topologies concerning advanced converters that have been developed by the authors. Some recently published topologies are also included. The prototypes presented here demonstrate novel approaches that the authors hope will be of great benefit to the area of power electronics.

Power electronics is the technology behind the conversion of electrical energy from a source to the requirements of the end-user. Although, it is of vital importance to both industry and the individual citizen, it is somewhat taken for granted in much the same way as the air we breathe and the water we drink. Energy conversion techniques are now a primary focus of the power electronics community with rapid advances being made in conversion technologies in recent years that are detailed in this book along with a look at the historical problems that have now been solved.

The necessary equipment for energy conversion can be divided into four groups: AC/DC rectifiers, DC/DC converters, DC/AC inverters, and AC/AC transformers. AC/DC rectifiers were the earliest converters to be developed and, consequently, most of the traditional circuits have now been widely published and discussed. However, some of those circuits have not been analyzed in any great detail with the single-phase diode rectifier with *R–C* load being a typical example. Recently, there has been a new approach to AC/DC rectifiers that involves power factor correction (PFC) and unity power factor (UPF), the techniques of which are introduced in this book.

The technology of DC/DC conversion is making rapid progress and, according to incomplete statistics, there are more than 600 topologies of DC/DC converters in existence with new ones being created every year. It would be an immense task to try and examine all of these approaches. However, in 2001, the authors were able to systematically sort and categorize the DC/DC converters into six groups. Our main contribution in this field involves voltage-lift and super-lift techniques for which more than 100 topologies are introduced in this book.

DC/AC inverters can be divided into two groups: pulse-width-modulation (PWM) inverters and multilevel inverters. People will be more familiar with PWM inverters as the voltage source inverter (VSI) and current source inverter (CSI). In 2003, details of the impedance-source inverter (ZSI) first appeared and a great deal of interest was created from power electronics experts. With its advantages so obvious in research and industrial applications, hundreds of papers concerning ZSI have been published in the ensuing years. Multilevel inverters were invented in the early 1980s and developed quickly. Many new topologies have been designed and applied to industrial applications, especially in renewable energy systems. Typical circuits include diode-clamped inverters, capacitor-clamped inverters, and hybrid H-bridge multilevel inverters. Multilevel inverters overcame the drawbacks of the PWM inverter and paved the way for industrial applications.

Traditional AC/AC converters are divided into three groups: voltage-modulation AC/AC converters, cycloconverters, and matrix converters. All traditional AC/AC converters can only convert a high voltage to a low voltage with adjustable amplitude and frequency. Their drawbacks are limited output voltage and poor total harmonic distortion (THD). Therefore, new types of AC/AC converters, such as sub-envelope-modulated (SEM) AC/AC converters and DC-modulated AC/AC converters have been created. These techniques successfully overcome the disadvantage of high THD. Also, DC-modulated AC/AC converters have other advantages, for instance, multiphase outputs.

Due to the world's increasing problem of energy resource shortage, the development of renewable energy sources, energy-saving techniques, and power supply quality has become an urgent issue. There is no time for delay. Renewable energy source systems require a large number of converters. For example, new AC/DC/AC converters are necessary in wind-turbine power systems, and DC/AC/DC converters are necessary in solar panel power systems.

The book consists of 12 chapters. The general knowledge on converters is introduced in Chapter 1. Traditional AC/DC diode rectifiers, controlled AC/DC rectifiers, and power factor correction and unity power factor techniques are discussed in Chapters 2 through 4. Classic DC/DC converters, voltage-lift and super-lift techniques are introduced in Chapters 5 through 7. Pulse-width-modulated DC/AC inverters are investigated in Chapter 8 and multilevel DC/AC inverters in Chapter 9. Traditional and improved AC/AC converters are introduced in Chapters 10 and 11. AC/DC/AC and DC/AC/DC converters used in renewable energy source systems are presented in Chapter 12.

As a textbook, there are many examples and homework questions in each chapter, which will help the reader thoroughly understand all aspects of research and application. This book can be both a textbook for university students studying power electronics and a reference book for practicing engineers involved in the design and application of power electronics.

Dr. Fang Lin Luo and Dr. Hong Ye
Nanyang Technological University
Singapore

Authors

Dr. Fang Lin Luo is an associate professor with the School of Electrical and Electronic Engineering, Nanyang Technological University (NTU), Singapore. He received his BSc degree, first class with honors, in Radio-Electronic Physics from Sichuan University, Chengdu, China, in 1968 and his PhD degree in Electrical Engineering and Computer Science (EE & CS) from Cambridge University, UK, in 1986.

After his graduation from Sichuan University, he joined the Chinese Automation Research Institute of Metallurgy (CARIM), Beijing, China, as Senior Engineer. From there, he then went to Entreprises Saunier Duval, Paris, France, as a project engineer in 1981–1982, and subsequently to Hocking NDT Ltd, Allen-Bradley IAP Ltd, and Simplatroll Ltd in England as senior engineer after he received his PhD degree from Cambridge University. He is Fellow of the Cambridge philosophical society and a senior member of IEEE. He has published nine books and 300 technical papers in *IEEE Transactions*, *IEE/IET Proceedings* and other international journals, and in various international conferences. His present research interests include power electronics and DC & AC motor drives with computerized artificial intelligent control (AIC) and digital signal processing (DSP), and digital power electronics.

He is currently the associate editor of both *IEEE Transactions on Power Electronics* and *IEEE Transactions on Industrial Electronics*. He is also an international editor for the international journal *Advanced Technology of Electrical Engineering and Energy*. Dr. Luo was chief editor of the international journal *Power Supply Technologies and Applications* in 1998–2003. He is general chairman of the First IEEE Conference on Industrial Electronics and Applications (ICIEA 2006) and of the Third IEEE Conference on Industrial Electronics and Applications (ICIEA 2008).

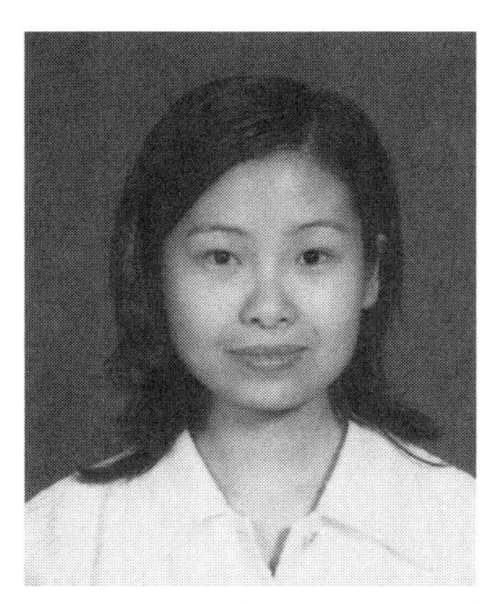

Dr. Hong Ye received her bachelor's degree, first class, in 1995, a master engineering degree from Xi'an JiaoTong University, China, in 1999, and her PhD degree from Nanyang Technological University (NTU), Singapore, in 2005. She was with the R&D Institute, XIYI Company Ltd, China, as a research engineer from 1995 to 1997. She was with NTU as a research associate in 2003–2004 and has been a research fellow from 2005.

Dr. Ye is an IEEE member and has co-authored nine books. She has published more than 60 technical papers in *IEEE Transactions*, *IEE Proceedings* and other international journals and various international conferences. Her research interests are power electronics and conversion technologies, signal processing, operations research, and structural biology.

1

Introduction

Power electronics is the technology of processing and controlling the flow of electric energy by supplying voltages and currents in a form that is optimally suited to the end-user's requirements [1]. A typical block diagram is given in Figure 1.1 [2]. The input power can be either AC and DC sources. A general example is one in which the AC input power is from the electric utility. The output power to the load can be either AC and DC voltages. The power processor in the block diagram is usually called a converter. Conversion technologies are used to construct converters. There are four types of converters [3]:

- AC/DC converters/rectifiers (AC to DC)
- DC/DC converters (DC to DC)
- DC/AC inverters/converters (DC to AC)
- AC/AC converters (AC to AC).

We will use *converter* as a generic term to refer to a single power conversion stage that may perform any of the functions listed above. To be more specific, during AC to DC and DC to AC conversion, the term *rectifier* refers to a converter in which the average power flow is from the AC to the DC side. The term *inverter* refers to a converter in which the average power flow is from the DC to the AC side. If the power flow through the converter is reversible, as shown in Figure 1.2 [2], we refer to the converter in terms of its rectifier and inverter modes of operation.

1.1 Symbols and Factors Used in This Book

In this chapter, we list the factors and symbols used in this book. If no specific description is given, the parameters follow the meaning stated here.

1.1.1 Symbols Used in Power Systems

For instantaneous values of variables such as voltage, current, and power, which are functions of time, lowercase letters v, i, and p are, respectively, used. They are functions of time performing in the time domain. We may or may not explicitly show that they are functions of time, for example, using v rather than $v(t)$. Uppercase symbols V and I refer to their computed values from their instantaneous waveforms. They generally refer to an average value in DC quantities and a root-mean-square (rms) value in AC quantities. If there is a

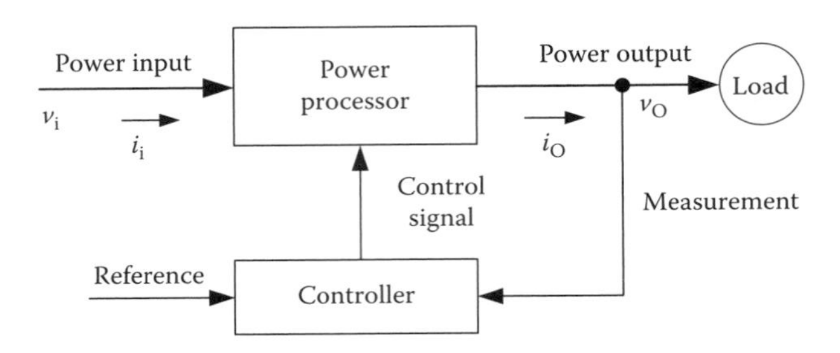

FIGURE 1.1 Block diagram of a power electronics system.

possibility of confusion, the subscript avg or rms is added explicitly. The average power is always indicated by P.

Usually, the input voltage and current are represented by v_{in} and i_{in} (or v_1 and i_1), and the output voltage and current are represented by v_O and i_O (or v_2 and i_2). The input and output powers are represented by P_{in} and P_O. The power transfer efficiency (η) is defined as $\eta = P_O/P_{in}$.

Passive loads such as resistor R, inductor L, and capacitor C are generally used in circuits. We use R, L, and C to indicate their symbols and values as well. All these three parameters and their combination Z are linear loads since the performance of the circuit constructed by these components is described by a linear differential equation. Z is used as the impedance of a linear load. If the circuit consists of a resistor R, an inductor L, and a capacitor C in series connection, the impedance Z is represented by

$$Z = R + j\omega L - j\frac{1}{\omega C} = |Z| \angle \phi, \tag{1.1}$$

where R is the resistance measured in units of Ω, L is the inductance measured in H, C is the capacitance measured in F, ω is the AC supply angular frequency measured in rad/s, and $\omega = 2\pi f$ where f is the AC supply frequency measured in Hz. For the calculation of Z, if there is no capacitor in the circuit, $j(1/\omega C)$ is omitted (do not take $c = 0$ and $j(1/\omega C) = >\infty$). The absolute impedance $|Z|$ and the phase angle ϕ are

$$\begin{aligned} |Z| &= \sqrt{R^2 + [\omega L - (1/\omega C)]^2}, \\ \phi &= \tan^{-1}\frac{\omega L - (1/\omega C)}{R}. \end{aligned} \tag{1.2}$$

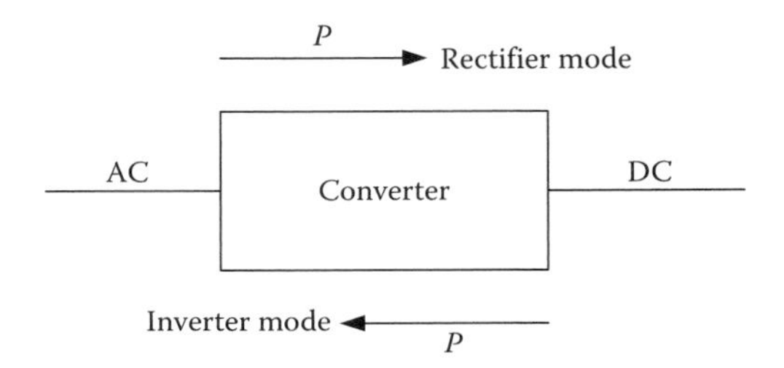

FIGURE 1.2 AC to DC converters.

Example 1.1

A circuit has a load with a resistor $R = 20\,\Omega$, an inductor $L = 20$ mH, and a capacitor $C = 200\,\mu$F in series connection. The voltage supply frequency $f = 60$ Hz. Calculate the load impedance and the phase angle.

SOLUTION

From Equation 1.1, the impedance Z is

$$Z = R + \mathrm{j}\omega L - \mathrm{j}\frac{1}{\omega C} = 20 + \mathrm{j}120\pi \times 0.02 - \mathrm{j}\frac{1}{120\pi \times 0.0002}$$
$$= 20 + \mathrm{j}(7.54 - 13.26) = 20 - \mathrm{j}5.72 = |Z|\angle\phi.$$

From Equation 1.2, the absolute impedance $|Z|$ and the phase angle ϕ are

$$|Z| = \sqrt{R^2 + \left(\omega L - \frac{1}{\omega C}\right)^2} = \sqrt{20^2 + 5.72^2} = 20.8\,\Omega,$$
$$\phi = \tan^{-1}\frac{\omega L - (1/\omega C)}{R} = \tan^{-1}\frac{-5.72}{20} = -17.73°.$$

If a circuit consists of a resistor R and an inductor L in series connection, the corresponding impedance Z is given by

$$Z = R + \mathrm{j}\omega L = |Z|\angle\phi. \tag{1.3}$$

The absolute impedance $|Z|$ and the phase angle ϕ are

$$|Z| = \sqrt{R^2 + (\omega L)^2}, \quad \phi = \tan^{-1}\frac{\omega L}{R}. \tag{1.4}$$

We define the circuit time constant τ as

$$\tau = \frac{L}{R}. \tag{1.5}$$

If a circuit consists of a resistor R and a capacitor C in series connection, the impedance Z is given by

$$Z = R - \mathrm{j}\frac{1}{\omega C} = |Z|\angle\phi. \tag{1.6}$$

The absolute impedance $|Z|$ and the phase angle ϕ are

$$|Z| = \sqrt{R^2 + \left(\frac{1}{\omega C}\right)^2}, \quad \phi = -\tan^{-1}\frac{1}{\omega CR}. \tag{1.7}$$

We define the circuit time constant τ as

$$\tau = RC. \tag{1.8}$$

Summary of the Symbols

Symbol	Explanation (measuring unit)
C	Capacitance (F)
f	Frequency (Hz)
i, I	Instantaneous current, Average/rms current (A)
L	Inductance (H)
R	Resistance (Ω)
p, P	Instantaneous power, Rated/real power (W)
q, Q	Instantaneous reactive power, Rated reactive power (VAR)
s, S	Instantaneous apparent power, Rated apparent power (VA)
v, V	Instantaneous voltage, Average/rms voltage (V)
Z	Impedance (Ω)
ϕ	Phase angle (° or rad)
η	Efficiency (%)
τ	Time constant (s)
ω	Angular frequency (rad/s), $\omega = 2\pi f$

1.1.2 Factors and Symbols Used in AC Power Systems

The input AC voltage can be either single-phase or three-phase voltages. They are usually a pure sinusoidal wave function. A single-phase input voltage $v(t)$ can be expressed as [4]

$$v(t) = \sqrt{2}V \sin \omega t = V_m \sin \omega t, \tag{1.9}$$

where v is the instantaneous input voltage, V the rms value, V_m the amplitude, and ω the angular frequency, $\omega = 2\pi f$ (f is the supply frequency). Usually, the input current may not be a pure sinusoidal wave that depends on load. If the input voltage supplies a linear load (resistive, inductive, capacitive loads or their combination), the input current $i(t)$ is not distorted, but may be delayed in a phase angle ϕ. In this case, it can be expressed as

$$i(t) = \sqrt{2}I \sin(\omega t - \phi) = I_m \sin(\omega t - \phi), \tag{1.10}$$

where i is the instantaneous input current, I the rms value, I_m the amplitude, and ϕ the phase-delay angle. We define the power factor (PF) as

$$\text{PF} = \cos \phi. \tag{1.11}$$

PF is the ratio of real power (P) to apparent power (S). We have the relation $S = P + jQ$, where Q is the reactive power. The power vector diagram is shown in Figure 1.3. We have

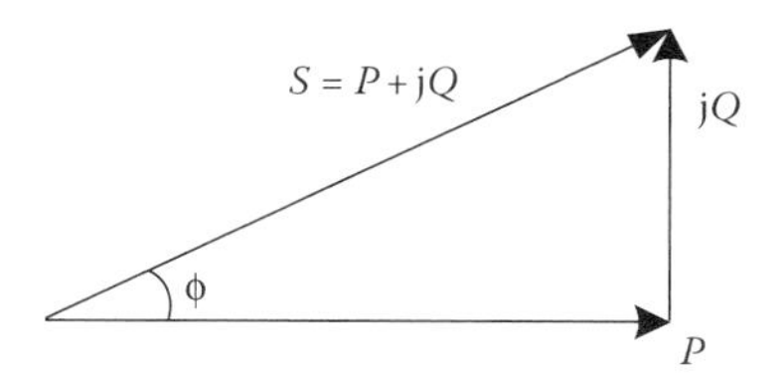

FIGURE 1.3 Power vector diagram.

the relations between the powers as follows:

$$S = VI^* = \frac{V^2}{Z^*} = P + jQ = |S| \angle \phi, \quad (1.12)$$

$$|S| = \sqrt{P^2 + Q^2}, \quad (1.13)$$

$$\phi = \tan^{-1} \frac{Q}{P}, \quad (1.14)$$

$$P = S \cos \phi, \quad (1.15)$$

$$Q = S \sin \phi. \quad (1.16)$$

If the input current is distorted, it consists of harmonics. Its fundamental harmonic can be expressed as

$$i_1 = \sqrt{2} I_1 \sin(\omega t - \phi_1) = I_{m1} \sin(\omega t - \phi_1), \quad (1.17)$$

where i_1 is the fundamental harmonic instantaneous value, I_1 the rms value, I_{m1} the amplitude, and ϕ_1 the phase angle. In this case, the displacement power factor (DPF) is defined as

$$\text{DPF} = \cos \phi_1. \quad (1.18)$$

Correspondingly, PF is defined as

$$\text{PF} = \frac{\text{DPF}}{\sqrt{1 + \text{THD}^2}}, \quad (1.19)$$

where THD is the total harmonic distortion. It can be used to measure both voltage and current waveforms. It is defined as

$$\text{THD} = \frac{\sqrt{\sum_{n=2}^{\infty} I_n^2}}{I_1} \quad \text{or} \quad \text{THD} = \frac{\sqrt{\sum_{n=2}^{\infty} V_n^2}}{V_1}, \quad (1.20)$$

where I_n or V_n is the amplitude of the nth-order harmonic.

The harmonic factor (HF) is a variable that describes the weighted percent of the nth-order harmonic referring to the amplitude of the fundamental harmonic V_1. It is defined as

$$\text{HF}_n = \frac{I_n}{I_1} \quad \text{or} \quad \text{HF}_n = \frac{V_n}{V_1}, \quad (1.21)$$

where $n = 1$ corresponds to the fundamental harmonic. Therefore, $HF_1 = 1$. THD can be written as

$$\mathrm{THD} = \sqrt{\sum_{n=2}^{\infty} \mathrm{HF}_n^2}. \tag{1.22}$$

A pure sinusoidal waveform has THD = 0.

The weighted total harmonic distortion (WTHD) is a variable that describes the waveform distortion. It is defined as

$$\mathrm{WTHD} = \frac{\sqrt{\sum_{n=2}^{\infty}(V_n^2/n)}}{V_1}. \tag{1.23}$$

Note that THD gives an immediate measure of the inverter output voltage waveform distortion. WTHD is often interpreted as the normalized current ripple expected in an inductive load when fed from the inverter output voltage.

Example 1.2

A load with a resistor $R = 20\,\Omega$, an inductor $L = 20\,\mathrm{mH}$, and a capacitor $C = 200\,\mu\mathrm{F}$ in series connection is supplied by an AC voltage of 240V (rms) with frequency $f = 60\,\mathrm{Hz}$. Calculate the circuit current, and the corresponding apparent power S, real power P, reactive power Q, and PF.

SOLUTION

From Example 1.1, the impedance Z is

$$Z = R + \mathrm{j}\omega L - \mathrm{j}\frac{1}{\omega C} = 20 + \mathrm{j}120\pi \times 0.02 - \mathrm{j}\frac{1}{120\pi \times 0.0002}$$
$$= 20 + \mathrm{j}(7.54 - 13.26) = 20 - \mathrm{j}5.72 = 20.8\angle{-17.73^\circ}\ \Omega.$$

The circuit current I is

$$I = \frac{V}{Z} = \frac{240}{20.8\angle{-17.73^\circ}} = 11.54\angle 17.73^\circ\ \mathrm{A}.$$

The apparent power S is

$$S = VI^* = 240 \times 11.54\angle{-17.73^\circ} = 2769.23\angle{-17.73^\circ}\ \mathrm{VA}.$$

The real power P is

$$P = |S|\cos\phi = 2769.23 \times \cos 17.73^\circ = 2637.7\ \mathrm{W}.$$

The reactive power Q is

$$Q = |S|\sin\phi = 2769.23 \times \sin -17.73^\circ = -843.3\ \mathrm{VAR}.$$

PF is

$$\mathrm{PF} = \cos\phi = 0.9525\ \text{leading}.$$

Summary of the Symbols

Symbol	Explanation (measuring unit)
DPF	Displacement power factor (%)
HF_n	nth-order harmonic factor
i_1, I_1	Instantaneous fundamental current, Average/rms fundamental current (A)
i_n, I_n	Instantaneous nth-order harmonic current, Average/rms nth-order harmonic current (A)
I_m	Current amplitude (A)
PF	Power factor (leading/lagging %)
q, Q	Instantaneous reactive power, Rated reactive power (VAR)
s, S	Instantaneous apparent power, Rated apparent power (VA)
t	Time (s)
THD	Total harmonic distortion (%)
v_1, V_1	Instantaneous fundamental voltage, Average/rms fundamental voltage (V)
v_n, V_n	Instantaneous nth-order harmonic voltage, Average/rms nth-order harmonic voltage (V)
WTHD	Weighted total harmonic distortion (%)
ϕ_1	Phase angle of the fundamental harmonic (° or rad)

1.1.3 Factors and Symbols Used in DC Power Systems

We define the output DC voltage instantaneous value as v_d and the average value as V_d (or V_{d0}) [5]. A pure DC voltage has no ripple; hence it is called ripple-free DC voltage. Otherwise, a DC voltage is distorted, and consists of DC components and AC harmonics. Its rms value is V_{d-rms}. For a distorted DC voltage, the rms value V_{d-rms} is constantly higher than the average value V_d. The ripple factor (RF) is defined as

$$\mathrm{RF} = \frac{\sqrt{\sum_{n=1}^{\infty} V_n^2}}{V_d}, \tag{1.24}$$

where V_n is the nth-order harmonic. The form factor (FF) is defined as

$$\mathrm{FF} = \frac{V_{d-rms}}{V_d} = \frac{\sqrt{\sum_{n=0}^{\infty} V_n^2}}{V_d}, \tag{1.25}$$

where V_0 is the 0th-order harmonic, that is, the average component V_d. Therefore, we obtain $\mathrm{FF} > 1$, and the relation

$$\mathrm{RF} = \sqrt{\mathrm{FF}^2 - 1}. \tag{1.26}$$

FF and RF are used to describe the quality of a DC waveform (voltage and current parameters). For a pure DC voltage, FF = 1 and RF = 0.

Summary of the Symbols

Symbol	Explanation (measuring unit)
FF	Form factor (%)
RF	Ripple factor (%)
v_d, V_d	Instantaneous DC voltage, Average DC voltage (V)
V_{d-rms}	rms DC voltage (V)
v_n, V_n	Instantaneous nth-order harmonic voltage, Average/rms nth-order harmonic voltage (V)

1.1.4 Factors and Symbols Used in Switching Power Systems

Switching power systems, such as power DC/DC converters, power PWM DC/AC inverters, soft-switching converters, and resonant converters, are widely used in power transfer equipment. In general, a switching power system has a pumping circuit and several energy-storage elements. It is likely an energy container to store some energy during performance. The input energy does not smoothly flow through the switching power system from the input source to the load. The energy is quantified by the switching circuit, and then pumped through the switching power system from the input source to the load [6–8].

We assume that the switching frequency is f and that the corresponding period is $T = 1/f$. The pumping energy (PE) is used to count the input energy in a switching period T. Its calculation formula is

$$\mathrm{PE} = \int_0^T P_{\mathrm{in}}(t)\,\mathrm{d}t = \int_0^T V_{\mathrm{in}} i_{\mathrm{in}}(t)\,\mathrm{d}t = V_{\mathrm{in}} I_{\mathrm{in}} T, \tag{1.27}$$

where

$$I_{\mathrm{in}} = \int_0^T i_{\mathrm{in}}(t)\,\mathrm{d}t \tag{1.28}$$

is the average value of the input current if the input voltage V_1 is constant. Usually, the input average current I_1 depends on the conduction duty cycle.

Energy storage in switching power systems has received much attention in the past. Unfortunately, there is still no clear concept to describe the phenomena and reveal the relationship between the stored energy (SE) and its characteristics.

The SE in an inductor is

$$W_{\mathrm{L}} = \frac{1}{2} L I_{\mathrm{L}}^2. \tag{1.29}$$

The SE across a capacitor is

$$W_{\mathrm{C}} = \frac{1}{2} C V_{\mathrm{C}}^2. \tag{1.30}$$

Therefore, if there are n_{L} inductors and n_{C} capacitors, the total SE in a DC/DC converter is

$$\mathrm{SE} = \sum_{j=1}^{n_{\mathrm{L}}} W_{\mathrm{L}j} + \sum_{j=1}^{n_{\mathrm{C}}} W_{\mathrm{C}j}. \tag{1.31}$$

Usually, the SE is independent of switching frequency f (as well as switching period T). Since inductor currents and capacitor voltages rely on the conduction duty cycle k, the SE *also relies on* k. We use SE as a new parameter in further descriptions.

Most switching power systems consist of inductors and capacitors. Therefore, we can define the capacitor–inductor stored energy ratio (CIR) as

$$\mathrm{CIR} = \frac{\sum_{j=1}^{n_{\mathrm{C}}} W_{\mathrm{C}j}}{\sum_{j=1}^{n_{\mathrm{C}}} W_{\mathrm{L}j}}. \tag{1.32}$$

As described in the previous sections, the input energy in a period T is the PE $= P_{in} \times T = V_{in}I_{in} \times T$. We now define the energy factor (EF), that is, the ratio of SE to PE, as

$$\mathrm{EF} = \frac{\mathrm{SE}}{\mathrm{PE}} = \frac{\mathrm{SE}}{V_{in}I_{in}T} = \frac{\sum_{j=1}^{m} W_{Lj} + \sum_{j=1}^{n} W_{Cj}}{V_{in}I_{in}T}. \tag{1.33}$$

EF is a very important factor of a switching power system. It is usually independent of the conduction duty cycle and inversely proportional to switching frequency f since PE is proportional to switching period T.

The *time constant* τ of a switching power system is a new concept that describes the transient process. If there are no power losses in the system, it is defined as

$$\tau = \frac{2T \times \mathrm{EF}}{1 + \mathrm{CIR}}. \tag{1.34}$$

This time constant τ is independent of switching frequency f (or period $T = 1/f$). It is available to estimate the system responses for a unit-step function and impulse interference.

If there are power losses and $\eta < 1$, τ is defined as

$$\tau = \frac{2T \times \mathrm{EF}}{1 + \mathrm{CIR}}\left(1 + \mathrm{CIR}\frac{1-\eta}{\eta}\right). \tag{1.35}$$

If there are no power losses, $\eta = 1$, Equation 1.35 becomes Equation 1.34. Usually, if the power losses (lower efficiency η) are higher, the time constant τ is larger since CIR > 1.

The *damping time constant* τ_d of a switching power system is a new concept that describes the transient process. If there are no power losses, it is defined as

$$\tau_d = \frac{2T \times \mathrm{EF}}{1 + \mathrm{CIR}}\mathrm{CIR}. \tag{1.36}$$

This damping time constant τ_d is independent of switching frequency f (or period $T = 1/f$). It is available to estimate the oscillation responses for a unit-step function and impulse interference.

If there are power losses and $\eta < 1$, τ_d is defined as

$$\tau_d = \frac{2T \times \mathrm{EF}}{1 + \mathrm{CIR}} \frac{\mathrm{CIR}}{\eta + \mathrm{CIR}(1-\eta)}. \tag{1.37}$$

If there are no power losses, $\eta = 1$, Equation 1.37 becomes Equation 1.36. Usually, if the power losses (lower efficiency η) are higher, the damping time constant τ_d is smaller since CIR > 1.

The *time constant ratio* ξ of a switching power system is a new concept that describes the transient process. If there are no power losses, it is defined as

$$\xi = \frac{\tau_d}{\tau} = \mathrm{CIR}. \tag{1.38}$$

This time constant ratio is independent of switching frequency f (or period $T = 1/f$). It is available to estimate the oscillation responses for a unit-step function and impulse interference.

If there are power losses and $\eta < 1, \xi$ is defined as

$$\xi = \frac{\tau_d}{\tau} = \frac{\text{CIR}}{\eta[1 + \text{CIR}(1 - \eta/\eta)]^2}. \tag{1.39}$$

If there are no power losses, $\eta = 1$, Equation 1.39 becomes Equation 1.38. Usually, if the power losses (the lower efficiency η) are higher, the time constant ratio ξ is smaller since CIR > 1. From this analysis, most switching power systems with lower power losses possess larger output voltage oscillation when the converter operation state changes. On the other hand, switching power systems with high power losses will possess smoothening output voltage when the converter operation state changes.

By cybernetic theory, we can estimate the unit-step function response using the ratio ξ. If the ratio ξ is equal to or smaller than 0.25, the corresponding unit-step function response has no oscillation and overshot. However, if the ratio ξ is greater than 0.25, the corresponding unit-step function response has oscillation and overshot. The higher the value of the ratio ξ, the heavier the oscillation with higher overshot.

Summary of the Symbols

Symbol	Explanation (measuring unit)
CIR	Capacitor–inductor stored energy ratio
EF	Energy factor
f	Switching frequency (Hz)
k	Conduction duty cycle
PE	Pumping energy (J)
SE	Total stored energy (J)
W_L, W_C	SE in an inductor/capacitor (J)
T	Switching period (s)
τ	Time constant (s)
τ_d	Damping time constant (s)
ξ	Time constant ratio

1.1.5 Other Factors and Symbols

A transfer function is the mathematical modeling of a circuit and a system. It describes the dynamic characteristics of the circuit and the system. Using the transfer function, we can easily obtain the system step and impulse responses by applying an input signal. A typical second-order transfer function is [6–8]

$$G(s) = \frac{M}{1 + s\tau + s^2\tau\tau_d} = \frac{M}{1 + s\tau + s^2\xi\tau^2}, \tag{1.40}$$

where M is the voltage transfer gain ($M = V_O/V_{in}$), τ the time constant (Equation 1.35), τ_d the damping time constant (Equation 1.37), $\tau_d = \xi\tau$ (Equation 1.39), and s the Laplace operator in the s-domain.

Using this mathematical model of a switching power system, it is significantly easier to describe the characteristics of the transfer function. In order to appreciate the characteristics of the transfer function more fully, a few situations are given below.

1.1.5.1 Very Small Damping Time Constant

If the damping time constant is very small (i.e., $\tau_d \ll \tau$, $\xi \ll 1$) and it can be ignored, the value of the damping time constant τ_d is omitted (i.e., $\tau_d = 0$, $\xi = 0$). The transfer function (Equation 1.40) is downgraded to first order as

$$G(s) = \frac{M}{1 + s\tau}. \tag{1.41}$$

The unit-step function response in the time domain is

$$g(t) = M(1 - e^{-t/\tau}). \tag{1.42}$$

The transient process (settling time) is nearly three times the time constant (3τ), to produce $g(t) = g(3\tau) = 0.95\,M$. The response in the time domain is shown in Figure 1.4 with $\tau_d = 0$.

The impulse interference response is

$$\Delta g(t) = U \cdot e^{-t/\tau}, \tag{1.43}$$

where U is the interference signal. The interference recovering progress is nearly three times the time constant (3τ), and is shown in Figure 1.5 with $\tau_d = 0$.

1.1.5.2 Small Damping Time Constant

If the damping time constant is small (i.e., $\tau_d < \tau/4, \xi < 0.25$) and cannot be ignored, the value of the damping time constant τ_d is not omitted. The transfer function (Equation 1.40) is retained as a second-order function with two real poles ($-\sigma_1$ and $-\sigma_2$) as

$$G(s) = \frac{M}{1 + s\tau + s^2\tau\tau_d} = \frac{M/\tau\tau_d}{(s + \sigma_1)(s + \sigma_2)}, \tag{1.44}$$

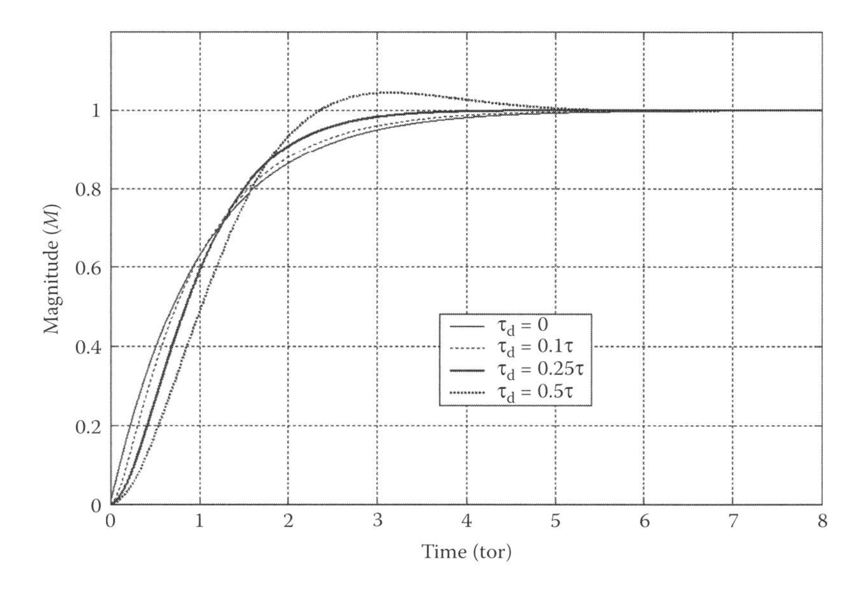

FIGURE 1.4 Unit-step function responses ($\tau_d = 0, 0.1\tau, 0.25\tau$, and 0.5τ).

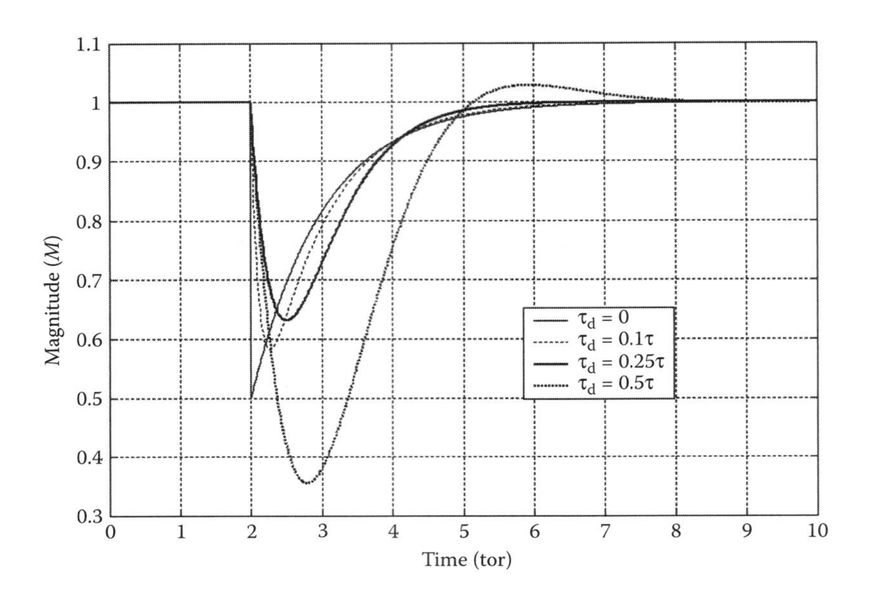

FIGURE 1.5 Impulse responses ($\tau_d = 0, 0.1\tau, 0.25\tau$, and 0.5τ).

where

$$\sigma_1 = \frac{\tau + \sqrt{\tau^2 - 4\tau\tau_d}}{2\tau\tau_d} \quad \text{and} \quad \sigma_2 = \frac{\tau - \sqrt{\tau^2 - 4\tau\tau_d}}{2\tau\tau_d}.$$

There are two real poles in the transfer function, assuming $\sigma_1 > \sigma_2$. The unit-step function response in the time domain is

$$g(t) = M(1 + K_1 e^{-\sigma_1 t} + K_2 e^{-\sigma_2 t}), \tag{1.45}$$

where

$$K_1 = -\frac{1}{2} + \frac{\tau}{2\sqrt{\tau^2 - 4\tau\tau_d}} \quad \text{and} \quad K_2 = -\frac{1}{2} - \frac{\tau}{2\sqrt{\tau^2 - 4\tau\tau_d}}.$$

The transient process is nearly three times the time value $1/\sigma_1, 3/\sigma_1 < 3\tau$. The response process is quick without oscillation. The corresponding waveform in the time domain is shown in Figure 1.4 with $\tau_d = 0.1\tau$.

The impulse interference response is

$$\Delta g(t) = \frac{U}{\sqrt{1 - 4\tau_d/\tau}}(e^{-\sigma_2 t} - e^{-\sigma_1 t}), \tag{1.46}$$

where U is the interference signal. The transient process is nearly three times the time value $1/\sigma_1, 3/\sigma_1 < 3\tau$. The response waveform in the time domain is shown in Figure 1.5 with $\tau_d = 0.1\tau$.

1.1.5.3 Critical Damping Time Constant

If the damping time constant is equal to the critical value (i.e., $\tau_d = \tau/4$), the transfer function (Equation 1.40) is retained as a second-order function with two poles $\sigma_1 = \sigma_2 =$

σ as

$$G(s) = \frac{M}{1 + s\tau + s^2\tau\tau_d} = \frac{M/\tau\tau_d}{(s+\sigma)^2}, \tag{1.47}$$

where

$$\sigma = \frac{1}{2\tau_d} = \frac{2}{\tau}.$$

There are two folded real poles in the transfer function. This expression describes the characteristics of the DC/DC converter. The unit-step function response in the time-domain is

$$g(t) = M\left[1 - \left(1 + \frac{2t}{\tau}\right)e^{-(2t/\tau)}\right]. \tag{1.48}$$

The transient process is nearly 2.4 times the time constant $\tau(2.4\tau)$. The response process is quick without oscillation. The response waveform in the time domain is shown in Figure 1.4 with $\tau_d = 0.25\tau$.

The impulse interference response is

$$\Delta g(t) = \frac{4U}{\tau} t e^{-(2t/\tau)}, \tag{1.49}$$

where U is the interference signal. The transient process is still nearly 2.4 times the time constant, 2.4τ. The response waveform in the time domain is shown in Figure 1.5 with $\tau_d = 0.25\tau$.

1.1.5.4 Large Damping Time Constant

If the damping time constant is large (i.e., $\tau_d > \tau/4, \xi > 0.25$), the transfer function 1.40 is a second-order function with a couple of conjugated complex poles $-s_1$ and $-s_2$ in the left-hand half plane (LHHP) in the s-domain:

$$G(s) = \frac{M}{1 + s\tau + s^2\tau\tau_d} = \frac{M/\tau\tau_d}{(s+s_1)(s+s_2)}, \tag{1.50}$$

where $s_1 = \sigma + j\omega$ and $s_2 = \sigma - j\omega$,

$$\sigma = \frac{1}{2\tau_d} \quad \text{and} \quad \omega = \frac{\sqrt{4\tau\tau_d - \tau^2}}{2\tau\tau_d}.$$

There are a couple of conjugated complex poles $-s_1$ and $-s_2$ in the transfer function. This expression describes the characteristics of the DC/DC converter. The unit-step function response in the time domain is

$$g(t) = M\left[1 - e^{-t/2\tau_d}\left(\cos\omega t - \frac{1}{\sqrt{4\tau_d/\tau - 1}}\sin\omega t\right)\right]. \tag{1.51}$$

The transient response has an oscillation progress with a damping factor σ and the frequency ω. The corresponding waveform in the time domain is shown in Figure 1.4 with $\tau_d = 0.5\tau$, and in Figure 1.6 with $\tau, 2\tau, 5\tau$, and 10τ.

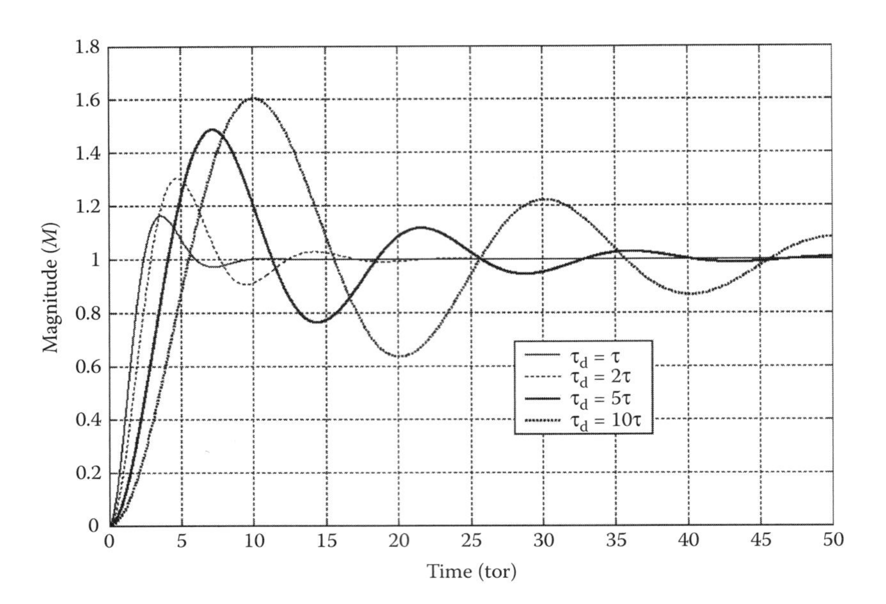

FIGURE 1.6 Unit-step function responses ($\tau_d = \tau, 2\tau, 5\tau$, and 10τ).

The impulse interference response is

$$\Delta g(t) = \frac{U}{\sqrt{(\tau_d/\tau) - (1/4)}} e^{-t/2\tau_d} \sin \omega t, \tag{1.52}$$

where U is the interference signal. The recovery process is a curve with damping factor σ and frequency ω. The response waveform in the time domain is shown in Figure 1.5 with $\tau_d = 0.5\tau$, and in Figure 1.7 with $\tau, 2\tau, 5\tau$, and 10τ.

1.1.6 Fast Fourier Transform

Fast Fourier transform (FFT) [9] is a very versatile method to analyze waveforms. A periodical function with radian frequency ω can be represented by a series of sinusoidal functions:

$$f(t) = \frac{a_0}{2} + \sum_{n=1}^{\infty} (a_n \cos n\omega t + b_n \sin n\omega t), \tag{1.53}$$

where the Fourier coefficients are

$$a_n = \frac{1}{\pi} \int_0^{2\pi} f(t) \cos(n\omega t)\, d(\omega t), \quad n = 0, 1, 2, \ldots, \infty \tag{1.54}$$

and

$$b_n = \frac{1}{\pi} \int_0^{2\pi} f(t) \sin(n\omega t)\, d(\omega t), \quad n = 0, 1, 2, \ldots, \infty. \tag{1.55}$$

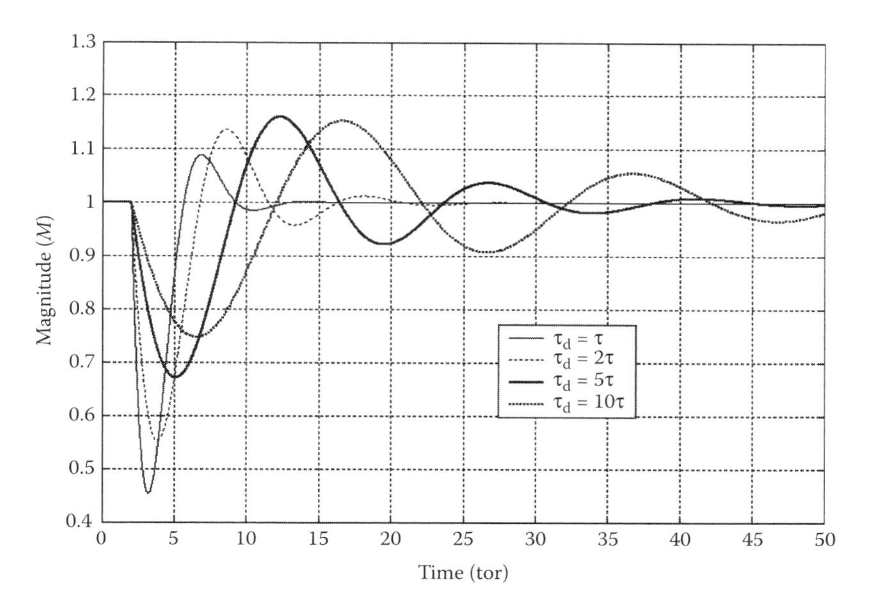

FIGURE 1.7 Impulse responses ($\tau_d = \tau, 2\tau, 5\tau$, and 10τ).

In this case, we call the item with radian frequency ω the fundamental harmonic and the items with radian frequency $n\omega$ ($n > 1$) higher-order harmonics. Draw the amplitudes of all harmonics in the frequency domain. We obtain the spectrum in an individual peak. The item $a_0/2$ is the DC component.

1.1.6.1 Central Symmetrical Periodical Function

If the periodical function is a central symmetrical periodical function, then all the items with cosine function disappear. The FFT remains as

$$f(t) = \sum_{n=1}^{\infty} b_n \sin n\omega t, \tag{1.56}$$

where

$$b_n = \frac{1}{\pi} \int_0^{2\pi} f(t) \sin(n\omega t)\, \mathrm{d}(\omega t), \quad n = 1, 2, \ldots, \infty. \tag{1.57}$$

We usually call this function an odd function. In this case, we call the item with radian frequency ω the fundamental harmonic and the items with radian frequency $n\omega$ ($n > 1$) the higher-order harmonics. Draw the amplitudes of all harmonics in the frequency domain. We obtain the spectrum in an individual peak. Since it is an odd function, the DC component is zero.

1.1.6.2 Axial (Mirror) Symmetrical Periodical Function

If the periodical function is an axial symmetrical periodical function, then all the items with sine function disappear. The FFT remains as

$$f(t) = \frac{a_0}{2} + \sum_{n=1}^{\infty} a_n \cos n\omega t, \tag{1.58}$$

where $a_0/2$ is the DC component and

$$a_n = \frac{1}{\pi} \int_0^{2\pi} f(t) \cos(n\omega t)\, \mathrm{d}(\omega t), \quad n = 0, 1, 2, \ldots, \infty. \tag{1.59}$$

The item $a_0/2$ is the DC component. We usually call this function an even function. In this case, we call the item with radian frequency ω the fundamental harmonic and the items with radian frequency $n\omega$ $(n > 1)$ higher-order harmonics. Draw the amplitudes of all harmonics in the frequency domain. We obtain the spectrum in an individual peak. Since it is an even function, the DC component is usually not zero.

1.1.6.3 Nonperiodical Function

The spectrum of a periodical function in the time domain is a discrete function in the frequency domain. If a function is a nonperiodical function in the time domain, it is possibly represented by Fourier integration. The spectrum is a continuous function in the frequency domain.

1.1.6.4 Useful Formulae and Data

Some trigonometric formulae are useful for FFT:

$$\sin^2 x + \cos^2 x = 1, \quad \sin x = \cos\left(\frac{\pi}{2} - x\right),$$

$$\sin x = -\sin(-x), \quad \sin x = \sin(\pi - x),$$

$$\cos x = \cos(-x), \quad \cos x = -\cos(\pi - x),$$

$$\frac{\mathrm{d}}{\mathrm{d}x}\sin x = \cos x, \quad \frac{\mathrm{d}}{\mathrm{d}x}\cos x = -\sin x,$$

$$\int \sin x\, \mathrm{d}x = -\cos x, \quad \int \cos x\, \mathrm{d}x = \sin x,$$

$$\sin(x \pm y) = \sin x \cos y \pm \cos x \sin y,$$

$$\cos(x \pm y) = \cos x \cos y \mp \sin x \sin y,$$

$$\sin 2x = 2 \sin x \cos x,$$

$$\cos 2x = \cos^2 x - \sin^2 x.$$

Some values corresponding to the special angles are usually used:

$$\sin\frac{\pi}{12} = \sin 15^\circ = 0.2588, \qquad \cos\frac{\pi}{12} = \cos 15^\circ = 0.9659,$$

$$\sin\frac{\pi}{8} = \sin 22.5^\circ = 0.3827, \qquad \cos\frac{\pi}{8} = \cos 22.5^\circ = 0.9239,$$

$$\sin\frac{\pi}{6} = \sin 30^\circ = 0.5, \qquad \cos\frac{\pi}{6} = \cos 30^\circ = \frac{\sqrt{3}}{2} = 0.866,$$

$$\sin\frac{\pi}{4} = \sin 45^\circ = \frac{\sqrt{2}}{2} = 0.7071, \qquad \cos\frac{\pi}{4} = \cos 45^\circ = \frac{\sqrt{2}}{2} = 0.7071,$$

$$\tan\frac{\pi}{12} = \tan 15^\circ = 0.2679, \qquad \tan\frac{\pi}{8} = \tan 22.5^\circ = 0.4142,$$

$$\tan\frac{\pi}{6} = \tan 30^\circ = \frac{\sqrt{3}}{3} = 0.5774, \qquad \tan\frac{\pi}{4} = \tan 45^\circ = 1,$$

$$\tan x = \frac{1}{\text{co-}\tan x}, \qquad \tan x = \text{co-}\tan\left(\frac{\pi}{2} - x\right).$$

1.1.6.5 Examples of FFT Applications

Example 1.3

An odd-square waveform is shown in Figure 1.8. Find FFT, HF up to seventh order, THD, and WTHD.

SOLUTION

The function $f(t)$ is

$$f(t) = \begin{cases} 1, & 2n\pi \le \omega t < (2n+1)\pi, \\ -1, & (2n+1)\pi \le \omega t < 2(n+1)\pi. \end{cases} \tag{1.60}$$

The Fourier coefficients are

$$b_n = \frac{1}{\pi}\int_0^{2\pi} f(t)\sin(n\omega t)\,\mathrm{d}(\omega t) = \frac{2}{n\pi}\int_0^{n\pi}\sin\theta\,\mathrm{d}\theta = 2\frac{1-(-1)^n}{n\pi}$$

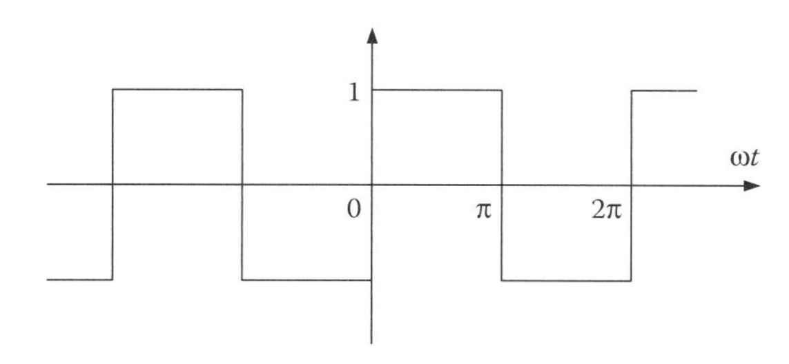

FIGURE 1.8 A waveform.

and

$$b_n = \frac{4}{n\pi}, \quad n = 1, 3, 5, \ldots, \infty. \tag{1.61}$$

Finally, we obtain

$$F(t) = \frac{4}{\pi}\sum_{n=1}^{\infty}\frac{\sin n\omega t}{n}, \quad n = 1, 3, 5, \ldots, \infty. \tag{1.62}$$

The fundamental harmonic has amplitude $4/\pi$. If we consider the higher-order harmonics up to the seventh order, that is, $n = 3, 5, 7$, the HFs are

$$\mathrm{HF}_3 = \frac{1}{3}, \quad \mathrm{HF}_5 = \frac{1}{5}, \quad \text{and} \quad \mathrm{HF}_7 = \frac{1}{7}.$$

The THD is

$$\mathrm{THD} = \frac{\sqrt{\sum_{n=2}^{\infty} V_n^2}}{V_1} = \sqrt{\left(\frac{1}{3}\right)^2 + \left(\frac{1}{5}\right)^2 + \left(\frac{1}{7}\right)^2} = 0.41415. \tag{1.63}$$

The WTHD is

$$\mathrm{WTHD} = \frac{\sqrt{\sum_{n=2}^{\infty}(V_n^2/n)}}{V_1} = \sqrt{\left(\frac{1}{3}\right)^3 + \left(\frac{1}{5}\right)^3 + \left(\frac{1}{7}\right)^3} = 0.219. \tag{1.64}$$

Example 1.4

An even-square waveform is shown in Figure 1.9. Find FFT, HF up to the seventh order, THD, and WTHD.

The function $f(t)$ is

$$f(t) = \begin{cases} 1, & (2n - 0.5)\pi \le \omega t < (2n + 0.5)\pi, \\ -1, & (2n + 0.5)\pi \le \omega t < (2n + 1.5)\pi. \end{cases} \tag{1.65}$$

The Fourier coefficients are

$$a_0 = 0,$$

$$a_n = \frac{1}{\pi}\int_0^{2\pi} f(t)\cos(n\omega t)\,\mathrm{d}(\omega t) = \frac{4}{n\pi}\int_0^{n\pi/2}\cos\theta\,\mathrm{d}\theta = \frac{4\sin(n\pi/2)}{n\pi},$$

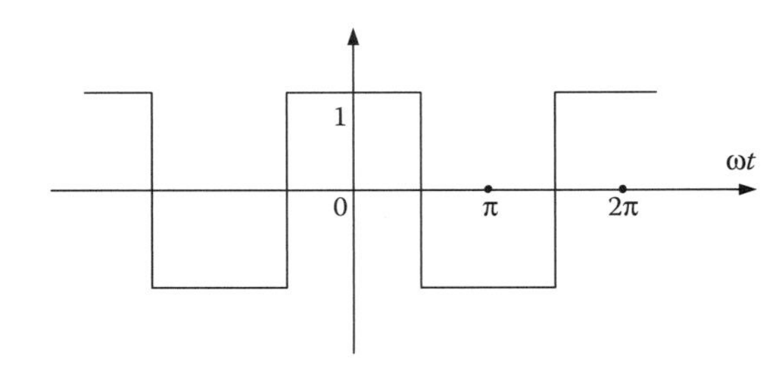

FIGURE 1.9 Even-square waveform.

and

$$a_n = \frac{4}{n\pi} \sin \frac{n\pi}{2}, \quad n = 1, 3, 5, \ldots, \infty. \tag{1.66}$$

The item $\sin(n\pi/2)$ is used to define the sign. Finally, we obtain

$$F(t) = \frac{4}{\pi} \sum_{n=1}^{\infty} \sin \frac{n\pi}{2} \cos(n\omega t), \quad n = 1, 3, 5, \ldots, \infty. \tag{1.67}$$

The fundamental harmonic has amplitude $4/\pi$. If we consider the higher-order harmonics up to the seventh order, that is, $n = 3, 5, 7$, the HFs are

$$\mathrm{HF}_3 = \frac{1}{3}, \quad \mathrm{HF}_5 = \frac{1}{5}, \quad \text{and} \quad \mathrm{HF}_7 = \frac{1}{7}.$$

The THD is

$$\mathrm{THD} = \frac{\sqrt{\sum_{n=2}^{\infty} V_n^2}}{V_1} = \sqrt{\left(\frac{1}{3}\right)^2 + \left(\frac{1}{5}\right)^2 + \left(\frac{1}{7}\right)^2} = 0.41415. \tag{1.68}$$

The WTHD is

$$\mathrm{WTHD} = \frac{\sqrt{\sum_{n=2}^{\infty} (V_n^2/n)}}{V_1} = \sqrt{\left(\frac{1}{3}\right)^3 + \left(\frac{1}{5}\right)^3 + \left(\frac{1}{7}\right)^3} = 0.219. \tag{1.69}$$

Example 1.5

An odd-waveform pulse with pulse width x is shown in Figure 1.10. Find FFT, HF up to the seventh order, THD, and WTHD.

The function $f(t)$ is in the period $-\pi$ to $+\pi$:

$$f(t) = \begin{cases} 1, & \dfrac{\pi - x}{2} \leq \omega t < \dfrac{\pi + x}{2}, \\ -1, & -\dfrac{\pi + x}{2} \leq \omega t < -\dfrac{\pi - x}{2}. \end{cases} \tag{1.70}$$

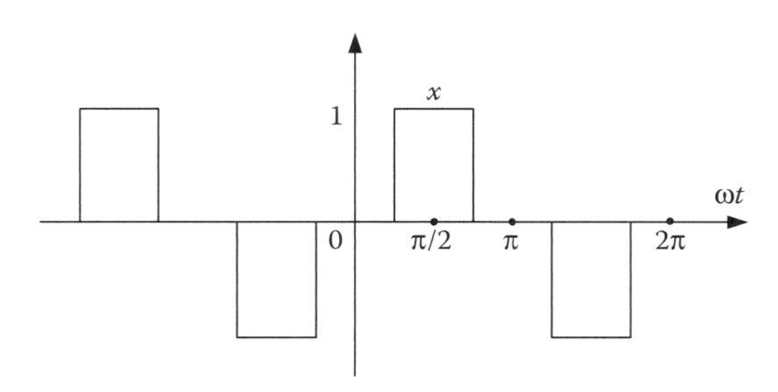

FIGURE 1.10 Odd-waveform pulse.

The Fourier coefficients are

$$b_n = \frac{1}{\pi}\int_0^{2\pi} f(t)\sin(n\omega t)\,d(\omega t) = \frac{2}{n\pi}\int_{n[(\pi-x)/2]}^{n[(\pi+x)/2]} \sin\theta\,d\theta = 2\frac{\cos[n(\pi-x)/2]-\cos[n(\pi+x)/2]}{n\pi}$$

$$= 2\frac{2\cos[n(\pi-x)/2]}{n\pi} = \frac{4\sin(n\pi/2)\sin(nx/2)}{n\pi},$$

or

$$b_n = \frac{4}{n\pi}\sin\frac{n\pi}{2}\sin\frac{nx}{2}, \quad n = 1,3,5,\ldots,\infty. \tag{1.71}$$

Finally, we obtain

$$F(t) = \frac{4}{\pi}\sum_{n=1}^{\infty}\frac{\sin(n\omega t)}{n}\sin\frac{n\pi}{2}\sin\frac{nx}{2}, \quad n = 1,3,5,\ldots,\infty. \tag{1.72}$$

The fundamental harmonic has amplitude $(4/\pi)\sin(x/2)$. If we consider the higher-order harmonics up to the seventh order, that is, $n = 3, 5, 7$, the HFs are

$$\text{HF}_3 = \frac{\sin(3x/2)}{3\sin(x/2)}, \quad \text{HF}_5 = \frac{\sin(5x/2)}{5\sin(x/2)}, \quad \text{and} \quad \text{HF}_7 = \frac{\sin(7x/2)}{7\sin(x/2)}.$$

The values of the HFs should be absolute values.
If $x = \pi$, the THD is

$$\text{THD} = \frac{\sqrt{\sum_{n=2}^{\infty} V_n^2}}{V_1} = \sqrt{\left(\frac{1}{3}\right)^2+\left(\frac{1}{5}\right)^2+\left(\frac{1}{7}\right)^2} = 0.41415. \tag{1.73}$$

The WTHD is

$$\text{WTHD} = \frac{\sqrt{\sum_{n=2}^{\infty}(V_n^2/n)}}{V_1} = \sqrt{\left(\frac{1}{3}\right)^3+\left(\frac{1}{5}\right)^3+\left(\frac{1}{7}\right)^3} = 0.219. \tag{1.74}$$

Example 1.6

A five-level odd waveform is shown in Figure 1.11. Find FFT, HF up to the seventh order, THD, and WTHD.

The function $f(t)$ is in the period $-\pi$ to $+\pi$:

$$f(t) = \begin{cases} 2, & \dfrac{\pi}{3} \le \omega t < \dfrac{2\pi}{3}, \\ 1, & \dfrac{\pi}{6} \le \omega t < \dfrac{\pi}{3}, \dfrac{2\pi}{3} \le \omega t < \dfrac{5\pi}{6}, \\ 0, & \text{other}, \\ -1, & -\dfrac{5\pi}{6} \le \omega t < -\dfrac{2\pi}{3}, -\dfrac{\pi}{3} \le \omega t < -\dfrac{\pi}{6}, \\ -2, & -\dfrac{2\pi}{3} \le \omega t < -\dfrac{\pi}{3}. \end{cases} \tag{1.75}$$

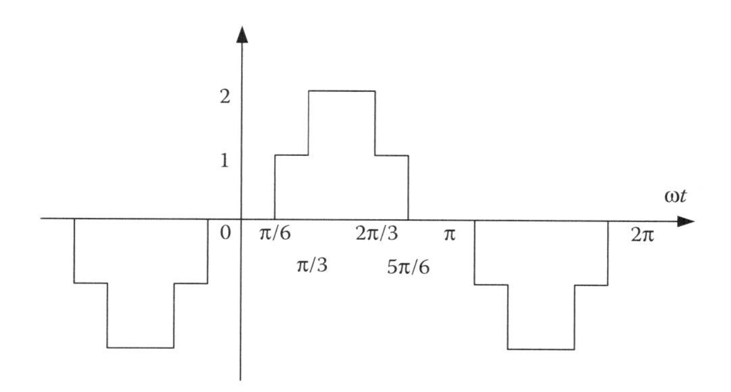

FIGURE 1.11 Five-level odd waveform.

The Fourier coefficients are

$$b_n = \frac{1}{\pi}\int_0^{2\pi} f(t)\sin(n\omega t)\,d(\omega t) = \frac{2}{n\pi}\left(\int_{n\pi/6}^{5n\pi/6}\sin\theta\,d\theta + \int_{n\pi/3}^{2n\pi/3}\sin\theta\,d\theta\right)$$

$$= \frac{2}{n\pi}\left[\left(\cos\frac{n\pi}{6} - \cos\frac{5n\pi}{6}\right) + \left(\cos\frac{n\pi}{3} - \cos\frac{2n\pi}{3}\right)\right] = \frac{4}{n\pi}\left(\cos\frac{n\pi}{6} + \cos\frac{n\pi}{3}\right)$$

or

$$b_n = \frac{4}{n\pi}\left(\cos\frac{n\pi}{6} + \cos\frac{n\pi}{3}\right), \quad n = 1, 3, 5, \ldots, \infty. \tag{1.76}$$

Finally, we obtain

$$F(t) = \frac{4}{\pi}\sum_{n=1}^{\infty}\frac{\sin(n\omega t)}{n}\left(\cos\frac{n\pi}{6} + \cos\frac{n\pi}{3}\right), \quad n = 1, 3, 5, \ldots, \infty. \tag{1.77}$$

The fundamental harmonic has amplitude $2/\pi(1+\sqrt{3})$. If we consider the higher-order harmonics up to the seventh order, that is, $n = 3, 5, 7$, the HFs are

$$\mathrm{HF}_3 = \frac{2}{3(1+\sqrt{3})} = 0.244, \quad \mathrm{HF}_5 = \frac{\sqrt{3}-1}{5(1+\sqrt{3})} = 0.0536, \quad \text{and}$$

$$\mathrm{HF}_7 = \frac{\sqrt{3}-1}{7(1+\sqrt{3})} = 0.0383.$$

The values of the HFs should be absolute values.

The THD is

$$\mathrm{THD} = \frac{\sqrt{\sum_{n=2}^{\infty} V_n^2}}{V_1} = \sqrt{\sum_{n=2}^{\infty}\mathrm{HF}_n^2} = \sqrt{0.244^2 + 0.0536^2 + 0.0383^2} = 0.2527. \tag{1.78}$$

The WTHD is

$$\text{WTHD} = \frac{\sqrt{\sum_{n=2}^{\infty}(V_n^2/n)}}{V_1} = \sqrt{\sum_{n=2}^{\infty}\frac{\text{HF}_n^2}{n}} = \sqrt{\frac{0.244^2}{3}+\frac{0.0536^2}{5}+\frac{0.0383^2}{7}} = 0.1436. \quad (1.79)$$

1.2 AC/DC Rectifiers

AC/DC rectifiers [3] have been used in industrial applications for a long time now. Before the 1960s, most power AC/DC rectifiers were constructed using mercury-arc rectifiers. Then the large power silicon diode and the thyristor (or SCR—silicon-controlled rectifier) were successfully developed in the 1960s. Since then, all power AC/DC rectifiers have been constructed using power silicon diodes and thyristors.

Using a power silicon diode, we can construct uncontrolled diode rectifiers. Using a power thyristor, we can construct controlled SCR rectifiers since the thyristor is usually triggered at firing angle α, which is variable. If the firing angle $\alpha = 0$, the characteristics of the controlled SCR rectifier will return to those of the uncontrolled diode rectifier. Research on the characteristics of the uncontrolled diode rectifier enables designers to get an idea of the characteristics of the controlled SCR rectifier.

A single-phase half-wave diode rectifier is shown in Figure 1.12. The load can be a resistive load, inductive load, capacitive load, or back electromotive force (emf) load. The diode can be conducting when current flows from the anode to the cathode, and the corresponding voltage applied across the diode is defined as positive. However, the diode is blocked when the voltage applied across the diode is negative, and no current flows through it. Therefore, the single-phase half-wave diode rectifier supplying different load has different output voltage waveform.

There are three important aims for this book:

- Clearing up the historic problems.
- Introducing updated circuits.
- Investigating PFC methods.

1.2.1 Historic Problems

Rectifier circuits are easily understood. The input power supply can be single-phase, three-phase, and multiphase sine-wave voltages. Usually, the more phases that an input power supplies to a circuit, the simpler the circuit operation. The most difficult analysis occurs in

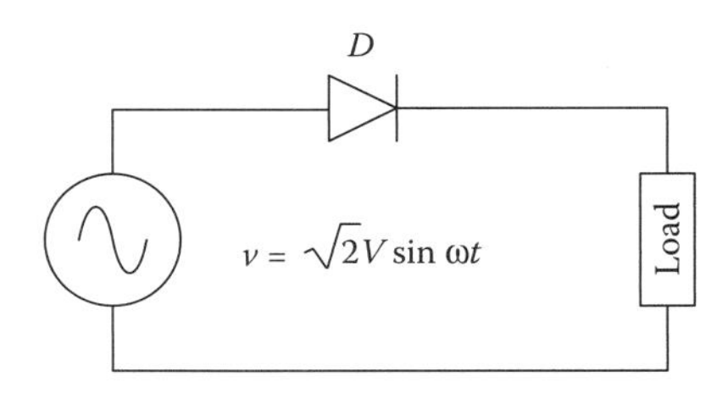

FIGURE 1.12 Single-phase half-wave diode rectifier.

the simplest circuit. Although a single-phase diode rectifier circuit is the simplest circuit, analysis of it has not been discussed in any great detail. Infact, the results presented in many recently published papers and books have given the wrong idea.

1.2.2 Updated Circuits

Many updated circuits and control methods have been developed in the last ten years. However, most of these updated circuits and control methods have yet to appear in dedicated books.

1.2.3 Power Factor Correction Methods

Power factor correction (PFC) methods have attracted the most attention in recent years. Many papers on PFC have been published but, as above, there is a distinct lack of dedicated textbooks on this subject.

1.3 DC/DC Converters

DC/DC conversion technology [5] is a vast subject area. It developed very fast and achieved much. There are believed to be more than 500 existing topologies of DC/DC converters according to current statistics. DC/DC converters have been widely used in industrial applications such as DC motor drives, communication equipment, mobile phones, and digital cameras. Many new topologies have been developed in the recent decade. They will be systematically introduced in this book.

Mathematical modeling is the historic problem accompanying the development of DC/DC conversion technology. From the 1940s onward, many scholars conducted research in this area and offered various mathematical modelings and control methods. We will discuss these problems in detail.

Most DC/DC converters have at least one pump circuit. For example, the buck–boost converter shown in Figure 1.13 has the pump circuit *S*–*L*. When the switch *S* is on, the inductor *L* absorbs energy from the source V_1. When the switch *S* is off, the inductor *L* releases the stored energy to the load and to charge the capacitor *C*.

From the example, we recognize that all energy obtained by the load must be a part of the energy stored in the inductor *L*. Theoretically, the energy transferred to the load looks no limit. In any particular operation, the energy rate cannot be very high. Consequently, power losses will increase sharply and the power transfer efficiency will largely decrease.

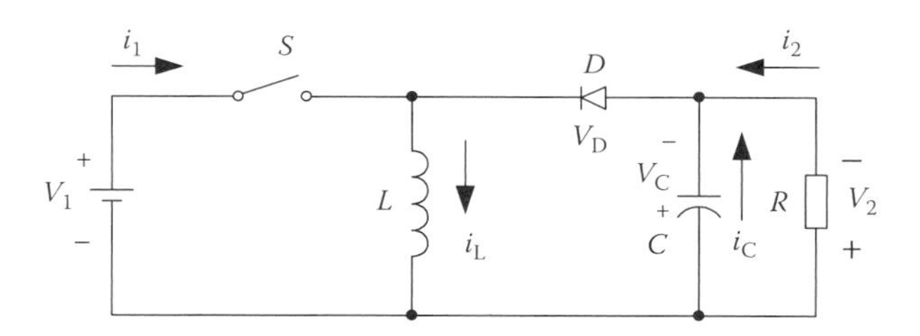

FIGURE 1.13 Buck–boost converter.

The following three important points will be emphasized in this book:

- The introduction of updated circuits.
- The introduction of new concepts and mathematical modeling.
- Checking the power rates.

1.3.1 Updated Converter

The voltage-lift (VL) conversion technique is widely used in electronic circuit design. Using this technique opened a the flood gates for designing DC/DC converters with in new topologies being developed in the last decade.

Furthermore, the super-lift (SL) technique and the ultralift (UL) technique have also been created. Both techniques facilitate on increase in the voltage transfer gains of DC/DC converters with the SL technique being the most outstanding with regard to the DC/DC conversion technology.

1.3.2 New Concepts and Mathematical Modeling

DC/DC converters are an element in an energy control system. In order to obtain satisfactory performance of the energy control system, it is necessary to know the mathematical modeling of the DC/DC converter used. Traditionally, the modeling of power DC/DC converters was derived from the impedance voltage-division method. The idea is that the inductor impedance is sL and the inductor impedance is $1/sC$, where s is the Laplace operator. The output voltage is the voltage divided by the impedance calculation. Actually, it successfully solves the problem of fundamental DC/DC converters. The transfer function of a DC/DC converter has an order number equal to the number of energy-storage elements. A DC/DC converter with two inductors and two capacitors has a fourth-order transfer function. Even more, a DC/DC converter with four inductors and four capacitors must have an eighth-order transfer function. It is hard to believe that it can be used for industrial applications.

1.3.3 Power Rate Checking

How can a large power be used in an energy system with DC/DC converters? This represents a very sensitive problem for industrial applications. DC/DC converters are quite different from transformers and AC/DC rectifiers. Their output power is limited by the pump circuit power rate.

The power rate of an inductor pump circuit depends on the inductance, applied current and current ripple, and switching frequency. The energy transferred by the inductor pump circuit in a cycle $T = 1/f$ is

$$\Delta E = \frac{L}{2}(I_{\text{max}}^2 - I_{\text{min}}^2). \tag{1.80}$$

The maximum power that can be transferred is

$$P_{\text{max}} = f\Delta E = \frac{fL}{2}(I_{\text{max}}^2 - I_{\text{min}}^2). \tag{1.81}$$

Therefore, when designing an energy system with a DC/DC converter, we have to estimate the power rate.

1.4 DC/AC Inverters

DC/AC inverters [1,2] were not widely used in industrial applications before the 1960s because of their complexity and cost. However, they were used in most fractional horsepower AC motor drives in the 1970s because AC motors have the advantage of lower cost when compared to DC motors, were smaller in size, and were maintenance free. In the 1980s, because of semiconductor development, more effective devices such as IGBT and MOSFET were produced, and DC/AC inverters started to be widely applied in industrial applications. To date, DC/AC conversion techniques can be sorted into two categories: pulse-width modulation (PWM) and multilevel modulation (MLM). Each category has many circuits to implement the modulation. Using PWM, we can design various inverters such as voltage-source inverters (VSI), current-source inverters (CSI), impedance-source inverters (ZSI), and multistage PWM inverters. A single-phase half-wave PWM is shown in Figure 1.14.

The PWM method is suitable for DC/AC conversion since the input voltage is usually a constant DC voltage (DC link). The pulse-phase modulation (PPM) method is also possible, but is less convenient. The pulse-amplitude modulation (PAM) method is not suitable for DC/AC conversion since the input voltage is usually a constant DC voltage. PWM operation has all the pulses' leading edge starting from the beginning of the pulse period, and their trailer edge is adjustable. The PWM method is a fundamental technique for many types of PWM DC/AC inverters such as VSI, CSI, ZSI, and multistage PWM inverters.

Another group of DC/AC inverters are the multilevel inverters (MLI). These inverters were invented in the late 1970s. The early MLIs are constructed using diode-clamped and capacitor-clamped circuits. Later, various MLIs were developed.

Three important points will be examined in this book:

- Sorting the existing inverters.
- Introducing updated circuits.
- Investigating soft-switching methods.

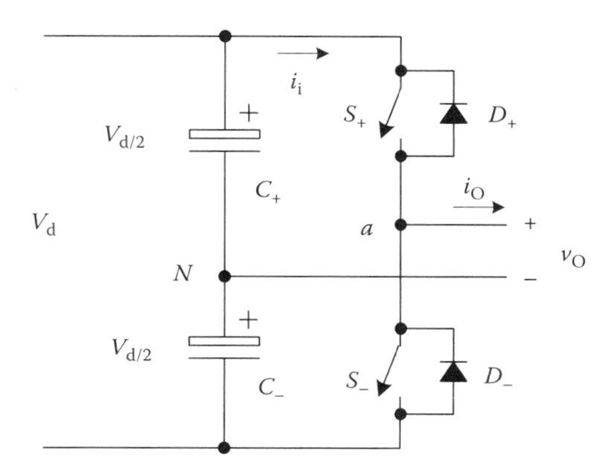

FIGURE 1.14 Single-phase half-wave PWM VSI.

1.4.1 Sorting Existing Inverters

Since a large number of inverters exist, we have to sort them systematically. Some circuits have not been defined with an exact title and thus mislead readers' understanding as to the particular function.

1.4.2 Updated Circuits

Many updated DC/AC inverters have been developed in the last decade but have not yet been introduced into textbooks. This book seeks to redress that point and show students the new methods.

1.4.3 Soft-Switching Methods

The soft-switching technique has been widely used in switching circuits for a long time now. It effectively reduce the power losses of equipment and increases the power transfer efficiency. A few soft-switching technique methods will be introduced into this book.

1.5 AC/AC Converters

AC/AC converters [10] were not very widely used in industrial applications before the 1960s because of their complexity and cost. They were used in heating systems for temperature control and in light dimmers in cinemas, theaters, and nightclubs, or in bedroom night dimmers for light color and brightening control. The early AC/AC converters were designed by the voltage-regulation (VR) method. A typical single-phase VR AC/AC converter is shown in Figure 1.15.

VR AC/AC converters have been successfully used in heating and light-dimming systems. One disadvantage is that the output AC voltage of VR AC/AC converters is a heavily distorted waveform with a poor THD and PF. Other disadvantages are that the

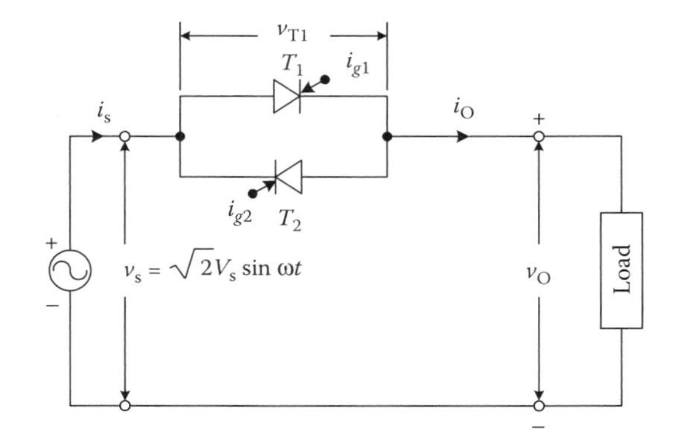

FIGURE 1.15 Single-phase VR AC/AC converter.

output voltage is constantly lower than the input voltage and the output frequency is not adjustable.

Cycloconverters and matrix converters can change the output frequency, but the output voltage is also constantly lower than the input voltage. Their THD and PF are also very poor.

DC-modulated (DM) AC/AC converters can easily give an output voltage higher than the input voltage, which will be discussed in this book. In addition, the DM method can successfully improve THD and PF.

1.6 AC/DC/AC and DC/AC/DC Converters

AC/DC/AC and DC/AC/DC converters are designed for special applications. In recent years, it has been realised that renewable energy sources and distributed generations (DG) need to be developed rapidly because fossil energy sources (coal, oil, gas, and so on) will soon be exhausted. Sources such as solar panels, photovoltaic cells, fuel cells, and wind turbines have unstable DC and/or AC output voltages. They are usually part of a microgrid. It is necessary to use special AC/DC/AC and DC/AC/DC converters to link these sources to the general buss inside the microgrid.

Wind turbines have single-phase or multiphase AC output voltages with variable amplitude and frequency since the wind speed varies constantly. As it is difficult to use these unstable AC voltages for any application, we need to use an AC/DC/AC converter to convert them to a suitable AC voltage (single-phase or multiphase) with stable amplitude and frequency.

Solar panels have DC output voltages with variable amplitude due to the variations of available sunlight. As it is difficult to use these unstable DC voltages for any application, we need to use a DC/AC/DC converter to convert them to a suitable DC voltage with stable amplitude and frequency.

Homework

1.1. A load Z with a resistance $R = 10\,\Omega$, an inductance $L = 10\,\text{mH}$, and a capacitance $C = 1000\,\mu\text{F}$ in series connection is supplied by a single-phase AC voltage with frequency $f = 60\,\text{Hz}$. Calculate the impedance Z and the phase angle ϕ.

1.2. A load Z with resistance $R = 10\,\Omega$ and inductance $L = 10\,\text{mH}$ in series connection is supplied by a single-phase AC voltage with frequency $f = 60\,\text{Hz}$. Calculate the impedance Z, the phase angle ϕ, and the time constant τ.

1.3. A load Z with resistance $R = 10\,\Omega$ and capacitance $C = 1000\,\mu\text{F}$ in series connection is supplied by a single-phase AC voltage with frequency $f = 60\,\text{Hz}$. Calculate the impedance Z, the phase angle ϕ, and the time constant τ.

1.4. Refer to Question 1.1. If the AC supply voltage is 240 V (rms) with $f = 60\,\text{Hz}$, calculate the circuit current, and the corresponding apparent power S, real power P, reactive power Q, and PF.

1.5. A five-level odd-waveform is shown in Figure 1.16.

The central symmetrical function $f(t)$ is in the period $-\pi$ to $+\pi$:

$$f(t) = \begin{cases} 2E, & \frac{3\pi}{8} \le \omega t < \frac{5\pi}{8}, \\ E, & \frac{\pi}{8} \le \omega t < \frac{3\pi}{8}, \ \frac{5\pi}{8} \le \omega t < \frac{7\pi}{8}, \\ 0, & \text{other}, \\ -E, & -\frac{7\pi}{8} \le \omega t < -\frac{5\pi}{8}, \ -\frac{3\pi}{8} \le \omega t < -\frac{\pi}{8}, \\ -2E, & -\frac{5\pi}{8} \le \omega t < -\frac{3\pi}{8}. \end{cases}$$

Consider the harmonics up to the seventh order and calculate the HFs, THD, and WTHD.

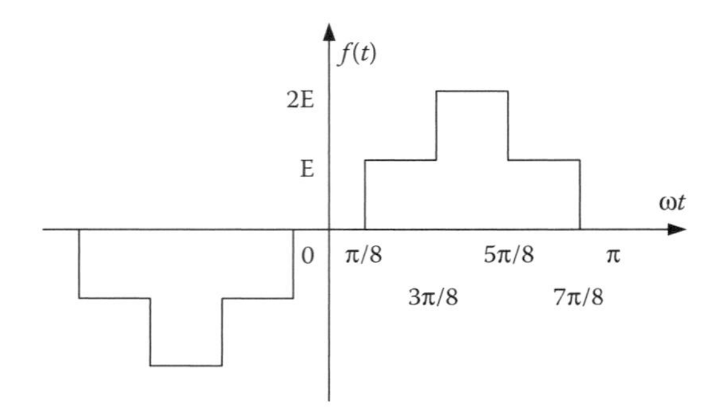

FIGURE 1.16 Five-level odd waveform.

References

1. Luo, F. L., Ye, H., and Rashid, M. H. 2005. *Digital Power Electronics and Applications.* Boston: Academic Press.
2. Mohan, N., Undeland, T. M., and Robbins, W. P. 2003. *Power Electronics: Converters, Applications and Design* (3rd edition). New York: Wiley.
3. Rashid, M. H. 2004. *Power Electronics: Circuits, Devices and Applications* (3rd edition). Englewood Cliffs, NJ: Prentice-Hall.
4. Luo, F. L. and Ye, H. 2007. DC-modulated single-stage power factor correction AC/AC converters. *Proceedings of ICIEA 2007* Harbin, China, pp. 1477–1483.
5. Luo, F. L. and Ye, H. 2004. *Advanced DC/DC Converters.* Boca Raton: CRC Press.
6. Luo, F. L. and Ye, H. 2005. Energy factor and mathematical modeling for power DC/DC converters. *IEE EPA Proceedings*, 152(2), 191–198.
7. Luo, F. L. and Ye, H. 2007. Small signal analysis of energy factor and mathematical modeling for power DC/DC converters. *IEEE Transactions on Power Electronics*, 22(1), 69–79.
8. Luo, F. L. and Ye, H. 2006. *Synchronous and Resonant DC/DC Conversion Technology, Energy Factor and Mathematical Modeling.* Boca Raton: Taylor & Francis.
9. Carlson A. B. 2000. *Circuits.* Pacific Grove, CA: Brooks/Cole.
10. Rashid, M. H. 2007. *Power Electronics Handbook* (2nd edition). Boston: Academic Press.

2

Uncontrolled AC/DC Converters

Most electronic equipment and circuits require DC sources for their operation. Dry cells and rechargeable batteries can be used for these applications but they only offer limited power and unstable voltage. The most useful DC sources are AC/DC converters [1]. The technology of AC/DC conversion is a wide subject area covering research investigation and industrial applications. AC/DC converters (usually called rectifiers) convert an AC power supply source voltage to a DC voltage load. Uncontrolled AC/DC converters usually consist of diode circuits. They can be sorted into the following groups [2]:

- Single-phase half-wave rectifiers
- Single-phase full-wave rectifiers
- Three-phase rectifiers
- Multipulse rectifiers
- PFC rectifiers
- Pulse-width-modulated boost-type rectifiers.

Since some of the theoretical analysis and calculation results in this book are different from that of some published papers and books, the associated underlying historical problems will be brought to the attention of the reader by way of ADVICE sections.

2.1 Introduction

The input of a diode rectifier is AC voltage, which can be either a single- or three-phase voltage, and is usually a pure sinusoidal wave. A single-phase input voltage $v(t)$ can be expressed as

$$v(t) = \sqrt{2}V \sin \omega t = V_{\mathrm{m}} \sin \omega t, \tag{2.1}$$

where $v(t)$ is the instantaneous input voltage, V the rms value, V_{m} the amplitude, and ω the angular frequency where $\omega = 2\pi f$ (f is the supply frequency). Usually, the input current $i(t)$ is a pure sinusoidal wave with a phase shift angle ϕ if it is not distorted, and is expressed as

$$i(t) = \sqrt{2}I \sin(\omega t - \phi) = I_{\mathrm{m}} \sin(\omega t - \phi), \tag{2.2}$$

where $i(t)$ is the instantaneous input current, I the rms value, I_{m} the amplitude, and ϕ the phase shift angle. In this case, we define the PF as

$$\mathrm{PF} = \cos \phi. \tag{2.3}$$

If the input current is distorted, it consists of harmonics. Its fundamental harmonic can be expressed as in Equation 1.17 and the DPF is defined in Equation 1.18. PF is measured as shown in Equation 1.19 and the THD is defined as in Equation 1.20 [3,4].

When a pure DC voltage has no ripple, it is called a ripple-free DC voltage. Otherwise, DC voltage is distorted and its rms value is V_{d-rms}. For a distorted DC voltage, its rms value V_{d-rms} is constantly higher than its average value V_d. The RF is defined in Equation 1.24 and the FF is defined in Equation 1.25.

2.2 Single-Phase Half-Wave Converters

A single-phase half-wave diode rectifier consists of a single-phase AC input voltage and one diode [5]. While it is the simplest rectifier, its analysis is the most complex. This rectifier can supply various loads as described in the following subsections.

2.2.1 *R* Load

A single-phase half-wave diode rectifier with R load is shown in Figure 2.1a, and the input voltage, input current, and output voltage waveforms are shown in Figures 2.1b–d, respectively. The output voltage is similar to the input voltage in the positive half-cycle and zero in the negative half-cycle.

The output average voltage is

$$V_d = \frac{1}{2\pi}\int_0^{\pi} \sqrt{2}V \sin\omega t \, d(\omega t) = \frac{2\sqrt{2}}{2\pi}V = 0.45\,V. \tag{2.4}$$

The output rms voltage is

$$V_{d-rms} = \sqrt{\frac{1}{2\pi}\int_0^{\pi}(\sqrt{2}V\sin\omega t)^2 \, d(\omega t)} = V\sqrt{\frac{1}{\pi}\int_0^{\pi}(\sin\alpha)^2 \, d\alpha} = \frac{1}{\sqrt{2}}V = 0.707\,V. \tag{2.5}$$

The output average and rms currents are

$$I_d = \frac{V_d}{R} = \frac{\sqrt{2}}{\pi}\frac{V}{R} = 0.45\frac{V}{R}, \tag{2.6}$$

$$I_{d-rms} = \frac{V_{d-rms}}{R} = \frac{1}{\sqrt{2}}\frac{V}{R} = 0.707\frac{V}{R}. \tag{2.7}$$

The FF, RF, and PF of the output voltage are

$$\mathrm{FF} = \frac{V_{d-rms}}{V_d} = \frac{1/\sqrt{2}}{\sqrt{2}/\pi} = \frac{\pi}{2} = 1.57, \tag{2.8}$$

$$\mathrm{RF} = \sqrt{\mathrm{FF}^2 - 1} = \sqrt{\left(\frac{\pi}{2}\right)^2 - 1} = 1.21, \tag{2.9}$$

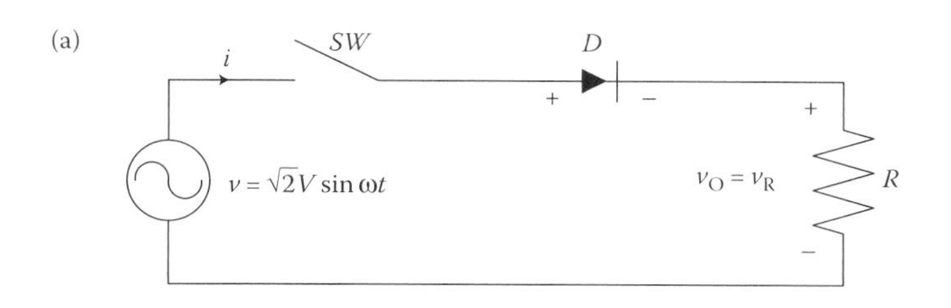

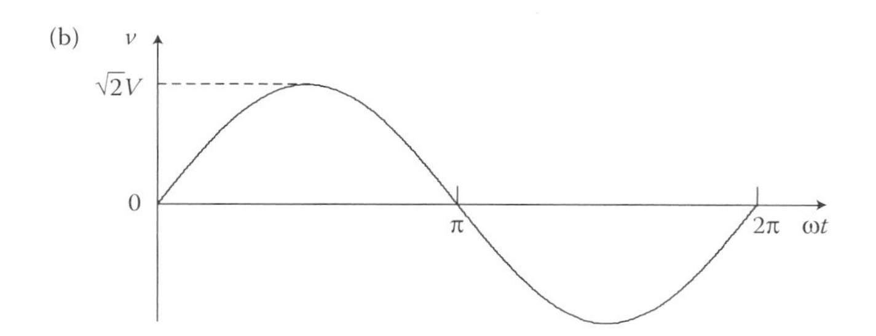

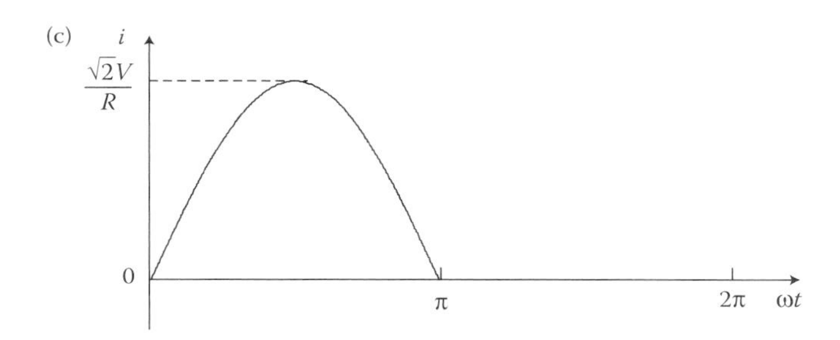

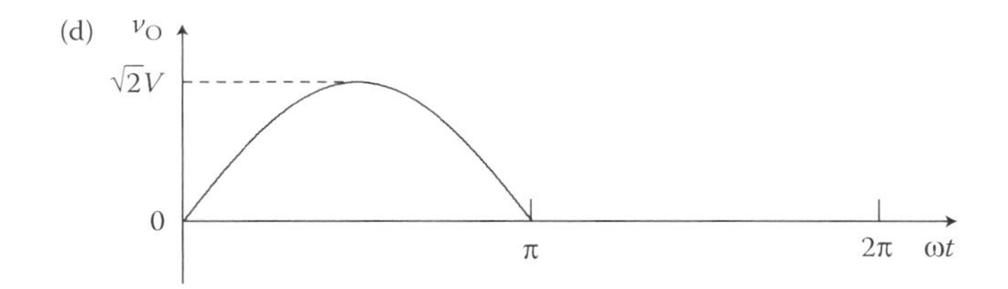

FIGURE 2.1 Single-phase half-wave diode rectifier with R load: (a) circuit, (b) input voltage, (c) input current, and (d) output voltage.

and

$$\text{PF} = \frac{1}{\sqrt{2}} = 0.707. \tag{2.10}$$

2.2.2 *R–L* Load

A single-phase half-wave diode rectifier with R–L load is shown in Figure 2.2a, while various circuit waveforms are shown in Figures 2.2b–d.

It can be seen that the load current flows not only in the positive half-cycle of the supply voltage, but also in a portion of the negative half-cycle of the supply voltage [6]. The load inductor SE maintains the load current, and the inductor's terminal voltage changes so as to overcome the negative supply and keep the diode forward biased and conducting. Area A is equal to area B in Figure 2.2c. During diode conduction, the following equation is available:

$$L\frac{\mathrm{d}i}{\mathrm{d}t} + Ri = \sqrt{2}V \sin \omega t \tag{2.11}$$

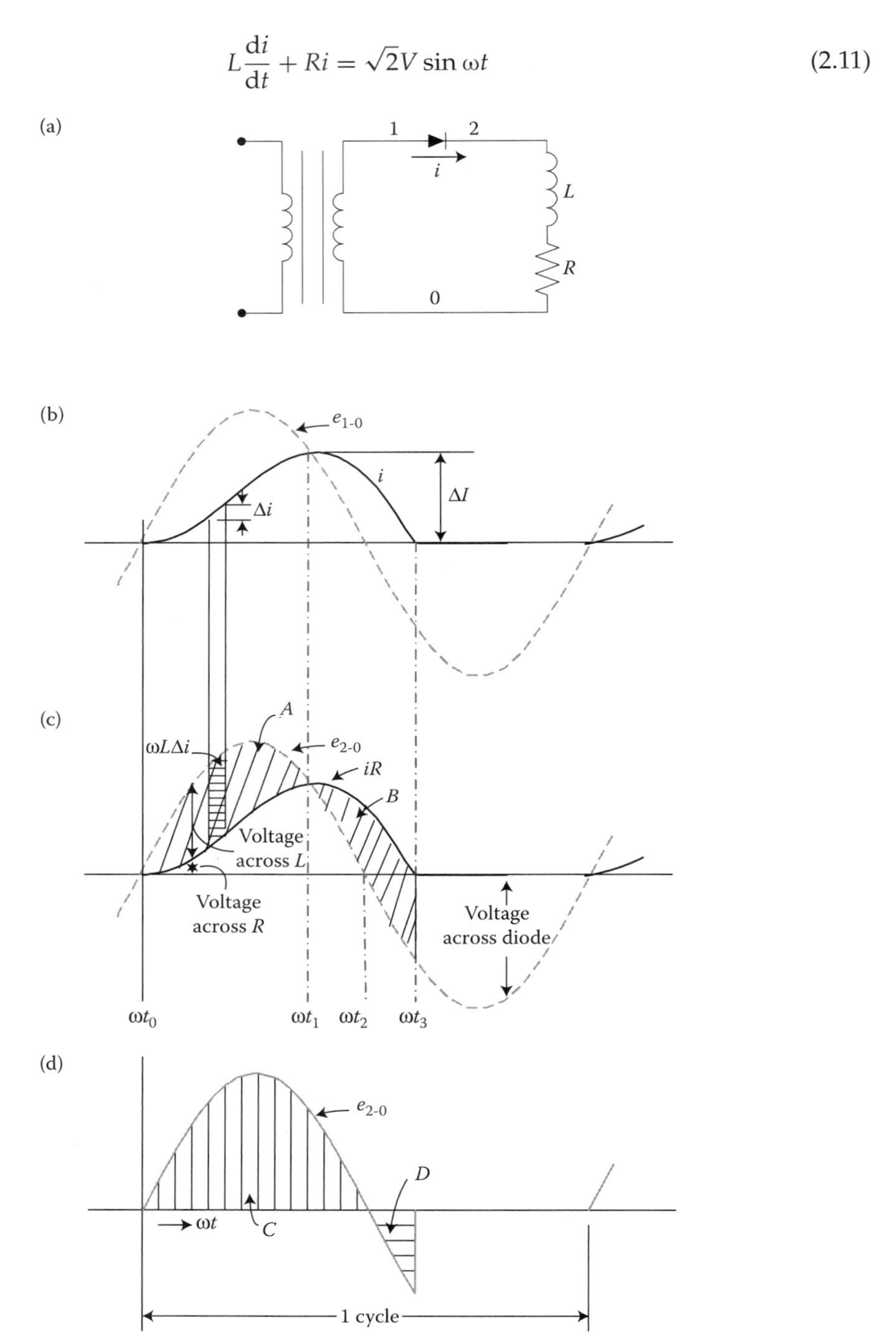

FIGURE 2.2 Half-wave rectifier with R–L load: (a) circuit, (b) input voltage and current, (c) analysis of input voltage and current, (d) output voltage.

or

$$\frac{\mathrm{d}i}{\mathrm{d}t} + \frac{R}{L}i = \frac{\sqrt{2}V}{L}\sin\omega t.$$

This is a nonnormalized differential equation. The solution has two parts. The forced component is determined by

$$i_{\mathrm{F}} = \mathrm{e}^{-(R/L)t}\int\left(\frac{\sqrt{2}V}{L}\sin\omega t\right)\mathrm{e}^{(R/L)t}\,\mathrm{d}t. \tag{2.12}$$

If the circuit is blocked during the negative half-cycle, then by sinusoidal steady-state circuit analysis the forced component of the current is

$$i_{\mathrm{F}} = \frac{\sqrt{2}V\sin(\omega t - \phi)}{\sqrt{R^2 + (\omega L)^2}}, \tag{2.13}$$

where

$$\phi = \tan^{-1}\left(\frac{\omega L}{R}\right). \tag{2.14}$$

The natural response of such a circuit is given by

$$i_{\mathrm{N}} = A\mathrm{e}^{-(R/L)t} = A\mathrm{e}^{-(t/\tau)} \quad \text{with} \quad \tau = \frac{L}{R}. \tag{2.15}$$

Thus,

$$i = i_{\mathrm{F}} + i_{\mathrm{N}} = \frac{\sqrt{2}V}{Z}\sin(\omega t - \phi) + A\mathrm{e}^{-(R/L)t}, \tag{2.16}$$

where

$$Z = \sqrt{R^2 + (\omega L)^2}. \tag{2.17}$$

The constant A is determined by substitution in Equation 2.16 of the initial condition $i = 0$ at $t = 0$, giving

$$A = \frac{\sqrt{2}V}{Z}\sin\phi.$$

Thus,

$$i = \frac{\sqrt{2}V}{Z}\left[\sin(\omega t - \phi) + \mathrm{e}^{-(R/L)t}\sin\phi\right]. \tag{2.18}$$

We define the *extinction angle* β where the current becomes zero. Therefore,

$$i = 0, \quad \beta \leq \omega t < 2\pi. \tag{2.19}$$

The current extinction angle β is determined by the load impedance and can be solved from Equation 2.18 when $i = 0$ and $\omega t = \beta$,

$$\sin(\beta - \phi) = -\mathrm{e}^{-(R\beta/\omega L)}\sin\phi. \tag{2.20}$$

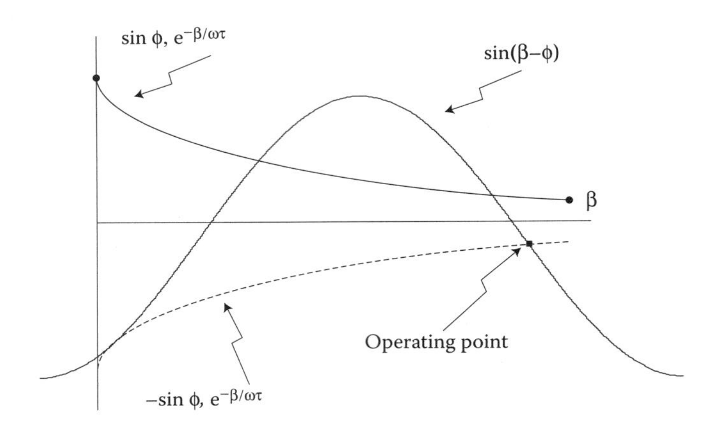

FIGURE 2.3 Determination of extinction angle β.

This is a transcendental equation with an unknown value of β (see Figure 2.3). The term $\sin(\beta - \phi)$ is a sinusoidal function and the term $e^{-(R\beta/\omega L)} \sin\phi$ is an exponentially decaying function; the operating point of β is the intersection of $\sin(\beta - \phi)$ and those terms. The value of β can be obtained by using MATLAB® simulation and can be solved by numerical techniques such as iterative methods.

2.2.2.1 *Graphical Method*

Using MATLAB® to solve Equation 2.20, the resultant values of β for the corresponding values of ϕ are plotted as a graph shown in Figure 2.4. It can be observed that the graph commences at 180° (or π radians) on the β (x) axis and, for small values of ϕ, the characteristic is linear,

$$\beta \approx \pi + \phi$$

However, for large values of ϕ, the corresponding value of β tends to be

$$\beta > \pi + \phi$$

with a terminal value of 2π (or 360°) for purely inductive load.

ADVICE

If $L > 0$, $\beta > \pi + \phi$. Using the graph in Figure 2.4, a highly accurate result cannot be obtained. (*Historic problem*: $\beta = \pi + \phi$.)

2.2.2.2 *Iterative Method 1*

The operating point setting: If $\beta \geq \pi + \phi$. Let starting point $\beta = \pi + \phi$.

L1: Calculate $x = \sin(\beta - \phi)$.
Calculate $y = -e^{-(R\beta/\omega L)} \sin\phi$.
If $x = y$, then β is the correct value, END.

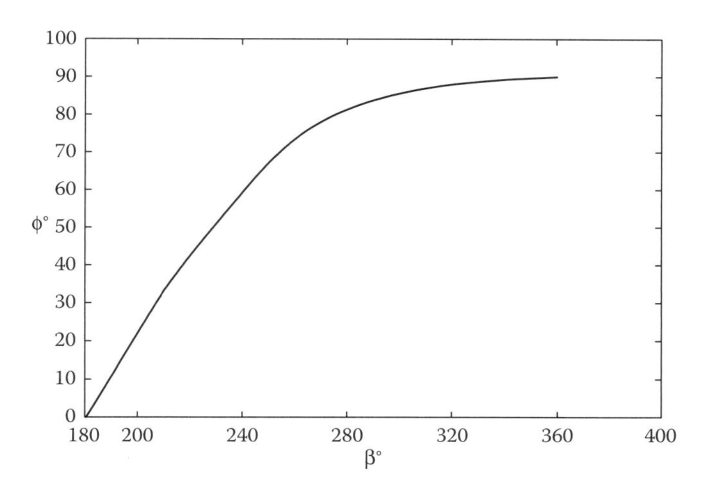

FIGURE 2.4 β versus φ.

If $|x| < y$, then increment β and return to L1.
If $|x| > y$, then decrement β and return to L1.

Example 2.1

A single-phase half-wave diode rectifier operates from a supply of $V = 240\,\text{V}$, 50 Hz to a load of $R = 10\,\Omega$ and $L = 0.1$ H. Determine the extinction angle β using iterative method 1.

Solution

From Equation 2.20, $\phi = \tan^{-1}(\omega L/R) = 72.34°$.
Then, letting $\beta_1 = \pi + \phi = 252.34°$:

Step	β (deg)	$x = \sin(\beta - \phi)$	$y = e^{-(R\beta/\omega L)} \sin\phi$	$\lvert x\rvert : y$
1	252.34	0	0.2345	$<$
2↑	260	−0.1332	0.2248	$<$
3↑	270	−0.3033	0.2126	$>$
4↓	265	−0.2191	0.2186	$\approx$
5↓	264	−0.2020	0.2198	$<$
6↓	266	−0.2360	0.2174	$>$

Therefore, to satisfy Equation 2.20, the best value is $\beta = 265°$.

2.2.2.3 *Iterative Method 2*

Let $\beta_n = \pi + \phi$.

L1: Calculate $x = \sin(\beta - \phi)$.
Calculate $y = e^{-(R\beta/\omega L)} \sin \phi$.
Let $x = y$ and $\beta_{n+1} = (\sin^{-1} y) + \pi + \phi$.
If $\beta_{n+1} = \beta_n$ then END. Else
Choose $\beta_n = \beta_{n+1}$ and return to L1.

The reader is referred to Homework Question 2.2.
The average value of the rectified current can be obtained by

$$
\begin{aligned}
v_{\mathrm{d}} &= v_{\mathrm{R}} + v_{\mathrm{L}} = \sqrt{2}V \sin \omega t, \\
\int_0^\beta v_{\mathrm{R}}\, \mathrm{d}(\omega t) + \int_0^\beta v_{\mathrm{L}}\, \mathrm{d}(\omega t) &= \int_0^\beta \sqrt{2}V \sin \omega t\, \mathrm{d}(\omega t), \\
R\int_0^\beta i(t)\, \mathrm{d}(\omega t) &= \sqrt{2}V(1 - \cos\beta), \\
I_{\mathrm{d}} = \frac{1}{2\pi}\int_0^\beta i(t)\, \mathrm{d}(\omega t) &= \frac{\sqrt{2}V}{2\pi R}(1 - \cos\beta).
\end{aligned}
\tag{2.21}
$$

The average output voltage is given by

$$
V_{\mathrm{d}} = \frac{\sqrt{2}V}{2\pi}(1 - \cos\beta). \tag{2.22}
$$

The output rms voltage is given by

$$
\begin{aligned}
V_{\mathrm{d-rms}} &= \sqrt{\frac{1}{2\pi}\int_0^\beta (\sqrt{2}V \sin \omega t)^2\, \mathrm{d}(\omega t)} = V\sqrt{\frac{1}{\pi}\int_0^\beta (\sin\alpha)^2\, \mathrm{d}\alpha} \\
&= V\sqrt{\frac{1}{\pi}\int_0^\beta \left(\frac{1 - \cos 2\alpha}{2}\right) \mathrm{d}\alpha} = V\sqrt{\frac{1}{\pi}\left(\frac{\beta}{2} - \frac{\sin 2\beta}{4}\right)}.
\end{aligned}
\tag{2.23}
$$

The FF and RF of the output voltage are

$$
\mathrm{FF} = \frac{V_{\mathrm{d-rms}}}{V_{\mathrm{d}}} = \frac{\sqrt{(1/\pi)[(\beta/2) - (\sin 2\beta/4)]}}{(\sqrt{2}/2\pi)(1 - \cos\beta)} = \sqrt{\frac{\pi}{2}}\frac{\sqrt{2\beta - \sin 2\beta}}{1 - \cos\beta}, \tag{2.24}
$$

$$
\mathrm{RF} = \sqrt{\mathrm{FF}^2 - 1} = \sqrt{\frac{\pi}{2}\frac{2\beta - \sin 2\beta}{(1 - \cos\beta)^2} - 1}. \tag{2.25}
$$

2.2.3 *R–L* Circuit with Freewheeling Diode

The circuit in Figure 2.2a, which has an *R–L* load, is characterized by discontinuous and high ripple current. Continuous load current can result when a diode is added across the load as shown in Figure 2.5a.

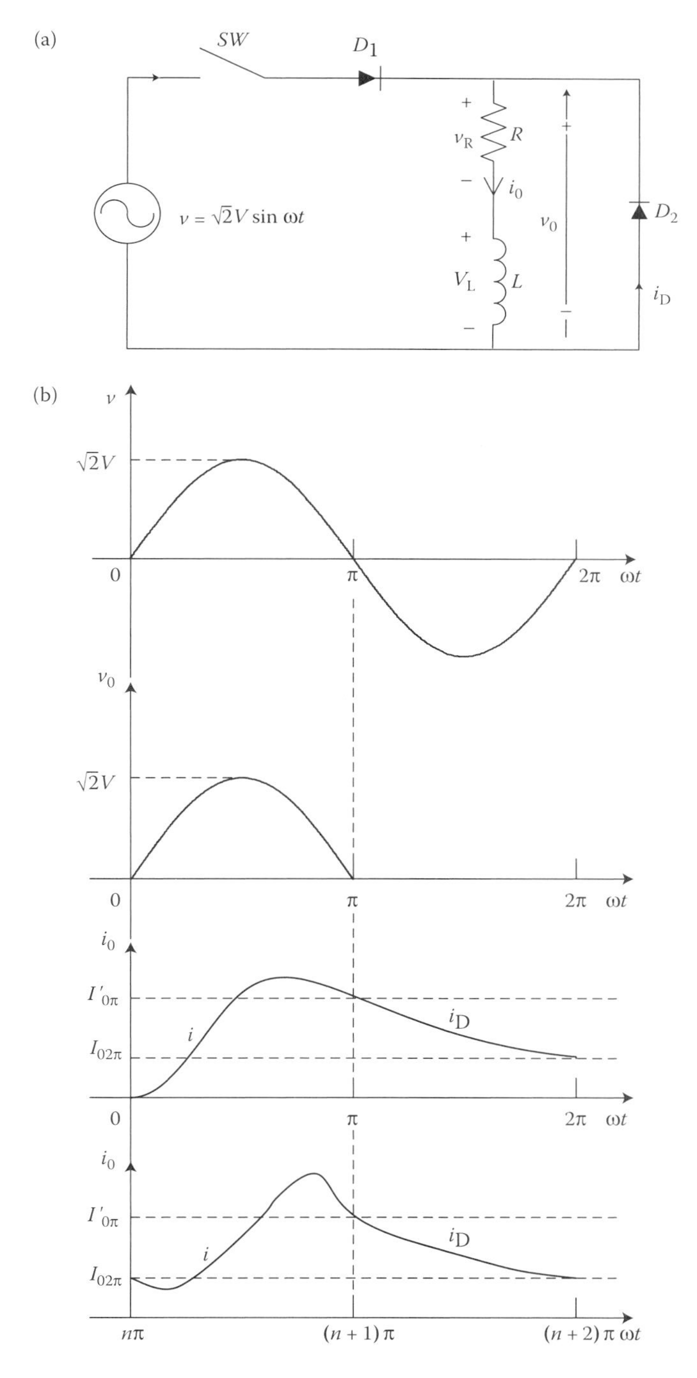

FIGURE 2.5 Half-wave rectifier with *R–L* load plus freewheeling diode. (a) Circuit and (b) waveforms.

The diode prevents the voltage across the load from reversing during the negative half-cycle of the supply voltage. When diode D_1 ceases to conduct at zero volts, diode D_2 provides an alternative freewheeling path as indicated by the waveforms in Figure 2.5b.

After a large number of supply cycles, steady-state load current conditions are established, and the load current is given by

$$i_0 = \frac{\sqrt{2}V}{Z}\sin(\omega t - \phi) + Ae^{-(R/L)t}. \tag{2.26}$$

Also,

$$i_0\,|_{t=0} = I_0\,|_{t=2\pi}. \tag{2.27}$$

Substitution of the initial conditions of Equation 2.27 into Equation 2.26 yields

$$i_0 = \frac{\sqrt{2}V}{Z}\sin(\omega t - \phi) + \left(I_{0-2\pi} + \frac{\sqrt{2}V}{Z}\sin\phi\right)e^{-(R/L)t}. \tag{2.28}$$

At $\omega t = \pi$, diode D_2 begins to conduct, the input current i falls instantaneously to zero, and from Equation 2.28,

$$I_{0-\pi} = i_0\,\big|_{\,t=\pi/\omega} = \frac{\sqrt{2}V}{Z}\sin(\pi - \phi) + \left(I_{0-2\pi} + \frac{\sqrt{2}V}{Z}\sin\phi\right)e^{-(\pi R/\omega L)}. \tag{2.29}$$

During the succeeding half-cycle, v_0 is zero. The SE in the inductor is dissipated by current i_D flowing in the $R-L-D_2$ mesh. Thus

$$i_0 = i_D = I_{0-\pi}e^{-(R/L)(t-\pi/\omega)} \tag{2.30}$$

at $\omega t = 2\pi$. Therefore, v, and hence v_0, becomes positive.

$$i_0\,\big|_{t=2\pi/\omega} = I_{0-\pi}e^{-(R/L)(t-\pi/\omega)} = I_0\,|_{\omega t=2\pi}\,. \tag{2.31}$$

Thus, from Equations 2.29 and 2.31,

$$\frac{\sqrt{2}V}{Z}\sin\phi + \left(I_{0-2\pi} + \frac{\sqrt{2}V}{Z}\sin\phi\right)e^{-(\pi R/\omega L)} = I_{0-2\pi}e^{\pi R/\omega L} \tag{2.32}$$

so that

$$I_{0-2\pi} = \frac{(\sqrt{2}V/Z)\sin\phi(1 + e^{-(\pi R/\omega L)})}{e^{\pi R/\omega L} - e^{-(\pi R/\omega L)}}. \tag{2.33}$$

2.2.4 An *R–L* Load Circuit with a Back Emf

A single-phase half-wave rectifier to supply an R–L load with a back emf V_c is shown in Figure 2.6a. The corresponding waveforms are shown in Figure 2.6b.

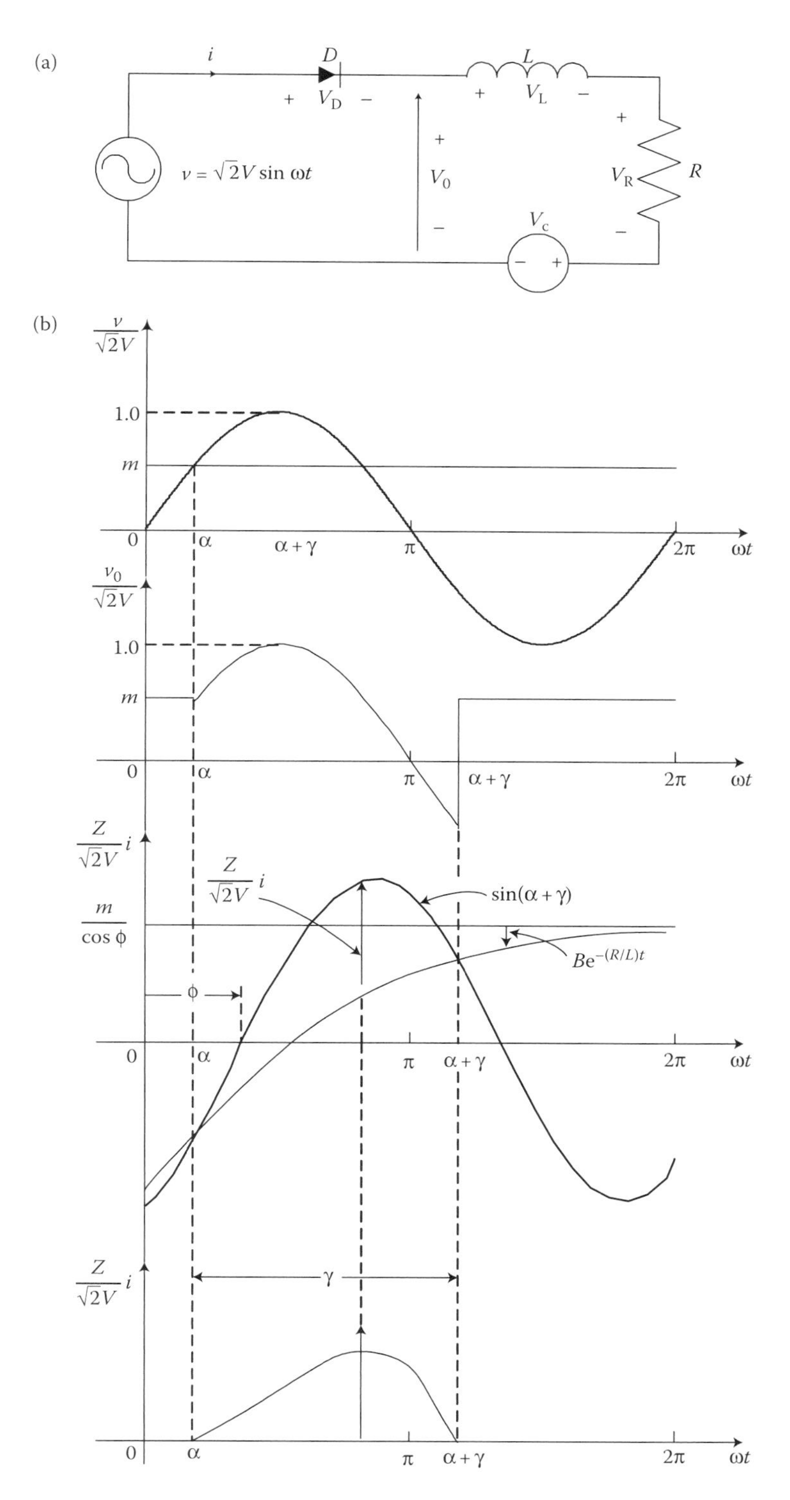

FIGURE 2.6 Half-wave rectifier with R–L load plus a back emf. (a) Circuit and (b) waveforms.

The effect of introducing a back electromotive force V_c into the load circuit of a half-wave rectifier is investigated in this section. This is the situation that would arise if such a circuit were employed to charge a battery or to excite a DC motor armature circuit.

The current component due to the AC source is

$$i_{SF} = \frac{\sqrt{2}V}{Z}\sin(\omega t - \phi). \tag{2.34}$$

The component due to the direct emf is

$$i_{cF} = \frac{-V_c}{R}. \tag{2.35}$$

The natural component is

$$i_N = Ae^{-(R/L)t}. \tag{2.36}$$

The total current in the circuit is the sum of these three components:

$$i = \frac{\sqrt{2}V}{Z}\sin(\omega t - \phi) - \frac{V_c}{R} + Ae^{-(R/L)t}, \quad \alpha < \omega t < \alpha + \gamma, \tag{2.37}$$

where α is the angle at which conduction begins and γ is the *conduction* angle. As may be seen from the voltage curve in Figure 2.6b,

$$\sin\alpha = \frac{V_c}{\sqrt{2}V} = m. \tag{2.38}$$

At $\omega t = \alpha$, $i = 0$ so that from Equation 2.37

$$A = \left[\frac{V_c}{R} - \frac{\sqrt{2}V}{Z}\sin(\alpha - \phi)\right]e^{\alpha R/\omega L}. \tag{2.39}$$

Also

$$R = Z\cos\phi. \tag{2.40}$$

Substituting Equations 2.38 through 2.40 into Equation 2.37 yields

$$\frac{Z}{\sqrt{2}V}i = \sin(\omega t - \phi) - \left[\frac{m}{\cos\phi} - Be^{-(R/L)t}\right], \quad \alpha < \omega t < \alpha + \gamma, \tag{2.41}$$

where

$$B = \left[\frac{m}{\cos\phi} - \sin(\alpha - \phi)\right]e^{\alpha R/\omega L}, \quad \omega t = \alpha. \tag{2.42}$$

The terms on the right-hand side of Equation 2.41 may be represented separately as shown in Figure 2.6b. At the end of the conduction period,

$$i = 0, \quad \omega t = \alpha + \gamma. \tag{2.43}$$

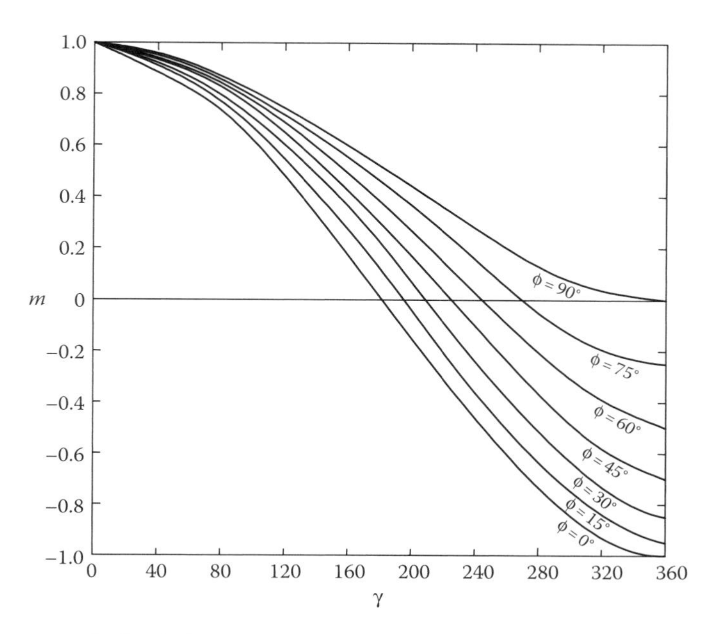

FIGURE 2.7 m versus γ referring to ϕ.

Substituting Equation 2.43 in Equation 2.41 yields

$$\frac{(m/\cos\phi) - \sin(\alpha + \gamma - \phi)}{(m/\cos\phi) - \sin(\alpha - \phi)} = e^{-\gamma/\tan\phi}. \tag{2.44}$$

We obtain

$$e^{-\gamma/\omega\tau} = \frac{(m/\cos\phi) - \sin(\eta + \gamma - \phi)}{(m/\cos\phi) - \sin(\eta - \phi)}. \tag{2.45}$$

Solve for conduction angle γ using suitable iterative techniques. For practicing design engineers, a quick reference graph of m–ϕ–γ is given in Figure 2.7.

Example 2.2

A single-phase half-wave diode rectifier operates from a supply of $V = 240\,\text{V}$, 50 Hz to a load of $R = 10\,\Omega$, $L = 0.1$ H, and an emf $V_C = 200\,\text{V}$. Determine the conduction angle γ and the total current $i(t)$.

SOLUTION

From Equation 2.20, $\phi = \tan^{-1}(\omega L/R) = 72.34°$.

Therefore,

$$\tau = \frac{L}{R} = 10\ \text{ms}, \quad Z = \sqrt{R^2 + (\omega L)^2} = \sqrt{100 + 986.96} = 32.969\,\Omega.$$

From Equation 2.38

$$m = \sin\alpha = \frac{200}{240\sqrt{2}} = 0.589.$$

Therefore, $\alpha = \sin^{-1} 0.589 = 36.1° = 0.63$ rad.
Checking the graph in Figure 2.7, we obtain $\gamma = 156°$.
From Equation 2.39

$$A = \left[\frac{V_c}{R} - \frac{\sqrt{2}V}{Z}\sin(\alpha - \phi)\right] e^{\alpha R/\omega L} = [20 - 10.295\sin(-36.24)]e^{0.2}$$
$$= 26.086 \times 1.2214 = 31.86.$$

Therefore, $i(t) = 10.295\sin(314.16t - 72.34°) - 20 + 31.86e^{-100t}$ A in $36.1° < \omega t < 192.1°$.

2.2.4.1 Negligible Load-Circuit Inductance

From Equation 2.37, if $L = 0$, we obtain

$$i = \frac{\sqrt{2}V}{R}\sin\omega t - \frac{V_c}{R} \tag{2.46}$$

or

$$\frac{R}{\sqrt{2}V} i = \sin\omega t - m. \tag{2.47}$$

The current $(R/\sqrt{2}V)i$ is shown in Figure 2.8, from which it may be seen that

$$\gamma = \pi - 2\alpha. \tag{2.48}$$

The average current is

$$\begin{aligned} I_0 &= \frac{1}{2\pi}\int_{\alpha}^{\pi-\alpha} \frac{\sqrt{2}V}{R}(\sin\omega t - m)\, d(\omega t) \\ &= \frac{\sqrt{2}V}{\pi R}[\cos\alpha - m(\pi/2 - \alpha)] = \frac{\sqrt{2}V}{\pi R}\left[\sqrt{1 - m^2} - m\cos^{-1} m\right]. \end{aligned} \tag{2.49}$$

2.2.5 Single-Phase Half-Wave Rectifier with a Capacitive Filter

The single-phase half-wave rectifier shown in Figure 2.9 has a parallel R–C load. The purpose of the capacitor is to reduce the variation in the output voltage, making it more like a pure DC voltage.

Assuming the rectifier works in steady-state, the capacitor is initially charged in a certain DC voltage and the circuit is energized at $\omega t = 0$; the diode becomes forward biased at the angle $\omega t = \alpha$ as the source becomes positive. As the source decreases after $\omega t = \pi/2$, the capacitor discharges from the discharging angle θ into the load resistor. From this point, the voltage of the source becomes less than the output voltage, reverse biasing the diode and isolating the load from the source. The output voltage is a decaying exponential with time constant RC while the diode is off.

(a)

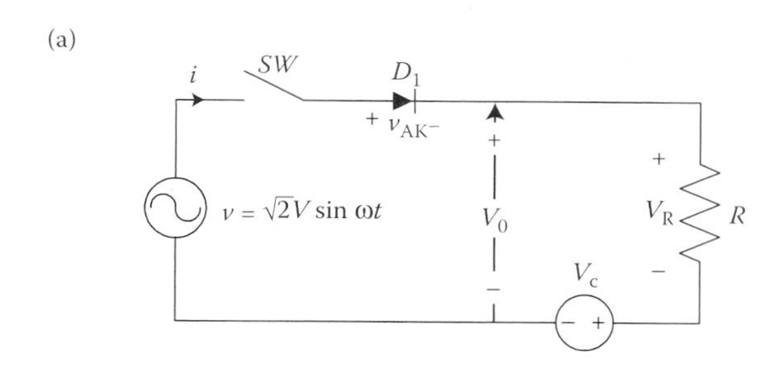

(b)

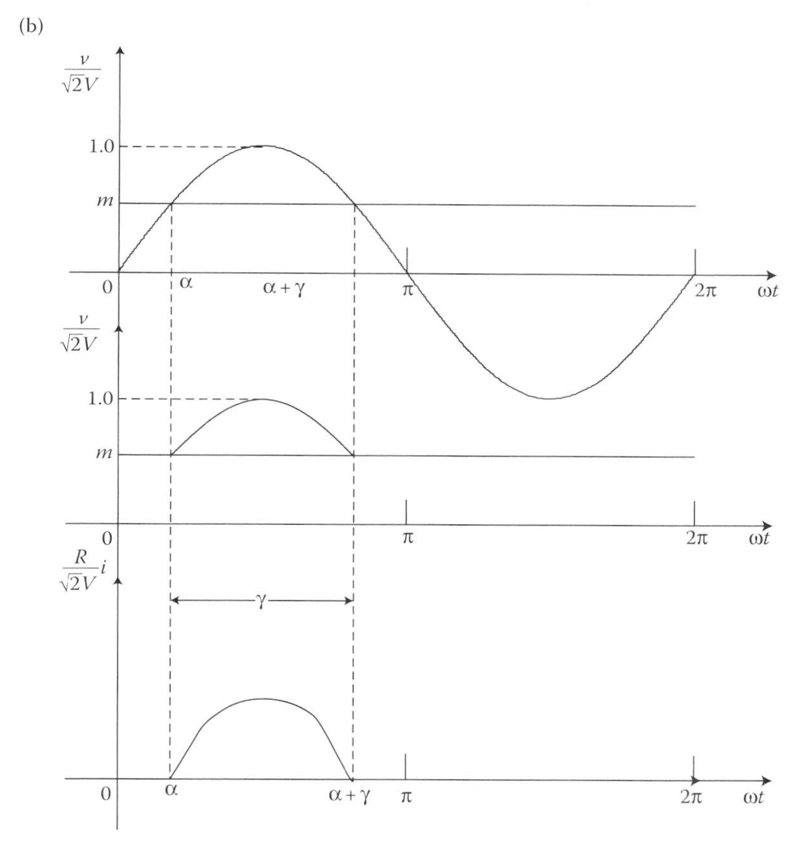

FIGURE 2.8 Half-wave rectifier with R load plus back emf. (a) Circuit and (b) waveforms.

The output voltage is described by

$$v_{\mathrm{d}}(\omega t)=\begin{cases}\sqrt{2}V\sin\omega t, & \text{diode on},\\ V_{\theta}\mathrm{e}^{-(\omega t-\theta)/\omega RC}, & \text{diode off},\end{cases} \tag{2.50}$$

where

$$V_{\theta}=\sqrt{2}V\sin\theta. \tag{2.51}$$

At $\omega t=\theta$, the slopes of the voltage functions are equal to

$$\sqrt{2}V\cos\theta=\frac{\sqrt{2}V\sin\theta}{-\omega RC}\mathrm{e}^{-(\theta-\theta)/\omega RC}.$$

(a)

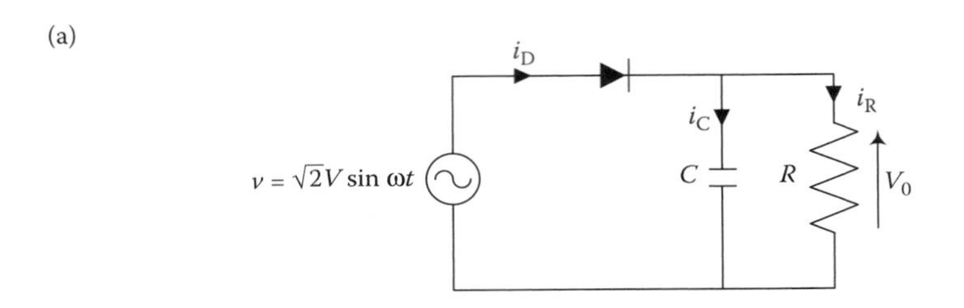

(b)

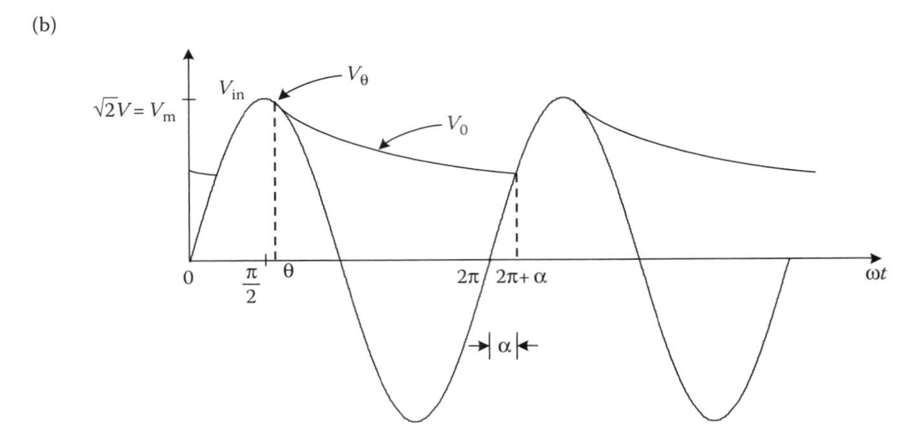

FIGURE 2.9 Half-wave rectifier with an R–C load: (a) circuit and (b) input and output voltage.

Hence

$$\frac{1}{\tan \theta} = \frac{-1}{\omega RC}. \tag{2.52}$$

Thus

$$\theta = \pi - \tan^{-1}(\omega RC).$$

ADVICE

The discharging angle θ must be $>\pi/2$. (*Historic problem*: $\theta = \pi/2$.)

The angle at which the diode turns on in the second period, $\omega t = 2\pi + \alpha$, is the point at which the sinusoidal source reaches the same value as the decaying exponential output.

$$\sqrt{2}V \sin(2\pi + \alpha) = (\sqrt{2}V \sin \theta)\mathrm{e}^{-(2\pi+\alpha-\theta)/\omega RC}$$

or

$$\sin \alpha - (\sin \theta)\mathrm{e}^{-(2\pi+\alpha-\theta)/\omega RC} = 0. \tag{2.53}$$

The preceding equation must be solved numerically.

Peak capacitor current occurs when the diode turns on at $\omega t = 2\pi + \alpha$,

$$i_{C-\text{peak}} = \omega C\sqrt{2}V \cos(2\pi + \alpha) = \omega C\sqrt{2}V \cos \alpha. \tag{2.54}$$

ADVICE

The capacitor peak current locates at $\omega t = \alpha$, which is usually much smaller than $\pi/2$. (*Historic problem*: $\alpha \approx \pi/2$.)

Resistor current $i_R(t)$ is

$$i_R(t) = \begin{cases} \dfrac{\sqrt{2}V}{R} \sin \omega t & \text{diode on,} \\ \dfrac{V_\theta}{R} e^{-(\omega t - \theta)/\omega RC} & \text{diode off,} \end{cases}$$

where $V_\theta = \sqrt{2}V \sin\theta$.

Its peak current at $\omega t = \pi/2$ is,

$$i_{R-peak} = \frac{\sqrt{2}V}{R}.$$

Its current at $\omega t = 2\pi + \alpha$ (and $\omega t = \alpha$) is

$$i_R(2\pi + \alpha) = \frac{\sqrt{2}V}{R} \sin(2\pi + \alpha) = \frac{\sqrt{2}V}{R} \sin\alpha. \tag{2.55}$$

Usually, the capacitive reactance is smaller than the resistance R; the main component of the source current is capacitor current. Therefore, the peak diode (source) current is

$$i_{D-peak} = \omega C\sqrt{2}V \cos\alpha + \frac{\sqrt{2}V}{R} \sin\alpha. \tag{2.56}$$

ADVICE

The source peak current locates at $\omega t = \alpha$, which is usually much smaller than $\pi/2$. (*Historic problem*: The source peak current locates at $\omega t = \pi/2$.)

The peak-to-peak ripple of the output voltage is given by

$$\Delta V_d = \sqrt{2}V - \sqrt{2}V \sin\alpha = \sqrt{2}V(1 - \sin\alpha). \tag{2.57}$$

Example 2.3

A single-phase half-wave diode rectifier shown in Figure 2.9a operates from a supply of $V = 240$ V, 50 Hz to a load of $R = 100\,\Omega$ and $C = 100\,\mu$F in parallel. If $\alpha = 12.63°$ (see Question 2.5), determine the peak capacitor current and peak source current.

SOLUTION

From Equation 2.54, the peak capacitor current at $\omega t = \alpha$ is

$$i_{C-peak} = \omega C\sqrt{2}V \cos\alpha = 100\,\pi \times 0.0001 \times 240 \times \sqrt{2} \times \cos 12.63° = 10.4\text{ A}.$$

From Equation 2.56, the peak source current at $\omega t = \alpha$ is

$$i_{D-peak} = \omega C\sqrt{2}V \cos\alpha + \frac{\sqrt{2}V}{R} \sin\alpha = 10.4 + \frac{240\sqrt{2}}{100} \sin 12.63° = 11.14\text{ A}.$$

In order to help readers understand the current waveforms, the simulation results are presented below (Figure 2.10) for reference: $V_{in} = 340$ V/50 Hz, $C = 100\,\mu$F, and $R = 100\,\Omega$.

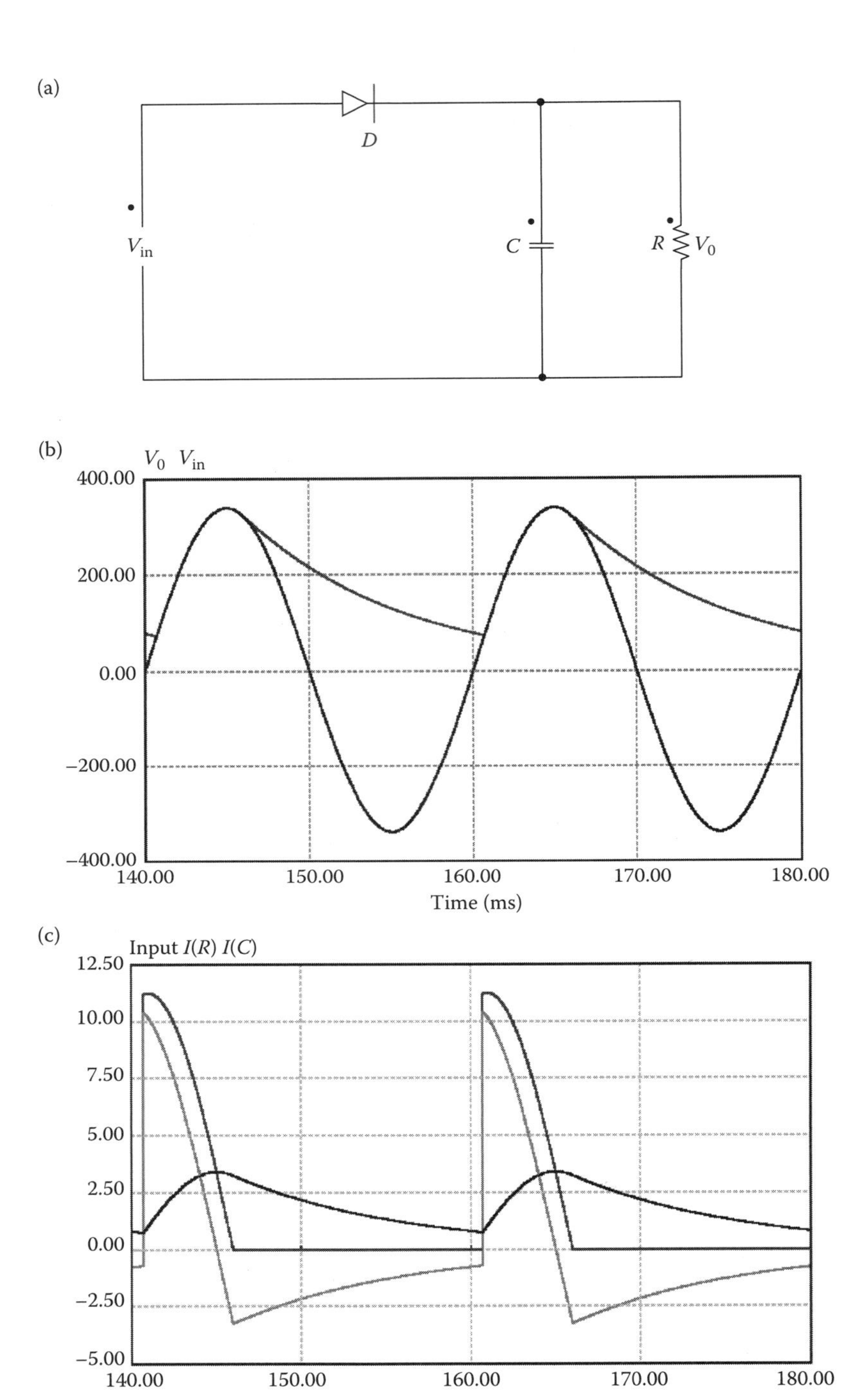

FIGURE 2.10 Simulation results: (a) circuit, (b) input (sine-wave) and output voltages, and (c) input (top), capacitor (middle), and resistor (lower) currents.

2.3 Single-Phase Full-Wave Converters

Single-phase uncontrolled full-wave bridge circuits are shown in Figures 2.11a and 2.12a. They are called the center-tap (midpoint) rectifier and the bridge (Graetz) rectifier,

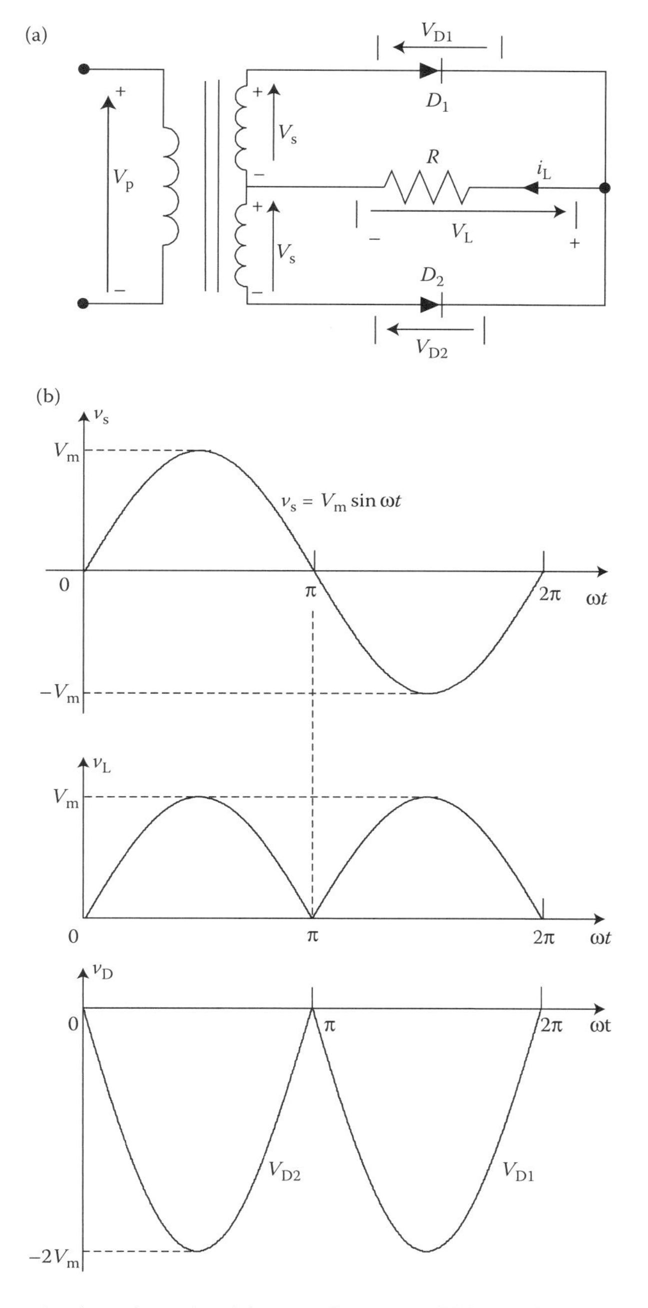

FIGURE 2.11 Center-tap (midpoint) rectifier: (a) circuit diagram and (b) waveforms.

respectively. Figures 2.11a and 2.12a appear identical as far as the load is concerned. It can be seen in Figure 2.11a that two less diodes are employed, but a center-tapped transformer is required. The rectifying diodes in Figure 2.11a experience twice the reverse voltage, as do the four diodes in the circuit of Figure 2.12a. As most industrial applications, use the bridge (Graetz) rectifier circuit, further analysis and discussion will be based on the bridge rectifier.

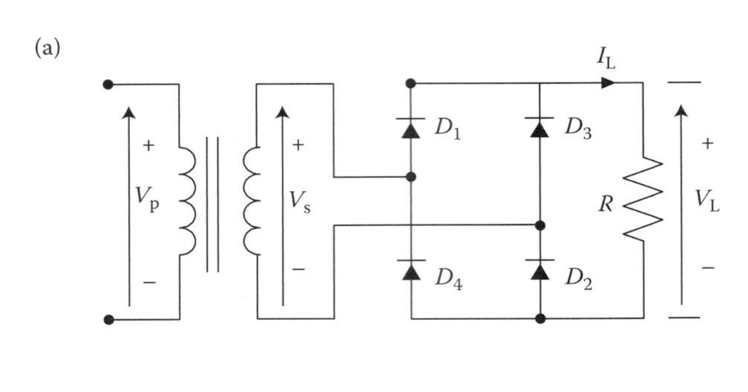

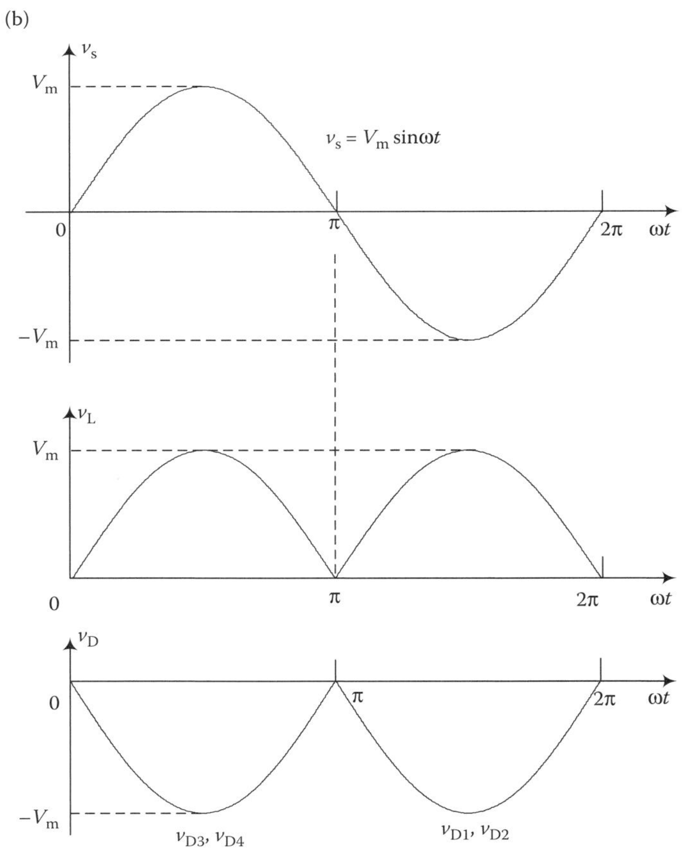

FIGURE 2.12 Bridge (Graetz) rectifier: (a) circuit diagram and (b) waveforms.

2.3.1 *R* Load

Referring to the bridge circuit shown in Figure 2.12, it is seen that the load is pure resistive, R. In Figure 2.12b, the bridge circuit voltage and current waveforms are shown. The output average voltage is

$$V_{\text{d}} = \frac{1}{\pi}\int_0^{\pi} \sqrt{2}V \sin \omega t \, \text{d}(\omega t) = \frac{2\sqrt{2}}{\pi} V = 0.9\,\text{V}. \tag{2.58}$$

The output rms voltage is

$$V_{\text{d-rms}} = \sqrt{\frac{1}{2\pi}\int_0^{\pi} \left(\sqrt{2}V \sin \omega t\right)^2 \text{d}(\omega t)} = V\sqrt{\frac{1}{\pi}\int_0^{\pi} (\sin \alpha)^2 \, \text{d}\alpha} = V. \tag{2.59}$$

The output average and rms currents are

$$I_{\text{d}} = \frac{V_{\text{d}}}{R} = \frac{2\sqrt{2}}{\pi}\frac{V}{R} = 0.9\frac{V}{R}, \tag{2.60}$$

$$I_{\text{d-rms}} = \frac{V_{\text{d-rms}}}{R} = \frac{V}{R}. \tag{2.61}$$

The FF and RF of the output voltage are

$$\text{FF} = \frac{V_{\text{d-rms}}}{V_{\text{d}}} = \frac{1}{2\sqrt{2}/\pi} = \frac{\pi}{2\sqrt{2}} = 1.11, \tag{2.62}$$

$$\text{RF} = \sqrt{\text{FF}^2 - 1} = \sqrt{(1.11)^2 - 1} = 0.48, \tag{2.63}$$

$$\text{PF} = \frac{1}{\sqrt{2}} = 0.707 \quad \text{for the mid-point circuit}, \tag{2.64}$$

$$\text{RF} = 1 \quad \text{for the Graetz circuit}. \tag{2.65}$$

ADVICE

For all diode rectifiers, only the Graetz (bridge) circuit has a unity power factor (UPF). (*Historic problem*: Multiphase full-wave rectifiers may have UPF.)

2.3.2 *R–C* Load

Linear and switch-mode DC power supplies require AC/DC rectification. To obtain a "smooth" output, capacitor C is connected as shown in Figure 2.13.

Neglecting diode forward voltage drop, the peak of the output voltage is $\sqrt{2}V$. During *each half-cycle*, the capacitor undergoes cyclic changes from $v_{\text{d(min)}}$ to $\sqrt{2}V$ in the period between $\omega t = \alpha$ and $\omega t = \pi/2$, and discharges from $\sqrt{2}V$ to $v_{\text{d(min)}}$ in the period between $\omega t = \theta$ and $\omega t = \pi + \alpha$. The resultant output of the diode bridge is unipolar, but time dependent.

$$v_{\text{d}}(\omega t) = \begin{cases} \sqrt{2}V \sin \omega t & \text{diode on,} \\ V_{\theta}\text{e}^{-(\omega t - \theta)/\omega RC} & \text{diode off,} \end{cases} \tag{2.66}$$

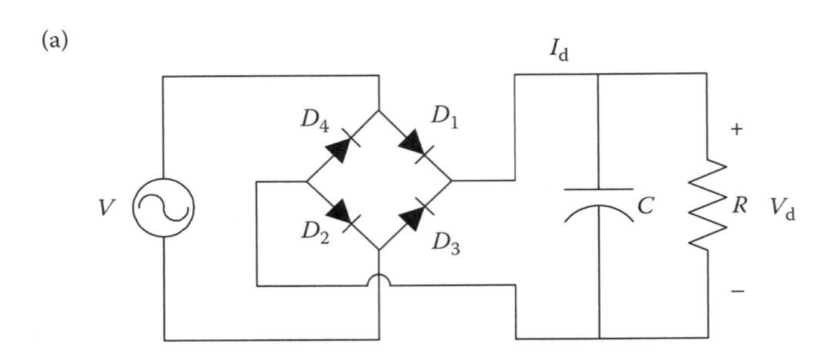

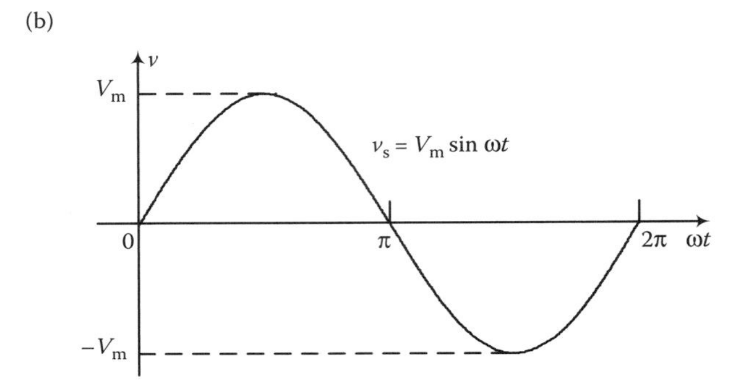

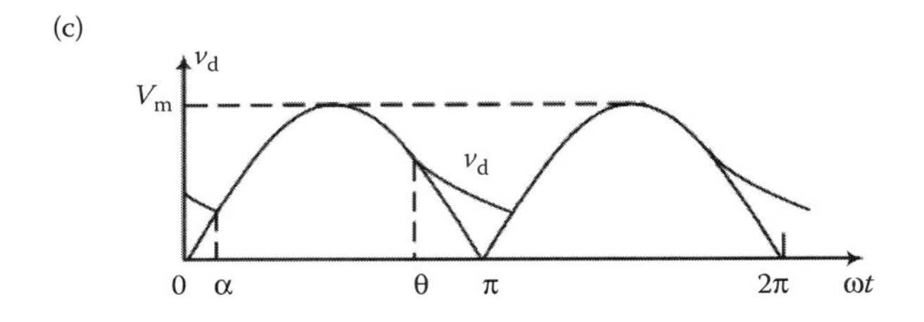

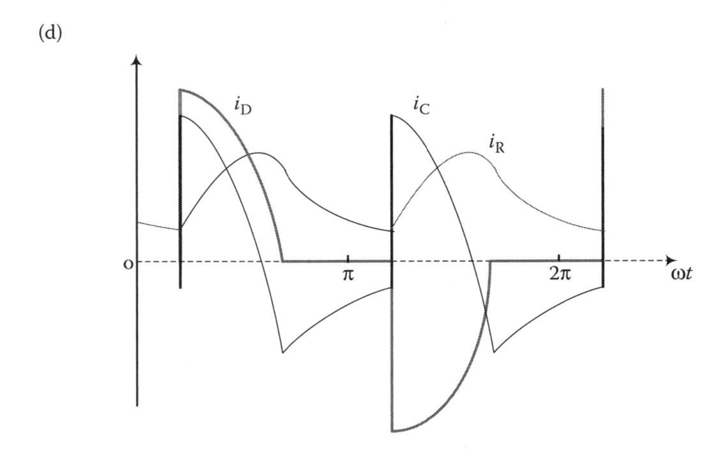

FIGURE 2.13 Single-phase full-wave bridge rectifier with R–C load: (a) circuit, (b) input voltage, (c) output voltage, and (d) current waveforms.

where

$$V_\theta = \sqrt{2}V \sin\theta. \tag{2.67}$$

At $\omega t = \theta$, the slopes of the voltage functions are equal to

$$\sqrt{2}V\cos\theta = \frac{\sqrt{2}V\sin\theta}{-\omega RC}\mathrm{e}^{-(\theta-\theta)/\omega RC}.$$

Therefore

$$\frac{1}{\tan\theta} = \frac{-1}{\omega RC}. \tag{2.68}$$

Thus

$$\theta = \pi - \tan^{-1}(\omega RC).$$

The angle at which the diode turns on in the second period, $\omega t = \pi + \alpha$, is the point at which the sinusoidal source reaches the same value as the decaying exponential output.

$$\sqrt{2}V\sin(\pi+\alpha) = (\sqrt{2}V\sin\theta)\mathrm{e}^{-(\pi+\alpha-\theta)/\omega RC}$$

or

$$\sin\alpha - (\sin\theta)\mathrm{e}^{-(\pi+\alpha-\theta)/\omega RC} = 0. \tag{2.69}$$

The preceding equation must be solved numerically.

The output average voltage is

$$\begin{aligned} V_\mathrm{d} &= \frac{1}{\pi}\int_\alpha^{\pi+\alpha} v_\mathrm{d}\,\mathrm{d}(\omega t) = \frac{\sqrt{2}V}{\pi}\left[\int_\alpha^{\theta}\sin\omega t\,\mathrm{d}(\omega t) + \int_\theta^{\pi+\alpha}\sin\theta\,\mathrm{e}^{-(t-\theta/\omega)/RC}\,\mathrm{d}(\omega t)\right] \\ &= \frac{\sqrt{2}V}{\pi}\left[(\cos\alpha - \cos\theta) + \omega RC\sin\theta\int_0^{(\pi+\alpha-\theta)/\omega}\mathrm{e}^{-t/RC}\,\mathrm{d}\left(\frac{t}{RC}\right)\right] \\ &= \frac{\sqrt{2}V}{\pi}\left[(\cos\alpha-\cos\theta) + \omega RC\sin\theta\left(1-\mathrm{e}^{-(\pi+\alpha-\theta)/\omega RC}\right)\right]. \end{aligned} \tag{2.70}$$

The output rms voltage is

$$\begin{aligned} V_{\mathrm{d-rms}} &= \sqrt{\frac{1}{\pi}\int_\alpha^{\pi+\alpha} v_\mathrm{d}^2\,\mathrm{d}(\omega t)} = \sqrt{\frac{2V^2}{\pi}\left[\int_\alpha^{\theta}(\sin\omega t)^2\,\mathrm{d}(\omega t) + \int_\theta^{\pi+\alpha}\sin^2\theta\,\mathrm{e}^{-2((t-\theta)\omega)/RC}\,\mathrm{d}(\omega t)\right]} \\ &= \sqrt{2}V\sqrt{\frac{1}{\pi}\left[\left(\frac{\theta-\alpha}{2} - \frac{\cos 2\alpha - \cos 2\theta}{4}\right) + \omega RC\sin^2\theta\left(1 - \frac{\mathrm{e}^{-2(\pi+\alpha-\theta)/\omega RC}}{2}\right)\right]}. \end{aligned} \tag{2.71}$$

Since the average capacitor current is zero, the output average current is

$$I_\mathrm{d} = \frac{V_\mathrm{d}}{R} = \frac{\sqrt{2}V}{\pi R}\left[(\cos\alpha - \cos\theta) + \omega RC\sin\theta\left(1 - \mathrm{e}^{-(\pi+\alpha-\theta)/\omega RC}\right)\right]. \quad (2.72)$$

The FF and RF of the output voltage are

$$\mathrm{FF} = \frac{V_\mathrm{d-rms}}{V_\mathrm{d}}$$

$$= \frac{\sqrt{2}V\sqrt{(1/\pi)[((\theta-\alpha)/2) - ((\cos 2\alpha - \cos 2\theta)/4) + \omega RC\sin^2\theta(1 - (\mathrm{e}^{-2(\pi+\alpha-\theta)/\omega RC}/2))]}}{\left(\sqrt{2}V/\pi\right)\left[(\cos\alpha - \cos\theta) + \omega RC\sin\theta(1 - \mathrm{e}^{-(\pi+\alpha-\theta)/\omega RC})\right]}$$

$$= \frac{\sqrt{\pi}\sqrt{((\theta-\alpha)/2) - ((\cos 2\alpha - \cos 2\theta)/4) + \omega RC\sin^2\theta(1 - (\mathrm{e}^{-2(\pi+\alpha-\theta)/\omega RC}/2))}}{\cos\alpha - \cos\theta + \omega RC\sin\theta\left(1 - \mathrm{e}^{-(\pi+\alpha-\theta)/\omega RC}\right)},$$

$$\mathrm{RF} = \sqrt{\mathrm{FF}^2 - 1}. \quad (2.73)$$

2.3.3 *R–L* Load

A single-phase full-wave diode rectifier with R–L load is shown in Figure 2.14a, while various circuit waveforms are shown in Figures 2.14b and c.

If the inductance L is large enough, the load current can be considered as a continuous constant current to simplify the analysis and calculations. It is accurate enough for theoretical analysis and engineering calculations. In this case, the load current is assumed to be a constant DC current.

The output average voltage is

$$V_\mathrm{d} = \frac{1}{\pi}\int_0^\pi \sqrt{2}V\sin\omega t\,\mathrm{d}(\omega t) = \frac{2\sqrt{2}}{\pi}V = 0.9\,\mathrm{V}. \quad (2.74)$$

The output rms voltage is

$$V_\mathrm{d-rms} = \sqrt{\frac{1}{2\pi}\int_0^\pi \left(\sqrt{2}V\sin\omega t\right)^2 \mathrm{d}(\omega t)} = V\sqrt{\frac{1}{\pi}\int_0^\pi (\sin\alpha)^2\,\mathrm{d}\alpha} = V. \quad (2.75)$$

The output current is a constant DC value; its average and rms currents are

$$I_\mathrm{d} = I_\mathrm{d-rms} = \frac{V_\mathrm{d}}{R} = \frac{2\sqrt{2}}{\pi}\frac{V}{R} = 0.9\frac{V}{R}. \quad (2.76)$$

The FF and RF of the output voltage are

$$\mathrm{FF} = \frac{V_\mathrm{d-rms}}{V_\mathrm{d}} = \frac{1}{2\sqrt{2}/\pi} = \frac{\pi}{2\sqrt{2}} = 1.11, \quad (2.77)$$

$$\mathrm{RF} = \sqrt{\mathrm{FF}^2 - 1} = \sqrt{(1.11)^2 - 1} = 0.48. \quad (2.78)$$

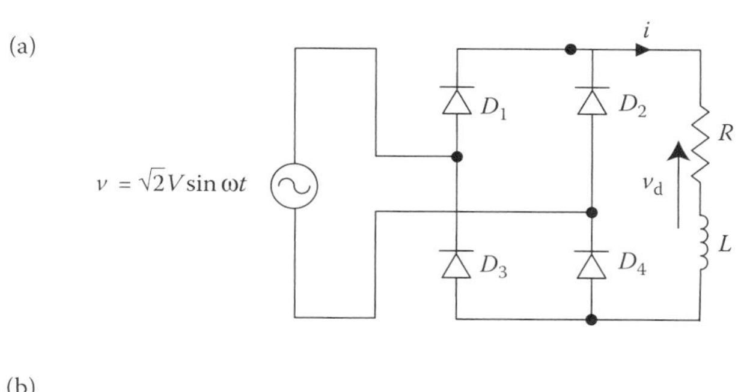

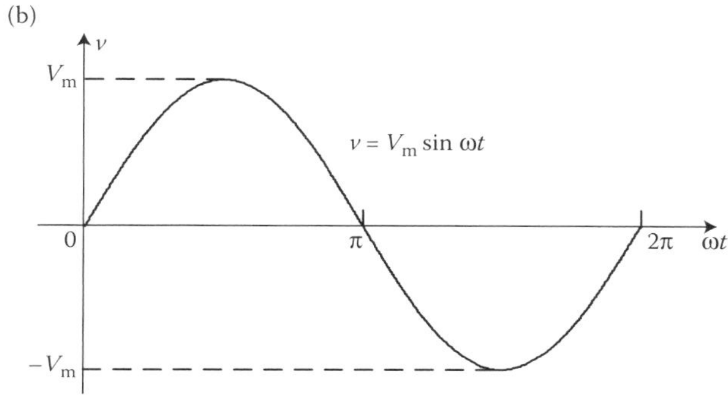

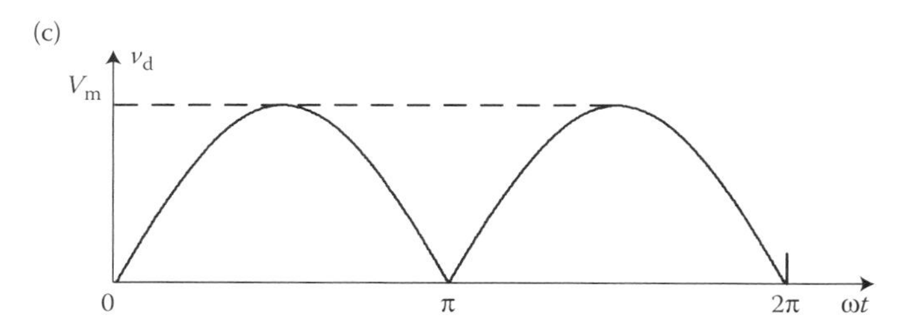

FIGURE 2.14 Single-phase full-wave bridge rectifier with R plus large L load: (a) circuit, (b) input voltage, and (c) output voltage.

2.4 Three-Phase Half-Wave Converters

If the AC power supply is from a transformer, four circuits can be used. The three-phase half-wave rectifiers are shown in Figure 2.15.

The first circuit is called a Y/Y circuit, shown in Figure 2.15a; the second circuit is called a Δ/Y circuit, shown in Figure 2.15b; the third circuit is called a Y/Y bending circuit, shown in Figure 2.15c; and the fourth circuit is called a Δ/Y bending circuit, shown in Figure 2.15d. Each diode is conducted in 120° a cycle. Some waveforms are shown in Figure 2.16 corresponding to $L = 0$. The three-phase voltages are balanced, so that

$$v_a(t) = \sqrt{2}V \sin \omega t, \tag{2.79}$$

$$v_b(t) = \sqrt{2}V \sin(\omega t - 120^\circ), \tag{2.80}$$

$$v_c(t) = \sqrt{2}V \sin(\omega t - 240^\circ). \tag{2.81}$$

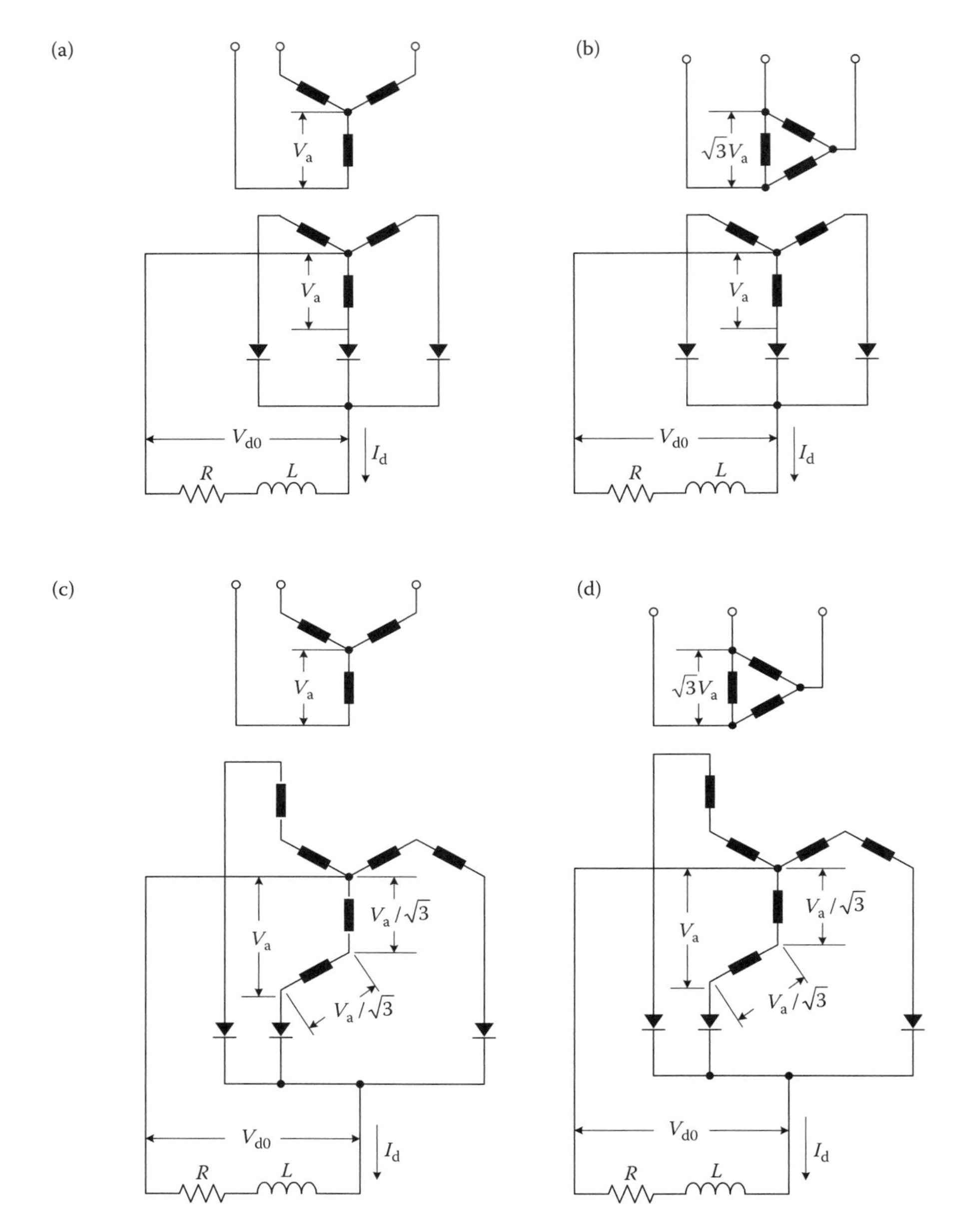

FIGURE 2.15 Three-phase half-wave diode rectifiers: (a) Y/Y circuit, (b) Δ/Y circuit, (c) Y/Y bending circuit, and (d) Δ/Y bending circuit.

2.4.1 *R* Load

Referring to the bridge circuit shown in Figure 2.15a, the load is pure resistive, R ($L = 0$). Figure 2.16 shows the voltage and current waveforms. The output average voltage is

$$V_{d0} = \frac{3}{2\pi}\int_{\pi/6}^{5\pi/6} \sqrt{2}V \sin \omega t \, d(\omega t) = \frac{3\sqrt{3}}{\sqrt{2}\pi}V = 1.17\,V. \tag{2.82}$$

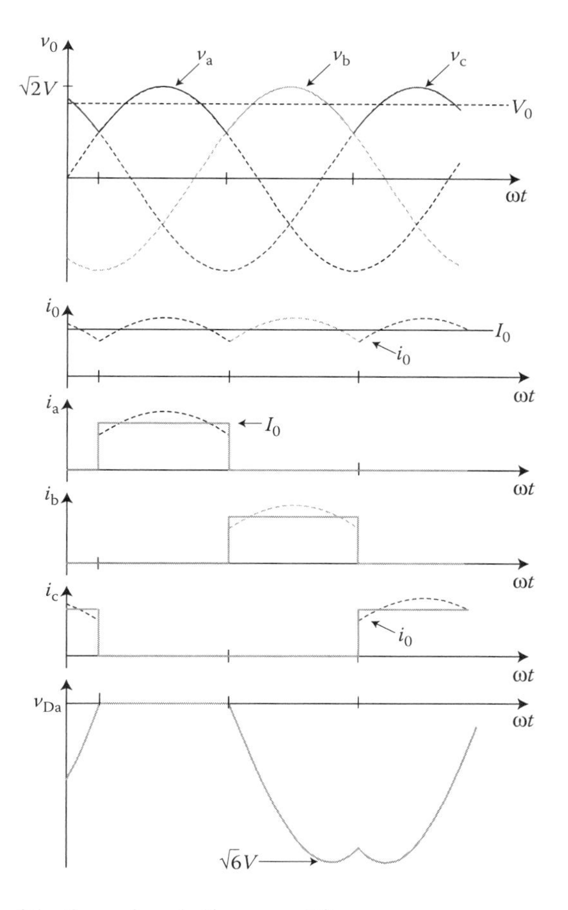

FIGURE 2.16 Waveforms of the three-phase half-wave rectifier.

The output rms voltage is

$$V_{\text{d-rms}} = \sqrt{\frac{3}{2\pi}\int_{\pi/6}^{5\pi/6}\left(\sqrt{2}V\sin\omega t\right)^2 \mathrm{d}(\omega t)} = V\sqrt{\frac{6}{\pi}\left(\frac{\pi}{6}+\frac{\sqrt{3}}{8}\right)} = 1.1889\,\mathrm{V}. \tag{2.83}$$

The output average and rms currents are

$$I_{\mathrm{d}} = \frac{V_{\mathrm{d}}}{R} = 1.17\frac{V}{R}, \tag{2.84}$$

$$I_{\text{d-rms}} = \frac{V_{\text{d-rms}}}{R} = 1.1889\frac{V}{R}. \tag{2.85}$$

The FF, RF, and PF of the output voltage are

$$\mathrm{FF} = \frac{V_{\text{d-rms}}}{V_{\mathrm{d}}} = \frac{1.1889}{1.17} = 1.016, \tag{2.86}$$

$$\mathrm{RF} = \sqrt{\mathrm{FF}^2 - 1} = \sqrt{(1.016)^2 - 1} = 0.18, \tag{2.87}$$

and

$$\mathrm{PF} = 0.686. \tag{2.88}$$

2.4.2 *R–L* Load

A three-phase half-wave diode rectifier with $R-L$ load is shown in Figure 2.15a. If the inductance L is large enough, the load current can be considered as a continuous constant current to simplify the analysis and calculations. It is accurate enough for theoretical analysis and engineering calculations. In this case, the load current is assumed to be a constant DC current. The output average voltage is

$$V_{\mathrm{d0}} = \frac{3}{2\pi} \int_{\pi/6}^{5\pi/6} \sqrt{2}V \sin \omega t \, \mathrm{d}(\omega t) = \frac{3\sqrt{3}}{\sqrt{2}\pi} V = 1.17\,\mathrm{V}. \tag{2.89}$$

The output rms voltage is

$$V_{\mathrm{d-rms}} = \sqrt{\frac{3}{2\pi} \int_{\pi/6}^{5\pi/6} \left(\sqrt{2}V \sin \omega t\right)^2 \mathrm{d}(\omega t)} = V\sqrt{\frac{6}{\pi}\left(\frac{\pi}{6} + \frac{\sqrt{3}}{8}\right)} = 1.1889\,\mathrm{V}. \tag{2.90}$$

The output current is nearly a constant DC value; its average and rms currents are

$$I_{\mathrm{d}} = I_{\mathrm{d-rms}} = \frac{V_{\mathrm{d}}}{R} = 1.17\frac{V}{R}. \tag{2.91}$$

The FF and RF of the output voltage are

$$\mathrm{FF} = \frac{V_{\mathrm{d-rms}}}{V_{\mathrm{d}}} = \frac{1.1889}{1.17} = 1.016, \tag{2.92}$$

$$\mathrm{RF} = \sqrt{\mathrm{FF}^2 - 1} = \sqrt{(1.016)^2 - 1} = 0.18. \tag{2.93}$$

2.5 Six-Phase Half-Wave Converters

Six-phase half-wave rectifiers have two constructions: six-phase with a neutral line circuit and double antistar with a balance-choke circuit. The following description is based on the R load or R plus large L load.

2.5.1 Six-Phase with a Neutral Line Circuit

If the AC power supply is from a transformer, four circuits can be used. The six-phase half-wave rectifiers are shown in Figure 2.17.

The first circuit is called a Y/star circuit, shown in Figure 2.17a; the second circuit is called a Δ/star circuit, shown in Figure 2.17b; the third circuit is called a Y/star bending circuit, shown in Figure 2.17c; and the fourth circuit is called a Δ/star bending circuit, shown in Figure 2.17d. Each diode is conducted in 60° a cycle. Since the load is an $R-L$ circuit, the output voltage average value is

$$V_{d0} = \frac{1}{\pi/3} \int_{\pi/3}^{2\pi/3} V_m \sin(\omega t)\, d(\omega t) = \frac{3\sqrt{2}}{\pi} V_a = 1.35 V_a, \tag{2.94}$$

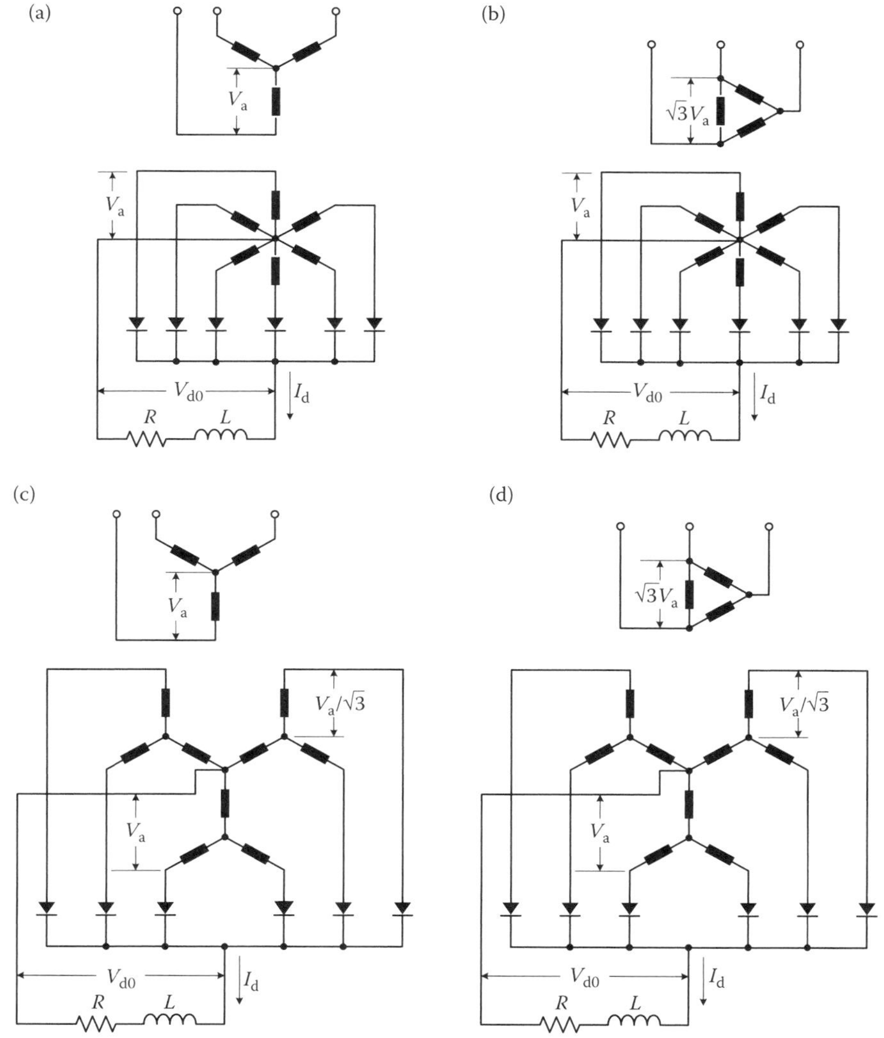

FIGURE 2.17 Six-phase half-wave diode rectifiers: (a) Y/star circuit, (b) Δ/star circuit, (c) Y/star bending circuit, and (d) Δ/star bending circuit.

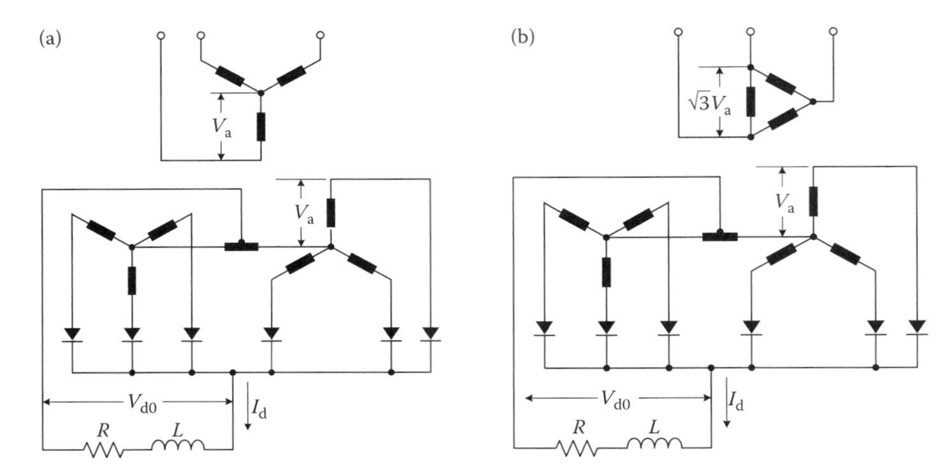

FIGURE 2.18 Three-phase double antistar with balance-choke diode rectifiers: (a) Y/Y-Y circuit and (b) Δ/Y-Y circuit.

$$\mathrm{FF} = 1.00088, \tag{2.95}$$

$$\mathrm{RF} = 0.042, \tag{2.96}$$

$$\mathrm{PF} = 0.552. \tag{2.97}$$

2.5.2 Double Antistar with Balance-Choke Circuit

If the AC power supply is from a transformer, two circuits can be used. The six-phase half-wave rectifiers are shown in Figure 2.18.

The first circuit is called a Y/Y-Y circuit, shown in Figure 2.18a, and the second circuit is called a Δ/Y-Y circuit, shown in Figure 2.18b. Each diode is conducted in 120° a cycle. Since the load is an R–L circuit, the output voltage average value is

$$V_{d0} = \frac{1}{2\pi/3} \int_{\pi/6}^{5\pi/6} V_m \sin(\omega t)\, d(\omega t) = \frac{3\sqrt{6}}{2\pi} V_a = 1.17 V_a, \tag{2.98}$$

$$\mathrm{FF} = 1.01615, \tag{2.99}$$

$$\mathrm{RF} = 0.18, \tag{2.100}$$

$$\mathrm{PF} = 0.686. \tag{2.101}$$

2.6 Three-Phase Full-Wave Converters

If the AC power supply is from a transformer, four circuits can be used. The three-phase full-wave diode rectifiers, shown in Figure 2.19, all consist of six diodes. The first circuit is called a Y/Y circuit, shown in Figure 2.19a; the second circuit is called a Δ/Y circuit, shown in Figure 2.19b; the third circuit is called a Y/Δ circuit, shown in Figure 2.19c; and

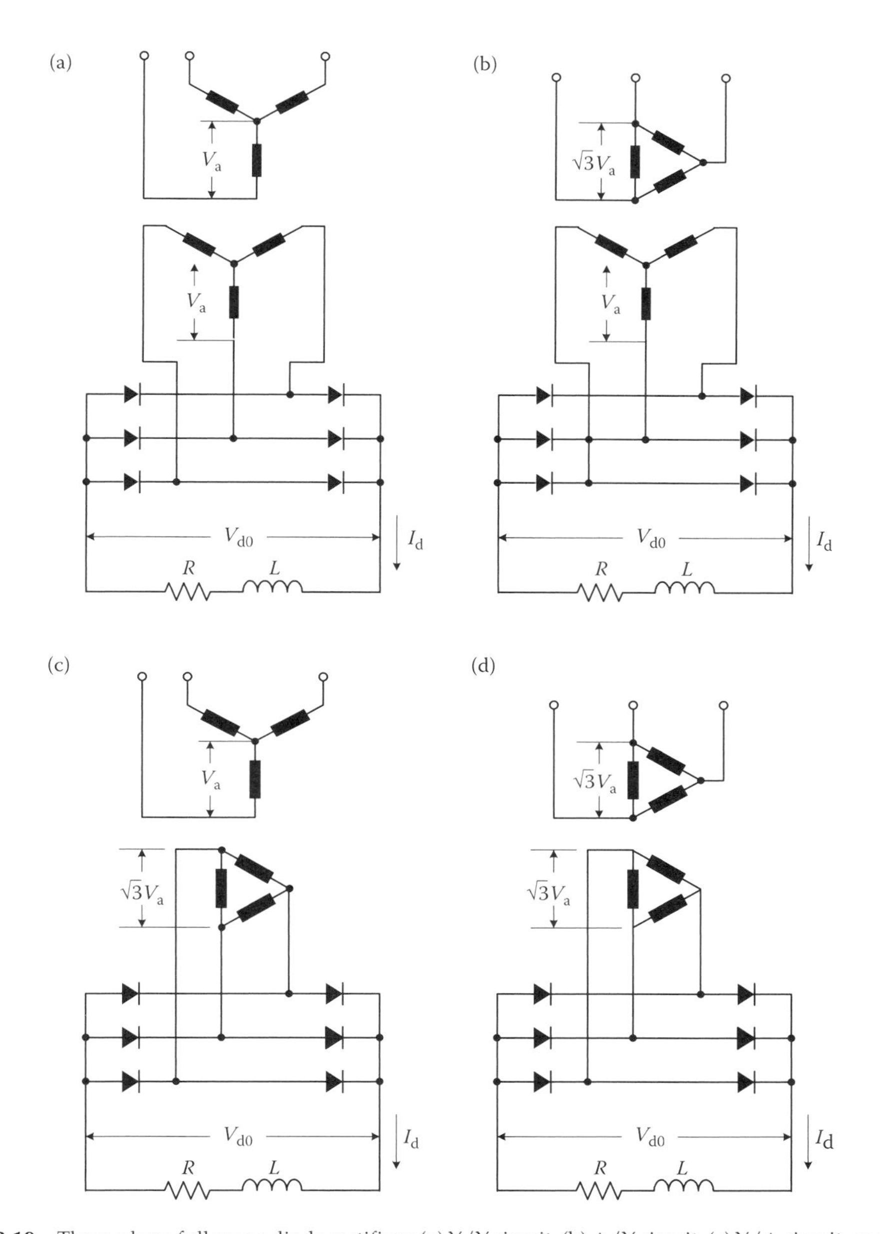

FIGURE 2.19 Three-phase full-wave diode rectifiers: (a) Y/Y circuit, (b) Δ/Y circuit, (c) Y/Δ circuit, and (d) Δ/Δ circuit.

the fourth circuit is called a Δ/Δ circuit, shown in Figure 2.19d. Each diode is conducted in 120° a cycle. Since the load is an R–L circuit, the output voltage average value is

$$V_{d0} = \frac{2}{2\pi/3}\int_{\pi/6}^{5\pi/6} V_m \sin(\omega t)\, d(\omega t) = \frac{3\sqrt{6}}{\pi}V_a = 2.34V_a, \tag{2.102}$$

$$FF = 1.00088, \tag{2.103}$$

$$\mathrm{RF} = 0.042, \tag{2.104}$$

$$\mathrm{PF} = 0.956. \tag{2.105}$$

Some waveforms are shown in Figure 2.20.

ADVICE

The three-phase full-wave bridge rectifier has high PF (although no UPF) and low RF = 4.2%. It is a proven circuit that can be used in most industrial applications.

2.7 Multiphase Full-Wave Converters

Usually, the more the phases, the smaller the output voltage ripple. In this section, several circuits with six-phase, twelve-phase, and eighteen-phase supply are investigated.

2.7.1 Six-Phase Full-Wave Diode Rectifiers

In Figure 2.21, two circuits of six-phase full-wave diode rectifiers, each all consisting of 12 diodes, are shown. The first circuit is called the six-phase bridge circuit (Figure 2.21a) and the second circuit is called the hexagon bridge circuit (Figure 2.21b). Each diode is conducted in 60° a cycle. Since the load is an R–L circuit, the average output voltage value is

$$V_{\mathrm{d0}} = \frac{2}{\pi/3}\int_{\pi/3}^{2\pi/3} V_{\mathrm{m}} \sin(\omega t)\, \mathrm{d}(\omega t) = \frac{6\sqrt{2}}{\pi} V_{\mathrm{a}} = 2.7 V_{\mathrm{a}}, \tag{2.106}$$

$$\mathrm{FF} = 1.00088, \tag{2.107}$$

$$\mathrm{RF} = 0.042, \tag{2.108}$$

$$\mathrm{PF} = 0.956. \tag{2.109}$$

2.7.2 Six-Phase Double-Bridge Full-Wave Diode Rectifiers

Figure 2.22 shows two circuits of the six-phase double-bridge full-wave diode rectifiers. The first circuit is called a Y/Y-Δ circuit (Figure 2.22a), and the second circuit is called a Δ/Y-Δ circuit (Figure 2.22b). Each diode is conducted in 120° a cycle. There are 12 pulses during each period and the phase shift is 30°. Since the load is an R–L circuit, the output voltage V_{d0} is nearly pure DC voltage.

$$V_{\mathrm{d0}} = \frac{4}{2\pi/3}\int_{\pi/6}^{5\pi/6} V_{\mathrm{m}} \sin(\omega t)\, \mathrm{d}(\omega t) = \frac{6\sqrt{6}}{\pi} V_{\mathrm{a}} = 4.678 V_{\mathrm{a}}, \tag{2.110}$$

$$\mathrm{FF} = 1.0000567, \tag{2.111}$$

$$\mathrm{RF} = 0.0106, \tag{2.112}$$

$$\mathrm{PF} = 0.956. \tag{2.113}$$

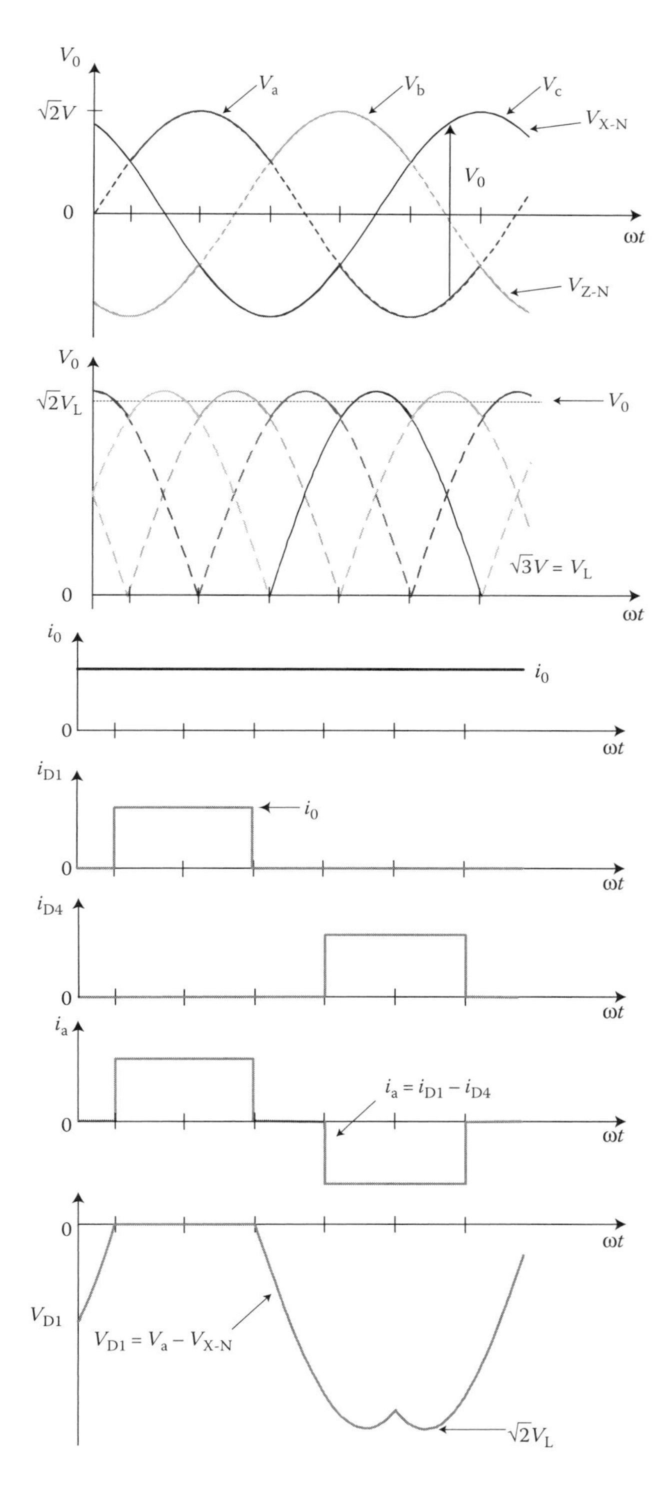

FIGURE 2.20 Waveforms of a three-phase full-wave bridge rectifier.

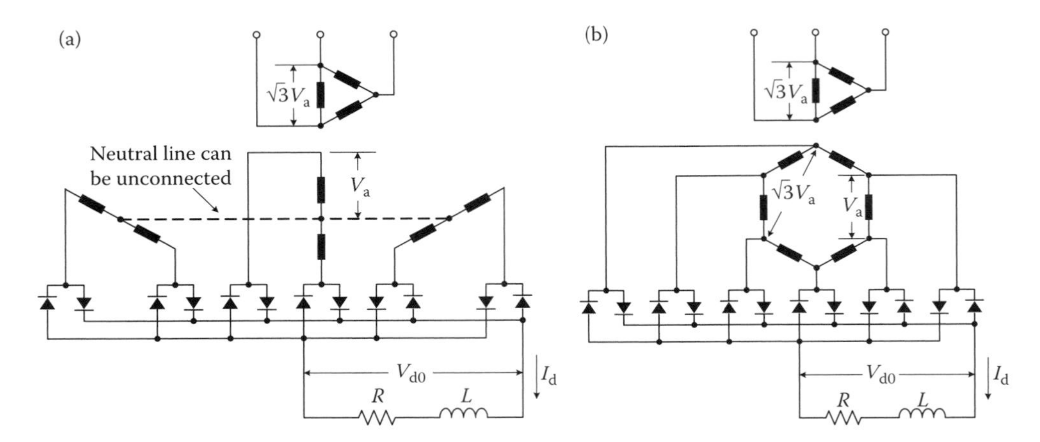

FIGURE 2.21 Six-phase full-wave diode rectifiers: (a) six-phase bridge circuit and (b) hexagon bridge circuit.

ADVICE

The six-phase double-bridge full-wave diode rectifier has high PF (although no UPF) and a low RF = 1.06%. It is a proven circuit that can be used in large power industrial applications.

2.7.3 Six-Phase Double-Transformer Double-Bridge Full-Wave Diode Rectifiers

Figure 2.23 shows the six-phase double-transformer double-bridge full-wave diode rectifier. The first transformer T_1 is called a Y/Y-Δ connection transformer, and the second transformer T_2 is called a bending Y/Y-Δ connection transformer with 15° phase shift. In total, there are 24 diodes involved in the rectifier. Each diode is conducted in 120° a cycle, There are 24 pulses a period and the phase shift is 15°. Since the load is an R–L circuit, the output voltage V_{d0} is nearly pure DC voltage.

$$V_{d0} = \frac{8}{2\pi/3} \int_{\pi/6}^{5\pi/6} V_m \sin(\omega t)\, d(\omega t) = \frac{12\sqrt{6}}{\pi} V_a = 9.356 V_a, \tag{2.114}$$

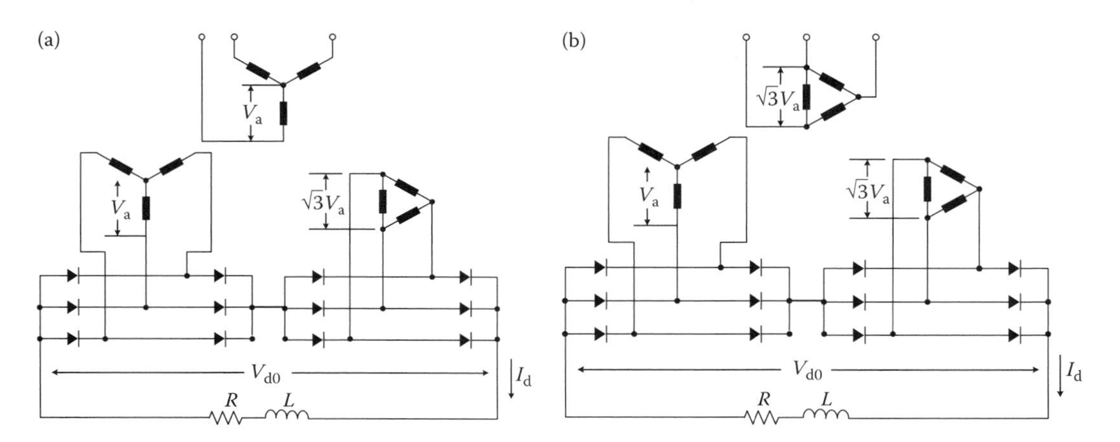

FIGURE 2.22 Six-phase double-bridge full-wave diode rectifiers: (a) Y/Y-Δ circuit and (b) Δ/Y-Δ circuit.

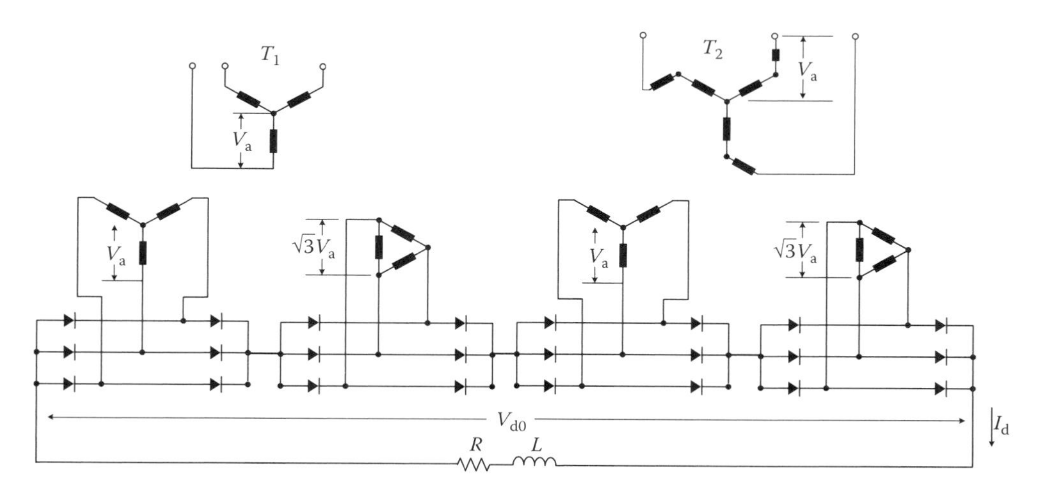

FIGURE 2.23 Six-phase double-transformer double-bridge full-wave diode rectifier.

$$\mathrm{FF} = 1.0000036, \tag{2.115}$$

$$\mathrm{RF} = 0.00267, \tag{2.116}$$

$$\mathrm{PF} = 0.956. \tag{2.117}$$

2.7.4 Six-Phase Triple-Transformer Double-Bridge Full-Wave Diode Rectifiers

Figure 2.24 shows the six-phase triple-transformer double-bridge full-wave diode rectifier. The first transformer T_1 is called a Y/Y-Δ connection transformer, the second transformer T_2 is called a positive-bending Y/Y-Δ connection transformer with a +10° phase shift, and the third transformer T_3 is called a negative-bending Y/Y-Δ connection transformer with a −10° phase shift. There are 36 diodes involved in the rectifier. Each diode is conducted in 120° a cycle. There are 36 pulses a period, and the phase shift is 10°. Since the load is an R–L circuit, the output voltage V_{d0} is nearly pure DC voltage.

$$V_{d0} = \frac{12}{2\pi/3}\int_{\pi/6}^{5\pi/6} V_m \sin(\omega t)\, d(\omega t) = \frac{18\sqrt{6}}{\pi}V_a = 14.035V_a, \tag{2.118}$$

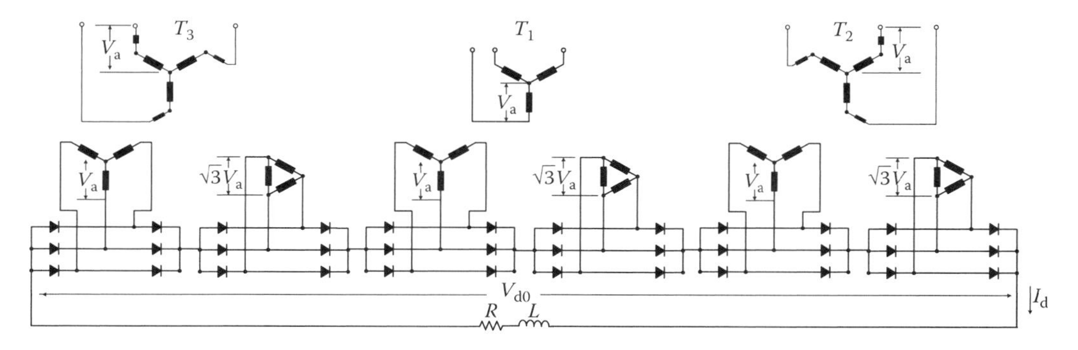

FIGURE 2.24 Six-phase triple-transformer double-bridge full-wave diode rectifier.

$$\text{FF} = 1.0000007, \tag{2.119}$$

$$\text{RF} = 0.00119, \tag{2.120}$$

$$\text{PF} = 0.956. \tag{2.121}$$

Homework

2.1. A single-phase half-wave diode rectifier operates from a supply of 200 V, 60 Hz to a load of $R = 15\,\Omega$ and $L = 0.2\,\text{H}$. Determine the extinction angle β using the graph in Figure 2.4.

2.2. A single-phase half-wave diode rectifier operates from a supply of 240 V, 50 Hz to a load of $R = 10\,\Omega$ and $L = 0.1\,\text{H}$. Determine the extinction angle β using iterative method 2 (see Section 2.2.2.3).

2.3. Referring to the single-phase half-wave rectifier with R–L load as shown in Figure 2.2a and given that $R = 100\,\Omega$, $L = 0.1\,\text{H}$, $\omega = 377\,\text{rad/s}$ ($f = 60\,\text{Hz}$), and $V = 100/\sqrt{2}\,\text{V}$, determine:

   a. Expression angle β for the current.

   b. Average current.

   c. Average output voltage.

2.4. In the circuit shown in Figure 2.8a, the source voltage $v(t) = 110\sqrt{2}\sin 120\,\pi t$, $R = 1\,\Omega$, and the load-circuit emf $V_c = 100\,\text{V}$. If the circuit is closed during the negative half-cycle of the source voltage, calculate:

   a. Angle α at which D starts to conduct.

   b. Conduction angle γ.

   c. Average value of current i.

   d. Rms value of current i.

   e. Power delivered by the AC source.

   f. The PF at the AC source.

2.5. A single-phase half-wave rectifier, as shown in Figure 2.9a, has an AC input of 240 V (rms) at $f = 50\,\text{Hz}$ with a load $R = 100\,\Omega$ and $C = 100\,\mu\text{F}$ in parallel. Determine angle α and angle θ within an accuracy of $0.1°$ using iterative method 1 (see Section 2.2.2.2).

2.6. A full-wave rectifier, as shown in Figure 2.12a, has an AC input of 240 V (rms) at 50 Hz with a load $R = 100\,\Omega$ and $C = 100\,\mu\text{F}$ in parallel. Determine angle α and angle θ within an accuracy of $0.1°$ using iterative method 1 (see Section 2.2.2.2). Calculate the average output voltage V_d and current I_d.

References

1. Rashid, M. H. 2007. *Power Electronics Handbook* (2nd edition). Boston: Academic Press.
2. Luo, F. L., Ye, H., and Rashid, M. H. 2005. *Digital Power Electronics and Applications*. New York: Academic Press.

3. Dorf, R. C. 2006. *The Electrical Engineering Handbook* (3rd edition). Boca Raton: Taylor & Francis.
4. Mohan, N., Undeland, T. M., and Robbins, W. P. 2003. *Power Electronics: Converters, Applications and Design* (3rd edition). New York: Wiley.
5. Rashid, M. H. 2003. *Power Electronics: Circuits, Devices and Applications* (3rd edition). New Jersey: Prentice-Hall, Inc.
6. Keown, J. 2001. *OrCAD PSpice and Circuit Analysis* (4th edition). New Jersey: Prentice-Hall.

3

Controlled AC/DC Converters

Controlled AC/DC converters are usually called controlled rectifiers. They convert an AC power supply source voltage to a controlled DC load voltage [1–3]. Controlled AC/DC conversion technology is a vast subject area spanning research investigation to industrial applications. Usually, such rectifier devices are thyristors (or SCRs—silicon-controlled rectifiers), gate-turn-off thyristors (GTOs), power transistors (PTs), insulated gate bipolar transistors (IGBTs), and so on. Generally, the device used most is the thyristor (or SCR). Controlled AC/DC converters consist of thyristor/diode circuits, which can be sorted into the following groups:

- Single-phase half-wave rectifiers.
- Single-phase full-wave rectifiers with half/full control.
- Three-phase rectifiers with half/full control.
- Multipulse rectifiers.

3.1 Introduction

As is the case of the diode rectifiers discussed in Chapter 2, the diode should be assumed that the diodes are replaced by thyristors or other semiconductor devices in controlled rectifiers, which are then supplied from an ideal AC source. Two conditions must be met before the thyristor can be conducting [4–10]:

1. The thyristor must be forward biased.
2. A current must be applied to the gate of the thyristor.

Only one condition must be met before the thyristor can be switched off: the current that flows through it should be lower than the latched value, irrespective of whether the thyristor is forward or reverse biased.

According to the above conditions, a firing pulse with a variable angle is then required to be applied to the gate of the thyristor. Usually, the firing angle is defined as α. If the firing angle $\alpha = 0$, the thyristor functions as a diode. The corresponding output DC voltage of the rectifier is its maximum value. Referring to the results in Chapter 2, properly controlled rectifiers can be designed that satisfy industrial application needs.

3.2 Single-Phase Half-Wave Controlled Converters

A single-phase half-wave controlled rectifier consists of a single-phase AC input voltage and one thyristor. It is the simplest rectifier. This rectifier can supply various loads as described in the following subsections.

3.2.1 *R* Load

A single-phase half-wave diode rectifier with *R* load is shown in Figure 3.1a; the input voltage, output voltage, and current waveforms are shown in Figures 3.1b–d. The output

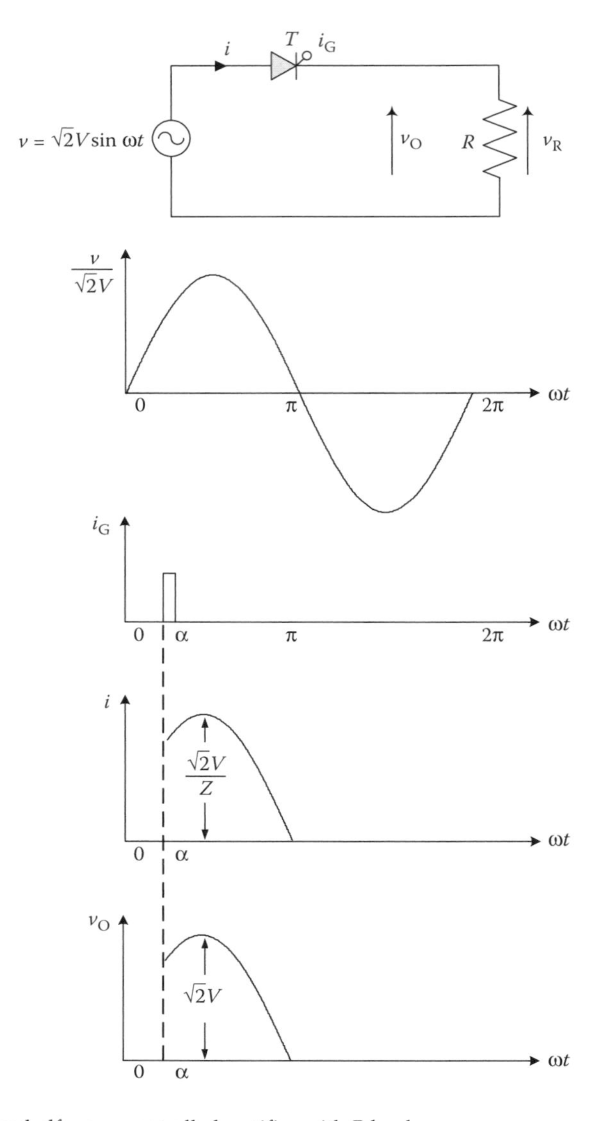

FIGURE 3.1 Single-phase half-wave controlled rectifier with *R* load.

voltage is the same as the input voltage in the positive half-cycle and zero in the negative half-cycle.

The output average voltage is

$$V_d = \frac{1}{2\pi}\int_{\alpha}^{\pi} \sqrt{2}V \sin \omega t \, d(\omega t) = \frac{\sqrt{2}}{2\pi}V(1+\cos\alpha) = 0.45V\frac{1+\cos\alpha}{2}. \tag{3.1}$$

Using the definition, we obtain

$$V_{dO} = \frac{1}{2\pi}\int_{0}^{\pi} \sqrt{2}V \sin \omega t \, d(\omega t) = \frac{\sqrt{2}}{2\pi}V. \tag{3.2}$$

We can rewrite Equation 3.1 as

$$V_d = \frac{1}{2\pi}\int_{\alpha}^{\pi} \sqrt{2}V \sin \omega t \, d(\omega t) = \frac{1+\cos\alpha}{2}V_{dO}. \tag{3.3}$$

The output rms voltage is

$$V_{d-rms} = \sqrt{\frac{1}{2\pi}\int_{\alpha}^{\pi} (\sqrt{2}V \sin \omega t)^2 \, d(\omega t)} = V\sqrt{\frac{1}{\pi}\int_{\alpha}^{\pi} (\sin x)^2 \, dx} = V\sqrt{\frac{1}{\pi}\left(\frac{\pi-\alpha}{2}+\frac{\sin 2\alpha}{4}\right)}. \tag{3.4}$$

The output average and rms currents are

$$I_d = \frac{V_d}{R} = \frac{\sqrt{2}}{\pi}\frac{V}{R}\frac{1+\cos\alpha}{2} = \frac{1+\cos\alpha}{2}\frac{V_{dO}}{R}, \tag{3.5}$$

$$I_{d-rms} = \frac{V_{d-rms}}{R} = \frac{V}{R}\sqrt{\frac{1}{\pi}\left(\frac{\pi-\alpha}{2}+\frac{\sin 2\alpha}{4}\right)}. \tag{3.6}$$

3.2.2 *R–L* Load

A single-phase half-wave diode rectifier with an *R–L* load is shown in Figure 3.2a, while various circuit waveforms are shown in Figures 3.2b–d.

It can be seen that load current flows not only during the positive part of the supply voltage but also during a portion of the negative supply voltage as well [11–21]. The load inductor SE maintains the load current, and the inductor's terminal voltage changes so as to overcome the negative supply and keep the diode forward biased and conducting. The load impedance Z is

$$Z = R + j\omega L = |Z|\angle\phi \quad \text{with } \phi = \tan^{-1}\frac{\omega L}{R},$$
$$|Z| = \sqrt{R^2 + (\omega L)^2}. \tag{3.7}$$

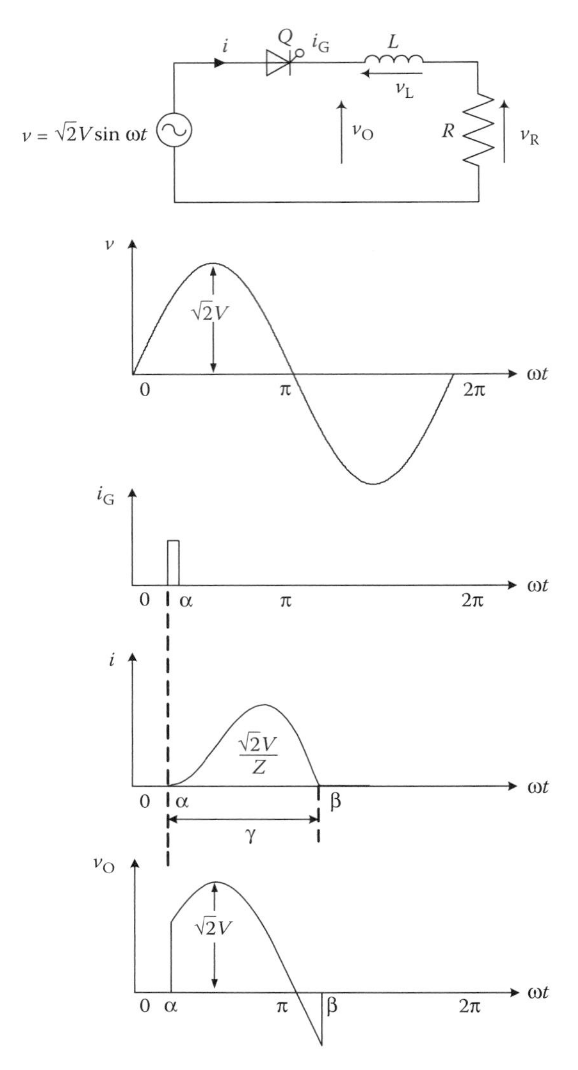

FIGURE 3.2 Half-wave controlled rectifier with *R–L* load.

When the thyristor is conducting, the dynamic equation is

$$L\frac{\mathrm{d}i}{\mathrm{d}t} + Ri = \sqrt{2}V\sin\omega t \quad \text{with } \alpha \le \omega t < \beta \tag{3.8}$$

or

$$\frac{\mathrm{d}i}{\mathrm{d}t} + \frac{R}{L}i = \frac{\sqrt{2}V}{L}\sin\omega t \quad \text{with } \alpha \le \omega t < \beta,$$

where α is the firing angle and β is the extinction angle. The thyristor conducts between α and β. Equation 3.8 is a non-normalized differential equation. The solution has two parts. The forced solution is determined by

$$i_F = \frac{\sqrt{2}V}{L}\sin(\omega t - \phi). \tag{3.9}$$

The natural response of such a circuit is given by

$$i_N = Ae^{-(R/L)t} = Ae^{-t/\tau} \quad \text{with } \tau = \frac{L}{R}. \tag{3.10}$$

The solution of Equation 3.8 is

$$i = i_F + i_N = \frac{\sqrt{2}V}{Z}\sin(\omega t - \phi) + Ae^{-(R/L)t}. \tag{3.11}$$

The constant A is determined by substitution in Equation 3.11 of the initial conditions $i = 0$ at $\omega t = \alpha$, which yields

$$i = \frac{\sqrt{2}V}{Z}\left[\sin(\omega t - \phi) - \sin(\alpha - \phi)e^{(R/L)((\alpha/\omega)-t)}\right]. \tag{3.12}$$

Also, $i = 0, \beta < \omega t < 2\pi$.

The current *extinction angle* β is determined by the load impedance and can be solved using Equation 2.12 when $i = 0$ and $\omega t = \beta$, that is,

$$\sin(\beta - \phi) = -e^{-(R\beta/\omega L)}\sin(\alpha - \phi), \tag{3.13}$$

which is a transcendental equation with an unknown value of β. The term $\sin(\beta - \phi)$ is a sinusoidal function. The term $e^{-(R\beta/\omega L)}\sin(\phi - \alpha)$ is an exponentially decaying function. The operating point of β is at the intersection of $\sin(\beta - \phi)$ and $e^{-(R\beta/\omega L)}\sin(\phi - \alpha)$, and its value can be determined by iterative methods and MATLAB®. The average output voltage is

$$V_O = \frac{1}{2\pi}\int_{\alpha}^{\beta}\sqrt{2}V\sin(\omega t)\,d(\omega t)$$

$$= \frac{V}{\sqrt{2}\pi}[\cos\alpha - \cos\beta]. \tag{3.14}$$

Example 3.1

A controlled half-wave rectifier has an AC input of 240V (rms) at 50 Hz with a load $R = 10\,\Omega$ and $L = 0.1$ H in series. The firing angle α is 45°, as shown in Figure 3.2. Determine the extinction angle β within an accuracy of 0.01° using iterative method 2 (see Section 2.2.2.3).

SOLUTION

Calculation of the extinction angle β using iterative method 2 (see Section 2.2.2.3).

$$\frac{\omega L}{R} = \pi \approx 3.14,$$

$$z = \sqrt{R^2 + \omega^2 L^2} = 33\,\Omega,$$

$$\Phi = \tan^{-1}\left(\frac{\omega L}{R}\right) = 72.34°,$$

$$\alpha = 45°, \quad V_m = \sqrt{2}V = 240\sqrt{2} = 340\,\text{V}.$$

At $\omega t = \beta$, the current is zero:

$$\sin(\beta - \phi) = e^{(\alpha-\beta)/\tan\phi}\sin(\alpha - \phi).$$

Using iterative method 2 (see Section 2.2.2.3), define

$$x = |\sin(\beta - \phi)|,$$

$$y = e^{(\alpha-\beta)/\tan\phi}\sin(\phi - \alpha) = \sin(72.34 - \alpha)e^{(\alpha-\beta)/\pi} = 0.46e^{(\alpha-\beta)/\pi}.$$

Make a table as follows:

β (deg)	x	y	$\sin^{-1} y$ (deg)	$\|x\|: y$
252.34	0	0.1454	8.36	$<$
260.7	0.1454	0.1388	7.977	$>$
260.32	0.1388	0.13907	7.994	$<$
260.33	0.13907	0.139066	7.994	$\approx$

From the above table, we can choose $\beta = 260.33°$.

3.2.3 *R–L* Load Plus Back Emf V_C

If the circuit involves an emf or battery V_C, the circuit diagram is shown in Figure 3.3. To guarantee that the thyristor is successfully fired on, the minimum firing angle is requested. If a firing angle is allowable to supply the load with an emf V_C, the minimum delay angle is

$$\alpha_{min} = \sin^{-1}\left(\frac{V_C}{\sqrt{2}V}\right). \tag{3.15}$$

This means that the firing pulse has to be applied to the thyristor when the supply voltage is higher than the emf V_C. Other characteristics can be derived as shown in Section 2.2.4.

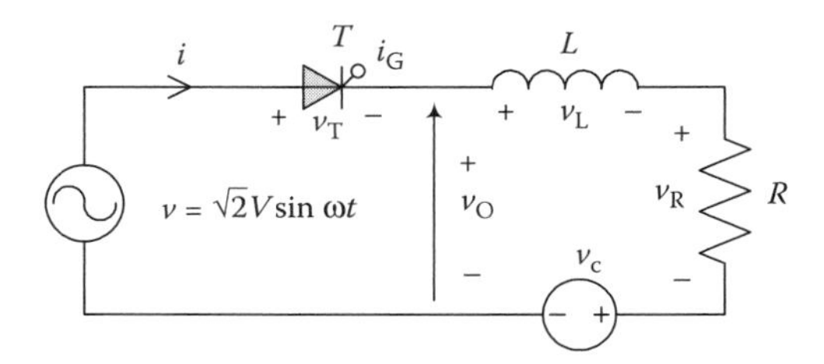

FIGURE 3.3 Half-wave controlled rectifier with *R–L* load plus an emf V_C.

Example 3.2

A controlled half-wave rectifier has an AC input of 120 V (rms) at 60 Hz, $R = 2\Omega$, $L = 20$ mH, and an emf of $V_C = 100$ V. The firing angle α is 45°. Determine

a. An expression for the current.
b. The power absorbed by the DC source V_C in the load.

Solution

From the parameters given,

$$z = \sqrt{R^2 + \omega^2 L^2} = 7.8\,\Omega,$$

$$\phi = \tan^{-1}\left(\frac{\omega L}{R}\right) = 1.312\text{ rad},$$

$$\frac{\omega L}{R} = 3.77,$$

$$\alpha = 45^\circ, \quad V_m = \sqrt{2}V = 120\sqrt{2} = 169.7\text{ V}.$$

a. First, use Equation 3.15 to determine the minimum delay angle, if $\alpha = 45^\circ$ is allowable. The minimum delay angle is

$$\alpha_{\min} = \sin^{-1}\left(\frac{100}{120\sqrt{2}}\right) = 36^\circ,$$

which indicates that $\alpha = 45^\circ$ is allowable. The equation

$$\frac{Z}{\sqrt{2}V} i = \sin(\omega t - \phi) - \left[\frac{m}{\cos\phi} - Be^{(\alpha - \omega t)/\tan\phi}\right], \quad \alpha < \omega t \le \beta,$$

$$B = \frac{m}{\cos\phi} - \sin(\alpha - \phi), \quad \omega t = \alpha,\ i = 0,$$

becomes

$$i = 21.8\sin(\omega t - 1.312) + 75e^{-\omega t/3.77} - 50 \quad \text{for } 0.785\text{ rad} \le \omega t \le 3.37\text{ rad}.$$

Here the extinction angle β is numerically found to be 3.37 rad from the equation $i(\beta) = 0$.

b. The power absorbed by the DC source V_C is

$$P_{DC} = IV_C = V_C \frac{1}{2\pi}\int_{\alpha}^{\beta} i(\omega t)\,d(\omega t) = 2.19 \times 100 = 219\text{ W}.$$

3.3 Single-Phase Full-Wave Controlled Converters

Full-wave voltage control is possible with the circuits with an R–L load shown in Figure 3.4a and b. The circuit in Figure 3.4a uses a center-tapped transformer and two thyristors, which experience a reverse bias of twice the supply. At high powers where a transformer may not be applicable, a four-thyristor configuration as in Figure 3.4b is suitable. The load current waveform becomes continuous when the (maximum) phase control angle α is given by

$$\alpha = \tan^{-1}\frac{\omega L}{R} = \phi \tag{3.16}$$

at which the output current is a rectifier sine wave.

For $\alpha > \phi$, discontinuous load current flows as shown in Figure 3.4c. At $\alpha = \phi$, the load current becomes continuous as shown in Figure 3.4d, where $\beta = \alpha + \pi$. A further decrease

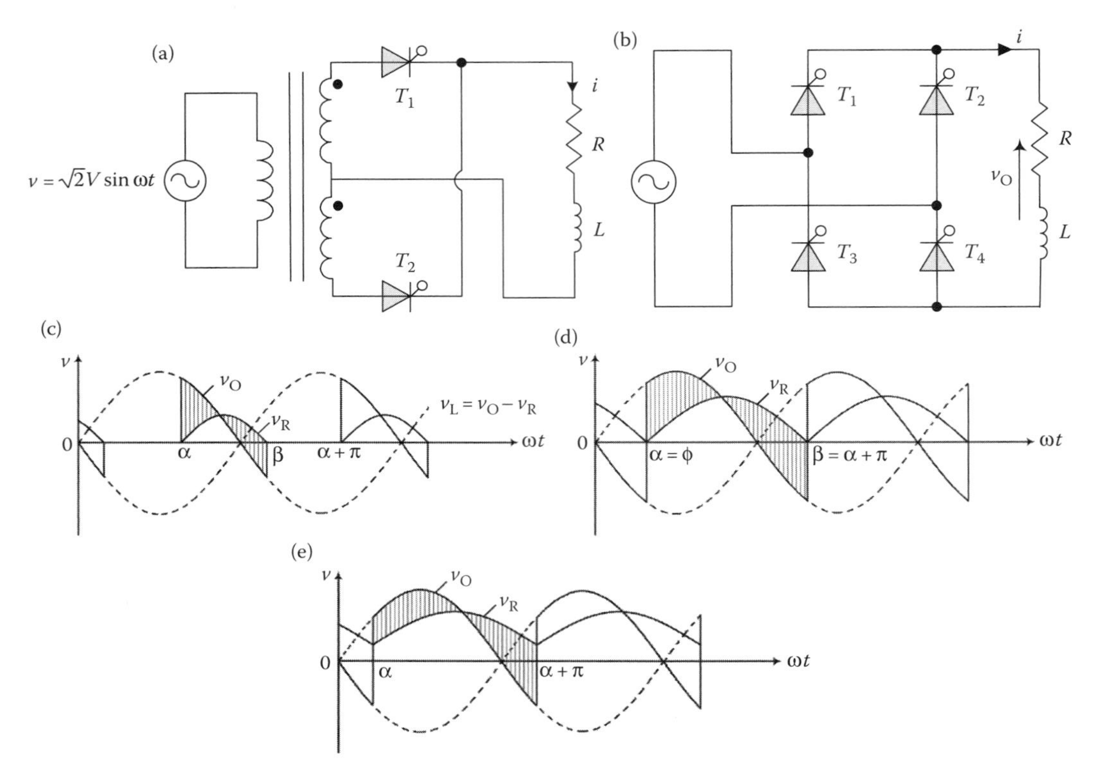

FIGURE 3.4 (a)–(e) Full-wave voltage-controlled circuit.

in α, that is, $\alpha < \phi$, results in continuous load current that is always greater than zero, as shown in Figure 3.4e.

3.3.1 $\alpha > \phi$, Discontinuous Load Current

The load current waveform is the same as for the half-wave situation considered in Section 3.2.2 by Equation 3.15, that is,

$$i = \frac{\sqrt{2}V}{Z}\left[\sin(\omega t - \phi) - \sin(\alpha - \phi)e^{(R/L)(\alpha/\omega - t)}\right]. \tag{3.17}$$

The average output voltage for this full-wave circuit will be twice that of the half-wave case in Section 3.2.2 by Equation 3.14.

$$V_O = \frac{1}{\pi}\int_{\alpha}^{\beta} \sqrt{2}V \sin(\omega t)\,d(\omega t)$$

$$= \frac{\sqrt{2}V}{\pi}[\cos\alpha - \cos\beta], \tag{3.18}$$

where β has to be found numerically.

Example 3.3

A full-wave controlled rectifier, shown in Figure 3.4, has an AC input of 240 V (rms) at 50 Hz with a load $R = 10\,\Omega$ and $L = 0.1$ H in series. The firing angle α is 80°.

a. Determine whether the load current is discontinuous. If it is, find the extinction angle β to within an accuracy of 0.01° using iterative method 2 (see Section 2.2.2.3).
b. Derive expressions for current i and output voltage v_O, and find the average output voltage V_O.

SOLUTION

a. The thyristor firing angle $\alpha = 80°$. Since the firing angle α is greater than the load phase angle $\phi = \tan^{-1}(\omega L/R) = 72.34°$, the load current is discontinuous. The extinction angle β is $>\pi$, but $<(\pi + \alpha) = 260°$. The output voltage becomes negative when $\pi \le \omega t \le \beta$.

   Calculation of the extinction angle β using iterative method 2 (see Section 2.2.2.3) is as follows:

$$\frac{\omega L}{R} = \pi \approx 3.14,$$

$$z = \sqrt{R^2 + \omega^2 L^2} = 33\,\Omega,$$

$$\Phi = \tan^{-1}\left(\frac{\omega L}{R}\right) = 72.34°,$$

$$\alpha = 80°, \quad V_m = \sqrt{2}V = 240\sqrt{2} = 340\,\text{V}.$$

   Since $\alpha > \phi$, the rectifier is working in the discontinuous current state.

   With $\omega t = \beta$ and the current is zero, we obtain the following equation

$$\sin(\beta - \phi) = e^{(\alpha-\beta)/\tan\phi}\sin(\alpha - \phi).$$

   Using iterative method 2 (see Section 2.2.2.3), we define

$$x = \sin(\beta - \phi),$$

$$y = e^{(\alpha-\beta)/\tan\phi}\sin(\alpha - \phi) = \sin(\alpha - 72.34)e^{(\alpha-\beta)/\pi} = 0.1333e^{(\alpha-\beta)/\pi}.$$

   Make a table as follows:

β (deg)	x	y	$\sin^{-1} y$ **(deg)**	$\|x\| >, =, < y$**?**
252.34	0	0.05117	2.933	<
255.273	0.05117	0.05034	2.886	>
255.226	0.05034	0.05036	2.8864	<
255.2264	0.05036	0.05036		≈

   From the above table, we choose $\beta = 255.23°$.

b. The equation of the current

$$i = \frac{\sqrt{2}V}{Z}\left[\sin(\omega t - \phi) - \sin(\alpha - \phi)e^{(R/L)(\alpha/\omega - t)}\right]$$

becomes

$$i = \frac{\sqrt{2}V}{Z}[\sin(\omega t - \phi) + 0.1333e^{(\alpha-\omega t)/\pi}] = 10.29\sin(\omega t - 72.34) + 1.37e^{(\alpha-\omega t)/\pi}.$$

The current expression is

$$i = 10.29\sin(\omega t - 72.34) + 2.138e^{-\omega t/\pi}.$$

The output voltage expression in a period is

$$v_O(t) = \begin{cases} 240\sqrt{2}\sin\omega t & \alpha \le \omega t \le \beta, (\pi + \alpha) \le \omega t \le (\pi + \beta), \\ 0 & \text{otherwise.} \end{cases}$$

The average output voltage V_O is

$$V_O = \frac{1}{\pi}\int_{\alpha}^{\beta} v\,d(\omega t) = \frac{240\sqrt{2}}{\pi}\int_{\alpha}^{\beta}\sin(\omega t)\,d(\omega t) = \frac{240\sqrt{2}}{\pi}(\cos\alpha - \cos\beta)$$

$$= \frac{240\sqrt{2}}{\pi}(0.1736 + 0.2549) = 46.3\text{ V}.$$

3.3.2 $\alpha < \phi$, Verge of Continuous Load Current

When $\alpha = \phi$, the load current is given by

$$i = \frac{\sqrt{2}V}{Z}\sin(\omega t - \varphi), \quad \varphi < \omega t < \varphi + \pi, \tag{3.19}$$

and the average output voltage is given by

$$V_O = \frac{2\sqrt{2}V}{\pi}\cos\alpha, \tag{3.20}$$

which is independent of the load.

3.3.3 $\alpha < \phi$, Continuous Load Current

Under these conditions, a thyristor is still conducting when another is forward biased and turned on. The first device is instantaneously reverse biased by the second device that has been turned on. The average output voltage is

$$V_O = \frac{2\sqrt{2}V}{\pi}\cos\alpha. \tag{3.21}$$

The rms output voltage is

$$V_r = V. \tag{3.22}$$

3.4 Three-Phase Half-Wave Controlled Rectifiers

A three-phase half-wave controlled rectifier is shown in Figure 3.5. The input three-phase voltages are:

$$\begin{aligned} v_a(t) &= \sqrt{2}V \sin \omega t, \\ v_b(t) &= \sqrt{2}V \sin(\omega t - 120°), \\ v_c(t) &= \sqrt{2}V \sin(\omega t + 120°). \end{aligned} \tag{3.23}$$

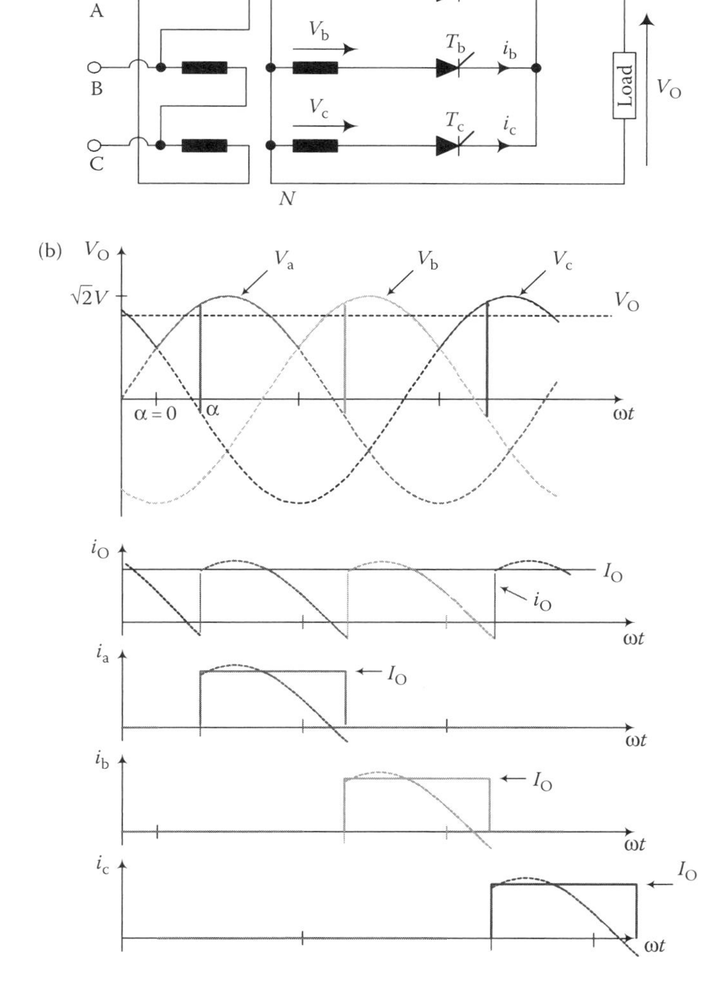

FIGURE 3.5 Three-phase half-wave controlled rectifier: (a) circuit and (b) waveforms.

Usually the load is an inductive load, that is, R–L load. If the inductance is large enough, the load current is continuous for most of the firing angle α, and the corresponding voltage and current waveforms are shown in Figure 3.5b. Each thyristor conducts for 120° a cycle. If the load is a pure resistive load and the firing angle is $0 < \alpha < \pi/6$, the output voltage and current are continuous and each thyristor is conducted in 120° a cycle. If the firing angle $\alpha > \pi/6$ (or 30°), the output voltage and current are discontinuous and each thyristor is conducting in the period between α to 150° a cycle.

3.4.1 *R* Load Circuit

If the load is a resistive load and the firing angle $\alpha \leq \pi/6$ ($\omega t = \alpha + \pi/6$), referring to Figure 3.5, the output voltage is

$$V_O = \frac{3}{2\pi}\int_{\alpha+(\pi/6)}^{\alpha+(5\pi/6)} \sqrt{2}V\sin(\omega t)\,\mathrm{d}(\omega t) = \frac{3V}{\sqrt{2}\pi}\left[\cos\left(\alpha+\frac{\pi}{6}\right) - \cos\left(\alpha+\frac{5\pi}{6}\right)\right]$$
$$= \frac{3\sqrt{3}V}{\sqrt{2}\pi}\cos\alpha = V_{dO}\cos\alpha. \tag{3.24}$$

Here V_{dO} is the output voltage corresponding to the firing angle $\alpha = 0$,

$$V_{dO} = \frac{3\sqrt{3}V}{\sqrt{2}\pi} = 1.17\,\mathrm{V}. \tag{3.25}$$

For $\alpha = \pi/6$, the output current is

$$I_O = \frac{V_O}{R} = \frac{V_{dO}}{R}\cos\alpha = 1.17\frac{V}{R}\cos\alpha. \tag{3.26}$$

If the load is a resistive load and the firing angle $\pi/6 < \alpha < 5\pi/6$ ($\omega t = \alpha + \pi/6$), the output voltage is

$$V_O = \frac{3}{2\pi}\int_{\alpha+(\pi/6)}^{\pi} \sqrt{2}V\sin(\omega t)\,\mathrm{d}(\omega t) = \frac{3V}{\sqrt{2}\pi}\left[\cos\left(\alpha+\frac{\pi}{6}\right)+1\right]$$
$$= \frac{3V}{\sqrt{2}\pi}\left(\frac{\sqrt{3}}{2}\cos\alpha - \frac{\sin\alpha}{2} + 1\right) = 0.675V\left(\frac{\sqrt{3}}{2}\cos\alpha - \frac{\sin\alpha}{2} + 1\right). \tag{3.27}$$

The output current is

$$I_0 = \frac{V_O}{R} = \frac{0.675V}{R}\left(\frac{\sqrt{3}}{2}\cos\alpha - \frac{\sin\alpha}{2} + 1\right). \tag{3.28}$$

Since $\pi/6 < \alpha < 5\pi/6$, the output current is always positive.

When $\alpha \geq 5\pi/6$, both the output voltage and current are zero. In this case, all thyristors are reversely biased when firing pulses are applied. Therefore, all thyristors cannot be conducting.

Example 3.4

A three-phase half-wave controlled rectifier shown in Figure 3.5 has an AC input of 200 V (rms) at 50 Hz with a load $R = 10\,\Omega$. The firing angle α is

a. 20°.
b. 60°.

Calculate the output voltage and current.

Solution

a. The firing angle $\alpha = 20°$, and the output voltage and current are continuous. Referring to Equations 3.24 through 3.26, the output voltage and current are

$$V_O = 1.17 V_{in} \cos\alpha = 1.17 \times 200 \times \cos 20° = 220\,\text{V},$$

$$I_O = \frac{V_O}{R} = \frac{220}{10} = 22\,\text{A}.$$

b. The firing angle $\alpha = 60°$, which is $>\pi/6 = 30°$. The output voltage and current are discontinuous. Referring to Equations 3.27 and 3.28, the output voltage and current are

$$V_O = 0.675V\left(\frac{\sqrt{3}}{2}\cos\alpha - \frac{\sin\alpha}{2} + 1\right)$$

$$= 0.675 \times 200(0.433 - 0.433 + 1) = 135\,\text{V},$$

$$I_O = \frac{V_O}{R} = \frac{135}{10} = 13.5\,\text{A}.$$

3.4.2 *R–L* Load Circuit

Figure 3.6 shows four circuit diagrams for an *R–L* load. If the inductance is large enough and can maintain current continuity, the output voltage is

$$V_O = V_{dO}\cos\alpha = 1.17V\cos\alpha. \tag{3.29}$$

For ($\alpha < \pi/2$), the output current is

$$I_O = \frac{V_O}{R} = \frac{V_{dO}}{R}\cos\alpha = 1.17\frac{V}{R}\cos\alpha. \tag{3.30}$$

When the firing angle α is $>\pi/2$, the output voltage can have negative values, but the output current can only have positive values. This situation corresponds to the regenerative state.

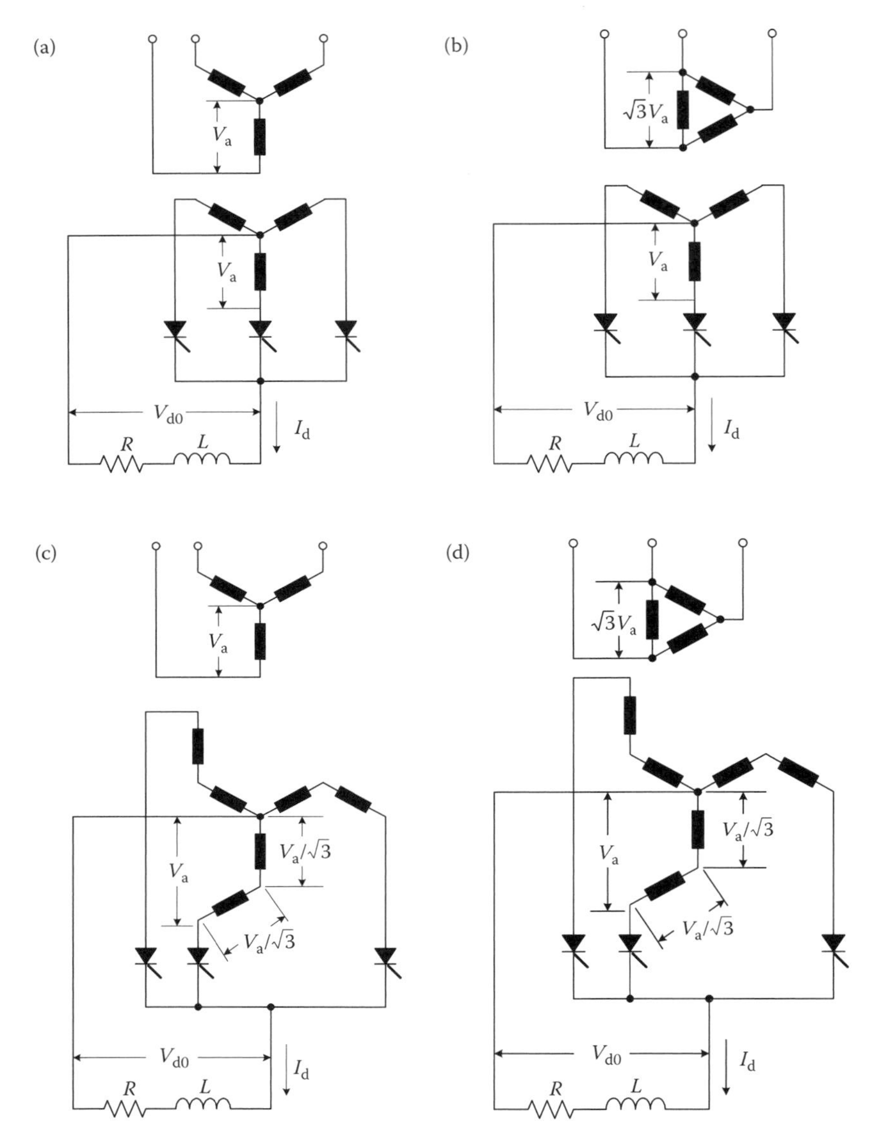

FIGURE 3.6 Three-phase half-wave controlled rectifiers: (a) Y/Y circuit, (b) Δ/Y circuit, (c) Y/Y bending circuit, and (d) Δ/Y bending circuit.

Example 3.5

A three-phase half-wave controlled rectifier shown in Figure 3.5 has an AC input of 200 V (rms) at 50 Hz with a load $R = 10\,\Omega$ plus a large inductance that can maintain the continuous output current. The firing angle α is

a. 20°.
b. 100°.

Calculate the output voltage and current.

Solution

a. The firing angle $\alpha = 20°$, and the output voltage and current are continuous. Referring to Equations 3.24 through 3.26, the output voltage and current are

$$V_O = 1.17V_{in} \cos\alpha = 1.17 \times 200 \times \cos 20° = 220\,\text{V},$$

$$I_O = \frac{V_0}{R} = \frac{220}{10} = 22\,\text{A}.$$

b. The firing angle $\alpha = 100°$, but the large inductance can maintain the output current as continuous. The output voltage and current are continuous and have negative values. Referring to Equations 3.29 and 3.30, the output voltage and current are

$$V_O = 1.17V_{in} \cos\alpha = 1.17 \times 200 \times \cos 100° = -40.6\,\text{V},$$

$$I_O = \frac{V_0}{R} = \frac{-40.6}{10} = -4.06\,\text{A}.$$

3.5 Six-Phase Half-Wave Controlled Rectifiers

Six-phase half-wave controlled rectifiers have two constructions: six-phase with a neutral line circuit and double antistar with a balance-choke circuit. The following description is based on the R load or R plus large L load.

3.5.1 Six-Phase with a Neutral Line Circuit

If the AC power supply is from a transformer, four circuits can be used. The six-phase half-wave rectifiers are shown in Figure 3.7.

The power supply is a six-phase balanced voltage source. Each phase is shifted by 60°.

$$\begin{aligned}
v_a(t) &= \sqrt{2}V \sin \omega t,\\
v_b(t) &= \sqrt{2}V \sin(\omega t - 60°),\\
v_c(t) &= \sqrt{2}V \sin(\omega t - 120°),\\
v_d(t) &= \sqrt{2}V \sin(\omega t - 180°),\\
v_e(t) &= \sqrt{2}V \sin(\omega t - 240°),\\
v_f(t) &= \sqrt{2}V \sin(\omega t - 300°).
\end{aligned} \tag{3.31}$$

The first circuit is called a Y/star circuit, shown in Figure 3.7a; the second circuit is called a Δ/star circuit, shown in Figure 3.7b; the third circuit is called a Y/star bending circuit, shown in Figure 3.7c, and the fourth circuit is called a Δ/star bending circuit, shown in Figure 3.7d. Each diode is conducted in 60° a cycle. The firing angle $\alpha = \omega t - \pi/3$ in the

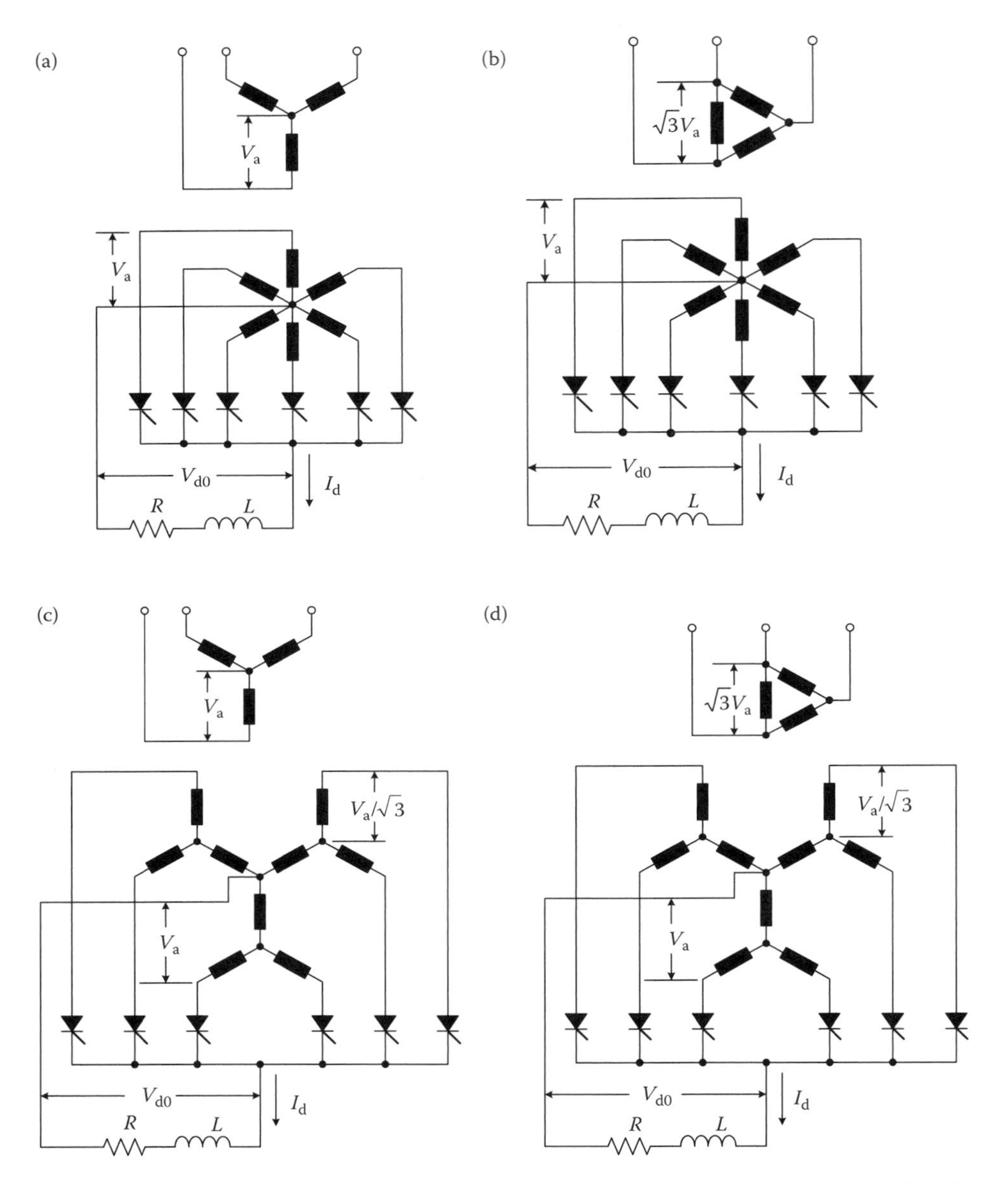

FIGURE 3.7 Six-phase half-wave controlled rectifiers: (a) Y/Star circuit, (b) Δ/Star circuit, (c) Y/Star bending circuit, and (d) Δ/Star bending circuit.

range of 0–2π/3. Since the load is an R–L circuit, the output voltage average value is

$$V_O = \frac{1}{\pi/3}\int_{\pi/3+\alpha}^{2\pi/3+\alpha} \sqrt{2}V\sin(\omega t)\,d(\omega t) = \frac{3\sqrt{2}V}{\pi}\left[\cos\left(\frac{\pi}{3}+\alpha\right) - \cos\left(\frac{2\pi}{3}+\alpha\right)\right]$$
$$= \frac{3\sqrt{2}}{\pi}V\cos\alpha = 1.35V\cos\alpha. \tag{3.32}$$

The output voltage can have positive ($\alpha < \pi/2$) and negative ($\alpha > \pi/2$) values. When $\alpha < \pi/2$, the output current is

$$I_O = \frac{V_O}{R} = \frac{3\sqrt{2}}{\pi R} V \cos\alpha = 1.35\frac{V}{R}\cos\alpha. \tag{3.33}$$

When the firing angle α is $> \pi/2$, the output voltage can have negative values, but the output current can only have positive values. This situation corresponds to the regenerative state.

3.5.2 Double Antistar with a Balance-Choke Circuit

If the AC power supply is from a transformer, two circuits can be used. Six-phase half-wave controlled rectifiers are shown in Figure 3.8. The three-phase double antistar with balance-choke controlled rectifiers is shown in Figure 3.8. The first circuit is called a Y/Y-Y circuit, shown in Figure 3.8a, and the second circuit is called a Δ/Y-Y circuit, shown in Figure 3.8b. Each device is conducted in 120° a cycle. The firing angle $\alpha = \omega t - \pi/6$. Since the load is an R–L circuit, the average output voltage value is

$$V_O = \frac{1}{2\pi/3}\int_{\pi/6+\alpha}^{5\pi/6+\alpha} \sqrt{2}V\sin(\omega t)\,d(\omega t) = \frac{3\sqrt{3}}{\sqrt{2}\pi}V\cos\alpha = 1.17V\cos\alpha. \tag{3.34}$$

The output voltage can have positive ($\alpha < \pi/2$) and negative ($\alpha > \pi/2$) value. The output current is

$$I_O = \frac{V_O}{R} = 1.17\frac{V}{R}\cos\alpha. \tag{3.35}$$

When the firing angle α is $> \pi/2$, the output voltage can have a negative value, but the output current can have only a positive value. This situation corresponds to the regenerative state.

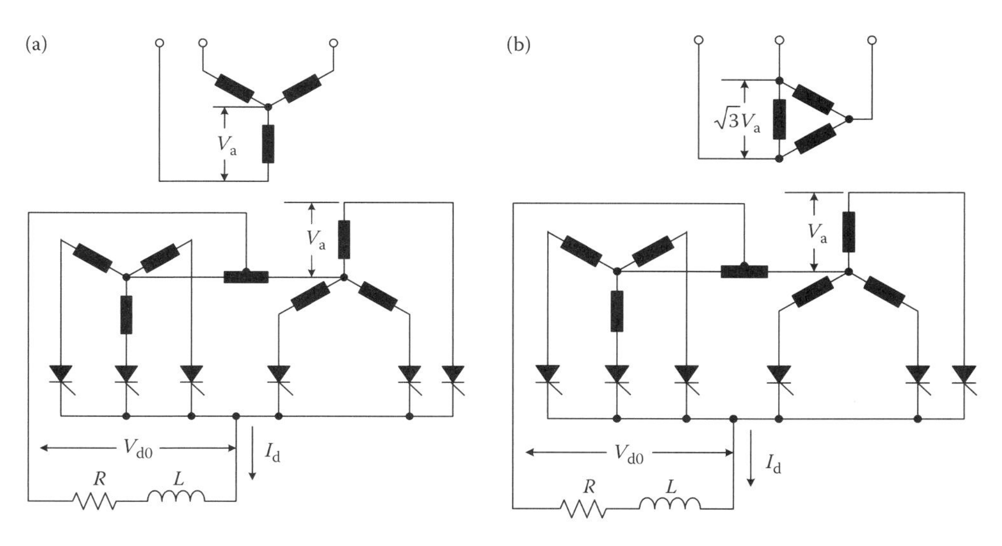

FIGURE 3.8 Three-phase double antistar with balance-choke controlled rectifiers: (a) Y/Y-Y circuit and (b) Δ/Y-Y circuit.

These circuits have the following advantages:

- A large output current can be obtained since there are two three-phase half-wave rectifiers.
- The output voltage has a lower ripple since each thyristor conducts at 120°.

3.6 Three-Phase Full-Wave Controlled Converters

A three-phase bridge is fully controlled when all six bridge devices are thyristors, as shown in Figure 3.9. The frequency of the output voltage ripple is six times the supply frequency. The average output voltage is given by

$$\begin{aligned} V_O &= \frac{3}{\pi}\int_{-\pi/3+\alpha}^{\alpha} v_{ry}\,\mathrm{d}(\omega t) = \frac{3}{\pi}\int_{-\pi/3+\alpha}^{\alpha} \sqrt{3}\sqrt{2}V\sin\left(\frac{\omega t+2\pi}{3}\right)\mathrm{d}(\omega t) \\ &= \frac{3\sqrt{3}}{\pi}\sqrt{2}V\ \cos\alpha = 2.34V\cos\alpha. \end{aligned} \tag{3.36}$$

The equation illustrates that the rectifier DC output voltage V_O is positive when the firing angle α is $<\pi/2$ and becomes negative for a firing angle $\alpha > \pi/2$. However, the DC current I_O is always positive irrespective of the polarity of the DC output voltage.

When the rectifier produces a positive DC voltage, the power is delivered from the supply to the load. With a negative DC voltage, the rectifier operates in an *inverter mode* and the power is fed from the load back to the supply. This phenomenon is usually used in electrical drive systems where the motor drive is allowed to decelerate and the kinetic energy of the motor and its mechanical load is converted to electrical energy and then sent back to the power supply by the thyristor rectifier for fast *dynamic braking*. The power flow in the thyristor rectifier is therefore *bidirectional*.

Figure 3.10 shows some waveforms corresponding to various firing angles. The shaded area A is the device conduction period and the corresponding rectified voltage.

The rms value of the output voltage is given by

$$V_{rms} = \sqrt{\frac{3}{\pi}\int_{-\pi/3+\alpha}^{\alpha}\left[\sqrt{3}\sqrt{2}V\sin\left(\frac{\omega t+2\pi}{3}\right)\right]^2\mathrm{d}(\omega t)} = \sqrt{2}\sqrt{6}V\left[\frac{1}{4}+\frac{3\sqrt{3}}{8\pi}\cos 2\alpha\right]^{1/2}. \tag{3.37}$$

The line current i_r can be expressed in a Fourier series as

$$\begin{aligned} i_r = \frac{2\sqrt{3}}{\pi}I_{DC}\Big[&\sin(\omega t-\varphi_1) - \frac{1}{5}\sin 5(\omega t-\varphi_1) \\ &-\frac{1}{7}\sin 7(\omega t-\varphi_1) + \frac{1}{11}\sin 11(\omega t-\varphi_1) + \frac{1}{13}\sin 13(\omega t-\varphi_1) - \cdots\Big], \end{aligned} \tag{3.38}$$

(a)

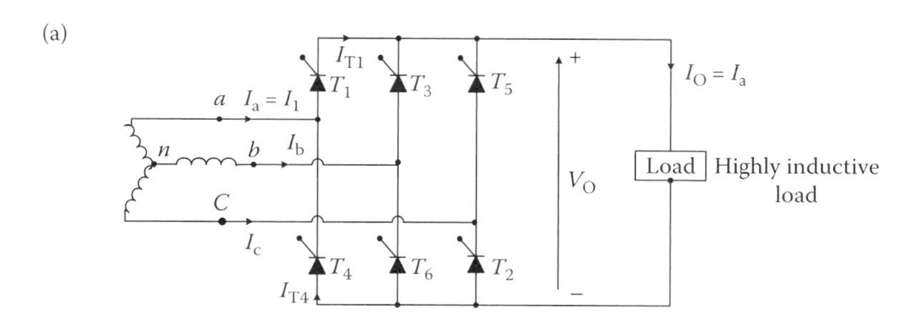

(b)

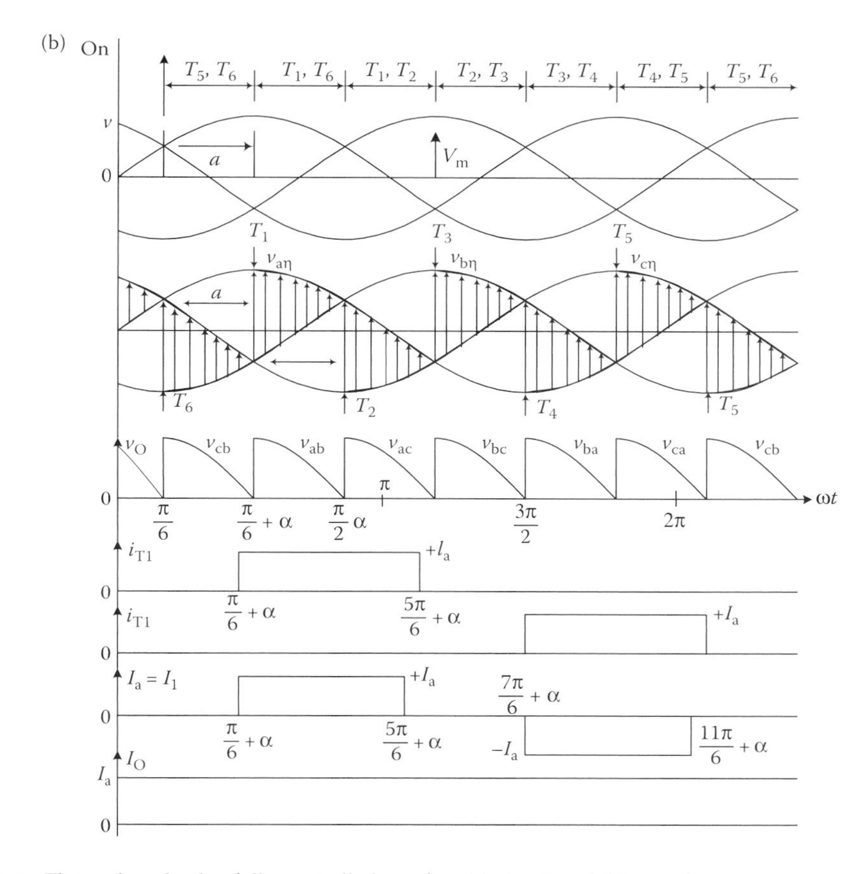

FIGURE 3.9 Three-phase bridge fully controlled rectifier: (a) circuit and (b) waveforms.

where ϕ_1 is the phase angle between the supply voltage v_r and the fundamental frequency line current i_{r1}. The rms value of i_r can be calculated using

$$I_r = \sqrt{\frac{1}{2\pi}\int_0^{2\pi} i_r \, d(\omega t)} = \sqrt{\frac{1}{2\pi}\left[\int_{-60+\alpha}^{60+\alpha} I_{DC}^2 \, d(\omega t) + \int_{120+\alpha}^{240+\alpha} I_{DC}^2 \, d(\omega t)\right]}$$

$$= \sqrt{\frac{2}{3}} I_{DC} = 0.816 I_{DC}. \tag{3.39}$$

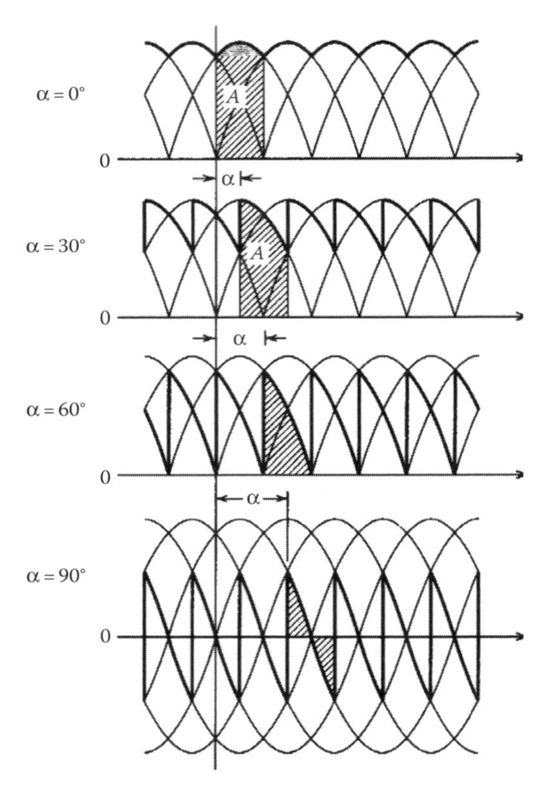

FIGURE 3.10 Rectified voltage waveforms for various firing angles.

from which the THD for the line current i_r is

$$\text{THD} = \frac{\sqrt{I_r^2 - I_{r1}^2}}{I_{r1}} = \frac{(0.816I_{DC})^2 - (0.78I_{DC})^2}{0.78I_{DC}} = 0.311, \tag{3.40}$$

where I_{r1} is the rms value of i_{r1} (i.e., $(\sqrt{6}/\pi)I_{DC}$).

Example 3.6

A three-phase full-wave controlled rectifier shown in Figure 3.9 has an AC input of 200 V (rms) at 50 Hz with a load $R = 10\,\Omega$ plus a large inductance that can maintain the continuous output current. Given that the firing angle α is (a) 30° and (b) 120°, calculate the output voltage and current.

Solution

a. With a firing angle $\alpha = 30°$ and the output voltage and current continuous, by referring to Equation 3.36, the output voltage and current are

$$V_O = 2.34V\cos\alpha = 2.34 \times 200\cos 30° = 234\text{V},$$

$$I_O = \frac{V_O}{R} = \frac{234}{10} = 23.4\,\text{A}.$$

b. With a firing angle $\alpha = 120°$ and the output voltage and current continuous and with negative values, by referring to Equation 3.36, the output voltage and current are

$$V_O = 2.34V \cos\alpha = 2.34 \times 200 \cos 120° = -234 \text{ V},$$

$$I_O = \frac{V_O}{R} = \frac{-234}{10} = -23.4 \text{ A}.$$

3.7 Multiphase Full-Wave Controlled Converters

Figure 3.11 shows the typical configuration of a 12-pulse series-type controlled rectifier. There are two identical three-phase controlled rectifiers to be used. Two six-pulse controlled rectifiers are powered by a phase-shifting transformer with two secondary windings in delta and star connections. Therefore, the phase angle between both secondary windings shifts 30°.

The DC outputs of the rectifiers are connected in series. To dominate lower-order harmonics in the line current i_A, the line-to-line voltage v_{a1b1} of the star-connected secondary winding is in phase with the primary voltage v_{AB}, while the delta-connected secondary winding voltage v_{a1b1} leads the primary voltage v_{AB} by

$$\delta = \angle v_{a2b2} - \angle V_{AB} = 30°. \tag{3.41}$$

The rms line-to-line voltage of each secondary winding is

$$V_{a1b1-rms} = V_{a2b2-rms} = \frac{V_{AB-rms}}{2}, \tag{3.42}$$

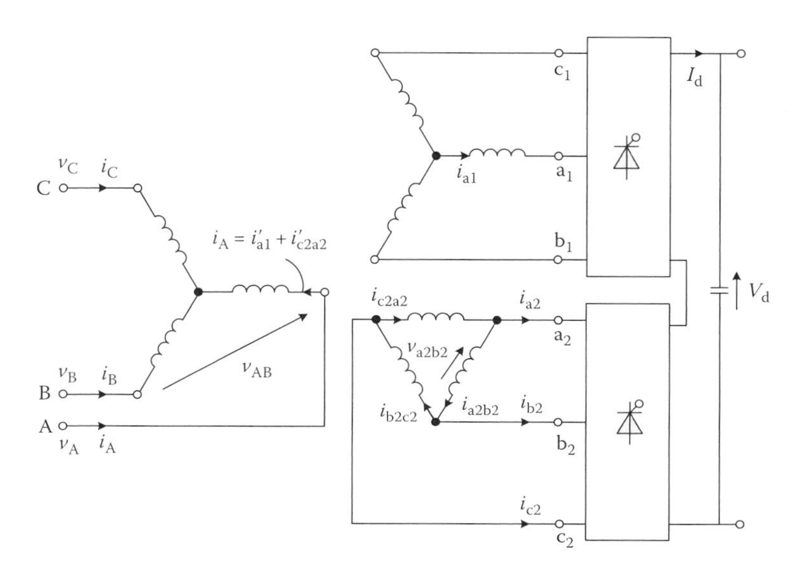

FIGURE 3.11 Twelve-pulse controlled rectifier.

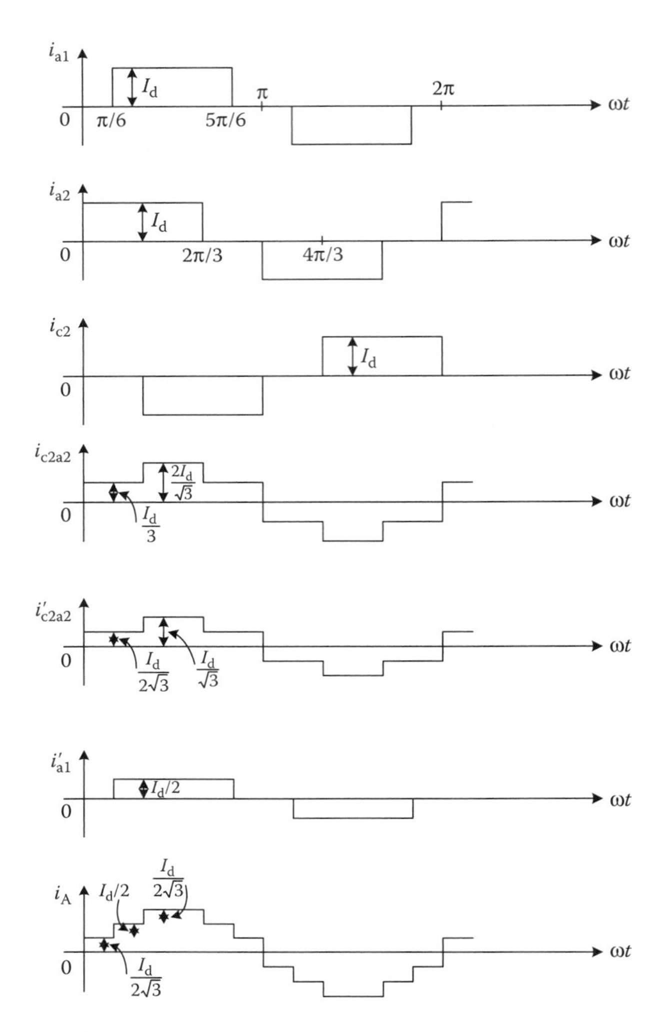

FIGURE 3.12 Current waveforms.

from which the turn's ratio of the transformer can be determined by

$$\begin{aligned} \frac{N_1}{N_2} &= 2 \qquad \text{for } Y/Y, \\ \frac{N_1}{N_3} &= \frac{2}{\sqrt{3}} \quad \text{for } Y/\Delta. \end{aligned} \tag{3.43}$$

Consider an idealized 12-pulse rectifier where the line inductance L_s and the total leakage inductance L_{lk} of the transformer are assumed to be zero. The current waveforms are illustrated in Figure 3.12, where i_{a1} and i_{c2a2} are the secondary line primary currents referred from the secondary side, and i_A is the primary line current given by $i_A = i'_{a1} + i'_{c2a2}$.

The secondary line current i_{a1} can be expressed as

$$i_{a1} = \frac{2\sqrt{3}}{\pi} I_d \left(\sin \omega t - \frac{1}{5} \sin 5\omega t - \frac{1}{7} \sin 7\omega t + \frac{1}{11} \sin 11\omega t + \frac{1}{13} \sin 13\omega t + \cdots \right), \tag{3.44}$$

where $\omega = 2\pi f$ is the angular frequency of the supply voltage. Since the waveform of current i_{a1} is of half-wave symmetry, it does not contain any even-order harmonics. Current i_A does not contain any triple harmonics either due to the balanced three-phase system.

Other secondary currents such as i_{a2} lead i_{a1} by 30°, and the Fourier expression is

$$i_{a2} = \frac{2\sqrt{3}}{\pi} I_d [\sin(\omega t + 30°) - \frac{1}{5}\sin 5(\omega t + 30°) - \frac{1}{7}\sin 7(\omega t + 30°) + \frac{1}{11}\sin 11(\omega t + 30°) + \frac{1}{13}\sin 13(\omega t + 30°) \cdots]. \tag{3.45}$$

The waveform for the referred current i'_{a1} in Figure 3.12 is identical to i_{a1} except that its magnitude is halved due to the turn's ratio of the Y/Y-connected windings. The current i'_{a1} can be expressed in Fourier series as

$$i'_{a1} = \frac{\sqrt{3}}{\pi} I_d \left(\sin \omega t - \frac{1}{5}\sin 5\,\omega t - \frac{1}{7}\sin 7\omega t + \frac{1}{11}\sin 11\omega t + \frac{1}{13}\sin 13\omega t \cdots \right). \tag{3.46}$$

The phase currents i_{b2a2}, i_{a2c2}, and i_{c2b2} can be derived from the line currents using the relationships in Equation 3.47:

$$\begin{pmatrix} i_{a2b2} \\ i_{b2c2} \\ i_{c2a2} \end{pmatrix} = \frac{1}{3} \begin{pmatrix} -1 & 1 & 0 \\ 0 & -1 & 1 \\ 1 & 0 & -1 \end{pmatrix} \begin{pmatrix} i_{a2} \\ i_{b2} \\ i_{c2} \end{pmatrix}. \tag{3.47}$$

These currents have a stepped waveform, each step being of 60° duration and the height of the steps being $I_d/3$ and $2I_d/3$. The currents i_{a2b2}, i_{b2a2}, and i_{c2a2} need to be multiplied by $\sqrt{3}/2$ when they are referred to the primary side. Using Equation 3.45 and similar equations for i_{b2} and i_{c2}, one can derive Fourier expressions for i_{a2b2}, i_{b2c2}, and i_{c2a2}. For example,

$$i_{a2b2} = \frac{1}{3}(i_{b2} - i_{a2}), \quad i_{b2c2} = \frac{1}{3}(i_{c2} - i_{b2}), \quad \text{and} \quad i_{c2a2} = \frac{1}{3}(i_{a2} - i_{c2}).$$

Therefore,

$$i_{c2a2} = \frac{1}{3}\frac{2\sqrt{3}}{\pi} I_d \left[\sin(\omega t + 30°) - \frac{1}{5}\sin 5(\omega t + 30°) - \frac{1}{7}\sin 7(\omega t + 30°) + \frac{1}{11}\sin 11(\omega t + 30°) \cdots + \sin(\omega t + 150°) - \frac{1}{5}\sin 5(\omega t + 150°) - \frac{1}{7}\sin 7(\omega t + 150°) + \frac{1}{11}\sin 11(\omega t + 150°) \cdots \right]. \tag{3.48}$$

By simplifying Equation 3.48 and multiplying with $\sqrt{3}/2$, we have

$$i'_{c2a2} = \frac{\sqrt{3}}{\pi} I_E \left(\sin \omega t + \frac{1}{5}\sin 5\omega t + \frac{1}{7}\sin 7\omega t + \frac{1}{11}\sin 11\,\omega t \cdots \right). \tag{3.49}$$

As can be seen from Equation 3.48, the phase angles of some harmonic currents are altered due to the Y/Δ-connected windings. As a result, the current i'_{c2a2} does not maintain the same wave shape as i'_{a1}. The line current i_A can be found from

$$i_A = i'_{a1} + i'_{c2a2} = \frac{2\sqrt{3}}{\pi} I_d \left(\sin \omega t + \frac{1}{11}\sin 11\omega t + \frac{1}{13}\sin 13\,\omega t \cdots \right),$$

where the two dominant current harmonics, the 5th and 7th, are cancelled in addition to the 17th and 19th.

The THD of the secondary and primary line currents i_{a1} and i_A can be determined by

$$\mathrm{THD}(i_{a1}) = \frac{\sqrt{I_{a1}^2 - I_{a1,1}^2}}{I_{a1,1}} = \frac{\sqrt{I_{a1,5}^2 + I_{a1,7}^2 + \cdots}}{I_{a1,1}} \tag{3.50}$$

and

$$\mathrm{THD}(i_A) = \frac{\sqrt{I_A^2 - I_{A,1}^2}}{I_{A,1}} = \frac{\sqrt{I_{A,11}^2 + I_{A,13}^2 + \cdots}}{I_{a1,1}}. \tag{3.51}$$

The THD of the primary line current i_A in the idealized 12-pulse rectifier is reduced by nearly 50% compared to that of i_{a1}.

3.7.1 Effect of Line Inductance on Output Voltage (Overlap)

We now investigate a three-phase fully-controlled rectifier as shown in Figure 3.9a. We partially redraw the circuit in Figure 3.13 (only show phase A and phase C). In practice, the cable length from phase A to A′ (or C to C′) has an inductance (L). The commutation process (e.g., for i_a to replace i_c) will take a certain time interval. This affects the voltage at point P to neutral point N and the final half output voltage is V_{PN}.

During the commutation process (e.g., for i_A to replace i_C), Kirchhoff's voltage law (KVL) for the commutation loop and Kirchhoff's current law (KCL) at point P give the output current I_O, which is filtered by a large inductance, and this implies that its change is much slower than that of i_C and i_A. We can write

$$v_{AN} - v_{CN} = L\frac{di_A}{dt} - L\frac{di_C}{dt}, \tag{3.52}$$

$$i_A + i_C = I_O \Rightarrow \frac{di_A}{dt} + \frac{di_C}{dt} = \frac{dI_O}{dt} \Rightarrow 0, \tag{3.53}$$

$$\frac{di_A}{dt} = -\frac{di_C}{dt}. \tag{3.54}$$

From Equations 3.52 and 3.54,

$$v_{AN} - v_{CN} = L\frac{di_A}{dt} - L\frac{di_C}{dt} \Rightarrow L\frac{di_A}{dt} = \frac{v_{AN} - v_{CN}}{2}. \tag{3.55}$$

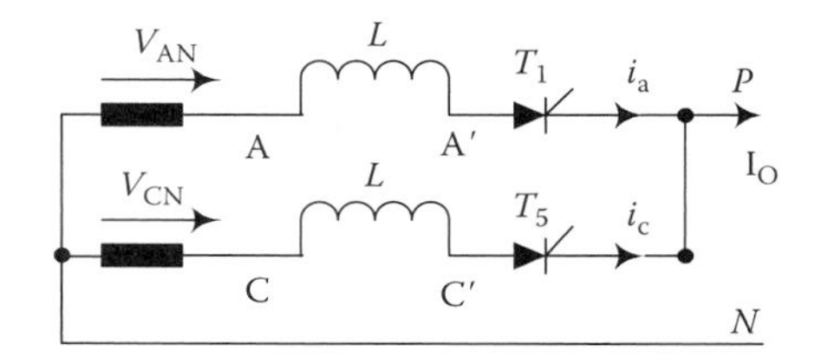

FIGURE 3.13 Effect of line inductance.

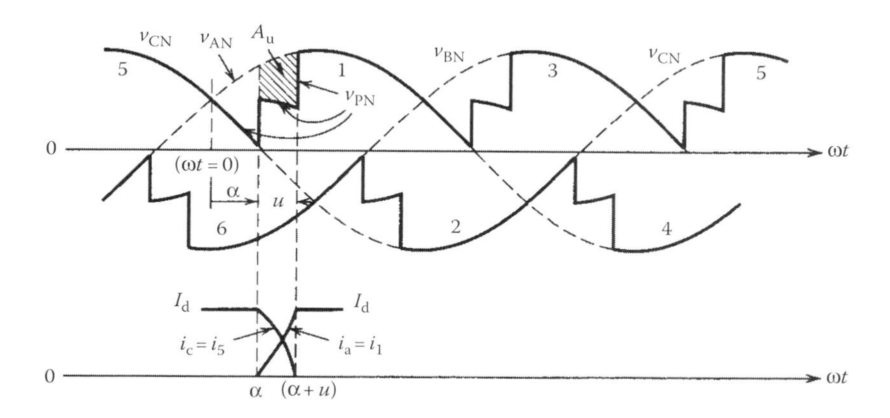

FIGURE 3.14 Waveforms affected by line inductance.

This allows one to derive V_{PN}. Thus, V_{PN} takes the midpoint value between V_{AN} and V_{CN} during commutation. The output voltage waveform is shown in Figure 3.14.

$$v_{PN} = v_{AN} - L\frac{di_A}{dt} = v_{AN} - \frac{v_{AN} - v_{CN}}{2} \Rightarrow v_{PN} = \frac{v_{AN} + v_{CN}}{2}. \tag{3.56}$$

Thus, the integral of V_{PN} will involve two parts: one from firing angle α to $(\alpha + u)$ where u is the overlap angle and, subsequently, the other from $(\alpha + u)$ to the next phase fired, where commutation happened and the v_{PN} is

$$v_{PN} = \frac{3}{2\pi}\left[\int_{\pi/6+\alpha}^{\pi/6+\alpha+u} \frac{v_{AN} + v_{CN}}{2}\,d(\omega t) + \int_{\pi/6+\alpha+u}^{\pi/6+\alpha+2\pi/3} v_{AN}\,d(\omega t)\right]. \tag{3.57}$$

Hence

$$v_{PN} = \frac{3}{2\pi}\left[\int_{\pi/6+\alpha}^{\pi/6+\alpha+u} \frac{v_{AN} + v_{CN}}{2}\,d(\omega t) + \int_{\pi/6+\alpha+u}^{\pi/6+\alpha+2\pi/3} v_{AN}\,d(\omega t)\right.$$
$$\left. + \int_{\pi/6+\alpha}^{\pi/6+\alpha+u} \frac{v_{AN} - v_{CN}}{2}\,d(\omega t) - \int_{\pi/6+\alpha}^{\pi/6+\alpha+u} \frac{v_{AN} - v_{CN}}{2}\,d(\omega t)\right]$$

Note that the first integral is the original integral involving V_{AN} for the full 120° interval. The second interval can be linked to the derivative of current i_A for the commutation interval

$$\int_{\pi/6+\alpha}^{\pi/6+\alpha+u} \frac{v_{AN} - v_{CN}}{2} \, d(\omega t) = \int_{\pi/6+\alpha}^{\pi/6+\alpha+u} \left(L\frac{di_A}{dt} \right) \frac{\omega}{\omega} \, d(\omega t)$$
$$= \int_{i_{start}}^{i_{end}} L\omega \, d(i_A) = L\omega[I_O + 0] = L\omega I_O. \tag{3.58}$$

Therefore, by an identical analysis for the bottom three thyristors, the output voltage is

$$V_O = 2v_{PN} = \frac{3\sqrt{6}}{\pi} V \cos\alpha - \frac{3}{\pi} L\omega I_O \quad \text{for } 0^\circ < \alpha < 180^\circ. \tag{3.59}$$

Thus, the commutation interval duration due to the line inductance modifies the output voltage waveform (finite time for current change) and this changes the average output voltage *by a reduction* of $(3/\pi)L\omega$. This can be compensated for by feedforward.

The above figure shows how V_{PN} is affected during the commutation interval u. It takes the midpoint value between the incoming phase (V_{AN}) and outgoing phase (V_{CN}) voltages. The corresponding currents i_A and i_C can be seen to rise and fall at finite rates. The rate of current change will be slower for high values of line interference (EMI) and certain standards limit this rise time.

Homework

3.1. A full-wave controlled rectifier shown in Figure 3.4 has a source of 120 V (rms) at 60 Hz, $R = 10\,\Omega$, $L = 20$ mH, and $\alpha = 60^\circ$.

   a. Determine an expression for load current.

   b. Determine the average load current.

   c. Determine the average output voltage.

3.2. A three-phase half-wave controlled rectifier shown in Figure 3.5 has an AC input of 240 V (rms) at 60 Hz with a load $R = 100\,\Omega$. Given that are firing angle α is (a) 15° and (b) 75°, calculate the output voltage and current.

3.3. A three-phase full-wave controlled rectifier shown in Figure 3.9 has an AC input of 240 V (rms) at 60 Hz with a load $R = 100\,\Omega$ with high inductance. Given that the firing angle α is (a) 20° and (b) 100°, calculate the output voltage and current.

References

1. Luo, F. L., Ye, H., and Rashid M. H. 2005. *Digital Power Electronics and Applications*. New York: Academic Press.
2. Rashid, M. H. 2007. *Power Electronics Handbook* (2nd edition). Boston: Academic Press.

3. Dorf, R. C. 2006. *The Electrical Engineering Handbook* (3rd edition). Boca Raton: Taylor & Francis.
4. Luo, F. L., Jackson, R. D., and Hill R. J. 1985. Digital controller for thyristor current source. *IEE-Proceedings Part B*, 132, 46–52.
5. Luo, F. L. and Hill, R. J. 1985. Disturbance response techniques for digital control systems. *IEEE Transactions on Industrial Electronics*, 32, 245–253.
6. Luo, F. L. and Hill, R. J. 1985. Minimisation of interference effects in thyristor converters by feedback feedforward control. *IEEE Transactions on Measurement and Control*, 7, 175–182.
7. Luo, F. L. and Hill, R. J. 1986. Influence of feedback filter on system stability area in digitally-controlled thyristor converters. *IEEE Transactions on Industry Applications*, 18–24.
8. Luo, F. L. and Hill, R. J. 1986. Fast response and optimum regulation in digitally-controlled thyristor converters. *IEEE Transactions on Industry Applications*, 22, 10–17.
9. Luo, F. L. and Hill, R. J. 1986. System analysis of digitally-controlled thyristor converters. *IEEE Transactions on Measurement and Control*, 8, 39–45.
10. Luo, F. L. and Hill, R. J. 1986. System optimisation—self-adaptive controller for digitally-controlled thyristor current controller. *IEEE Transactions on Industrial Electronics*, 33, 254–261.
11. Luo, F. L. and Hill, R. J. 1987. Stability analysis of thyristor current controllers. *IEEE Transactions on Industry Applications*, 23, 49–56.
12. Luo, F. L. and Hill, R. J. 1987. Current source optimisation in AC-DC GTO thyristor converters. *IEEE Transactions on Industrial Electronics*, 34, 475–482.
13. Luo, F. L. and Hill, R. J. 1989. Microprocessor-based control of steel rolling mill digital DC drives. *IEEE Transactions on Power Electronics*, 4, 289–297.
14. Luo, F. L. and Hill, R. J. 1990. Microprocessor-controlled power converter using single-bridge rectifier and GTO current switch. *IEEE Transactions on Measurement and Control*, 12, 2–8.
15. Muth, E. J. 1977. *Transform Method with Applications to Engineering and Operation Research*. New Jersey: Prentice-Hall.
16. Oliver, G., Stefanovic, R., and Jamil, A. 1979. Digitally controlled thyristor current source. *IEEE Transactions on Industrial Electronics and Control Instrumentation*, 185–191.
17. Fallside, F. and Jackson, R. D. 1969. Direct digital control of thyristor amplifiers. *IEE-Proceedings, Part B*, 873–878.
18. Arrillaga, J., Galanos, G., and Posner, E. T. 1970. Direct digital control of HVDC converters. *IEEE Transactions on Power Apparatus and Systems*, 2056–2065.
19. Daniels, A. R. and Lipczyski, R. T. 1969. Digital firing angle circuit for thyristor motor controllers. *IEE-Proceedings, Part B*, pp. 245–256.
20. Dewan, S. B. and Dunford, W. G. 1983. A microprocessor-based controller for a three-phase controlled bridge rectifiers. *IEEE Transactions on Industry Applications*, 113–119.
21. Cheung, W. N. 1971. The realisation of converter control using sampled-and-delay method. *IEE-Proceedings, Part B*, pp. 701–705.

4

Implementing Power Factor Correction in AC/DC Converters

Power factor correction (PFC) is the capacity for generating or absorbing the reactive power produced by a load [1–3]. Power quality issues and regulations require rectifier loads to be connected to the utility to achieve high PFs. This means that a PFC rectifier needs to draw close to a sinusoidal current in phase with the supply voltage, unlike phase-controlled rectifiers (making the PFC rectifier "look like" a resistive load to the utility).

4.1 Introduction

Refer to the following formula [3]:

$$\mathrm{PF} = \frac{\mathrm{DPF}}{\sqrt{1 + \mathrm{THD}^2}},$$

where DPF is the displacement power factor and THD is the total harmonic distortion. We can explain DPF as the fundamental harmonic of the current that has a delay angle θ (or ϕ), that is, $\mathrm{DPF} = \cos\theta$ (or $\cos\phi$). THD is calculated using Equation 1.20. Most AC/DC uncontrolled and controlled rectifiers have poor PFs, except for the single-phase full-wave uncontrolled bridge (Graetz) rectifier with *R* load. All three-phase uncontrolled and controlled rectifiers have the input current fundamental harmonic delaying its corresponding voltage by an angle 30° plus α, where α is the firing angle of the controlled rectifier. Consequently, AC/DC rectifiers naturally have poor PFs. In order to maintain power quality, PFC is necessary. Implementing the PFC means

- Reducing the phase difference between the line voltage and current ($\mathrm{DPF} = >1$).
- Shaping the line current to a sinusoidal waveform ($\mathrm{THD} = >0$).

The first condition requires that the fundamental harmonic of the current has a delay angle $\theta = >0°$. The second condition requires that the harmonic components are as small as possible. In recent research, the following methods have been used to implement PFC:

1. DC/DC converterized rectifiers.
2. Pulse-width modulation (PWM) boost-type rectifiers.
3. Tapped-transformer converters.

4. Single-stage PFC AC/DC converters.
5. VIENNA rectifiers.
6. Other methods.

4.2 DC/DC Converterized Rectifiers

A full-wave diode rectifier with R load has a high PF. If this rectifier supplies an R–C load, the PF is poor. Using a DC/DC converter in this circuit will improve the PF. The PFC rectifier circuit is shown in Figure 4.1.

The resistor emulation of the PFC rectifier is carried out by the DC/DC converter. The input to the DC/DC converter is a fully rectified sinusoidal voltage waveform. A constant DC voltage is maintained at the output of the PFC rectifier. The DC/DC converter is switched at a switching frequency f_s that is many times higher than the line frequency f. The input current waveform into the diode bridge is modified to contain a strong fundamental sinusoid at the line frequency but with harmonics at a frequency several times higher than the line frequency.

Since the switching frequency f_s is very high in comparison with the line frequency f, the input and output voltages of the PFC rectifier may be considered as constant throughout the switching period. Thus, the PFC rectifier can be analyzed like a regular DC/DC converter:

$$\begin{aligned} v_S &= V_S \sin\theta, \\ v_1 &= V_S|\sin\theta| \quad \text{with } \theta = 2\pi ft. \end{aligned} \tag{4.1}$$

The voltage transfer ratio of the PFC rectifier is required to vary with angle θ in a half supply period. The voltage transfer ratio of the DC/DC converter is

$$T_{VV}(\theta) = \frac{\overline{V_{DC}}}{V_1(\theta)} = \frac{\overline{V_{DC}}}{V_S \sin\theta} \quad \text{with } f_S \gg f, \tag{4.2}$$

where $\overline{V_{DC}}$ is the local average DC output voltage.

T_{VV} in a supply period is shown in Figure 4.2. The high voltage transfer ratio in the vicinity of $\omega t = 0°$ and 180° can be achieved by using converters such as boost, buck–boost, or fly-back converters.

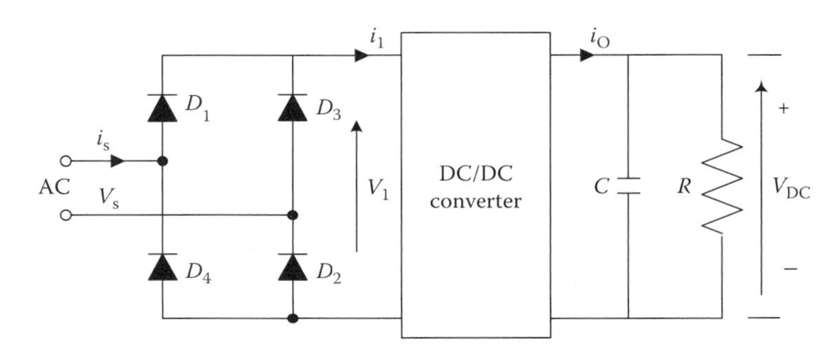

FIGURE 4.1 PFC rectifier.

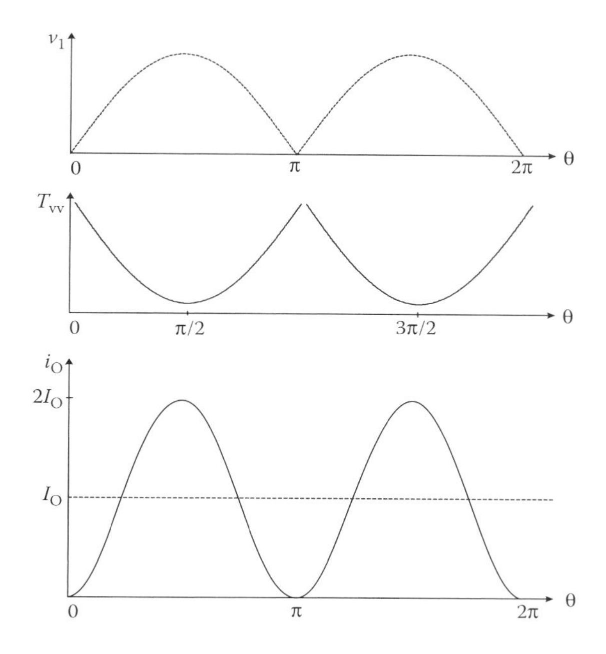

FIGURE 4.2 DC/DC converter output current required.

To prove this technique, a full-wave diode rectifier with *R–C* load ($R = 100\,\Omega$ and $C = 100\,\mu F$) plus a buck–boost converter is investigated. Before applying any converter, the input voltage and current waveforms are shown in Figure 4.3. The fundamental harmonic of the input current delays the input voltage by an angle $\theta = 33.45°$.

The harmonics (FFT spectrum) of the input current are shown in Figure 4.4.

The harmonics, values are listed in Table 4.1.

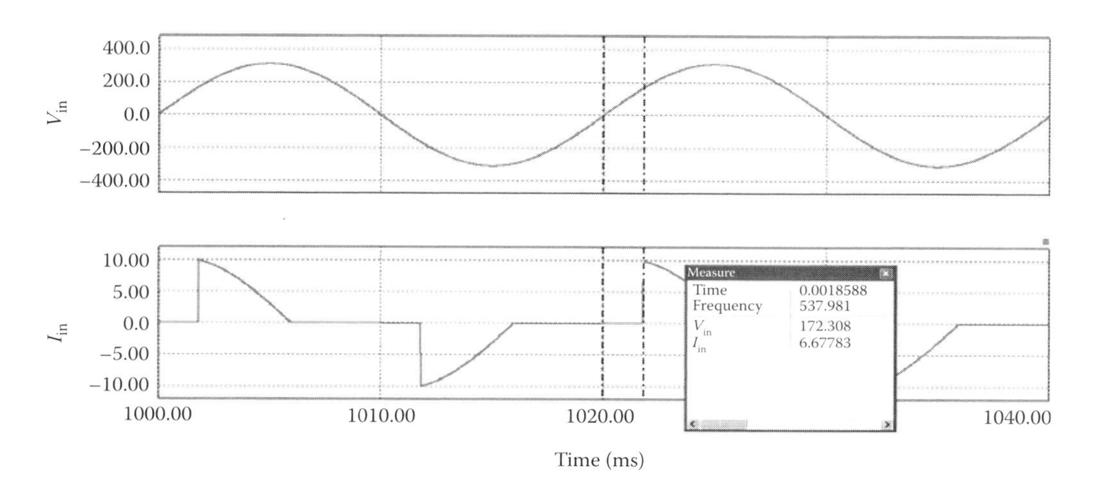

FIGURE 4.3 Input voltage and current waveforms.

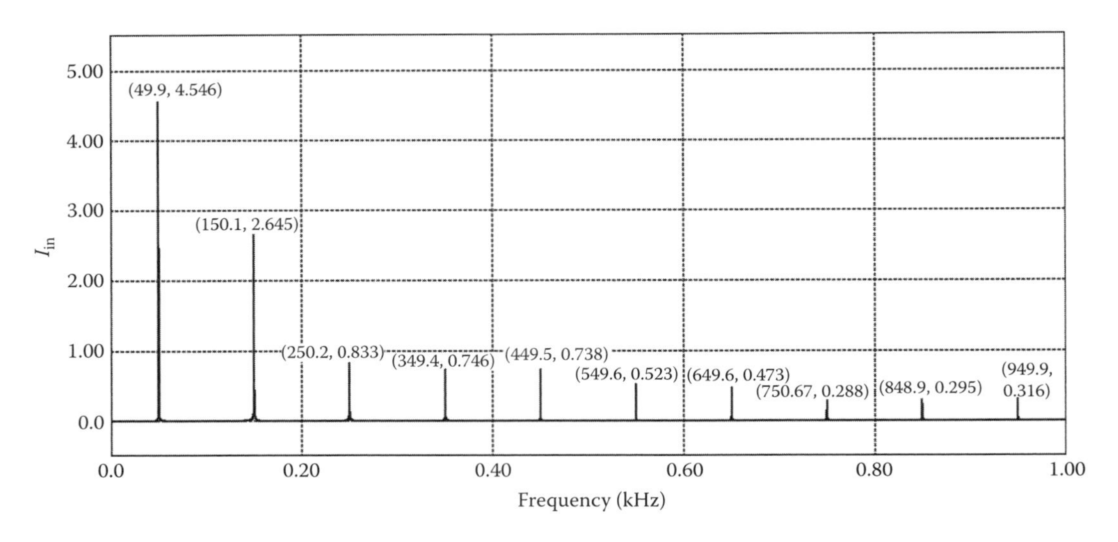

FIGURE 4.4 FFT spectrum of the input current.

The THD of the input current is obtained as $\text{THD}^2 = \sum_{i=2}^{\alpha}(I_i/I_1)^2 = 0.4625$ and the DPF is obtained as $\cos(33.45°) = 0.834$. Therefore,

$$\text{PF} = \frac{\text{DPF}}{\sqrt{1+\text{THD}^2}} = \frac{\cos 33.45}{\sqrt{1+0.4625}} = 0.689. \tag{4.3}$$

A buck–boost converter (see Figure 5.7 in Chapter 5) is used for this purpose. The circuit diagram is shown in Figure 4.5.

The input voltage is 311 V (peak)/50 Hz. The duty ratio k is calculated as 20 chopping periods for a half-cycle. For one cycle, there are 40 chopping periods (maintain the same duty ratio) corresponding to its frequency of 2 kHz. The inductance value was set as $L = 0.6$ mH and the capacitance value as 800 μF to maintain the output voltage at 200 V. The duty ratio k was calculated to set a constant DC output voltage of 200 V (see Table 4.2).

The duty ratio k waveform in two-and-a-half cycles is shown in Figure 4.6 and the switch (transistor) turn-on and turn-off in a half-cycle are shown in Figure 4.7.

TABLE 4.1

Harmonic Current Values of Normal AC to DC Converter

Current	Frequency (Hz)	Fourier Component
I1	50	4.546
I3	150	2.645
I5	250	0.833
I7	350	0.746
I9	450	0.738
I11	550	0.523
I13	650	0.473
I15	750	0.288
I17	850	0.295
I19	950	0.316

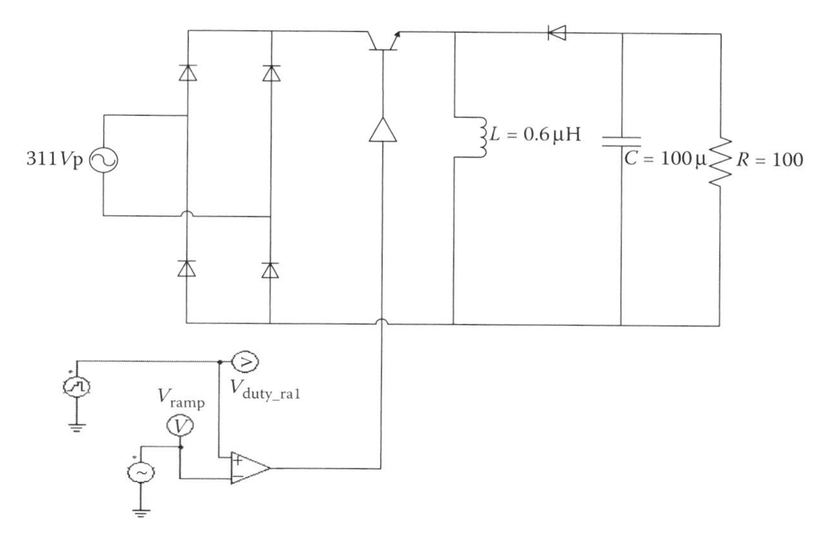

FIGURE 4.5 Buck–boost converter used for PFC with *R*–*C* load.

The input voltage and current waveforms are shown in Figure 4.8. From the waveform we can see that the fundamental harmonic delay angle θ is about 3.21°. The output voltage of the buck–boost converter is 200 V, as shown in Figure 4.9.

The FFT spectrum of the input current is shown in Figure 4.10 and the harmonic components are shown in Table 4.3.

TABLE 4.2

Duty Ratio *k* in the 20 Chopping Periods in a Half-Cycle

ωt (deg)	Input Voltage = $311\sin(\omega t)$ (V)	k
9	48.65	0.804
18	96.1	0.676
27	141.2	0.586
36	182.8	0.522
45	219.9	0.476
54	251.6	0.443
63	277.1	0.419
72	295.8	0.403
81	307.2	0.394
90	311	0.391
99	307.2	0.394
108	295.8	0.403
117	277.1	0.419
126	251.6	0.443
135	219.9	0.476
144	182.8	0.522
153	141.2	0.586
162	96.1	0.676
171	48.65	0.804
180	0	∞

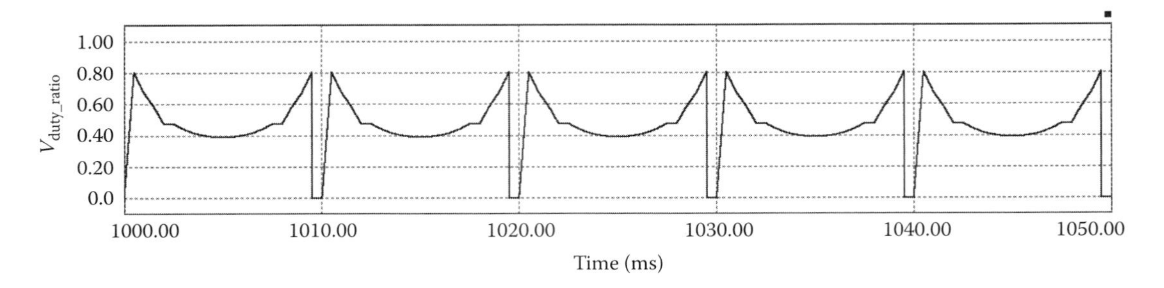

FIGURE 4.6 Duty ratio k waveform in two-and-a-half cycles.

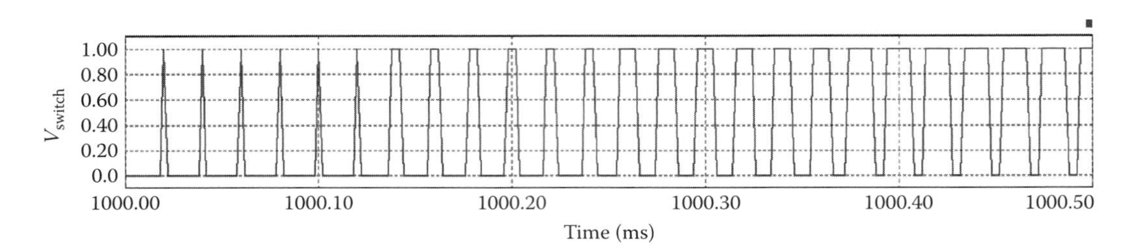

FIGURE 4.7 Switch turn-on and turn-off waveform in a half-cycle.

From the data in Table 4.3, the THD of the input current is obtained as $\text{THD}^2 = \sum_{i=2}^{\alpha}(I_i/I_1)^2 = 0.110062$ and DPF is obtained as $\cos(3.21°) = 0.998431$. Therefore,

$$\text{PF} = \frac{\text{DPF}}{\sqrt{1+\text{THD}^2}} = \frac{\cos 3.21}{\sqrt{1+0.110062}} = 0.95. \tag{4.4}$$

Using this technique, PF is significantly improved from 0.689 to 0.95.

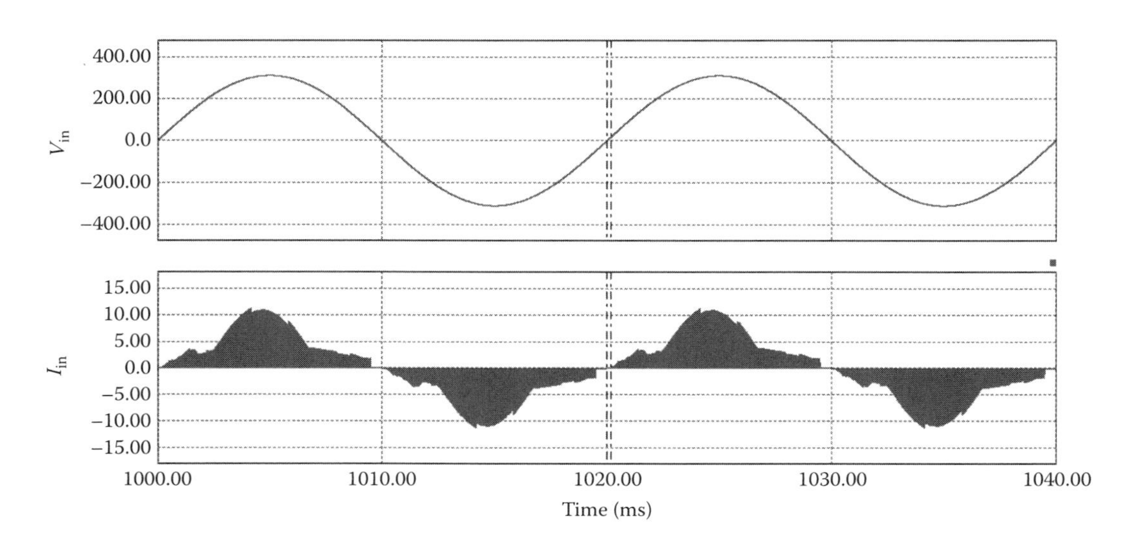

FIGURE 4.8 Input voltage and current waveforms.

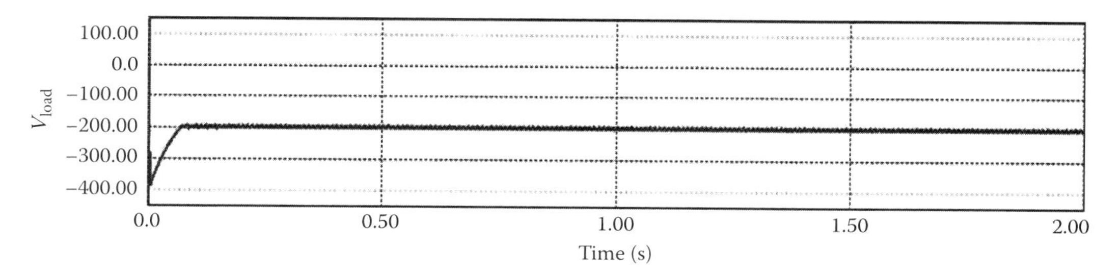

FIGURE 4.9 Output voltage of the buck–boost converter.

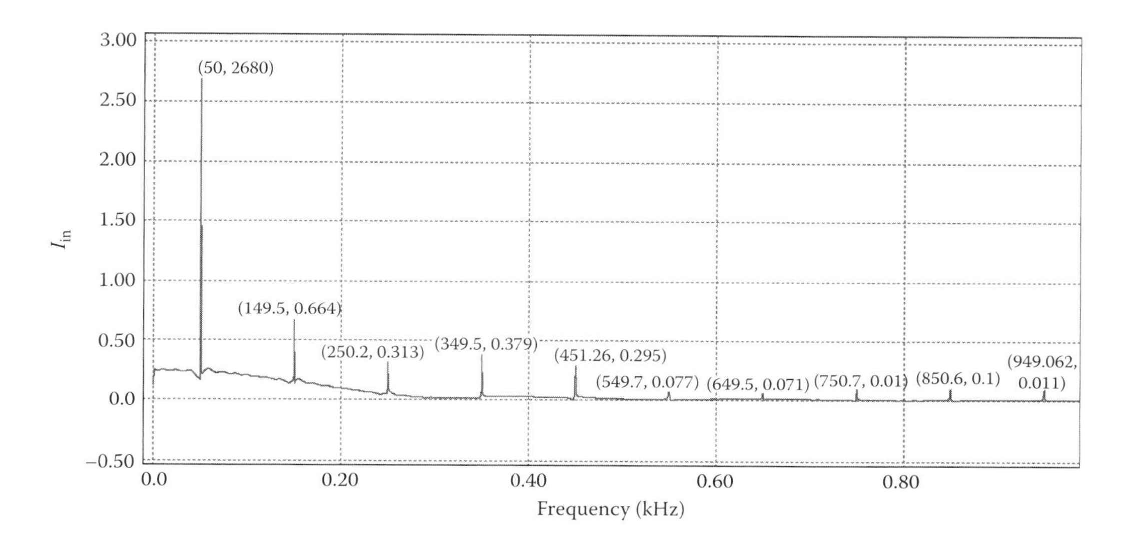

FIGURE 4.10 FFT spectrum of the input current.

TABLE 4.3

Harmonic Components of the Input Current

Current	Frequency (Hz)	Fourier Component
I1	50	2.680
I3	150	0.664
I5	250	0.313
I7	350	0.379
I9	450	0.295
I11	550	0.077
I13	650	0.071
I15	750	0.010
I17	850	0.100
I19	950	0.011

From the above investigation, we know that using a buck–boost converter to implement PFC can be successful, but the output voltage has a negative polarity. If a P/O Luo-converter or SEPIC or a P/O buck–boost converter is used, we can obtain the P/O voltage.

Example 4.1

A P/O Luo-converter (see Figure 5.11 in Chapter 5) is used to implement PFC in a single-phase diode rectifier with an R–C load. The AC supply voltage is 240 V/50 Hz and the required output voltage is 200 V. The switching frequency is 4 kHz. Determine the duty cycle k in a half supply period (10 ms). Other component values for reference are the following: $R = 100\,\Omega$, $C = C_O = 20\,\mu\text{F}$, and $L_1 = L_2 = 10\,\text{mH}$.

SOLUTION

Since the supply frequency is 50 Hz and the switching frequency is 4 kHz, there are 40 switching periods in a half supply period (10 ms). The voltage transfer gain of the P/O Luo-converter is

$$V_O = \frac{k}{1-k} V_{in},$$

$$k = \frac{V_O}{V_O + V_{in}} = \frac{200}{200 + 240\sqrt{2}\sin\omega t}.$$

Duty cycle k is listed in Table 4.4.

4.3 PWM Boost-Type Rectifiers

Using this method, we can obtain the UPF. In order to obtain UPF, that is, PF = 1, the current from the diode bridge must be identical in shape and in phase with the supply voltage waveform. Hence,

$$i_1 = I_s|\sin\theta|. \tag{4.5}$$

The input and output powers averaged over a switching period are

$$\begin{aligned} P_{in} &= V_s I_s \sin^2\theta, \\ P_O &= \overline{V_{DC}} i_O. \end{aligned} \tag{4.6}$$

Assuming a lossless rectifier, the output current requirement is determined as

$$i_O = \frac{V_s I_s}{\overline{V_{DC}}} \sin^2\theta. \tag{4.7}$$

The input and output powers averaged over a supply period are

$$\begin{aligned} P_{in} &= \frac{V_s I_s}{2}, \\ P_O &= \overline{V_{DC}} I_O, \end{aligned} \tag{4.8}$$

where I_O is the averaged DC output current.

TABLE 4.4

Duty Ratio k in the 40 Chopping Periods in a Half-Cycle

ωt (deg)	Input Voltage = $240\sqrt{2}\sin(\omega t)$ (V)	k
4.5	26.6	0.88
9	53.1	0.79
13.5	79.2	0.72
18	104.9	0.66
22.5	129.9	0.61
27	154.1	0.56
31.5	177.3	0.53
36	199.5	0.5
40.5	220.4	0.48
45	240	0.45
49.5	258.1	0.44
54	274.6	0.42
58.5	289.4	0.41
63	302.4	0.4
67.5	313.6	0.39
72	322.8	0.38
76.5	330	0.377
81	335.2	0.374
85.5	338.4	0.371
90	339.4	0.37
94.5	338.4	0.371
99	335.2	0.374
103.5	330	0.377
108	322.8	0.38
112.5	313.6	0.39
117	302.4	0.4
121.5	289.4	0.41
126	274.6	0.42
130.5	258.1	0.44
135	240	0.45
139.5	220.4	0.48
144	199.5	0.5
148.5	177.3	0.53
153	154.1	0.56
157.5	129.9	0.61
162	104.9	0.66
166.5	79.2	0.72
171	53.1	0.79
175.5	26.6	0.88
180	0	∞

The instantaneous output currents are

$$\begin{aligned} i_O &= \frac{V_s I_s}{V_{DC}} \sin^2 \theta = 2I_O \sin^2 \theta \\ &= I_O(1 - \cos 2\theta). \end{aligned} \tag{4.9}$$

The DC/DC converter output current required for a UPF, as a function of angle θ, is shown in Figure 4.2.

Because the input current to the DC/DC converter is to be shaped, the DC/DC converter is operated in a current-regulated mode.

4.3.1 DC-Side PWM Boost-Type Rectifier

The DC-side PWM boost-type rectifier is shown in Figure 4.11 where i_1^* is the reference of the desired value of the current i_1. Here i_1^* has the same waveform shape as $|v_s|$. The amplitude of i_1^* should be able to maintain the output voltage at a desired or reference level v_{dc}^*, in spite of the variation on load and the fluctuation of line voltage from its nominal value. The waveform of i_1^* is obtained by measuring $|v_s|$ and multiplying it by the amplified error between v_{dc}^* and v_{dc}. The actual current i_1 is measured. The status of the switch in the DC/DC converter is controlled by comparing the actual current with i_1^*.

Once i_1^* and i_1 are available, there are various ways of implementing the current-mode control of the DC/DC converter.

4.3.1.1 Constant-Frequency Control

Here, the switching frequency f_s is kept constant. When i_1 reaches i_1^*, the switch in the DC/DC converter is turned off. The switch is turned on by a clock period at a fixed frequency f_s. This method is likely an open-loop control. The operation indication is shown in Figure 4.12.

Example 4.2

A boost converter (see Figure 5.5 in Chapter 5) is used to implement PFC in the circuit shown in Figure 4.11a. The switching frequency is 2 kHz, $L = 10$ mH, $C_d = 20\,\mu$F, $R = 100\,\Omega$, and the output voltage $V_O = 400$V. The AC supply voltage is 240 V/50 Hz. Determine the duty cycle k in a half supply period (10 ms).

SOLUTION

Since the supply frequency is 50 Hz and the switching frequency is 2 kHz, there are 20 switching periods in a half supply cycle (10 ms). The voltage transfer gain of the boost converter is

$$V_O = \frac{1}{1-k} V_{in},$$

$$k = \frac{V_O - V_{in}}{V_O} = \frac{400 - 240\sqrt{2}\,\sin \omega t}{400}.$$

The duty ratio k is listed in Table 4.5.

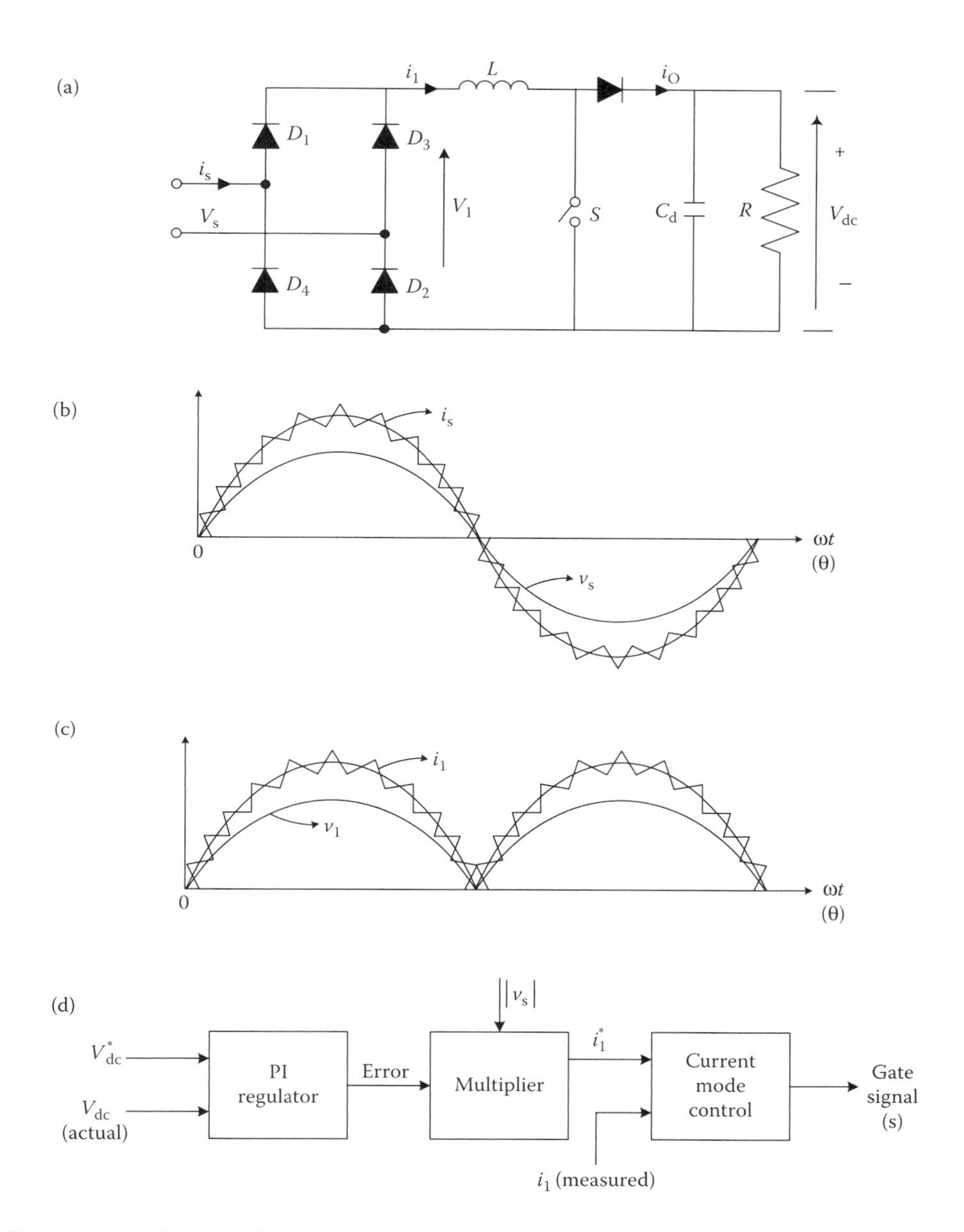

FIGURE 4.11 UPF diode rectifier with feedback control: (a) circuit, (b) input voltage and current, (c) output voltage and current of the diode rectifier, and (d) control block diagram.

4.3.1.2 Constant-Tolerance-Band (Hysteresis) Control

Here, the constant i_1 is controlled so that the peak-to-peak ripple (I_{rip}) in i_1 remains constant. With a preselected value of I_{rip}, i_1 is forced to be within the tolerance band ($i_1^* + I_{rip}/2$) and ($i_1^* - I_{rip}/2$) by controlling the switch status. This method is likely to be a closed-loop control. A current sensor is necessary to measure the particular current i_1 to determine switch-on and switch-off. The operation indication is shown in Figure 4.13.

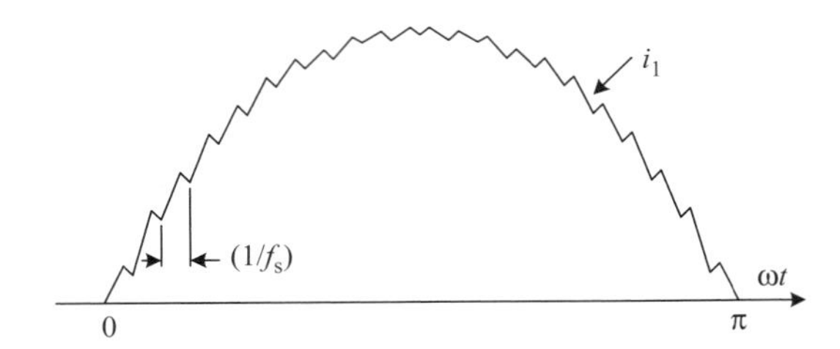

FIGURE 4.12 Operation indication of constant-frequency control.

TABLE 4.5

Duty Ratio k in the 20 Chopping Periods in a Half-Cycle (10 ms)

ωt (deg)	Input Current $= 240\sqrt{2}\sin(\omega t)$ (V)	k
9	53.1	0.867
18	104.9	0.738
27	154.1	0.615
36	199.5	0.501
45	240	0.4
54	274.6	0.314
63	302.4	0.244
72	322.8	0.193
81	335.2	0.162
90	339.4	0.152
99	335.2	0.162
108	322.8	0.193
117	302.4	0.244
126	274.6	0.314
135	240	0.4
144	199.5	0.501
153	154.1	0.615
162	104.9	0.738
171	53.1	0.867
180	0	∞

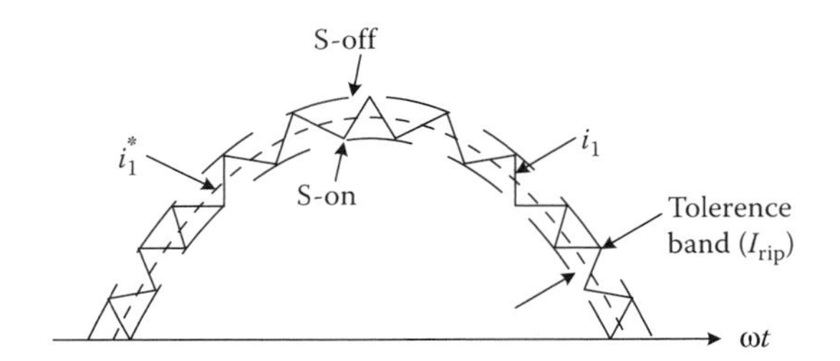

FIGURE 4.13 Operation indication of hysteresis control.

4.3.2 Source-Side PWM Boost-Type Rectifiers

In motor drive applications with regenerative braking, the power flow from the AC line is required to be bidirectional. A bidirectional converter can be designed using phase angle delay control but at the expense of poor input PF and high waveform distortion in the line current. It is possible to overcome these limitations by using a switch-mode converter, as shown in Figure 4.14.

The rectifier being the dominant mode of operation, i_s is defined with a direction. An inductance L_s (that augments the internal inductance of the utility source) is included to reduce the ripple in i_s at a finite switching frequency. The four switching devices (IGBTs or MOSFETs) are operated in PWM. Their switching frequency f_s is usually measured in kilohertz. From Figure 4.14, we have

$$v_s = v_{conv} + v_L. \tag{4.10}$$

Assuming v_s to be sinusoidal, the fundamental frequency components of v_{conv} and i_s in Figure 4.14 can be expressed as phasors $\overrightarrow{V_{conv1}}$ and $\overrightarrow{I_{s1}}$, respectively (subscript 1 denotes the fundamental component). Arbitrarily choosing the reference phasor to be $\overrightarrow{V_s} = V_s e^{j0°}$, at the line frequency $\omega = 2\pi f$

$$\overrightarrow{V_s} = \overrightarrow{V_{conv1}} + \overrightarrow{V_{L1}}, \tag{4.11}$$

where

$$\overrightarrow{V_{L1}} = i\omega L_s \overrightarrow{I_{s1}}. \tag{4.12}$$

A phasor diagram corresponding to Equations 4.11 and 4.12 is shown in Figure 4.15 where $\overrightarrow{I_{s1}}$ lags $\overrightarrow{V_s}$ by an arbitrary phase angle θ.

The real power P supplied by the AC source to the converter is

$$P = V_s I_{s1} \cos\theta = \frac{V_s^2}{\omega L_s} \frac{V_{conv1}}{V_s} \sin\delta. \tag{4.13}$$

From Figure 4.15a,

$$V_{L1} \cos\theta = \omega L_s I_{s1} \cos\theta = V_{conv1} \sin\delta. \tag{4.14}$$

In the phasor diagram of Figure 4.15a, the reactive power Q supplied by the AC source is positive. It can be expressed as

$$Q = V_s I_{s1} \sin\theta = \frac{V_s^2}{\omega L_s}\left(1 - \frac{V_{conv1}}{V_s}\cos\delta\right). \tag{4.15}$$

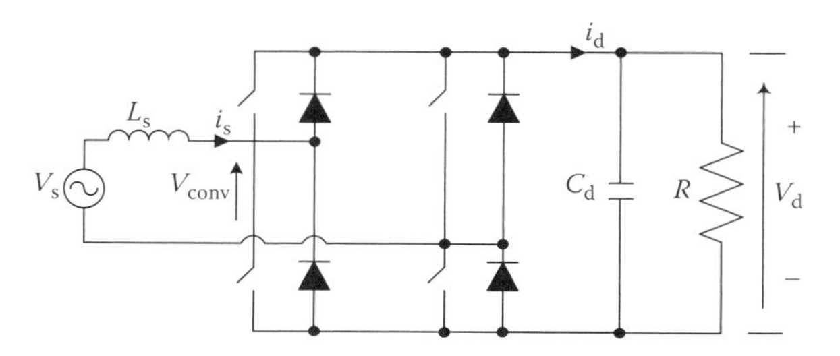

FIGURE 4.14 Switch-mode converter.

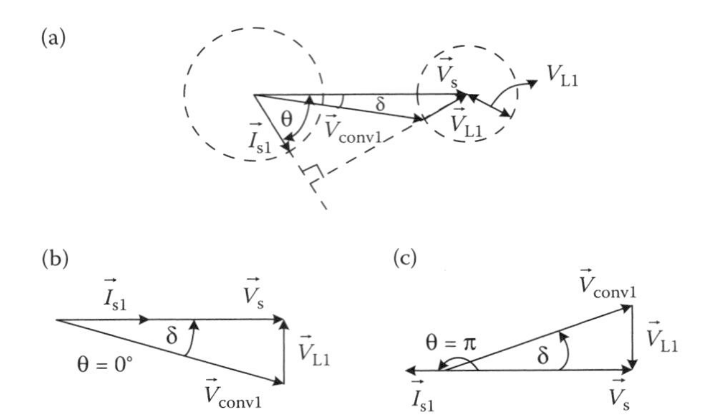

FIGURE 4.15 Phasor diagram: (a) overall diagram; (b) δ is negative; (c) δ is positive.

From Figure 4.15a, we also have

$$V_s - \omega L_s I_{s1} \sin\theta = V_{conv1} \cos\delta. \tag{4.16}$$

From these equations, it is clear that for a given line voltage v_s and the chosen inductance L_s, the desired values of P and Q can be obtained by controlling the magnitude and the phase of v_{conv1}.

Figure 4.15 shows how $\overrightarrow{V_{conv1}}$ can be varied, keeping the magnitude of $\overrightarrow{I_{s1}}$ constant. The two special cases of rectification and inversion at a UPF are shown in Figure 4.15b and c. In both cases

$$V_{conv1} = \sqrt{V_s^2 + (\omega L_s I_{s1})^2}. \tag{4.17}$$

In the circuit of Figure 4.14, V_d is established by charging the capacitor C_d through the switch-mode converter. The value of V_d should have a sufficiently large magnitude so that v_{conv1} at the AC side of the converter is produced by a PWM that corresponds to a PWM in a linear region. The control circuit to regulate V_d in Figure 4.14 is shown in Figure 4.16. The reference value V_d^* intends to achieve a UPF of operation. The amplified error between V_d and V_d^* is multiplied by the signal proportional to the input voltage v_s waveform to produce the reference signal i_s^*. A current-mode control such as a tolerance band control or a fixed-frequency control can be used to deliver i_s equal to i_s^*. The magnitude and direction of power flow are automatically controlled by regulating V_d at its desired value.

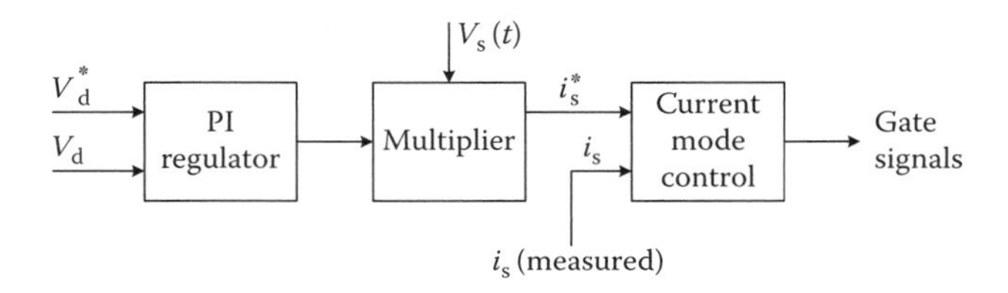

FIGURE 4.16 Block diagram of UPF operation.

4.4 Tapped-Transformer Converters

A simple method to improve the PF is to use tapped-transformer converters. DC motor variable speed control drive systems are widely used in industrial applications. Some applications require the DC motor to run at lower speeds. For example, winding machines and rolling mills mostly work at lower speeds (lower than their 50% rated speed). If DC motors are supplied by AC/DC rectifiers, the lower speed corresponds to lower armature voltage.

Assume that the DC motor rated voltage corresponding to the rectifier firing angle α is about 10°. The firing angle α will be about 60° if the motor runs at half rated speed. In the first case, the DPF is about ($\cos\alpha$), that is, DPF = 0.98. In the second case, the DPF is about 0.48. This means that the PF is very poor if the DC motor works at lower speed.

A tapped-transformer converter is shown in Figure 4.17a, which is a single-phase controlled rectifier. The original bridge consists of thyristors T_1–T_4. The transformer is tapped at 50% of the secondary winding. The third leg consists of thyristors T_5–T_6, which are linked at the tapped point at the middle point of the secondary winding. Since the DC motor armature circuit has enough inductance, the armature current is always continuous. The motor armature voltage is

$$V_O = V_{dO} \cos\alpha. \tag{4.18}$$

(a)

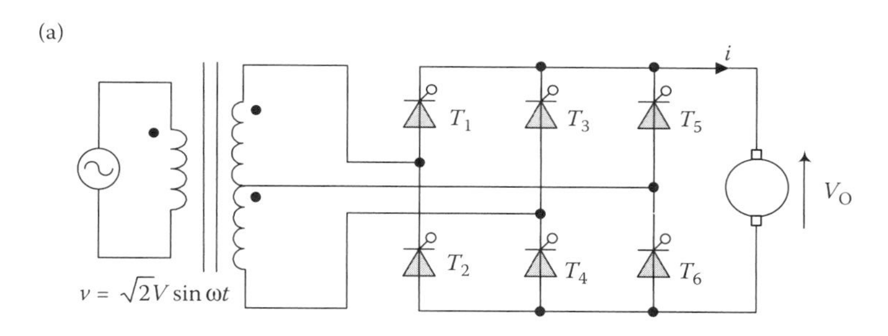

(b)

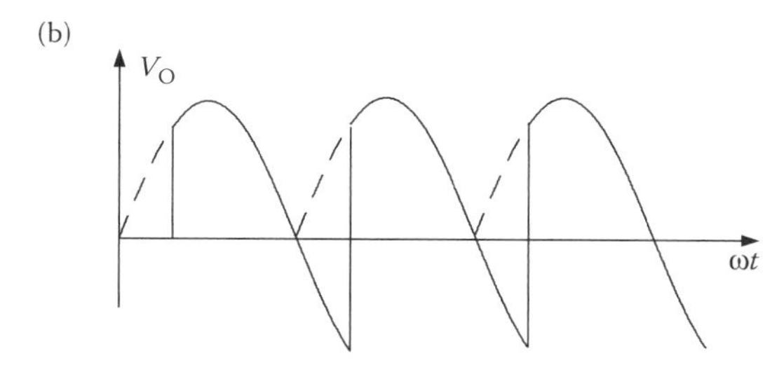

(c)

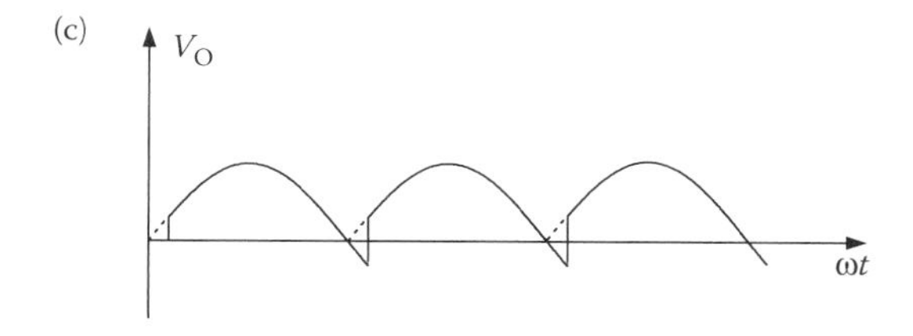

FIGURE 4.17 Tapped-transformer converter: (a) circuit diagram, (b) output voltage waveform from original bridge, and (c) output voltage waveform from new leg.

If the motor works at a lower speed, for example, at 45% of its rated speed, the corresponding firing angle α is about 64°. The output voltage waveform from the original bridge is shown in Figure 4.17b. The fundamental harmonic component sine wave must have the delay angle $\phi_1 = \alpha = 64°$ and DPF $= \cos\alpha$. After Fourier transform analysis and THD calculation, the voltage waveform in Figure 4.17b is 0.24. Therefore,

$$\mathrm{PF} = \frac{\mathrm{DPF}}{\sqrt{1+\mathrm{THD}^2}} = \frac{\cos 64°}{\sqrt{1+0.24^2}} = \frac{0.443}{1.028} = 0.43. \tag{4.19}$$

Keeping the same armature voltage, we obtain the voltage from legs 2 and 3, that is, thyristors T_1 and T_2 are idled. This means that the input voltage is reduced by half the supply voltage, and the firing angle α' is about 27.6°. The output voltage waveform from legs 2 and 3 is shown in Figure 4.17c. The fundamental harmonic component sine wave must have the delay angle $\phi_1 = \alpha' = 27.6°$ and DPF $= \cos\alpha'$. After Fourier transform analysis and THD calculation, the voltage waveform in Figure 4.17c is 0.07. Therefore,

$$\mathrm{PF} = \frac{\mathrm{DPF}}{\sqrt{1+\mathrm{THD}^2}} = \frac{\cos 27.6°}{\sqrt{1+0.07^2}} = \frac{0.8863}{1.0024} = 0.884. \tag{4.20}$$

In comparison with the PFs in Equations 4.19 and 4.20, it is obvious that the PF has been significantly corrected.

This method is very simple and straightforward. The tapped point can be shifted to any other percentage (not fixed at 50%), depending on the applications.

A test rig can be constructed for collecting the measured results. The circuit is shown in Figure 4.18. The secondary voltage of the transformer is 230/115 V. The requested output voltage is set as 80 V.

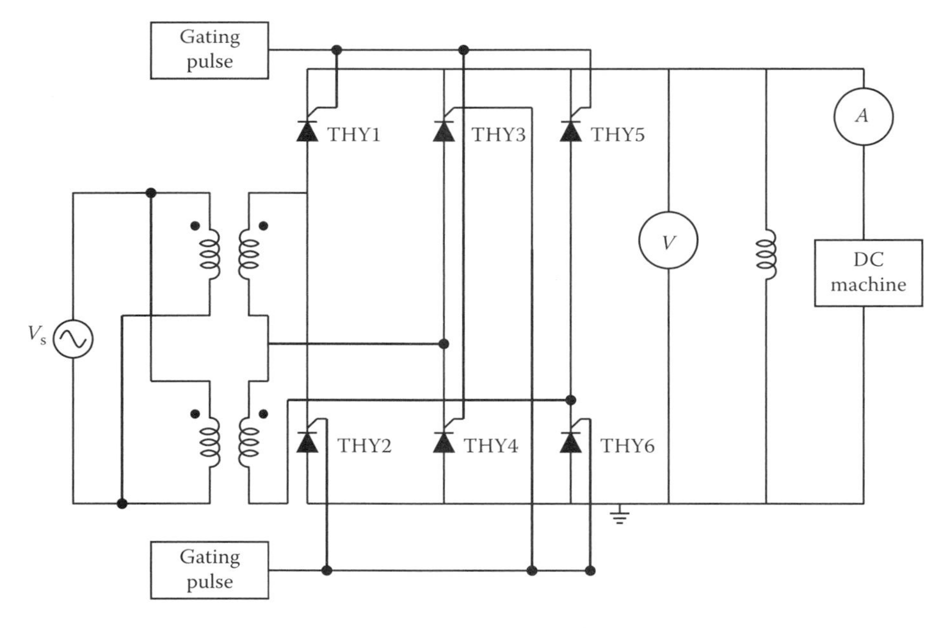

FIGURE 4.18 Single-phase controlled rectifier with a tapped transformer.

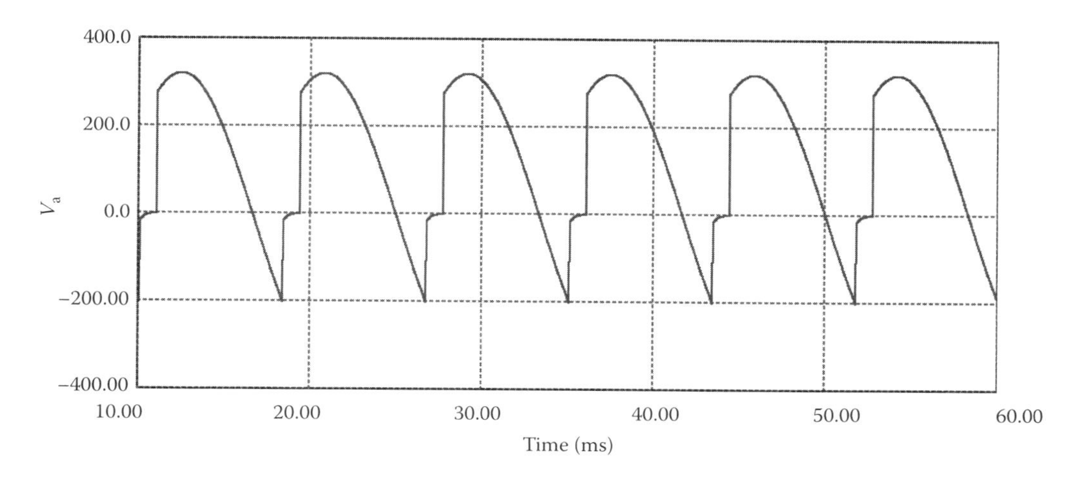

FIGURE 4.19 Output voltage 80 V with input voltage 230 V.

If the supply voltage is 230 V, the firing angle is approximately 67°. The output voltage is shown in Figure 4.19 and the measured record is shown in Figure 4.20. PF is indicated to be 0.64.

If the supply voltage is 115 V, the firing angle is approximately 39.4°. The output voltage is shown in Figure 4.21, and the measured record is shown in Figure 4.22. The indication of the PF in it is 0.87.

If the output voltage increases to 103 V and the supply voltage remains at 115 V, the firing angle is approximately 1°. The output voltage is shown in Figure 4.23, and the measurement record is shown in Figure 4.24. PF is indicated to be 0.98.

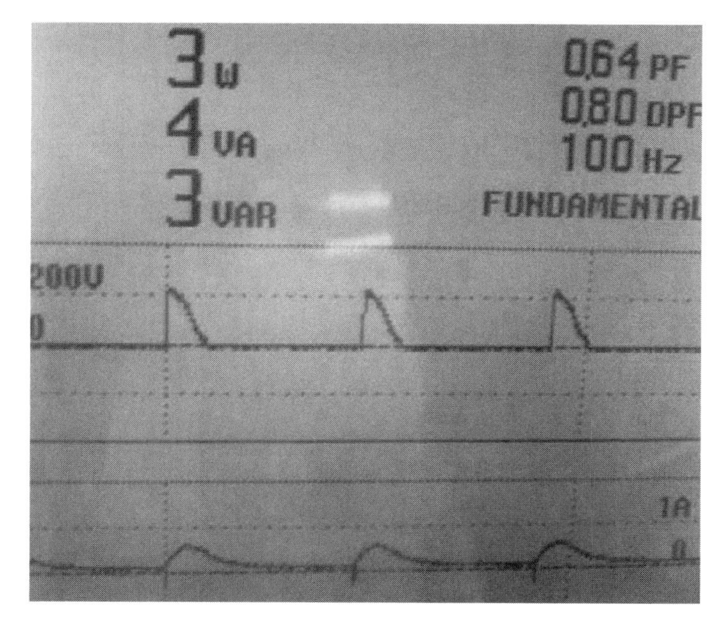

FIGURE 4.20 PF with input voltage 230 V and output voltage 80 V.

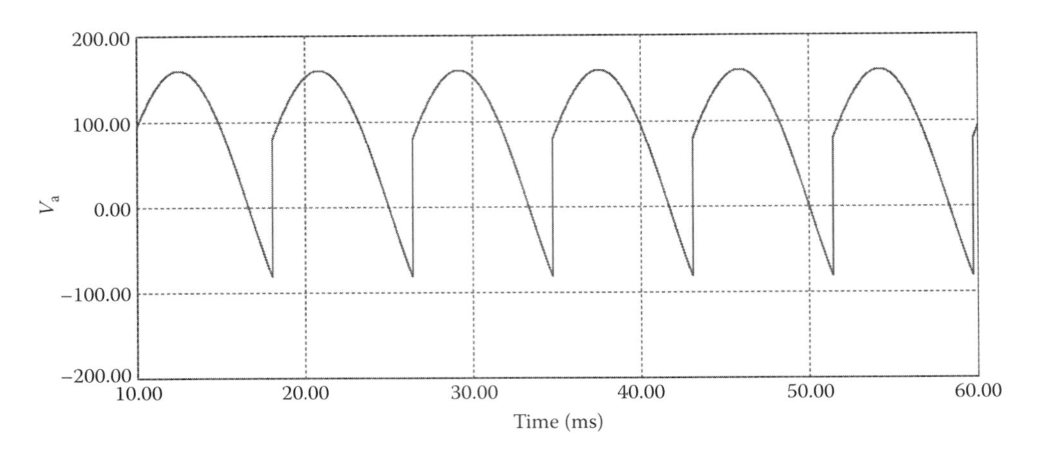

FIGURE 4.21 Output voltage 80 V with input voltage 115 V.

4.5 Single-Stage PFC AC/DC Converters

A double-current synchronous rectifier converter is a popular circuit that is used in computers [1,2]. Unfortunately, its PF is not high. However, the single-stage PFC double-current synchronous rectifier (DC-SR) converter is able to improve its PF nearly to unity. The circuit diagram is shown in Figure 4.25.

The system consists of an AC/DC diode rectifier and a DC-SR converter [1]. Suppose that the output inductors L_1 and L_2 are equal to each other, $L_1 = L_2 = L_O$. There are three switches: main switch S and two auxiliary synchronous switches S_1 and S_2. It inherently

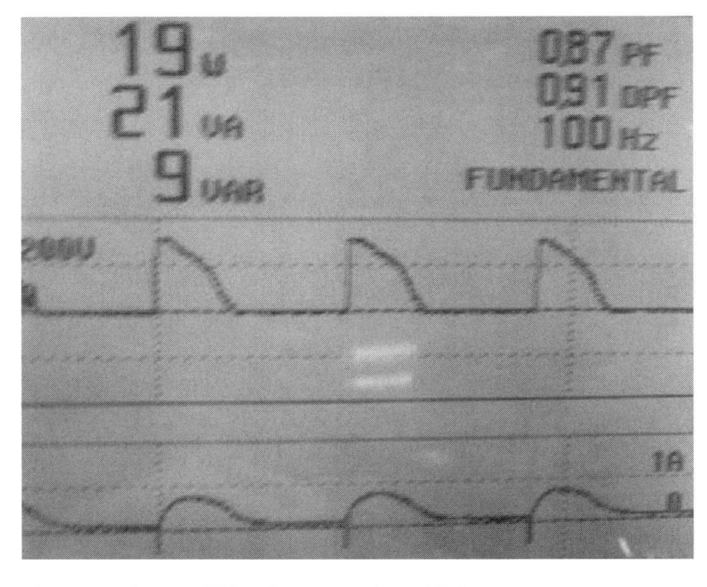

FIGURE 4.22 PF with input voltage 115 V and output voltage 80 V.

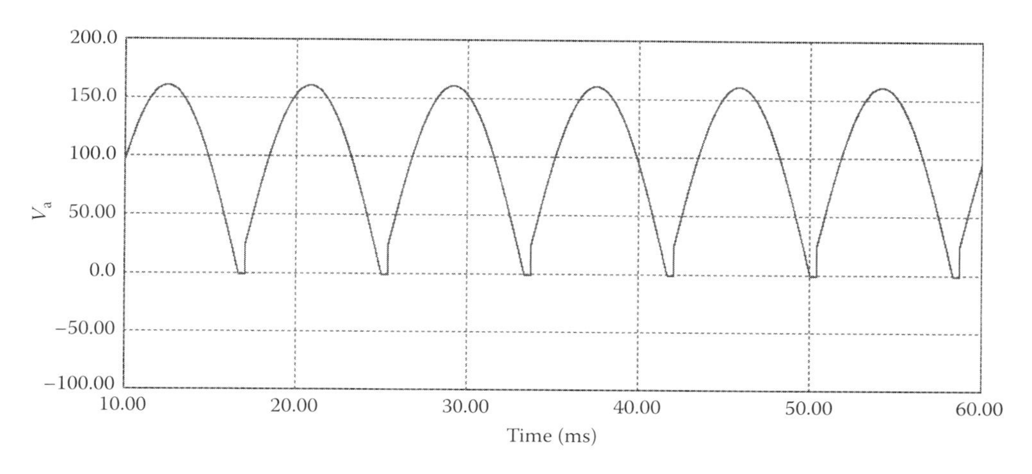

FIGURE 4.23 Output voltage 103 V with input voltage 115 V.

exhibits high PF because the PFC cell operates in continuous conduction mode (CCM). In addition, it is also free to have high voltage stress across the bulk capacitor at light loads. In order to investigate the dynamical behaviors, the averaging method is used to drive the DC operating point and the small-signal model. A proportional-integral-differential (PID) controller is designed to achieve output voltage regulation despite variations in line voltage and load resistance.

In power electronic equipment, the PFC circuits are usually added between the bridge rectifier and the loads to eliminate high harmonic distortion of the line current. In general, they can be divided into two categories, the two-stage approach and the single-stage

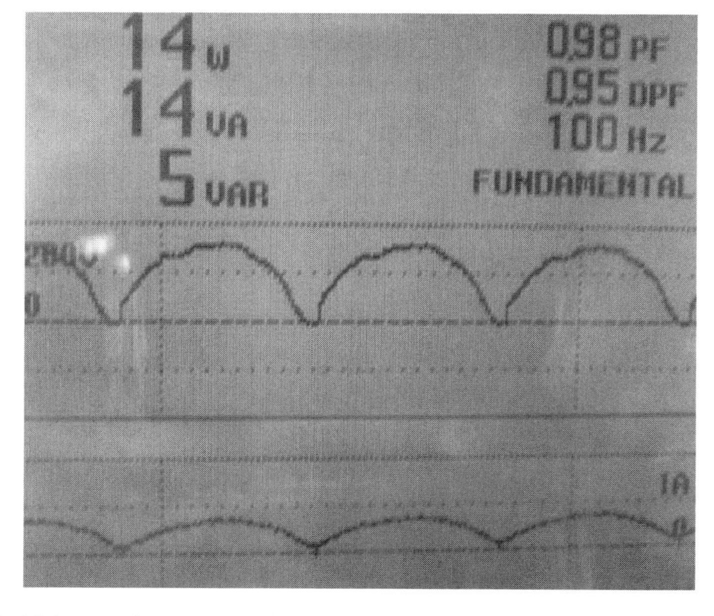

FIGURE 4.24 PF with input voltage 115 V and output voltage 103 V.

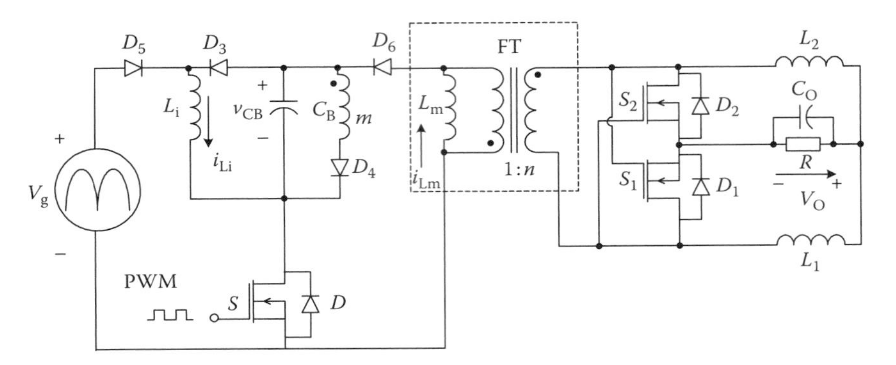

FIGURE 4.25 Proposed single-stage PFC DC-SR converter.

approach. The two-stage approach includes a PFC stage and a DC/DC regulation stage. This approach has good PFC and fast output regulations, but the size and cost increase. To overcome the drawbacks, the graft scheme is proposed in reference [4]. Many single-stage approaches have been proposed in the literature [5–8]. They integrate a PFC cell and a DC/DC conversion cell to form a single stage with a common switch. Therefore, the sinusoidal input current waveform and the output voltage regulation can be simultaneously achieved, thereby meeting the requirements of performance and cost.

However, a high voltage stress exists across the bulk capacitor C_B at light loads if a DC/DC cell operates in discontinuous current mode. To overcome this drawback, a negative magnetic feedback technique has been proposed in the literature [5–8]. However, the dead band exists in the input current and the PF is thereby degraded. To deal with this problem, the DC/DC cell operates in discontinuous current mode. The voltage across the bulk capacitor is independent of loads and the voltage stress is effectively reduced.

4.5.1 Operating Principles

Figure 4.25 depicts the proposed single-stage high PFC converter topology. A physical three-winding transformer has a turns ratio of 1:n:m. A tertiary transformer winding, in series with diode D_4, is added to the converter for transformer flux resetting. The magnetizing inductance L_m is parallel with the ideal transformer. In the proposed converter, both the PFC cell and the DC/DC conversion cell are operating in CCM. To simplify the analysis of the circuit, the following assumptions are made:

1. The large-valued bulk capacitor C_B and the output capacitor C_O are sufficiently large enough to allow the voltages across the bulk capacitor and the output capacitor to be approximately constant during one switching period T_S.
2. All switches and diodes of the converter are ideal. The switching time of the switch and the reverse recovery time of the diodes are negligible.
3. The inductors and the capacitors of the converter are considered to be ideal without parasitic components.

Based on the switching of the switch and diodes, the proposed converter operating in one switching period T_s can be divided into five linear stages as described below.

Stage 1 $[0,t_1]$ (S: on; D_1: on; D_2: off; D_3: off; D_4: off; D_5: on; D_6: on): In the first stage, the switch S is turned on. The diodes D_1, D_5, and D_6 are turned on and the diodes D_2, D_3, and D_4 are turned off. Power is transferred from the bulk capacitor C_B to the output via the transformer.

Stage 2 $[t_1,t_2]$ (S: off; D_1: off; D_2: on; D_3: on; D_4: on; D_5: off; D_6: off): The stage begins when the switch S is turned off. The diodes D_2, D_3, and D_4 are turned on and the diodes D_1, D_5, and D_6 are turned off. The current i_{Li} flows through diode D_3 and charges the bulk capacitor C_B. Diode D_4 is turned on for transformer flux resetting. In this stage, the output power is provided by the inductor L_O.

Stage 3 $[t_2,t_3]$ (S: off; D_1: off; D_2: on; D_3: off; D_4: on; D_5: off; D_6: off): The stage begins at t_2 when the input current i_{Li} falls to zero and thus diode D_3 is turned off. Switch S is still off. All diodes, except D_3, maintain their states as shown in the previous stage. During this stage, the voltages $-v_{CB}/m$ and $-v_O$ are applied across the inductors L_m and L_O, and thus the inductor currents continue to decrease linearly. The output power is also provided by the output inductor L_O.

Stage 4 $[t_3,t_4]$ (S: off; D_1: off; D_2: off; D_3: off; D_4: on; D_5: off; D_6: off): The stage begins when the current i_{LO} decreases to zero and thus diode D_2 is turned off. Switch S is still off. Diode D_4 is still turned on and diodes D_1, D_3, D_5, and D_6 are still turned off. During this stage, the voltage $-v_{CB}/m$ is applied across inductor L_m. The inductor current continues to decrease linearly. The output power is provided by the output capacitor C_O in this stage.

Stage 5 $[t_4,t_5]$ (S: off; D_1: off; D_2: off; D_3: off; D_4: off; D_5: off; D_6: off): The stage begins when the current i_{Lm} falls to zero and thus diode D_4 is turned off. Switch S is still off and all diodes are off. The output power is also provided by the output capacitor C_O. The operation of the converter returns to the first stage when switch S is turned on again.

According to the analysis of the proposed converter, the key waveforms over one switching period T_s are schematically depicted in Figure 4.26. The slopes of the waveforms $i_{CO}(t)$ and $i_{CB}(t)$ are defined as

$$m_{CO1} = \frac{nv_{CB} - v_{CO}}{L_O}, \quad m_{CO2} = -\frac{v_{CO}}{L_O}, \quad m_{CB1} = -\left[\frac{v_{CB}}{L_m} + \frac{n(nv_{CB} - v_{CO})}{L_O}\right],$$
$$m_{CB2} = -\left(\frac{v_{CB}}{L_i} + \frac{v_{CB}}{m^2 L_m}\right), \quad m_{CB2} = -\frac{v_{CB}}{m^2 L_m}. \tag{4.21}$$

4.5.2 Mathematical Model Derivation

In this section, the small-signal model of the proposed converter can be derived by the averaging method. The moving average of a variable, voltage or current, over one switching period T_s is defined as the area, encompassed by its waveform and time axis, divided by T_s.

4.5.2.1 Averaged Model over One Switching Period T_s

There are six storage elements in the proposed converter in Figure 4.25. The state variables of the converter are chosen as the current through the inductor and the voltage across the capacitor. Since both PFC cells and DC/DC cells operate in discontinuous current mode, the initial and final values of inductor currents vanish in each switching period T_s. From a system point of view, the inductor currents i_{Li}, i_{LO}, and i_{Lm} should not be considered as

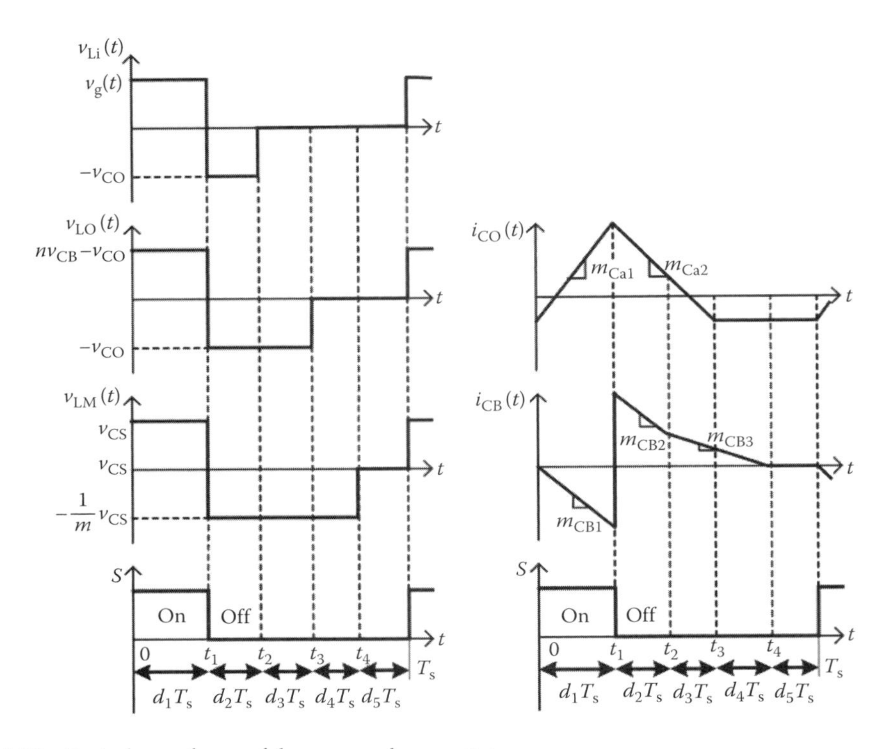

FIGURE 4.26 Typical waveforms of the proposed converter.

state variables. Only the bulk capacitor voltage v_{CB} and the output capacitor voltage v_{CO} are considered to be state variables of the proposed converter.

For notational brevity, a variable with an upper bar denotes its moving average over one switching period T_{S}. With the aid of this definition, the averaged state-variable description of the converter is given by

$$C_{\mathrm{B}}\frac{\mathrm{d}\bar{v}_{\mathrm{CB}}}{\mathrm{d}t} = \bar{i}_{\mathrm{CB}} \quad \text{and} \quad C_{\mathrm{O}}\frac{\mathrm{d}\bar{v}_{\mathrm{CO}}}{\mathrm{d}t} = \bar{i}_{\mathrm{CO}}. \tag{4.22}$$

Moreover, in discontinuous conduction, the averaged voltage across each inductor over one switching period is zero. Hence we have three constraints of the form

$$L_{\mathrm{i}}\frac{\mathrm{d}\bar{i}_{\mathrm{Li}}}{\mathrm{d}t} = \bar{v}_{\mathrm{Li}} = 0, \quad L_{\mathrm{O}}\frac{\mathrm{d}\bar{i}_{\mathrm{LO}}}{\mathrm{d}t} = \bar{v}_{\mathrm{LO}} = 0, \quad L_{\mathrm{m}}\frac{\mathrm{d}\bar{i}_{\mathrm{Lm}}}{\mathrm{d}t} = \bar{v}_{\mathrm{Lm}} = 0. \tag{4.23}$$

The output equation is expressed as

$$\bar{v}_{\mathrm{O}} = \bar{v}_{\mathrm{CO}}. \tag{4.24}$$

Based on the typical waveforms in Figure 4.26, the averaged variables are given by

$$\bar{i}_{CB} = \frac{1}{T_s}\sum_{j=1}^{5}\text{area}[i_{CB(j)}] = \frac{1}{T_s}\left[d_1^2T_s^2m_{CB1} + \frac{1}{2}d_2T_s^2, [d_2m_{CB2} + 2(d_3+d_4)m_{CB2}] + \frac{1}{2}(d_3+d_4)^2T_s^2m_{CB3}\right], \tag{4.25}$$

$$\bar{i}_{CO} = \frac{1}{T_s}\sum_{j=1}^{5}\text{area}[i_{CO(j)}] = \frac{1}{T_s}\left[d_1T_s^2(d_1+d_2+d_3)\frac{n\bar{v}_{CB}-\bar{v}_{CO}}{2L_O} - T_s\frac{\bar{v}_{CO}}{R}\right], \tag{4.26}$$

where the notation area $[i_{CB(j)}]$ denotes the area, encompassed by the waveform $i_{CB}(t)$ and time axis, during stage j. Similarly, we have

$$\begin{aligned}
\bar{v}_{Li} &= \frac{1}{T_s}\sum_{j=1}^{5}\text{area}[v_{Li(j)}] = \frac{1}{T_s}[d_1T_s\bar{v}_g(t) + d_2T_s(-\bar{v}_{CB})],\\
\bar{v}_{Lm} &= \frac{1}{T_s}\sum_{j=1}^{5}\text{area}[v_{Lm(j)}] = \frac{1}{T_s}\left[d_1T_s\bar{v}_{CB} + (d_2+d_3+d_4)T_s\left(-\frac{\bar{v}_{CB}}{m}\right)\right],\\
\bar{v}_{LO} &= \frac{1}{T_s}\sum_{j=1}^{5}\text{area}[v_{LO(j)}] = \frac{1}{T_s}[d_1T_s(n\bar{v}_{CB}-\bar{v}_{CO}) + (d_2+d_3)T_s(-\bar{v}_{CO})].
\end{aligned} \tag{4.27}$$

Substituting Equation 4.27 into the constraints given by Equation 4.23, and performing mathematical manipulations, gives

$$d_2 = \frac{\bar{v}_g(t)}{\bar{v}_{CB}}d_1, \quad d_3 = \left(\frac{n\bar{v}_{CB}}{\bar{v}_{CO}} - 1 - \frac{\bar{v}_g(t)}{\bar{v}_{CB}}\right)d_1, \quad d_4 = \left(m + 1 - \frac{n\bar{v}_{CB}}{\bar{v}_{CO}}\right)d_1. \tag{4.28}$$

Now, substituting Equations 4.21 and 4.28 into Equations 4.25 and 4.26, the averaged state Equation 4.22 can be rewritten as

$$\begin{aligned}
C_B\frac{d\bar{v}_{CB}}{dt} &= -d_1^2T_s\frac{n(n\bar{v}_{CB}-\bar{v}_{CO})}{2L_o} + \frac{d_1^2T_s\bar{v}_g^2(t)}{2L_i\bar{v}_{CB}} \quad \text{and}\\
C_O\frac{d\bar{v}_{CO}}{dt} &= -\frac{\bar{v}_{CO}}{R} + d_1^2T_s\frac{n(n\bar{v}_{CB}-\bar{v}_{CO})}{2L_O\bar{v}_{CO}}.
\end{aligned} \tag{4.29}$$

The averaged rectified line current is given by

$$\bar{i}_g(t) = \frac{1}{T_s}\{\text{area}[i_{Li(1)}]\} = \frac{1}{T_s}\left[\frac{1}{2}(d_1T_s)^2\frac{\bar{v}_g(t)}{L_i}\right]. \tag{4.30}$$

It is revealed from Equation 4.30 that $\bar{i}_g(t)$ is proportional to $\bar{v}_g(t)$.' Thus, the proposed converter is provided with an UPF.

4.5.2.2 Averaged Model over One Half Line Period T_L

Based on the derived averaged model described by Equation 4.30 over one switching period T_s, we now proceed to develop the averaged model over one half line period T_L. Since the bulk capacitance and the output capacitance are sufficiently large, both capacitor voltages can be considered as constants over T_L. Therefore, the state equations of the averaged model over one half line period T_L can be given by

$$\begin{aligned} C_B\frac{d\langle\bar{v}_{CB}\rangle_{T_L}}{dt} &= \left\langle\frac{d_1^2T_s}{2}\left[\frac{-(n^2\bar{v}_{CB}+n\bar{v}_{CO})}{L_O}+\frac{\bar{v}_g^2(t)}{L_i\bar{v}_{CB}}\right]\right\rangle_{T_L} \\ &= \frac{1}{\pi}\int_0^{\pi}\frac{d_1^2T_s}{2}\left[\frac{-(n^2\bar{v}_{CB}+n\bar{v}_{CO})}{L_O}+\frac{v_m^2\sin^2(\omega t)}{L_i\bar{v}_{CB}}\right]d(\omega t) \\ &= \frac{d_1^2T_s}{2}\left[\frac{-n^2\langle\bar{v}_{CB}\rangle_{T_L}+n\langle\bar{v}_{CO}\rangle_{T_L}}{L_O}+\frac{v_m^2}{2L_i\langle\bar{v}_{CB}\rangle_{T_L}}\right], \end{aligned} \tag{4.31}$$

$$\begin{aligned} C_O\frac{d\langle\bar{v}_{CO}\rangle_{T_L}}{dt} &= \left\langle-\frac{\bar{v}_{CO}}{R}+d_1^2T_s\frac{n^2\bar{v}_{CB}^2-n\bar{v}_{CB}\bar{v}_{CO}}{2L_O\bar{v}_{CO}}\right\rangle_{T_L} \\ &= \frac{1}{\pi}\int_0^{\pi}\left[-\frac{\bar{v}_{CO}}{R}+d_1^2T_s\frac{n^2\bar{v}_{CB}^2-n\bar{v}_{CB}\bar{v}_{CO}}{L_O\bar{v}_{CO}}\right]d(\omega t) \\ &= \frac{\langle\bar{v}_{CO}\rangle_{T_L}}{R}+\frac{d_1^2T_s\left[-n^2\langle\bar{v}_{CB}\rangle_{T_L}^2-n\langle\bar{v}_{CB}\rangle_{T_L}\langle\bar{v}_{CO}\rangle_{T_L}\right]}{2L_O\langle\bar{v}_{CO}\rangle_{T_L}}, \end{aligned} \tag{4.32}$$

and the output equation is given by

$$\langle\bar{v}_O\rangle_{T_L} = \langle\bar{v}_{CO}\rangle_{T_L}. \tag{4.33}$$

Notably, Equations 4.31 and 4.32 are nonlinear state equations that can be linearized around the DC operating point. The DC operating point can be determined by setting $d\langle\bar{v}_{CB}\rangle_{T_L}/dt = 0$ and $d\langle\bar{v}_{CO}\rangle_{T_L}/dt = 0$ in Equations 4.31 and 4.32. Mathematically, we then successively compute the bulk capacitor voltage V_{CB} and the output voltage V_O as

$$V_{CB} = \frac{1}{2n}\left(\sqrt{\frac{D_1^2RT_s}{4L_i}+\frac{2L_O}{L_i}}+\sqrt{\frac{D_1^2RT_s}{4L_i}}\right), \quad V_O = D_1\sqrt{\frac{RT_s}{4L_i}}V_m. \tag{4.34}$$

The design specifications and the component values of the proposed converter are listed in Table 4.6. In Table 4.6, it follows directly from Equation 4.34 that $V_{CB} = 146.6$ V and $V_O = 108$ V. Therefore, the proposed converter exhibits low voltage stress across the bulk capacitor for a VAC 110 input voltage.

TABLE 4.6

Design Specifications and Component Values of the Proposed Converter

Input peak voltage V_m	156 V	Duty ratio D_1	0.26
Input inductor L_i	75 μH	Switching period T_s	20 μs
Magnetizing inductor L_m	3.73 mH	Switching frequency f_s	50 kHz
Output inductor L_O	340 μH	Load resistance R	108 Ω
Bulk capacitor C_B	330 μF	Turns ratio $1{:}n{:}m$	1:2:1
Output capacitor C_O	1000 μF	PWM gain k_{PWM}	$1/12\,V^{-1}$
Bulk capacitor voltage V_{CB}	146.6 V	Output voltage V_O	108 V

After determining the DC operating point, we proceed to derive the small-signal model linearized around the operating point. To proceed, small perturbations

$$v_m = V_m + \tilde{v}_m, \quad d_1 = D_1 + \tilde{d}_1, \quad \langle \bar{v}_{CB} \rangle_{T_L} = V_{C_B} + \tilde{v}_{C_B}, \\ \langle \bar{v}_{CO} \rangle_{T_L} = V_{CO} + \tilde{v}_{CO}, \quad \langle \bar{v}_O \rangle_{T_L} = V_O + \tilde{v}_O, \tag{4.35}$$

with

$$V_m \gg \tilde{v}_m, \quad D_1 \gg \tilde{d}_1, \quad V_{CB} \gg \tilde{v}_{CB}, \quad V_{CO} \gg \tilde{v}_{CO}, \quad V_O \gg \tilde{v}_O, \tag{4.36}$$

are introduced into Equations 4.31 and 4.32 and high-order terms are neglected, yielding dynamical equations of the form

$$\begin{aligned} C_B \frac{d\tilde{v}_{CB}}{dt} &= \frac{D_1^2 T_s}{2}\left(-\frac{n^2}{L_O} - \frac{V_m^2}{2L_i V_{CB}^2}\right)\tilde{v}_{CB} + \frac{D_1^2 T_s}{2}\left(\frac{n}{L_O}\right)\tilde{v}_{CO} \\ &\quad + \frac{D_1^2 T_s}{2}\left(\frac{V_m}{L_i V_{CB}}\right)\tilde{v}_m + D_1 T_s\left(\frac{-n^2 V_{CB} + nV_{CO}}{L_O} + \frac{V_m^2}{2L_i V_{CB}}\right)\tilde{d}_1 \\ &= a_{11}\tilde{v}_{CB} + a_{12}\tilde{v}_{CO} + b_{11}\tilde{v}_m + b_{12}\tilde{d}_1, \end{aligned} \tag{4.37}$$

$$\begin{aligned} C_O \frac{d\tilde{v}_{CO}}{dt} &= \frac{D_1^2 T_s}{2}\left(\frac{2n^2 V_{CB}}{L_O V_{CO}} - \frac{n}{L_O}\right)\tilde{v}_{CB} + \left(-\frac{1}{R} - \frac{D_1^2 T_s}{2}\frac{n^2 V_{CB}^2}{L_O V_{CO}^2}\right)\tilde{v}_{CO} \\ &\quad + 0 \cdot \tilde{v}_m + D_1 T_s\left(\frac{n^2 V_{CB}^2}{L_O V_{CO}} - \frac{nV_{CB}}{L_O}\right)\tilde{d}_1 \\ &= a_{21}\tilde{v}_{CB} + a_{22}\tilde{v}_{CO} + b_{21}\tilde{v}_m + b_{22}\tilde{d}_1. \end{aligned} \tag{4.38}$$

The parameters are defined as

$$a_{11} = \frac{-D_1^2 T_s}{2}\left(\frac{n^2}{L_O} + \frac{V_m^2}{2L_i V_{CB}^2}\right), \quad a_{12} = \frac{D_1^2 T_s}{2}\left(\frac{n}{L_O}\right),$$

$$a_{21} = \frac{D_1^2 T_s}{2}\left(\frac{2n^2 V_{CB}}{L_O V_{CO}} - \frac{n}{L_O}\right), \quad a_{22} = -\left(\frac{1}{R} + \frac{D_1^2 T_s}{2}\frac{n^2 V_{CB}^2}{L_O V_{CO}^2}\right),$$

$$b_{11} = \frac{D_1^2 T_s}{2}\left(\frac{V_m}{L_i V_{CB}}\right), \quad b_{12} = D_1 T_s\left(\frac{-n^2 V_{CB} + nV_{CO}}{L_O} + \frac{V_m^2}{2L_i V_{CB}}\right),$$

$$b_{21} = 0, \quad b_{22} = D_1 T_s\left(\frac{n^2 V_{CB}^2}{L_O V_{C_O}} - \frac{nV_{CB}}{L_O}\right).$$

Mathematically, the dynamical equations in Equations 4.37 and 4.38 can be expressed in matrix form as

$$\begin{bmatrix} \dot{\tilde{v}}_{CB} \\ \dot{\tilde{v}}_{CO} \end{bmatrix} = \begin{bmatrix} \dfrac{a_{11}}{C_B} & \dfrac{a_{12}}{C_B} \\ \dfrac{a_{21}}{C_O} & \dfrac{a_{22}}{C_O} \end{bmatrix} \begin{bmatrix} \tilde{v}_{CB} \\ \tilde{v}_{C_O} \end{bmatrix} + \begin{bmatrix} \dfrac{b_{11}}{C_B} & \dfrac{b_{12}}{C_B} \\ \dfrac{b_{21}}{C_O} & \dfrac{b_{22}}{C_O} \end{bmatrix} \begin{bmatrix} \tilde{v}_m \\ \tilde{d}_1 \end{bmatrix}, \tag{4.39}$$

$$\tilde{v}_O = [0\ 1]\begin{bmatrix} \tilde{v}_{CB} \\ \tilde{v}_{CO} \end{bmatrix}. \tag{4.40}$$

Now taking the Laplace transform for the dynamical equation, the resulting transfer functions from line to output and duty ratio to output are given by

$$\begin{aligned} \frac{\tilde{v}_O(s)}{\tilde{v}_m(s)} &= \frac{b_{11}a_{21}/C_B C_O}{s^2 + [(-a_{11}/C_B) - (a_{22}/C_O)]s + (a_{11}a_{22} - a_{12}a_{21})/C_B C_O}, \\ \frac{\tilde{v}_o(s)}{\tilde{d}_1(s)} &= \frac{(b_{22}/C_O)s + (a_{21}b_{12} - a_{11}b_{22})/C_B C_O}{s^2 + [(-a_{11}/C_B) - (a_{22}/C_O)]s + (a_{11}a_{22} - a_{12}a_{21})/C_B C_O}. \end{aligned} \tag{4.41}$$

4.5.3 Simulation Results

The PSpice simulation results presented in Figure 4.27 demonstrate that both PFC and DC/DC cells are operating in discontinuous current mode. The input inductor current $i_{Li}(t)$ and the output inductor current $i_{LO}(t)$ both reach zero for the remainder of the switching period. Figure 4.28a presents the bulk capacitor voltage $V_{CB} = 149$ V and Figure 4.28b presents the output capacitor voltage $V_{CO} = 110$ V. They are close to the theoretical results $V_{CB} = 146.6$ V and $V_{CO} = 108$ V.

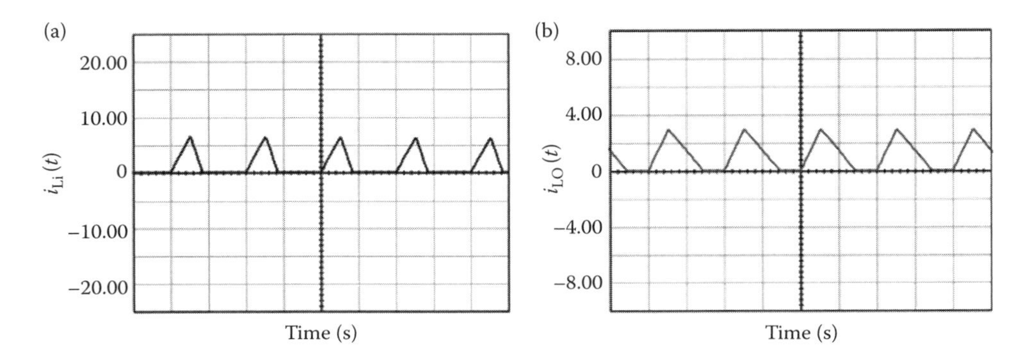

FIGURE 4.27 Current waveforms: (a) input inductor currents $i_{Li}(t)$ (horizontal: 10 μs/div) and (b) output inductor currents $i_{LO}(t)$ (horizontal: 10 μs/div).

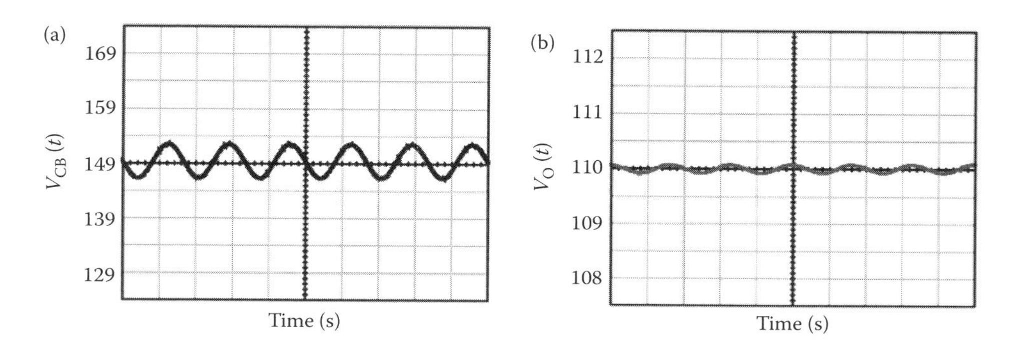

FIGURE 4.28 Ripples of (a) bulk capacitor voltage $V_{CB}(t)$ (vertical: 5 V/div; horizontal: 5 ms/div) and (b) output capacitor voltage $V_{CO}(t)$ (vertical: 0.5 V/div; horizontal: 5 ms/div)

4.5.4 Experimental Results

A prototype based on the topology depicted in Figure 4.25 was built and tested to verify the operating principle of the proposed converter. The experimental results are depicted in the following figures. Figure 4.29a presents the rectified line voltage and current. Figure 4.29b presents the input line voltage and current. This reveals that the proposed converter has a high PF. According to the THD obtained in the simulation results, PF = 0.999.

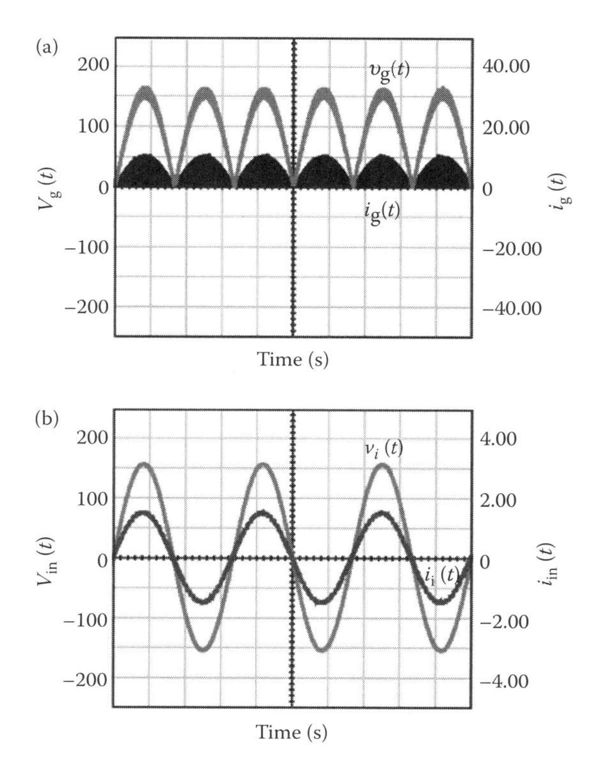

FIGURE 4.29 Line voltages and currents: (a) rectified line voltage and current (horizontal: 5 ms/div) and (b) input line voltage and current (horizontal: 5 ms/div).

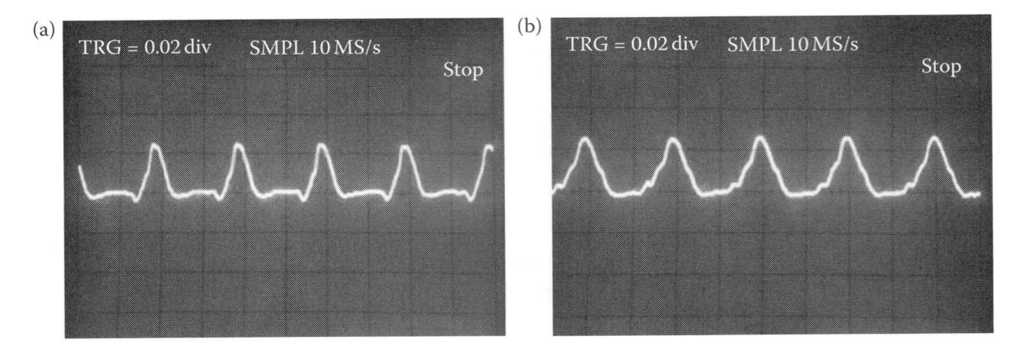

FIGURE 4.30 Inductor currents (horizontal: 10 μs/div): (a) input inductor currents $i_{Li}(t)$ (vertical: 5 A/div) and (b) output inductor currents $i_{LO}(t)$ (vertical: 2 A/div).

Figure 4.30 presents the waveform of the input inductor current $i_{Li}(t)$ and the output inductor current $i_{LO}(t)$. Figure 4.31 presents the voltage ripples of the bulk capacitor voltage $V_{CB}(t)$ and the output capacitor voltage $V_{CO}(t)$. Figure 4.32 presents the rectified line voltage and current and the input line voltage and current. The proposed converter exhibits low voltage stress and a high PF. The measured PF of the converter is 0.998. The efficiency of the proposed converter is about 72%.

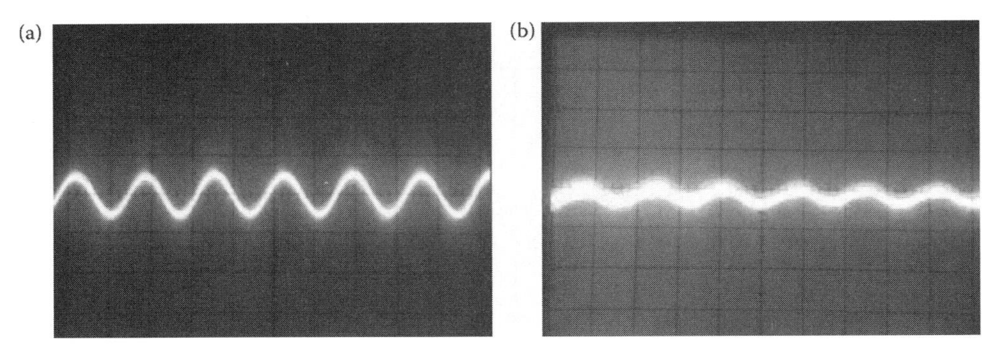

FIGURE 4.31 Ripples of (a) bulk capacitor voltage $V_{CB}(t)$ (vertical: 5 V/div; horizontal: 5 ms/div) and (b) output capacitor voltage $V_{CO}(t)$ (vertical: 0.5 V/div; horizontal: 5 ms/div).

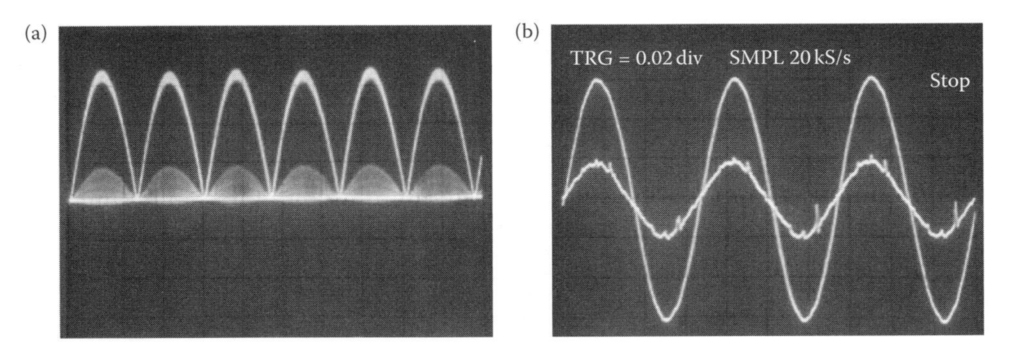

FIGURE 4.32 Line voltages and currents (horizontal: 5 ms/div): (a) rectified line voltage and current (vertical: 50 V/div, 10 A/div) and (b) input line voltage and current (vertical: 50 V/div, 2 A/div).

4.6 VIENNA Rectifiers

The VIENNA rectifier can be used to improve the PF of a three-phase rectifier. The "critical input inductor" is calculated for the nominal load condition, and both PF and THD are degraded in the low-output power region. A novel strategy implementing reference compensation current is proposed based on the operation principle of the VIENNA rectifier in this section. This strategy can realize a three-phase three-level UPF rectifier. With the proposed control algorithm, the converter draws high-quality sinusoidal supply currents and maintains good DC-link voltage regulation under wide load variation. Theoretical analysis is initially verified by digital simulation. Finally, experimental results of a 1-kVA laboratory prototype system confirm the feasibility and effectivity of the proposed technique.

Diode rectifiers with smoothing capacitors have been widely used in many three-phase power electronic systems such as DC motor drives and switch-mode power supplies. However, this topology injects large current harmonics into utilities, which result in the decrease of PF. Expressions of the current THD and the input PF are given as

$$\mathrm{THD} = 100 \times \frac{\sqrt{\sum_{\mathrm{h}=2}^{\infty} I_{\mathrm{sh}}^{2}}}{I_{\mathrm{s1}}}, \tag{4.42}$$

$$\mathrm{PF} = \frac{1}{\sqrt{1+\mathrm{THD}^{2}}}\mathrm{DPF}. \tag{4.43}$$

The international standards presented in IEC 1000-3-2 and EN61000-3-2 imposed harmonic restrictions to modern rectifiers that stimulated a focused research effort on the topic of UPF rectifiers. A slew of new topologies, including those based on three-level power conversion, have been proposed to realize high-quality input waveforms [9–20].

Among the reported three-phase rectifier topologies, the three-phase star-connected switch three-level (VIENNA) rectifier [21–25] is an attractive choice because its switch voltage stress is one-half the total output voltage. This rectifier with three bidirectional switches, three input inductors, and two series-connected capacitors is shown in Figure 4.33.

Each bidirectional switch is turned on when the corresponding phase voltage crosses the zero-volt point and conducts for 30° of the line voltage cycle. Thus, the input current waveform is well shaped and approximately sinusoidal. The input current THD can be as low as 6.6%, and the PF can be as high as 0.99. In addition, the bidirectional switches conduct at twice the line frequency; therefore, the switching losses are negligible.

However, the optimal input inductance required to obtain such a result is usually large, and this technique was proposed for the rectifier operating with a fixed load and a fixed optimal input inductor. Therefore, the DC-link voltage is sensitive to load variation and high performance is achieved within a very limited output power range.

In order to overcome these drawbacks, some control strategies have been proposed [26–31]. A control strategy that takes into account the actual load level on the rectifier is proposed in reference [27]. With this method, high performance can be achieved within a wide output power range. The required optimal input inductance for a prototype rated at 8 kW is about 4 mH. This method is especially suitable for medium-to high-power applications. However, for low-power application (i.e., 1–5 kW), the required optimal input inductance should be larger: for example, around 24 mH for a converter with rated power 1.5 kW. This can result in a bulky and impractical structure.

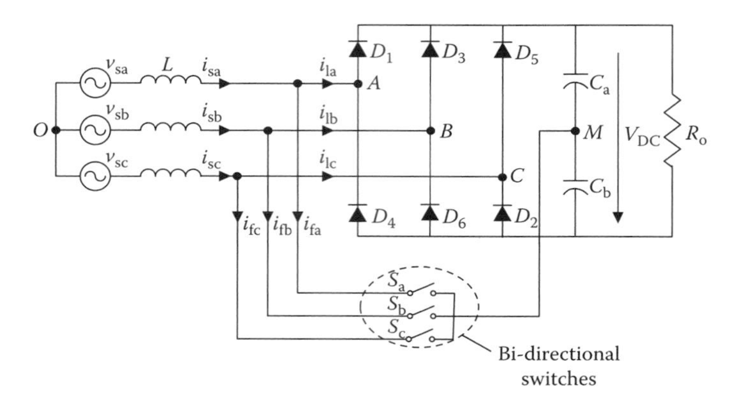

FIGURE 4.33 AC/DC converter with bidirectional switches—the VIENNA rectifier.

The ramp comparison current control presented in reference [26] derives the duty cycle by a comparison of the current error and the fixed-frequency carrier signal. The ripple current in the input inductor makes the current error noisy, although synchronization is carefully considered. Another approach that features constant switching frequency was proposed based on integration control [28]. The input voltage sensors were eliminated in the integration control. However, a significant low-frequency distortion can be observed in the input currents. Recently, a synchronous-reference-frame-based hysteresis current control (HCC) was adopted as the inner loop and DC-link voltage control as the outer loop [29], but a reference-frame transformation was required that increased the controller operation time (digital signal processor [DSP] [29]). A hysteresis current controller was proposed in references [30,31]. The switching signals are generated by the comparison of a reference current template (sinusoidal) and the measured main currents. Although this approach is easy to implement, one needs to measure the DC current and the equipment is costly.

The novel control method proposed in this chapter was based on the operation principle of the VIENNA rectifier. The VIENNA rectifier is composed of two parts: an active compensation circuit and a conventional rectifier circuit. The harmonics injected by a conventional rectifier can be compensated by the active compensation circuit, which enables the input PF can be increased. The average real power consumed by the load is supplied by the source and the active compensation circuit does not provide or consume any average real power. Then the reference compensational current can be obtained. The conduction period of bidirectional switches (S_a, S_b, and S_c) is controlled by using HCC. The idea is that a high switching frequency results in the input inductor size being effectively reduced. This control method does not need to measure the DC-link current and so results in the decrease of the equipment size and cost. Simulation and experimental results have shown that the input PF can be significantly improved and the input current harmonics can be effectively eliminated under wide load variation. The proposed control strategy also maintains good DC-link voltage.

4.6.1 Circuit Analysis and Principle of Operation

The AC/DC converter topology shown in Figure 4.33 is composed of a three-phase diode rectifier with two identical series-connected capacitors and three bidirectional switches

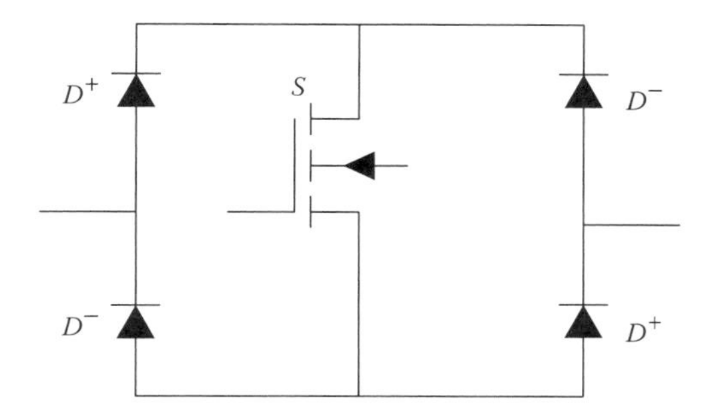

FIGURE 4.34 Construction of a bidirectional switch.

(S_a, S_b, and S_c). The switches consist of four diodes and a MOSFET to form a bidirectional switch (see Figure 4.34).

These bidirectional switches are controlled by using HCC to ensure good supply current waveform, constant DC-link voltage, and accurate voltage balance between the two capacitors. In Figure 4.33, the voltage sources v_{sa}, v_{sb}, and v_{sc} denote the three-phase AC system. The waveforms and the current of phase a (i_{sa}) are shown in Figure 4.35.

For the circuit analysis (Figure 4.35), six topological stages are presented, corresponding to a half-cycle (0° to 180°), which refer to the input voltage v_{sa} shown in Figure 4.35; for simplicity, only the components where current is present are pictured at each of those intervals.

In the interval between 0° and 30° (see Figure 4.36a and b), the polarities of the source voltages v_{sa} and v_{sc} are positive with that of v_{sb} negative. When the bidirectional switch S_a is on, the source current i_{sa} flows through S_a, and diodes D_5 and D_6 are on. The other diodes not shown in Figure 4.36a are off. When the bidirectional switch S_a is off, the current i_{sa} flowing through the input inductor is continued through diode D_1 and diodes D_5 and D_6 are still on. The other diodes not shown in Figure 4.36b are off. The current commutation from S_a to D_1 is at a certain moment determined by HCC. Diodes D_5 and D_6 offer a conventional rectifying wave. Switch S_a and diode D_1 turn on exclusively, and offer the active compensation current.

In the interval between 30° and 60° (see Figure 4.36c and d), the polarities of the source voltages v_{sa} and v_{sc} are positive with that of v_{sb} negative. When the bidirectional switch S_c is on, the source current i_{sc} flows through S_c, and diodes D_1 and D_6 are on. The other diodes not shown in Figure 4.36c are off. When the bidirectional switch S_c is off, the current i_{sc} flowing through the input inductor continues through diode D_5, and diodes D_1 and D_6 are still on.

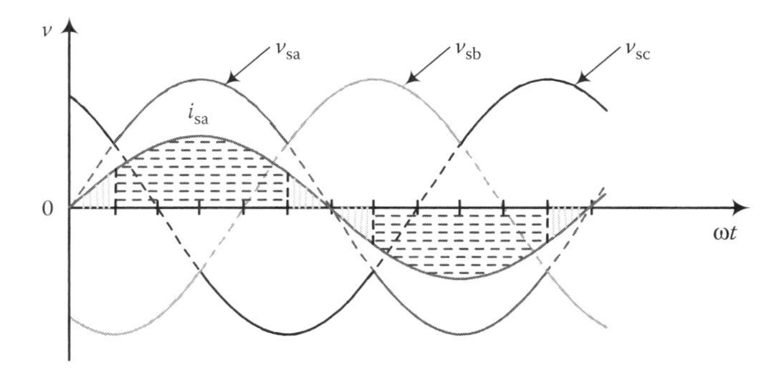

FIGURE 4.35 Waveforms of source voltages and current of phase a, i_{sa}.

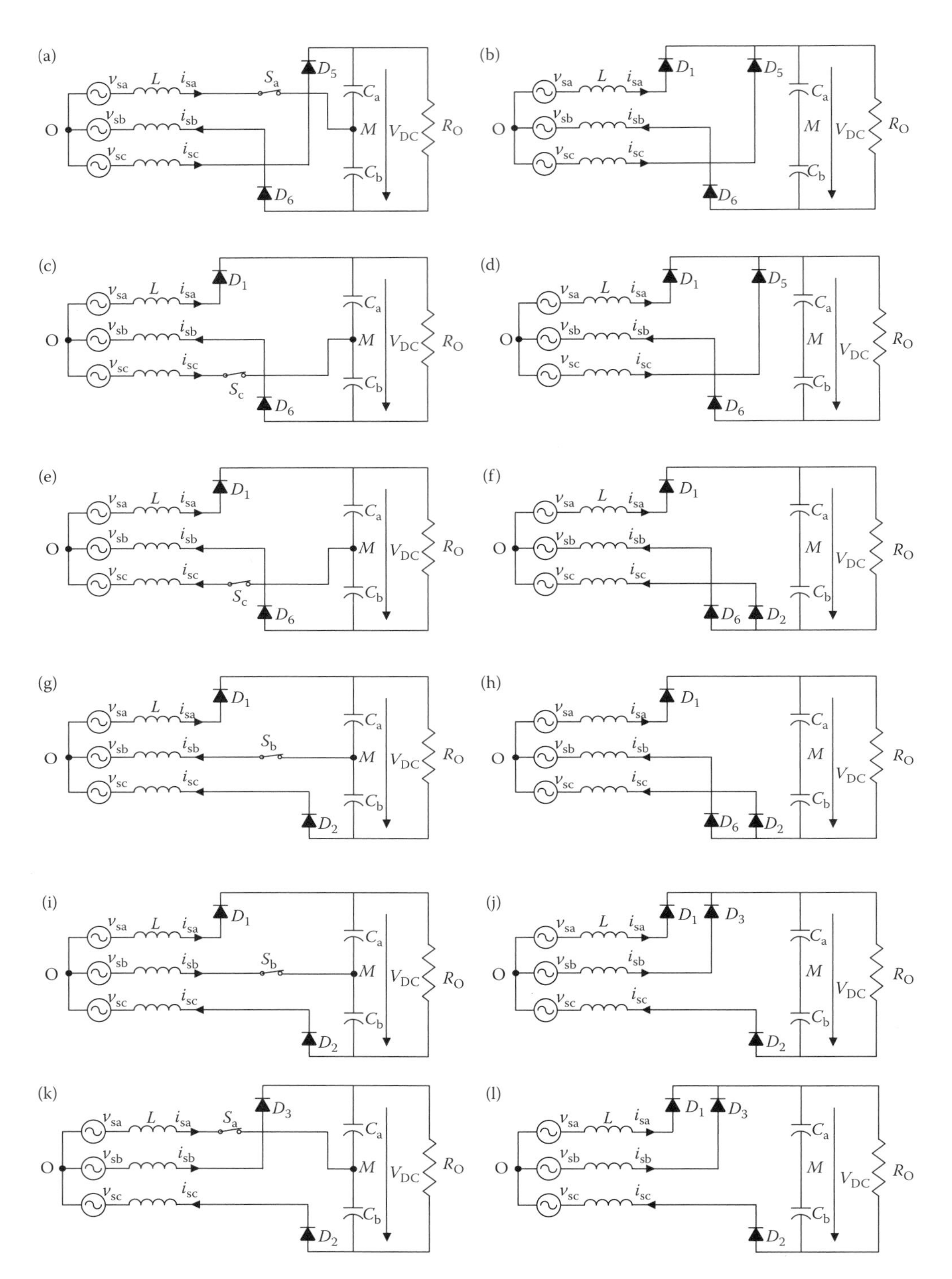

FIGURE 4.36 Topological stages for 0°–180° referring to the input voltage v_{sa} (a) 0°–30°; S_a is on; (b) 0°–30°; S_a is off, the current i_{sa} flowing through the input inductor is continued through the diode D_1, diodes D_5 and D_6 are still on. (c) 30°–60°; S_c is on; (d) 30°–60°; S_c is off, the current i_{sc} flowing through the input inductor is continued through the diode D_5, diodes D_1 and D_6 are still on. (e) 60°–90°; S_c is on; (f) 60°–90°; S_c is off, the current i_{sc} flowing through the input inductor is continued through the diode D_2, diodes D1 and D6 are still on. (g) 90°–120°; S_b is on; (h) 90°–120°; S_b is off, the current i_{sb} flowing through the input inductor is continued through the diode D_6, diodes D_1 and D_2 are still on. (i) 120°–150°; S_b is on; (j) 90°–120°; S_b is off, the current i_{sb} flowing through the input inductor is continued through the diode D_3, diodes D_1 and D_2 are still on. (k) 150°–180°; S_a is on; (l) 150°–180°; S_a is off, the current i_{sa} flowing through the input inductor is continued through the diode D_1, diodes D_3 and D_2 are still on.

The other diodes not shown in Figure 4.36d are off. The current commutation from S_c to D_5 is at a certain moment determined by HCC. Diodes D_1 and D_6 offer a conventional rectifying wave. Switch S_c and diode D_5 turn on exclusively, and offer the active compensation current.

In the interval between 60° and 90° (see Figure 4.36e and f), the polarity of the source voltage v_{sa} is positive with those of v_{sb} and v_{sc} negative. When the bidirectional switch S_c is on, the source current i_{sc} flows through S_c, and diodes D_1 and D_6 are on. The other diodes not shown in Figure 4.36e are off. When the bidirectional switch S_c is off, the current i_{sc} flowing through the input inductor is continued through diode D_2, and diodes D_1 and D_6 are still on. The other diodes not shown in Figure 4.36f are off. The current commutation from S_c to D_2 is at a certain moment determined by HCC. Diodes D_1 and D_6 offer a conventional rectifying wave. Switch S_c and diode D_2 turn on exclusively, and offer the active compensation current.

In the interval between 90° and 120° (see Figure 4.36g and h), the polarity of the source voltage v_{sa} is positive with those of v_{sb} and v_{sc} negative. When the bidirectional switch S_b is on, the source current i_b flows through S_b and diodes D_1 and D_2 are on. The other diodes not shown in Figure 4.36g are off. When the bidirectional switch S_b is off, the current i_b flowing through the input inductor continues through diode D_6, and diodes D_1 and D_2 are still on. The other diodes not shown in Figure 4.36h are off. The current commutation from S_b to D_6 is at a certain moment determined by HCC. Diodes D_1 and D_2 offer a conventional rectifying wave. Switch S_b and diode D_6 turn on exclusively, and offer the active compensation current.

In the interval between 120° and 150° (see Figure 4.36i and j), the polarities of the source voltages v_{sa} and v_{sb} are positive with that of v_{sc} negative. When the bidirectional switch S_b is on, the source current i_{sb} flows through S_b and diodes D_1 and D_2 are on. The other diodes not shown in Figure 4.36i are off. When the bidirectional switch S_b is off, the current i_{sb} flowing through the input inductor continues through diode D_3 and diodes D_1 and D_2 are still on. The other diodes not shown in Figure 4.36j are off. The current commutation from S_b to D_3 is at a certain moment determined by HCC. Diodes D_1 and D_2 offer a conventional rectifying wave. Switch S_b and diode D_3 turn on exclusively, and offer the active compensation current.

In the interval between 150° and 180° (see Figure 4.36k and l), the polarities of the source voltages v_{sa} and v_{sb} are positive with that of v_{sc} negative. When the bidirectional switch S_a is on, the source current i_{sa} flows through S_a and diodes D_3 and D_2 are on. The other diodes not shown in Figure 4.36k are off. When the bidirectional switch S_a is off, the current i_{sa} flowing through the input inductor continues through diode D_1 and diodes D_3 and D_2 are still on. The other diodes not shown in Figure 4.36l are off. The current commutation from S_a to D_1 is at a certain moment determined by HCC. Diodes D_3 and D_2 offer a conventional rectifying wave. Switch S_a and diode D_1 turn on exclusively, and offer the active compensation current.

An active compensation circuit is composed of one of the bidirectional switches and an off-diode in the rectifier bridge legs, but the other legs act as a conventional rectifier. So there are two circuits in the VIENNA rectifier, namely the conventional rectifier circuit and the active compensation circuit. Thus, the load average real power is supplied by the source (the same as a conventional rectifier) and the active compensation circuit does not provide or consume any real power.

4.6.2 Proposed Control Arithmetic

The proposed controller is based on the requirement that the source currents need to be balanced, undistorted, and in phase with the source voltages. The functions of the active compensation circuit are to (1) unitize supply PF, (2) minimize average real power consumed or supplied by the active compensation circuit, and (3) compensate harmonics and reactive currents. To carry out the functions, the desired three-phase source currents of Equation 4.44 must be in phase with the source voltages of Equation 4.45:

$$\begin{cases} i_{sa} = I_m \sin(\omega t + \phi), \\ i_{sb} = I_m \sin(\omega t + \phi - 120°), \\ i_{sc} = I_m \sin(\omega t + \phi + 120°), \end{cases} \tag{4.44}$$

$$\begin{cases} v_{sa} = V_m \sin(\omega t + \phi), \\ v_{sb} = V_m \sin(\omega t + \phi - 120°), \\ v_{sc} = V_m \sin(\omega t + \phi + 120°). \end{cases} \tag{4.45}$$

where V_m and ϕ are the voltage magnitude and the phase angle of the source voltages, respectively. Under the conditions that the load active power is supplied by the source and the active compensation circuit does not provide or consume any real power, the current magnitude I_m needs to be determined from the sequential instantaneous voltage and real power components supplied to the load. According to the symmetrical component transformation for the three-phase rms currents at each harmonic order, the three-phase instantaneous load currents can be expressed by

$$i_{lk} = \sum_{n=1}^{\infty} i_{lkn}^{+} + \sum_{n=1}^{\infty} i_{lkn}^{-} + \sum_{n=1}^{\infty} i_{lkn}^{0}, \quad k \in K. \tag{4.46}$$

In Equation 4.46, $K = \{a, b, c\}$; 0, +, and − stands for zero-, positive-, and negative-sequence components, respectively, and n represents the fundamental (i.e., $n = 1$) and the harmonic components. Since the average real power consumed by the load over one period of time T must be supplied by the source and requires that the active compensation circuit consumes or supplies null average real power, Equations 4.47 through 4.51 must hold

$$p_s = p_l + p_f, \tag{4.47}$$

$$\bar{p}_s = \frac{1}{T}\int_0^T \sum_{k \in K} v_{sk} i_{sk}\, dt, \tag{4.48}$$

$$\bar{p}_l = \frac{1}{T}\int_0^T \sum_{k \in K} v_{sk} i_{lk}\, dt, \tag{4.49}$$

$$\bar{p}_f = 0, \tag{4.50}$$

$$\bar{p}_s = \bar{p}_l. \tag{4.51}$$

Substituting Equation 4.46 into Equation 4.49 yields the sum of the fundamental and the harmonic power terms at the three sequential components

$$\bar{p}_l = \bar{p}_{l1}^{+} + \bar{p}_{l1}^{-} + \bar{p}_{l1}^{0} + \bar{p}_{lh}^{+} + \bar{p}_{lh}^{-} + \bar{p}_{lh}^{0}, \tag{4.52}$$

where

$$\bar{p}_{l1}^{+} = \frac{1}{T}\int_0^T \sum_{k \in K} v_{sk} i_{lk1}^{+}\, dt = \frac{1}{T}\int_0^T \sum_{k \in K} v_{sk} i_{sk}\, dt = \frac{3V_m I_m}{2} \tag{4.53}$$

and

$$\bar{p}_{l1}^{-} = \bar{p}_{l1}^{0} = \bar{p}_{lh}^{+} = \bar{p}_{lh}^{-} = \bar{p}_{lh}^{0} = 0. \tag{4.54}$$

Each power term in Equation 4.54 is determined based on the orthogonal theorem for a periodic sinusoidal function. Then, Equation 4.49 becomes

$$\bar{p}_s = \bar{p}_l = \bar{p}_{l1}^{+} = \frac{1}{T}\int_0^T \sum_{k\in K} v_{sk} i_{sk}\, dt. \tag{4.55}$$

Using Equations 4.51, 4.53, and 4.55, the desired source current magnitude at each phase is determined as

$$I_m = \frac{2\bar{p}_l}{3V_m} = \frac{2\int_0^T \sum_{k\in K} v_{sk} i_{lk}\, dt}{3TV_m} \tag{4.56}$$

and the source currents of Equation 4.44 can be expressed by

$$i_{sk} = I_m \frac{v_{sk}}{V_m} = \frac{2\bar{p}_l}{3(V_m)^2} v_{sk}, \quad k \in K. \tag{4.57}$$

The required current compensation at each phase by the active compensation circuit is then obtained by subtracting the desired source current from the load current as

$$i_{fk}^{*} = i_{lk} - i_{sk} = i_{lk} - \frac{2\bar{p}_l}{3(V_m)^2} v_{sk}, \quad k \in K. \tag{4.58}$$

The average real power consumed or supplied by the active compensation circuit is expressed as

$$\bar{p}_f = \frac{1}{T}\int_0^T \sum_{k\in K} v_{sk} i_{fk}\, dt. \tag{4.59}$$

Substituting Equation 4.58 into Equation 4.59 yields

$$\begin{aligned}\bar{p}_f &= \frac{1}{T}\int_0^T \sum_{k\in K} v_{sk} i_{lk}\, dt - \frac{2\bar{p}_l}{3(V_m)^2}\frac{1}{T}\int_0^T \sum_{k\in K} v_{sk}^2\, dt \\ &= \bar{p}_l - \frac{2\bar{p}_l}{3(V_m)^2}\frac{3(V_m)^2}{2} = \bar{p}_l - \bar{p}_l = 0.\end{aligned} \tag{4.60}$$

Therefore, the active compensation circuit does not consume or supply average real power.

4.6.3 Block Diagram of the Proposed Controller for the VIENNA Rectifier

Figure 4.37 depicts the block diagram of the control circuit based on the proposed approach to fulfill the function of the reference compensation current calculator. The source voltages

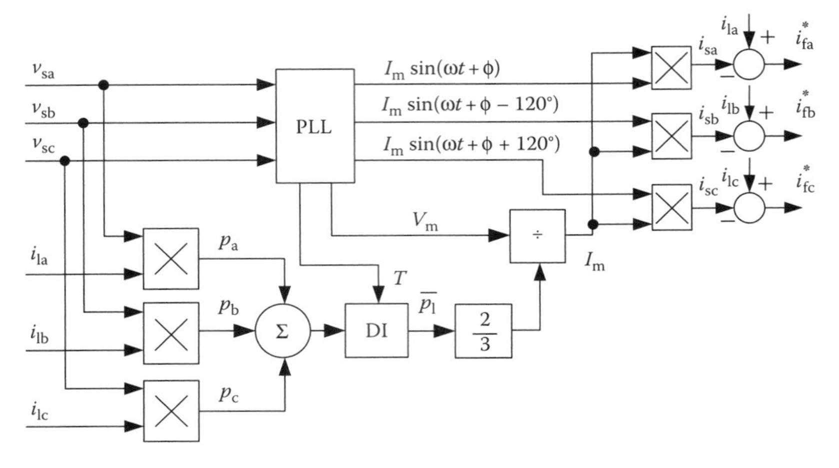

FIGURE 4.37 Block diagram of the controller.

are input to a phase locked-loop (PLL) where the peak voltage magnitude V_m, the unity voltages (i.e., v_{sk}/V_m), and the period T are generated. The average real power of the load consumed is calculated using Equation 4.55 and is input to a divider to obtain the desired source current amplitude I_m in Equation 4.56. DI denotes the calculation of definite integral. The desired source currents in Equation 4.57 and the reference compensation currents of the active compensation circuit in Equation 4.58 are computed by using the voltage magnitude and the unity voltages.

Once the reference compensation currents are determined, they are input to a current controller to produce control signals to the bidirectional switches. The block diagram of the proposed control scheme is shown in Figure 4.38. The bidirectional switches are controlled by the HCC technique to ensure sinusoidal input current with UPF and DC-link voltage. In addition, since the capacitor voltage must be maintained at a constant level, the power losses caused by switching and capacitor voltage variations are supplied by the source. The sum of the power losses, $\bar{p}_{sw}$, is controlled via a proportional-integral (PI) controller and is then input to the reference compensation current calculator. Since the rectifier provides continuous input currents, the current stresses on the switching devices are smaller and the critical input inductor size can be reduced.

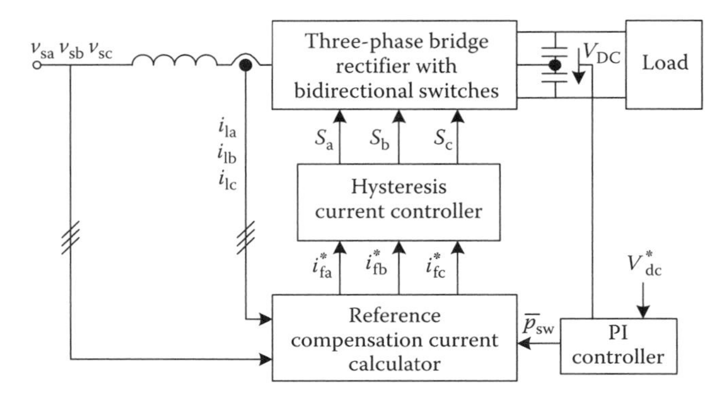

FIGURE 4.38 Block diagram of the control system.

4.6.4 Converter Design and Simulation Results

To verify the performance of the proposed control strategy, a MATLAB®–SIMULINK® prototype of the rectifier is developed. A sinusoidal PWM (SPWM) voltage source inverter, which is a very popular topology in industry, is used as the DC/AC inverter for the intended rectifier–inverter AC motor drive topology (see Figure 4.39).

To illustrate the design feasibility of the proposed converter, a prototype with the following specifications is chosen:

1. Input line-to-line voltage 220 V.
2. DC-link reference voltage 370 V.
3. Input inductance 5 mH.
4. Rated output power 1 kW.

A MATLAB®–SIMULINK® model for the proposed rectifier–inverter structure is developed to perform the digital simulation. Figure 4.40 shows the converter input phase current waveform and its harmonic spectrum at rated output power operation. The same waveform for a conventional converter is shown in Figure 4.41.

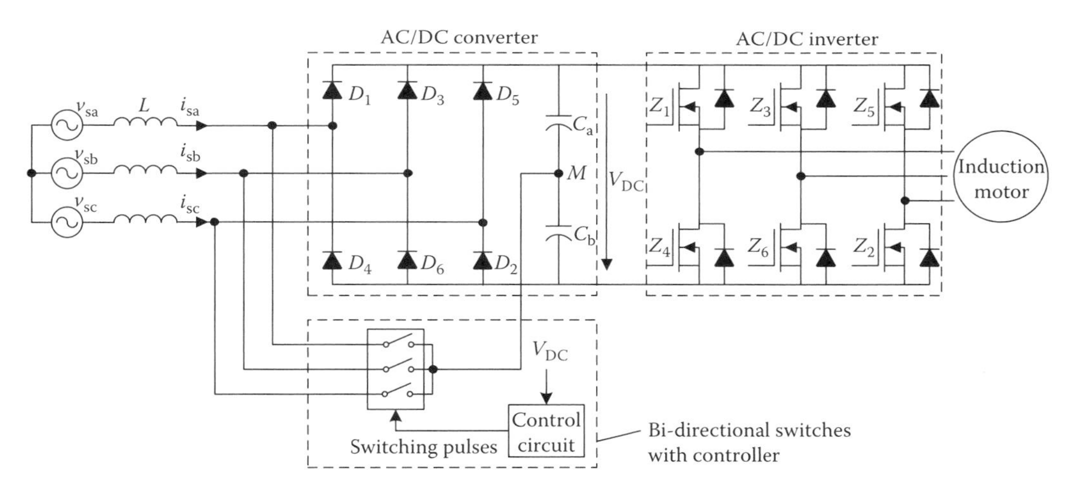

FIGURE 4.39 Complete diagram of the proposed UPF AC drive.

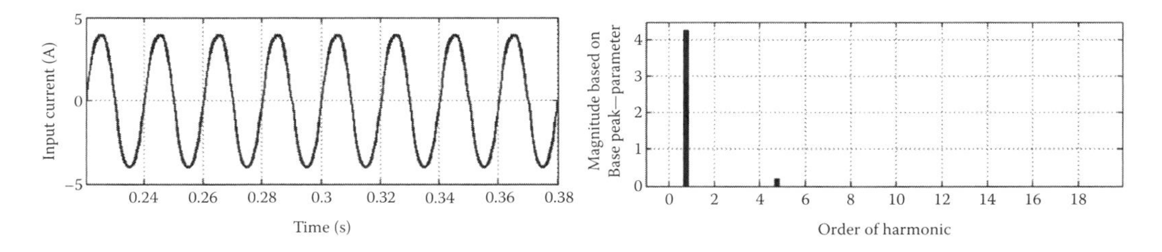

FIGURE 4.40 Input current and spectral composition of the proposed scheme at rated load.

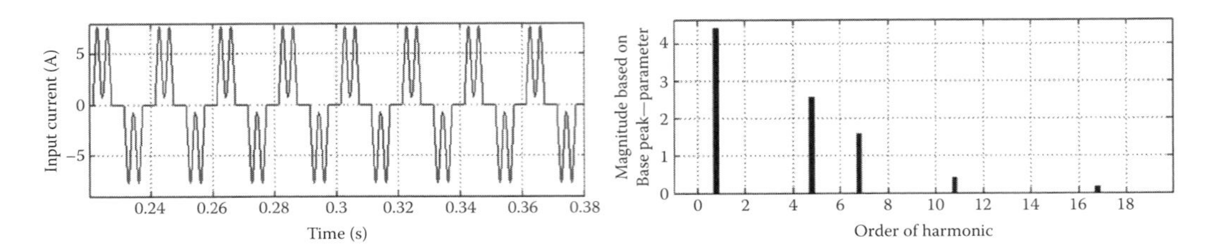

FIGURE 4.41 Input current and spectral composition of a typical commercial converter.

Before improvement, the THD of the rectifier input current was found to be 91.5% and the input PF was 0.72. After improvement, the input current THD was 3.8% and the input PF was 0.999. Thus, with the proposed reference compensation current strategy, the harmonics are effectively reduced and the PF is dramatically increased.

In order to show the performance of the converter under varying load conditions, it is operated below and above its rated value. The converter input phase current waveform and its harmonic spectrum at 50% rated output power are shown in Figure 4.42. The converter input PF is found to be 0.996 and the input current THD is 4.0%.

The converter input phase current waveform and its harmonic spectrum at 150% rated output power are shown in Figure 4.43. The converter input PF is found to be 0.999 and the input current THD is 3.7%. It is evident that the proposed control strategy has a good adaptability to different load conditions. This strategy can also be used for rectifiers operating at various rated power levels.

Figure 4.44 illustrates the input phase currents and DC-link voltage waveforms when the converter output power demand changes instantaneously from 50% to 100% of its rated value due to load disturbance. The load change was initiated at 0.26 s where the converter was in steady state. One can clearly see that the converter exhibits a good response to the

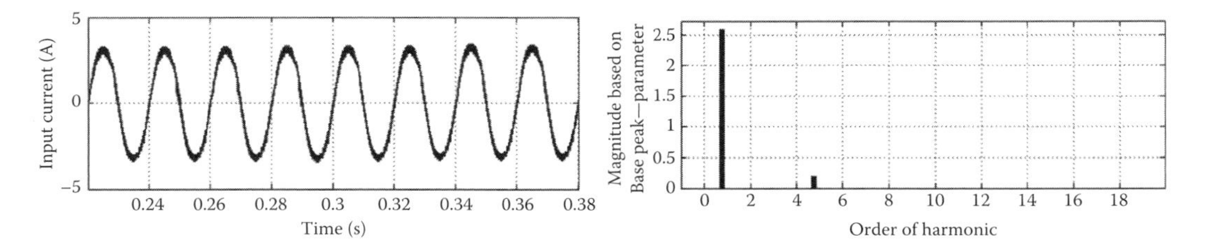

FIGURE 4.42 Input current and spectral composition of the proposed scheme at 50% rated load.

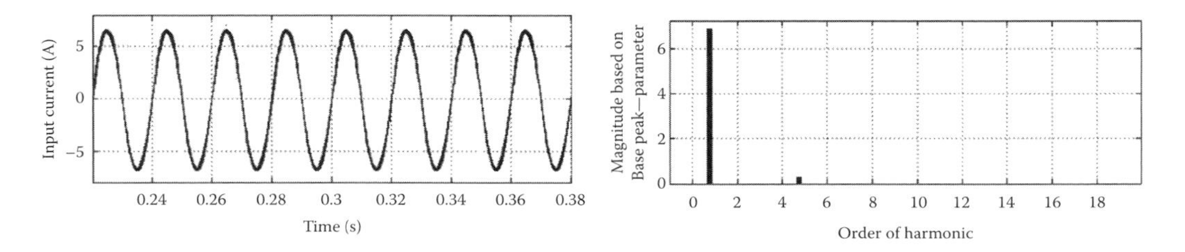

FIGURE 4.43 Input current and spectral composition of the proposed scheme at 150% rated load.

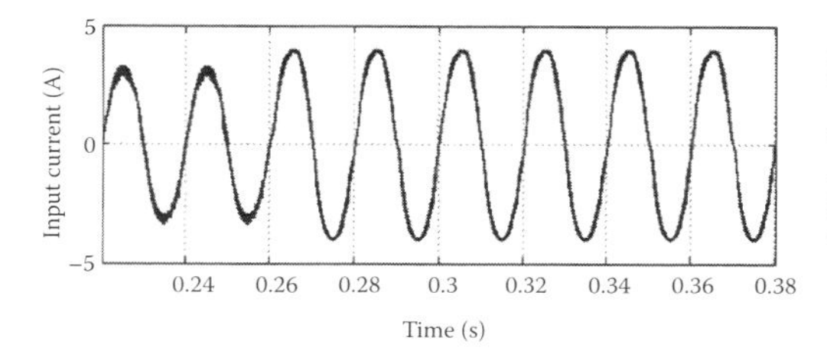

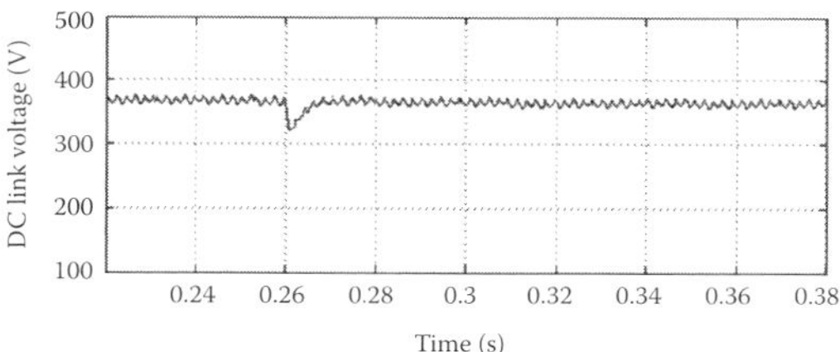

FIGURE 4.44 Converter response due to load change.

sudden load variation. From this figure, it can be seen that this proposed control technique has a good adaptability to load variation.

4.6.5 Experimental Results

The control system is implemented using a single-board dSPACE 1102 microprocessor and is developed under the integrated development of MATLAB®–SIMULINK® RTW provided by The Math Works. A 1-kW hardware prototype of the rectifier–inverter structure as shown in Figure 4.39 was constructed and its performance was observed.

The rectifier input current and voltage waveforms before and after improvements are shown in Figures 4.45 and 4.46, respectively. The fluke-43 spectrum analyzer with online numerical value illustration is used to monitor the waveforms. The input PF is shown online at the upper right-hand side of Figures 4.45 and 4.46. Prior to improvement, the input current THD and PF were 91.5% and 0.72, respectively.

The proposed scheme is able to improve the input current THD to 3.8% and the input PF to 0.99. There is a remarkable improvement in PF and THD. The experimental results are identical to the MATLAB® predicted ones calculated based on the waveforms in Figures 4.40 and 4.41. Figures 4.47 and 4.48 show the experimental input current fast-Fourier transform (FFT) spectrum for a typical conventional converter and the proposed converter, respectively.

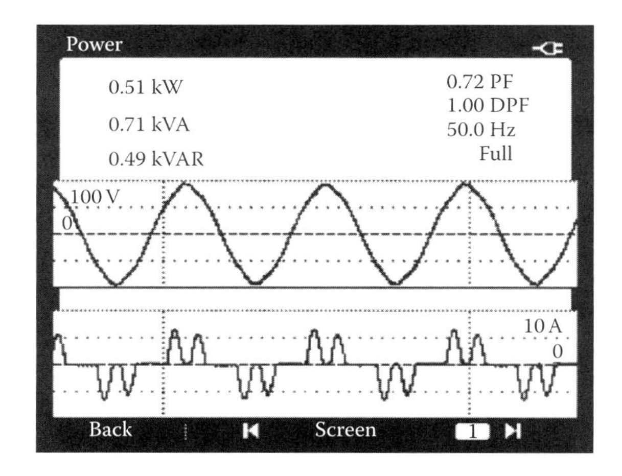

FIGURE 4.45 Input voltage and current of a typical conventional converter.

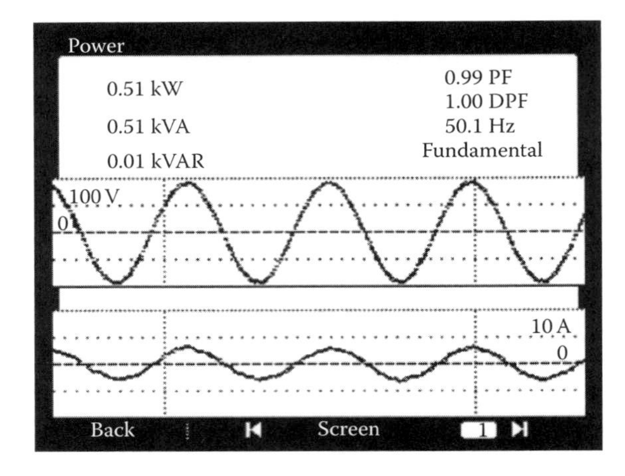

FIGURE 4.46 Input voltage and current of the proposed prototype.

At 50% rated output power, the converter input PF is found to be 0.99 and the input current THD has increased to 4.0%, as shown in Figure 4.49. At 150% rated output power, the converter input PF is found to be 0.99 and the input current THD is reduced to 3.7% (see Figure 4.50).

Figure 4.51 shows the DC-link voltage waveforms when the converter output power demand changes instantaneously from 50% to 100% of its rated value responding to load disturbance. One can see that with the proposed control strategy, the converter exhibits a good response to sudden load variation.

To investigate the effect of input inductance, this was varied as well. Under 3 and 7 mH input inductances, the converter input currents and voltages are shown in Figures 4.52 and 4.53, respectively. These results illustrate that the proposed converter with bidirectional switches coupled with the proposed strategy overcomes most of the shortcomings of the conventional converters such as change of input PF due to output power, input inductance, and load torque variations.

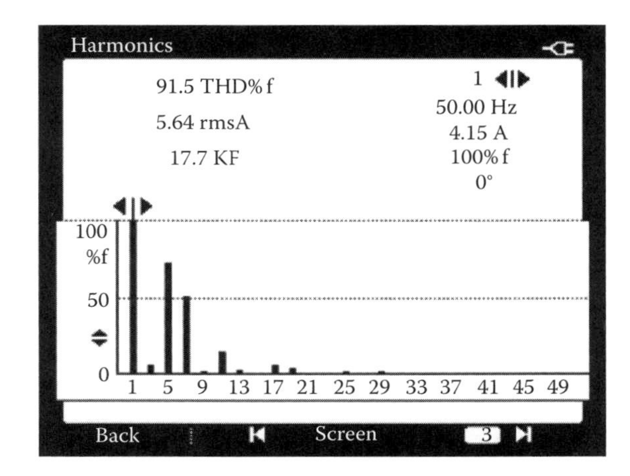

FIGURE 4.47 Input current FFT of a typical conventional converter.

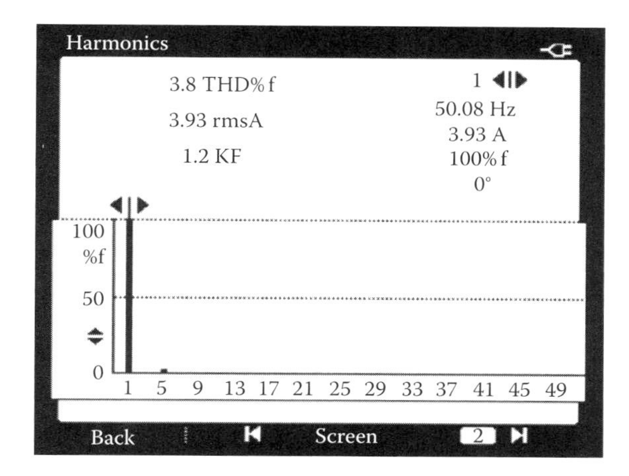

FIGURE 4.48 Input current FFT of the proposed prototype conventional converter.

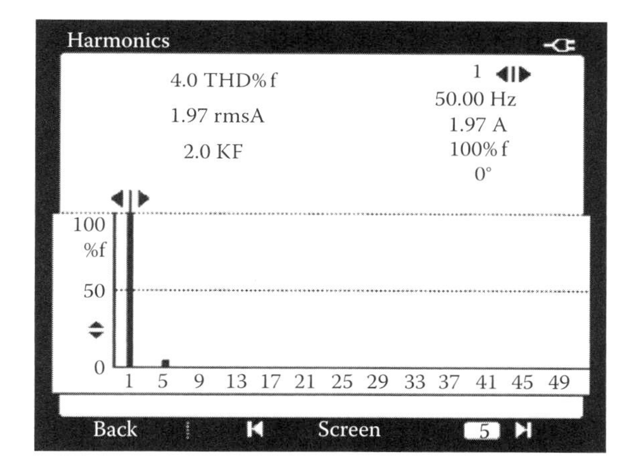

FIGURE 4.49 Input current FFT of the proposed prototype at 50% rated load.

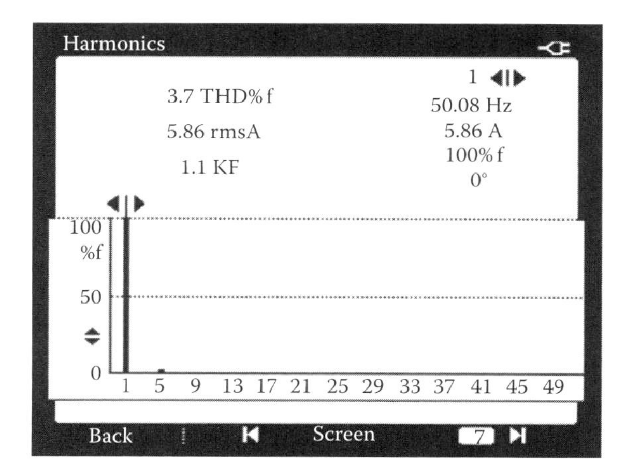

FIGURE 4.50 Input current FFT of the proposed prototype at 150% rated load.

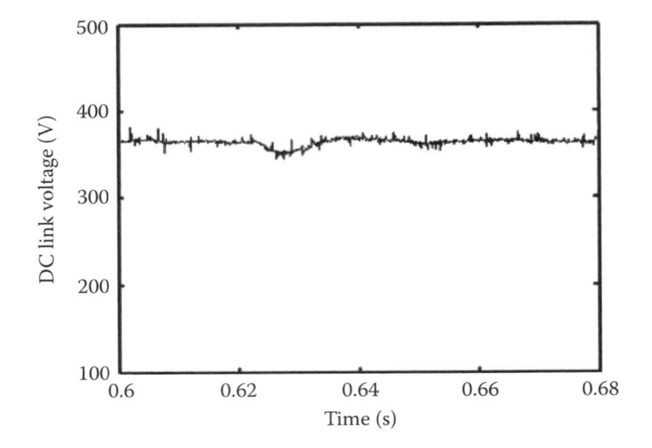

FIGURE 4.51 Converter response to a sudden load change in DC-link voltage.

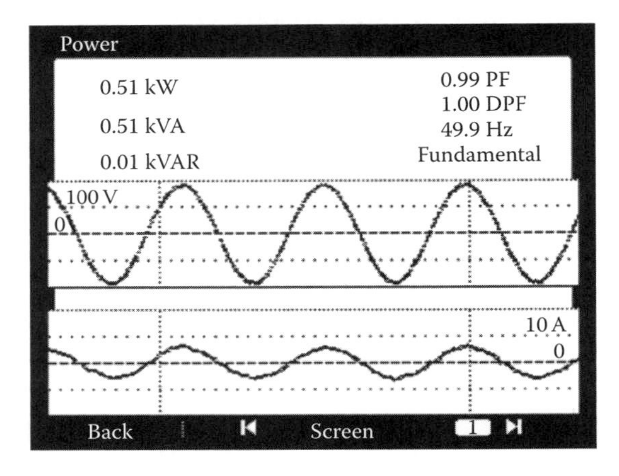

FIGURE 4.52 Converter input current and voltage for 3 mH input inductance.

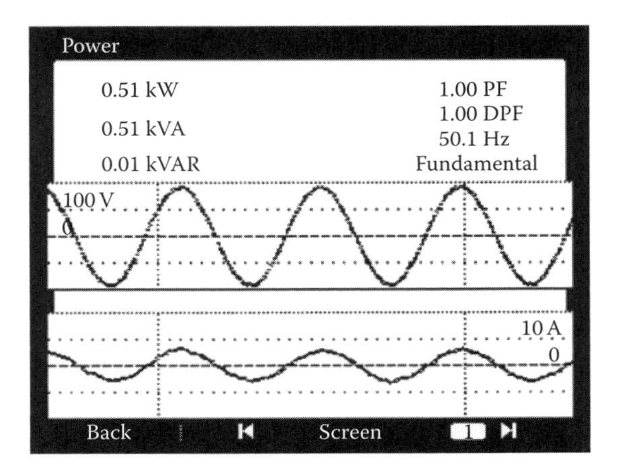

FIGURE 4.53 Converter input current and voltage for 7 mH input inductance.

Homework

4.1. A P/O self-lift Luo-converter (see Figure 6.4 in Chapter 6) is used to implement PFC in a single-phase diode rectifier with R–C load. The AC supply voltage is 200 V/60 Hz and the required output voltage is 400 V. The switching frequency is 2.4 kHz. Determine the duty cycle k in a half supply period (8.33 ms). Other component values for reference are $R = 100\,\Omega, L_1 = L_2 = 10$, and $C = C_1 = C_O = 20\,\mu F$.

4.2. A P/O super-lift Luo-converter (see Figure 7.1 in Chapter 7) is used to implement PFC in a single-phase diode rectifier with R–C load. The AC supply voltage is 200 V/60 Hz and the required output voltage is 600 V. The switching frequency is 3.6 kHz. Determine the duty cycle k in a half supply period (8.33 ms). Other component values for reference are $R = 100\,\Omega, L_1 = L_2 = 10$, and $C = C_1 = C_O = 20\,\mu F$.

References

1. Luo, F. L. and Ye, H. 2004. *Advanced DC/DC Converters*. Boca Raton: CRC Press.
2. Luo, F. L. 2005. A single-stage power factor correction AC/DC converter. *Proceedings of the International Conference IPEC* 2005, pp. 513–518.
3. Mohan, N., Undeland, T. M., and Robbins, W. P. 2003. *Power Electronics: Converters, Applications and Design* (3rd edition). New York: Wiley.
4. Wu, T. F. and Chen, Y. K. 1998. A systematic and unified approach to modeling PWM DC/DC converters based on the graft scheme. *IEEE Transactions on Industrial Electronics*, 45, 88–98.
5. Kheraluwala, M. H. 1991. Fast-response high power factor converter with a single power stage. *Proceedings of the IEEE-PESC*, pp. 769–779.
6. Lee, Y. S. and Siu, K. W. 1996. Single-switch fast-response switching regulators with unity power factor. *Proceedings of the IEEE-APEC*, pp. 791–796.
7. Shen, M. and Qian, Z. 2002. A novel high-efficiency single-stage PFC converter with reduced voltage stress. *IEEE Transactions on Industry Applications*, 49, 507–513.
8. Qiu, M. 1999. Analysis and design of a single stage power factor corrected full-bridge converter. *Proceedings of the IEEE-APEC*, pp. 119–125.
9. Zhang, S. and Luo, F. L. 2009. A novel reference compensation current strategy for three-phase three-level unity PF rectifier. *Proceedings of the IEEE-ICIEA* 2009, pp. 581–586.
10. Suryawanshi, H. M., Ramteke, M. R., Thakre, K. L., and Borghate, V. B. 2008. Unity-power-factor operation of three-phase AC-DC soft switched converter based on boost active clamp topology in modular approach. *IEEE Transactions on Industrial Electronics*, 55, 229–236.
11. Lu, D. D., Iu, H. H., and Jevalica, P. 2008. A single-stage AC/DC converter with high power factor, regulated bus voltage, and output voltage. *IEEE Transactions on Power Electronics*, 23, 218–228.
12. Chen, J. F., Chen, R. Y., and Liang, T. J. 2008. Study and implementation of a single-stage current-fed boost PFC converter with ZCS for high voltage applications. *IEEE Transactions on Power Electronics*, 23, 379–386.
13. Kong, P., Wang, S., and Lee, F. C. 2008. Common mode EMI noise suppression for bridgeless PFC converters. *IEEE Transactions on Power Electronics*, 23, 291–297.
14. Chen, M., Mathew, A., and Sun, J. 2007. Nonlinear current control of single-phase PFC converters. *IEEE Transactions on Power Electronics*, 22, 2187–2194.

15. Tutakne, D. R., Suryawanshi, H. M., and Tarnekar, S. G. 2007. Adaptive pulse synchronizing control for high-power-factor operation of variable speed DC-drive. *IEEE Transactions on Power Electronics*, 22, 2499–2510.
16. Greul, R., Round, S. D., and Kolar, J. W. 2007. Analysis and control of a three-phase, unity power factor Y-rectifier. *IEEE Transactions on Power Electronics*, 22, 1900–1911.
17. Bendre, A. and Venkataramanan, G. 2003. Modeling and design of a neutral point regulator for a three level diode clamped rectifier, *Proceedings of IEEE IAS* 2003, pp. 1758–1765.
18. Kolar, J. W. and Drofenik, U. 1999. A new switching loss reduced discontinuous PWM scheme for a unidirectional three-phase/switch/level boost type PWM (VIENNA) rectifier. *Proceedings of the 21st INTELEC*, Paper 29-2.
19. Kolar, J. W. and Zach, F. C. 1994. A novel three-phase utility interface minimizing line current harmonics of high-power telecommunications rectifier modules. *Proceedings of the 16th INTELEC*, pp. 367–374.
20. Youssef, N. B. H., Fnaiech, F., and Al-Haddad, K. 2003. Small signal modeling and control design of a three-phase AC/DC Vienna converter. *Proceedings of the 29th IEEE IECON*, pp. 656–661.
21. Mehl, E. L. M. and Barbi, I. 1997. An improved high power factor and low cost three-phase rectifier. *IEEE Transactions on Industry Applications*, pp. 485–492.
22. Salmon, J. 1995. Circuit topologies for PWM boost rectifiers operated from 1-phase and 3-phase AC supplies and using either single or split DC rail voltage outputs. *Proceedings of the IEEE Applied Power Electronics Conference*, pp. 473–479.
23. Kolar, J. W. and Zach, F. C. 1997. A novel three-phase utility interface minimizing line current harmonics of high power telecommunications rectifiers modules. *IEEE Transactions on Industrial Electronics*, 456–467.
24. Kolar, J. W., Ertl, H., and Zach, F. C. 1996. Design and experimental investigation of a three-phase high power density high efficiency unity-powerfactor PWM (VIENNA) rectifier employing a novel integrated power semiconductor module. *Proceedings of APEC* 96, pp. 514–523.
25. Maswood, A. I., Yusop, A. K., and Rahman, M. A. 2002. A novel suppressed-link rectifier–inverter topology with unity power factor. *IEEE Transactions on Power Electronics*, 692–700.
26. Drofenik, U. and Kolar, J. W. 1999. Comparison of not synchronized sawtooth carrier and synchronized triangular carrier phase current control for the VIENNA rectifier I. *Proceedings of IEEE ISIE*, pp. 13–18.
27. Maswood, A. I. and Liu, F. 2005. A novel unity power factor input stage for AC drive application. *IEEE Transactions on Power Electronics*, pp. 839–846.
28. Qiao, C. and Smedley, K. M. 2003. Three-phase unity-power-factor star-connected switch (VIENNA) rectifier with unified constant-frequency integration control. *IEEE Transactions on Power Electronics*, 952–957.
29. Liu, F. and Maswood, A. I. 2006. A novel variable hysteresis band current control of three-phase three-level unity PF rectifier with constant switching frequency. *IEEE Transactions on Power Electronics*, 1727–1734.
30. Maswood, A. I. and Liu, F. 2006. A unity power factor front-end rectifier with hysteresis current control. *IEEE Transactions on Energy Conversion*, 69–76.
31. Maswood, A. I. and Liu, F. 2007. A unity-power-factor converter using the synchronous-reference-frame-based hysteresis current control. *IEEE Transactions on Industry Applications*, 593–599.

5

Ordinary DC/DC Converters

According to certain statistics, there are more than 600 prototypes at present of DC/DC. In their book *Advanced DC/DC Converters* [1,2], the authors have systematically sorted them into six categories. According to the systematic categorization, the ordinary converters introduced in this book will fall under these generations.

5.1 Introduction

DC/DC conversion technology is an important area of research and has industrial applications. Since the last century, the DC/DC conversion technique has been extensively developed and there are now many new topologies of DC/DC converters. DC/DC converters are now widely used in communication equipment, cell phones and digital cameras, computer hardware circuits, dental apparatus, and other industrial applications. Since there are a lot of DC/DC converters, we have sorted them into six generations: first-generation (classical/traditional), second-generation (multiquadrant), third-generation (switched-component), fourth-generation (soft-switching), fifth-generation (synchronous rectifier), and sixth-generation (multielement resonant power).

The first-generation DC/DC converters are so-called classical or traditional converters. These converters operate in a single-quadrant mode and in a low power range (up to 100 W). Since there are a large number of prototype converters in this generation, they are further sorted into the following six categories [1–5]:

- Fundamental
- Transformer-type
- Developed
- VL
- SL
- UL

Fundamental converters such as the buck converter, the boost converter, and the buck–boost converter are named after their functions. These three prototypes perform basic functions and therefore will be investigated in detail. Because of the effects of parasitic elements, the output voltage and power transfer efficiency of these converters are restricted. As a consequence, transformer-type and developed converters were created.

The VL technique is a popular method that is widely applied in electronic circuit design. Applying this technique can effectively overcome the effects of parasitic elements and greatly increase the voltage transfer gain. Therefore, these DC/DC converters can convert the source voltage into a higher output voltage with a high power efficiency, a high power density, and a simple structure. The SL and UL techniques are even more powerful methods that are used to increase the voltage transfer gain in power series.

The second-generation converters perform two-quadrant or four-quadrant operation with output power in a medium range (say, 100–1000 W). These converters are usually used in industrial applications, for example, DC motor drives with multiquadrant operation. Since most second-generation converters are still made of capacitors and inductors, they are large in size.

The third-generation converters are called switched-component DC/DC converters; as they are made of either capacitors or inductors, they are called switched-capacitor converters or switched-inductor converters, respectively. They usually perform two-quadrant or four-quadrant operation with output power in a high range (say, 1000 W). Since they consist of only capacitors or inductors, they are small in size.

Switched-capacitor DC/DC converters consist of capacitors only. Since switched-capacitors can be integrated into power semiconductor integrated circuit (IC) chips, they have a limited size and work at a high switching frequency. They have been successfully employed in inductorless DC/DC converters and this has opened up the way for the construction of converters with a high power density. As a consequence, they have received a great deal of attention from research workers and manufacturers. However, most switched-capacitor converters in the literature perform single-quadrant operation and work in the push–pull status. In addition, their control circuit and topologies are very complex due to the large difference between input and output voltages.

Switched-inductor DC/DC converters consist of inductors only and have been derived from four-quadrant choppers. They usually perform multiquadrant operation with a very simple structure. Two advantages of these converters are simplicity and high power density. No matter how large the difference between the input and output voltages, only one inductor is required for each switched-inductor DC/DC converter. Consequently, they are widely used in industrial applications.

The fourth-generation converters are called soft-switching converters. The soft-switching technique involves many methods for implementing resonance characteristics with resonant switching a popular method. There are two main groups of fourth-generation converters: zero-current-switching (ZCS) and zero-voltage-switching (ZVS). As described in the literature, they usually perform in single-quadrant operation.

ZCS and ZVS converters have large current and voltage stresses. In addition, the conduction duty cycle k and switching frequency f are not individually adjusted. In order to overcome these drawbacks, zero-voltage-plus-zero-current-switching (ZV/ZCS) and zero-transition (ZT) converters were developed, which implement the ZVS and ZCS techniques in the operation. Since the switches turn on and off at the moment the voltage and/or current is equal to zero, the power losses during switching-on and switching-off become zero. As a consequence, these converters have a high power density and a high transfer efficiency. Usually, the repeating frequency is not very high and the converter works in the resonance state. As the components of higher-order harmonics are very low, the EMI is low and EMS and EMC should be reasonable.

The fifth-generation converters are called synchronous rectifier DC/DC converters. Corresponding to the development of microelectronics and computer science, power supplies with low output voltage and strong current are widely required in industrial applications. These power supplies provide very low voltages (5, 3.3, 2.5, and 1.8–1.5 V) and a strong current (30, 60, and 100–200 A) with a high power density and a high power transfer efficiency

(88%, 90–92%). Traditional diode bridge rectifiers are not available for this requirement. The new type of synchronous rectifier DC/DC converters can realize these technical features.

The sixth-generation converters are called multielement resonant power converters (RPC). There are eight topologies of two-element RPC, 38 topologies of three-element RPC, and 98 topologies of four-element RPC. They are widely applied in military equipment and industrial applications.

The DC/DC converter family tree is shown in Figure 5.1.

In this book, the input voltage is represented by V_1 and/or V_I (V_{in}), the output voltage by V_2 and/or V_O, the input current by I_1 and/or I_I (I_{in}), and the output current by I_2 and/or I_O. The switching frequency is represented by f and the switching period is represented by $T = 1/f$. The conduction duty cycle/ratio is represented by **k** and k is the ratio of the switching-on time over the period T. The value of k is in the range of $0 < k < 1$.

5.2 Fundamental Converters

Fundamental converters are exemplified by the buck converter, the boost converter, the buck–boost converter, and the P/O buck–boost converter. Considering the *input current continuity*, we can divide all DC/DC converters into two main modes: continuous input current mode (CICM) and discontinuous input current mode (DICM). The boost converter operates in CICM whereas the buck converter and the buck–boost converter operate in DICM [6–12].

5.2.1 Buck Converter

A buck converter is shown in Figure 5.2a. It converts the input voltage into output voltage that is less than the input voltage. Its switch-on and switch-off equivalent circuits are shown in Figure 5.2b and 5.2c.

5.2.1.1 Voltage Relations

When switch S is on, the inductor current increases. For easy analysis in the steady state, we assume that the capacitor C is large enough (the ripple can be negligible), namely $v_C = V_2$. Therefore, we have

$$V_1 = v_L + v_C = L\frac{di_L}{dt} + v_C, \tag{5.1}$$

$$\frac{di_L}{dt} = \frac{V_1 - v_C}{L} = \frac{V_1 - V_2}{L}. \tag{5.2}$$

For the period of time kT, the inductor current increases at a constant slope $(V_1 - V_2)/L$ (see Figure 5.3). The inductor current starts at the initial value I_{min} and changes to a top value I_{max} at the end of the switch-closure period.

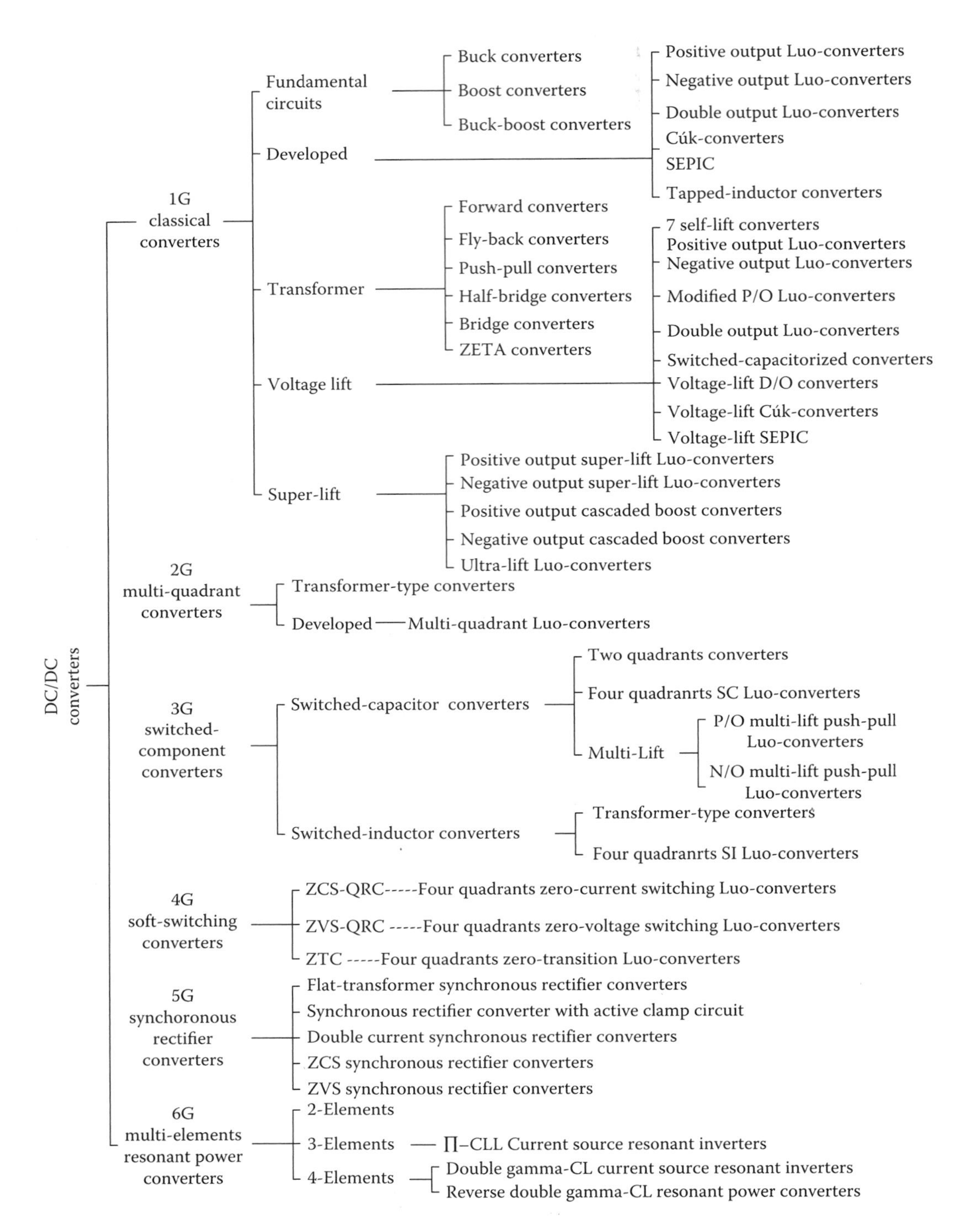

FIGURE 5.1 DC/DC converter family tree.

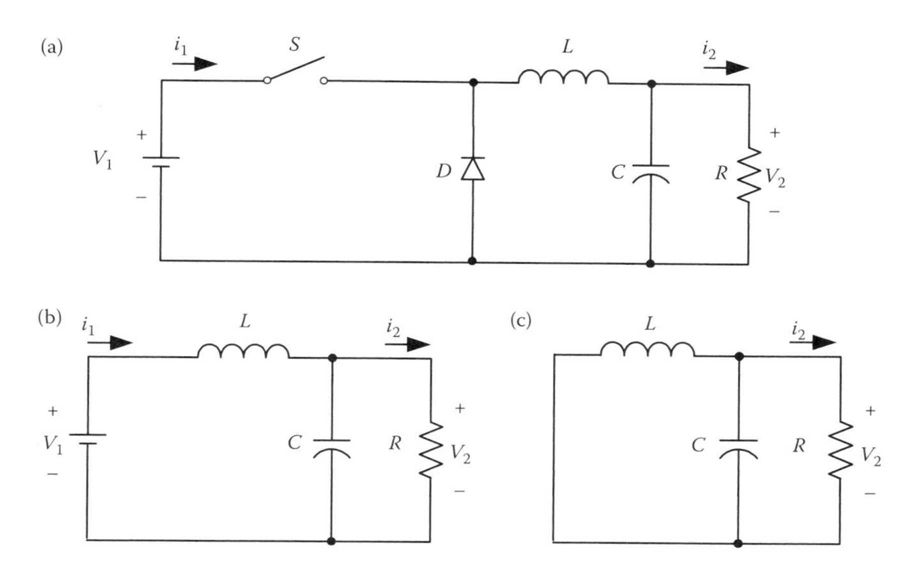

FIGURE 5.2 A buck converter and its equivalent circuits: (a) buck converter, (b) switch-on, and (c) switch-off. (Reprinted from Luo, F. L. and Ye, H. 2006. *Essential DC/DC Converters*. Boca Raton: Taylor & Francis Group LLC, p. 22. With permission.)

When the switch is off, the inductor current decreases and freewheels through the diode. We have the following equations:

$$0 = v_L + v_C, \tag{5.3}$$

$$\frac{di_L}{dt} = -\frac{v_C}{L} = -\frac{V_2}{L}. \tag{5.4}$$

When the switch is off in the time interval $(1-k)T$, the inductor current decreases with a constant slope $-V_2/L$ from I_{max} to I_{min}. The ending value I_{min} must be the same as that at the beginning of the period in the steady state. The current increment during switch-on is equal to the current decrement during switch-off:

$$I_{max} - I_{min} = \frac{V_1 - V_2}{L}kT, \tag{5.5}$$

$$I_{min} - I_{max} = \frac{-V_2}{L}(1-k)T. \tag{5.6}$$

Thus,

$$\frac{V_1 - V_2}{L}kT = \frac{V_2}{L}(1-k)T, \quad V_2 = kV_1. \tag{5.7}$$

The output voltage (capacitor voltage) depends solely on the duty cycle k and the input voltage. From Figure 5.3, it can be seen that the input source current i_1 (which is equal to switch current i_S) is discontinuous. Consequently, the buck converter operates in DICM.

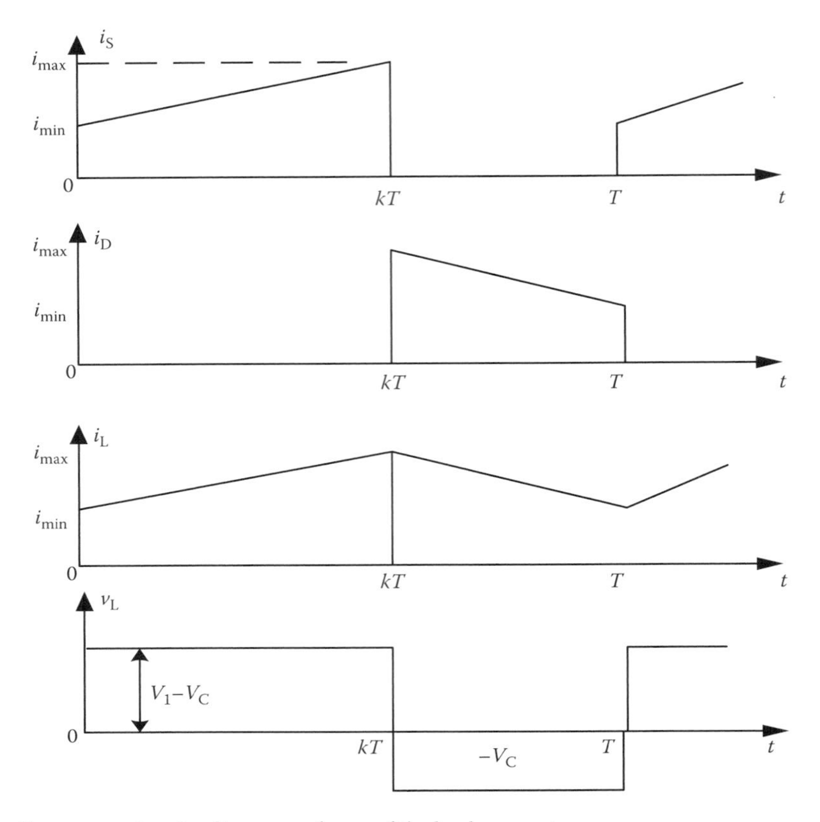

FIGURE 5.3 Some current and voltage waveforms of the buck converter.

5.2.1.2 Circuit Currents

From Figure 5.3, we can find the average value of inductor current easily by inspecting the waveform:

$$I_L = \frac{I_{max} + I_{min}}{2}. \tag{5.8}$$

Applying the Kirchhoff current law (KCL), we have

$$i_L = i_C + i_2. \tag{5.9}$$

Because the average capacitor current is zero in periodic operation, the result can be written by averaging values over one period of operation:

$$I_L = I_2. \tag{5.10}$$

By Ohm's law, the current I_2 is given by

$$I_2 = \frac{V_2}{R}. \tag{5.11}$$

Considering Equations 5.5, 5.10, and 5.11, we have

$$I_{max} + I_{min} = 2\frac{V_2}{R}, \tag{5.12}$$

$$I_{max} = kV_1\left(\frac{1}{R} + \frac{1-k}{2L}T\right), \tag{5.13}$$

$$I_{min} = kV_1\left(\frac{1}{R} - \frac{1-k}{2L}T\right). \tag{5.14}$$

5.2.1.3 Continuous Current Condition (Continuous Conduction Mode)

If I_{min} is zero, we obtain a relation for the minimum inductance that results in a continuous inductor current:

$$L_{min} = \frac{1-k}{2}TR \tag{5.15}$$

5.2.1.4 Capacitor Voltage Ripple

The condition that there are no ripples in the capacitor voltage is now relaxed to allow a small ripple. This has only a second-order effect on the currents calculated in the previous section, so the previous results can be used without change.

As noted previously, in order to have periodic operation, the capacitor current must be entirely alternating. The graph of the capacitor current needs to be as shown in Figure 5.4 for the continuous inductor current. The peak value of this triangular waveform is $(I_{max} - I_{min})/2$. The resulting ripple in the capacitor voltage depends on the area under the curve of the capacitor current versus time. The charge added to the capacitor in a half-cycle is given by the triangular area above the axis:

$$\Delta Q = \frac{1}{2}\frac{I_{max} - I_{min}}{2}\frac{T}{2} = \frac{I_{max} - I_{min}}{8}T. \tag{5.16}$$

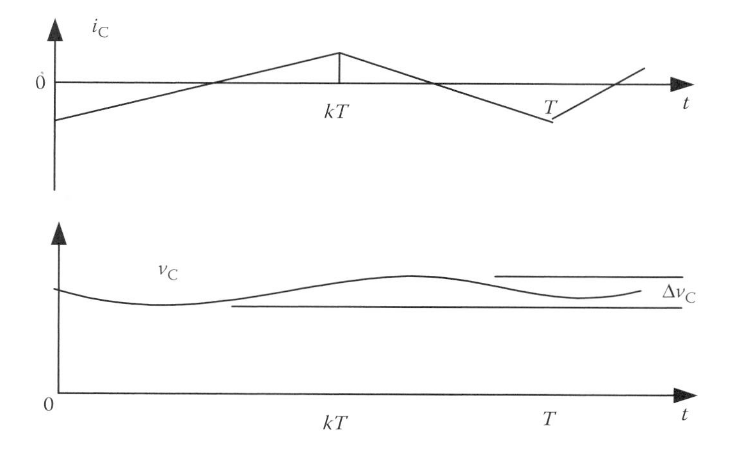

FIGURE 5.4 Waveforms of i_C and v_C.

The graph of the capacitor voltage is also shown in the lower graph of Figure 5.4. The ripple in the voltage is exaggerated to show its effect. Minimum and maximum capacitor voltage values occur at the time the capacitor current becomes zero. The peak-to-peak value of the capacitor voltage ripple is given by

$$\Delta v_C = \Delta Q/C = \frac{I_{max} - I_{min}}{8C}T = \frac{k(1-k)V_1}{8CL}T^2. \quad (5.17)$$

Example 5.1

A buck converter has the following components: $V_1 = 20\,\text{V}$, $L = 10\,\text{mH}$, $C = 20\,\mu\text{F}$, $R = 20\,\Omega$, switching frequency $f = 20\,\text{kHz}$, and conduction duty cycle $k = 0.6$. Calculate the output voltage and its ripple in the steady state. Does this converter work in CCM or discontinuous conduction mode (DCM)?

SOLUTION

1. From Equation 5.7, the output voltage is $V_2 = kV_1 = 0.6 \times 20 = 12\,\text{V}$.
2. From Equation 5.17, the output voltage ripple is

$$\Delta v_2 = \Delta v_C = \frac{k(1-k)V_1}{8CL}T^2 = \frac{0.6 \times 0.4 \times 20}{8 \times 20\,\mu\text{F} \times 10\,\text{mH} \times (20\,\text{k})^2} = 7.5\,\text{mV}.$$

3. From Equation 5.15, the inductor

$$L = 10\,\text{mH} > L_{min} = \left(\frac{1-k}{2}\right)TR = \left(\frac{0.4}{2 \times 20\,\text{k}}\right)20 = 0.2\,\text{mH}.$$

This converter works in CCM.

5.2.2 Boost Converter

If the three elements S, L, and D of the buck converter are rearranged as shown in Figure 5.5a, a boost converter is created. Its equivalent circuits during switch-on and switch-off are shown in Figure 5.5b and 5.5c.

5.2.2.1 Voltage Relations

When the switch S is on, the inductor current increases:

$$\frac{di_L}{dt} = \frac{V_1}{L}. \quad (5.18)$$

Since the diode is inversely biased, the capacitor supplies current to the load, and the capacitor current i_C is negative. Upon opening the switch, the inductor current must decrease so that the current at the end of the cycle can be the same as that at the start of the cycle in the steady state. For the inductor current to decrease, the value $V_C = V_2$ must be $>V_1$. For this interval with the switch open, the inductor current derivative is given by

$$\frac{di_L}{dt} = \frac{V_1 - V_C}{L} = \frac{V_1 - V_2}{L}. \quad (5.19)$$

A graph of the inductor current versus time is shown in Figure 5.6.

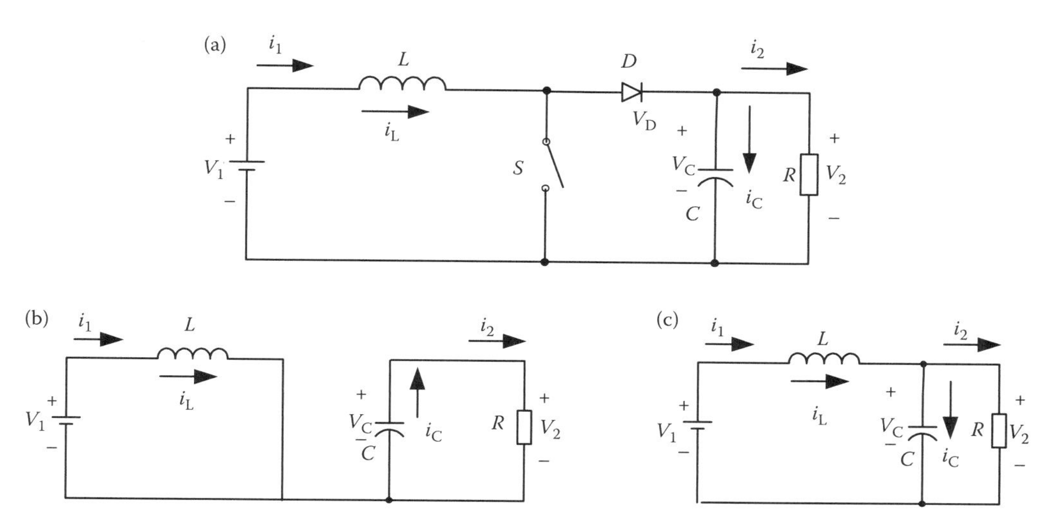

FIGURE 5.5 Boost converter: (a) circuit, (b) switch-on, and (c) switch-off. (Reprinted from Luo, F. L. and Ye, H. 2006. *Essential DC/DC Converters*. Boca Raton: Taylor & Francis Group LLC, p. 23. With permission.)

The increment of i_L during switch-on must be equal to its decrement during switch-off:

$$I_{max} - I_{min} = \frac{V_1}{L} kT \tag{5.20}$$

and

$$I_{min} - I_{max} = \frac{V_1 - V_C}{L}(1-k)T; \tag{5.21}$$

$$V_2 = V_C = \frac{V_1}{1-k}. \tag{5.22}$$

From Equation 5.22, we can see that if k is large, the output voltage V_2 can be very large. In fact, as k approaches unity, the output voltage decreases rather than increasing because of the effect of circuit parasitic elements. The value of k must be limited within a certain upper limit (say 0.9) to prevent such a problem. Practical limits to this also become important for an increase in the voltage transfer gain, for example, 10. The switch may be open for only a very short time ($0.1\,T$ since $k = 0.9$).

5.2.2.2 Circuit Currents

The I_{max} and I_{min} values can be found via the input average power and the load average power, if there are no power losses:

$$P_{in} = \frac{I_{max} + I_{min}}{2} V_1 \quad \text{(input power)} \tag{5.23}$$

and

$$P_O = \frac{V_2^2}{R} \quad \text{(output power)}. \tag{5.24}$$

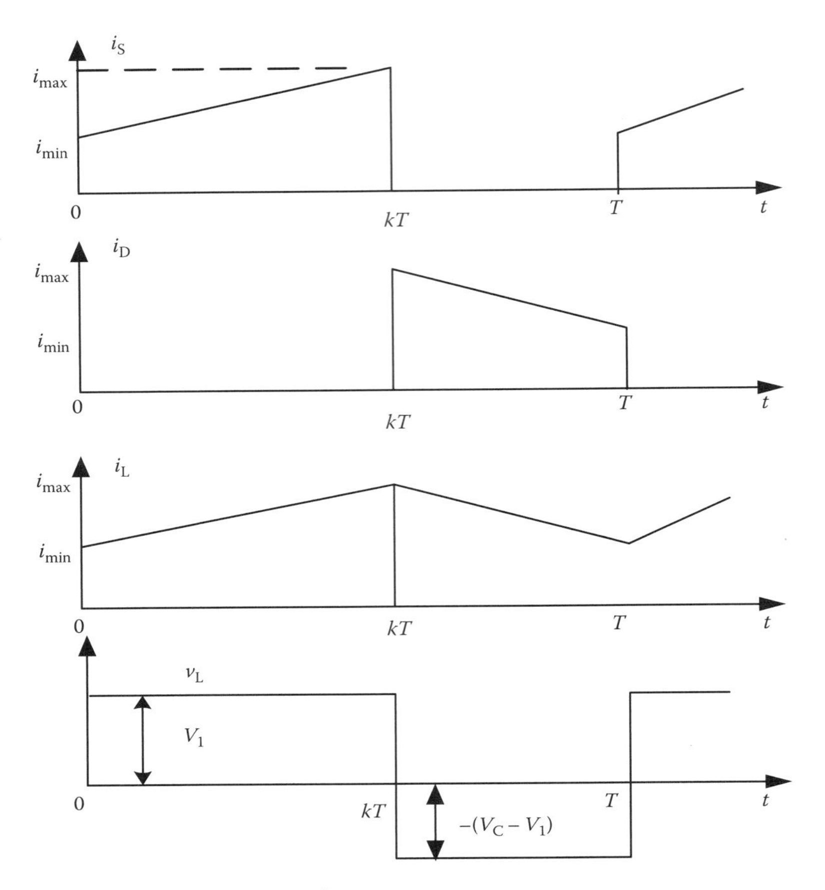

FIGURE 5.6 Some current and voltage waveforms.

Considering Equation 5.22, we have

$$I_{max} + I_{min} = 2\frac{V_1}{R(1-k)^2}. \tag{5.25}$$

From Equations 5.21 and 5.25

$$I_{min} = \frac{V_1}{R(1-k)^2} - \frac{V_1}{2L}kT, \tag{5.26}$$

$$I_{max} = \frac{V_1}{R(1-k)^2} + \frac{V_1}{2L}kT. \tag{5.27}$$

The load current value I_2 is given by $I_2 = V_2/R$, and the average current flowing through the capacitor is zero. The instantaneous capacitor current is likely a triangular waveform, which is approximately $(i_L - I_2)$ during switch-off and $-I_2$ during switch-on. From Figure 5.6, the input source current $i_1 = i_S = i_L$ is continuous. Hence, the buck converter operates in CICM.

5.2.2.3 *Continuous Current Condition*

When the I_{min} is equal to zero, the minimum inductance can be determined to ensure a continuous inductor current. Using Equation 5.26 and solving it, we obtain

$$L_{min} = \frac{k(1-k)^2}{2} TR. \tag{5.28}$$

5.2.2.4 *Output Voltage Ripple*

The change of the charge across the capacitor C is

$$\Delta Q = kTI_2 = kT\frac{V_2}{R} = \frac{kTV_1}{(1-k)R}.$$

Therefore, the ripple voltage Δv_C across the capacitor C is

$$\Delta v_C = \frac{\Delta Q}{C} = \frac{kTV_2}{RC} = \frac{kTV_1}{(1-k)RC}. \tag{5.29}$$

5.2.3 Buck–Boost Converter

If the three elements S, D, and L in a boost converter are rearranged as shown in Figure 5.7a, a buck–boost-type converter is created. Applying a similar analysis to this converter, we can easily obtain all the characteristics of a buck–boost converter under steady-state operating conditions.

5.2.3.1 *Voltage and Current Relations*

With the switch closed, the inductor current changes:

$$\frac{\mathrm{d}i_L}{\mathrm{d}t} = \frac{V_1}{L} \tag{5.30}$$

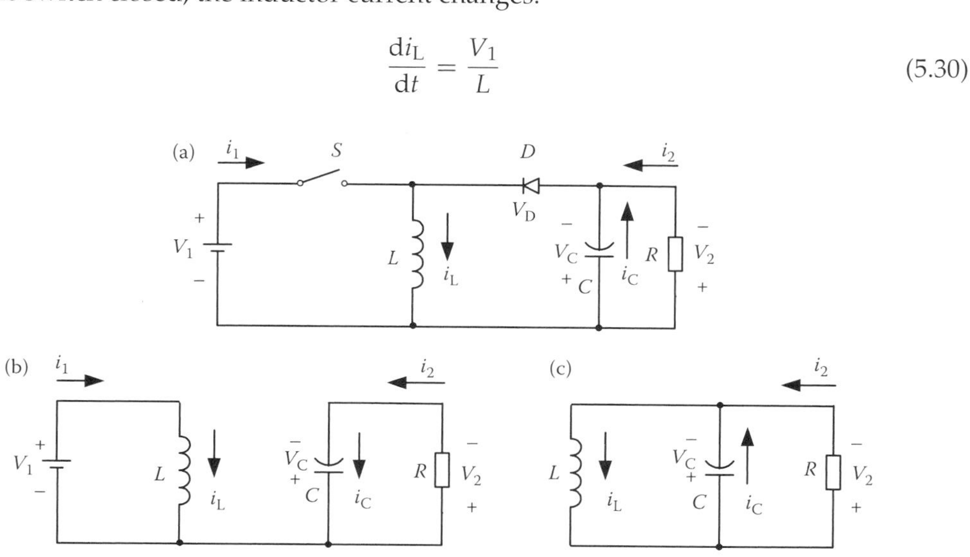

FIGURE 5.7 Buck–boost converter: (a) circuit, (b) switch-on, and (c) switch-off. (Reprinted from Luo, F. L. and Ye, H. 2006. *Essential DC/DC Converters*. Boca Raton: Taylor & Francis Group LLC, p. 151. With permission.)

and

$$I_{max} - I_{min} = \frac{V_1}{L} kT. \tag{5.31}$$

With the switch open,

$$\frac{di_L}{dt} = -\frac{V_C}{L} \tag{5.32}$$

and

$$I_{min} - I_{max} = -\frac{V_C}{L}(1-k)T. \tag{5.33}$$

Equating these two changes in i_L gives the result

$$V_2 = V_C = \frac{k}{1-k} V_1. \tag{5.34}$$

5.2.3.2 CCM Operation and Circuit Currents

Some waveforms are shown in Figure 5.8. The input source current $i_1 = i_S$ is discontinuous during switch-off. Hence, the buck–boost converter operates in DICM. The input average power is then found from

$$P_{in} = \frac{I_{max} + I_{min}}{2} kV_1 \quad \text{(input power)}, \tag{5.35}$$

and

$$P_O = \frac{V_2^2}{R} \quad \text{(output power)}. \tag{5.36}$$

Other parameters are listed below:

$$I_{max} + I_{min} = \frac{2kV_1}{R(1-k)^2}, \tag{5.37}$$

$$I_{min} = \frac{kV_1}{R(1-k)^2} - \frac{V_1}{2L} kT, \tag{5.38}$$

$$I_{max} = \frac{kV_1}{R(1-k)^2} + \frac{V_1}{2L} kT. \tag{5.39}$$

The boundary for a continuous current is found by setting I_{min} to zero; this defines a minimum inductance to ensure a continuous inductor current. Using Equation 5.38 and solving it, we obtain

$$L_{min} = \frac{(1-k)^2}{2} TR. \tag{5.40}$$

The ripple voltage Δv_C across the capacitor C is

$$\Delta v_C = \frac{\Delta Q}{C} = \frac{kTI_2}{C} = \frac{kTV_2}{RC} = \frac{k^2 TV_1}{(1-k)RC}. \tag{5.41}$$

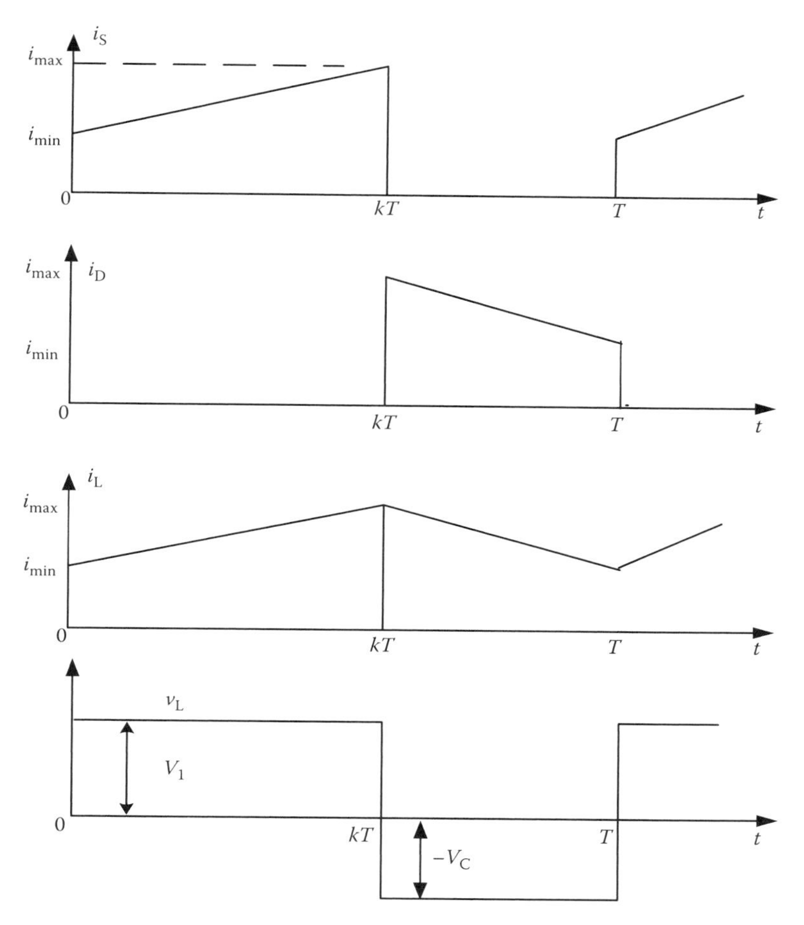

FIGURE 5.8 Some current and voltage waveforms.

Example 5.2

A buck–boost converter has the following components: $V_1 = 20\text{V}$, $L = 10\,\text{mH}$, $C = 20\,\mu\text{F}$, $R = 20\,\Omega$, switching frequency $f = 50\,\text{kHz}$, and conduction duty cycle $k = 0.6$. Calculate the output voltage and its ripple in the steady state. Does this converter work in CCM or DCM?

SOLUTION

1. From Equation 5.34, the output voltage is

$$V_2 = V_C = \frac{k}{1-k}V_1 = \frac{0.6}{0.4}20 = 30\,\text{V}.$$

2. From Equation 5.41, the output voltage ripple is

$$\Delta v_2 = \Delta v_C = \frac{kV_2}{fRC} = \frac{0.6 \times 20}{50\,\text{k} \times 20 \times 20\,\mu} = 0.6\,\text{V}.$$

3. From Equation 5.40, the inductor

$$L = 10\,\text{mH} > L_{\text{min}} = \frac{1-k}{2}TR = \frac{0.4}{2 \times 50\,\text{k}}20 = 0.08\,\text{mH}.$$

This converter works in CCM.

5.3 P/O Buck–Boost Converter

Traditional buck–boost converters have negative output (N/O) voltage. In some applications, changing the voltage polarity is not allowed. For example, the Li-ion battery is the common choice for most portable applications such as mobile phones and digital cameras. With the increasing use of low-voltage portable devices and increasing requirements of functionalities embedded into such devices, efficient power management techniques are needed for a longer battery life. The voltage of a single Li-ion battery varies from 4.2 to 2.7 V. A DC/DC converter is needed to maintain the varying voltage of the Li-ion battery at a constant value of 3.3 V. This converter needs to operate in both the step-up and step-down conditions. Smooth transition from the buck mode to the boost mode is the most desired criteria for a longer battery life. A P/O buck–boost converter with two independent controlled switches is shown in Figure 5.9.

There are three operation modes shown in Figure 5.10:

- Buck operation mode, if V_1 is higher than V_2
- Boost operation mode, if V_1 is lower than V_2
- Buck–boost operation mode, if V_1 is similar to V_2.

Here $V_2 = 3.3\,\text{V}$ for this application.

This converter can work as a buck converter or a boost converter depending on input–output voltages. The problem of output regulation with guaranteed transient performances for noninverting buck–boost converter topology is discussed. Various digital control techniques are addressed, which can smoothly perform the transition job. In the first two modes, the operation principles are the same as those of the buck converter and the boost converter described in the previous section. The third operation needs to be described here.

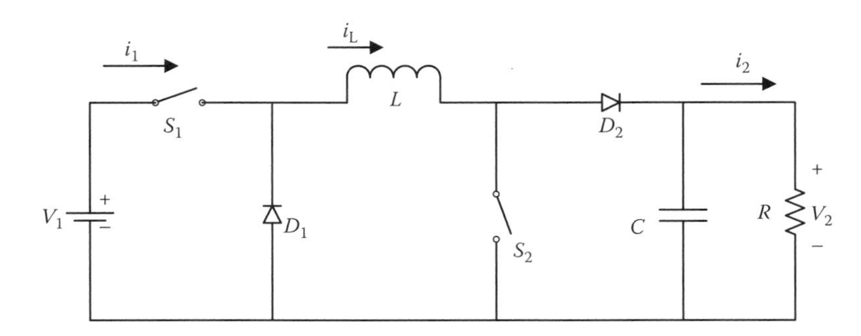

FIGURE 5.9 Circuit diagram of a P/O buck–boost converter.

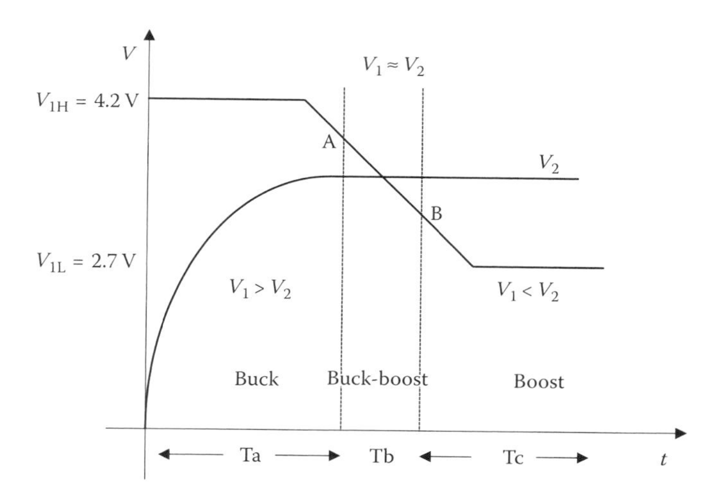

FIGURE 5.10 Input and output characteristics curves of the P/O buck–boost converter.

5.3.1 Buck Operation Mode

When the input voltage V_1 is higher than the output voltage V_2 (e.g., $V_1 > 1.03V_2$, say 3.4 V), the positive buck–boost converter can be operated in the "Buck Operation Mode." In this case, the switch S_2 is constantly open, and the diode D_2 will be constantly on. The remaining components are the same as those of a buck converter.

5.3.2 Boost Operation Mode

When the input voltage V_1 is lower than the output voltage V_2 (e.g., $V_1 > 0.97V_2$, say 3.2 V), the positive buck–boost converter can be operated in the "Boost Operation Mode." In this case, the switch S_1 is constantly on, and the diode D_1 will be constantly blocked. The remaining components are the same as those of a boost converter.

5.3.3 Buck–Boost Operation Mode

When the input voltage V_1 is nearly equal to the output voltage V_2 (e.g., 3.2 V< V_1 < 3.4 V), the positive buck–boost converter can be operated in the "buck–boost operation mode." In this case, both the switches S_1 and S_2 switch on and switch off simultaneously. When the switches are on, the inductor current increases:

$$\Delta i_L = \frac{V_1}{L} kT. \tag{5.42}$$

When the switches are off, the inductor current decreases:

$$\Delta i_L = \frac{V_2}{L}(1-k)T. \tag{5.43}$$

Hence

$$V_2 = \frac{k}{1-k} V_1. \tag{5.44}$$

The other parameters can be determined by the corresponding formulae of the normal buck–boost converter. Therefore, the positive buck–boost converter operates in "buck–boost operation mode," and the output voltage keeps positive polarity.

When this converter works in "buck operation mode" and "buck–boost operation mode," its input current is discontinuous, that is, it works in DICM.

5.3.4 Operation Control

The general control block diagram is shown in Figure 5.11. It implements two functions: logic control to select the operation mode and voltage closed-loop control to keep the output voltage constant.

Refer to Figure 5.11. When the input voltage V_1 is higher than the upper limit voltage, for example, $1.03V_{ref}$ (here the upper limit voltage is set as 3.4 V) as the point A in Figure 5.10, the P/O buck–boost converter operates in the buck mode. When the input voltage V_1 is lower than the lower limit voltage, for example, $0.97V_{ref}$ (the upper limit voltage is set as 3.2 V) as the point B in Figure 5.10, the P/O buck–boost converter operates in the boost mode. When the input voltage V_1 is that between the upper and lower limit voltages, for example, $0.97V_{ref} < V_1 < 1.03V_{ref}$, the P/O buck–boost converter operates in the buck–boost mode.

The output voltage feedback signal compares with the $V_{ref} = 3.3$ V to regulate the duty cycle k in order to keep the output voltage $V_2 = 3.3$ V. In order to analyze the performance of the system during operation in the buck and boost modes and the behavior of the system in transition, the typical parameters of the converter are shown in Table 5.1. The voltage source is modeled to act as a single-cell Li-ion battery, whose voltage varies from $V_{1H} = 4.2$ V when it is fully charged to $V_{1L} = 2.7$ V when it is not charged.

A proportional-integral (PI) controller is used for voltage closed-loop control. All logic operations and the voltage feedback control diagram of the P/O buck–boost converter are shown in Figure 5.12.

The simulation results are shown in Figure 5.13.

A test rig is constructed for experimental testing. The measured results are shown in Table 5.2.

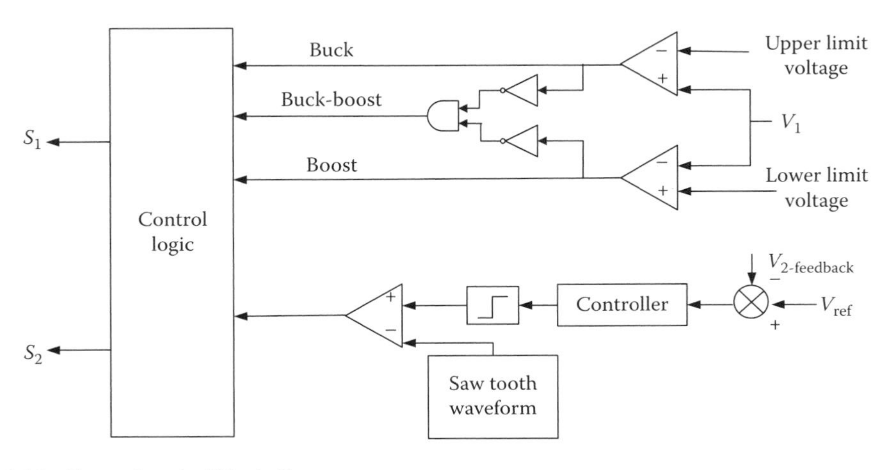

FIGURE 5.11 General control block diagram.

TABLE 5.1

Circuit Parameters of the P/O Buck–Boost Converter

Variable	Parameter	Value
L	Magnetizing inductance	220 μH
C	Output filter capacitance	500 μF
V_1	Input voltage	4.2–2.7 V
	Upper limit voltage	3.4 V
V_{ref}	Output voltage reference	3.3 V
	Lower limit voltage	3.2 V
R	Load resistance	7 Ω
f	Switching frequency	20 kHz

5.4 Transformer-Type Converters

Transformer-type converters consist of transformers and other parts. They can isolate the input and output circuits, and have additional voltage transfer gain corresponding to the winding turn's ratio n. After reviewing popular topologies, a few new circuits will be introduced.

- Forward converter
- Fly-back converter

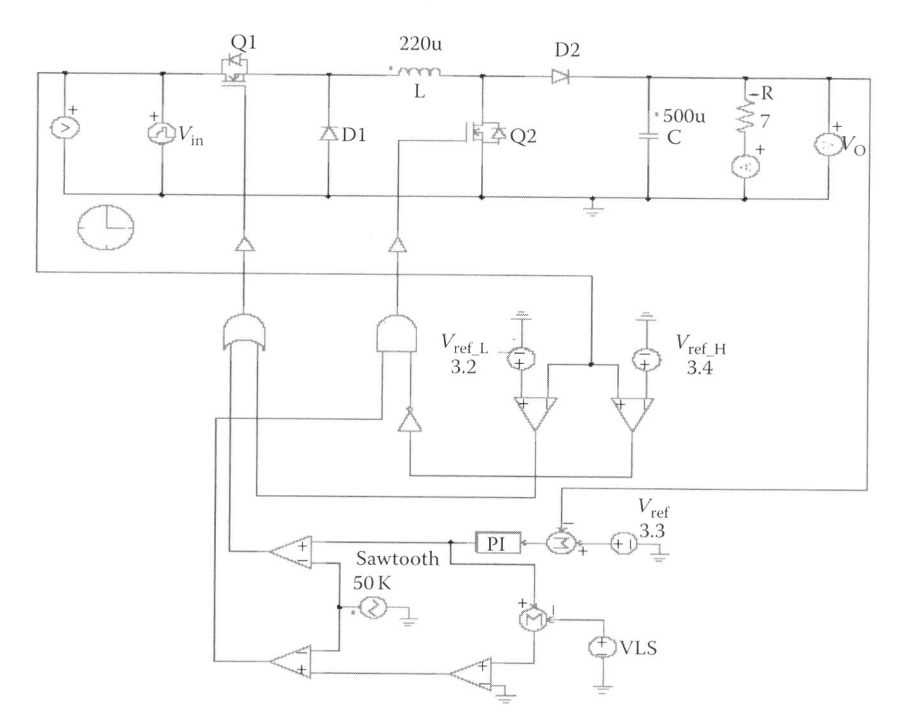

FIGURE 5.12 Simulation diagram of the P/O buck–boost converter.

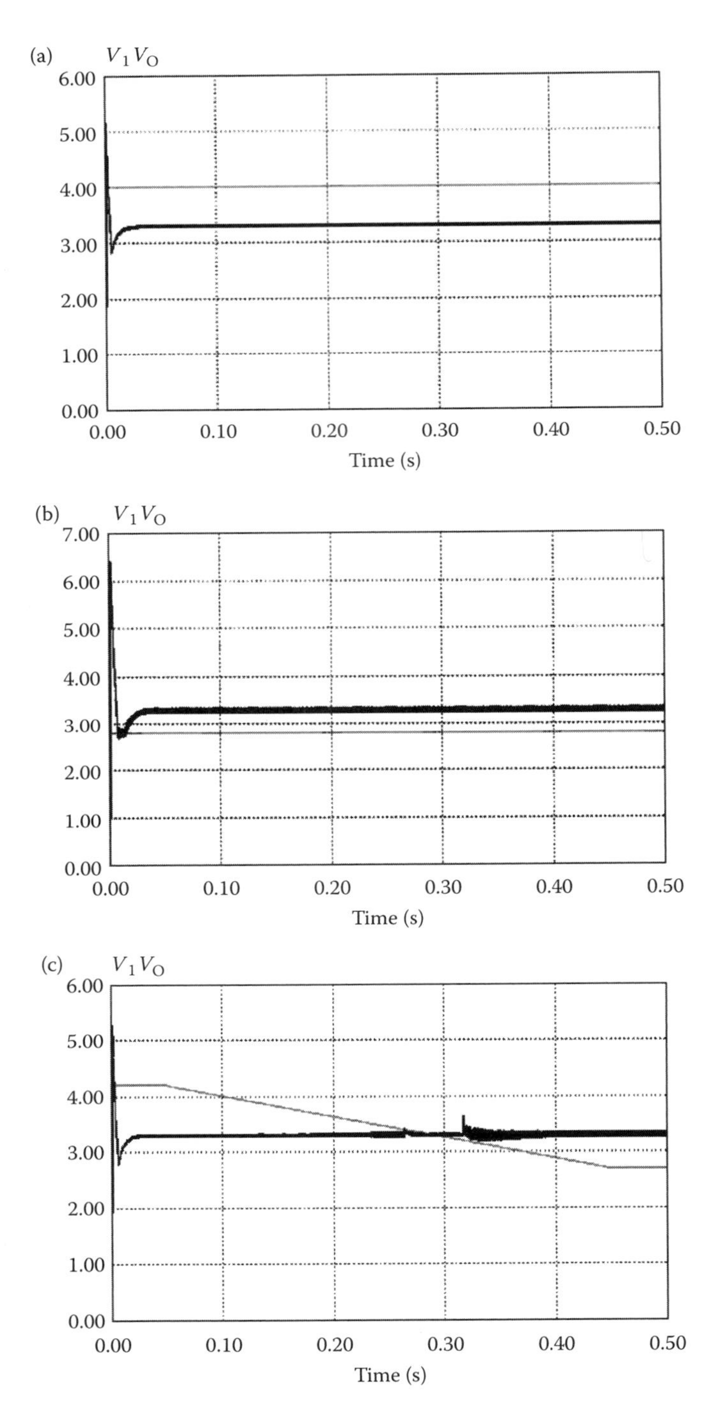

FIGURE 5.13 Simulation results: (a) buck mode operation with $V_1 = 4.0\,\text{V}$, (b) boost mode operation with $V_1 = 2.8\,\text{V}$, and (c) overall operation with $V_1 = 2.7$–$4.2\,\text{V}$.

TABLE 5.2

Measured Simulation Results

Step	V_{in}	V_{out}
1	4.20000	3.30
2	4.15909	3.30
3	3.99091	3.30
4	3.75748	3.30
5	3.54412	3.30
6	3.44875	3.30
7	3.18519	3.30
8	3.08228	3.30
9	2.95426	3.30
10	2.82877	3.30
11	2.70000	3.30

- Push–pull converters
- Half-bridge converters
- Bridge converters
- Zeta converter.

5.4.1 Forward converter

A forward converter is the first transformer-type converter, and is widely applied in industrial applications.

5.4.1.1 Fundamental Forward Converter

The forward converter shown in Figure 5.14 is a transformer-type topology, which consists of a transformer and other parts in the circuits. This converter insolates the input and output circuitry. Therefore, the output voltage can be applied in any floating circuit. Furthermore, since the secondary winding polarity is reversible, it is very convenient to perform N/O and multiquadrant operation. In this text explanation, the polarity is shown in Figure 5.14, which means that the output voltage is positive.

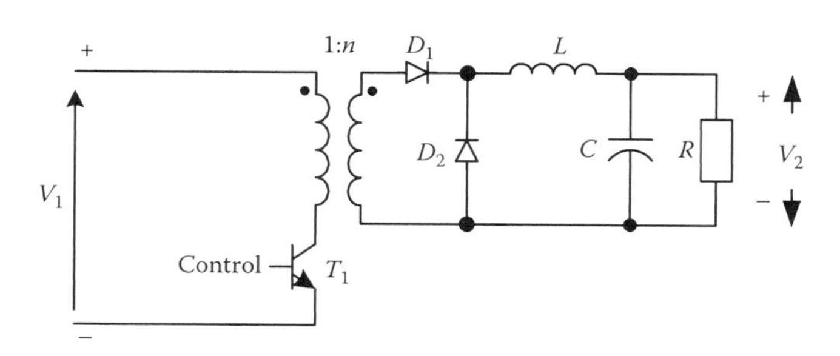

FIGURE 5.14 Forward converter. (Reprinted from Luo, F. L. and Ye, H. 2006. *Essential DC/DC Converters*. Boca Raton: Taylor & Francis Group LLC, p. 24. With permission.)

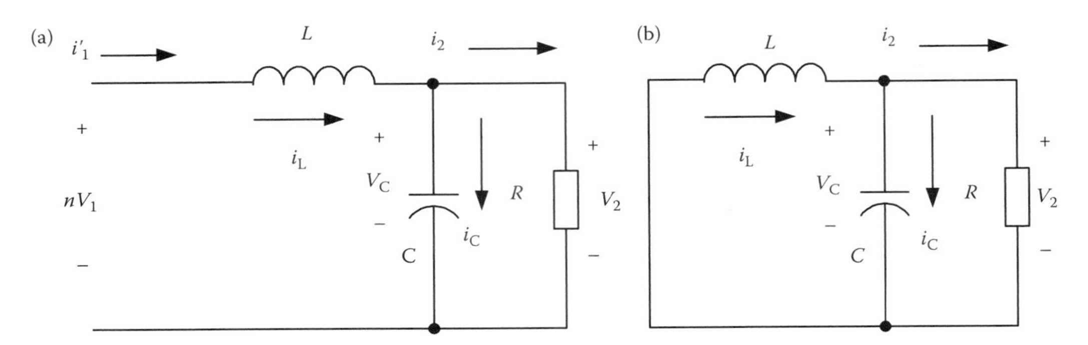

FIGURE 5.15 Equivalent circuits: (a) switching-on and (b) switching-off.

In Figure 5.14, n is the transformer turn's ratio and k is the conduction duty cycle. The turn's ratio n can be any value greater or smaller than unity; the conduction duty cycle k is definitely smaller than unity.

The equivalent circuits during switching-on and switching-off are shown in Figure 5.15a and 5.15b. During switching-on, we have the following equations:

$$nV_1 = v_L + v_C, \quad nV_1 = L\frac{di_L}{dt} + V_C,$$

$$\frac{di_L}{dt} = \frac{nV_1 - V_C}{L}. \tag{5.45}$$

During switching-off, we have the following equations:

$$0 = v_L + v_C, \quad 0 = L\frac{di_L}{dt} + V_C,$$

$$\frac{di_L}{dt} = \frac{-V_C}{L}. \tag{5.46}$$

Some voltage and current waveforms are shown in Figure 5.16.

In the steady state, the current increment ($I_{max} - I_{min}$) during switching-on is equal to the current decrement ($I_{min} - I_{max}$) during switching-off. We have obtained the following Equations to determine the voltage transfer gain:

$$I_{max} - I_{min} = \frac{nV_1 - V_C}{L}kT, \tag{5.47}$$

$$I_{min} - I_{max} = \frac{-V_C}{L}(1-k)T. \tag{5.48}$$

Thus,

$$\frac{nV_1 - V_C}{L}kT = \frac{V_C}{L}(1-k)T, \tag{5.49}$$

$$(nV_1 - V_C)kT = V_C(1-k)T,$$

$$V_2 = V_C = nkV_1. \tag{5.50}$$

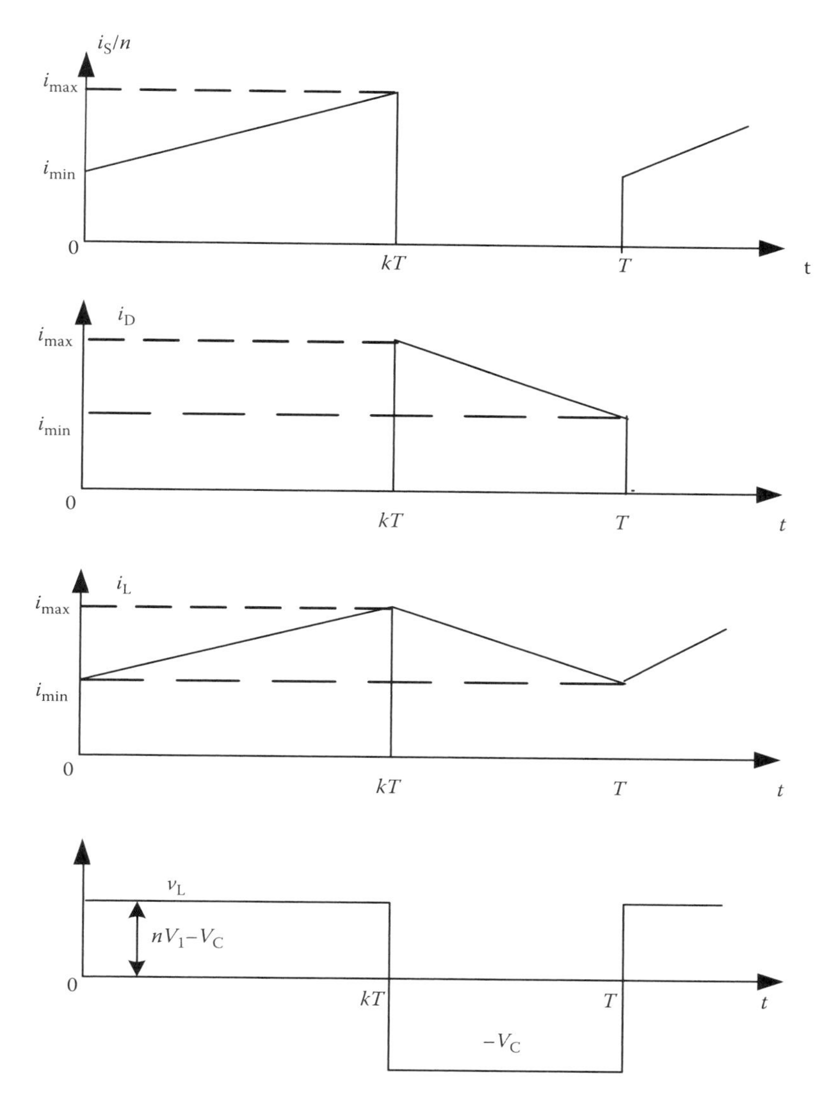

FIGURE 5.16 Some voltage and current waveforms.

From Figure 5.16, we can find the average value of the inductor current easily by inspecting the waveform.

$$I_L = I_2 = \frac{V_2}{R} = \frac{I_{max} + I_{min}}{2}. \tag{5.51}$$

The values of I_{max} and I_{min} are expressed below:

$$I_{max} = V_2 \left(\frac{1}{R} + \frac{1-k}{2L} T \right), \tag{5.52}$$

$$I_{min} = V_2 \left(\frac{1}{R} - \frac{1-k}{2L} T \right). \tag{5.53}$$

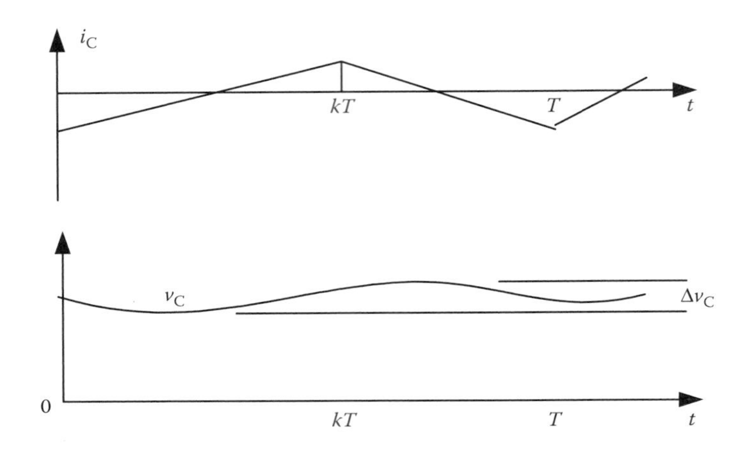

FIGURE 5.17 Waveforms of i_C and v_C.

If the I_{min} is greater than zero, we call the operation the CCM, and vice versa, the DCM. Solving Equation 5.53 for a zero value of I_{min} yields a relation for the minimum value of circuit inductance, which results in continuous inductor current:

$$L_{min} = \frac{1-k}{2} TR. \tag{5.54}$$

The ripple-less condition in the capacitor voltage is now relaxed to allow a small ripple. This has only a second-order effect on the currents calculated in the previous section; so the previous results can be used without change.

As noted previously, the capacitor current must be entirely alternating to have periodic operation. The graph of the capacitor current must be as shown in Figure 5.17 for a continuous inductor current. The peak value of this triangular waveform is located at $(I_{max} - I_{min})/2$. The resulting ripple in the capacitor voltage depends on the area under the curve of capacitor current versus time. The charge added to the capacitor in a half-cycle is given by the triangular area above the axis:

$$\Delta Q = \frac{1}{2}\frac{I_{max} - I_{min}}{2}\frac{T}{2} = \frac{I_{max} - I_{min}}{8}T. \tag{5.55}$$

The graph of capacitor voltage is also shown as part of Figure 5.17. The ripple in the voltage is exaggerated to show its effect. Minimum and maximum capacitor voltage values occur at the time the capacitor current becomes zero. The peak-to-peak value of the capacitor voltage ripple is given by

$$\Delta v_2 = \Delta v_C = \frac{\Delta Q}{C} = \frac{I_{max} - I_{min}}{8C}T = \frac{(1-k)V_2}{8CL}T^2 = \frac{nk(1-k)V_1}{8CL}T^2. \tag{5.56}$$

5.4.1.2 *Forward Converter with Tertiary Winding*

In order to exploit the magnetizing characteristics ability, a tertiary winding is applied in a forward converter. The circuit diagram is shown in Figure 5.18.

The tertiary winding very much exploits the core magnetization ability and reduces the transformer size largely.

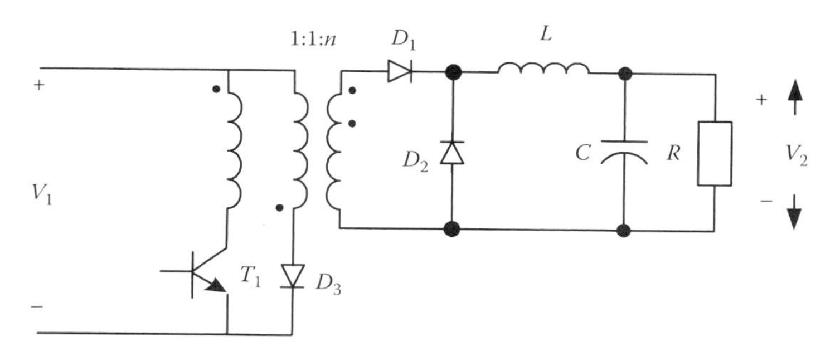

FIGURE 5.18 Forward converter with tertiary winding. (From Luo, F. L. and Ye, H. 2006. *Essential DC/DC Converters*. Boca Raton: Taylor & Francis Group LLC, p. 25. With permission.)

5.4.1.3 Switch Mode Power Supplies with Multiple Outputs

In many applications, more than one output is required, with each output likely to have different voltage and current specifications. A forward converter with three outputs is shown in Figure 5.19. Each output voltage will be determined by the turn's ratio n_1, n_2, or n_3. The three output voltages are

$$\begin{aligned} V_{O1} &= n_1 k V_1, \\ V_{O2} &= n_2 k V_1, \\ V_{O3} &= n_3 k V_1. \end{aligned} \tag{5.50a}$$

However, multiple outputs can be readily obtained using any of the converters that have an isolating transformer, by employing a separate secondary winding for each output, as shown in the forward converter in Figure 5.19.

5.4.2 Fly-Back Converter

A fly-back converter is a transformer-type converter using the demagnetizing effect. Its circuit diagram is shown in Figure 5.20. The output voltage is calculated by

$$V_O = \frac{k}{1-k} n V_{in}, \tag{5.57}$$

where n is the transformer turn's ratio and k is the conduction duty cycle, $k = t_{on}/T$.

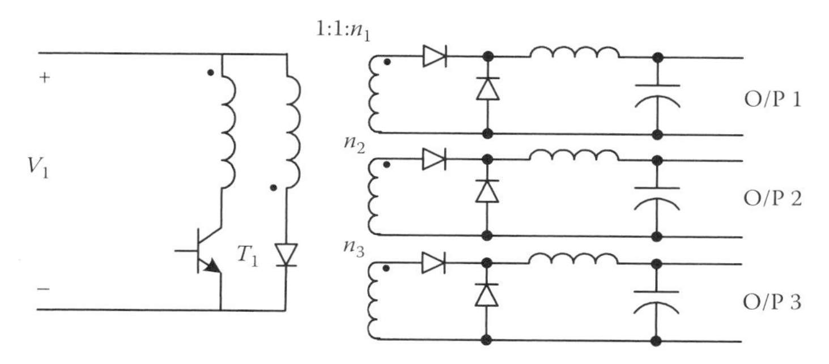

FIGURE 5.19 Forward converter with three outputs. (Reprinted from Luo, F. L. and Ye, H. 2006. *Essential DC/DC Converters*. Boca Raton: Taylor & Francis Group LLC, p. 25. With permission.)

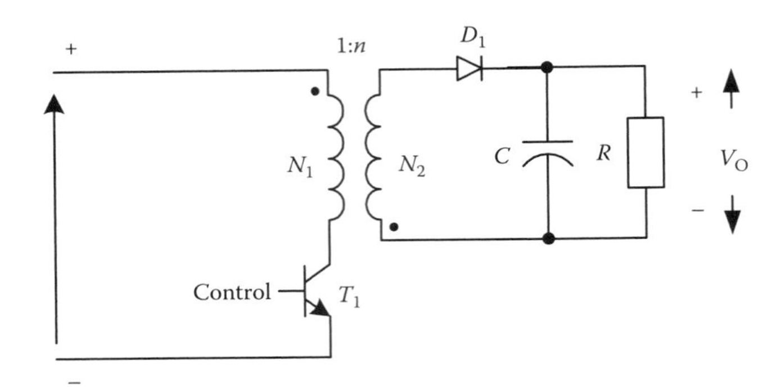

FIGURE 5.20 Fly-back converter. (Reprinted from Luo, F. L. and Ye, H. 2006. *Essential DC/DC Converters*. Boca Raton: Taylor & Francis Group LLC, p. 26. With permission.)

5.4.3 Push–Pull Converter

A push–pull converter works in the push–pull state, which effectively avoids the iron core saturation. Its circuit diagram is shown in Figure 5.21. Since there are two switches working alternatively, the output voltage is doubled. The output voltage is calculated by

$$V_O = 2nkV_{in}, \tag{5.58}$$

where n is the transformer turn's ratio and k is the conduction duty cycle, $k = t_{on}/T$.

5.4.4 Half-Bridge Converter

In order to reduce the primary side in one winding, a half-bridge converter was constructed. Its circuit diagram is shown in Figure 5.22. The output voltage is calculated by

$$V_O = nkV_{in}, \tag{5.59}$$

where n is the transformer turn's ratio and k is the conduction duty cycle, $k = t_{on}/T$.

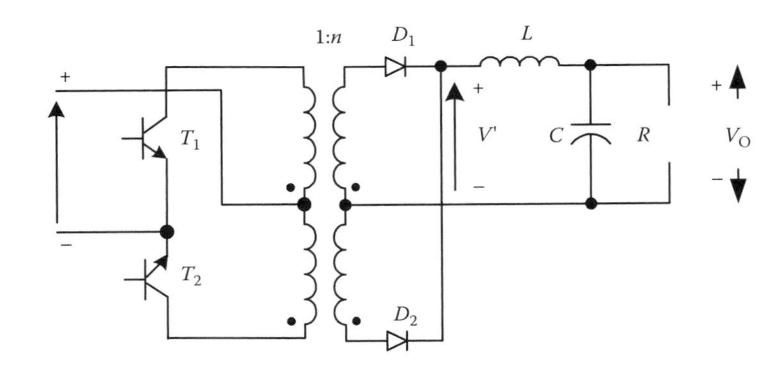

FIGURE 5.21 Push–pull converter. (Reprinted from Luo, F. L. and Ye, H. 2006. *Essential DC/DC Converters*. Boca Raton: Taylor & Francis Group LLC, p. 25. With permission.)

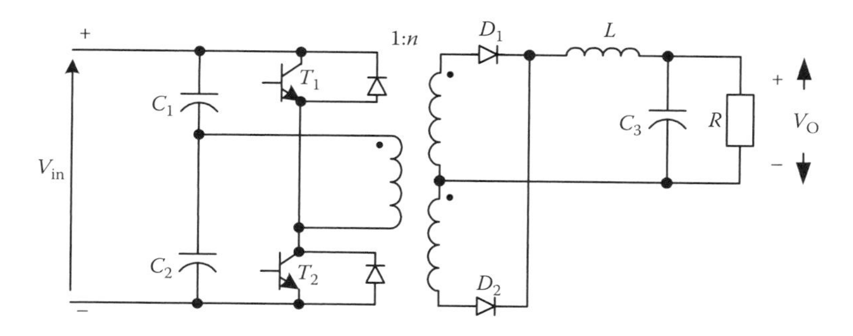

FIGURE 5.22 Half-bridge converter. (Reprinted from Luo, F. L. and Ye, H. 2006. *Essential DC/DC Converters*. Boca Raton: Taylor & Francis Group LLC, p. 151. With permission.)

5.4.5 Bridge Converter

A bridge converter is shown in Figure 5.23. The transformer has a couple of identical secondary windings. The primary circuit is a bridge inverter; hence it is called a bridge converter. Since the two pairs of the switches work symmetrically with 180° phase angle shift, the transformer iron core is not saturated, and the magnetizing characteristics have been fully exploited. No tertiary winding is required. The secondary side contains an antiparalleled diode full-wave rectifier. It is likely that the two antiparalleled forward converters work together.

To avoid short circuit, each pair of the switches can be switched on only in the phase angle 0–180°; usually it is set at 18–162°. The corresponding conduction duty cycle k is in the range of 0.05–0.45.

The circuit analysis is also similar to the forward converter. Some voltage and current waveforms are shown in Figure 5.24. The repeating period is $T/2$ in bridge converter operation, while it is T in forward converter operation.

The voltage transfer gain is

$$V_2 = 2nkV_1. \tag{5.60}$$

Analogously, the average current is

$$I_L = I_2 = \frac{V_2}{R} = \frac{I_{max} + I_{min}}{2}. \tag{5.61}$$

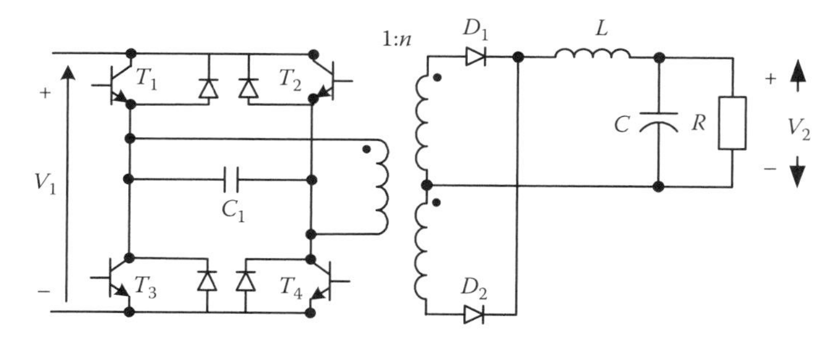

FIGURE 5.23 Bridge converter. (Reprinted from Luo, F. L. and Ye, H. 2006. *Essential DC/DC Converters*. Boca Raton: Taylor & Francis Group LLC, p. 27. With permission.)

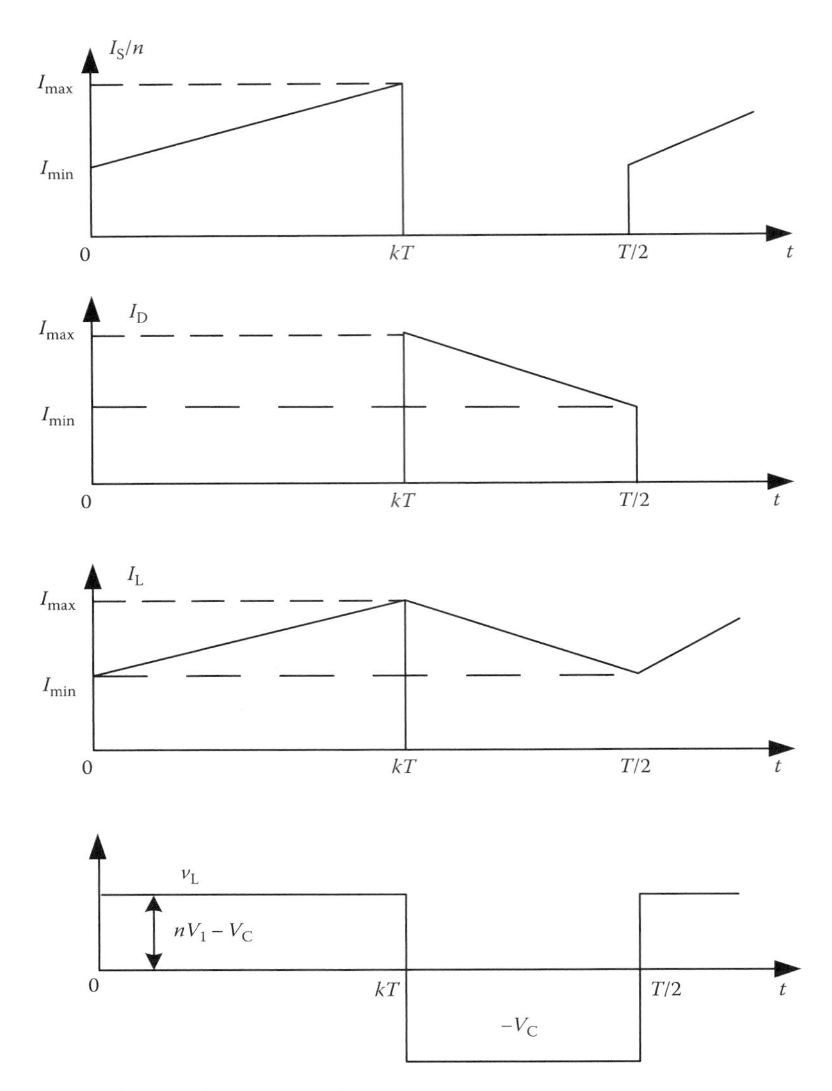

FIGURE 5.24 Some voltage and current waveforms.

The currents I_{max} and I_{min} are

$$I_{max} = V_2\left(\frac{1}{R} + \frac{0.5-k}{2L}T\right), \tag{5.62}$$

$$I_{min} = V_2\left(\frac{1}{R} - \frac{0.5-k}{2L}T\right). \tag{5.63}$$

The minimum inductor to retain CCM is

$$L_{min} = \frac{0.5-k}{2}TR. \tag{5.64}$$

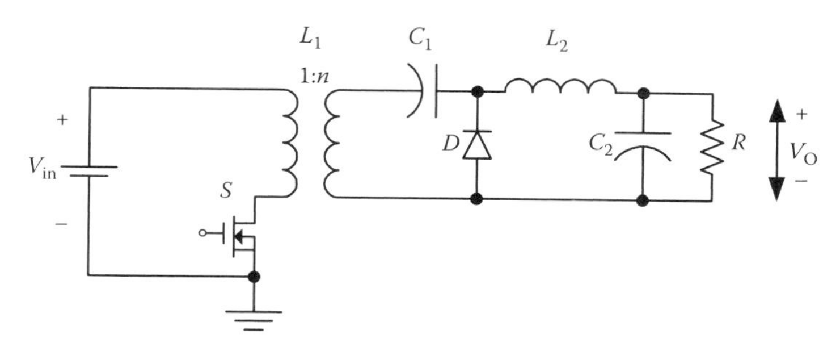

FIGURE 5.25 Zeta converter. (Reprinted from Luo, F. L. and Ye, H. 2006. *Essential DC/DC Converters*. Boca Raton: Taylor & Francis Group LLC, p. 27. With permission.)

The peak-to-peak value of the capacitor voltage ripple is

$$\Delta v_2 = \Delta v_C = \frac{\Delta Q}{C} = \frac{I_{\max} - I_{\min}}{8C}T = \frac{(0.5-k)V_2}{8CL}T^2 = \frac{nk(0.5-k)V_1}{4CL}T^2. \tag{5.65}$$

5.4.6 Zeta Converter

A zeta converter is a transformer-type converter with a low-pass filter. Its circuit diagram is shown in Figure 5.25. Many people do not know its original circuit and call a P/O Luo-converter as a zeta converter. The output voltage ripple of the zeta converter is small. The output voltage is calculated by

$$V_O = \frac{k}{1-k} n v_{in} \tag{5.66}$$

where n is the transformer turn's ratio and k is the conduction duty cycle, $k = t_{on}/T$.

5.5 Developed Converters

All the developed converters are derived from fundamental converters. Since there are more components, the output voltage ripple is smaller. Five types of developed converters are introduced in this section.

- P/O Luo-converter
- N/O Luo-converter
- Double output (D/O) Luo-converter
- Cúk-converter
- Single-ended primary inductance converter (SEPIC).

5.5.1 P/O Luo-Converter (Elementary Circuit)

A P/O Luo-converter (elementary circuit) is shown in Figure 5.26a. The capacitor C acts as the primary means of storing and transferring energy from the input source to the output

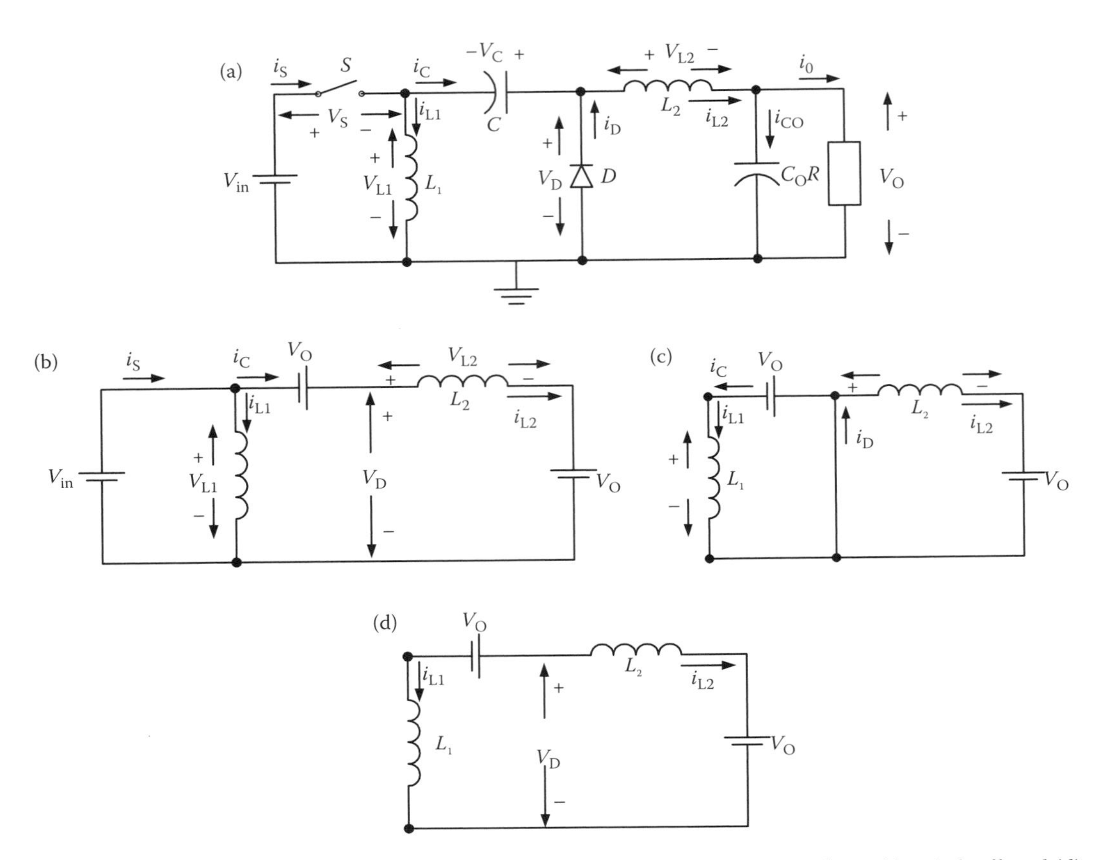

FIGURE 5.26 P/O Luo-converter (elementary circuit): (a) circuit diagram, (b) switch-on, (c) switch off, and (d) discontinuous conduction mode. (Reprinted from Luo, F. L. and Ye, H. 2006. *Essential DC/DC Converters*. Boca Raton: Taylor & Francis Group LLC, p. 29. With permission.)

load via the pump inductor L_1. Assuming the capacitor C to be sufficiently large, the variation of the voltage across the capacitor C from its average value V_C can be neglected in the steady state, that is, $v_C(t) \approx V_C$, even though it stores and transfers energy from the input to the output.

When the switch S is on, the source current $i_I = i_{L1} + i_{L2}$. The inductor L_1 absorbs energy from the source. In the meantime, the inductor L_2 absorbs energy from the source and the capacitor C, and both currents i_{L1} and i_{L2} increase. When the switch S is off, source current $i_I = 0$. Current i_{L1} flows through the freewheeling diode D to the charge capacitor C. The inductor L_1 transfers its SE to the capacitor C. In the meantime, the inductor current i_{L2} flows through the $(C_O - R)$ circuit and freewheeling diode D to keep itself continuous. Both currents i_{L1} and i_{L2} decrease. In order to analyze the progress in the circuit's working, the equivalent circuits in switching-on and switching-off states are shown in Figure 5.26b–d.

Actually, the variations of currents i_{L1} and i_{L2} are small so that $i_{L1} \approx I_{L1}$ and $i_{L2} \approx I_{L2}$. The charge on the capacitor C increases during switch-off:

$$Q+ = (1 - k)TI_{L1}.$$

It decreases during switch-on: $Q- = kTI_{L2}$.

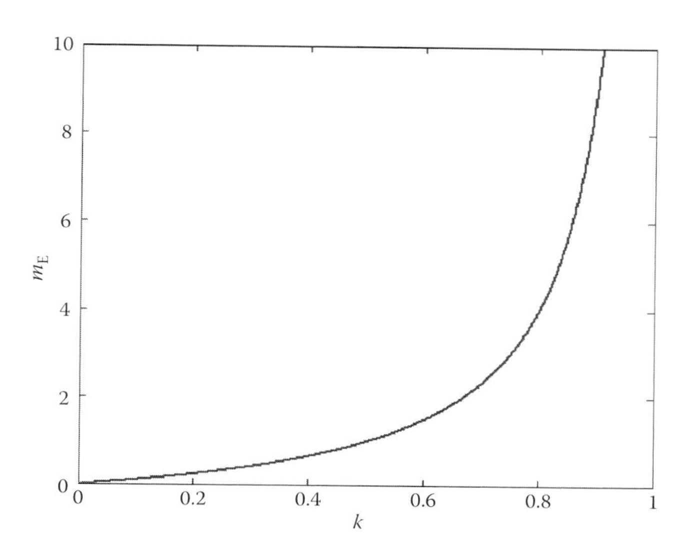

FIGURE 5.27 Voltage transfer gain M_E versus k.

In the whole period of the investigation, $Q+ = Q-$. Thus,

$$I_{L2} = \frac{1-k}{k} I_{L1}.$$

Since the capacitor C_O performs as a low-pass filter, the output current

$$I_{L2} = I_O. \tag{5.67}$$

Equations 5.66 and 5.67 are available for all P/O Luo-converters.

The source current $i_I = i_{L1} + i_{L2}$ during the switch-on period, and $i_I = 0$ during the switch-off period. Thus, the average source current I_I is

$$I_I = k \times i_I = k(i_{L1} + i_{L2}) = k\left(1 + \frac{1-k}{k}\right) I_{L1} = I_{L1}. \tag{5.68}$$

Therefore, the output current is

$$I_O = \frac{1-k}{k} I_I. \tag{5.69}$$

Hence, the output voltage is

$$V_O = \frac{k}{1-k} V_I. \tag{5.70}$$

The voltage transfer gain in continuous mode is

$$M_E = \frac{V_O}{V_I} = \frac{k}{1-k}. \tag{5.71}$$

The curve of M_E versus k is shown in Figure 5.27.

The current i_{L1} increases and is supplied by V_I during switch-on. It decreases and is inversely biased by $-V_C$ during switch-off. Therefore, $kTV_I = (1-k)TV_C$. The average voltage across the capacitor C is

$$V_C = \frac{k}{1-k}V_I = V_O. \tag{5.72}$$

The current i_{L1} increases and is supplied by V_I during switch-on. It decreases and is inversely biased by $-V_C$ during switch-off. Therefore, its peak-to-peak variation is

$$\Delta i_{L1} = \frac{kTV_I}{L_1}.$$

Considering Equation 5.68, the variation ratio of the current i_{L1} is

$$\xi_1 = \frac{\Delta i_{L1}/2}{I_{L1}} = \frac{kTV_I}{2L_1I_I} = \frac{1-k}{2M_E}\frac{R}{fL_1}. \tag{5.73}$$

The current i_{L2} increases and is supplied by the voltage $(V_I + V_C - V_O) = V_I$ during switch-on. It decreases and is inversely biased by $-V_O$ during switch-off. Therefore its peak-to-peak variation is

$$\Delta i_{L2} = \frac{kTV_I}{L_2}. \tag{5.74}$$

Considering Equation 5.66, the variation ratio of the current i_{L2} is

$$\xi_2 = \frac{\Delta i_{L2}/2}{I_{L2}} = \frac{kTV_I}{2L_2I_O} = \frac{k}{2M_E}\frac{R}{fL_2}. \tag{5.75}$$

When the switch is off, the freewheeling diode current $i_D = i_{L1} + i_{L2}$ and

$$\Delta i_D = \Delta i_{L1} + \Delta i_{L2} = \frac{kTV_I}{L_1} + \frac{kTV_I}{L_2} = \frac{kTV_I}{L} = \frac{(1-k)TV_O}{L}. \tag{5.76}$$

Considering Equations 5.66 and 5.67, the average current in the switch-off period is

$$I_D = I_{L1} + I_{L2} = I_O/(1-k).$$

The variation ratio of current i_D is

$$\zeta = \frac{\Delta i_D/2}{I_D} = \frac{(1-k)^2TV_O}{2LI_O} = \frac{k(1-k)R}{2M_E fL} = \frac{k^2}{M_E^2}\frac{R}{2fL}. \tag{5.77}$$

The peak-to-peak variation of v_C is

$$\Delta v_C = \frac{Q+}{C} = \frac{1-k}{C}TI_I.$$

Considering Equation 5.72, the variation ratio of v_C is

$$\rho = \frac{\Delta v_C/2}{V_C} = \frac{(1-k)TI_I}{2CV_O} = \frac{k}{2}\frac{1}{fCR}. \tag{5.78}$$

In order to investigate the variation of output voltage v_O, we have to calculate the charge variation on the output capacitor C_O, because $Q = C_O V_O$ and $\Delta Q = C_O \Delta v_O$. Here, ΔQ is caused by Δi_{L2} and corresponds to the *area* of the triangle with the *height* of half of Δi_{L2} and the *width* of half of the repeating period $T/2$. Considering Equation 5.74,

$$\Delta Q = \frac{1}{2}\frac{\Delta i_{L2}}{2}\frac{T}{2} = \frac{T}{8}\frac{kTV_I}{L_2}.$$

Thus, the half peak-to-peak variation of output voltage v_O and v_{CO} is

$$\frac{\Delta v_O}{2} = \frac{\Delta Q}{2C_O} = \frac{kT^2 V_I}{16 C_O L_2}.$$

The variation ratio of output voltage v_O is

$$\varepsilon = \frac{\Delta v_O/2}{V_O} = \frac{kT^2}{16C_O L_2}\frac{V_I}{V_O} = \frac{k}{16M_E}\frac{1}{f^2 C_O L_2}. \tag{5.79}$$

For analysis in DCM, referring to Figure 5.26d, we can see that the diode current i_D becomes zero during switch-off before the next period switch-on. The condition for DCM is $\zeta \geq 1$, that is,

$$\frac{k^2}{M_E^2}\frac{R}{2fL} \geq 1, \quad M_E \leq k\sqrt{\frac{R}{2fL}} = k\sqrt{\frac{z_N}{2}}. \tag{5.80}$$

The graph of the boundary curve versus the normalized load $z_N = R/fL$ is shown in Figure 5.28. It can be seen that the boundary curve is a monorising function of the parameter k.

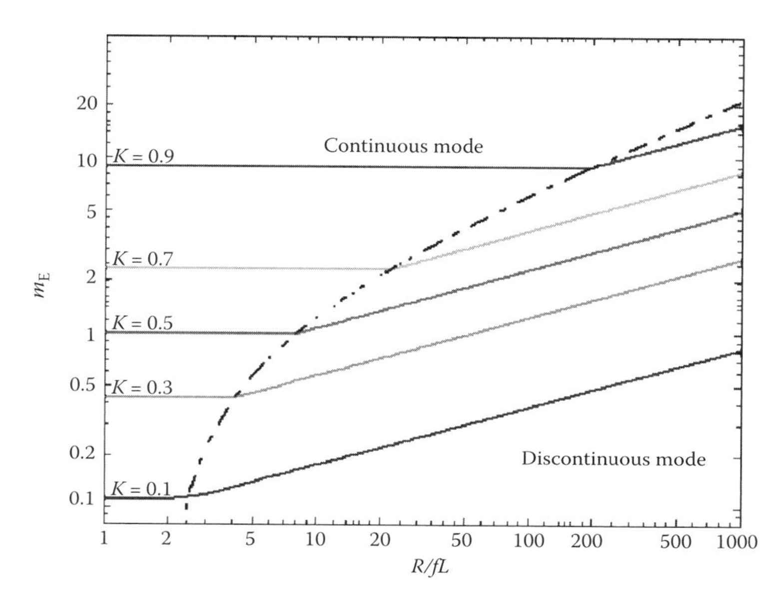

FIGURE 5.28 The boundary between continuous and discontinuous modes and the output voltage versus the normalized load $z_N = R/fL$.

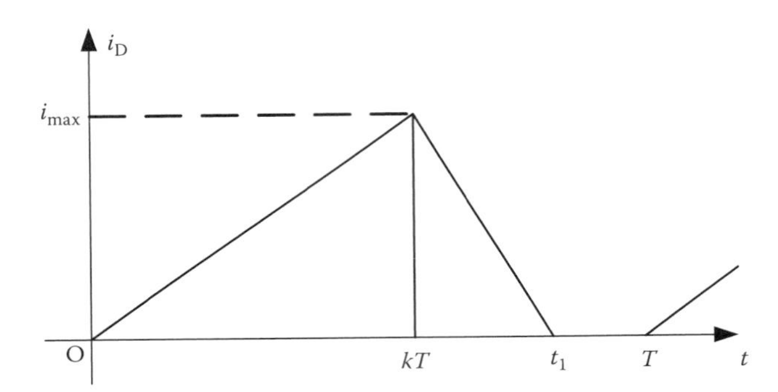

FIGURE 5.29 The discontinuous current waveform. (Reprinted from Luo, F. L. and Ye, H. 2006. *Essential DC/DC Converters*. Boca Raton: Taylor & Francis Group LLC. With permission.)

In the DCM case, the current i_D exists in the period between kT and $t_1 = [k + (1 - k)m_E]T$, where m_E is the *filling efficiency* and is defined as

$$m_E = \frac{1}{\zeta} = \frac{M_E^2}{k^2(R/2fL)}. \tag{5.81}$$

The diode current i_D decreases to zero at $t = t_1 = kT + (1 - k)m_E T$; therefore, $0 < m_E < 1$ (Figure 5.29).

For the current i_L, we have

$$kTV_I = (1 - k)m_E TV_C$$

or

$$V_C = \frac{k}{(1 - k)m_E} V_I = k(1 - k)\frac{R}{2fL} V_I \quad \text{with} \quad \sqrt{\frac{R}{2fL}} \geq \frac{1}{1 - k}.$$

For the current i_{LO}, we have $kT(V_I + V_C - V_O) = (1 - k)m_E TV_O$.

Therefore, the output voltage in discontinuous mode is

$$V_O = \frac{k}{(1 - k)m_E} V_I = k(1 - k)\frac{R}{2fL} V_I \quad \text{with} \quad \sqrt{\frac{R}{2fL}} \geq \frac{1}{1 - k}. \tag{5.82}$$

The output voltage increases linearly with an increase in the load resistance R. The output voltage versus the normalized load $z_N = R/fL$ is shown in Figure 5.28. It can be seen that larger load resistance R may cause higher output voltage in DCM.

Example 5.3

A P/O Luo-converter has the following components: $V_I = 20$V, $L_1 = L_2 = 10$ mH, $C = C_O = 20\,\mu$F, $R = 20\,\Omega$, switching frequency $f = 50$ kHz, and conduction duty cycle $k = 0.6$. Calculate the output voltage, its variation ratio, and the variation ratio of the inductor currents i_{L1} and i_{L2} in steady state.

SOLUTION

1. From Equation 5.70, the output voltage is $V_O = kV_I/(1-k) = 0.6 \times 20/0.4 = 30\text{V}$.
2. From Equation 5.79, the variation ratio of v_O is

$$\varepsilon = \frac{k}{16M_E}\frac{1}{f^2C_OL_2} = \frac{0.6}{16\times 1.5}\frac{1}{(50\,\text{k})^2\times 20\,\mu\times 10\,\text{m}} = 0.00005.$$

3. From Equation 5.73, the variation ratio of the current i_{L1} is

$$\xi_1 = \frac{1-k}{2M_E}\frac{R}{fL_1} = \frac{0.4}{2\times 1.5}\frac{20}{50\,\text{k}\times 10\,\text{m}} = 0.0053.$$

4. From Equation 5.75, the variation ratio of the current i_{L2} is

$$\xi_2 = \frac{k}{2M_E}\frac{R}{fL_2} = \frac{0.6}{2\times 1.5}\frac{20}{50\,\text{k}\times 10\,\text{m}} = 0.008.$$

5.5.2 N/O Luo-Converter (Elementary Circuit)

The N/O Luo-converter (elementary circuit) and its switch-on and switch-off equivalent circuits are shown in Figure 5.30. This circuit can be considered as a combination of an electronic pump *S-L-D-(C)* and a "Π"-type low-pass filter C-L_O-C_O. The electronic pump injects certain energy to the low-pass filter in each cycle. The capacitor C in Figure 5.30 acts as the primary means of storing and transferring energy from the input source to the output load. Assuming the capacitor C to be sufficiently large, the variation of the voltage across the capacitor C from its average value V_C can be neglected in the steady state, that is, $V_C(t) \approx V_C$, even though it stores and transfers energy from the input to the output.

The voltage transfer gain in CCM is

$$M_E = \frac{V_O}{V_I} = \frac{I_I}{I_O} = \frac{k}{1-k}. \tag{5.83}$$

The transfer gain is shown in Figure 5.27. The current i_L increases and is supplied by V_I during switch-on. Thus, its peak-to-peak variation is $\Delta i_L = kTV_I/L$. The inductor current I_L is

$$I_L = I_{C-\text{off}} + I_O = \frac{I_O}{1-k}. \tag{5.84}$$

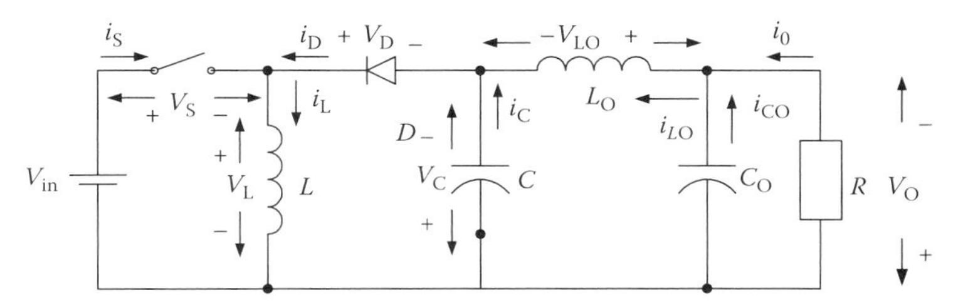

FIGURE 5.30 N/O Luo-converter (elementary circuit). (Reprinted from Luo, F. L. and Ye, H. 2006. *Essential DC/DC Converters*. Boca Raton: Taylor & Francis Group LLC, p. 29. With permission.)

Considering $R = V_O/I_O$, the variation ratio of the current i_L is

$$\zeta = \frac{\Delta i_L/2}{I_L} = \frac{k(1-k)V_I T}{2LI_O} = \frac{k(1-k)R}{2M_E fL} = \frac{k^2}{M_E^2}\frac{R}{2fL}. \quad (5.85)$$

The peak-to-peak variation of the voltage v_C is

$$\Delta v_C = \frac{Q-}{C} = \frac{k}{C}TI_O. \quad (5.86)$$

The variation ratio of the voltage v_C is

$$\rho = \frac{\Delta v_C/2}{V_C} = \frac{kI_O T}{2CV_O} = \frac{k}{2}\frac{1}{fCR}. \quad (5.87)$$

The peak-to-peak variation of current i_{LO} is

$$\Delta i_{LO} = \frac{k}{8f^2 CL_O} I_O. \quad (5.88)$$

Considering $I_{LO} = I_O$,

$$\xi = \frac{\Delta i_{LO}/2}{I_{LO}} = \frac{k}{16}\frac{1}{f^2 CL_O}. \quad (5.89)$$

The variation of the voltage v_{CO} is

$$\Delta v_{CO} = \frac{A}{C_O} = \frac{1}{2}\frac{T}{2}\frac{k}{16f^2 CC_O L_O} I_O = \frac{k}{64f^3 CC_O L_O} I_O. \quad (5.90)$$

The variation ratio of the output voltage v_{CO} is

$$\varepsilon = \frac{\Delta v_{CO}/2}{V_{CO}} = \frac{k}{128f^3 CC_O L_O}\frac{I_O}{V_O} = \frac{k}{128}\frac{1}{f^3 CC_O L_O R}. \quad (5.91)$$

In DCM, the diode current i_D becomes zero during switch-off before the next period switch-on. The condition for DCM is $\zeta \geq 1$, that is,

$$\frac{k^2}{M_E^2}\frac{R}{2fL} \geq 1$$

or

$$M_E \leq k\sqrt{\frac{R}{2fL}} = k\sqrt{\frac{z_N}{2}}. \quad (5.92)$$

The graph of the boundary curve versus the normalized load $z_N = R/fL$ is shown in Figure 5.28. It can be seen that the boundary curve is a monorising function of the parameter k.

In the DCM case, the current i_D exists in the period between kT and $t_1 = [k + (1-k)m_E]T$, where m_E is the *filling efficiency* and is defined as

$$m_E = \frac{1}{\zeta} = \frac{M_E^2}{k^2(R/2fL)}. \tag{5.93}$$

Considering $\zeta > 1$ for DCM operation, therefore $0 < m_E < 1$. The diode current i_D becomes zero at $t = t_1 = kT + (1-k)m_E T$.

For the current i_L, we have

$$TV_I = (1-k)m_E TV_C$$

or

$$V_C = \frac{k}{(1-k)m_E}V_I = k(1-k)\frac{R}{2fL}V_I \quad \text{with} \quad \sqrt{\frac{R}{2fL}} \geq \frac{1}{1-k}.$$

For the current i_{LO}, we have $kT(V_I + V_C - V_O) = (1-k)m_E TV_O$.
Therefore, the output voltage in discontinuous mode is

$$V_O = \frac{k}{(1-k)m_E}V_I = k(1-k)\frac{R}{2fL}V_I \quad \text{with} \quad \sqrt{\frac{R}{2fL}} \geq \frac{1}{1-k}. \tag{5.94}$$

That is, the output voltage increases linearly with an increase in the load resistance R. Larger load resistance R may cause higher output voltage in DCM.

5.5.3 D/O Luo-Converter (Elementary Circuit)

Combining P/O and N/O elementary Luo-converters together, we obtain the D/O elementary Luo-converter that is shown in Figure 5.31. For all the analyses, refer to the previous two sections on P/O and N/O output elementary Luo-converters. The voltage transfer gains are calculated by

$$\frac{V_{O+}}{V_I} = \frac{V_{O-}}{V_I} = \frac{k}{1-k}. \tag{5.95}$$

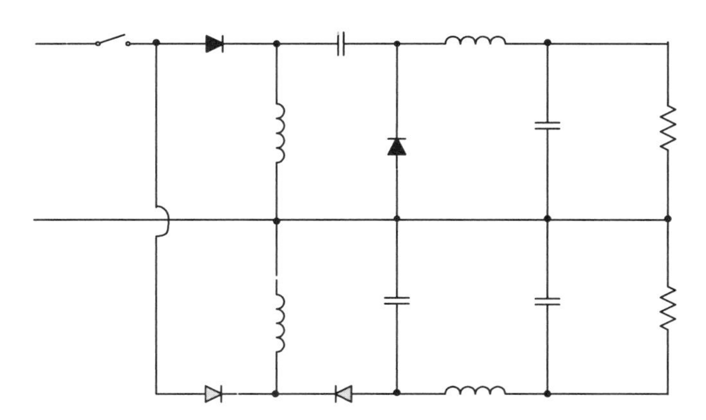

FIGURE 5.31 D/O elementary Luo-converter. (Reprinted from Luo, F. L. and Ye, H. 2006. *Essential DC/DC Converters*. Boca Raton: Taylor & Francis Group LLC, p. 30. With permission.)

5.5.4 Cúk-Converter

The Cúk-converter is derived from the boost converter. Its circuit diagram is shown in Figure 5.32. The Cúk-converter was published in 1977 as the boost–buck converter, and was renamed by Cúk's students afterwards in 1986–1990.

The inductor current i_L increases with slope $+V_1/L$ during switch-on and decreases with slope $-(V_C - V_1)/L$ during switch-off. Thus

$$\frac{V_I}{L}kT = \frac{V_C - V_I}{L}(1-k)T,$$

$$V_C = \frac{1}{1-k}V_I.$$

Since $L_O - C_O$ is a low-pass filter, the output voltage is calculated by

$$V_O = V_C - V_I = \frac{k}{1-k}V_I. \tag{5.96}$$

The voltage transfer gain is

$$M = \frac{V_O}{V_I} = \frac{k}{1-k}, \tag{5.97}$$

and also

$$M = \frac{I_I}{I_O} = \frac{k}{1-k}.$$

Since the inductor L is connected in series to the source voltage and the inductor L_O is connected in series to the output circuit R–C_O, we have the relations

$$I_L = I_I \quad \text{and} \quad I_{LO} = I_O.$$

The variation of the current i_L is

$$\Delta i_L = \frac{V_I}{L}kT.$$

Therefore, the variation ratio of the current i_L is

$$\xi = \frac{\Delta i_L/2}{I_L} = \frac{V_I}{2I_I L}kT = \frac{k}{2M^2}\frac{R}{fL}. \tag{5.98}$$

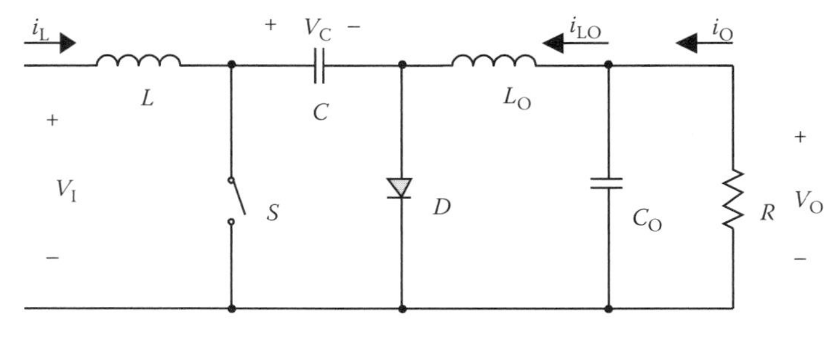

FIGURE 5.32 Cúk-converter. (Reprinted from Luo, F. L. and Ye, H. 2006. *Essential DC/DC Converters*. Boca Raton: Taylor & Francis Group LLC, p. 30. With permission.)

The variation of the current i_{LO} is

$$\Delta i_{LO} = \frac{V_O}{L_O}(1-k)T.$$

Therefore, the variation ratio of the current i_{LO} is

$$\xi_O = \frac{\Delta i_{LO}/2}{I_{LO}} = \frac{V_O}{2I_O L_O}(1-k)T = \frac{1-k}{2}\frac{R}{fL_O}. \quad (5.99)$$

The variation of the diode current i_D is

$$\Delta i_D = \Delta i_L + \Delta i_{LO} = \left(\frac{V_O}{L} + \frac{V_O}{L_O}\right)(1-k)T.$$

We can define $L_{/} = L//L_O$.

$$\Delta i_D = \Delta i_L + \Delta i_{LO} = \frac{V_O}{L_{//}}(1-k)T$$

and

$$I_D = I_L + I_{LO} = I_I + I_O = (M+1)I_O = \frac{1}{1-k}I_O.$$

Therefore, the variation ratio of the diode current i_D is

$$\zeta = \frac{\Delta i_D/2}{I_D} = \frac{V_O}{2I_O L_{//}}(1-k)^2 T = \frac{(1-k)^2}{2}\frac{R}{fL_{//}}. \quad (5.100)$$

The variation of the voltage v_C is

$$\Delta v_C = \frac{\Delta Q}{C} = \frac{I_I}{C}(1-k)T.$$

Therefore, the variation ratio of the voltage v_C is

$$\rho = \frac{\Delta v_C/2}{V_C} = \frac{I_I}{2CV_C}(1-k)T = \frac{k(1-k)M}{2}\frac{1}{fRC}. \quad (5.101)$$

The variation of the voltage v_{CO} is

$$\Delta v_{CO} = \frac{\Delta Q_O}{C_O} = \frac{T}{8C_O}\Delta i_{LO} = \frac{V_O}{8f^2 C_O L_O}(1-k).$$

Therefore, the variation ratio of the voltage v_{CO} is

$$\varepsilon = \frac{\Delta v_{CO}/2}{V_O} = \frac{1-k}{16f^2 C_O L_O}. \quad (5.102)$$

The boundary is determined by the condition

$$\zeta = 1$$

or

$$\zeta = \frac{(1-k)^2}{2}\frac{R}{fL_{//}} = \frac{1}{2(1+M)^2}Z_N = 1 \quad \text{with} \quad Z_N = \frac{R}{fL_{//}}.$$

Therefore, the boundary between CCM and DCM is

$$M = \sqrt{\frac{Z_N}{2}} - 1. \tag{5.103}$$

If $(M+1) > \sqrt{Z_N/2}$, the converter works in CCM; if $(M+1) < \sqrt{Z_N/2}$, the converter works in DCM.

Example 5.4

A Cúk-converter has the following components: $V_1 = 20\,\text{V}$, $L = L_O = 10\,\text{mH}$, $C = C_O = 20\,\mu\text{F}$, $R = 20\,\Omega$, switching frequency $f = 50\,\text{kHz}$, and conduction duty cycle $k = 0.6$. Calculate the output voltage and its ripple in the steady state. Does this converter work in CCM or DCM?

Solution

1. From Equation 5.59, the output voltage is

$$V_2 = V_C = \frac{k}{1-k}V_1 = \frac{0.6}{0.4}20 = 30\,\text{V}.$$

2. From Equation 5.102, the output voltage ripple is

$$\varepsilon = \frac{1-k}{16f^2C_OL_O} = \frac{0.4}{16(50\,\text{k})^2 \times 20\,\mu \times 10\,\text{m}} = 0.00005.$$

3. We have $M + 1 = 2.5$, which is greater than $\sqrt{Z_N/2} = \sqrt{20/(2 \times 5\,\text{m} \times 50\,\text{k})} = 0.2$. Referring to Equation 5.103, we know that this converter works in CCM.

5.5.5 SEPIC

The SEPIC is derived from the boost converter. Its circuit diagram is shown in Figure 5.33. The SEPIC was created immediately after the Cúk-converter, and is also called the P/O Cúk-converter.

The inductor current i_{L1} increases with slope $+V_C/L_1$ during switching-on and decreases with slope $-V_O/L_1$ during switching-off.

Thus

$$\frac{V_C}{L_1}kT = \frac{V_O}{L_1}(1-k)T,$$
$$V_C = \frac{1-k}{k}V_O. \tag{5.104}$$

The inductor current i_L increases with slope $+V_I/L$ during switching-on and decreases with slope $-(V_C + V_O - V_I)/L$ during switching-off.

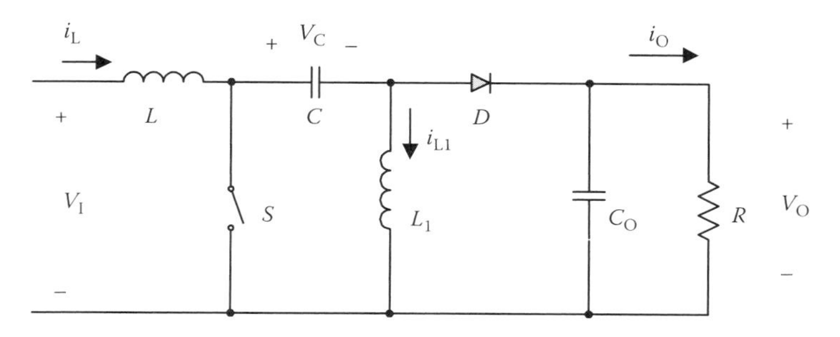

FIGURE 5.33 Single-ended primary inductance converter (SEPIC). (Reprinted from Luo, F. L. and Ye, H. 2006. *Essential DC/DC Converters*. Boca Raton: Taylor & Francis Group LLC, p. 30. With permission.)

Thus

$$\frac{V_I}{L}kT = \frac{V_C + V_O - V_I}{L}(1-k)T, \qquad V_O = \frac{k}{1-k}V_I, \tag{5.105}$$

that is,

$$M = \frac{V_O}{V_I} = \frac{k}{1-k}.$$

Since the inductor L is connected in series to the source voltage, the inductor average current I_L is

$$I_L = I_I.$$

Since the inductor L_1 is connected in parallel to the capacitor C during switching-off, the inductor average current I_{L1} is ($I_{CO-on} = I_O$ and $I_{CO-off} = I_I$), $I_{L1} = I_O$.

The variation of the current i_L is

$$\Delta i_L = \frac{V_I}{L}kT.$$

Therefore, the variation ratio of the current i_L is

$$\xi = \frac{\Delta i_L/2}{I_L} = \frac{V_I}{2I_I L}kT = \frac{k}{2M^2}\frac{R}{fL}. \tag{5.106}$$

The variation of the current i_{L1} is

$$\Delta i_{L1} = \frac{V_C}{L_1}kT.$$

Therefore, the variation ratio of the current i_{L1} is

$$\xi_1 = \frac{\Delta i_{L1}/2}{I_{L1}} = \frac{V_C}{2I_O L_1}kT = \frac{1-k}{2}\frac{R}{fL_1}. \tag{5.107}$$

The variation of the diode current i_D is

$$\Delta i_D = \Delta i_L + \Delta i_{L1} = \left(\frac{V_O}{L} + \frac{V_O}{L_1}\right)(1-k)T.$$

We can define $L_{//} = L//L_1$.
Hence

$$\Delta i_D = \Delta i_L + \Delta i_{L1} = \frac{V_O}{L_{//}}(1-k)T$$

and

$$I_D = I_L + I_{LO} = I_I + I_O = (M+1)I_O = \frac{1}{1-k}I_O.$$

Therefore, the variation ratio of the diode current i_D is

$$\zeta = \frac{\Delta i_D/2}{I_D} = \frac{V_O}{2I_O L_{//}}(1-k)^2 T = \frac{(1-k)^2}{2}\frac{R}{fL_{//}}. \tag{5.108}$$

The variation of the voltage v_C is

$$\Delta v_C = \frac{\Delta Q}{C} = \frac{I_I}{C}(1-k)T.$$

Therefore, the variation ratio of the voltage v_C is

$$\rho = \frac{\Delta v_C/2}{V_C} = \frac{I_I}{2CV_C}(1-k)T = \frac{kM}{2}\frac{1}{fRC}. \tag{5.109}$$

The variation of the voltage v_{CO} is

$$\Delta v_{CO} = \frac{\Delta Q_O}{C_O} = \frac{kTI_O}{C_O} = \frac{kI_O}{fC_O}.$$

Therefore, the variation ratio of the voltage v_{CO} is

$$\varepsilon = \frac{\Delta v_{CO}/2}{V_O} = \frac{kI_O}{2fC_O V_O} = \frac{k}{2fRC_O}. \tag{5.110}$$

The boundary is determined by the condition

$$\zeta = 1$$

or

$$\zeta = \frac{(1-k)^2}{2}\frac{R}{fL_{//}} = \frac{1}{2(1+M)^2}Z_N = 1 \quad \text{with} \quad Z_N = \frac{R}{fL_{//}}.$$

Therefore, the boundary between CCM and DCM is

$$M = \sqrt{\frac{Z_N}{2}} - 1. \tag{5.111}$$

TABLE 5.3

Circuit Diagrams of the Tapped-Inductor Fundamental Converters

	Standard Converter	Switch Tap	Diode to Tap	Rail to Tap
Buck				
Boost				
Buck–boost				

Source: Data from Luo, F. L. and Ye, H. 2006. *Essential DC/DC Converters*,. Boca Raton: Taylor & Francis Group LLC, p. 31.

TABLE 5.4

Voltage Transfer Gains of the Tapped-Inductor Fundamental Converters

Converter	No Tap	Switched to Tap	Diode to Tap	Rail to Tap
Buck	k	$\frac{k}{n+k(1-n)}$	$\frac{nk}{1+k(n-1)}$	$\frac{k-n}{k(1-n)}$
Boost	$\frac{1}{1-k}$	$\frac{n+k(1-n)}{n(1-k)}$	$\frac{1+k(n-1)}{1-k}$	$\frac{n-k}{n(1-k)}$
Buck-boost	$\frac{k}{1-k}$	$\frac{k}{n(1-k)}$	$\frac{nk}{1-k}$	$\frac{k}{1-k}$

Source: Data from Luo, F. L. and Ye, H. 2006. *Essential DC/DC Converters*, Boca Raton: Taylor & Francis Group LLC, p. 32.

5.6 Tapped-Inductor Converters

These converters have been derived from fundamental converters, whose circuit diagrams are shown in Table 5.3. The voltage transfer gains are presented in Table 5.4. Here the tapped-inductor ratio is $n = n_1/(n_1 + n_2)$.

Homework

5.1. A boost converter has the following components: $V_1 = 20\,\text{V}$, $L = 10\,\text{mH}$, $C = 20\,\mu\text{F}$, $R = 20\,\Omega$, switching frequency $f = 50\,\text{kHz}$, and conduction duty cycle $k = 0.6$. Calculate the output voltage and its ripple in the steady state. Does this converter work in CCM or DCM?

5.2. A P/O buck–boost converter working in "buck–boost operation mode" has the following components: $V_1 = 20\,\text{V}$, $L = 10\,\text{mH}$, $C = 20\,\mu\text{F}$, $R = 20\,\Omega$, switching frequency $f = 20\,\text{kHz}$, and conduction duty cycle $k = 0.6$. Calculate the output voltage and its ripple in the steady state. Does this converter work in CCM or DCM?

5.3. A multiple charger is required to offer three output voltages at 6, 9, and 12 V to charge mobile phones, digital cameras, and GPS. A forward converter with multiple outputs in Figure 5.19 can be used for this purpose. It has the following components: $V_1 = 20\,\text{V}$, all $L = 10\,\text{mH}$, all $C = 20\,\mu\text{F}$, all R are about $20\,\Omega$, switching frequency $f = 20\,\text{kHz}$, and conduction duty cycle $k = 0.5$. Calculate the turn's ratio for each secondary winding of the transformer. If the primary winding has 600 turns, what are the three secondary winding's turns?

5.4. A Zeta converter in Figure 5.25 is used to provide high output voltage $V_O = 1500\,\text{V}$. It has the following components: $V_{in} = 50\,\text{V}$, $L_1 = L_2 = 10\,\text{mH}$, $C_1 = C_2 = 20\,\mu\text{F}$, $R = 100\,\Omega$, switching frequency $f = 50\,\text{kHz}$, and conduction duty cycle $k = 0.8$. If the primary winding has 200 turns, calculate the transformer turn's ratio and the particular turns of the secondary winding.

5.5. A negative output Luo-converter has the following components: $V_I = 20\,V$, $L = L_O = 10\,mH$, $C = C_O = 20\,\mu F$, $R = 3000\,\Omega$, switching frequency $f = 20\,kHz$, and conduction duty cycle $k = 0.6$. Calculate the output voltage and its variation ratio in the steady state.

References

1. Luo, F. L. and Ye, H. 2004. *Advanced DC/DC Converters*. Boca Raton: CRC Press.
2. Luo, F. L. and Ye, H. 2006. *Essential DC/DC Converters*. Boca Raton: Taylor & Francis Group LLC.
3. Luo, F. L. 1999. Positive output Luo-converters: Voltage lift technique. *IEE-EPA Proceedings*, vol. 146, pp. 415–432.
4. Luo, F. L. 1999. Negative output Luo-converters: Voltage lift technique. *IEE-EPA Proceedings*, vol. 146, pp. 208–224.
5. Luo, F. L. 2000. Double output Luo-converters: Advanced voltage lift technique. *IEE-EPA Proceedings*, vol. 147, pp. 469–485.
6. Erickson, R. W. and Maksimovic, D. 1999. *Fundamentals of Power Electronics*. Norwell, MA: Kluwer and Academic Publishers.
7. Middlebrook, R. D. and Cuk, S. 1981. *Advances in Switched-Mode Power Conversion*. Pasadena: TESLAco.
8. Maksimovic, D. and Cuk, S. 1991. Switching converters with wide DC conversion range. *IEEE Transactions on Power Electronics*, 151–159.
9. Smedley, K. M. and Cuk, S. 1995. One-cycle control of switching converters. *IEEE Transactions on Power Electronics*, 10, 625–634.
10. Redl, R., Molnar, B., and Sokal, N. O. 1986. Class-E resonant DC-DC power converters: Analysis of operations, and experimental results at 1.5 MHz. *IEEE Transactions on Power Electronics*, 1, 111–121.
11. Kazimierczuk, M. K. and Bui, X. T. 1989. Class-E DC–DC converters with an inductive impedance inverter. *IEEE Transactions on Power Electronics*, 4, 124–133.
12. Liu, Y. and Sen, P. C. 1996. New Class-E DC-DC converter topologies with constant switching frequency. *IEEE Transactions on Industry Applications*, 32, 961–972.

6

Voltage Lift Converters

The ordinary DC/DC converter has limited voltage transfer gain. Considering the effects of the component called parasitic elements, the conduction duty cycle k can only be $0.1 < k < 0.9$. This restriction blocks ordinary DC/DC converter voltage transfer gain increase. The VL technique is a common method used in electronics circuitry design to amplify output voltage. Using this technique in DC/DC conversion technology, we can design, stage by stage, VL power converters with high voltage transfer gains in arithmetic progression. It opens the way to significantly increase the voltage transfer gain of DC/DC converters. Using this technique, the following series of VL converters are designed [1,2]:

- P/O Luo-converters
- N/O Luo-converters
- D/O Luo-converters
- VL Cúk-converters
- VL SEPIC
- Other VL D/O converters
- Switched-capacitorized (SC) converters.

6.1 Introduction

The VL technique is applied to the periodical switching circuit. Usually, a capacitor is charged, during switch-on, by a certain voltage, for example, the source voltage. This charged capacitor voltage can be arranged on top-up to some parameter, for example, output voltage during switch-off. Therefore, the output voltage can be lifted higher. Consequently, this circuit is called a self-lift circuit. A typical example is the sawtooth wave generator with a self-lift circuit.

Repeating this operation, another capacitor can be charged by a certain voltage, which is possibly the input voltage or other equivalent voltages. The second capacitor-charged voltage can also be arranged on top-up to some parameter, especially the output voltage. Therefore, the output voltage can be higher than that of a self-lift circuit. Usually, this circuit is called a re-lift circuit.

Analogously, this operation can be repeated many times. Consequently, the series circuits are called a triple-lift circuit, a quadruple-lift circuit, and so on.

Because of the effect of parasitic elements, the output voltage and power transfer efficiency of DC–DC converters are limited. The VL technique offers a good way of improving circuit characteristics. After long-term research, this technique has been successfully applied to DC–DC converters. Three series of Luo-converters have now been developed from prototypes using the VL technique. These converters perform DC–DC voltage increasing conversion with high power density, high efficiency, and cheap topology in a simple structure. They are different from other DC–DC step-up converters and possess many advantages, including the high output voltage with small ripples. Therefore, these converters will be widely used in computer peripheral equipment and industrial applications, especially for high-output-voltage projects. The contents of this chapter are arranged as follows:

1. Seven types of self-lift converters
2. P/O Luo-converters
3. N/O Luo-converters
4. Modified P/O Luo-converters
5. D/O Luo-converters.

Using the VL technique, we can easily obtain the other series of VL converters. For example, VL Cúk-converters, VL SEPICs, other types of D/O converters, and switched-capacitorized converters.

6.2 Seven Self-Lift Converters

All self-lift converters introduced here are derived from developed converters such as Luo-converters, Cúk-converters, and SEPICs, which were described in Section 5.5. Since all circuits are simple, usually only one more capacitor and diode are required; the output voltage is higher than the input voltage [3–5]. The output voltage is calculated by

$$V_O = \left(\frac{k}{1-k} + 1\right) V_{in} = \frac{1}{1-k} V_{in}. \tag{6.1}$$

Seven circuits were developed:

- Self-lift Cúk-converter
- Self-lift P/O Luo-converter
- Reverse self-lift P/O Luo-converter
- Self-lift N/O Luo-converter
- Reverse self-lift Luo-converter
- Self-lift SEPIC
- Enhanced self-lift P/O Luo-converter.

These converters perform DC–DC voltage increasing conversion in simple structures. In these circuits, the switch S is a semiconductor device (MOSFET, BJT, IGBT, and so on). It is driven by a PWM switching signal with variable frequency f and conduction duty cycle k. For all circuits, the load is usually resistive, that is, $R = V_O/I_O$.

The normalized impedance Z_N is

$$Z_N = \frac{R}{f L_{eq}}, \tag{6.2}$$

where L_{eq} is the equivalent inductance.

We concentrate on the absolute values rather than polarity in the description and calculations given below. The directions of all voltages and currents are defined and shown in the corresponding figures. We also assume that the semiconductor switch and the passive components are all ideal. All capacitors are assumed to be large enough that the ripple voltage across the capacitors can be negligible in one switching cycle, for the average value discussions.

For any component X (e.g., C, L, and so on), its instantaneous current and voltage are expressed as i_X and v_X. Its average current and voltage values are expressed as I_x and V_x. The output voltage and current are V_O and I_O; the input voltage and current are V_I and I_I. T and f are the switching period and frequency.

The voltage transfer gain for the CCM is as follows:

$$M = \frac{V_O}{V_I} = \frac{I_I}{I_O}. \tag{6.3}$$

$$\text{Variation of current } i_L\text{: } \zeta_1 = \frac{\Delta i_L/2}{I_L}. \tag{6.4}$$

$$\text{Variation of current } i_{LO}\text{: } \zeta_2 = \frac{\Delta i_{LO}/2}{I_{LO}}. \tag{6.5}$$

$$\text{Variation of current } i_D\text{: } \xi = \frac{\Delta i_D/2}{I_D}. \tag{6.6}$$

$$\text{Variaton of voltage } v_C\text{: } \rho = \frac{\Delta v_C/2}{V_C}. \tag{6.7}$$

$$\text{Variation of voltage } v_{C1}\text{: } \sigma_1 = \frac{\Delta v_{C1}/2}{v_{C1}}. \tag{6.8}$$

$$\text{Variation of voltage } v_{C2}\text{: } \sigma_2 = \frac{\Delta v_{C2}/2}{v_{C2}}. \tag{6.9}$$

$$\text{Variation of output voltage } v_O\text{: } \varepsilon = \frac{\Delta V_O/2}{V_O}. \tag{6.10}$$

Here, I_D refers to the average current i_D that flows through the diode D during the switch-off period, and not its average current over the whole period.

A detailed analysis of the seven self-lift DC–DC converters is given in the following sections. Due to the limit on the length of the book, only the simulation and experimental results of the self-lift Cúk-converter are given. However, the results and conclusions of other self-lift converters should be quite similar to those of the self-lift Cúk-converter.

6.2.1 Self-Lift Cúk-Converter

The self-lift Cúk-converter and its equivalent circuits during the switch-on and switch-off periods are shown in Figure 6.1. It is derived from the Cúk-converter. During the switch-on

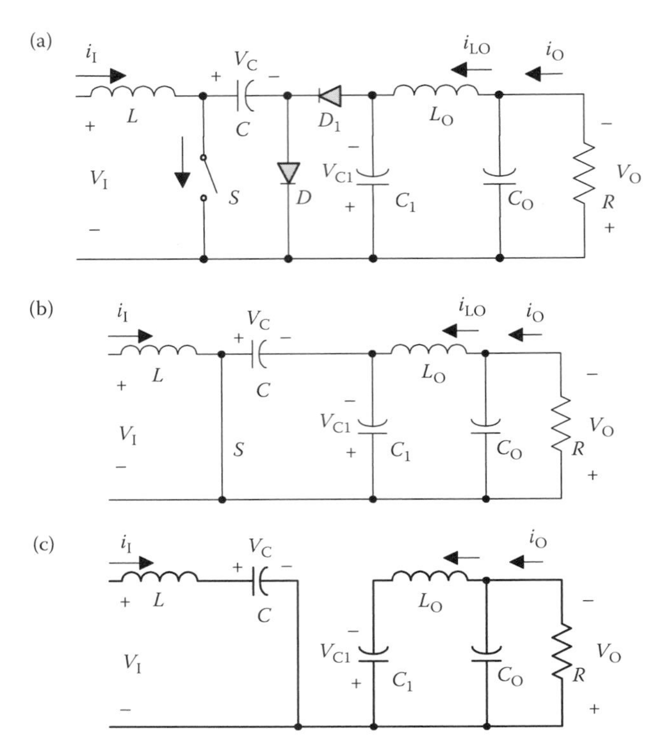

FIGURE 6.1 (a) Self-lift Cúk-converter circuit and its equivalent circuits during (b) switch-on, and (c) switch-off. (Reprinted from Luo, F. L. and Ye, H. 2006. *Essential DC/DC Converters*. Boca Raton: Taylor & Francis Group LLC, p. 45. With permission.)

period, S and D_1 are on and D is off. During the switch-off period, D is on and S and D_1 are off.

6.2.1.1 Continuous Conduction Mode

In steady state, the average of inductor voltages over a period is zero. Thus

$$V_{C1} = V_{CO} = V_O. \tag{6.11}$$

During the switch-on period, the voltages across capacitors C and C_1 are equal. Since we assume that C and C_1 are sufficiently large,

$$V_C = V_{C1} = V_O. \tag{6.12}$$

The inductor current i_L increases during switch-on and decreases during switch-off. The corresponding voltages across L are V_I and $-(V_C - V_I)$.

Therefore, $kTV_I = (1 - k)T(V_C - V_I)$.

Hence,

$$V_O = V_C = V_{C1} = V_{CO} = \frac{1}{1-k}V. \tag{6.13}$$

The voltage transfer gain in the CCM is

$$M = \frac{V_O}{V_I} = \frac{I_I}{I_O} = \frac{1}{1-k}. \tag{6.14}$$

The characteristics of M versus conduction duty cycle k are shown in Figure 6.2.

Since all the components are considered ideal, the power loss associated with all the circuit elements is neglected. Therefore the output power P_O is considered to be equal to the input power P_{IN}: $V_O I_O = V_I I_I$.

Thus,

$$I_L = I_I = \frac{1}{(1-k)} I_O.$$

During switch-off,

$$i_D = i_L, \quad I_D = \frac{1}{1-k} I_O. \tag{6.15}$$

The capacitor C_O acts as a low-pass filter, so that $I_{LO} = I_O$.

The current i_L increases during switch-on. The voltage across it during switch-on is V_I; therefore its peak-to-peak current variation is $\Delta I_L = kTV_I/L$.

The variation ratio of current i_L is

$$\zeta_1 = \frac{\Delta i_L/2}{I_L} = \frac{kTV_I}{2I_L} = \frac{k(1-k)^2 R}{2fL} = \frac{kR}{2M^2 fL}. \tag{6.16}$$

The variation of current i_D is

$$\xi = \zeta_1 = \frac{kR}{2M^2 fL}. \tag{6.17}$$

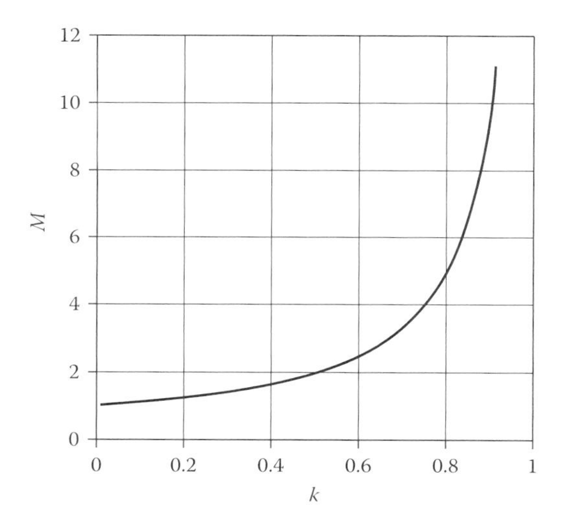

FIGURE 6.2 Voltage transfer gain M versus k. (Reprinted from Luo, F. L. and Ye, H. 2006. *Essential DC/DC Converters*. Boca Raton: Taylor & Francis Group LLC, p. 46. With permission.)

The peak-to-peak variation of voltage v_C is

$$\Delta v_C = \frac{I_L(1-k)T}{C} = \frac{I_O}{fC}. \tag{6.18}$$

The variation ratio of voltage v_C is

$$\rho = \frac{\Delta v_C/2}{V_C} = \frac{I_O}{2fCV_O} = \frac{1}{2fRC}. \tag{6.19}$$

The peak-to-peak variation of voltage v_{C1} is

$$\Delta v_{C1} = \frac{I_{LO}(1-k)T}{C_1} = \frac{I_O(1-k)}{fC_1}. \tag{6.20}$$

The variation ratio of voltage v_{C1} is

$$\sigma_1 = \frac{\Delta v_{C1}/2}{V_{C1}} = \frac{I_O(1-k)}{2fC_1V_O} = \frac{1}{2MfRC_1}. \tag{6.21}$$

The peak-to-peak variation of current i_{LO} is approximately

$$\Delta i_{LO} = \frac{(1/2)(\Delta v_{C1}/2)(T/2)}{L_O} = \frac{I_O(1-k)}{8f^2L_OC_1}. \tag{6.22}$$

The variation ratio of current i_{LO} is approximately

$$\zeta_2 = \frac{\Delta i_{LO}/2}{I_{LO}} = \frac{I_O(1-k)}{16f^2L_OC_1I_O} = \frac{1}{16Mf^2L_OC_1}. \tag{6.23}$$

The peak-to-peak variation of voltages v_O and v_{CO} is

$$\Delta v_O = \Delta v_{CO} = \frac{(1/2)(\Delta i_{LO}/2)(T/2)}{C_O} = \frac{I_O(1-k)}{64f^3L_OC_1C_O}. \tag{6.24}$$

The variation ratio of the output voltage is

$$\varepsilon = \frac{\Delta v_O/2}{V_O} = \frac{I_O(1-k)}{128f^3L_OC_1C_OV_O} = \frac{1}{128Mf^3L_OC_1C_OR}. \tag{6.25}$$

The voltage transfer gain of the self-lift Cúk-converter is the same as the original boost converter. However, the output current of the self-lift Cúk-converter is continuous, with small ripples.

The output voltage of the self-lift Cúk-converter is higher than the corresponding Cúk-converter by an input voltage. It retains one of the merits of the Cúk-converter. They both have continuous input and output currents in the CCM. As for component stress, it can be seen that the self-lift Cúk-converter has a smaller voltage and current stresses than the original Cúk-converter.

6.2.1.2 Discontinuous Conduction Mode

The self-lift Cúk-converter operates in the DCM if the current i_D decreases to zero during switch-off. A special case is seen when i_D decreases to zero at $t = T$, then, the circuit operates at the boundary of CCM and DCM. The variation ratio of current i_D is 1 when the circuit works in the boundary state:

$$\xi = \frac{k}{2}\frac{R}{M^2 f L} = 1. \tag{6.26}$$

Therefore the boundary between CCM and DCM is

$$M_B = \sqrt{k}\sqrt{\frac{R}{2fL}} = \sqrt{\frac{kz_N}{2}}, \tag{6.27}$$

where z_N is the normalized load $R/(fL)$. The boundary between CCM and DCM is shown in Figure 6.3a. The curve that describes the relationship between M_B and z_N has the minimum value $M_B = 1.5$ and $k = 1/3$ when the normalized load z_N is 13.5.

When $M > M_B$, the circuit operates in the DCM. In this case, the diode current i_D decreases to zero at $t = t_1 = [k + (1-k)m]T$, where $kT < t_1 < T$ and $0 < m < 1$.

Define m as the current filling factor. After mathematical manipulation,

$$m = \frac{1}{\xi} = \frac{M^2}{k(R/2fL)}. \tag{6.28}$$

From the above equation, we can see that the DCM is caused by the following factors:

- Switching frequency f is too low
- Duty cycle k is too small
- Inductance L is too small
- Load resistor R is too big.

In the DCM, current i_L increases during switch-on and decreases in the period from kT to $(1-k)mT$. The corresponding voltages across L are V_I and $-(V_C - V_I)$. Therefore,

$$kTV_I = (1-k)mT(V_C - V_I).$$

Hence,

$$V_C = \left[1 + \frac{k}{(1-k)m}\right]V_I. \tag{6.29}$$

Since we assume that C, C_1, and C_O are large enough,

$$V_O = V_C = V_{CO} = \left[1 + \frac{k}{(1-k)m}\right]V_I \tag{6.30}$$

or

$$V_O = \left[1 + k^2(1-k)\frac{R}{2fL}\right]V_I. \tag{6.31}$$

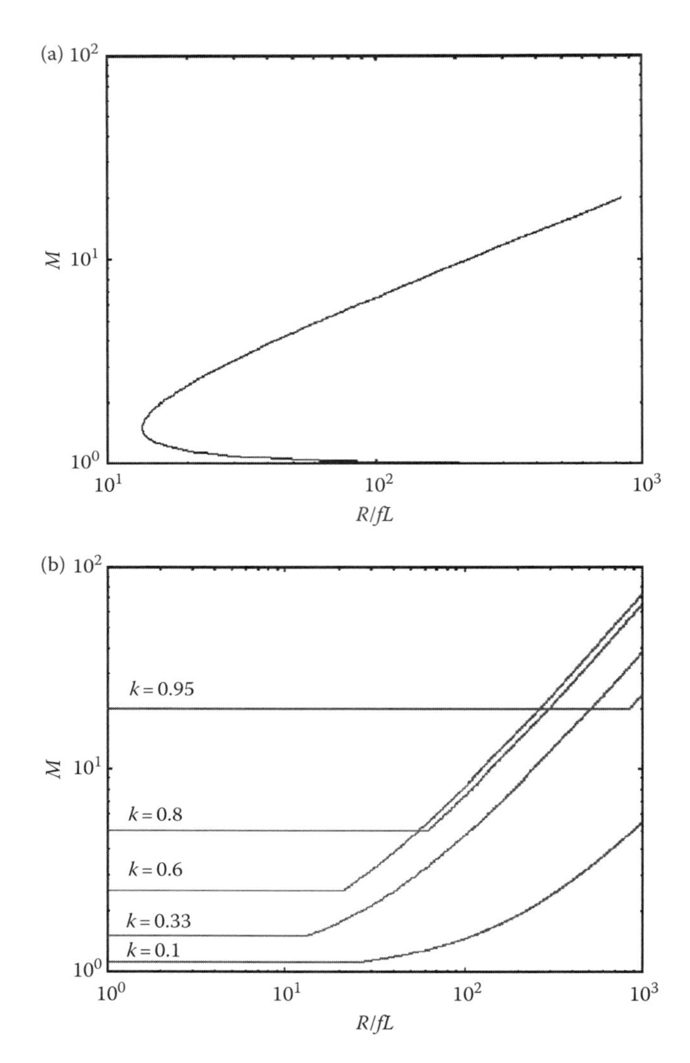

FIGURE 6.3 Output voltage characteristics of the self-lift Cúk-converter: (a) boundary between CCM and DCM and (b) Voltage transfer gain M versus the normalized load at various k. (Reprinted from Luo, F. L. and Ye, H. 2006. *Essential DC/DC Converters*. Boca Raton: Taylor & Francis Group LLC, p. 49. With permission.)

The voltage transfer gain in the DCM is

$$M_{DCM} = 1 + k^2(1 - k)\frac{R}{2fL}. \tag{6.32}$$

The relationship between DC voltage transfer gain M and the normalized load at various k in the DCM is also shown in Figure 6.3b. It can be seen that in the DCM, the output voltage increases as the load resistance R increases.

6.2.2 Self-Lift P/O Luo-Converter

A self-lift P/O Luo-converter and its equivalent circuits during the switch-on and switch-off periods are shown in Figure 6.4. It is the self-lift circuit of the P/O Luo-converter. It is derived from the elementary circuit of the P/O Luo-converter. During the switch-on period, S and D_1 are switched on and D is switched off. During the switch-off period, D is on, and S and D_1 are off.

6.2.2.1 Continuous Conduction Mode

In steady state, the average of inductor voltages over a period is zero. Thus

$$V_C = V_{CO} = V_O.$$

During the switch-on period, the voltage across capacitor C_1 is equal to the source voltage. Since we assume that C and C_1 are sufficiently large, $V_{C1} = V_I$.

The inductor current i_L increases in the switch-on period and decreases in the switch-off period. The corresponding voltages across L are V_I and $-(V_C - V_{C1})$. Therefore, $kTV_I = (1-k)T(V_C - V_{C1})$. Hence, $V_O = (1/(1-k))V_I$.

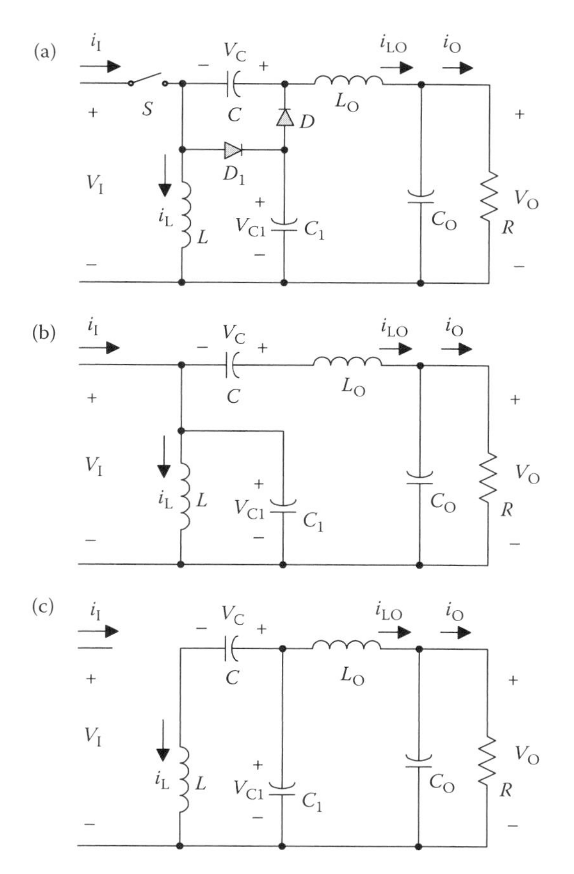

FIGURE 6.4 (a) Self-lift P/O Luo-converter circuit and its equivalent circuits during (b) switch-on, and (c) switch-off. (Reprinted from Luo, F. L. and Ye, H. 2006. *Essential DC/DC Converters*. Boca Raton: Taylor & Francis Group LLC, p. 51. With permission.)

The voltage transfer gain in the CCM is

$$M = \frac{V_O}{V_I} = \frac{1}{1-k}. \tag{6.33}$$

Since all the components are considered to be ideal, the power loss associated with all the circuit elements is neglected. Therefore, the output power P_O is considered to be equal to the input power P_{IN} : $V_O I_O = V_I I_I$. Thus, $I_I = (1/1-k)I_O$.

The capacitor C_O acts as a low-pass filter so that $I_{LO} = I_O$.

The charge of capacitor C increases during switch-on and decreases during switch-off:

$$Q_+ = I_{C-on}kT = I_O kT, \quad Q_- = I_{C-off}(1-k)T = I_L(1-k)T.$$

In a switching period, $Q_+ = Q_-$, $\quad I_L = (k/(1-k))I_O$.

During the switch-off period, $i_D = i_L + i_{LO}$.

Therefore, $I_D = I_L + I_{LO} = (1/(1-k))I_O$.

For the current and voltage variations and boundary condition, we can obtain the following equations using a similar method to that used in the analysis of the self-lift Cúk-converter.

$$\text{Current variations: } \zeta_1 = \frac{1}{2M^2}\frac{R}{fL}, \quad \zeta_2 = \frac{k}{2M}\frac{R}{fL_O}, \quad \xi = \frac{k}{2M^2}\frac{R}{fL_{eq}},$$

where L_{eq} refers to $L_{eq} = LL_O/(L+L_O)$.

$$\text{Voltage variations: } \rho = \frac{k}{2}\frac{1}{fCR}, \quad \sigma_1 = \frac{M}{2}\frac{1}{fC_1R}, \quad \varepsilon = \frac{k}{8M}\frac{1}{f^2L_OC_O}.$$

6.2.2.2 Discontinuous Conduction Mode

The self-lift P/O Luo-converter operates in the DCM if the current i_D decreases to zero during switch-off. In the critical case when i_D decreases to zero at $t = T$, the circuit operates at the boundary of CCM and DCM.

The variation ratio of current i_D is 1 when the circuit works in the boundary state:

$$\xi = \frac{k}{2M^2}\frac{R}{fL_{eq}} = 1.$$

Therefore, the boundary between CCM and DCM is

$$M_B = \sqrt{k}\sqrt{\frac{R}{2fL_{eq}}} = \sqrt{\frac{kz_N}{2}}, \tag{6.34}$$

where z_N is the normalized load $R/(fL_{eq})$, and L_{eq} refers to $L_{eq} = LL_O/(L+L_O)$.

When $M > M_B$, the circuit operates in the DCM. In this case, the diode current i_D decreases to zero at $t = t_1 = [k + (1-k)m]T$, where $KT < t_1 < T$ and $0 < m < 1$. Here, m is the current filling factor. We define m as

$$m = \frac{1}{\xi} = \frac{M^2}{k(R/2fL_{eq})}. \tag{6.35}$$

In the DCM, the current i_L increases in the switch-on period kT and decreases in the period from kT to $(1-k)mT$. The corresponding voltages across L are V_I and $-(V_C - V_{C1})$. Therefore, $kTV_I = (1-k)mT(V_C - V_{C1})$ and $V_C = V_{CO} = V_O \ V_{C1} = V_I$. Hence,

$$V_O = \left[1 + \frac{k}{(1-k)m}\right] V_I \quad \text{or} \quad V_O = \left[1 + k^2(1-k)\frac{R}{2fL_{eq}}\right] V_I. \tag{6.36}$$

So the real DC voltage transfer gain in the DCM is

$$M_{DCM} = 1 + k^2(1-k)\frac{R}{2fL_{eq}}. \tag{6.37}$$

In DCM, the output voltage increases as the load resistance R increases.

Example 6.1

A P/O self-lift Luo-converter has the following components: $V_I = 20$V, $L = L_O = 1$ mH, $C = C_1 = C_O = 20\,\mu$F, $R = 40\,\Omega$, $f = 50$ kHz, and $k = 0.5$. Calculate the output voltage, and the variation ratios ζ_1, ζ_2, ξ, ρ, σ_1, and ε in steady state.

SOLUTION

1. From Equation 6.33, the output voltage is $V_O = V_I/(1-k) = 20/0.5 = 40$V, that is, $M = 2$.
2. From the formulae we can obtain the following ratios:

$$\zeta_1 = \frac{1}{2M^2}\frac{R}{fL} = \frac{1}{2 \times 2^2}\frac{40}{50\,\text{k} \times 1\,\text{m}} = 0.1,$$

$$\zeta_2 = \frac{k}{2M}\frac{R}{fL_O} = \frac{1}{2 \times 2^2}\frac{40}{50\,\text{k} \times 1\,\text{m}} = 0.1,$$

$$\xi = \frac{k}{2M^2}\frac{R}{fL_{eq}} = \frac{1}{2 \times 2^2}\frac{40}{50\,\text{k} \times 0.5\,\text{m}} = 0.2,$$

$$\rho = \frac{k}{2}\frac{1}{fCR} = \frac{0.5}{2}\frac{1}{50\,\text{k} \times 20\,\mu \times 40} = 0.00625,$$

$$\sigma_1 = \frac{M}{2}\frac{1}{fC_1R} = \frac{2}{2}\frac{1}{50\,\text{k} \times 20\,\mu \times 40} = 0.025,$$

$$\varepsilon = \frac{k}{8M}\frac{1}{f^2L_OC_O} = \frac{0.5}{8 \times 2}\frac{1}{(50\,\text{k})^2 \times 20\,\mu \times 1\,\text{m}} = 0.000625.$$

From the calculations, the variations of i_{L1}, i_{L2}, v_C, and v_{C1} are small. The output voltage v_O (also v_{C1}) is almost a real DC voltage with very small ripples. Because of the resistive load, the output current $i_O (i_O = v_O/R)$ is almost a real DC waveform with very small ripples as well.

6.2.3 Reverse Self-Lift P/O Luo-Converter

The reverse self-lift P/O Luo-converter and its equivalent circuits during the switch-on and switch-off periods are shown in Figure 6.5. It is derived from the elementary circuit of P/O Luo-converters. During the switch-on period, S and D_1 are on and D is off. During the switch-off period, D is on, and S and D_1 are off.

6.2.3.1 Continuous Conduction Mode

In steady state, the average of inductor voltages over a period is zero. Thus $V_{C1} = V_{CO} = V_O$.

During the switch-on period, the voltage across capacitor C is equal to the source voltage plus the voltage across C_1. Since we assume that C and C_1 are sufficiently large, $V_{C1} = V_I + V_C$.

Therefore,

$$V_{C1} = V_I + \frac{k}{1-k}V_I = \frac{1}{1-k}V_I, \quad V_O = V_{CO} = V_{C1} = \frac{1}{1-k}V_I. \tag{6.38}$$

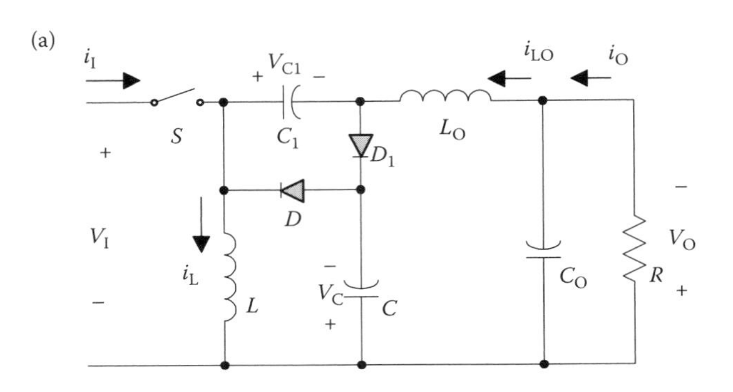

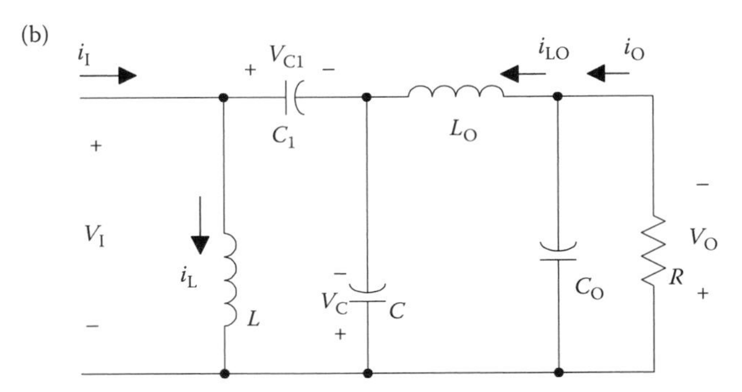

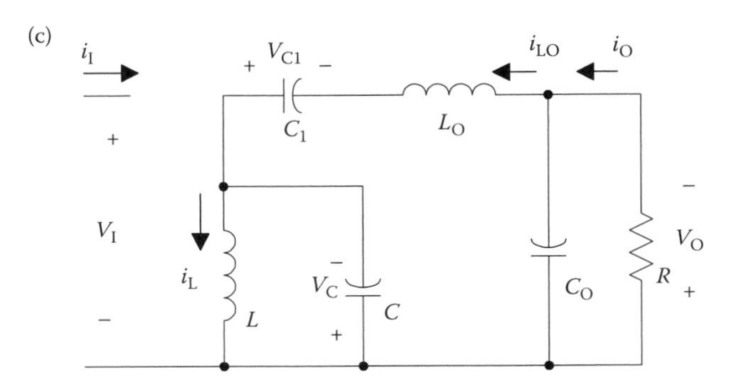

FIGURE 6.5 (a) Reverse self-lift P/O Luo-converter circuit and its equivalent circuits during (b) switch-on, and (c) switch-off. (Reprinted from Luo, F. L. and Ye, H. 2006. *Essential DC/DC Converters*. Boca Raton: Taylor & Francis Group LLC, p. 55. With permission.)

The voltage transfer gain in the CCM is

$$M = \frac{V_O}{V_I} = \frac{1}{1-k}. \tag{6.39}$$

Since all the components are considered to be ideal, the power losses on all the circuit elements are neglected. Therefore, the output power P_O is considered to be equal to the input power P_{IN}:

$$V_O I_O = V_I I_I.$$

Thus, $I_I = (1/(1-k))I_O$.

The capacitor C_O acts as a low-pass filter, so that $I_{LO} = I_O$.

The charge of capacitor C_1 increases during switch-on and decreases during switch-off:

$$Q_+ = I_{C1-on}kT,$$
$$Q_- = I_{LO}(1-k)T = I_O(1-k)T.$$

In a switching period,

$$Q_+ = Q_-,$$
$$I_{C1-on} = ((1-k)/k)I_O,$$
$$I_{C-on} = I_{LO} + I_{C1-on} = I_O + \frac{1-k}{k}I_O = \frac{1}{k}I_O. \tag{6.40}$$

The charge on the capacitor C increases during switch-off and decreases during switch-on.

$$Q_+ = I_{C-off}(1-k)T, \quad Q_- = I_{C-on}kT = \frac{1}{k}I_O kT.$$

In a switching period,

$$Q_+ = Q_-, \quad I_{C-off} = \frac{1-k}{k}I_{C-on} = \frac{1}{1-k}I_O. \tag{6.41}$$

Therefore,

$$I_L = I_{LO} + I_{C-off} = I_O + \frac{1}{1-k}I_O = \frac{2-k}{1-k}I_O = I_O + I_I.$$

During switch-off, $i_D = i_L - i_{LO}$.

Therefore, $I_D = I_L - I_{LO} = I_O$.

The following equations are used for current and voltage variations and boundary condition:

$$\text{Current variations: } \zeta_1 = \frac{k}{(2-k)M^2}\frac{R}{fL}, \quad \zeta_2 = \frac{k}{2M}\frac{R}{fL_O}, \quad \xi = \frac{1}{2M^2}\frac{R}{fL_{eq}},$$

where L_{eq} refers to $L_{eq} = (LL_O/L + L_O)$.

$$\text{Voltage variations: } \rho = \frac{1}{2k}\frac{1}{fCR}, \quad \sigma_1 = \frac{1}{2M}\frac{1}{fC_1R}, \quad \varepsilon = \frac{k}{16M}\frac{1}{f^2C_OL_O}.$$

6.2.3.2 *Discontinuous Conduction Mode*

The reverse self-lift P/O Luo-converter operates in the DCM; if the current i_D decreases to zero during switch-off at $t = T$, then the circuit operates at the boundary of CCM and DCM. The variation ratio of current i_D is 1 when the circuit works in the boundary state:

$$\xi = \frac{k}{2M^2}\frac{R}{fL_{eq}} = 1.$$

Therefore, the boundary between CCM and DCM is

$$M_B = \sqrt{k}\sqrt{\frac{R}{2fL_{eq}}} = \sqrt{\frac{kz_N}{2}}, \quad (6.42)$$

where z_N is the normalized load $R/(fL_{eq})$, and L_{eq} refers to $L_{eq} = LL_O/(L + L_O)$.

When $M > M_B$, the circuit operates in the DCM. In this case, the diode current i_D decreases to zero at $t = t_1 = [k + (1 - k)m]T$, where $kT < t_1 < T$ and $0 < m < 1$. Here, m is the current filling factor and is defined as

$$m = \frac{1}{\xi} = \frac{M^2}{k(R/2fL_{eq})}. \quad (6.43)$$

In the DCM, current i_L increases during switch-on and decreases in the period from kT to $(1 - k)mT$. The corresponding voltages across L are V_I and $-V_C$.

Therefore, $kTV_I = (1 - k)mTV_C$ and $V_{C1} = V_{CO} = V_O, \quad V_{C1} = V_I + V_C$.
Hence,

$$V_O = \left[1 + \frac{k}{(1-k)m}\right]V_I \quad \text{or} \quad V_O = \left(1 + k^2(1-k)\frac{R}{2fL_{eq}}\right)V_I. \quad (6.44)$$

So the real DC voltage transfer gain in the DCM is

$$M_{DCM} = 1 + k^2(1-k)\frac{R}{2fL}. \quad (6.45)$$

In DCM, the output voltage increases as the load resistance R increases.

6.2.4 Self-Lift N/O Luo-Converter

The self-lift N/O Luo-converter and its equivalent circuits during the switch-on and switch-off periods are shown in Figure 6.6. It is the self-lift circuit of the N/O Luo-converter. The function of capacitor C_1 is to lift the voltage V_C to a level higher than the source voltage V_I. S and D_1 are on and D is off during the switch-on period. D is on and S and D_1 are off during the switch-off period.

6.2.4.1 *Continuous Conduction Mode*

In steady state, the average of inductor voltages over a period is zero. Thus $V_C = V_{CO} = V_O$. During the switch-on period, the voltage across capacitor C_1 is equal to the source voltage. Since we assume that C and C_1 are sufficiently large, $V_{C1} = V_I$.

The inductor current i_L increases in the switch-on period and decreases in the switch-off period. The corresponding voltages across L are V_I and $-(V_C - V_{C1})$.

Therefore, $kTV_I = (1-k)T(V_C - V_{C1})$.

Hence,

$$V_O = V_C = V_{CO} = \frac{1}{1-k}V_I. \tag{6.46}$$

The voltage transfer gain in the CCM is

$$M = \frac{V_O}{V_I} = \frac{1}{1-k}. \tag{6.47}$$

Since all the components are considered to be ideal, the power loss associated with all the circuit elements is neglected. Therefore, the output power P_O is considered to be equal to the input power P_{IN}: $V_O I_O = V_I I_I$. Thus, $I_I = (1/(1-k))I_O$. The capacitor C_O acts as a low-pass filter so that $I_{LO} = I_O$.

For the current and voltage variations and boundary condition, the following equations can be obtained using a similar method to that used in the analysis of the self-lift

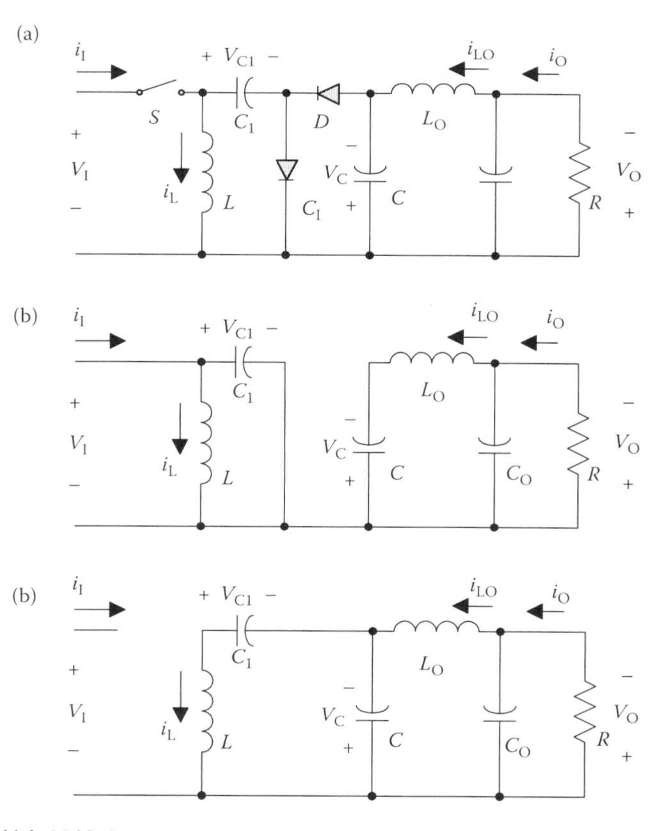

FIGURE 6.6 (a) Self-lift N/O Luo-converter circuit and its equivalent circuits during (b) switch-on, and (c) switch-off. (Reprinted from Luo, F. L. and Ye, H. 2006. *Essential DC/DC Converters*. Boca Raton: Taylor & Francis Group LLC, p. 59. With permission.)

Cúk-converter:

$$\text{Current variations: } \zeta_1 = \frac{k}{2M^2}\frac{R}{fL}, \quad \zeta_2 = \frac{k}{16}\frac{1}{f^2 L_O C}, \quad \xi = \zeta_1 = \frac{k}{2M^2}\frac{R}{fL}.$$

$$\text{Voltage variations: } \rho = \frac{k}{2}\frac{1}{fCR}, \quad \sigma_1 = \frac{M}{2}\frac{1}{fC_1 R}, \quad \varepsilon = \frac{k}{128}\frac{1}{f^3 L_O C C_O R}.$$

6.2.4.2 *Discontinuous Conduction Mode*

The self-lift N/O Luo-converter operates in the DCM; if the current i_D decreases to zero at $t = T$, then the circuit operates at the boundary of CCM and DCM. The variation ratio of current i_D is 1 when the circuit works at the boundary state:

$$\xi = \frac{k}{2M^2}\frac{R}{fL} = 1.$$

Therefore, the boundary between CCM and DCM is

$$M_B = \sqrt{k}\sqrt{\frac{R}{2fL_{eq}}} = \sqrt{\frac{kz_N}{2}}, \tag{6.48}$$

where L_{eq} refers to $L_{eq} = L$ and z_N is the normalized load $R/(fL)$.

When $M > M_B$, the circuit operates in the DCM. In this case, the diode current i_D decreases to zero at $t = t_1 = [k + (1 - k)m]T$, where $KT < t_1 < T$ and $0 < m < 1$. Here, m is the current filling factor and is defined as

$$m = \frac{1}{\xi} = \frac{M^2}{k(R/2fL)}. \tag{6.49}$$

In the DCM, current i_L increases during switch-on and decreases during the period from kT to $(1 - k)mT$. The voltages across L are V_I and $-(V_C - V_{C1})$.

Therefore,

$$kTV_I = (1 - k)mT(V_C - V_{C1})$$

and $V_{C1} = V_I$, $V_C = V_{CO} = V_O$. Hence,

$$V_O = \left[1 + \frac{k}{(1-k)m}\right]V_I \quad \text{or} \quad V_O = \left[1 + k^2(1-k)\frac{R}{2fL}\right]V_I.$$

So the real DC voltage transfer gain in the DCM is

$$M_{DCM} = 1 + k^2(1-k)\frac{R}{2fL}. \tag{6.50}$$

We can see that in DCM, the output voltage increases as the load resistance R increases.

6.2.5 Reverse Self-Lift N/O Luo-Converter

The reverse self-lift N/O Luo-converter and its equivalent circuits during the switch-on and switch-off periods are shown in Figure 6.7. During the switch-on period, S and D_1 are on and D is off. During the switch-off period, D is on and S and D_1 are off.

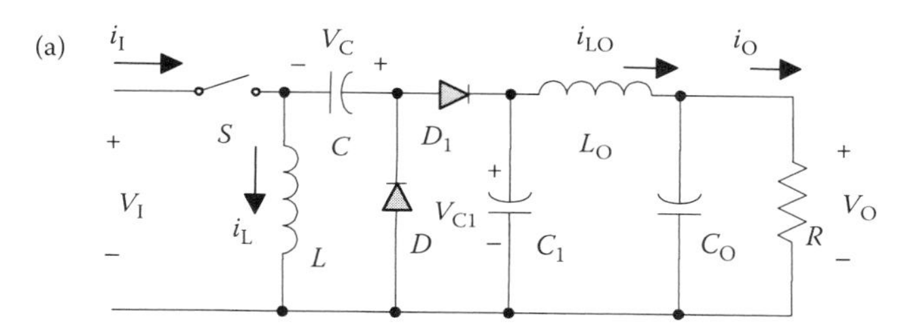

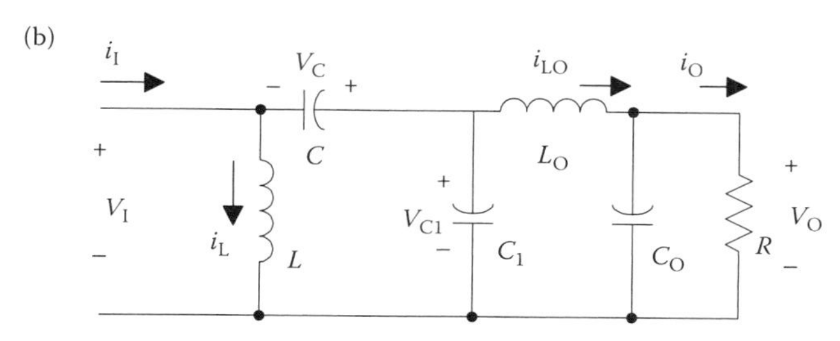

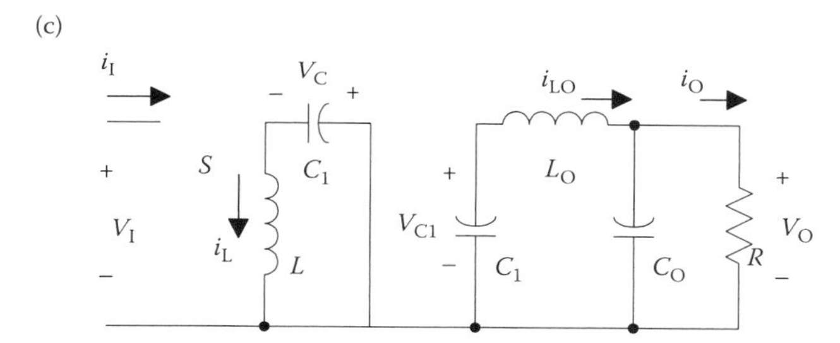

FIGURE 6.7 (a) Reverse self-lift N/O Luo-converter circuit and its equivalent circuits during (b) switch-on, and (c) switch-off. (Reprinted from Luo, F. L. and Ye, H. 2006. *Essential DC/DC Converters*. Boca Raton: Taylor & Francis Group LLC, p. 62. With permission.)

6.2.5.1 Continuous Conduction Mode

In steady state, the average of inductor voltages over a period is zero. Thus

$$V_{C1} = V_{CO} = V_O.$$

The inductor current i_L increases in the switch-on period and decreases in the switch-off period. The corresponding voltages across L are V_I and $-V_C$.

Therefore, $kTV_I = (1-k)TV_C$.

Hence,

$$V_C = \frac{k}{1-k}V_I \tag{6.51}$$

is the voltage across C. Since we assume that C and C_1 are sufficiently large, $V_{C1} = V_I + V_C$.

Therefore,

$$V_{C1} = V_I + \frac{k}{1-k}V_I = \frac{1}{1-k}V_I, \quad V_O = V_{CO} = V_{C1} = \frac{1}{1-k}V_I.$$

The voltage transfer gain in the CCM is

$$M = \frac{V_O}{V_I} = \frac{1}{1-k}. \tag{6.52}$$

Since all the components are considered ideal, the power loss associated with all the circuit elements is neglected. Therefore, the output power P_O is considered to be equal to the input power P_{IN}: $V_O I_O = V_I I_I$. Thus, $I_I = (1/(1-k))I_O$. The capacitor C_O acts as a low-pass filter so that $I_{LO} = I_O$.

The charge of capacitor C_1 increases during switch-on and decreases during switch-off:

$$Q_+ = I_{C1-on}kT, \quad Q_- = I_{C1-off}(1-k)T = I_O(1-k)T.$$

In a switching period,

$$Q_+ = Q_-, \quad I_{C1-on} = \frac{1-k}{k}I_{C-off} = \frac{1-k}{k}I_O.$$

The charge of capacitor C increases during switch-on and decreases during switch-off:

$$Q_+ = I_{C-on}kT, \quad Q_- = I_{C-off}(1-k)T.$$

In a switching period, $Q_+ = Q_-$.

$$I_{C-on} = I_{C1-on} + I_{LO} = \frac{1-k}{k}I_O + I_O = \frac{1}{k}I_O,$$

$$I_{C-off} = \frac{k}{1-k}I_{C-on} = \frac{k}{1-k}\frac{1}{k}I_O = \frac{1}{1-k}I_O.$$

Therefore,

$$I_L = I_{C-off} = \frac{1}{1-k}I_O.$$

During the switch-off period,

$$i_D = i_L, \quad I_D = I_L = \frac{1}{1-k}I_O.$$

For the current and voltage variations and the boundary condition, we can obtain the following equations using a similar method to that used in the analysis of the self-lift Cúk-converter.

$$\text{Current variations: } \zeta_1 = \frac{k}{2M^2}\frac{R}{fL}, \quad \zeta_2 = \frac{1}{16M}\frac{R}{f^2 L_O C_1}, \quad \xi = \frac{k}{2M^2}\frac{R}{fL}.$$

$$\text{Voltage variations: } \rho = \frac{1}{2k}\frac{1}{fCR}, \quad \sigma_1 = \frac{1}{2M}\frac{1}{fC_1 R}, \quad \varepsilon = \frac{1}{128M}\frac{1}{f^3 L_O C_1 C_O R}.$$

6.2.5.2 Discontinuous Conduction Mode

The reverse self-lift N/O Luo-converter operates in the DCM if the current i_D decreases to zero during switch-off. In the special case when i_D decreases to zero at $t = T$, the circuit operates at the boundary of CCM and DCM.

The variation ratio of current i_D is 1 when the circuit works in the boundary state:

$$\xi = \frac{k}{2M^2}\frac{R}{fL_{eq}} = 1.$$

The boundary between CCM and DCM is

$$M_B = \sqrt{k}\sqrt{\frac{R}{2fL_{eq}}} = \sqrt{\frac{kz_N}{2}},$$

where z_N is the normalized load $R/(fL_{eq})$ and L_{eq} refers to $L_{eq} = L$.

When $M > M_B$, the circuit operates at the DCM. In this case, diode current i_D decreases to zero at $t = t_1 = [k + (1 - k)m]T$, where $KT < t_1 < T$ and $0 < m < 1$ with m being the current filling factor:

$$m = \frac{1}{\xi} = \frac{M^2}{k(R/2fL_{eq})}. \tag{6.53}$$

In the DCM, current i_L increases in the switch-on period kT and decreases in the period from kT to $(1 - k)mT$. The corresponding voltages across L are V_I and $-V_C$.

Therefore,

$$kTV_I = (1 - k)mTV_C$$

and $V_{C1} = V_{CO} = V_O$, $V_{C1} = V_I + V_C$. Hence,

$$V_O = \left[1 + \frac{k}{(1 - k)m}\right]V_I \quad \text{or} \quad V_O = \left(1 + k^2(1 - k)\frac{R}{2fL}\right)V_I. \tag{6.54}$$

The voltage transfer gain in the DCM is

$$M_{DCM} = 1 + k^2(1 - k)\frac{R}{2fL}. \tag{6.55}$$

It can be seen that in DCM, the output voltage increases as the load resistance R increases.

6.2.6 Self-Lift SEPIC

The self-lift SEPIC and its equivalent circuits during the switch-on and switch-off periods are shown in Figure 6.8. It is derived from the SEPIC (with output filter). S and D_1 are on and D is off during the switch-on period, whereas D is on and S and D_1 are off during the switch-off period.

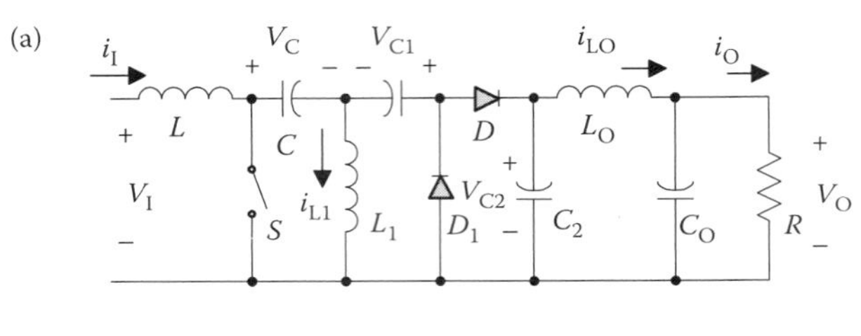

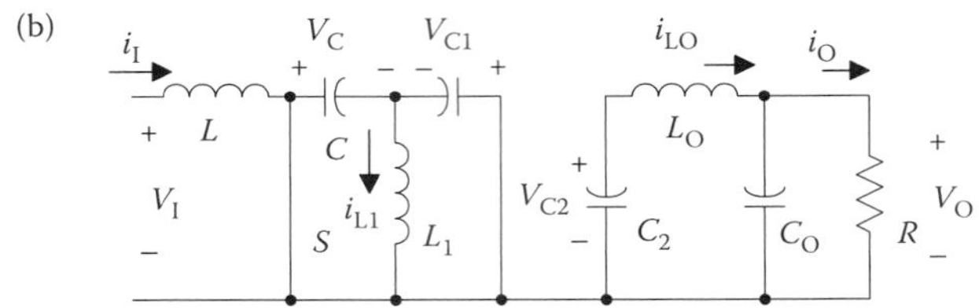

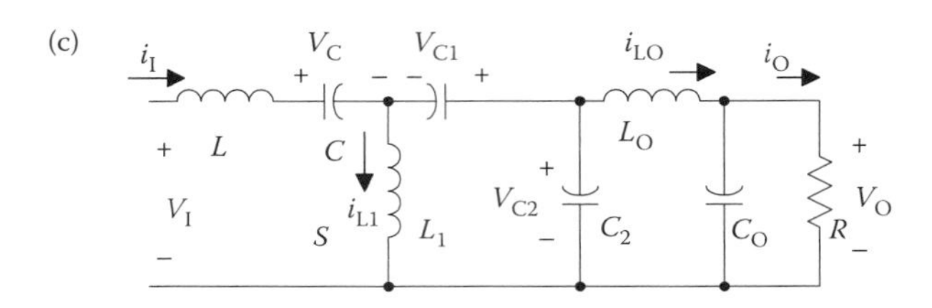

FIGURE 6.8 (a) Self-lift SEPIC converter and its equivalent circuits during (b) switch-on, and (c) switch-off. (Reprinted from Luo, F. L. and Ye, H. 2006. *Essential DC/DC Converters*. Boca Raton: Taylor & Francis Group LLC, p. 67. With permission.)

6.2.6.1 Continuous Conduction Mode

In the steady state, the average voltage across inductor L over a period is zero. Thus $V_C = V_I$.

During the switch-on period, the voltage across capacitor C_1 is equal to the voltage across C. Since we assume that C and C_1 are sufficiently large, $V_{C1} = V_C = V_I$.

In the steady state, the average voltage across inductor L_O over a period is also zero. Thus $V_{C2} = V_{CO} = V_O$.

The inductor current i_L increases in the switch-on period and decreases in the switch-off period. The corresponding voltages across L are V_I and $-(V_C - V_{C1} + V_{C2} - V_I)$.

Therefore

$$kTV_I = (1-k)T(V_C - V_{C1} + V_{C2} - V_I)$$

or

$$kTV_I = (1-k)T(V_O - V_I).$$

Hence,

$$V_O = \frac{1}{1-k}V_I = V_{CO} = V_{C2}. \tag{6.56}$$

The voltage transfer gain in the CCM is

$$M = \frac{V_O}{V_I} = \frac{1}{1-k}. \tag{6.57}$$

Since all the components are considered to be ideal, the power loss associated with all the circuit elements is neglected. Therefore, the output power P_O is considered to be equal to the input power P_{IN}: $V_O I_O = V_I I_I$. Thus,

$$I_I = \frac{1}{1-k} I_O = I_L.$$

The capacitor C_O acts as a low-pass filter so that $I_{LO} = I_O$.

The charge of capacitor C increases during switch-off and decreases during switch-on:

$$Q_- = I_{C-on} kT, \quad Q_+ = I_{C-off}(1-k)T = I_I(1-k)T.$$

In a switching period,

$$Q_+ = Q_-, \quad I_{C-on} = \frac{1-k}{k} I_{C-off} = \frac{1-k}{k} I_I.$$

The charge of capacitor C_2 increases during switch-off and decreases during switch-on:

$$Q_- = I_{C2-on} kT = I_O kT, \quad Q_+ = I_{C2-off}(1-k)T.$$

In a switching period,

$$Q_+ = Q_-, \quad I_{C2-off} = \frac{k}{1-k} I_{C-N} = \frac{k}{1-k} I_O.$$

The charge of capacitor C_1 increases during switch-on and decreases during switch-off:

$$Q_+ = I_{C1-on} kT, \quad Q_- = I_{C1-off}(1-k)T.$$

In a switching period,

$$Q_+ = Q_-, \quad I_{C1-off} = I_{C2-off} + I_{LO} = \frac{k}{1-k} I_O + I_O = \frac{1}{1-k} I_O.$$

Therefore,

$$I_{C1-on} = \frac{1-k}{k} I_{C1-off} = \frac{1}{k} I_O, \quad I_{L1} = I_{C1-on} - I_{C-on} = 0.$$

During switch-off, $i_D = i_L - i_{L1}$.

Therefore,

$$I_D = I_I = \frac{1}{1-k} I_O.$$

For the current and voltage variations and the boundary condition, we can obtain the following equations using a similar method to that used in the analysis of the self-lift Cúk-converter:

$$\text{Current variations: } \zeta_1 = \frac{k}{2M^2} \frac{R}{fL}, \quad \zeta_2 = \frac{k}{16} \frac{R}{f^2 L_O C_2}, \quad \xi = \frac{k}{2M^2} \frac{R}{f L_{eq}},$$

where L_{eq} refers to $L_{eq} = L L_O/(L + L_O)$.

$$\text{Voltage variations: } \rho = \frac{M}{2} \frac{1}{fCR} \quad \sigma_1 = \frac{M}{2} \frac{1}{fC_1 R}, \quad \sigma_2 = \frac{k}{2} \frac{1}{fC_2 R}, \quad \varepsilon = \frac{k}{128} \frac{1}{f^3 L_O C_2 C_O R}.$$

6.2.6.2 Discontinuous Conduction Mode

The self-lift SEPIC converter operates in the DCM if the current i_D decreases to zero during switch-off. As a special case, when i_D decreases to zero at $t = T$, the circuit operates at the boundary of CCM and DCM.

The variation ratio of current i_D is 1 when the circuit works in the boundary state:

$$\xi = \frac{k}{2M^2}\frac{R}{fL_{eq}} = 1.$$

Therefore, the boundary between CCM and DCM is

$$M_B = \sqrt{k}\sqrt{\frac{R}{2fL_{eq}}} = \sqrt{\frac{kz_N}{2}}, \tag{6.58}$$

where z_N is the normalized load $R/(fL_{eq})$ and L_{eq} refers to $L_{eq} = LL_O/(L + L_O)$.

When $M > M_B$, the circuit operates in the DCM. In this case, the diode current i_D decreases to zero at $t = t_1 = [k + (1-k)m]T$, where $KT < t_1 < T$ and $0 < m < 1$. Here, m is defined as

$$m = \frac{1}{\xi} = \frac{M^2}{k(R/2fL_{eq})}. \tag{6.59}$$

In the DCM, current i_L increases during switch-on and decreases in the period from kT to $(1-k)mT$. The corresponding voltages across L are V_I and $-(V_C - V_{C1} + V_{C2} - V_I)$. Thus,

$$kTV_I = (1-k)T(V_C - V_{C1} + V_{C2} - V_I)$$

and $V_C = V_I$, $V_{C1} = V_C = V_I$, $V_{C2} = V_{CO} = V_O$.

Hence,

$$V_O = \left[1 + \frac{k}{(1-k)m}\right]V_I \quad \text{or} \quad V_O = \left(1 + k^2(1-k)\frac{R}{2fL_{eq}}\right)V_I.$$

So the real DC voltage transfer gain in the DCM is

$$M_{DCM} = 1 + k^2(1-k)\frac{R}{2fL_{eq}}. \tag{6.60}$$

In DCM, the output voltage increases as the load resistance R increases.

6.2.7 Enhanced Self-Lift P/O Luo-Converter

Enhanced self-lift P/O Luo-converter circuit and the equivalent circuits during the switch-on and switch-off periods are shown in Figure 6.9. It is derived from the self-lift P/O Luo-converter in Figure 6.4 with swapping of the positions of switch S and inductor L.

During the switch-on period, S and D_1 are on and D is off. We obtain

$$V_C = V_{C1} \quad \text{and} \quad \Delta i_L = \frac{V_I}{L}kT.$$

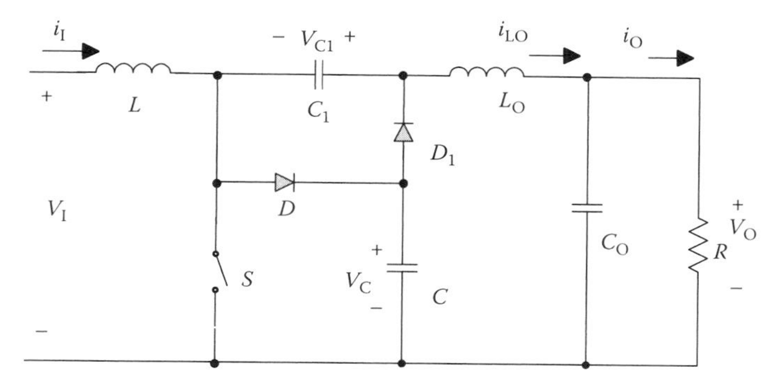

FIGURE 6.9 Enhanced self-lift P/O Luo-converter. (Reprinted from Luo, F. L. and Ye, H. 2006. *Essential DC/DC Converters*. Boca Raton: Taylor & Francis Group LLC, p. 71. With permission.)

During the switch-off period, D is on and S and D_1 are off:

$$\Delta i_L = \frac{V_C - V_I}{L}(1-k)T,$$

so that

$$V_C = \frac{1}{1-k}V_I.$$

The output voltage and current and the voltage transfer gain are

$$V_O = V_I + V_{C1} = \left(1 + \frac{1}{1-k}\right)V_I, \tag{6.61}$$

$$I_O = \frac{1-k}{2-k}I_I, \tag{6.62}$$

$$M = 1 + \frac{1}{1-k} = \frac{2-k}{1-k}. \tag{6.63}$$

Average voltages are

$$V_C = \frac{1}{1-k}V_I, \tag{6.64}$$

$$V_{C1} = \frac{1}{1-k}V_I. \tag{6.65}$$

Average currents are

$$I_{LO} = I_O, \tag{6.66}$$

$$I_L = \frac{2-k}{1-k}I_O = I_I. \tag{6.67}$$

Therefore,

$$\frac{V_O}{V_I} = \frac{1}{1-k} + 1 = \frac{2-k}{1-k}. \tag{6.68}$$

6.3 P/O Luo-Converters

P/O Luo-converters perform the voltage conversion from positive to positive voltages using the VL technique. They work in the first quadrant with large voltage amplification. Five circuits have been introduced in the literature [6–11]:

- Elementary circuit
- Self-lift circuit
- Re-lift circuit
- Triple-lift circuit
- Quadruple-lift circuit

The elementary circuit is discussed in Section 5.5.1 and the self-lift circuit is discussed in Section 6.2.2.

6.3.1 Re-Lift Circuit

The re-lift circuit and its equivalent switch-on and switch-off circuits are shown in Figure 6.10, which is derived from the self-lift circuit. Capacitors C_1 and C_2 perform characteristics to lift the capacitor voltage V_C to a level 2 times higher than the source voltage V_I. L_3 performs the function of a *ladder joint* to link the two capacitors C_1 and C_2 and lift the capacitor voltage V_C up.

When switches S and S_1 are turned on, the source's instantaneous current $i_I = i_{L1} + i_{L2} + i_{C1} + i_{L3} + i_{C2}$. Inductors L_1 and L_3 absorb energy from the source. In the meantime, inductor L_2 absorbs energy from the source and capacitor C. Three currents i_{L1}, i_{L3}, and i_{L2} increase. When switches S and S_1 turn off, the source current $i_I = 0$. Current i_{L1} flows through capacitor C_1, inductor L_3, capacitor C_2, and diode D to charge capacitor C. Inductor L_1 transfers its SE to capacitor C. In the meantime, current i_{L2} flows through the $(C_O - R)$ circuit, capacitor C_1, inductor L_3, capacitor C_2, and diode D to keep itself continuous. Both currents i_{L1} and i_{L2} decrease. In order to analyze the progress of the working of the circuit, the equivalent circuits in switch-on and switch-off states are shown in Figure 6.10b–d. Assume that capacitors C_1 and C_2 are sufficiently large, and the voltages V_{C1} and V_{C2} across them are equal to V_I in steady state.

Voltage v_{L3} is equal to V_I during switch-on. The peak-to-peak variation of current i_{L3} is

$$\Delta i_{L3} = \frac{V_I kT}{L_3}. \tag{6.69}$$

This variation is equal to the current reduction during switch-off. Suppose that its voltage is $-V_{L3-off}$, then

$$\Delta i_{L3} = \frac{V_{L3-off}(1-k)T}{L_3}.$$

Thus, during switch-off, the voltage-drop across inductor L_3 is

$$V_{L3-off} = \frac{k}{1-k} V_I. \tag{6.70}$$

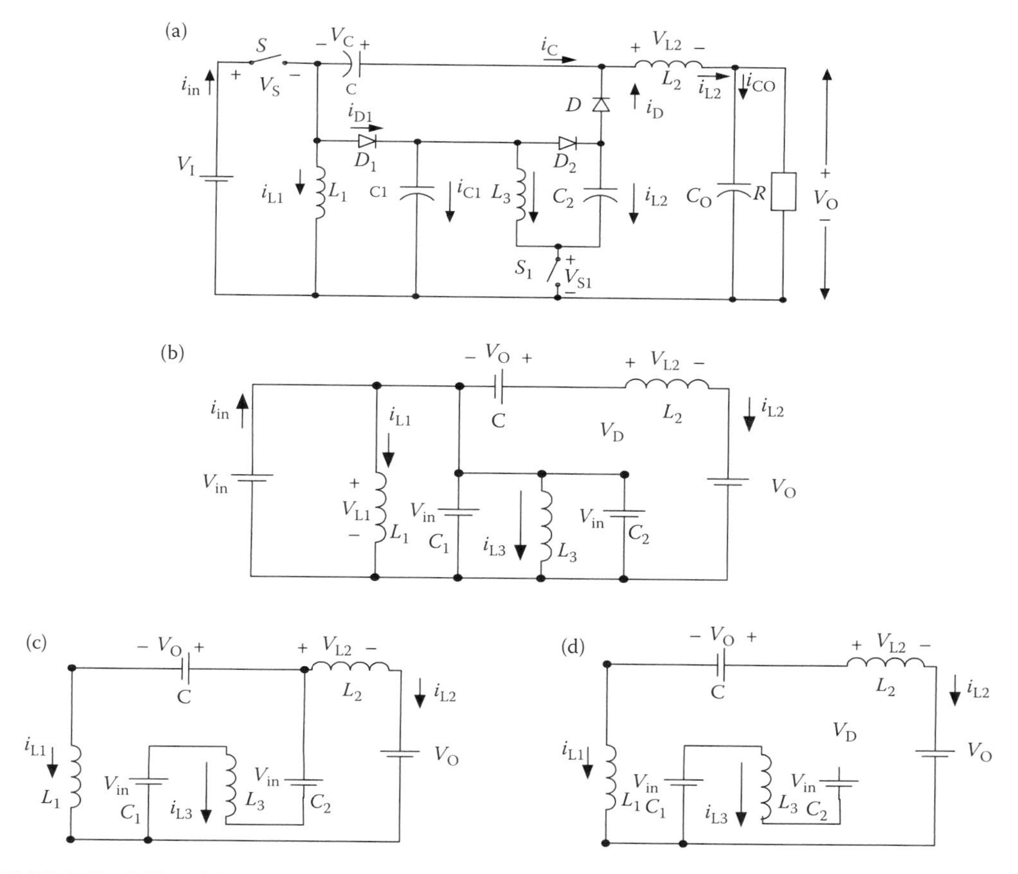

FIGURE 6.10 P/O re-lift circuit (a) circuit diagram, (b) switch-on (c) switch-off, and (d) discontinuous mode. (Reprinted from Luo, F. L. and Ye, H. 2006. *Essential DC/DC Converters*. Boca Raton: Taylor & Francis Group LLC, p. 97. With permission.)

Current i_{L1} increases in the switch-on period kT, and decreases in the switch-off period $(1-k)T$. The corresponding voltages applied across L_1 are V_I and $-(V_C - 2V_I - V_{L3-off})$. Therefore,

$$kTV_I = (1-k)T(V_C - 2V_I - V_{L3-off}).$$

Hence,

$$V_C = \frac{2}{1-k}V_I. \tag{6.71}$$

Current i_{L2} increases in the switch-on period kT, and it decreases in the switch-off period $(1-k)T$. The corresponding voltages applied across L_2 are $(V_I + V_C - V_O)$ and $-(V_O - 2V_I - V_{L3-off})$. Therefore,

$$kT(V_C + V_I - V_O) = (1-k)T(V_O - 2V_I - V_{L3-off}).$$

Hence,

$$V_O = \frac{2}{1-k}V_I \tag{6.72}$$

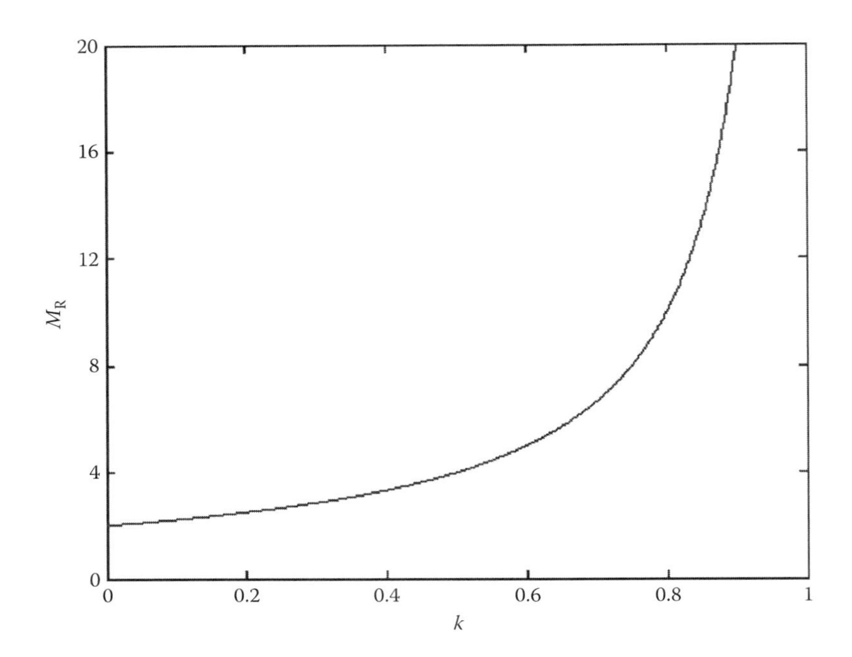

FIGURE 6.11 Voltage transfer gain M_R versus k. (Reprinted from Luo, F. L. and Ye, H. 2006. *Essential DC/DC Converters*. Boca Raton: Taylor & Francis Group LLC, p. 99. With permission.)

and the output current is

$$I_O = \frac{1-k}{2} I_I. \tag{6.73}$$

The voltage transfer gain in the continuous mode is

$$M_R = \frac{V_O}{V_I} = \frac{2}{1-k}. \tag{6.74}$$

The curve of M_R versus k is shown in Figure 6.11.

Other average currents are

$$I_{L1} = \frac{k}{1-k} I_O = \frac{k}{2} I_I \tag{6.75}$$

and

$$I_{L3} = I_{L1} + I_{L2} = \frac{1}{1-k} I_O. \tag{6.76}$$

Currents i_{C1} and i_{C2} are equal to $(i_{L1} + i_{L2})$ during the *switch-off* period $(1-k)T$, and the charges on capacitors C_1 and C_2 decrease, that is,

$$i_{C1} = i_{C2} = (i_{L1} + i_{L2}) = \frac{1}{1-k} I_O.$$

The charges increase during the *switch-on* period kT, so their average currents are

$$I_{C1} = I_{C2} = \frac{1-k}{k}(I_{L1} + I_{L2}) = \frac{1-k}{k}\left(\frac{k}{1-k} + 1\right) I_O = \frac{I_O}{k}. \tag{6.77}$$

During switch-off, the source current i_I is 0, and in the switch-on period kT, it is

$$i_I = i_{L1} + i_{L2} + i_{C1} + i_{L3} + i_{C2}.$$

Hence,

$$\begin{aligned} I_I = ki_I &= k(I_{L1} + I_{L2} + I_{C1} + I_{L3} + I_{C2}) = k[2(I_{L1} + I_{L2}) + 2I_{C1}] \\ &= 2k(I_{L1} + I_{L2})\left(1 + \frac{1-k}{k}\right) = 2k\frac{I_{L2}}{1-k}\frac{1}{k} = \frac{2}{1-k}I_O. \end{aligned} \quad (6.78)$$

6.3.1.1 Variations of Currents and Voltages

Current i_{L1} increases and is supplied by V_I during the switch-on period kT. It decreases and is inversely biased by $-(V_C - 2V_I - V_{L3})$ during the switch-off period $(1-k)T$. Therefore, its peak-to-peak variation is

$$\Delta i_{L1} = \frac{kTV_I}{L_1}.$$

The variation ratio of current i_{L1} is

$$\xi_1 = \frac{\Delta i_{L1}/2}{I_{L1}} = \frac{kV_I T}{kL_1 I_I} = \frac{1-k}{2M_R}\frac{R}{fL_1}. \quad (6.79)$$

Current i_{L2} increases and is supplied by the voltage $(V_I + V_C - V_O) = V_I$ during the switch-on period kT. It decreases and is inversely biased by $-(V_O - 2V_I - V_{L3})$ during switch-off. Therefore, its peak-to-peak variation is

$$\Delta i_{L_2} = \frac{kTV_I}{L_2}.$$

The variation ratio of current i_{L2} is

$$\xi_2 = \frac{\Delta i_{L2}/2}{I_{L2}} = \frac{kTV_I}{2L_2 I_O} = \frac{k}{2M_R}\frac{R}{fL_2}. \quad (6.80)$$

When the switch is off, the freewheeling diode current $i_D = i_{L1} + i_{L2}$ and

$$\Delta i_D = \Delta i_{L3} = \Delta i_{L1} + \Delta i_{L2} = \frac{kTV_I}{L} = \frac{k(1-k)V_O}{2L}T. \quad (6.81)$$

Since $I_D = I_{L1} + I_{L2} = I_O/1-k$, the variation ratio of current i_D is

$$\zeta = \frac{\Delta i_D/2}{I_D} = \frac{k(1-k)^2 TV_O}{4LI_O} = \frac{k(1-k)R}{2M_R fL} = \frac{k}{M_R^2}\frac{R}{fL}. \quad (6.82)$$

The variation ratio of current i_{L3} is

$$\chi_1 = \frac{\Delta i_{L3}/2}{I_{L3}} = \frac{kV_I T}{2L_3(1/1-k)I_O} = \frac{k}{M_R^2}\frac{R}{fL_3}. \quad (6.83)$$

The peak-to-peak variation of v_C is

$$\Delta v_C = \frac{Q+}{C} = \frac{1-k}{C} TI_{L1} = \frac{k(1-k)}{2C} TI_I.$$

Considering Equation 6.71, the variation ratio is

$$\rho = \frac{\Delta v_C/2}{V_C} = \frac{k(1-k)TI_I}{4CV_O} = \frac{k}{2fCR}. \tag{6.84}$$

The charges on capacitors C_1 and C_2 increase during the switch-on period kT, and decrease during the switch-off period $(1-k)T$ because of the current $(I_{L1}+I_{L2})$. Therefore, their peak-to-peak variations are

$$\Delta v_{C1} = \frac{(1-k)T(I_{L1}+I_{L2})}{C_1} = \frac{(1-k)I_I}{2C_1 f},$$

$$\Delta v_{C2} = \frac{(1-k)T(I_{L1}+I_{L2})}{C_2} = \frac{(1-k)I_I}{2C_2 f}.$$

Considering $V_{C1} = V_{C2} = V_I$, the variation ratios of voltages v_{C1} and v_{C2} are

$$\sigma_1 = \frac{\Delta v_{C1}/2}{V_{C1}} = \frac{(1-k)I_I}{4fC_1V_I} = \frac{M_R}{2fC_1R}, \tag{6.85}$$

$$\sigma_2 = \frac{\Delta v_{C2}/2}{V_{C2}} = \frac{(1-k)I_I}{4V_IC_2f} = \frac{M_R}{2fC_2R}. \tag{6.86}$$

Analogously, the variation ratio of output voltage v_O is

$$\varepsilon = \frac{\Delta v_O/2}{V_O} = \frac{kT^2}{16C_OL_2}\frac{V_I}{V_O} = \frac{k}{16M_R}\frac{1}{f^2C_OL_2}. \tag{6.87}$$

Example 6.2

A P/O re-lift Luo-converter has the following components: $V_I = 20$V, $L_1 = L_2 = 1$ mH, $L_3 = 0.5$ mH, and all capacitors have 20 μF, $R = 160\,\Omega$, $f = 50$ kHz, and $k = 0.5$. Calculate the output voltage and the variation ratios $\xi_1, \xi_2, \zeta, \chi_1, \rho, \sigma_1, \sigma_2$, and ε in steady state.

SOLUTION

From Equation 6.72, we obtain the output voltage as

$$V_O = \frac{2}{1-k}V_I = \frac{2}{1-0.5}20 = 80\,\text{V}.$$

The variation ratios are $\xi_1 = 0.2$, $\xi_2 = 0.2$, $\zeta = 0.1$, $\chi_1 = 0.1$, $\rho = 0.0016$ $\sigma_1 = 0.0125$, $\sigma_2 = 0.0125$, and $\varepsilon = 1.56 \times 10^{-4}$. Therefore, the variations are small.

From the example, we know the variations are small. Therefore, the output voltage v_O is almost a real DC voltage with very small ripples. Because of the resistive load, the output current $i_O(t)$ is almost a real DC waveform with very small ripples as well, and $I_O = V_O/R$.

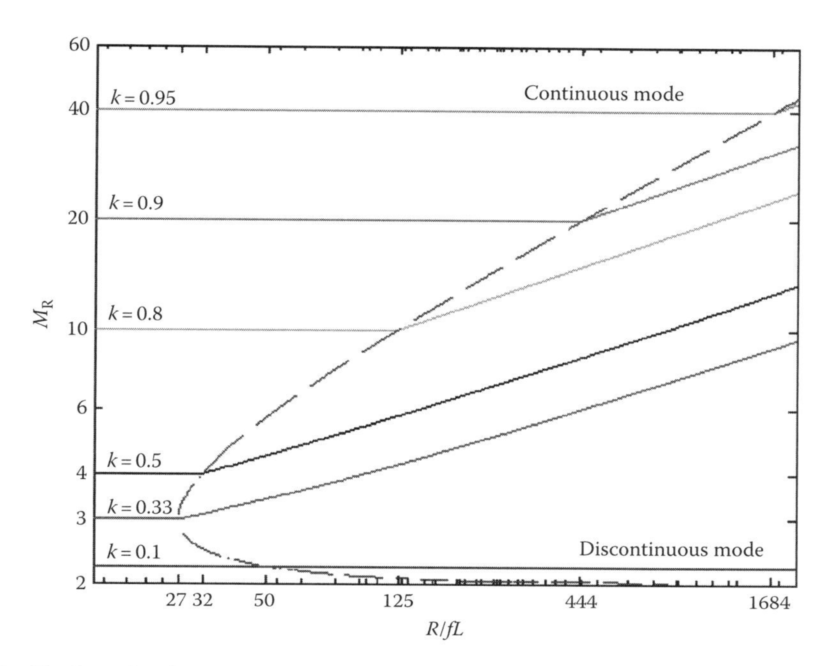

FIGURE 6.12 The boundary between continuous and discontinuous modes and the output voltage versus the normalized load $z_N = R/fL$. (Reprinted from Luo, F. L. and Ye, H. 2006. *Essential DC/DC Converters*. Boca Raton: Taylor & Francis Group LLC, p. 106. With permission.)

For DCM, referring to Figure 6.10d, we can see that the diode current i_D becomes zero during switch-off before the next period switch-on. The condition for the DCM is $\zeta \geq 1$, that is,

$$\frac{k}{M_R^2}\frac{R}{fL} \geq 1,$$

or

$$M_R \leq \sqrt{k}\sqrt{\frac{R}{fL}} = \sqrt{k}\sqrt{z_N}. \tag{6.88}$$

The graph of the boundary curve versus the normalized load $z_N = R/fL$ is shown in Figure 6.12. It can be seen that the boundary curve has a minimum value of 3.0 at $k = 1/3$.

In this case, the current i_D exists in the period between kT and $t_1 = [k + (1-k)m_R]T$, where m_R is the *filling efficiency* and is defined as

$$m_R = \frac{1}{\zeta} = \frac{M_R^2}{k(R/fL)}. \tag{6.89}$$

Therefore, $0 < m_R < 1$. Since the diode current i_D becomes zero at $t = t_1 = kT + (1-k)m_R T$, for the current i_L

$$kTV_I = (1-k)m_R T(V_C - 2V_I - V_{L3-off})$$

or

$$V_C = \left[2 + \frac{k}{1-k} + \frac{k}{(1-k)m_R}\right]V_I = \left[2 + \frac{k}{1-k} + k^2(1-k)\frac{R}{4fL}\right]V_I \quad \text{with } \sqrt{k}\sqrt{\frac{R}{fL}} \geq \frac{2}{1-k},$$

and for the current i_{LO}

$$kT(V_I + V_C - V_O) = (1-k)m_R T(V_O - 2V_I - V_{L3-off}).$$

Therefore, the output voltage in the discontinuous mode is

$$V_O = \left[2 + \frac{k}{1-k} + \frac{k}{(1-k)m_R}\right]V_I = \left[2 + \frac{k}{1-k} + k^2(1-k)\frac{R}{4fL}\right]V_I$$
$$\text{with} \quad \sqrt{k}\sqrt{\frac{R}{fL}} \geq \frac{2}{1-k}. \tag{6.90}$$

That is, the output voltage linearly increases as the load resistance R increases. The output voltage versus the normalized load $z_N = R/fL$ is shown in Figure 6.12. Larger load resistance R may cause higher output voltage in the discontinuous mode.

6.3.2 Triple-Lift Circuit

The triple-lift circuit, shown in Figure 6.13, consists of two static switches S and S_1, four inductors L_1, L_2, L_3, and L_4, five capacitors C, C_1, C_2, C_3, and C_O, and five diodes. Capacitors C_1, C_2, and C_3 perform characteristics to lift the capacitor voltage V_C to a level 3 times higher than the source voltage V_I. L_3 and L_4 perform the function of ladder joints to link the capacitors C_1, C_2, and C_3 and lift the capacitor voltage V_C up. Currents $i_{C1}(t)$, $i_{C2}(t)$, and $i_{C3}(t)$ are exponential functions. They have large values at the moment of switching power on, but they are small because $v_{C1} = v_{C2} = v_{C3} = V_I$ in steady state.

The output voltage and current are

$$V_O = \frac{3}{1-k}V_I \tag{6.91}$$

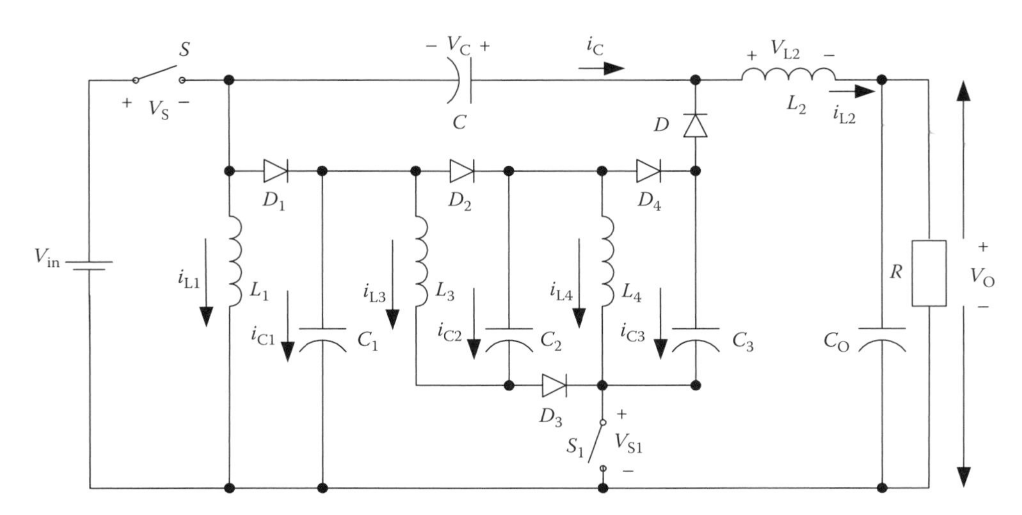

FIGURE 6.13 Triple-lift circuit. (Reprinted from Luo, F. L. and Ye, H. 2006. *Essential DC/DC Converters*. Boca Raton: Taylor & Francis Group LLC, p. 110. With permission.)

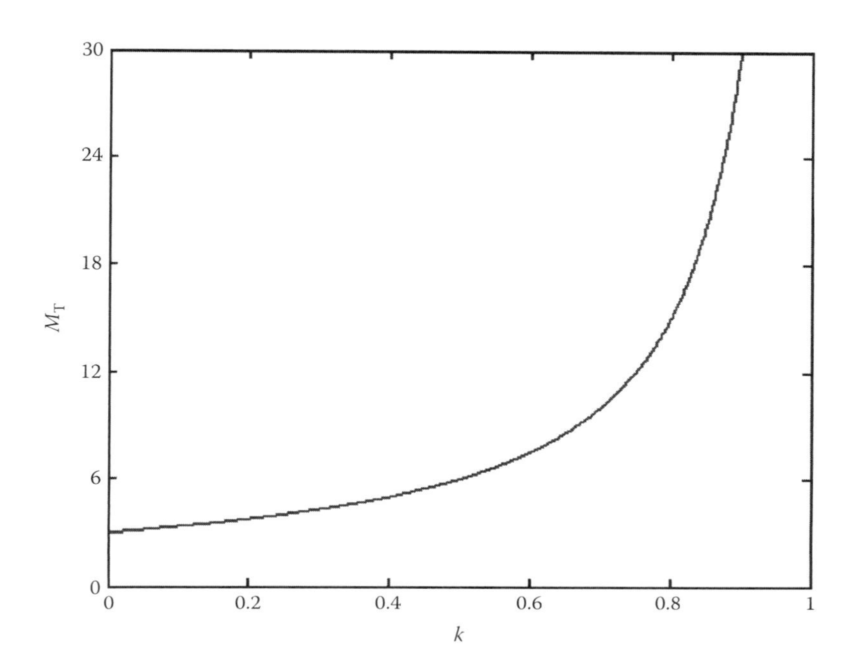

FIGURE 6.14 Voltage transfer gain M_T versus k. (Reprinted from Luo, F. L. and Ye, H. 2006. *Essential DC/DC Converters*. Boca Raton: Taylor & Francis Group LLC, p. 111. With permission.)

and

$$I_O = \frac{1-k}{3} I_I. \tag{6.92}$$

The voltage transfer gain in the continuous mode is

$$M_T = \frac{V_O}{V_I} = \frac{3}{1-k}. \tag{6.93}$$

The curve of M_T versus k is shown in Figure 6.14.

Other average voltages: $V_C = V_O, \quad V_{C1} = V_{C2} = V_{C3} = V_I$.

Other average currents: $I_{L2} = I_O, \quad I_{L1} = \frac{k}{1-k} I_O, \quad I_{L3} = I_{L4} = I_{L1} + I_{L2} = \frac{1}{1-k} I_O$.

Current variations: $\xi_1 = \frac{1-k}{2M_T} \frac{R}{fL_1}, \quad \xi_2 = \frac{k}{2M_T} \frac{R}{fL_2}, \quad \zeta = \frac{k(1-k)R}{2M_T fL} = \frac{k}{M_T^2} \frac{3R}{2fL},$

$$\chi_1 = \frac{k}{M_T^2} \frac{3R}{2fL_3}, \quad \chi_2 = \frac{k}{M_T^2} \frac{3R}{2fL_4}.$$

Voltage variations: $\rho = \frac{k}{2fCR}, \quad \sigma_1 = \frac{M_T}{2fC_1R}, \quad \sigma_2 = \frac{M_T}{2fC_2R}, \quad \sigma_3 = \frac{M_T}{2fC_3R}.$

The variation ratio of output voltage v_C is

$$\varepsilon = \frac{k}{16M_T} \frac{1}{f^2 C_O L_2}. \tag{6.94}$$

The output voltage ripple is very small.

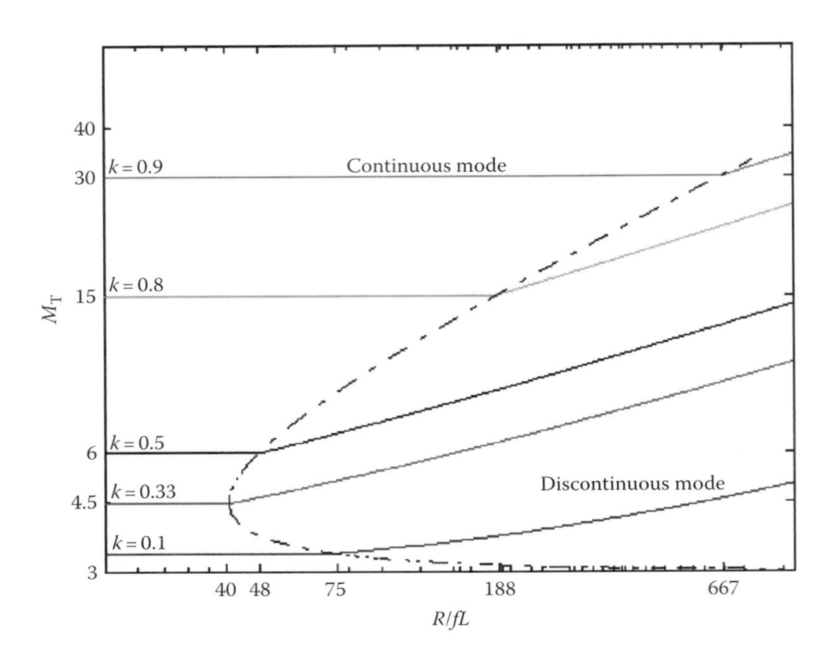

FIGURE 6.15 The boundary between continuous and discontinuous modes and the output voltage versus the normalized load $z_N = R/fL$. (Reprinted from Luo, F. L. and Ye, H. 2006. *Essential DC/DC Converters*. Boca Raton: Taylor & Francis Group LLC, p. 113. With permission.)

The boundary between CCM and DCM is

$$M_T \le \sqrt{k}\sqrt{\frac{3R}{2fL}} = \sqrt{\frac{3kz_N}{2}}. \tag{6.95}$$

This boundary curve is shown in Figure 6.15. It can be seen that the boundary curve has a minimum value of M_T that is equal to 4.5, corresponding to $k = 1/3$.

In the discontinuous mode, the current i_D exists in the period between kT and $t_1 = [k + (1-k)m_T]T$, where m_T is the filling efficiency, that is,

$$m_T = \frac{1}{\zeta} = \frac{M_T^2}{k(3R/2fL)}. \tag{6.96}$$

The diode current i_D becomes zero at $t = t_1 = kT + (1-k)m_T T$; therefore, $0 < m_T < 1$.

For the current i_{L1},

$$kTV_I = (1-k)m_T T(V_C - 3V_I - V_{L3-off} - V_{L4-off})$$

or

$$V_C = \left[3 + \frac{2k}{1-k} + \frac{k}{(1-k)m_T}\right]V_I = \left[3 + \frac{2k}{1-k} + k^2(1-k)\frac{R}{6fL}\right]V_I$$

$$\text{with} \quad \sqrt{k}\sqrt{\frac{3R}{2fL}} \ge \frac{3}{1-k},$$

and for the current i_{L2}, $\quad kT(V_I + V_C - V_O) = (1-k)m_T T(V_O - 2V_I - V_{L3-off} - V_{L4-off})$.

Therefore, the output voltage in the discontinuous mode is

$$V_O = \left[3 + \frac{2k}{1-k} + \frac{k}{(1-k)m_T}\right] V_I = \left[3 + \frac{2k}{1-k} + k^2(1-k)\frac{R}{6fL}\right] V_I$$
$$\text{with} \quad \sqrt{k}\sqrt{\frac{3R}{2fL}} \geq \frac{3}{1-k}. \tag{6.97}$$

That is, the output voltage linearly increases as the load resistance R increases, as shown in Figure 6.15.

6.3.3 Quadruple-Lift Circuit

The quadruple-lift circuit, shown in Figure 6.16, consists of two static switches S and S_1, five inductors L_1, L_2, L_3, L_4, and L_5, six capacitors C, C_1, C_2, C_3, C_4, and C_O, and seven diodes. Capacitors C_1, C_2, C_3, and C_4 perform characteristics to lift the capacitor voltage V_C to a level 4 times higher than the source voltage V_I. L_3, L_4, and L_5 perform the function of ladder joints to link the capacitors C_1, C_2, C_3, and C_4, and lift the output capacitor voltage V_C up. Current $i_{C1}(t)$, $i_{C2}(t)$, $i_{C3}(t)$, and $i_{C4}(t)$ are exponential functions. They have large values at the moment of power on, but they are small because $v_{C1} = v_{C2} = v_{C3} = v_{C4} = V_I$ in steady state.

The output voltage and current are

$$V_O = \frac{4}{1-k} V_I \tag{6.98}$$

and

$$I_O = \frac{1-k}{4} I_I. \tag{6.99}$$

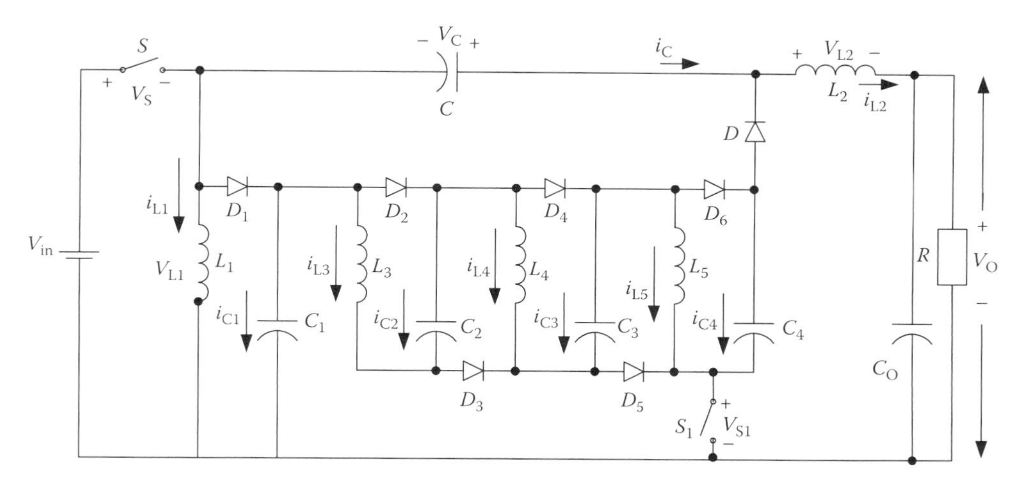

FIGURE 6.16 Quadruple-lift circuit. (Reprinted from Luo, F. L. and Ye, H. 2006. *Essential DC/DC Converters*. Boca Raton: Taylor & Francis Group LLC, p. 114. With permission.)

The voltage transfer gain in the continuous mode is

$$M_Q = \frac{V_O}{V_I} = \frac{4}{1-k}. \tag{6.100}$$

The curve of M_Q versus k is shown in Figure 6.17.

Other average voltages: $V_C = V_O, \quad V_{C1} = V_{C2} = V_{C3} = V_{C4} = V_I.$

Other average currents: $I_{L2} = I_O, \quad I_{L1} = \frac{k}{1-k} I_O,$

$$I_{L3} = I_{L4} = L_{L5} = I_{L1} + I_{L2} = \frac{1}{1-k} I_O.$$

Inductor current variations: $\xi_1 = \frac{1-k}{2M_Q}\frac{R}{fL_1}, \quad \xi_2 = \frac{k}{2M_Q}\frac{R}{fL_2},$

$$\zeta = \frac{k(1-k)R}{2M_Q fL} = \frac{k}{M_Q^2}\frac{2R}{fL} \quad \chi_1 = \frac{k}{M_Q^2}\frac{2R}{fL_3},$$

$$\chi_2 = \frac{k}{M_Q^2}\frac{2R}{fL_4}, \quad \chi_3 = \frac{k}{M_Q^2}\frac{2R}{fL_5}.$$

Capacitor voltage variations: $\rho = \frac{k}{2fCR} \quad \sigma_1 = \frac{M_Q}{2fC_1R} \quad \sigma_2 = \frac{M_Q}{2fC_2R}$

$$\sigma_3 = \frac{M_Q}{2fC_3R} \quad \sigma_4 = \frac{M_Q}{2fC_4R}.$$

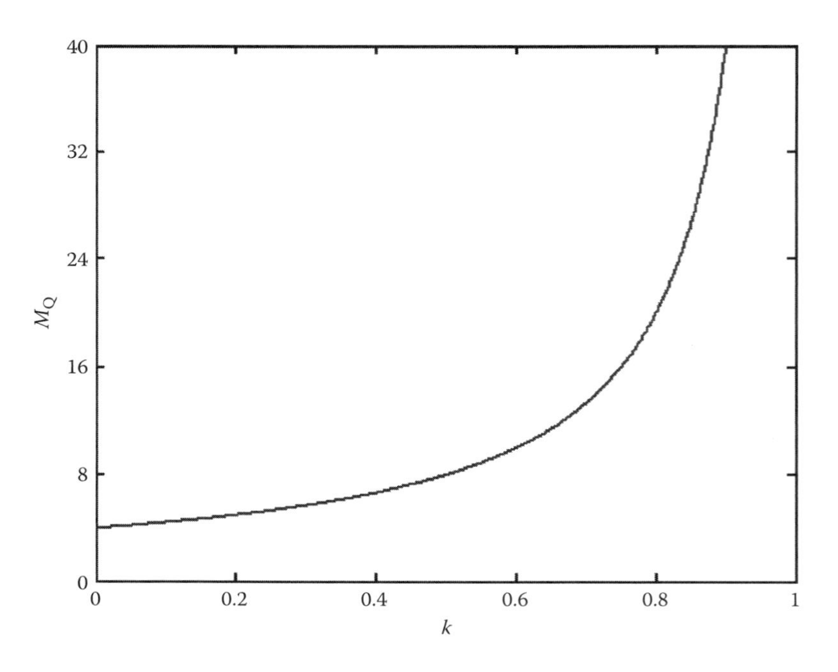

FIGURE 6.17 Voltage transfer gain M_Q versus k. (Reprinted from Luo, F. L. and Ye, H. 2006. *Essential DC/DC Converters*. Boca Raton: Taylor & Francis Group LLC, p. 115. With permission.)

The variation ratio of output voltage V_C is

$$\varepsilon = \frac{k}{16M_Q}\frac{1}{f^2C_OL_2}. \tag{6.101}$$

The output voltage ripple is very small.

The boundary between continuous and discontinuous modes is

$$M_Q \le \sqrt{k}\sqrt{\frac{2R}{fL}} = \sqrt{2kz_N}. \tag{6.102}$$

This boundary curve is shown in Figure 6.18. It can be seen that it has a minimum value of M_Q that is equal to 6.0, corresponding to $k = 1/3$.

In the discontinuous mode, the current i_D exists in the period between kT and $t_1 = [k + (1 - k)m_Q]T$, where m_Q is the filling efficiency, that is,

$$m_Q = \frac{1}{\zeta} = \frac{M_Q^2}{k(2R/fL)}. \tag{6.103}$$

The current i_D becomes zero at $t = t_1 = kT + (1 - k)m_QT$; therefore, $0 < m_Q < 1$. For the current i_{L1}, we have

$$kTV_I = (1 - k)m_QT(V_C - 4V_I - V_{L3-off} - V_{L4-off} - V_{L5-off}),$$

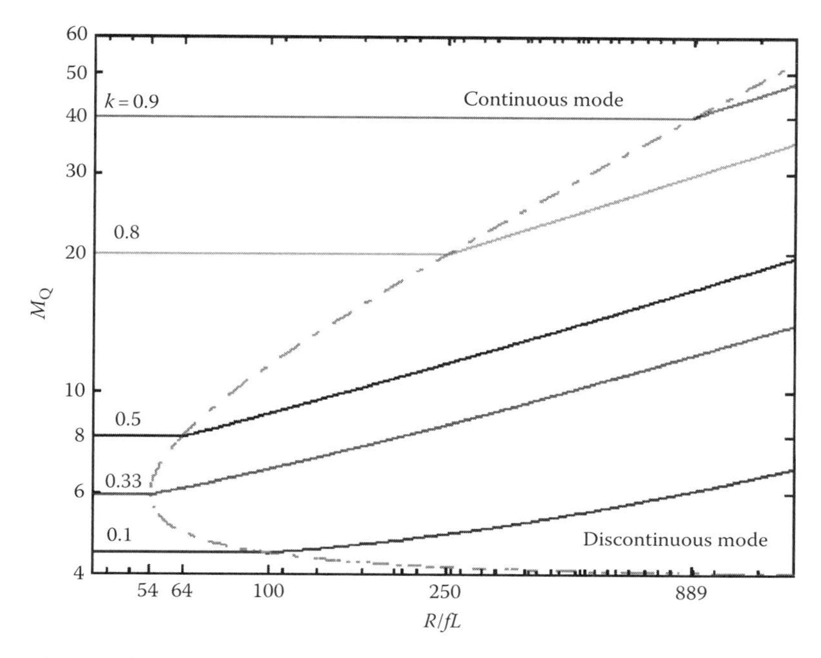

FIGURE 6.18 The boundary between continuous and discontinuous modes and the output voltage versus the normalized load $z_N = R/fL$. (Reprinted from Luo, F. L. and Ye, H. 2006. *Essential DC/DC Converters*. Boca Raton: Taylor & Francis Group LCC, p. 116. With permission.)

or

$$V_C = \left[4 + \frac{3k}{1-k} + \frac{k}{(1-k)m_Q}\right] V_I = \left[4 + \frac{3k}{1-k} + k^2(1-k)\frac{R}{8fL}\right] V_I$$

$$\text{with} \quad \sqrt{k}\sqrt{\frac{2R}{fL}} \geq \frac{4}{1-k},$$

and for current i_{L2}, we have

$$kT(V_I + V_C - V_O) = (1-k)m_Q T(V_O - 2V_I - V_{L3-off} - V_{L4-off} - V_{L5-off}).$$

Therefore, the output voltage in the discontinuous mode is

$$V_O = \left[4 + \frac{3k}{1-k} + \frac{k}{(1-k)m_Q}\right] V_I = \left[4 + \frac{3k}{1-k} + k^2(1-k)\frac{R}{8fL}\right] V_I$$

$$\text{with} \quad \sqrt{k}\sqrt{\frac{2R}{fL}} \geq \frac{4}{1-k}. \tag{6.104}$$

That is, the output voltage increases linearly as the load resistance R increases, as shown in Figure 6.18.

6.3.4 Summary

From the analysis and calculation in previous sections, the common formulae for all circuits can be obtained:

$$M = \frac{V_O}{V_I} = \frac{I_I}{I_O}, \quad L = \frac{L_1 L_2}{L_1 + L_2}, \quad z_N = \frac{R}{fL}, \quad R = \frac{V_O}{I_O}.$$

$$\text{Inductor current variations: } \xi_1 = \frac{1-k}{2M}\frac{R}{fL_1}, \quad \xi_2 = \frac{k}{2M}\frac{R}{fL_2}, \quad \chi_i = \frac{k}{M^2}\frac{n}{2}\frac{R}{fL_{i+2}},$$

where i is the component number ($i = 1, 2, 3, \ldots, n-1$), and n the stage number.

$$\text{Capacitor voltage variations: } \rho = \frac{k}{2fCR}; \quad \varepsilon = \frac{k}{16M}\frac{1}{f^2 C_O L_2};$$

$$\sigma_i = \frac{M}{2fC_i R}, \quad i = 1, 2, 3, 4, \ldots, n.$$

In order to write common formulae for the boundaries between continuous and discontinuous modes and output voltage for all circuits, the circuits can be numbered. The definition is that subscript $n = 0$ denotes the elementary circuit, 1 denotes the self-lift circuit, 2 denotes the re-lift circuit, 3 denotes the triple-lift circuit, 4 denotes the quadruple-lift circuit, and so on. The voltage transfer gain is

$$M_n = \frac{n + kh(n)}{1-k}, \quad n = 0, 1, 2, 3, 4, \ldots. \tag{6.105}$$

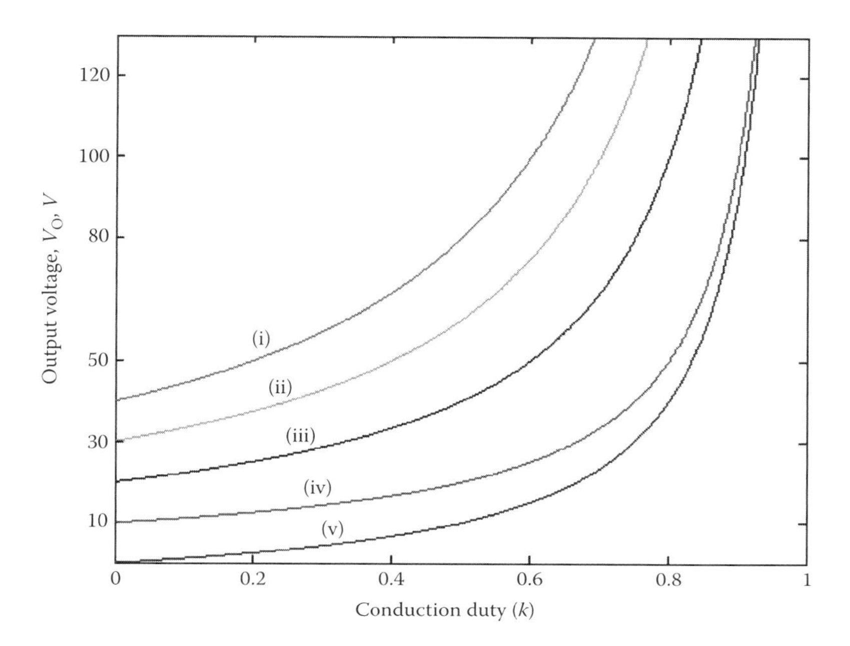

FIGURE 6.19 Output voltages of all P/O Luo-converters ($V_I = 10\,V$). (i) Quadruple-lift circuit; (ii) triple-lift circuit; (iii) re-lift circuit; (iv) self-lift circuit; and (v) elementary circuit. (Reprinted from Luo, F. L. and Ye, H. 2006. *Essential DC/DC Converters*. Boca Raton: Taylor & Francis Group LLC, p. 120. With permission.)

Assuming that $f = 50\,\text{kHz}, L_1 = L_2 = 1\,\text{mH}, L_2 = L_3 = L_4 = L_5 = 0.5\,\text{mH}, C = C_1 = C_2 = C_3 = C_4 = C_O = 20\,\mu\text{F}$, and the source voltage $V_I = 10\,V$, the values of the output voltage V_O with various conduction duty cycles k in the continuous mode are shown in Figure 6.19. The variation of freewheeling diode current i_D is given by

$$\zeta_n = \frac{k^{[1+h(n)]}}{M_n^2}\frac{n+h(n)}{2}z_N. \tag{6.106}$$

The boundaries are determined by the condition:

$$\zeta_n \geq 1$$

or

$$\frac{k^{[1+h(n)]}}{M_n^2}\frac{n+h(n)}{2}z_N \geq 1, \quad n = 0,1,2,3,4,\ldots. \tag{6.107}$$

Therefore, the boundaries between continuous and discontinuous modes for all circuits are

$$M_n = k^{(1+h(n))/2}\sqrt{\frac{n+h(n)}{2}z_N}, \quad n = 0,1,2,3,4,\ldots. \tag{6.108}$$

The filling efficiency is

$$m_n = \frac{1}{\zeta_n} = \frac{M_n^2}{k^{[1+h(n)]}}\frac{2}{n+h(n)}\frac{1}{z_N}. \tag{6.109}$$

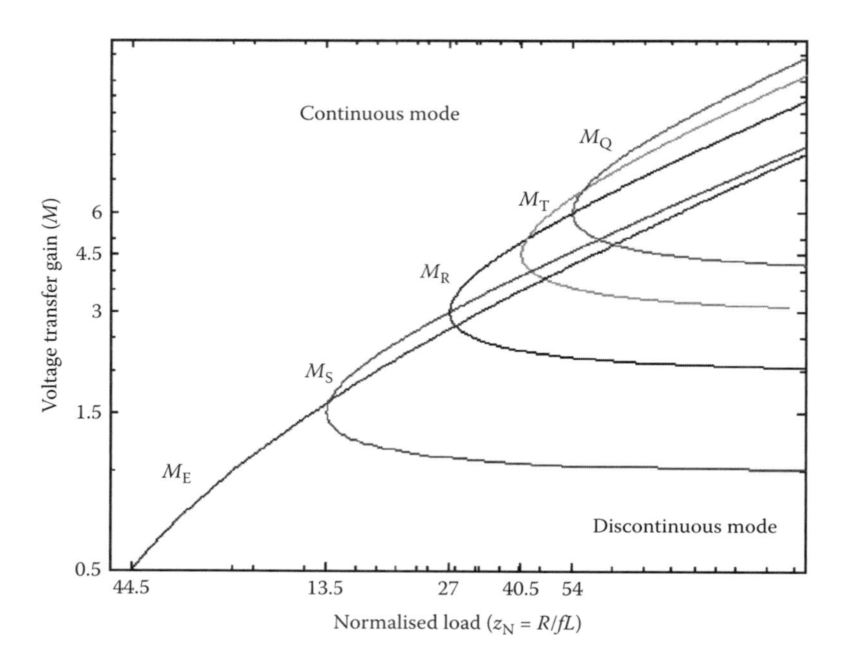

FIGURE 6.20 Boundaries between CCM and DCM of P/O Luo-converters. (Reprinted from Luo, F. L. and Ye, H. 2006. *Essential DC/DC Converters*. Boca Raton: Taylor & Francis Group LLC, p. 121. With permission.)

The output voltage in the DCM for all circuits is

$$V_{O-n} = \left[n + \frac{n + h(n) - 1}{1 - k} + k^{[2-h(n)]} \frac{1 - k}{2[n + h(n)]} z_N \right] V_I, \quad n = 0, 1, 2, 3, 4, \ldots, \qquad (6.110)$$

where

$$h(n) = \begin{cases} 0 & \text{if} \quad n \geq 1 \\ 1 & \text{if} \quad n = 0 \end{cases} \quad \text{is the Hong function.} \qquad (6.111)$$

The boundaries between continuous and discontinuous modes of all circuits are shown in Figure 6.20. The curves of all M versus z_N suggest that the continuous mode area increases from M_E via M_S, M_R, M_T to M_Q. The boundary of an elementary circuit is a monorising curve, but other curves are not monorising. There are minimum values of the boundaries of other circuits which for M_S, M_R, M_T, and M_Q correspond at $k = 1/3$.

6.4 N/O Luo-Converters

N/O Luo-converters perform the voltage conversion from positive to negative voltages using the VL technique. They work in the second quadrant with large voltage amplification. Five circuits have been introduced in the literature [12,13]:

- Elementary circuit
- Self-lift circuit

- Re-lift circuit
- Triple-lift circuit
- Quadruple-lift circuit.

The elementary circuit was discussed in Section 5.5.2 and the self-lift circuit was discussed in Section 6.2.4. Therefore, further circuits will be discussed in this section.

6.4.1 Re-Lift Circuit

Figure 6.21 shows the N/O re-lift circuit, which is derived from the self-lift circuit. It consists of one static switch S, three inductors L, L_1, and L_O, four capacitors C, C_1, C_2, and C_O, and diodes. It can be seen that one capacitor C_2, one inductor L_1, and two diodes D_2 and D_{11} have been added into the re-lift circuit. Circuit C_1-D_1-D_{11}-L_1-C_2-D_2 is the lift circuit. Capacitors C_1 and C_2 perform characteristics to lift the capacitor voltage V_C to a level 2 times higher than the source voltage $2V_I$. Inductor L_1 performs the function as a ladder joint to link the two capacitors C_1 and C_2 and lift the capacitor voltage V_C. Currents $i_{C1}(t)$ and $i_{C2}(t)$ are exponential functions $\delta_1(t)$ and $\delta_2(t)$. They have large values at the moment of power switching on, but they are small because $v_{C1} = v_{C2} \cong V_I$ in steady state.

When switch S is on, the source current $i_I = i_L + i_{C1} + i_{C2}$. Inductor L absorbs energy from the source, and current i_L linearly increases with slope V_I/L. In the meantime the diodes D_1 and D_2 are conducted so that capacitors C_1 and C_2 are charged by the currents i_{C1} and i_{C2}. Inductor L_O keeps the output current I_O continuous and transfers energy from capacitor C to the load R, that is, $i_{C-on} = i_{LO}$. When switch S is off, the source current $i_I = 0$. Current i_L flows through the freewheeling diode D, capacitors C_1 and C_2, and inductor L_1 to charge capacitor C and enhance current i_{LO}. Inductor L transfers its SE to capacitor C and load R via inductor L_O, that is, $i_L = i_{C1-off} = i_{C2-off} = i_{L1-off} = i_{C-off} + i_{LO}$. Thus, the current i_L decreases.

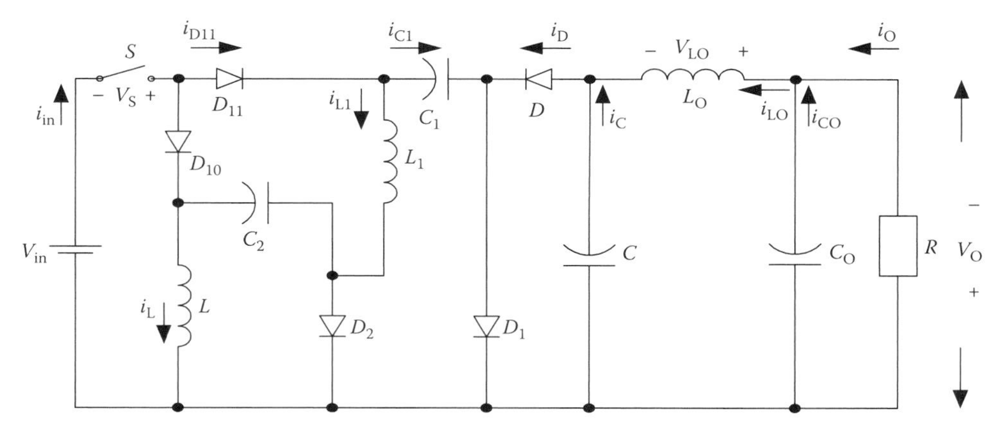

FIGURE 6.21 N/O re-lift circuit. (Reprinted from Luo, F. L. and Ye, H. 2006. *Essential DC/DC Converters*. Boca Raton: Taylor & Francis Group LLC, p. 142. With permission.)

The output current $I_O = I_{LO}$ because the capacitor C_O does not consume any energy in the steady state. The average output current is

$$I_O = I_{LO} = I_{C-on}. \tag{6.112}$$

The charge of capacitor C increases during switch-off:

$$Q+ = (1-k)TI_{C-off}.$$

It decreases during switch-on: $Q- = kTI_{C-on}$.
In the whole repeating period T, $Q+ = Q-$. Thus,

$$I_{C-off} = \frac{k}{1-k}I_{C-on} = \frac{k}{1-k}I_O.$$

Therefore, the inductor current I_L is

$$I_L = I_{C-off} + I_O = \frac{I_O}{1-k}. \tag{6.113}$$

We know that

$$I_{C1-off} = I_{C2-off} = I_{L1} = I_L = \frac{1}{1-k}I_O, \tag{6.114}$$

$$I_{C1-on} = \frac{1-k}{k}I_{C1-off} = \frac{1}{k}I_O, \tag{6.115}$$

and

$$I_{C2-on} = \frac{1-k}{k}I_{C2-off} = \frac{1}{k}I_O. \tag{6.116}$$

In the steady state, we can use

$$V_{C1} = V_{C2} = V_I$$

and

$$V_{L1-on} = V_I, \quad V_{L1-off} = \frac{k}{1-k}V_I.$$

Considering current i_L, it increases during switch-on with slope V_I/L and decreases during switch-off with slope $-(V_O - V_{C1} - V_{C2} - V_{L1-off})/L = -[V_O - 2V_I - kV_I/(1-k)]/L$.
Therefore,

$$kTV_I = (1-k)T\left(V_O - 2V_I - \frac{k}{1-k}V_I\right)$$

or

$$V_O = \frac{2}{1-k}V_I, \tag{6.117}$$

and

$$I_O = \frac{1-k}{2} I_I. \quad (6.118)$$

The voltage transfer gain in the continuous mode is

$$M_R = \frac{V_O}{V_I} = \frac{I_I}{I_O} = \frac{2}{1-k}. \quad (6.119)$$

The curve of M_R versus k is shown in Figure 6.11. The circuit (C-L_O-C_O) is a "Π"-type low-pass filter. Therefore,

$$V_C = V_O = \frac{2}{1-k} V_I. \quad (6.120)$$

Current i_L increases and is supplied by V_I during switch-on. Thus, its peak-to-peak variation is

$$\Delta i_L = \frac{kTV_I}{L}.$$

The variation ratio of current i_L is

$$\zeta = \frac{\Delta i_L/2}{I_L} = \frac{k(1-k)V_I T}{2LI_O} = \frac{k(1-k)R}{2M_R f L} = \frac{k}{M_R^2} \frac{R}{fL}. \quad (6.121)$$

The peak-to-peak variation of current i_{L1} is

$$\Delta i_{L1} = \frac{k}{L_1} TV_I.$$

The variation ratio of current i_{L1} is

$$\chi_1 = \frac{\Delta i_{L1}/2}{I_{L1}} = \frac{kTV_I}{2L_1 I_O}(1-k) = \frac{k(1-k)}{2M_R} \frac{R}{fL_1}. \quad (6.122)$$

The peak-to-peak variation of voltage v_C is

$$\Delta v_C = \frac{Q-}{C} = \frac{k}{C} TI_O.$$

The variation ratio of voltage v_C is

$$\rho = \frac{\Delta v_C/2}{V_C} = \frac{kI_O T}{2CV_O} = \frac{k}{2} \frac{1}{fCR}. \quad (6.123)$$

The peak-to-peak variation of voltage v_{C1} is

$$\Delta v_{C1} = \frac{kT}{C_1} I_{C1-on} = \frac{1}{fC} I_O.$$

The variation ratio of voltage v_{C1} is

$$\sigma_1 = \frac{\Delta v_{C1}/2}{V_{C1}} = \frac{I_O}{2fC_1V_I} = \frac{M_R}{2}\frac{1}{fC_1R}. \tag{6.124}$$

Using the same operation, the variation ratio of voltage v_{C2} is

$$\sigma_2 = \frac{\Delta v_{C2}/2}{V_{C2}} = \frac{I_O}{2fC_2V_I} = \frac{M_R}{2}\frac{1}{fC_2R}. \tag{6.125}$$

Since

$$\Delta i_{LO} = \frac{1}{2}\frac{T}{2}\frac{k}{2CL_O}TI_O = \frac{k}{8f^2CL_O}I_O,$$

the variation ratio of current i_{LO} is

$$\xi = \frac{\Delta i_{LO}/2}{I_{LO}} = \frac{k}{16}\frac{1}{f^2CL_O}. \tag{6.126}$$

Since

$$\Delta v_{CO} = \frac{B}{C_O} = \frac{1}{2}\frac{T}{2}\frac{k}{16f^2CC_OL_O}I_O = \frac{k}{64f^3CC_OL_O}I_O,$$

the variation ratio of current v_{CO} is

$$\varepsilon = \frac{\Delta v_{CO}/2}{V_{CO}} = \frac{k}{128f^3CC_OL_O}\frac{I_O}{V_O} = \frac{k}{128}\frac{1}{f^3CC_OL_OR}. \tag{6.127}$$

Example 6.3

An N/O re-lift Luo-converter has the following components: $V_I = 20$V, $L = L_1 = L_O = 1$ mH, all capacitances are equal to 20 μF, $R = 160\ \Omega$, $f = 50$ kHz, and $k = 0.5$. Calculate the output voltage and the variation ratios ξ, ζ, χ_1, ρ, σ_1, σ_2, and ε in steady state.

Solution

From Equation 6.127, we obtain the output voltage as

$$V_O = \frac{2}{1-k}V_I = \frac{2}{1-0.5}20 = 80\,\text{V}.$$

The variation ratios are $\xi = 6.25 \times 10^{-4}$, $\zeta = 0.04$, $\chi_1 = 0.1$, $\rho = 0.0016$, $\sigma_1 = 0.04$, $\sigma_2 = 0.04$, and $\varepsilon = 7.8 \times 10^{-5}$. Therefore, the variations are small.

In the DCM, the diode current i_D becomes zero during switch-off before the next period switch-on. The condition for DCM is $\zeta \geq 1$, that is,

$$\frac{k}{M_R^2}\frac{R}{fL} \geq 1$$

or

$$M_R \le \sqrt{k}\sqrt{\frac{R}{fL}} = \sqrt{k}\sqrt{z_N}. \quad (6.128)$$

The graph of the boundary curve versus the normalized load $z_N = R/fL$ is shown in Figure 6.12. It can be seen that the boundary curve has a minimum value of 3.0 at $k = 1/3$.

In this case, the current i_D exists in the period between kT and $t_1 = [k + (1-k)m_R]T$, where m_R is the *filling efficiency* and it is defined as

$$m_R = \frac{1}{\zeta} = \frac{M_R^2}{k(R/fL)}. \quad (6.129)$$

Therefore, $0 < m_R < 1$. Because inductor current $i_{L1} = 0$ at $t = t_1$,

$$V_{L1-off} = \frac{k}{(1-k)m_R}V_I.$$

Since the current i_D becomes zero at $t = t_1 = [k + (1-k)m_R]T$, for the current i_L,

$$kTV_I = (1-k)m_R T(V_C - 2V_I - V_{L1-off})$$

or

$$V_C = \left[2 + \frac{2k}{(1-k)m_R}\right]V_I = \left[2 + k^2(1-k)\frac{R}{2fL}\right]V_I \quad \text{with} \quad \sqrt{k}\sqrt{\frac{R}{fL}} \ge \frac{2}{1-k},$$

and for the current i_{LO}, $kT(V_I + V_C - V_O) = (1-k)m_R T(V_O - 2V_I - V_{L1-off})$.

Therefore, the output voltage in the discontinuous mode is

$$V_O = \left[2 + \frac{2k}{(1-k)m_R}\right]V_I = \left[2 + k^2(1-k)\frac{R}{2fL}\right]V_I \quad \text{with} \quad \sqrt{k}\sqrt{\frac{R}{fL}} \ge \frac{2}{1-k}. \quad (6.130)$$

That is, the output voltage linearly increases as the load resistance R increases. Larger load resistance R may cause higher output voltage in the discontinuous mode.

6.4.2 N/O Triple-Lift Circuit

An N/O triple-lift circuit is shown in Figure 6.22. It consists of one static switch S, four inductors L, L_1, L_2, and L_O, five capacitors C, C_1, C_2, C_3, and C_O, and diodes. The circuit C_1-D_1-L_1-C_2-D_2-D_{11}-L_2-C_3-D_3-D_{12} is the lift circuit. Capacitors C_1, C_2, and C_3 perform characteristics to lift the capacitor voltage V_C to a level 3 times higher than the source voltage V_I. L_1 and L_2 perform the function as ladder joints to link the three capacitors C_1, C_2, and C_3 and lift the capacitor voltage V_C up. Currents $i_{C1}(t)$, $i_{C2}(t)$, and $i_{C3}(t)$ are exponential functions. They have large values at the moment of power switching on, but they are small because $v_{C1} = v_{C2} = v_{C3} \cong V_I$ in steady state.

The output voltage and current are

$$V_O = \frac{3}{1-k}V_I \quad (6.131)$$

and

$$I_O = \frac{1-k}{3}I_I. \quad (6.132)$$

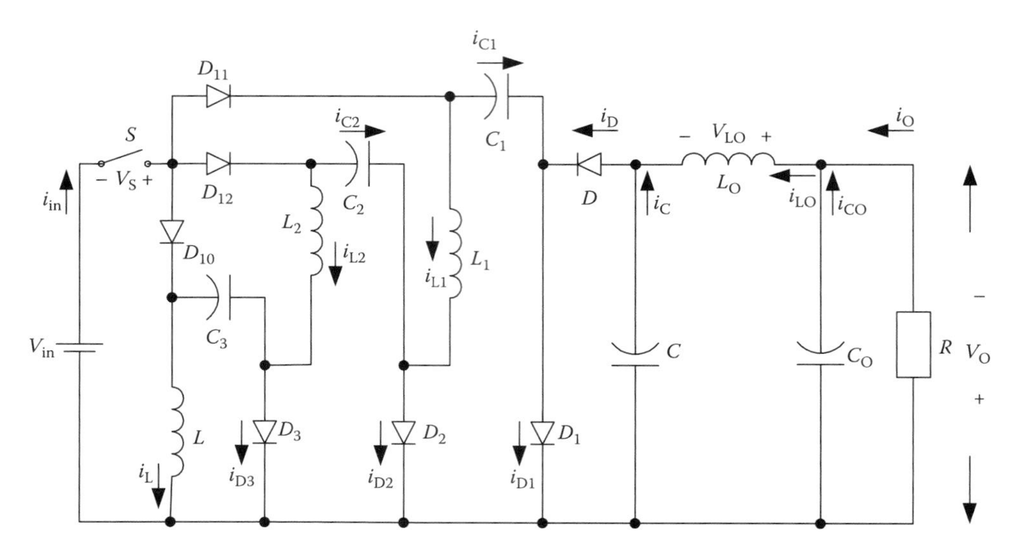

FIGURE 6.22 N/O triple-lift circuit. (Reprinted from Luo, F. L. and Ye, H. 2006. *Essential DC/DC Converters*. Boca Raton: Taylor & Francis Group LLC, p. 151. With permission.)

The voltage transfer gain in the continuous mode is

$$M_T = \frac{V_O}{V_I} = \frac{3}{1-k}. \tag{6.133}$$

The curve of M_T versus k is shown in Figure 6.14.

$$\text{Other average voltages: } V_C = V_O; \quad V_{C1} = V_{C2} = V_{C3} = V_I.$$

$$\text{Other average currents: } I_{LO} = I_O; \quad I_L = I_{L1} = I_{L2} = \frac{1}{1-k} I_O.$$

$$\text{Current variation ratios: } \zeta = \frac{k}{M_T^2}\frac{3R}{2fL}; \quad \xi = \frac{k}{16}\frac{1}{f^2 C L_O}; \quad \chi_1 = \frac{k(1-k)}{2M_T}\frac{R}{fL_1};$$

$$\chi_2 = \frac{k(1-k)}{2M_T}\frac{R}{fL_2}.$$

$$\text{Voltage variation ratios: } \rho = \frac{k}{2}\frac{1}{fCR}; \quad \sigma_1 = \frac{M_T}{2}\frac{1}{fC_1R}; \quad \sigma_2 = \frac{M_T}{2}\frac{1}{fC_2R}; \quad \sigma_3 = \frac{M_T}{2}\frac{1}{fC_3R}.$$

The variation ratio of output voltage V_C is

$$\varepsilon = \frac{k}{128}\frac{1}{f^3 C C_O L_O R}. \tag{6.134}$$

The boundary between continuous and discontinuous modes is

$$M_T \leq \sqrt{k}\sqrt{\frac{3R}{2fL}} = \sqrt{\frac{3kz_N}{2}}. \tag{6.135}$$

It can be seen that the boundary curve has a minimum value of M_T that is equal to 4.5, corresponding to $k = 1/3$. The boundary curve versus the normalized load $z_N = R/fL$ is shown in Figure 6.15.

In the discontinuous mode, the current i_D exists in the period between kT and $t_1 = [k + (1-k)m_T]T$, where m_T is the filling efficiency, that is,

$$m_T = \frac{1}{\zeta} = \frac{M_T^2}{k(3R/2fL)}. \tag{6.136}$$

Because inductor current $i_{L1} = i_{L2} = 0$ at $t = t_1$; therefore $0 < m_T < 1$:

$$V_{L1-off} = V_{L2-off} = \frac{k}{(1-k)m_T} V_I.$$

Since the current i_D becomes zero at $t = t_1 = [k + (1-k)m_T]T$, for the current i_L, we have

$$kTV_I = (1-k)m_T T(V_C - 3V_I - V_{L1-off} - V_{L2-off})$$

or

$$V_C = \left[3 + \frac{3k}{(1-k)m_T}\right] V_I = \left[3 + k^2(1-k)\frac{R}{2fL}\right] V_I \quad \text{with} \quad \sqrt{k}\sqrt{\frac{3R}{2fL}} \geq \frac{3}{1-k},$$

and for the current i_{LO}, we have

$$kT(V_I + V_C - V_O) = (1-k)m_T T(V_O - 2V_I - V_{L1-off} - V_{L2-off}).$$

Therefore, output voltage in the discontinuous mode is

$$V_O = \left[3 + \frac{3k}{(1-k)m_T}\right] V_I = \left[3 + k^2(1-k)\frac{R}{2fL}\right] V_I \quad \text{with} \quad \sqrt{k}\sqrt{\frac{3R}{2fL}} \geq \frac{3}{1-k}. \tag{6.137}$$

That is, the output voltage increases linearly as the load resistance R increases. We can see that the output voltage increases as the load resistance R increases.

6.4.3 N/O Quadruple-Lift Circuit

An N/O quadruple-lift circuit is shown in Figure 6.23. It consists of one static switch S, five inductors L, L_1, L_2, L_3, and L_O, and six capacitors C, C_1, C_2, C_3, C_4, and C_O. Capacitors C_1, C_2, C_3, and C_4 perform characteristics to lift the capacitor voltage V_C to a level 4 times higher than the source voltage V_I. L_1, L_2, and L_3 perform the function of ladder joints to link the four capacitors C_1, C_2, C_3, and C_4 and lift the output capacitor voltage V_C. Currents $i_{C1}(t)$, $i_{C2}(t)$, $i_{C3}(t)$, and $i_{C4}(t)$ are exponential functions. They have large values at the moment of power switching on, but they are small because $v_{C1} = v_{C2} = v_{C3} = v_{C4} \cong V_I$ in steady state.

The output voltage and current are

$$V_O = \frac{4}{1-k} V_I \tag{6.138}$$

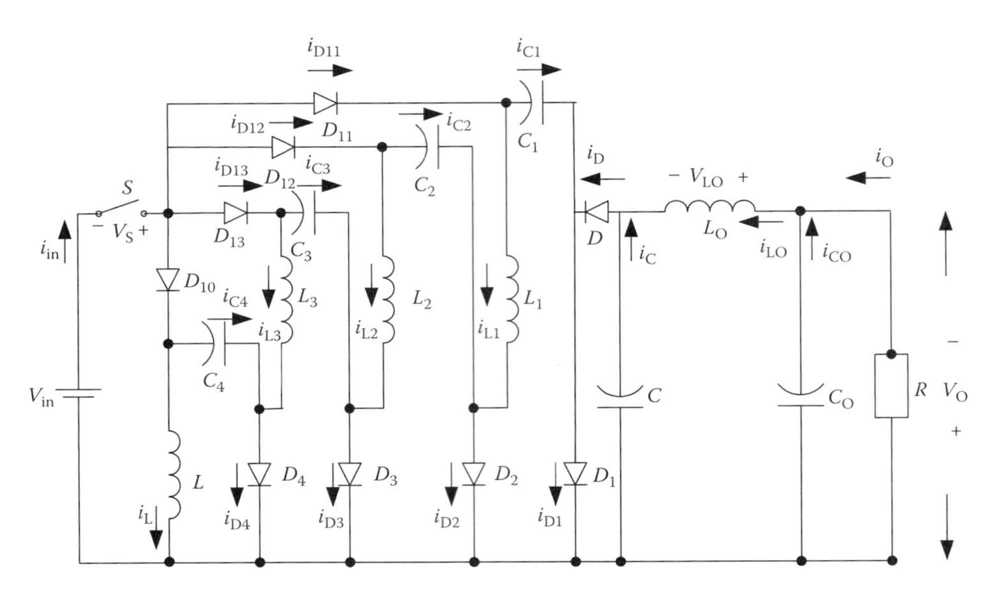

FIGURE 6.23 N/O quadruple-lift circuit. (Reprinted from Luo, F. L. and Ye, H. 2006. *Essential DC/DC Converters*. Boca Raton: Taylor & Francis Group LLC, p. 155. With permission.)

and

$$I_O = \frac{1-k}{4} I_I. \tag{6.139}$$

The voltage transfer gain in the continuous mode is

$$M_Q = \frac{V_O}{V_I} = \frac{4}{1-k}. \tag{6.140}$$

The curve of M_Q versus k is shown in Figure 6.17.

Other average voltages: $V_C = V_O$; $V_{C1} = V_{C2} = V_{C3} = V_{C4} = V_I$.

Other average currents: $I_{LO} = I_O$; $I_L = I_{L1} = I_{L2} = I_{L3} = \frac{1}{1-k} I_O$.

Current variation ratios: $\zeta = \frac{k}{M_Q^2} \frac{2R}{fL}$; $\xi = \frac{k}{16} \frac{1}{f^2 C L_O}$;

$$\chi_1 = \frac{k(1-k)}{2M_Q} \frac{R}{fL_1}; \quad \chi_2 = \frac{k(1-k)}{2M_Q} \frac{R}{fL_2}; \quad \chi_3 = \frac{k(1-k)}{2M_Q} \frac{R}{fL_3}.$$

Voltage variation ratios: $\rho = \frac{k}{2} \frac{1}{fCR}$; $\sigma_1 = \frac{M_Q}{2} \frac{1}{fC_1 R}$;

$$\sigma_2 = \frac{M_Q}{2} \frac{1}{fC_2 R}; \quad \sigma_3 = \frac{M_Q}{2} \frac{1}{fC_3 R}; \quad \sigma_4 = \frac{M_Q}{2} \frac{1}{fC_4 R}.$$

The variation ratio of output voltage V_C is

$$\varepsilon = \frac{k}{128} \frac{1}{f^3 C C_O L_O R}. \tag{6.141}$$

The output voltage ripple is very small.

The boundary between CCM and DCM is

$$M_Q \le \sqrt{k}\sqrt{\frac{2R}{fL}} = \sqrt{2kz_N}. \quad (6.142)$$

It can be seen that the boundary curve has a minimum value of M_Q that is equal to 6.0, corresponding to $k = 1/3$. The boundary curve is shown in Figure 6.18.

In the discontinuous mode, the current i_D exists in the period between kT and $t_1 = [k + (1-k)m_Q]T$, where m_Q is the filling efficiency, that is,

$$m_Q = \frac{1}{\zeta} = \frac{M_Q^2}{k(2R/fL)}. \quad (6.143)$$

Because inductor current $i_{L1} = i_{L2} = i_{L3} = 0$ at $t = t_1$; therefore $0 < m_Q < 1$:

$$V_{L1-off} = V_{L2-off} = V_{L3-off} = \frac{k}{(1-k)m_Q}V_I.$$

Since the current i_D becomes zero at $t = t_1 = kT + (1-k)m_QT$, for the current i_L, we have

$$kTV_I = (1-k)m_QT(V_C - 4V_I - V_{L1-off} - V_{L2-off} - V_{L3-off})$$

or with

$$V_C = \left[4 + \frac{4k}{(1-k)m_Q}\right]V_I = \left[4 + k^2(1-k)\frac{R}{2fL}\right]V_I \quad \text{with} \quad \sqrt{k}\sqrt{\frac{2R}{fL}} \ge \frac{4}{1-k},$$

and for current i_{LO}, we have

$$kT(V_I + V_C - V_O) = (1-k)m_QT(V_O - 2V_I - V_{L1-off} - V_{L2-off} - V_{L3-off}).$$

Therefore, the output voltage in the discontinuous mode is

$$V_O = \left[4 + \frac{4k}{(1-k)m_Q}\right]V_I = \left[4 + k^2(1-k)\frac{R}{2fL}\right]V_I \quad \text{with} \quad \sqrt{k}\sqrt{\frac{2R}{fL}} \ge \frac{4}{1-k}. \quad (6.144)$$

That is, the output voltage linearly increases as the load resistance R increases. We can see that the output voltage increases as load resistance R increases.

6.4.4 Summary

From the analysis and calculation in previous sections, the common formulae for all these circuits can be obtained:

$$M = \frac{V_O}{V_I} = \frac{I_I}{I_O}; \quad z_N = \frac{R}{fL}; \quad R = \frac{V_O}{I_O}.$$

Inductor current variation ratios: $$\zeta = \frac{k(1-k)R}{2MfL}; \quad \xi = \frac{k}{16f^2CL_O};$$

$$\chi_i = \frac{k(1-k)R}{2MfL_i}, \quad i = 1, 2, 3, \ldots, n-1 \quad \text{with} \quad n \geq 2.$$

Capacitor voltage variation ratios: $$\rho = \frac{k}{2fCR}; \quad \varepsilon = \frac{k}{128f^3CC_OL_OR};$$

$$\sigma_i = \frac{M}{2fC_iR}, \quad i = 1, 2, 3, 4, \ldots, n \quad \text{with} \quad n \geq 1.$$

Here i is the component number and n is the stage number. In order to write common formulae for the boundaries between continuous and discontinuous modes and the output voltage for all circuits, the circuits can be numbered. The definition is that subscript $n = 0$ denotes the elementary circuit, 1 the self-lift circuit, 2 the re-lift circuit, 3 the triple-lift circuit, 4 the quadruple-lift circuit, and so on. Therefore, the voltage transfer gain in the continuous

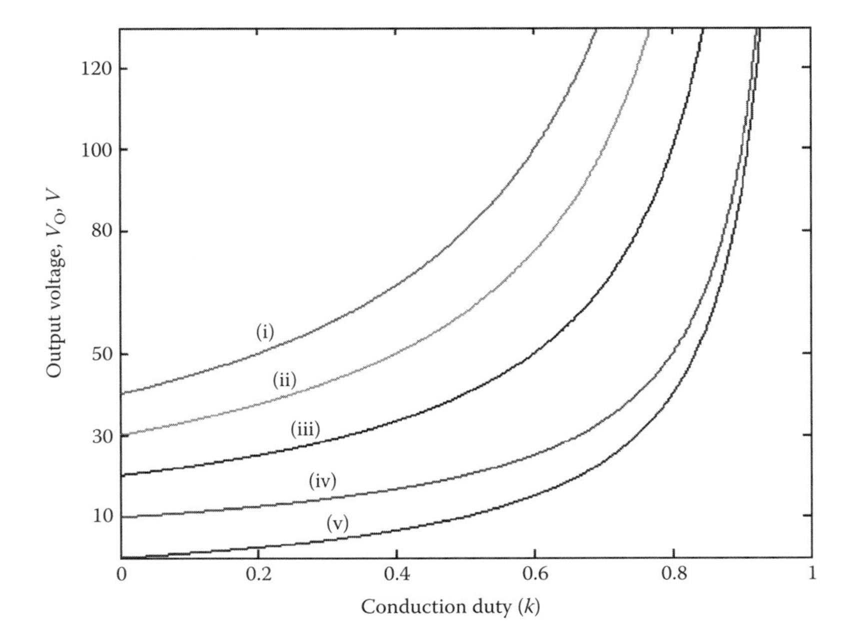

FIGURE 6.24 Output voltages of N/O Luo-converters ($V_I = 10$ V). (i) Quadruple-lift circuit; (ii) triple-lift circuit; (iii) re-lift circuit; (iv) self-lift circuit; and (v) elementary circuit. (Reprinted from Luo, F. L. and Ye, H. 2006. *Essential DC/DC Converters*. Boca Raton: Taylor & Francis Group LLC, p. 160. With permission.)

mode for all circuits is (Figure 6.24)

$$M_n = \frac{n + kh(n)}{1 - k}, \quad n = 0, 1, 2, 3, 4, \ldots. \tag{6.145}$$

The variation of freewheeling diode current i_D is

$$\zeta_n = \frac{k^{[1+h(n)]}}{M_n^2} \frac{n + h(n)}{2} z_N. \tag{6.146}$$

The boundaries are determined by the condition:

$$\zeta_n \geq 1$$

or

$$\frac{k^{[1+h(n)]}}{M_n^2} \frac{n + h(n)}{2} z_N \geq 1, \quad n = 0, 1, 2, 3, 4, \ldots. \tag{6.147}$$

Therefore, the boundaries between continuous and discontinuous modes for all circuits are

$$M_n = k^{(1+h(n))/2} \sqrt{\frac{n + h(n)}{2} z_N}, \quad n = 0, 1, 2, 3, 4, \ldots. \tag{6.148}$$

For DCM, the filling efficiency is

$$m_n = \frac{1}{\zeta_n} = \frac{M_n^2}{k^{[1+h(n)]}} \frac{2}{n + h(n)} \frac{1}{z_N}. \tag{6.149}$$

The voltage across capacitor C in the discontinuous mode for all circuits is

$$V_{C-n} = \left[n + k^{[2-h(n)]} \frac{1 - k}{2} z_N \right] V_I, \quad n = 0, 1, 2, 3, 4, \ldots. \tag{6.150}$$

The output voltage in the discontinuous mode for all circuits is

$$V_{O-n} = \left[n + k^{[2-h(n)]} \frac{1 - k}{2} z_N \right] V_I, \quad n = 0, 1, 2, 3, 4, \ldots, \tag{6.151}$$

where

$$h(n) = \begin{cases} 0 & \text{if} \quad n \geq 1 \\ 1 & \text{if} \quad n = 0 \end{cases}$$

is the Hong function.

The voltage transfer gains in CCM for all circuits are shown in Figure 6.24. The boundaries between continuous and discontinuous modes of all circuits are shown in Figure 6.25. The

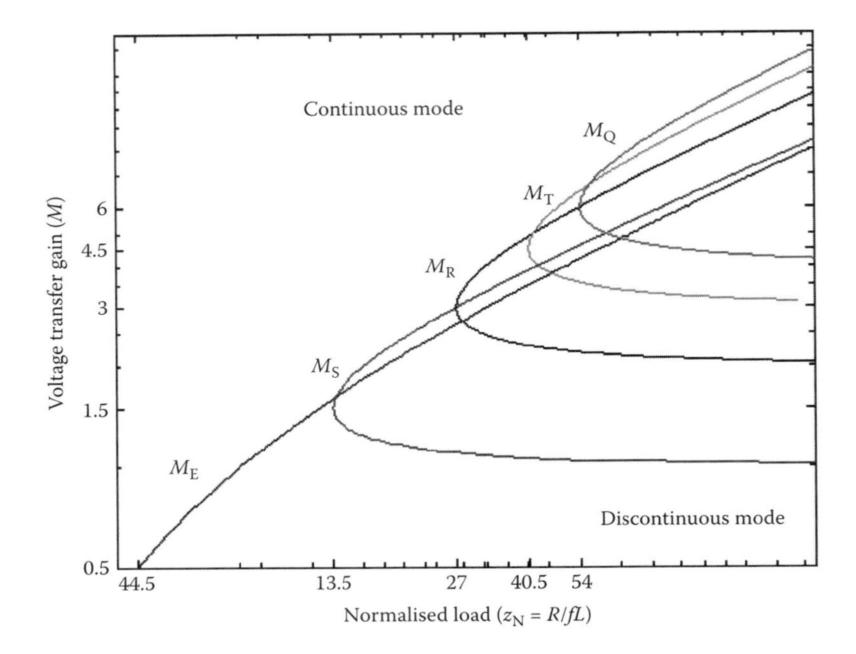

FIGURE 6.25 Boundaries between CCM and DCM of N/O Luo-converters. (Reprinted from Luo, F. L. and Ye, H. 2006. *Essential DC/DC Converters*.Boca Raton: Taylor & Francis Group LLC, p. 161. With permission.)

curves of all M versus z_N suggest that the continuous mode area increases from M_E via M_S, M_R, and M_T to M_Q. The boundary of the elementary circuit is a monorising curve, but other curves are not monorising. There are minimum values of the boundaries of other circuits, which for M_S, M_R, M_T, and M_Q correspond at $k = 1/3$.

6.5 Modified P/O Luo-Converters

N/O Luo-converters perform the voltage conversion from positive to negative voltages using the VL technique with only one switch S. This section introduces the technique to modify P/O Luo-converters that can employ only *one* switch for all circuits. Five circuits have been introduced in the literature [14]:

- Elementary circuit
- Self-lift circuit
- Re-lift circuit
- Triple-lift circuit
- Quadruple-lift circuit.

The elementary circuit is the original P/O Luo-converter. We will introduce the self-lift circuit, re-lift circuit, and multiple-lift circuit in this section.

6.5.1 Self-Lift Circuit

The self-lift circuit is shown in Figure 6.26. It is derived from the elementary circuit of the P/O Luo-converter. In steady state, the average of inductor voltages in a period is zero. Thus

$$V_{C1} = V_{CO} = V_O. \tag{6.152}$$

The inductor current i_L increases in the switch-on period and decreases in the switch-off period. The corresponding voltages across L are V_I and $-V_C$.

Therefore, $kTV_I = (1 - k)TV_C$. Hence,

$$V_C = \frac{k}{1-k}V_I. \tag{6.153}$$

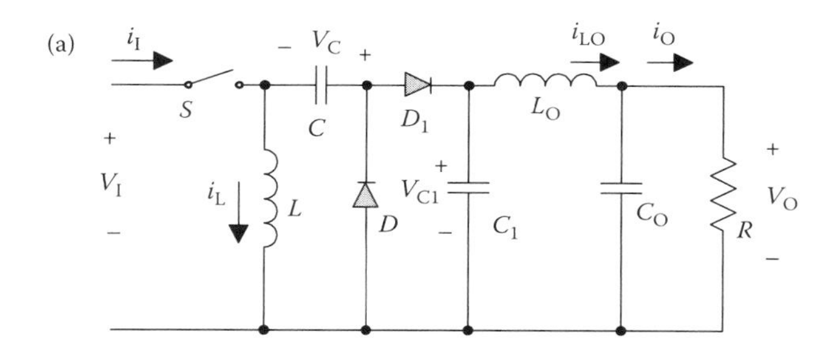

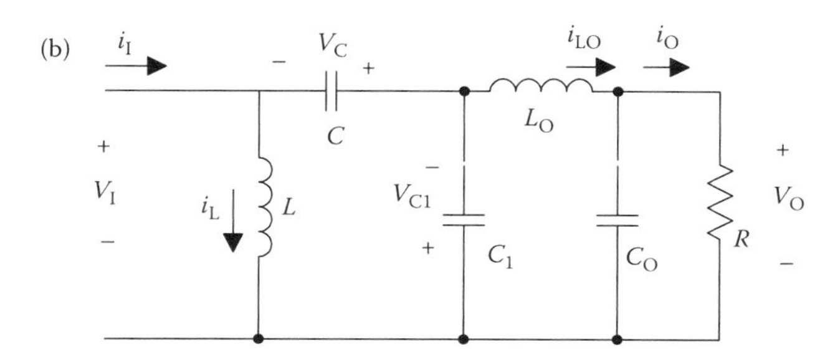

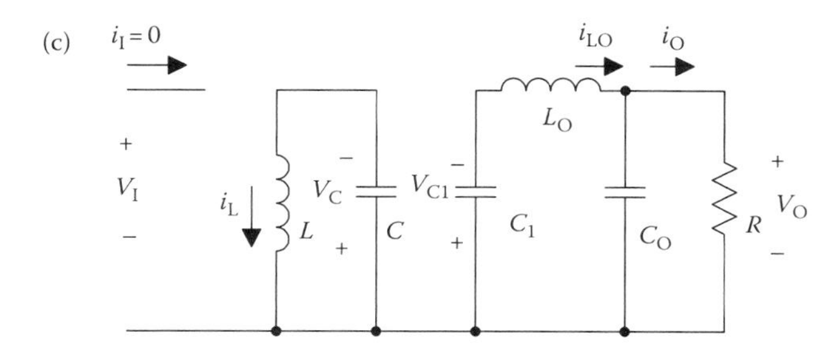

FIGURE 6.26 (a) Self-lift circuit of modified P/O Luo-converters and its equivalent circuit during (b) switch-on, and (c) switch-off. (Reprinted from Luo, F. L. and Ye, H. 2006. *Essential DC/DC Converters*. Boca Raton: Taylor & Francis Group LLC, p. 163. With permission.)

During the switch-on period, the voltage across capacitor C_1 is equal to the source voltage plus the voltage across C. Since we assume that C and C_1 are sufficiently large,

$$V_{C1} = V_I + V_C.$$

Therefore,

$$V_{C1} = V_I + \frac{k}{1-k} V_I = \frac{1}{1-k} V_I.$$

$$V_O = V_{CO} = V_{C1} = \frac{1}{1-k} V_I.$$

The voltage transfer gain of CCM is

$$M = \frac{V_O}{V_I} = \frac{1}{1-k}.$$

The output voltage and current and the voltage transfer gain are

$$\begin{aligned} V_O &= \frac{1}{1-k} V_I, \\ I_O &= (1-k) I_I, \\ M_S &= \frac{1}{1-k}. \\ \text{Average voltages: } V_C &= k V_O, \\ V_{C1} &= V_O. \\ \text{Average currents: } I_{LO} &= I_O, \\ I_L &= \frac{1}{1-k} I_O. \end{aligned} \tag{6.154}$$

We also implement the breadboard prototype of the proposed self-lift circuit. NMOS IRFP460 is used as the semiconductor switch. The diode is MR824. The other parameters are

$$V_I = 0\text{–}30\,\text{V}, \quad R = 30\text{–}340, \quad k = 0.1\text{–}0.9,$$
$$C = C_O = 100\,\text{mF}, \quad \text{and} \quad L = 470\,\mu\text{H}.$$

6.5.2 Re-Lift Circuit

The re-lift circuit and its equivalent circuits are shown in Figure 6.27. It is derived from the self-lift circuit. The function of capacitor C_2 is to lift the voltage v_C to a level higher than the source voltage V_I; inductor L_1 performs the function of the hinge of a foldable ladder (capacitor C_2) to lift the voltage v_C during switch-off.

In steady state, the average of inductor voltages over a period is zero. Thus

$$V_{C1} = V_{CO} = V_O.$$

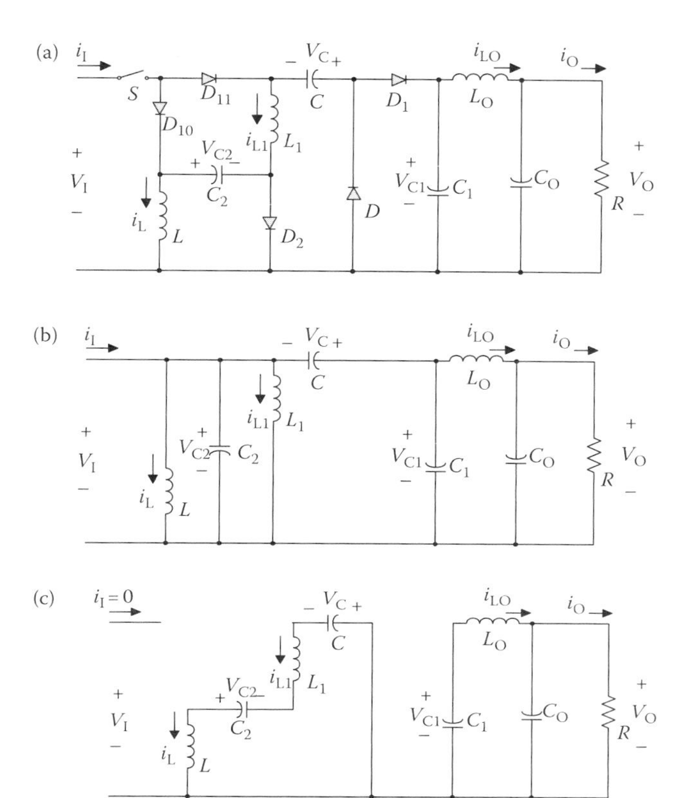

FIGURE 6.27 (a) Re-lift circuit and its equivalent circuit during (b) switch-on, and (c) switch-off. (Reprinted from Luo, F. L. and Ye, H. 2006. *Essential DC/DC Converters*. Boca Raton: Taylor & Francis Group LLC, p. 166. With permission.)

Since we assume that C_2 is large enough and C_2 is biased by the source voltage V_I during the switch-on period, $V_{C2} = V_I$.

From the switch-on equivalent circuit, another capacitor voltage equation can also be derived since we assume all the capacitors to be large enough,

$$V_O = V_{C1} = V_C + V_I.$$

The inductor current i_L increases in the switch-on period and decreases in the switch-off period. The corresponding voltages across L_1 are V_I and $-V_{L-off}$. Therefore, $kTV_I = (1-k)TV_{L-off}$. Hence,

$$V_{L-off} = \frac{k}{1-k}V_I.$$

The inductor current i_{L1} increases in the switch-on period and decreases in the switch-off period. The corresponding voltages across L_1 are V_I and $-V_{L1-off}$. Therefore, $kTV_I = (1-k)TV_{L1-off}$.
Hence,

$$V_{L1-off} = \frac{k}{1-k}V_I.$$

From the switch-off period equivalent circuit,

$$V_C = V_{C-off} = V_{L-off} + V_{L1-off} + V_{C2}.$$

Therefore,

$$V_C = \frac{k}{1-k}V_I + \frac{k}{1-k}V_I + V_I = \frac{1+k}{1-k}V_I, \tag{6.155}$$

$$V_O = \frac{1+k}{1-k}V_I + V_I = \frac{2}{1-k}V_I.$$

Then we get the voltage transfer ratio in the CCM,

$$M = M_R = \frac{2}{1-k}. \tag{6.156}$$

The following is a brief summary of the main equations for the re-lift circuit. The output voltage and current and the voltage transfer gain are

$$V_O = \frac{2}{1-k}V_I,$$

$$I_O = \frac{1-k}{2}I_I,$$

$$M_R = \frac{2}{1-k}.$$

$$\text{Average voltages: } V_C = \frac{1+k}{1-k}V_I,$$

$$V_{C1} = V_{CO} = V_O,$$

$$V_{C2} = V_I.$$

$$\text{Average currents: } I_{LO} = I_O,$$

$$I_L = I_{L1} = \frac{1}{1-k}I_O.$$

6.5.3 Multiple-Lift Circuit

Multiple-lift circuits are derived from re-lift circuits by repeating the section of L_1-C_1-D_1 multiple times. For example, a triple-lift circuit is shown in Figure 6.28. The function of capacitors C_2 and C_3 is to lift the voltage V_C across capacitor C to a level 2 times higher than the source voltage $2V_I$, and the inductors L_1 and L_2 perform the function of the hinges of a foldable ladder (capacitors C_2 and C_3) to lift the voltage V_C during switch-off.

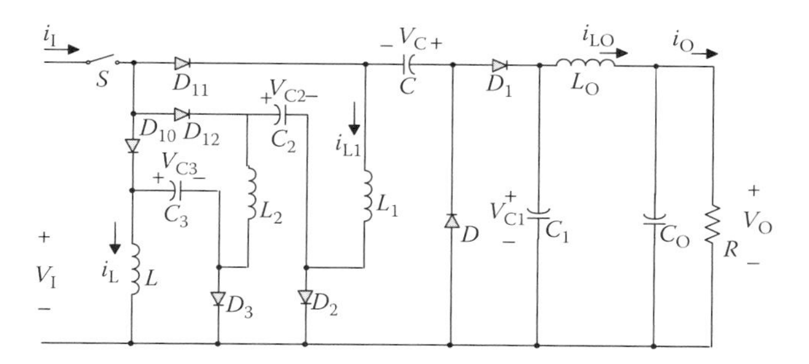

FIGURE 6.28 Triple-lift circuit. (Reprinted from Luo, F. L. and Ye, H. 2006. *Essential DC/DC Converters*. Boca Raton: Taylor & Francis Group LLC, p. 168. With permission.)

The output voltage and current and the voltage transfer gain are

$$V_O = \frac{3}{1-k}V_I,$$
$$I_O = \frac{1-k}{3}I_I, \quad (6.157)$$
$$M_T = \frac{3}{1-k}.$$

Other average voltages: $V_C = \dfrac{2+k}{1-k}V_I,$

$$V_{C1} = V_O,$$
$$V_{C2} = V_{C3} = V_I.$$

Other average currents: $I_{LO} = I_O,$

$$I_{L1} = I_{L2} = I_L = \frac{1}{1-k}I_O.$$

The quadruple-lift circuit is shown in Figure 6.29. The function of capacitors C_2, C_3, and C_4 is to lift the voltage V_C across capacitor C to a level 3 times higher than the source voltage

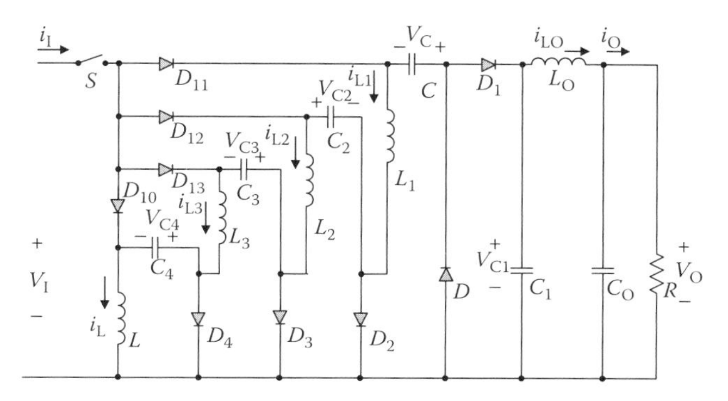

FIGURE 6.29 Quadruple-lift circuit. (Reprinted from Luo, F. L. and Ye, H. 2006. *Essential DC/DC Converters*. Boca Raton: Taylor & Francis Group LLC, p. 169. With permission.)

$3V_I$. The inductors L_1, L_2, and L_3 perform the function of the hinges of a foldable ladder (capacitors C_2, C_3, and C_4) to lift the voltage V_C during switch-off. The output voltage and current and voltage transfer gain are

$$V_O = \frac{4}{1-k}V_I, \quad I_O = \frac{1-k}{4}I_I, \tag{6.158}$$

$$M_Q = \frac{4}{1-k}.$$

$$\text{Average voltages: } V_C = \frac{3+k}{1-k}V_I, \quad V_{C1} = V_O,$$

$$V_{C2} = V_{C3} = V_{C4} = V_I.$$

$$\text{Average currents: } I_{LO} = I_O, \quad I_L = \frac{k}{1-k}I_O,$$

$$I_{L1} = I_{L2} = I_{L3} = I_L + I_{LO} = \frac{1}{1-k}I_O.$$

6.6 D/O Luo-Converters

Mirror-symmetrical D/O voltages are specially required in industrial applications and computer periphery circuits. The D/O DC–DC Luo-converter can convert positive input source voltage to P/O and N/O voltages. It consists of two conversion paths. It performs increasing conversion from positive to positive and negative DC–DC voltages with high power density, high efficiency, and cheap topology in a simple structure [15,16]. Like P/O and N/O Luo-converters, there are five circuits in this series:

- Elementary circuit
- Self-lift circuit
- Re-lift circuit
- Triple-lift circuit
- Quadruple-lift circuit.

The elementary circuit is the original D/O Luo-converter introduced in Section 5.53. We will introduce the self-lift circuit, re-lift circuit, triple-lift circuit, and quadruple-lift circuit in this section.

6.6.1 Self-Lift Circuit

The self-lift circuit shown in Figure 6.30 is derived from the elementary circuit. The positive conversion path consists of a pump circuit S-L_1-D_0-C_1, a filter (C_2)-L_2-C_O, and a lift circuit D_1-C_2. The negative conversion path consists of a pump circuit S-L_{11}-D_{10}-(C_{11}), an "Π"-type filter C_{11}-L_{12}-C_{10}, and a lift circuit D_{11}-C_{12}.

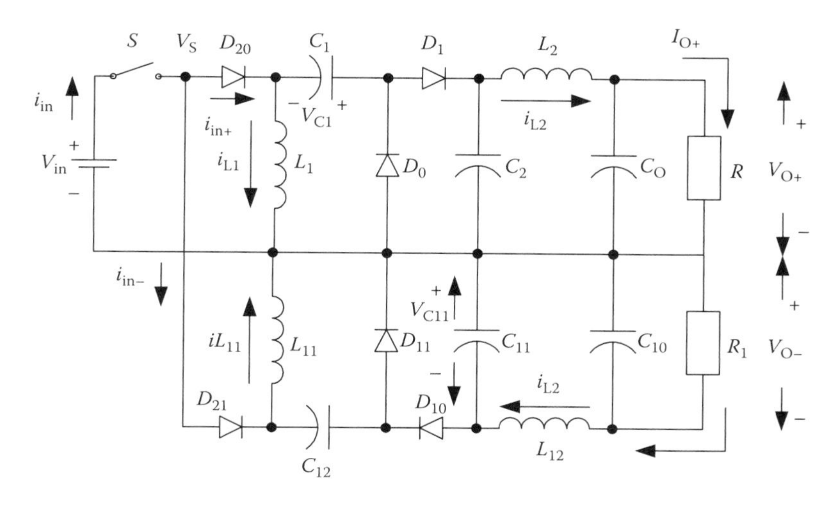

FIGURE 6.30 D/O self-lift circuit. (Reprinted from Luo, F. L. and Ye, H. 2006. *Essential DC/DC Converters*. Boca Raton: Taylor & Francis Group LLC, p. 181. With permission.)

6.6.1.1 Positive Conversion Path

The equivalent circuit during switch-on is shown in Figure 6.31a and its equivalent circuit during switch-off is shown in Figure 6.31b. The voltage across inductor L_1 is equal to V_I during switch-on and $-V_{C1}$ during switch-off. We have the relation:

$$V_{C1} = \frac{k}{1-k}V_I.$$

Hence,

$$V_O = V_{CO} = V_{C2} = V_I + V_{C1} = \frac{1}{1-k}V_I$$

and $V_{O+} = (1/(1-k))V_I$. The output current is $I_{O+} = (1-k)I_{I+}$.

Other relations are

$$I_{I+} = ki_{I+}, \quad i_{I+} = I_{L1} + i_{C1-on}, \quad i_{C1-off} = \frac{k}{1-k}i_{C1-on},$$

and

$$I_{L1} = i_{C1-off} = ki_{I+} = I_{I+}. \tag{6.159}$$

Therefore, the voltage transfer gain in the continuous mode is

$$M_{S+} = \frac{V_{O+}}{V_I} = \frac{1}{1-k}. \tag{6.160}$$

The variation ratios of the parameters are

$$\xi_{2+} = \frac{\Delta i_{L2}/2}{I_{L2}} = \frac{k}{16}\frac{1}{f^2C_2L_2}, \quad \rho_+ = \frac{\Delta v_{C1}/2}{V_{C1}} = \frac{(1-k)I_{I+}}{2fC_1(k/1-k)V_I} = \frac{1}{2kfC_1R},$$

$$\text{and} \quad \sigma_{1+} = \frac{\Delta v_{C2}/2}{V_{C2}} = \frac{k}{2fC_2R}.$$

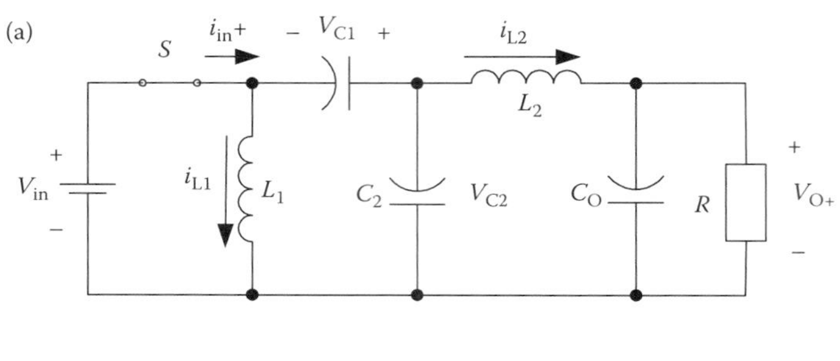

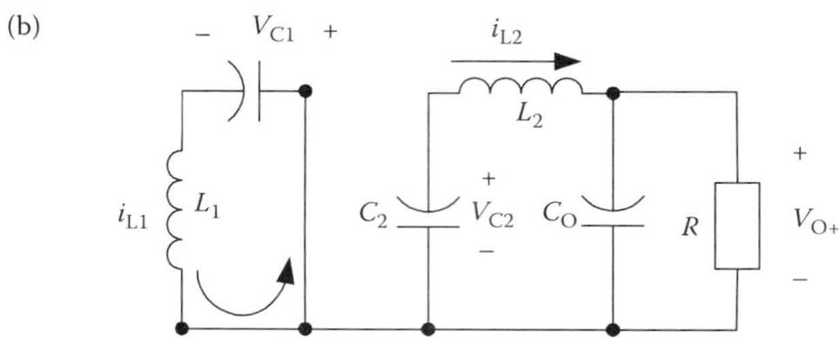

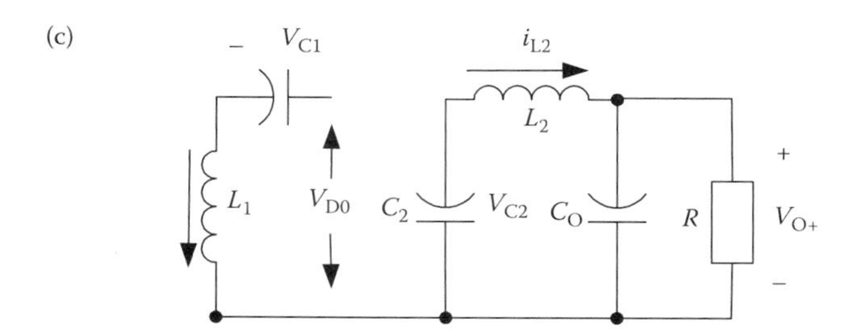

FIGURE 6.31 Equivalent circuits positive path of the D/O self-lift circuit: (a) switch-on, (b) switch-off, and (c) discontinuous conduction mode. (Reprinted from Luo, F. L. and Ye, H. 2006. *Essential DC/DC Converters*. Boca Raton: Taylor & Francis Group LLC, p. 182. With permission.)

The variation ratio of currents i_{D0} and i_{L1} is

$$\zeta_+ = \xi_{1+} = \frac{\Delta i_{L1}/2}{I_{L1}} = \frac{kV_I T}{2L_1 I_{I+}} = \frac{k}{M_S^2}\frac{R}{2fL_1}. \tag{6.161}$$

The variation ratio of output voltage v_{O+} is

$$\varepsilon_+ = \frac{\Delta v_{O+}/2}{V_{O+}} = \frac{k}{128 f^3 C_2 C_O L_2 R}. \tag{6.162}$$

6.6.1.2 Negative Conversion Path

The equivalent circuit during switch-on is shown in Figure 6.32a, and its equivalent circuit during switch-off is shown in Figure 6.32b. The relations of the average currents and

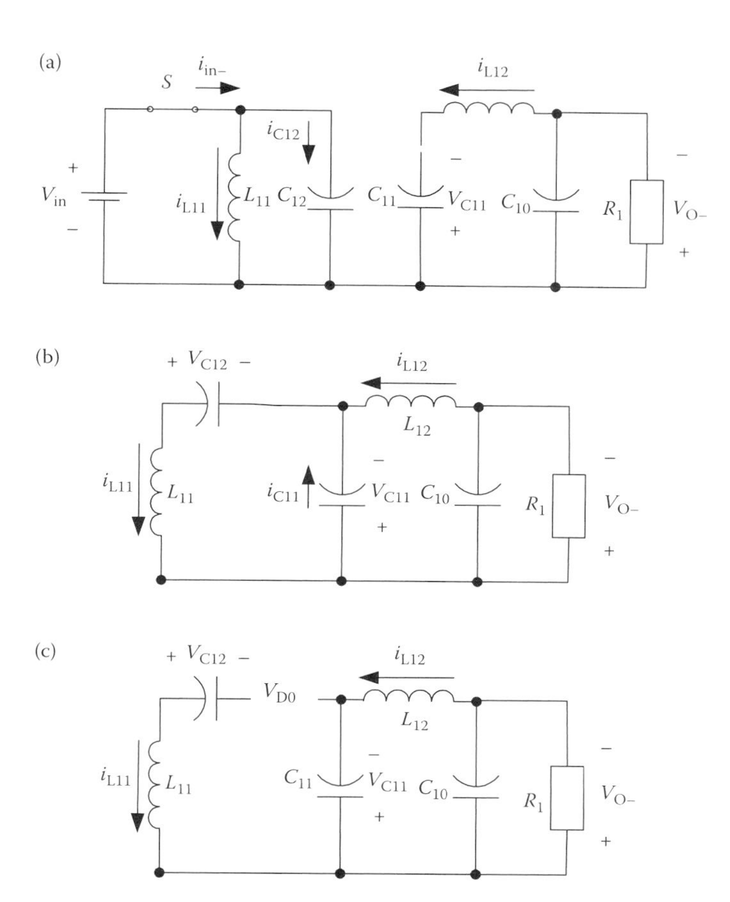

FIGURE 6.32 Equivalent circuits negative path of the D/O self-lift circuit: (a) switch-on, (b) switch-off, and (c) discontinuous conduction mode. (Reprinted from Luo, F. L. and Ye, H. 2006. *Essential DC/DC Converters*. Boca Raton: Taylor & Francis Group LLC, p. 184. With permission.)

voltages are

$$I_{O-} = I_{L12} = I_{C11-on}, \quad I_{C11-off} = \frac{k}{1-k} I_{C11-on} = \frac{k}{1-k} I_{O-},$$
$$\text{and} \quad I_{L11} = I_{C11-off} + I_{O-} = \frac{I_{O-}}{1-k}. \tag{6.163}$$

We know that $I_{C12-off} = I_{L11} = (1/(1-k))I_{O-}$ and $I_{C12-on} = ((1-k)/k)I_{C12-off} = (1/k)I_{O-}$, so that $V_{O-} = (1/(1-k))V_I$ and $I_{O-} = (1-k)I_I$.

The voltage transfer gain in the continuous mode is

$$M_{S-} = \frac{V_{O-}}{V_I} = \frac{1}{1-k}. \tag{6.164}$$

The circuit (C_{11}-L_{12}-C_{10}) is a "Π"-type low-pass filter. Therefore, $V_{C11} = V_{O-} = (k/(1-k))V_I$. From Equations 6.160 and 6.161, define $M_S = M_{S+} = M_{S-}$. The curve of M_S versus k is shown in Figure 6.33.

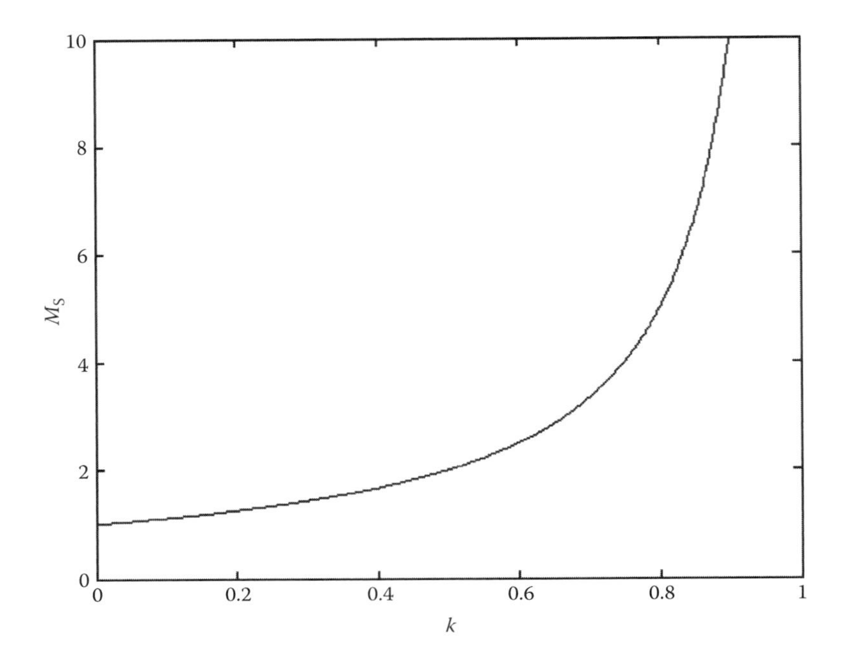

FIGURE 6.33 Voltage transfer gain M_S versus k. (Reprinted from Luo, F. L. and Ye, H. 2006. *Essential DC/DC Converters*. Boca Raton: Taylor & Francis Group LLC, p. 185. With permission.)

The variation ratios of the parameters are

$$\xi_- = \frac{\Delta i_{L12}/2}{I_{L12}} = \frac{k}{16} \frac{1}{f^2 C_{10} L_{12}}; \quad \rho_- = \frac{\Delta v_{C11}/2}{V_{C11}} = \frac{kI_{O-}T}{2C_{11}V_{O-}} = \frac{k}{2} \frac{1}{fC_{11}R_1};$$

$$\sigma_{1-} = \frac{\Delta v_{C12}/2}{V_{C12}} = \frac{I_{O-}}{2fC_{12}V_I} = \frac{M_S}{2} \frac{1}{fC_{12}R_1}.$$

The variation ratio of currents i_{D10} and i_{L11} is

$$\zeta_- = \frac{\Delta i_{L11}/2}{I_{L11}} = \frac{k(1-k)V_I T}{2L_{11}I_{O-}} = \frac{k(1-k)R_1}{2M_S f L_{11}} = \frac{k}{M_S^2} \frac{R_1}{2fL_{11}}. \tag{6.165}$$

The variation ratio of current v_{C10} is

$$\varepsilon_- = \frac{\Delta v_{C10}/2}{V_{C10}} = \frac{k}{128f^3 C_{11} C_{10} L_{12}} \frac{I_{O-}}{V_{O-}} = \frac{k}{128} \frac{1}{f^3 C_{11} C_{10} L_{12} R_1}. \tag{6.166}$$

Example 6.4

A D/O self-lift Luo-converter has the following components: $V_I = 20\text{V}$, all inductances are 1 mH, all capacitances are equal to $20\,\mu\text{F}$, $R = R_1 = 160\,\Omega$, $f = 50\,\text{kHz}$, and $k = 0.5$. Calculate the output voltage and the variation ratios, and ε in steady state.

Solution

From Equations 6.160 and 6.164, we obtain the output voltage as

$$V_{O+} = V_{O-} = \frac{1}{1-k}V_I = \frac{1}{1-0.5}20 = 40\,\text{V}.$$

The variation ratios:

$$\xi_{2+} = 6.25 \times 10^{-4}, \quad \xi_{1+} = \zeta_{1+} = 0.2, \quad \rho_+ = 0.05, \quad \sigma_{1+} = 0.00625, \quad \text{and} \quad \varepsilon_+ = 2 \times 10^{-6}.$$

$$\xi_- = 6.25 \times 10^{-4}, \quad \zeta_- = 0.05, \quad \rho_- = 0.00625, \quad \sigma_{1-} = 0.025, \quad \text{and} \quad \varepsilon_+ = 2 \times 10^{-6}.$$

Therefore, the variations are small.

6.6.1.3 Discontinuous Conduction Mode

The equivalent circuits of the DCM's operation are shown in Figures 6.31c and 6.32c. Since we select $z_N = z_{N+} = z_{N-}$, $M_S = M_{S+} = M_{S-}$, and $\zeta = \zeta_+ = \zeta_-$, the boundary between CCM and DCM is: $\zeta \geq 1$ or

$$\frac{k}{M_S^2}\frac{z_N}{2} \geq 1.$$

Hence,

$$M_S \leq \sqrt{k}\sqrt{\frac{z}{2}} = \sqrt{\frac{kz_N}{2}}. \tag{6.167}$$

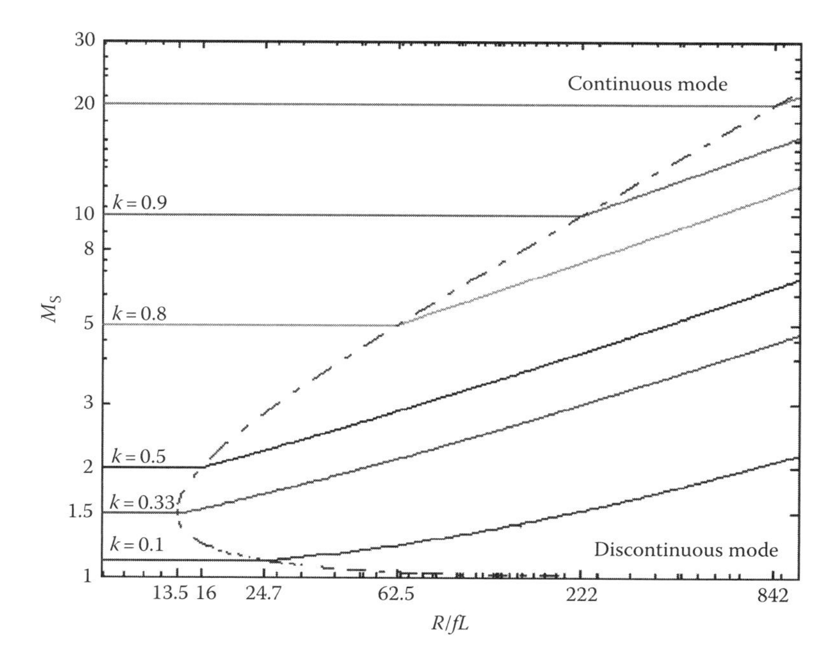

FIGURE 6.34 The boundary between continuous and discontinuous modes and the output voltage versus the normalized load $z_N = R/fL$ (D/O self-lift circuit). (Reprinted from Luo, F. L. and Ye, H. 2006. *Essential DC/DC Converters*. Boca Raton: Taylor & Francis Group LLC, p. 187. With permission.)

This boundary curve is shown in Figure 6.34. This curve has a minimum value of M_S that is equal to 1.5 at $k = 1/3$.

The filling efficiency is defined as

$$m_S = \frac{1}{\zeta} = \frac{2M_S^2}{kz_N}. \tag{6.168}$$

For the current i_{L1}, we have

$$TV_I = (1-k)m_{S+}TV_{C1}$$

or

$$V_{C1} = \frac{k}{(1-k)m_S}V_I = k^2(1-k)\frac{z_N}{2}V_I \quad \text{with} \quad \sqrt{\frac{kz_N}{2}} \geq \frac{1}{1-k}. \tag{6.169}$$

Therefore, the P/O voltage in the DCM is

$$V_{O+} = V_{C1} + V_I = \left[1 + \frac{k}{(1-k)m_S}\right]V_I = \left[1 + k^2(1-k)\frac{z_N}{2}\right]V_I \quad \text{with} \quad \sqrt{\frac{kz_N}{2}} \geq \frac{1}{1-k}. \tag{6.170}$$

For the current i_{L11}, we have

$$kTV_I = (1-k)m_S T(V_{C11} - V_I)$$

or

$$V_{C11} = \left[1 + \frac{k}{(1-k)m_S}\right]V_I = \left[1 + k^2(1-k)\frac{z_N}{2}\right]V_I \quad \text{with} \quad \sqrt{\frac{kz_N}{2}} \geq \frac{1}{1-k} \tag{6.171}$$

and for the current i_{L12}, we have $kT\ (V_I + V_{C11} - V_{O-}) = (1-k)m_{S-}T(V_{O-} - V_I)$.

Therefore, the N/O voltage in the DCM is

$$V_{O-} = \left[1 + \frac{k}{(1-k)m_S}\right]V_I = \left[1 + k^2(1-k)\frac{z_N}{2}\right]V_I \quad \text{with} \quad \sqrt{\frac{kz_N}{2}} \geq \frac{1}{1-k}. \tag{6.172}$$

Then we have $V_O = V_{O+} = V_{O-} = [1 + k^2(1-k)(z_N/2)]V_I$; that is, the output voltage linearly increases as the load resistance R increases. Larger load resistance causes higher output voltage in the DCM, as shown in Figure 6.34.

6.6.2 Re-Lift Circuit

The re-lift circuit shown in Figure 6.35 is derived from the self-lift circuit. The positive conversion path consists of a pump circuit S-L_1-D_0-C_1, a filter (C_2)-L_2-C_O, and a lift circuit D_1-C_2-D_3-L_3-D_2-C_3. The negative conversion path consists of a pump circuit S-L_{11}-D_{10}-(C_{11}), an "Π"- type filter C_{11}-L_{12}-C_{10}, and a lift circuit D_{11}-C_{12}-L_{13}-D_{22}-C_{13}-D_{12}.

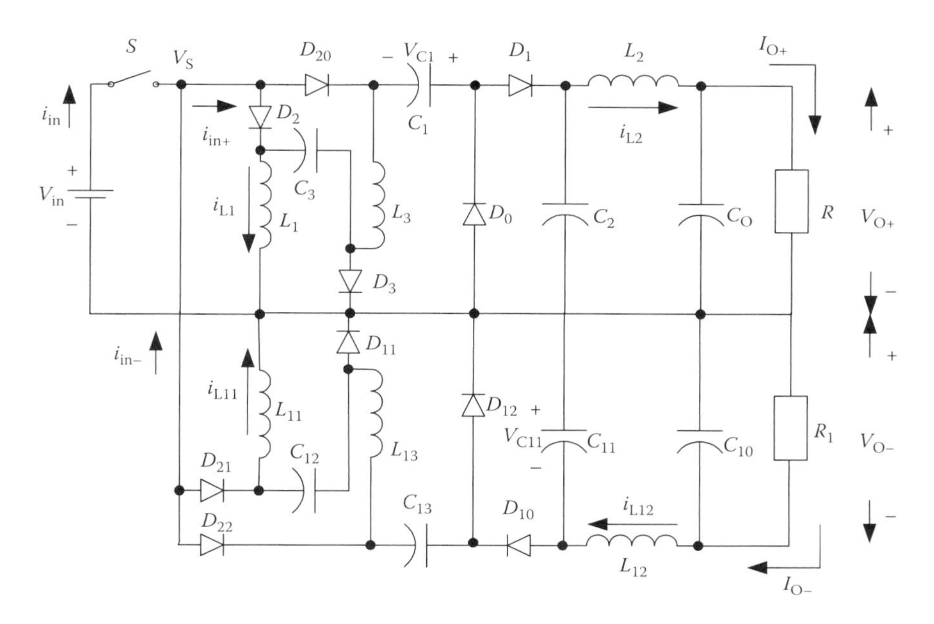

FIGURE 6.35 D/O re-lift circuit. (Reprinted from Luo, F. L. and Ye, H. 2006. *Essential DC/DC Converters*. Boca Raton: Taylor & Francis Group LLC, p. 189. With permission.)

6.6.2.1 Positive Conversion Path

The equivalent circuit during switch-on is shown in Figure 6.36a, and its equivalent circuit during switch-off is shown in Figure 6.36b.

The voltage across inductors L_1 and L_3 is equal to V_I during switch-on, and $-(V_{C1} - V_I)$ during switch-off. We have the following relations:

$$V_{C1} = \frac{1+k}{1-k}V_I \quad \text{and} \quad V_O = V_{CO} = V_{C2} = V_I + V_{C1} = \frac{2}{1-k}V_I.$$

Thus,

$$V_{O+} = \frac{2}{1-k}V_I \quad \text{and} \quad I_{O+} = \frac{1-k}{2}I_{I+}.$$

The other relations are $I_{I+} = ki_{I+}$, $i_{I+} = I_{L1} + I_{L3} + i_{C3-on} + i_{C1-on}$, $i_{C1-off} = k/(1-k)i_{C1-on}$ and

$$I_{L1} = I_{L3} = i_{C1-off} = i_{C3-off} = \frac{k}{2}i_{I+} = \frac{1}{2}I_{I+}. \tag{6.173}$$

The voltage transfer gain in the continuous mode is

$$M_{R+} = \frac{V_{O+}}{V_I} = \frac{2}{1-k}. \tag{6.174}$$

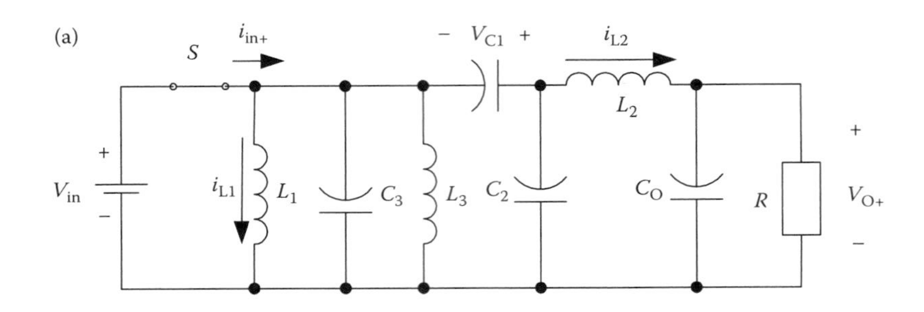

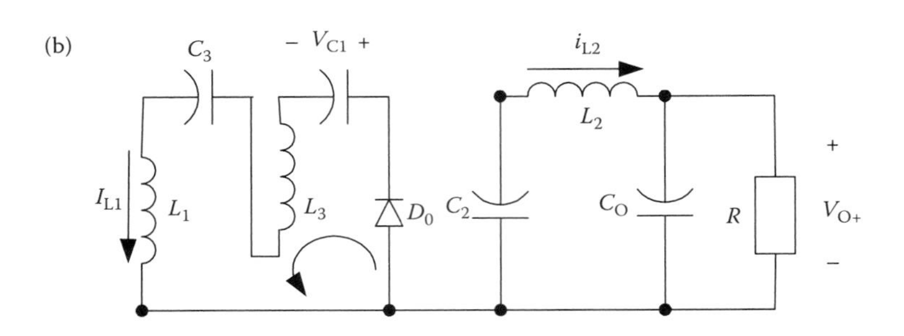

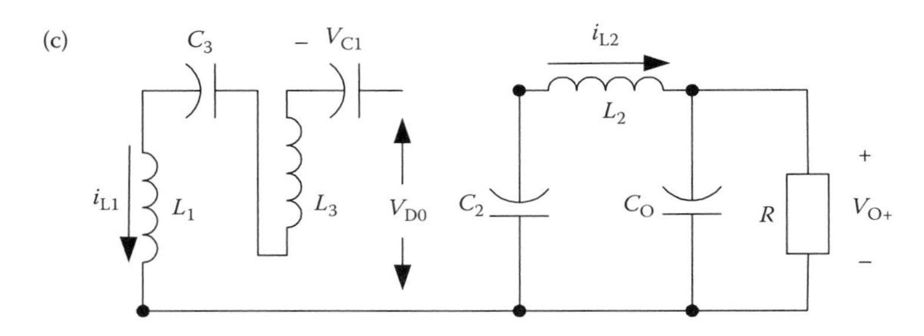

FIGURE 6.36 Equivalent circuits positive path of the D/O re-lift circuit: (a) switch-on, (b) switch-off, and (c) discontinuous conduction mode. (Reprinted from Luo, F. L. and Ye, H. 2006. *Essential DC/DC Converters*. Boca Raton: Taylor & Francis Group LLC, p. 151. With permission.)

The variation ratios of the parameters are

$$\xi_{2+} = \frac{\Delta i_{L2}/2}{I_{L2}} = \frac{k}{16} \frac{1}{f^2 C_2 L_2}; \quad \chi_{1+} = \frac{\Delta i_{L3}/2}{I_{L3}} = \frac{k V_I T}{2L_3(1/2)I_{I+}} = \frac{k}{M_R^2} \frac{R}{f L_3};$$

and

$$\rho_{+} = \frac{\Delta v_{C1}/2}{V_{C1}} = \frac{(1-k)TI_I}{4C_1(1+k/1-k)V_I} = \frac{1}{(1+k)fC_1R}; \quad \sigma_{1+} = \frac{\Delta v_{C2}/2}{V_{C2}} = \frac{k}{2fC_2R};$$

$$\sigma_{2+} = \frac{\Delta v_{C3}/2}{V_{C3}} = \frac{1-k}{4fC_3} \frac{I_{I+}}{V_I} = \frac{M_R}{2fC_3R}.$$

The variation ratio of currents i_{D0} and i_{L1} is

$$\zeta_+ = \xi_{1+} = \frac{\Delta i_{D0}/2}{I_{D0}} = \frac{kV_I T}{L_1 I_{I+}} = \frac{k}{M_R^2}\frac{R}{fL_1}, \tag{6.175}$$

and the variation ratio of output voltage v_{O+} is

$$\varepsilon_+ = \frac{\Delta v_{O+}/2}{V_{O+}} = \frac{k}{128}\frac{1}{f^3 C_2 C_O L_2 R}. \tag{6.176}$$

6.6.2.2 Negative Conversion Path

The equivalent circuit during switch-on is shown in Figure 6.37a, and its equivalent circuit during switch-off is shown in Figure 6.37b.

The relations of the average currents and voltages are

$$I_{O-} = I_{L12} = I_{C11-on} \quad I_{C11-off} = \frac{k}{1-k} I_{C11-on} = \frac{k}{1-k} I_{O-}$$

(a)

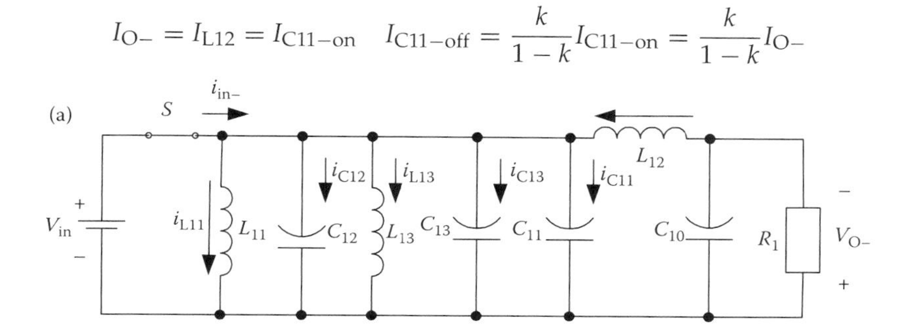

(b)

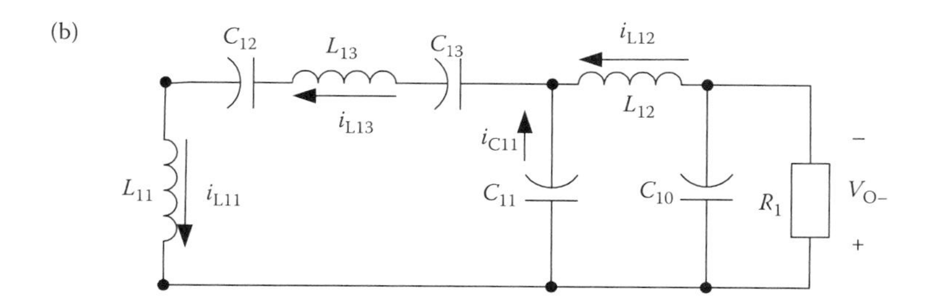

(c)

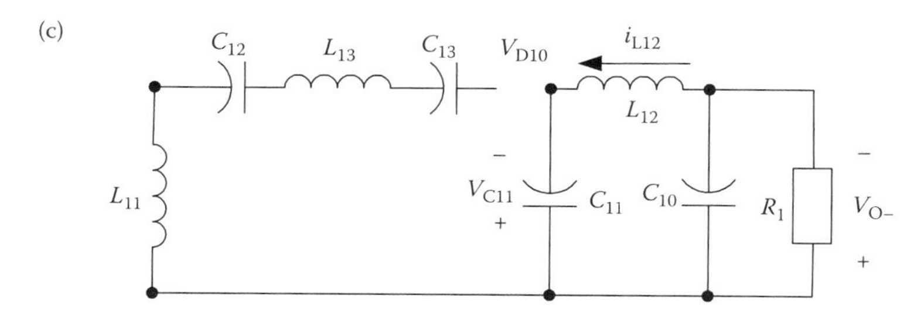

FIGURE 6.37 Equivalent circuits negative path of the D/O re-lift circuit: (a) switch-on, (b) switch-off, and (c) discontinuous conduction mode. (Reprinted from Luo, F. L. and Ye, H. 2006. *Essential DC/DC Converters*. Boca Raton: Taylor & Francis Group LLC, p. 192. With permission.)

and

$$I_{L11} = I_{C11-off} + I_{O-} = \frac{I_{O-}}{1-k}. \tag{6.177}$$

$$I_{C12-off} = I_{C13-off} = I_{L11} = \frac{1}{1-k} I_{O-}; \quad I_{C12-on} = \frac{1-k}{k} I_{C12-off} = \frac{1}{k} I_{O-};$$

$$I_{C13-on} = \frac{1-k}{k} I_{C13-off} = \frac{1}{k} I_{O-}.$$

In the steady state, we have: $V_{C12} = V_{C13} = V_I$, $V_{L13-on} = V_I$, and $V_{L13-off} = (k/1-k)V_I$.

$$V_{O-} = \frac{2}{1-k} V_I \quad \text{and} \quad I_{O-} = \frac{1-k}{2} I_{I-}.$$

The voltage transfer gain in the continuous mode is

$$M_{R-} = \frac{V_{O-}}{V_I} = \frac{I_{I-}}{I_{O-}} = \frac{2}{1-k}. \tag{6.178}$$

The circuit (C_{11}-L_{12}-C_{10}) is a "Π"-type low-pass filter.

Therefore, $V_{C11} = V_{O-} = (2/(1-k))V_I$.

From Equations 6.174 and 6.178, we define $M_R = M_{R+} = M_{R-}$. The curve of M_R versus k is shown in Figure 6.38.

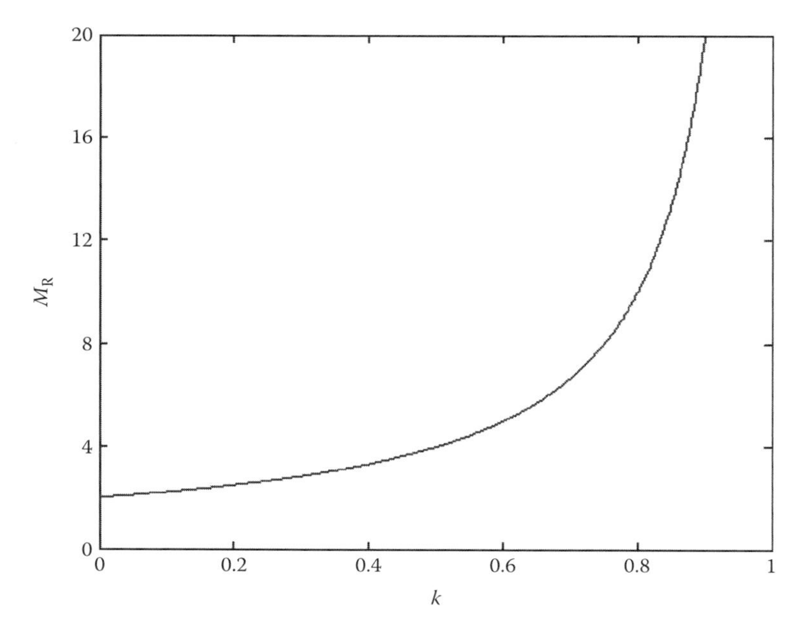

FIGURE 6.38 Voltage transfer gain M_R versus k. (Reprinted from Luo, F. L. and Ye, H. 2006. *Essential DC/DC Converters*. Boca Raton: Taylor & Francis Group LLC, p. 193. With permission.)

The variation ratios of the parameters are

$$\xi_- = \frac{\Delta i_{L12}/2}{I_{L12}} = \frac{k}{16} \frac{1}{f^2 C_{10} L_{12}}; \quad \chi_{1-} = \frac{\Delta i_{L13}/2}{I_{L13}} = \frac{kTV_I}{2L_{13} I_{O-}}(1-k) = \frac{k(1-k)}{2M_R} \frac{R_1}{fL_{13}};$$

and

$$\rho_- = \frac{\Delta v_{C11}/2}{V_{C11}} = \frac{kI_{O-}T}{2C_{11}V_{O-}} = \frac{k}{2} \frac{1}{fC_{11}R_1}; \quad \sigma_{1-} = \frac{\Delta v_{C12}/2}{V_{C12}} = \frac{I_{O-}}{2fC_{12}V_I} = \frac{M_R}{2} \frac{1}{fC_{12}R_1};$$

$$\sigma_{2-} = \frac{\Delta v_{C13}/2}{V_{C13}} = \frac{I_{O-}}{2fC_{13}V_I} = \frac{M_R}{2} \frac{1}{fC_{13}R_1}.$$

The variation ratio of currents i_{D10} and i_{L11} is

$$\zeta_- = \frac{\Delta i_{L11}/2}{I_{L11}} = \frac{k(1-k)V_I T}{2L_{11} I_{O-}} = \frac{k(1-k)R_1}{2M_R f L_{11}} = \frac{k}{M_R^2} \frac{R_1}{fL_{11}}. \quad (6.179)$$

The variation ratio of current v_{C10} is

$$\varepsilon_- = \frac{\Delta v_{C10}/2}{V_{C10}} = \frac{k}{128 f^3 C_{11} C_{10} L_{12}} \frac{I_{O-}}{V_{O-}} = \frac{k}{128} \frac{1}{f^3 C_{11} C_{10} L_{12} R_1}. \quad (6.180)$$

6.6.2.3 Discontinuous Conduction Mode

The equivalent circuits of the DCM are shown in Figures 6.36c and 6.37c. In order to obtain the mirror-symmetrical D/O voltages, we purposely select $z_N = z_{N+} = z_{N-}$ and $\zeta = \zeta_+ = \zeta_-$. The freewheeling diode currents i_{D0} and i_{D10} become zero during switch-off before the next switch-on period. The boundary between CCM and DCM is

$$\zeta \geq 1$$

or

$$\frac{k}{M_R^2} z_N \geq 1.$$

Hence,

$$M_R \leq \sqrt{kz_N}. \quad (6.181)$$

This boundary curve is shown in Figure 6.39. It can be seen that the boundary curve has a minimum value of M_R that is equal to 3.0, corresponding to $k = 1/3$.

The filling efficiency m_R is

$$m_R = \frac{1}{\zeta} = \frac{M_R^2}{kz_N}. \quad (6.182)$$

So

$$V_{C1} = \left[1 + \frac{2k}{(1-k)m_R}\right] V_I = \left[1 + k^2(1-k)\frac{z_N}{2}\right] V_I \quad \text{with} \quad \sqrt{kz_N} \geq \frac{2}{1-k}. \quad (6.183)$$

Therefore, the P/O voltage in the DCM is

$$V_{O+} = V_{C1} + V_I = \left[2 + \frac{2k}{(1-k)m_R}\right] V_I = \left[2 + k^2(1-k)\frac{z_N}{2}\right] V_I \quad \text{with} \quad \sqrt{kz_N} \geq \frac{2}{1-k}. \tag{6.184}$$

For the current i_{L11}, because inductor current i_{L13}=0 at $t = t_1$, $V_{L13-off} = (k/(1-k)m_R)V_I$. For the current i_{L11}, we have

$$kTV_I = (1-k)m_R T(V_{C11} - 2V_I - V_{L13-off})$$

or

$$V_{C11} = \left[2 + \frac{2k}{(1-k)m_R}\right] V_I = \left[2 + k^2(1-k)\frac{z_N}{2}\right] V_I \quad \text{with} \quad \sqrt{kz_N} \geq \frac{2}{1-k}, \tag{6.185}$$

and for the current i_{L12} we have $kT(V_I + V_{C11} - V_{O-}) = (1-k)m_R T(V_{O-} - 2V_I - V_{L13-off})$. Therefore, the N/O voltage in the DCM is

$$V_{O-} = \left[2 + \frac{2k}{(1-k)m_R}\right] V_I = \left[2 + k^2(1-k)\frac{z_N}{2}\right] V_I \quad \text{with} \quad \sqrt{kz_N} \geq \frac{2}{1-k}. \tag{6.186}$$

So

$$V_O = V_{O+} = V_{O-} = \left[2 + k^2(1-k)\frac{z_N}{2}\right] V_I.$$

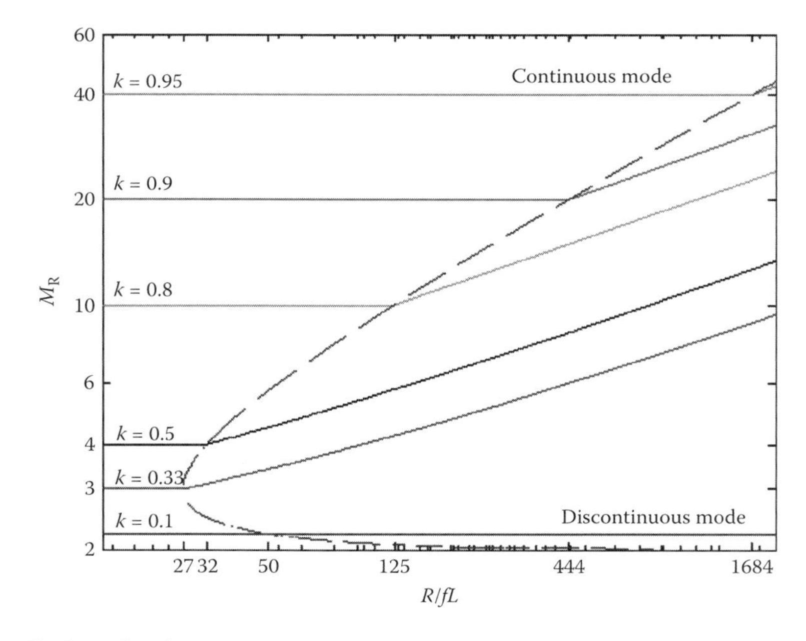

FIGURE 6.39 The boundary between continuous and discontinuous modes and the output voltage versus the normalized load $z_N = R/fL$ (D/O re-lift circuit). (Reprinted from Luo, F. L. and Ye, H. 2006. *Essential DC/DC Converters*. Boca Raton: Taylor & Francis Group LLC, p. 196. With permission.)

That is, the output voltage linearly increases as the load resistance R increases. Larger load resistance may cause higher output voltage in the discontinuous mode as shown in Figure 6.39.

6.6.3 Triple-Lift Circuit

The triple-lift circuit is shown in Figure 6.40.

The positive conversion path consists of a pump circuit S-L_1-D_0-C_1, a filter (C_2)-L_2-C_O, and a lift circuit D_1-C_2-D_2-C_3-D_3-L_3-D_4-C_4-D_5-L_4. The negative conversion path consists of a pump circuit S-L_{11}-D_{10}-(C_{11}), an "Π"-type filter C_{11}-L_{12}-C_{10}, and a lift circuit D_{11}-C_{12}-D_{22}-C_{13}-L_{13}-D_{12}-D_{23}-L_{14}-C_{14}-D_{13}.

6.6.3.1 *Positive Conversion Path*

The lift circuit is D_1-C_2-D_2-C_3-D_3-L_3-D_4-C_4-D_5-L_4. Capacitors C_2, C_3, and C_4 perform characteristics to lift the capacitor voltage V_{C1} to a level 3 times higher than the source voltage V_I. L_3, and L_4 perform the function of ladder joints to link the three capacitors C_3 and C_4 and lift the capacitor voltage V_{C1} up. Current $i_{C2}(t)$, $i_{C3}(t)$, and $i_{C4}(t)$ are exponential functions. They have large values at the moment of power switching on, but they are small because $v_{C3} = v_{C4} = V_I$ and $v_{C2} = V_{O+}$ in the steady state.

The output voltage and current are

$$V_{O+} = \frac{3}{1-k}V_I \quad \text{and} \quad I_{O+} = \frac{1-k}{3}I_{I+}$$

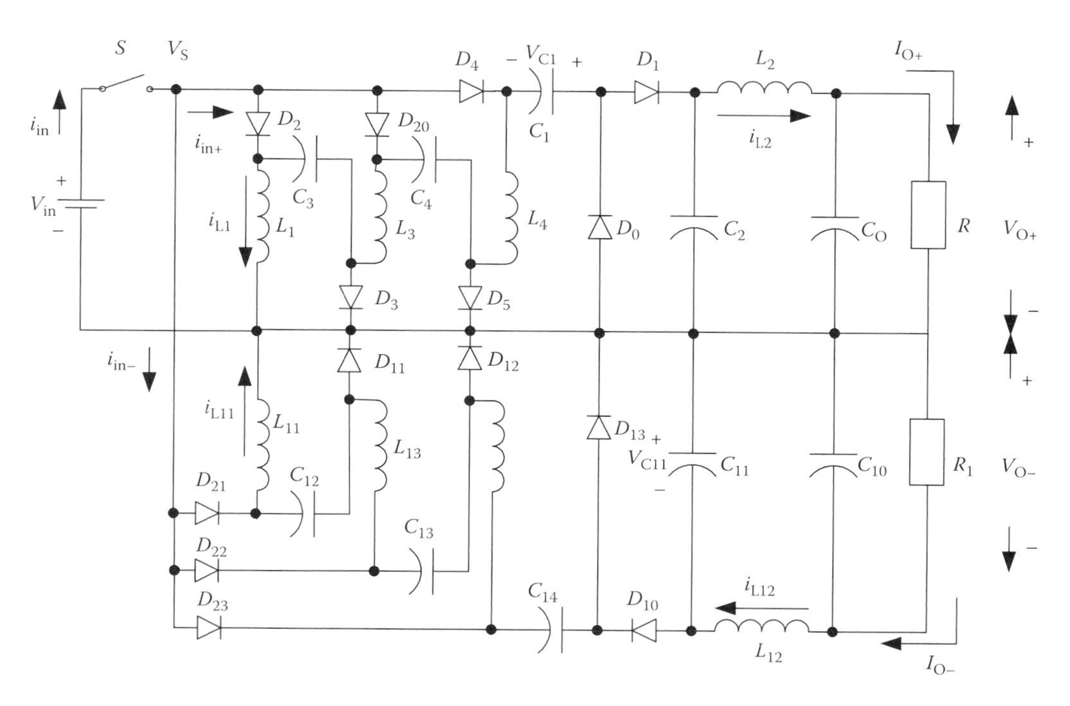

FIGURE 6.40 D/O triple-lift circuit. (Reprinted from Luo, F. L. and Ye, H. 2006. *Essential DC/DC Converters*. Boca Raton: Taylor & Francis Group LLC, p. 197. With permission.)

The voltage transfer gain in the continuous mode is

$$M_{T+} = \frac{V_{O+}}{V_I} = \frac{3}{1-k}. \tag{6.187}$$

$$\text{Other average voltages: } V_{C1} = \frac{2+k}{1-k}V_I; \quad V_{C3} = V_{C4} = V_I; \quad V_{CO} = V_{C2} = V_{O+}.$$

$$\text{Other average currents: } I_{L2} = I_{O+}; \quad I_{L1} = I_{L3} = I_{L4} = \frac{1}{3}I_{I+} = \frac{1}{1-k}I_{O+}.$$

$$\text{Current variations: } \xi_{1+} = \zeta_+ = \frac{k(1-k)R}{2M_T fL} = \frac{k}{M_T^2}\frac{3R}{2fL}; \quad \xi_{2+} = \frac{k}{16}\frac{1}{f^2C_2L_2};$$

$$\chi_{1+} = \frac{k}{M_T^2}\frac{3R}{2fL_3}; \quad \chi_{2+} = \frac{k}{M_T^2}\frac{3R}{2fL_4}.$$

$$\text{Voltage variations: } \rho_+ = \frac{3}{2(2+k)fC_1R}; \quad \sigma_{1+} = \frac{k}{2fC_2R}; \quad \sigma_{2+} = \frac{M_T}{2fC_3R};$$

$$\sigma_{3+} = \frac{M_T}{2fC_4R}.$$

The variation ratio of the output voltage V_{C0} is

$$\varepsilon_+ = \frac{k}{128}\frac{1}{f^3C_2C_OL_2R}. \tag{6.188}$$

6.6.3.2 Negative Conversion Path

The circuit C_{12}-D_{11}-L_{13}-D_{22}-C_{13}-D_{12}-L_{14}-D_{23}-C_{14}-D_{13} is the lift circuit. Capacitors C_{12}, C_{13}, and C_{14} perform characteristics to lift the capacitor voltage V_{C11} to a level 3 times higher than the source voltage V_I. L_{13} and L_{14} perform the function of ladder joints to link the three capacitors C_{12}, C_{13}, and C_{14} and lift the capacitor voltage V_{C11} up. Currents $i_{C12}(t)$, $i_{C13}(t)$, and $i_{C14}(t)$ are exponential functions. They have large values at the moment of power switching on, but they are small because $v_{C12} = v_{C13} = v_{C14} \cong V_I$ in the steady state.

The output voltage and current are

$$V_{O-} = \frac{3}{1-k}V_I \quad \text{and} \quad I_{O-} = \frac{1-k}{3}I_{I-}.$$

The voltage transfer gain in the continuous mode is

$$M_{T-} = \frac{V_{O-}}{V_I} = \frac{3}{1-k}. \tag{6.189}$$

From Equations 6.187 and 6.189, we define $M_T = M_{T+} = M_{T-}$. The curve of M_T versus k is shown in Figure 6.41.

$$\text{Other average voltages: } V_{C11} = V_{O-}; \quad V_{C12} = V_{C13} = V_{C14} = V_I.$$

$$\text{Other average currents: } I_{L12} = I_{O-}; \quad I_{L11} = I_{L13} = I_{L14} = \frac{1}{1-k} I_{O-}.$$

$$\text{Current variation ratios: } \zeta_- = \frac{k}{M_T^2} \frac{3R_1}{2fL_{11}}; \quad \xi_{2-} = \frac{k}{16} \frac{1}{f^2 C_{10} L_{12}};$$

$$\chi_{1-} = \frac{k(1-k)}{2M_T} \frac{R_1}{fL_{13}}; \quad \chi_{2-} = \frac{k(1-k)}{2M_T} \frac{R_1}{fL_{14}}.$$

$$\text{Voltage variation ratios: } \rho_- = \frac{k}{2} \frac{1}{fC_{11}R_1}; \quad \sigma_{1-} = \frac{M_T}{2} \frac{1}{fC_{12}R_1};$$

$$\sigma_{2-} = \frac{M_T}{2} \frac{1}{fC_{13}R_1}; \quad \sigma_{3-} = \frac{M_T}{2} \frac{1}{fC_{14}R_1}.$$

The variation ratio of output voltage V_{C10} is

$$\varepsilon_- = \frac{k}{128} \frac{1}{f^3 C_{11} C_{10} L_{12} R_1}. \tag{6.190}$$

6.6.3.3 Discontinuous Mode

To obtain the mirror-symmetrical D/O voltages, we purposely select: $L_1 = L_{11}$ and $R = R_1$.

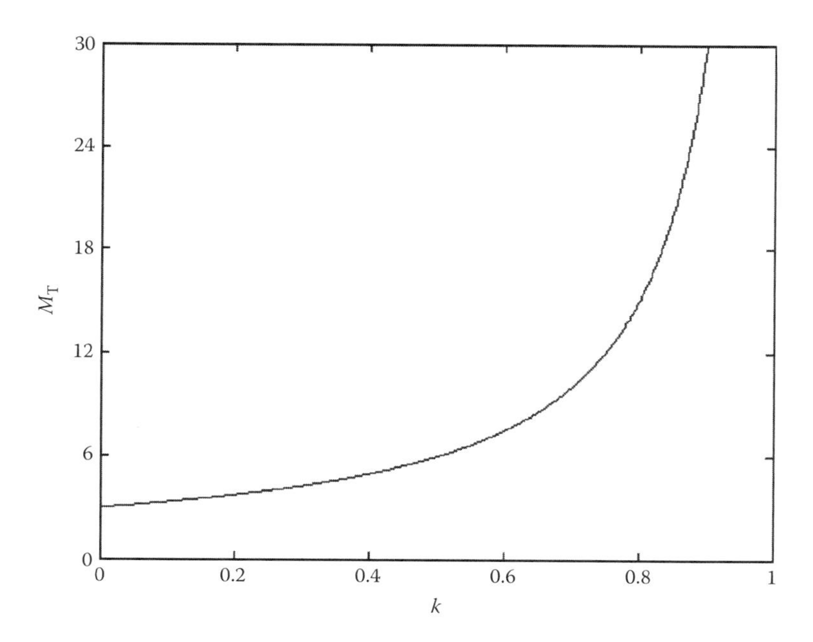

FIGURE 6.41 Voltage transfer gain M_T versus k. (Reprinted from Luo, F. L. and Ye, H. 2006. *Essential DC/DC Converters*. Boca Raton: Taylor & Francis Group LLC, p. 199. With permission.)

Define:

$$V_O = V_{O+} = V_{O-}, \quad M_T = M_{T+} = M_{T-} = V_O/V_I = (3/(1-k)), \quad z_N = z_{N+} = z_{N-},$$

and $\zeta = \zeta_+ = \zeta_-$

The freewheeling diode currents i_{D0} and i_{D10} become zero during switch-off before the next switch-on period. The boundary between continuous and discontinuous modes is $\zeta \geq 1$. The boundary between continuous and discontinuous modes is

$$M_T \leq \sqrt{\frac{3kz_N}{2}}. \quad (6.191)$$

This boundary curve is shown in Figure 6.42. It can be seen that the boundary curve has a minimum value of M_T that is equal to 4.5, corresponding to $k = 1/3$.

In the discontinuous mode, the currents i_{D0} and i_{D10} exist in the period between kT and $[k + (1-k)m_T]T$, where m_T is the filling efficiency, that is,

$$m_T = \frac{1}{\zeta} = \frac{2M_T^2}{3kz_N}. \quad (6.192)$$

Considering Equation 6.191, therefore, $0 < m_T < 1$. Since the current i_{D0} becomes zero at $t = t_1 = [k + (1-k)m_T]T$, for the current i_{L1}, i_{L3}, and i_{L4}, we have

$$3kTV_I = (1-k)m_T T(V_{C1} - 2V_I)$$

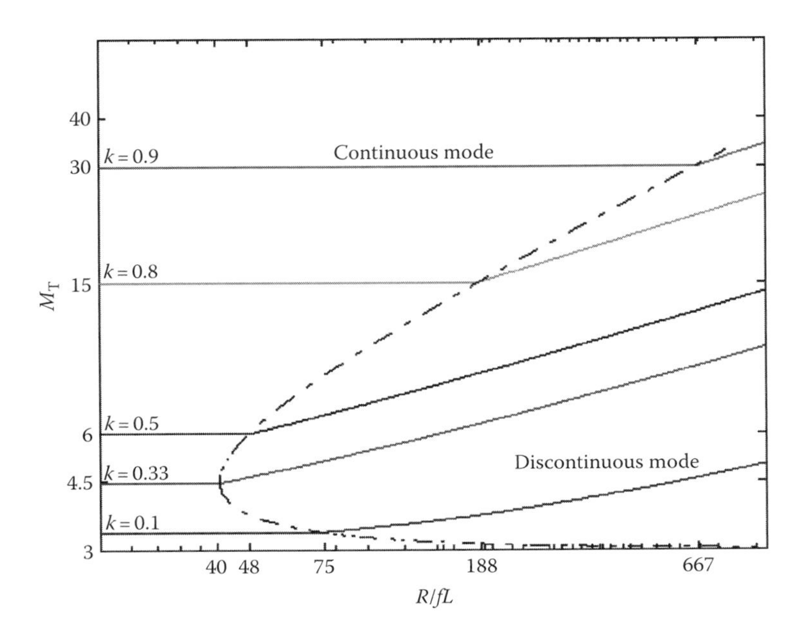

FIGURE 6.42 The boundary between continuous and discontinuous modes and the output voltage versus the normalized load $z_N = R/fL$ (D/O triple-lift circuit). (Reprinted from Luo, F. L. and Ye, H. 2006. *Essential DC/DC Converters*. Boca Raton: Taylor & Francis Group LLC, p. 201. With permission.)

or

$$V_{C1} = \left[2 + \frac{3k}{(1-k)m_T}\right] V_I = \left[2 + k^2(1-k)\frac{z_N}{2}\right] V_I \quad \text{with} \quad \sqrt{\frac{3kz_N}{2}} \geq \frac{3}{1-k}. \tag{6.193}$$

Therefore, the P/O voltage in the discontinuous mode is

$$V_{O+} = V_{C1} + V_I = \left[3 + \frac{3k}{(1-k)m_T}\right] V_I = \left[3 + k^2(1-k)\frac{z_N}{2}\right] V_I \quad \text{with} \quad \sqrt{\frac{3kz_N}{2}} \geq \frac{3}{1-k}. \tag{6.194}$$

Because inductor current $i_{L11} = 0$ at $t = t_1$,

$$V_{L13-\text{off}} = V_{L14-\text{off}} = \frac{k}{(1-k)m_T} V_I.$$

Since i_{D10} becomes 0 at $t_1 = [k + (1-k)m_T]T$, for the current i_{L11}, we have $kTV_I = (1-k)m_T - T(V_{C11} - 3V_I - V_{L13-\text{off}} - V_{L14-\text{off}})$
or

$$V_{C11} = \left[3 + \frac{3k}{(1-k)m_T}\right] V_I = \left[3 + k^2(1-k)\frac{z_N}{2}\right] V_I \quad \text{with} \quad \sqrt{\frac{3kz_N}{2}} \geq \frac{3}{1-k}, \tag{6.195}$$

and for the current i_{L12}, we have $kT(V_I + V_{C14} - V_{O-}) = (1-k)m_T - T(V_{O-} - 2V_I - V_{L13-\text{off}} - V_{L14-\text{off}})$.

Therefore, the N/O voltage in discontinuous mode is

$$V_{O-} = \left[3 + \frac{3k}{(1-k)m_T}\right] V_I = \left[3 + k^2(1-k)\frac{z_N}{2}\right] V_I \quad \text{with} \quad \sqrt{\frac{3kz_N}{2}} \geq \frac{3}{1-k}. \tag{6.196}$$

So $V_O = V_{O+} = V_{O-} = [3 + k^2(1-k)(z_N/2)]V_I$ that is, the output voltage linearly increases as the load resistance R increases. The output voltage increases as the load resistance R increases, as shown in Figure 6.42.

6.6.4 Quadruple-Lift Circuit

The quadruple-lift circuit is shown in Figure 6.43.

The positive conversion path consists of a pump circuit S-L_1-D_0-C_1 and a filter (C_2)-L_2-C_O, and a lift circuit D_1-C_2-L_3-D_2-C_3-D_3-L_4-D_4-C_4-D_5-L_5-D_6-C_5-S_1. The negative conversion path consists of a pump circuit S-L_{11}-D_{10}-(C_{11}) and an "Π"-type filter C_{11}-L_{12}-C_{10}, and a lift circuit D_{11}-C_{12}-D_{22}-L_{13}-C_{13}-D_{12}-D_{23}-L_{14}-C_{14}-D_{13}-D_{24}-L_{15}-C_{15}-D_{14}.

6.6.4.1 *Positive Conversion Path*

Capacitors C_2, C_3, C_4, and C_5 perform characteristics to lift the capacitor voltage V_{C1} to a level 4 times higher than the source voltage V_I. L_3, L_4, and L_5 perform the function as ladder joints to link the four capacitors C_2, C_3, C_4, and C_5, and lift the output capacitor voltage V_{C1}

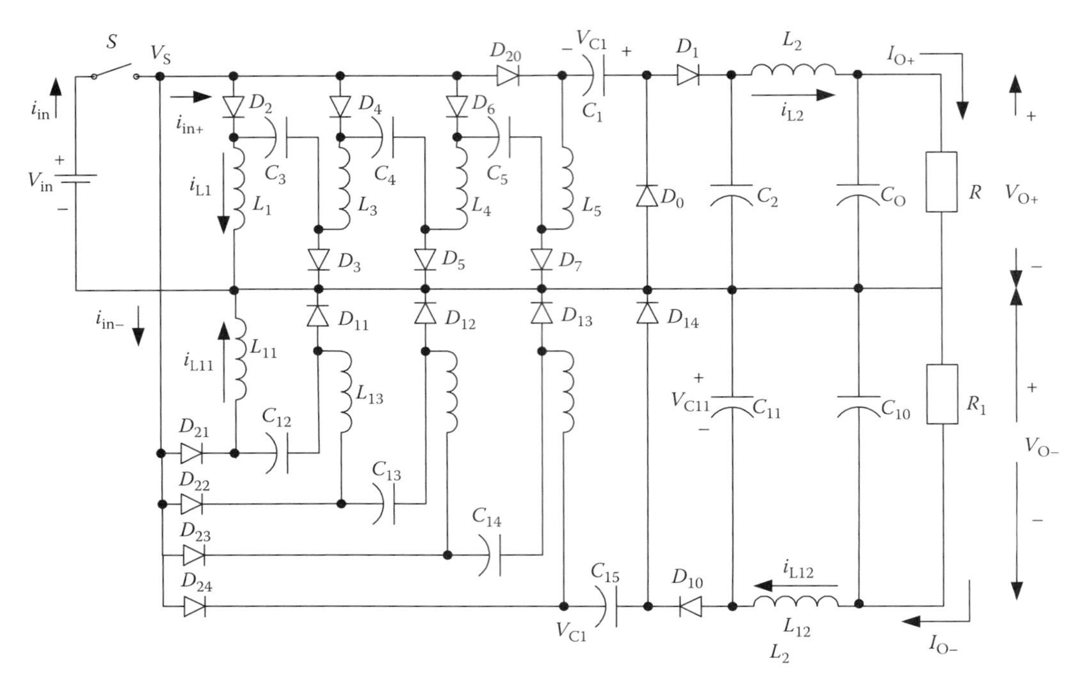

FIGURE 6.43 D/O quadruple-lift circuit. (Reprinted from Luo, F. L. and Ye, H. 2006. *Essential DC/DC Converters*. Boca Raton: Taylor & Francis Group LLC, p. 203. With permission.)

up. Current $i_{C2}(t)$, $i_{C3}(t)$, $i_{C4}(t)$, and $i_{C5}(t)$ are exponential functions. They have large values at the moment of power switching on, but they are small because $v_{C3} = v_{C4} = v_{C5} = V_I$ and $v_{C2} = V_{O+}$ in steady state.

The output voltage and current are

$$V_{O+} = \frac{4}{1-k}V_I \quad \text{and} \quad I_{O+} = \frac{1-k}{4}I_{I+}.$$

The voltage transfer gain in the continuous mode is

$$M_{Q+} = \frac{V_{O+}}{V_I} = \frac{4}{1-k}. \tag{6.197}$$

Other average voltages: $V_{C1} = \frac{3+k}{1-k}V_I; \quad V_{C3} = V_{C4} = V_{C5} = V_I; \quad V_{CO} = V_{C2} = V_O.$

Other average currents: $I_{L2} = I_{O+}; \quad I_{L1} = I_{L3} = I_{L4} = I_{L5} = \frac{1}{4}I_{I+} = \frac{1}{1-k}I_{O+}.$

Current variations: $\xi_{1+} = \zeta_+ = \frac{k(1-k)R}{2M_Q fL} = \frac{k}{M_Q^2}\frac{2R}{fL}; \quad \xi_{2+} = \frac{k}{16}\frac{1}{f^2C_2L_2};$

$$\chi_{1+} = \frac{k}{M_Q^2}\frac{2R}{fL_3}; \quad \chi_{2+} = \frac{k}{M_Q^2}\frac{2R}{fL_4}; \quad \chi_{3+} = \frac{k}{M_Q^2}\frac{2R}{fL_5}.$$

$$\text{Voltage variations: } \rho_+ = \frac{2}{(3+2k)fC_1R}; \quad \sigma_{1+} = \frac{M_Q}{2fC_2R};$$

$$\sigma_{2+} = \frac{M_Q}{2fC_3R}; \quad \sigma_{3+} = \frac{M_Q}{2fC_4R}; \quad \sigma_{4+} = \frac{M_Q}{2fC_5R}.$$

The variation ratio of output voltage V_{C0} is

$$\varepsilon_+ = \frac{k}{128}\frac{1}{f^3C_2C_OL_2R}. \tag{6.198}$$

6.6.4.2 Negative Conversion Path

Capacitors C_{12}, C_{13}, C_{14}, and C_{15} perform characteristics to lift the capacitor voltage V_{C11} to a level 4 times higher than the source voltage V_I. L_{13}, L_{14}, and L_{15} perform the function of ladder joints to link the four capacitors C_{12}, C_{13}, C_{14}, and C_{15}, and lift the output capacitor voltage V_{C11} up. Currents $i_{C12}(t)$, $i_{C13}(t)$, $i_{C14}(t)$, and $i_{C15}(t)$ are exponential functions. They have large values at the moment of power switching on, but they are small because $v_{C12} = v_{C13} = v_{C14} = v_{C15} \cong V_I$ in the steady state.

The output voltage and current are

$$V_{O-} = \frac{4}{1-k}V_I \quad \text{and} \quad I_{O-} = \frac{1-k}{4}I_{I-}.$$

The voltage transfer gain in the continuous mode is

$$M_{Q-} = \frac{V_{O-}}{V_I} = \frac{4}{1-k}. \tag{6.199}$$

From Equations 6.197 and 6.199, we define $M_Q = M_{Q+} = M_{Q-}$. The curve of M_Q versus k is shown in Figure 6.44.

$$\text{Other average voltages: } V_{C10} = V_{O-}; \quad V_{C12} = V_{C13} = V_{C14} = V_{C15} = V_I.$$

$$\text{Other average currents: } I_{L12} = I_{O-}; \quad I_{L11} = I_{L13} = I_{L14} = I_{L15} = \frac{1}{1-k}I_{O-}.$$

$$\text{Current variation ratios: } \zeta_- = \frac{k}{M_Q^2}\frac{2R_1}{fL_{11}}; \quad \xi_- = \frac{k}{16}\frac{1}{f^2CL_{12}};$$

$$\chi_{1-} = \frac{k(1-k)}{2M_Q}\frac{R_1}{fL_{13}}; \quad \chi_{2-} = \frac{k(1-k)}{2M_Q}\frac{R_1}{fL_{14}}; \quad \chi_{3-} = \frac{k(1-k)}{2M_Q}\frac{R_1}{fL_{15}}.$$

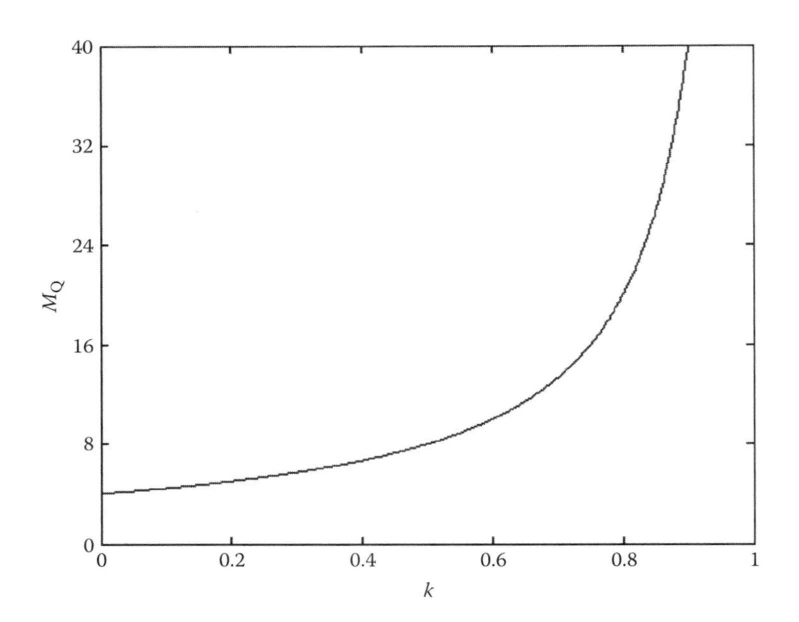

FIGURE 6.44 Voltage transfer gain M_Q versus k. (Reprinted from Luo, F. L. and Ye, H. 2006. *Essential DC/DC Converters*. Boca Raton: Taylor & Francis Group LLC, p. 205. With permission.)

$$\text{Voltage variation ratios: } \rho_- = \frac{k}{2}\frac{1}{fC_{11}R_1}; \quad \sigma_{1-} = \frac{M_Q}{2}\frac{1}{fC_{12}R_1};$$

$$\sigma_{2-} = \frac{M_Q}{2}\frac{1}{fC_{13}R_1}; \quad \sigma_{3-} = \frac{M_Q}{2}\frac{1}{fC_{14}R_1}; \quad \sigma_{4-} = \frac{M_Q}{2}\frac{1}{fC_{15}R_1}.$$

The variation ratio of output voltage V_{C10} is

$$\varepsilon_- = \frac{k}{128}\frac{1}{f^3C_{11}C_{10}L_{12}R_1}. \tag{6.200}$$

6.6.4.3 Discontinuous Conduction Mode

In order to obtain the mirror-symmetrical D/O voltages, we purposely select $L_1 = L_{11}$ and $R = R_1$. Therefore, we may define

$$V_O = V_{O+} = V_{O-}, \quad M_Q = M_{Q+} = M_{Q-} = \frac{V_O}{V_I} = \frac{4}{1-k},$$

$$z_N = z_{N+} = z_{N-}, \quad \text{and} \quad \zeta = \zeta_+ = \zeta_-.$$

The freewheeling diode currents i_{D0} and i_{D10} become zero during switch-off before the next switch-on period. The boundary between CCM and DCM is

$$\zeta \geq 1$$

or

$$M_Q \leq \sqrt{2kz_N}. \tag{6.201}$$

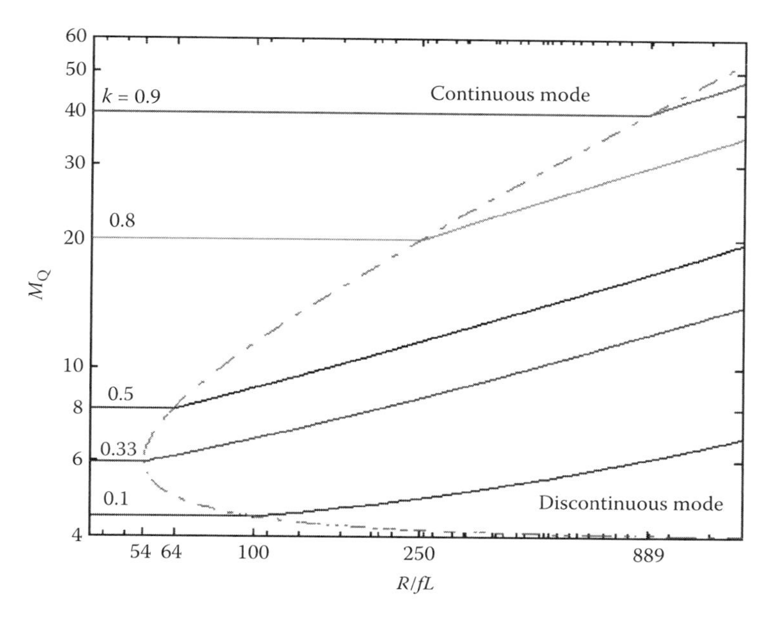

FIGURE 6.45 The boundary between continuous and discontinuous modes and the output voltage versus the normalized load $z_N = R/fL$ (D/O quadruple-lift circuit). (Reprinted from Luo, F. L. and Ye, H. 2006. *Essential DC/DC Converters*. Boca Raton: Taylor & Francis Group LLC, p. 206. With permission.)

This boundary curve is shown in Figure 6.45. It can be seen that it has a minimum value of M_Q that is equal to 6.0, corresponding to $k = 1/3$.

In the discontinuous mode, the currents i_{D0} and i_{D10} exist in the period between kT and $[k + (1-k)m_Q]T$, where m_Q is the filling efficiency, that is,

$$m_Q = \frac{1}{\zeta} = \frac{M_Q^2}{2kz_N}. \tag{6.202}$$

Considering Equation 6.201, therefore, $0 < m_Q < 1$. Since the current i_{D0} becomes zero at $t = t_1 = kT + (1-k)m_QT$, for the currents i_{L1}, i_{L3}, i_{L4}, and i_{L5}, we have

$$4kTV_I = (1-k)m_QT(V_{C1} - 3V_I)$$

or

$$V_{C1} = \left[3 + \frac{4k}{(1-k)m_Q}\right]V_I = \left[3 + k^2(1-k)\frac{z_N}{2}\right]V_I \quad \text{with} \quad \sqrt{2kz_N} \geq \frac{4}{1-k}. \tag{6.203}$$

Therefore, the P/O voltage in the DCM is

$$V_{O+} = V_{C1} + V_I = \left[4 + \frac{4k}{(1-k)m_Q}\right]V_I = \left[4 + k^2(1-k)\frac{z_N}{2}\right]V_I \quad \text{with} \quad \sqrt{2kz_N} \geq \frac{4}{1-k}. \tag{6.204}$$

Because inductor current $i_{L11} = 0$ at $t = t_1$,

$$V_{L13-\text{off}} = V_{L14-\text{off}} = V_{L15-\text{off}} = \frac{k}{(1-k)m_Q}V_I.$$

Since the current i_{D10} becomes zero at $t = t_1 = kT + (1-k)m_Q T$, for the current i_{L11}, we have

$$kTV_I = (1-k)m_Q - T(V_{C11} - 4V_I - V_{L13-off} - V_{L14-off} - V_{L15-off}).$$

So, with

$$V_{C11} = \left[4 + \frac{4k}{(1-k)m_Q}\right]V_I = \left[4 + k^2(1-k)\frac{z_N}{2}\right]V_I \quad \text{with} \quad \sqrt{2kz_N} \geq \frac{4}{1-k}. \tag{6.205}$$

For the current i_{L12}, we have $kT(V_I + V_{C15} - V_{O-}) = (1-k)m_Q T(V_{O-} - 2V_I - V_{L13-off} - V_{L14-off} - V_{L15-off})$.

Therefore, the N/O voltage in the DCM is

$$V_{O-} = \left[4 + \frac{4k}{(1-k)m_Q}\right]V_I = \left[4 + k^2(1-k)\frac{z_N}{2}\right]V_I \quad \text{with} \quad \sqrt{2kz_N} \geq \frac{4}{1-k}. \tag{6.206}$$

So $V_O = V_{O+} = V_{O-} = [4 + k^2(1-k)(z_N/2)]V_I$, that is, the output voltage linearly increases as the load resistance R increases. It can be seen that the output voltage increases as the load resistance R increases, as shown in Figure 6.45.

6.6.5 Summary

6.6.5.1 Positive Conversion Path

From the analysis and calculation in previous sections, the common formulae for all circuits can be obtained:

$$M = \frac{V_{O+}}{V_I} = \frac{I_{I+}}{I_{O+}}; \quad z_N = \frac{R}{fL}; \quad R = \frac{V_{O+}}{I_{O+}};$$

$$L = \frac{L_1 L_2}{L_1 + L_2} \quad \text{for the elementary circuit only;}$$

$$L = L_1 \quad \text{for other lift circuits.}$$

Current variations: $\xi_{1+} = \dfrac{1-k}{2M_E}\dfrac{R}{fL_1}$ and $\xi_{2+} = \dfrac{k}{2M_E}\dfrac{R}{fL_2}$ for the elementary circuit only;

$$\xi_{1+} = \zeta_+ = \frac{k(1-k)R}{2MfL} \quad \text{and} \quad \xi_{2+} = \frac{k}{16}\frac{1}{f^2C_2L_2} \quad \text{for other lift circuits;}$$

$$\zeta_+ = \frac{k(1-k)R}{2MfL}; \quad \chi_{j+} = \frac{k}{M^2}\frac{R}{fL_{j+2}}, \quad j = 1, 2, 3, \ldots.$$

$$\text{Voltage variations: } \rho_+ = \frac{k}{2fC_1R}; \quad \varepsilon_+ = \frac{k}{8M_E}\frac{1}{f^2C_OL_2} \quad \text{for the elementary circuit only;}$$

$$\rho_+ = \frac{M}{M-1}\frac{1}{2fC_1R}; \quad \varepsilon_+ = \frac{k}{128}\frac{1}{f^3C_2C_OL_2R} \quad \text{for other lift circuits;}$$

$$\sigma_{1+} = \frac{k}{2fC_2R}; \quad \sigma_{j+} = \frac{M}{2fC_{j+1}R}, \quad j = 2, 3, 4, \ldots.$$

6.6.5.2 Negative Conversion Path

From the analysis and calculation in previous sections, the common formulae for all circuits can be obtained:

$$M = \frac{V_{O-}}{V_I} = \frac{I_{I-}}{I_{O-}}; \quad z_{N-} = \frac{R_1}{fL_{11}}; \quad R_1 = \frac{V_{O-}}{I_{O-}}.$$

$$\text{Current variation ratios: } \zeta_- = \frac{k(1-k)R_1}{2MfL_{11}}; \quad \xi_- = \frac{k}{16f^2C_{11}L_{12}}; \quad \chi_{j-} = \frac{k(1-k)R_1}{2MfL_{j+2}},$$

$$j = 1, 2, 3, \ldots.$$

$$\text{Voltage variation ratios: } \rho_- = \frac{k}{2fC_{11}R_1}; \quad \varepsilon_- = \frac{k}{128f^3C_{11}C_{10}L_{12}R_1}; \quad \sigma_{j-} = \frac{M}{2fC_{j+11}R_1},$$

$$j = 1, 2, 3, 4, \ldots.$$

6.6.5.3 Common Parameters

Usually, we select the loads $R = R_1$, $L = L_{11}$, so that we obtain $z_N = z_{N+} = z_{N-}$. In order to write common formulae for the boundaries between continuous and discontinuous modes and output voltage for all circuits, the circuits can be numbered. The definition is that subscript $j = 0$ denotes the elementary circuit, 1 the self-lift circuit, 2 the re-lift circuit, 3 the triple-lift circuit, 4 the quadruple-lift circuit, and so on.

The voltage transfer gain is

$$M_j = \frac{k^{h(j)}[j + h(j)]}{1-k}, \quad j = 0, 1, 2, 3, 4, \ldots.$$

The characteristics of output voltage of all circuits are shown in Figure 6.46.

The freewheeling diode current's variation is given by

$$\zeta_j = \frac{k^{[1+h(j)]}}{M_j^2}\frac{j + h(j)}{2}z_N.$$

The boundaries are determined by the condition:

$$\zeta \geq 1$$

or

$$\frac{k^{[1+h(j)]}}{M_j^2}\frac{j + h(j)}{2}z_N \geq 1, \quad j = 0, 1, 2, 3, 4, \ldots.$$

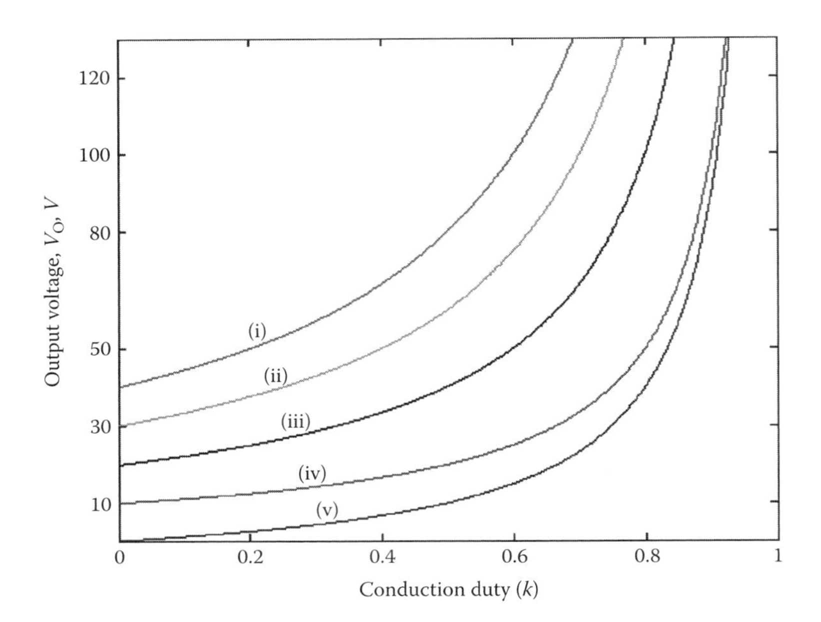

FIGURE 6.46 Output voltages of all D/O Luo-converters ($V_I = 10$ V). (i) Quadruple-lift circuit; (ii) triple-lift circuit; (iii) re-lift circuit; (iv) self-lift circuit; and (v) elementary circuit. (Reprinted from Luo, F. L. and Ye, H. 2006. *Essential DC/DC Converters*. Boca Raton: Taylor & Francis Group LLC, p. 211. With permission.)

Therefore, the boundaries between continuous and discontinuous modes for all circuits are

$$M_j = k^{(1+h(j))/2}\sqrt{\frac{j+h(j)}{2}z_N}, \quad j = 0,1,2,3,4,\ldots.$$

The filling efficiency is

$$m_j = \frac{1}{\zeta_j} = \frac{M_j^2}{k^{[1+h(j)]}}\frac{2}{j+h(j)}\frac{1}{z_N}, \quad j = 0,1,2,3,4,\ldots.$$

The output voltage in the discontinuous mode for all circuits is

$$V_{O-j} = \left[j + k^{[2-h(j)]}\frac{1-k}{2}z_N\right]V_I,$$

where

$$h(j) = \begin{cases} 0 & \text{if} \quad j \geq 1, \\ 1 & \text{if} \quad j = 0, \end{cases} \quad j = 0,1,2,3,4\ldots;$$

where $h(j)$ is the Hong function.

The boundaries between continuous and discontinuous modes of all circuits are shown in Figure 6.47. The curves of all M versus z_N suggest that the continuous mode area increases from M_E via M_S, M_R, and M_T to M_Q. The boundary of the elementary circuit is a monorising curve, but other curves are not monorising. There are minimum values of the boundaries of other circuits, which for M_S, M_R, M_T, and M_Q correspond at $k = 1/3$.

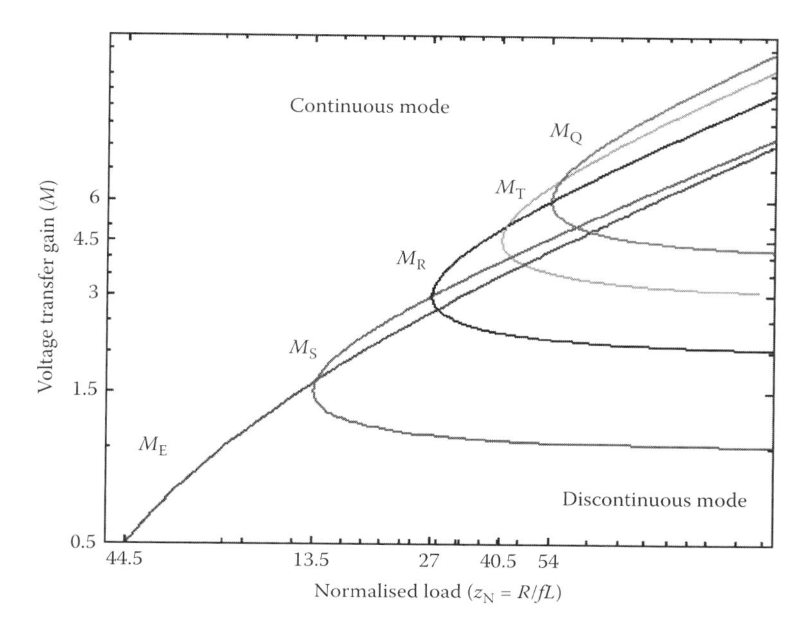

FIGURE 6.47 Boundaries between continuous and discontinuous modes of all D/O Luo-converters. (Reprinted from Luo, F. L. and Ye, H. 2006. *Essential DC/DC Converters*. Boca Raton: Taylor & Francis Group LCC, p. 212. With permission.)

6.7 VL Cúk-Converters

The proposed N/O Cúk-converters were developed from the Cúk-converter, as shown in Figure 5.32. They are as follows:

- Elementary self-lift circuit
- Developed self-lift circuit
- Re-lift circuit
- Multiple-lift circuits (e.g., triple-lift and quadruple-lift circuits).

These converters perform positive to negative DC–DC voltage is increasing conversion with higher voltage transfer gains, power density, small ripples, high efficiency, and cheap topology in a simple structure [17–19].

6.7.1 Elementary Self-Lift Cúk Circuit

The elementary self-lift circuit is derived from the Cúk-converter by adding the components ($D_1 - C_1$). The circuit diagram is shown in Figure 6.48. The lift circuit consists of L_1-D_1-C_1, and it is a basic VL cell. When switch S turns on, D_1 is on and D_o is off. When switch S turns off, D_1 is off and D_o is on. The capacitor C_1 performs characteristics to lift the output capacitor voltage V_{Co} to a level higher than the capacitor voltage V_{Cs}.

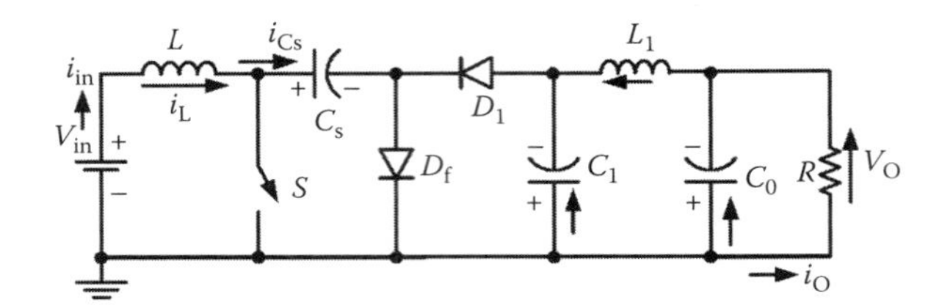

FIGURE 6.48 Elementary self-lift Cúk-converter.

In the steady state, the average voltage across inductor L_1 over a period is zero. Thus

$$V_{C1} = V_{Co} = V_O.$$

During the switch-on period, the voltages across capacitor C_1 is equal to the voltage across C_s. Since C_s and C_1 are sufficiently large, we have $V_{C1} = V_{Cs} = V_O$.

The inductor current i_L increases during switch-on and decreases during switch-off. The corresponding voltages across L are V_{in} and $-(V_{Cs} - V_{in})$. Therefore, $kTV_{in} = (1-k)T(V_{Cs} - V_{in})$.

Hence, the voltage transfer gain of the elementary self-lift circuit is

$$M_S = \frac{V_O}{V_{in}} = \frac{1}{1-k}. \tag{6.207}$$

6.7.2 Developed Self-Lift Cúk Circuit

The developed self-lift circuit is derived from the elementary self-lift Cúk circuit by adding the components $(D_o - S_1)$ and redesigning the connection of L_1. Static switches S and S_1 are switched on simultaneously. The circuit diagram is shown in Figure 6.49. The lift circuit consists of C_1-L_1-S_1-D_1. When switches S and S_1 turn on, D_1 is on and D_f and D_o are off. When S and S_1 turn off, D_1 is off and D_f and D_o are on. The capacitor C_1 performs characteristics to lift the output capacitor voltage V_{CO} to a level higher than the capacitor voltage V_{Cs}.

During the switch-on period, the voltage across capacitor C_1 is equal to the voltage across C_s. Since C_s and C_1 are sufficiently large, we have $V_{C1} = V_{Cs} = (1/(1-k))V_{in}$.

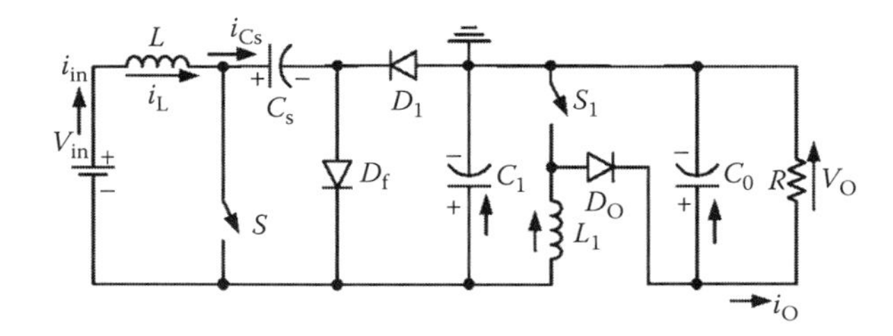

FIGURE 6.49 Developed self-lift Cúk circuit.

The inductor current i_{L1} increases during switch-on and decreases during switch-off. The corresponding voltages across L are V_{Cs} and $-(V_O - V_{C1})$. Therefore,

$$kTV_{Cs} = (1-k)T(V_O - V_{C1}).$$

Hence, the voltage transfer gain of the developed self-lift circuit is

$$M'_S = \frac{V_O}{V_{in}} = \frac{1}{(1-k)^2}. \tag{6.208}$$

6.7.3 Re-Lift Cúk Circuit

The re-lift circuit is derived from the developed self-lift Cúk circuit by adding the components (D_2-C_2-L_2-D_3). Static switches S and S_1 are switched on simultaneously. The circuit diagram is shown in Figure 6.50. The lift circuit consists of L_1-D_1-C_1-D_2-C_2-L_2-D_3-S_1 and it can be divided into two basic VL cells. When switches S and S_1 turn on, D_1, D_2, and D_3 are on, and D_O is off. When S and S_1 turn off, D_1, D_2, and D_3 are off and D_O is on. Capacitors C_1 and C_2 perform characteristics to lift the output capacitor voltage V_{Co} to a level 2 times higher than the capacitor voltage V_{Cs}. L_2 performs the function of a ladder joint to link the two capacitors C_1 and C_2 and lift V_{Co}. To avoid the abnormal phenomena of diodes during switch-off, it is assumed that L_1 and L_2 are the same to simplify the theoretical analysis.

During the switch-on period, both the voltages across capacitors C_1 and C_2 are equal to the voltage across C_s. Since C_s, C_1, and C_2 are sufficiently large, we have

$$V_{C1} = V_{C2} = V_{Cs} = V_{in} = \frac{1}{1-k}V_{in}.$$

The voltage across L_1 is equal to V_{Cs} during switch-on. With the second voltage balance, we have $V_{L1-off} = (k/1-k)V_{Cs}$.

The inductor current i_{L2} increases during switch-on and decreases during switch-off. The corresponding voltages across L_2 are V_{Cs} and $-(V_O - V_{C1} - V_{C2} - V_{L1-off})$. Therefore,

$$kTV_{Cs} = (1-k)T(V_O - V_{C1} - V_{C2} - V_{L1-off}).$$

Hence, the voltage transfer gain of the re-lift circuit is

$$M_R = \frac{V_O}{V_{in}} = \frac{2}{(1-k)^2}. \tag{6.209}$$

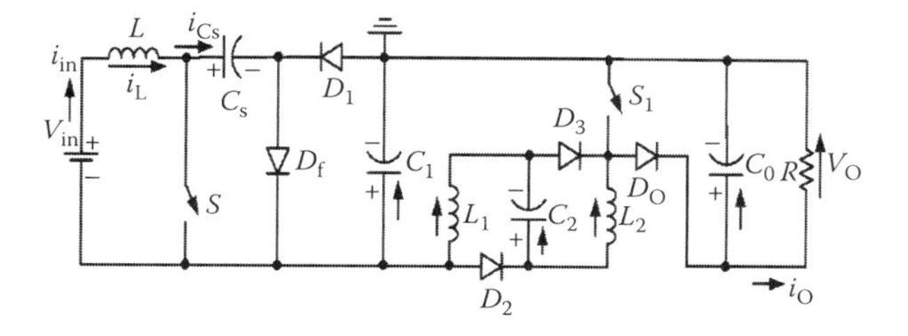

FIGURE 6.50 Re-lift Cúk circuit.

6.7.4 Multiple-Lift Cúk Circuit

It is possible to construct a multiple-lift circuit by adding the components (D_2-C_2-L_2-D_3). Assuming that there are n VL cells, the generalized representation of multiple-lift circuits is shown in Figure 6.51. Only two synchronous switches S and S_1 are required for each complex multiple-lift circuit, which simplifies the control scheme and decreases the cost significantly. Hence, each circuit has two switches, $(n+1)$ inductors, $(n+1)$ capacitors, and $(2n-1)$ diodes. It is noted that all inductors existing in the VL cells are the same here for the reasons explained in the re-lift circuit. All the capacitors are sufficiently large. From the foregoing analysis and calculation, the general formulae for all multiple-lift circuits can be obtained according to similar steps.

The generalized voltage transfer gain is

$$M = \frac{n}{(1-k)^{h(n)}}, \quad n = 1,2,3,4,\ldots, \tag{6.210}$$

where

$$h(n) = \begin{cases} 1 & \text{self-lift} \\ 2 & \text{others} \end{cases}$$

If the generalized circuit possesses three VL cells, it is termed the triple-lift circuit. If the generalized circuit possesses four VL cells, it is termed the quadruple-lift circuit.

6.7.5 Simulation and Experimental Verification of an Elementary and a Developed Self-Lift Circuit

Referring to Figures 6.48 and 6.49, we set these two circuits to have the same conditions: $V_{in} = 10$ V, $R = 100\ \Omega$, $L = 1$ mH, $L_1 = 500\ \mu$H, $C_s = 110\ \mu$F, $C_1 = 22\ \mu$F, $C_O = 47\ \mu$F, $k = 0.5$, and $f = 100$ kHz. According to Equation 6.207, the theoretical value V_O of the elementary self-lift circuit is equal to 20 V. According to Equation 6.208, the theoretical value V_O of the developed self-lift circuit is equal to 40 V. The simulation results of Psim are shown in Figure 6.52, where curve 1 is for the v_O of the elementary self-lift circuit and curve 2 is for the v_O of the developed self-lift circuit. The steady-state values in the simulation identically match the theoretical analysis.

Similar parameters are chosen to construct the corresponding testing hardware circuits. A single n-channel MOSFET is used in the elementary self-lift circuit. Two n-channel MOSFETs are used in the developed self-lift circuit. The corresponding experimental curves in the

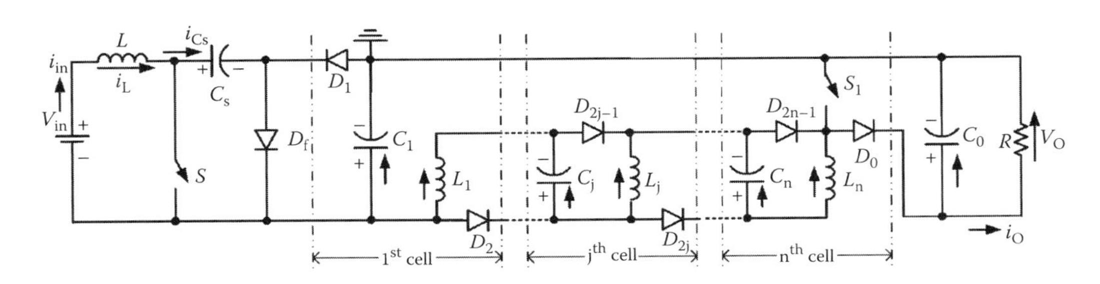

FIGURE 6.51 Generalized representation of N/O Cúk-converters.

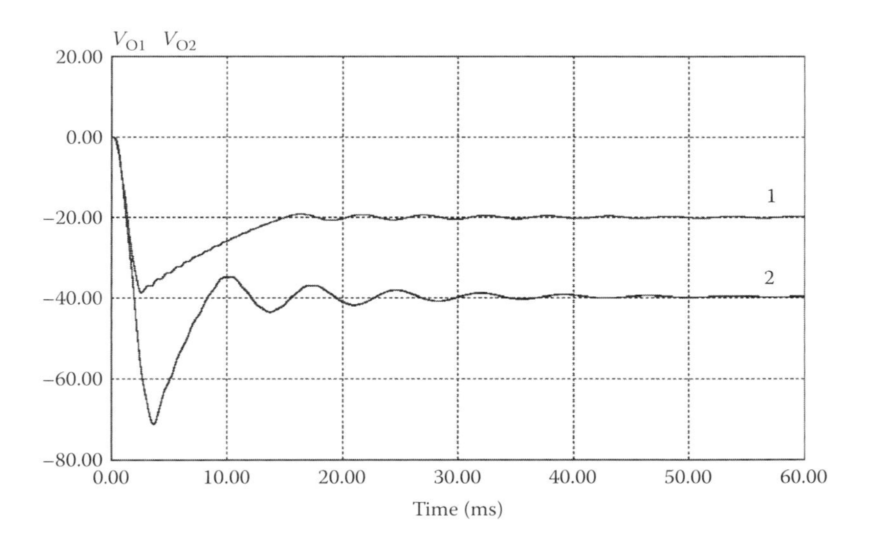

FIGURE 6.52 Simulation results of the elementary and developed self-lift circuits.

steady state are shown in Figure 6.53. The curve shown in Channel 1 with 10 V/Div corresponds to the output voltage of the elementary self-lift circuit, which is about 19 V. The curve shown in Channel 2 with 10 V/Div corresponds to the output voltage of the developed self-lift circuit, which is about 37 V. Considering the effects caused by the parasitic parameters, we can see that the measured results are very close to the theoretical analysis and simulation results.

6.8 VL SEPICs

The proposed P/O SEPICs are developed from SEPIC as shown in Figure 5.33. They are as follows:

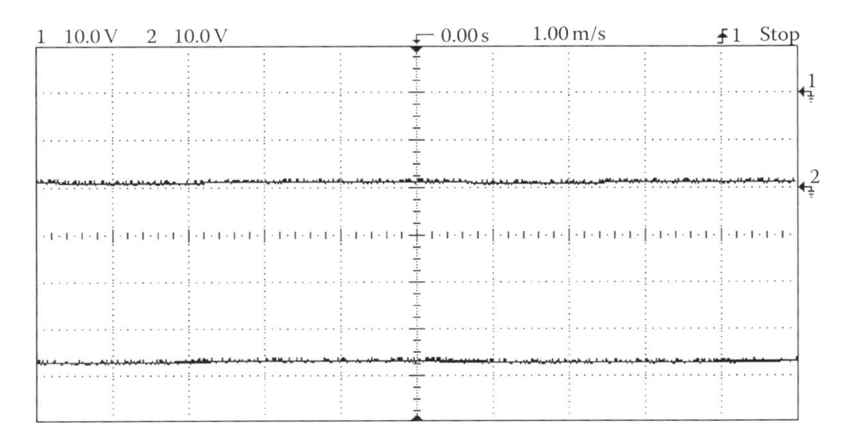

FIGURE 6.53 Experimental results of the elementary and developed self-lift circuits.

- Self-lift circuit
- Re-lift circuit
- Multiple circuits (e.g., triple-lift and quadruple-lift circuits).

These converters perform positive-to-positive DC–DC voltage-increasing conversion with higher voltage transfer gains, power density, small ripples, high efficiency, and cheap topology in a simple structure [18–21].

6.8.1 Self-Lift SEPIC

The self-lift circuit is derived from the SEPIC converter by adding the components D_1-C_1. The circuit diagram is shown in Figure 6.54. The lift circuit consists of L_1-D_1-C_1 and is a basic VL cell. When switch S turns on, D_1 is on and D_o is off. When switch S turns off, D_1 is off and D_o is on. Capacitor C_1 performs characteristics to lift the output capacitor voltage V_{Co} to a level higher than the capacitor voltage V_{Cs}.

In the steady state, the average voltage across inductor L over a period is zero. Thus

$$V_{Cs} = V_{in}.$$

During the switch-on period, the voltage across capacitor C_1 is equal to the voltage across C_s. Since C and C_1 are sufficiently large, we have $V_{C1} = V_{Cs} = V_{in}$.

The inductor current i_L increases during switch-on and decreases during switch-off. The corresponding voltages across L are V_{Cs} and $-(V_{Co} - V_{C1} - V_{in} + V_{Cs})$. Therefore,

$$kTV_{Cs} = (1-k)T(V_{Co} - V_{C1} - V_{in} + V_{Cs}).$$

Hence, the voltage transfer gain of the self-lift circuit is

$$M_S = \frac{V_O}{V_{in}} = \frac{1}{1-k}. \tag{6.211}$$

6.8.2 Re-Lift SEPIC

The re-lift circuit is derived from the self-lift circuit by adding the components L_2-D_2-C_2-S_1. Static switches S and S_1 are switched on simultaneously. The circuit diagram and equivalent circuits during switch-on and switch-off are shown in Figure 6.55. The lift circuit consists of L_1-D_1-C_1-L_2-D_2-C_2-S_1 and can be divided into two basic VL cells. When switches S and S_1

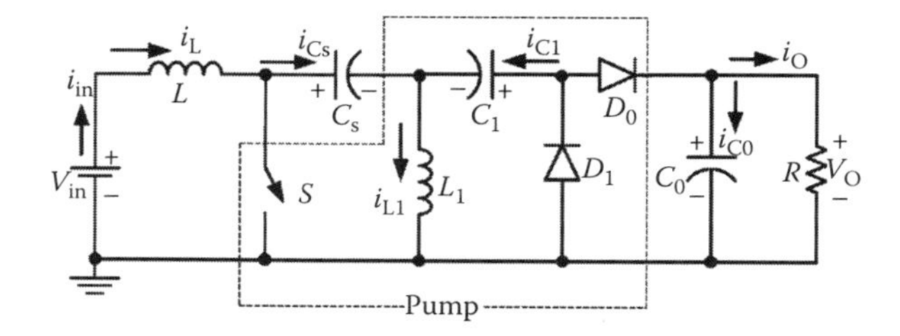

FIGURE 6.54 Self-lift SEPIC.

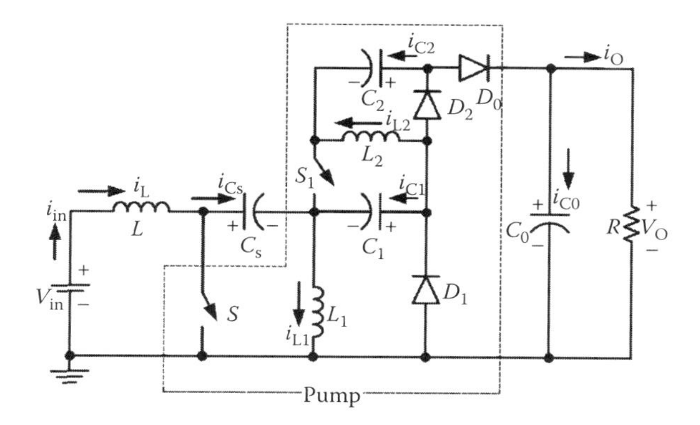

FIGURE 6.55 Re-lift SEPIC.

turn on, D_1 and D_2 are on and D_o is off. When S and S_1 turn off, D_1 and D_2 are off and D_o is on. Capacitors C_1 and C_2 perform characteristics to lift the output capacitor voltage V_{Co} to a level 2 times higher than the capacitor voltage V_{Cs}. L_2 performs the function of a ladder joint to link the two capacitors C_1 and C_2 and lift V_{Co}. To avoid the abnormal phenomena of diodes during switch-off [11], it is assumed that L_1 and L_2 are the same, which simplifies the theoretical analysis.

In steady state, both the average voltages across inductors L and L_1 over a period equal zero. Thus $V_{Cs} = V_{in}$.

During the switch-on period, both the voltages across capacitors C_1 and C_2 are equal to the voltage across C_s. Since C, C_1, and C_2 are sufficiently large, we have

$$V_{C1} = V_{C2} = V_{Cs} = V_{in}.$$

The voltage across L_1 is equal to V_{Cs} during switch-on. With the second voltage balance, we have $V_{L1-off} = (k/(1-k))V_{in}$.

The inductor current i_{L2} increases during switch-on and decreases during switch-off. The corresponding voltages across L_2 are V_{Cs} and $-(V_{Co} - V_{C1} - V_{C2} - V_{L1-off})$. Therefore,

$$kTV_{Cs} = (1-k)T(V_{Co} - V_{C1} - V_{C2} - V_{L1-off}).$$

Hence, the voltage transfer gain of the re-lift circuit is

$$M_R = \frac{V_O}{V_{in}} = \frac{2}{1-k}. \tag{6.212}$$

6.8.3 Multiple-Lift SEPICs

It is possible to construct a multiple-lift circuit by adding the components L_2-D_2-C_2-S_1. Assuming that there are n VL cells, the generalized representation of multiple-lift circuits is shown in Figure 6.56.

All future active switches can be replaced by passive diodes. According to this principle, only two synchronous switches S and S_1 are required for each complex multiple-lift circuit, which simplifies the control scheme and decreases the cost significantly. Hence, each circuit has two switches, $(n+1)$ inductors, $(n+1)$ capacitors, and $(2n-1)$ diodes. It is noted that all

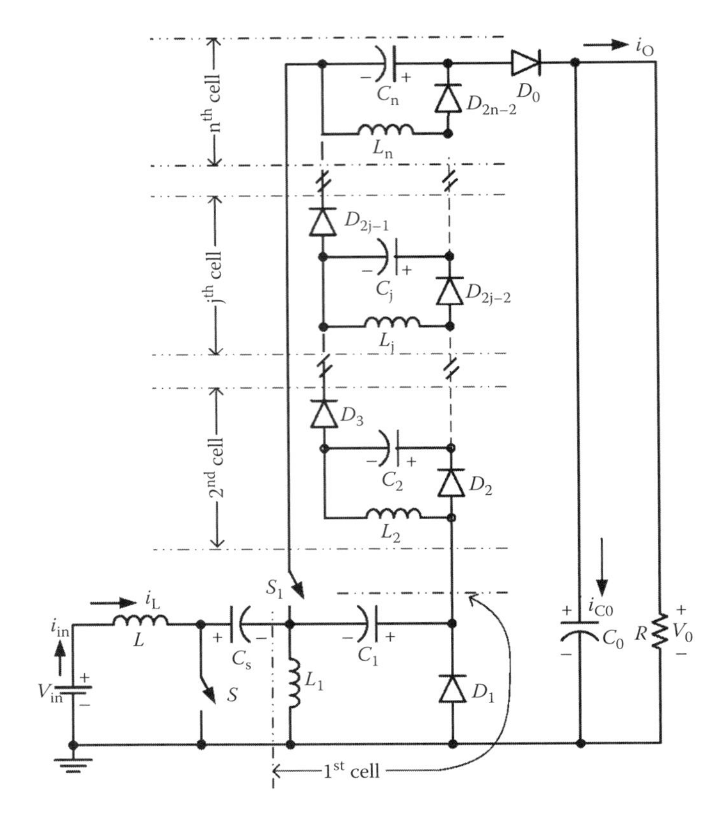

FIGURE 6.56 Multi-lift SEPIC.

inductors existing in the VL cells are the same here for the reasons explained in the re-lift circuit. All the capacitors are sufficiently large. From the foregoing analysis and calculation, the general formulae for all multiple-lift circuits can be obtained according to similar steps. The generalized voltage transfer gain is

$$M = \frac{n}{1-k}, \quad n = 1, 2, 3, 4, \ldots. \tag{6.213}$$

If the generalized circuit possesses three VL cells, it is termed the triple-lift circuit. If the generalized circuit possesses four VL cells, it is termed the quadruple-lift circuit.

6.8.4 Simulation and Experimental Results of a Re-Lift SEPIC

The circuit parameters for simulation are $V_{in} = 10\,\text{V}$, $R = 100\,\Omega$, $L = 1\,\text{mH}$, $L_1 = L_2 = 500\,\mu\text{H}$, $C_s = 110\,\mu\text{F}$, $C_1 = C_2 = 22\,\mu\text{F}$, $C_O = 110\,\mu\text{F}$, and $k = 0.6$. The switching frequency f is 100 kHz. According to Equation 6.212, we obtain the theoretical value V_O, which is equal to 50 V. The simulation results of Psim are shown in Figure 6.57, where curves 1–3 are for v_O, i_{L2}, and i_{L1}, respectively. The steady-state performance in the simulation identically matches the theoretical analysis.

Similar parameters are chosen to construct a testing hardware circuit. Two n-channel MOSFETs 2SK2267 are selected. We obtained the output voltage value of $V_O = 46.2\,\text{V}$

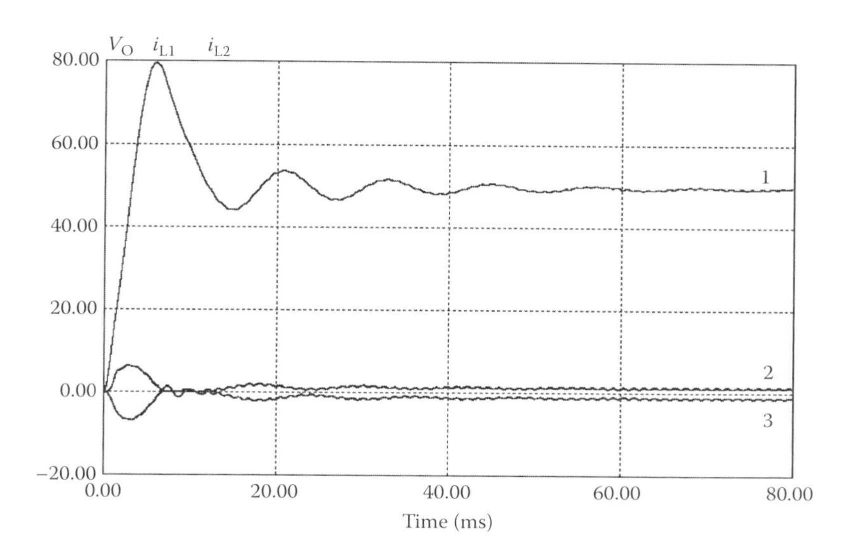

FIGURE 6.57 Simulation result of a re-lift SEPIC.

(shown in Channel 1 with 10 V/Div) and the capacitor value of $V_{Cs} = 9.9$ V (shown in Channel 1 with 10 V/Div). The corresponding experimental curves in the steady state are shown in Figure 6.58. The practical output voltage is smaller than the theoretical values due to the effects caused by parasitic parameters. It is seen that the measured results are very close to the theoretical analysis and simulation results.

6.9 Other D/O Voltage-Lift Converters

For all the above-mentioned converters, each topology is divided into two sections: the source section including voltage source, inductor L, and active switch S, and the pump

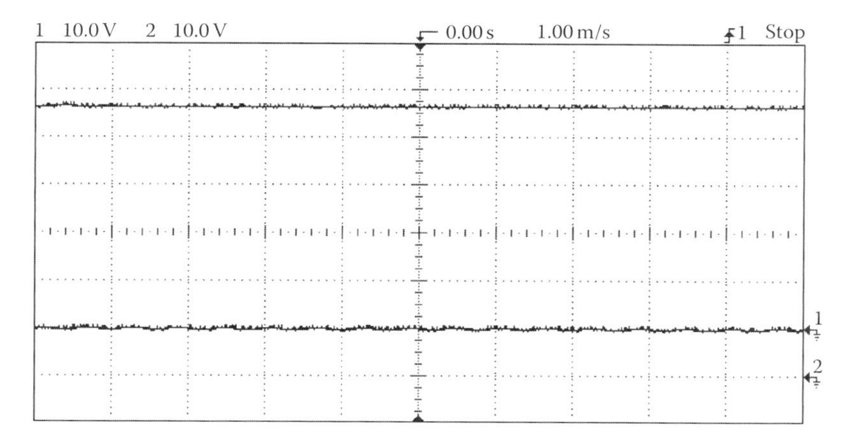

FIGURE 6.58 Experimental result of a re-lift SEPIC.

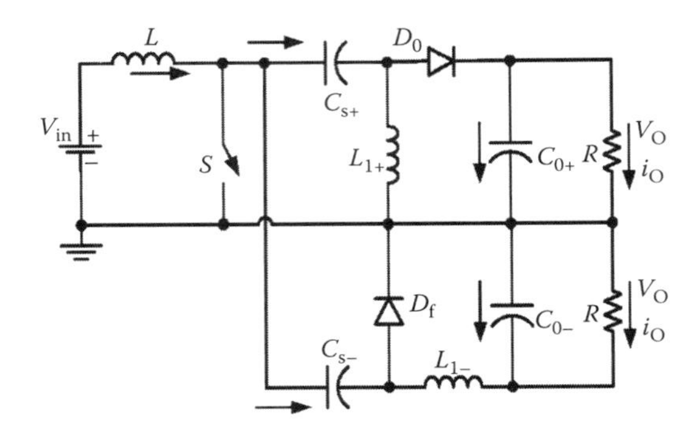

FIGURE 6.59 Novel elementary D/O converter.

section consisting of the rest of the components. Each topology can be considered as a special cascade connection of these two sections.

We compare the SEPIC converter to the Cúk-converter; both converters have the same source sections and the same voltage transfer gains with opposite polarities. Hence, a series of novel D/O converters based on the SEPIC and Cúk-converters can be constructed by combining the two converters at the input side. They are the elementary circuit, the self-lift circuit, and the corresponding enhanced series [18,19].

6.9.1 Elementary Circuit

Combining the prototypes of the SEPIC and Cúk-converters, we obtain the elementary circuit of novel D/O converters, which is shown in Figure 6.59. The positive conversion path is the same as that of the SEPIC converter. The negative conversion path is the same as that of the Cúk-converter. Hence, from the foregoing analysis and calculation, the voltage transfer gains are obtained as

$$M_{E+} = \frac{V_{O+}}{V_{in}} = \frac{k}{1-k},$$
$$M_{E-} = \frac{V_{O-}}{V_{in}} = -\frac{k}{1-k}. \tag{6.214}$$

6.9.2 Self-Lift D/O Circuit

The self-lift circuit is a derivative of the elementary circuit shown in Figure 6.60.

The positive conversion path is the same as that of the self-lift SEPIC converter. The negative conversion path is the same as that of the self-lift Cúk-converter. Hence, from the foregoing analysis and calculation, the voltage transfer gains are obtained as

$$M_{S+} = \frac{V_{O+}}{V_{in}} = \frac{1}{1-k},$$
$$M_{S-} = \frac{V_{O-}}{V_{in}} = -\frac{1}{1-k}. \tag{6.215}$$

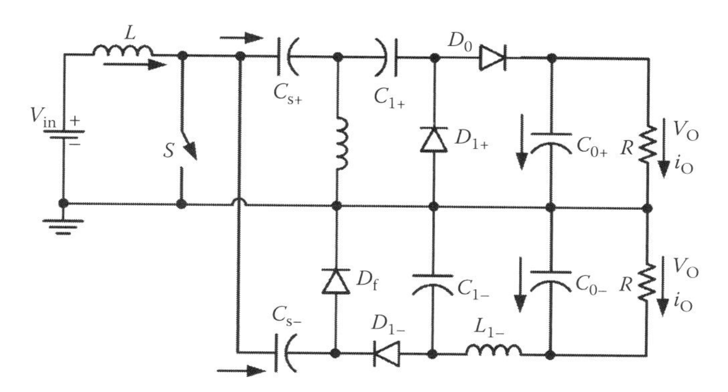

FIGURE 6.60 Novel self-lift D/O converter.

6.9.3 Enhanced Series D/O Circuits

Since the positive and negative conversion paths share a common source section that can be regarded as a boost converter circuit, we can construct the corresponding enhanced series using the VL technique. A series of novel boost circuits is applied into the source section, which transfers much more energy to C_{s+} and C_{s-} in each cycle and increases V_{Cs+} and V_{Cs-} stage-by-stage along geometric progression.

As shown in Figure 6.61, the source section is redesigned by adding the components L_{s1}-D_{s1}-D_{s2}-C_{s1}, which form a basic VL cell and are expressed by *boost*^{1}. The newly derived topology provides a single boost circuit enhancement using supplementary components. When switch S turns on, D_{s2} is on and D_{s1} is off. When switch S turns off, D_{s2} is off and D_{s1} is on. Capacitor C_{s1} performs characteristics to lift the source voltage V_{in}. The energy is transferred to C_{s+} and C_{s-} in each cycle from C_{s1} and increases V_{Cs+} and V_{Cs-}. We obtain

$$V_{Cs+} = V_{Cs1} = \frac{1}{1-k} V_{in},$$
$$V_{Cs-} = \frac{1}{1-k} V_{Cs1} = \frac{1}{(1-k)^2} V_{in}. \tag{6.216}$$

Therefore, from the foregoing analysis and calculation, the voltage transfer gains of this enhanced D/O self-lift DC–DC converters are

$$M_{boost^1-S+} = \frac{V_{O+}}{V_{in}} = \frac{1}{(1-k)^2},$$
$$M_{boost^1-S-} = \frac{V_{O-}}{V_{in}} = -\frac{1}{(1-k)^2}. \tag{6.217}$$

Referring to Figure 6.61, it is possible to realize multiple boost circuits enhancement in the source section by repeating the components L_{s1}-D_{s1}-D_{s2}-C_{s1} stage-by-stage. Assuming that there are n VL cells (denoted by *boost*^{M}), the generalized representation of the enhanced series for the D/O self-lift DC–DC converter is shown in Figure 6.62. All circuits share the same power switch S, which simplifies the control scheme and decreases the cost significantly. Hence, each circuit has one switch, $(n+3)$ inductors, $(n+5)$ capacitors, and $(2n+4)$ diodes. It is noted that all inductors existing in the VL cells are the same here for the same reasons as explained in foregoing sections. All the capacitors are sufficiently large. The

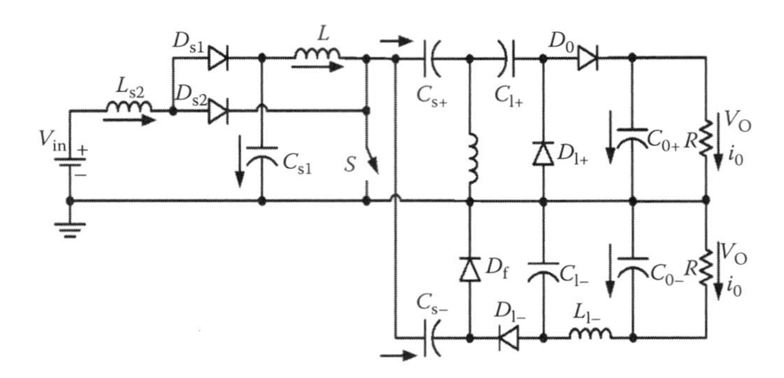

FIGURE 6.61 Enhanced D/O self-lift DC–DC converter (single boost circuit enhancement).

energy is transferred to C_{s+} and C_{s-} in each cycle from C_{sn}, and increases by V_{Cs+} and V_{Cs-}. We obtain

$$V_{Cs+} = V_{Csn} = \frac{1}{(1-k)^n} V_{in},$$
$$V_{Cs-} = \frac{1}{1-k} V_{Csn} = \frac{1}{(1-k)^{n+1}} V_{in}. \tag{6.218}$$

Therefore, from the foregoing analysis and calculation, the general voltage transfer gains of enhanced D/O self-lift DC–DC converters are

$$M_{boostM-S+} = \frac{V_{O+}}{V_{in}} = \frac{1}{(1-k)^{n+1}},$$
$$M_{boostM-S-} = \frac{V_{O-}}{V_{in}} = -\frac{1}{(1-k)^{n+1}}. \tag{6.219}$$

Analogically, we can also develop a series of enhanced D/O elementary circuits using the same source section. The general voltage transfer gains of enhanced D/O elementary

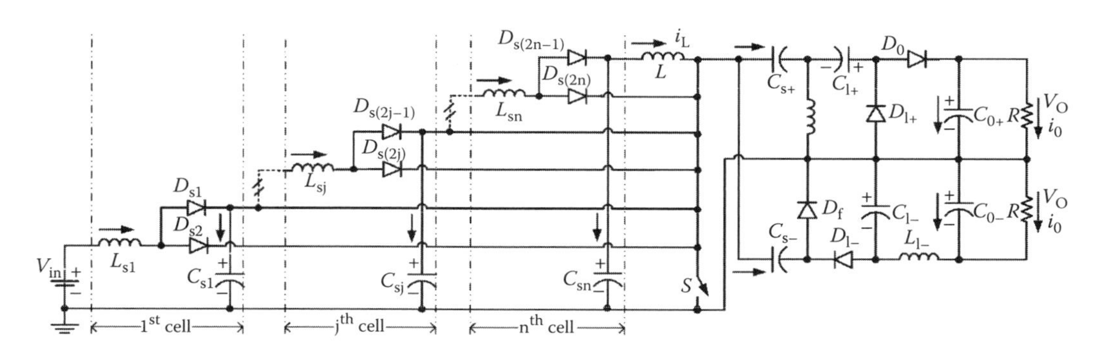

FIGURE 6.62 Generalized representation of enhanced D/O self-lift DC–DC converters (multiple boost circuits enhancement).

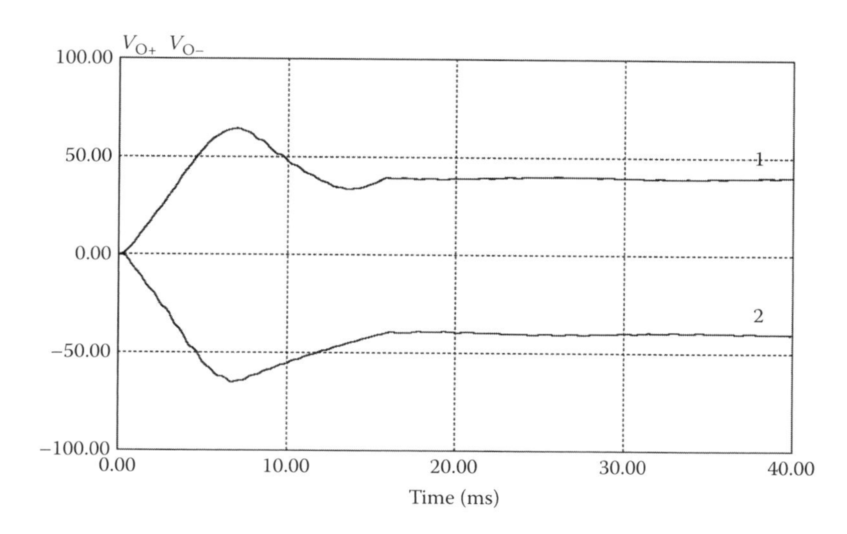

FIGURE 6.63 Simulation result for an enhanced D/O self-lift circuit (single boost circuit enhancement).

DC–DC converters are also given here for ready reference.

$$\begin{aligned} M_{boost^{M}-E+} &= \frac{V_{O+}}{V_{in}} = \frac{k}{(1-k)^{n+1}}, \\ M_{boost^{M}-E-} &= \frac{V_{O-}}{V_{in}} = -\frac{k}{(1-k)^{n+1}}. \end{aligned} \tag{6.220}$$

6.9.4 Simulation and Experimental Verification of an Enhanced D/O Self-Lift Circuit

Referring to Figure 6.61, the circuit parameters for simulation are $V_{in} = 10\,\text{V}$, $R = 100\,\Omega$, $L_{s1} = L = 1\,\text{mH}$, $C_{1+} = C_{1-} = C_{s1} = 22\,\mu\text{F}$, $C_{s+} = C_{s-} = 110\,\mu\text{F}$, $C_{o+} = C_{o-} = 47\,\mu\text{F}$, $C_O = 110\,\mu\text{F}$, $k = 0.5$, and $f = 100\,\text{kHz}$. According to Equation 6.219, we obtain the theoretical values of D/O voltages V_{O+} and V_{O-}, which are equal to 40 and −40 V, respectively. The simulation results of Psim are shown in Figure 6.63, where curve 1 is for the v_{O+} of the positive conversion path and curve 2 is for the v_{O-} of the negative conversion path. The steady-state values in the simulation identically match the theoretical analysis.

Similar parameters are chosen to construct the testing hardware circuit. Only a single n-channel MOSFET is used in the circuit. The corresponding experimental curves in the steady state are shown in Figure 6.64. The curve shown in Channel 1 with 20 V/Div corresponds to P/O v_{O+}, which is about 37 V. The curve shown in Channel 2 with 20 V/Div corresponds to N/O v_{O-}, which is also about 37 V. Considering the effects caused by the parasitic parameters, we can see that the measured results are very close to the theoretical analysis and simulation results.

6.10 SC Converters

A switched capacitor is an improved component used in power electronics. Switched capacitors can be used to construct a new type of DC–DC converter called the switched-

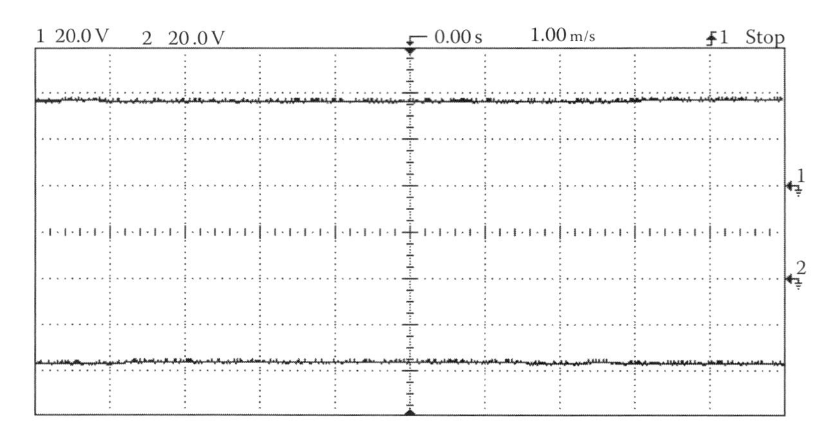

FIGURE 6.64 Experimental result for an enhanced D/O self-lift circuit (single boost circuit enhancement).

capacitor DC–DC converter. Switched capacitors can be integrated into a power IC chip. By using this manufacturing technology, we have the advantages of small size and low power losses. Consequently, switched-capacitor DC–DC converters have a small size, a high power density, a high power transfer efficiency, and a high voltage transfer gain [22–27].

Current is supplied to DC–DC converters by a DC voltage source. The input source current can be continuous or discontinuous. In some converters such as buck converters and buck–boost converters, the input current is discontinuous. This is called working in the DICM. In other converters such as boost converters, the input current is continuous. This is called working in the CICM. The VL technique can be applied to the switched capacitor to construct DC/DC converters. The idea is that for converters to operate in the DICM, switched capacitors can be charged with the source voltage and energy can be stored during the input current discontinuous period (when the main switch is off). They will join the conversion operation during the time the main switch is on, and their SE will be delivered through the DICM converters to the load. These converters are called SC DC–DC converters.

It is easy to construct SC DC–DC converters. Depending on how many switched capacitors need to be used, they are called one-stage SC converters, two-stage SC converters, three-stage SC converters, and n-stage SC converters. The corresponding circuits are shown in Figures 6.65 through 6.67.

The one-stage SC converter circuit is shown in Figure 6.65a. The input source voltage is V_{in} and the output voltage is V_O. To simplify the description, we assume that the load is resistive load R. The auxiliary switches S_1 and S_2 are switched on (the auxiliary switch S_3 is off) during the switch-off period. The switched capacitor C_1 is charged with the source voltage V_{in}. The auxiliary switches S_1 and S_2 are switched off, and the auxiliary switch S_3 is on during the switch-on period. The equivalent circuit is shown in Figure 6.65b. Therefore, the equivalent input voltage supplied to the DICM converter is $2V_{in}$ [28–32]. In other words, the equivalent input voltage has been lifted by using the switched capacitor.

Analogously, the circuit diagram of the two-stage SC converter is shown in Figure 6.66a, and the corresponding equivalent circuit when the main switch is on is shown in Figure 6.66b. It supplies $3V_{in}$ to the DICM converter. The equivalent input voltage is lifted to a level 2 times higher than the supplied voltage V_{in}.

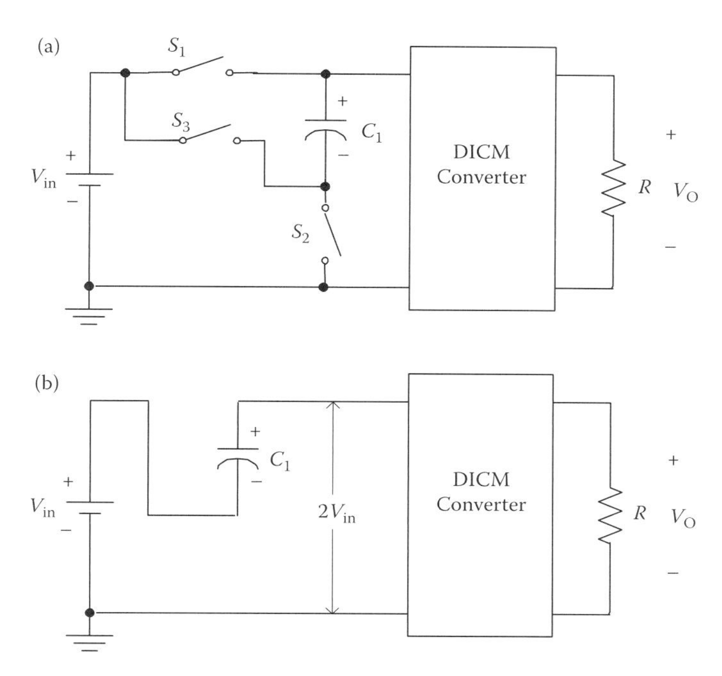

FIGURE 6.65 One-stage SC converter: (a) circuit diagram and (b) equivalent circuit during main switch-on.

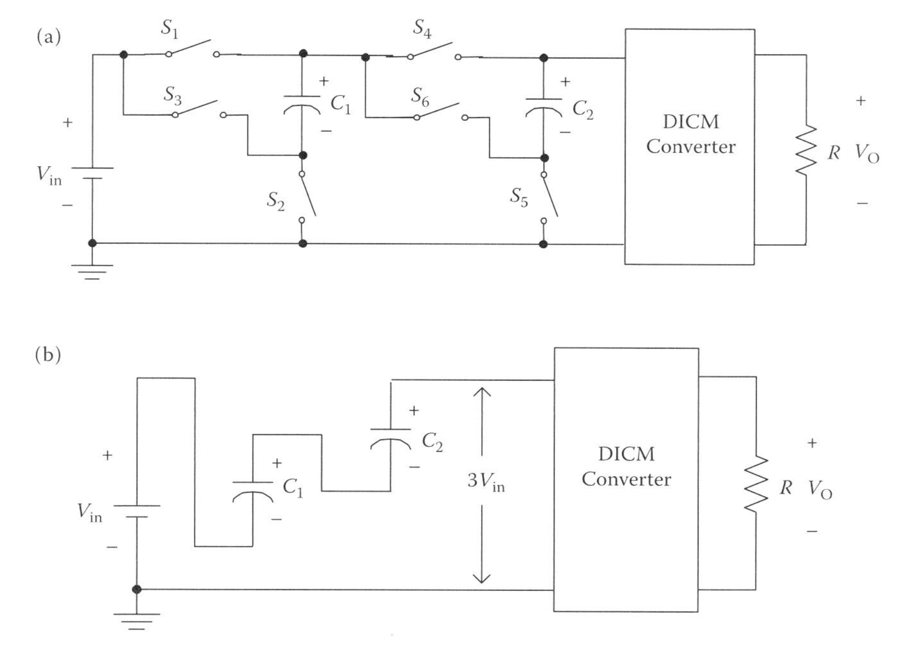

FIGURE 6.66 Two-stage SC converter: (a) circuit diagram and (b) equivalent circuit during main switch-on.

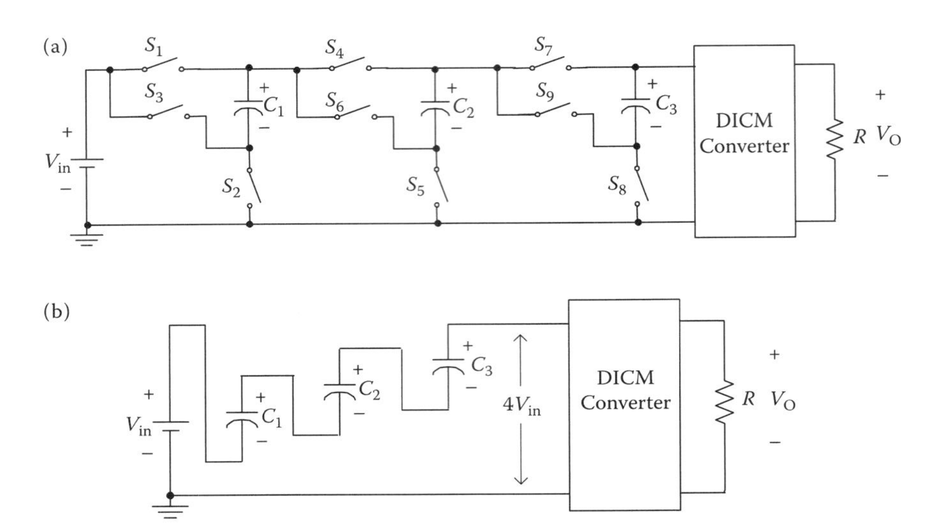

FIGURE 6.67 Three-stage SC converter: (a) circuit and (b) equivalent circuit during main switch-on.

The circuit diagram of the three-stage SC-converter is shown in Figure 6.67a, and the corresponding equivalent circuit when the main switch is on is shown in Figure 6.67b. It supplies $4V_{in}$ to the DICM converter. The equivalent input voltage is lifted to a level 3 times higher than the supplied voltage V_{in}.

Several circuits will be introduced in this chapter:

- SC buck converters
- SC buck–boost converters
- SC P/O Luo-converters
- SC N/O Luo-converters.

Assume that the stage number is n and the voltage transfer gain of the DICM converter is M. Then, in the ideal condition, we obtain the output voltage as

$$V_O = (n + 1)MV_{in}. \tag{6.221}$$

The ideal condition means that the voltage drop across all switches and diodes is zero and the voltage across all the SCs has no drop-down when the main switch is off. This assumption is reasonable for the investigation. We will discuss the unideal condition operation in Section 6.10.5 [33–38].

There is another advantage in the input current being continuous. The input current of the original DICM converter is zero when the main switch is off. For example, the input current of the one-stage SC DC–DC converter flows through the auxilliary switches S_1 and S_2 to the charge capacitor C_1 when the main switch is off. For the n-stage SC DC–DC converter, each switched capacitor is discharged by the discharging current I_D shown in Figure 6.68a. The charging current of each switched capacitor should be I_d in the switch-off period since the average current of each switched capacitor is zero in the steady state. Therefore, the source input average current should be

$$I_{in} = (n + 1)I_d. \tag{6.222}$$

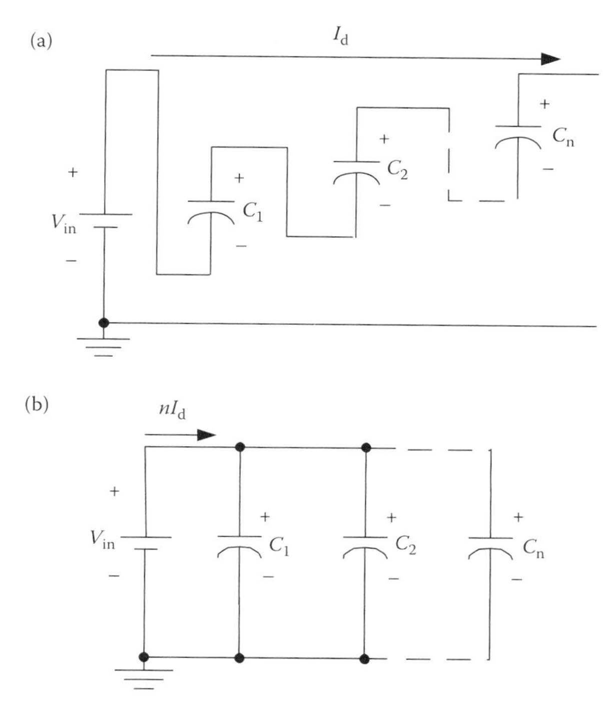

FIGURE 6.68 Discharging and charging currents of switched-capacitors: (a) discharging current during switch-on and (b) charging current during switch-off.

6.10.1 One-Stage SC Buck Converter

The one-stage SC buck converter is shown in Figure 6.69. The main switch S and the auxiliary switch S_3 are on and off simultaneously. The auxiliary switches S_1 and S_2 are off and on separately.

6.10.1.1 Operation Analysis

We assume that the converter works in the steady state and the switched capacitor C_1 is fully charged. The main switch S is on during the switch-on period, and the auxiliary

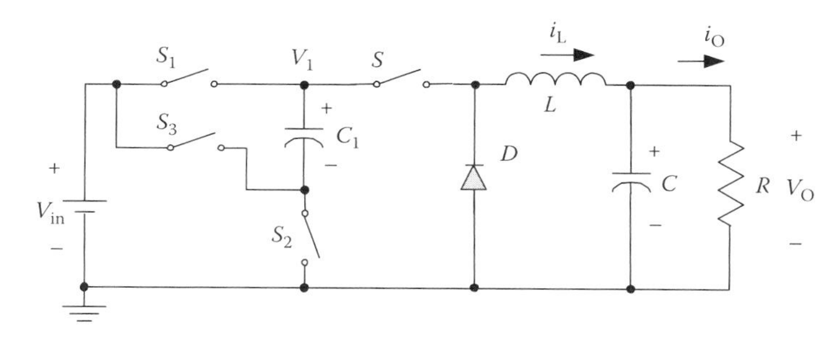

FIGURE 6.69 One-stage SC buck converter.

switch S_3 is on simultaneously. The voltage V_1 is about $2V_{in}$ when the main switch S is on. This is the equivalent input voltage of $2V_{in}$ for supply to the buck converter. Referring to the buck converter voltage transfer gain $M = k$, we can easily obtain the output voltage as

$$V_O = 2kV_{in}. \tag{6.223}$$

Using this technique, we can obtain an output voltage that is higher than the input voltage if the conduction duty cycle k is >0.5. The output voltage of the original buck converter is always lower than the input voltage.

6.10.1.2 Simulation and Experimental Results

In order to verify the design and analysis, the simulation result is shown in Figure 6.70. The simulation condition is that $V_{in} = 20\,V$, $L = 10\,mH$, $C = C_1 = 20\,\mu F$, $f = 50\,kHz$, $R = 100\,\Omega$, and conduction duty cycle $k = 0.8$. The voltage at the top end of the switched capacitor C_1 varies from 20 to 40 V. The output voltage $V_O = 32\,V$, which is the same as the calculation result.

$$V_O = 2kV_{in} = 2 \times 0.8 \times 20 = 32\,V. \tag{6.224}$$

The experimental result is shown in Figure 6.71. The test condition is the same: $V_{in} = 20\,V$ (Channel 1 in Figure 6.71), $L = 10\,mH$, $C = C_1 = 20\,\mu F$, $f = 50kHz$, $R = 100\,\Omega$ and conduction duty cycle $k = 0.8$. The output voltage $V_O = 32\,V$ (Channel 2 in Figure 6.71), which is the same as the calculation and simulation results.

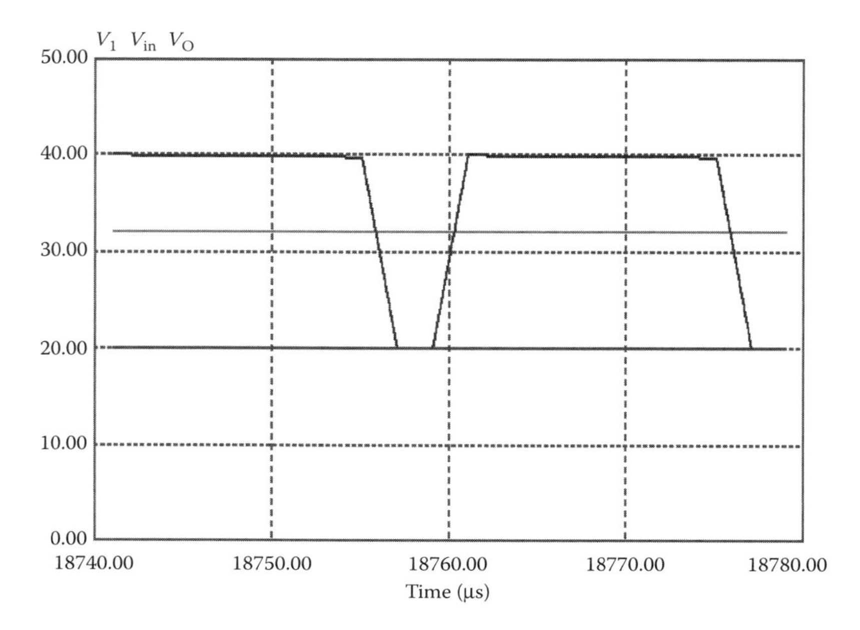

FIGURE 6.70 Simulation result.

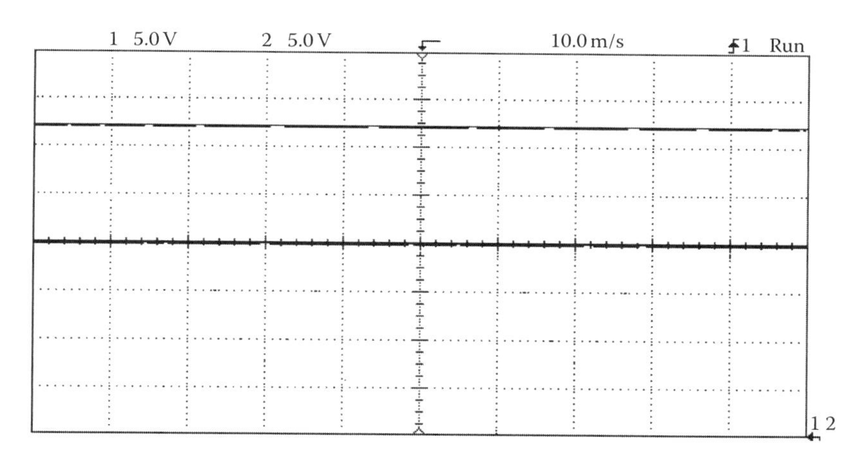

FIGURE 6.71 Experimental result.

6.10.2 Two-Stage SC Buck–Boost Converter

The two-stage SC buck–boost converter is shown in Figure 6.72. The main switch S and the auxiliary switches S_3 and S_6 are on and off simultaneously. The auxiliary switches S_1, S_2, S_4, and S_5 are off and on separately.

6.10.2.1 Operation Analysis

We assume that the converter works in the steady state and the switched capacitors C_1 and C_2 are fully charged. The main switch S is on during the switch-on period and the auxiliary switches S_3 and S_6 are on simultaneously. The voltage V_1 is about $2V_{in}$ and the voltage V_2 is about $3V_{in}$ when the main switch S is on. This is the equivalent input voltage of $3V_{in}$ for supply to the buck–boost converter. Referring to the buck–boost converter voltage transfer gain $M = -k/(1-k)$, we easily obtain the output voltage as

$$V_O = -\frac{3k}{1-k}V_{in}. \tag{6.225}$$

Using this technique, we effortlessly obtain a higher output voltage. For example, if $k = 0.5$, the output voltage of the original buck–boost converter is equal to the input source

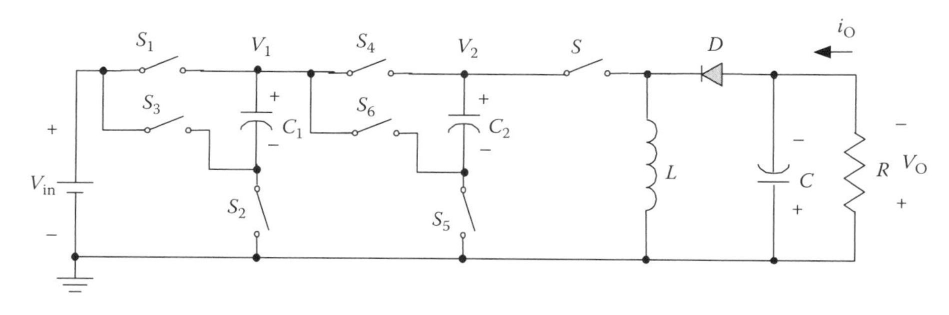

FIGURE 6.72 Two-stage SC buck–boost converter.

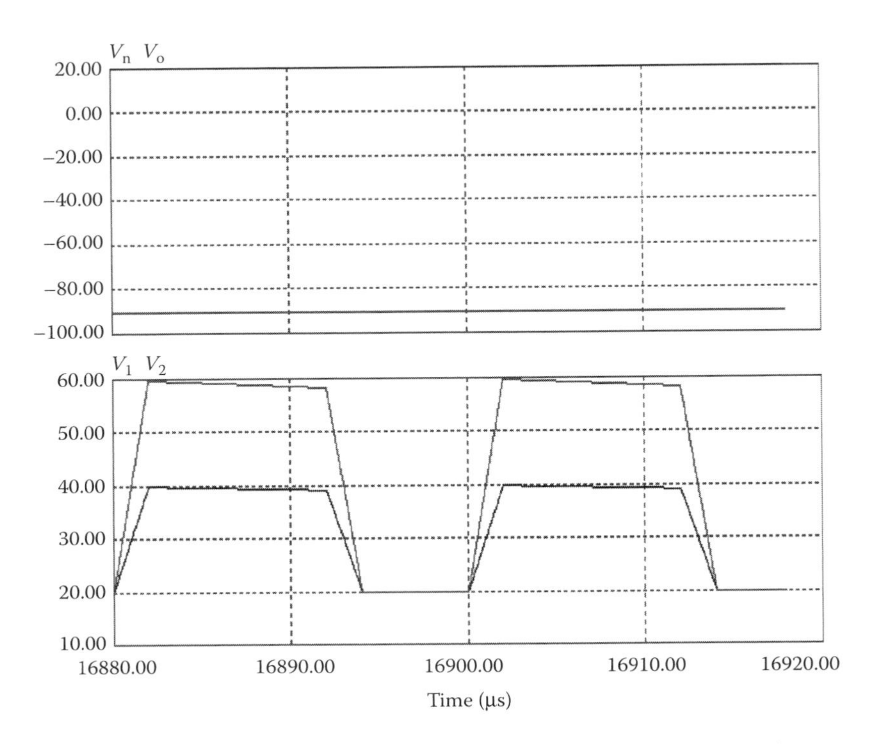

FIGURE 6.73 Simulation result. (a) Waveforms of V_{in} and V_O and (b) waveforms of V_1 and V_2.

voltage V_{in}. The value of the output voltage of the two-stage SC buck–boost converter is 6 times the value of the source voltage.

6.10.2.2 Simulation and Experimental Results

In order to verify the design, the simulation result is shown in Figure 6.73. The simulation condition is that $V_{in} = 20\,\text{V}$, $L = 10\,\text{mH}$, $C = C_1 = C_2 = 20\,\mu\text{F}$, $f = 50\,\text{kHz}$, $R = 200\,\Omega$, and conduction duty cycle $k = 0.6$. The voltage at the top end of the switched capacitor C_1 in Figure 6.72 varies from 20 to 40 V. The voltage at the top end of the switched capacitor C_2 varies from 20 to 60 V. The output voltage $V_O = -90\,\text{V}$, which is similar to the calculation result.

$$V_O = -\frac{3k}{1-k}V_{in} = -\frac{3 \times 0.6}{1-0.6} \times 20 = -90\,\text{V}. \tag{6.226}$$

The experimental result is shown in Figure 6.74. The test condition is that $V_{in} = 20\,\text{V}$ (Channel 1 in Figure 6.74), $L = 10\,\text{mH}$, $C = C_1 = C_2 = 20\,\mu\text{F}$, $f = 50\,\text{kHz}$, $R = 200\,\Omega$ and conduction duty cycle $k = 0.6$. The output voltage $V_O = -90\,\text{V}$ (Channel 2 in Figure 6.74), which is similar to the simulation result and the calculation result.

6.10.3 Three-Stage SC P/O Luo-Converter

The three-stage SC P/O Luo-converter is shown in Figure 6.75. The main switch S and the auxiliary switches S_3, S_6, and S_9 are on and off simultaneously. The auxiliary switches S_1, S_2, S_4, S_5, S_7, and S_8 are off and on separately.

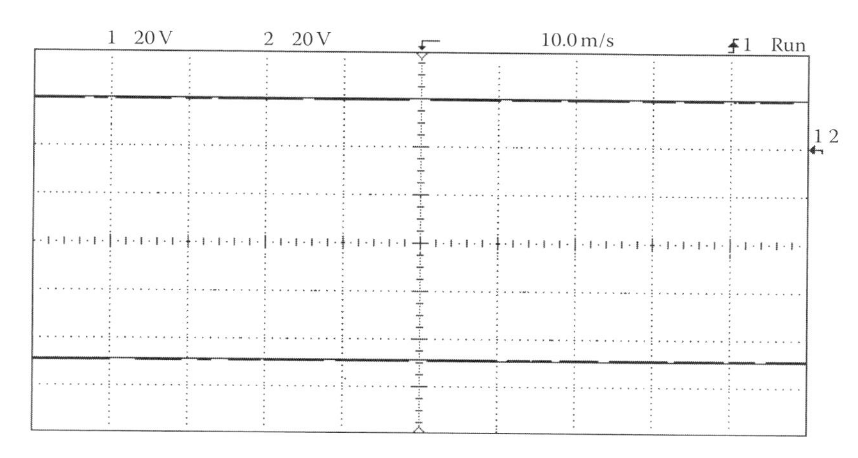

FIGURE 6.74 Experimental result.

6.10.3.1 Operation Analysis

We assume that the converter works in the steady state, and the switched capacitors C_1, C_2, and C_3 are fully charged. The main switch S is on during the switch-on period, and the auxiliary switches S_3, S_6, and S_9 are on simultaneously. The voltage V_1 is about $2V_{in}$, the voltage V_2 is about $3V_{in}$, and the voltage V_3 is about $4V_{in}$ when the main switch S is on. This is the equivalent input voltage of $4V_{in}$ for supply to the P/O Luo-converter. Referring to the P/O Luo-converter voltage transfer gain $M = k/(1-k)$, we can easily obtain the output voltage as

$$V_O = \frac{4k}{1-k} V_{in}. \tag{6.227}$$

6.10.3.2 Simulation and Experimental Results

In order to verify the design, the simulation result is shown in Figure 6.76. The simulation condition is that $V_{in} = 20$ V, $L = L_O = 10$ mH, $C = C_1 = C_2 = C_3 = 20\,\mu$F, $f = 50$ kHz, $R = 400\,\Omega$ and conduction duty cycle $k = 0.6$. The voltage on the top end of the switched capacitor C_1 varies from 20 to 40 V. The voltage on the top end of the switched capacitor C_2 varies from 20 to 60 V. The voltage on the top end of the switched capacitor C_3 varies from

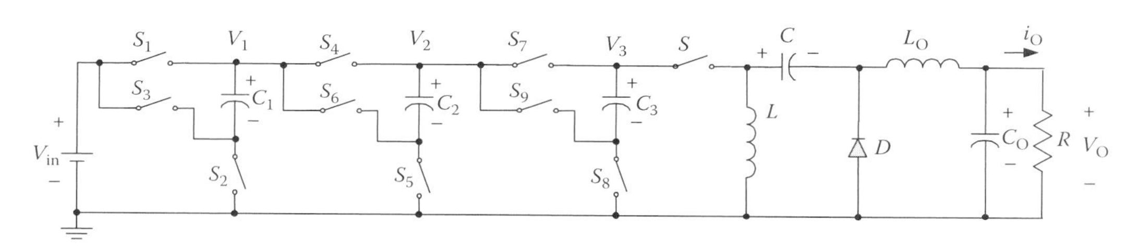

FIGURE 6.75 Three-stage SC P/O Luo-converter.

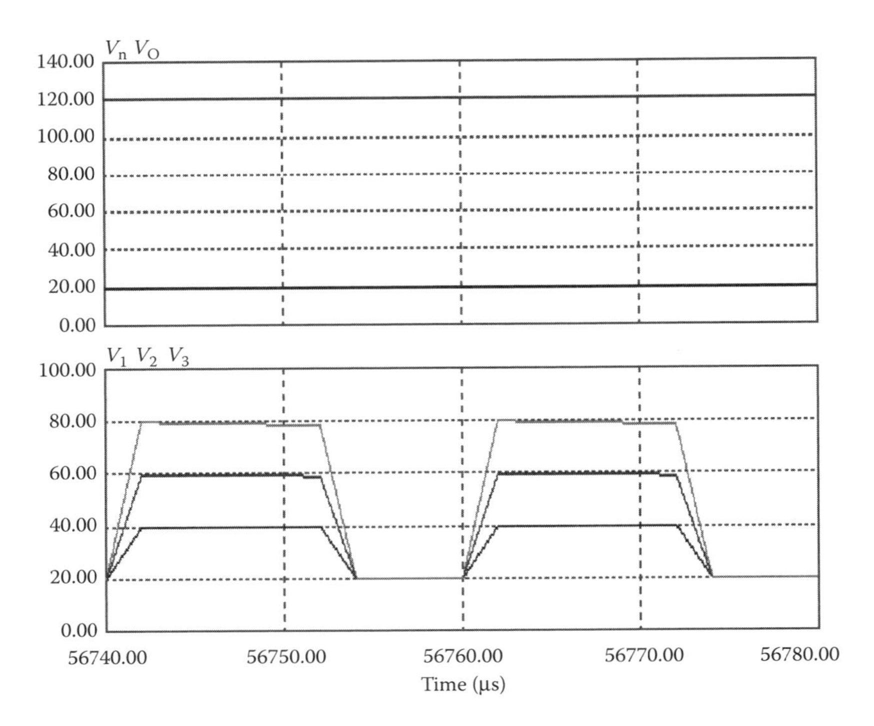

FIGURE 6.76 Simulation result.

20 to 80 V. The output voltage $V_O = 120$ V, which is the same as the calculation result.

$$V_O = \frac{4k}{1-k} V_{in} = \frac{4 \times 0.6}{1-0.6} \times 20 = 120\,\text{V}. \tag{6.228}$$

The experimental result is shown in Figure 6.77. The test condition is the same: $V_{in} = 20$ V (Channel 1 in Figure 6.77), $L = L_O = 10$ mH, $C = C_O = C_1 = C_2 = C_3 = 20\,\mu$F, $f = 50$ kHz, $R = 400\,\Omega$ and conduction duty cycle $k = 0.6$. The output voltage $V_O = 120$ V (Channel 2 in Figure 6.77), which is the same as the simulation and calculation results.

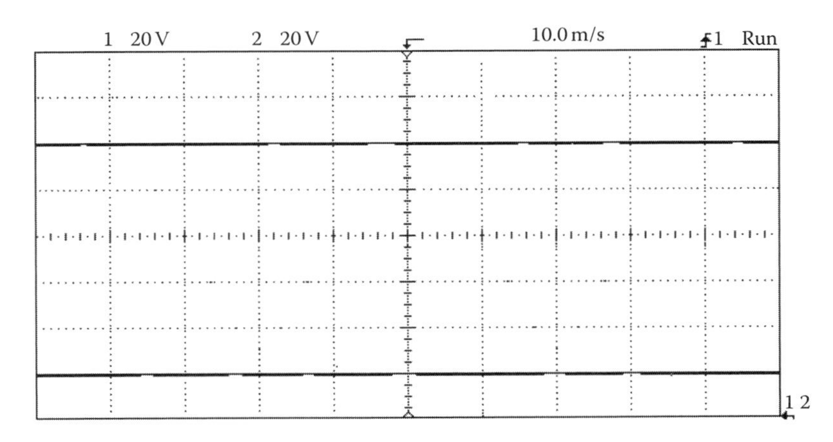

FIGURE 6.77 Experimental result.

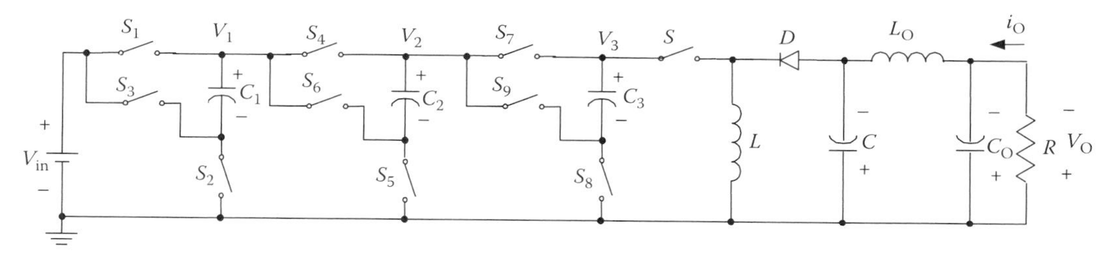

FIGURE 6.78 Three-stage SC N/O Luo-converter.

6.10.4 Three-Stage SC N/O Luo-Converter

The three-stage SC N/O Luo-converter is shown in Figure 6.78. The main switch S and the auxiliary switches S_3, S_6, and S_9 are on and off simultaneously. The auxiliary switches S_1, S_2, S_4, S_5, S_7, and S_8 are off and on separately.

6.10.4.1 *Operation Analysis*

We assume that the converter works in the steady state, and the switched capacitors C_1, C_2, and C_3 are fully charged. The main switch S is on during the switch-on period and the auxiliary switches S_3, S_6, and S_9 are on simultaneously. The voltage V_1 is about $2V_{in}$, V_2 is about $3V_{in}$, and V_3 is about $4V_{in}$ when the main switch S is on. This is the equivalent input voltage of $4V_{in}$ for supply to the N/O Luo-converter. Referring to the N/O Luo-converter voltage transfer gain $M = -k/(1-k)$, we can easily obtain the output voltage as

$$V_O = -\frac{4k}{1-k}V_{in}. \qquad (6.229)$$

6.10.4.2 *Simulation and Experimental Results*

In order to verify the design, the simulation result is shown in Figure 6.79. The simulation condition is that $V_{in} = 20$ V, $L = L_O = 10$ mH, $C = C_1 = C_2 = C_3 = 20\,\mu$F, $f = 50$ kHz, $R = 400\,\Omega$ and conduction duty cycle $k = 0.6$. The voltage at the top end of the switched capacitor C_1 varies from 20 to 40 V. The voltage at the top end of the switched capacitor C_2 varies from 20 to 60 V. The voltage at the top end of the switched capacitor C_3 varies from 20 to 80 V. The output voltage $V_O = -120$ V, which is the same as the calculation result.

$$V_O = -\frac{4k}{1-k}V_{in} = -\frac{4 \times 0.6}{1-0.6} \times 20 = -120\text{ V}. \qquad (6.230)$$

The experimental result is shown in Figure 6.80. The test condition is the same: $V_{in} = 20$ V (Channel 1 in Figure 6.80), $L = L_O = 10$ mH, $C = C_O = C_1 = C_2 = C_3 = 20\,\mu$F, $f = 50$ kHz, $R = 400\,\Omega$ and conduction duty cycle $k = 0.6$. The output voltage $V_O = 120$ V (Channel 2 in Figure 6.80), which is the same as the simulation and calculation results.

6.10.5 Discussion

In this section, we will discuss several factors of this technique for converter design consideration and industrial applications.

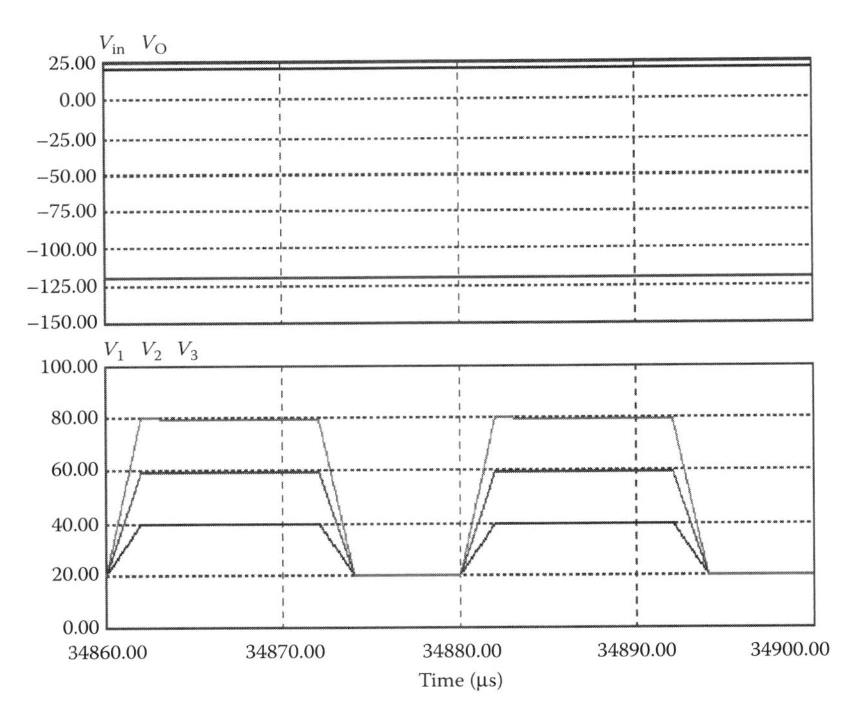

FIGURE 6.79 Simulation result.

6.10.5.1 *Voltage Drop across Switched Capacitors*

Referring to the waveform in Figures 6.72, 6.75, and 6.78, we can clearly see the voltage drop across the switched capacitors. For an n-stage SC converter, n switched capacitors need to be used. In the ideal condition, the total voltage across all switched capacitors should be

$$V_n = nV_{in}. \tag{6.231}$$

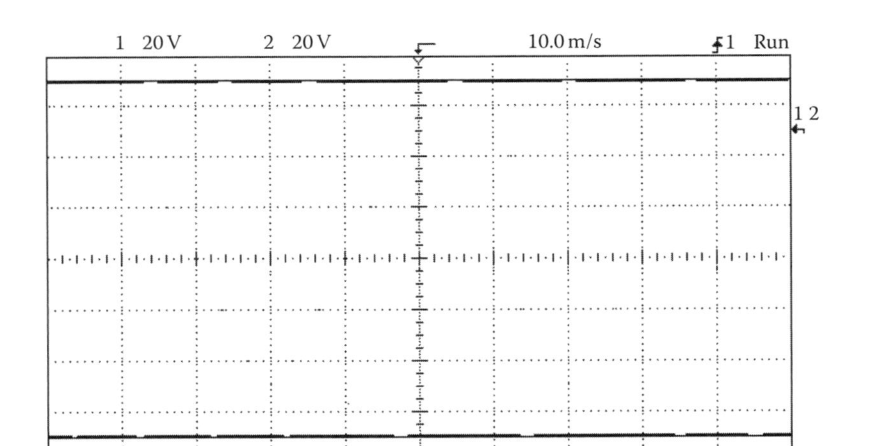

FIGURE 6.80 Experimental result.

If all switched capacitors have the same capacitance C, the equivalent capacitance in the switch-on period is C/n. We assume that the discharging current during the switch-on period is a constant value I_d, the conduction duty cycle is k, the switching frequency is f, and the switch-on period is $kT = k/f$. Then we calculate the voltage drop of the last switched capacitor as

$$\Delta V_n = \frac{1}{C/n}\int_0^{kT} i_d\,dt = \frac{nkT}{C} I_d. \tag{6.232}$$

The average current flowing through switched capacitors in a period T is zero in the steady state. The average input current from the source is $I_{in} = (n+1)I_D$. Current I_D is the input current of the DICM converter. If there are no energy losses inside the DICM converter, we can obtain it as

$$I_{in}V_{in} = (n+1)I_d V_{in} = V_O I_O = \frac{V_O^2}{R}. \tag{6.233}$$

Considering Equation 6.221, we have

$$I_d = \frac{V_O}{(n+1)V_{in}} I_O = MI_O = M\frac{V_O}{R}, \tag{6.234}$$

$$\Delta V_n = \frac{nkT}{C} I_d = \frac{nk}{fC} MI_O = \frac{nkM}{fC}\frac{V_O}{R}. \tag{6.235}$$

From Equation 6.235, we can see that the voltage drop is directly proportional to stages n, duty cycle k, and output voltage V_O. It is inversely proportional to switching frequency f, capacitance C of the used switched capacitors, and load R. In order to reduce the voltage drop for our design, one of the following ways can be used:

- Increase the switching frequency f
- Increase the capacitance C
- Increase the load R
- Decrease the duty cycle k.

Correspondingly, the voltage drop across each switched capacitor is

$$\Delta V_{each} = \frac{\Delta V_n}{n} = \frac{k}{fC} I_d = \frac{kM}{fC}\frac{V_O}{R}. \tag{6.236}$$

6.10.5.2 Necessity of the Voltage Drop across Switched-Capacitors and Energy Transfer

Voltage drops across switched capacitors are necessary for energy transfer from the source to the DICM converter. Switched capacitors absorb energy from the supply source during the switch-off period and release the SE to the DICM converter during the switch-on period. In the steady state, the energy transferred by the switched capacitors in a period T is

$$\Delta E = \frac{1}{2}\frac{C}{n}\left[V_n^2 - (V_n - \Delta V_n)^2\right] = \frac{C}{2n}\left(2V_n\Delta V_n - \Delta V_n^2\right) = \frac{C}{2n}(2V_n - \Delta V_n)\,\Delta V_n. \tag{6.237}$$

Considering that $2V_n \gg \Delta V_n$, Equation 6.237 can be rewritten as

$$\Delta E \approx \frac{C}{n} V_n \Delta V_n. \tag{6.238}$$

Substituting Equations 6.231 and 6.235 into Equation 6.238, the total power transferred by the switched capacitors is

$$P = f\Delta E = \frac{fC}{n} V_n \Delta V_n = \frac{fC}{n}(nV_{in})\left(\frac{nkM}{fC} I_O\right) = nkMV_{in}I_O. \tag{6.239}$$

If we would like to obtain the power transferred to the DICM converter as high, increasing the switching frequency f and capacitance C is necessary. From Equation 6.239, helpful methods are the following:

- Increase the duty cycle k
- Increase the stage number n
- Increase the transfer gain M.

6.10.5.3 Inrush Input Current

Inrush input current is large for all SC DC–DC converters, since the charging current to the switched-capacitors is high during the main switch-off period. As an example, the simulation result of the inrush input current of a three-stage SC P/O Luo-converter is shown in Figure 6.81.

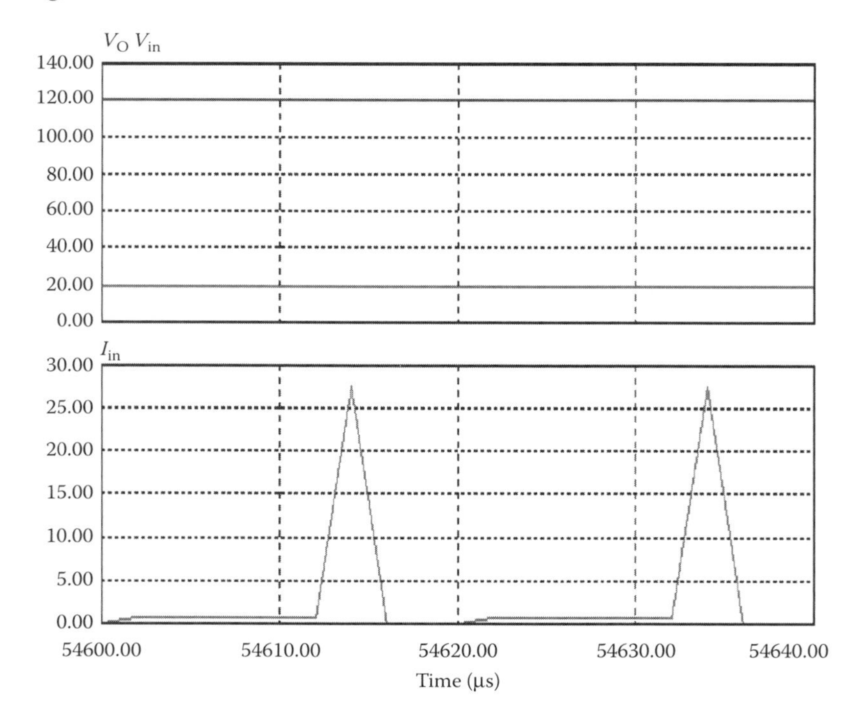

FIGURE 6.81 Simulation result (inrush input current).

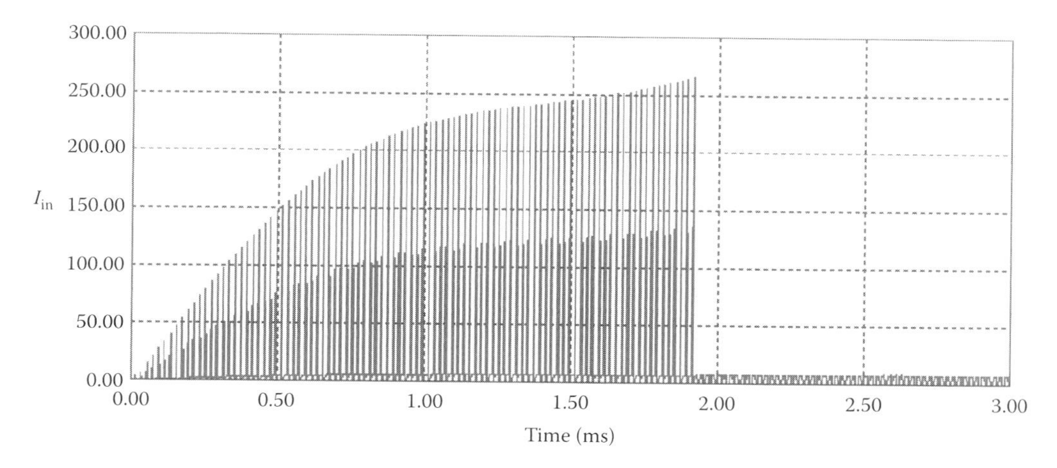

FIGURE 6.82 Simulation result (power-on surge input current).

The load current is very small, namely $I = 120/400 = 0.3\,\text{A}$, but the peak value of the input inrush current is about 27.3 A. Another phenomenon is that the input inrush current usually does not fully occupy the switch-off period. We will discuss how to overcome this phenomenon in Section 6.10.5.5.

6.10.5.4 Power Switch-On Process

Surge input current is large for all SC DC–DC converters during the power switch-on process since all switched capacitors are not precharged. For example, we show the simulation result of the power-on surge input current of a three-stage SC P/O Luo-converter in Figure 6.82.

The peak value of the power-on surge input current is very high, namely about 262 A.

6.10.5.5 Suppression of the Inrush and Surge Input Currents

From Figures 6.81 and 6.82, we can see that the peak inrush input current can be 90 times the normal load current, and the peak power-on surge input current can be about 880 times the normal load current. This is a serious problem for industrial applications of the SC DC/DC converters. In order to suppress the large inrush input current and the peak power-on surge

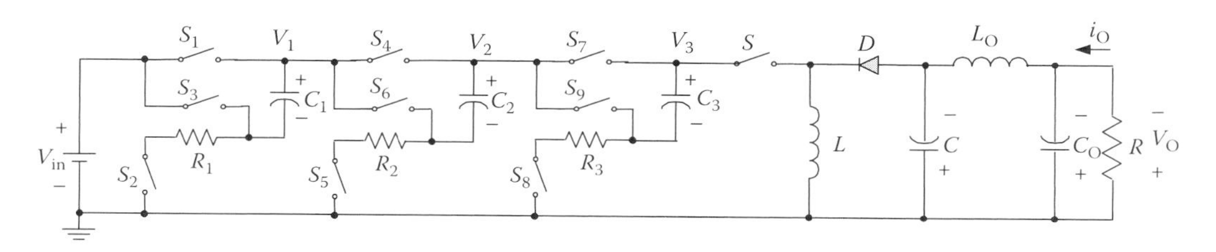

FIGURE 6.83 Improved three-stage SC P/O Luo-converter.

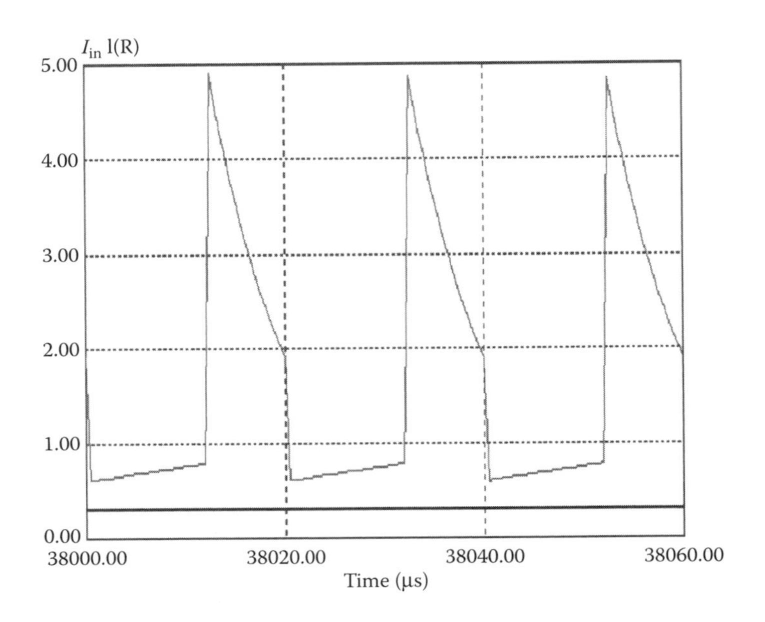

FIGURE 6.84 Simulation result (inrush input current) with R_S.

input current, we set a small resistor (the so-called suppression resistor R_S) in series with each switched capacitor. The circuit of such a three-stage SC P/O Luo-converter is shown in Figure 6.83. The resistance R_S is designed to have the time constant of the RC circuit compete with the switch-off period.

$$R_S = \frac{1-k}{C}T = \frac{1-k}{fC}. \tag{6.240}$$

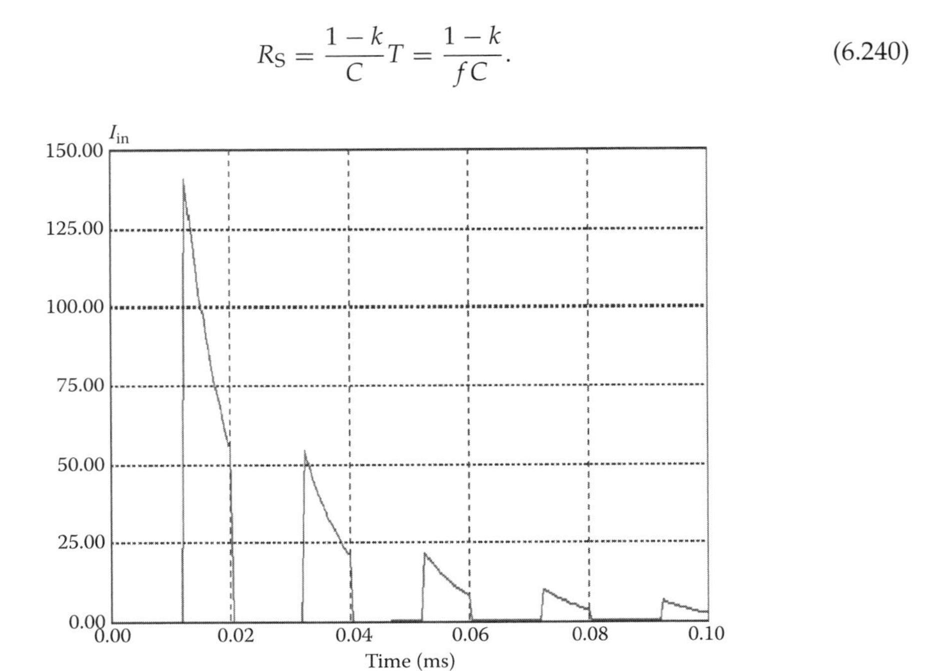

FIGURE 6.85 Simulation result (power-on surge input current) with R_S.

The same conditions as those mentioned in the previous section were used: $f = 50\,\text{kHz}$, all capacitances are $C = 20\,\mu\text{F}$, and conduction duty cycle $k = 0.6$. We can choose $R_1 = R_2 = R_3 = 0.4\,\Omega$. The inrush input current and the load current are shown in Figure 6.84.

By comparison with Figure 6.81, we can see that the peak inrush input current is largely reduced to 4.8 A and the input current becomes continuous in the switch-off period.

The power-on surge input current waveform is shown in Figure 6.85. The peak power-on surge input current is about 138 A, which is largely reduced.

Homework

6.1. An N/O self-lift Luo-converter shown in Figure 6.6a has the following components: $V_I = 20\,\text{V}$, $L = L_O = 1\,\text{mH}$, $C = C_1 = C_O = 20\,\mu\text{F}$, $R = 40\,\Omega$, $f = 50\,\text{kHz}$, and $k = 0.5$. Calculate the output voltage and the variation ratios $\zeta_1, \zeta_2, \rho, \sigma_1$, and ε in the steady state.

6.2. An N/O self-lift Luo-converter shown in Figure 6.6a has the following components: $V_I = 20\,\text{V}$, all inductances are 1 mH, all capacitances are $20\,\mu\text{F}$, $R = 1000\,\Omega$, $f = 50\,\text{kHz}$, and $k = 0.5$. Calculate the output voltage in the steady state.

6.3. An enhanced self-lift P/O Luo-converter shown in Figure 6.9 has the following components: $V_I = 20\,\text{V}$, all inductances are 1 mH, all capacitances are $20\,\mu\text{F}$, $R = 100\,\Omega$, $f = 50\,\text{kHz}$, and $k = 0.5$. Calculate the output voltage in the steady state.

6.4. An N/O triple-lift Luo-converter shown in Figure 6.22 has the following components: $V_I = 20\,\text{V}$, $L_1 = L_2 = 0.5\,\text{mH}$, $L = L_O = 1\,\text{mH}$, all capacitors have $20\,\mu\text{F}$, $R = 300\,\Omega$, $f = 50\,\text{kHz}$, and $k = 0.5$. Calculate the output voltage and the variation ratios $\zeta, \xi, \chi_1, \chi_2, \rho, \sigma_1, \sigma_2, \sigma_3$, and ε in the steady state.

6.5. An enhanced D/O self-lift DC–DC converter shown in Figure 6.61 has the following components: $V_I = 20\,\text{V}$, all inductances are 1 mH, all capacitances are $20\,\mu\text{F}$, $R = R_1 = 300\,\Omega$, $f = 50\,kHz$, and $k = 0.5$. Calculate the output voltage in the steady state.

6.6. A three-stage SC P/O Luo-converter shown in Figure 6.75 has the following components: $V_{in} = 20\,\text{V}$, all inductances are 1 mH, all capacitances are $20\,\mu\text{F}$, $R = 300\,\Omega$, $f = 50\,\text{kHz}$, and k varies from 0.1 to 0.9 with an increment of 0.1. Calculate the output voltage in the steady state.

References

1. Luo, F. L. and Ye, H. 2004. *Advanced DC/DC Converters*. Boca Raton: CRC Press.
2. Luo, F. L. and Ye, H. 2006. *Essential DC/DC Converters*. Boca Raton: Taylor & Francis Group LLC.
3. Luo, F. L. 2001. Seven self-lift DC/DC converters: Voltage lift technique. *IEE-Proceedings on EPA*, vol. 148, pp. 329–338.
4. Luo, F. L. 2001. Six self-lift DC/DC converters: Voltage lift technique. *IEEE Transactions on Industrial Electronics*, 48, 1268–1272.
5. Luo, F. L. and Chen X. F. 1998. Self-lift DC–DC converters. *Proceedings of the 2nd IEEE International Conference PEDES'98*, pp. 441–446.

6. Luo, F. L. 1999. Positive output Luo-converters: Voltage lift technique. *IEE-EPA Proceedings*, vol. 146, pp. 415–432.
7. Luo, F. L. 1998. Re-lift converter: Design, test, simulation and stability analysis. *IEE-EPA Proceedings*, vol. 145, pp. 315–325.
8. Luo, F. L. 1997. Re-lift circuit: A new DC–DC step-up (boost) converter. *IEE Electronics Letters*, 33, 5–7.
9. Luo, F. L. 1998. Luo-converters—voltage lift technique. *Proceedings of the IEEE Power Electronics Special Conference IEEE-PESC'98*, pp. 1783–1789.
10. Luo, F. L. 1997. Luo-converters, a series of new DC–DC step-up (boost) conversion circuits. *Proceedings of the IEEE International Conference on Power Electronics and Drive Systems—1997*, pp. 882–888.
11. Massey, R.P. and Snyder, E.C. 1977. High voltage single ended DC–DC converter. *Record of IEEE PESC*, 156–159.
12. Luo, F. L. 1999. Negative output Luo-converters: Voltage lift technique. *IEE-EPA Proceedings*, vol. 146, pp. 208–224.
13. Luo, F. L. 1998. Negative output Luo-converters, implementing the voltage lift technique. *Proceedings of the Second World Energy System International Conference'98*, pp. 253–260.
14. Luo, F. L. and Ye, H. 1999. Modified positive output Luo converters. *Proceedings of the IEEE International Conference PEDS'99*, pp. 450–455.
15. Luo, F. L. 2000. Double output Luo-converters: Advanced voltage lift technique. *Proceedings of IEE-EPA*, vol. 147, pp. 469–485.
16. Luo, F. L. 1999. Double output Luo-converters. *Proceedings of the International Conference IEE-IPEC'99*, pp. 647–652.
17. Cuk, S. and Middlebrook, R. D. 1977. A new optimum topology switching DC-to-DC converter. *Proceedings of IEEE PESC*, pp. 160–179.
18. Zhu, M. and Luo, F. L. 2007. Implementing of developed voltage lift technique on SEPIC, Cúk and double-output DC/DC converters. *Proceedings of IEEE-ICIEA 2007*, pp. 674–681.
19. Zhu, M. and Luo, F. L. 2007. Implementing of development of voltage lift technique on double-output transformerless DC–DC converters. *Proceedings of IECON 2007*, pp. 1983–1988.
20. Jozwik, J. J. and Kazimerczuk, M. K. 1989. Dual SEPIC PWM switching-mode DC/DC power converter. *IEEE Transactions on Industrial Electronics*, 36, 64–70.
21. Adar, D., Rahav, G., and Ben-Yaakov, S. 1996. Behavioural average model of SEPIC converters with coupled inductors. *IEE Electronics Letters*, 32, 1525–1526.
22. Luo, F. L. 2009. Switched-capacitorized DC–DC converters. *Proceedings of IEEE-ICIEA 2009*, pp. 385–389.
23. Luo, F. L. 2009. Investigation of switched-capacitorized DC–DC converters. *Proceedings of IEEE-IPEMC 2009*, pp. 1283–1288.
24. Luo, F. L. and Ye, H. 2004. Positive output multiple-lift push-pull switched-capacitor Luo-converters. *IEEE-Transactions on Industrial Electronics*, 51, 594–602.
25. Gao, Y. and Luo, F. L. 2001. Theoretical analysis on performance of a 5 V/12 V push-pull switched capacitor DC/DC converter. *Proceedings of the International Conference IPEC 2001*, pp. 711–715.
26. Luo, F. L. and Ye, H. 2003. Negative output multiple-lift push-pull switched-capacitor Luo-converters. *Proceedings of IEEE International Conference PESC 2003*, pp. 1571–1576.
27. Luo, F. L., Ye, H., and Rashid, M. H. 1999. Switched capacitor four-quadrant Luo-converter. *Proceedings of the IEEE-IAS Annual Meeting*, pp. 1653–1660.
28. Makowski, M. S. 1997. Realizability conditions and bounds on synthesis of switched capacitor DC–DC voltage multiplier circuits. *IEEE Transactions on Circuits and Systems*, 45, 684–691.
29. Cheong, S. V., Chung, H., and Ioinovici, A. 1994. Inductorless DC–DC converter with high power density. *IEEE Transactions on Industrial Electronics*, 42, 208–215.
30. Midgley, D. and Sigger, M. 1974. Switched-capacitors in power control. *IEE Proceedings*, 124, 703–704.
31. Mak, O. C., Wong, Y. C., and Ioinovici, A. 1995. Step-up DC power supply based on a switched-capacitor circuit. *IEEE Transactions on Industrial Electronics*, 43, 90–97.

32. Chung, H. S., Hui, S. Y. R., Tang, S. C., and Wu, A. 2000. On the use of current control scheme for switched-capacitor DC/DC converters. *IEEE Transactions on Industrial Electronics*, 47, 238–244.
33. Pan, C. T. and Liao, Y. H. 2007. Modeling and coordinate control of circulating currents in parallel three-phase boost rectifiers. *IEEE Transactions on Industrial Electronics*, 54, 825–838.
34. Mazumder, S. K., Tahir, M., and Acharya, K. 2008. Master–slave current-sharing control of a parallel DC–DC converter system over an RF communication interface. *IEEE Transactions on Industrial Electronics*, 55, 59–66.
35. Asiminoaei, L., Aeloiza, E., Enjeti, P., and Blaabjerg, F. 2008. Shunt active-power-filter topology based on parallel interleaved inverters. *IEEE Transactions on Industrial Electronics*, 55, 1175–1189.
36. Chen, W. and Ruan, X. 2008. Zero-voltage-switching PWM hybrid full-bridge three-level converter with secondary-voltage clamping scheme. *IEEE Transactions on Industrial Electronics*, 55, 644–654.
37. Wang, C. M. 2006. New family of zero-current-switching PWM converters using a new zero-current-switching PWM auxiliary circuit. *IEEE Transactions on Industrial Electronics*, 53, 768–777.
38. Ye, Z., Jain, P. K., and Sen, P. C. 2007. Circulating current minimization in high-frequency AC power distribution architecture with multiple inverter modules operated in parallel. *IEEE Transactions on Industrial Electronics*, 54, 2673–2687.

7

Super-Lift Converters and Ultralift Converter

The VL technique has been successfully employed in the design of DC/DC converters, and effectively enhances the voltage transfer gains of the VL converters. However, the output voltage increases in arithmetic progression stage by stage. The SL technique is more powerful than the VL technique; its voltage transfer gain can be a very large value. The SL technique implements the output voltage increasing in geometric progression stage by stage. It effectively enhances the voltage transfer gain in power series [1–6].

7.1 Introduction

The SL technique is the most important contribution to DC/DC conversion technology. By applying this technique, a large number of SL converters can be designed. The following series of VL converters are introduced in this chapter:

- P/O SL Luo-converters
- N/O SL Luo-converters
- P/O cascaded boost converters
- N/O cascaded boost converters
- UL Luo-converters.

Each series of converters has several subseries. For example, the P/O SL Luo-converters have five subseries:

- *The main series*: Each circuit of the main series has only one switch S, n inductors for the nth stage circuit, $2n$ capacitors, and $(3n-1)$ diodes.
- *Additional series*: Each circuit of the additional series has one switch S, n inductors for the nth stage circuit, $2(n+1)$ capacitors, and $(3n+1)$ diodes.
- *Enhanced series*: Each circuit of the enhanced series has one switch S, n inductors for the nth stage circuit, $4n$ capacitors, and $(5n-1)$ diodes.
- *Re-enhanced series*: Each circuit of the re-enhanced series has one switch S, n inductors for the nth stage circuit, $6n$ capacitors, and $(7n-1)$ diodes.
- *Multiple* (j)*-enhanced series*: Each circuit of the multiple (j times)-enhanced series has one switch S, n inductors for the nth stage circuit, $2(1+j)n$ capacitors, and $[(3+2j)n-1]$ diodes.

In order to concentrate the voltage enhancement, assume that the converters are working in the steady state in the CCM. The conduction duty ratio is k, the switching frequency is f, the switching period is $T = 1/f$, and the load is resistive load R. The input voltage and current are V_{in} and I_{in}, and the output voltage and current are V_O and I_O. Assuming that there are no power losses during the conversion process, $V_{in} \times I_{in} = V_O \times I_O$. The voltage transfer gain G is given by

$$G = \frac{V_O}{V_{in}}.$$

7.2 P/O SL Luo-Converters

We introduce here only three circuits from each subseries. Once the readers grasp the clue, they can design the other circuits easily [1–4].

7.2.1 Main Series

The first three stages of P/O SL Luo-converters, namely the main series, are shown in Figures 7.1 through 7.3. To make it easy to explain, they are called the elementary circuit, the re-lift circuit, and the triple-lift circuit, respectively, and are numbered $n = 1, 2$, and 3, respectively.

7.2.1.1 Elementary Circuit

The elementary circuit and its equivalent circuits during switch-on and switch-off periods are shown in Figure 7.1.

The voltage across capacitor C_1 is charged with V_{in}. The current i_{L1} flowing through inductor L_1 increases with V_{in} during the switch-on period kT and decreases with $-(V_O - 2V_{in})$ during the switch-off period $(1 - k)T$. Therefore, the ripple of the inductor current i_{L1} is

$$\Delta i_{L1} = \frac{V_{in}}{L_1} kT = \frac{V_O - 2V_{in}}{L_1}(1 - k)T, \tag{7.1}$$

$$V_O = \frac{2 - k}{1 - k} V_{in}. \tag{7.2}$$

The voltage transfer gain is

$$G = \frac{V_O}{V_{in}} = \frac{2 - k}{1 - k}. \tag{7.3}$$

The input current I_{in} is equal to $(i_{L1} + i_{C1})$ during switch-on, and only i_{L1} during switch-off. The capacitor current i_{C1} is equal to i_{L1} during switch-off. In the steady state, the average charge across capacitor C_1 should not change. The following relations are obtained:

$$i_{in-off} = i_{L1-off} = i_{C1-off}, \quad i_{in-on} = i_{L1-on} + i_{C1-on}, \quad kTi_{C1-on} = (1 - k)Ti_{C1-off}.$$

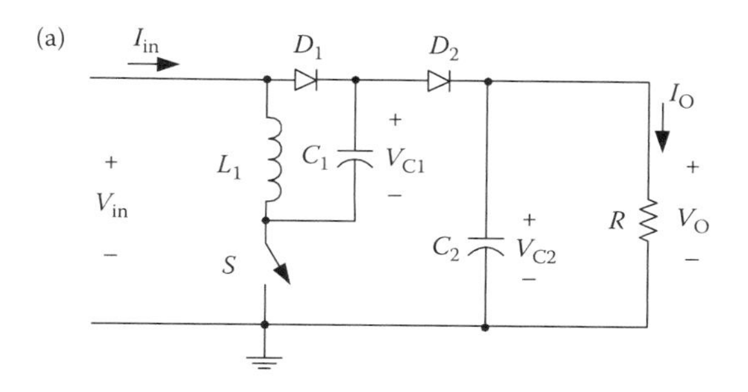

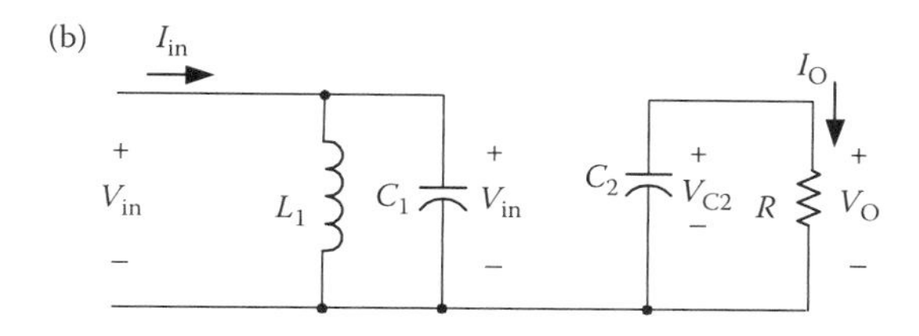

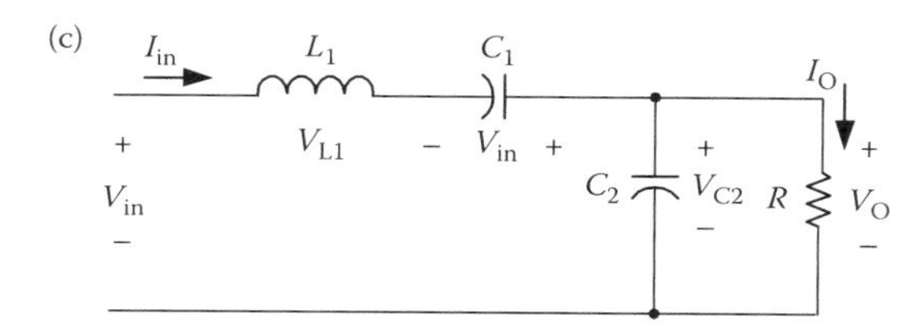

FIGURE 7.1 Elementary circuit of P/O SL Luo-converters—main series: (a) circuit diagram, (b) equivalent circuit during switch-on, and (c) equivalent circuit during switch-off. (Reprinted from Luo, F. L. and Ye, H. 2006. *Essential DC/DC Converters*. Boca Raton: Taylor & Francis Group LLC, p. 217. With permission.)

If inductance L_1 is large enough, i_{L1} is nearly equal to its average current I_{L1}. Therefore,

$$i_{in-off} = i_{C1-off} = I_{L1}, \quad i_{in-on} = I_{L1} + \frac{1-k}{k}I_{L1} = \frac{I_{L1}}{k}, \quad i_{C1-on} = \frac{1-k}{k}I_{L1},$$

and the average input current is

$$I_{in} = ki_{in-on} + (1-k)i_{in-off} = I_{L1} + (1-k)I_{L1} = (2-k)I_{L1}. \tag{7.4}$$

Considering $V_{in}/I_{in} = ((1-k)/(2-k))^2 V_O/I_O = ((1-k)/(2-k))^2 R$, the variation ratio of current i_{L1} through inductor L_1 is

$$\xi_1 = \frac{\Delta i_{L1}/2}{I_{L1}} = \frac{k(2-k)TV_{in}}{2L_1 I_{in}} = \frac{k(1-k)^2}{2(2-k)}\frac{R}{fL_1}. \tag{7.5}$$

Usually ξ_1 is small (much lower than unity); this means that this converter normally works in the continuous mode.

The ripple voltage of output voltage v_O is

$$\Delta v_O = \frac{\Delta Q}{C_2} = \frac{I_O kT}{C_2} = \frac{k}{fC_2}\frac{V_O}{R}.$$

Therefore, the variation ratio of output voltage v_O is

$$\varepsilon = \frac{\Delta v_O/2}{V_O} = \frac{k}{2RfC_2}. \tag{7.6}$$

Example 7.1

A P/O SL Luo-converter in Figure 7.1a has $V_{in} = 20V$, $L_1 = 10\,mH$, $C_1 = C_2 = 20\,\mu F$, $R = 100\,\Omega$, $f = 50\,kHz$, and conduction duty cycle $k = 0.6$. Calculate the variation ratio of current i_{L1}, and the output voltage and its variation ratio.

SOLUTION

From Equation 7.5, we can obtain the variation ratio of current i_{L1},

$$\xi_1 = \frac{k(1-k)^2}{2(2-k)}\frac{R}{fL_1} = \frac{0.6(1-0.6)^2}{2(2-0.6)}\frac{100}{50\,k \times 10\,m} = 0.00686.$$

From Equation 7.2, we can obtain the output voltage

$$V_O = \frac{2-k}{1-k}V_{in} = \frac{2-0.6}{1-0.6}20 = 70\,V.$$

From Equation 7.6, its variation ratio is

$$\varepsilon = \frac{k}{2RfC_2} = \frac{0.6}{2 \times 100 \times 50\,k \times 20\,\mu} = 0.003.$$

7.2.1.2 Re-Lift Circuit

The re-lift circuit is derived from the elementary circuit by adding the parts (L_2-D_3-D_4-D_5-C_3-C_4). Its circuit diagram and equivalent circuits during switch-on and switch-off periods are shown in Figure 7.2. The voltage across capacitor C_1 is charged with V_{in}. As described in the previous section, the voltage V_1 across capacitor C_2 is $V_1 = ((2-k)/(1-k))V_{in}$.

The voltage across capacitor C_3 is charged with V_1. The current flowing through inductor L_2 increases with V_1 during the switch-on period kT and decreases with $-(V_O - 2V_1)$ during the switch-off period $(1-k)T$. Therefore, the ripple of the inductor current i_{L2} is

$$\Delta i_{L2} = \frac{V_1}{L_2}kT = \frac{V_O - 2V_1}{L_2}(1-k)T, \tag{7.7}$$

$$V_O = \frac{2-k}{1-k}V_1 = \left(\frac{2-k}{1-k}\right)^2 V_{in}. \tag{7.8}$$

The voltage transfer gain is

$$G = \frac{V_O}{V_{in}} = \left(\frac{2-k}{1-k}\right)^2. \tag{7.9}$$

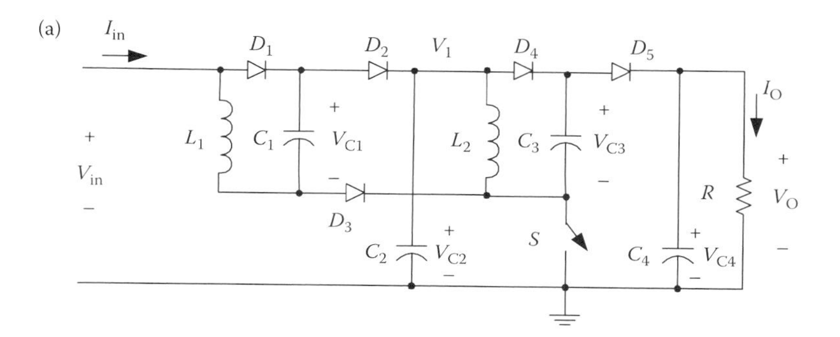

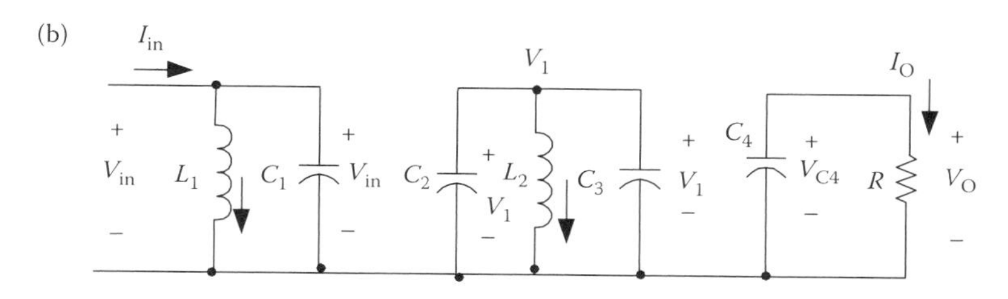

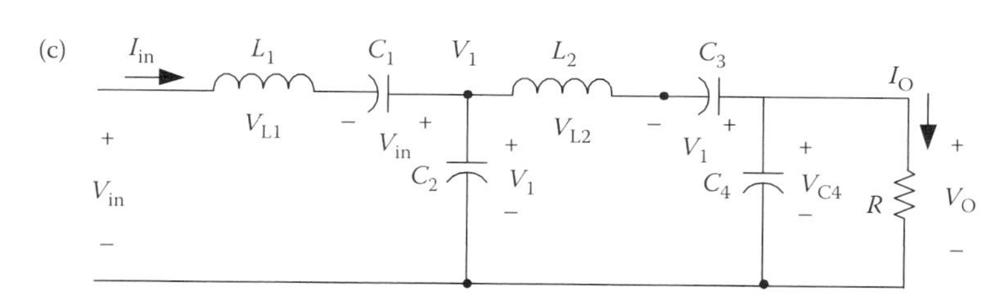

FIGURE 7.2 Re-lift circuit of P/O SL Luo-converters—main series: (a) circuit diagram, (b) equivalent circuit during switch-on, and (c) equivalent circuit during switch-off. (Reprinted from Luo, F. L. and Ye, H. 2006. *Essential DC/DC Converters*. Boca Raton: Taylor & Francis Group LLC, p. 218. With permission.)

Analogously, the following relations are obtained:

$$\Delta i_{L1} = \frac{V_{in}}{L_1}kT, \qquad I_{L1} = \frac{I_{in}}{2-k},$$

$$\Delta i_{L2} = \frac{V_1}{L_2}kT, \qquad I_{L2} = \left(\frac{2-k}{1-k} - 1\right) I_O = \frac{I_O}{1-k}.$$

Therefore, the variation ratio of current i_{L1} through inductor L_1 is

$$\xi_1 = \frac{\Delta i_{L1}/2}{I_{L1}} = \frac{k(2-k)TV_{in}}{2L_1 I_{in}} = \frac{k(1-k)^4}{2(2-k)^3}\frac{R}{fL_1}. \tag{7.10}$$

The variation ratio of current i_{L2} through inductor L_2 is

$$\xi_2 = \frac{\Delta i_{L2}/2}{I_{L2}} = \frac{k(1-k)TV_1}{2L_2 I_O} = \frac{k(1-k)^2 TV_O}{2(2-k)L_2 I_O} = \frac{k(1-k)^2}{2(2-k)}\frac{R}{fL_2}, \tag{7.11}$$

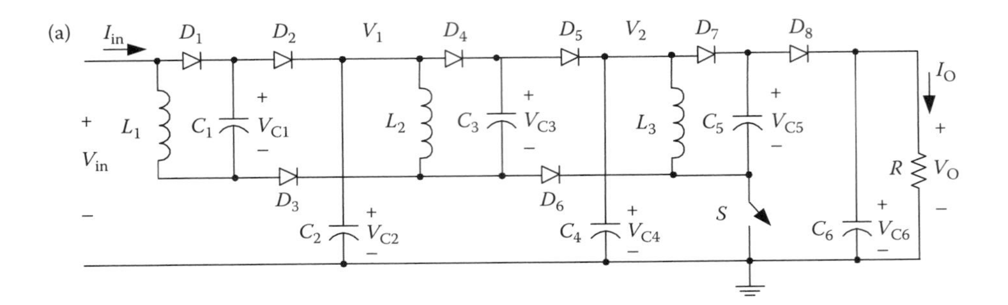

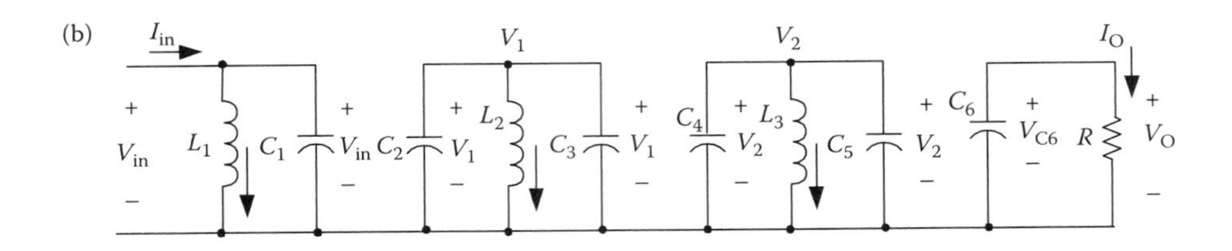

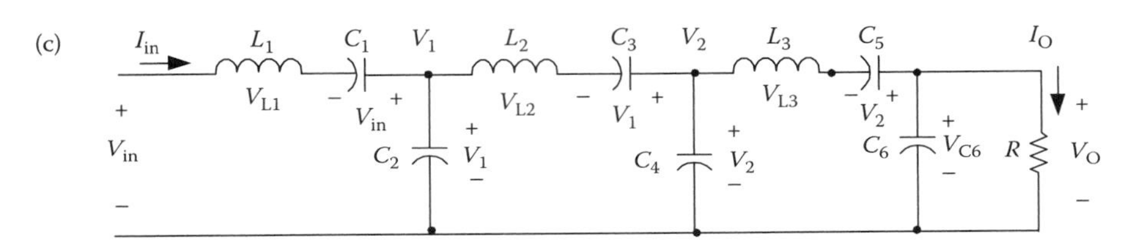

FIGURE 7.3 Triple-lift circuit of P/O SL Luo-converters—main series: (a) circuit diagram, (b) equivalent circuit during switch-on, and (c) equivalent circuit during switch-off. (Reprinted from Luo, F. L. and Ye, H. 2006. *Essential DC/DC Converters*. Boca Raton: Taylor & Francis Group LLC, p. 219. With permission.)

and the variation ratio of output voltage v_O is

$$\varepsilon = \frac{\Delta v_O/2}{V_O} = \frac{k}{2RfC_4}. \tag{7.12}$$

7.2.1.3 Triple-Lift Circuit

The triple-lift circuit is derived from the re-lift circuit by twice repeating the parts (L_2-D_3-D_4-D_5-C_3-C_4). Its circuit diagram and equivalent circuits during switch-on and switch-off periods are shown in Figure 7.3. The voltage across capacitor C_1 is charged with V_{in}. As described in the previous section, the voltage V_1 across capacitor C_2 is $V_1 = ((2-k)/(1-k))V_{in}$, and the voltage V_2 across capacitor C_4 is $V_2 = ((2-k)/(1-k))^2 V_{in}$.

The voltage across capacitor C_5 is charged with V_2. The current flowing through inductor L_3 increases with V_2 during the switch-on period kT and decreases with $-(V_O - 2V_2)$ during the switch-off period $(1-k)T$. Therefore, the ripple of the inductor current i_{L2} is

$$\Delta i_{L3} = \frac{V_2}{L_3} kT = \frac{V_O - 2V_2}{L_3}(1-k)T, \tag{7.13}$$

$$V_O = \frac{2-k}{1-k}V_2 = \left(\frac{2-k}{1-k}\right)^2 V_1 = \left(\frac{2-k}{1-k}\right)^3 V_{in}. \tag{7.14}$$

The voltage transfer gain is

$$G = \frac{V_O}{V_{in}} = \left(\frac{2-k}{1-k}\right)^3. \tag{7.15}$$

Analogously,

$$\Delta i_{L1} = \frac{V_{in}}{L_1}kT, \quad I_{L1} = \frac{I_{in}}{2-k},$$

$$\Delta i_{L2} = \frac{V_1}{L_2}kT, \quad I_{L2} = \frac{2-k}{(1-k)^2}I_O,$$

$$\Delta i_{L3} = \frac{V_2}{L_3}kT, \quad I_{L3} = \frac{I_O}{1-k}.$$

Therefore, the variation ratio of current i_{L1} through inductor L_1 is

$$\xi_1 = \frac{\Delta i_{L1}/2}{I_{L1}} = \frac{k(2-k)TV_{in}}{2L_1 I_{in}} = \frac{k(1-k)^6}{2(2-k)^5}\frac{R}{fL_1}. \tag{7.16}$$

The variation ratio of current i_{L2} through inductor L_2 is

$$\xi_2 = \frac{\Delta i_{L2}/2}{I_{L2}} = \frac{k(1-k)^2 TV_1}{2(2-k)L_2 I_O} = \frac{kT(2-k)^4 V_O}{2(1-k)^3 L_2 I_O} = \frac{k(2-k)^4}{2(1-k)^3}\frac{R}{fL_2}. \tag{7.17}$$

The variation ratio of current i_{L3} through inductor L_3 is

$$\xi_3 = \frac{\Delta i_{L3}/2}{I_{L3}} = \frac{k(1-k)TV_2}{2L_3 I_O} = \frac{k(1-k)^2 TV_O}{2(2-k)L_2 I_O} = \frac{k(1-k)^2}{2(2-k)}\frac{R}{fL_3}, \tag{7.18}$$

and the variation ratio of output voltage v_O is

$$\varepsilon = \frac{\Delta v_O/2}{V_O} = \frac{k}{2RfC_6}. \tag{7.19}$$

Example 7.2

A triple-lift circuit of the P/O SL Luo-converter in Figure 7.3a has $V_{in} = 20$V, all inductors have 10 mH, all capacitors have 20 μF, $R = 1000\,\Omega$, $f = 50$ kHz, and conduction duty cycle $k = 0.6$. Calculate the variation ratio of current i_{L1}, and the output voltage and its variation ratio.

Solution

From Equation 7.16, we can obtain the variation ratio of current i_{L1},

$$\xi_1 = \frac{k(1-k)^6}{2(2-k)^5}\frac{R}{fL_1} = \frac{0.6(1-0.6)^6}{2(2-0.6)^5}\frac{1000}{50\,\text{k} \times 10\,\text{m}} = 0.00046.$$

From Equation 7.14, we can obtain the output voltage

$$V_O = \left(\frac{2-k}{1-k}\right)^3 V_{in} = \left(\frac{2-0.6}{1-0.6}\right)^3 20 = 857.5\text{V}.$$

From Equation 7.19, its variation ratio is

$$\varepsilon = \frac{k}{2RfC_6} = \frac{0.6}{2\times 1000\times 50\,\text{k}\times 20\,\mu} = 0.0003.$$

7.2.1.4 Higher-Order Lift Circuit

The higher-order lift circuit can be designed by just multiple repeating of the parts (L_2-D_3-D_4-D_5-C_3-C_4). For the nth-order lift circuit, the final output voltage across capacitor C_{2n} is

$$V_O = \left(\frac{2-k}{1-k}\right)^n V_{in}.$$

The voltage transfer gain is

$$G = \frac{V_O}{V_{in}} = \left(\frac{2-k}{1-k}\right)^n. \tag{7.20}$$

The variation ratio of current i_{Li} through inductor L_i ($i = 1, 2, 3, \ldots, n$) is

$$\xi_i = \frac{\Delta i_{Li}/2}{I_{Li}} = \frac{k(1-k)^{2(n-i+1)}}{2(2-k)^{2(n-i)+1}}\frac{R}{fL_i}, \tag{7.21}$$

and the variation ratio of output voltage v_O is

$$\varepsilon = \frac{\Delta v_O/2}{V_O} = \frac{1-k}{2RfC_{2n}}. \tag{7.22}$$

7.2.2 Additional Series

By using two diodes and two capacitors (D_{11}-D_{12}-C_{11}-C_{12}), a circuit called "double/enhance circuit" (DEC) can be constructed, which is shown in Figure 7.4. If the input voltage is V_{in}, the output voltage V_O can be $2V_{in}$ or another value higher than V_{in}. The DEC is very useful to enhance the DC/DC converter's voltage transfer gain.

All circuits of P/O SL Luo-converters—additional series—are derived from the corresponding circuits of the main series by adding a DEC. The first three stages of this series are shown in Figures 7.5 through 7.7. For ease of understanding, they are called the elementary additional circuit, the re-lift additional circuit, and the triple-lift additional circuit, respectively, and are numbered as $n = 1$, 2, and 3, respectively.

7.2.2.1 Elementary Additional Circuit

The elementary additional circuit is derived from the elementary circuit by adding a DEC. Its circuit and switch-on and switch-off equivalent circuits are shown in Figure 7.5.

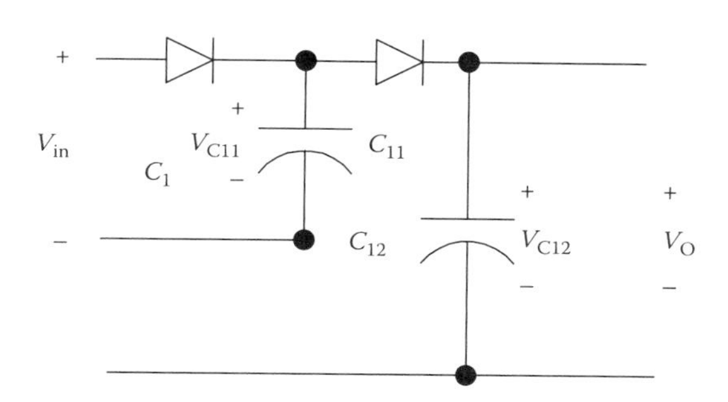

FIGURE 7.4 DEC. (Reprinted from Luo, F. L. and Ye, H. 2006. *Essential DC/DC Converters*. Boca Raton: Taylor & Francis Group LLC, p. 223. With permission.)

The voltage across capacitor C_1 is charged with V_{in} and the voltage across capacitors C_2 and C_{11} is charged with V_1. The current i_{L1} flowing through inductor L_1 increases with V_{in} during the switch-on period kT and decreases with $-(V_O - 2V_{in})$ during the switch-off period $(1 - k)T$. Therefore,

$$V_1 = \frac{2-k}{1-k} V_{in} \tag{7.23}$$

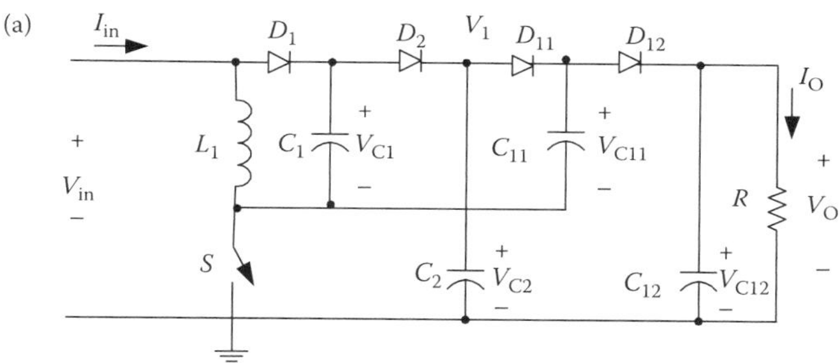

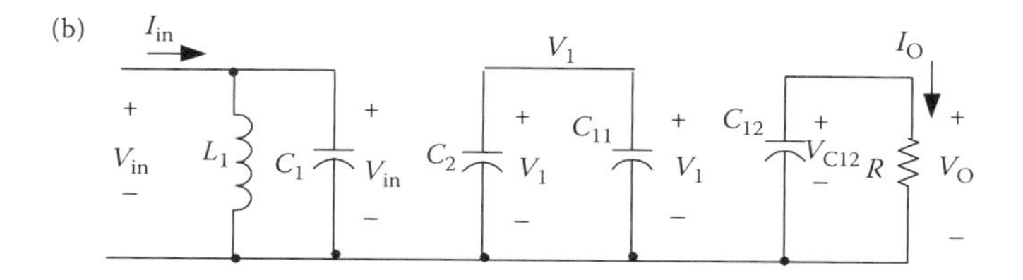

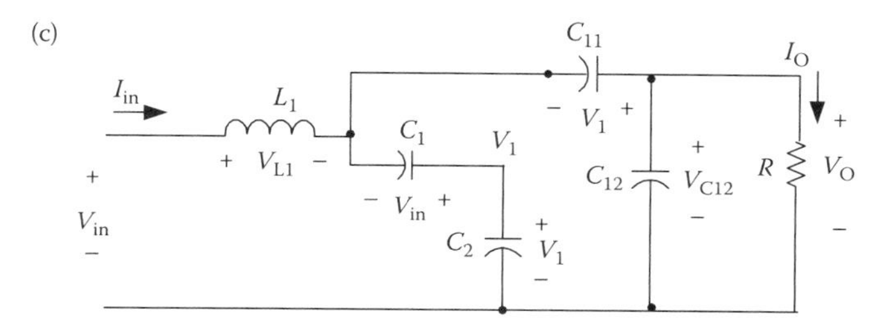

FIGURE 7.5 Elementary additional circuit of P/O SL Luo-converters: (a) circuit diagram, (b) equivalent circuit during switch-on, and (c) equivalent circuit during switch-off. (Reprinted from Luo, F. L. and Ye, H. 2006. *Essential DC/DC Converters*, p. 224. Boca Raton: Taylor & Francis Group LLC. With permission.)

and

$$V_{L1} = \frac{k}{1-k} V_{in}. \tag{7.24}$$

The output voltage is

$$V_O = V_{in} + V_{L1} + V_1 = \frac{3-k}{1-k} V_{in}. \tag{7.25}$$

The voltage transfer gain is

$$G = \frac{V_O}{V_{in}} = \frac{3-k}{1-k}. \tag{7.26}$$

The following relations are derived:

$$i_{in-off} = I_{L1} = i_{C11-off} + i_{C1-off} = \frac{2I_O}{1-k}, \quad i_{in-on} = i_{L1-on} + i_{C1-on} = I_{L1} + \frac{I_O}{k},$$

$$i_{C1-on} = \frac{1-k}{k} i_{C1-off} = \frac{I_O}{k}, \quad i_{C1-off} = i_{C2-off} = \frac{I_O}{1-k},$$

$$i_{C2-off} = \frac{k}{1-k} i_{C2-on} = \frac{k}{1-k} i_{C11-on} = \frac{I_O}{1-k}, \quad i_{C11-on} = \frac{1-k}{k} i_{C11-off} = \frac{I_O}{k},$$

$$i_{C11-off} = I_O + i_{C12-off} = I_O + \frac{k}{1-k} i_{C12-on} = \frac{I_O}{1-k}, \quad i_{C12-off} = \frac{k}{1-k} i_{C12-on} = \frac{kI_O}{1-k}.$$

If inductance L_1 is large enough, i_{L1} is nearly equal to its average current I_{L1}. Therefore,

$$i_{in-off} = I_{L1} = \frac{2I_O}{1-k}, \quad i_{in-on} = I_{L1} + \frac{I_O}{k} = \left(\frac{2}{1-k} + \frac{1}{k}\right) I_O = \frac{1+k}{k(1-k)} I_O.$$

$$\text{Verification: } I_{in} = k i_{in-on} + (1-k) i_{in-off} = \left(\frac{1+k}{1-k} + 2\right) I_O = \frac{3-k}{1-k} I_O.$$

Considering $(V_{in}/I_{in}) = ((1-k)/(2-k))^2 (V_O/I_O) = ((1-k)/(2-k))^2 R$, the variation of current i_{L1} is $\Delta i_{L1} = kTV_{in}/L_1$.

Therefore, the variation ratio of current i_{L1} through inductor L_1 is

$$\xi_1 = \frac{\Delta i_{L1}/2}{I_{L1}} = \frac{k(1-k)TV_{in}}{4L_1 I_O} = \frac{k(1-k)^2}{4(3-k)} \frac{R}{fL_1}. \tag{7.27}$$

The ripple voltage of output voltage v_O is

$$\Delta v_O = \frac{\Delta Q}{C_{12}} = \frac{I_O kT}{C_{12}} = \frac{k}{fC_{12}} \frac{V_O}{R}.$$

Therefore, the variation ratio of output voltage v_O is

$$\varepsilon = \frac{\Delta v_O/2}{V_O} = \frac{k}{2RfC_{12}}. \tag{7.28}$$

7.2.2.2 Re-Lift Additional Circuit

This circuit is derived from the re-lift circuit by adding a DEC. Its circuit diagram and switch-on and switch-off equivalent circuits are shown in Figure 7.6. The voltage across capacitor C_1 is charged with V_{in}. As described in the previous section, the voltage across C_2 is $V_1 = ((2-k)/(1-k))V_{in}$.

The voltage across capacitor C_3 is charged with V_1 and the voltage across capacitors C_4 and C_{11} is charged with V_2. The current flowing through inductor L_2 increases with V_1 during the switch-on period kT and decreases with $-(V_O - 2V_1)$ during the switch-off period $(1-k)T$. Therefore,

$$V_2 = \frac{2-k}{1-k}V_1 = \left(\frac{2-k}{1-k}\right)^2 V_{in} \tag{7.29}$$

(a)

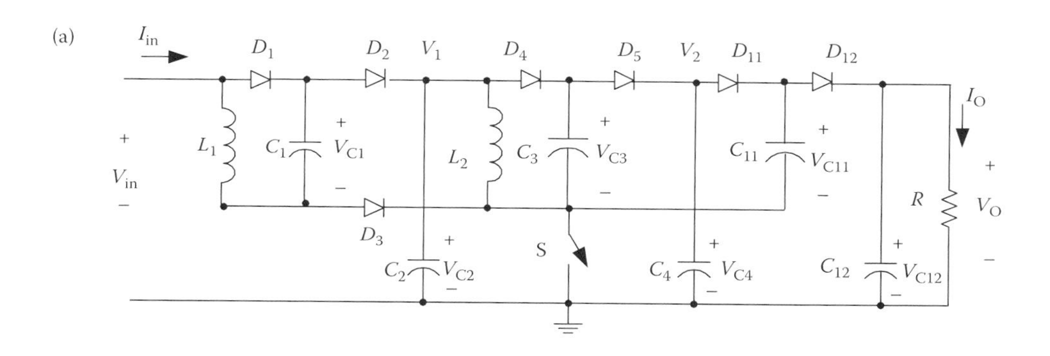

(b)

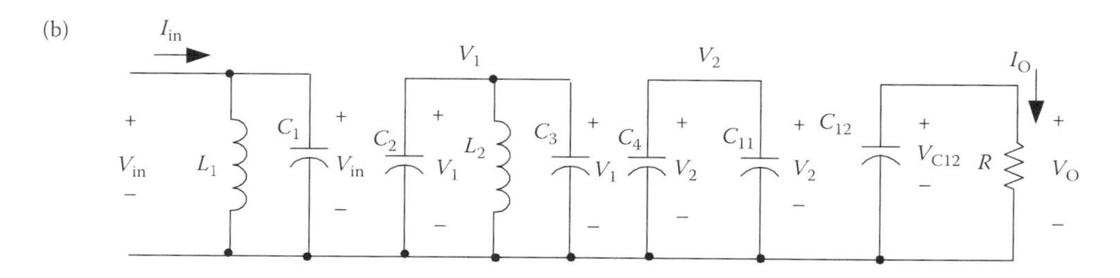

(c)

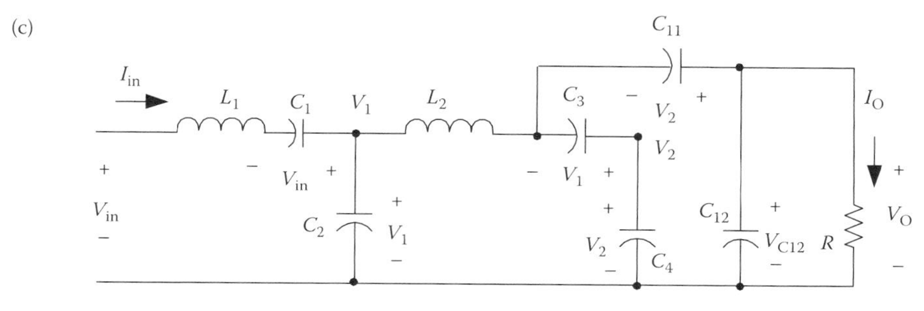

FIGURE 7.6 Re-lift additional circuit of P/O SL Luo-converters: (a) circuit diagram, (b) equivalent circuit during switch-on, and (c) equivalent circuit during switch-off. (Reprinted from Luo, F. L. and Ye, H. 2006. *Essential DC/DC Converters*. Boca Raton: Taylor & Francis Group LLC, p. 225. With permission.)

and

$$V_{L2} = \frac{k}{1-k} V_1. \tag{7.30}$$

The output voltage is

$$V_O = V_1 + V_{L2} + V_2 = \frac{2-k}{1-k} \frac{3-k}{1-k} V_{in}. \tag{7.31}$$

The voltage transfer gain is

$$G = \frac{V_O}{V_{in}} = \frac{2-k}{1-k} \frac{3-k}{1-k}. \tag{7.32}$$

Analogously,

$$\Delta i_{L1} = \frac{V_{in}}{L_1} kT, \qquad I_{L1} = \frac{3-k}{(1-k)^2} I_O,$$

$$\Delta i_{L2} = \frac{V_1}{L_2} kT, \qquad I_{L2} = \frac{2I_O}{1-k}.$$

Therefore, the variation ratio of current i_{L1} through inductor L_1 is

$$\xi_1 = \frac{\Delta i_{L1}/2}{I_{L1}} = \frac{k(1-k)^2 T V_{in}}{2(3-k) L_1 I_O} = \frac{k(1-k)^4}{2(2-k)(3-k)^2} \frac{R}{f L_1}, \tag{7.33}$$

and the variation ratio of current i_{L2} through inductor L_2 is

$$\xi_2 = \frac{\Delta i_{L2}/2}{I_{L2}} = \frac{k(1-k) T V_1}{4 L_2 I_O} = \frac{k(1-k)^2}{4(3-k)} \frac{R}{f L_2}. \tag{7.34}$$

The ripple voltage of output voltage v_O is

$$\Delta v_O = \frac{\Delta Q}{C_{12}} = \frac{I_O kT}{C_{12}} = \frac{k}{f C_{12}} \frac{V_O}{R}.$$

Therefore, the variation ratio of output voltage v_O is

$$\varepsilon = \frac{\Delta v_O/2}{V_O} = \frac{k}{2 R f C_{12}}. \tag{7.35}$$

7.2.2.3 Triple-Lift Additional Circuit

This circuit is derived from the triple-lift circuit by adding a DEC. Its circuit diagram and equivalent circuits during switch-on and switch-off periods are shown in Figure 7.7. The voltage across capacitor C_1 is charged with V_{in}. As described in the previous section, the voltage across C_2 is $V_1 = ((2-k)/(1-k))V_{in}$ and the voltage across C_4 is

$$V_2 = \frac{2-k}{1-k} V_1 = \left(\frac{2-k}{1-k}\right)^2 V_{in}.$$

(a)

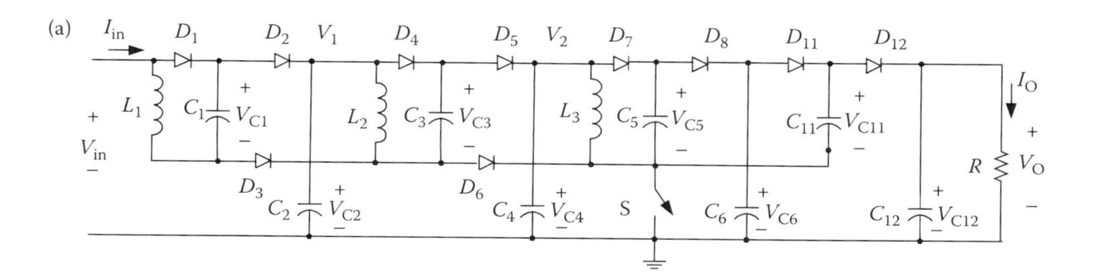

(b)

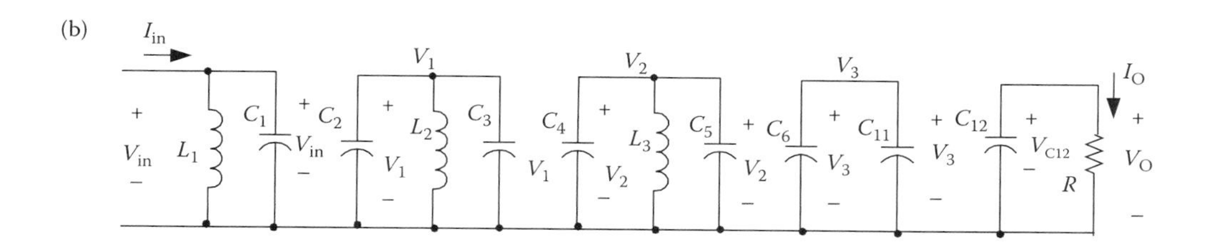

(c)

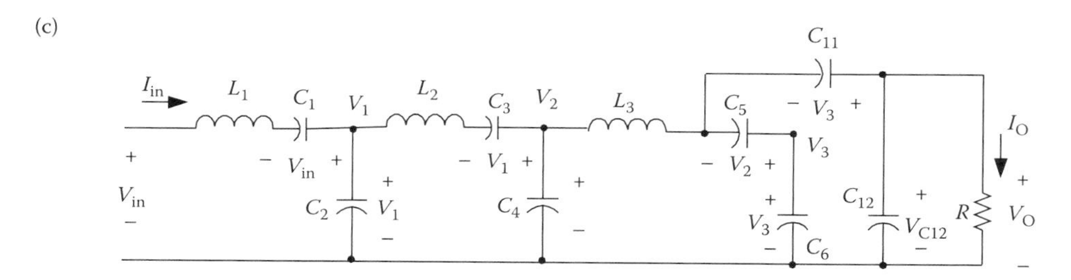

FIGURE 7.7 Triple-lift additional circuit of P/O SL Luo-converters: (a) circuit diagram, (b) equivalent circuit during switch-on, and (c) equivalent circuit during switch-off. (Reprinted from Luo, F. L. and Ye, H. 2006. *Essential DC/DC Converters*. Boca Raton: Taylor & Francis Group LLC, p. 226. With permission.)

The voltage across capacitor C_5 is charged with V_2 and the voltage across capacitors C_6 and C_{11} is charged with V_3. The current flowing through inductor L_3 increases with V_2 during the switch-on period kT and decreases with $-(V_O - 2V_2)$ during the switch-off period $(1-k)T$. Therefore,

$$V_3 = \frac{2-k}{1-k}V_2 = \left(\frac{2-k}{1-k}\right)^2 V_1 = \left(\frac{2-k}{1-k}\right)^3 V_{in} \tag{7.36}$$

and

$$V_{L3} = \frac{k}{1-k}V_2. \tag{7.37}$$

The output voltage is

$$V_O = V_2 + V_{L3} + V_3 = \left(\frac{2-k}{1-k}\right)^2 \frac{3-k}{1-k}V_{in}. \tag{7.38}$$

The voltage transfer gain is

$$G = \frac{V_O}{V_{in}} = \left(\frac{2-k}{1-k}\right)^2 \frac{3-k}{1-k}. \tag{7.39}$$

Analogously,

$$\Delta i_{L1} = \frac{V_{in}}{L_1} kT, \quad I_{L1} = \frac{(2-k)(3-k)}{(1-k)^3} I_O,$$

$$\Delta i_{L2} = \frac{V_1}{L_2} kT, \quad I_{L2} = \frac{3-k}{(1-k)^2} I_O,$$

$$\Delta i_{L3} = \frac{V_2}{L_3} kT, \quad I_{L3} = \frac{2I_O}{1-k}.$$

Considering

$$\frac{V_{in}}{I_{in}} = \left(\frac{1-k}{2-k}\right)^2 \frac{V_O}{I_O} = \left(\frac{1-k}{2-k}\right)^2 R,$$

the variation ratio of current i_{L1} through inductor L_1 is

$$\xi_1 = \frac{\Delta i_{L1}/2}{I_{L1}} = \frac{k(1-k)^3 T V_{in}}{2(2-k)(3-k)L_1 I_O} = \frac{k(1-k)^3 T}{2(2-k)(3-k)L_1 I_O} \frac{(1-k)^3}{(2-k)^2(3-k)} V_O$$

$$= \frac{k(1-k)^6}{2(2-k)^3(3-k)^2} \frac{R}{fL_1}, \tag{7.40}$$

the variation ratio of current i_{L2} through inductor L_2 is

$$\xi_2 = \frac{\Delta i_{L2}/2}{I_{L2}} = \frac{k(1-k)^2 T V_1}{2(3-k)L_2 I_O} = \frac{k(1-k)^2 T}{2(3-k)L_2 I_O} \frac{(1-k)^2}{(2-k)(3-k)} V_O = \frac{k(1-k)^4}{2(2-k)(3-k)^2} \frac{R}{fL_2}, \tag{7.41}$$

and the variation ratio of current i_{L3} through inductor L_3 is

$$\xi_3 = \frac{\Delta i_{L3}/2}{I_{L3}} = \frac{k(1-k)TV_2}{4L_3 I_O} = \frac{k(1-k)T}{4L_3 I_O} \frac{1-k}{3-k} V_O = \frac{k(1-k)^2}{4(3-k)} \frac{R}{fL_3}. \tag{7.42}$$

The ripple voltage of output voltage v_O is

$$\Delta v_O = \frac{\Delta Q}{C_{12}} = \frac{I_O kT}{C_{12}} = \frac{k}{fC_{12}} \frac{V_O}{R}.$$

Therefore, the variation ratio of output voltage v_O is

$$\varepsilon = \frac{\Delta v_O/2}{V_O} = \frac{k}{2RfC_{12}}. \tag{7.43}$$

7.2.2.4 Higher-Order Lift Additional Circuit

The higher-order lift additional circuit is derived from the corresponding circuits of the main series by adding a DEC. For the nth-order lift additional circuit, the final output voltage is

$$V_O = \left(\frac{2-k}{1-k}\right)^{n-1}\frac{3-k}{1-k}V_{in}.$$

The voltage transfer gain is

$$G = \frac{V_O}{V_{in}} = \left(\frac{2-k}{1-k}\right)^{n-1}\frac{3-k}{1-k}. \quad (7.44)$$

Analogously, the variation ratio of current i_{Li} through inductor L_i $(i = 1, 2, 3, \ldots, n)$ is

$$\xi_i = \frac{\Delta i_{Li}/2}{I_{Li}} = \frac{k(1-k)^{2(n-i+1)}}{2[2(2-k)]^{h(n-i)}(2-k)^{2(n-i)+1}(3-k)^{2u(n-i-1)}}\frac{R}{fL_i}, \quad (7.45)$$

where

$$h(x) = \begin{cases} 0 & x > 0 \\ 1 & x \le 0 \end{cases} \quad \text{is the Hong function}$$

and

$$u(x) = \begin{cases} 1 & x \ge 0 \\ 0 & x < 0 \end{cases} \quad \text{is the unit-step function,}$$

and the variation ratio of output voltage v_O is

$$\varepsilon = \frac{\Delta v_O/2}{V_O} = \frac{k}{2RfC_{12}}. \quad (7.46)$$

7.2.3 Enhanced Series

All circuits of P/O SL Luo-converters—enhanced series—are derived from the corresponding circuits of the main series by adding the DEC in circuits of each stage. The first three stages of this series are shown in Figures 7.5, 7.8, and 7.9. For ease of understanding, they are called the elementary-enhanced circuit, the re-lift enhanced circuit, and the triple-lift enhanced circuit, respectively, and numbered $n = 1, 2$, and 3, respectively.

7.2.3.1 Elementary Enhanced Circuit

This circuit is the same as the elementary additional circuit shown in Figure 7.5.

The output voltage is

$$V_O = V_{in} + V_{L1} + V_1 = \frac{3-k}{1-k}V_{in}. \quad (7.25)$$

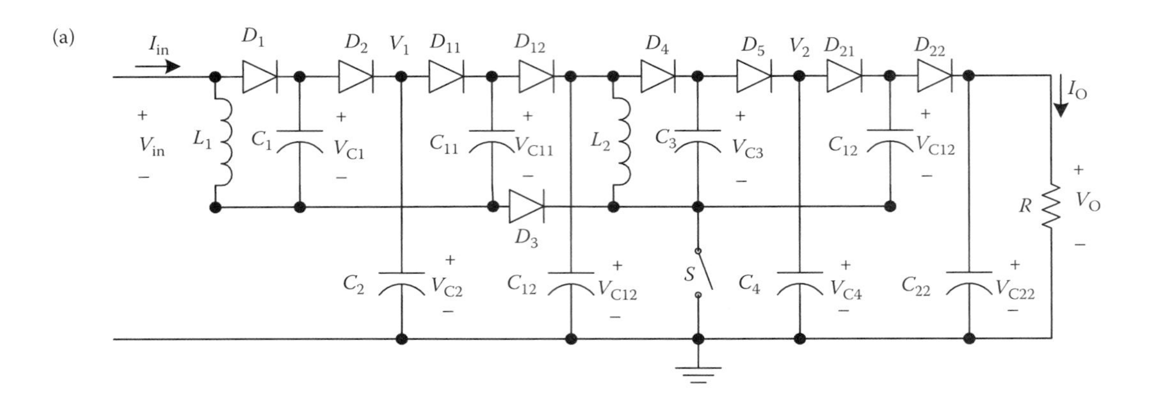

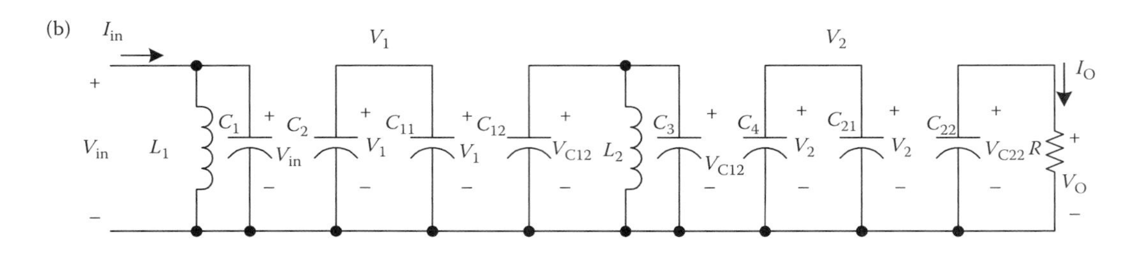

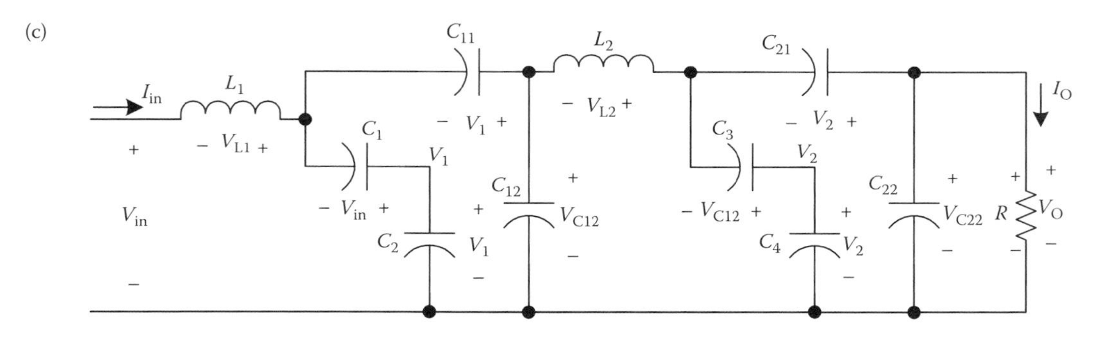

FIGURE 7.8 Re-lift enhanced circuit of P/O SL Luo-converters: (a) circuit diagram, (b) equivalent circuit during switch-on, and (c) equivalent circuit during switch-off. (Reprinted from Luo, F. L. and Ye, H. 2006. *Essential DC/DC Converters*. Boca Raton: Taylor & Francis Group LLC, p. 232. With permission.)

The voltage transfer gain is

$$G = \frac{V_O}{V_{in}} = \frac{3-k}{1-k}. \tag{7.26}$$

The variation of current i_{L1} is $\Delta i_{L1} = kTV_{in}/L_1$.

Therefore, the variation ratio of current i_{L1} through inductor L_1 is

$$\xi_1 = \frac{\Delta i_{L1}/2}{I_{L1}} = \frac{k(1-k)TV_{in}}{4L_1 I_O} = \frac{k(1-k)^2}{4(3-k)} \frac{R}{fL_1}. \tag{7.27}$$

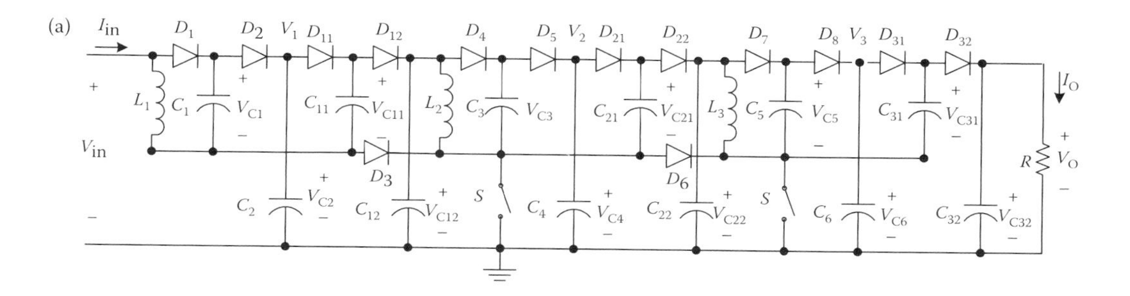

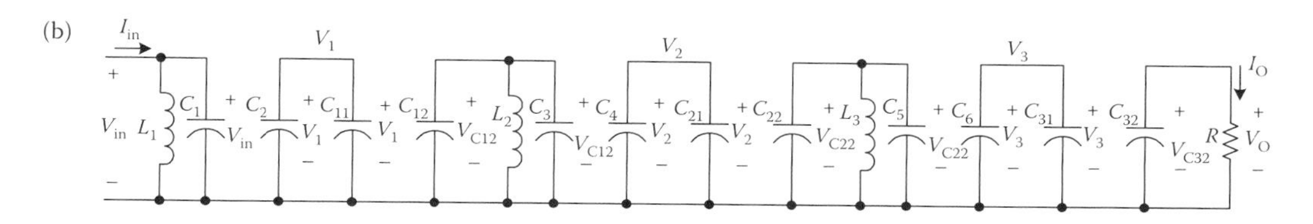

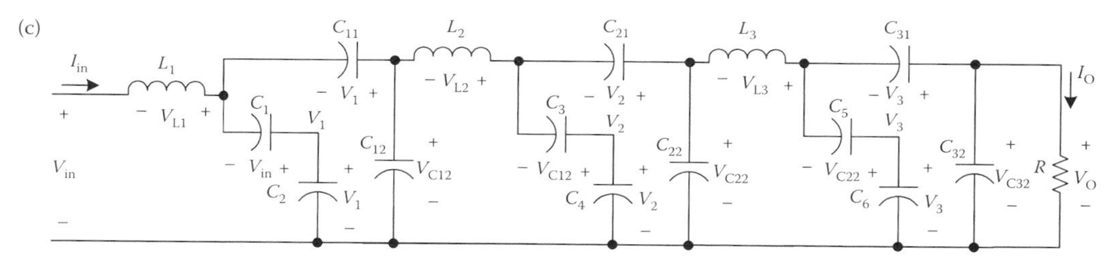

FIGURE 7.9 Triple-lift enhanced circuit of P/O SL Luo-converters: (a) circuit diagram, (b) equivalent circuit during switch-on, and (c) equivalent circuit during switch-off. (Reprinted from Luo, F. L. and Ye, H. 2006. *Essential DC/DC Converters*. Boca Raton: Taylor & Francis Group LLC, p. 233. With permission.)

The ripple voltage of output voltage v_O is

$$\Delta v_O = \frac{\Delta Q}{C_{12}} = \frac{I_O kT}{C_{12}} = \frac{k}{fC_{12}}\frac{V_O}{R}.$$

Therefore, the variation ratio of output voltage v_O is

$$\varepsilon = \frac{\Delta v_O/2}{V_O} = \frac{k}{2RfC_{12}}. \quad (7.28)$$

7.2.3.2 Re-Lift Enhanced Circuit

The re-lift enhanced circuit is derived from the re-lift circuit of the main series by adding the DEC in each stage circuit. Its circuit diagram and switch-on and switch-off equivalent circuits are shown in Figure 7.8. As described in the previous section, the voltage across capacitor C_{12} is charged with $V_{C12} = ((3-k)/(1-k))V_{in}$.

The voltage across capacitor C_3 is charged with V_{C12} and the voltage across capacitors C_4 and C_{21} is charged with V_{C4},

$$V_{C4} = \frac{2-k}{1-k}V_{C12} = \frac{2-k}{1-k}\frac{3-k}{1-k}V_{in}. \quad (7.47)$$

The current flowing through inductor L_2 increases with V_{C12} during the switch-on period kT and decreases with $-(V_O - V_{C4} - V_{C12})$ during the switch-off period $(1-k)T$. Therefore,

$$\Delta i_{L2} = \frac{k}{L_2}V_{C12} = \frac{1-k}{L_2}(V_O - V_{C4} - V_{C12}), \tag{7.48}$$

$$V_O = \frac{3-k}{1-k}V_{C12} = \left(\frac{3-k}{1-k}\right)^2 V_{in}. \tag{7.49}$$

The voltage transfer gain is

$$G = \frac{V_O}{V_{in}} = \left(\frac{3-k}{1-k}\right)^2. \tag{7.50}$$

Analogously,

$$\Delta i_{L1} = \frac{V_{in}}{L_1}kT, \quad I_{L1} = \frac{3-k}{(1-k)^2}I_O,$$

$$\Delta i_{L2} = \frac{V_1}{L_2}kT, \quad I_{L2} = \frac{2I_O}{1-k}.$$

Therefore, the variation ratio of current i_{L1} through inductor L_1 is

$$\xi_1 = \frac{\Delta i_{L1}/2}{I_{L1}} = \frac{k(1-k)^2TV_{in}}{2(3-k)L_1I_O} = \frac{k(1-k)^4}{2(2-k)(3-k)^2}\frac{R}{fL_1} \tag{7.51}$$

and the variation ratio of current i_{L2} through inductor L_2 is

$$\xi_2 = \frac{\Delta i_{L2}/2}{I_{L2}} = \frac{k(1-k)TV_1}{4L_2I_O} = \frac{k(1-k)^2}{4(3-k)}\frac{R}{fL_2}. \tag{7.52}$$

The ripple voltage of output voltage v_O is

$$\Delta v_O = \frac{\Delta Q}{C_{22}} = \frac{I_OkT}{C_{22}} = \frac{k}{fC_{22}}\frac{V_O}{R}.$$

Therefore, the variation ratio of output voltage v_O is

$$\varepsilon = \frac{\Delta v_O/2}{V_O} = \frac{k}{2RfC_{22}}. \tag{7.53}$$

7.2.3.3 Triple-Lift Enhanced Circuit

The triple-lift enhanced circuit is derived from the triple-lift circuit of the main series by adding the DEC in each stage circuit. Its circuit diagram and equivalent circuits during switch-on and switch-off periods are shown in Figure 7.9. As described in the previous section, the voltage across capacitor C_{12} is charged with $V_{C12} = ((3-k)/(1-k))V_{in}$, and the voltage across capacitor C_{22} is charged with $V_{C22} = ((3-k)/(1-k))^2V_{in}$.

The voltage across capacitor C_5 is charged with V_{C22} and the voltage across capacitors C_6 and C_{31} is charged with V_{C6},

$$V_{C6} = \frac{2-k}{1-k} V_{C22} = \frac{2-k}{1-k} \left(\frac{3-k}{1-k} \right)^2 V_{in}. \tag{7.54}$$

The current flowing through inductor L_3 increases with V_{C22} during the switch-on period kT and decreases with $-(V_O - V_{C6} - V_{C22})$ during the switch-off period $(1-k)T$. Therefore,

$$\Delta i_{L3} = \frac{k}{L_3} V_{C22} = \frac{1-k}{L_3} (V_O - V_{C6} - V_{C22}), \tag{7.55}$$

$$V_O = \frac{3-k}{1-k} V_{C22} = \left(\frac{3-k}{1-k} \right)^3 V_{in}. \tag{7.56}$$

The voltage transfer gain is

$$G = \frac{V_O}{V_{in}} = \left(\frac{3-k}{1-k} \right)^3. \tag{7.57}$$

Analogously,

$$\Delta i_{L1} = \frac{V_{in}}{L_1} kT, \quad I_{L1} = \frac{(2-k)(3-k)}{(1-k)^3} I_O,$$

$$\Delta i_{L2} = \frac{V_1}{L_2} kT, \quad I_{L2} = \frac{3-k}{(1-k)^2} I_O,$$

$$\Delta i_{L3} = \frac{V_2}{L_3} kT, \quad I_{L3} = \frac{2 I_O}{1-k}.$$

Considering

$$\frac{V_{in}}{I_{in}} = \left(\frac{1-k}{2-k} \right)^2 \frac{V_O}{I_O} = \left(\frac{1-k}{2-k} \right)^2 R,$$

the variation ratio of current i_{L1} through inductor L_1 is

$$\xi_1 = \frac{\Delta i_{L1}/2}{I_{L1}} = \frac{k(1-k)^3 T V_{in}}{2(2-k)(3-k) L_1 I_O} = \frac{k(1-k)^3 T}{2(2-k)(3-k) L_1 I_O} \frac{(1-k)^3}{(2-k)^2(3-k)} V_O$$
$$= \frac{k(1-k)^6}{2(2-k)^3(3-k)^2} \frac{R}{f L_1}. \tag{7.58}$$

The variation ratio of current i_{L2} through inductor L_2 is

$$\xi_2 = \frac{\Delta i_{L2}/2}{I_{L2}} = \frac{k(1-k)^2 T V_1}{2(3-k) L_2 I_O} = \frac{k(1-k)^2 T}{2(3-k) L_2 I_O} \frac{(1-k)^2}{(2-k)(3-k)} V_O = \frac{k(1-k)^4}{2(2-k)(3-k)^2} \frac{R}{f L_2}, \tag{7.59}$$

and the variation ratio of current i_{L3} through inductor L_3 is

$$\xi_3 = \frac{\Delta i_{L3}/2}{I_{L3}} = \frac{k(1-k)TV_2}{4L_3 I_O} = \frac{k(1-k)T}{4L_3 I_O}\frac{1-k}{3-k}V_O = \frac{k(1-k)^2}{4(3-k)}\frac{R}{fL_3}. \quad (7.60)$$

The ripple voltage of output voltage v_O is

$$\Delta v_O = \frac{\Delta Q}{C_{32}} = \frac{I_O kT}{C_{32}} = \frac{k}{fC_{32}}\frac{V_O}{R}.$$

Therefore, the variation ratio of output voltage v_O is

$$\varepsilon = \frac{\Delta v_O/2}{V_O} = \frac{k}{2RfC_{32}}. \quad (7.61)$$

7.2.3.4 Higher-Order Lift Enhanced Circuit

The higher-order lift enhanced circuit is derived from the corresponding circuits of the main series by adding the DEC in each stage circuit. For the nth-order lift enhanced circuit, the final output voltage is $V_O = ((3-k)/(1-k))^n V_{in}$.

The voltage transfer gain is

$$G = \frac{V_O}{V_{in}} = \left(\frac{3-k}{1-k}\right)^n. \quad (7.62)$$

Analogously, the variation ratio of current i_{Li} through inductor L_i ($i = 1, 2, 3, \ldots, n$) is

$$\xi_i = \frac{\Delta i_{Li}/2}{I_{Li}} = \frac{k(1-k)^{2(n-i+1)}}{2[2(2-k)]^{h(n-i)}(2-k)^{2(n-i)+1}(3-k)^{2u(n-i-1)}}\frac{R}{fL_i}, \quad (7.63)$$

where

$$h(x) = \begin{cases} 0 & x > 0 \\ 1 & x \le 0 \end{cases} \quad \text{is the Hong function}$$

and

$$u(x) = \begin{cases} 1 & x \ge 0 \\ 0 & x < 0 \end{cases} \quad \text{is the unit-step function,}$$

and the variation ratio of output voltage v_O is

$$\varepsilon = \frac{\Delta v_O/2}{V_O} = \frac{k}{2RfC_{n2}}. \quad (7.64)$$

7.2.4 Re-Enhanced Series

All circuits of P/O SL Luo-converters—re-enhanced series—are derived from the corresponding circuits of the main series by adding the DEC twice in each stage circuit.

(a)

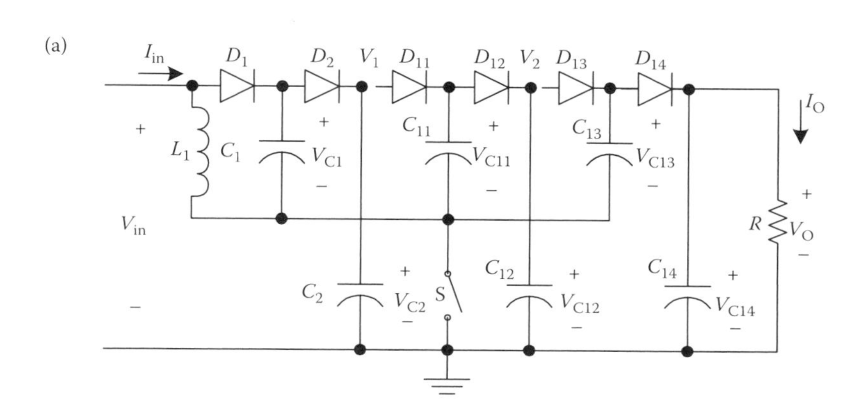

(b)

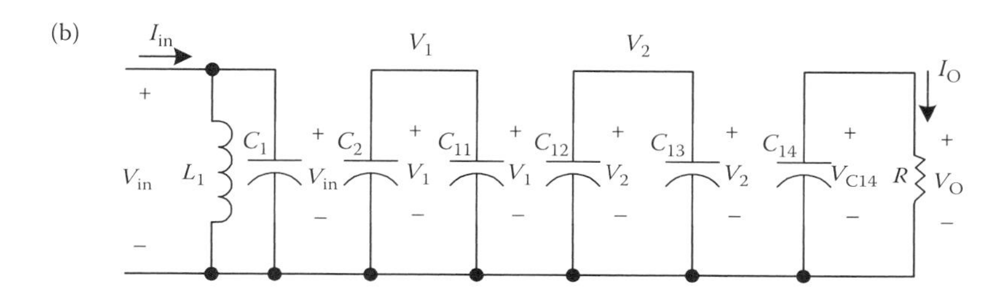

(c)

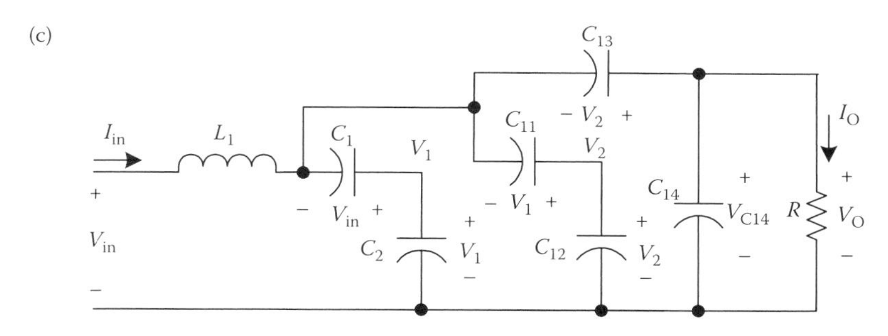

FIGURE 7.10 Elementary re-enhanced circuit of P/O SL Luo-converters: (a) circuit diagram, (b) equivalent circuit during switch-on, and (c) equivalent circuit during switch-off. (Reprinted from Luo, F. L. and Ye, H. 2006. *Essential DC/DC Converters*. Boca Raton: Taylor & Francis Group LLC, p. 238. With permission.)

The first three stages of this series are shown in Figures 7.10 through 7.12. For ease of understanding, they are called the elementary re-enhanced circuit, the re-lift re-enhanced circuit, and the triple-lift re-enhanced circuit, respectively, and numbered $n = 1$, 2, and 3, respectively.

7.2.4.1 Elementary Re-Enhanced Circuit

This circuit is derived from the elementary circuit by adding the DEC twice in each stage circuit. Its circuit and switch-on and switch-off equivalent circuits are shown in Figure 7.10.

The output voltage is

$$V_O = V_{in} + V_{L1} + V_{C12} = \frac{4-k}{1-k}V_{in}. \tag{7.65}$$

The voltage transfer gain is

$$G = \frac{V_O}{V_{in}} = \frac{4-k}{1-k}, \tag{7.66}$$

where

$$V_{C2} = \frac{2-k}{1-k}V_{in}, \tag{7.67}$$

$$V_{C12} = \frac{3-k}{1-k}V_{in}, \tag{7.68}$$

and

$$V_{L1} = \frac{k}{1-k}V_{in}. \tag{7.69}$$

The following relations are obtained:

$$i_{in-off} = I_{L1} = i_{C11-off} + i_{C1-off} = \frac{2I_O}{1-k}, \qquad i_{in-on} = i_{L1-on} + i_{C1-on} = I_{L1} + \frac{I_O}{k},$$

$$i_{C1-on} = \frac{1-k}{k}i_{C1-off} = \frac{I_O}{k}, \qquad i_{C1-off} = i_{C2-off} = \frac{I_O}{1-k},$$

$$i_{C2-off} = \frac{k}{1-k}i_{C2-on} = \frac{k}{1-k}i_{C11-on} = \frac{I_O}{1-k}, \qquad i_{C11-on} = \frac{1-k}{k}i_{C11-off} = \frac{I_O}{k},$$

$$i_{C11-off} = I_O + i_{C12-off} = I_O + \frac{k}{1-k}i_{C12-on} = \frac{I_O}{1-k}, \quad i_{C12-off} = \frac{k}{1-k}i_{C12-on} = \frac{kI_O}{1-k}.$$

If inductance L_1 is large enough, i_{L1} is nearly equal to its average current I_{L1}. Therefore,

$$i_{in-off} = I_{L1} = \frac{2I_O}{1-k}, \quad i_{in-on} = I_{L1} + \frac{I_O}{k} = \left(\frac{2}{1-k} + \frac{1}{k}\right) I_O = \frac{1+k}{k(1-k)}I_O.$$

$$\text{Verification: } I_{in} = ki_{in-on} + (1-k)i_{in-off} = \left(\frac{1+k}{1-k} + 2\right) I_O = \frac{3-k}{1-k}I_O.$$

Considering

$$\frac{V_{in}}{I_{in}} = \left(\frac{1-k}{2-k}\right)^2 \frac{V_O}{I_O} = \left(\frac{1-k}{2-k}\right)^2 R,$$

the variation of current i_{L1} is $\Delta i_{L1} = kTV_{in}/L_1$.

Therefore, the variation ratio of current I_{L1} through inductor L_1 is

$$\xi_1 = \frac{\Delta i_{L1}/2}{I_{L1}} = \frac{k(1-k)TV_{in}}{4L_1 I_O} = \frac{k(1-k)^2}{4(3-k)}\frac{R}{fL_1}. \tag{7.70}$$

The ripple voltage of output voltage v_O is

$$\Delta v_O = \frac{\Delta Q}{C_{14}} = \frac{I_O kT}{C_{14}} = \frac{k}{fC_{14}} \frac{V_O}{R}.$$

Therefore, the variation ratio of output voltage v_O is

$$\varepsilon = \frac{\Delta v_O / 2}{V_O} = \frac{k}{2RfC_{14}}. \tag{7.71}$$

7.2.4.2 Re-Lift Re-Enhanced Circuit

This circuit is derived from the re-lift circuit of the main series by adding the DEC twice in each stage circuit. Its circuit and switch-on and switch-off equivalent circuits are shown in Figure 7.11.

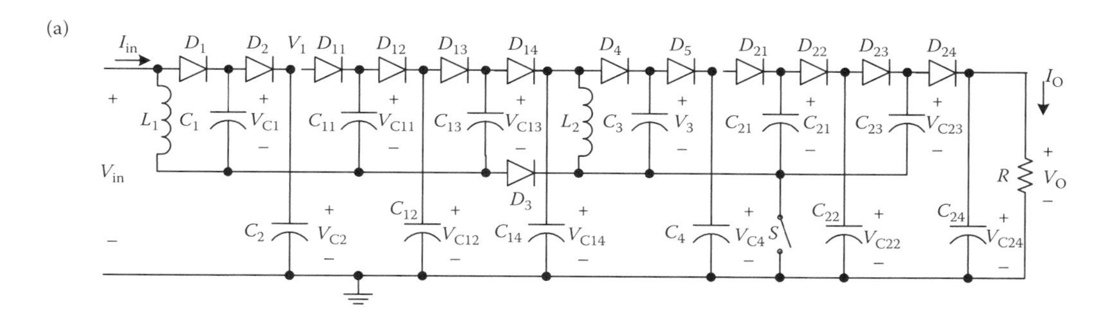

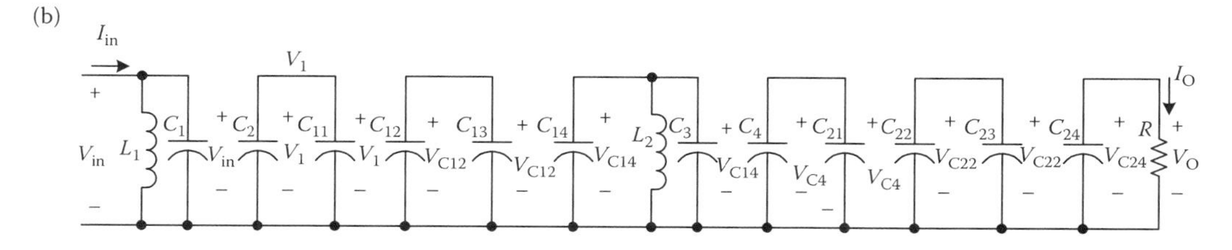

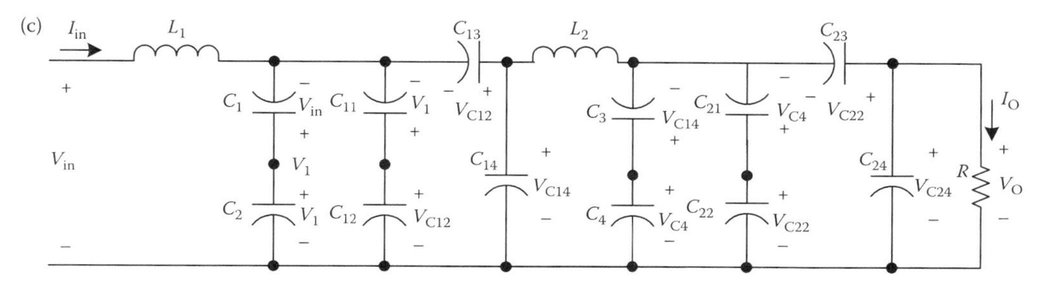

FIGURE 7.11 Re-lift re-enhanced circuit of P/O SL Luo-converters: (a) circuit diagram, (b) equivalent circuit during switch-on, and (c) equivalent circuit during switch-off. (Reprinted from Luo, F. L. and Ye, H. 2006. *Essential DC/DC Converters*. Boca Raton: Taylor & Francis Group LLC, p. 239. With permission.)

The voltage across capacitor C_{14} is

$$V_{C14} = \frac{4-k}{1-k}V_{in}. \tag{7.72}$$

By the same analysis

$$V_O = \frac{4-k}{1-k}V_{C14} = \left(\frac{4-k}{1-k}\right)^2 V_{in}. \tag{7.73}$$

The voltage transfer gain is

$$G = \frac{V_O}{V_{in}} = \left(\frac{4-k}{1-k}\right)^2. \tag{7.74}$$

Analogously,

$$\Delta i_{L1} = \frac{V_{in}}{L_1}kT, \qquad I_{L1} = \frac{3-k}{(1-k)^2}I_O,$$

$$\Delta i_{L2} = \frac{V_1}{L_2}kT, \qquad I_{L2} = \frac{2I_O}{1-k}.$$

Therefore, the variation ratio of current i_{L1} through inductor L_1 is

$$\xi_1 = \frac{\Delta i_{L1}/2}{I_{L1}} = \frac{k(1-k)^2 TV_{in}}{2(3-k)L_1 I_O} = \frac{k(1-k)^4}{2(2-k)(3-k)^2}\frac{R}{fL_1}. \tag{7.75}$$

The variation ratio of current i_{L2} through inductor L_2 is

$$\xi_2 = \frac{\Delta i_{L2}/2}{I_{L2}} = \frac{k(1-k)TV_1}{4L_2 I_O} = \frac{k(1-k)^2}{4(3-k)}\frac{R}{fL_2}. \tag{7.76}$$

The ripple voltage of output voltage v_O is

$$\Delta v_O = \frac{\Delta Q}{C_{24}} = \frac{I_O kT}{C_{24}} = \frac{k}{fC_{24}}\frac{V_O}{R}.$$

Therefore, the variation ratio of output voltage v_O is

$$\varepsilon = \frac{\Delta v_O/2}{V_O} = \frac{k}{2RfC_{24}}. \tag{7.77}$$

7.2.4.3 Triple-Lift Re-Enhanced Circuit

This circuit is derived from the triple-lift circuit of the main series by adding the DEC twice in each stage circuit. Its circuit and switch-on and switch-off equivalent circuits are shown in Figure 7.12.

(a)

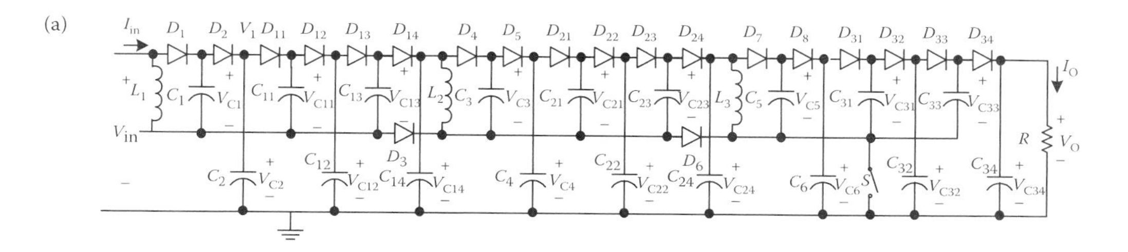

(b)

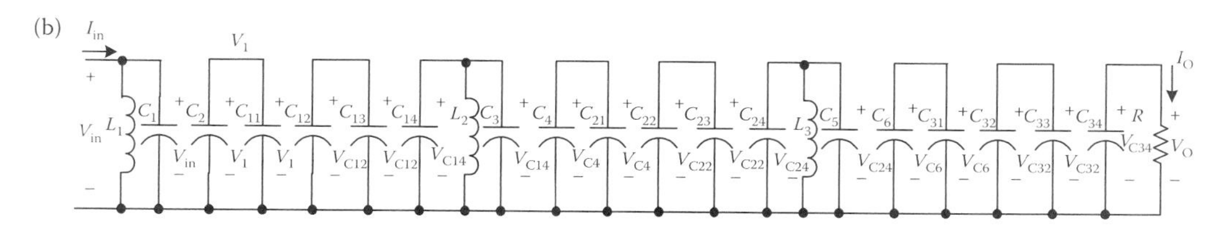

(c)

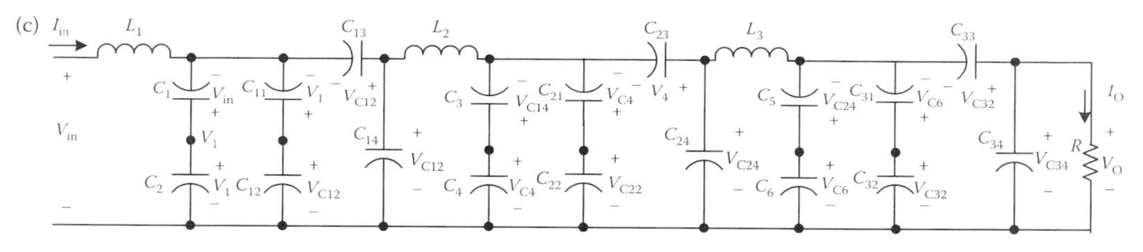

FIGURE 7.12 Triple-lift re-enhanced circuit of P/O SL Luo-converters: (a) circuit diagram, (b) equivalent circuit during switch-on, and (c) equivalent circuit during switch-off. (Reprinted from Luo, F. L. and Ye, H. 2006. *Essential DC/DC Converters*. Boca Raton: Taylor & Francis Group LLC, p. 240. With permission.)

The voltage across capacitor C_{14} is

$$V_{C14} = \frac{4-k}{1-k}V_{in}. \tag{7.78}$$

The voltage across capacitor C_{24} is

$$V_{C24} = \left(\frac{4-k}{1-k}\right)^2 V_{in}. \tag{7.79}$$

By the same analysis

$$V_O = \frac{4-k}{1-k}V_{C24} = \left(\frac{4-k}{1-k}\right)^3 V_{in}. \tag{7.80}$$

The voltage transfer gain is

$$G = \frac{V_O}{V_{in}} = \left(\frac{4-k}{1-k}\right)^3. \tag{7.81}$$

Analogously,

$$\Delta i_{L1} = \frac{V_{in}}{L_1}kT, \qquad I_{L1} = \frac{(2-k)(3-k)}{(1-k)^3}I_O,$$

$$\Delta i_{L2} = \frac{V_1}{L_2}kT, \qquad I_{L2} = \frac{3-k}{(1-k)^2}I_O,$$

$$\Delta i_{L3} = \frac{V_2}{L_3}kT, \qquad I_{L3} = \frac{2I_O}{1-k}.$$

Considering

$$\frac{V_{in}}{I_{in}} = \left(\frac{1-k}{2-k}\right)^2 \frac{V_O}{I_O} = \left(\frac{1-k}{2-k}\right)^2 R,$$

the variation ratio of current i_{L1} through inductor L_1 is

$$\xi_1 = \frac{\Delta i_{L1}/2}{I_{L1}} = \frac{k(1-k)^3 TV_{in}}{2(2-k)(3-k)L_1 I_O} = \frac{k(1-k)^3 T}{2(2-k)(3-k)L_1 I_O}\frac{(1-k)^3}{(2-k)^2(3-k)}V_O$$
$$= \frac{k(1-k)^6}{2(2-k)^3(3-k)^2}\frac{R}{fL_1}. \tag{7.82}$$

The variation ratio of current i_{L2} through inductor L_2 is

$$\xi_2 = \frac{\Delta i_{L2}/2}{I_{L2}} = \frac{k(1-k)^2 TV_1}{2(3-k)L_2 I_O} = \frac{k(1-k)^2 T}{2(3-k)L_2 I_O}\frac{(1-k)^2}{(2-k)(3-k)}V_O = \frac{k(1-k)^4}{2(2-k)(3-k)^2}\frac{R}{fL_2}. \tag{7.83}$$

The variation ratio of current i_{L3} through inductor L_3 is

$$\xi_3 = \frac{\Delta i_{L3}/2}{I_{L3}} = \frac{k(1-k)TV_2}{4L_3 I_O} = \frac{k(1-k)T}{4L_3 I_O}\frac{1-k}{3-k}V_O = \frac{k(1-k)^2}{4(3-k)}\frac{R}{fL_3}. \tag{7.84}$$

The ripple voltage of output voltage v_O is

$$\Delta v_O = \frac{\Delta Q}{C_{34}} = \frac{I_O kT}{C_{34}} = \frac{k}{fC_{34}}\frac{V_O}{R}.$$

Therefore, the variation ratio of output voltage v_O is

$$\varepsilon = \frac{\Delta v_O/2}{V_O} = \frac{k}{2RfC_{34}}. \tag{7.85}$$

7.2.4.4 Higher-Order Lift Re-Enhanced Circuit

The higher-order lift additional circuit is derived from the corresponding circuits of the main series by adding the DEC twice in each stage circuit. For the nth-order lift additional circuit, the final output voltage is $V_O = ((4-k)/(1-k))^n V_{in}$.

The voltage transfer gain is

$$G = \frac{V_O}{V_{in}} = \left(\frac{4-k}{1-k}\right)^n. \tag{7.86}$$

Analogously, the variation ratio of current i_{Li} through inductor $L_i (i = 1, 2, 3, \ldots, n)$ is

$$\xi_i = \frac{\Delta i_{Li}/2}{I_{Li}} = \frac{k(1-k)^{2(n-i+1)}}{2[2(2-k)]^{h(n-i)}(2-k)^{2(n-i)+1}(3-k)^{2u(n-i-1)}} \frac{R}{fL_i}, \tag{7.87}$$

where

$$h(x) = \begin{cases} 0 & x > 0 \\ 1 & x \le 0 \end{cases} \quad \text{is the Hong function}$$

and

$$u(x) = \begin{cases} 1 & x \ge 0 \\ 0 & x < 0 \end{cases} \quad \text{is the unit-step function,}$$

and the variation ratio of output voltage v_O is

$$\varepsilon = \frac{\Delta v_O/2}{V_O} = \frac{k}{2RfC_{n4}}. \tag{7.88}$$

7.2.5 Multiple-Enhanced Series

All circuits of P/O SL Luo-converters—multiple-enhanced series—are derived from the corresponding circuits of the main series by adding the DEC multiple (j) times in circuits of each stage. The first three stages of this series are shown in Figures 7.13 through 7.15. For ease of understanding, they are called the elementary multiple-enhanced circuit, the re-lift multiple-enhanced circuit, and the triple-lift multiple-enhanced circuit, respectively, and numbered $n = 1, 2$, and 3, respectively.

7.2.5.1 Elementary Multiple-Enhanced Circuit

This circuit is derived from the elementary circuit of the main series by adding the DEC multiple (j) times. Its circuit and switch-on and switch-off equivalent circuits are shown in Figure 7.13.

The output voltage is

$$V_O = \frac{j+2-k}{1-k} V_{in}. \tag{7.89}$$

The voltage transfer gain is

$$G = \frac{V_O}{V_{in}} = \frac{j+2-k}{1-k}. \tag{7.90}$$

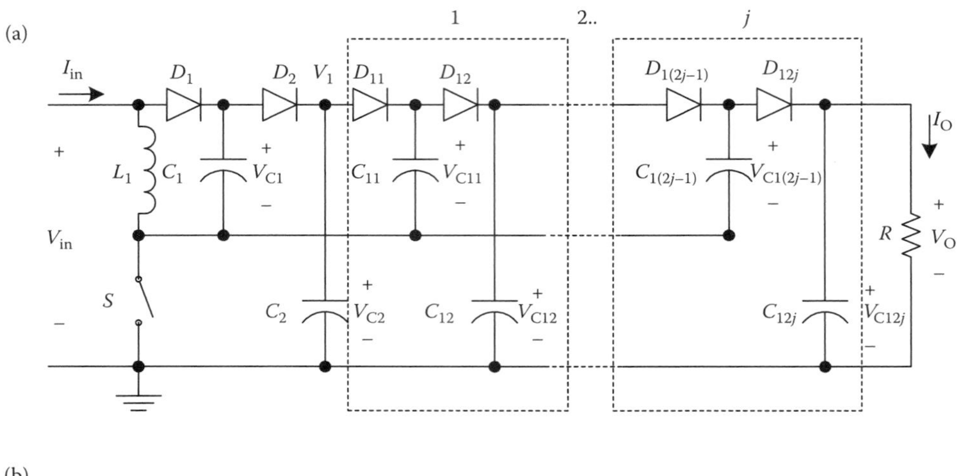

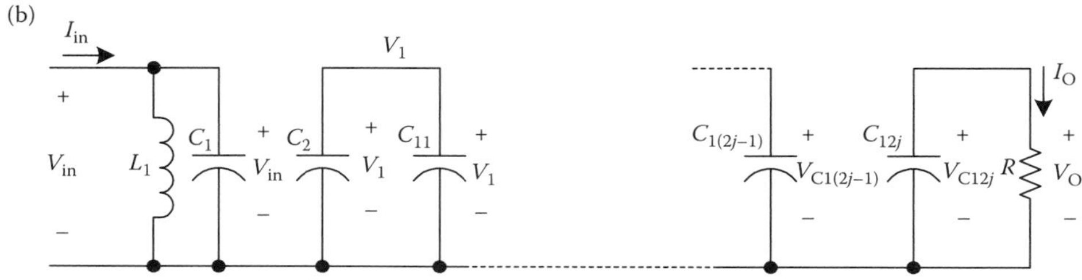

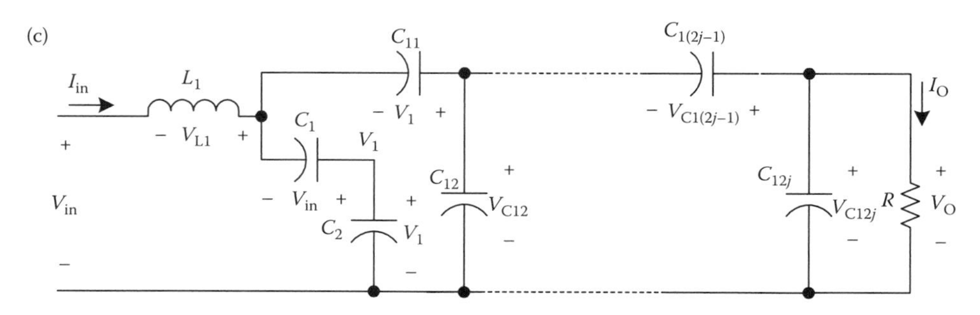

FIGURE 7.13 Elementary multiple-enhanced circuit of P/O SL Luo-converters: (a) circuit diagram, (b) equivalent circuit during switch-on, and (c) equivalent circuit during switch-off. (Reprinted from Luo, F. L. and Ye, H. 2006. *Essential DC/DC Converters*. Boca Raton: Taylor & Francis Group LLC, p. 246. With permission.)

The following relations are obtained:

$$i_{\text{in-off}} = I_{L1} = i_{\text{C11-off}} + i_{\text{C1-off}} = \frac{2I_O}{1-k}, \qquad i_{\text{in-on}} = i_{\text{L1-on}} + i_{\text{C1-on}} = I_{L1} + \frac{I_O}{k},$$

$$i_{\text{C1-on}} = \frac{1-k}{k} i_{\text{C1-off}} = \frac{I_O}{k}, \qquad i_{\text{C1-off}} = i_{\text{C2-off}} = \frac{I_O}{1-k},$$

$$i_{\text{C2-off}} = \frac{k}{1-k} i_{\text{C2-on}} = \frac{k}{1-k} i_{\text{C11-on}} = \frac{I_O}{1-k}, \qquad i_{\text{C11-on}} = \frac{1-k}{k} i_{\text{C11-off}} = \frac{I_O}{k},$$

$$i_{\text{C11-off}} = I_O + i_{\text{C12-off}} = I_O + \frac{k}{1-k} i_{\text{C12-on}} = \frac{I_O}{1-k}, \qquad i_{\text{C12-off}} = \frac{k}{1-k} i_{\text{C12-on}} = \frac{kI_O}{1-k}.$$

If inductance L_1 is large enough, i_{L1} is nearly equal to its average current I_{L1}. Therefore,

$$i_{in-off} = I_{L1} = \frac{2I_O}{1-k}, \quad i_{in-on} = I_{L1} + \frac{I_O}{k} = \left(\frac{2}{1-k} + \frac{1}{k}\right) I_O = \frac{1+k}{k(1-k)} I_O.$$

$$\text{Verification: } I_{in} = k i_{in-on} + (1-k) i_{in-off} = \left(\frac{1+k}{1-k} + 2\right) I_O = \frac{3-k}{1-k} I_O.$$

Considering

$$\frac{V_{in}}{I_{in}} = \left(\frac{1-k}{2-k}\right)^2 \frac{V_O}{I_O} = \left(\frac{1-k}{2-k}\right)^2 R,$$

the variation of current i_{L1} is $\Delta i_{L1} = kTV_{in}/L_1$.

Therefore, the variation ratio of current i_{L1} through inductor L_1 is

$$\xi_1 = \frac{\Delta i_{L1}/2}{I_{L1}} = \frac{k(1-k)TV_{in}}{4L_1 I_O} = \frac{k(1-k)^2}{4(3-k)} \frac{R}{fL_1}. \tag{7.91}$$

The ripple voltage of output voltage v_O is

$$\Delta v_O = \frac{\Delta Q}{C_{12j}} = \frac{I_O kT}{C_{12j}} = \frac{k}{fC_{12j}} \frac{V_O}{R}.$$

Therefore, the variation ratio of output voltage v_O is

$$\varepsilon = \frac{\Delta v_O/2}{V_O} = \frac{k}{2RfC_{12j}}. \tag{7.92}$$

7.2.5.2 Re-Lift Multiple-Enhanced Circuit

This circuit is derived from the re-lift circuit of the main series by adding the DEC multiple (j) times in each stage circuit. Its circuit diagram and switch-on and switch-off equivalent circuits are shown in Figure 7.14.

The voltage across capacitor C_{12j} is

$$V_{C12j} = \frac{j+2-k}{1-k} V_{in}. \tag{7.93}$$

The output voltage across capacitor C_{22j} is

$$V_O = V_{C22j} = \left(\frac{j+2-k}{1-k}\right)^2 V_{in}. \tag{7.94}$$

The voltage transfer gain is

$$G = \frac{V_O}{V_{in}} = \left(\frac{j+2-k}{1-k}\right)^2. \tag{7.95}$$

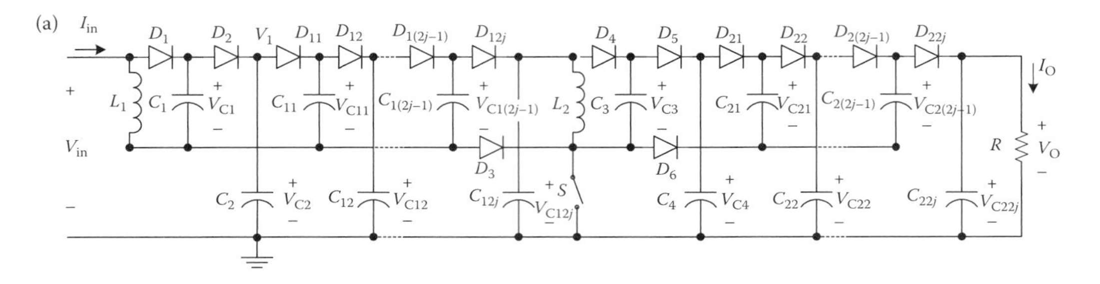

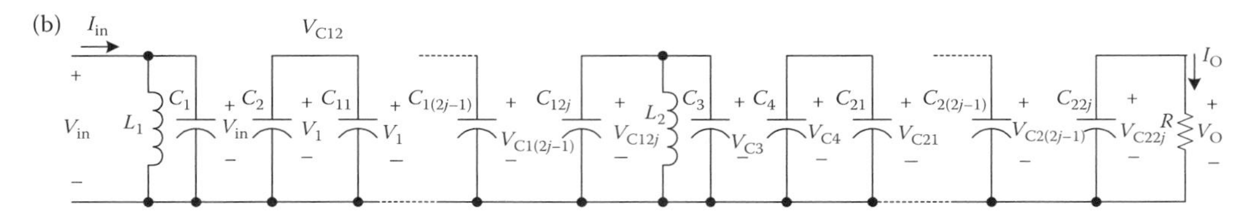

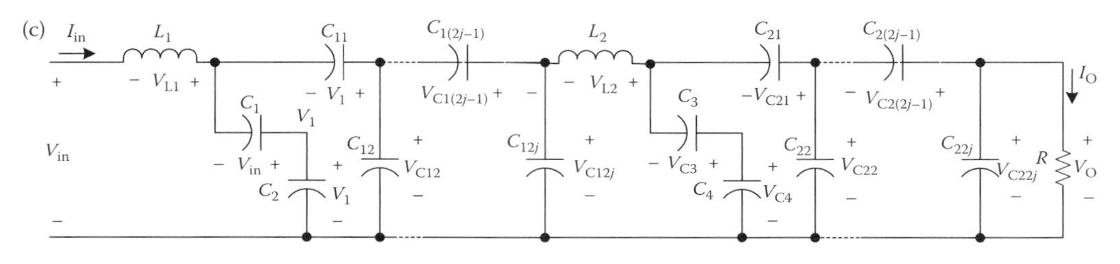

FIGURE 7.14 Re-lift multiple-enhanced circuit of P/O SL Luo-converters: (a) circuit diagram, (b) equivalent circuit during switch-on, and (c) equivalent circuit during switch-off. (Reprinted from Luo, F. L. and Ye, H. 2006. *Essential DC/DC Converters*. Boca Raton: Taylor & Francis Group LLC, p. 247. With permission.)

Analogously,

$$\Delta i_{L1} = \frac{V_{in}}{L_1}kT, \quad I_{L1} = \frac{3-k}{(1-k)^2}I_O,$$

$$\Delta i_{L2} = \frac{V_1}{L_2}kT, \quad I_{L2} = \frac{2I_O}{1-k}.$$

Therefore, the variation ratio of current i_{L1} through inductor L_1 is

$$\xi_1 = \frac{\Delta i_{L1}/2}{I_{L1}} = \frac{k(1-k)^2 TV_{in}}{2(3-k)L_1 I_O} = \frac{k(1-k)^4}{2(2-k)(3-k)^2}\frac{R}{fL_1} \tag{7.96}$$

and the variation ratio of current i_{L2} through inductor L_2 is

$$\xi_2 = \frac{\Delta i_{L2}/2}{I_{L2}} = \frac{k(1-k)TV_1}{4L_2 I_O} = \frac{k(1-k)^2}{4(3-k)}\frac{R}{fL_2}. \tag{7.97}$$

The ripple voltage of output voltage v_O is

$$\Delta v_O = \frac{\Delta Q}{C_{22j}} = \frac{I_O kT}{C_{22j}} = \frac{k}{fC_{22j}}\frac{V_O}{R}.$$

Therefore, the variation ratio of output voltage v_O is

$$\varepsilon = \frac{\Delta v_O/2}{V_O} = \frac{k}{2RfC_{22j}}. \tag{7.98}$$

7.2.5.3 Triple-Lift Multiple-Enhanced Circuit

This circuit is derived from the triple-lift circuit of the main series by adding the DEC multiple (j) times in each stage circuit. Its circuit and switch-on and switch-off equivalent circuits are shown in Figure 7.15.

The voltage across capacitor C_{12j} is

$$V_{C12j} = \frac{j+2-k}{1-k} V_{in}. \tag{7.99}$$

The voltage across capacitor C_{22j} is

$$V_{C22j} = \left(\frac{j+2-k}{1-k}\right)^2 V_{in}. \tag{7.100}$$

By the same analysis,

$$V_O = \frac{j+2-k}{1-k} V_{C22j} = \left(\frac{j+2-k}{1-k}\right)^3 V_{in}. \tag{7.101}$$

(a)

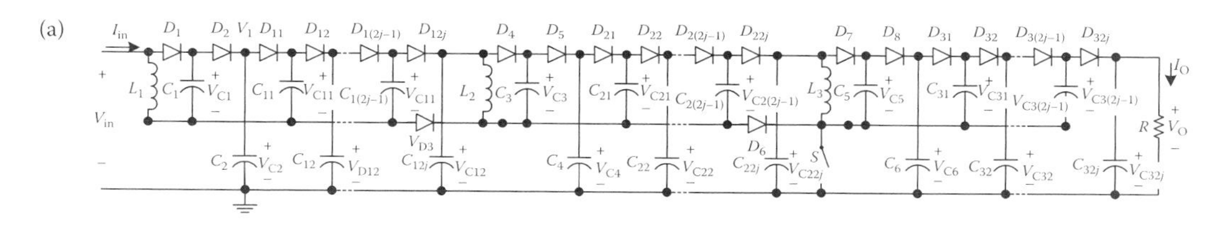

(b)

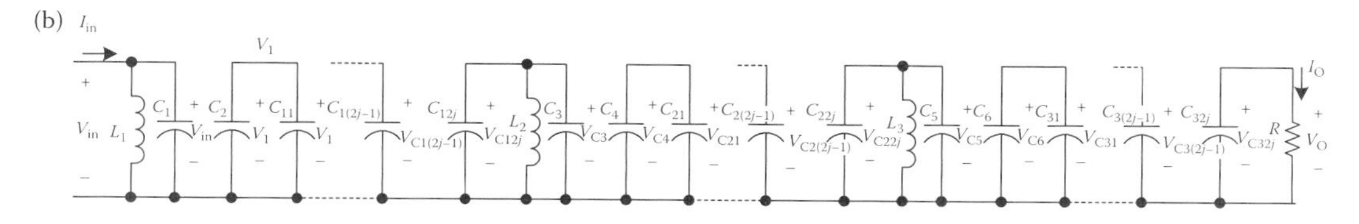

(c)

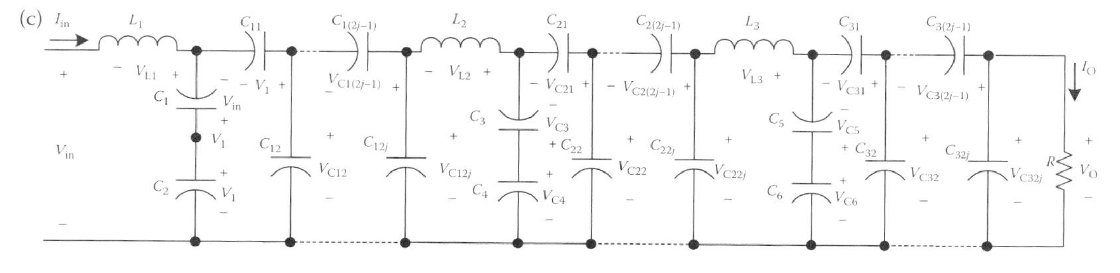

FIGURE 7.15 Triple-lift multiple-enhanced circuit of P/O SL Luo-converters: (a) circuit diagram, (b) equivalent circuit during switch-on, and (c) equivalent circuit during switch-off. (Reprinted from Luo, F. L. and Ye, H. 2006. *Essential DC/DC Converters*. Boca Raton: Taylor & Francis Group LLC, p. 248. With permission.)

The voltage transfer gain is

$$G = \frac{V_O}{V_{in}} = \left(\frac{j+2-k}{1-k}\right)^3. \tag{7.102}$$

Analogously,

$$\Delta i_{L1} = \frac{V_{in}}{L_1}kT, \qquad I_{L1} = \frac{(2-k)(3-k)}{(1-k)^3}I_O,$$

$$\Delta i_{L2} = \frac{V_1}{L_2}kT, \qquad I_{L2} = \frac{3-k}{(1-k)^2}I_O.$$

$$\Delta i_{L3} = \frac{V_2}{L_3}kT, \qquad I_{L3} = \frac{2I_O}{1-k}.$$

Considering

$$\frac{V_{in}}{I_{in}} = \left(\frac{1-k}{2-k}\right)^2 \frac{V_O}{I_O} = \left(\frac{1-k}{2-k}\right)^2 R,$$

the variation ratio of current i_{L1} through inductor L_1 is

$$\xi_1 = \frac{\Delta i_{L1}/2}{I_{L1}} = \frac{k(1-k)^3 T V_{in}}{2(2-k)(3-k)L_1 I_O} = \frac{k(1-k)^3 T}{2(2-k)(3-k)L_1 I_O}\frac{(1-k)^3}{(2-k)^2(3-k)}V_O$$

$$= \frac{k(1-k)^6}{2(2-k)^3(3-k)^2}\frac{R}{fL_1}. \tag{7.103}$$

The variation ratio of current i_{L2} through inductor L_2 is

$$\xi_2 = \frac{\Delta i_{L2}/2}{I_{L2}} = \frac{k(1-k)^2 T V_1}{2(3-k)L_2 I_O} = \frac{k(1-k)^2 T}{2(3-k)L_2 I_O}\frac{(1-k)^2}{(2-k)(3-k)}V_O$$

$$= \frac{k(1-k)^4}{2(2-k)(3-k)^2}\frac{R}{fL_2}. \tag{7.104}$$

The variation ratio of current i_{L3} through inductor L_3 is

$$\xi_3 = \frac{\Delta i_{L3}/2}{I_{L3}} = \frac{k(1-k)TV_2}{4L_3 I_O} = \frac{k(1-k)T}{4L_3 I_O}\frac{1-k}{3-k}V_O = \frac{k(1-k)^2}{4(3-k)}\frac{R}{fL_3}. \tag{7.105}$$

The ripple voltage of output voltage v_O is

$$\Delta v_O = \frac{\Delta Q}{C_{32j}} = \frac{I_O kT}{C_{32j}} = \frac{k}{fC_{32j}}\frac{V_O}{R}.$$

Therefore, the variation ratio of output voltage v_O is

$$\varepsilon = \frac{\Delta v_O/2}{V_O} = \frac{k}{2RfC_{32j}}. \tag{7.106}$$

7.2.5.4 Higher-Order Lift Multiple-Enhanced Circuit

The higher-order lift multiple-enhanced circuit can be derived from the corresponding circuits of the main series converters by adding the DEC multiple (j) times in each stage circuit. For the nth-order lift additional circuit, the final output voltage is

$$V_O = \left(\frac{j+2-k}{1-k}\right)^n V_{in}.$$

The voltage transfer gain is

$$G = \frac{V_O}{V_{in}} = \left(\frac{j+2-k}{1-k}\right)^n. \quad (7.107)$$

Analogously, the variation ratio of current i_{Li} through inductor $L_i (i = 1, 2, 3, \ldots, n)$ is

$$\xi_i = \frac{\Delta i_{Li}/2}{I_{Li}} = \frac{k(1-k)^{2(n-i+1)}}{2[2(2-k)]^{h(n-i)}(2-k)^{2(n-i)+1}(3-k)^{2u(n-i-1)}} \frac{R}{fL_i}, \quad (7.108)$$

where

$$h(x) = \begin{cases} 0 & x > 0 \\ 1 & x \le 0 \end{cases} \quad \text{is the Hong function}$$

and

$$u(x) = \begin{cases} 1 & x \ge 0 \\ 0 & x < 0 \end{cases} \quad \text{is the unit-step function.}$$

The variation ratio of output voltage v_O is

$$\varepsilon = \frac{\Delta v_O/2}{V_O} = \frac{k}{2RfC_{n2j}}. \quad (7.109)$$

7.2.6 Summary of P/O SL Luo-Converters

All circuits of P/O SL Luo-converters can be shown in Figure 7.16 as the family tree.

From the analysis in previous sections, the common formula to calculate the output voltage can be presented as

$$V_O = \begin{cases} \left(\frac{2-k}{1-k}\right)^n V_{in} & \text{main series,} \\ \left(\frac{2-k}{1-k}\right)^{n-1} \left(\frac{3-k}{1-k}\right) V_{in} & \text{additional series,} \\ \left(\frac{3-k}{1-k}\right)^n V_{in} & \text{enhanced series,} \\ \left(\frac{4-k}{1-k}\right)^n V_{in} & \text{re-enhanced series,} \\ \left(\frac{j+2-k}{1-k}\right)^n V_{in} & \text{multiple-enhanced series.} \end{cases} \quad (7.110)$$

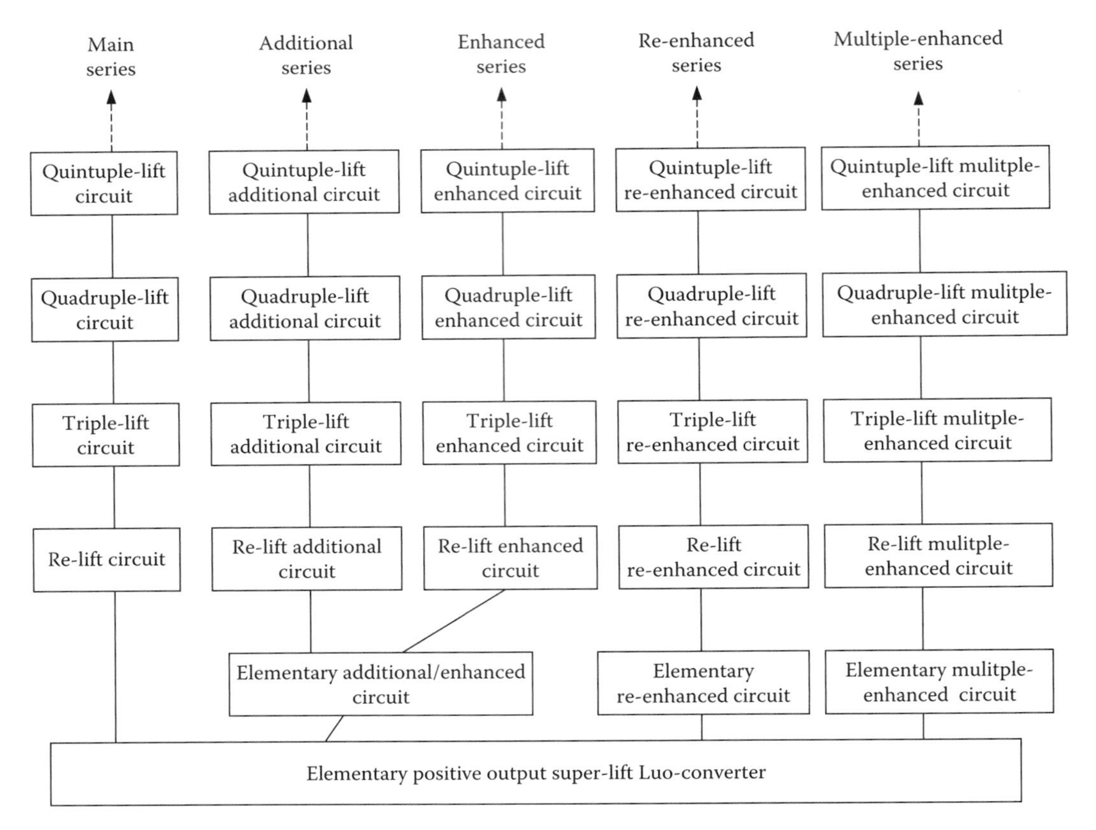

FIGURE 7.16 The family of P/O SL Luo-converters. (Reprinted from Luo, F. L. and Ye, H. 2006. *Essential DC/DC Converters*. Boca Raton: Taylor & Francis Group LLC, p. 255. With permission.)

The voltage transfer gain is

$$G = \frac{V_O}{V_{in}} = \begin{cases} \left(\frac{2-k}{1-k}\right)^n & \text{main series,} \\ \left(\frac{2-k}{1-k}\right)^{n-1}\left(\frac{3-k}{1-k}\right) & \text{additional series,} \\ \left(\frac{3-k}{1-k}\right)^n & \text{enhanced series,} \\ \left(\frac{4-k}{1-k}\right)^n & \text{re-enhanced series,} \\ \left(\frac{j+2-k}{1-k}\right)^n & \text{multiple-enhanced series.} \end{cases} \tag{7.111}$$

In order to show the advantages of the SL Luo-converters, their voltage transfer gains can be compared with that of other converters:

Buck converter: $G = V_O/V_{in} = k$.

Forward converter: $G = V_O/V_{in} = kN$, where N is the transformer turn's ratio.

Cúk-converter: $G = V_O/V_{in} = k/(1-k)$.

Fly-back converter: $G = V_O/V_{in} = (k/(1-k))N$, where N is the transformer turn's ratio.

Boost converter: $G = V_O/V_{in} = 1/(1-k)$.

P/O Luo-converters:

$$G = \frac{V_O}{V_{in}} = \frac{n}{1-k}. \tag{7.112}$$

Assume that the conduction duty cycle k is 0.2; the output voltage transfer gains are listed in Table 7.1. Similarly, for $k = 0.5$ and 0.8, the output voltage transfer gains are listed in Tables 7.2 and 7.3.

7.3 N/O SL Luo-Converters

The N/O SL Luo-converters were developed at the same time as the P/O SL Luo-converters. They too perform the SL technique. Only three circuits from each subseries will be introduced in this section [1,2,5,6].

TABLE 7.1

Voltage Transfer Gains of Converters in the Condition $k = 0.2$

Stage No. (n)	1	2	3	4	5	n
Buck converter				0.2		
Forward converter				0.2N (N is the transformer turn's ratio)		
Cúk-converter				0.25		
Fly-back converter				0.25N (N is the transformer turn's ratio)		
Boost converter				1.25		
P/O Luo-converters	1.25	2.5	3.75	5	6.25	$1.25n$
P/O SL Luo-converters—main series	2.25	5.06	11.39	25.63	57.67	2.25^n
P/O SL Luo-converters—additional series	3.5	7.88	17.72	39.87	89.7	$3.5 \times 2.25^{(n-1)}$
P/O SL Luo-converters—enhanced series	3.5	12.25	42.88	150	525	3.5^n
P/O SL Luo-converters—re-enhanced series	4.75	22.56	107.2	509	2418	4.75^n
P/O SL Luo-converters—multiple ($j = 4$)-enhanced series	7.25	52.56	381	2762	20,030	7.25^n

Source: Data from Luo, F. L. and Ye, H. 2006. *Essential DC/DC Converters*, p. 256. Boca Raton: Taylor & Francis Group LLC.

TABLE 7.2

Voltage Transfer Gains of Converters in the Condition $k = 0.5$

Stage No. (n)	1	2	3	4	5	n
Buck converter				0.5		
Forward converter				$0.5N$ (N is the transformer turn's ratio)		
Cúk-converter				1		
Fly-back converter				N (N is the transformer turn's ratio)		
Boost converter				2		
P/O Luo-converters	2	4	6	8	10	$2n$
P/O SL Luo-converters—main series	3	9	27	81	243	3^n
P/O SL Luo-converters—additional series	5	15	45	135	405	$5 \times 3^{(n-1)}$
P/O SL Luo-converters—enhanced series	5	25	125	625	3125	5^n
P/O SL Luo-converters—re-enhanced series	7	49	343	2401	16,807	7^n
P/O SL Luo-converters—multiple ($j = 4$)-enhanced series	11	121	1331	14,641	16×10^4	11^n

Source: Data from Luo, F. L. and Ye, H. 2006. *Essential DC/DC Converters*, p. 257. Boca Raton: Taylor & Francis Group LLC.

7.3.1 Main Series

The first three stages of N/O SL Luo-converters—main series—are shown in Figures 7.17 through 7.19. For ease of understanding, they are called the elementary circuit, the re-lift circuit, and the triple-lift circuit, respectively, and are numbered $n = 1, 2$, and 3, respectively.

7.3.1.1 N/O Elementary Circuit

The N/O elementary circuit and its equivalent circuits during switch-on and switch-off periods are shown in Figure 7.17.

The voltage across capacitor C_1 is charged with V_{in}. The current flowing through inductor L_1 increases along the slope V_{in}/L_1 during the switch-on period kT and decreases along the slope $-(V_O - V_{in})/L_1$ during the switch-off period $(1-k)T$. Therefore, the variation of current i_{L1} is

$$\Delta i_{L1} = \frac{V_{in}}{L_1}kT = \frac{V_O - V_{in}}{L_1}(1-k)T, \quad (7.113)$$

$$V_O = \frac{1}{1-k}V_{in} = \left(\frac{2-k}{1-k} - 1\right)V_{in}. \quad (7.114)$$

TABLE 7.3

Voltage Transfer Gains of Converters in the Condition $k = 0.8$

Stage No. (n)	1	2	3	4	5	n
Buck converter				0.8		
Forward converter				0.8N (N is the transformer turn's ratio)		
Cúk-converter				4		
Fly-back converter				4N (N is the transformer turn's ratio)		
Boost converter				5		
P/O Luo-converters	5	10	15	20	25	$5n$
P/O SL Luo-converters—main series	6	36	216	1296	7776	6^n
P/O SL Luo-converters—additional series	11	66	396	2376	14,256	$11 \times 6(n-1)$
P/O SL Luo-converters—enhanced series	11	121	1331	14,641	16×10^4	11^n
P/O SL Luo-converters—re-enhanced series	16	256	4096	65,536	104×10^4	16^n
P/O SL Luo-converters—multiple ($j = 4$)-enhanced series	26	676	17,576	46×10^4	12×10^6	26^n

Source: Data from Luo, F. L. and Ye, H. 2006. *Essential DC/DC Converters*, p. 257. Boca Raton: Taylor & Francis Group LLC.

The voltage transfer gain is

$$G_1 = \frac{V_O}{V_{in}} = \frac{2-k}{1-k} - 1. \tag{7.115}$$

In the steady state, the average charge across the capacitor C_1 in a period should be zero. The following relations are available:

$$kTi_{C1-on} = (1-k)Ti_{C1-off} \quad \text{and} \quad i_{C1-on} = \frac{1-k}{k} i_{C1-off}.$$

These relations are available for all the capacitors' current in switch-on and switch-off periods.

The input current i_{in} is equal to $(i_{L1} + i_{C1})$ during the switch-on period and zero during the switch-off period. The capacitor current i_{C1} is equal to i_{L1} during switch-off.

$$i_{in-on} = i_{L1-on} + i_{C1-on}, \quad i_{L1-off} = i_{C1-off} = I_{L1}.$$

If inductance L_1 is large enough, i_{L1} is nearly equal to its average current I_{L1}. Therefore,

$$i_{in-on} = i_{L1-on} + i_{C1-on} = i_{L1-on} + \frac{1-k}{k} i_{C1-off} = \left(1 + \frac{1-k}{k}\right) I_{L1} = \frac{1}{k} I_{L1}$$

and

$$I_{in} = ki_{in-on} = I_{L1}. \tag{7.116}$$

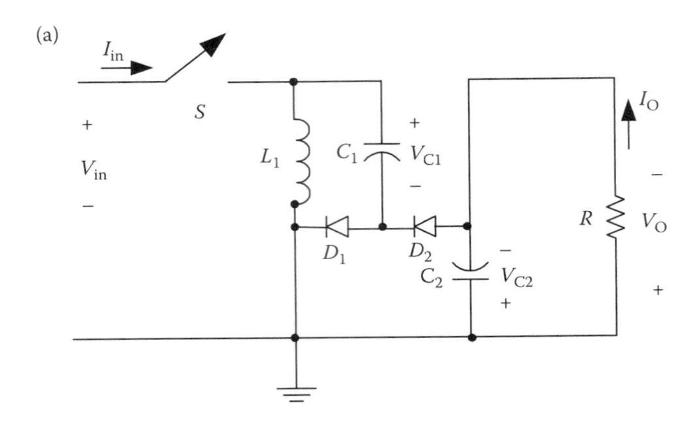

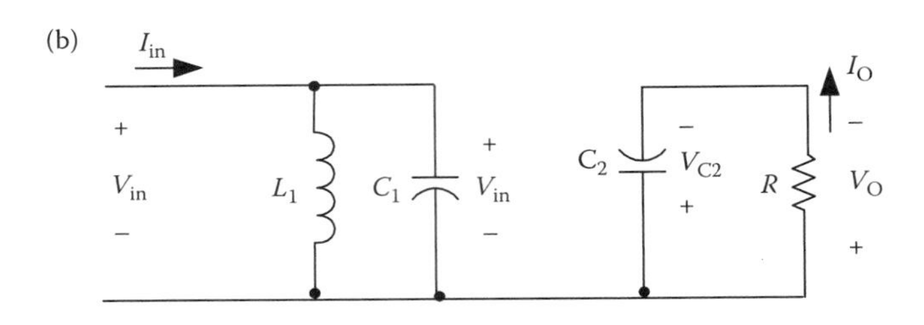

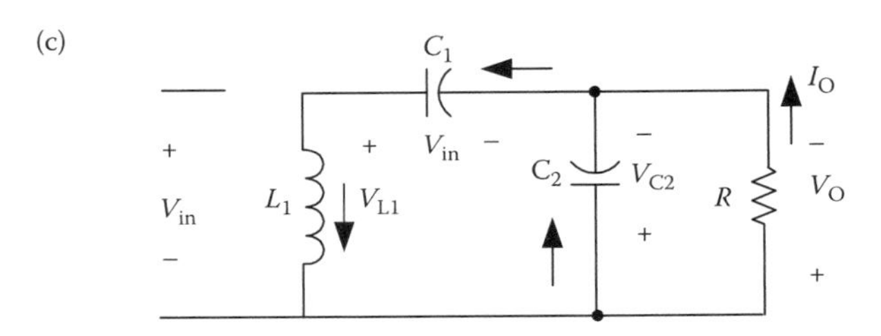

FIGURE 7.17 Elementary circuit of N/O SL Luo-converters—main series: (a) circuit diagram, (b) equivalent circuit during switch-on, and (c) equivalent circuit during switch-off. (Reprinted from Luo, F. L. and Ye, H. 2006. *Essential DC/DC Converters*. Boca Raton: Taylor & Francis Group LLC, p. 264. With permission.)

Further,

$$i_{C2\text{-on}} = I_O, \quad i_{C2\text{-off}} = \frac{k}{1-k} I_O,$$

$$I_{L1} = i_{C2\text{-off}} + I_O = \frac{k}{1-k} i_{C2\text{-on}} + I_O = \frac{1}{1-k} I_O.$$

The variation ratio of inductor current i_{L1} is

$$\xi_1 = \frac{\Delta i_{L1}/2}{I_{L1}} = \frac{k(1-k)TV_{in}}{2L_1 I_O} = \frac{k(1-k)}{G_1} \frac{R}{2fL_1}. \tag{7.117}$$

Usually ξ_1 is small (much lower than unity); this means that this converter works in the CCM.

The ripple voltage of output voltage v_O is

$$\Delta v_O = \frac{\Delta Q}{C_2} = \frac{I_O kT}{C_2} = \frac{k}{fC_2}\frac{V_O}{R}.$$

Therefore, the variation ratio of output voltage v_O is

$$\varepsilon = \frac{\Delta v_O/2}{V_O} = \frac{k}{2RfC_2}. \tag{7.118}$$

7.3.1.2 N/O Re-Lift Circuit

The N/O re-lift circuit is derived from the N/O elementary circuit by adding the parts (L_2-D_3-D_4-D_5-C_3-C_4). Its circuit diagram and equivalent circuits during switch-on and switch-off periods are shown in Figure 7.18.

The voltage across capacitor C_1 is charged to V_{in}. As described in the previous section, the voltage V_1 across capacitor C_2 is $V_1 = (1/(1-k))V_{in}$.

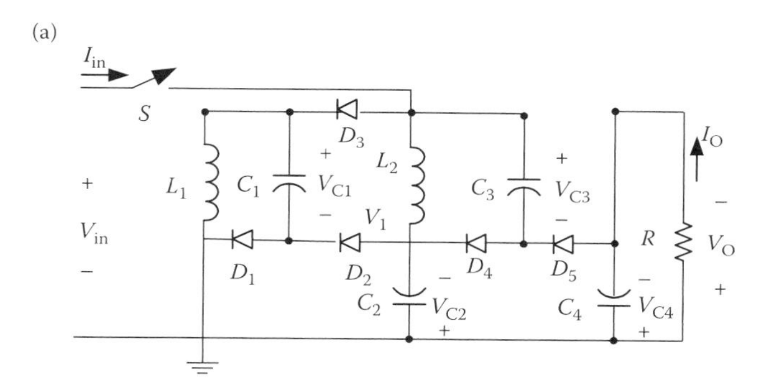

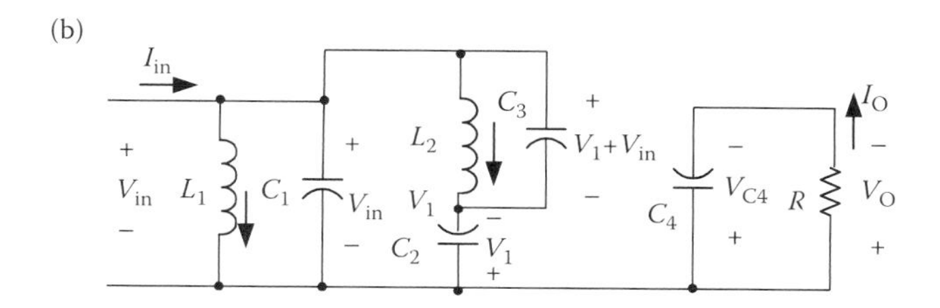

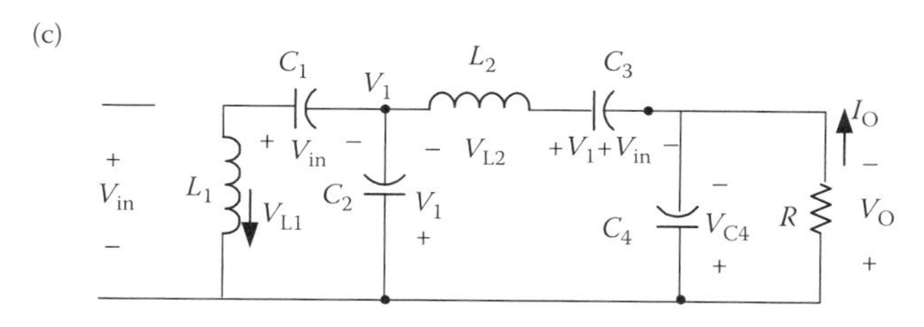

FIGURE 7.18 Re-lift circuit of N/O SL Luo-converters—main series: (a) circuit diagram, (b) equivalent circuit during switch-on, and (c) equivalent circuit during switch-off. (Reprinted from Luo, F. L. and Ye, H. 2006. *Essential DC/DC Converters*. Boca Raton: Taylor & Francis Group LLC, p. 265. With permission.)

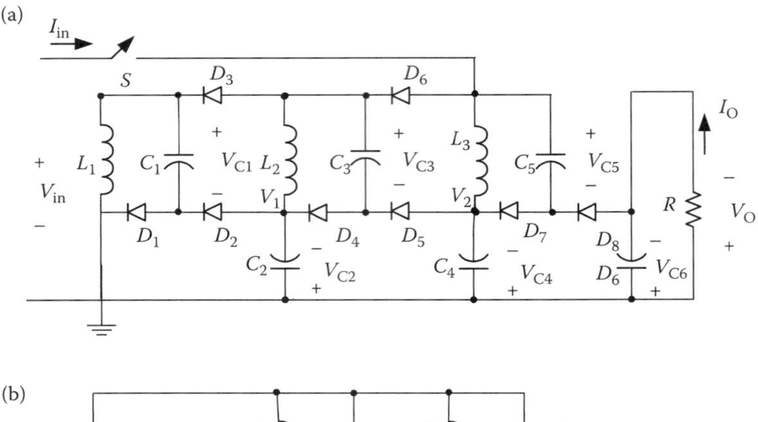

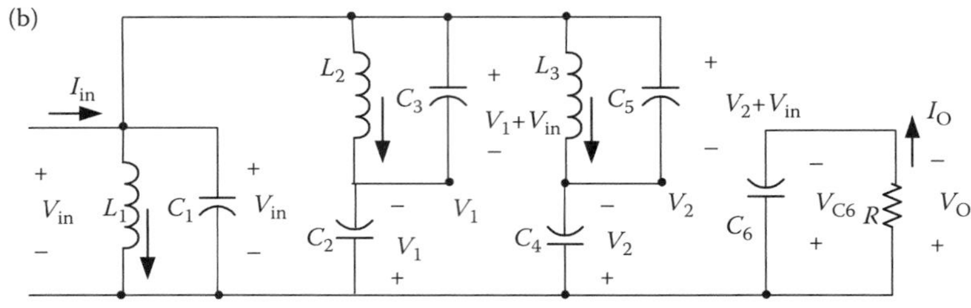

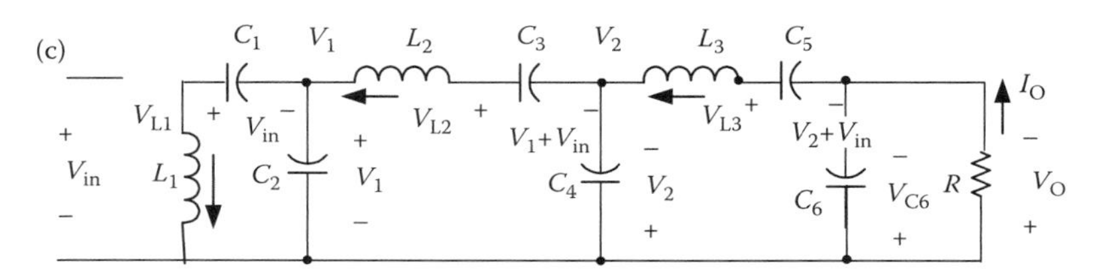

FIGURE 7.19 Triple-lift circuit of N/O SL Luo-converters—main series: (a) circuit diagram, (b) equivalent circuit during switch-on, and (c) equivalent circuit during switch-off. (Reprinted from Luo, F. L. and Ye, H. 2006. *Essential DC/DC Converters*. Boca Raton: Taylor & Francis Group LLC, p. 266. With permission.)

The voltage across capacitor C_3 is charged with $(V_1 + V_{in})$. The current flowing through inductor L_2 increases along the slope $(V_1 + V_{in})/L_2$ during the switch-on period kT and decreases along the slope $-(V_O-2V_1 - V_{in})/L_2$ during the switch-off period $(1-k)T$. Therefore, the variation of current i_{L2} is

$$\Delta i_{L2} = \frac{V_1 + V_{in}}{L_2} kT = \frac{V_O - 2V_1 - V_{in}}{L_2}(1-k)T, \tag{7.119}$$

$$V_O = \frac{(2-k)V_1 + V_{in}}{1-k} = \left[\left(\frac{2-k}{1-k}\right)^2 - 1\right] V_{in}. \tag{7.120}$$

The voltage transfer gain is

$$G_2 = \frac{V_O}{V_{in}} = \left(\frac{2-k}{1-k}\right)^2 - 1. \tag{7.121}$$

The input current i_{in} is equal to $(i_{L1} + i_{C1} + i_{L2} + i_{C3})$ during the switch-on period and is zero during the switch-off period. In the steady state, the following relations are available:

$$i_{in-on} = i_{L1-on} + i_{C1-on} + i_{L2-on} + i_{C3-on},$$

$$i_{C4-on} = I_O, \qquad i_{C4-off} = \frac{k}{1-k} I_O,$$

$$i_{C3-off} = I_{L2} = I_O + i_{C4-off} = \frac{I_O}{1-k}, \qquad i_{C3-on} = \frac{I_O}{k},$$

$$i_{C2-on} = I_{L2} + i_{C3-on} = \frac{I_O}{1-k} + \frac{I_O}{k} = \frac{I_O}{k(1-k)}, \qquad i_{C2-off} = \frac{I_O}{(1-k)^2},$$

$$i_{C1-off} = I_{L1} = I_{L2} + i_{C2-off} = \frac{I_O}{1-k} + \frac{I_O}{(1-k)^2} = \frac{2-k}{(1-k)^2} I_O, \quad i_{C1-on} = \frac{2-k}{k(1-k)} I_O.$$

Thus,

$$i_{in-on} = i_{L1-on} + i_{C1-on} + i_{L2-on} + i_{C3-on} = \frac{1}{k}(I_{L1} + I_{L2}) = \frac{3-2k}{k(1-k)^2} I_O.$$

Therefore,

$$I_{in} = k i_{in-on} = \frac{3-2k}{(1-k)^2} I_O.$$

Since

$$\Delta i_{L1} = \frac{V_{in}}{L_1} kT, \quad I_{L1} = \frac{2-k}{(1-k)^2} I_O,$$

$$\Delta i_{L2} = \frac{V_1 + V_{in}}{L_2} kT = \frac{2-k}{1-k} \frac{kT}{L_2} V_{in}, \quad I_{L2} = \frac{1}{1-k} I_O.$$

Therefore, the variation ratio of current I_{L1} through inductor L_1 is

$$\xi_1 = \frac{\Delta i_{L1}/2}{I_{L1}} = \frac{kTV_{in}}{(2-k/(1-k)^2)2L_1 I_O} = \frac{k(1-k)^2}{(2-k)G_2} \frac{R}{2fL_1}. \tag{7.122}$$

The variation ratio of current i_{L2} through inductor L_2 is

$$\xi_2 = \frac{\Delta i_{L2}/2}{I_{L2}} = \frac{k(2-k)TV_{in}}{2L_2 I_O} = \frac{k(2-k)}{G_2} \frac{R}{2fL_2}. \tag{7.123}$$

The ripple voltage of output voltage v_O is

$$\Delta v_O = \frac{\Delta Q}{C_4} = \frac{I_O kT}{C_4} = \frac{k}{fC_4} \frac{V_O}{R}.$$

Therefore, the variation ratio of output voltage v_O is

$$\varepsilon = \frac{\Delta v_O/2}{V_O} = \frac{k}{2RfC_4}. \tag{7.124}$$

Example 7.3

An N/O re-lift circuit in Figure 7.18a has $V_{in} = 20$V, all inductors have 10 mH, all capacitors have 20 μ F, $R = 200\,\Omega$, $f = 50$ kHz, and conduction duty cycle $k = 0.6$. Calculate the variation ratio of current I_{L1}, and the output voltage and its variation ratio.

SOLUTION

From Equation 7.122, we can obtain the variation ratio of current i_{L1}:

$$\xi_1 = \frac{k(1-k)^3}{(2-k)^2}\frac{R}{2fL_1} = \frac{0.6(1-0.6)^3}{(2-0.6)^2}\frac{200}{2\times 50\,\text{k}\times 10\,\text{m}} = 0.0039.$$

From Equation 7.120, we can obtain the output voltage:

$$V_O = \left[\left(\frac{2-k}{1-k}\right)^2 - 1\right]V_{in} = \left[\left(\frac{2-0.6}{1-0.6}\right)^2 - 1\right]\times 20 = 225\text{V}.$$

From Equation 7.124, its variation ratio is

$$\varepsilon = \frac{k}{2RfC_4} = \frac{0.6}{2\times 200\times 50\,\text{k}\times 20\,\mu} = 0.0015.$$

7.3.1.3 N/O Triple-Lift Circuit

The N/O triple-lift circuit is derived from the N/O re-lift circuit by twice repeating the parts (L_2-D_3-D_4-D_5-C_3-C_4). Its circuit diagram and equivalent circuits during switch-on and switch-off periods are shown in Figure 7.19.

The voltage across capacitor C_1 is charged with V_{in}. As described in the previous section, the voltage V_1 across capacitor C_2 is $V_1 = (((2-k)/(1-k)) - 1)V_{in} = (1/(1-k))V_{in}$, and the voltage V_2 across capacitor C_4 is

$$V_2 = \left[\left(\frac{2-k}{1-k}\right)^2 - 1\right]V_{in} = \frac{3-2k}{(1-k)^2}V_{in}.$$

The voltage across capacitor C_5 is charged with $(V_2 + V_{in})$. The current flowing through inductor L_3 increases along the slope $(V_2 + V_{in})/L_3$ during the switch-on period kT and decreases along the slope $-(V_O - 2V_2 - V_{in})/L_3$ during the switch-off period $(1-k)T$. Therefore, the variation of current i_{L3} is

$$\Delta i_{L3} = \frac{V_2 + V_{in}}{L_3}kT = \frac{V_O - 2V_2 - V_{in}}{L_3}(1-k)T, \tag{7.125}$$

$$V_O = \frac{(2-k)V_2 + V_{in}}{1-k} = \left[\left(\frac{2-k}{1-k}\right)^3 - 1\right]V_{in}. \tag{7.126}$$

The voltage transfer gain is

$$G_3 = \frac{V_O}{V_{in}} = \left(\frac{2-k}{1-k}\right)^3 - 1. \tag{7.127}$$

The input current i_{in} is equal to $(i_{L1} + i_{C1} + i_{L2} + i_{C3} + i_{L3} + i_{C5})$ during the switch-on period and is zero during the switch-off period. In the steady state, the following relations are available:

$$i_{in\text{-}on} = i_{L1\text{-}on} + i_{C1\text{-}on} + i_{L2\text{-}on} + i_{C3\text{-}on} + i_{L3\text{-}on} + i_{C5\text{-}on},$$

$$i_{C6\text{-}on} = I_O, \qquad i_{C6\text{-}off} = \frac{k}{1-k}I_O,$$

$$i_{C5\text{-}off} = I_{L3} = I_O + i_{C6\text{-}off} = \frac{I_O}{1-k}, \qquad i_{C5\text{-}on} = \frac{I_O}{k},$$

$$i_{C4\text{-}on} = I_{L3} + i_{C5\text{-}on} = \frac{I_O}{1-k} + \frac{I_O}{k} = \frac{I_O}{k(1-k)}, \qquad i_{C4\text{-}off} = \frac{I_O}{(1-k)^2},$$

$$i_{C3\text{-}off} = I_{L2} = I_{L3} + i_{C4\text{-}off} = \frac{2-k}{(1-k)^2}I_O, \qquad i_{C3\text{-}on} = \frac{2-k}{k(1-k)}I_O,$$

$$i_{C2\text{-}on} = I_{L2} + i_{C3\text{-}on} = \frac{2-k}{k(1-k)^2}I_O, \qquad i_{C2\text{-}off} = \frac{2-k}{(1-k)^3}I_O,$$

$$i_{C1\text{-}off} = I_{L1} = I_{L2} + i_{C2\text{-}off} = \frac{(2-k)^2}{(1-k)^3}I_O, \qquad i_{C1\text{-}on} = \frac{(2-k)^2}{k(1-k)^2}I_O.$$

Thus,

$$i_{in\text{-}on} = i_{L1\text{-}on} + i_{C1\text{-}on} + i_{L2\text{-}on} + i_{C3\text{-}on} + i_{L3\text{-}on} + i_{C5\text{-}on} = \frac{1}{k}(I_{L1} + I_{L2} + I_{L3})$$
$$= \frac{7 - 9k + 3k^2}{k(1-k)^3}I_O.$$

Therefore,

$$I_{in} = ki_{in\text{-}on} = \frac{7 - 9k + 3k^2}{(1-k)^3}I_O = \left[\left(\frac{2-k}{1-k}\right)^3 - 1\right]I_O.$$

Analogously,

$$\Delta i_{L1} = \frac{V_{in}}{L_1}kT, \qquad I_{L1} = \frac{(2-k)^2}{(1-k)^3}I_O,$$

$$\Delta i_{L2} = \frac{V_1 + V_{in}}{L_2}kT = \frac{2-k}{(1-k)L_2}kTV_{in}, \qquad I_{L2} = \frac{2-k}{(1-k)^2}I_O,$$

$$\Delta i_{L3} = \frac{V_2 + V_{in}}{L_3}kT = \left(\frac{2-k}{1-k}\right)^2\frac{kT}{L_3}V_{in}, \qquad I_{L3} = \frac{I_O}{1-k}.$$

Therefore, the variation ratio of current i_{L1} through inductor L_1 is

$$\xi_1 = \frac{\Delta i_{L1}/2}{I_{L1}} = \frac{k(1-k)^3TV_{in}}{2(2-k)^2L_1I_O} = \frac{k(1-k)^3}{(2-k)^2G_3}\frac{R}{2fL_1}. \tag{7.128}$$

The variation ratio of current i_{L2} through inductor L_2 is

$$\xi_2 = \frac{\Delta i_{L2}/2}{I_{L2}} = \frac{k(1-k)TV_{in}}{2L_2 I_O} = \frac{k(1-k)}{G_3}\frac{R}{2fL_2}. \quad (7.129)$$

The variation ratio of current i_{L3} through inductor L_3 is

$$\xi_3 = \frac{\Delta i_{L3}/2}{I_{L3}} = \frac{k(2-k)^2 TV_{in}}{2(1-k)L_3 I_O} = \frac{k(2-k)^2}{(1-k)G_3}\frac{R}{2fL_3}. \quad (7.130)$$

The ripple voltage of output voltage v_O is

$$\Delta v_O = \frac{\Delta Q}{C_6} = \frac{I_O kT}{C_6} = \frac{k}{fC_6}\frac{V_O}{R}.$$

Therefore, the variation ratio of output voltage v_O is

$$\varepsilon = \frac{\Delta v_O/2}{V_O} = \frac{k}{2RfC_6}. \quad (7.131)$$

7.3.1.4 N/O Higher-Order Lift Circuit

The N/O higher-order lift circuit can be designed by just multiple repeating of the parts (L_2-D_3-D_4-D_5-C_3-C_4). For the nth-order lift circuit, the final output voltage across capacitor C_{2n} is

$$V_O = \left[\left(\frac{2-k}{1-k}\right)^n - 1\right] V_{in}. \quad (7.132)$$

The voltage transfer gain is

$$G_n = \frac{V_O}{V_{in}} = \left(\frac{2-k}{1-k}\right)^n - 1. \quad (7.133)$$

The variation ratio of current i_{Li} through inductor L_i ($i = 1, 2, 3, \ldots, n$) is

$$\xi_1 = \frac{\Delta i_{L1}/2}{I_{L1}} = \frac{k(1-k)^n}{(2-k)^{(n-1)}G_n}\frac{R}{2fL_i}, \quad (7.134)$$

$$\xi_2 = \frac{\Delta i_{L2}/2}{I_{L2}} = \frac{k(2-k)^{(3-n)}}{(1-k)^{(n-3)}G_n}\frac{R}{2fL_i}, \quad (7.135)$$

$$\xi_3 = \frac{\Delta i_{L3}/2}{I_{L3}} = \frac{k(2-k)^{(n-i+2)}}{(1-k)^{(n-i+1)}G_n}\frac{R}{2fL_3}. \quad (7.136)$$

The variation ratio of the output voltage v_O is

$$\varepsilon = \frac{\Delta v_O/2}{V_O} = \frac{k}{2RfC_{2n}}. \quad (7.137)$$

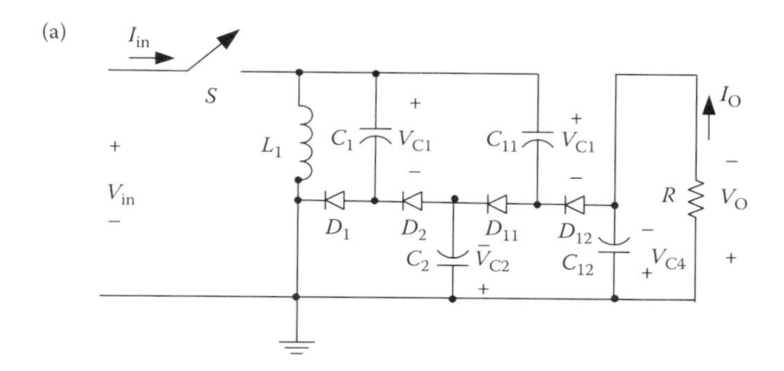

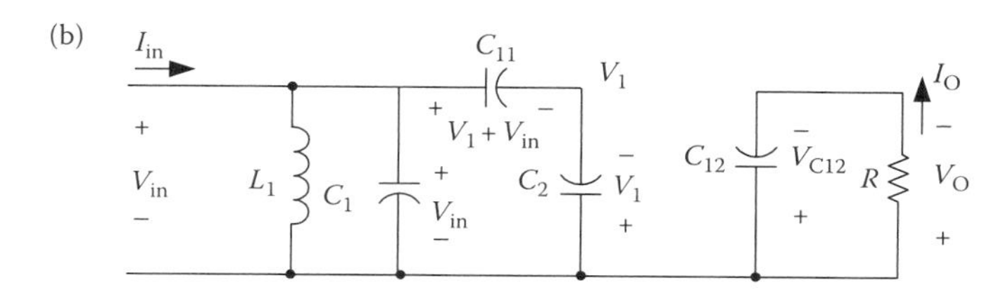

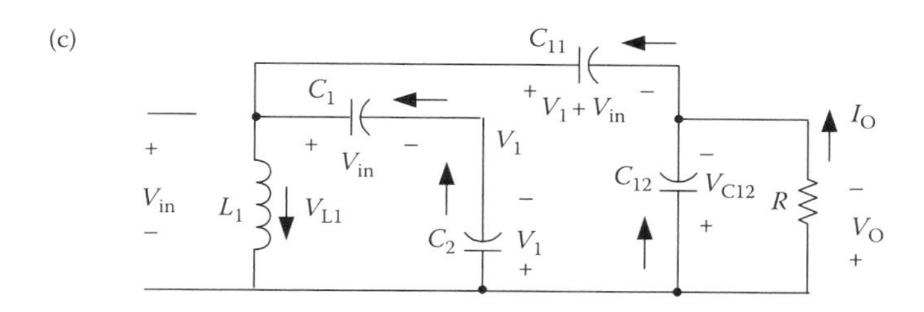

FIGURE 7.20 Elementary additional circuit of N/O SL Luo-converters: (a) circuit diagram, (b) equivalent circuit during switch-on, and (c) equivalent circuit during switch-off. (Reprinted from Luo, F. L. and Ye, H. 2006. *Essential DC/DC Converters*. Boca Raton: Taylor & Francis Group LLC, p. 274. With permission.)

7.3.2 N/O Additional Series

All circuits of the N/O SL Luo-converters—additional series—are derived from the corresponding circuits of the main series by adding a DEC. The first three stages of this series are shown in Figures 7.20 through 7.22. For ease of understanding, they are called the elementary additional circuit, the re-lift additional circuit, and the triple-lift additional circuit, respectively, and are numbered $n = 1, 2$, and 3, respectively.

7.3.2.1 N/O Elementary Additional Circuit

This circuit is derived from the N/O elementary circuit by adding a DEC. Its circuit and switch-on and switch-off equivalent circuits are shown in Figure 7.20.

The voltage across capacitor C_1 is charged with V_{in}. The voltage across capacitor C_2 is charged with V_1 and C_{11} is charged with $(V_1 + V_{\text{in}})$. The current I_{L1} flowing through inductor L_1 increases with the slope V_{in}/L_1 during the switch-on period kT and decreases with the slope $-(6.V_1 - V_{\text{in}})/L_1$ during the switch-off period $(1 - k)T$.

Therefore,

$$\Delta i_{L1} = \frac{V_{\text{in}}}{L_1} kT = \frac{V_1 - V_{\text{in}}}{L_1}(1 - k)T, \tag{7.138}$$

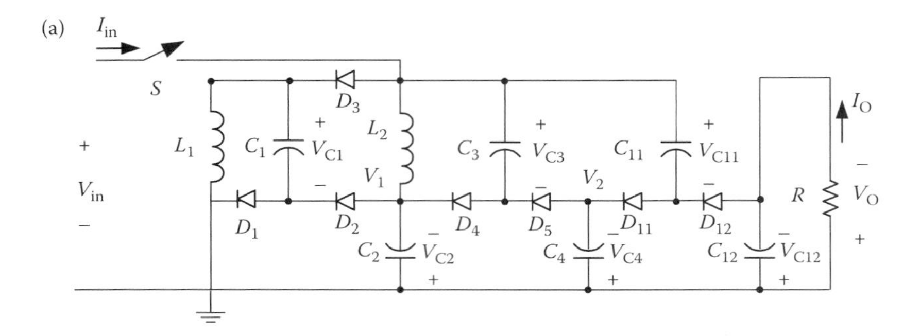

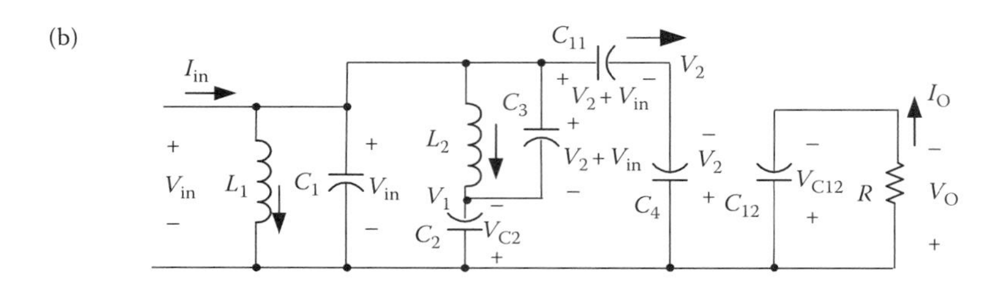

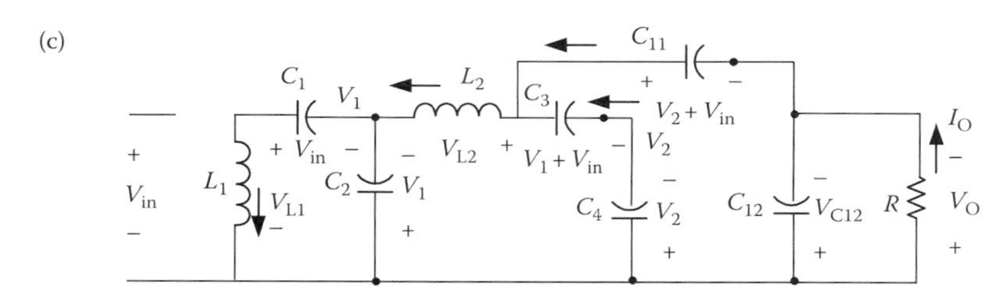

FIGURE 7.21 Re-lift additional circuit of N/O SL Luo-converters: (a) circuit diagram, (b) equivalent circuit during switch-on, and (c) equivalent circuit during switch-off. (Reprinted from Luo, F. L. and Ye, H. 2006. *Essential DC/DC Converters*. Boca Raton: Taylor & Francis Group LLC, p. 275. With permission.)

$$V_1 = \frac{1}{1-k}V_{in} = \left(\frac{2-k}{1-k} - 1\right)V_{in},$$

$$V_{L1\text{-off}} = \frac{k}{1-k}V_{in}.$$

The output voltage is

$$V_O = V_{in} + V_{L1} + V_1 = \frac{2}{1-k}V_{in} = \left[\frac{3-k}{1-k} - 1\right]V_{in}. \tag{7.139}$$

The voltage transfer gain is

$$G_1 = \frac{V_O}{V_{in}} = \frac{3-k}{1-k} - 1. \tag{7.140}$$

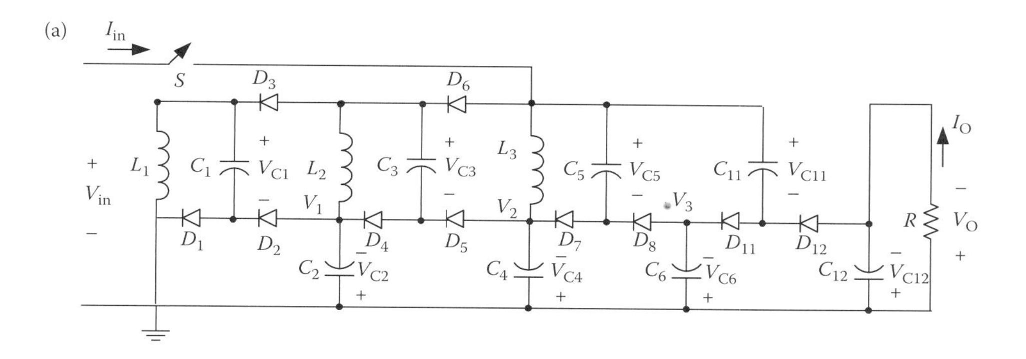

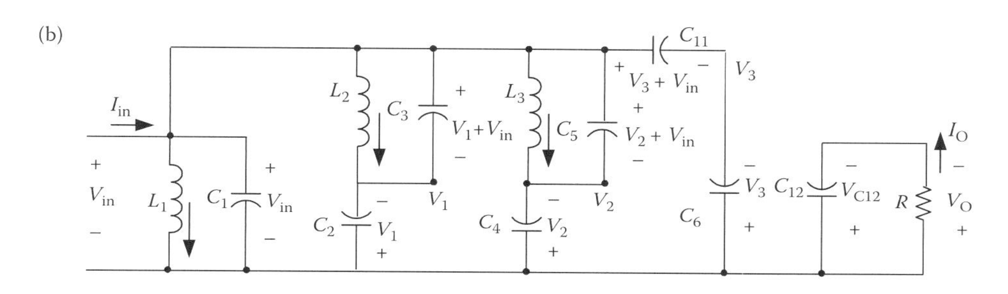

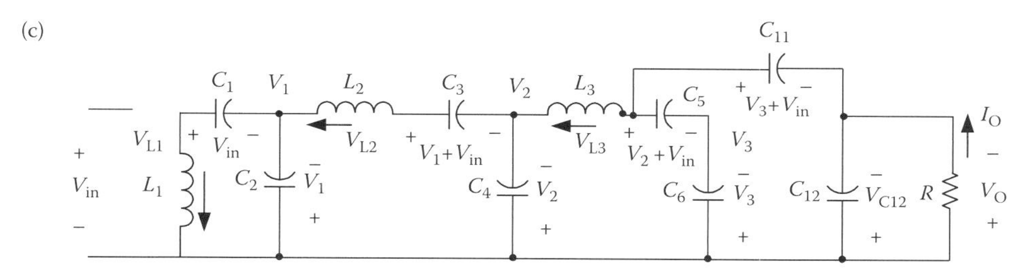

FIGURE 7.22 Triple-lift additional circuit of N/O SL Luo-converters: (a) circuit diagram, (b) equivalent circuit during switch-on, and (c) equivalent circuit during switch-off. (Reprinted from Luo, F. L. and Ye, H. 2006. *Essential DC/DC Converters*. Boca Raton: Taylor p. 276. With permission.)

The following relations are obtained:

$$i_{C12-on} = I_O, \qquad i_{C12-off} = \frac{kI_O}{1-k},$$

$$i_{C11-off} = I_O + i_{C12-off} = \frac{I_O}{1-k}, \qquad i_{C11-on} = i_{C2-on} = \frac{I_O}{k},$$

$$i_{C2-off} = i_{C1-off} = \frac{I_O}{1-k}, \qquad i_{C1-on} = \frac{I_O}{k},$$

$$I_{L1} = i_{C1-off} + i_{C11-on} = \frac{2I_O}{1-k},$$

$$i_{in} = I_{L1} + i_{C1-on} + i_{C11-on} = \left(\frac{2}{1-k} + \frac{1}{k} + \frac{1}{k}\right) I_O = \frac{2}{k(1-k)} I_O.$$

Therefore,

$$I_{in} = ki_{in} = \frac{2}{1-k}I_O = \left[\frac{3-k}{1-k} - 1\right]I_O.$$

The variation ratio of current I_{L1} through inductor L_1 is

$$\xi_1 = \frac{\Delta i_{L1}/2}{I_{L1}} = \frac{k(1-k)TV_{in}}{4L_1 I_O} = \frac{k(1-k)}{2G_1}\frac{R}{2fL_1}. \tag{7.141}$$

The ripple voltage of output voltage v_O is

$$\Delta v_O = \frac{\Delta Q}{C_{12}} = \frac{I_O kT}{C_{12}} = \frac{k}{fC_{12}}\frac{V_O}{R}.$$

Therefore, the variation ratio of output voltage v_O is

$$\varepsilon = \frac{\Delta v_O/2}{V_O} = \frac{k}{2RfC_{12}}. \tag{7.142}$$

7.3.2.2 N/O Re-Lift Additional Circuit

The N/O re-lift additional circuit is derived from the N/O re-lift circuit by adding a DEC. Its circuit diagram and switch-on and switch-off equivalent circuits are shown in Figure 7.21.

The voltage across capacitor C_1 is charged with V_{in}. As described in the previous section, the voltage across C_2 is $V_1 = (1/(1-k))V_{in}$.

The voltage across capacitor C_3 is charged with $(V_1 + V_{in})$, the voltage across capacitor C_4 is charged with V_2, and the voltage across capacitor C_{11} is charged with $(V_2 + V_{in})$. The current flowing through inductor L_2 increases with $(V_1 + V_{in})$ during the switch-on period kT and decreases with $-(V_2 - 2V_1 - V_{in})$ during the switch-off period $(1-k)T$. Therefore,

$$\Delta i_{L2} = \frac{V_1 + V_{in}}{L_2}kT = \frac{V_2 - 2V_1 - V_{in}}{L_2}(1-k)T, \tag{7.143}$$

$$V_2 = \frac{(2-k)V_1 + V_{in}}{1-k} = \frac{3-2k}{(1-k)^2} = \left[\left(\frac{2-k}{1-k}\right)^2 - 1\right]V_{in},$$

and

$$V_{L2\text{–}off} = V_2 - 2V_1 - V_{in} = \frac{k(2-k)}{(1-k)^2}V_{in}. \tag{7.144}$$

The output voltage is

$$V_O = V_2 + V_{in} + V_{L2} + V_1 = \frac{5-3k}{(1-k)^2}V_{in} = \left[\frac{3-k}{1-k}\frac{2-k}{1-k} - 1\right]V_{in}. \tag{7.145}$$

The voltage transfer gain is

$$G_2 = \frac{V_O}{V_{in}} = \frac{2-k}{1-k}\frac{3-k}{1-k} - 1. \tag{7.146}$$

The following relations are obtained:

$$i_{C12-on} = I_O, \qquad i_{C12-off} = \frac{kI_O}{1-k},$$

$$i_{C11-off} = I_O + i_{C12-off} = \frac{I_O}{1-k}, \qquad i_{C11-on} = i_{C4-on} = \frac{I_O}{k},$$

$$i_{C4-off} = i_{C3-off} = \frac{I_O}{1-k}, \qquad i_{C3-on} = \frac{I_O}{k},$$

$$I_{L2} = i_{C11-off} + i_{C3-off} = \frac{2I_O}{1-k},$$

$$i_{C2-on} = I_{L2} + i_{C3-on} = \frac{1+k}{k(1-k)}I_O, \qquad i_{C2-off} = \frac{1+k}{(1-k)^2}I_O.$$

$$I_{L1} = i_{C1-off} = I_{L2} + i_{C2-off} = \frac{3-k}{(1-k)^2}I_O, \qquad i_{C1-on} = \frac{3-k}{k(1-k)}I_O,$$

$$i_{in} = I_{L1} + i_{C1-on} + i_{C2-on} + i_{C4-on} = \left[\frac{3-k}{(1-k)^2} + \frac{3-k}{k(1-k)} + \frac{1+k}{k(1-k)} + \frac{1}{k}\right]I_O$$
$$= \frac{5-3k}{k(1-k)^2}I_O.$$

Therefore,

$$I_{in} = ki_{in} = \frac{5-3k}{(1-k)^2}I_O = \left[\frac{3-k}{1-k}\frac{2-k}{1-k} - 1\right]I_O.$$

Analogously,

$$\Delta i_{L1} = \frac{V_{in}}{L_2}kT, \quad I_{L1} = \frac{3-k}{(1-k)^2}I_O,$$

$$\Delta i_{L2} = \frac{V_1 + V_{in}}{L_2}kT = \frac{2-k}{(1-k)L_2}kTV_{in}, \quad I_{L2} = \frac{2I_O}{1-k}.$$

Therefore, the variation ratio of current i_{L1} through inductor L_1 is

$$\xi_1 = \frac{\Delta i_{L1}/2}{I_{L1}} = \frac{k(1-k)^2 TV_{in}}{2(3-k)L_1 I_O} = \frac{k(1-k)^2}{(3-k)G_2}\frac{R}{2fL_1}. \tag{7.147}$$

The variation ratio of current i_{L2} through inductor L_2 is

$$\xi_2 = \frac{\Delta i_{L2}/2}{I_{L2}} = \frac{k(2-k)TV_{in}}{4L_2 I_O} = \frac{k(2-k)}{2G_2}\frac{R}{2fL_2}. \tag{7.148}$$

The ripple voltage of output voltage v_O is

$$\Delta v_O = \frac{\Delta Q}{C_{12}} = \frac{I_O kT}{C_{12}} = \frac{k}{fC_{12}}\frac{V_O}{R}.$$

Therefore, the variation ratio of output voltage v_O is

$$\varepsilon = \frac{\Delta v_O/2}{V_O} = \frac{k}{2RfC_{12}}. \tag{7.149}$$

7.3.2.3 Triple-Lift Additional Circuit

This circuit is derived from the N/O triple-lift circuit by adding a DEC. Its circuit diagram and equivalent circuits during switch-on and switch-off periods are shown in Figure 7.22.

The voltage across capacitor C_1 is charged with V_{in}. As described in the previous section, the voltage across C_2 is $V_1 = (1/(1-k))V_{in}$ and the voltage across C_4 is

$$V_2 = \frac{3-2k}{1-k}V_1 = \frac{3-2k}{(1-k)^2}V_{in}.$$

The voltage across capacitor C_5 is charged with $(V_2 + V_{in})$, the voltage across capacitor C_6 is charged with V_3, and the voltage across capacitor C_{11} is charged with $(V_3 + V_{in})$. The current flowing through inductor L_3 increases with $(V_2 + V_{in})$ during the switch-on period kT and decreases with $-(V_3 - 2V_2 - V_{in})$ during the switch-off period $(1-k)T$. Therefore,

$$\Delta i_{L3} = \frac{V_2 + V_{in}}{L_3}kT = \frac{V_3 - 2V_2 - V_{in}}{L_3}(1-k)T, \tag{7.150}$$

$$V_3 = \frac{(2-k)V_2 + V_{in}}{1-k} = \frac{7 - 9k + 3k^2}{(1-k)^3}V_{in} = \left[\left(\frac{2-k}{1-k}\right)^3 - 1\right]V_{in},$$

and

$$V_{L3\text{-}off} = V_3 - 2V_2 - V_{in} = \frac{k(2-k)^2}{(1-k)^3}V_{in}. \tag{7.151}$$

The output voltage is

$$V_O = V_3 + V_{in} + V_{L3} + V_2 = \frac{11 - 13k + 4k^2}{(1-k)^3}V_{in} = \left[\frac{3-k}{1-k}\left(\frac{2-k}{1-k}\right)^2 - 1\right]V_{in}. \tag{7.152}$$

The voltage transfer gain is

$$G_3 = \frac{V_O}{V_{in}} = \left(\frac{2-k}{1-k}\right)^2\frac{3-k}{1-k} - 1. \tag{7.153}$$

The following relations are available:

$$i_{C12\text{-}on} = I_O, \qquad i_{C12\text{-}off} = \frac{kI_O}{1-k},$$

$$i_{C11\text{-}off} = I_O + i_{C12\text{-}off} = \frac{I_O}{1-k}, \qquad i_{C11\text{-}on} = i_{C6\text{-}on} = \frac{I_O}{k},$$

$$i_{C6-off} = i_{C5-off} = \frac{I_O}{1-k}, \qquad i_{C5-on} = \frac{I_O}{k},$$

$$I_{L3} = i_{C11-off} + i_{C5-off} = \frac{2I_O}{1-k},$$

$$i_{C4-on} = I_{L3} + i_{C5-on} = \frac{1+k}{k(1-k)}I_O, \qquad i_{C4-off} = \frac{1+k}{(1-k)^2}I_O.$$

$$I_{L2} = i_{C3-off} = I_{L3} + i_{C4-off} = \frac{3-k}{(1-k)^2}I_O, \qquad i_{C3-on} = \frac{3-k}{k(1-k)}I_O,$$

$$i_{C2-on} = I_{L2} + i_{C3-on} = \frac{3-k}{k(1-k)^2}I_O, \qquad i_{C2-off} = \frac{3-k}{(1-k)^3}I_O,$$

$$I_{L1} = i_{C1-off} = I_{L2} + i_{C2-off} = \frac{(3-k)(2-k)}{(1-k)^3}I_O, \qquad i_{C1-on} = \frac{(3-k)(2-k)}{k(1-k)^2}I_O,$$

$$\begin{aligned} i_{in} &= I_{L1} + i_{C1-on} + i_{C2-on} + i_{C4-on} + i_{C6-on} \\ &= \left[\frac{(3-k)(2-k)}{(1-k)^3} + \frac{(3-k)(2-k)}{k(1-k)^2} + \frac{3-k}{k(1-k)^2} + \frac{1+k}{k(1-k)} + \frac{1}{k}\right]I_O \\ &= \frac{11-13k+4k^2}{k(1-k)^3}I_O. \end{aligned}$$

Therefore,

$$I_{in} = ki_{in} = \frac{11-13k+4k^2}{(1-k)^3}I_O = \left[\frac{3-k}{1-k}\left(\frac{2-k}{1-k}\right)^2 - 1\right]I_O.$$

Analogously,

$$\Delta i_{L1} = \frac{V_{in}}{L_2}kT, \qquad I_{L1} = \frac{(2-k)(3-k)}{(1-k)^3}I_O,$$

$$\Delta i_{L2} = \frac{V_1 + V_{in}}{L_2}kT = \frac{2-k}{(1-k)L_2}kTV_{in}, \qquad I_{L2} = \frac{3-k}{(1-k)^2}I_O,$$

$$\Delta i_{L3} = \frac{V_2 + V_{in}}{L_3}kT = \frac{(2-k)^2}{(1-k)^2L_3}kTV_{in}, \qquad I_{L3} = \frac{2I_O}{1-k}.$$

Therefore, the variation ratio of current i_{L1} through inductor L_1 is

$$\xi_1 = \frac{\Delta i_{L1}/2}{I_{L1}} = \frac{k(1-k)^3TV_{in}}{2(2-k)(3-k)L_1I_O} = \frac{k(1-k)^3}{(2-k)(3-k)G_3}\frac{R}{2fL_1}, \tag{7.154}$$

the variation ratio of current i_{L2} through inductor L_2 is

$$\xi_2 = \frac{\Delta i_{L2}/2}{I_{L2}} = \frac{k(1-k)(2-k)TV_1}{2(3-k)L_2I_O} = \frac{k(1-k)(2-k)}{(3-k)G_3}\frac{R}{2fL_2}, \tag{7.155}$$

and the variation ratio of current i_{L3} through inductor L_3 is

$$\xi_3 = \frac{\Delta i_{L3}/2}{I_{L3}} = \frac{k(2-k)^2TV_{in}}{4(1-k)L_3I_O} = \frac{k(2-k)^2}{2(1-k)G_3}\frac{R}{2fL_3}. \tag{7.156}$$

The ripple voltage of output voltage v_O is

$$\Delta v_O = \frac{\Delta Q}{C_{12}} = \frac{I_O kT}{C_{12}} = \frac{k}{fC_{12}} \frac{V_O}{R}.$$

Therefore, the variation ratio of output voltage v_O is

$$\varepsilon = \frac{\Delta v_O/2}{V_O} = \frac{k}{2RfC_{12}}. \tag{7.157}$$

7.3.2.4 N/O Higher-Order Lift Additional Circuit

The higher-order N/O lift additional circuit can be derived from the corresponding circuits of the main series by adding a DEC. Each stage voltage $V_i (i = 1, 2, \ldots, n)$ is

$$V_i = \left[\left(\frac{2-k}{1-k}\right)^i - 1\right] V_{in}. \tag{7.158}$$

It means that V_1 is the voltage across capacitor C_2, V_2 is the voltage across capacitor C_4, and so on.

For the nth-order lift additional circuit, the final output voltage is

$$V_O = \left[\frac{3-k}{1-k}\left(\frac{2-k}{1-k}\right)^{n-1} - 1\right] V_{in}. \tag{7.159}$$

The voltage transfer gain is

$$G_n = \frac{V_O}{V_{in}} = \frac{3-k}{1-k}\left(\frac{2-k}{1-k}\right)^{n-1} - 1. \tag{7.160}$$

Analogously, the variation ratio of current i_{Li} through inductor $L_i (i = 1, 2, 3, \ldots, n)$ is

$$\xi_1 = \frac{\Delta i_{L1}/2}{I_{L1}} = \frac{k(1-k)^n}{2^{h(1-n)}[(2-k)^{(n-2)}(3-k)]^{u(n-2)} G_n} \frac{R}{fL_1}, \tag{7.161}$$

$$\xi_2 = \frac{\Delta i_{L2}/2}{I_{L2}} = \frac{k(1-k)^{(n-2)}(2-k)}{2^{h(n-2)}(3-k)^{(n-2)} G_n} \frac{R}{2fL_2}, \tag{7.162}$$

$$\xi_3 = \frac{\Delta i_{L3}/2}{I_{L3}} = \frac{k(2-k)^{(n-1)}}{2^{h(n-3)}(1-k)^{(n-2)} G_n} \frac{R}{2fL_3}, \tag{7.163}$$

where

$$h(x) = \begin{cases} 0 & x > 0 \\ 1 & x \leq 0 \end{cases} \quad \text{is the Hong function}$$

and

$$u(x) = \begin{cases} 1 & x \geq 0 \\ 0 & x < 0 \end{cases} \quad \text{is the unit-step function,}$$

and the variation ratio of output voltage v_O is

$$\varepsilon = \frac{\Delta v_O/2}{V_O} = \frac{k}{2RfC_{12}}. \quad (7.164)$$

7.3.3 Enhanced Series

All circuits of the N/O SL Luo-converters—enhanced series—are derived from the corresponding circuits of the main series by adding the DEC in each stage circuit of all series converters.

The first three stages of this series are shown in Figures 7.20, 7.23, and 7.24. For ease of understanding, they are called the elementary enhanced circuit, the re-lift enhanced circuit, and the triple-lift enhanced circuit, respectively, and are numbered $n = 1, 2$, and 3, respectively.

7.3.3.1 N/O Elementary Enhanced Circuit

This circuit is derived from the N/O elementary circuit by adding a DEC. Its circuit and switch-on and switch-off equivalent circuits are shown in Figure 7.20.

The output voltage is

$$V_O = V_{in} + V_{L1} + V_1 = \frac{2}{1-k}V_{in} = \left[\frac{3-k}{1-k} - 1\right]V_{in}. \quad (7.139)$$

The voltage transfer gain is

$$G_1 = \frac{V_O}{V_{in}} = \frac{3-k}{1-k} - 1. \quad (7.140)$$

7.3.3.2 N/O Re-Lift Enhanced Circuit

The N/O re-lift enhanced circuit is derived from the N/O re-lift circuit of the main series by adding the DEC in each stage circuit. Its circuit diagram and switch-on and switch-off equivalent circuits are shown in Figure 7.23.

The voltage across capacitor C_{12} is charged to

$$V_{C12} = \frac{3}{1-k}V_{in}. \quad (7.165)$$

The voltage across capacitor C_3 is charged with V_{C12}, and the voltage across capacitors C_4 and C_{12} is charged with V_{C4}

$$V_{C4} = \frac{2-k}{1-k}V_{C12} = \frac{2-k}{1-k}\frac{3-k}{1-k}V_{in}. \quad (7.166)$$

(a)

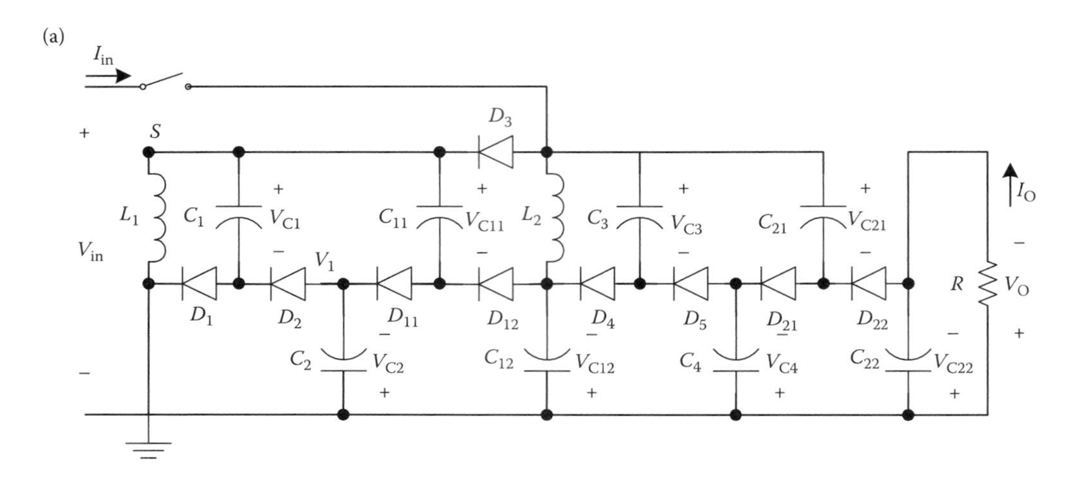

(b)

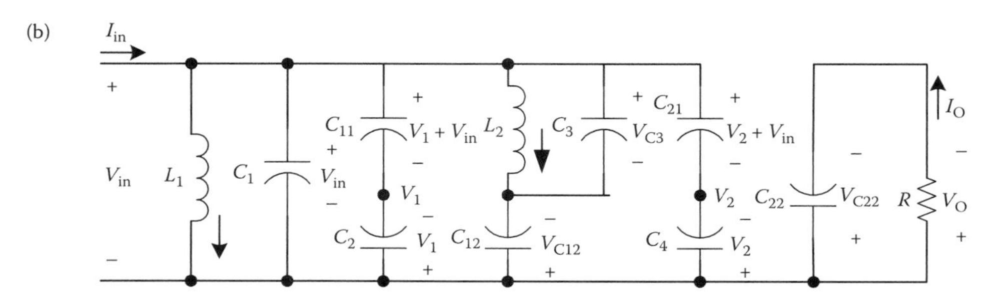

(c)

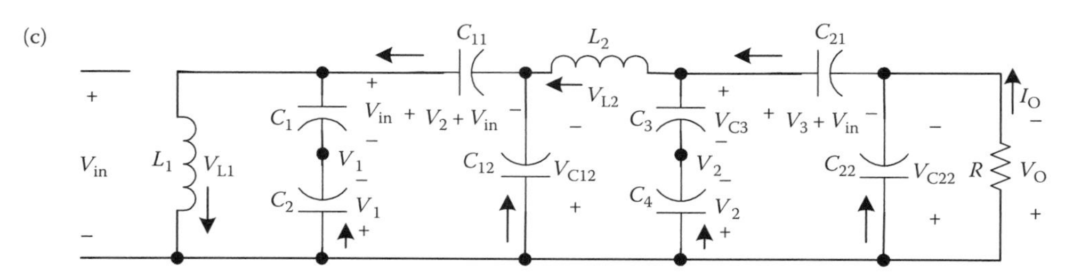

FIGURE 7.23 Re-lift enhanced circuit of N/O SL Luo-converters: (a) circuit diagram, (b) equivalent circuit during switch-on, and (c) equivalent circuit during switch-off. (Reprinted from Luo, F. L. and Ye, H. 2006. *Essential DC/DC Converters*. Boca Raton: Taylor & Francis Group LLC, p. 284. With permission.)

The current flowing through inductor L_2 increases with V_{C12} during the switch-on period kT and decreases with $-(V_{C21} - V_{C4} - V_{C12})$ during the switch-off period $(1-k)T$.

Therefore,

$$\Delta i_{L2} = \frac{kT}{L_2}(V_{C12} - V_{in}) = \frac{V_{C21} - V_{C4} - V_{C12}}{L_2}(1-k)T, \tag{7.167}$$

$$V_{C21} = \left(\frac{3-k}{1-k}\right)^2 V_{in}.$$

The output voltage is

$$V_O = V_{C21} - V_{in} = \left[\left(\frac{3-k}{1-k}\right)^2 - 1\right] V_{in}. \tag{7.168}$$

The voltage transfer gain is

$$G_2 = \frac{V_O}{V_{in}} = \left(\frac{3-k}{1-k}\right)^2 - 1. \tag{7.169}$$

The following relations are obtained:

$$i_{C22\text{–}on} = I_O, \qquad i_{C22\text{–}off} = \frac{kI_O}{1-k},$$

$$i_{C21\text{–}off} = I_O + i_{C22\text{–}off} = \frac{I_O}{1-k}, \qquad i_{C21\text{–}on} = i_{C4\text{–}on} = \frac{I_O}{k},$$

$$i_{C4\text{–}off} = i_{C3\text{–}off} = \frac{I_O}{1-k}, \qquad i_{C3\text{—}on} = \frac{I_O}{k},$$

$$I_{L2} = i_{C21\text{–}off} + i_{C3\text{–}off} = \frac{2I_O}{1-k},$$

$$i_{C12\text{–}on} = I_{L2} + i_{C3\text{–}on} = \frac{1+k}{k(1-k)} I_O, \qquad i_{C12\text{–}off} = \frac{1+k}{(1-k)^2} I_O,$$

$$i_{C11\text{–}off} = I_{L2} + i_{C12\text{–}off} = \frac{3-k}{(1-k)^2} I_O, \qquad i_{C2\text{–}off} = \frac{3-k}{(1-k)^2} I_O,$$

$$i_{C11\text{–}on} = i_{C2\text{–}on} = \frac{3-k}{k(1-k)} I_O,$$

$$I_{L1} = i_{C11\text{–}off} + i_{C2\text{–}off} = 2\frac{3-k}{(1-k)^2} I_O, \qquad i_{C1\text{–}on} = \frac{3-k}{k(1-k)} I_O,$$

$$i_{in} = I_{L1} + i_{C1\text{–}on} + i_{C11\text{–}on} + i_{C12\text{–}on} + i_{C21\text{–}on} = \frac{4(2-k)}{k(1-k)^2} I_O.$$

Therefore,

$$I_{in} = k i_{in} = \frac{4(2-k)}{(1-k)^2} I_O = \left[\frac{(3-k)^2}{(1-k)^2} - 1\right] I_O.$$

Analogously,

$$\Delta i_{L1} = \frac{V_{in}}{L_2} kT, \qquad I_{L1} = 2\frac{3-k}{(1-k)^2} I_O,$$

$$\Delta i_{L2} = \frac{V_{C12} - V_{in}}{L_2} kT = \frac{2+k}{(1-k)L_2} kTV_{in}, \qquad I_{L2} = \frac{2I_O}{1-k}.$$

Therefore, the variation ratio of current i_{L1} through inductor L_1 is

$$\xi_1 = \frac{\Delta i_{L1}/2}{I_{L1}} = \frac{k(1-k)^2 TV_{in}}{4(3-k)L_1 I_O} = \frac{k(1-k)^2}{2(3-k)G_2} \frac{R}{2fL_1}. \tag{7.170}$$

The variation ratio of current i_{L2} through inductor L_2 is

$$\xi_2 = \frac{\Delta i_{L2}/2}{I_{L2}} = \frac{k(2+k)TV_{in}}{4L_2 I_O} = \frac{k(2+k)}{2G_2}\frac{R}{2fL_2}. \tag{7.171}$$

The ripple voltage of output voltage v_O is

$$\Delta v_O = \frac{\Delta Q}{C_{22}} = \frac{I_O kT}{C_{22}} = \frac{k}{fC_{22}}\frac{V_O}{R}.$$

Therefore, the variation ratio of output voltage v_O is

$$\varepsilon = \frac{\Delta v_O/2}{V_O} = \frac{k}{2RfC_{22}}. \tag{7.172}$$

7.3.3.3 N/O Triple-Lift Enhanced Circuit

This circuit is derived from the N/O triple-lift circuit of the main series by adding the DEC in each stage circuit. Its circuit diagram and equivalent circuits during switch-on and switch-off periods are shown in Figure 7.24.

The voltage across capacitor C_{12} is charged with V_{C12}. As described in the previous section, the voltage across C_{C12} is $V_{C12} = ((3-k)/(1-k))V_{in}$ and the voltage across C_4 and C_{C22} is

$$V_{C22} = \frac{3-k}{1-k}V_{C12} = \left(\frac{3-k}{1-k}\right)^2 V_{in}.$$

The voltage across capacitor C_5 is charged with V_{C22}, and the voltage across capacitor C_6 is charged with V_{C6}

$$V_{C6} = \frac{2-k}{1-k}V_{C22} = \frac{2-k}{1-k}\left(\frac{3-k}{1-k}\right)^2 V_{in}.$$

The current flowing through inductor L_3 increases with V_{C22} during the switch-on period kT and decreases with $-(V_{C32} - V_{C6} - V_{C22})$ during the switch-off period $(1-k)T$.

Therefore,

$$\Delta i_{L3} = \frac{kT}{L_3}(V_{C22} - V_{in}) = \frac{V_{C31} - V_{C6} - V_{C22}}{L_3}(1-k)T, \tag{7.173}$$

$$V_{C31} = \left(\frac{3-k}{1-k}\right)^3 V_{in},$$

and

$$V_O = V_{C31} - V_{in} = \left[\left(\frac{3-k}{1-k}\right)^3 - 1\right]V_{in}. \tag{7.174}$$

The voltage transfer gain is

$$G_3 = \frac{V_O}{V_{in}} = \left(\frac{3-k}{1-k}\right)^2 - 1. \tag{7.175}$$

(a)

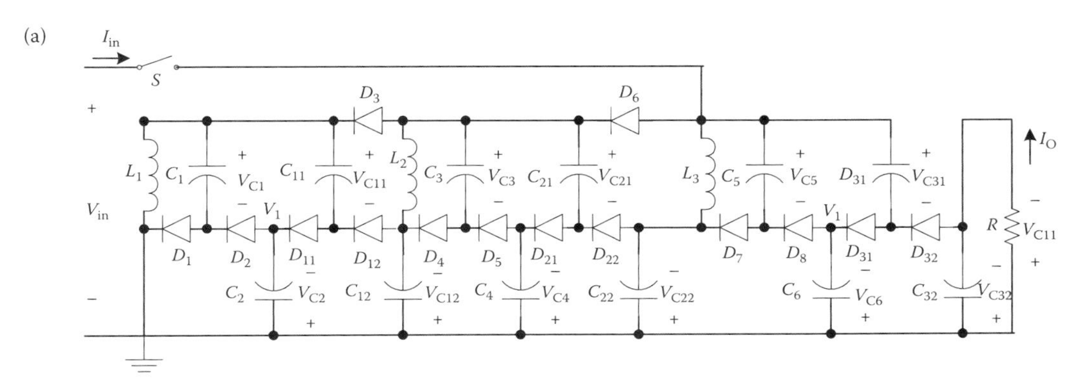

(b)

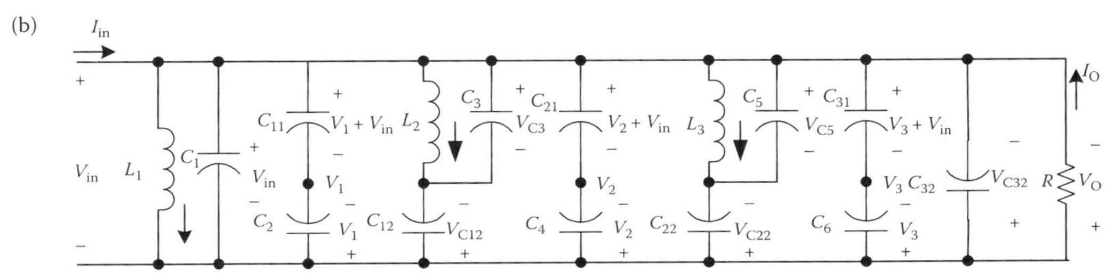

(c)

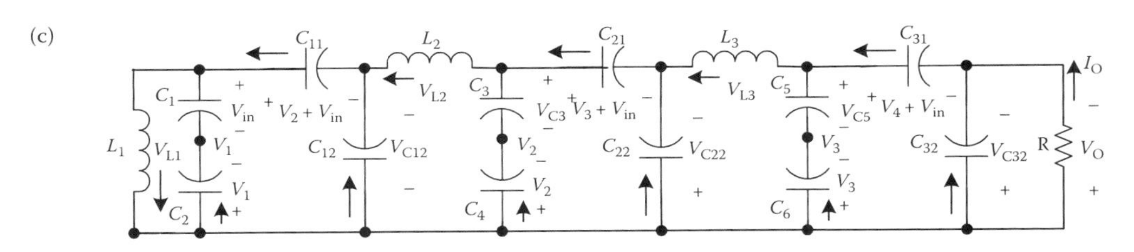

FIGURE 7.24 Triple-lift enhanced circuit of N/O SL Luo-converters: (a) circuit diagram, (b) equivalent circuit during switch-on, and (c) equivalent circuit during switch-off. (Reprinted from Luo, F. L. and Ye, H. 2006. *Essential DC/DC Converters*. Boca Raton: Taylor & Francis Group LLC, p. 285. With permission.)

The following relations are obtained:

$$i_{C32\text{-on}} = I_O, \qquad i_{C32\text{-off}} = \frac{kI_O}{1-k},$$

$$i_{C31\text{-off}} = I_O + i_{C32\text{-off}} = \frac{I_O}{1-k}, \qquad i_{C31\text{-on}} = i_{C6\text{-on}} = \frac{I_O}{k},$$

$$i_{C6\text{-off}} = i_{C5\text{-off}} = \frac{I_O}{1-k}, \qquad i_{C6\text{-on}} = i_{C5\text{-on}} = \frac{I_O}{k},$$

$$I_{L3} = i_{C31\text{-off}} + i_{C5\text{-off}} = \frac{2I_O}{1-k},$$

$$i_{C22\text{-on}} = I_{L3} + i_{C5\text{-on}} = \frac{1+k}{k(1-k)}I_O, \qquad i_{C22\text{-off}} = \frac{1+k}{(1-k)^2}I_O,$$

$$i_{C21\text{-off}} = i_{C4\text{-off}} = I_{L3} + i_{C22\text{-off}} = \frac{3-k}{(1-k)^2}I_O, \qquad i_{C4\text{-on}} = \frac{3-k}{k(1-k)}I_O,$$

$$I_{L2} = i_{C4\text{–off}} + i_{C21\text{–off}} = 2\frac{3-k}{(1-k)^2}I_O, \qquad i_{C3\text{–on}} = \frac{3-k}{k(1-k)}I_O,$$

$$i_{C12\text{–on}} = I_{L2} + i_{C3\text{–on}} = \frac{(3-k)(2-k)}{k(1-k)^2}I_O, \qquad i_{C12\text{–off}} = \frac{(3-k)(2-k)}{(1-k)^3}I_O,$$

$$i_{C11\text{–off}} = I_{L2} + i_{C12\text{–off}} = \frac{(3-k)(4-3k)}{(1-k)^3}I_O, \qquad i_{C11\text{–on}} = \frac{(3-k)(4-k)}{k(1-k)^2}I_O,$$

$$I_{L1} = i_{C11\text{–off}} + i_{C1\text{–off}} = 2\frac{(3-k)(4-k)}{(1-k)^3}I_O, \qquad i_{C1\text{–on}} = i_{C2\text{–on}} = \frac{(3-k)(4-k)}{k(1-k)^2}I_O,$$

$$i_{in} = I_{L1} + i_{C1\text{–on}} + i_{C2\text{–on}} + i_{C12\text{–on}} + i_{C4\text{–on}} + i_{C22\text{–on}} + i_{C6\text{–on}}$$
$$= \frac{2(13-12k+3k^2)}{k(1-k)^3}I_O.$$

Therefore,

$$I_{in} = ki_{in} = 2\frac{13-12k+3k^2}{(1-k)^3}I_O = \left[\left(\frac{3-k}{1-k}\right)^3 - 1\right]I_O.$$

Analogously,

$$\Delta i_{L1} = \frac{V_{in}}{L_2}kT, \qquad I_{L1} = \frac{2(4-k)(3-k)}{(1-k)^3}I_O,$$

$$\Delta i_{L2} = \frac{V_1 + V_{in}}{L_2}kT = \frac{2-k}{(1-k)L_2}kTV_{in}, \qquad I_{L2} = 2\frac{3-k}{(1-k)^2}I_O,$$

$$\Delta i_{L3} = \frac{V_2 + V_{in}}{L_3}kT = \frac{(2-k)^2}{(1-k)^2L_3}kTV_{in}, \qquad I_{L3} = \frac{2I_O}{1-k}.$$

Therefore, the variation ratio of current i_{L1} through inductor L_1 is

$$\xi_1 = \frac{\Delta i_{L1}/2}{I_{L1}} = \frac{k(1-k)^3TV_{in}}{4(4-k)(3-k)L_1I_O} = \frac{k(1-k)^3}{2(4-k)(3-k)G_3}\frac{R}{2fL_1}, \tag{7.176}$$

the variation ratio of current i_{L2} through inductor L_2 is

$$\xi_2 = \frac{\Delta i_{L2}/2}{I_{L2}} = \frac{k(1-k)(2-k)TV_1}{4(3-k)L_2I_O} = \frac{k(1-k)(2-k)}{2(3-k)G_3}\frac{R}{2fL_2}, \tag{7.177}$$

and the variation ratio of current i_{L3} through inductor L_3 is

$$\xi_3 = \frac{\Delta i_{L3}/2}{I_{L3}} = \frac{k(2-k)^2TV_{in}}{4(1-k)L_3I_O} = \frac{k(2-k)^2}{2(1-k)G_3}\frac{R}{2fL_3}. \tag{7.178}$$

The ripple voltage of output voltage v_O is

$$\Delta v_O = \frac{\Delta Q}{C_{32}} = \frac{I_OkT}{C_{32}} = \frac{k}{fC_{32}}\frac{V_O}{R}.$$

Therefore, the variation ratio of output voltage v_O is

$$\varepsilon = \frac{\Delta v_O/2}{V_O} = \frac{k}{2RfC_{32}}. \tag{7.179}$$

7.3.3.4 N/O Higher-Order Lift Enhanced Circuit

The higher-order N/O lift enhanced circuit is derived from the corresponding circuits of the main series by adding the DEC in each stage of the circuit. At each stage, the final voltage $V_{Ci1} (i = 1, 2, \ldots, n)$ is

$$V_{Ci1} = \left(\frac{3-k}{1-k}\right)^i V_{in}. \tag{7.180}$$

For the nth-order lift enhanced circuit, the final output voltage is

$$V_O = \left[\left(\frac{3-k}{1-k}\right)^n - 1\right] V_{in}. \tag{7.181}$$

The voltage transfer gain is

$$G_n = \frac{V_O}{V_{in}} = \left(\frac{3-k}{1-k}\right)^n - 1. \tag{7.182}$$

The variation ratio of output voltage v_O is

$$\varepsilon = \frac{\Delta v_O/2}{V_O} = \frac{k}{2RfC_{n2}}. \tag{7.183}$$

7.3.4 Re-Enhanced Series

All circuits of the N/O SL Luo-converters—re-enhanced series—are derived from the corresponding circuits of the main series by adding the DEC *twice* in circuits of each stage.

The first three stages of this series are shown in Figures 7.25 through 7.27. For ease of understanding, they are called the elementary re-enhanced circuit, the re-lift re-enhanced circuit, and the triple-lift re-enhanced circuit, respectively, and are numbered $n = 1, 2$, and 3, respectively.

7.3.4.1 N/O Elementary Re-Enhanced Circuit

This circuit is derived from the N/O elementary circuit by adding the DEC twice in each stage circuit. Its circuit and switch-on and switch-off equivalent circuits are shown in Figure 7.25.

The voltage across capacitor C_1 is charged with V_{in}. The voltage across capacitor C_{12} is charged with V_{C12}.

The voltage across capacitor C_{13} is charged with V_{C13}.

$$V_{C13} = \frac{4-k}{1-k} V_{in}. \tag{7.184}$$

(a)

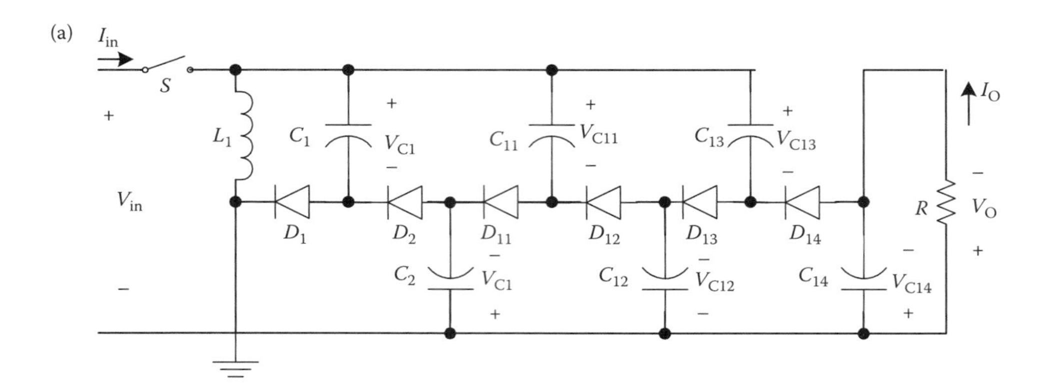

(b)

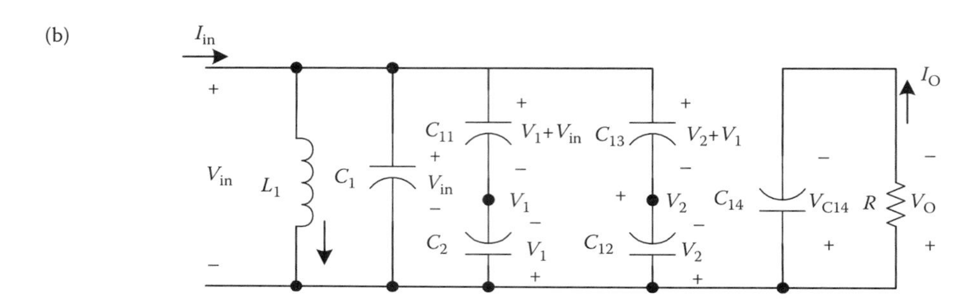

(c)

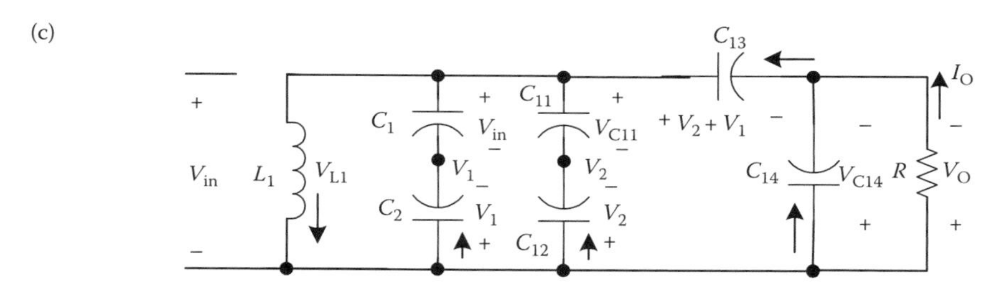

FIGURE 7.25 Elementary re-enhanced circuit of N/O SL Luo-converters: (a) circuit diagram, (b) equivalent circuit during switch-on, and (c) equivalent circuit during switch-off. (Reprinted from Luo, F. L. and Ye, H. 2006. *Essential DC/DC Converters*. Boca Raton: Taylor & Francis Group LLC, p. 292. With permission.)

The output voltage is

$$V_O = V_{C13} - V_{in} = \left[\frac{4-k}{1-k} - 1\right] V_{in}. \tag{7.185}$$

The voltage transfer gain is

$$G_1 = \frac{V_O}{V_{in}} = \frac{4-k}{1-k} - 1. \tag{7.186}$$

The ripple voltage of output voltage v_O is

$$\Delta v_O = \frac{\Delta Q}{C_{14}} = \frac{I_O k T}{C_{14}} = \frac{k}{f C_{14}} \frac{V_O}{R}.$$

Therefore, the variation ratio of output voltage v_O is

$$\varepsilon = \frac{\Delta v_O/2}{V_O} = \frac{k}{2RfC_{14}}. \tag{7.187}$$

7.3.4.2 N/O Re-Lift Re-Enhanced Circuit

The N/O re-lift re-enhanced circuit is derived from the N/O re-lift circuit by adding the DEC twice in circuits of each stage. Its circuit diagram and switch-on and switch-off equivalent circuits are shown in Figure 7.26.

The voltage across capacitor C_{13} is charged with V_{C13}. As described in the previous section, the voltage across C_{13} is

$$V_{C13} = \frac{4-k}{1-k}V_{in}.$$

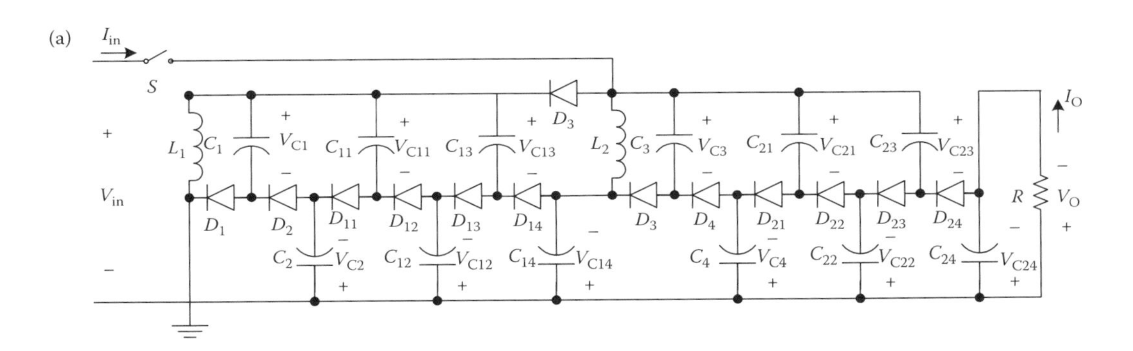

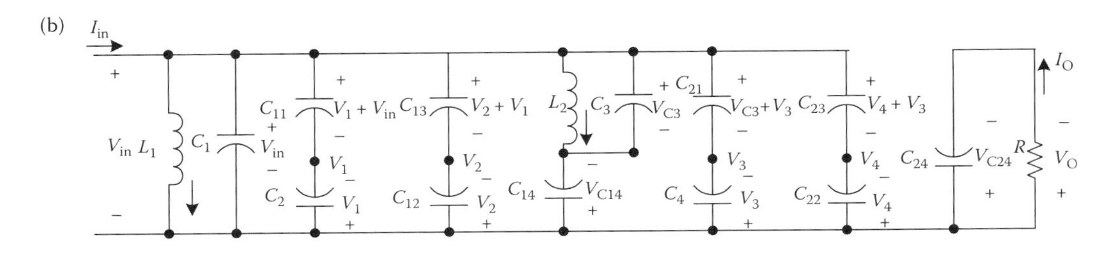

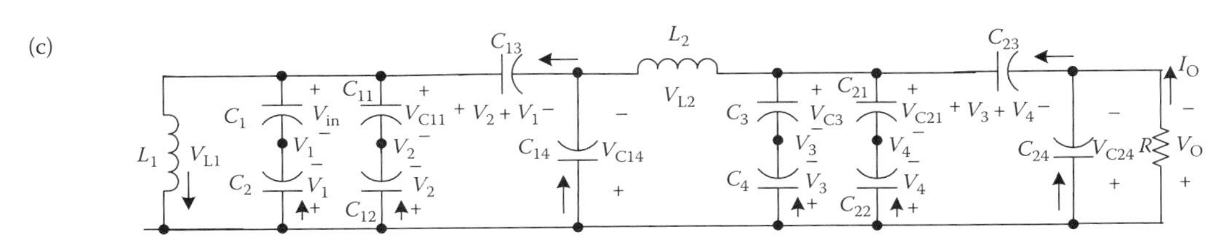

FIGURE 7.26 Re-lift re-enhanced circuit of N/O SL Luo-converters: (a) circuit diagram, (b) equivalent circuit during switch-on, and (c) equivalent circuit during switch-off. (Reprinted from Luo, F. L. and Ye, H. 2006. *Essential DC/DC Converters*. Boca Raton: Taylor & Francis Group LLC, p. 293. With permission.)

Analogously,

$$V_{C23} = \left(\frac{4-k}{1-k}\right)^2 V_{in}. \tag{7.188}$$

The output voltage is

$$V_O = V_{C23} - V_{in} = \left[\left(\frac{4-k}{1-k}\right)^2 - 1\right] V_{in}. \tag{7.189}$$

The voltage transfer gain is

$$G_2 = \frac{V_O}{V_{in}} = \left(\frac{4-k}{1-k}\right)^2 - 1. \tag{7.190}$$

The ripple voltage of output voltage v_O is

$$\Delta v_O = \frac{\Delta Q}{C_{24}} = \frac{I_O kT}{C_{24}} = \frac{k}{fC_{24}} \frac{V_O}{R}.$$

Therefore, the variation ratio of output voltage v_O is

$$\varepsilon = \frac{\Delta v_O/2}{V_O} = \frac{k}{2RfC_{24}}. \tag{7.191}$$

7.3.4.3 N/O Triple-Lift Re-Enhanced Circuit

This circuit is derived from the N/O triple-lift circuit by adding the DEC twice in each stage circuit. Its circuit diagram and equivalent circuits during switch-on and switch-off periods are shown in Figure 7.27.

The voltage across capacitor C_{13} is

$$V_{C13} = \frac{4-k}{1-k} V_{in}.$$

The voltage across capacitor C_{23} is

$$V_{C23} = \left(\frac{4-k}{1-k}\right)^2 V_{in}.$$

Analogously, the voltage across capacitor C_{33} is

$$V_{C33} = \left(\frac{4-k}{1-k}\right)^3 V_{in}. \tag{7.192}$$

The output voltage is

$$V_O = V_{C33} - V_{in} = \left[\left(\frac{4-k}{1-k}\right)^3 - 1\right] V_{in}. \tag{7.193}$$

(a)

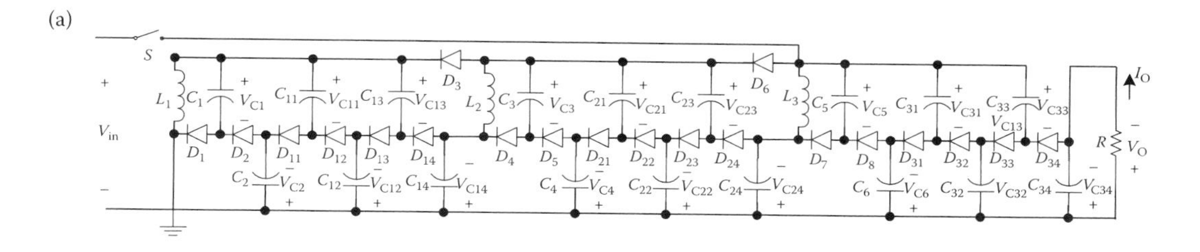

(b)

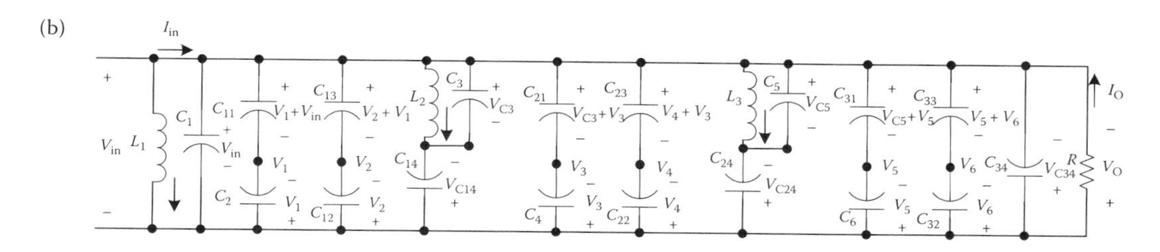

(c)

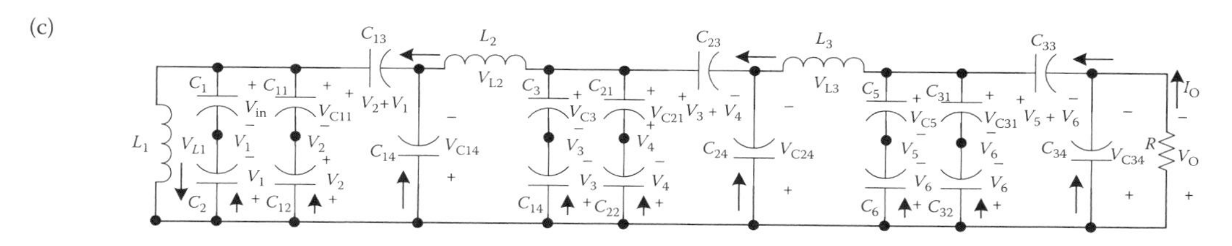

FIGURE 7.27 Triple-lift re-enhanced circuit of N/O SL Luo-converters: (a) circuit diagram, (b) equivalent circuit during switch-on, and (c) equivalent circuit during switch-off. (Reprinted from Luo, F. L. and Ye, H. 2006. *Essential DC/DC Converters*. Boca Raton: Taylor & Francis Group LLC, p. 294. With permission.)

The voltage transfer gain is

$$G_3 = \frac{V_O}{V_{in}} = \left(\frac{4-k}{1-k}\right)^3 - 1. \tag{7.194}$$

The ripple voltage of output voltage v_O is

$$\Delta v_O = \frac{\Delta Q}{C_{34}} = \frac{I_O kT}{C_{34}} = \frac{k}{fC_{34}}\frac{V_O}{R}.$$

Therefore, the variation ratio of output voltage v_O is

$$\varepsilon = \frac{\Delta v_O/2}{V_O} = \frac{k}{2RfC_{34}}. \tag{7.195}$$

7.3.4.4 N/O Higher-Order Lift Re-Enhanced Circuit

The higher-order N/O lift re-enhanced circuit can be derived from the corresponding circuits of the main series by adding the DEC twice in each stage circuit. At each stage, the final voltage $V_{Ci3} (i = 1, 2, \ldots, n)$ is

$$V_{Ci3} = \left(\frac{4-k}{1-k}\right)^i V_{in}. \quad (7.196)$$

For the nth-order lift additional circuit, the final output voltage is

$$V_O = V_{Cn3} - V_{in} = \left[\left(\frac{4-k}{1-k}\right)^n - 1\right] V_{in}. \quad (7.197)$$

The voltage transfer gain is

$$G_n = \frac{V_O}{V_{in}} = \left(\frac{4-k}{1-k}\right)^n - 1. \quad (7.198)$$

The variation ratio of output voltage v_O is

$$\varepsilon = \frac{\Delta v_O/2}{V_O} = \frac{k}{2RfC_{n4}}. \quad (7.199)$$

7.3.5 N/O Multiple-Enhanced Series

All circuits of the N/O SL Luo-converters—multiple-enhanced series—are derived from the corresponding circuits of the main series by adding the DEC multiple (j) times in each stage circuit.

The first three stages of this series are shown in Figures 7.28 through 7.30. For ease of understanding, they are called the elementary multiple-enhanced circuit, the re-lift multiple-enhanced circuit, and the triple-lift multiple-enhanced circuit, respectively, and are numbered $n = 1$, 2, and 3, respectively.

7.3.5.1 N/O Elementary Multiple-Enhanced Circuit

This circuit is derived from the N/O elementary circuit by adding the DEC multiple (j) times. Its circuit and switch-on and switch-off equivalent circuits are shown in Figure 7.28.

The voltage across capacitor C_{12j-1} is

$$V_{C12j-1} = \frac{j+2-k}{1-k} V_{in}. \quad (7.200)$$

The output voltage is

$$V_O = V_{C12j-1} - V_{in} = \left[\frac{j+2-k}{1-k} - 1\right] V_{in}. \quad (7.201)$$

The voltage transfer gain is

$$G_1 = \frac{V_O}{V_{in}} = \frac{j+2-k}{1-k} - 1. \quad (7.202)$$

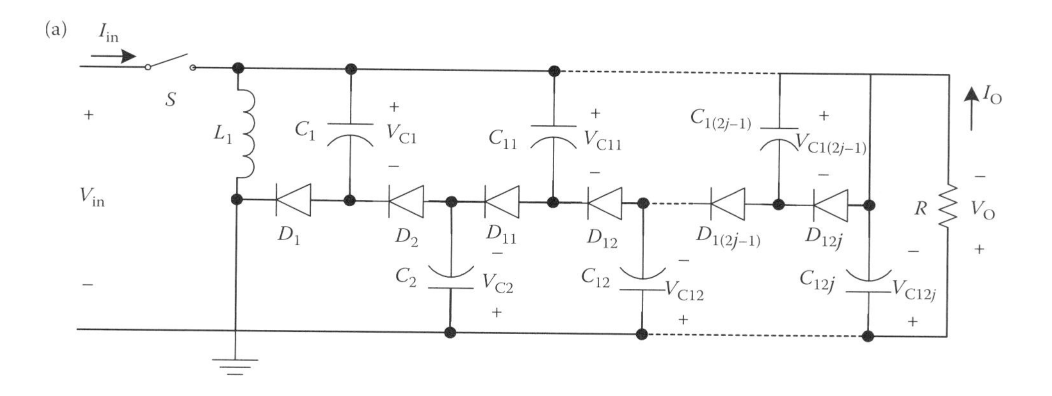

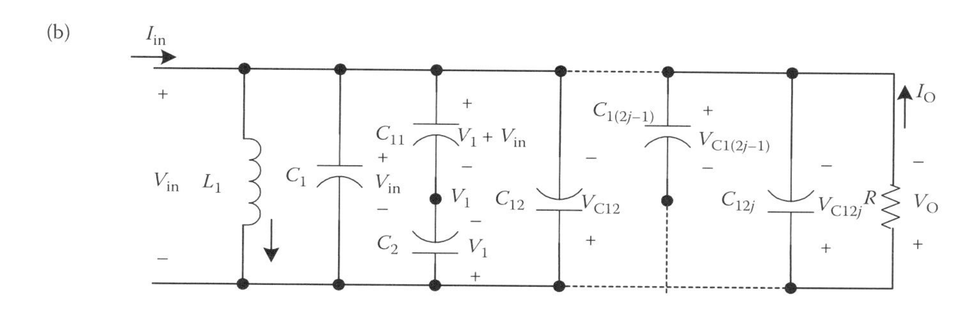

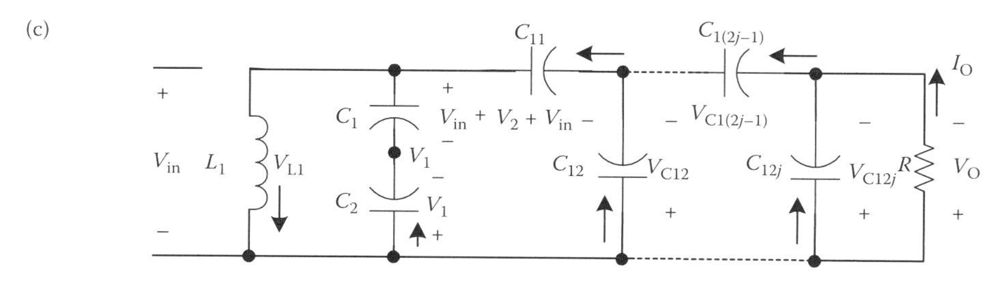

FIGURE 7.28 Elementary multiple-enhanced circuit of N/O SL Luo-converters: (a) circuit diagram, (b) equivalent circuit during switch-on, and (c) equivalent circuit during switch-off. (Reprinted from Luo, F. L. and Ye, H. 2006. *Essential DC/DC Converters*. Boca Raton: Taylor & Francis Group LLC, p. 298. With permission.)

The ripple voltage of output voltage v_O is

$$\Delta v_O = \frac{\Delta Q}{C_{12j}} = \frac{I_O kT}{C_{12j}} = \frac{k}{fC_{12j}} \frac{V_O}{R}.$$

Therefore, the variation ratio of output voltage v_O is

$$\varepsilon = \frac{\Delta v_O / 2}{V_O} = \frac{k}{2RfC_{12j}}. \tag{7.203}$$

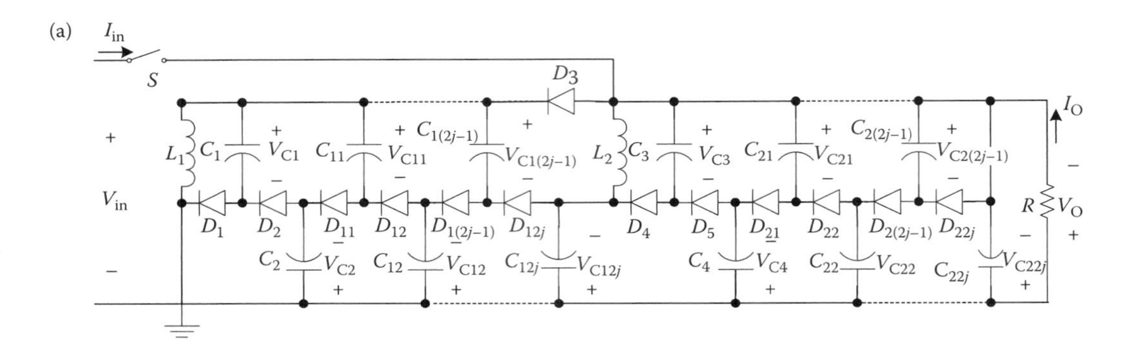

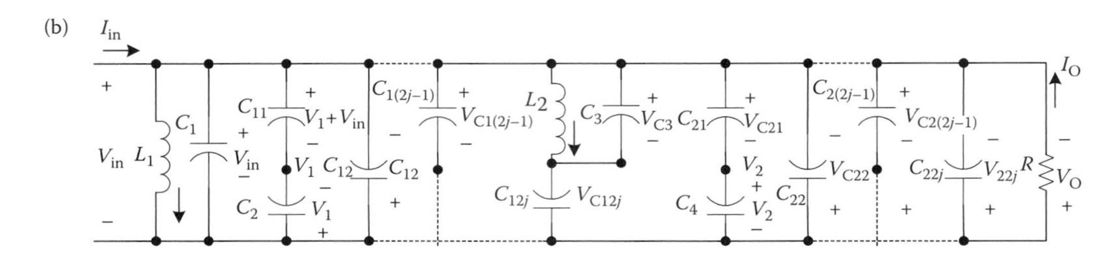

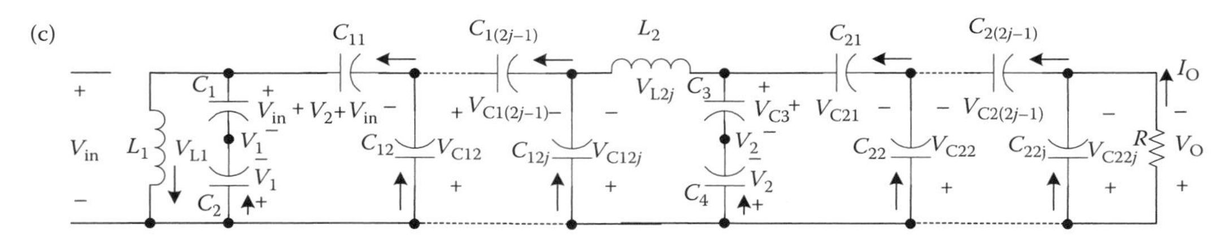

FIGURE 7.29 Re-lift multiple-enhanced circuit of N/O SL Luo-converters: (a) circuit diagram, (b) equivalent circuit during switch-on, and (c) equivalent circuit during switch-off. (Reprinted from Luo, F. L. and Ye, H. 2006. *Essential DC/DC Converters*. Boca Raton: Taylor & Francis Group LLC, p. 299. With permission.)

7.3.5.2 N/O Re-Lift Multiple-Enhanced Circuit

The N/O re-lift multiple-enhanced circuit is derived from the N/O re-lift circuit by adding the DEC multiple (j) times in circuits of each stage. Its circuit diagram and switch-on and switch-off equivalent circuits are shown in Figure 7.29.

The voltage across capacitor C_{22j-1} is

$$V_{C22j-1} = \left(\frac{j+2-k}{1-k}\right)^2 V_{in}. \tag{7.204}$$

The output voltage is

$$V_O = V_{C22j-1} - V_{in} = \left[\left(\frac{j+2-k}{1-k}\right)^2 - 1\right] V_{in}. \tag{7.205}$$

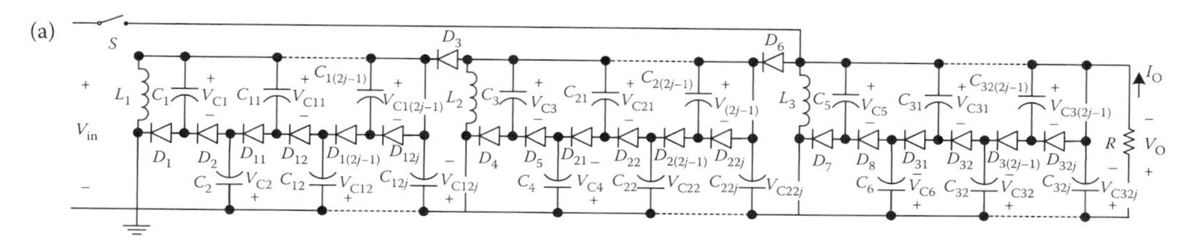

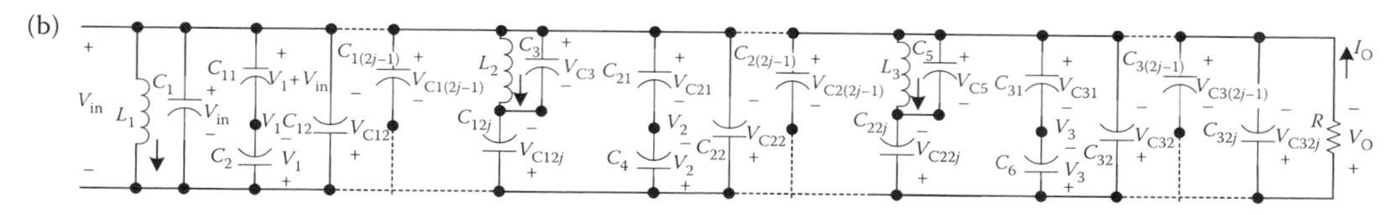

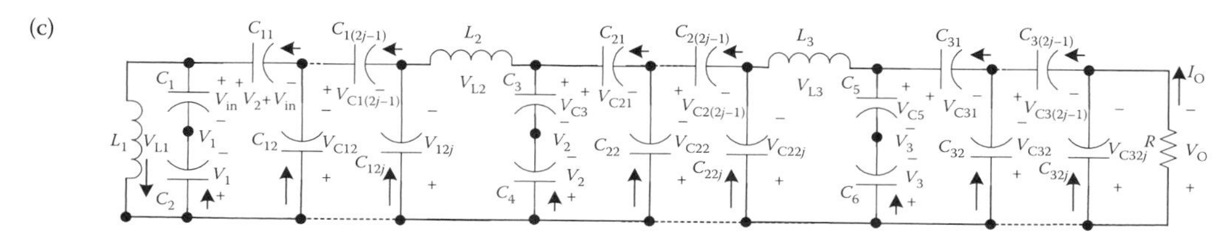

FIGURE 7.30 Triple-lift multiple-enhanced circuit of N/O SL Luo-converters: (a) circuit diagram, (b) equivalent circuit during switch-on, and (c) equivalent circuit during switch-off. (Reprinted from Luo, F. L. and Ye, H. 2006. *Essential DC/DC Converters*. Boca Raton: Taylor & Francis Group LLC, p. 300. With permission.)

The voltage transfer gain is

$$G_2 = \frac{V_O}{V_{in}} = \left(\frac{j+2-k}{1-k}\right)^2 - 1. \tag{7.206}$$

The ripple voltage of output voltage v_O is

$$\Delta v_O = \frac{\Delta Q}{C_{22j}} = \frac{I_O kT}{C_{22j}} = \frac{k}{fC_{22j}}\frac{V_O}{R}.$$

Therefore, the variation ratio of output voltage v_O is

$$\varepsilon = \frac{\Delta v_O/2}{V_O} = \frac{k}{2RfC_{22j}}. \tag{7.207}$$

7.3.5.3 N/O Triple-Lift Multiple-Enhanced Circuit

This circuit is derived from the N/O triple-lift circuit by adding the DEC multiple (j) times in each stage circuit. Its circuit diagram and equivalent circuits during switch-on and switch-off periods are shown in Figure 7.30.

The voltage across capacitor C_{32j-1} is

$$V_{C32j-1} = \left(\frac{j+2-k}{1-k}\right)^3 V_{in}. \tag{7.208}$$

The output voltage is

$$V_O = V_{C32j-1} - V_{in} = \left[\left(\frac{j+2-k}{1-k}\right)^3 - 1\right] V_{in}. \tag{7.209}$$

The voltage transfer gain is

$$G_3 = \frac{V_O}{V_{in}} = \left(\frac{j+2-k}{1-k}\right)^3 - 1. \tag{7.210}$$

The ripple voltage of output voltage v_O is

$$\Delta v_O = \frac{\Delta Q}{C_{32j}} = \frac{I_O kT}{C_{32j}} = \frac{k}{fC_{32j}} \frac{V_O}{R}.$$

Therefore, the variation ratio of output voltage v_O is

$$\varepsilon = \frac{\Delta v_O/2}{V_O} = \frac{k}{2RfC_{32j}}. \tag{7.211}$$

7.3.5.4 N/O Higher-Order Lift Multiple-Enhanced Circuit

The higher-order N/O lift multiple-enhanced circuit is derived from the corresponding circuits of the main series by adding the DEC multiple (j) times in each stage circuit. At each stage, the final voltage $V_{Ci2j-1} (i = 1, 2, \ldots, n)$ is

$$V_{Ci2j-1} = \left(\frac{j+2-k}{1-k}\right)^i V_{in}. \tag{7.212}$$

For the nth-order lift multiple-enhanced circuit, the final output voltage is

$$V_O = \left[\left(\frac{j+2-k}{1-k}\right)^n - 1\right] V_{in}. \tag{7.213}$$

The voltage transfer gain is

$$G_n = \frac{V_O}{V_{in}} = \left(\frac{j+2-k}{1-k}\right)^n - 1. \tag{7.214}$$

The variation ratio of output voltage v_O is

$$\varepsilon = \frac{\Delta v_O/2}{V_O} = \frac{k}{2RfC_{n2j}}. \tag{7.215}$$

7.3.6 Summary of N/O SL Luo-Converters

All circuits of the N/O SL Luo-converters can be shown in Figure 7.31 as the family tree.

From the analysis in previous sections, the common formula to calculate the output voltage can be presented as

$$V_O = \begin{cases} \left[\left(\frac{2-k}{1-k}\right)^n - 1\right]V_{in} & \text{main series,} \\ \left[\left(\frac{2-k}{1-k}\right)^{n-1}\left(\frac{3-k}{1-k}\right) - 1\right]V_{in} & \text{additional series,} \\ \left[\left(\frac{3-k}{1-k}\right)^n - 1\right]V_{in} & \text{enhanced series,} \\ \left[\left(\frac{4-k}{1-k}\right)^n - 1\right]V_{in} & \text{re-enhanced series,} \\ \left[\left(\frac{j+2-k}{1-k}\right)^n - 1\right]V_{in} & \text{multiple-enhanced series.} \end{cases} \tag{7.216}$$

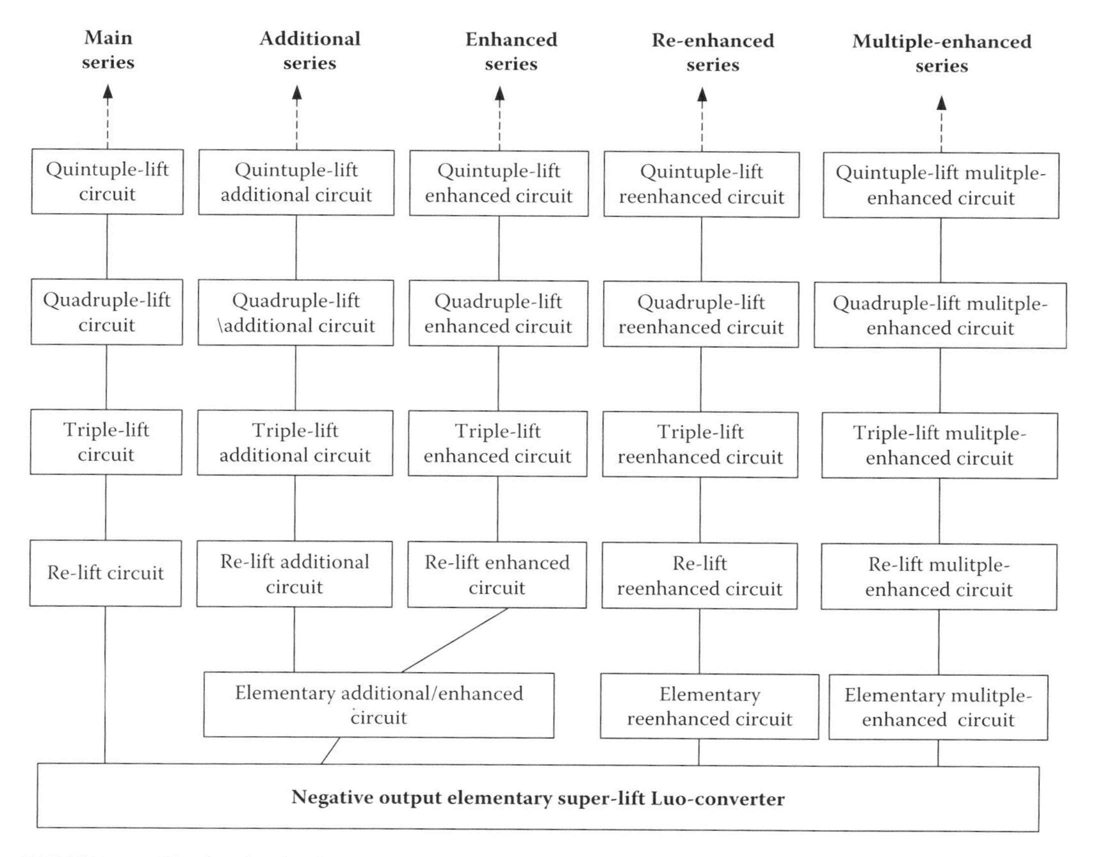

FIGURE 7.31 The family of N/O SL Luo-converters. (Reprinted from Luo, F. L. and Ye, H. 2006. *Essential DC/DC Converters*. Boca Raton: Taylor & Francis Group LLC, p. 304. With permission.)

The corresponding voltage transfer gain is

$$G = \frac{V_O}{V_{in}} = \begin{cases} \left(\frac{2-k}{1-k}\right)^n - 1 & \text{main series,} \\ \left(\frac{2-k}{1-k}\right)^{n-1}\left(\frac{3-k}{1-k}\right) - 1 & \text{additional series,} \\ \left(\frac{3-k}{1-k}\right)^n - 1 & \text{enhanced series,} \\ \left(\frac{4-k}{1-k}\right)^n - 1 & \text{re-enhanced series,} \\ \left(\frac{j+2-k}{1-k}\right)^n - 1 & \text{multiple-enhanced series.} \end{cases} \tag{7.217}$$

In order to show the advantages of N/O SL converters, their voltage transfer gains can be compared with that of the buck converter, $G = V_O/V_{in} = k$.

$$\text{Forward converter: } G = \frac{V_O}{V_{in}} = kN \quad (N \text{ is the transformer turn's ratio}),$$

$$\text{Cúk converter: } G = \frac{V_O}{V_{in}} = \frac{k}{1-k},$$

$$\text{Fly-back converter: } G = \frac{V_O}{V_{in}} = \frac{kN}{1-k} \quad (N \text{ is the transformer turn's ratio}),$$

$$\text{Boost converter: } G = \frac{V_O}{V_{in}} = \frac{1}{1-k},$$

and

$$\text{N/O Luo-converter: } G = \frac{V_O}{V_{in}} = \frac{n}{1-k}. \tag{7.218}$$

Assume that the conduction duty cycle k is 0.2; the output voltage transfer gains are listed in Table 7.4. Assume that the conduction duty cycle k is 0.5; the output voltage transfer gains are listed in Table 7.5. Assume that the conduction duty cycle k is 0.8; the output voltage transfer gains are listed in Table 7.6.

7.4 P/O Cascaded Boost-Converters

SL Luo-converters largely increase the voltage transfer gain in geometric progression. However, their circuits are a bit complex. We introduce a novel approach—P/O cascaded boost converters—that implement the output voltage increasing in geometric progression as well, but with a simpler structure. They also effectively enhance the voltage transfer gain in power law. There are several subseries. As described in previous sections, only three circuits of each subseries are introduced [1,2,7–9].

TABLE 7.4

Voltage Transfer Gains of Converters in the Condition $k = 0.2$

Stage No. (n)	1	2	3	4	5	n
Buck converter				0.2		
Forward converter				0.2N (N is the transformer turn's ratio)		
Cúk-converter				0.25		
Fly-back converter				0.25N (N is the transformer turn's ratio)		
Boost converter				1.25		
N/O Luo-converters	1.25	2.5	3.75	5	6.25	$1.25n$
N/O SL converters—main series	1.25	4.06	10.39	24.63	56.67	$2.25^n - 1$
N/O SL converters—additional series	2.5	6.88	16.72	38.87	88.7	$3.5 \times 2.25^{(n-1)} - 1$

Source: Data from Luo, F. L. and Ye, H. 2006. *Essential DC/DC Converters*, p. 305. Boca Raton: Taylor & Francis Group LLC.

TABLE 7.5

Voltage Transfer Gains of Converters in the Condition $k = 0.5$

Stage No. (n)	1	2	3	4	5	n
Buck converter				0.5		
Forward converter				0.5N (N is the transformer turn's ratio)		
Cúk-converter				1		
Fly-back converter				N (N is the transformer turn's ratio)		
Boost converter				2		
N/O Luo-converters	2	4	6	8	10	$2n$
N/O SL converters—main series	2	8	26	80	242	$3^n - 1$
N/O SL converters—additional series	4	14	44	134	404	$5 \times 3^{(n-1)} - 1$

Source: Data from Luo, F. L. and Ye, H. 2006. *Essential DC/DC Converters*, p. 305. Boca Raton: Taylor & Francis Group LLC.

TABLE 7.6

Voltage Transfer Gains of Converters in the Condition $k = 0.8$

Stage No. (n)	1	2	3	4	5	n
Buck converter				0.8		
Forward converter				0.8N (N is the transformer turn's ratio)		
Cúk-converter				4		
Fly-back converter				4N (N is the transformer turn's ratio)		
Boost converter				5		
N/O Luo-converters	5	10	15	20	25	$5n$
N/O SL converters—main series	5	35	215	1295	7775	$6^n - 1$
N/O SL converters—additional series	10	65	395	2375	14,255	$11 \times 6^{(n-1)} - 1$

Source: Data from Luo, F. L. and Ye, H. 2006. *Essential DC/DC Converters*, p. 305. Boca Raton: Taylor & Francis Group LLC.

7.4.1 Main Series

The first three stages of P/O cascaded boost converters—main series—are shown in Figures 5.5, 7.32, and 7.33. For ease of understanding, they are called the elementary boost converter, two-stage circuit, and three-stage circuit, respectively, and are numbered $n = 1$, 2, and 3, respectively.

7.4.1.1 *Elementary Boost Circuit*

The elementary boost converter is the fundamental boost converter; it is also introduced in Section 5.2.2. Its circuit diagram and its equivalent circuits during switch-on and switch-off periods are shown in Figure 5.5. The output voltage is

$$V_O = \frac{1}{1-k} V_{in}.$$

The voltage transfer gain is $G = V_O/V_{in} = 1/(1-k)$.

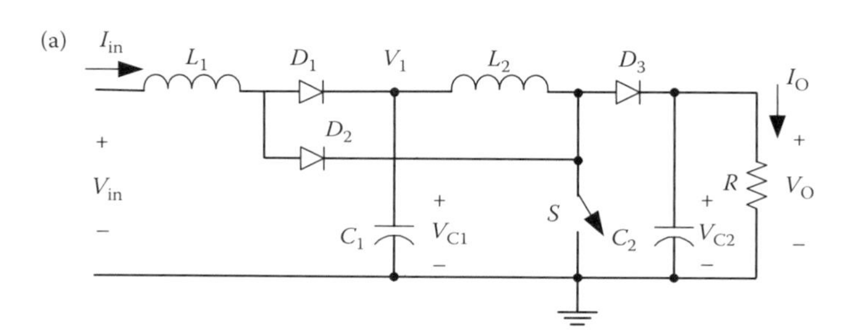

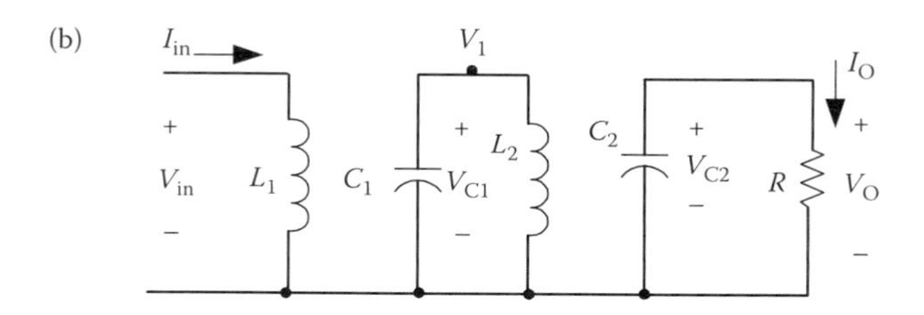

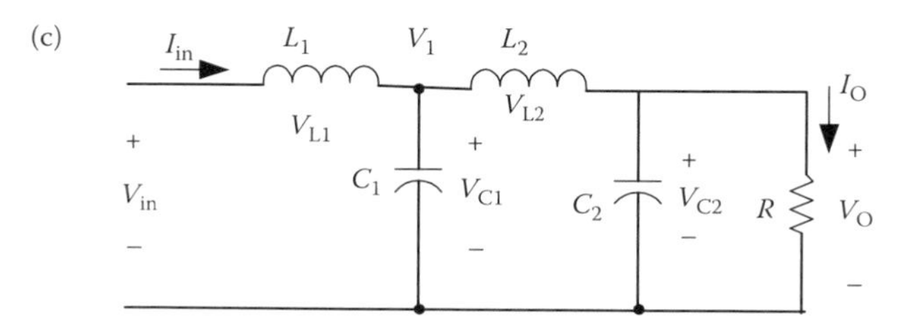

FIGURE 7.32 Two-stage boost circuit: (a) circuit diagram, (b) equivalent circuit during switch-on, and (c) equivalent circuit during switch-off. (Reprinted from Luo, F. L. and Ye, H. 2006. *Essential DC/DC Converters*. Boca Raton: Taylor & Francis Group LLC, p. 314. With permission.)

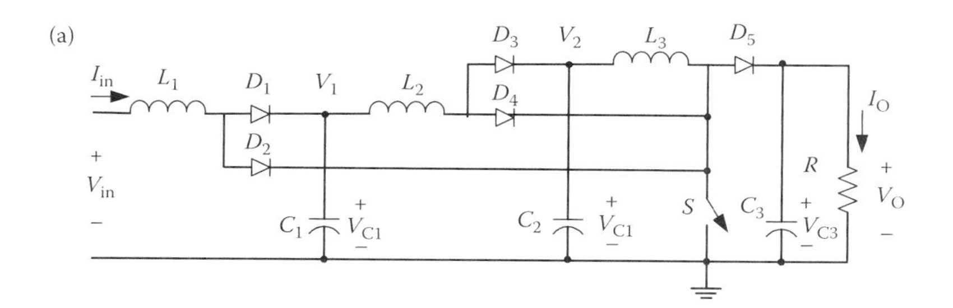

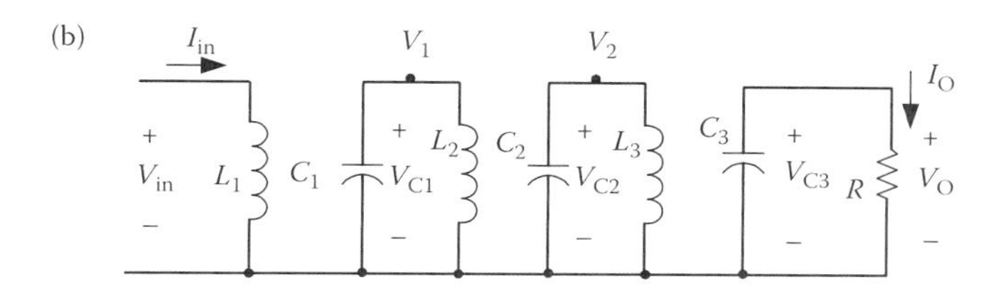

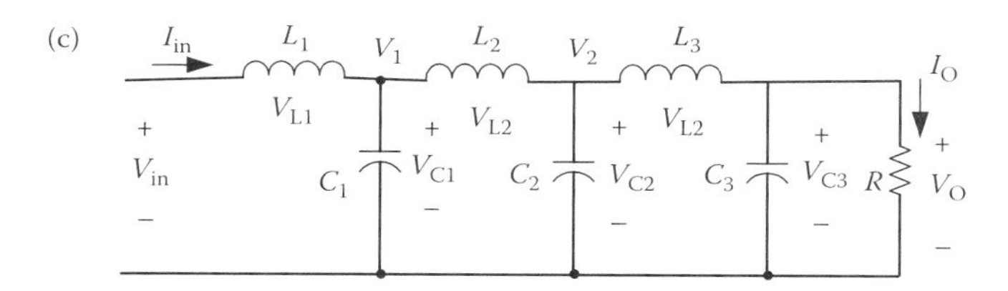

FIGURE 7.33 Three-stage boost circuit: (a) circuit diagram, (b) equivalent circuit during switch-on, and (c) equivalent circuit during switch-off. (Reprinted from Luo, F. L. and Ye, H. 2006. *Essential DC/DC Converters*. Boca Raton: Taylor & Francis Group LLC, p. 316. With permission.)

Therefore, the variation ratio of output voltage v_O is

$$\varepsilon = \frac{\Delta v_O/2}{V_O} = \frac{k}{2RfC}.$$

7.4.1.2 Two-Stage Boost Circuit

The two-stage boost circuit is derived from the elementary boost converter by adding the parts (L_2-D_2-D_3-C_2). Its circuit diagram and equivalent circuits during switch-on and switch-off periods are shown in Figure 7.32.

The voltage across capacitor C_1 is charged with V_1. As described in the previous section, the voltage V_1 across capacitor C_1 is $V_1 = (1/(1-k))V_{in}$.

The voltage across capacitor C_2 is charged with V_O. The current flowing through inductor L_2 increases with V_1 during the switch-on period kT and decreases with $-(V_O - V_1)$ during the switch-off period $(1-k)T$. Therefore, the ripple of the inductor current i_{L2} is

$$\Delta i_{L2} = \frac{V_1}{L_2}kT = \frac{V_O - V_1}{L_2}(1-k)T, \tag{7.219}$$

$$V_O = \frac{1}{1-k}V_1 = \left(\frac{1}{1-k}\right)^2 V_{in}. \tag{7.220}$$

The voltage transfer gain is

$$G = \frac{V_O}{V_{in}} = \left(\frac{1}{1-k}\right)^2. \quad (7.221)$$

Analogously,

$$\Delta i_{L1} = \frac{V_{in}}{L_1}kT, \qquad I_{L1} = \frac{I_O}{(1-k)^2},$$

$$\Delta i_{L2} = \frac{V_1}{L_2}kT, \qquad I_{L2} = \frac{I_O}{1-k}.$$

Therefore, the variation ratio of current i_{L1} through inductor L_1 is

$$\xi_1 = \frac{\Delta i_{L1}/2}{I_{L1}} = \frac{k(1-k)^2 TV_{in}}{2L_1 I_O} = \frac{k(1-k)^4}{2}\frac{R}{fL_1}, \quad (7.222)$$

the variation ratio of current i_{L2} through inductor L_2 is

$$\xi_2 = \frac{\Delta i_{L2}/2}{I_{L2}} = \frac{k(1-k)TV_1}{2L_2 I_O} = \frac{k(1-k)^2}{2}\frac{R}{fL_2}, \quad (7.223)$$

and the variation ratio of output voltage v_O is

$$\varepsilon = \frac{\Delta v_O/2}{V_O} = \frac{k}{2RfC_2}. \quad (7.224)$$

7.4.1.3 Three-Stage Boost Circuit

The three-stage boost circuit is derived from the two-stage boost circuit by twice repeating the parts (L_2-D_2-D_3-C_2). Its circuit diagram and equivalent circuits during switch-on and switch-off periods are shown in Figure 7.33.

The voltage across capacitor C_1 is charged with V_1. As described in the previous section, the voltage V_1 across capacitor C_1 is $V_1 = (1/(1-k))V_{in}$, and the voltage V_2 across capacitor C_2 is $V_2 = (1/(1-k))^2 V_{in}$.

The voltage across capacitor C_3 is charged with V_O. The current flowing through inductor L_3 increases with V_2 during the switch-on period kT and decreases with $-(V_O - V_2)$ during the switch-off period $(1-k)T$. Therefore, the ripple of the inductor current i_{L3} is

$$\Delta i_{L3} = \frac{V_2}{L_3}kT = \frac{V_O - V_2}{L_3}(1-k)T, \quad (7.225)$$

$$V_O = \frac{1}{1-k}V_2 = \left(\frac{1}{1-k}\right)^2 V_1 = \left(\frac{1}{1-k}\right)^3 V_{in}. \quad (7.226)$$

The voltage transfer gain is

$$G = \frac{V_O}{V_{in}} = \left(\frac{1}{1-k}\right)^3. \quad (7.227)$$

Analogously,

$$\Delta i_{L1} = \frac{V_{in}}{L_1} kT, \qquad I_{L1} = \frac{I_O}{(1-k)^3},$$

$$\Delta i_{L2} = \frac{V_1}{L_2} kT, \qquad I_{L2} = \frac{I_O}{(1-k)^2},$$

$$\Delta i_{L3} = \frac{V_2}{L_3} kT, \qquad I_{L3} = \frac{I_O}{1-k}.$$

Therefore, the variation ratio of current i_{L1} through inductor L_1 is

$$\xi_1 = \frac{\Delta i_{L1}/2}{I_{L1}} = \frac{k(1-k)^3 T V_{in}}{2L_1 I_O} = \frac{k(1-k)^6}{2} \frac{R}{fL_1}, \tag{7.228}$$

the variation ratio of current i_{L2} through inductor L_2 is

$$\xi_2 = \frac{\Delta i_{L2}/2}{I_{L2}} = \frac{k(1-k)^2 T V_1}{2L_2 I_O} = \frac{k(1-k)^4}{2} \frac{R}{fL_2}, \tag{7.229}$$

the variation ratio of current i_{L3} through inductor L_3 is

$$\xi_3 = \frac{\Delta i_{L3}/2}{I_{L3}} = \frac{k(1-k) T V_2}{2L_3 I_O} = \frac{k(1-k)^2}{2} \frac{R}{fL_3}, \tag{7.230}$$

and the variation ratio of output voltage v_O is

$$\varepsilon = \frac{\Delta v_O/2}{V_O} = \frac{k}{2RfC_3}. \tag{7.231}$$

Example 7.4

A three-stage boost converter in Figure 7.33a has $V_{in} = 20$V, all inductors have 10 mH, all capacitors have 20 μF, $R = 400\,\Omega$, $f = 50$ kHz, and conduction duty cycle $k = 0.6$. Calculate the variation ratio of current I_{L1}, and the output voltage and its variation ratio.

SOLUTION

From Equation 7.228, we can obtain the variation ratio of current i_{L1}:

$$\xi_1 = \frac{k(1-k)^6}{2} \frac{R}{fL_1} = \frac{0.6(1-0.6)^6}{2} \frac{400}{50\,\text{k} \times 10\,\text{m}} = 0.00098.$$

From Equation 7.226, we can obtain the output voltage:

$$V_O = \left(\frac{1}{1-k}\right)^3 V_{in} = \left(\frac{1}{1-0.6}\right)^3 \times 20 = 312.5\text{V}.$$

From Equation 7.231, its variation ratio is

$$\varepsilon = \frac{k}{2RfC_3} = \frac{0.6}{2 \times 400 \times 50\,\text{k} \times 20\,\mu} = 0.00075.$$

7.4.1.4 Higher-Stage Boost Circuit

A higher-stage boost circuit can be designed by just multiple repeating of the parts (L_2-D_2-D_3-C_2). For the nth stage boost circuit, the final output voltage across capacitor C_n is

$$V_O = \left(\frac{1}{1-k}\right)^n V_{in}.$$

The voltage transfer gain is

$$G = \frac{V_O}{V_{in}} = \left(\frac{1}{1-k}\right)^n, \tag{7.232}$$

the variation ratio of current i_{Li} through inductor L_i ($i = 1, 2, 3, \ldots, n$) is

$$\xi_i = \frac{\Delta i_{Li}/2}{I_{Li}} = \frac{k(1-k)^{2(n-i+1)}}{2}\frac{R}{fL_i}, \tag{7.233}$$

and the variation ratio of output voltage v_O is

$$\varepsilon = \frac{\Delta v_O/2}{V_O} = \frac{k}{2RfC_n}. \tag{7.234}$$

7.4.2 Additional Series

All circuits of P/O cascaded boost converters—additional series—are derived from the corresponding circuits of the main series by adding a DEC.

The first three stages of this series are shown in Figures 7.34 through 7.36. For ease of understanding, they are called the elementary additional circuit, the two-stage additional circuit, and the three-stage additional circuit, respectively, and are numbered $n = 1$, 2, and 3, respectively.

7.4.2.1 Elementary Boost Additional (Double) Circuit

The elementary boost additional circuit is derived from the elementary boost converter by adding a DEC. Its circuit and switch-on and switch-off equivalent circuits are shown in Figure 7.34.

The voltage across capacitors C_1 and C_{11} is charged with V_1 and the voltage across capacitor C_{12} is charged with $V_O = 2V_1$. The current i_{L1} flowing through inductor L_1 increases with V_{in} during the switch-on period kT and decreases with $-(V_1 - V_{in})$ during the switch-off period $(1-k)T$. Therefore,

$$\Delta i_{L1} = \frac{V_{in}}{L_1}kT = \frac{V_1 - V_{in}}{L_1}(1-k)T \tag{7.235}$$

$$V_1 = \frac{1}{1-k}V_{in}.$$

The output voltage is

$$V_O = 2V_1 = \frac{2}{1-k}V_{in}. \tag{7.236}$$

(a)

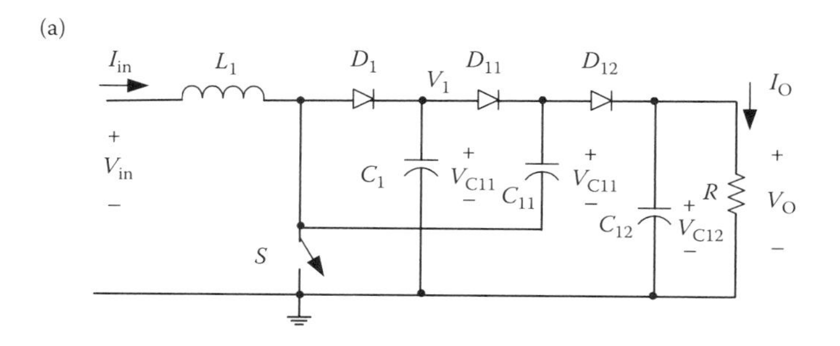

(b)

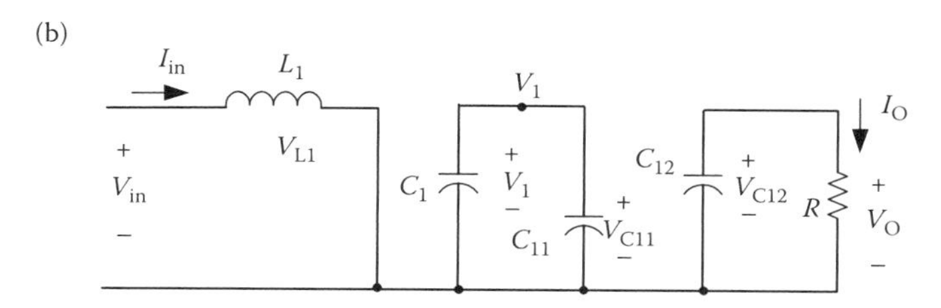

(c)

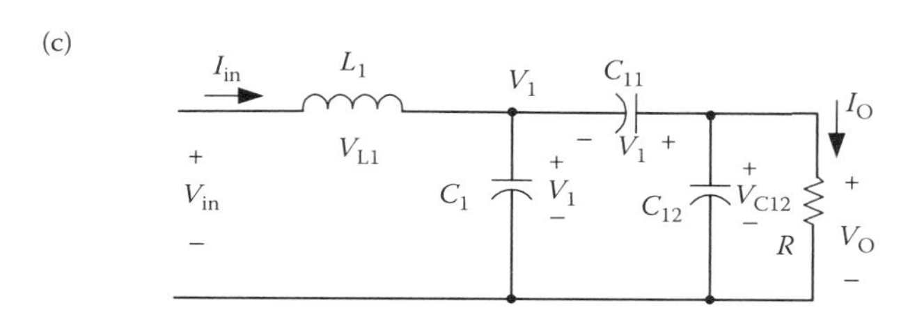

FIGURE 7.34 Elementary boost additional circuit: (a) circuit diagram, (b) equivalent circuit during switch-on, and (c) equivalent circuit during switch-off. (Reprinted from Luo, F. L. and Ye, H. 2006. *Essential DC/DC Converters*. Boca Raton: Taylor & Francis Group LLC, p. 319. With permission.)

The voltage transfer gain is

$$G = \frac{V_O}{V_{in}} = \frac{2}{1-k} \tag{7.237}$$

and

$$i_{in} = I_{L1} = \frac{2}{1-k} I_O. \tag{7.238}$$

The variation ratio of current i_{L1} through inductor L_1 is

$$\xi_1 = \frac{\Delta i_{L1}/2}{I_{L1}} = \frac{k(1-k)TV_{in}}{4L_1 I_O} = \frac{k(1-k)^2}{8}\frac{R}{fL_1}. \tag{7.239}$$

The ripple voltage of output voltage v_O is

$$\Delta v_O = \frac{\Delta Q}{C_{12}} = \frac{I_O kT}{C_{12}} = \frac{k}{fC_{12}}\frac{V_O}{R}.$$

Therefore, the variation ratio of output voltage v_O is

$$\varepsilon = \frac{\Delta v_O/2}{V_O} = \frac{k}{2RfC_{12}}. \tag{7.240}$$

7.4.2.2 Two-Stage Boost Additional Circuit

The two-stage additional boost circuit is derived from the two-stage boost circuit by adding a DEC. Its circuit diagram and switch-on and switch-off equivalent circuits are shown in Figure 7.35.

The voltage across capacitor C_1 is charged with V_1. As described in the previous section, the voltage V_1 across capacitor C_1 is $V_1 = (1/(1-k))V_{in}$.

The voltage across capacitors C_2 and C_{11} is charged with V_2 and the voltage across capacitor C_{12} is charged with V_O. The current flowing through inductor L_2 increases with V_1 during the switch-on period kT and decreases with $-(V_2 - V_1)$ during the switch-off period $(1-k)T$. Therefore, the ripple of the inductor current i_{L2} is

$$\Delta i_{L2} = \frac{V_1}{L_2}kT = \frac{V_2 - V_1}{L_2}(1-k)T, \tag{7.241}$$

$$V_2 = \frac{1}{1-k}V_1 = \left(\frac{1}{1-k}\right)^2 V_{in}. \tag{7.242}$$

The output voltage is

$$V_O = 2V_2 = \frac{2}{1-k}V_1 = 2\left(\frac{1}{1-k}\right)^2 V_{in}. \tag{7.243}$$

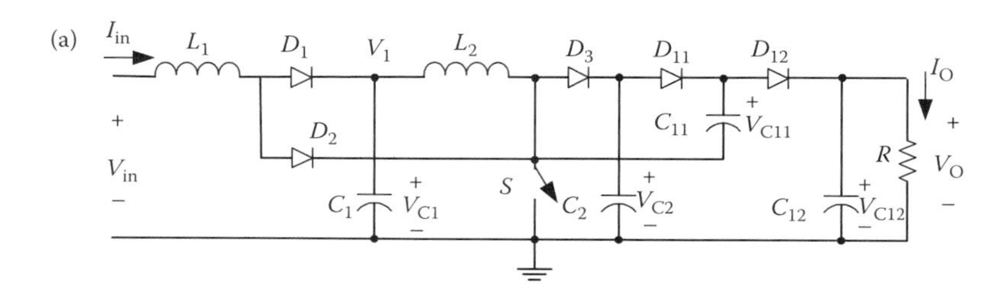

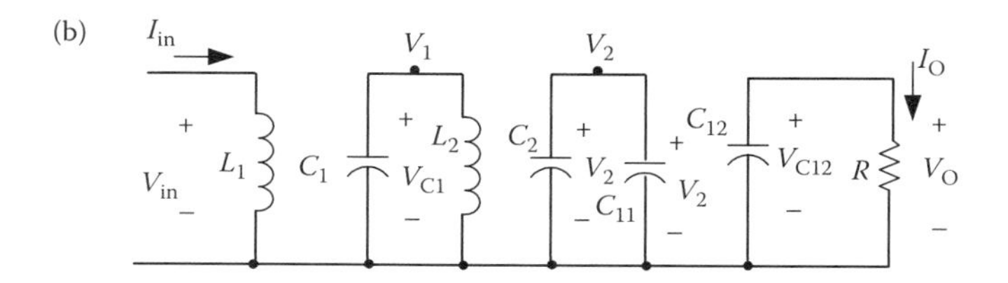

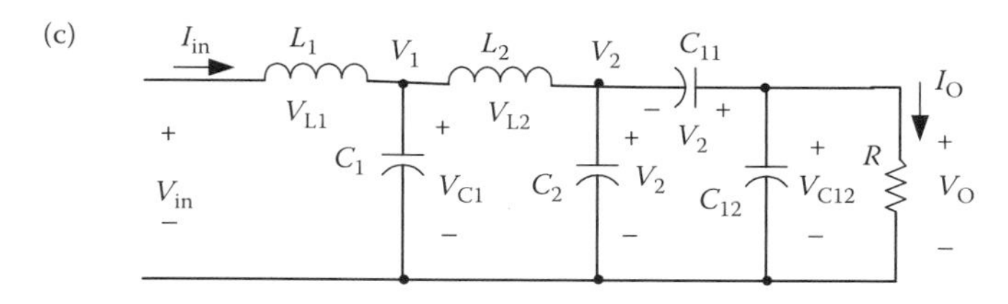

FIGURE 7.35 Two-stage additional boost circuit: (a) circuit diagram, (b) equivalent circuit during switch-on, and (c) equivalent circuit during switch-off. (Reprinted from Luo, F. L. and Ye, H. 2006. *Essential DC/DC Converters*. Boca Raton: Taylor & Francis Group LLC, p. 321. With permission.)

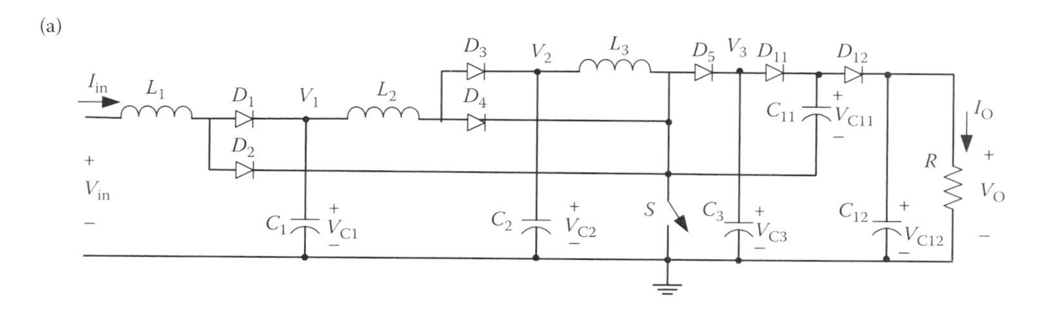

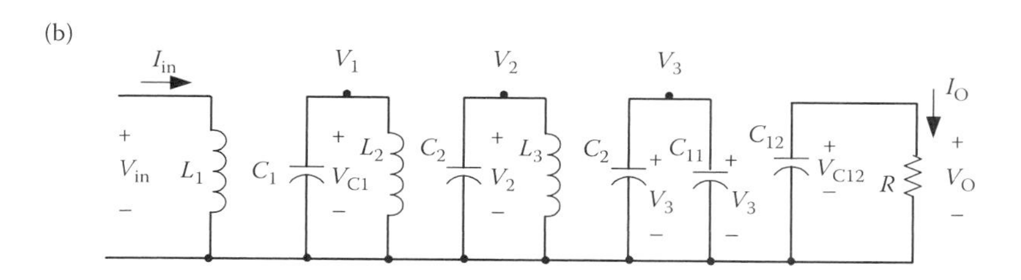

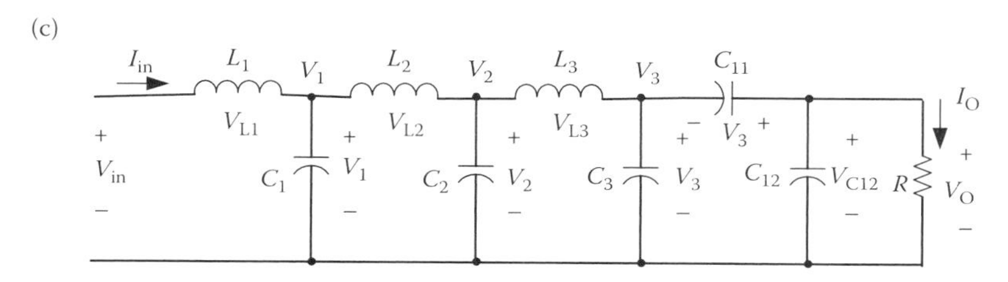

FIGURE 7.36 Three-stage additional boost circuit: (a) circuit diagram, (b) equivalent circuit during switch-on, and (c) equivalent circuit during switch-off. (Reprinted from Luo, F. L. and Ye, H. 2006. *Essential DC/DC Converters*. Boca Raton: Taylor & Francis Group LLC, p. 323. With permission.)

The voltage transfer gain is

$$G = \frac{V_O}{V_{in}} = 2\left(\frac{1}{1-k}\right)^2. \tag{7.244}$$

Analogously,

$$\Delta i_{L1} = \frac{V_{in}}{L_1}kT, \qquad I_{L1} = \frac{2}{(1-k)^2}I_O,$$

$$\Delta i_{L2} = \frac{V_1}{L_2}kT, \qquad I_{L2} = \frac{2I_O}{1-k}.$$

Therefore, the variation ratio of current i_{L1} through inductor L_1 is

$$\xi_1 = \frac{\Delta i_{L1}/2}{I_{L1}} = \frac{k(1-k)^2 TV_{in}}{4L_1 I_O} = \frac{k(1-k)^4}{8}\frac{R}{fL_1} \tag{7.245}$$

and the variation ratio of current i_{L2} through inductor L_2 is

$$\xi_2 = \frac{\Delta i_{L2}/2}{I_{L2}} = \frac{k(1-k)TV_1}{4L_2 I_O} = \frac{k(1-k)^2}{8}\frac{R}{fL_2}. \quad (7.246)$$

The ripple voltage of output voltage v_O is

$$\Delta v_O = \frac{\Delta Q}{C_{12}} = \frac{I_O kT}{C_{12}} = \frac{k}{fC_{12}}\frac{V_O}{R}.$$

Therefore, the variation ratio of output voltage v_O is

$$\varepsilon = \frac{\Delta v_O/2}{V_O} = \frac{k}{2RfC_{12}}. \quad (7.247)$$

7.4.2.3 Three-Stage Boost Additional Circuit

This circuit is derived from the three-stage boost circuit by adding a DEC. Its circuit diagram and equivalent circuits during switch-on and switch-off periods are shown in Figure 7.36.

The voltage across capacitor C_1 is charged with V_1. As described in the previous section, the voltage V_1 across capacitor C_1 is $V_1 = (1/(1-k))V_{in}$, and the voltage V_2 across capacitor C_2 is $V_2 = (1/(1-k))^2 V_{in}$.

The voltage across capacitors C_3 and C_{11} is charged with V_3. The voltage across capacitor C_{12} is charged with V_O. The current flowing through inductor L_3 increases with voltage V_2 during the switch-on period kT and decreases with $-(V_3 - V_2)$ during the switch-off period $(1-k)T$. Therefore,

$$\Delta i_{L3} = \frac{V_2}{L_3}kT = \frac{V_3 - V_2}{L_3}(1-k)T, \quad (7.248)$$

and

$$V_3 = \frac{1}{1-k}V_2 = \left(\frac{1}{1-k}\right)^2 V_1 = \left(\frac{1}{1-k}\right)^3 V_{in}. \quad (7.249)$$

The output voltage is

$$V_O = 2V_3 = 2\left(\frac{1}{1-k}\right)^3 V_{in}. \quad (7.250)$$

The voltage transfer gain is

$$G = \frac{V_O}{V_{in}} = 2\left(\frac{1}{1-k}\right)^3. \quad (7.251)$$

Analogously,

$$\Delta i_{L1} = \frac{V_{in}}{L_1}kT, \quad I_{L1} = \frac{2}{(1-k)^3}I_O,$$

$$\Delta i_{L2} = \frac{V_1}{L_2}kT, \quad I_{L2} = \frac{2}{(1-k)^2}I_O,$$

$$\Delta i_{L3} = \frac{V_2}{L_3}kT, \quad I_{L3} = \frac{2I_O}{1-k}.$$

Therefore, the variation ratio of current i_{L1} through inductor L_1 is

$$\xi_1 = \frac{\Delta i_{L1}/2}{I_{L1}} = \frac{k(1-k)^3 T V_{in}}{4L_1 I_O} = \frac{k(1-k)^6}{8}\frac{R}{fL_1}, \tag{7.252}$$

the variation ratio of current i_{L2} through inductor L_2 is

$$\xi_2 = \frac{\Delta i_{L2}/2}{I_{L2}} = \frac{k(1-k)^2 T V_1}{4L_2 I_O} = \frac{k(1-k)^4}{8}\frac{R}{fL_2}, \tag{7.253}$$

and the variation ratio of current i_{L3} through inductor L_3 is

$$\xi_3 = \frac{\Delta i_{L3}/2}{I_{L3}} = \frac{k(1-k) T V_2}{4L_3 I_O} = \frac{k(1-k)^2}{8}\frac{R}{fL_3}. \tag{7.254}$$

The ripple voltage of output voltage v_O is

$$\Delta v_O = \frac{\Delta Q}{C_{12}} = \frac{I_O kT}{C_{12}} = \frac{k}{fC_{12}}\frac{V_O}{R}.$$

Therefore, the variation ratio of output voltage v_O is

$$\varepsilon = \frac{\Delta v_O/2}{V_O} = \frac{k}{2RfC_{12}}. \tag{7.255}$$

7.4.2.4 Higher-Stage Boost Additional Circuit

A higher-stage boost additional circuit can be designed by just multiple repeating of the parts (L_2-D_2-D_3-C_2). For the nth stage additional circuit, the final output voltage is

$$V_O = 2\left(\frac{1}{1-k}\right)^n V_{in}.$$

The voltage transfer gain is

$$G = \frac{V_O}{V_{in}} = 2\left(\frac{1}{1-k}\right)^n. \tag{7.256}$$

Analogously, the variation ratio of current i_{Li} through inductor L_i ($i = 1, 2, 3, \ldots, n$) is

$$\xi_i = \frac{\Delta i_{Li}/2}{I_{Li}} = \frac{k(1-k)^{2(n-i+1)}}{8}\frac{R}{fL_i} \tag{7.257}$$

and the variation ratio of output voltage v_O is

$$\varepsilon = \frac{\Delta v_O/2}{V_O} = \frac{k}{2RfC_{12}}. \tag{7.258}$$

7.4.3 Double Series

All circuits of the P/O cascaded boost converters—double series—are derived from the corresponding circuits of the main series by adding the DEC in each stage of the circuit. The first three stages of this series are shown in Figures 7.34, 7.37, and 7.38. For ease of understanding, they are called the elementary double circuit, the two-stage double circuit, and the three-stage double circuit, respectively, and are numbered $n = 1$, 2 and 3, respectively.

7.4.3.1 *Elementary Double Boost Circuit*

From the construction principle, the elementary double boost circuit is derived from the elementary boost converter by adding a DEC. Its circuit and switch-on and switch-off equivalent circuits are shown in Figure 7.34, which is the same as the elementary boost additional circuit.

7.4.3.2 *Two-Stage Double Boost Circuit*

The two-stage double boost circuit is derived from the two-stage boost circuit by adding the DEC in circuits of each stage. Its circuit diagram and switch-on and switch-off equivalent circuits are shown in Figure 7.37.

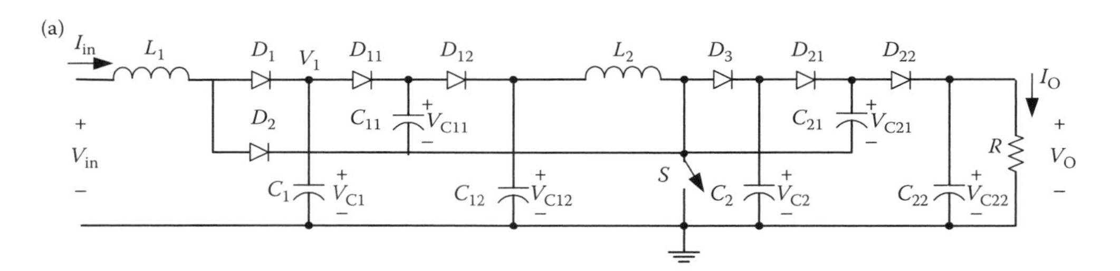

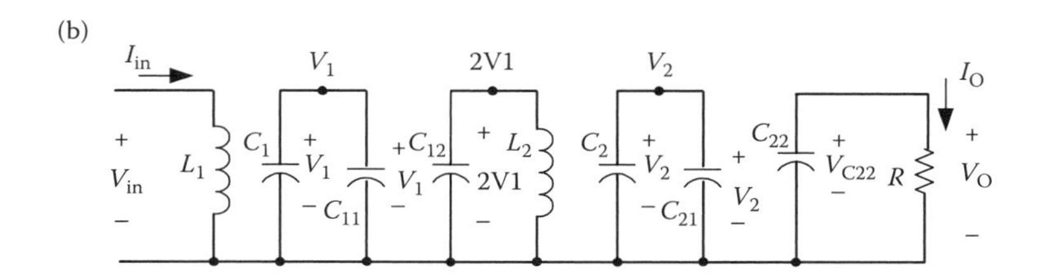

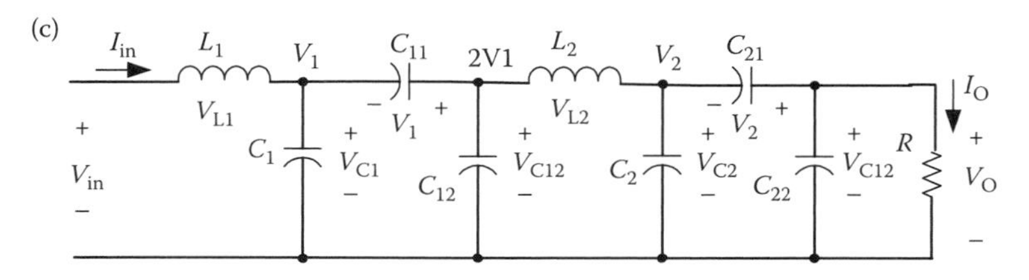

FIGURE 7.37 Two-stage double boost circuit: (a) circuit diagram, (b) equivalent circuit during switch-on, and (c) equivalent circuit during switch-off. (Reprinted from Luo, F. L. and Ye, H. 2006. *Essential DC/DC Converters*. Boca Raton: Taylor & Francis Group LLC, p. 326. With permission.)

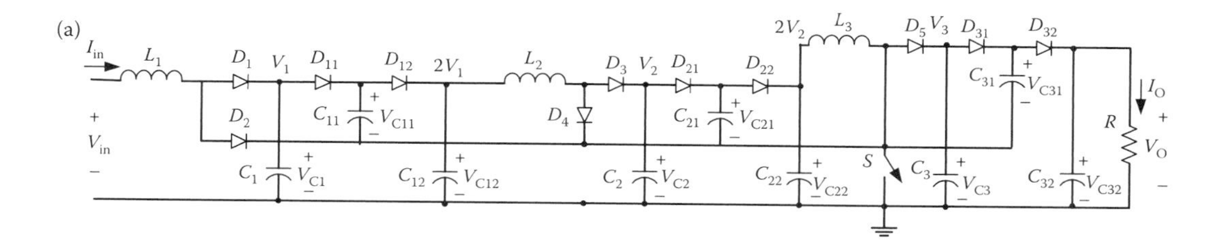
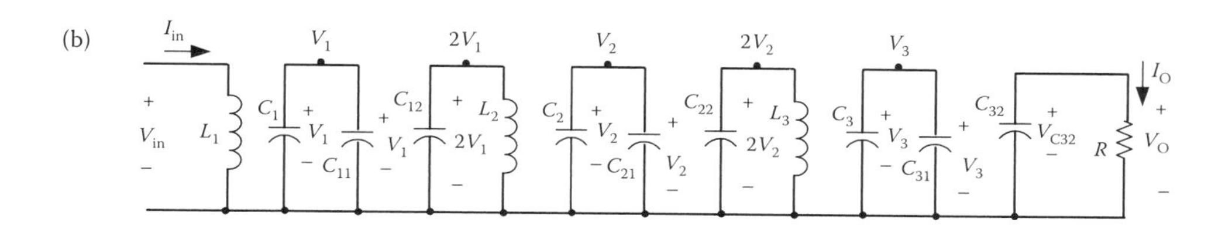
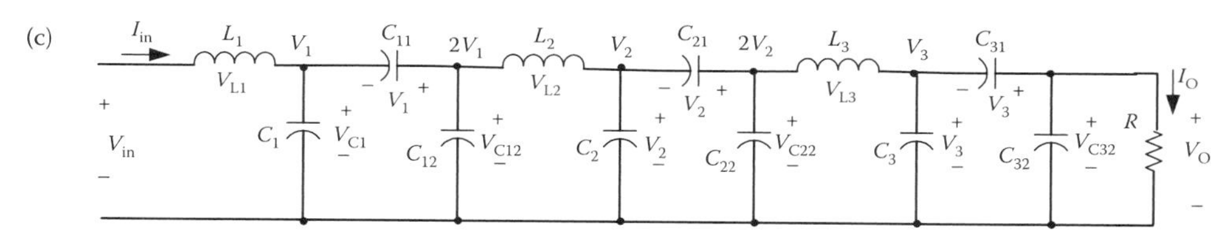

FIGURE 7.38 Three-stage double boost circuit: (a) circuit diagram, (b) equivalent circuit during switch-on, and (c) equivalent circuit during switch-off. (Reprinted from Luo, F. L. and Ye, H. 2006. *Essential DC/DC Converters*. Boca Raton: Taylor & Francis Group LLC, p. 328. With permission.)

The voltage across capacitors C_1 and C_{11} is charged with V_1. As described in the previous section, the voltage V_1 across capacitors C_1 and C_{11} is $V_1 = (1/(1-k))V_{in}$. The voltage across capacitor C_{12} is charged with $2V_1$.

The current flowing through inductor L_2 increases with $2V_1$ during the switch-on period kT and decreases with $-(V_2 - 2V_1)$ during the switch-off period $(1-k)T$. Therefore, the ripple of the inductor current i_{L2} is

$$\Delta i_{L2} = \frac{2V_1}{L_2}kT = \frac{V_2 - 2V_1}{L_2}(1-k)T, \tag{7.259}$$

$$V_2 = \frac{2}{1-k}V_1 = 2\left(\frac{1}{1-k}\right)^2 V_{in}. \tag{7.260}$$

The output voltage is

$$V_O = 2V_2 = \left(\frac{2}{1-k}\right)^2 V_{in}. \tag{7.261}$$

The voltage transfer gain is

$$G = \frac{V_O}{V_{in}} = \left(\frac{2}{1-k}\right)^2. \tag{7.262}$$

Analogously,

$$\Delta i_{L1} = \frac{V_{in}}{L_1} kT, \qquad I_{L1} = \left(\frac{2}{1-k}\right)^2 I_O,$$

$$\Delta i_{L2} = \frac{V_1}{L_2} kT, \qquad I_{L2} = \frac{2I_O}{1-k}.$$

Therefore, the variation ratio of current i_{L1} through inductor L_1 is

$$\xi_1 = \frac{\Delta i_{L1}/2}{I_{L1}} = \frac{k(1-k)^2 TV_{in}}{8L_1 I_O} = \frac{k(1-k)^4}{16} \frac{R}{fL_1} \tag{7.263}$$

and the variation ratio of current i_{L2} through inductor L_2 is

$$\xi_2 = \frac{\Delta i_{L2}/2}{I_{L2}} = \frac{k(1-k) TV_1}{4L_2 I_O} = \frac{k(1-k)^2}{8} \frac{R}{fL_2}. \tag{7.264}$$

The ripple voltage of output voltage v_O is

$$\Delta v_O = \frac{\Delta Q}{C_{22}} = \frac{I_O kT}{C_{22}} = \frac{k}{fC_{22}} \frac{V_O}{R}.$$

Therefore, the variation ratio of output voltage v_O is

$$\varepsilon = \frac{\Delta v_O/2}{V_O} = \frac{k}{2RfC_{22}}. \tag{7.265}$$

7.4.3.3 Three-Stage Double Boost Circuit

This circuit is derived from the three-stage boost circuit by adding the DEC in each stage circuit. Its circuit diagram and equivalent circuits during switch-on and switch-off periods are shown in Figure 7.38.

The voltage across capacitors C_1 and C_{11} is charged with V_1. As described in the previous section, the voltage V_1 across capacitors C_1 and C_{11} is $V_1 = (1/(1-k))V_{in}$, and the voltage V_2 across capacitors C_2 and C_{12} is $V_2 = 2(1/(1-k))^2 V_{in}$.

The voltage across capacitor C_{22} is $2V_2 = (2/(1-k))^2 V_{in}$. The voltage across capacitors C_3 and C_{31} is charged with V_3. The voltage across capacitor C_{12} is charged with V_O. The current flowing through inductor L_3 increases with V_2 during the switch-on period kT and decreases with $-(V_3 - 2V_2)$ during the switch-off period $(-k)T$. Therefore,

$$\Delta i_{L3} = \frac{2V_2}{L_3} kT = \frac{V_3 - 2V_2}{L_3} (1-k)T \tag{7.266}$$

and

$$V_3 = \frac{2V_2}{(1-k)} = \frac{4}{(1-k)^3} V_{in}. \tag{7.267}$$

The output voltage is

$$V_O = 2V_3 = \left(\frac{2}{1-k}\right)^3 V_{in}. \tag{7.268}$$

The voltage transfer gain is

$$G = \frac{V_O}{V_{in}} = \left(\frac{2}{1-k}\right)^3. \tag{7.269}$$

Analogously,

$$\Delta i_{L1} = \frac{V_{in}}{L_1}kT, \qquad I_{L1} = \frac{8}{(1-k)^3}I_O$$
$$\Delta i_{L2} = \frac{V_1}{L_2}kT, \qquad I_{L2} = \frac{4}{(1-k)^2}I_O$$
$$\Delta i_{L3} = \frac{V_2}{L_3}kT, \qquad I_{L3} = \frac{2I_O}{1-k}.$$

Therefore, the variation ratio of current i_{L1} through inductor L_1 is

$$\xi_1 = \frac{\Delta i_{L1}/2}{I_{L1}} = \frac{k(1-k)^3 TV_{in}}{16L_1 I_O} = \frac{k(1-k)^6}{128}\frac{R}{fL_1}, \tag{7.270}$$

the variation ratio of current i_{L2} through inductor L_2 is

$$\xi_2 = \frac{\Delta i_{L2}/2}{I_{L2}} = \frac{k(1-k)^2 TV_1}{8L_2 I_O} = \frac{k(1-k)^4}{32}\frac{R}{fL_2}, \tag{7.271}$$

and the variation ratio of current i_{L3} through inductor L_3 is

$$\xi_3 = \frac{\Delta i_{L3}/2}{I_{L3}} = \frac{k(1-k) TV_2}{4L_3 I_O} = \frac{k(1-k)^2}{8}\frac{R}{fL_3}. \tag{7.272}$$

The ripple voltage of output voltage v_O is

$$\Delta v_O = \frac{\Delta Q}{C_{32}} = \frac{I_O kT}{C_{32}} = \frac{k}{fC_{32}}\frac{V_O}{R}.$$

Therefore, the variation ratio of output voltage v_O is

$$\varepsilon = \frac{\Delta v_O/2}{V_O} = \frac{k}{2RfC_{32}}. \tag{7.273}$$

7.4.3.4 Higher-Stage Double Boost Circuit

A higher-stage double boost circuit can be derived from the corresponding main series circuit by adding the DEC in each stage circuit. For the nth stage additional circuit, the final output voltage is

$$V_O = \left(\frac{2}{1-k}\right)^n V_{in}.$$

The voltage transfer gain is

$$G = \frac{V_O}{V_{in}} = \left(\frac{2}{1-k}\right)^n. \tag{7.274}$$

Analogously, the variation ratio of current i_{Li} through inductor L_i ($i = 1, 2, 3, \ldots, n$) is

$$\xi_i = \frac{\Delta i_{Li}/2}{I_{Li}} = \frac{k(1-k)^{2(n-i+1)}}{2 \times 2^{2n}} \frac{R}{fL_i}. \tag{7.275}$$

The variation ratio of output voltage v_O is

$$\varepsilon = \frac{\Delta v_O/2}{V_O} = \frac{k}{2RfC_{n2}}. \tag{7.276}$$

7.4.4 Triple Series

All circuits of the P/O cascaded boost converters—triple series—are derived from the corresponding circuits of the double series by adding the DEC twice in circuits of each stage. The first three stages of this series are shown in Figures 7.39 through 7.41. To make it easy to explain, they are called the elementary triple boost circuit, the two-stage triple boost circuit, and the three-stage triple boost circuit, respectively, and are numbered $n = 1, 2,$ and 3, respectively.

7.4.4.1 Elementary Triple Boost Circuit

From the construction principle, the elementary triple boost circuit is derived from the elementary double boost circuit by adding another DEC. Its circuit and switch-on and switch-off equivalent circuits are shown in Figure 7.39.

The output voltage of the first-stage boost circuit is V_1, $V_1 = V_{in}/(1-k)$.

The voltage across capacitors C_1 and C_{11} is charged with V_1 and the voltage across capacitors C_{12} and C_{13} is charged with $V_{C13} = 2V_1$ The current i_{L1} flowing through inductor L_1 increases with V_{in} during the switch-on period kT and decreases with $-(V_1 - V_{in})$ during the switch-off period $(1-k)T$. Therefore,

$$\Delta i_{L1} = \frac{V_{in}}{L_1}kT = \frac{V_1 - V_{in}}{L_1}(1-k)T, \tag{7.277}$$

$$V_1 = \frac{1}{1-k}V_{in}.$$

The output voltage is

$$V_O = V_{C1} + V_{C13} = 3V_1 = \frac{3}{1-k}V_{in}. \tag{7.278}$$

The voltage transfer gain is

$$G = \frac{V_O}{V_{in}} = \frac{3}{1-k}. \tag{7.279}$$

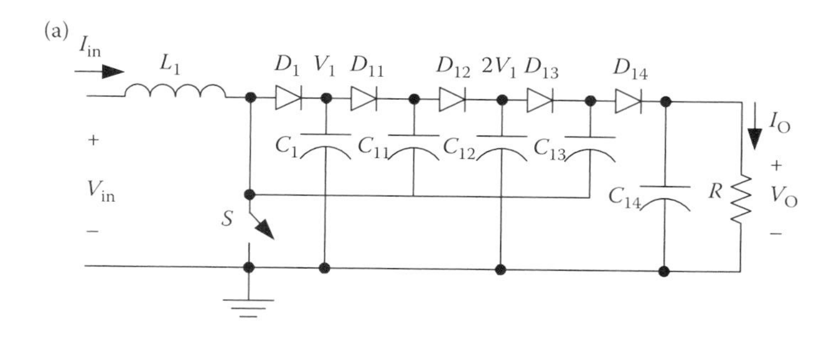

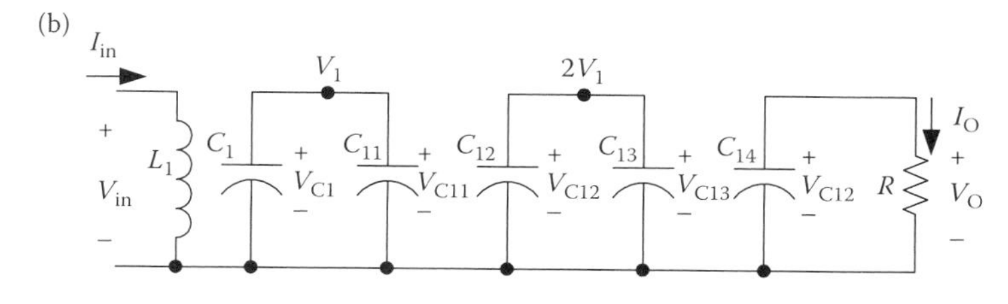

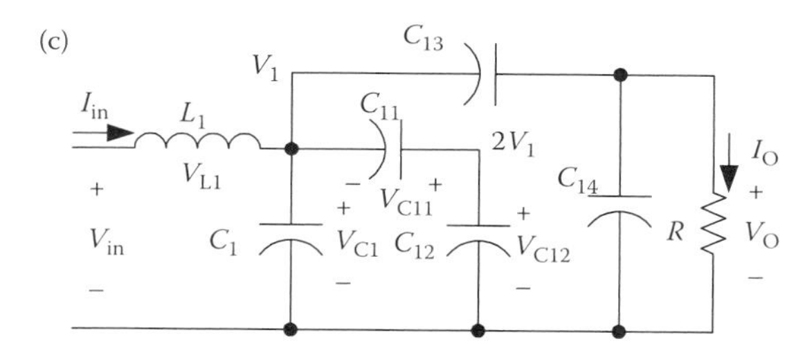

FIGURE 7.39 Elementary triple boost circuit: (a) circuit diagram, (b) equivalent circuit during switch-on, and (c) equivalent circuit during switch-off. (Reprinted from Luo, F. L. and Ye, H. 2006. *Essential DC/DC Converters*. Boca Raton: Taylor & Francis Group LLC, p. 331. With permission.)

7.4.4.2 Two-Stage Triple Boost Circuit

The two-stage triple boost circuit is derived from the two-stage double boost circuit by adding another DEC in circuits of each stage. Its circuit diagram and switch-on and switch-off equivalent circuits are shown in Figure 7.40.

As described in the previous section, the voltage V_1 across capacitors C_1 and C_{11} is $V_1 = (1/(1-k))V_{in}$. The voltage across capacitor C_{14} is charged with $3V_1$.

The voltage across capacitors C_2 and C_{21} is charged with V_2 and the voltage across capacitors C_{22} and C_{23} is charged with $V_{C23} = 2V_2$. The current flowing through inductor L_2 increases with $3V_1$ during the switch-on period kT, and decreases with $-(V_2 - 3V_1)$ during the switch-off period $(1-k)T$. Therefore, the ripple of the inductor current i_{L2} is

$$\Delta i_{L2} = \frac{3V_1}{L_2}kT = \frac{V_2 - 3V_1}{L_2}(1-k)T, \tag{7.280}$$

$$V_2 = \frac{3}{1-k}V_1 = 3\left(\frac{1}{1-k}\right)^2 V_{in}. \tag{7.281}$$

(a)

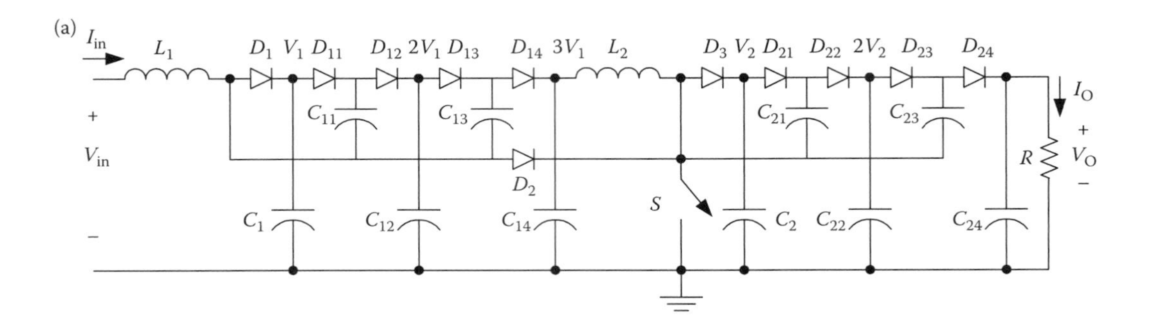

(b)

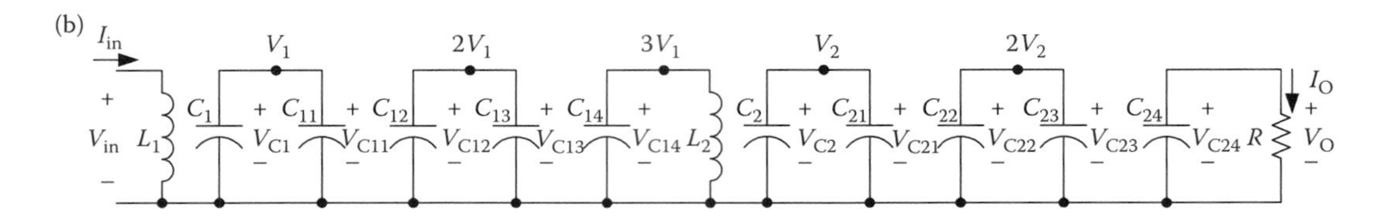

(c)

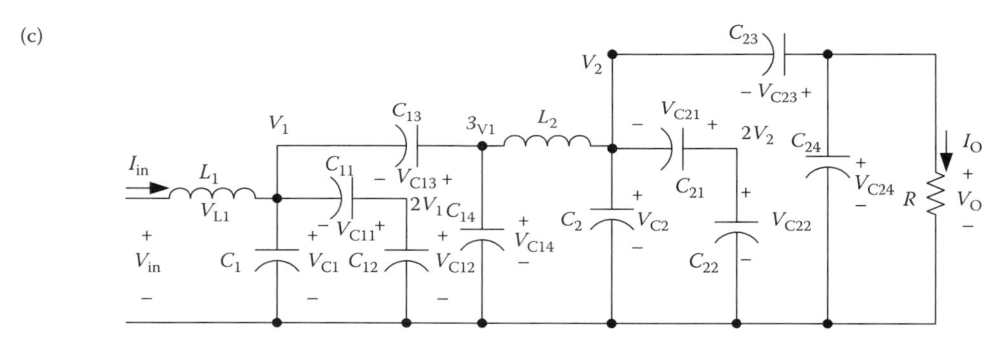

FIGURE 7.40 Two-stage triple boost circuit: (a) circuit diagram, (b) equivalent circuit during switch-on, and (c) equivalent circuit during switch-off. (Reprinted from Luo, F. L. and Ye, H. 2006. *Essential DC/DC Converters*. Boca Raton: Taylor & Francis Group LLC, p. 333. With permission.)

The output voltage is

$$V_O = V_{C2} + V_{C23} = 3V_2 = \left(\frac{3}{1-k}\right)^2 V_{in}. \tag{7.282}$$

The voltage transfer gain is

$$G = \frac{V_O}{V_{in}} = \left(\frac{3}{1-k}\right)^2. \tag{7.283}$$

Analogously,

$$\Delta i_{L1} = \frac{V_{in}}{L_1}kT, \qquad I_{L1} = \left(\frac{2}{1-k}\right)^2 I_O,$$

$$\Delta i_{L2} = \frac{V_1}{L_2}kT, \qquad I_{L2} = \frac{2I_O}{1-k}.$$

Therefore, the variation ratio of current i_{L1} through inductor L_1 is

$$\xi_1 = \frac{\Delta i_{L1}/2}{I_{L1}} = \frac{k(1-k)^2 T V_{in}}{8 L_1 I_O} = \frac{k(1-k)^4}{16} \frac{R}{f L_1} \tag{7.284}$$

and the variation ratio of current i_{L2} through inductor L_2 is

$$\xi_2 = \frac{\Delta i_{L2}/2}{I_{L2}} = \frac{k(1-k) T V_1}{4 L_2 I_O} = \frac{k(1-k)^2}{8} \frac{R}{f L_2}. \tag{7.285}$$

The ripple voltage of output voltage v_O is

$$\Delta v_O = \frac{\Delta Q}{C_{22}} = \frac{I_O k T}{C_{22}} = \frac{k}{f C_{22}} \frac{V_O}{R}.$$

Therefore, the variation ratio of output voltage v_O is

$$\varepsilon = \frac{\Delta v_O/2}{V_O} = \frac{k}{2 R f C_{22}}. \tag{7.286}$$

7.4.4.3 Three-Stage Triple Boost Circuit

This circuit is derived from the three-stage double boost circuit by adding another DEC in each stage circuit. Its circuit diagram and equivalent circuits during switch-on and switch-off periods are shown in Figure 7.41.

As described in the previous section, the voltage V_2 across capacitors C_2 and C_{11} is $V_2 = 3V_1 = (3/(1-k))V_{in}$, and the voltage across capacitor C_{24} is charged with $3V_2$.

The voltage across capacitors C_3 and C_{31} is charged with V_3 and the voltage across capacitors C_{32} and C_{33} is charged with $V_{C33} = 2V_3$. The current flowing through inductor L_3 increases with $3V_2$ during the switch-on period kT and decreases with $-(V_3 - 3V_2)$ during the switch-off period $(1-k)T$. Therefore, the ripple of the inductor current i_{L3} is

$$\Delta i_{L3} = \frac{3V_2}{L_3} kT = \frac{V_3 - 3V_2}{L_3}(1-k)T \tag{7.287}$$

and

$$V_3 = \frac{3}{1-k} V_2 = 9\left(\frac{1}{1-k}\right)^3 V_{in}. \tag{7.288}$$

The output voltage is

$$V_O = V_{C3} + V_{C33} = 3V_3 = \left(\frac{3}{1-k}\right)^3 V_{in}. \tag{7.289}$$

The voltage transfer gain is

$$G = \frac{V_O}{V_{in}} = \left(\frac{3}{1-k}\right)^3. \tag{7.290}$$

(a)

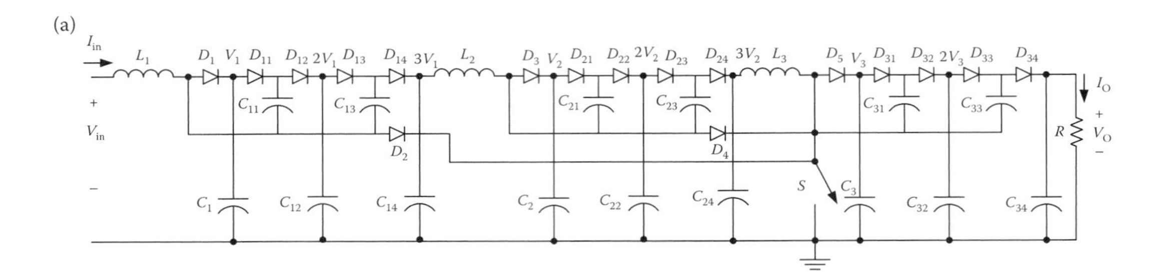

(b)

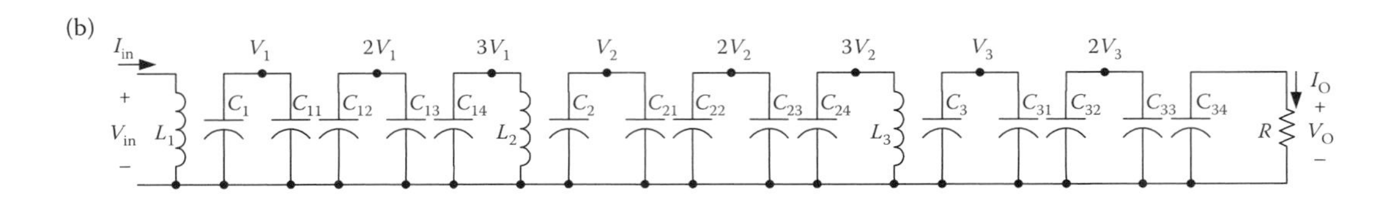

(c)

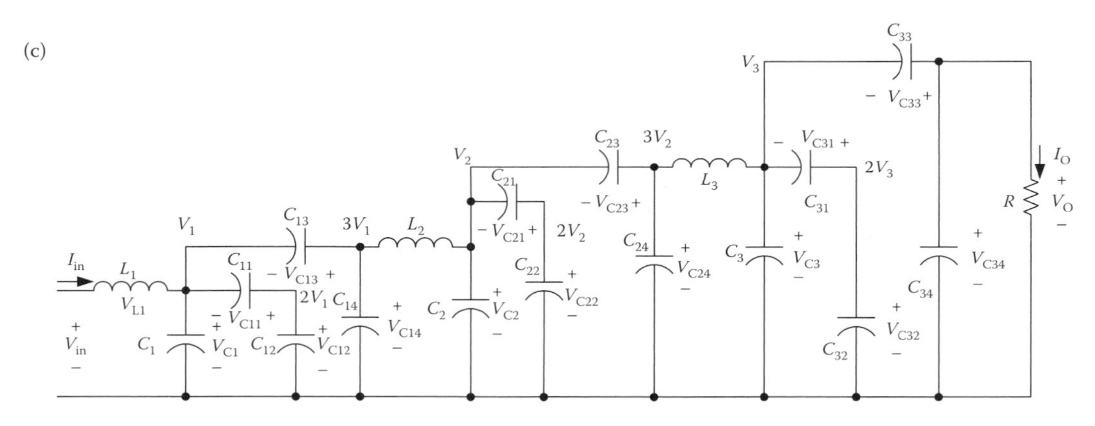

FIGURE 7.41 Three-stage triple boost circuit: (a) circuit diagram, (b) equivalent circuit during switch-on, and (c) equivalent circuit during switch-off. (Reprinted from Luo, F. L. and Ye, H. 2006. *Essential DC/DC Converters*. Boca Raton: Taylor & Francis Group LLC, p. 335. With permission.)

Analogously,

$$\Delta i_{L1} = \frac{V_{in}}{L_1}kT, \qquad I_{L1} = \frac{32}{(1-k)^3}I_O,$$

$$\Delta i_{L2} = \frac{V_1}{L_2}kT, \qquad I_{L2} = \frac{8}{(1-k)^2}I_O,$$

$$\Delta i_{L3} = \frac{V_2}{L_3}kT, \qquad I_{L3} = \frac{2}{1-k}I_O.$$

Therefore, the variation ratio of current i_{L1} through inductor L_1 is

$$\xi_1 = \frac{\Delta i_{L1}/2}{I_{L1}} = \frac{k(1-k)^3 TV_{in}}{64L_1 I_O} = \frac{k(1-k)^6}{12^3}\frac{R}{fL_1}, \tag{7.291}$$

the variation ratio of current i_{L2} through inductor L_2 is

$$\xi_2 = \frac{\Delta i_{L2}/2}{I_{L2}} = \frac{k(1-k)^2 T V_1}{16 L_2 I_O} = \frac{k(1-k)^4}{12^2}\frac{R}{f L_2}, \tag{7.292}$$

and the variation ratio of current i_{L3} through inductor L_3 is

$$\xi_3 = \frac{\Delta i_{L3}/2}{I_{L3}} = \frac{k(1-k) T V_2}{4 L_3 I_O} = \frac{k(1-k)^2}{12}\frac{R}{f L_3}. \tag{7.293}$$

The ripple voltage of output voltage v_O is

$$\Delta v_O = \frac{\Delta Q}{C_{32}} = \frac{I_O k T}{C_{32}} = \frac{k}{f C_{32}}\frac{V_O}{R}.$$

Therefore, the variation ratio of output voltage v_O is

$$\varepsilon = \frac{\Delta v_O/2}{V_O} = \frac{k}{2 R f C_{32}}. \tag{7.294}$$

7.4.4.4 Higher-Stage Triple Boost Circuit

A higher-stage triple boost circuit can be derived from the corresponding circuits of the double boost series by adding another DEC in each stage circuit. For the nth stage additional circuit, the final output voltage is

$$V_O = \left(\frac{3}{1-k}\right)^n V_{in}.$$

The voltage transfer gain is

$$G = \frac{V_O}{V_{in}} = \left(\frac{3}{1-k}\right)^n. \tag{7.295}$$

Analogously, the variation ratio of current i_{Li} through inductor L_i ($i = 1, 2, 3, \ldots, n$) is

$$\xi_i = \frac{\Delta i_{Li}/2}{I_{Li}} = \frac{k(1-k)^{2(n-i+1)}}{12^{(n-i+1)}}\frac{R}{f L_i} \tag{7.296}$$

and the variation ratio of output voltage v_O is

$$\varepsilon = \frac{\Delta v_O/2}{V_O} = \frac{k}{2 R f C_{n2}}. \tag{7.297}$$

7.4.5 Multiple Series

All circuits of P/O cascaded boost converters—multiple series—are derived from the corresponding circuits of the main series by adding the DEC multiple (j) times in circuits of each stage. The first three stages of this series are shown in Figures 7.42 through 7.44. For ease of understanding, they are called the elementary multiple boost circuit, the two-stage multiple boost circuit, and the three-stage multiple boost circuit, respectively, and are numbered $n = 1$, 2, and 3, respectively.

7.4.5.1 Elementary Multiple Boost Circuit

From the construction principle, the elementary multiple boost circuit is derived from the elementary boost converter by adding the DEC multiple (j) times in the circuit. Its circuit and switch-on and switch-off equivalent circuits are shown in Figure 7.42.

The voltage across capacitors C_1 and C_{11} is charged with V_1 and the voltage across capacitors C_{12} and C_{13} is charged with $V_{C13} = 2V_1$. The voltage across capacitors $C_{1(2j-2)}$ and $C_{1(2j-1)}$ is charged with $V_{C1}(2j-1) = jV_1$. The current i_{L1} flowing through inductor L_1 increases with V_{in} during the switch-on period kT and decreases with $-(V_1 - V_{in})$ during the switch-off period $(1-k)T$. Therefore,

$$\Delta i_{L1} = \frac{V_{in}}{L_1}kT = \frac{V_1 - V_{in}}{L_1}(1-k)T, \quad (7.298)$$

$$V_1 = \frac{1}{1-k}V_{in}. \quad (7.299)$$

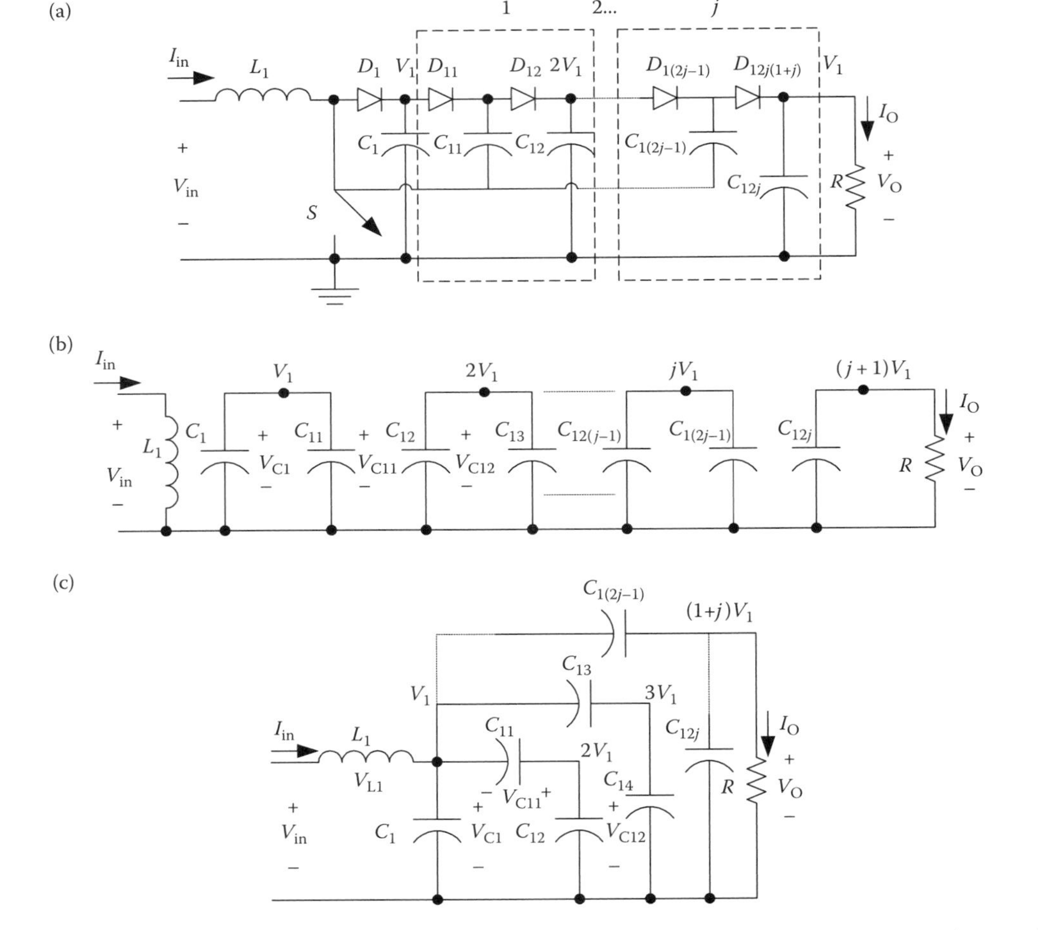

FIGURE 7.42 Elementary multiple boost circuit: (a) circuit diagram, (b) equivalent circuit during switch-on, and (c) equivalent circuit during switch-off. (Reprinted from Luo, F. L. and Ye, H. 2006. *Essential DC/DC Converters*. Boca Raton: Taylor & Francis Group LLC, p. 338. With permission.)

The output voltage is

$$V_O = V_{C1} + V_{C1(2j-1)} = (1+j)V_1 = \frac{1+j}{1-k}V_{in}. \tag{7.300}$$

The voltage transfer gain is

$$G = \frac{V_O}{V_{in}} = \frac{1+j}{1-k}. \tag{7.301}$$

7.4.5.2 Two-Stage Multiple Boost Circuit

The two-stage multiple boost circuit is derived from the two-stage boost circuit by adding multiple (j) DECs in each stage of the circuit. Its circuit diagram and switch-on and switch-off equivalent circuits are shown in Figure 7.43.

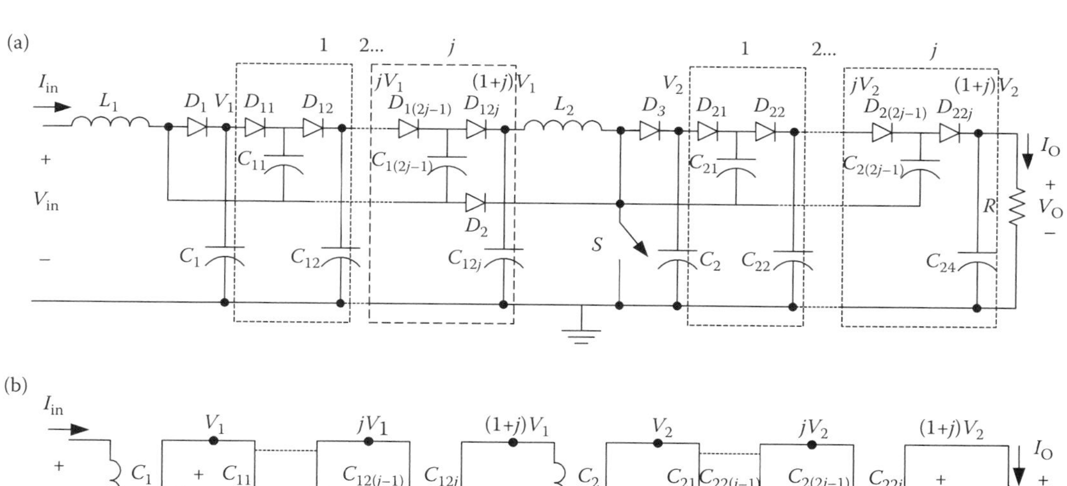

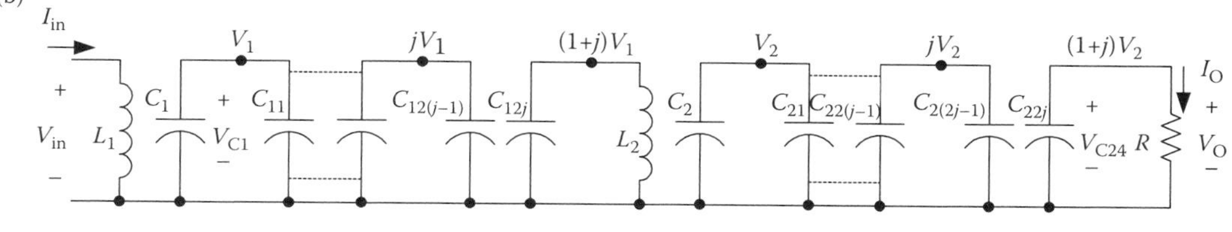

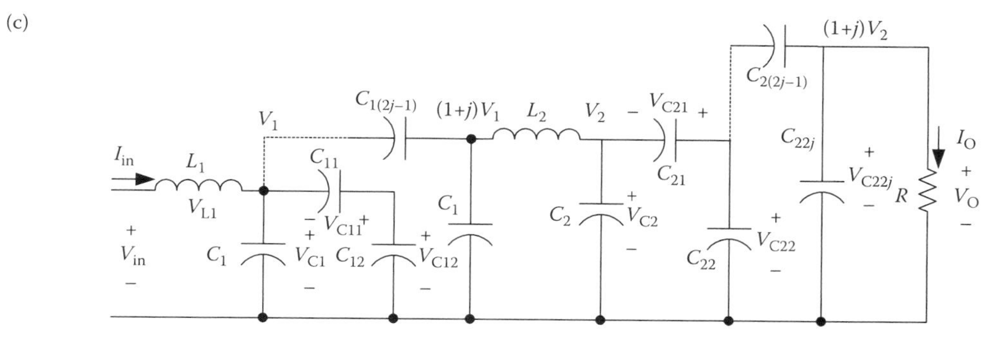

FIGURE 7.43 Two-stage multiple boost circuit: (a) circuit diagram, (b) equivalent circuit during switch-on, and (c) equivalent circuit during switch-off. (Reprinted from Luo, F. L. and Ye, H. 2006. *Essential DC/DC Converters*. Boca Raton: Taylor & Francis Group LLC, p. 340. With permission.)

The voltage across capacitors C_1 and C_{11} is charged with $V_1 = (1/(1-k))V_{in}$. The voltage across capacitor $C_{1(2j)}$ is charged with $(1+j)V_1$.

The current flowing through inductor L_2 increases with $(1+j)V_1$ during the switch-on period kT and decreases with $-[V_2 - (1+j)V_1]$ during the switch-off period $(1-k)T$. Therefore, the ripple of the inductor current i_{L2} is

$$\Delta i_{L2} = \frac{1+j}{L_2}kTV_1 = \frac{V_2 - (1+j)V_1}{L_2}(1-k)T, \tag{7.302}$$

$$V_2 = \frac{1+j}{1-k}V_1 = (1+j)\left(\frac{1}{1-k}\right)^2 V_{in}. \tag{7.303}$$

The output voltage is

$$V_O = V_{C1} + V_{C1(2j-1)} = (1+j)V_2 = \left(\frac{1+j}{1-k}\right)^2 V_{in}. \tag{7.304}$$

The voltage transfer gain is

$$G = \frac{V_O}{V_{in}} = \left(\frac{1+j}{1-k}\right)^2. \tag{7.305}$$

The ripple voltage of output voltage v_O is

$$\Delta v_O = \frac{\Delta Q}{C_{22j}} = \frac{I_O kT}{C_{22j}} = \frac{k}{fC_{22j}}\frac{V_O}{R}.$$

Therefore, the variation ratio of output voltage v_O is

$$\varepsilon = \frac{\Delta v_O/2}{V_O} = \frac{k}{2RfC_{22j}}. \tag{7.306}$$

7.4.5.3 Three-Stage Multiple Boost Circuit

This circuit is derived from the three-stage boost circuit by adding multiple (j) DECs in circuits of each stage. Its circuit diagram and equivalent circuits during switch-on and switch-off periods are shown in Figure 7.44.

The voltage across capacitors C_1 and C_{11} is charged with $V_1 = (1/(1-k))V_{in}$. The voltage across capacitor $C_{1(2j)}$ is charged with $(1+j)V_1$. The voltage V_2 across capacitors C_2 and $C_{2(2j)}$ is charged with $(1+j)V_2$.

The current flowing through inductor L_3 increases with $(1+j)V_2$ during the switch-on period kT and decreases with $-[V_3 - (1+j)V_2]$ during the switch-off period $(1-k)T$. Therefore,

$$\Delta i_{L3} = \frac{1+j}{L_3}kTV_2 = \frac{V_3 - (1+j)V_2}{L_3}(1-k)T \tag{7.307}$$

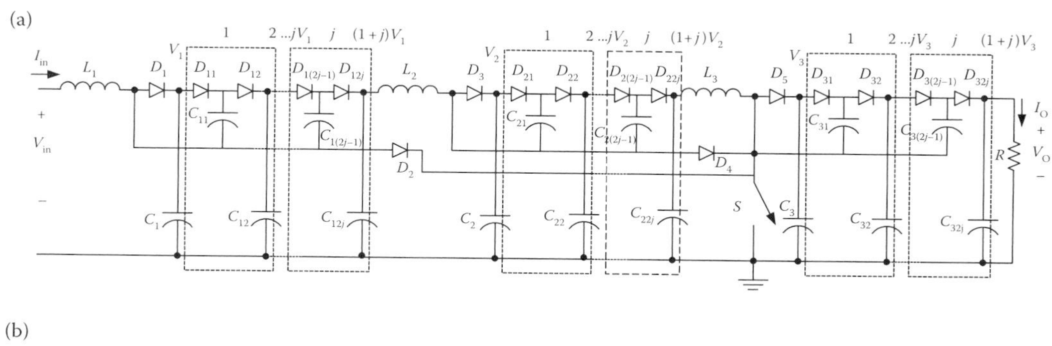

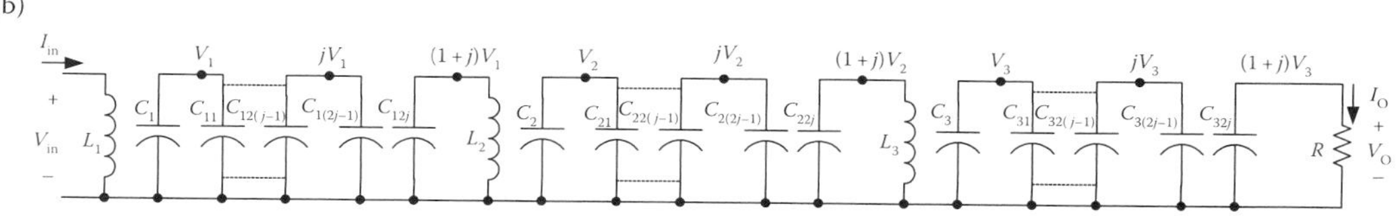

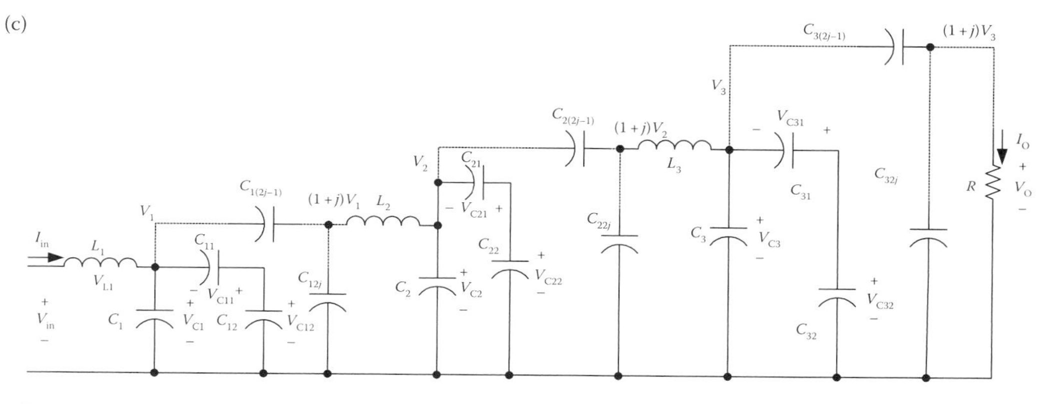

FIGURE 7.44 Three-stage multiple boost circuit: (a) circuit diagram, (b) equivalent circuit during switch-on, and (c) equivalent circuit during switch-off. (Reprinted from Luo, F. L. and Ye, H. 2006. *Essential DC/DC Converters*. Boca Raton: Taylor & Francis Group LLC, p. 342. With permission.)

and

$$V_3 = \frac{(1+j)V_2}{(1-k)} = \frac{(1+j)^2}{(1-k)^3} V_{in}. \tag{7.308}$$

The output voltage is

$$V_O = V_{C3} + V_{C3(2j-1)} = (1+j)V_3 = \left(\frac{1+j}{1-k}\right)^3 V_{in}. \tag{7.309}$$

The voltage transfer gain is

$$G = \frac{V_O}{V_{in}} = \left(\frac{1+j}{1-k}\right)^3. \tag{7.310}$$

The ripple voltage of output voltage v_O is

$$\Delta v_O = \frac{\Delta Q}{C_{32j}} = \frac{I_O kT}{C_{32j}} = \frac{k}{fC_{32j}}\frac{V_O}{R}.$$

Therefore, the variation ratio of output voltage v_O is

$$\varepsilon = \frac{\Delta v_O/2}{V_O} = \frac{k}{2RfC_{32j}}. \tag{7.311}$$

7.4.5.4 Higher-Stage Multiple Boost Circuit

A higher-stage multiple boost circuit is derived from the corresponding circuits of the main series by adding multiple (j) DECs in circuits of each stage. For the nth stage additional circuit, the final output voltage is

$$V_O = \left(\frac{1+j}{1-k}\right)^n V_{in}.$$

The voltage transfer gain is

$$G = \frac{V_O}{V_{in}} = \left(\frac{1+j}{1-k}\right)^n. \tag{7.312}$$

Analogously, the variation ratio of output voltage v_O is

$$\varepsilon = \frac{\Delta v_O/2}{V_O} = \frac{k}{2RfC_{n2j}}. \tag{7.313}$$

7.4.6 Summary of P/O Cascaded Boost Converters

All circuits of the P/O cascaded boost converters can be shown in Figure 7.45 as the family tree.

From the analysis of the previous two sections, the common formula to calculate the output voltage can be presented as

$$V_O = \begin{cases} \left(\frac{1}{1-k}\right)^n V_{in} & \text{main series,} \\ 2*\left(\frac{1}{1-k}\right)^n V_{in} & \text{additional series,} \\ \left(\frac{2}{1-k}\right)^n V_{in} & \text{double series,} \\ \left(\frac{3}{1-k}\right)^n V_{in} & \text{triple series,} \\ \left(\frac{j+1}{1-k}\right)^n V_{in} & \text{multiple } (j) \text{ series.} \end{cases} \tag{7.314}$$

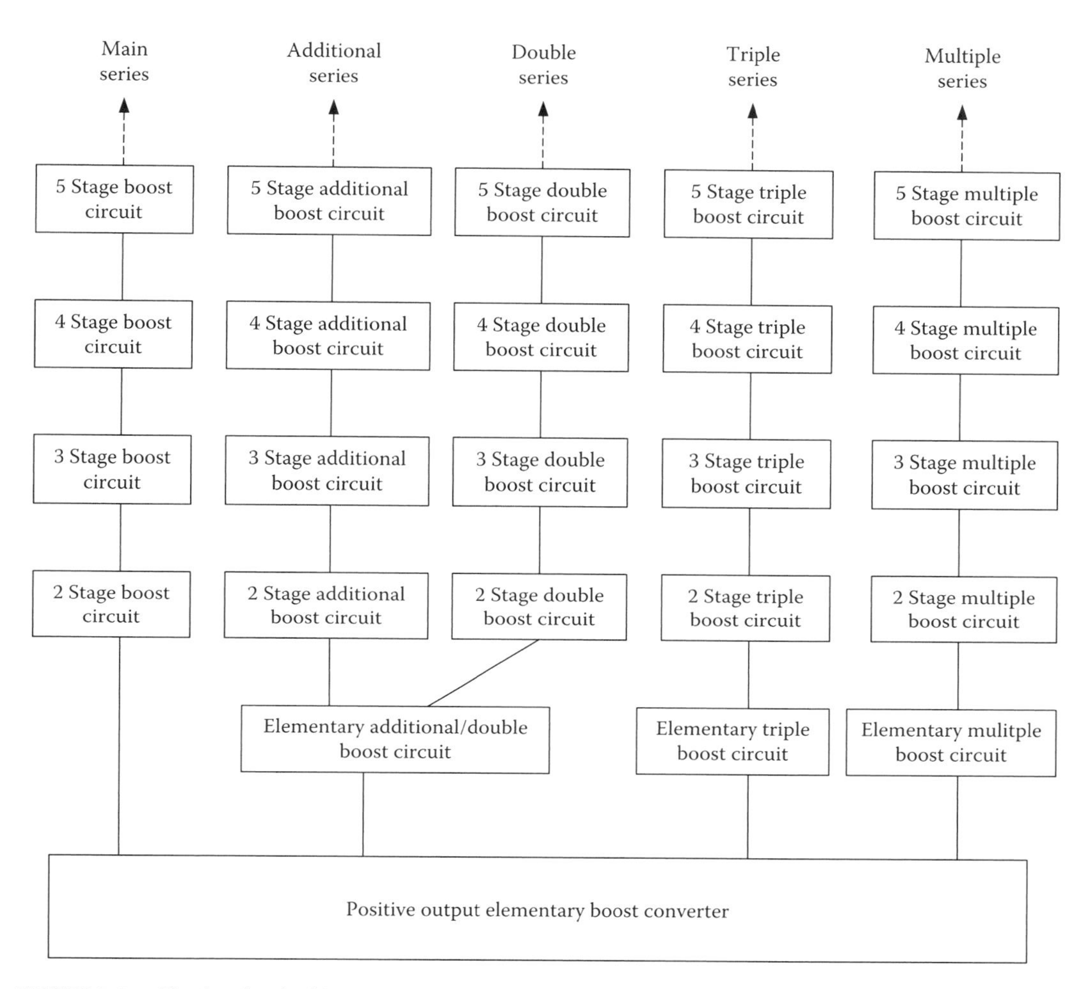

FIGURE 7.45 The family of P/O cascaded boost converters. (Reprinted from Luo, F. L. and Ye, H. 2006. *Essential DC/DC Converters*. Boca Raton: Taylor & Francis Group LLC, p. 344. With permission.)

The voltage transfer gain is

$$V_O = \frac{V_O}{V_{in}} = \begin{cases} \left(\frac{1}{1-k}\right)^n & \text{main series,} \\ 2 * \left(\frac{1}{1-k}\right)^n & \text{additional series,} \\ \left(\frac{2}{1-k}\right)^n & \text{double series,} \\ \left(\frac{3}{1-k}\right)^n & \text{triple series,} \\ \left(\frac{j+1}{1-k}\right)^n & \text{multiple } (j) \text{ series.} \end{cases} \quad (7.315)$$

7.5 N/O Cascaded Boost Converters

This section introduces N/O cascaded boost converters. Like P/O cascaded boost converters, these converters implement the SL technique [1,2].

7.5.1 Main Series

The first three stages of the N/O cascaded boost converters—main series—are shown in Figures 7.46 through 7.48. For ease of understanding, they are called the elementary boost converter, the two-stage boost circuit, and the three-stage boost circuit, respectively, and are numbered $n = 1, 2$, and 3, respectively.

7.5.1.1 N/O Elementary Boost Circuit

The N/O elementary boost converter and its equivalent circuits during switch-on and switch-off periods are shown in Figure 7.46.

The voltage across capacitor C_1 is charged with V_{C1}. The current i_{L1} flowing through inductor L_1 increases with V_{in} during the switch-on period kT and decreases with

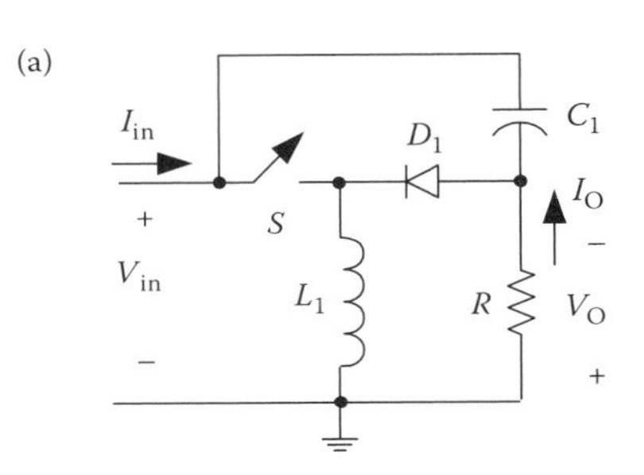

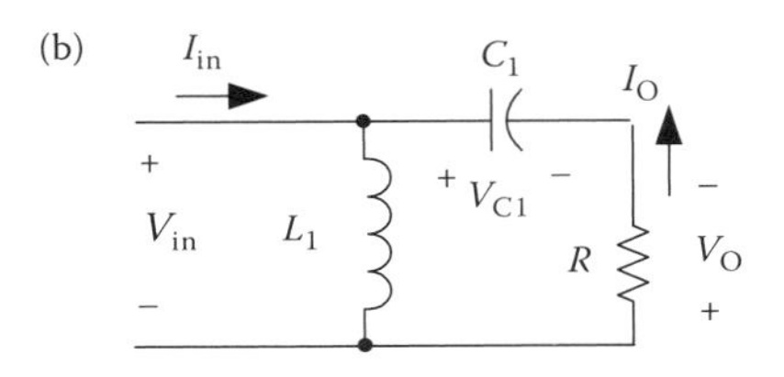

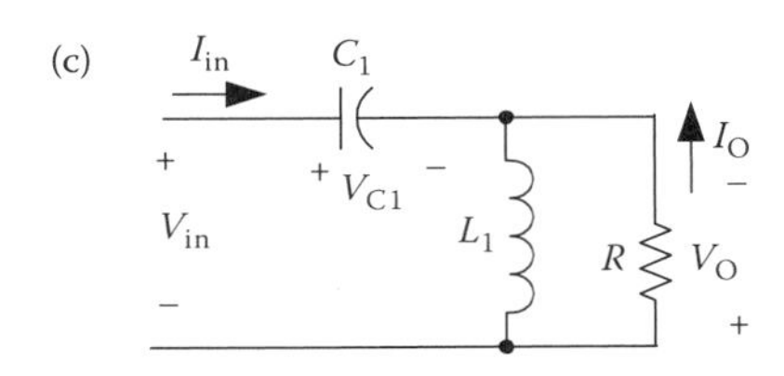

FIGURE 7.46 N/O elementary boost converter: (a) circuit diagram, (b) equivalent circuit during switch-on, and (c) equivalent circuit during switch-off. (Reprinted from Luo, F. L. and Ye, H. 2006. *Essential DC/DC Converters*. Boca Raton: Taylor & Francis Group LLC, p. 352. With permission.)

$-(V_{C1} - V_{in})$ during the switch-off period $(1-k)T$. Therefore, the ripple of the inductor current i_{L1} is

$$\Delta i_{L1} = \frac{V_{in}}{L_1}kT = \frac{V_{C1} - V_{in}}{L_1}(1-k)T, \quad (7.316)$$

$$V_{C1} = \frac{1}{1-k}V_{in},$$

$$V_O = V_{C1} - V_{in} = \frac{k}{1-k}V_{in}. \quad (7.317)$$

The voltage transfer gain is

$$G = \frac{V_O}{V_{in}} = \frac{1}{1-k} - 1. \quad (7.318)$$

The inductor average current is

$$I_{L1} = \frac{1}{1-k}\frac{V_O}{R}. \quad (7.319)$$

The variation ratio of current i_{L1} through inductor L_1 is

$$\xi_1 = \frac{\Delta i_{L1}/2}{I_{L1}} = \frac{kTV_{in}}{2L_1V_O/(1-k)R} = \frac{k(1-k)}{2G}\frac{R}{fL_1} = \frac{(1-k)^2}{2}\frac{R}{fL_1}. \quad (7.320)$$

Usually ξ_1 is small (much lower than unity); this means that this converter works in the continuous mode.

The ripple voltage of output voltage v_O is

$$\Delta v_O = \frac{\Delta Q}{C_1} = \frac{I_O kT}{C_1} = \frac{k}{fC_1}\frac{V_O}{R},$$

since $\Delta Q = I_O kT$.

Therefore, the variation ratio of output voltage v_O is

$$\varepsilon = \frac{\Delta v_O/2}{V_O} = \frac{k}{2RfC_1}. \quad (7.321)$$

7.5.1.2 N/O Two-Stage Boost Circuit

The N/O two-stage boost circuit is derived from the N/O elementary boost converter by adding the parts (L_2-D_2-D_3-C_2). Its circuit diagram and equivalent circuits during switch-on and switch-off periods are shown in Figure 7.47.

The voltage across capacitor C_1 is charged with V_1. As described in the previous section, the voltage V_1 across capacitor C_1 is $V_1 = (1/(1-k))V_{in}$.

The voltage across capacitor C_2 is charged with V_{C2}. The current flowing through inductor L_2 increases with V_1 during the switch-on period kT and decreases with $-(V_{C2} - V_1)$ during

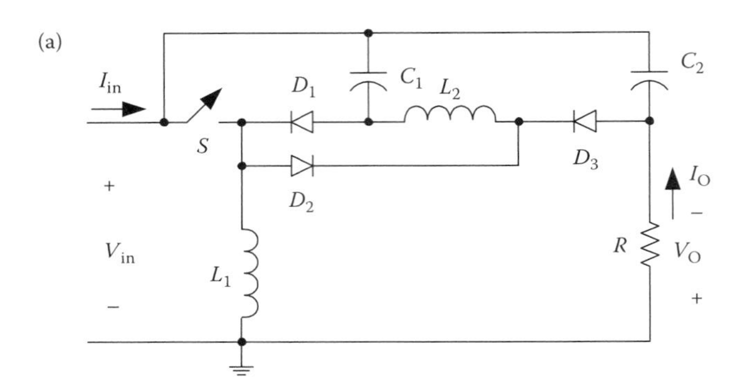

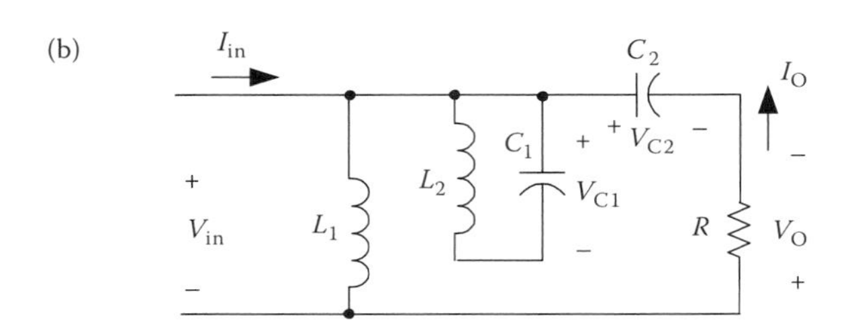

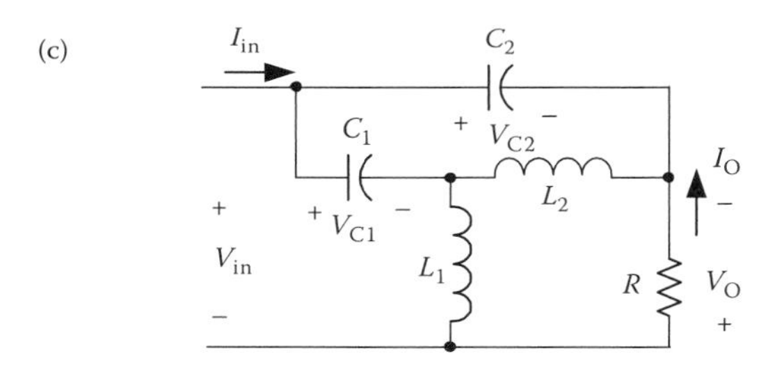

FIGURE 7.47 N/O two-stage boost circuit: (a) circuit diagram, (b) equivalent circuit during switch-on, and (c) equivalent circuit during switch-off. (Reprinted from Luo, F. L. and Ye, H. 2006. *Essential DC/DC Converters*. Boca Raton: Taylor & Francis Group LLC, p. 354. With permission.)

the switch-off period $(1-k)T$. Therefore, the ripple of the inductor current i_{L2} is

$$\Delta i_{L2} = \frac{V_1}{L_2}kT = \frac{V_{C2}-V_1}{L_2}(1-k)T, \tag{7.322}$$

$$V_{C2} = \frac{1}{1-k}V_1 = \left(\frac{1}{1-k}\right)^2 V_{in},$$

$$V_O = V_{C2} - V_{in} = \left[\left(\frac{1}{1-k}\right)^2 - 1\right]V_{in}. \tag{7.323}$$

The voltage transfer gain is

$$G = \frac{V_O}{V_{in}} = \left(\frac{1}{1-k}\right)^2 - 1. \tag{7.324}$$

Analogously,

$$\Delta i_{L1} = \frac{V_{in}}{L_1}kT, \qquad I_{L1} = \frac{I_O}{(1-k)^2},$$

$$\Delta i_{L2} = \frac{V_1}{L_2}kT, \qquad I_{L2} = \frac{I_O}{1-k}.$$

Therefore, the variation ratio of current i_{L1} through inductor L_1 is

$$\xi_1 = \frac{\Delta i_{L1}/2}{I_{L1}} = \frac{k(1-k)^2 TV_{in}}{2L_1 I_O} = \frac{k(1-k)^4}{2}\frac{R}{fL_1}, \tag{7.325}$$

the variation ratio of current i_{L2} through inductor L_2 is

$$\xi_2 = \frac{\Delta i_{L2}/2}{I_{L2}} = \frac{k(1-k)TV_1}{2L_2 I_O} = \frac{k(1-k)^2}{2}\frac{R}{fL_2}, \tag{7.326}$$

and the variation ratio of output voltage v_O is

$$\varepsilon = \frac{\Delta v_O/2}{V_O} = \frac{k}{2RfC_2}. \tag{7.327}$$

7.5.1.3 N/O Three-Stage Boost Circuit

The N/O three-stage boost circuit is derived from the N/O two-stage boost circuit by twice repeating the parts (L_2-D_2-D_3-C_2). Its circuit diagram and equivalent circuits during switch-on and switch-off periods are shown in Figure 7.48.

The voltage across capacitor C_1 is charged with V_1. As described in the previous section, the voltage V_{C1} across capacitor C_1 is $V_{C1} = (1/(1-k))V_{in}$, and the voltage V_{C2} across capacitor C_2 is $V_{C2} = (1/(1-k))^2 V_{in}$.

The voltage across capacitor C_3 is charged with V_O. The current flowing through inductor L_3 increases with V_{C2} during the switch-on period kT and decreases with $-(V_{C3} - V_{C2})$ during the switch-off period $(1-k)T$. Therefore, the ripple of the inductor current i_{L3} is

$$\Delta i_{L3} = \frac{V_{C2}}{L_3}kT = \frac{V_{C3} - V_{C2}}{L_3}(1-k)T, \tag{7.328}$$

$$V_{C3} = \frac{1}{1-k}V_{C2} = \left(\frac{1}{1-k}\right)^2 V_{C1} = \left(\frac{1}{1-k}\right)^3 V_{in},$$

$$V_O = V_{C3} - V_{in} = \left[\left(\frac{1}{1-k}\right)^3 - 1\right]V_{in}. \tag{7.329}$$

The voltage transfer gain is

$$G = \frac{V_O}{V_{in}} = \left(\frac{1}{1-k}\right)^3 - 1. \tag{7.330}$$

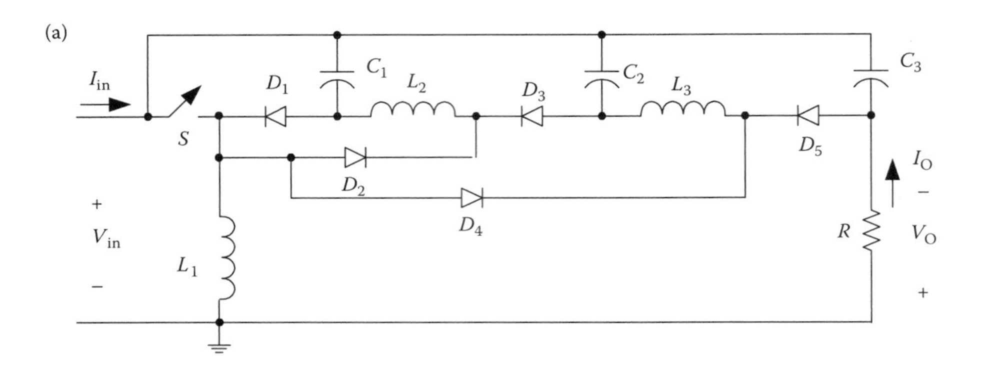

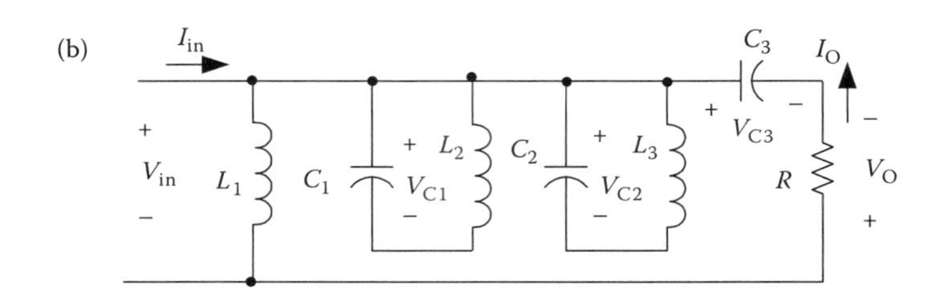

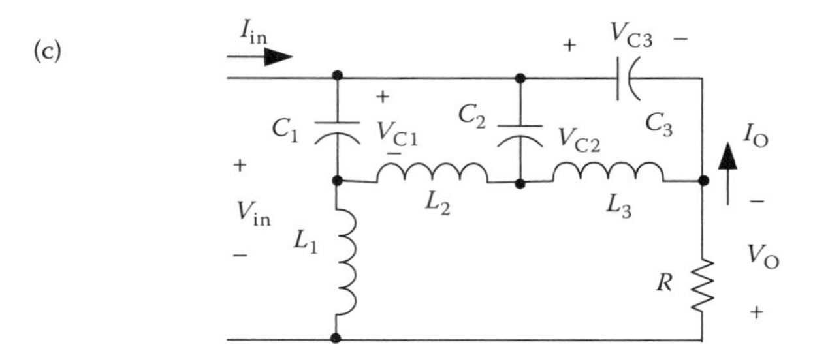

FIGURE 7.48 N/O three-stage boost circuit: (a) circuit diagram, (b) equivalent circuit during switch-on, and (c) equivalent circuit during switch-off. (Reprinted from Luo, F. L. and Ye, H. 2006. *Essential DC/DC Converters*. Boca Raton: Taylor & Francis Group LLC, p. 356. With permission.)

Analogously,

$$\Delta i_{L1} = \frac{V_{in}}{L_1}kT, \qquad I_{L1} = \frac{I_O}{(1-k)^3},$$

$$\Delta i_{L2} = \frac{V_1}{L_2}kT, \qquad I_{L2} = \frac{I_O}{(1-k)^2},$$

$$\Delta i_{L3} = \frac{V_2}{L_3}kT, \qquad I_{L3} = \frac{I_O}{1-k}.$$

Therefore, the variation ratio of current i_{L1} through inductor L_1 is

$$\xi_1 = \frac{\Delta i_{L1}/2}{I_{L1}} = \frac{k(1-k)^3 T V_{in}}{2L_1 I_O} = \frac{k(1-k)^6}{2}\frac{R}{fL_1}, \tag{7.331}$$

the variation ratio of current i_{L2} through inductor L_2 is

$$\xi_2 = \frac{\Delta i_{L2}/2}{I_{L2}} = \frac{k(1-k)^2 T V_1}{2L_2 I_O} = \frac{k(1-k)^4}{2}\frac{R}{fL_2}, \tag{7.332}$$

the variation ratio of current i_{L3} through inductor L_3 is

$$\xi_3 = \frac{\Delta i_{L3}/2}{I_{L3}} = \frac{k(1-k) T V_2}{2L_3 I_O} = \frac{k(1-k)^2}{2}\frac{R}{fL_3}, \tag{7.333}$$

and the variation ratio of output voltage v_O is

$$\varepsilon = \frac{\Delta v_O/2}{V_O} = \frac{k}{2RfC_3}. \tag{7.334}$$

7.5.1.4 N/O Higher-Stage Boost Circuit

An N/O higher-stage boost circuit can be designed by just multiple repeating of the parts (L_2-D_2-D_3-C_2). For the nth stage boost circuit; the final output voltage across capacitor C_n is

$$V_O = \left[\left(\frac{1}{1-k}\right)^n - 1\right] V_{in}.$$

The voltage transfer gain is

$$G = \frac{V_O}{V_{in}} = \left(\frac{1}{1-k}\right)^n - 1, \tag{7.335}$$

the variation ratio of current i_{Li} through inductor L_i $(i = 1, 2, 3, \ldots, n)$ is

$$\xi_i = \frac{\Delta i_{Li}/2}{I_{Li}} = \frac{k(1-k)^{2(n-i+1)}}{2}\frac{R}{fL_i}, \tag{7.336}$$

and the variation ratio of output voltage v_O is

$$\xi_i = \frac{\Delta i_{Li}/2}{I_{Li}} = \frac{k(1-k)^{2(n-i+1)}}{2}\frac{R}{fL_i}. \tag{7.337}$$

7.5.2 N/O Additional Series

All circuits of N/O cascaded boost converters—additional series—are derived from the corresponding circuits of the main series by adding a DEC.

The first three stages of this series are shown in Figures 7.49 through 7.51. For ease of understanding, they are called the elementary additional boost circuit, the two-stage additional boost circuit, and the three-stage additional boost circuit, respectively, and are numbered $n = 1, 2$ and 3, respectively.

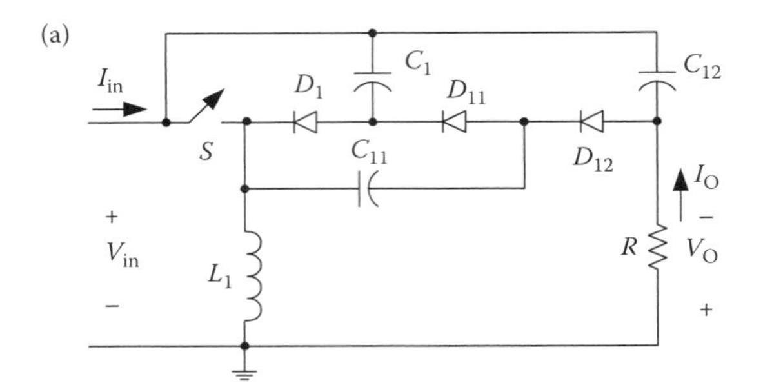

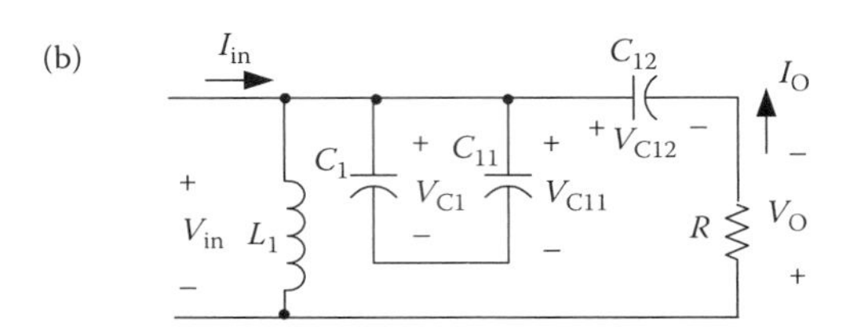

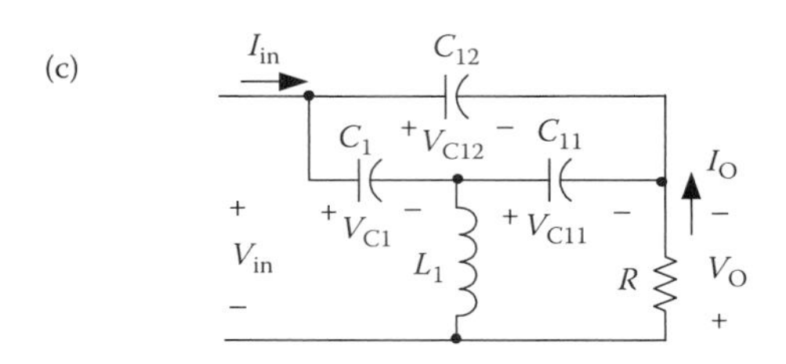

FIGURE 7.49 N/O elementary additional (double) boost circuit: (a) circuit diagram, (b) equivalent circuit during switch-on, and (c) equivalent circuit during switch-off. (Reprinted from Luo, F. L. and Ye, H. 2006. *Essential DC/DC Converters*. Boca Raton: Taylor & Francis Group LLC, p. 359. With permission.)

7.5.2.1 N/O Elementary Additional Boost Circuit

The N/O elementary additional boost circuit is derived from the N/O elementary boost converter by adding a DEC. Its circuit and switch-on and switch-off equivalent circuits are shown in Figure 7.49.

The voltage across capacitors C_1 and C_{11} is charged with V_{C1} and the voltage across capacitor C_{12} is charged with $V_{C12} = 2V_{C1}$. The current i_{L1} flowing through inductor L_1 increases with V_{in} during the switch-on period kT and decreases with $-(V_{C1} - V_{in})$ during the switch-off period $(1-k)T$. Therefore,

$$\Delta i_{L1} = \frac{V_{in}}{L_1}kT = \frac{V_{C1} - V_{in}}{L_1}(1-k)T, \tag{7.338}$$

$$V_{C1} = \frac{1}{1-k}V_{in}.$$

The voltage V_{C12} is

$$V_{C12} = 2V_{C1} = \frac{2}{1-k}V_{in}. \tag{7.339}$$

The output voltage is

$$V_O = V_{C12} - V_{in} = \left[\frac{2}{1-k} - 1\right] V_{in}. \tag{7.340}$$

The voltage transfer gain is

$$G = \frac{V_O}{V_{in}} = \frac{2}{1-k} - 1. \tag{7.341}$$

The variation ratio of current i_{L1} through inductor L_1 is

$$\xi_1 = \frac{\Delta i_{L1}/2}{I_{L1}} = \frac{k(1-k)TV_{in}}{4L_1 I_O} = \frac{k(1-k)^2}{8}\frac{R}{fL_1}. \tag{7.342}$$

The ripple voltage of output voltage v_O is

$$\Delta v_O = \frac{\Delta Q}{C_{12}} = \frac{I_O kT}{C_{12}} = \frac{k}{fC_{12}}\frac{V_O}{R}.$$

Therefore, the variation ratio of output voltage v_O is

$$\varepsilon = \frac{\Delta v_O/2}{V_O} = \frac{k}{2RfC_{12}}. \tag{7.343}$$

7.5.2.2 N/O Two-Stage Additional Boost Circuit

The N/O two-stage additional boost circuit is derived from the N/O two-stage boost circuit by adding a DEC. Its circuit diagram and switch-on and switch-off equivalent circuits are shown in Figure 7.50.

The voltage across capacitor C_1 is charged with V_{C1}. As described in the previous section, the voltage V_{C1} across capacitor C_1 is $V_{C1} = (1/(1-k))V_{in}$.

The voltage across capacitors C_2 and C_{11} is charged with V_{C2} and the voltage across the capacitor C_{12} is charged with V_{C12}. The current flowing through the inductor L_2 increases with V_{C1} during the switch-on period kT and decreases with $-(V_{C2} - V_{C1})$ during the switch-off period $(1-k)T$. Therefore, the ripple of the inductor current i_{L2} is

$$\Delta i_{L2} = \frac{V_{C1}}{L_2}kT = \frac{V_{C2} - V_{C1}}{L_2}(1-k)T, \tag{7.344}$$

$$V_{C2} = \frac{1}{1-k}V_{C1} = \left(\frac{1}{1-k}\right)^2 V_{in}, \tag{7.345}$$

$$V_{C12} = 2V_{C2} = \frac{2}{1-k}V_{C1} = 2\left(\frac{1}{1-k}\right)^2 V_{in}.$$

The output voltage is

$$V_O = V_{C12} - V_{in} = \left[2\left(\frac{1}{1-k}\right)^2 - 1\right] V_{in}. \tag{7.346}$$

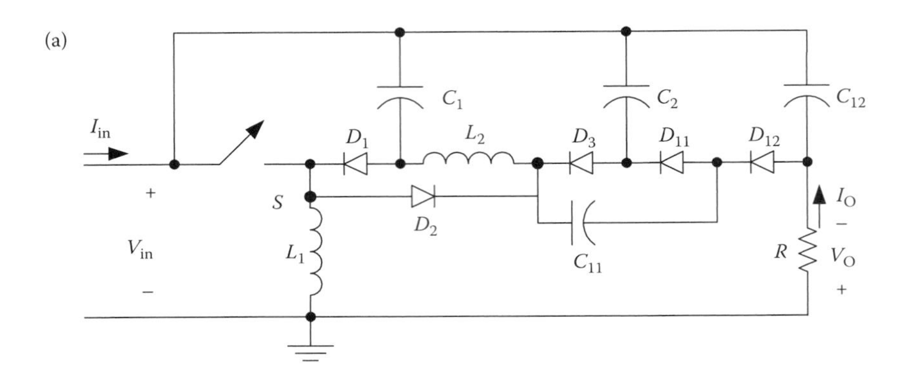

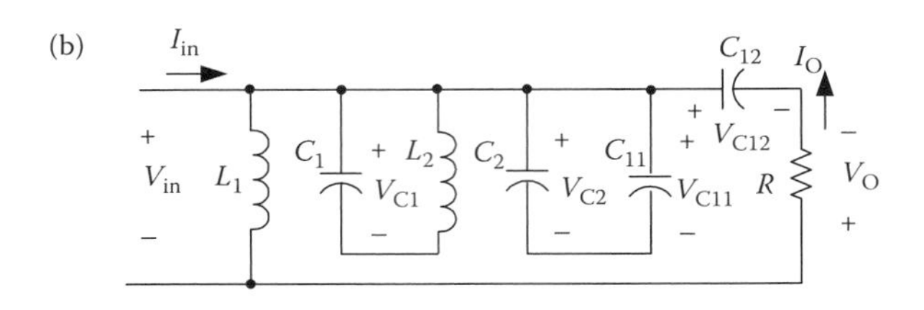

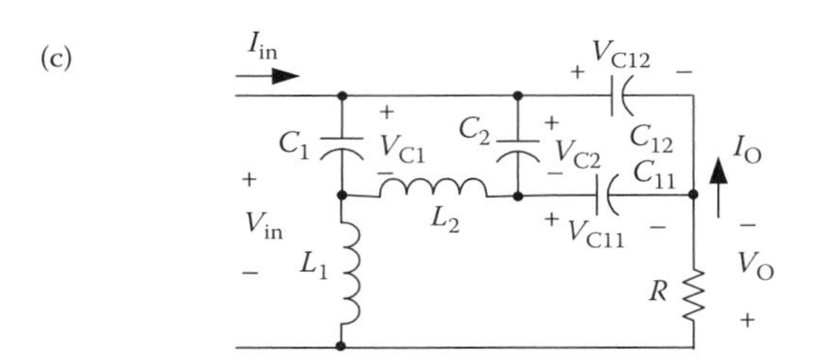

FIGURE 7.50 N/O two-stage additional boost circuit: (a) circuit diagram, (b) equivalent circuit during switch-on, and (c) equivalent circuit during switch-off. (Reprinted from Luo, F. L. and Ye, H. 2006. *Essential DC/DC Converters*. Boca Raton: Taylor & Francis Group LLC, p. 361. With permission.)

The voltage transfer gain is

$$G = \frac{V_O}{V_{in}} = 2\left(\frac{1}{1-k}\right)^2 - 1. \tag{7.347}$$

Analogously,

$$\Delta i_{L1} = \frac{V_{in}}{L_1}kT, \qquad I_{L1} = \frac{2}{(1-k)^2}I_O,$$

$$\Delta i_{L2} = \frac{V_1}{L_2}kT, \qquad I_{L2} = \frac{2I_O}{1-k}.$$

Therefore, the variation ratio of current i_{L1} through inductor L_1 is

$$\xi_1 = \frac{\Delta i_{L1}/2}{I_{L1}} = \frac{k(1-k)^2 TV_{in}}{4L_1 I_O} = \frac{k(1-k)^4}{8}\frac{R}{fL_1} \tag{7.348}$$

and the variation ratio of current i_{L2} through inductor L_2 is

$$\xi_2 = \frac{\Delta i_{L2}/2}{I_{L2}} = \frac{k(1-k)TV_1}{4L_2 I_O} = \frac{k(1-k)^2}{8}\frac{R}{fL_2}. \quad (7.349)$$

The ripple voltage of output voltage v_O is

$$\Delta v_O = \frac{\Delta Q}{C_{12}} = \frac{I_O kT}{C_{12}} = \frac{k}{fC_{12}}\frac{V_O}{R}.$$

Therefore, the variation ratio of output voltage v_O is

$$\varepsilon = \frac{\Delta v_O/2}{V_O} = \frac{k}{2RfC_{12}}. \quad (7.350)$$

7.5.2.3 N/O Three-Stage Additional Boost Circuit

The N/O three-stage additional boost circuit is derived from the three-stage boost circuit by adding a DEC. Its circuit diagram and equivalent circuits during switch-on and switch-off periods are shown in Figure 7.51.

The voltage across capacitor C_1 is charged with V_{C1}. As described in the previous section, the voltage V_{C1} across capacitor C_1 is $V_{C1} = (1/(1-k))V_{in}$, and the voltage V_2 across capacitor C_2 is $V_{C2} = (1/(1-k))^2 V_{in}$.

The voltage across capacitors C_3 and C_{11} is charged with V_{C3}. The voltage across capacitor C_{12} is charged with V_{C12}. The current flowing through inductor L_3 increases with V_{C2} during the switch-on period kT and decreases with $-(V_{C3} - V_{C2})$ during the switch-off period $(1-k)T$. Therefore,

$$\Delta i_{L3} = \frac{V_{C2}}{L_3}kT = \frac{V_{C3} - V_{C2}}{L_3}(1-k)T \quad (7.351)$$

and

$$V_{C3} = \frac{1}{1-k}V_{C2} = \left(\frac{1}{1-k}\right)^2 V_{C1} = \left(\frac{1}{1-k}\right)^3 V_{in}. \quad (7.352)$$

The voltage V_{C12} is $V_{C12} = 2V_{C3} = 2(1/(1-k))^3 V_{in}$.

The output voltage is

$$V_O = V_{C12} - V_{in} = \left[2\left(\frac{1}{1-k}\right)^3 - 1\right]V_{in}. \quad (7.353)$$

The voltage transfer gain is

$$G = \frac{V_O}{V_{in}} = 2\left(\frac{1}{1-k}\right)^3 - 1. \quad (7.354)$$

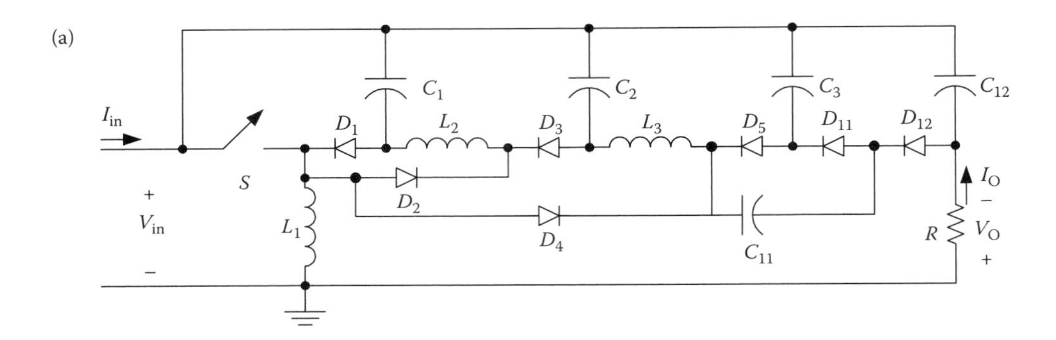

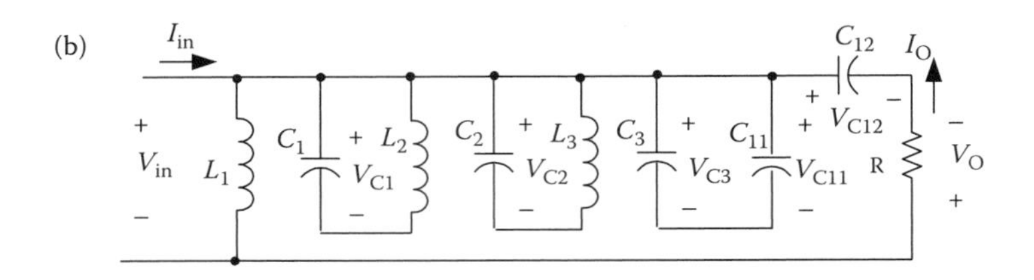

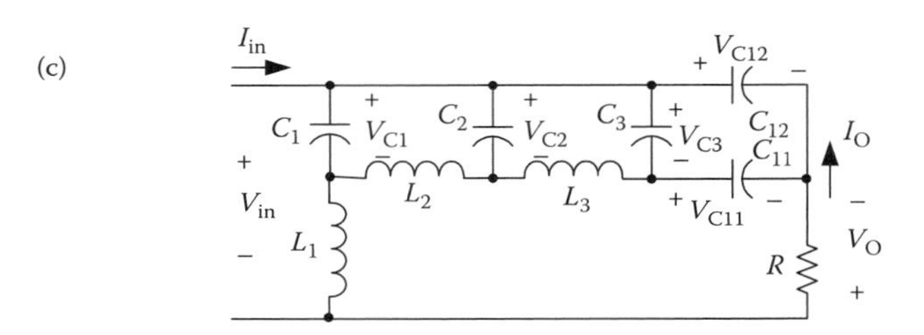

FIGURE 7.51 N/O three-stage additional boost circuit: (a) circuit diagram, (b) equivalent circuit during switch-on, and (c) equivalent circuit during switch-off. (Reprinted from Luo, F. L. and Ye, H. 2006. *Essential DC/DC Converters*. Boca Raton: Taylor & Francis Group LLC, p. 363. With permission.)

Analogously,

$$\Delta i_{L1} = \frac{V_{in}}{L_1}kT, \qquad I_{L1} = \frac{2}{(1-k)^3}I_O,$$

$$\Delta i_{L2} = \frac{V_1}{L_2}kT, \qquad I_{L2} = \frac{2}{(1-k)^2}I_O,$$

$$\Delta i_{L3} = \frac{V_2}{L_3}kT, \qquad I_{L3} = \frac{2I_O}{1-k}.$$

Therefore, the variation ratio of current i_{L1} through inductor L_1 is

$$\xi_1 = \frac{\Delta i_{L1}/2}{I_{L1}} = \frac{k(1-k)^3 TV_{in}}{4L_1 I_O} = \frac{k(1-k)^6}{8}\frac{R}{fL_1}, \tag{7.355}$$

the variation ratio of current i_{L2} through inductor L_2 is

$$\xi_2 = \frac{\Delta i_{L2}/2}{I_{L2}} = \frac{k(1-k)^2 TV_1}{4L_2 I_O} = \frac{k(1-k)^4}{8} \frac{R}{fL_2}, \tag{7.356}$$

and the variation ratio of current i_{L3} through inductor L_3 is

$$\xi_3 = \frac{\Delta i_{L3}/2}{I_{L3}} = \frac{k(1-k)TV_2}{4L_3 I_O} = \frac{k(1-k)^2}{8} \frac{R}{fL_3}. \tag{7.357}$$

The ripple voltage of output voltage v_O is

$$\Delta v_O = \frac{\Delta Q}{C_{12}} = \frac{I_O kT}{C_{12}} = \frac{k}{fC_{12}} \frac{V_O}{R}.$$

Therefore, the variation ratio of output voltage v_O is

$$\varepsilon = \frac{\Delta v_O/2}{V_O} = \frac{k}{2RfC_{12}}. \tag{7.358}$$

7.5.2.4 N/O Higher-Stage Additional Boost Circuit

The N/O higher-stage boost additional circuit is derived from the corresponding circuits of the main series by adding a DEC. For the nth stage additional circuit, the final output voltage is

$$V_O = \left[2\left(\frac{1}{1-k}\right)^n - \right] V_{in}.$$

The voltage transfer gain is

$$G = \frac{V_O}{V_{in}} = 2\left(\frac{1}{1-k}\right)^n - 1. \tag{7.359}$$

Analogously, the variation ratio of current i_{Li} through inductor L_i ($i = 1, 2, 3, \ldots, n$) is

$$\xi_i = \frac{\Delta i_{Li}/2}{I_{Li}} = \frac{k(1-k)^{2(n-i+1)}}{8} \frac{R}{fL_i}, \tag{7.360}$$

and the variation ratio of output voltage v_O is

$$\varepsilon = \frac{\Delta v_O/2}{V_O} = \frac{k}{2RfC_{12}}. \tag{7.361}$$

7.5.3 Double Series

All circuits of the N/O cascaded boost converters—double series—are derived from the corresponding circuits of the main series by adding the DEC in each stage circuit. The first three stages of this series are shown in Figures 7.49, 7.52, and 7.53. For ease of understanding, they are called the elementary double circuit, the two-stage double circuit, and three-stage double boost circuit, respectively, and are numbered $n = 1$, 2, and 3, respectively.

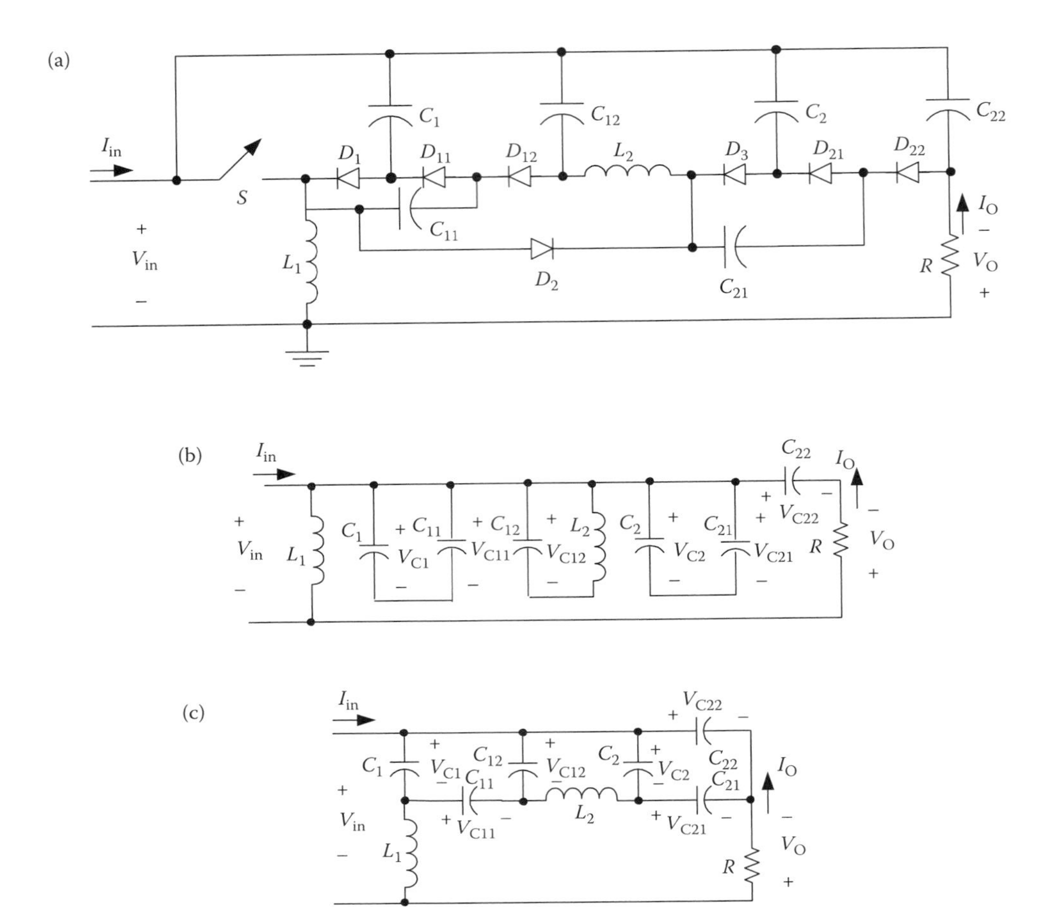

FIGURE 7.52 N/O two-stage double boost circuit: (a) circuit diagram, (b) equivalent circuit during switch-on, and (c) equivalent circuit during switch-off. (Reprinted from Luo, F. L. and Ye, H. 2006. *Essential DC/DC Converters*. Boca Raton: Taylor & Francis Group LLC, p. 366. With permission.)

7.5.3.1 N/O Elementary Double Boost Circuit

The N/O elementary double boost circuit is derived from the elementary boost converter by adding a DEC. Its circuit and switch-on and switch-off equivalent circuits are shown in Figure 7.49, which is the same as the elementary boost additional circuit.

7.5.3.2 N/O Two-Stage Double Boost Circuit

The N/O two-stage double boost circuit is derived from the two-stage boost circuit by adding the DEC in each stage circuit. Its circuit diagram and switch-on and switch-off equivalent circuits are shown in Figure 7.52.

The voltage across capacitors C_1 and C_{11} is charged with V_1. As described in the previous section, the voltage V_{C1} across capacitors C_1 and C_{11} is $V_{C1} = (1/(1-k))V_{in}$. The voltage across capacitor C_{12} is charged with $2V_{C1}$.

The current flowing through inductor L_2 increases with $2V_{C1}$ during the switch-on period kT and decreases with $-(V_{C2} - 2V_{C1})$ during the switch-off period $(1-k)T$. Therefore, the

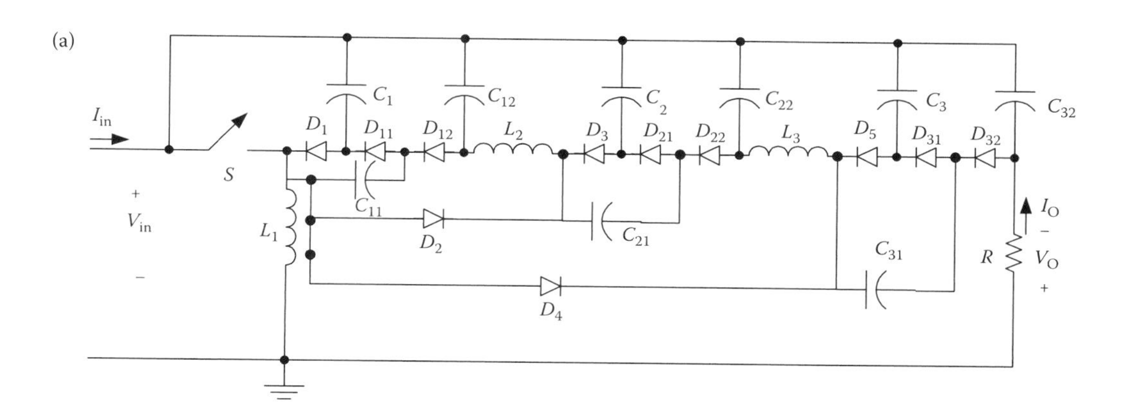

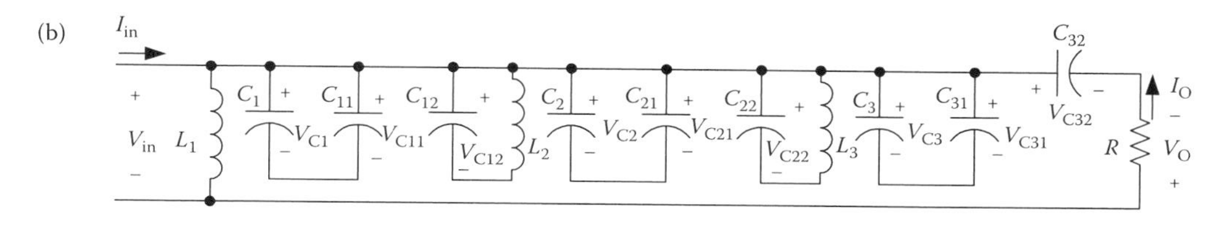

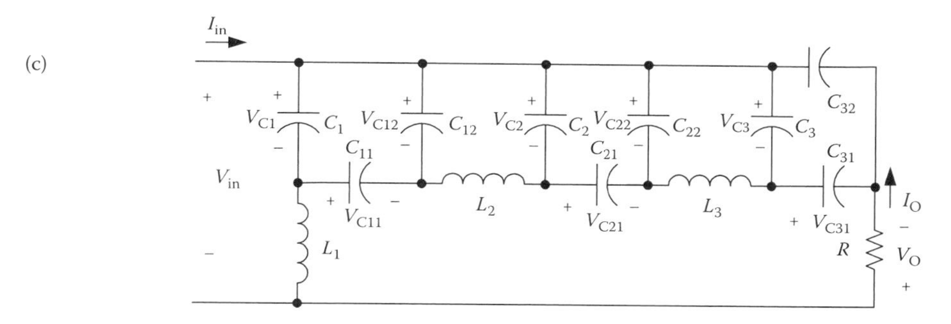

FIGURE 7.53 N/O three-stage double boost circuit: (a) circuit diagram, (b) equivalent circuit during switch-on, and (c) Equivalent circuit during switch-off. (Reprinted from Luo, F. L. and Ye, H. 2006. *Essential DC/DC Converters*. Boca Raton: Taylor & Francis Group LLC, p. 368. With permission.)

ripple of the inductor current i_{L2} is

$$\Delta i_{L2} = \frac{2V_{C1}}{L_2}kT = \frac{V_{C2} - 2V_{C1}}{L_2}(1-k)T, \tag{7.362}$$

$$V_{C2} = \frac{2}{1-k}V_{C1} = 2\left(\frac{1}{1-k}\right)^2 V_{in}. \tag{7.363}$$

The voltage V_{C22} is

$$V_{C22} = 2V_{C2} = \left(\frac{2}{1-k}\right)^2 V_{in}.$$

The output voltage is

$$V_O = V_{C22} - V_{in} = \left[\left(\frac{2}{1-k}\right)^2 - 1\right] V_{in}. \quad (7.364)$$

The voltage transfer gain is

$$G = \frac{V_O}{V_{in}} = \left(\frac{2}{1-k}\right)^2 - 1 \quad (7.365)$$

Analogously,

$$\Delta i_{L1} = \frac{V_{in}}{L_1} kT, \quad I_{L1} = \left(\frac{2}{1-k}\right)^2 I_O,$$

$$\Delta i_{L2} = \frac{V_1}{L_2} kT, \quad I_{L2} = \frac{2I_O}{1-k}.$$

Therefore, the variation ratio of current i_{L1} through inductor L_1 is

$$\xi_1 = \frac{\Delta i_{L1}/2}{I_{L1}} = \frac{k(1-k)^2 T V_{in}}{8 L_1 I_O} = \frac{k(1-k)^4}{16} \frac{R}{f L_1} \quad (7.366)$$

and the variation ratio of current i_{L2} through inductor L_2 is

$$\xi_2 = \frac{\Delta i_{L2}/2}{I_{L2}} = \frac{k(1-k) T V_1}{4 L_2 I_O} = \frac{k(1-k)^2}{8} \frac{R}{f L_2}. \quad (7.367)$$

The ripple voltage of output voltage v_O is

$$\Delta v_O = \frac{\Delta Q}{C_{22}} = \frac{I_O kT}{C_{22}} = \frac{k}{f C_{22}} \frac{V_O}{R}.$$

Therefore, the variation ratio of output voltage v_O is

$$\varepsilon = \frac{\Delta v_O/2}{V_O} = \frac{k}{2 R f C_{22}}. \quad (7.368)$$

7.5.3.3 N/O Three-Stage Double Boost Circuit

The N/O three-stage double boost circuit is derived from the three-stage boost circuit by adding the DEC in each stage circuit. Its circuit diagram and equivalent circuits during switch-on and switch-off periods are shown in Figure 7.53.

The voltage across capacitors C_1 and C_{11} is charged with V_{C1}. As described in the previous section, the voltage V_{C1} across capacitors C_1 and C_{11} is $V_{C1} = (1/(1-k))V_{in}$, and the voltage V_{C2} across capacitors C_2 and C_{12} is $V_{C2} = 2(1/(1-k))^2 V_{in}$.

The voltage across capacitor C_{22} is $2V_{C2} = (2/(1-k))^2 V_{in}$. The voltage across capacitors C_3 and C_{31} is charged with V_3. The voltage across capacitor C_{12} is charged with V_O. The current flowing through inductor L_3 increases with V_2 during the switch-on period kT and decreases with $-(V_{C3} - 2V_{C2})$ during the switch-off period $(1-k)T$. Therefore,

$$\Delta i_{L3} = \frac{2V_{C2}}{L_3}kT = \frac{V_{C3} - 2V_{C2}}{L_3}(1-k)T, \tag{7.369}$$

$$V_{C3} = \frac{2V_{C2}}{(1-k)} = \frac{4}{(1-k)^3}V_{in}. \tag{7.370}$$

The voltage V_{C32} is

$$V_{C32} = 2V_{C3} = \left(\frac{2}{1-k}\right)^3 V_{in}.$$

The output voltage is

$$V_O = V_{C32} - V_{in} = \left[\left(\frac{2}{1-k}\right)^3 - 1\right] V_{in}. \tag{7.371}$$

The voltage transfer gain is

$$G = \frac{V_O}{V_{in}} = \left(\frac{2}{1-k}\right)^3 - 1. \tag{7.372}$$

Analogously,

$$\Delta i_{L1} = \frac{V_{in}}{L_1}kT, \qquad I_{L1} = \frac{8}{(1-k)^3}I_O,$$

$$\Delta i_{L2} = \frac{V_1}{L_2}kT, \qquad I_{L2} = \frac{4}{(1-k)^2}I_O,$$

$$\Delta i_{L3} = \frac{V_2}{L_3}kT, \qquad I_{L3} = \frac{2I_O}{1-k}.$$

Therefore, the variation ratio of current i_{L1} through inductor L_1 is

$$\xi_1 = \frac{\Delta i_{L1}/2}{I_{L1}} = \frac{k(1-k)^3 TV_{in}}{16L_1 I_O} = \frac{k(1-k)^6}{128}\frac{R}{fL_1}, \tag{7.373}$$

the variation ratio of current i_{L2} through inductor L_2 is

$$\xi_2 = \frac{\Delta i_{L2}/2}{I_{L2}} = \frac{k(1-k)^2 TV_1}{8L_2 I_O} = \frac{k(1-k)^4}{32}\frac{R}{fL_2}, \tag{7.374}$$

and the variation ratio of current i_{L3} through inductor L_3 is

$$\xi_3 = \frac{\Delta i_{L3}/2}{I_{L3}} = \frac{k(1-k)TV_2}{4L_3 I_O} = \frac{k(1-k)^2}{8}\frac{R}{fL_3}. \tag{7.375}$$

The ripple voltage of output voltage v_O is

$$\Delta v_O = \frac{\Delta Q}{C_{32}} = \frac{I_O k T}{C_{32}} = \frac{k}{f C_{32}} \frac{V_O}{R}.$$

Therefore, the variation ratio of output voltage v_O is

$$\varepsilon = \frac{\Delta v_O/2}{V_O} = \frac{k}{2 R f C_{32}}. \quad (7.376)$$

7.5.3.4 N/O Higher-Stage Double Boost Circuit

The N/O higher-stage double boost circuit is derived from the corresponding circuits of the main series by adding the DEC in each stage circuit. For the nth stage additional circuit, the final output voltage is

$$V_O = \left[\left(\frac{2}{1-k} \right)^n - 1 \right] V_{in}.$$

The voltage transfer gain is

$$G = \frac{V_O}{V_{in}} = \left(\frac{2}{1-k} \right)^n - 1. \quad (7.377)$$

Analogously, the variation ratio of current i_{Li} through inductor L_i $(i = 1, 2, 3, \ldots, n)$ is

$$\xi_i = \frac{\Delta i_{Li}/2}{I_{Li}} = \frac{k(1-k)^{2(n-i+1)}}{2 \times 2^{2n}} \frac{R}{f L_i}, \quad (7.378)$$

and the variation ratio of output voltage v_O is

$$\varepsilon = \frac{\Delta v_O/2}{V_O} = \frac{k}{2 R f C_{n2}}. \quad (7.379)$$

7.5.4 Triple Series

All circuits of the N/O cascaded boost converters—triple series—are derived from the corresponding circuits of the main series by adding the DEC twice in circuits of each stage. The first three stages of this series are shown in Figures 7.54 through 7.56. For ease of understanding, they are called the elementary double (or additional) circuit, the two-stage double circuit, and the three-stage double circuit, respectively, and are numbered $n = 1, 2,$ and 3, respectively.

7.5.4.1 N/O Elementary Triple Boost Circuit

The N/O elementary triple boost circuit is derived from the elementary boost converter by adding the DEC twice in each stage circuit. Its circuit and switch-on and switch-off equivalent circuits are shown in Figure 7.54. The output voltage of the first stage boost circuit is V_{C1}, $V_{C1} = V_{in}/(1-k)$.

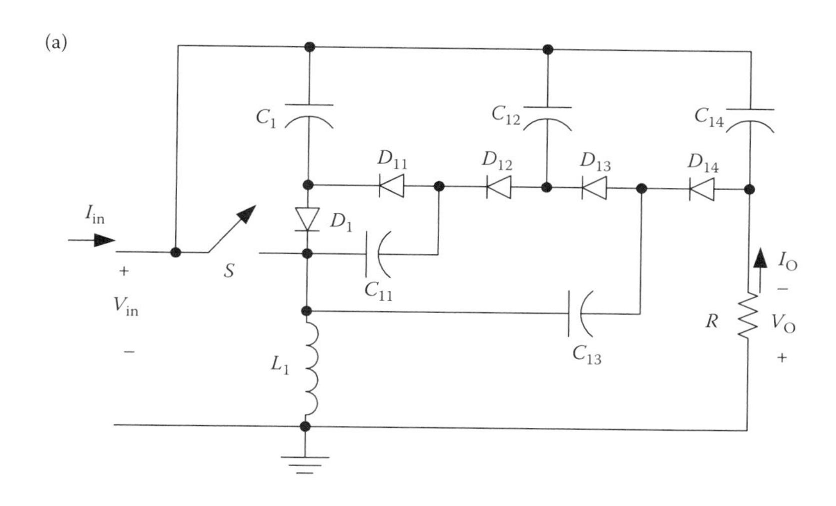

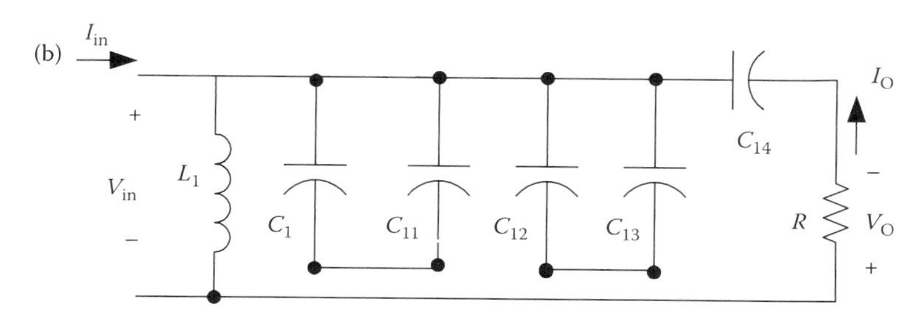

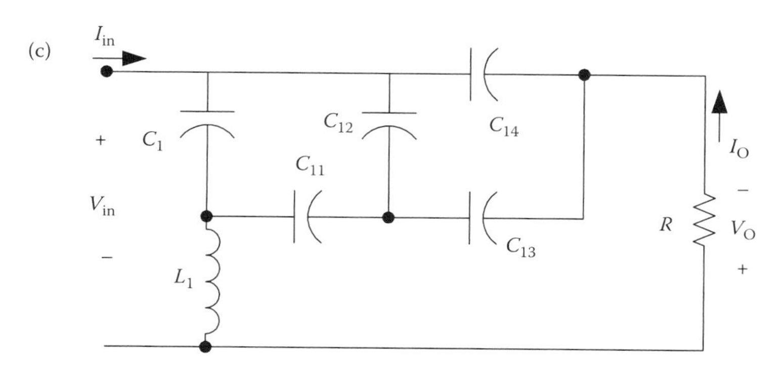

FIGURE 7.54 N/O elementary triple boost circuit: (a) circuit diagram, (b) equivalent circuit during switch-on, and (c) equivalent circuit during switch-off. (Reprinted from Luo, F. L. and Ye, H. 2006. *Essential DC/DC Converters*. Boca Raton: Taylor & Francis Group LLC, p. 372. With permission.)

After the first DEC, the voltage (across capacitor C_{12}) increases to

$$V_{C12} = 2V_{C1} = \frac{2}{1-k}V_{in}. \tag{7.380}$$

After the second DEC, the voltage (across capacitor C_{14}) increases to

$$V_{C14} = V_{C12} + V_{C1} = \frac{3}{1-k}V_{in}. \tag{7.381}$$

The final output voltage V_O is equal to

$$V_O = V_{C14} - V_{in} = \left[\frac{3}{1-k} - 1\right] V_{in}. \tag{7.382}$$

The voltage transfer gain is

$$G = \frac{V_O}{V_{in}} = \frac{3}{1-k} - 1. \tag{7.383}$$

7.5.4.2 N/O Two-Stage Triple Boost Circuit

The N/O two-stage triple boost circuit is derived from the two-stage boost circuit by adding the DEC twice in circuits of each stage. Its circuit diagram and switch-on and switch-off equivalent circuits are shown in Figure 7.55.

As described in the previous section, the voltage across capacitor C_{14} is $V_{C14} = (3/(1-k))V_{in}$. Analogously, the voltage across capacitor C_{24} is

$$V_{C24} = \left(\frac{3}{1-k}\right)^2 V_{in}. \tag{7.384}$$

The final output voltage V_O is equal to

$$V_O = V_{C24} - V_{in} = \left[\left(\frac{3}{1-k}\right)^2 - 1\right] V_{in}. \tag{7.385}$$

The voltage transfer gain is

$$G = \frac{V_O}{V_{in}} = \left(\frac{3}{1-k}\right)^2 - 1. \tag{7.386}$$

Analogously,

$$\Delta i_{L1} = \frac{V_{in}}{L_1} kT, \qquad I_{L1} = \left(\frac{2}{1-k}\right)^2 I_O,$$

$$\Delta i_{L2} = \frac{V_1}{L_2} kT, \qquad I_{L2} = \frac{2I_O}{1-k}.$$

Therefore, the variation ratio of current i_{L1} through inductor L_1 is

$$\xi_1 = \frac{\Delta i_{L1}/2}{I_{L1}} = \frac{k(1-k)^2 T V_{in}}{8L_1 I_O} = \frac{k(1-k)^4}{16} \frac{R}{fL_1} \tag{7.387}$$

and the variation ratio of current i_{L2} through inductor L_2 is

$$\xi_2 = \frac{\Delta i_{L2}/2}{I_{L2}} = \frac{k(1-k) T V_1}{4L_2 I_O} = \frac{k(1-k)^2}{8} \frac{R}{fL_2}. \tag{7.388}$$

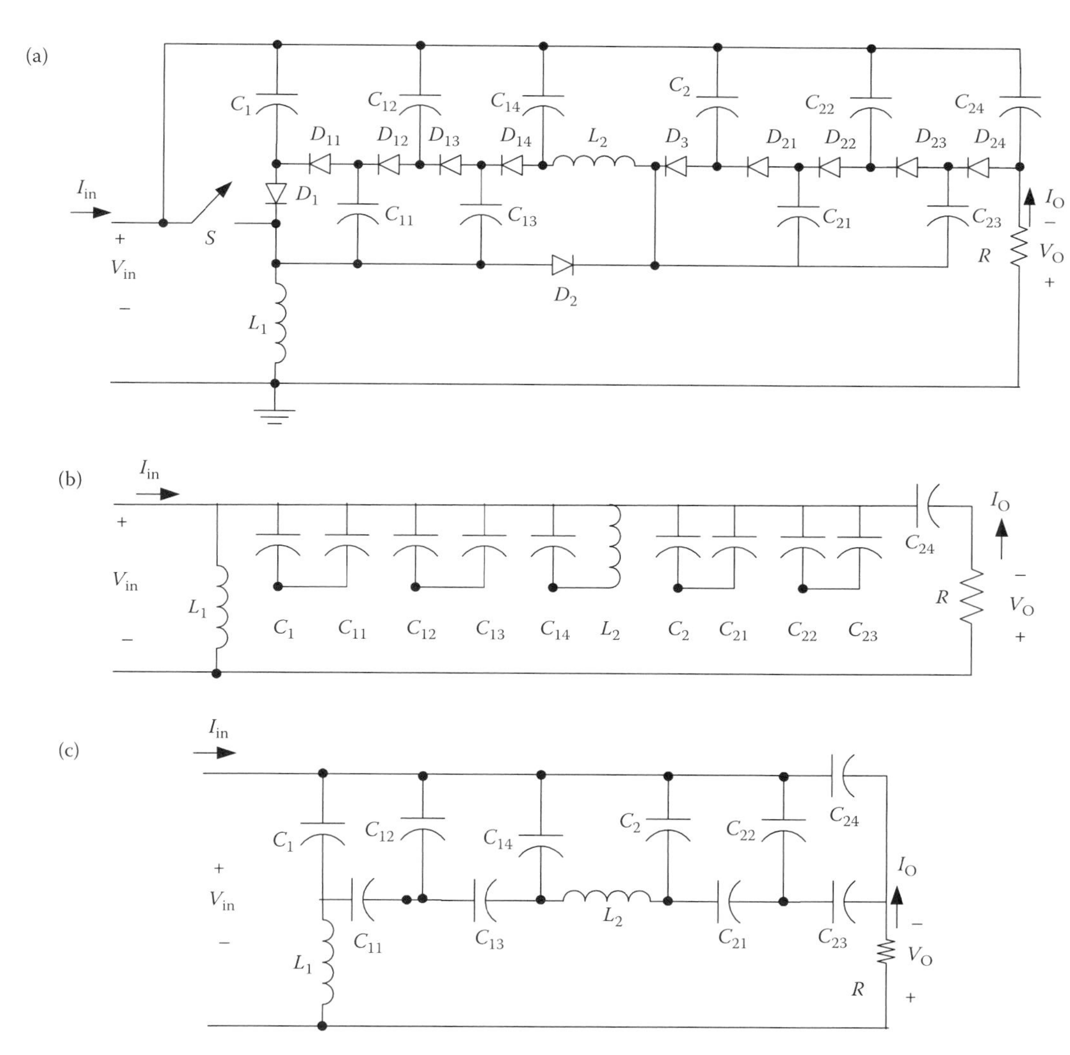

FIGURE 7.55 N/O two-stage triple boost circuit: (a) circuit diagram, (b) equivalent circuit during switch-on, and (c) equivalent circuit during switch-off. (Reprinted from Luo, F. L. and Ye, H. 2006. *Essential DC/DC Converters*. Boca Raton: Taylor & Francis Group LLC, p. 373. With permission.)

The ripple voltage of output voltage v_O is

$$\Delta v_O = \frac{\Delta Q}{C_{22}} = \frac{I_O kT}{C_{22}} = \frac{k}{fC_{22}}\frac{V_O}{R}.$$

Therefore, the variation ratio of output voltage v_O is

$$\varepsilon = \frac{\Delta v_O/2}{V_O} = \frac{k}{2RfC_{22}}. \tag{7.389}$$

7.5.4.3 N/O Three-Stage Triple Boost Circuit

This N/O three-stage triple boost circuit is derived from the three-stage boost circuit by adding the DEC twice in each stage circuit. Its circuit diagram and equivalent circuits during switch-on and switch-off periods are shown in Figure 7.56.

As described in the previous section, the voltage across capacitor C_{14} is $V_{C14} = (3/(1-k))V_{in}$ and the voltage across capacitor C_{24} is $V_{C24} = (3/(1-k))^2 V_{in}$. Analogously, the voltage across capacitor C_{34} is

$$V_{C34} = \left(\frac{3}{1-k}\right)^3 V_{in}. \tag{7.390}$$

The final output voltage V_O is

$$V_O = V_{C34} - V_{in} = \left[\left(\frac{3}{1-k}\right)^3 - 1\right] V_{in}. \tag{7.391}$$

(a)

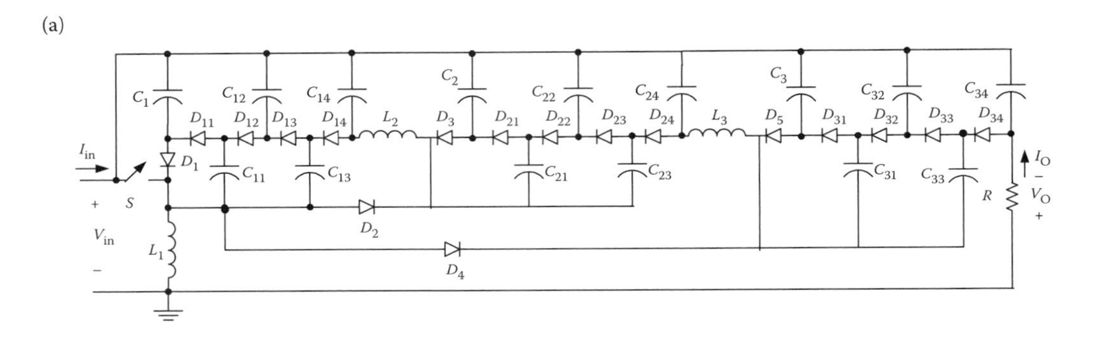

(b)

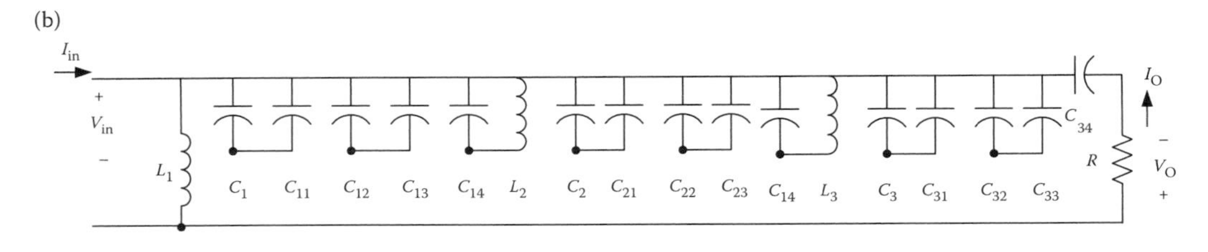

(c)

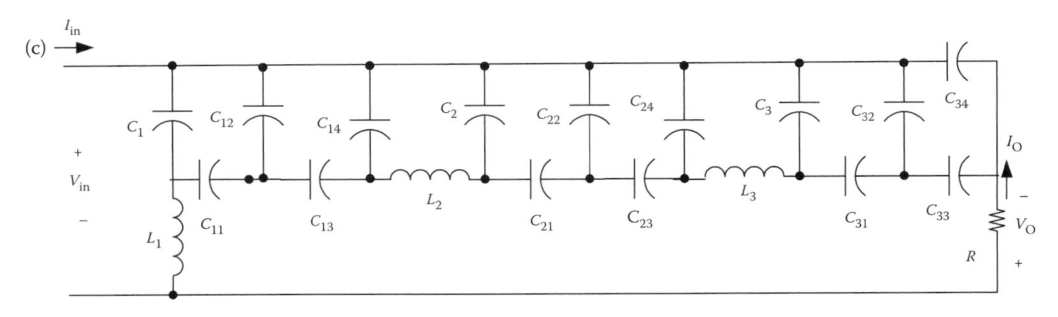

FIGURE 7.56 N/O three-stage triple boost circuit: (a) circuit diagram, (b) equivalent circuit during switch-on, and (c) equivalent circuit during switch-off. (Reprinted from Luo, F. L. and Ye, H. 2006. *Essential DC/DC Converters*. Boca Raton: Taylor & Francis Group LLC, p. 375. With permission.)

The voltage transfer gain is

$$G = \frac{V_O}{V_{in}} = \left(\frac{3}{1-k}\right)^3 - 1. \tag{7.392}$$

Analogously,

$$\Delta i_{L1} = \frac{V_{in}}{L_1}kT, \qquad I_{L1} = \frac{32}{(1-k)^3}I_O,$$

$$\Delta i_{L2} = \frac{V_1}{L_2}kT, \qquad I_{L2} = \frac{8}{(1-k)^2}I_O,$$

$$\Delta i_{L3} = \frac{V_2}{L_3}kT, \qquad I_{L3} = \frac{2}{1-k}I_O.$$

Therefore, the variation ratio of current i_{L1} through inductor L_1 is

$$\xi_1 = \frac{\Delta i_{L1}/2}{I_{L1}} = \frac{k(1-k)^3 TV_{in}}{64L_1 I_O} = \frac{k(1-k)^6}{12^3}\frac{R}{fL_1}, \tag{7.393}$$

the variation ratio of current i_{L2} through inductor L_2 is

$$\xi_2 = \frac{\Delta i_{L2}/2}{I_{L2}} = \frac{k(1-k)^2 TV_1}{16L_2 I_O} = \frac{k(1-k)^4}{12^2}\frac{R}{fL_2}, \tag{7.394}$$

and the variation ratio of current i_{L3} through inductor L_3 is

$$\xi_3 = \frac{\Delta i_{L3}/2}{I_{L3}} = \frac{k(1-k) TV_2}{4L_3 I_O} = \frac{k(1-k)^2}{12}\frac{R}{fL_3}. \tag{7.395}$$

Usually ξ_1, ξ_2, and ξ_3 are small; this means that this converter works in the continuous mode.

The ripple voltage of output voltage v_O is

$$\Delta v_O = \frac{\Delta Q}{C_{32}} = \frac{I_O kT}{C_{32}} = \frac{k}{fC_{32}}\frac{V_O}{R}.$$

Therefore, the variation ratio of output voltage v_O is

$$\varepsilon = \frac{\Delta v_O/2}{V_O} = \frac{k}{2RfC_{32}}. \tag{7.396}$$

Usually R is expressed in kΩ, f in 10 kHz, and C_{34} in μF; this ripple is much smaller than 1%.

7.5.4.4 N/O Higher-Stage Triple Boost Circuit

An N/O higher-stage triple boost circuit is derived from the corresponding circuits of the main series by adding the DEC twice in circuits of each stage. For the nth stage additional circuit, the voltage across capacitor C_{n4} is

$$V_{Cn4} = \left(\frac{3}{1-k}\right)^n V_{in}.$$

The output voltage is

$$V_O = V_{Cn4} - V_{in} = \left[\left(\frac{3}{1-k}\right)^n - 1\right] V_{in}. \quad (7.397)$$

The voltage transfer gain is

$$G = \frac{V_O}{V_{in}} = \left(\frac{3}{1-k}\right)^n - 1. \quad (7.398)$$

Analogously, the variation ratio of current i_{Li} through inductor L_i $(i = 1, 2, 3, \ldots, n)$ is

$$\xi_i = \frac{\Delta i_{Li}/2}{I_{Li}} = \frac{k(1-k)^{2(n-i+1)}}{12^{(n-i+1)}} \frac{R}{fL_i}, \quad (7.399)$$

and the variation ratio of output voltage v_O is

$$\varepsilon = \frac{\Delta v_O/2}{V_O} = \frac{k}{2RfC_{n2}}. \quad (7.400)$$

7.5.5 Multiple Series

All circuits of the N/O cascaded boost converters—multiple series—are derived from the corresponding circuits of the main series by adding the DEC multiple (j) times in each stage circuit. The first three stages of this series are shown in Figures 7.57 through 7.59. To make it easy to explain, they are called the elementary multiple boost circuit, the two-stage multiple boost circuit, and the three-stage multiple boost circuit, respectively, and are numbered as $n = 1$, 2, and 3, respectively.

7.5.5.1 N/O Elementary Multiple Boost Circuit

The N/O elementary multiple boost circuit is derived from the elementary boost converter by adding the DEC multiple (j) times. Its circuit and switch-on and switch-off equivalent circuits are shown in Figure 7.57.

The output voltage of the first DEC (across capacitor C_{12j}) increases to

$$V_{C12j} = \frac{j+1}{1-k} V_{in}. \quad (7.401)$$

The final output voltage V_O is

$$V_O = V_{C12j} - V_{in} = \left[\frac{j+1}{1-k} - 1\right] V_{in}. \quad (7.402)$$

The voltage transfer gain is

$$G = \frac{V_O}{V_{in}} = \frac{j+1}{1-k} - 1. \quad (7.403)$$

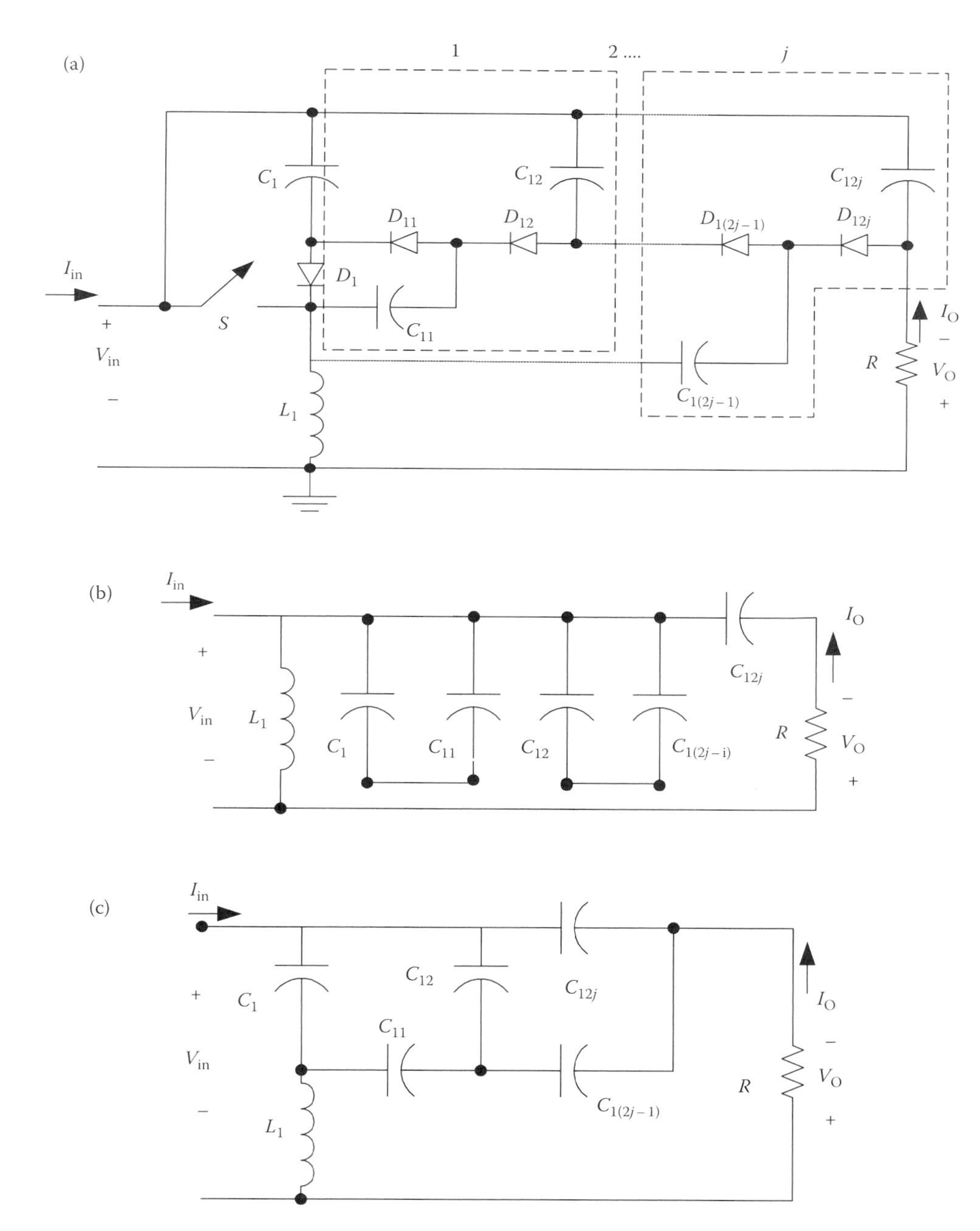

FIGURE 7.57 N/O elementary multiple boost circuit: (a) circuit diagram, (b) equivalent circuit during switch-on, and (c) equivalent circuit during switch-off. (Reprinted from Luo, F. L. and Ye, H. 2006. *Essential DC/DC Converters*. Boca Raton: Taylor & Francis Group LLC, p. 379. With permission.)

7.5.5.2 N/O Two-Stage Multiple Boost Circuit

The N/O two-stage multiple boost circuit is derived from the two-stage boost circuit by adding the DEC multiple (j) times in each stage circuit. Its circuit diagram and switch-on and switch-off equivalent circuits are shown in Figure 7.58.

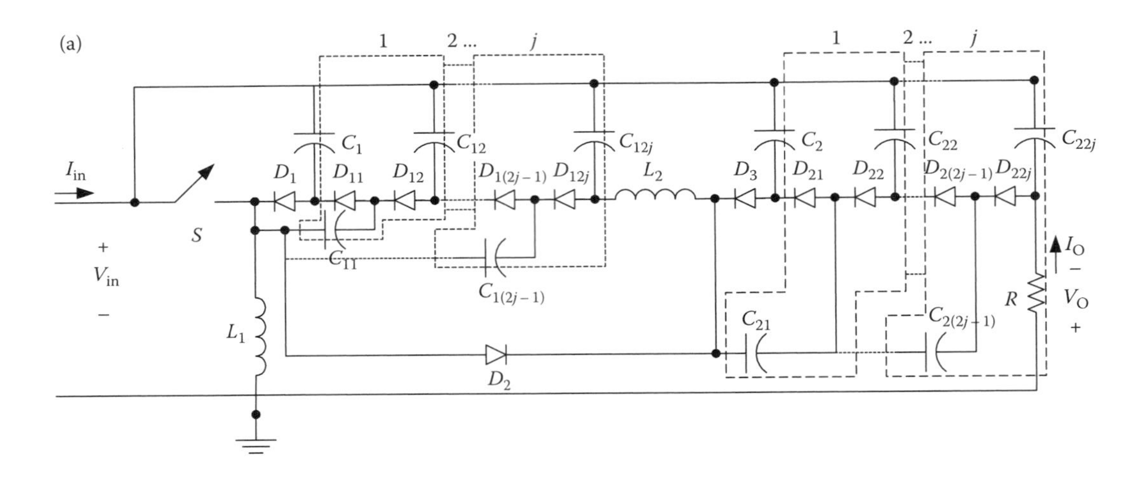

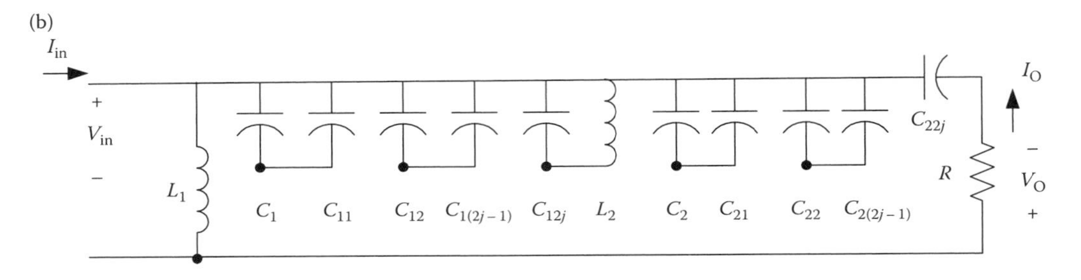

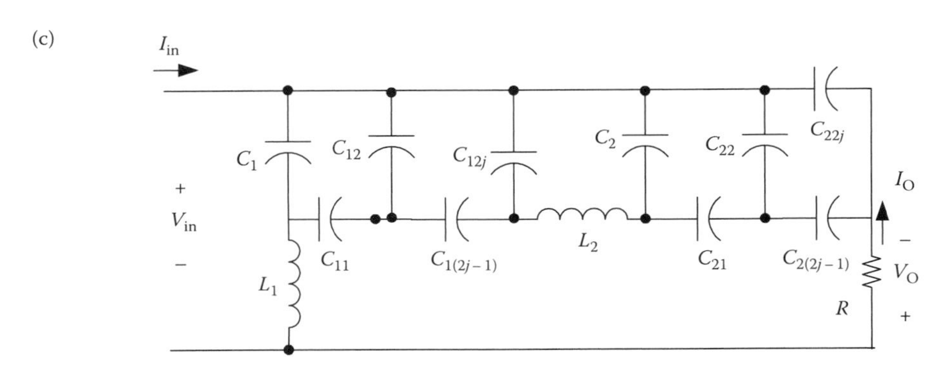

FIGURE 7.58 N/O two-stage multiple boost circuit: (a) circuit diagram, (b) equivalent circuit during switch-on, and (c) equivalent circuit during switch-off. (Reprinted from Luo, F. L. and Ye, H. 2006. *Essential DC/DC Converters*. Boca Raton: Taylor & Francis Group LLC, p. 380. With permission.)

As described in the previous section, the voltage across capacitor C_{12j} is $V_{C12j} = ((j+1)/(1-k))V_{in}$.

Analogously, the voltage across capacitor C_{22j} is

$$V_{C22j} = \left(\frac{j+1}{1-k}\right)^2 V_{in}. \tag{7.404}$$

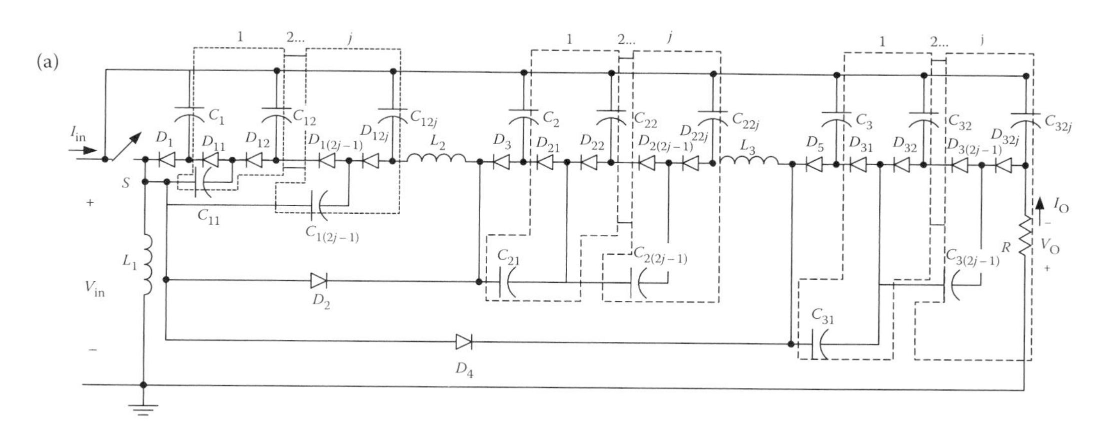

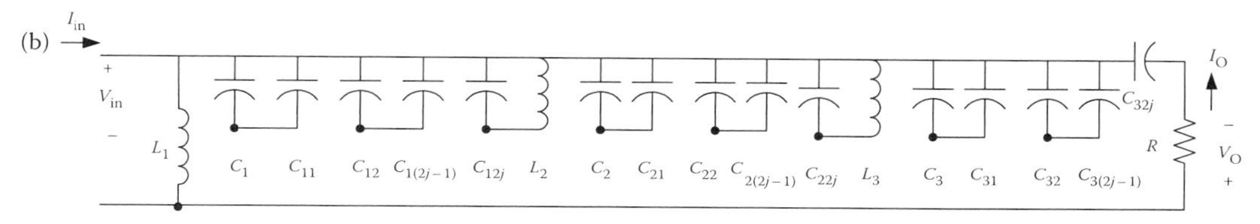

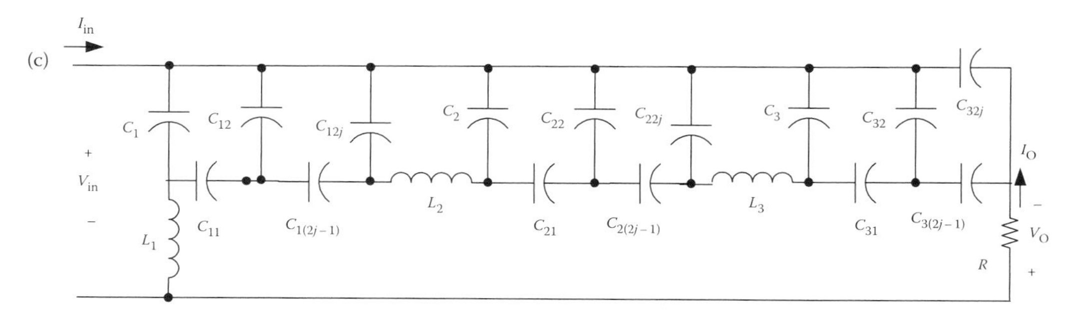

FIGURE 7.59 N/O three-stage multiple boost circuit: (a) circuit diagram, (b) equivalent circuit during switch-on, and (c) equivalent circuit during switch-off. (Reprinted from Luo, F. L. and Ye, H. 2006. *Essential DC/DC Converters*. Boca Raton: Taylor & Francis Group LLC, p. 382. With permission.)

The final output voltage V_O is

$$V_O = V_{C22j} - V_{in} = \left[\left(\frac{j+1}{1-k}\right)^2 - 1\right] V_{in}. \tag{7.405}$$

The voltage transfer gain is

$$G = \frac{V_O}{V_{in}} = \left(\frac{j+1}{1-k}\right)^2 - 1. \tag{7.406}$$

The ripple voltage of output voltage v_O is

$$\Delta v_O = \frac{\Delta Q}{C_{22j}} = \frac{I_O kT}{C_{22j}} = \frac{k}{fC_{22j}} \frac{V_O}{R}.$$

Therefore, the variation ratio of output voltage V_O is

$$\varepsilon = \frac{\Delta v_O/2}{V_O} = \frac{k}{2RfC_{22j}}. \quad (7.407)$$

Example 7.5

An N/O two-stage multiple ($j = 4$) boost converter in Figure 7.58a has $V_{in} = 20$V, all inductors have 10 mH, all capacitors have 20 μF, $R = 10\,k\Omega$, $f = 50$ kHz, and conduction duty cycle $k = 0.6$. Calculate the output voltage and its variation ratio.

SOLUTION

From Equation 7.405, we can obtain the output voltage

$$V_O = \left[\left(\frac{j+1}{1-k}\right)^2 - 1\right] V_{in} = \left[\left(\frac{4+1}{1-0.6}\right)^2 - 1\right] \times 20 = 605\,V.$$

From Equation 7.407, its variation ratio is

$$\varepsilon = \frac{k}{2RfC_{28}} = \frac{0.6}{2 \times 10{,}000 \times 50\,k \times 20\,\mu} = 0.00003.$$

7.5.5.3 N/O Three-Stage Multiple Boost Circuit

The N/O three-stage multiple boost circuit is derived from the three-stage boost circuit by adding the DEC multiple (j) times in circuits of each stage. Its circuit diagram and equivalent circuits during switch-on and switch-off periods are shown in Figure 7.59.

As described in the previous section, the voltage across capacitor C_{12j} is $V_{C12j} = ((j+1)/(1-k))V_{in}$ and the voltage across capacitor C_{22j} is $V_{C22j} = ((j+1)/(1-k))^2 V_{in}$. Analogously, the voltage across capacitor C_{32j} is

$$V_{C32j} = \left(\frac{j+1}{1-k}\right)^3 V_{in}. \quad (7.408)$$

The final output voltage V_O is

$$V_O = V_{C32j} - V_{in} = \left[\left(\frac{j+1}{1-k}\right)^3 - 1\right] V_{in}. \quad (7.409)$$

The voltage transfer gain is

$$G = \frac{V_O}{V_{in}} = \left(\frac{j+1}{1-k}\right)^3 - 1. \quad (7.410)$$

The ripple voltage of output voltage v_O is

$$\Delta v_O = \frac{\Delta Q}{C_{32j}} = \frac{I_O kT}{C_{32j}} = \frac{k}{fC_{32j}} \frac{V_O}{R}.$$

Therefore, the variation ratio of output voltage v_O is

$$\varepsilon = \frac{\Delta v_O/2}{V_O} = \frac{k}{2RfC_{32j}}. \quad (7.411)$$

7.5.5.4 N/O Higher-Stage Multiple Boost Circuit

An N/O higher-stage multiple boost circuit is derived from the corresponding circuits of the main series by adding the DEC multiple (j) times in circuits of each stage. For the nth stage multiple boost circuit, the voltage across capacitor C_{n2j} is

$$V_{Cn2j} = \left(\frac{j+1}{1-k}\right)^n V_{in}.$$

The output voltage is

$$V_O = V_{Cn2j} - V_{in} = \left[\left(\frac{j+1}{1-k}\right)^n - 1\right] V_{in}. \quad (7.412)$$

The voltage transfer gain is

$$G = \frac{V_O}{V_{in}} = \left(\frac{j+1}{1-k}\right)^n - 1. \quad (7.413)$$

The variation ratio of output voltage v_O is

$$\varepsilon = \frac{\Delta v_O/2}{V_O} = \frac{k}{2RfC_{n2j}}. \quad (7.414)$$

7.5.6 Summary of N/O Cascaded Boost Converters

All circuits of the N/O cascaded boost converters can be shown in Figure 7.60 as the family tree.

From the analysis of the previous two sections, the common formula to calculate the output voltage can be presented as

$$V_O = \begin{cases} \left[\left(\frac{1}{1-k}\right)^n - 1\right] V_{in} & \text{main series,} \\ \left[2 * \left(\frac{1}{1-k}\right)^n - 1\right] V_{in} & \text{additional series,} \\ \left[\left(\frac{2}{1-k}\right)^n - 1\right] V_{in} & \text{double series,} \\ \left[\left(\frac{3}{1-k}\right)^n - 1\right] V_{in} & \text{triple series,} \\ \left[\left(\frac{j+1}{1-k}\right)^n - 1\right] V_{in} & \text{multiple } (j) \text{ series.} \end{cases} \quad (7.415)$$

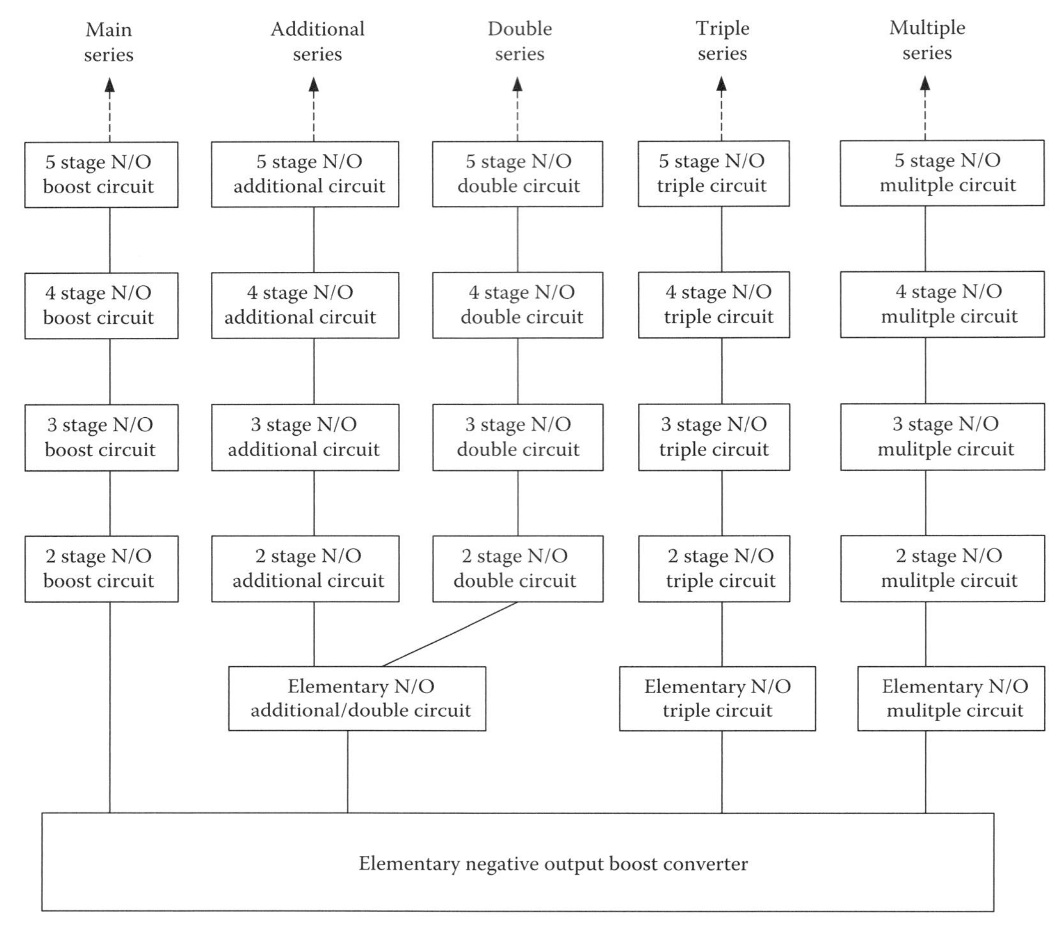

FIGURE 7.60 The family of N/O cascaded boost converters. (Reprinted from Luo, F. L. and Ye, H. 2006. *Essential DC/DC Converters*. Boca Raton: Taylor & Francis Group LLC, p. 384. With permission.)

The voltage transfer gain is

$$G = \frac{V_O}{V_{in}} = \begin{cases} \left(\frac{1}{1-k}\right)^n - 1 & \text{main series,} \\ 2 * \left(\frac{1}{1-k}\right)^n - 1 & \text{additional series,} \\ \left(\frac{2}{1-k}\right)^n - 1 & \text{double series,} \\ \left(\frac{3}{1-k}\right)^n - 1 & \text{triple series,} \\ \left(\frac{j+1}{1-k}\right)^n - 1 & \text{multiple } (j) \text{ series.} \end{cases} \tag{7.416}$$

7.6 UL Luo-Converter

The VL technique has been widely applied in the design of electronic circuits. Since the last century, it has been successfully applied in the design of power DC/DC converters. Good examples are the three series Luo-converters. Using the VL technique, one can obtain the converter's voltage transfer gain stage by stage in arithmetic series, which is higher than that of classical converters such as the buck converter, the boost converter and the buck–boost converter. Assume that the input voltage and current of a DC/DC converter are V_1 and I_1, the output voltage and current are V_2 and I_2, and the conduction duty cycle is k. To compare the transfer gains of these converters, we list the formulae below:

$$\text{Buck converter: } G = \frac{V_2}{V_1} = k,$$

$$\text{Boost converter: } G = \frac{V_2}{V_1} = \frac{1}{1-k},$$

$$\text{Buck-boost converter: } G = \frac{V_2}{V_1} = \frac{k}{1-k},$$

$$\text{Luo-converter: } G = \frac{V_2}{V_1} = \frac{k^{h(n)}[n + h(n)]}{1-k}, \tag{7.417}$$

where n is the stage number, and $h(n)$ is the Hong function:

$$h(n) = \begin{cases} 1 & n = 0 \\ 0 & n > 0 \end{cases},$$

and $n = 0$ for the elementary circuit with the voltage transfer gain

$$G = \frac{V_2}{V_1} = \frac{k}{1-k}. \tag{7.418}$$

The SL technique has been paid much more attention since it yields higher voltage transfer gain. Good examples are the SL Luo-converters. Using this technique, one can obtain the converter's voltage transfer gain stage by stage in geometrical series. The gain calculation formula is

$$G = \frac{V_2}{V_1} = \left(\frac{j+2-k}{1-k}\right)^n, \tag{7.419}$$

where n is the stage number and j is the multiple-enhanced number. Note that $n = 1$ and $j = 0$ for the elementary circuit with

$$G = \frac{V_2}{V_1} = \frac{2-k}{1-k}. \tag{7.420}$$

We introduce the UL Luo-converter as a novel approach of the new technology called the UL technique, which produces even higher voltage transfer gains [1,2,10,11]. Simulation results verified our analysis and calculation, and illustrated the advanced characteristics of this converter.

7.6.1 Operation of the UL Luo-Converter

The circuit diagram is shown in Figure 7.61a, which consists of one switch S, two inductors L_1 and L_2, two capacitors C_1 and C_2, three diodes, and the load R. Its switch-on equivalent circuit is shown in Figure 7.61b. Its switch-off equivalent circuit for the CCM is shown in Figure 7.61c and the switch-off equivalent circuit for the DCM is shown in Figure 7.61d.

It is a converter with a very simple structure when compared with other converters. As usual, the input voltage and current of the UL Luo-converter are V_1 and I_1, the output voltage and current are V_2 and I_2, the conduction duty cycle is k, and the switching frequency is f. Consequently, the repeating period $T = 1/f$, the switch-on period is kT, and the switch-off period is $(1 - k)T$. To concentrate the operation process, we assume that all components except load R are ideal ones. Therefore, no power losses are considered during power transformation, that is, $P_{in} = P_O$ or $V_1 \times I_1 = V_2 \times I_2$.

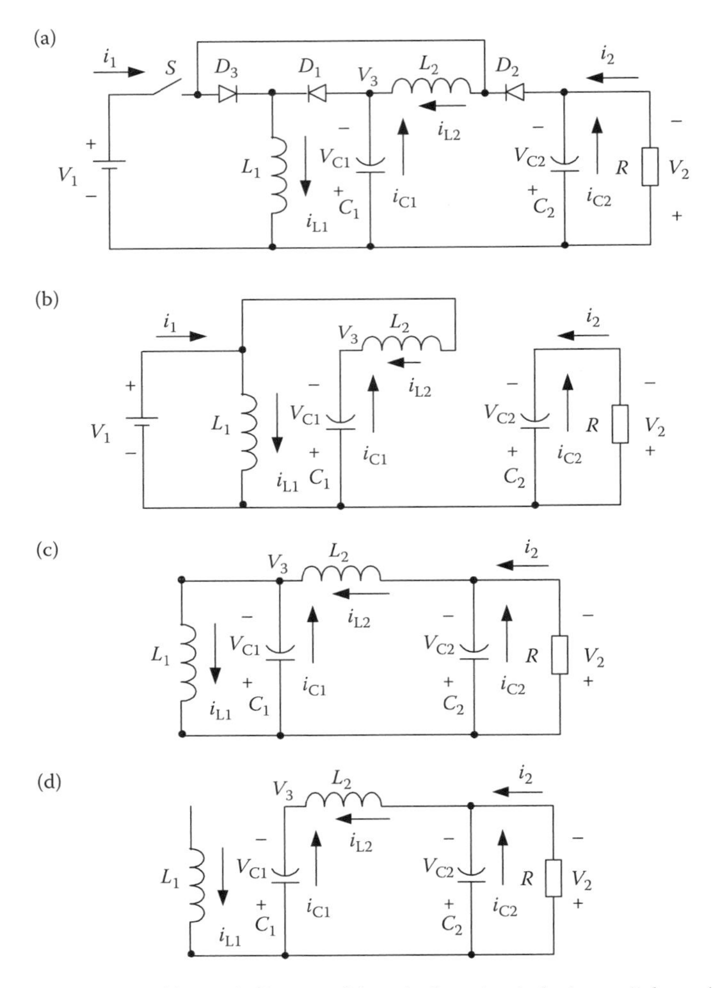

FIGURE 7.61 UL Luo-converter: (a) circuit diagram, (b) equivalent circuit during switch-on, (c) equivalent circuit during switch-off (CCM), and (d) equivalent circuit during switch-off (DCM). (Reprinted from Luo, F. L. and Ye, H. 2006. *Essential DC/DC Converters*. Boca Raton: Taylor & Francis Group LLC. With permission.)

7.6.1.1 Continuous Conduction Mode

Referring to Figures 7.61b and c, we have obtained the result that the current i_{L1} increases along the slope $+V_1/L_1$ during the switch-on period and decreases along the slope $-V_3/L_1$ during the switch-off period. In the steady state, the current increment is equal to the current decrement in the whole period T. The following relation is obtained:

$$kT\frac{V_1}{L_1} = (1-k)T\frac{V_3}{L_1}.$$

Thus,

$$V_{C1} = V_3 = \frac{k}{1-k}V_1. \tag{7.421}$$

The current i_{L2} increases with the slope $+(V_1 - V_3)/L_2$ during the switch-on period and decreases with the slope $-(V_3 - V_2)/L_2$ during the switch-off period. In the steady state, the current increment is equal to the current decrement in the whole period T. We obtain the following relation:

$$kT\frac{V_1 + V_3}{L_2} = (1-k)T\frac{V_2 - V_3}{L_2},$$

$$V_2 = V_{C2} = \frac{2-k}{1-k}V_3 = \frac{k}{1-k}\frac{2-k}{1-k}V_1 = \frac{k(2-k)}{(1-k)^2}V_1 \tag{7.422}$$

The voltage transfer gain is

$$G = \frac{V_2}{V_1} = \frac{k}{1-k}\frac{2-k}{1-k} = \frac{k(2-k)}{(1-k)^2}. \tag{7.423}$$

It is much higher than the voltage transfer gains of the VL Luo-converter and SL Luo-converter in Equations 7.418 and 7.420. Actually, the gain in Equation 7.423 is the consequence of those in Equations 7.418 and 7.420. Another advantage is the starting output voltage of 0 V. The curve of the voltage transfer gain M versus the conduction duty cycle k is shown in Figure 7.62.

The relation between input and output average currents is

$$I_2 = \frac{(1-k)^2}{k(2-k)}I_1. \tag{7.424}$$

The relation between average currents I_{L2} and I_{L1} is

$$I_{L2} = (1-k)I_{L1}. \tag{7.425}$$

The other relations are

$$I_{L2} = \left(1 + \frac{k}{1-k}\right)I_2 = \frac{1}{1-k}I_2, \tag{7.426}$$

$$I_{L1} = \frac{1}{1-k}I_{L2} = \left(\frac{1}{1-k}\right)^2 I_2. \tag{7.427}$$

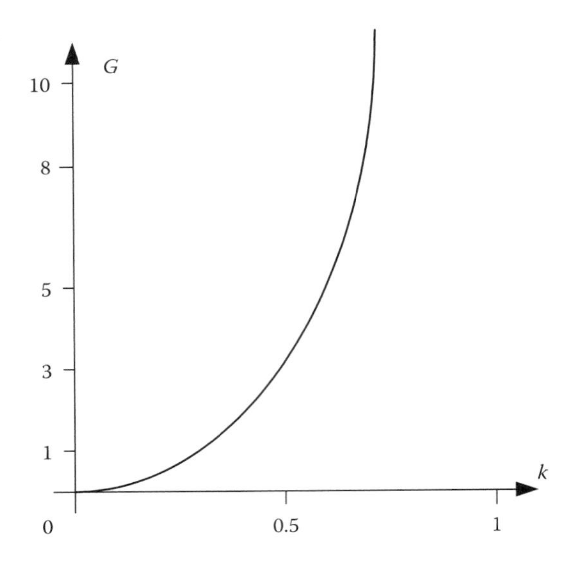

FIGURE 7.62 Voltage transfer gain G versus conduction duty cycle k. (Reprinted from Luo, F. L. and Ye, H. 2006. *Essential DC/DC Converters*. Boca Raton: Taylor & Francis Group LLC, p. 395. With permission.)

The variation of inductor current i_{L1} is

$$\Delta i_{L1} = kT\frac{V_1}{L_1} \tag{7.428}$$

and its variation ratio is

$$\xi_1 = \frac{\Delta i_{L1}/2}{I_{L1}} = \frac{k(1-k)^2 TV_1}{2L_1 I_2} = \frac{k(1-k)^2 TR}{2L_1 M} = \frac{(1-k)^4 R}{2(2-k)fL_1}. \tag{7.429}$$

The diode current i_{D1} is the same as the inductor current i_{L1} during the switch-off period. For the CCM operation, both currents do not descend to zero, that is, $\xi_1 \geq 1$.

The variation of inductor current i_{L2} is

$$\Delta i_{L2} = \frac{kTV_1}{(1-k)L_2} \tag{7.430}$$

and its variation ratio is

$$\xi_2 = \frac{\Delta i_{L2}/2}{I_{L2}} = \frac{kTV_1}{2L_2 I_2} = \frac{kTR}{2L_2 M} = \frac{(1-k)^2 R}{2(2-k)fL_2}. \tag{7.431}$$

The variation of capacitor voltage v_{C1} is

$$\Delta v_{C1} = \frac{\Delta Q_{C1}}{C_1} = \frac{kTI_{L2}}{C_1} = \frac{kTI_2}{(1-k)C_1} \tag{7.432}$$

and its variation ratio is

$$\sigma_1 = \frac{\Delta v_{C1}/2}{V_{C1}} = \frac{kTI_2}{2(1-k)V_3 C_1} = \frac{k(2-k)}{2(1-k)^2 fC_1 R}. \tag{7.433}$$

The variation of capacitor voltage v_{C2} is

$$\Delta v_{C2} = \frac{\Delta Q_{C2}}{C_2} = \frac{kTI_2}{C_2} \tag{7.434}$$

and its variation ratio is

$$\varepsilon = \sigma_2 = \frac{\Delta v_{C2}/2}{V_{C2}} = \frac{kTI_2}{2V_2C_2} = \frac{k}{2fC_2R}. \tag{7.435}$$

Example 7.6

An UL Luo-converter, shown in Figure 7.61a, has $V_1 = 20\,\text{V}$, all inductors have 10 mH, all capacitors have 20 μF, $R = 500\,\Omega$, $f = 50\,\text{kHz}$, and conduction duty cycle $k = 0.6$. Calculate the variation ratios of current i_{L1}, current i_{L2} and voltage v_{C1}, and the output voltage and its variation ratio.

SOLUTION

From Equation 7.429, we can obtain the variation ratio of current i_{L1}:

$$\xi_1 = \frac{(1-k)^4 R}{2(2-k)fL_1} = \frac{(1-0.6)^4 \times 500}{2(2-0.6) \times 50\,\text{k} \times 10\,\text{m}} = 0.0091.$$

From Equation 7.431, we can obtain the variation ratio of current i_{L2}:

$$\xi_2 = \frac{(1-k)^2 R}{2(2-k)fL_2} = \frac{(1-0.6)^2 \times 500}{2(2-0.6) \times 50\,\text{k} \times 10\,\text{m}} = 0.057.$$

From Equation 7.433, we can obtain the variation ratio of voltage v_{C1}:

$$\sigma_1 = \frac{k(2-k)}{2(1-k)^2 fC_1R} = \frac{0.6(2-0.6)}{2(1-0.6)^2 \times 50\,\text{k} \times 20\,\mu \times 500} = 0.00525.$$

This converter works in the CCM. From Equation 7.422, we can obtain the output voltage

$$V_2 = \frac{k(2-k)}{(1-k)^2}V_1 = \frac{0.6(2-0.6)}{(1-0.6)^2}20 = 105\,\text{V}.$$

From Equation 7.435, its variation ratio is

$$\varepsilon = \frac{k}{2fC_2R} = \frac{0.6}{2 \times 50\,\text{k} \times 20\,\mu \times 500} = 0.0006.$$

7.6.1.2 Discontinuous Conduction Mode

Referring to Figures 7.61b–d, we have obtained the result that the current i_{L1} increases along the slope $+V_1/L_1$ during the switch-on period and decreases along the slope $-V_3/L_1$ during the switch-off period. The inductor current i_{L1} decreases to zero before $t = T$, that is, the current becomes zero before the switch turns on once again.

The current waveform is shown in Figure 7.63. The DCM operation condition is defined as

$$\xi_1 \geq 1$$

or

$$\xi_1 = \frac{k(1-k)^2 TR}{2L_1 M} = \frac{(1-k)^4 R}{2(2-k) f L_1} \geq 1. \tag{7.436}$$

Taking the equal sign, we obtain the boundary between CCM and DCM operations. Here we define the normalized impedance Z_N as

$$Z_N = \frac{R}{f L_1}. \tag{7.437}$$

The boundary equation is

$$G = \frac{k(1-k)^2}{2} Z_N \tag{7.438}$$

or

$$\frac{G}{Z_N} = \frac{k(1-k)^2}{2}.$$

The corresponding Z_N is

$$Z_N = \frac{k(2-k)/(1-k)^2}{k(1-k)^2/2} = \frac{2(2-k)}{(1-k)^4}. \tag{7.439}$$

The curve is shown in Figure 7.64 and Table 7.7.

We define the filling factor m to describe the current's survival time. For DCM operation,

$$0 < m \leq 1.$$

In the steady state, the current increment is equal to the current decrement in the whole period T. The following relation is obtained:

$$kT\frac{V_1}{L_1} = (1-k)mT\frac{V_3}{L_1}.$$

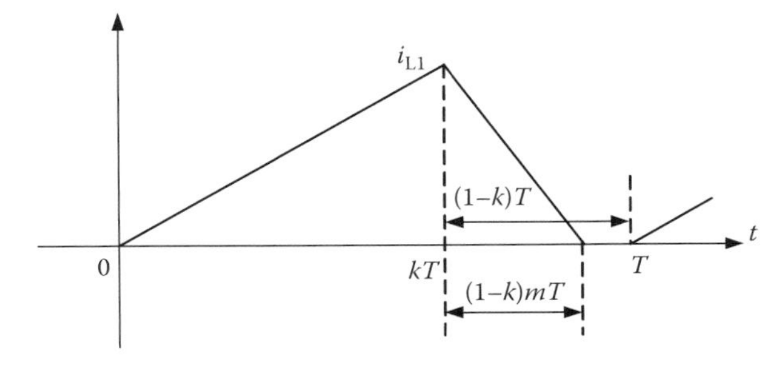

FIGURE 7.63 Discontinuous inductor current i_{L1}. (Reprinted from Luo, F. L. and Ye, H. 2006. *Essential DC/DC Converters*. Boca Raton: Taylor & Francis Group LLC, p. 397. With permission.)

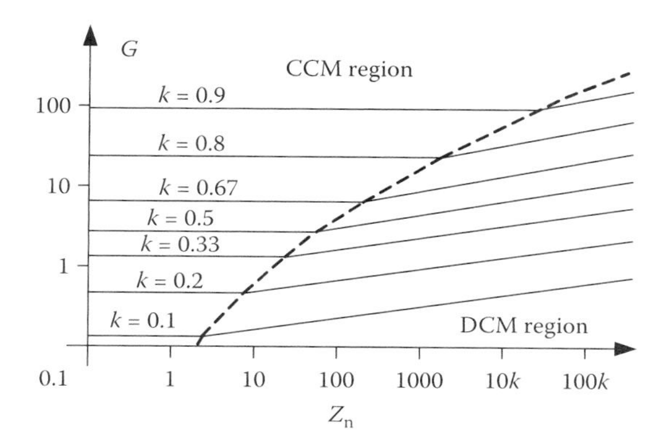

FIGURE 7.64 Boundary between CCM and DCM. (Reprinted from Luo, F. L. and Ye, H. 2006. *Essential DC/DC Converters*. Boca Raton: Taylor & Francis Group LLC, p. 398. With permission.)

TABLE 7.7

Boundary between CCM and DCM

k	0.2	0.33	0.5	0.67	0.8	0.9
G	0.5625	1.25	3	8	24	99
G/Z_N	0.064	2/27	1/16	1/27	0.016	0.0045
Z_N	8.8	16.9	48	216	1500	22,000

Source: Data from Luo, F. L. and Ye, H. 2006. *Essential DC/DC Converters*, p. 398. Boca Raton: Taylor & Francis Group LLC.

Thus,

$$V_{C1} = V_3 = \frac{k}{(1-k)m} V_1. \tag{7.440}$$

Comparing Equations 7.421 and 7.440, we found that the voltage V_3 is higher during DCM operation since the filling factor $m < 1$. Its expression is

$$m = \frac{1}{\xi_1} = \frac{2L_1 G}{k(1-k)^2 TR} = \frac{2(2-k)}{(1-k)^4 Z_N}. \tag{7.441}$$

The current i_{L2} increases along the slope $+(V_1 - V_3)/L_2$ during the switch-on period and decreases along the slope $-(V_3 - V_2)/L_2$ during the switch-off period. In the steady state, the current increment is equal to the current decrement in the whole period T. We obtain the following relation:

$$kT\frac{V_1 + V_3}{L_2} = (1-k)T\frac{V_2 - V_3}{L_2},$$

$$V_2 = V_{C2} = \frac{2-k}{1-k} V_3 = \frac{k(2-k)}{m(1-k)^2} V_1. \tag{7.442}$$

The voltage transfer gain in the DCM is

$$G_{DCM} = \frac{V_2}{V_1} = \frac{k(2-k)}{m(1-k)^2} = \frac{k(1-k)^2}{2} Z_N. \tag{7.443}$$

It is higher than the voltage transfer gain during CCM operation since $m < 1$. It can be seen that the voltage transfer gain G_{DCM} is directly proportional to the normalized impedance Z_N, and this is shown in Figure 7.64.

7.6.2 Instantaneous Values

Instantaneous values of the voltage and current of each component are very important to describe the converter operation. Referring to Figure 7.61, we have obtained the components' values in CCM and DCM operations.

7.6.2.1 Continuous Conduction Mode

Referring to Figure 7.61b and c, we have obtained the instantaneous values of the voltage and current of each component in CCM operation, as listed below:

$$i_{L1}(t) = \begin{cases} I_{L1-min} + \dfrac{V_1}{L_1}t, & 0 \le t \le kT, \\ I_{L1-max} - \dfrac{V_3}{L_1}t, & kT \le t \le T, \end{cases} \tag{7.444}$$

$$i_{L2}(t) = \begin{cases} I_{L2-min} + \dfrac{V_1 - V_3}{L_2}t, & 0 \le t \le kT, \\ I_{L2-max} - \dfrac{V_2 - V_1}{L_2}t, & kT \le t \le T, \end{cases} \tag{7.445}$$

$$i_1(t) = i_s = \begin{cases} I_{1-min} + \left(\dfrac{V_1}{L_1} + \dfrac{V_1 - V_3}{L_2}\right)t, & 0 \le t \le kT, \\ 0, & kT \le t \le T, \end{cases} \tag{7.446}$$

$$i_{D1}(t) = \begin{cases} 0, & 0 \le t \le kT, \\ I_{L1-max} - \dfrac{V_3}{L_1}t, & kT \le t \le T, \end{cases} \tag{7.447}$$

$$i_{C1}(t) = \begin{cases} -\left(I_{L2-min} + \dfrac{V_1 - V_3}{L_2}t\right), & 0 \le t \le kT, \\ I_{C1}, & kT \le t \le T, \end{cases} \tag{7.448}$$

$$i_{C2}(t) = \begin{cases} -I_2, & 0 \le t \le kT, \\ I_{C2}, & kT \le t \le T, \end{cases} \tag{7.449}$$

$$v_{L1}(t) = \begin{cases} V_1, & 0 \le t \le kT, \\ V_3, & kT \le t \le T, \end{cases} \tag{7.450}$$

$$v_{L2}(t) = \begin{cases} V_1 - V_3, & 0 \le t \le kT, \\ V_2 - V_3, & kT \le t \le T, \end{cases} \tag{7.451}$$

$$v_s = \begin{cases} 0, & 0 \le t \le kT, \\ V_1 - V_3, & kT \le t \le T, \end{cases} \tag{7.452}$$

$$v_{D1}(t) = \begin{cases} V_1 - V_3, & 0 \le t \le kT, \\ 0, & kT \le t \le T, \end{cases} \tag{7.453}$$

$$v_{C1}(t) = \begin{cases} V_3 - \dfrac{I_{L2}}{C_1}t, & 0 \le t \le kT, \\ V_3 + \dfrac{I_{C1}}{C_1}t, & kT \le t \le T, \end{cases} \tag{7.454}$$

$$v_{C2}(t) = \begin{cases} V_2 - \dfrac{I_2}{C_2}t, & 0 \le t \le kT, \\ V_2 + \dfrac{I_{C2}}{C_2}t, & kT \le t \le T. \end{cases} \tag{7.455}$$

7.6.2.2 Discontinuous Conduction Mode

Referring to Figure 7.61b–d, we have obtained the instantaneous values of the voltage and current of each component in DCM operation. Since the inductor current i_{L1} is discontinuous, some parameters have three states with $T' = kT + (1-k)mT < T$.

$$i_{L1}(t) = \begin{cases} \dfrac{V_1}{L_1}t, & 0 \le t \le kT, \\ I_{L1-\max} - \dfrac{V_3}{L_1}t, & kT \le t \le T', \\ 0, & T' \le t \le kT, \end{cases} \tag{7.456}$$

$$i_{L2}(t) = \begin{cases} I_{L2-\min} + \dfrac{V_1 - V_3}{L_2}t, & 0 \le t \le kT, \\ I_{L2-\max} - \dfrac{V_2 - V_1}{L_2}t, & kT \le t \le T, \end{cases} \tag{7.457}$$

$$i_1(t) = i_s = \begin{cases} I_{1-\min} + \left(\dfrac{V_1}{L_1} + \dfrac{V_1 - V_3}{L_2}\right)t, & 0 \le t \le kT, \\ 0, & kT \le t \le T, \end{cases} \tag{7.458}$$

$$i_{D1}(t) = \begin{cases} 0, & 0 \le t \le kT, \\ I_{L1-\max} - \dfrac{V_3}{L_1}t, & kT \le t \le T', \\ 0, & T' \le t \le kT, \end{cases} \tag{7.459}$$

$$i_{C1}(t) = \begin{cases} -\left(I_{L2-\min} + \dfrac{V_1 - V_3}{L_2}t\right), & 0 \le t \le kT, \\ I_{C1}, & kT \le t \le T, \end{cases} \tag{7.460}$$

$$i_{C2}(t) = \begin{cases} -I_2, & 0 \le t \le kT, \\ I_{C2}, & kT \le t \le T, \end{cases} \tag{7.461}$$

$$v_{L1}(t) = \begin{cases} V_1, & 0 \le t \le kT, \\ V_3, & kT \le t \le T', \\ 0, & T' \le t \le kT, \end{cases} \tag{7.462}$$

$$v_{L2}(t) = \begin{cases} V_1 - V_3, & 0 \le t \le kT, \\ V_2 - V_3, & kT \le t \le T, \end{cases} \tag{7.463}$$

$$v_{s}(t) = \begin{cases} 0, & 0 \le t \le kT, \\ V_1 - V_3, & kT \le t \le T', \\ V_1, & T' \le t \le kT, \end{cases} \tag{7.464}$$

$$v_{D1}(t) = \begin{cases} V_1 - V_3, & 0 \le t \le kT, \\ 0, & kT \le t \le T', \\ -V_3, & T' \le t \le kT, \end{cases} \tag{7.465}$$

$$v_{C1}(t) = \begin{cases} V_3 - \dfrac{I_{L2}}{C_1}t, & 0 \le t \le kT, \\ V_3 + \dfrac{I_{C1}}{C_1}t, & kT \le t \le T, \end{cases} \tag{7.466}$$

$$v_{C2}(t) = \begin{cases} V_2 - \dfrac{I_2}{C_2}t, & 0 \le t \le kT, \\ V_2 + \dfrac{I_{C2}}{C_2}t, & kT \le t \le T. \end{cases} \tag{7.467}$$

TABLE 7.8

Comparison of Various Converters Gains

k	0.2	0.33	0.5	0.67	0.8	0.9
Buck	0.2	0.33	0.5	0.67	0.8	0.9
Boost	1.25	1.5	2	3	5	10
Buck–boost	0.25	0.5	1	2	4	9
Luo-converter	0.25	0.5	1	2	4	9
SL Luo-converter	2.25	2.5	3	4	6	11
UL Luo-converter	0.56	1.25	3	8	24	99

Source: Data from Luo, F. L. and Ye, H. 2006. *Essential DC/DC Converters*. Boca Raton: Taylor & Francis Group LLC, p. 403.

7.6.3 Comparison of the Gain to Other Converters' Gains

The UL Luo-converter has been successfully developed using a novel approach of the new technology called UL. Table 7.8 lists the voltage transfer gains of various converters at $k = 0.2, 0.33, 0.5, 0.67, 0.8$, and 0.9. The outstanding characteristics of the UL Luo-converter

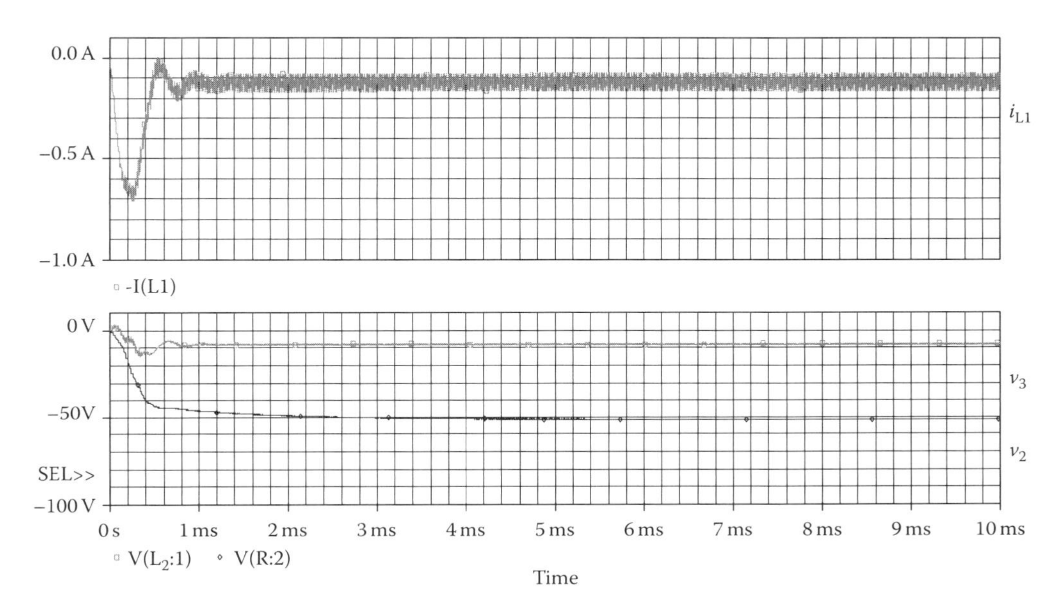

FIGURE 7.65 Simulation results for $k = 0.6$. (Reprinted from Luo, F. L. and Ye, H. 2006. *Essential DC/DC Converters*. Boca Raton: Taylor & Francis Group LLC, p. 404. With permission.)

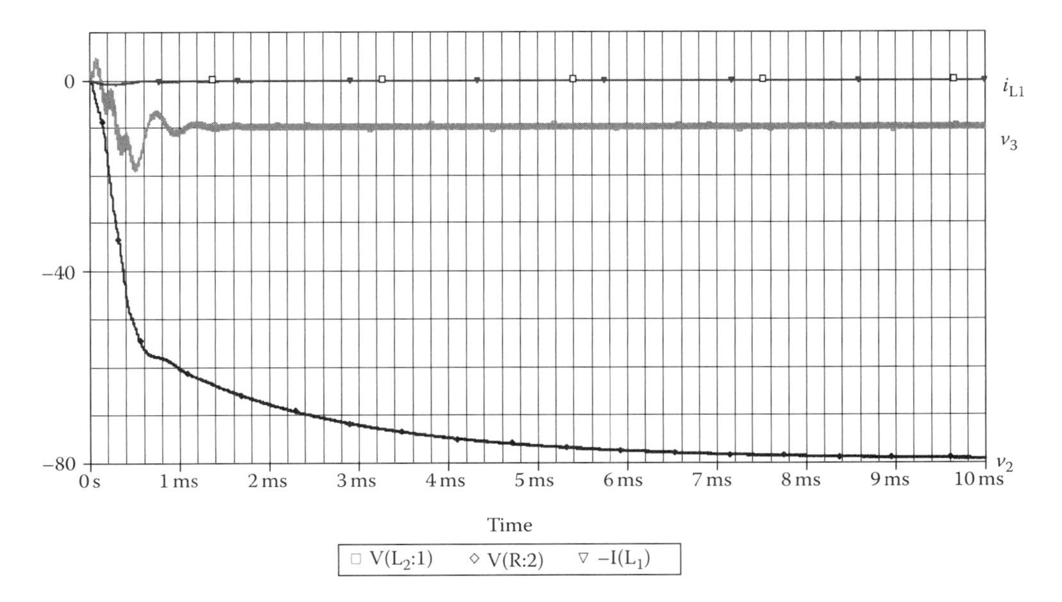

FIGURE 7.66 Simulation results for $k = 0.66$. (Reprinted from Luo, F. L. and Ye, H. 2006. *Essential DC/DC Converters*. Boca Raton: Taylor & Francis Group LLC, p. 404. With permission.)

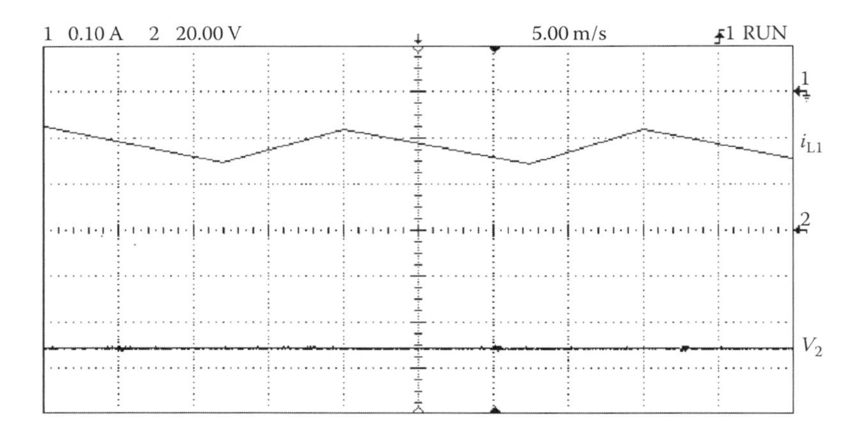

FIGURE 7.67 Experimental results for $k = 0.6$. (Reprinted from Luo, F. L. and Ye, H. 2006. *Essential DC/DC Converters*. Boca Raton: Taylor & Francis Group LLC, p. 405. With permission.)

are very well presented. From the comparison, we can clearly see that the UL Luo-converter has very high voltage transfer gain: $G(k)|_{k=0.5} = 3$, $G(k)|_{k=0.67} = 8, G(k)|_{k=0.8} = 24$, and $G(k)|_{k=0.9} = 99$.

7.6.4 Simulation Results

To verify the advantages of the UL Luo-converter, a PSpice simulation method was applied. We choose the following parameters: $V_1 = 10\,\text{V}, L_1 = L_2 = 1\,\text{mH}, C_1 = C_2 = 1\,\mu\text{F}, R = 3\,\text{k}\Omega$, $f = 50\,\text{kHz}$, and conduction duty cycle $k = 0.6$ and 0.66. The output voltage is $V_2 = 52.5$ and 78 V, correspondingly. The first waveform is the inductor current i_{L1}, which flows through the inductor L_1. The second and third waveforms are the voltage V_3 and the output voltage V_2. These simulation results are identical to the calculation results. The results are shown in Figures 7.65 and 7.66, respectively.

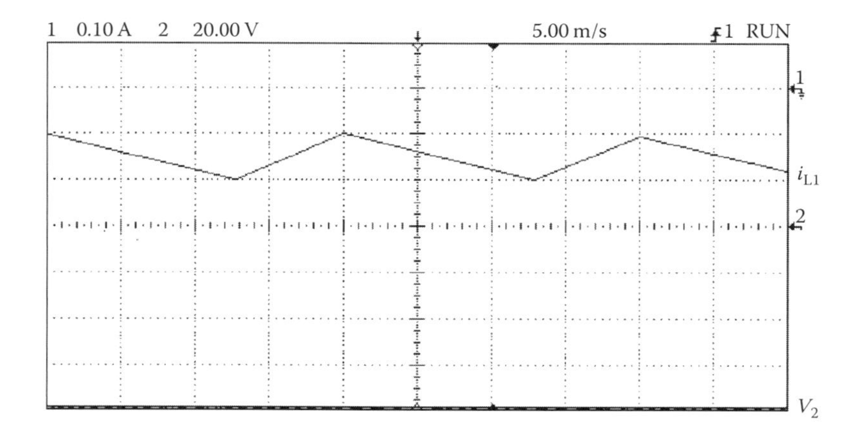

FIGURE 7.68 Experimental results for $k = 0.66$. (Reprinted from Luo, F. L. and Ye, H. 2006. *Essential DC/DC Converters*. Boca Raton: Taylor & Francis Group LLC, p. 405. With permission.)

7.6.5 Experimental Results

To verify the advantages and design of the UL Luo-converter and compare them with the simulation results, we constructed a test rig with the following components: $V_1 = 10\,\text{V}$, $L_1 = L_2 = 1\,\text{mH}$, $C_1 = C_2 = 1\,\mu\text{F}$, $R = 3\,\text{k}\Omega$, $f = 50\,\text{kHz}$, and conduction duty cycle $k = 0.6$ and 0.66. The output voltage is $V_2 = 52$ and 78 V, correspondingly. The first waveform is the inductor current i_{L1}, which flows through the inductor L_1. The second waveform is the output voltage V_2. The experimental results are shown in Figures 7.67 and 7.68, respectively. The test results are identical to those of the simulation results shown in Figures 7.65 and 7.66, and confirm the calculation results and our design.

7.6.6 Summary

The UL Luo-converter has been successfully developed using a novel approach of the new technology called the UL technique, that produces even higher voltage transfer gain. The voltage transfer gain of the UL Luo-converter is much higher than that of VL Luo-converter and the SL Luo-converter. This chapter introduced the operation and characteristics of this converter in detail. The converter will be applied in industrial applications with high output voltages.

Homework

7.1. A re-lift circuit of the P/O SL Luo-converter, shown in Figure 7.2a, has $V_{in} = 20\,\text{V}$, $L_1 = 10\,\text{mH}$, $C_2 = 20\,\mu\text{F}$, $R = 100\,\Omega$, $f = 50\,\text{kHz}$, and conduction duty cycle $k = 0.6$. Calculate the variation ratio of current i_{L1}, and the output voltage and its variation ratio.

7.2. An elementary additional circuit of the P/O SL Luo-converter, shown in Figure 7.5a, has $V_{in} = 20\,\text{V}$, all inductors have 10 mH, all capacitors have $20\,\mu\text{F}$, $R = 1000\,\Omega$, $f = 50\,\text{kHz}$, and conduction duty cycle $k = 0.6$. Calculate the variation ratio of current i_{L1}, and the output voltage and its variation ratio.

7.3. An N/O triple-lift circuit, shown in Figure 7.19a, has $V_{in} = 20\,\text{V}$, all inductors have 10 mH, all capacitors have $20\,\mu\text{F}$, $R = 200\,\Omega$, $f = 50\,\text{kHz}$, and conduction duty cycle $k = 0.6$. Calculate the variation ratio of current i_{L1}, and the output voltage and its variation ratio.

7.4. An elementary boost additional circuit, shown in Figure 7.34a, has $V_{in} = 20\,\text{V}$, $L_1 = 10\,\text{mH}$, all capacitors have $20\,\mu\text{F}$, $R = 400\,\Omega$, $f = 50\,\text{kHz}$, and conduction duty cycle $k = 0.6$. Calculate the variation ratio of current i_{L1}, and the output voltage and its variation ratio.

7.5. An N/O three-stage multiple ($j = 5$) boost converter, shown in Figure 7.59a, has $V_{in} = 20\,\text{V}$, all inductors have 10 mH, all capacitors have $20\,\mu\text{F}$, $R = 10\,\text{k}\Omega$, $f = 50\,\text{kHz}$, and conduction duty cycle $k = 0.6$. Calculate the output voltage and its variation ratio.

7.6. An UL Luo-converter, shown in Figure 7.61a, has $V_1 = 20\,\text{V}$, all inductors have 1 mH, all capacitors have $2\,\mu\text{F}$, $R = 10\,\text{k}\Omega$, $f = 50\,\text{kHz}$, and conduction duty cycle $k = 0.6$. Calculate the output voltage.

References

1. Luo, F. L. and Ye, H. 2004. *Advanced DC/DC Converters*. Boca Raton: CRC Press.
2. Luo, F. L. and Ye, H. 2006. *Essential DC/DC Converters*. Boca Raton: Taylor & Francis Group LLC.
3. Luo, F. L. and Ye, H. 2002. Super-lift Luo-converters. *Proceedings of the IEEE International Conference PESC 2002*, pp. 425–430.
4. Luo, F. L. and Ye, H. 2003. Positive output super-lift converters. *IEEE Transactions on Power Electronics*, 18, 105–113.
5. Luo, F. L. and Ye, H. 2003. Negative output super-lift Luo-converters. *Proceedings of the IEEE International Conference PESC 2003*, pp. 1361–1366.
6. Luo, F. L. and Ye, H. 2003. Negative output super-lift converters. *IEEE Transactions on Power Electronics*, 18, 1113–1121.
7. Luo, F. L. and Ye, H. 2004. Positive output cascaded boost converters. *IEE-Proceedings on Electric Power Applications*, 151, pp. 590–606.
8. Zhu, M. and Luo, F. L. 2006. Steady-state performance analysis of cascaded boost converters. *Proceeding of IEEE Asia Pacific Conference on Circuits and Systems*, pp. 659–662.
9. Zhu, M. and Luo, F. L. 2006. Generalized steady-state analysis on developed series of cascaded boost converters. *Proceedings of IEEE Asia Pacific Conference on Circuits and Systems, APCCAS 2006*, pp. 1399–1402.
10. Luo, F. L. and Ye, H. 2005. Ultra-lift Luo-converter. *IEE-Proceedings on Electric Power Applications*, 152, pp. 27–32.
11. Luo, F. L. and Ye, H. 2004. Investigation of ultra-lift Luo-converter. *Proceedings of the IEEE International Conference POWERCON 2004*, pp. 13–18.

8

Pulse-Width-Modulated DC/AC Inverters

DC/AC inverters are used to quickly develop knowledge of the power switching circuits applied in industrial applications in comparison with other power switching circuits. In the 20th century, a number of topologies of DC/AC inverters were created. Generally, DC/AC inverters are mainly used in an AC motor adjustable speed drive (ASD). Power DC/AC inverters have been widely used in other industrial applications since the late 1980s. Semiconductor manufacture development resulted in power devices such as GTO, triac, BT, IGBT, MOSFET, and so on in higher switching frequencies (say from tens of kHz up to a few MHz). Because of devices such as thyristors (SCRs) with low switching frequency and high power rate, the above-mentioned devices have low power rate and high switching frequency [1,2].

Square-waveform DC/AC inverters were used before the 1980s. Among this equipment, thyristor, GTO, and triac can be used in low-frequency switching operations. Also, high-frequency devices such as power BT and IGBT were produced. Corresponding equipment implementing the PWM technique has a large range of output voltage and frequency, and low THD.

Today, two DC/AC inversion techniques are popular: the PWM technique and the MLM technique. Most DC/AC inverters continue to be different prototypes of PWM DC/AC inverters. We introduce PWM inverters in this chapter and MLM inverters in the next.

8.1 Introduction

DC/AC inverters are used for converting a DC power source into an AC power application. They are generally used in the following applications:

1. Variable voltage/variable frequency AC supplies in an ASD, such as induction motor drives, synchronous machine drives, and so on.
2. Constant regulated voltage AC power supplies, such as uninterruptible power supplies (UPSs).
3. Static var (reactive power) compensations.
4. Passive/active series/parallel filters.
5. Flexible AC transmission systems (FACTSs).
6. Voltage compensations.

Adjustable-speed induction motor drive systems are widely used in industrial applications. These systems require DC/AC power supply with variable frequency, usually from

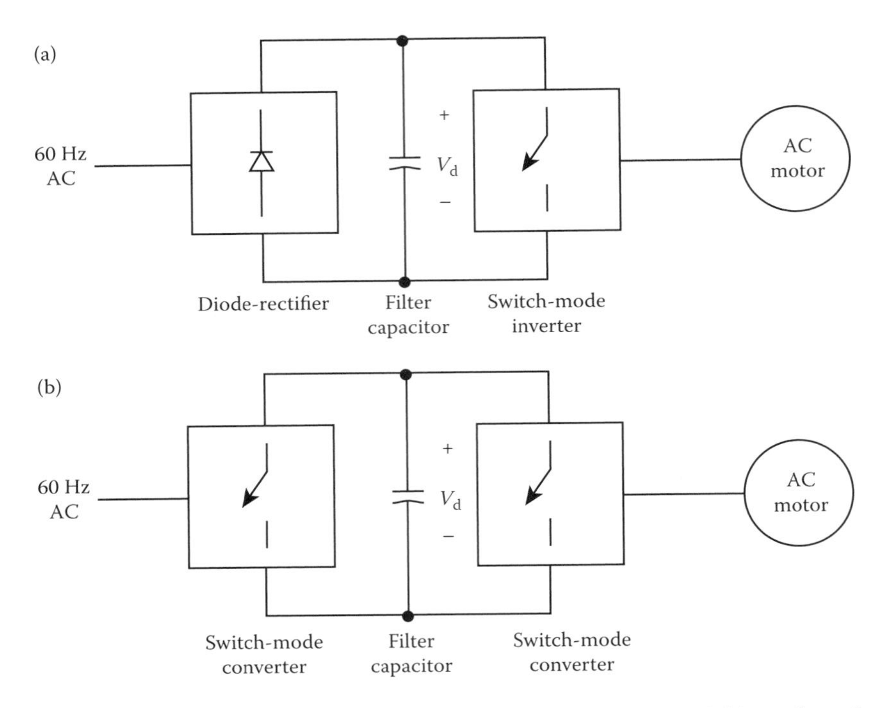

FIGURE 8.1 Standard ASD scheme: (a) switch-mode inverter in AC motor drive and (b) switch-mode converters for motoring/regenerative braking.

0 to 400 Hz in fractional horsepower (hp) to hundreds of HP. Today, there are a large number of DC/AC inverters in the world market. The typical block circuit of an ASD is shown in Figure 8.1. From this block diagram, we see that the power DC/AC inverter produces variable frequency and voltage to implement ASD.

The PWM technique is different from the PAM and PPM techniques. By implementing this technique, all pulses have adjustable pulse width with constant amplitude and phase. The corresponding circuit is called the pulse-width modulator. Typical input and output waveforms of a pulse-width modulator are shown in Figure 8.2. The output pulse train has pulses with the same amplitude and different widths, which correspond to the input signal at the sampling instants.

8.2 Parameters Used in PWM Operations

Some parameters are specially used in PWM operations.

8.2.1 Modulation Ratios

The modulation ratio is usually yielded by a uniformed-amplitude triangle (carrier) signal with amplitude V_{tri-m}. The maximum amplitude of the input signal is assumed to be V_{in-m}.

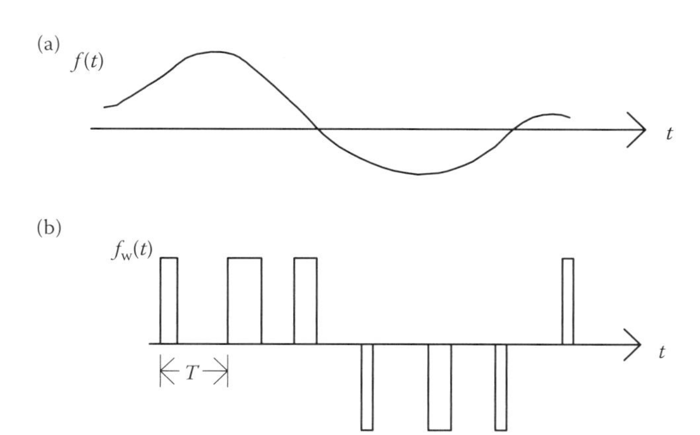

FIGURE 8.2 Typical (a) input and (b) output waveforms of a pulse-width modulator.

We define the amplitude modulation ratio m_a for a single-phase inverter as

$$m_a = \frac{V_{in-m}}{V_{tri-m}}. \tag{8.1}$$

We also define the frequency modulation ratio m_f as

$$m_f = \frac{f_{tri-m}}{f_{in-m}}. \tag{8.2}$$

A one-leg switch-mode inverter is shown in Figure 8.3. The DC-link voltage is V_d. Two large capacitors are used to establish the neutral point N. The AC output voltage from point a to N is V_{AO} and its fundamental component is $(V_{AO})_1$. We mark $(\hat{V}_{AO})_1$ to show the maximum amplitude of $(V_{AO})_1$. The waveforms of the input (control) signal and the triangle signal, and the spectrum of the PWM pulse train are shown in Figure 8.4.

If the maximum amplitude $(\hat{V}_{AO})_1$ of the input signal is smaller than and/or equal to half the DC-link voltage $V_d/2$, the modulation ratio m_a is smaller than and/or equal to unity. In this case, the fundamental component $(V_{AO})_1$ of the output AC voltage is proportional to the input voltage. The voltage control by varying m_a for a single-phase PWM is split in to three areas, as shown in Figure 8.5.

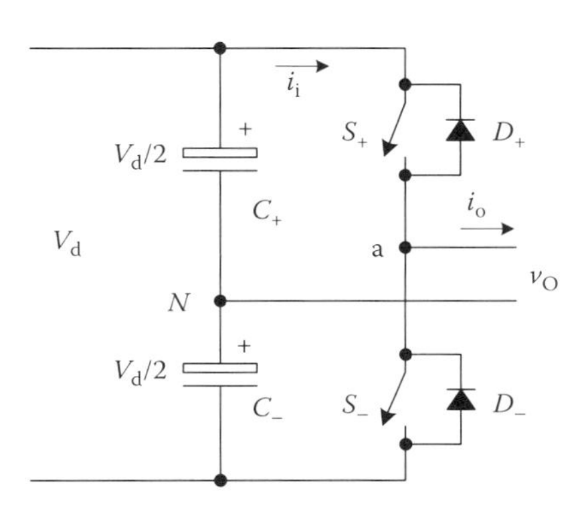

FIGURE 8.3 One-leg switch-mode inverter.

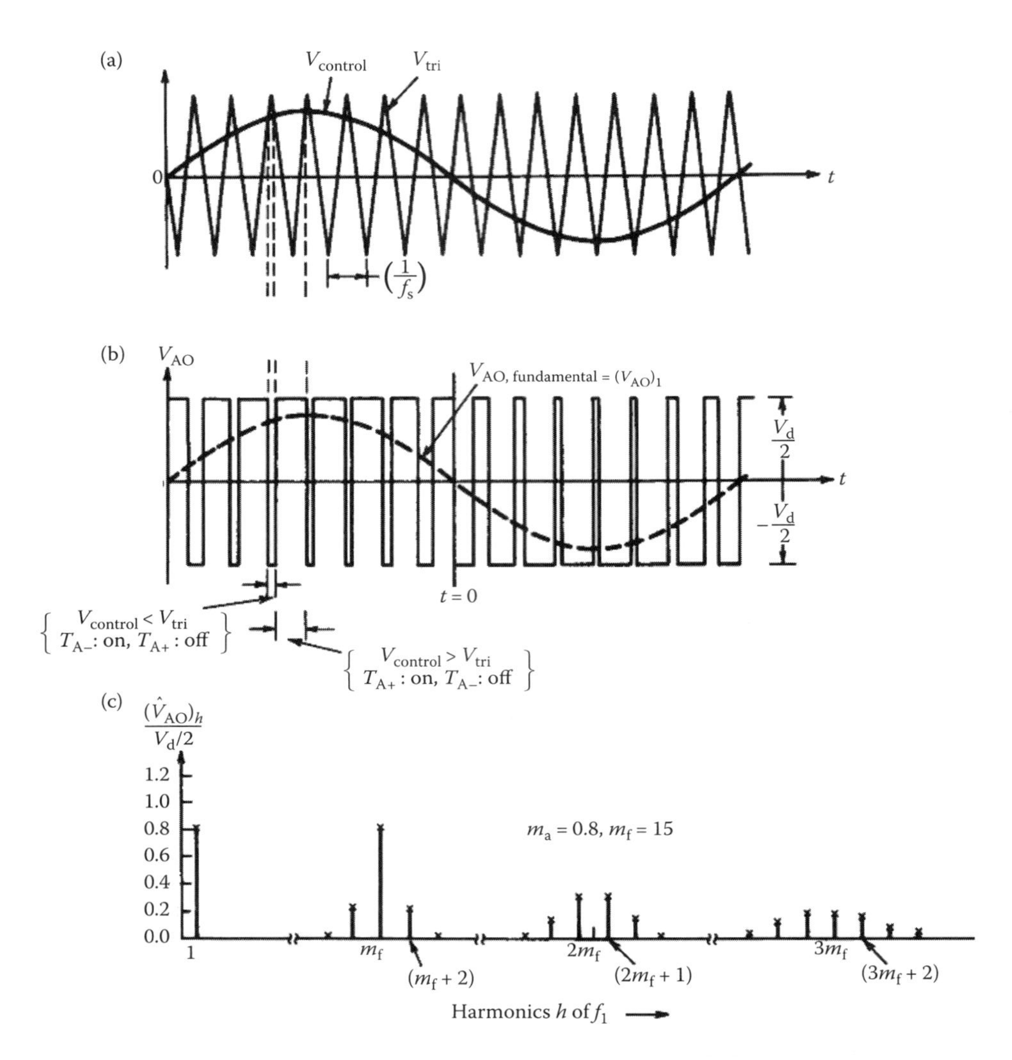

FIGURE 8.4 Pulse-width modulation. (a) Control and triangle waveforms, (b) inverter output waveform and its fundamental wave, and (c) spectrum of the inverter output waveform.

8.2.1.1 Linear Range ($m_a \leq 1.0$)

The condition $(\hat{V}_{AO})_1 = m_a(V_d/2)$ determines the linear region. It is a sinusoidal PWM where the amplitude of the fundamental frequency voltage varies linearly with the amplitude modulation ratio m_a. The PWM pushes the harmonics into a high frequency range around the switching frequency and its multiples. However, the maximum available amplitude of the fundamental frequency component may not be as high as desired.

8.2.1.2 Overmodulation ($1.0 < m_a \leq 1.27$)

The condition $(V_d/2) < (\hat{V}_{AO})_1 \leq (4/\pi)(V_d/2)$ determines the overmodulation region. When the amplitude of the fundamental frequency component in the output voltage

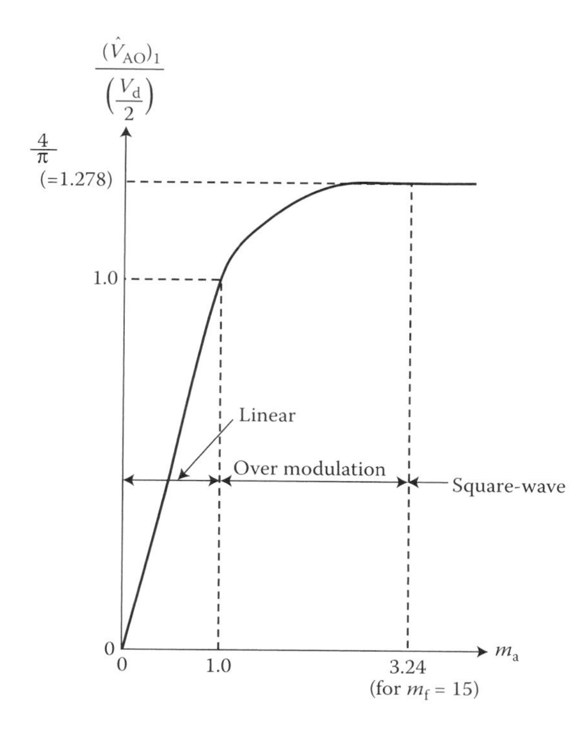

FIGURE 8.5 Voltage control by varying m_a.

increases beyond 1.0, it reaches overmodulation. In the overmodulation range, the amplitude of the fundamental frequency voltage no longer varies linearly with m_a.

Overmodulation causes the output voltage to contain many more harmonics in the sidebands as compared with the linear range. The harmonics with dominant amplitudes in the linear range may not be dominant during overmodulation.

8.2.1.3 Square Wave (Sufficiently Large $m_a > 1.27$)

The condition $(\hat{V}_{AO})_1 > (4/\pi)(V_d/2)$ determines the square-wave region. The inverter voltage waveform degenerates from a pulse-width-modulated waveform into a square wave. Each switch of the inverter leg in Figure 8.3 is on for one half-cycle (180°) of the desired output frequency.

8.2.1.4 Small m_f ($m_f \leq 21$)

Usually the triangle waveform frequency is much larger than the input signal frequency to obtain small THD. For the situation with a small $m_f \leq 21$, two points have to be mentioned:

- *Synchronous PWM*: For a small value of m_f, the triangle waveform signal and the input signal should be synchronized to each other (synchronous PWM). This synchronous PWM requires that m_f be an integer. The reason for using synchronous PWM is that asynchronous PWM (where m_f is not an integer) results in subharmonics (of the fundamental frequency) that are very undesirable in most applications. This implies that the triangle waveform frequency varies with the desired inverter frequency (e.g., if the inverter output frequency and hence the input signal frequency is 65.42 Hz and $m_f = 15$, the triangle wave frequency should be exactly $15 \times 65.42 = 981.3$ Hz).

- $m_f \leq 21$ *should be an odd integer*: As discussed previously, m_f should be an odd integer except in single-phase inverters with PWM unipolar voltage switching, which will be discussed in Section 8.7.1.

8.2.1.5 Large m_f ($m_f > 21$)

The amplitudes of subharmonics due to asynchronous PWM are small at large values of m_f. Therefore, at large values of m_f, asynchronous PWM can be used where the frequency of the triangle waveform is kept constant, whereas the input signal frequency varies, resulting in noninteger values of m_f (so long as they are large). However, if the inverter is supplying a load such as an AC motor, the subharmonics at zero or close to zero frequency, even though small in amplitude, will result in large currents, which will be highly undesirable. Therefore, asynchronous PWM should be avoided.

It is extremely important to determine the harmonic components of the output voltage. Referring to Figure 8.4c, we have the FFT spectrum and the harmonics. Choosing the frequency modulation ratio m_f as an odd integer and the amplitude modulation ratio $m_a < 1$, we obtain the generalized harmonics of the output voltage shown in Table 8.1.

The rms voltages of the output voltage harmonics are calculated by

$$(V_O)_h = \frac{V_d}{\sqrt{2}} \frac{(\hat{V}_{AO})_h}{V_d/2}, \tag{8.3}$$

TABLE 8.1

Generalized Harmonics of V_O (or V_{AO}) for a Large Value of m_f

	m_a				
h	**0.2**	**0.4**	**0.6**	**0.8**	**1.0**
1 (Fundamental)	0.2	0.4	0.6	0.8	1.0
m_f	1.242	1.15	1.006	0.818	0.601
$m_f \pm 2$	0.016	0.061	0.131	0.220	0.318
$m_f \pm 4$					0.018
$2m_f \pm 1$	0.190	0.326	0.370	0.314	0.181
$2m_f \pm 3$		0.024	0.071	0.139	0.212
$2m_f \pm 5$				0.013	0.033
$3m_f$	0.335	0.123	0.083	0.171	0.113
$3m_f \pm 2$	0.044	0.139	0.203	0.176	0.062
$3m_f \pm 4$		0.012	0.047	0.104	0.157
$3m_f \pm 6$				0.016	0.044
$4m_f \pm 1$	0.163	0.157	0.008	0.105	0.068
$4m_f \pm 3$	0.012	0.070	0.132	0.115	0.009
$4m_f \pm 5$			0.034	0.084	0.119
$4m_f \pm 7$				0.017	0.050

Note: $(\hat{V}_{AO})_h/(V_d/2)$ or $(\hat{V}_{AO})_h/(V_d/2)$ is tabulated as a function of m_a.

where $(V_O)_h$ is the hth harmonic rms voltage of the output voltage, V_d is the DC-link voltage, and $(\hat{V}_{AO})_1/(V_d/2)$ or $(\hat{V}_{AO})_h/(V_d/2)$ is tabulated as a function of m_a.

If the input (control) signal is a sinusoidal wave, we usually call this inversion SPWM. The typical waveforms of an SPWM are also shown in Figure 8.4a and b.

Example 8.1

A single-phase half-bridge DC/AC inverter is shown in Figure 8.3 to implement SPWM with $V_d = 200\,\text{V}$, $m_a = 0.8$, and $m_f = 27$. The fundamental frequency is 50 Hz. Determine the rms value of the fundamental frequency and some of the harmonics in the output voltage using Table 8.1.

SOLUTION

From Equation 8.3, we have the general rms values

$$(V_O)_h = \frac{V_d}{\sqrt{2}} \frac{(\hat{V}_{AO})_h}{V_d/2} = \frac{200}{\sqrt{2}} \frac{(\hat{V}_{AO})_h}{V_d/2} = 141.42 \frac{(\hat{V}_{AO})_h}{V_d/2}\ \text{V}. \tag{8.4}$$

Checking the data from Table 8.1, we obtained the following rms values.
Fundamental:

$$(V_O)_1 = 141.42 \times 0.8 = 113.14\text{V at } 50\,\text{Hz},$$
$$(V_O)_{23} = 141.42 \times 0.818 = 115.68\text{V at } 1150\,\text{Hz},$$
$$(V_O)_{25} = 141.42 \times 0.22 = 31.11\text{V at } 1250\,\text{Hz},$$
$$(V_O)_{27} = 141.42 \times 0.818 = 115.68\text{V at } 1350\,\text{Hz},$$
$$(V_O)_{51} = 141.42 \times 0.139 = 19.66\text{V at } 2550\,\text{Hz},$$
$$(V_O)_{53} = 141.42 \times 0.314 = 44.41\text{V at } 2650\,\text{Hz},$$
$$(V_O)_{55} = 141.42 \times 0.314 = 44.41\text{V at } 2750\,\text{Hz},$$
$$(V_O)_{57} = 141.42 \times 0.139 = 19.66\text{V at } 2850\,\text{Hz, and so on.}$$

8.2.2 Harmonic Parameters

Refering to Figure 8.4c, various harmonic parameters were introduced in Chapter 1, which are used in PWM operation.
Harmonic factor:

$$\text{HF}_n = \frac{V_n}{V_1}. \tag{1.21}$$

Total harmonic distortion:

$$\text{THD} = \frac{\sqrt{\sum_{n=2}^{\infty} V_n^2}}{V_1}. \tag{1.22}$$

Weighted total harmonic distortion:

$$\text{WTHD} = \frac{\sqrt{\sum_{n=2}^{\infty} (V_n^2/n)}}{V_1}. \tag{1.23}$$

8.3 Typical PWM Inverters

DC/AC inverters have three typical supply methods:

- VSI
- CSI
- Impedance Source Inverter (zZ-source inverter or ZSI).

Generally speaking, the circuits of various PWM inverters can be the same. The difference between them is the type of power supply sources or network, which are voltage source, current source, or impedance source.

8.3.1 Voltage Source Inverter

A VSI is supplied by a voltage source. The source is a DC voltage power supply. In an ASD, the DC source is usually an AC/DC rectifier. A large capacitor is used to keep the DC-link voltage stable. Usually, a VSI has buck operation function. Its output voltage peak value is lower than the DC-link voltage.

It is necessary to avoid a *short circuit* across the DC voltage source during operation. If a VSI takes bipolar operation, that is, the upper switch and the lower switch in a leg work to provide a PWM output waveform, the control circuit and interface have to be designed to leave small gaps between switching signals to the upper switch and the lower switch in the same leg. For example, the output voltage frequency is in the range of 0–400 Hz, and the PWM carrying frequency is in the range of 2–20 kHz; the gaps are usually set at 20–100 ns. This requirement is not very convenient for the control circuit and interface design. Therefore, the unipolar operation is implemented in most industrial applications.

8.3.2 Current Source Inverter

A CSI is supplied by a DC current source. In an ASD, the DC current source is usually an AC/DC rectifier with a large inductor to keep the current supply stable. Usually, a CSI has a boost operation function. Its output voltage peak value can be higher than the DC-link voltage.

Since the source is a DC current source, it is necessary to avoid the *open circuit* across the inverter during operation. The control circuit and interface have to be designed to have small overlaps between switching signals to the upper and lower switches at least in one leg. For example, the output voltage frequency is in the range of 0–400 Hz, and the PWM carrying frequency is in the range of 2–20 kHz; the overlaps are usually set at 20–100 ns. This requirement is easy for the control circuit and interface design.

8.3.3 Impedance Source Inverter (Z-SI)

A ZSI is supplied by a voltage source or current source via an "X"-shaped impedance network formed by two capacitors and two inductors, which is called a Z-network. In an ASD, the DC impedance source is usually an AC/DC rectifier. An Z-network is located between the rectifier and the inverter. Since there are two inductors and two capacitors to be set in front of the chopping legs, there is no restriction to avoid the opened or short-circuited legs. A ZSI has the buck–boost operation function. Its output voltage peak value can be higher or lower than the DC-link voltage.

8.3.4 Circuits of DC/AC Inverters

The commonly used DC/AC inverters are introduced below:

1. Single-phase half-bridge VSI
2. Single-phase full-bridge VSI
3. Three-phase full-bridge VSI
4. Three-phase full-bridge CSI
5. Multistage PWM inverters
6. Soft-switching inverters
7. Impedance-source Inverters (ZSI).

8.4 Single-Phase VSI

Single-phase VSIs can be implemented using the half-bridge circuit and the full-bridge circuit.

8.4.1 Single-Phase Half-Bridge VSI

A single-phase half-bridge VSI is shown in Figure 8.6. The carrier-based PWM technique is applied in this inverter. Two large capacitors are required to provide a neutral point N; therefore, each capacitor keeps half of the input DC voltage. Since the output voltage refers to the neutral point N, the maximum output voltage is smaller than half of the DC-link voltage if it is operating in linear modulation. The modulation operations are shown in Figure 8.5. Two switches S_+ and S_- in one chopping leg are switched by the PWM signal. Two switches S_+ and S_- operate in an exclusive state with a short dead time to avoid a short circuit.

In general, linear modulation operation is considered, so that m_a is usually smaller than unity, for example, $m_a = 0.8$. Generally, in order to obtain low THD, m_f is usually taken as a large number. For description convenience, we choose $m_f = 9$. In order to understand each inverter, we show some typical waveforms in Figure 8.7.

How to determine whether the pulse width is the clue to the PWM. If the control signal v_C is a sine-wave function as shown in Figure 8.7a, the modulation is called an SPWM. Figure 8.7b offers the switching signal. When it is positive to switch on the upper switch S_+, and switch off the lower switch S_-; vice versa it is to switch off the upper switch S_+, and the lower switch S_- on. Assume that the amplitude of the triangle wave is unity, and the amplitude of the sine wave is 0.8. Referring to Figure 8.7a, the sine-wave function is

$$f(t) = m_a \sin \omega t = 0.8 \sin 100\pi t, \tag{8.5}$$

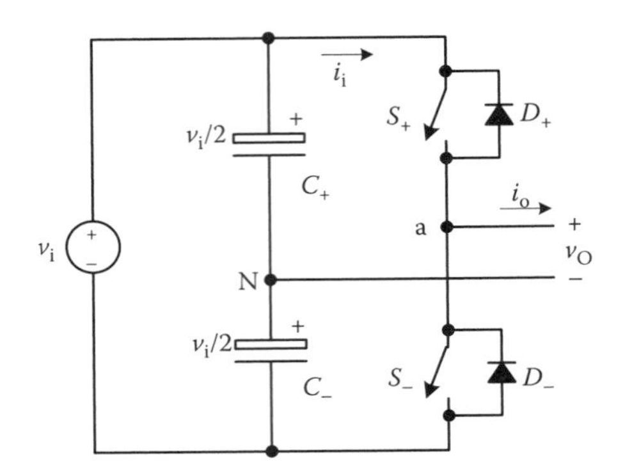

FIGURE 8.6 Single-phase half-bridge VSI.

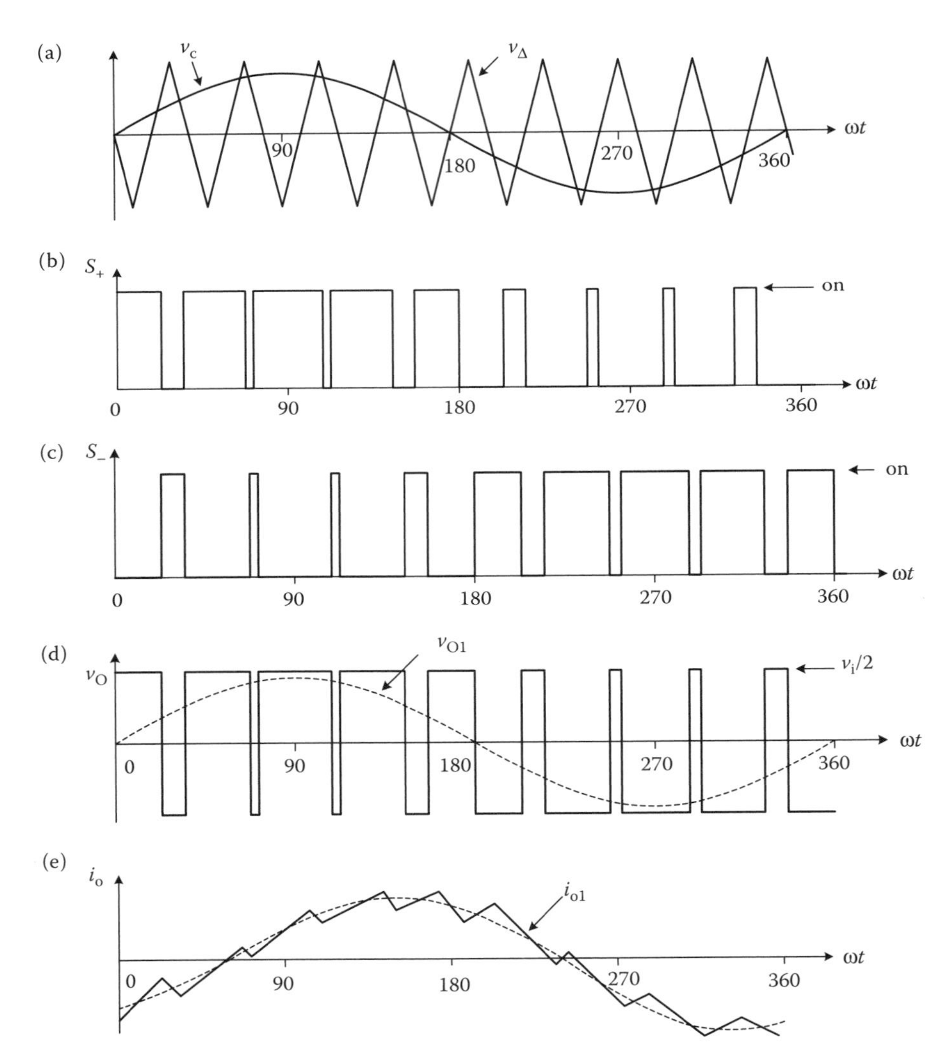

FIGURE 8.7 Single-phase half-bridge VSI ($m_a = 0.8$, $m_f = 9$): (a) carrier and modulating signals, (b) switch S_+ state, (c) switch S_- state, (d) AC output voltage, and (e) AC output current.

where $\omega = 2\pi f$ and $f = 50\,\text{Hz}$. The triangle functions are lines

$$f_{\Delta 1}(t) = -4fm_{\text{f}}t = -1800t, \quad f_{\Delta 2}(t) = 4fm_{\text{f}}t - 2 = 1800t - 2,$$
$$f_{\Delta 3}(t) = 4 - 4fm_{\text{f}}t = 4 - 1800t, \quad f_{\Delta 4}(t) = 4fm_{\text{f}}t - 6 = 1800t - 6,$$

......

$$f_{\Delta(2n-1)}(t) = 4(n-1) - 4fm_{\text{f}}t, \quad f_{\Delta 2n}(t) = 4fm_{\text{f}}t - (4n-2), \tag{8.6}$$

......

$$f_{\Delta 17}(t) = 32 - 1800t, \quad f_{\Delta 18}(t) = 1800t - 34, \quad f_{\Delta 19}(t) = 36 - 1800t.$$

Example 8.2

A single-phase half-bridge DC/AC inverter is shown in Figure 8.6 to implement SPWM with $m_{\text{a}} = 0.8$ and $m_{\text{f}} = 9$. Determine the first pulse width of the pulse shown in Figure 8.7a.

SOLUTION

The leading edge of the first pulse is at $t = 0$. Referring to the triangle formulae, the first pulse width (time or degree) is determined by

$$0.8 \sin 100\pi t = 1800t - 2. \tag{8.7}$$

This is a transcendental equation with the unknown parameter t. Using an iterative method to solve the equation, let $x = 0.8 \sin 100\pi t$ and $y = 1800t - 2$. We can choose the initial $t_0 = 1.38889\,\text{ms} = 25°$.

t(*ms*/°)	*x*	*y*	|*x*| : *y*	Remarks
1.38889/25°	0.338	0.5	$<$	Decrease *t*
1.27778/23°	0.3126	0.3	$>$	Increase *t*
1.2889/23.2°	0.3152	0.32	$<$	Decrease *t*
1.2861/23.15°	0.3145	0.315	$\approx$	

Note: The first pulse width to switch-on and switch-off the switch S_+ is 1.2861 ms (or 23.15°).

Other pulse widths can be determined from other equations using the iterative method. For a PWM operation with large values of m_{f}, readers can refer to Figure 8.8.

Figure 8.7 shows the ideal waveforms associated with the half-bridge VSI. We can find the phase delay between the output current and voltage. For a large m_{f}, we see the cross points demonstrated in Figure 8.8 with smaller phase delay between the output current and voltage.

8.4.2 Single-Phase Full-Bridge VSI

A single-phase full-bridge VSI is shown in Figure 8.9. The carrier-based PWM technique is applied in this inverter. Two large capacitors may be used to provide a neutral point N, but not necessarily. Since the output voltage is not referring to the neutral point N, the maximum output voltage is possibly greater than half the DC-link voltage. If it is operating in linear modulation, the output voltage is smaller than the DC-link voltage. The modulation

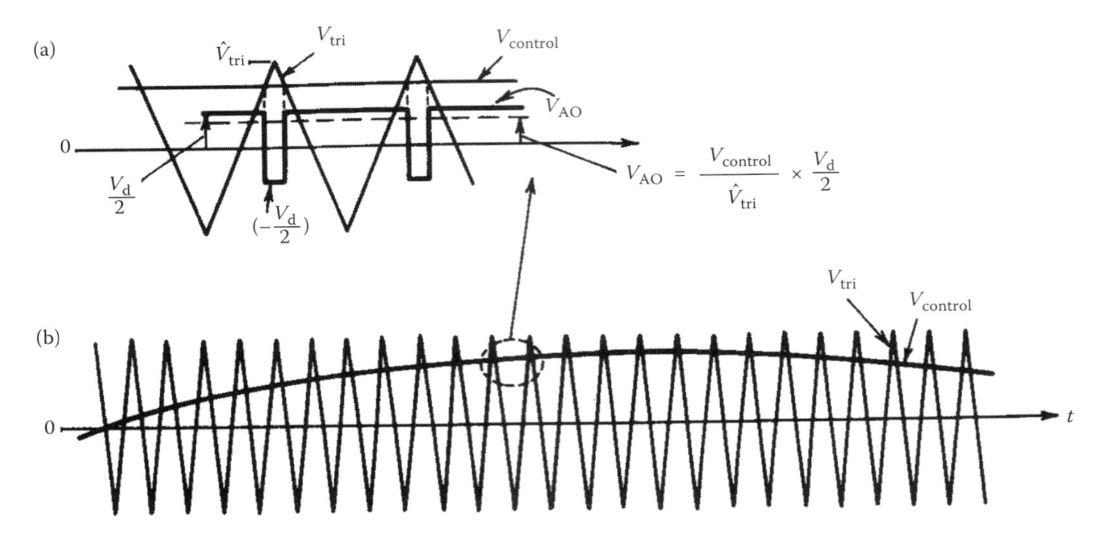

FIGURE 8.8 Sinusoidal PWM. (a) Enlarged the partial waveform and (b) original waveform.

operation is different from that of the single-phase half-bridge VSI described in the previous subsection. This is shown in Figure 8.13. Four switches S_{1+}/S_{1-} and S_{2+}/S_{2-} in two legs are applied and switched by the PWM signal.

Figure 8.10 shows the ideal waveforms associated with the full-bridge VSI. Two sine-waves are used in Figure 8.10a, corresponding to the operation of two legs. We can find the phase delay between the output current and voltage.

The method to determine the pulse widths is the same as that introduced in the previous section. Referring to Figure 8.10a, we find that there are two sine-wave functions:

$$f_+(t) = m_a \sin \omega t = 0.8 \sin 100\pi t \tag{8.8}$$

and

$$f_-(t) = -m_a \sin \omega t = -0.8 \sin 100\pi t. \tag{8.9}$$

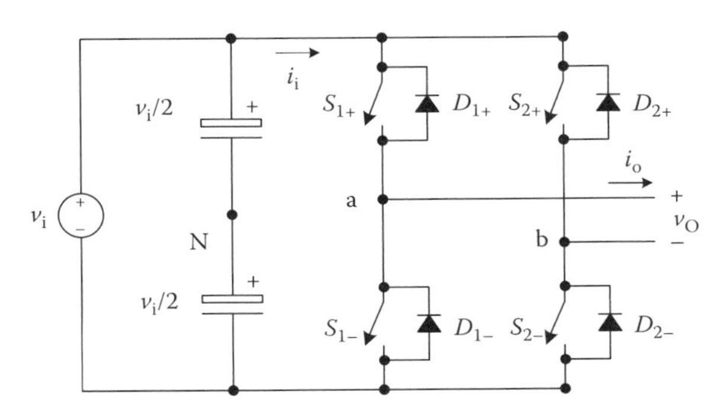

FIGURE 8.9 Single-phase full-bridge VSI.

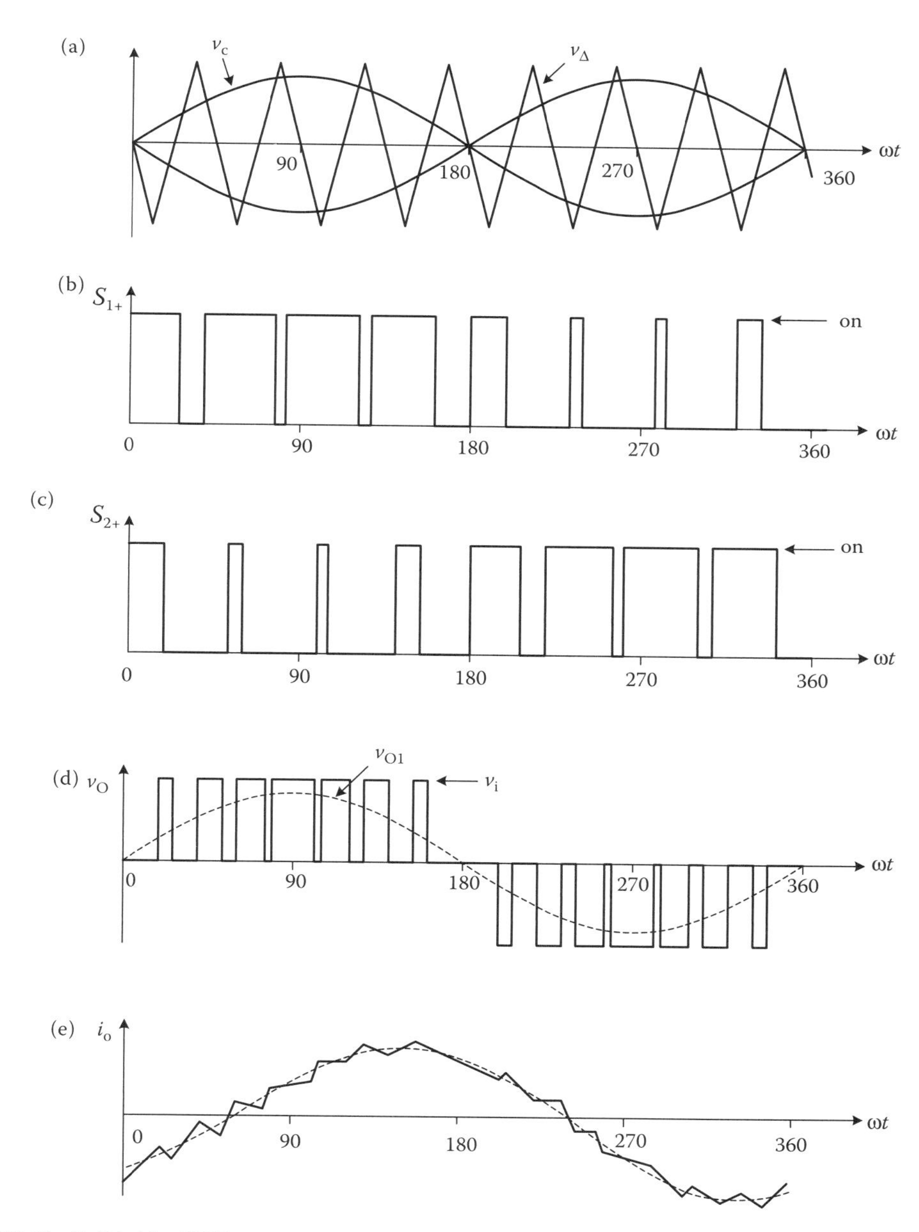

FIGURE 8.10 Full-bridge VSI ($m_a = 0.8, m_f = 8$): (a) carrier and modulating signals, (b) switch S_{1+} and S_{1-} state, (c) switch S_{2+} and S_{2-} state, (d) AC output voltage, and (e) AC output current.

The triangle functions are

$$f_{\Delta 1}(t) = -4fm_f t = -1600t, \quad f_{\Delta 2}(t) = 4fm_f t - 2 = 1600t - 2,$$
$$f_{\Delta 3}(t) = 4 - 4fm_f t = 4 - 1600t, \quad f_{\Delta 4}(t) = 4fm_f t - 6 = 1600t - 6,$$

......

$$f_{\Delta(2n-1)}(t) = 4(n-1) - 4fm_f t, \quad f_{\Delta 2n}(t) = 4fm_f t - (4n-2), \tag{8.10}$$

......

$$f_{\Delta 15}(t) = 28 - 1600t, \quad f_{\Delta 16}(t) = 1600t - 30,$$
$$f_{\Delta 17}(t) = 32 - 1600t.$$

The first pulse width to switch-on and switch-off switches S_{1+} and S_{1-} is determined by

$$0.8 \sin 100\pi t = 1600t - 2. \tag{8.11}$$

The first pulse width to switch-on and switch-off switches S_{2+} and S_{2-} is determined by

$$-0.8 \sin 100\pi t = 1600t - 2$$

or

$$0.8 \sin 100\pi t = 2 - 1600t. \tag{8.12}$$

In the output voltage between leg-to-leg, the rms voltages of the output voltage harmonics are calculated by

$$(V_{\mathrm{O}})_h = \frac{2V_{\mathrm{d}}}{\sqrt{2}} \frac{(\hat{V}_{\mathrm{AO}})_h}{V_{\mathrm{d}}/2}, \tag{8.13}$$

where $(V_{\mathrm{O}})_h$ is the hth harmonic rms voltage of the output voltage, V_{d} is the DC-link voltage, and $(\hat{V}_{\mathrm{AO}})_h/(V_{\mathrm{d}}/2)$ is tabulated as a function of m_{a}, which can be obtained from Table 8.1.

Example 8.3

A single-phase full-bridge DC/AC inverter is shown in Figure 8.9 to implement SPWM with $V_{\mathrm{d}} = 300\,\mathrm{V}$, $m_{\mathrm{a}} = 1.0$, and $m_{\mathrm{f}} = 31$. The fundamental frequency is 50 Hz. Determine the rms value of the fundamental frequency and some of the harmonics in the output voltage using Table 8.1.

SOLUTION

From Equation 8.13, we have the general rms values

$$(V_{\mathrm{O}})_h = \frac{2V_{\mathrm{d}}}{\sqrt{2}} \frac{(\hat{V}_{\mathrm{AO}})_h}{V_{\mathrm{d}}/2} = \frac{600}{\sqrt{2}} \frac{(\hat{V}_{\mathrm{AO}})_h}{V_{\mathrm{d}}/2} = 424.26 \frac{(\hat{V}_{\mathrm{AO}})_h}{V_{\mathrm{d}}/2}\,\mathrm{V}.$$

Checking the data from Table 8.1, we obtain the following rms values:

Fundamental:

$$
\begin{aligned}
(V_O)_1 &= 424.26 \times 1.0 = 424.26\,\text{V at } 50\,\text{Hz},\\
(V_O)_{27} &= 424.26 \times 0.018 = 7.64\,\text{V at } 1350\,\text{Hz},\\
(V_O)_{29} &= 424.26 \times 0.318 = 134.92\,\text{V at } 1450\,\text{Hz},\\
(V_O)_{31} &= 424.26 \times 0.601 = 254.98\,\text{V at } 1550\,\text{Hz},\\
(V_O)_{33} &= 424.26 \times 0.318 = 134.92\,\text{V at } 1650\,\text{Hz},\\
(V_O)_{35} &= 424.26 \times 0.018 = 7.64\,\text{V at } 1750\,\text{Hz},\\
(V_O)_{57} &= 424.26 \times 0.033 = 14\,\text{V at } 2850\,\text{Hz},\\
(V_O)_{59} &= 424.26 \times 0.212 = 89.94\,\text{V at } 2950\,\text{Hz},\\
(V_O)_{61} &= 424.26 \times 0.181 = 76.79\,\text{V at } 3050\,\text{Hz},\\
(V_O)_{63} &= 424.26 \times 0.181 = 76.79\,\text{V at } 3150\,\text{Hz},\\
(V_O)_{65} &= 424.26 \times 0.212 = 89.94\,\text{V at } 3250\,\text{Hz},\\
(V_O)_{67} &= 424.26 \times 0.033 = 14\,V \text{ at } 3350\,\text{Hz, and so on.}
\end{aligned}
$$

8.5 Three-Phase Full-Bridge VSI

A three-phase full-bridge VSI is shown in Figure 8.11. The carrier-based PWM technique is applied in this single-phase full-bridge VSI. Two large capacitors may be used to provide a neutral point N, but not necessarily. Six switches, S_1–S_6, are applied in three legs and switched by the PWM signal.

Figure 8.12 shows the ideal waveforms associated with the full-bridge VSI. We can find out the phase delay between output current and voltage.

Since the three-phase waveform in Figure 8.12a does not refer to the neutral point N, the operation conditions are different from the single-phase half-bridge VSI. The maximum output line-to-line voltage is possibly greater than half the DC-link voltage. If it is operating in linear modulation, the output voltage is smaller than the DC-link voltage. The modulation indication of a three-phase VSI is different from that of a single-phase half-bridge VSI in Section 8.4.1, as shown in Figure 8.13.

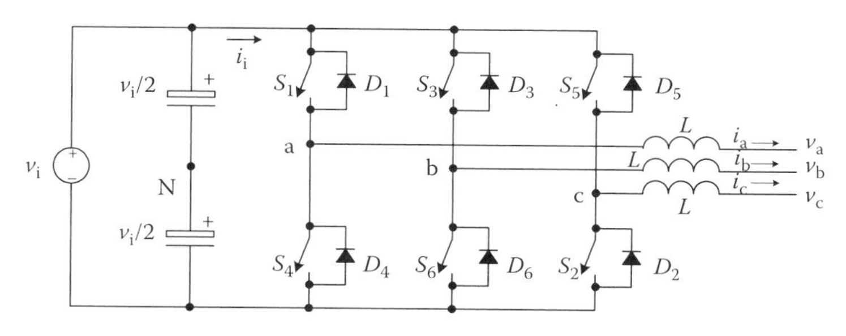

FIGURE 8.11 Three-phase full-bridge VSI.

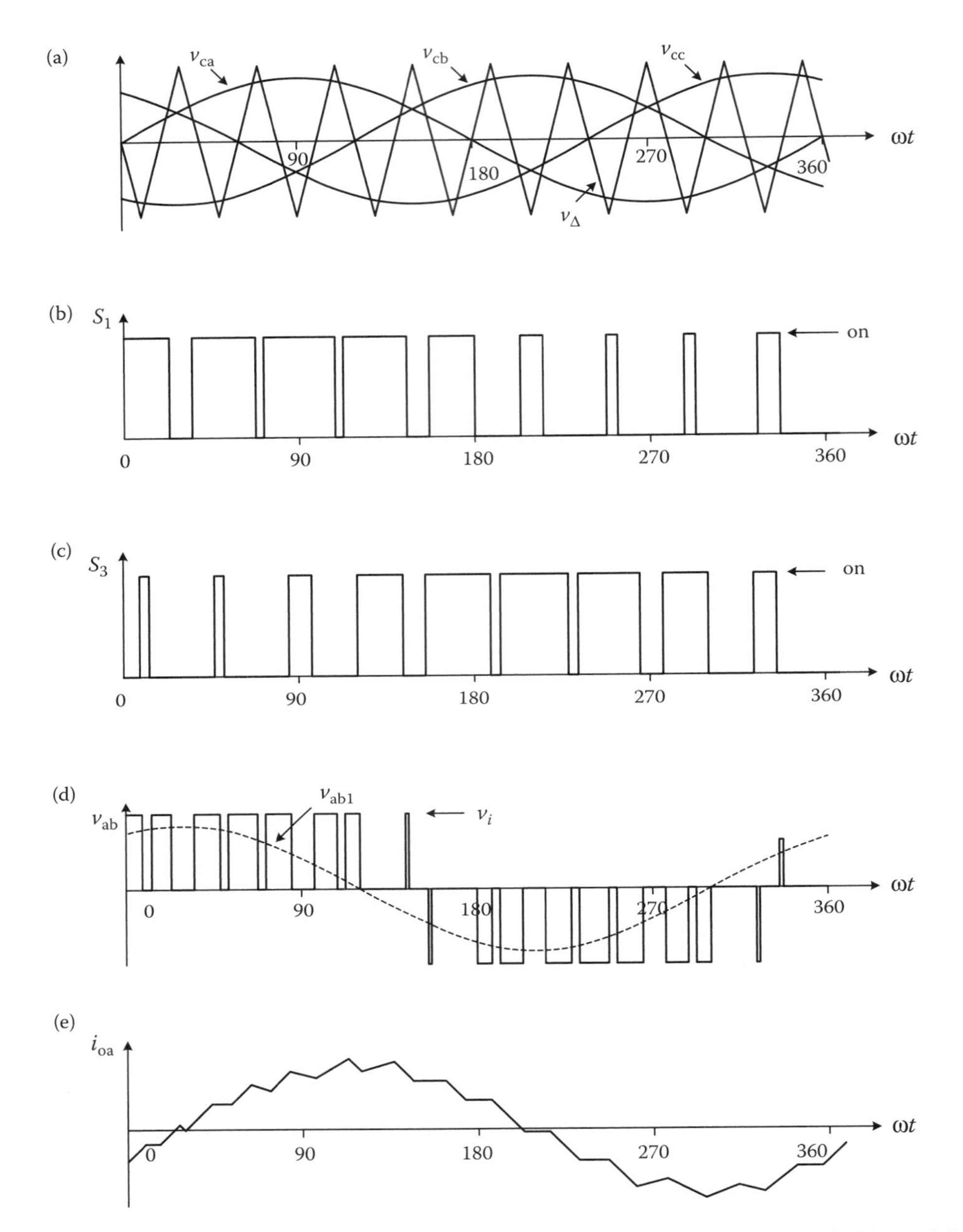

FIGURE 8.12 Three-phase full-bridge VSI ($m_a = 0.8, m_f = 9$): (a) carrier and modulating signals, (b) switch S_1/S_4 state, (c) switch S_3/S_4 state, (d) AC output voltage, and (e) AC output current.

8.6 Three-Phase Full-Bridge CSI

A three-phase full-bridge CSI is shown in Figure 8.14.

The carrier-based PWM technique is applied in this three-phase full-bridge CSI. The main objective of these static power converters is to produce AC output current waveforms from

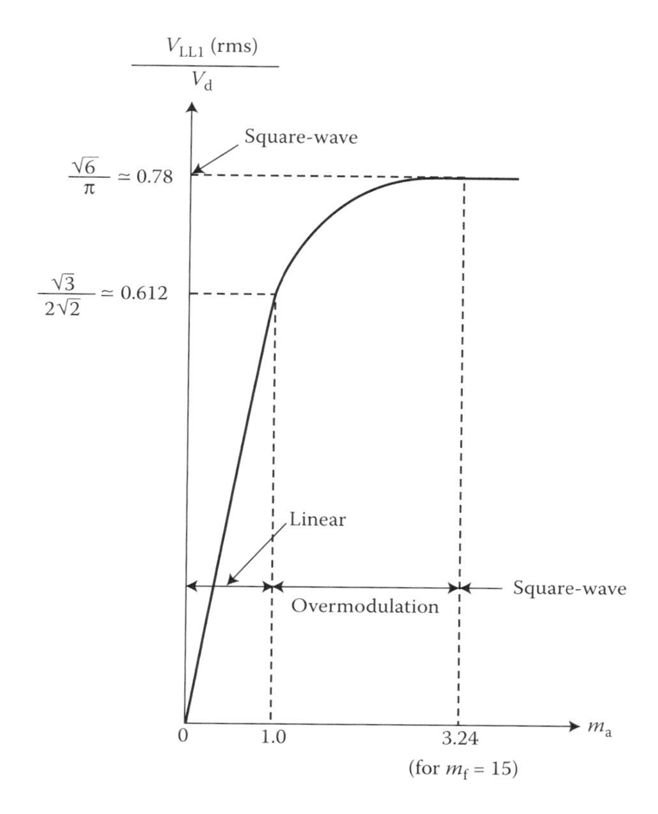

FIGURE 8.13 Function of m_a for a three-phase inverter.

a DC current power supply. Six switches, $S_1 - S_6$, are applied and switched by the PWM signal. Figure 8.15 shows the ideal waveforms associated with the full-bridge CSI.

The CSI has a boost function. Usually, the output voltage can be higher than the input voltage. We can find the phase ahead between the output voltage and current.

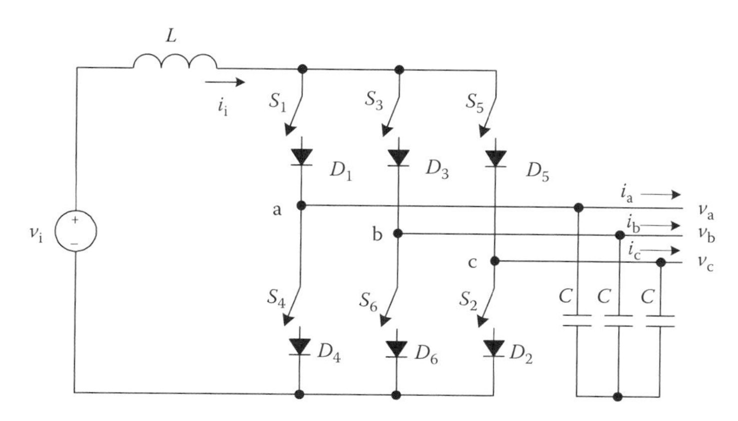

FIGURE 8.14 Three-phase CSI.

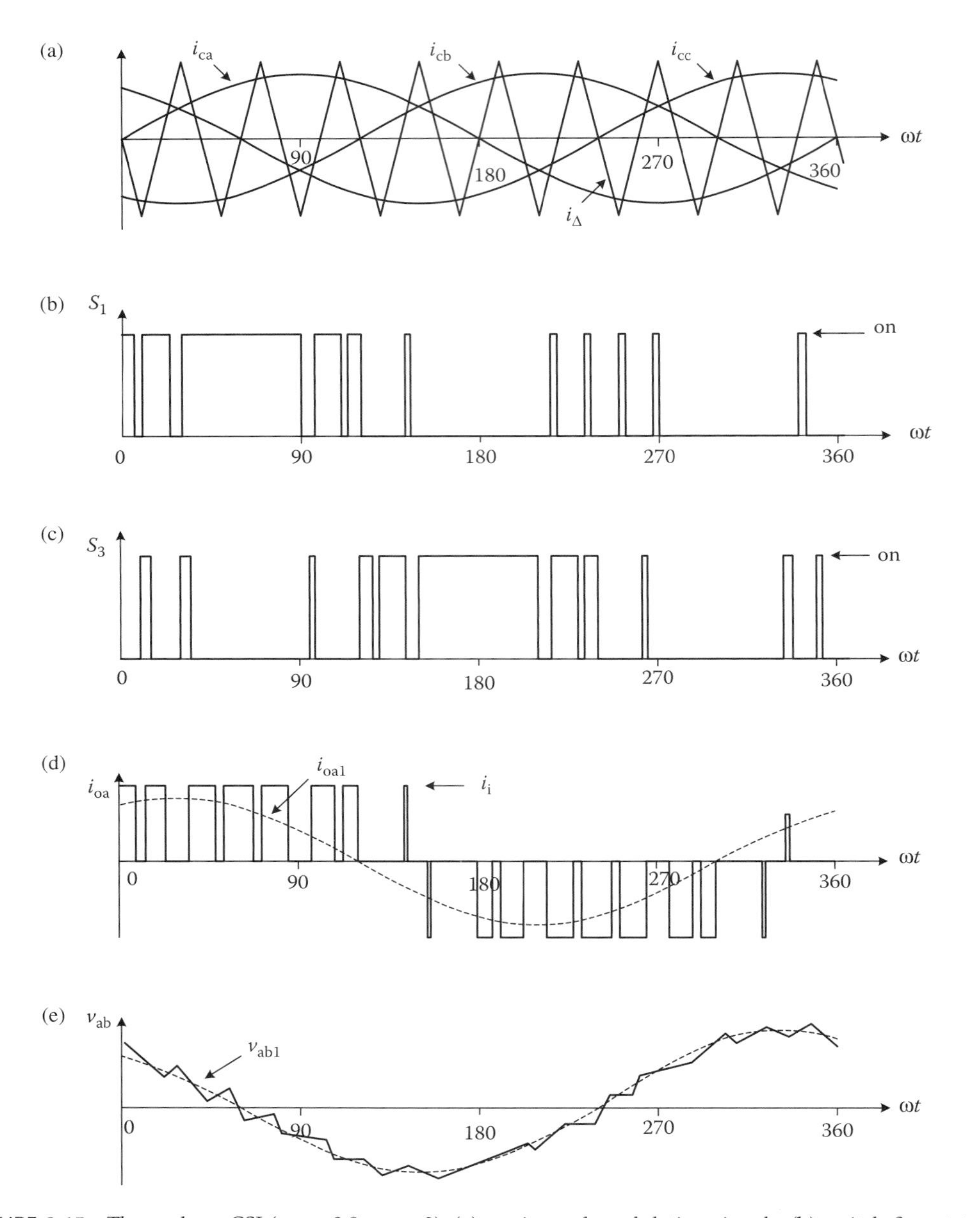

FIGURE 8.15 Three-phase CSI ($m_a = 0.8$, $m_f = 9$): (a) carrier and modulating signals, (b) switch S_{1+} state, (c) switch S_3 state, (d) AC output current, and (e) AC output voltage.

8.7 Multistage PWM Inverter

Multistage PWM inverters can be constructed by two methods: multicell and multilevel. Unipolar modulation PWM inverters can be considered as multistage inverters.

8.7.1 Unipolar PWM VSI

In Section 8.4, we introduced the single-phase source inverter operating in the *bipolar modulation*. Referring to the circuit in Figure 8.6, the upper switch S_+ and the lower switch S_- work together. The carrier and modulating signals are shown in Figure 8.7a, and the switching signals for upper switch S_+ and lower switch S_- are shown in Figures 8.7b and 8.7c. The output voltage of the inverter is the pulse train with both polarities, as shown in Figure 8.7d.

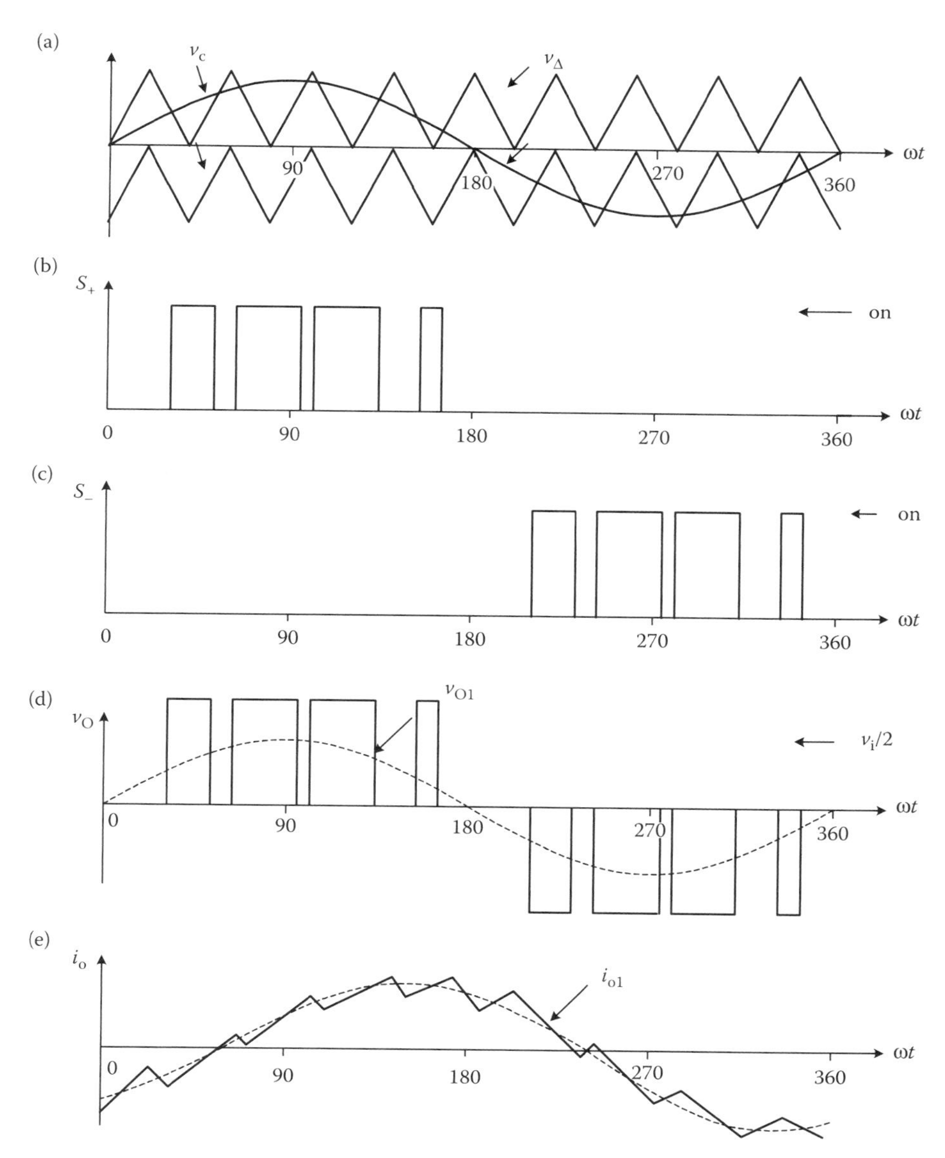

FIGURE 8.16 Three-phase unipolar regulation inverter ($m_a = 0.8$, $m_f = 9$). (a) Control and triangle waveforms, (b) positive half-cycle pulse waveform, (c) negative half-cycle pulse waveform, (d) inverter output waveform and its fundamental wave, and (e) output voltage and current waveforms after filters.

There are some drawbacks to using bipolar modulation: (1) if the inverter is VSI, a dead time has to be set to avoid short circuit; (2) the zero output voltage corresponds to the equal pulse width of positive and negative pulses; (3) power losses are high since two devices work, and hence efficiency is lower; and (4) two devices should be controlled simultaneously.

In most industrial applications, unipolar modulation is widely used. The regulation and corresponding waveforms are shown in Figure 8.16 with $m_a = 0.8$ and $m_f = 9$. For unipolar regulation, m_a is measured by

$$m_a = \frac{V_{in-m}}{2V_{tri-m}}. \tag{8.14}$$

This regulation method is like a two-stage PWM inverter. If the output voltage is positive, only the upper device works and the lower device idles. Therefore, the output voltage only remains as the positive polarity pulse train. On the other hand, if the output voltage is negative, only the lower device works and the upper device idles. Therefore, the output voltage only remains as the negative polarity pulse train. The advantages of implementing unipolar regulation are as follows:

- No need to set a dead time
- The pulses are narrow, for example, the zero output voltage requires zero pulse width
- Power losses are low and hence the efficiency is high
- Only one device should be controlled in a half-cycle.

8.7.2 Multicell PWM VSI

Multistage PWM inverters can consist of many cells. Each cell can be a single-phase or three-phase input plus a single-phase output VSI, which is shown in Figure 8.17. If the three-phase AC supply is a secondary winding of a main transformer, it is floating and isolated from other cells and a common ground point. Therefore, all cells can be linked in series or in parallel.

A three-stage PWM inverter is shown in Figure 8.18. Each phase consists of three cells with a difference phase-angle shift of 20° to each other.

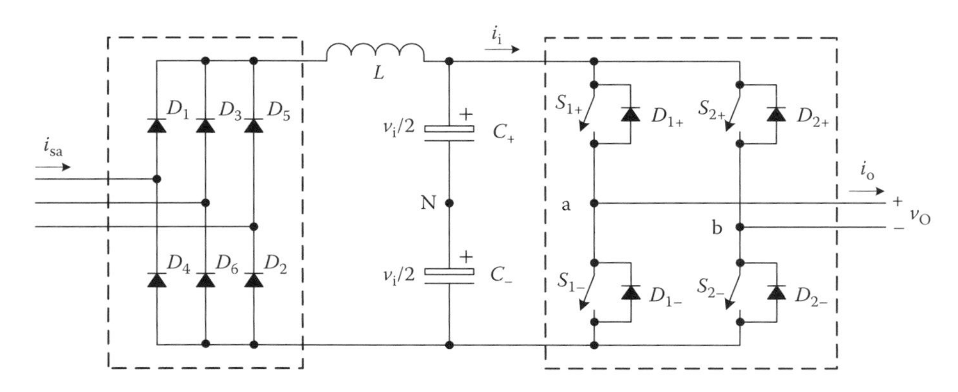

FIGURE 8.17 Three-phase input single-phase output cell.

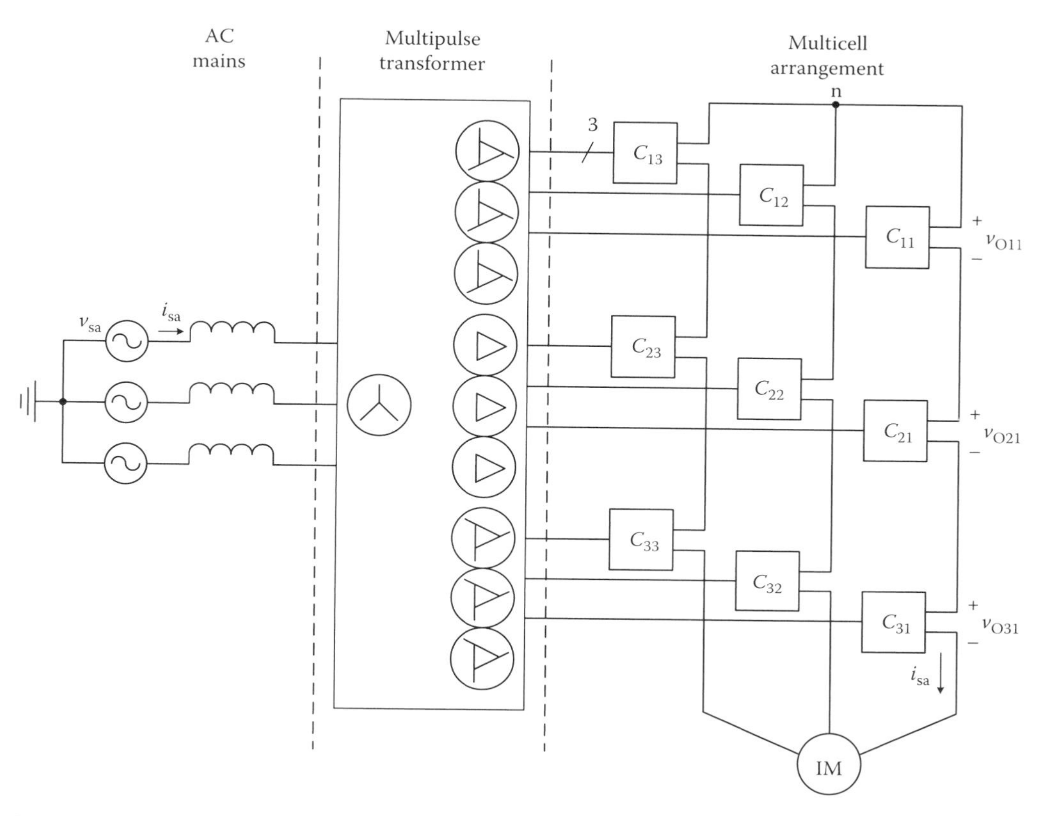

FIGURE 8.18 Multistage converter based on a multicell arrangement.

The carrier-based PWM technique is applied in this three-phase multistage PWM inverter. Figure 8.19 shows the ideal waveforms associated with the full-bridge VSI. We can calculate the output voltage, and the current phase delayed beyond the output voltage.

8.7.3 Multilevel PWM Inverter

A three-level PWM inverter is shown in Figure 8.20. The carrier-based PWM technique is applied in this multilevel PWM inverter. Figure 8.21 shows the ideal waveforms associated with the multilevel PWM inverter. We can find the output and the phase delayed between the output current and voltage.

8.8 Impedance-Source Inverters

ZSI is a new approach of DC/AC conversion technology. It was published by Peng in 2003 [3–5]. The ZSI circuit diagram shown in Figure 8.22 consists of an "X"-shaped impedance network formed by two capacitors and two inductors, and it provides unique buck–boost characteristics. Moreover, unlike VSI, the need for dead time would not arise with this

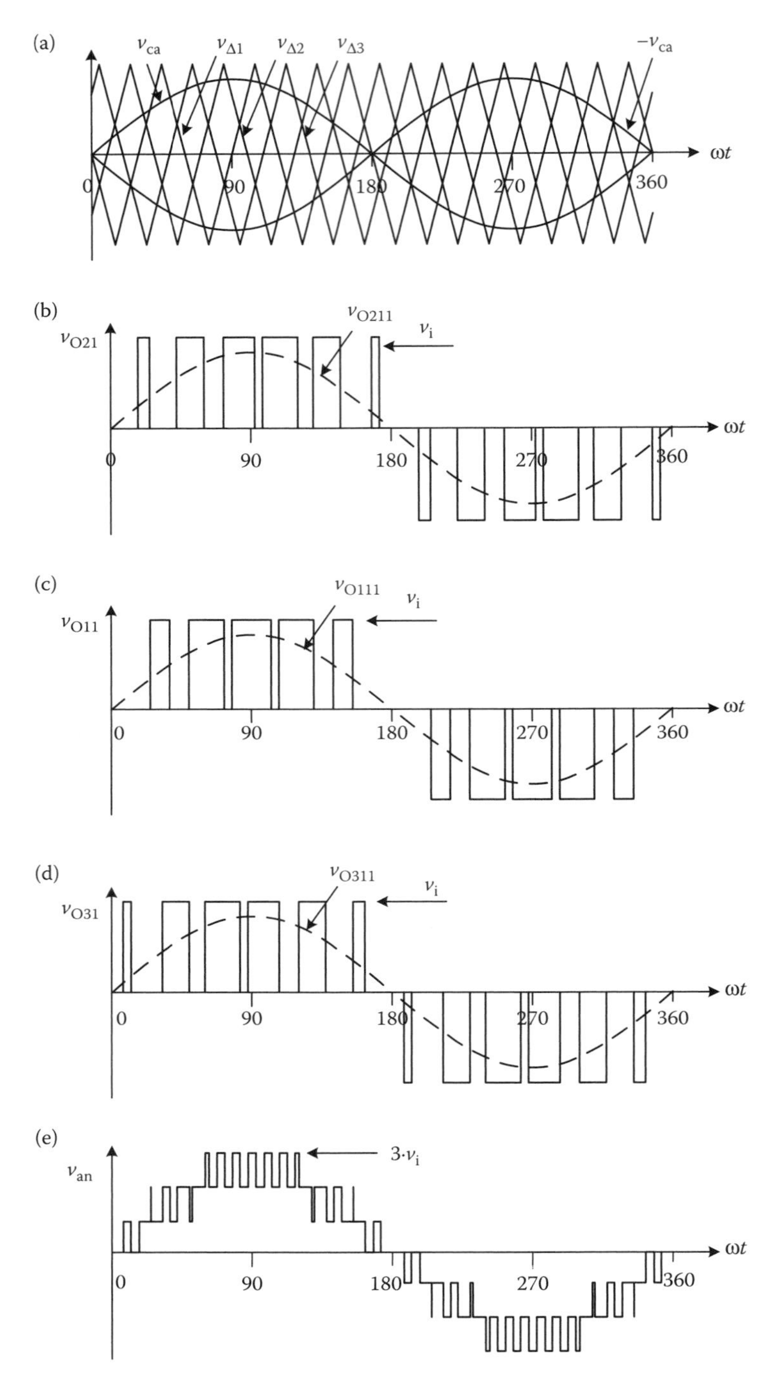

FIGURE 8.19 Multicell PWM inverter (three stages, $m_a = 0.8$, $m_f = 6$): (a) carrier and modulating signals, (b) cell C_{11} AC output voltage, (c) cell C_{21} AC output voltage, (d) cell C_{31} AC output voltage, and (e) phase a load voltage.

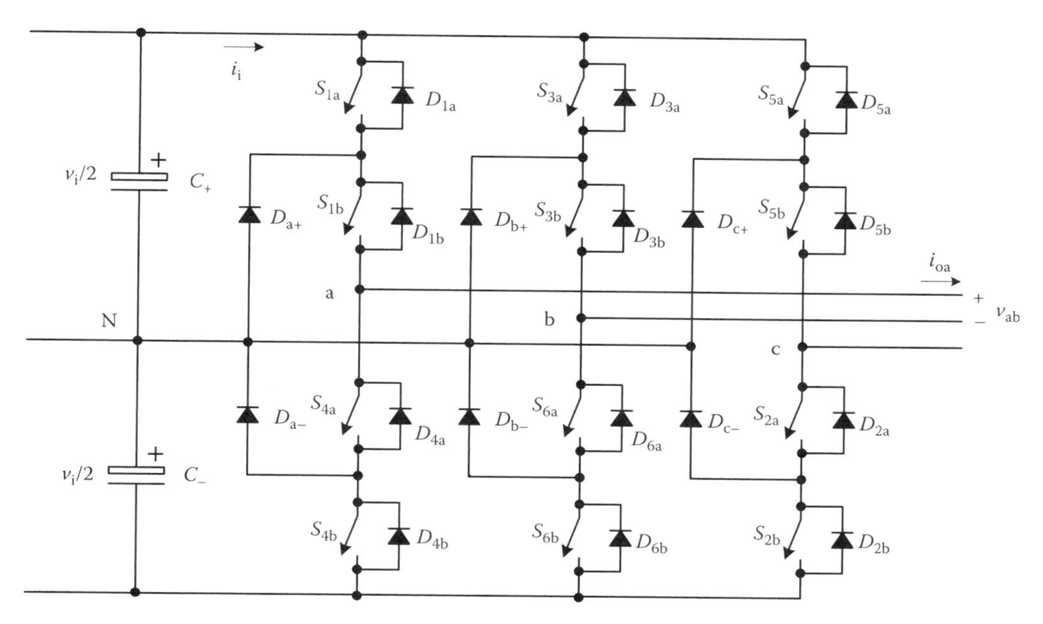

FIGURE 8.20 Three-phase three-level VSI.

topology. Due to these attractive features, it has found use in numerous industrial applications, including variable speed drives and DG. However, it has not been widely researched as a DG topology. Moreover, all these industrial applications require proper closed-loop controlling to adjust the operating conditions subjected to changes in both input and output conditions. On the other hand, the presence of the "X"-shaped impedance network and the need for short-circuiting of the inverter arm to boost the voltage would complicate the controlling of ZSI.

8.8.1 Comparison with VSI and CSI

ZSI is a new inverter that is different from traditional VSIs and CSIs. In order to express the advantages of ZSI, it is necessary to compare it with VSI and CSI.

A three-phase VSI is shown in Figure 8.11. A DC voltage source supported by a relatively large capacitor feeds the main converter circuit, a three-phase bridge. The V-source inverter has the following conceptual and theoretical barriers and limitations.

1. The AC output voltage is limited below, and cannot exceed, the DC link. Therefore, the VSI is a buck (step-down) inverter for DC/AC power conversion. For applications where an overdrive is desirable and the available DC voltage is limited, an additional DC/DC boost converter is needed to obtain a desired AC output. The additional power converter stage increases system cost and lowers efficiency.
2. The upper and lower devices of each phase leg cannot be gated simultaneously either by purpose or by EMI noise. Otherwise, a shoot-through would occur and destroy the devices. The shoot-through problem by EMI noise's misgating-on is a major killer to the converter's reliability. Dead time to block both the upper and lower devices has to be provided in the VSI, which causes waveform distortion, and so on.

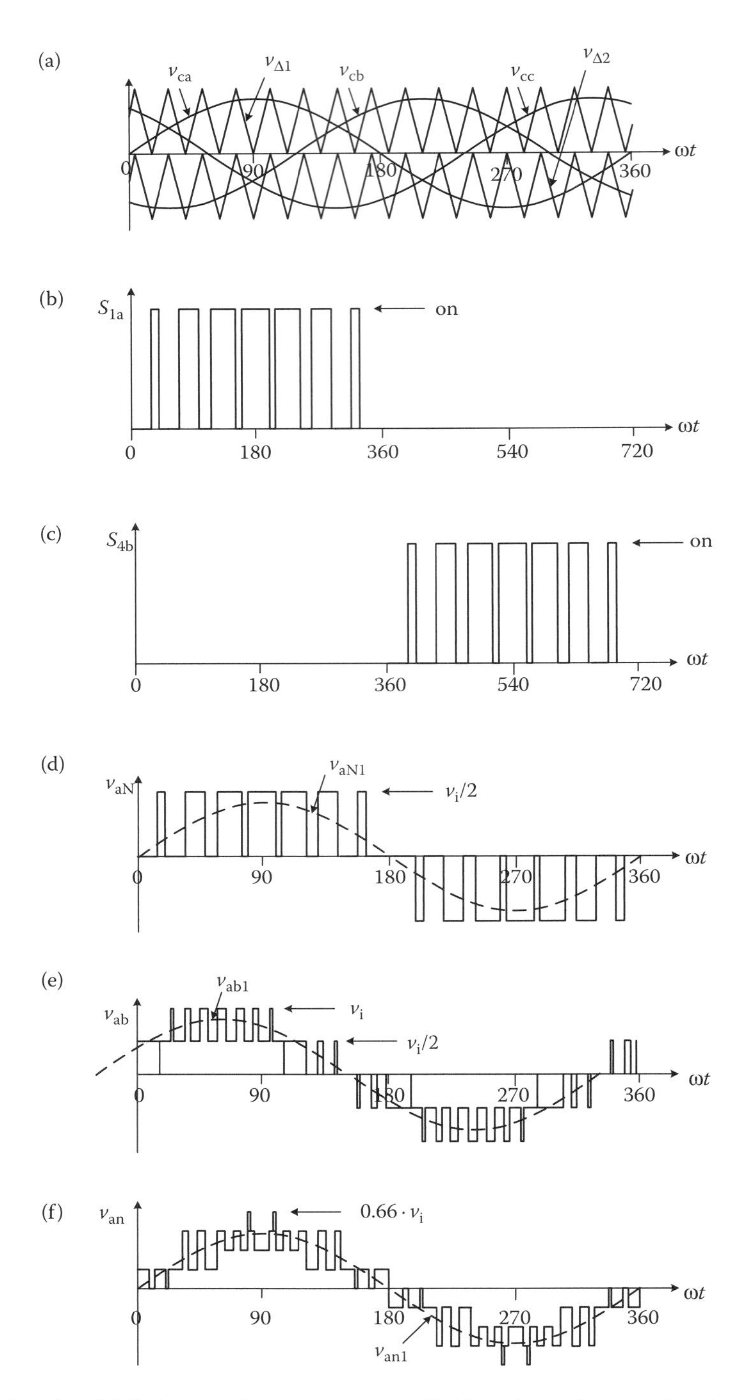

FIGURE 8.21 Three-level VSI (three levels, $m_a = 0.8$, $m_f = 15$): (a) carrier and modulating signals, (b) switch S_{1a} status, (c) switch S_{4b} status, (d) inverter phase a-N voltage, (e) AC output line voltage, and (f) AC output phase voltage.

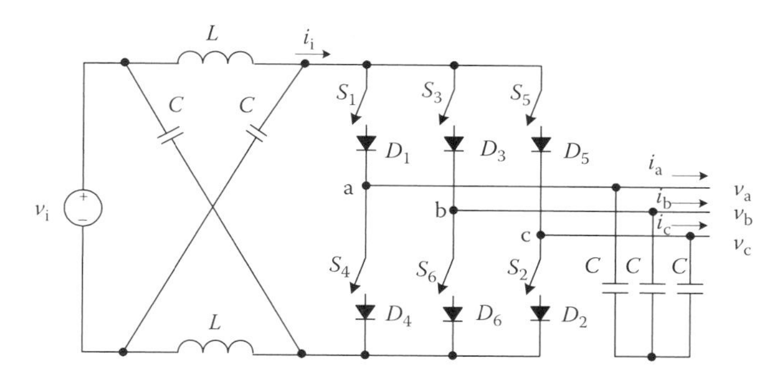

FIGURE 8.22 Impedance source inverter.

3. An output LC filter is needed for providing a sinusoidal voltage, contrasted with the CSI, which causes additional power loss and control complexity.

A three-phase CSI is shown in Figure 8.14. A DC voltage source feeds the main inverter circuit, a three-phase bridge. The DC current source can be a relatively large DC inductor fed by a voltage source such as a battery, fuel-cell stack, diode rectifier, or thyristor converter. The CSI has the following conceptual and theoretical barriers and limitations.

1. The AC output voltage has to be greater than the original DC voltage that feeds the DC inductor, or the DC voltage produced is always smaller than the AC input voltage. Therefore, the CSI is a boost inverter for DC/AC power conversion. For applications where a wide voltage range is desirable, an additional DC/DC buck (or boost) converter is needed. The additional power conversion stage increases system cost and lowers efficiency.
2. At least one of the upper devices and one of the lower devices have to be gated on and maintained at any time. Otherwise, an open circuit of the DC inductor would occur and destroy the devices. The open-circuit problem caused by the misgating-off of the EMI noise is a major concern of the converter's reliability. Overlap time for safe current commutation is needed in the I-source converter, which also causes waveform distortion, and so on.
3. The main switches of the I-source converter have to block the reverse voltage that requires a series diode to be used in combination with high-speed and high-performance transistors such as IGBTs. This prevents the direct use of low-cost and high-performance IGBT modules and intelligent power modules (IPMs).

In addition, both the VSI and the CSI have the following common problems:

1. They are either a boost or a buck converter and cannot be a buck–boost converter. That is, their obtainable output voltage range is limited to being either greater or smaller than the input voltage.
2. Their main circuits cannot be interchangeable. That is, the VSI main circuit cannot be used for the CSI, and vice versa.
3. They are vulnerable to EMI noise in terms of reliability.

To overcome these problems of the traditional VSI and CSI, ZST was designed, as shown in Figure 8.22. It employs a unique impedance network to couple the converter main circuit to the power source. ZSI overcomes the above-mentioned conceptual and theoretical barriers and limitations of the traditional VSI and CSI and provides a novel power conversion concept.

In Figure 8.22, a two-port network that consists of split inductors L_1 and L_2 and capacitors C_1 and C_2 connected in X shape is employed to provide an impedance source (Z-source) coupling the converter (or inverter) to the DC source. Switches used in ZSI can be a combination of switching devices and diodes such as those shown in Figures 8.11 and 8.14. If the two inductors have zero inductance, the ZSI becomes a VSI. On the other hand, if the two capacitors have zero capacitance, the ZSI becomes a CSI. The advantages of the ZSI are listed below:

1. The AC output voltage is not fixed lower or higher than the DC-link (or DC source) voltage. Therefore, the ZSI is a buck–boost inverter for DC/AC power conversion. For applications where overdrive is desirable and the available DC voltage is not limited, there is no need for an additional DC/DC boost converter to obtain a desired AC output. Therefore, the system cost is low and efficiency is high.
2. The Z-circuit consists of two inductors and two capacitors and can restrict the overvoltage and overcurrent. Therefore, the legs in the main bridge can operate in short circuit and open circuit in a short time. There are restrictions for the main bridge such as dead time for VSI and overlap-time for CSI.
3. ZSI has a function to suppress EMI noise. The shoot-through problem by EMI noise's misgating-on will not damage the devices and the converter's reliability.

8.8.2 Equivalent Circuit and Operation

A three-phase ZSI used for fuel-cell application is shown in Figure 8.23. It has nine permissible switching states (vectors): six active vectors, as a traditional VSI has, and three zero vectors when the load terminals are shorted through both the upper and lower devices of any one phase leg (i.e., both devices are gated on), any two phase legs, or all three phase legs. This shoot-through zero state (or vector) is forbidden in the traditional VSI, because it would cause a shoot-through. We call this third zero state (vector) the shoot-through zero state (or vector), which can be generated in seven different ways: shoot-through via any one

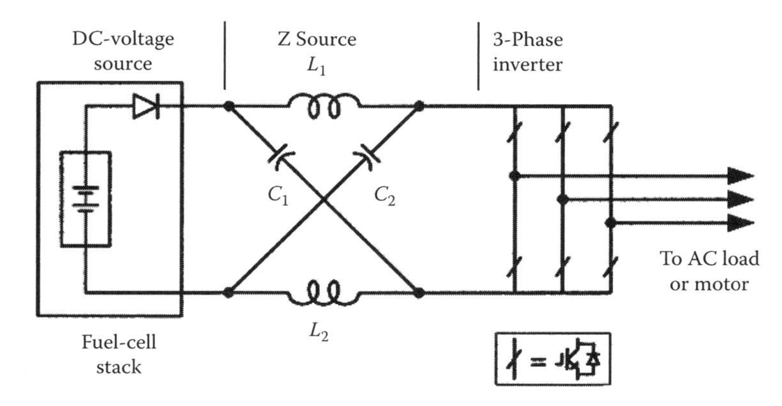

FIGURE 8.23 ZSI for fuel-cell applications. [Reprinted from Peng, F. Z. 2003. *IEEE Transactions on Industry Applications*, 504–510. (©2003 IEEE). With permission.]

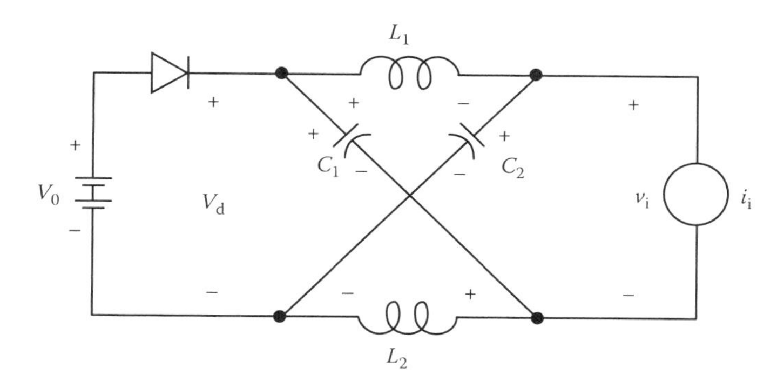

FIGURE 8.24 Equivalent circuit of the ZSI viewed from the DC link. [Reprinted from Peng, F. Z. 2003. *IEEE Transactions on Industry Applications*, 504–510. (©2003 IEEE). With permission.]

phase leg, combinations of any two phase legs, and all three phase legs. The Z-source network makes the shoot-through zero state possible. This shoot-through zero state provides the unique buck–boost feature to the inverter.

Figure 8.24 shows the equivalent circuit of the ZSI shown in Figure 8.23 when viewed from the DC link. The inverter bridge is equivalent to a short circuit when the inverter bridge is in the shoot-through zero state, as shown in Figure 8.25, whereas the inverter bridge becomes an equivalent current source as shown in Figure 8.26 when the inverter bridge is in one of the six active states. Note that the inverter bridge can also be represented by a current source with zero value (i.e., an open circuit) when it is in one of the two traditional zero states. Therefore, Figure 8.26 shows the equivalent circuit of the ZSI viewed from the DC link when the inverter bridge is in one of the eight non-shoot-through switching states.

All the traditional PWM schemes can be used to control the ZSI, and their theoretical input–output relationships still hold. Figure 8.27 shows the traditional PWM switching sequence based on the triangular carrier method. In every switching cycle, the two non-shoot-through zero states are used along with two adjacent active states to synthesize the desired voltage. When the DC voltage is high enough to generate the desired AC voltage, the traditional PWM of Figure 8.27 is used. While the DC voltage is not enough to directly generate a desired output voltage, a modified PWM with shoot-through zero states will

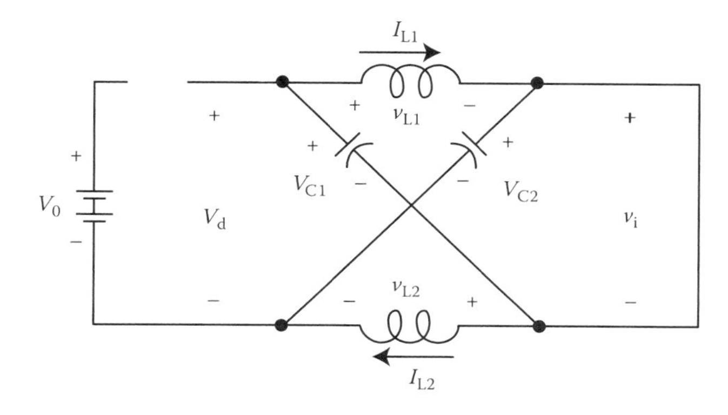

FIGURE 8.25 Equivalent circuit of the ZSI viewed from the DC link when the inverter bridge is in the shoot-through zero state. [From Peng, F. Z. 2003. *IEEE Transactions on Industry Applications*, 504–510. (©2003 IEEE). With permission.]

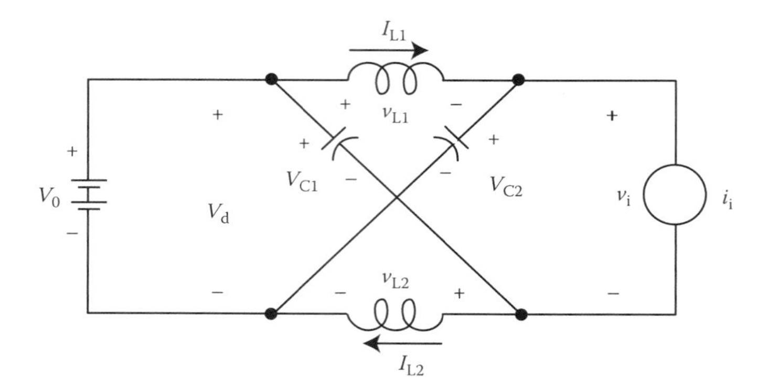

FIGURE 8.26 Equivalent circuit of the ZSI viewed from the DC link when the inverter bridge is in one of the eight non-shoot-through switching states. [Reprinted from Peng, F. Z. 2003. *IEEE Transactions on Industry Applications*, 504–510. (©2003 IEEE).With permission.]

be used, as shown in Figure 8.28, to boost voltage. It should be noted that each phase leg still switches on and off once per switching cycle. Without changing the total zero-state time interval, the shoot-through zero states are evenly allocated into each phase. That is, the active states are unchanged. However, the equivalent DC-link voltage to the inverter is

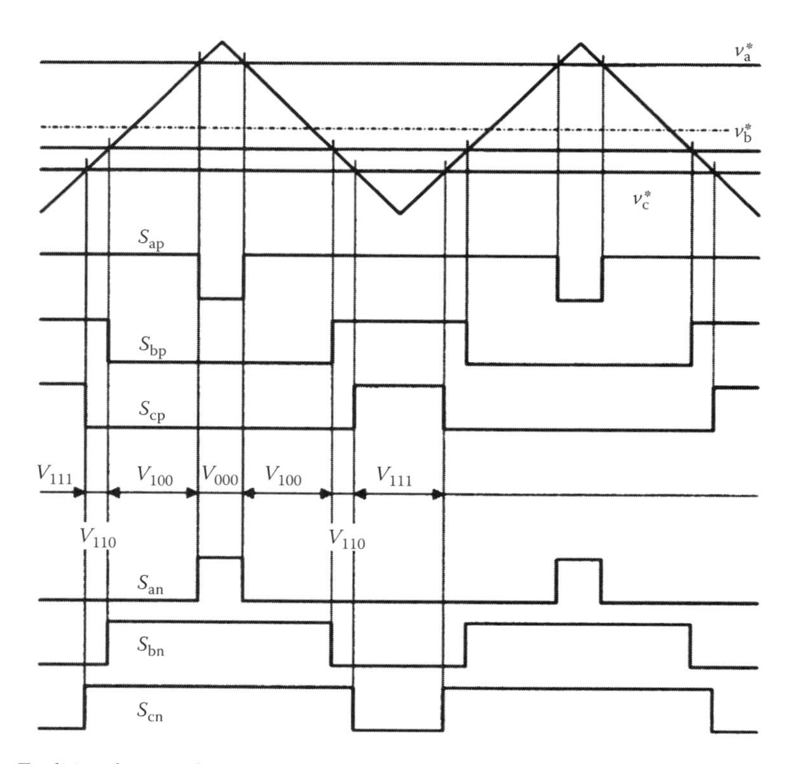

FIGURE 8.27 Traditional carrier-based PWM control without shoot-through zero states, where the traditional zero states (vectors) V_{111} and V_{000} are generated in every switching cycle and determined by the references. [Reprinted from Peng, F. Z. 2003. *IEEE Transactions on Industry Applications*, 504–510. (©2003 IEEE). With permission.]

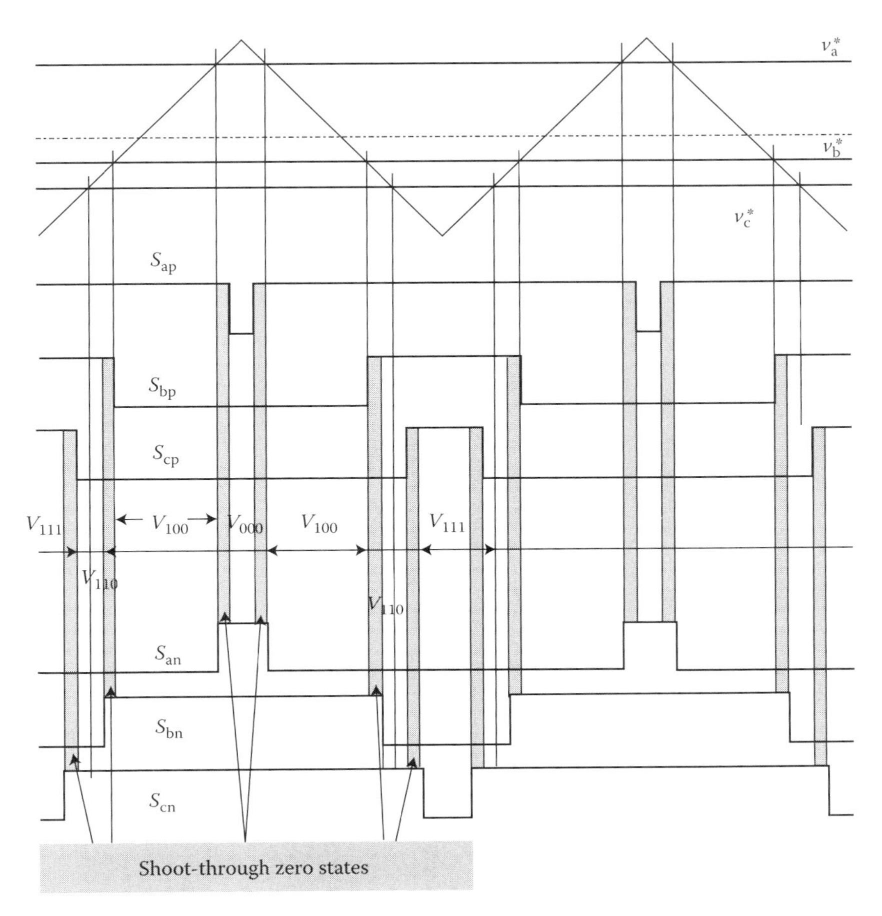

FIGURE 8.28 Modified carrier-based PWM control with shoot-through zero states that are evenly distributed among the three phase legs, while the equivalent active vectors are unchanged. [Reprinted from Peng, F. Z. 2003. *IEEE Transactions on Industry Applications*, 504–510. (©2003 IEEE). With permission.]

boosted because of the shoot-through states. The detailed relationship will be analyzed in the next section. It is noted here that the equivalent switching frequency viewed from the Z-source network is six times the switching frequency of the main inverter, which greatly reduces the required inductance of the Z-source network.

8.8.3 Circuit Analysis and Calculations

Assuming that the inductors L_1 and L_2 and capacitors C_1 and C_2 have the same inductance L and capacitance C, respectively, the Z-source network becomes symmetrical. From the symmetry and equivalent circuits, we have

$$V_{C1} = V_{C2} = V_C, \quad v_{L1} = v_{L2} = v_L. \tag{8.15}$$

Given that the inverter bridge is in the shoot-through *zero state* for an interval of T_0 during a switching cycle T, from the equivalent circuit in Figure 8.25 one has

$$v_L = V_C, \quad V_d = 2V_C, \quad v_i = 0. \tag{8.16}$$

Now, consider that the inverter bridge is in one of the eight non-shoot-through states for an interval of T_1 during the switching cycle T. From the equivalent circuit in Figure 8.25, one has

$$v_L = V_0 - V_C, \quad V_d = V_0, \quad v_i = V_C - v_L = 2V_C - V_0, \tag{8.17}$$

where V_0 is the DC voltage source and $T = T_0 + T_1$. The switching duty cycle $k = T_1/T$.

The average voltage of the inductors over one switching period should be zero in steady state, from Equations 8.16 and 8.17; thus, we have

$$V_L = \bar{v}_L = \frac{T_0 V_C + T_1(V_0 - V_C)}{T} = 0 \tag{8.18}$$

or

$$\frac{V_C}{V_0} = \frac{T_1}{T_1 - T_0}. \tag{8.19}$$

Similarly, the average DC-link voltage across the inverter bridge can be found as follows:

$$V_i = \bar{v}_i = \frac{T_0 \times 0 + T_1(2V_C - V_0)}{T} = \frac{T_1}{T_1 - T_0} V_0 = V_C. \tag{8.20}$$

The peak DC-link voltage across the inverter bridge is expressed in Equation 8.17 and can be rewritten as

$$\hat{v}_i = V_C - v_L = 2V_C - V_0 = \frac{T}{T_1 - T_0} V_0 = BV_0, \tag{8.21}$$

where

$$B = \frac{T}{T_1 - T_0} = \frac{1}{1 - 2(T_0/T)} \geq 1. \tag{8.22}$$

B is the boost factor resulting from the shoot-through zero state. Usually, T_1 is greater than T_0, that is, $T_0 < T/2$. The peak DC-link voltage $\hat{v}_i$ is the equivalent DC-link voltage of the inverter. On the other hand, the output peak phase voltage from the inverter can be expressed as

$$\hat{v}_{AC} = M\frac{\hat{v}_i}{2}, \tag{8.23}$$

where M is the modulation index. Using Equation 8.21, Equation 8.23 can be further expressed as

$$\hat{v}_{AC} = MB\frac{V_0}{2}. \tag{8.24}$$

For the traditional VSI, we have the well-known relationship $\hat{v}_{AC} = M(V_0/2)$. Equation 8.24 shows that the output voltage can be stepped up and down by choosing an appropriate buck–boost factor MB.

$$MB = \frac{T}{T_1 - T_0} M. \tag{8.25}$$

MB is changeable from 0 to ∞. From Equations 8.15, 8.19, and 8.22, the capacitor voltage can be expressed as

$$V_C = \frac{1 - (T_1/T)}{1 - 2(T_0/T)} V_0. \tag{8.26}$$

The buck–boost factor *MB* is determined by the modulation index *M* and the boost factor *B*. The boost factor *B* as expressed in Equation 8.22 can be controlled by the duty cycle (i.e., interval ratio) of the shoot-through zero state over the non-shoot-through states of the inverter PWM.

Note that the shoot-through zero state does not affect the PWM control of the inverter, because it produces, similarly zero voltage to the load terminal. The available shoot-through period is limited by the zero-state period that is determined by the modulation index.

8.9 Extended Boost ZSIs

In recent years, many researchers have focused, in many directions, on developing ZSIs in order to achieve different objectives [6–13]. Some have worked on developing different kinds of topological variations whereas others have worked on developing ZSIs into different applications where controller design, modeling and analyzing its operating modes, and developing modulation methods are addressed. Theoretically, ZSI can produce infinite gain like many other DC–DC boosting topologies; however, in practice this cannot be achieved because of the effects of parasitic components where the gain tends to drop drastically [6]. Conversely, high boost could increase power losses and instability. On the other hand, the shoot-through inverter can change its variables to respond the increasing gain, which is interdependent with the other variable modulation index that controls the output of the ZSI, and also imposes limitation on variability and thereby the boosting of output voltage. That is, an increase in the boosting factor would compromise the modulation index and result in a lower modulation index [7]. Also, the voltage stress on the switches would be high due to the pulsating nature of the output voltage.

Unlike in the case of DC–DC converters, so far researchers of ZSIs have not focused on improving the gain of the converter. This results in a significant research gap in the field of ZSI development. Particularly, some applications like solar and fuel cells, where generated power is integrated into the grid, may require high voltage gain to match the voltage difference and also to compensate the voltage variations. The effect is significant when such sources are connected to 415 V three-phase systems. In the case of fuel and solar cells, although it is possible to increase the number of cells to increase the voltage, there are other influencing factors that need to be taken into account. Sometimes the available number of cells is limited, or environmental factors could come into play due to the shading of some cells from light, which could result in poor overall energy catchment. Then with fuel cells, some manufacturers produce fuel cells with a lower voltage to achieve a faster response. Such factors could demand power converters with a larger boosting ratio. This cannot be realized with a single ZSI. Hence this chapter focuses on developing a new family of ZSIs that would realize extended boosting capability.

8.9.1 Introduction to ZSI and Basic Topologies

The basic topology of ZSI was originally proposed in reference [3]. This is a single-stage buck–boost topology due to the presence of the X-shaped impedance network, as shown in Figure 8.29a, which allows the safe shoot-through of inverter arms and avoids the need for dead time (which was needed in the traditional VSI). However, unlike the VSI, the original ZSI does not share the ground point of the DC source with the converter and also, the current drawn from the source will be discontinuous. These are disadvantages in some applications, and a decoupling capacitor bank at the front end may be required to avoid current discontinuity. Subsequently, the ZSI was modified as shown in Figures 8.29b and

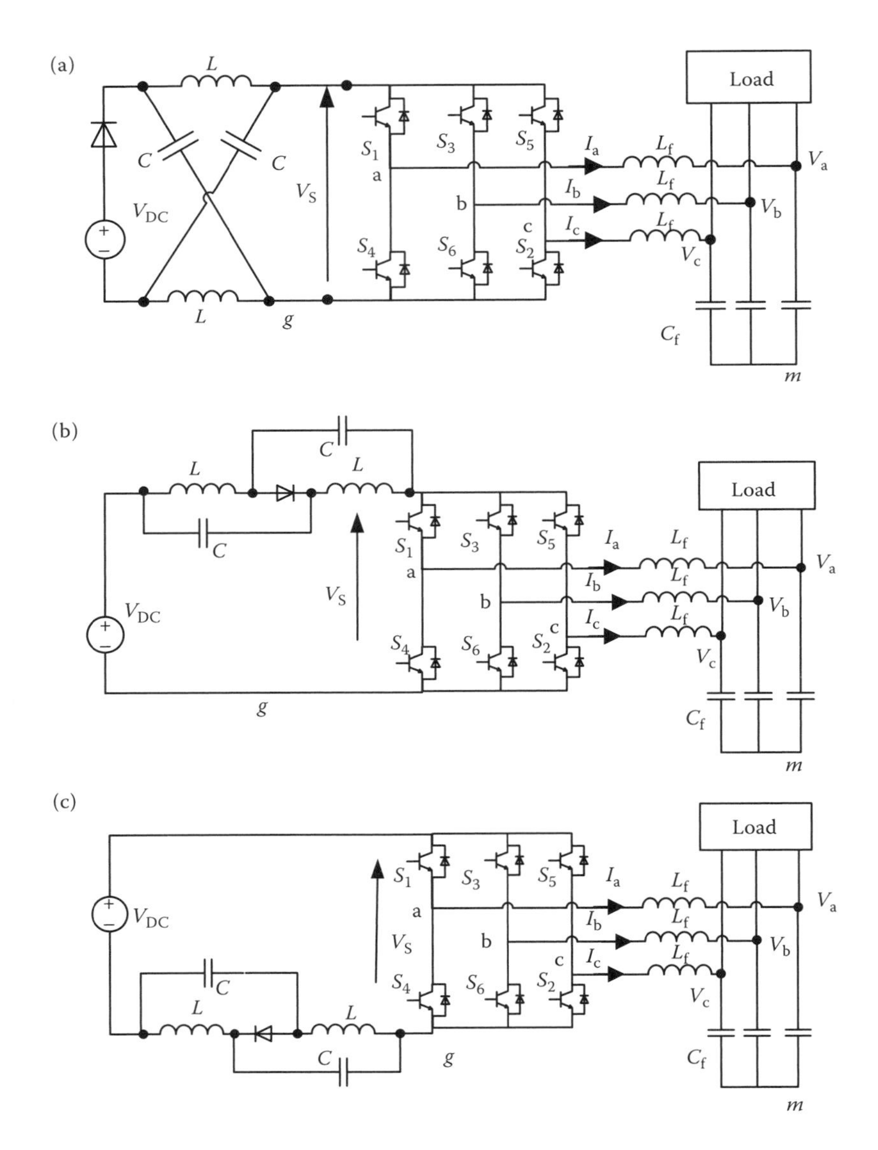

FIGURE 8.29 Various ZSIs: (a) original ZSI, (b) discontinuous current quasi Z-source inverter with shared ground, and (c) discontinuous current qZSI with low voltage level at components.

8.29c, where now an impedance network is placed at the bottom or top arm of the inverter. The advantage of this topology is that in one topology the ground point can be shared and in both cases the voltage stress on the component is much lower compared with that of the traditional ZSI. However, the current discontinuity still prevails; an alternative continuous current quasi-ZSI (qZSI) is proposed, but this continuous current circuit is not considered in developing new converters. In terms of topology, the qZSI has no disadvantage over the traditional topology. In this chapter, a discontinuous current qZSI inverter is used to extend the boosting capability. In summary, the proposed qZSIs operate similarly to the original ZSI, and the same modulation schemes can be applied.

8.9.2 Extended Boost qZSI Topologies

In this chapter, four new converter topologies have been proposed. These topologies can be mainly categorized into diode-assisted boost or capacitor-assisted boost topologies, and can be further divided into continuous current and discontinuous current topologies. Their operation is extensively described in subsequent sections. All these topologies can be modulated using the modulation methods proposed for the original ZSI. The other advantage of the proposed new topologies is their expandability. This was not possible with the original ZSI, that is, if one needs additional boosting, another stage can be cascaded at the front end. The new topology would operate with the same number of active switches. The only addition would be one inductor, one capacitor, and two diodes for the diode-assisted case, and one inductor, two capacitors, and one diode for the capacitor-assisted case for each new stage added. By defining the shoot-through duty ratio (D_S) for each new added stage, the boosting factor can be increased by a factor of $1/(1 - D_S)$ in the case of diode-assisted topology. Then the capacitor-assisted topology would have a boosting factor of $1/(1 - 3D_S)$ compared with $1/(1 - 2D_S)$ in the traditional topology. However, similar to the other boosting topologies, it is not advisable to operate with very high or very low shoot-through values. Also, a careful consideration is required when selecting the boosting factor modulation index for suitable topology to achieve high efficiency. These aspects need further research and will be addressed in a future paper.

8.9.2.1 Diode-Assisted Extended Boost qZSI Topologies

In this category, two new families of topologies are proposed, namely the continuous current-type topology and the discontinuous current-type topology. Figure 8.30 shows the continuous current-type topology, which can be extended to have very high boost by cascading more stages as shown in Figure 8.31. This new topology comprises an additional inductor, a capacitor, and two diodes. The operating principle of this additional impedance network is similar to that found in the cascaded boost and Luo-converters [9–12]. The added impedance network provides the boosting function without disturbing the operation inverter.

Considering the continuous current topology and its steady-state operation, we know that this converter has three operating states similar to those of traditional ZSI topology. It can be simplified into shoot-through and non-shoot-through states. Then the inverter's action is replaced by a current source and a single switch. First consider the non-shoot-through state, which is represented by an open switch. Also, diodes D_1 and D_2 are in the conducting state and D_3 is in the blocking state; therefore, the inductors discharge, and the capacitors

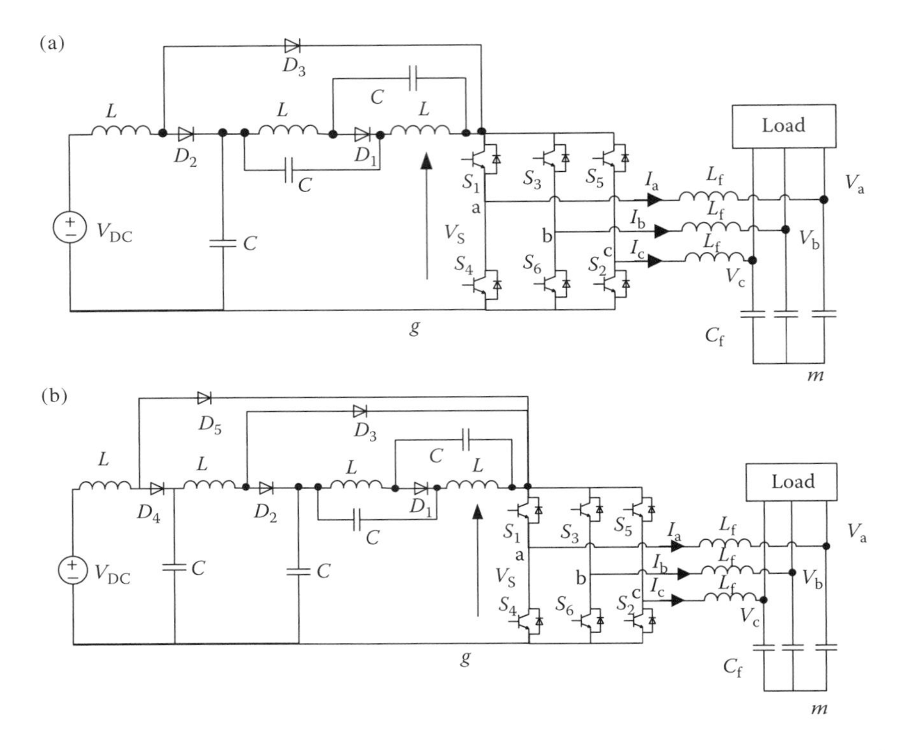

FIGURE 8.30 Diode-assisted extended boost continuous current qZSI: (a) first extension and (b) second extension.

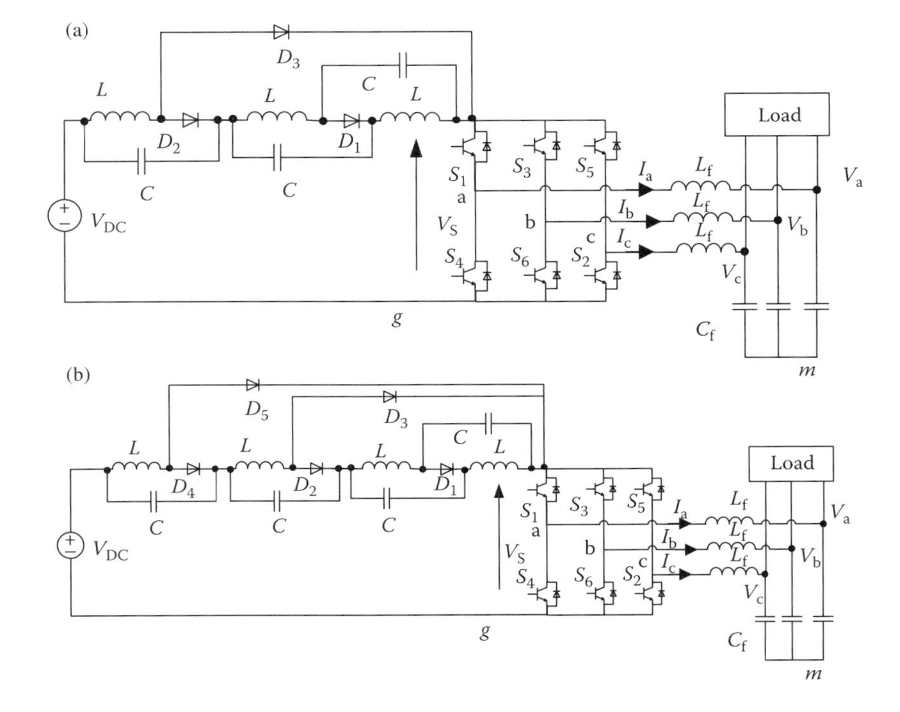

FIGURE 8.31 Diode-assisted extended boost discontinuous current qZSI: (a) first extension and (b) second extension.

get charged. Figure 8.32b shows the equivalent circuit diagram for the non-shoot-through state.

By applying KVL, the following steady-state relationships can be observed: $V_{DC} + V_{L3} = V_{C3}$, $V_{L1} = V_{C1}$, $V_{L2} = V_{C2}$, and $V_S = V_{C3} + V_{C2} + V_{L1}$. Figure 8.32c shows the equivalent circuit diagram for the shoot-through state where it is represented by the closed switch, and D_3 is in the conducting state and D_1 and D_2 diodes are in the blocking state where all the inductors get charged. Energy is transferred from the source to the inductor or from the capacitor to the inductor while the capacitors are getting discharged. Similar relationships can be derived as $V_{DC} + V_{L3} = 0$, $V_{C3} + V_{L2} + V_{C1} = 0$, $V_{C3} + V_{C2} + V_{L1} = 0$, $V_S = 0$, and $V_{C3} + V_{C2} = V_{L1}$. Considering that the average voltage across the inductors is zero and by defining the shoot-through duty ratio as D_S and the non-shoot-through duty ratio as D_A,

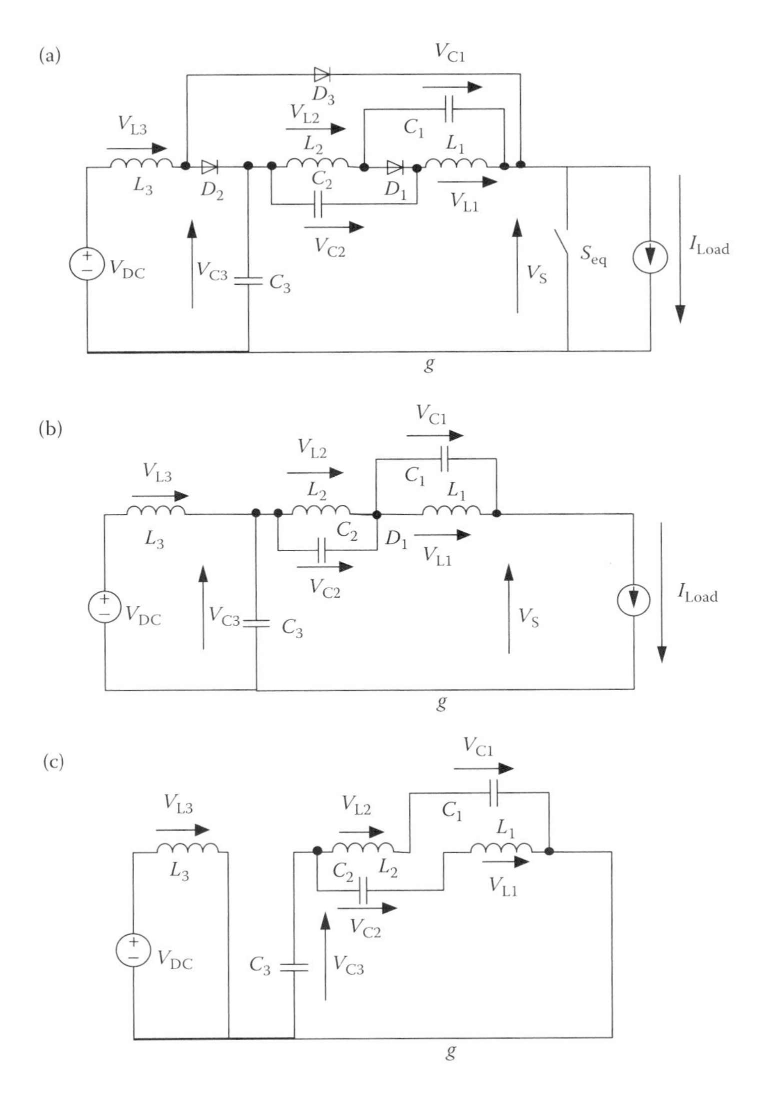

FIGURE 8.32 Simplified diagram of diode-assisted extended boost continuous current qZSI: (a) simplified circuit, (b) non-shoot-through state, and (c) shoot-through state.

where $D_A + D_S = 1$, the following relations can be derived:

$$V_{C3} = \frac{1}{1 - D_S} V_{DC} \quad \text{and} \quad V_{C1} = V_{C2} = \frac{D_S}{1 - 2D_S} V_{C3} = \frac{D_S}{(1 - 2D_S)(1 - D_S)} V_{DC}. \tag{8.27}$$

From the above equations, the peak voltage across the inverter $\hat{v}_S$ and the peak AC output voltage $\hat{v}_x$ can be obtained as

$$\hat{v}_S = \frac{1}{(1 - 2D_S)(1 - D_S)} V_{DC} \quad \text{and} \quad \hat{v}_x = M\frac{\hat{v}_S}{2}. \tag{8.28}$$

Define $B = 1/[(1 - 2D_S)(1 - D_S)]$, the boost factor in the DC side; then the peak in the AC side can be written as

$$\hat{v}_x = B\left(M\frac{V_{DC}}{2}\right). \tag{8.29}$$

Now the boosting factor has increased by a factor of $1/(1 - D_S)$ compared with that of the original ZSI. Similarly, steady-state equations can be derived for the diode-assisted extended boost discontinuous current qZSI. Then it is possible to prove that this converter also has the same boosting factor as that of continuous current topology. Also, the voltage stresses on the capacitors are similar, except for the voltage across C_3; this can be written as $V_{C3} = D_S/(1 - D_S) \times V_{DC}$. By studying these two topologies, it can be noted that with the discontinuous current topology, capacitors are subjected to a small voltage stress, and if there is no boosting then the voltage across them is zero. Also, it is possible to derive the boost factor for the topologies shown in Figures 8.30b and 8.31b as $B = 1/[(1 - 2D_S)(1 - D_S)^2]$.

8.9.2.2 Capacitor-Assisted Extended Boost qZSI Topologies

Similar to the previous family of extended boost qZSIs, this section proposes another family of converters. The difference is that now a much higher boost is achieved with only a simple structural change to the previous topology. Now D_3 is replaced by a capacitor, as shown in Figure 8.36. In this context also, two topological variations are derived as continuous current or discontinuous current forms, as shown in Figure 8.33.

In the previous scenario, the steady-state relations are derived using continuous current topology; therefore, in this context, the discontinuous current topology is considered. In this case also, the converter's three operating states are simplified into shoot-through and non-shoot-through states.

The simplified circuit diagram is shown in Figure 8.34a. First consider the non-shoot-through state shown in Figure 8.34b, which is represented by an open switch. As diodes D_1 and D_2 are conducting the inductors discharge, and the capacitors get charged. Then by applying KVL, the following steady-state relationships can be observed. $V_{DC} + V_{C3} + V_{C2} + V_{C1} = V_S$ and $V_{DC} + V_{C3} + V_{C4} + V_{C1} = V_S$, $V_{C1} = V_{L1}$, $V_{C2} = V_{L2}$, $V_{C3} = V_{L3}$, $V_{DC} + V_{C3} = V_d$, $V_{C2} = V_{C4}$. Figure 8.34c shows the equivalent circuit diagram for the shoot-through state, where it is represented by the closed switch. Diodes D_1 and D_2 are in the blocking state, where all the inductors get charged and energy is transferred from

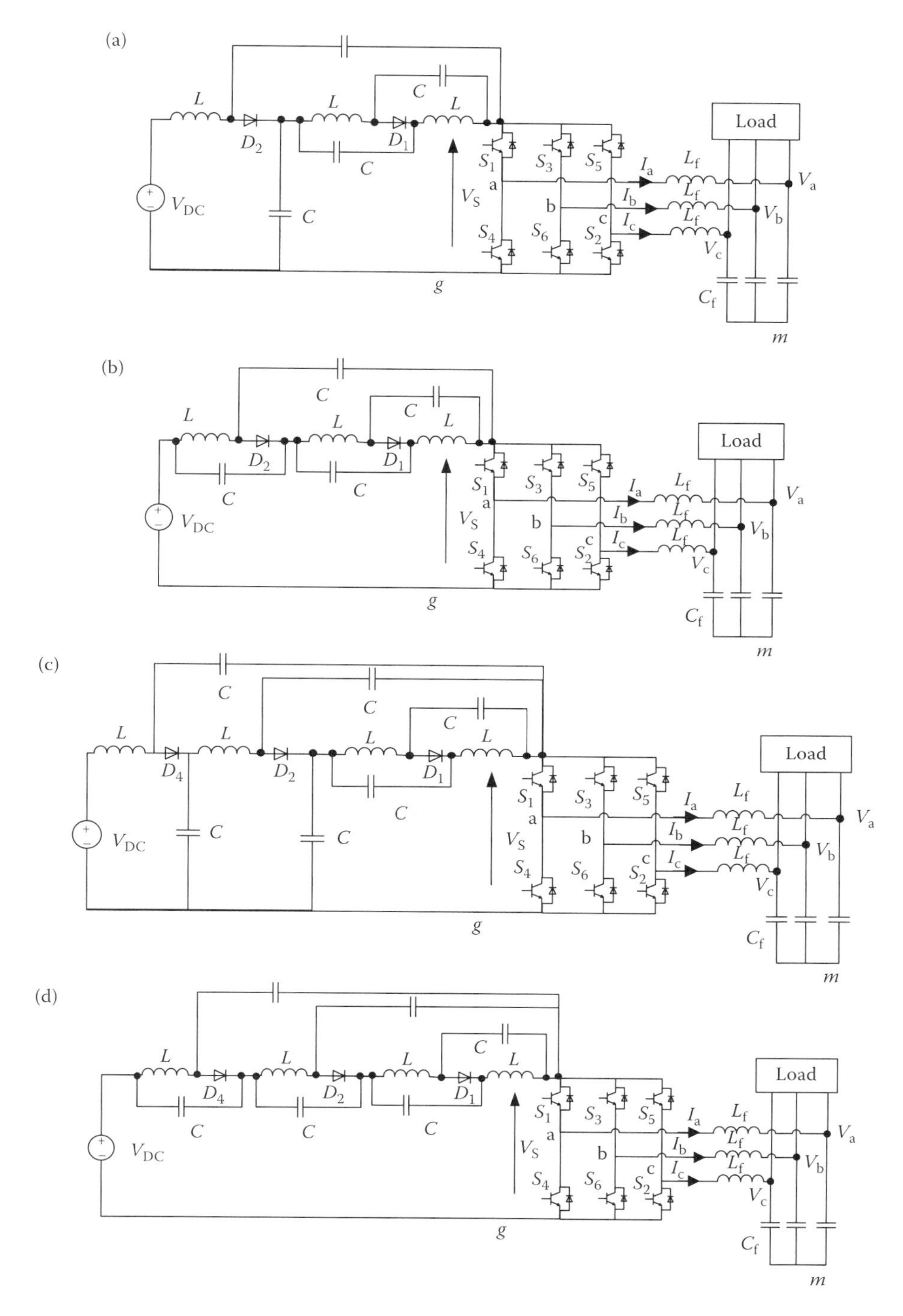

FIGURE 8.33 Capacitor-assisted extended boost qZSIs: (a) continuous current, (b) discontinuous current, (c) high extended continuous current, and (d) discontinuous current.

(a)

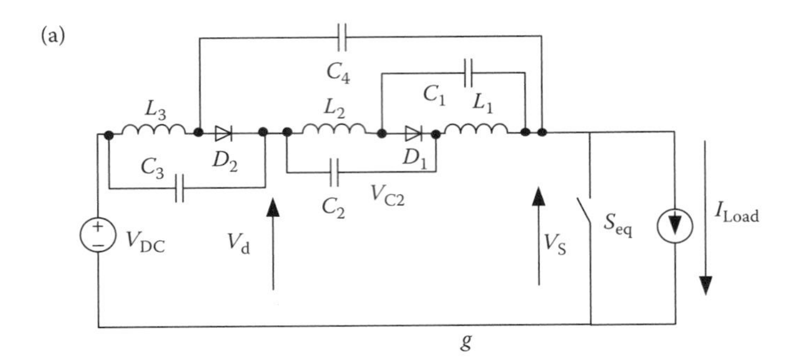

(b)

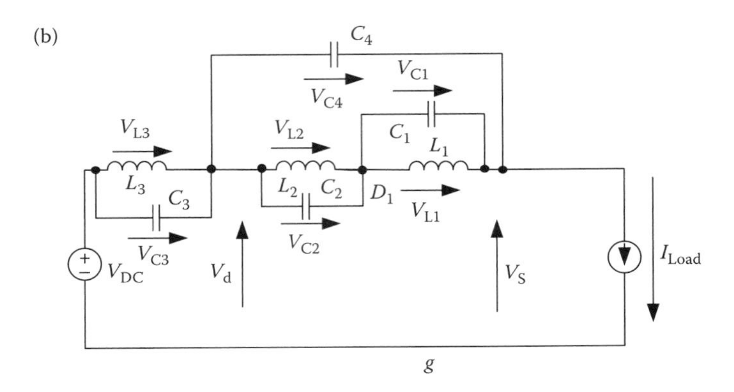

(c)

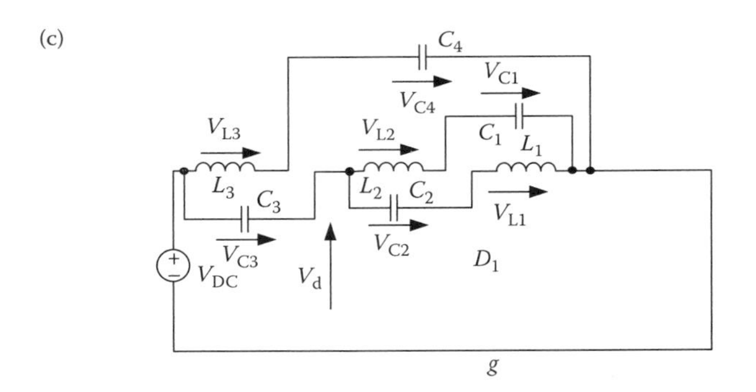

FIGURE 8.34 Simplified diagram of capacitor-assisted extended boost continuous current qZSI: (a) simplified circuit, (b) non-shoot-through state, and (c) shoot-through state.

the source to the inductors or from the capacitor to the inductors while the capacitors are getting discharged. Similar relationships can be derived as $V_{DC} + v_{L3} + V_{C4} + V_{C1} = 0$, $V_{DC} + V_{C3} = V_d$, $V_d + V_{L1} + V_{C2} = 0$, $V_d + V_{L2} + V_{C1} = 0$, and $V_S = 0$. Considering the fact that the average voltage across the inductors is zero, the following relations can be derived:

$$V_d = \frac{1 - 2D_S}{1 - 3D_S} V_{DC} \quad \text{and} \quad V_{C_1} = V_{C2} = V_{C_3} = V_{C_4} = \frac{D_S}{1 - 2D_S} V_d = \frac{D_S}{1 - 3D_S} V_{DC}. \tag{8.30}$$

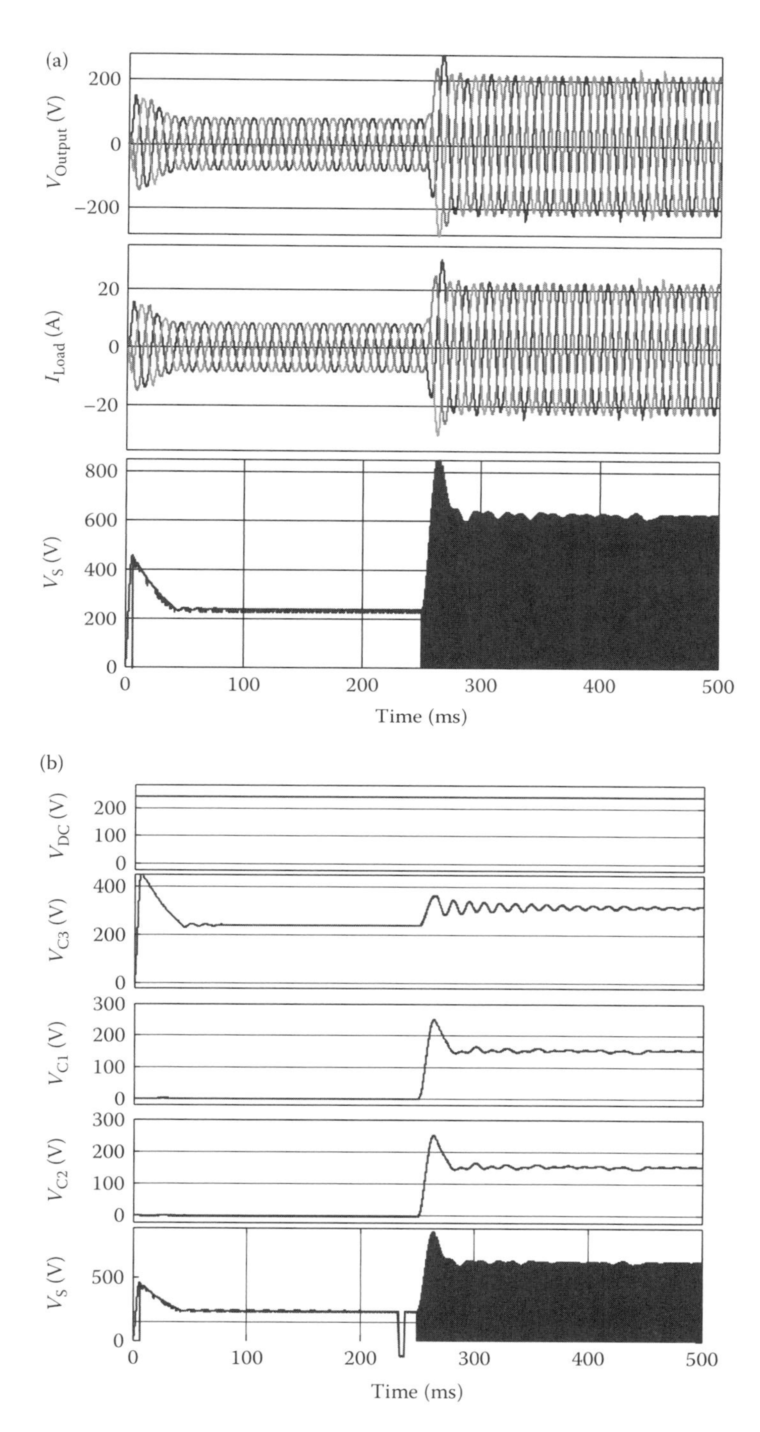

FIGURE 8.35 Simulation results for diode-assisted extended boost continuous current qZSI: (a) waveforms of V_O, I_{load}, and V_S; (b) waveforms of V_{DC}, V_{C3}, V_{C1}, V_{C2}, and V_S.

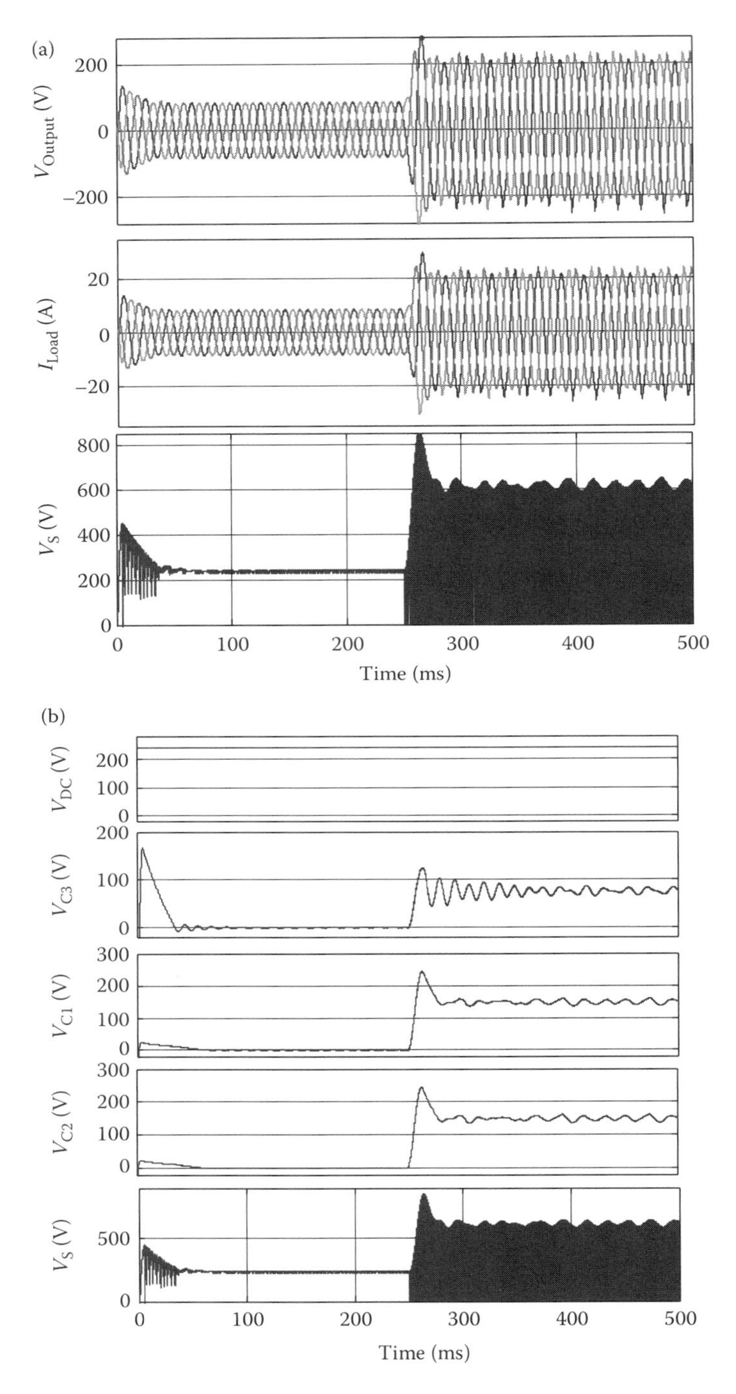

FIGURE 8.36 Simulation results for capacitor-assisted extended boost discontinuous current qZSI: (a) waveforms of V_O, I_{load}, and V_S; (b) waveforms of V_{DC}, V_{C3}, V_{C1}, V_{C2}; and V_S.

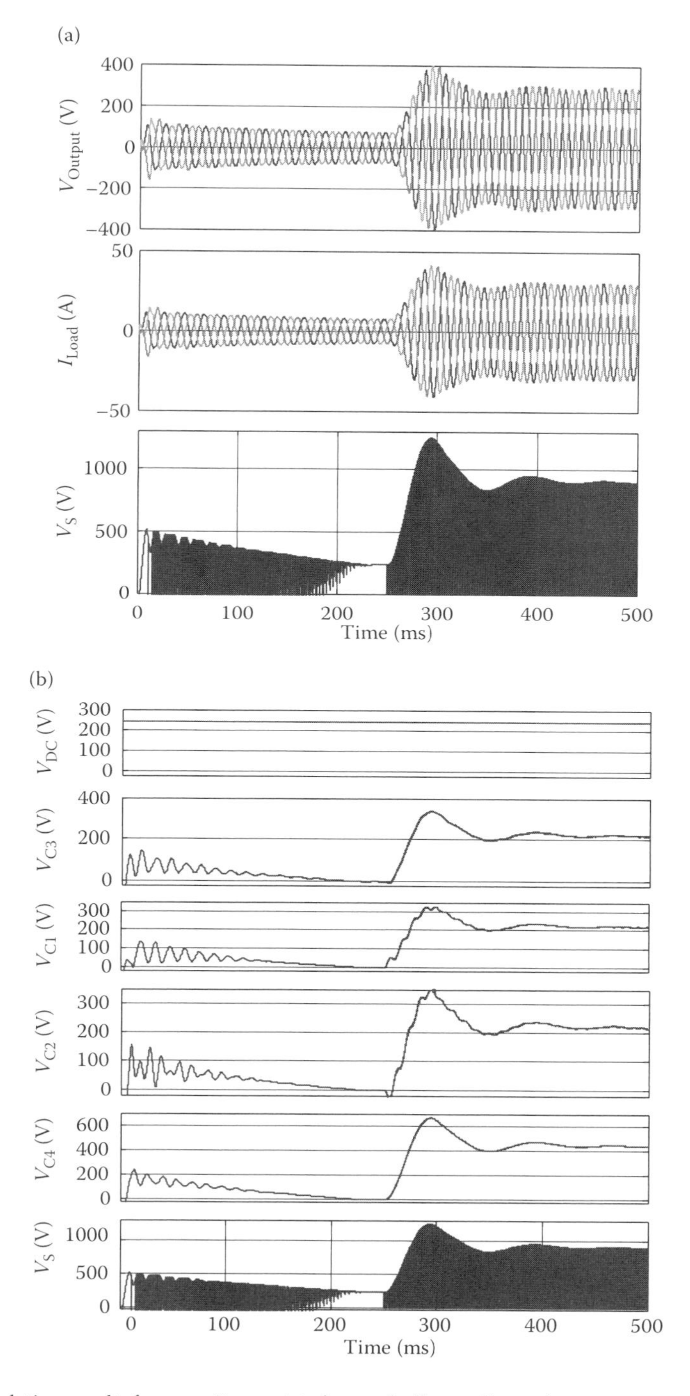

FIGURE 8.37 Simulation results for capacitor-assisted extended boost discontinuous current qZSI: (a) waveforms of V_O, I_{load}, and V_S; (b) waveforms of V_{DC}, V_{C3}, V_{C1}, V_{C2}, and V_S.

Then, from the above equations, the peak voltage across the inverter $\hat{v}_S$ can be obtained as

$$\hat{v}_S = \frac{1}{1 - 3D_S} V_{DC}. \tag{8.31}$$

Similar equations can be derived for the continuous current topology. The only difference would be the continuity of source current and the difference in voltage across the C_3 now it is equal to V_d where the voltage across the capacitor is much larger than other topology. Similarly, it is possible to derive the boost factor for topologies shown in Figures 8.33c and 8.33d as $B = 1/(1 - 4D_S)$.

8.9.3 Simulation Results

Extensive simulation studies are performed on the open-loop configuration of all proposed topologies in MATLAB®/SIMULINK® using the modulation method proposed in reference [5]. However, due to space limitations, only a few results are presented. This would validate the operation of diode-assisted and capacitor-assisted topologies as well as continuous current and discontinuous current topologies. Here, three cases are simulated. In all three cases, the input voltage is kept constant at 240 V and a three-phase load of 9.7 Ω resistor bank is used. All DC-side capacitors are 1000 μF and inductors are 3.5 mH. The AC-side second-order filter is used with a 10- μF capacitor and a 7-mH inductor. In all three cases the converter is operated with zero boosting in the beginning, and at $t = 250$ ms the shoot-through is increased to 0.25 while the modulation index is kept constant at 0.7. Figures 8.35 through 8.37 show the simulation results corresponding to the topologies shown in Figures 8.30a, 8.31a, and 8.33b. From these figures, it is possible to note that in the first two cases equal boosting is achieved and the difference is the voltage across V_{C3}. This complies with the theoretical finding. From Figure 8.37 it can be noted that with the capacitor-assisted topology a much higher boosting can be achieved with the same shoot-through value; also, the voltage across all four capacitors is equal and complies with the equations derived in Section 8.9.2. A comprehensive set of simulation results will be presented in the full paper.

Homework

8.1. A single-phase half-bridge DC/AC inverter is shown in Figure 8.3 to implement SPWM with $V_d = 400$ V, $m_a = 0.8$, and $m_f = 35$. The fundamental frequency is 50 Hz. Determine the rms value of the fundamental frequency and some of the harmonics in the output voltage using Table 8.1.

8.2. A single-phase full-bridge VSI with amplitude modulation ratio $(m_a) = 0.8$ and frequency modulation ratio $(m_f) = 8$ is shown in Figure 8.8. The SPWM technique is applied in this VSI. The required frequency of the output voltage is 50 Hz. Calculate the pulse widths (times or angles) of the first pulses to turn on and turn off the two pairs of switches.

8.3. A three-phase full-bridge DC/AC inverter is shown in Figure 8.11 to implement SPWM with $V_d = 500$ V, $m_a = 1.0$, and $m_f = 41$. The fundamental frequency is 50 Hz. Determine the rms value of the fundamental frequency and some of the harmonics in the output voltage using Table 8.1.

References

1. Mohan, N., Undeland, T. M., and Robbins, W. P. 2003. *Power Electronics: Converters, Applications and Design* (3rd edition). New York: Wiley.
2. Holtz, J. 1992. Pulsewidth modulation—a survey. *IEEE Transactions on Industrial Electronics*, 28, 410–420.
3. Peng, F. Z. 2003. Z-source inverter. *IEEE Transactions on Industry Applications*, 39, 504–510.
4. Trzynadlowski, A. M. 1998. *Introduction to Modern Power Electronics*. New York: Wiley.
5. Middlebrook, R. D. and Cúk, S. 1981. *Advances in Switched-Mode Power Conversion* (Vols. I and II). Pasadena, CA: TESLAco.
6. Gajanayake, C. J. and Luo, F. L. 2009. Extended boost Z-source inverters. *Proceedings of IEEE ECCE 2009*, pp. 368–373.
7. Gajanayake, C. J., Vilathgamuwa, D. M., and Loh, P. C. 2007. Development of a comprehensive model and a multiloop controller for Z-source inverter DG systems. *IEEE Transactions on Industrial Electronics*, 54, 2352–2359.
8. Anderson, J. and Peng, F. Z. 2008. Four quasi-Z-source inverters. *Proceedings of IEEE PESC 2008*, pp. 2743–2749.
9. Luo, F. L. and Ye, H. 2005. *Advanced DC/DC Converters*. Boca Raton: CRC Press.
10. Luo, F. L. and Ye, H. 2005. *Essential DC/DC Converters*. Boca Raton: Taylor & Francis Group LLC.
11. Luo, F. L. 1999. Positive output Luo-converters: Voltage lift technique. *IEE-Proceedings on Electric Power Applications*, 146, pp. 415–432.
12. Luo, F. L. 1999. Negative output Luo-converters: Voltage lift technique. *IEE-Proceedings on Electric Power Applications*, 146, pp. 208–224.
13. Ortiz-Lopez, M. G., Leyva-Ramos, J. E., Carbajal-Gutierrez, E., and Morales-Saldana, J. A. 2008. Modelling and analysis of switch-mode cascade converters with a single active switch. *Power Electronics, IET*, 155, 478–487.

9

Multilevel and Soft-Switching DC/AC Inverters

Multilevel inverters represent a novel method of constructing DC/AC inverters. This idea was published by Nabae in 1980 in an IEEE international conference IEEE APEC'80 [1], and the same idea was published in 1981 in *IEEE Transactions on Industry Applications* [2]. Actually, multilevel inverters represent a different technique from the PWM method, which consists of vertically chopping a reference waveform to achieve a similar output waveform (e.g., sine wave). The multilevel inverting technique consists of accumulating the levels horizontally to achieve the waveform (e.g., sine wave). The soft-switching technique was implemented in DC/DC conversion in the 1980s. We would like to introduce this technique in DC/AC inverters as well, in this chapter.

9.1 Introduction

Although PWM inverters have been used in industrial applications, they have many drawbacks:

1. The carrier frequency must be very high. Mr. Mohan nominated $m_f > 21$, which means that $f_\Delta > 1$ kHz if the frequency of the output waveform is 50 Hz. Usually, in order to keep the THD small, f_Δ is chosen to be 2–20 kHz [3].
2. The pulse height is very high. In a normal PWM waveform (not multistage PWM), the height of all pulses is the DC-link voltage. The output voltage of this PWM inverter has a large jumping span. For example, if the DC-link voltage is 400 V, all pulses have the peak value of 400 V. Usually, this causes a large dv/dt and a strong EMI.
3. The pulse width would be very narrow when the output voltage has a low value. For example, if the DC-link voltage is 400 V, the output is 10 V; the corresponding pulse width should be 2.5% of the full pulse period.
4. Items 2 and 3 induce a number of harmonics to produce poor THD.
5. Items 2 and 3 offer very rigorous switching conditions. The switching devices have large switching power losses.
6. The inverter control circuitry is complex and the devices are costly. Therefore the whole inverter is costly.

The multilevel inverter accumulates the output voltage in horizontal levels (layers). Therefore, using this technique, the above drawbacks of the PWM technique can be overcome because of the following features of multilevel inverters:

1. The switching frequencies of most switching devices are low and are equal to or only few times the output signal frequency.
2. The pulse heights are quite low. For an m-level inverter with output amplitude V_m, the pulse heights are V_m/m or only few times of it. Usually, this causes a low dv/dt and an ignorable EMI.
3. The pulse widths of all pulses have reasonable values to be comparable to the output signal.
4. Items 2 and 3 cannot induce enough harmonics to produce lower THD.
5. Items 2 and 3 offer smooth switching conditions. The switching devices have small switching power losses.
6. The inverter control circuitry is comparatively simple and the devices are not costly. Therefore the whole inverter is economical.

Multilevel inverters contain several power switches and capacitors [4]. The output voltages of multilevel inverters are the additions of the voltages due to the commutation of the switches. Figure 9.1 shows a schematic diagram of one phase leg of inverters with different level numbers. A two-level inverter, as shown in Figure 9.1a, generates an output voltage with two levels with respect to the negative terminal of the capacitor. The three-level inverter shown in Figure 9.1b generates a three-level voltage, and the m-level inverter shown in Figure 9.1c generates an m-level voltage. Thus, the output voltages of multilevel inverters have several levels. Moreover, they can reach high voltage, whereas the power semiconductors must withstand only reduced voltages.

Multilevel inverters have been receiving increasing attention in recent decades, since they have many attractive features as described before. Various kinds of multilevel inverters have been proposed, tested, and installed.

- Diode-clamped (neutral-clamped) multilevel inverters (DCMI)
- Capacitors-clamped (flying capacitors) multilevel inverters
- Cascaded multilevel inverters (CMIs) with separate DC sources
- Hybrid multilevel inverters
- Generalized multilevel inverters (GMIs)

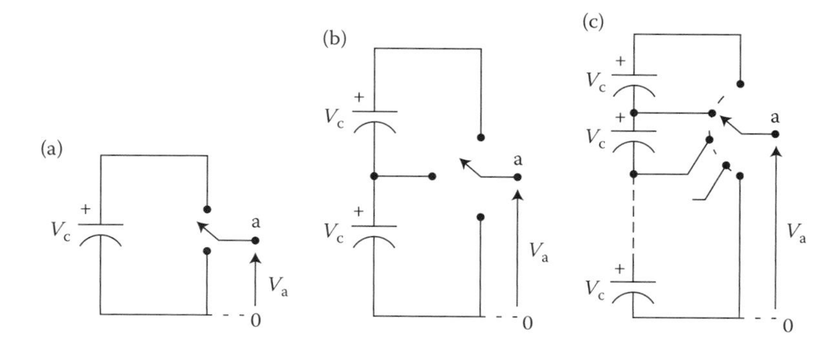

FIGURE 9.1 One phase leg of an inverter: (a) two levels, (b) three levels, and (c) m levels.

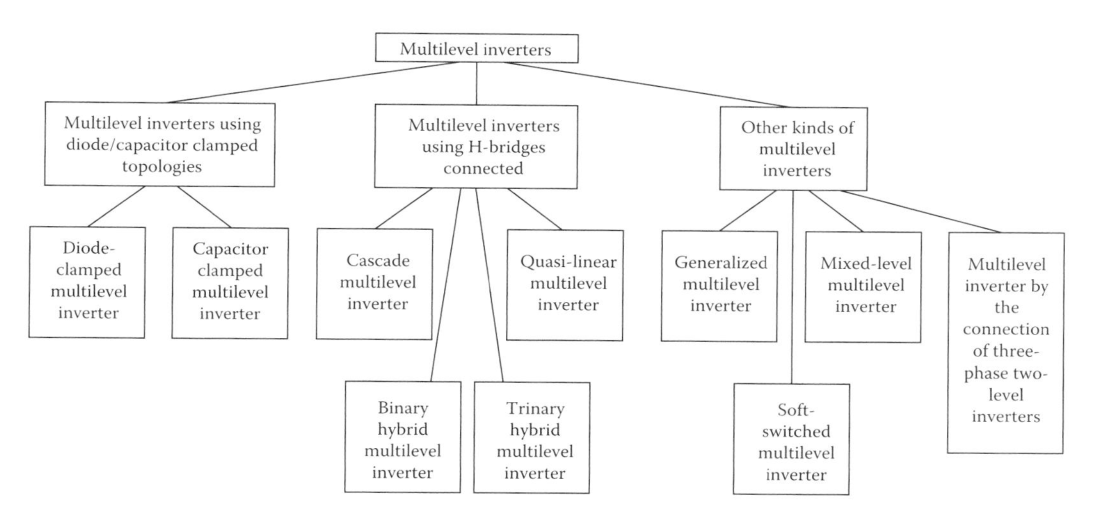

FIGURE 9.2 Family tree of multilevel inverters.

- Mixed-level multilevel inverters
- Multilevel inverters through the connection of three-phase two-level inverters
- Soft-switched multilevel inverters.

The family tree of multilevel inverters is shown in Figure 9.2.

The family of multilevel inverters has emerged as the solution for high-power application, since implementation via a single power semiconductor switch directly in a medium-voltage network is hard work. Multilevel inverters have been applied to different high-power applications, such as large motor drives, railway traction applications, high-voltage DC transmissions (HVDC), unified power flow controllers (UPFC), static var compensators (SVC), and static synchronous compensators (STATCOM). The output voltage of the multilevel inverter has many levels, synthesized from several DC voltage sources. The quality of the output voltage is improved as the number of voltage levels increases; hence the effort of output filters can be decreased. The transformers can be eliminated due to the reduced voltage that the switch endures. Moreover, as cost-effective solutions, the applications of multilevel inverters have also been extended to medium- and low-power applications such as electrical vehicle propulsion systems, active power filters (APF), voltage sag compensations, photovoltaic systems, and distributed power systems.

Multilevel inverter circuits have been investigated for nearly 30 years. Separate DC-sourced full-bridge cells were connected in series to synthesize a staircase AC output voltage. The diode-clamped inverter, also called the neutral-point clamped (NPC) inverter, was presented in 1980 by Nabae. Because the NPC inverter effectively doubles the device voltage level without requiring precise voltage matching, the circuit topology prevailed in the 1980s. The capacitor-clamped multilevel inverter (CCMI) appeared in the 1990s. Although the CMI was invented earlier, its application did not prevail until the mid-1990s. The advantages of CMIs were indicated for motor drives and utility applications. The cascaded inverter has drawn great interest due to the high demand for medium-voltage high-power inverters.

The cascaded inverter is also used in regenerative-type motor drive applications. Recently, some new topologies of multilevel inverters have emerged, such as GMIs, mixed multilevel inverters, hybrid multilevel inverters, and soft-switched multilevel inverters. Today, multilevel inverters are extensively used in high-power applications with medium-voltage levels such as laminators, mills, conveyors, pumps, fans, blowers, compressors, and so on. Moreover, as a cost-effective solution, the applications of multilevel inverters are also extended to low-power applications, such as photovoltaic systems, hybrid electrical vehicles, and voltage sag compensation, in which the effort of the output filter components can become much decreased due to low harmonics distortions of the output voltages of the multilevel inverters.

9.2 Diode-Clamped Multilevel Inverters

In this category, the switching devices are connected in series to make up the desired voltage rating and output levels. The inner voltage points are clamped by either two extra diodes or one high-frequency capacitor. The switching devices of an m-level inverter are required to block a voltage level of $V_{DC}/(m-1)$. The clamping diode needs to have different voltage ratings for different inner voltage levels. In summary, for an m-level diode-clamped inverter,

- Number of power electronic switches $= 2(m-1)$
- Number of DC-link capacitors $= (m-1)$
- Number of clamped-diodes $= 2\ (m-2)$
- The voltage across each DC-link capacitor $= V_{DC}/(m-1)$.

where V_{DC} is the DC-link voltage. A three-level diode-clamped inverter is shown in Figure 9.3a with $V_{DC} = 2E$. In this circuit, the DC-bus voltage is split into three levels by two series-connected bulk capacitors, C_1 and C_2. The middle point of the two capacitors, n, can be defined as the neutral point. The output voltage v_{an} has three states: E, 0, and $-E$. For voltage level E, switches S_1 and S_2 need to be turned on; for $-E$, switches $S_{1'}$ and $S_{2'}$ need to be turned on; and for the 0 level, switches S_2 and $S_{2'}$, need to be turned on.

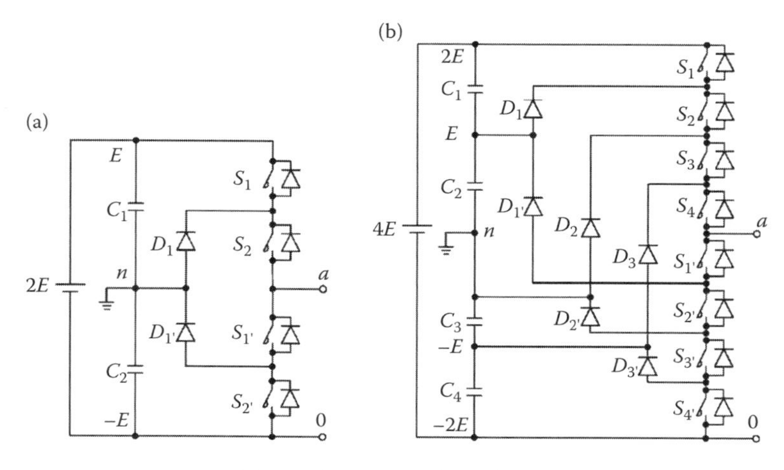

FIGURE 9.3 DCMI circuit topologies: (a) three levels and (b) five levels.

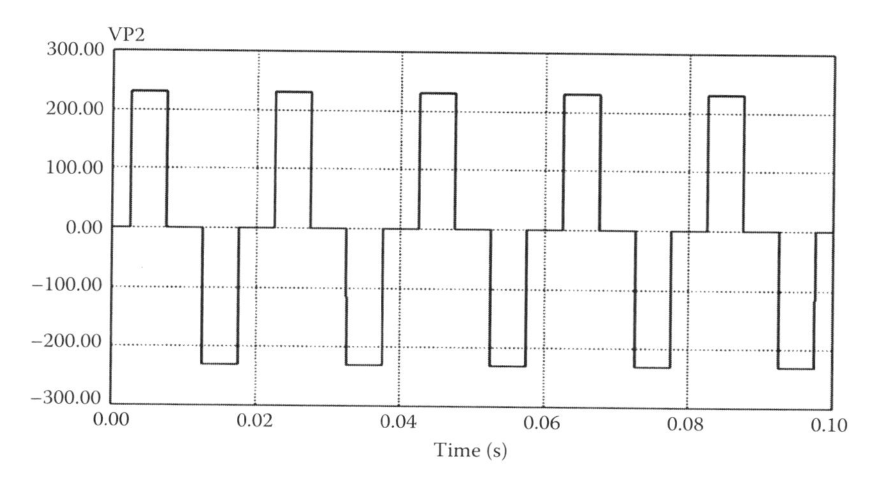

FIGURE 9.4 Output waveform of a three-level inverter.

The key components that distinguish this circuit from a conventional two-level inverter are D_1 and $D_{1'}$. These two diodes clamp the switch voltage to half the level of the DC-bus voltage. When both S_1 and S_2 turn on, the voltage across a and 0 is $2E$, that is, $v_{a0} = 2E$. In this case, $D_{1'}$, balances out the voltage sharing between $S_{1'}$ and $S_{2'}$ with $S_{1'}$ blocking the voltage across C_1 and $S_{2'}$ blocking the voltage across C_2. Note that the output voltage v_{an} is AC, and v_{a0} is DC. The difference between v_{an} and v_{a0} is the voltage across C_2, which is E. If the output is removed between a and 0, then the circuit becomes a DC/DC converter, which has three output voltage levels: E, 0, and $-E$. The simulation waveform is shown in Figure 9.4.

Usually, the higher the number of levels, the lower the THD of the output voltage. The switching angle decides the THD of the output voltage as well. The THD of the three-level diode-clamped inverter is shown in Table 9.1.

Figure 9.3b shows a five-level diode-clamped inverter in which the DC bus consists of four capacitors, C_1, C_2, C_3, and C_4. For DC-bus voltage $4E$, the voltage across each capacitor is E, and each device voltage stress will be limited to one capacitor voltage level E through clamping diodes.

To explain how the staircase voltage is synthesized, the neutral point n is considered as the output phase voltage reference point. There are five switch combinations to synthesize a five-level voltage across a and n.

- For voltage level $v_{an} = 2E$, turn on all upper switches S_1–S_4.
- For voltage level $v_{an} = E$, turn on three upper switches S_2–S_4 and one lower switch $S_{1'}$.

TABLE 9.1

THD Content for Different Switching Angle

Switching Angle	THD (%)
15	31.76
30	30.9

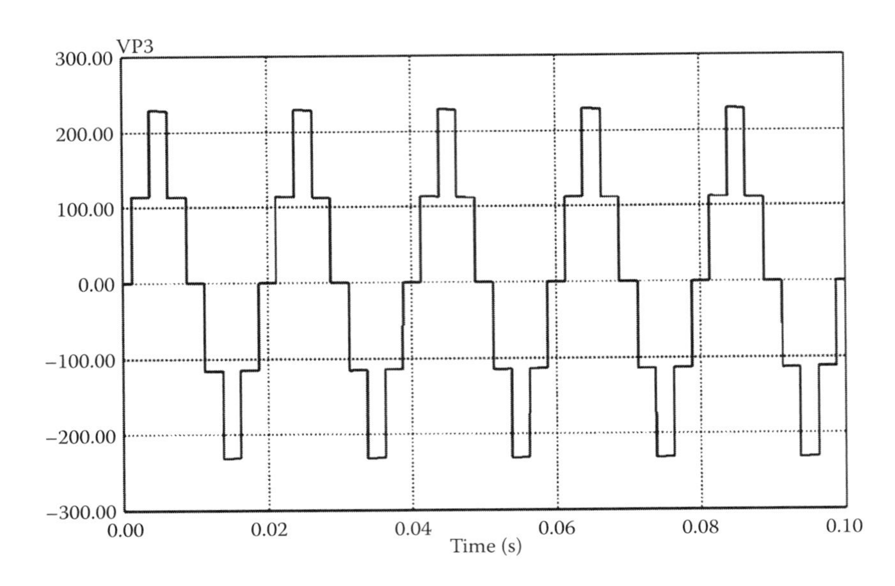

FIGURE 9.5 Output waveform of a five-level inverter.

- For voltage level $v_{an} = 0$, turn on two upper switches S_3 and S_4 and two lower switches $S_{1'}$ and $S_{2'}$.
- For voltage level $v_{an} = -E$, turn on one upper switch S_4 and three lower switches $S_{1'}$–$S_{3'}$.
- For voltage level $v_{an} = -2E$, turn on all lower switches $S_{1'}$–$S_{4'}$.

For a diode-clamped inverter, each output level has only one combination to implement its output voltage. Four complementary switch pairs exist in each phase. The complementary switch pair is defined such that turning on one of the switches will exclude the other from being turned on. In this example, the four complementary pairs are (S_1, $S_{1'}$), (S_2, $S_{2'}$), (S_3, $S_{3'}$), and (S_4, $S_{4'}$). Although each active switching device is only required to block a voltage level of E, the clamping diodes must have different voltage ratings for reverse voltage blocking. Using $D_{1'}$ of Figure 9.3b as an example, when lower devices $S_{2'}$–$S_{4'}$ are turned on, $D_{1'}$ needs to block three capacitor voltages, or $3E$. Similarly, D_2 and $D_{2'}$ need to block $2E$, and D_1 needs to block $3E$.

The simulation waveform is shown in Figure 9.5.

A seven-level diode-clamped inverter has the waveform shown in Figure 9.6.

From Figures 9.4 through 9.6, the THD is reduced when the number of levels of the inverter is increased. Hence, higher levels of the inverter will be considered to produce the output with less harmonic content. For each inverter, by carefully setting the firing angles, the best THD can be obtained. Table 9.2 shows various inverters' best firing angles to produce the lowest THD.

By applying MATLAB® graph fitting tool, the relationship between the lowest THD and the number of levels of the inverter can be estimated as

$$\mathrm{THD}_{\mathrm{Lowest}} = 72.42\mathrm{e}^{-0.4503m} + 11.86\mathrm{e}^{-0.05273m}, \tag{9.1}$$

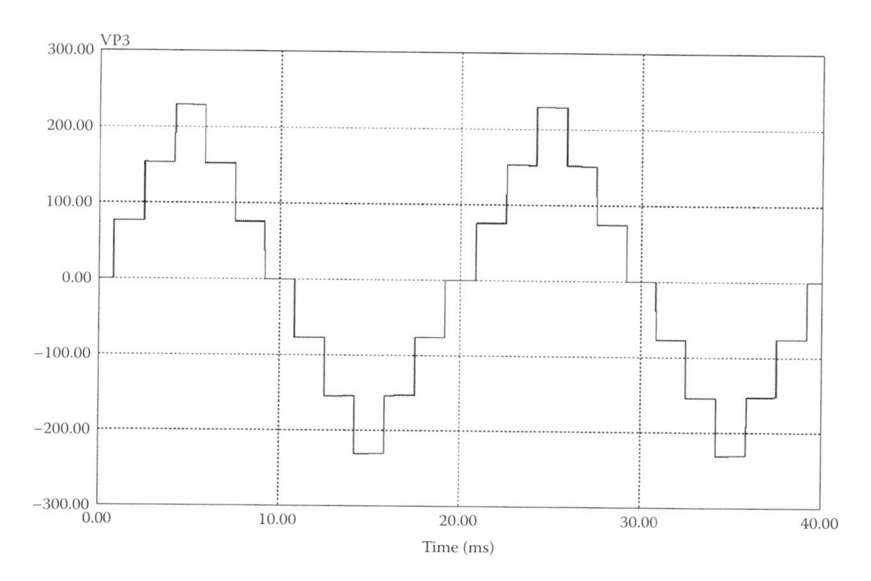

FIGURE 9.6 Output waveform of a seven-level inverter.

where m is the level number of the inverter. The corresponding figure for the THD versus m is shown in Figure 9.7.

Example 9.1

A diode-clamped three-level inverter shown in Figure 9.3a operates in the state with the best THD. Determine the corresponding switching angles, switch status, and THD.

Solution

Refer to Table 9.2; the best switching angles in a cycle are

$$\alpha_1 = 0.2332\text{rad} = 13.36^\circ.$$

$$\alpha_2 = \pi - \alpha_1 = 180^\circ - 13.36^\circ = 166.64^\circ.$$

$$\alpha_3 = \pi + \alpha_1 = 180^\circ + 13.36^\circ = 193.36^\circ.$$

$$\alpha_4 = 2\pi - \alpha_1 = 360^\circ - 13.36^\circ = 346.64^\circ.$$

The switches referring to Figure 9.3a operate in a cycle (0° to 360°) as follows:

Turn on the upper switches S_2 and the lower switches $S_{1'}$ in $0^\circ - \alpha_1$.
Turn on all upper switches S_1 and S_2 in $\alpha_1 - \alpha_2$.
Turn on the upper switches S_2 and the lower switches $S_{1'}$ in $\alpha_2 - \alpha_3$.
Turn on all lower switches $S_{1'}$–$S_{2'}$ in $\alpha_3 - \alpha_4$.
Turn on the upper switches S_2 and the lower switches $S_{1'}$ in $\alpha_4 - 360^\circ$.

The best THD = 28.96%.

TABLE 9.2
Best Switching Angle

Number of Level (m)	α_1	α_2	α_3	α_4	α_5	α_6	α_7	α_8	α_9	α_{10}	α_{11}	α_{12}	α_{13}	α_{14}	α_{15}	THD (%)
3	0.2332	—	—	—	—	—	—	—	—	—	—	—	—	—	—	28.96
5	0.2242	0.7301	—	—	—	—	—	—	—	—	—	—	—	—	—	16.42
7	0.155	0.4817	0.8821	—	—	—	—	—	—	—	—	—	—	—	—	11.53
9	0.1185	0.3625	0.6323	0.9744	—	—	—	—	—	—	—	—	—	—	—	8.90
11	0.0958	0.2912	0.4989	0.7341	1.3078	—	—	—	—	—	—	—	—	—	—	7.26
13	0.0804	0.2436	0.4136	0.5976	0.8088	1.0848	—	—	—	—	—	—	—	—	—	6.13
15	0.0693	0.2094	0.3538	0.5064	0.6733	0.8666	1.1214	—	—	—	—	—	—	—	—	5.31
17	0.0609	0.1836	0.3093	0.4402	0.5798	0.7337	0.913	1.1509	—	—	—	—	—	—	—	4.68
19	0.0544	0.1635	0.275	0.3897	0.5105	0.6400	0.7834	0.9513	1.1754	—	—	—	—	—	—	4.19
21	0.049	0.1475	0.2474	0.3500	0.4565	0.569	0.6902	0.8252	0.9839	1.1961	—	—	—	—	—	3.79
23	0.0466	0.1342	0.225	0.3176	0.4132	0.513	0.6187	0.7331	0.8609	1.0116	1.2137	—	—	—	—	3.46
25	0.0412	0.1233	0.2063	0.2909	0.3777	0.4675	0.5616	0.6619	0.7705	0.8921	1.0359	1.2294	—	—	—	3.18
27	0.0379	0.1138	0.1905	0.2683	0.3478	0.4297	0.5147	0.6042	0.6995	0.8032	0.9195	1.0573	1.243	—	—	2.95
29	0.0353	0.1058	0.1769	0.2491	0.3224	0.3977	0.4754	0.5563	0.6416	0.7328	0.8320	0.9437	1.0761	1.2551	—	2.74
31	0.0329	0.0988	0.1652	0.2324	0.3005	0.3703	0.4419	0.516	0.5934	0.6751	0.7625	0.8580	0.9655	1.0933	1.266	2.57

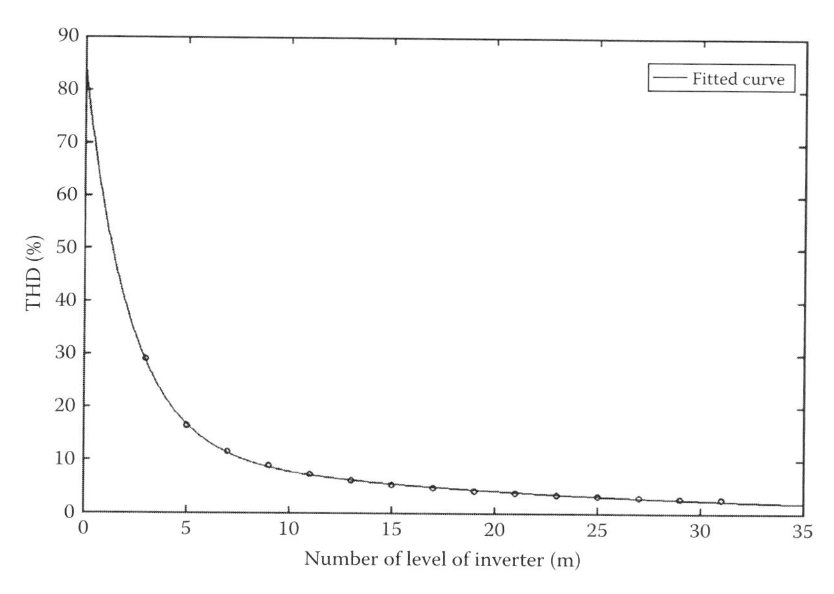

FIGURE 9.7 THD versus m.

9.3 Capacitor-Clamped Multilevel Inverters (Flying Capacitor Inverters)

Figure 9.8 illustrates the fundamental building block of a phase-leg capacitor-clamped inverter. The circuit has been called the flying capacitor inverter with dependent capacitors clamping the device voltage to one capacitor voltage level. The inverter in Figure 9.8a provides a three-level output across a and n, that is, $v_{an} = E, 0$, or $-E$. For the voltage level E, switches S_1 and S_2 need to be turned on; for $-E$, switches $S_{1'}$ and $S_{2'}$ need to be turned on; and for the 0 level, either pair (S_1, $S_{1'}$) or (S_2, $S_{2'}$) needs to be turned on. Clamping capacitor C_1 is charged when S_1 and $S_{1'}$ are turned on, and is discharged when S_2 and $S_{2'}$ are turned on. The charge of C_1 can be balanced by proper selection of the zero-level switch combination.

The voltage synthesis in a five-level capacitor-clamped inverter has more flexibility than a diode-clamped inverter. Using Figure 9.8b as an example, the voltage of the five-level phase leg a output with respect to the neutral point n, v_{an}, can be synthesized by the following switching combinations.

- For voltage level $v_{an} = 2E$, turn on all upper switches S_1–S_4.
- For voltage level $v_{an} = E$, there are three combinations:
  - $S_1, S_2, S_3, S_{1'} : v_{an} = 2E$ (upper C_4) $- E$ (C_1).
  - $S_2, S_3, S_4, S_{4'} : v_{an} = 3E$ (C_3) $- 2E$ (lower C_4).
  - $S_1, S_3, S_4, S_{3'} : v_{an} = 2E$ (upper C_4) $- 3E$ (C_3) $+ 2E$ (C_2).
- For voltage level $v_{an} = 0$, there are six combinations:
  - $S_1, S_2, S_{1'}, S_{4'} : v_{an} = 2E$ (upper C_4) $- 2E(C_2)$.
  - $S_3, S_4, S_{3'}, S_{4'} : v_{an} = 2E(C_2) - 2E$ (lower C_4).

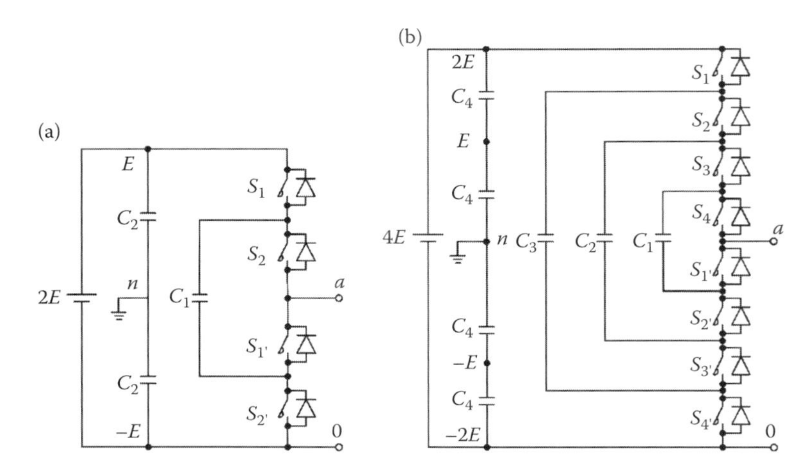

FIGURE 9.8 Capacitor-clamped multilevel inverter circuit topologies: (a) three levels and (b) five levels.

- ∘ $S_1, S_3, S_{1'}, S_{3'} : v_{an} = 2E$ (upper C_4) $- 3E(C_3) + 2E(C_2) - E(C_1)$.
- ∘ $S_1, S_4, S_{2'}, S_{3'} : v_{an} = 2E$ (upper C_4) $- 3E(C_3) + E(C_1)$.
- ∘ $S_2, S_4, S_{2'}, S_{4'} : v_{an} = 3E(C_3) - 2E(C_2) + E(C_1) - 2E$ (lower C_4).
- ∘ $S_2, S_3, S_{1'}, S_{4'} : v_{an} = 3E(C_3) - E(C_1) - 2E$ (lower C_4).
- For voltage level $V_{an} = -E$, there are three combinations:
  - ∘ $S_1, S_{1'}, S_{2'}, S_{3'} : v_{an} = 2E$ (upper C_4) $- 3E(C_3)$.
  - ∘ $S_4, S_{2'}, S_{3'}, S_{4'} : v_{an} = E(C_1) - 2E$ (lower C_4).
  - ∘ $S_3, S_{1'}, S_{3'}, S_{4'} : v_{an} = 2E(C_2) - E(C_1) - 2E$ (lower C_4).
- For voltage level $v_{an} = -2E$, turn on all lower switches, $S_{1'}$–$S_{4'}$.

Usually the positive top level and the negative top level have only one combination to implement their output values. Other levels have various combinations to implement their output values. In the preceding description, the capacitors with positive signs are in the discharging mode, while those with negative sign are in the charging mode. By proper selection of capacitor combinations, it is possible to balance the capacitor charge.

Example 9.2

A capacitor-clamped three-level inverter is shown in Figure 9.8a. It operates in the equal-angle state, that is, the operation time in each level is 90°. Determine the switches' status and the corresponding THD.

Solution

Refer to Figure 9.4; the switching angles in a cycle are

$$\alpha_1 = 45°$$

$$\alpha_2 = 135°$$

$$\alpha_3 = 225°$$
$$\alpha_4 = 315°.$$

The switches referring to Figure 9.8a operate in a cycle $(0 - 360°)$ as follows:

Turn on the upper switches S_2 and the lower switches $S_{2'}$ in $0° - \alpha_1$. (Or turn on the upper switches S_1 and the lower switches S_1, in $0° - \alpha_1$.)
Turn on all upper switches S_1 and S_2 in $\alpha_1 - \alpha_2$.
Turn on the upper switches S_2 and the lower switches S_2, in $\alpha_2 - \alpha_3$. (Or turn on the upper switches S_1 and the lower switches S_1, in $\alpha_2 - \alpha_3$.)
Turn on all lower switches $S_1, -S_2$, in $\alpha_3 - \alpha_4$.
Turn on the upper switches S_2 and the lower switches S_2, in $\alpha_4 - 360°$. (Or turn on the upper switches S_1 and the lower switches S_1, in $\alpha_4 - 360°$.)

Refer to Example 1.6; the fundamental harmonic has the amplitude $(4/\pi)\sin(x/2)$, where $x = 90°$ in this example. Therefore, $(4/\pi)\sin(x/2) = 0.9$. If we consider the higher-order harmonics until the seventh order, that is, $n = 3, 5, 7$, then the HFs are;;

$$HF_3 = \frac{\sin(3x/2)}{3\sin(x/2)} = \frac{1}{3}; \quad HF_5 = \frac{\sin(5x/2)}{5\sin(x/2)} = -\frac{1}{5}; \quad HF_7 = \frac{\sin(7x/2)}{7\sin(x/2)} = -\frac{1}{7}.$$

The values of the HFs should be absolute values.

$$THD = \frac{\sqrt{\sum_{n=2}^{\infty} V_n^2}}{V_1} = \sqrt{\left(\frac{1}{3}\right)^2 + \left(\frac{1}{5}\right)^2 + \left(\frac{1}{7}\right)^2} = 0.41415.$$

9.4 Multilevel Inverters Using H-Bridge Converters

The basic structure is based on the connection of H-bridges (HBs). Figure 9.9 shows the power circuit for one phase leg of a multilevel inverter with three HBs (HB_1, HB_2, and HB_3) in each phase. Each HB is supplied by a separate DC source. The resulting phase voltage is synthesized by the addition of the voltages generated by the different HBs. If the DC-link voltages of HBs are identical, the multilevel inverter is called the CMI. However, it is possible to have different values among the DC-link voltages of HBs, and the circuit can be called as the hybrid multilevel inverter.

Example 9.3

A three-HB multilevel inverter is shown in Figure 9.9. The output voltage is v_{an}. It implements as a binary hybrid multilevel inverter (BHMI). Explain the inverter working operation, draw the corresponding waveforms, and indicate the source voltages arrangement and how many levels can be implemented.

Solution

The DC-link voltages of HB_i (the ith HB), V_{DCi}, are $2^{i-1}E$. In a three-HB one phase leg,

$$V_{DC1} = E, \quad V_{DC2} = 2E, \quad V_{DC3} = 4E.$$

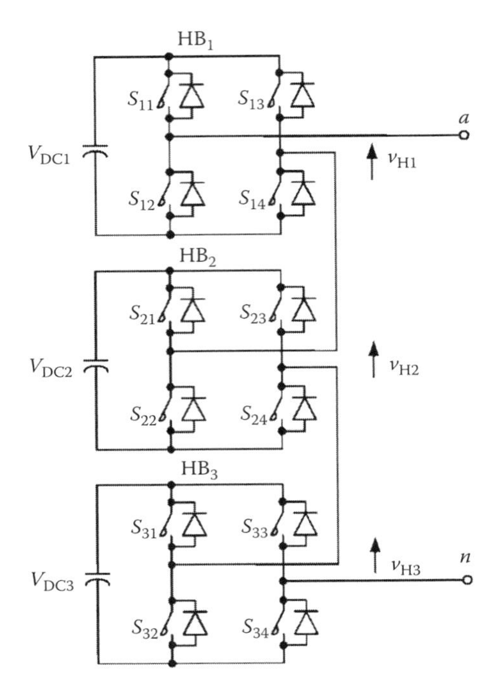

FIGURE 9.9 Multilevel inverter based on the connection of HBs.

The operation is listed below:

$+0$: $v_{H1} = 0$, $v_{H2} = 0$, $v_{H3} = 0$.

$+1E$: $v_{H1} = E$, $v_{H2} = 0$, $v_{H3} = 0$.

$+2E$: $v_{H1} = 0$, $v_{H2} = 2E$, $v_{H3} = 0$.

$+3E$: $v_{H1} = E$, $v_{H2} = 2E$, $v_{H3} = 0$.

$+4E$: $v_{H1} = 0$, $v_{H2} = 0$, $v_{H3} = 4E$.

$+5E$: $v_{H1} = E$, $v_{H2} = 0$, $v_{H3} = 4E$.

$+6E$: $v_{H1} = 0$, $v_{H2} = 2E$, $v_{H3} = 4E$.

$+7E$: $v_{H1} = E$, $v_{H2} = 2E$, $v_{H3} = 4E$.

$-E$: $v_{H1} = -E$, $v_{H2} = 0$, $v_{H3} = 0$.

$-2E$: $v_{H1} = 0$, $v_{H2} = -2E$, $v_{H3} = 0$.

$-3E$: $v_{H1} = -E$, $v_{H2} = -2E$, $v_{H3} = 0$.

$-4E$: $v_{H1} = 0$, $v_{H2} = 0$, $v_{H3} = -4E$.

$-5E$: $v_{H1} = -E$, $v_{H2} = 0$, $v_{H3} = -4E$.

$-6E$: $v_{H1} = 0$, $v_{H2} = -2E$, $v_{H3} = -4E$.

$-7E$: $v_{H1} = -E$, $v_{H2} = -2E$, $v_{H3} = -4E$.

As shown in the above figure, the output waveform, v_{an}, has 15 levels. One of the advantages is that the HB with higher DC-link voltage has a lower number of commutations, thereby reducing the associated switching losses. The higher switching frequency components, for example, IGBT, are used to construct the HB with lower DC-link voltages.

9.4.1 Cascaded Equal-Voltage Multilevel Inverters

In a cascaded equal-voltage multilevel inverter (CEMI), the DC-link voltages of HBs are identical, as shown in Figure 9.9.

$$V_{DC1} = V_{DC2} = V_{DC3} = E, \tag{9.2}$$

where E is the unit voltage. Each HB generates three voltages at the output: $+E$, 0, and $-E$. This is made possible by connecting the capacitors sequentially to the AC side via the three power switches. The resulting output AC voltage swings from $-3E$ to $3E$ with seven levels, as shown in Figure 9.10.

9.4.2 Binary Hybrid Multilevel Inverter

In a BHMI, the DC-link voltages of HB_i (the ith HB), V_{DCi}, are $2^{i-1}E$. In a three-HB one phase leg,

$$V_{DC1} = E, \qquad V_{DC2} = 2E, \qquad V_{DC3} = 4E. \tag{9.3}$$

As shown in Figure 9.11, the output waveform, v_{an}, has 15 levels. One of the advantages is that the HB with higher DC-link voltage has a lower number of commutations, thereby reducing the associated switching losses. The BHMI illustrates a seven-level (in half-cycle) inverter using this hybrid topology. The HB with higher DC-link voltage consists of a lower switching frequency component, for example, IGBT. The higher switching frequency components, for example, IGBT, are used to construct the HB with lower DC-link voltages.

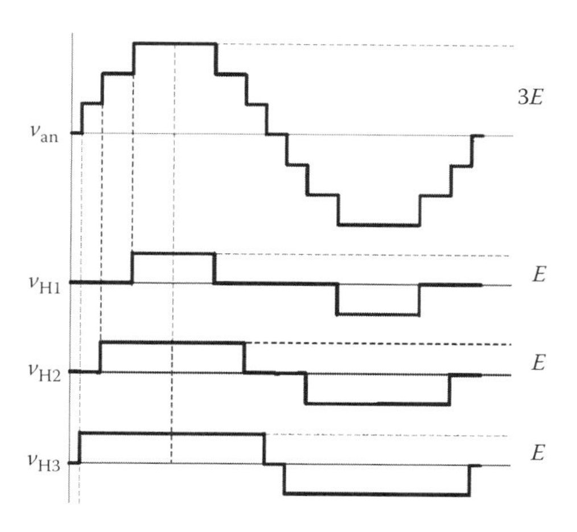

FIGURE 9.10 Waveforms of a CMI.

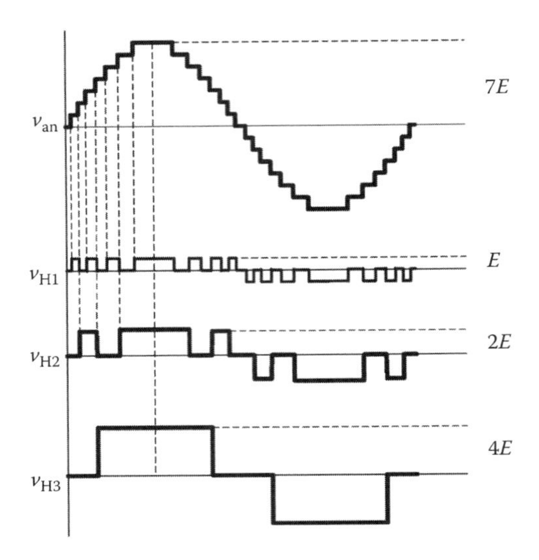

FIGURE 9.11 Waveforms of a BHMI.

9.4.3 Quasi-Linear Multilevel Inverter

In a quasi-linear multilevel inverter (QLMI), the DC-link voltages of HB_i, V_{DCi}, can be expressed as

$$V_{DCi} = \begin{cases} E, & i = 1, \\ 2 \times 3^{i-2}E, & i \geq 2. \end{cases} \tag{9.4}$$

In a three-HB one phase leg,

$$V_{DC1} = E, \quad V_{DC2} = 2E, \quad V_{DC3} = 6E. \tag{9.5}$$

As shown in Figure 9.12, the output waveform, v_{an}, has 19 levels.

9.4.4 Trinary Hybrid Multilevel Inverter

In a trinary hybrid multilevel inverter (THMI), the DC-link voltages of HB_i, V_{DCi}, are $3^{i-1}E$. In a three-HB one phase leg,

$$V_{DC1} = E, \quad V_{DC2} = 3E, \quad V_{DC3} = 9E. \tag{9.6}$$

As shown in Figure 9.13, the output waveform, v_{an}, has 27 levels. To the best of the author's knowledge, this circuit has the greatest level number for a given number of HBs among existing multilevel inverters.

9.5 Investigation of THMI

THMI has many advantages. Therefore, we would like to analyze carefully its characteristics in this section [5].

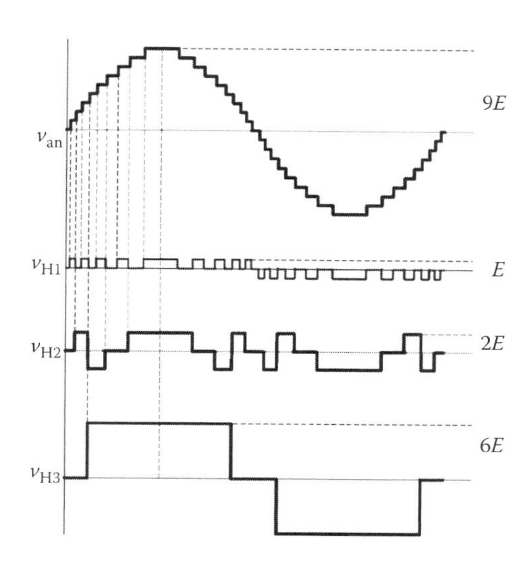

FIGURE 9.12 Waveforms of a QLMI.

9.5.1 Topology and Operation

A single-phase THMI with h HBs connected in series is shown in Figure 9.14. The key feature of the THMI is that the ratio of DC-link voltage is $1 : 3 : \cdots : 3^{h-1}$, where h is the number of HBs. The maximum number of synthesized voltage levels is 3^h.

As shown in Figure 9.14, v_{Hi} represents the output voltage of the ith HB. V_{DCi} represents the DC-link voltage of the ith HB. A switching function, F_i, is used to relate V_{Hi} and V_{DCi} as shown in the following equation:

$$v_{Hi} = F_i \cdot V_{DCi}. \tag{9.7}$$

The value of F_i can be either 1 or −1 or 0. For the value 1, switches S_{i1} and S_{i4} need to be turned on. For the value −1, switches S_{i2} and S_{i3} need to be turned on. For the value 0, switches S_{i1} and S_{i3} need to be turned on or S_{i2} and S_{i4} need to be turned on. Table 9.3

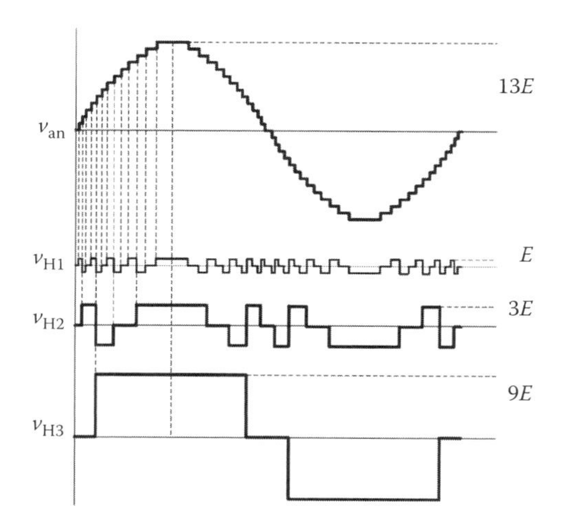

FIGURE 9.13 Waveforms of a 27-level THMI.

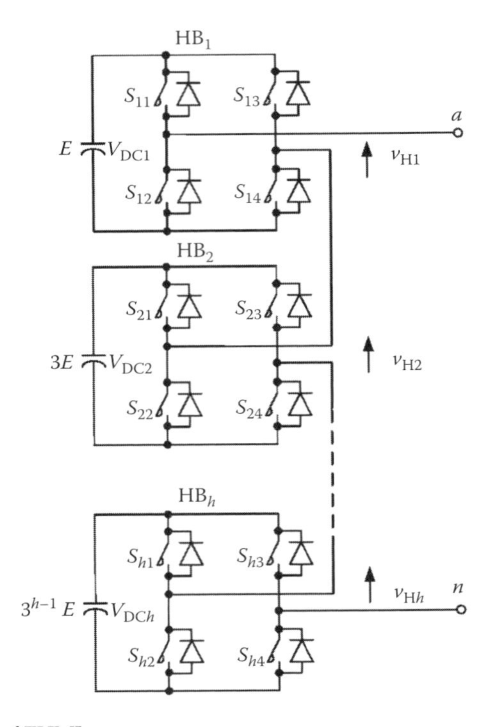

FIGURE 9.14 Configuration of THMI.

represents the relationship between the switching function, the output voltage of an HB, and states of switches.

The output voltage of the THMI, v_{an}, is the summation of the output voltages of HBs.

$$v_{an} = \sum_{i=1}^{h} v_{Hi}. \tag{9.8}$$

From Equations 9.7 and 9.8, we obtain

$$v_{an} = \sum_{i=1}^{h} \mathrm{F}_i \cdot V_{DCi}. \tag{9.9}$$

TABLE 9.3

Relationship between the Switching Function, Output Voltage of an HB, and States of Switches

F_i	v_{Hi}	S_{i1}	S_{i2}	S_{i3}	S_{i4}
1	V_{DCi}	Conduct	Block	Block	Conduct
−1	$-V_{DCi}$	Block	Conduct	Conduct	Block
0	0	Conduct	Conduct	Block	Block
0	0	Block	Block	Conduct	Conduct

In a single-phase h-HB THMI, the ratio of DC-link voltage is $1:3:\cdots:3^{h-1}$. Suppose that E is the unit voltage, then the DC-link voltage can be expressed as

$$V_{DCi} = 3^{i-1}E. \tag{9.10}$$

From Equations 9.8 and 9.9, we obtain

$$v_{an} = \sum_{i=1}^{h} F_i \cdot 3^{i-1}E. \tag{9.11}$$

Suppose that l is the ordinal of the expected voltage level that the inverter outputs. If l is not negative, the inverter outputs the positive lth voltage level. If l is negative, the inverter outputs the negative $(-l)$th voltage level. In a single-phase THMI with h HBs, given the value of l, the value of F_i can be determined by

$$\begin{aligned}
F_h &= \frac{\mathrm{ABS}(l)}{l}B_b\left(\mathrm{ABS}(l) - \frac{3^{h-1}-1}{2}\right),\\
F_{h-1} &= \frac{\mathrm{ABS}(l)}{l}B_b\left(\mathrm{ABS}(l) - \mathrm{ABS}(F_h)\cdot 3^{h-1} - \frac{3^{h-2}-1}{2}\right),\\
&\vdots\\
F_i &= \frac{\mathrm{ABS}(l)}{l}B_b\left(\mathrm{ABS}(l) - \sum_{k=i+1}^{h}(\mathrm{ABS}(F_k)\cdot 3^{k-1}) - \frac{3^{i-1}-1}{2}\right),\\
&\vdots\\
F_2 &= \frac{\mathrm{ABS}(l)}{l}B_b\left(\mathrm{ABS}(l) - \sum_{k=3}^{h}(\mathrm{ABS}(F_k)\cdot 3^{k-1}) - 1\right),\\
F_1 &= \frac{\mathrm{ABS}(l)}{l}B_b\left(\mathrm{ABS}(l) - \sum_{k=2}^{h}(\mathrm{ABS}(F_k)\cdot 3^{k-1})\right),
\end{aligned} \tag{9.12}$$

where ABS is the function of the absolute value, and the bipolar binary function, B_b, is defined as

$$B_b(\tau) = \begin{cases} 1, & \tau > 0,\\ 0, & \tau = 0,\\ -1, & \tau < 0. \end{cases} \tag{9.13}$$

From Equation 9.13, we obtain the relationship between the output voltage of the inverter, v_{an}, and the values of the switching functions in the THMI with different numbers of HBs. In the case of a two-HB THMI, Table 9.4 shows the relationship between the output voltage of the inverter and the values of switching functions. The waveforms of a single-phase two-HB THMI are shown in Figure 9.15.

TABLE 9.4

Relationship between the Output Voltage of the Inverter and the Values of Switching Functions in a Single-Phase Two-HB THMI

v_{an}	$-4E$	$-3E$	$-2E$	$-E$	0
F_1	−1	0	1	−1	0
F_2	−1	−1	−1	0	0
v_{an}	$4E$	$3E$	$2E$	E	
F_1	1	0	−1	1	
F_2	1	1	1	0	

The output voltage of a single-phase three-HB THMI has 27 levels. v_{H1}, v_{H2}, and v_{H3} can be negative when v_{an} is positive. Table 9.5 shows the relationship between the output voltage of the inverter and the values of switching functions in a single-phase three-HB THMI. From Equation 9.12, we obtain

$$v_{an} = -v'_{an} \Leftrightarrow F_i = F'_i, \quad i = 1, \ldots, h. \tag{9.14}$$

The conclusions about the cases of negative value of v_{an} can be deduced from Table 9.5.

9.5.2 Proof that the THMI has the Greatest Number of Output Voltage Levels

Among the existing multilevel levels, THMI has the greatest levels of output voltage using the same number of components. In this section, first, the theoretical proof for this conclusion is specified; then the comparison of various kinds of multilevel inverters is given.

9.5.2.1 Theoretical Proof

This section proves that the THMI has the greatest levels of output voltage using the same number of HBs among the multilevel inverters using HBs connected. A phase voltage waveform is obtained by summing the output voltages of h HBs as shown in Equation 9.8.

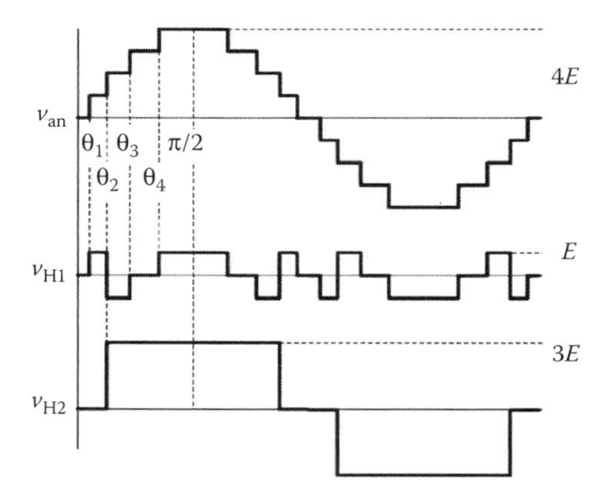

FIGURE 9.15 Waveforms of a single-phase two-HB THMI.

TABLE 9.5

Relationship between the Output Voltage of the Inverter and the Values of Switching Functions in a Single-Phase Three-HB THMI

v_{an}	13E	12E	11E	10E	9E	8E	7E
F_1	1	0	−1	1	0	−1	1
F_2	1	1	1	0	0	0	−1
F_3	1	1	1	1	1	1	1
v_{an}	6E	5E	4E	3E	2E	E	0
F_1	0	−1	1	0	−1	1	0
F_2	−1	−1	1	1	1	0	0
F_3	1	1	0	0	0	0	0

If the DC-link sources of all HB cells are equal, the multilevel inverter is called the CMI and the maximum number of levels of phase voltage is given by

$$m = 1 + 2h. \tag{9.15}$$

On the other hand, if at least one of the DC-link sources is different from the others, the multilevel inverter is called the hybrid multilevel inverter. In Section 9.4, the BHMI, the QLMI, and the THMI are introduced. Thus, considering that the lowest DC-link source E is chosen as the base value for the p.u. notation, the normalized values of all DC-link voltages must be natural numbers to obtain a uniform step multilevel inverter, that is,

$$V_{DCi^*} \in E, \quad i = 1, 2, \ldots, h. \tag{9.16}$$

Moreover, to obtain a uniform step multilevel inverter, the DC-link voltage of the HB cells must also satisfy the following relation:

$$V_{DCi^*} \leq 1 + 2\sum_{k=1}^{i-1} V_{DCk^*}, \quad i = 2, 3, \ldots, h, \tag{9.17}$$

where it is also considered that the DC-link voltages are arranged in an increasing order, that is,

$$V_{DC1^*} \leq V_{DC2^*} \leq V_{DC3^*} \leq \cdots \leq V_{DCh^*}. \tag{9.18}$$

Therefore, the maximum number of levels of the output phase voltage waveform can be given by

$$m = 1 + 2\sigma_{max}, \tag{9.19}$$

where σ_{max} is the maximum number of positive/negative voltage levels and can be expressed as

$$\sigma_{max} = \sum_{i=1}^{h} V_{DCi^*}. \tag{9.20}$$

TABLE 9.6
First Comparison between Multilevel Inverters

Converter Type	DCMI	CCMI	GMI	CMI	BHMI	THMI
Main switching devices	$2m-2$	$2m-2$	$2m-2$	$2m-2$	$2m-2$	$2m-2$
Diodes	$m(m-1)$	$m-1$	$2m-2$	$2m-2$	$2m-2$	$2m-2$
Capacitors	$m-1$	$0.5m(m-1)$	$m-1$	$(m-1)/2$	$(m-1)/2$	$(m-1)/2$
Total components	$(m-1)(m+1)$	$(m-1)(0.5m+3)$	$2^{m+1}+m-5$	$4.5(m-1)$	$4.5(m-1)$	$4.5(m-1)$

From Equations 9.15, 9.19, and 9.20, it is possible to verify that hybrid multilevel inverters can generate a large number of levels with the same number of cells. Moreover, in the THMI, the DC-link voltages satisfy

$$V_{DCi^*} = 1 + 2\sum_{k=1}^{i-1} V_{DCk^*}, \quad i = 2, 3, \ldots, h. \tag{9.21}$$

Therefore, the THMI has the greatest levels of output voltages using the same number of HBs among the multilevel inverters using HBs connected.

9.5.2.2 Comparison of Various Kinds of Multilevel Inverters

Two kinds of comparisons are presented in this section. In the first comparison, the components are considered to have the same voltage rating, E. This comparison is for high-power and high-voltage applications, in which the devices connected in series are used to satisfy the requirement of high-voltage ratings. Table 9.6 shows the comparison between multilevel inverters: DCMI, CCMI, CMI, GMI, BHMI, and THMI; m is the number of steps of phase voltage. From Table 9.6, we can find that CMI, BHMI, and THMI use fewer components. The CMI, BHMI, and THMI use the same number of components. However, in practical systems, the redundancy requirement must be satisfied. THMI uses fewer components than BHMI and CMI in practical systems since THMI uses less redundant components. Moreover, the THMI uses fewer DC sources than the CMI and BHMI.

The second comparison is for medium- and low-power applications, in which the voltage rating of the main switching components, diodes, and capacitors can be researched easily. Therefore, the numbers of the main switching components, diodes, and capacitors are the minimal required values. Table 9.7 shows the comparison results of DCMI, CCMI, CMI, GMI, BHMI, and THMI. From Table 9.7, we can find that the THMI uses the fewest components among these multilevel inverters.

TABLE 9.7
Second Comparison between Multilevel Inverters

Converter Type	DCMI	CCMI	CMI	GMI	BHMI	THMI
Main switching devices	$2m-2$	$2m-2$	$2m-2$	2^m-2	$4\times\log_2[(m+1)/2]$	$4\times\log_3 m$
Diodes	$4m-6$	$2m-2$	$2m-2$	2^m-2	$4\times\log_2[(m+1)/2]$	$4\times\log_3 m$
Capacitors	$m-1$	$2m-3$	$0.5m-0.5$	$m-1$	$\log_2[(m+1)/2]$	$\log_3 m$
Total components	$7m-9$	$6m-7$	$4.5m-4.5$	$2^{m+1}+m-5$	$9\times\log_2[(m+1)/2]$	$9\times\log_3 m$

9.5.2.3 Modulation Strategies for THMI

Five modulation strategies for the THMI are investigated: the step modulation strategy, the virtual stage modulation strategy, the hybrid modulation strategy, the subharmonics PWM strategy, and the simple modulation strategy. Since multilevel inverters are used in three-phase systems generally, only modulation strategies for the three-phase systems will be investigated here. In the three-phase systems, the triple-order harmonic components of voltages need not be eliminated by the modulation strategies since they can be eliminated by proper connection of three-phase voltage sources and loads. In other words, only 5th, 7th, 11th, 13th, 17th, 19th ... harmonic components should be eliminated by the modulation strategies. In addition, the amplitude of the fundamental component should be controlled. The list can be expressed by

$$\eta_i = \begin{cases} 3i - 2 & \forall i = \text{odd} \\ 3i - 1 & \forall i = \text{even} \end{cases}, \quad i > 0. \tag{9.22}$$

The step modulation strategy, the virtual stage modulation strategy, and the simple modulation strategy belong to low-frequency modulation strategies. The high-frequency modulation strategies used in the hybrid multilevel inverters include the hybrid modulation strategy and the subharmonic PWM strategy.

9.5.2.3.1 *Step Modulation Strategy*

Figure 9.16 shows a general quarter-wave symmetric stepped voltage waveform synthesized by a THMI, where E indicates unit voltage of the DC source. Consider that ς is the number of switching angles in a quarter wave of v_{an} and σ is the number of positive/negative levels of v_{an}. In the step modulation strategy,

$$\varsigma = \sigma. \tag{9.23}$$

By applying Fourier series analysis, the amplitude of any odd jth harmonic of v_{an} can be expressed as

$$|v_{\text{an}}|_j = \frac{4}{j\pi} \sum_{i=1}^{\varsigma} [E \cos(j\theta_i)], \tag{9.24}$$

where j is an odd harmonic order and θ_i is the ith switching angle. The amplitudes of all even harmonics are zero. According to Figure 9.16, θ_1 to θ_ς must satisfy

$$0 < \theta_1 < \theta_2 < \ldots < \theta_\varsigma < \frac{\pi}{2}. \tag{9.25}$$

The switching angles controlled by the step modulation technique are derived from Equation 9.26. Up to ($\varsigma-1$), harmonic contents can be removed from the voltage waveform and the amplitude of the fundamental component can be controlled.

$$\begin{cases} \sum_{i=1}^{\varsigma} \cos(\eta_1 \theta_i) = \sigma \cdot MR \\ \sum_{i=1}^{\varsigma} \cos(\eta_2 \theta_i) = 0 \\ \vdots \\ \sum_{i=1}^{\varsigma} \cos(\eta_\varsigma \theta_i) = 0 \end{cases}, \tag{9.26}$$

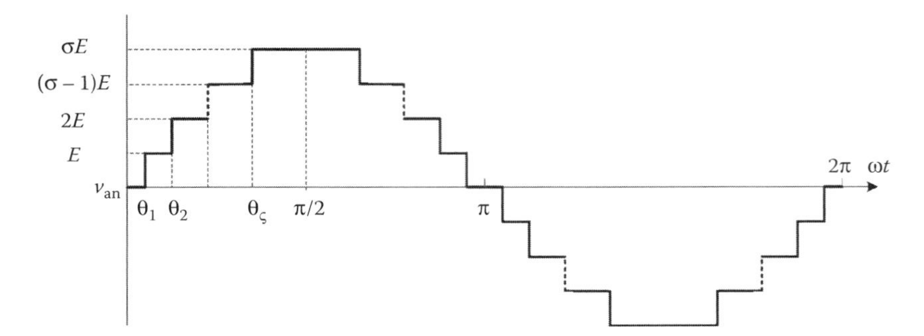

FIGURE 9.16 Step modulation strategy of THMI.

where *MR* is the relative modulation index and is expressed as

$$MR = \frac{\pi \, |v_{an}|_1}{4\sigma E}, \tag{9.27}$$

where $|v_{an}|_1$ is the amplitude of the fundamental component of the output voltage of the inverter.

The set of Equations 9.26 from which the switching angles can be derived are nonlinear and transcendental. For example, in a two-HB THMI, with the step modulation technique, the set of equations are expressed as Equations 9.28 when the *MR* is 0.83. The correct solution must satisfy the inequational condition as shown in Equation 9.25.

$$\begin{aligned}
&\cos(\theta_1) + \cos(\theta_2) + \cos(\theta_3) + \cos(\theta_4) = 0.83 \times 4, \\
&\cos(5\theta_1) + \cos(5\theta_2) + \cos(5\theta_3) + \cos(5\theta_4) = 0, \\
&\cos(7\theta_1) + \cos(7\theta_2) + \cos(7\theta_3) + \cos(7\theta_4) = 0, \\
&\cos(11\theta_1) + \cos(11\theta_2) + \cos(11\theta_3) + \cos(11\theta_4) = 0.
\end{aligned} \tag{9.28}$$

The constrained optimization approach can be used to solve the nonlinear and transcendental set of equations. Each equation is regarded as an equational constraint. However, the computational problems of constrained optimization do not converge easily. Since in the actual electric system, there are always mismatches and parameter tolerances, lower-order harmonics will be small but not exactly zero. This gives rise to the idea of transforming the constraint optimization model to a nonconstraint one. The nonconstraint optimization is expected to have a better convergence property.

The target function of the new scheme of optimization without equational constraints can be written as

$$\mathrm{FT} = p_1\left[\sum_{i=1}^{\varsigma}\cos(\eta_1\theta_i) - \sigma\cdot M\right]^2 + p_2\left[\sum_{i=1}^{\varsigma}\cos(\eta_2\theta_i)\right]^2 + \cdots + p_\varsigma\left[\sum_{i=1}^{\varsigma}\cos(\eta_\varsigma\theta_i)\right]^2. \tag{9.29}$$

Here $p_1 - p_\varsigma$ are the penalty factors. The penalty factors were selected as

$$p_i = \frac{4}{2i-1}. \tag{9.30}$$

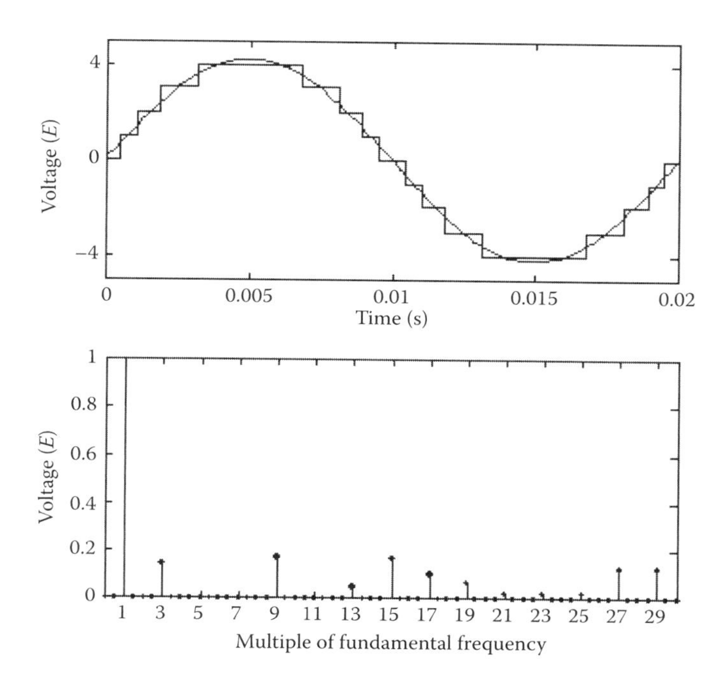

FIGURE 9.17 Synthesized phase leg voltage waveform and frequency spectrum of a two-HB THMI with step modulation technique.

Thus, the penalty factors place more emphasis on elimination of lower-order harmonics. The function f_{mincon} in the MATLAB® optimization toolbox was used to solve this minimization problem.

The two-HB THMI can synthesize nine-level output voltage. Figures 9.17 and 9.18 show the typical synthesized waveform of the phase leg voltage, line-to-line voltage waveform, and their frequency spectrums, as *MR* is equal to 0.83. The switching angles are 0.1478, 0.3232, 0.5738, and 0.9970. According to Equation 9.26, the fifth, seventh, and eleventh harmonics of the phase leg voltage can be eliminated in the two-HB THMI as shown in Figure 9.17. The THD of the phase leg voltage is 9.66%. The triple-order harmonic components do not exist in the line-to-line voltage as shown in Figure 9.18. The THD of the line-to-line voltage is 5.91%.

According to Equation 9.26, all switching angles must satisfy the constraint (Equation 9.25). If switching angles do not satisfy the constraint, this scheme no longer exists. The theoretical maximum amplitude of the fundamental component is $4\varsigma E/\pi$, which occurs as $\theta_1 - \theta_h$ is equal to zero. Because of the internal restriction of switching angles, the relative modulation index has upper and lower limitations. The limitation of the relative modulation index can be explained using Figures 9.19 and 9.20.

As shown in Figure 9.19, when the relative modulation index is less than a certain value, denoted by *MR* (min), θ_ς approaches $\pi/2$ and the limitation of the minimum modulation index occurs. Similarly, when the relative modulation index is greater than *MR*(max), θ_1 approaches 0 and the limitation of the maximum modulation index occurs as shown in Figure 9.20.

For a THMI with h HBs, the maximum number of levels of the phase leg voltage is m, which is equal to 3^h. The maximum number of the positive/negative phase leg voltage levels is $\sigma_{\max}$, which is equal to $(m-1)/2$. As mentioned above, the relative modulation

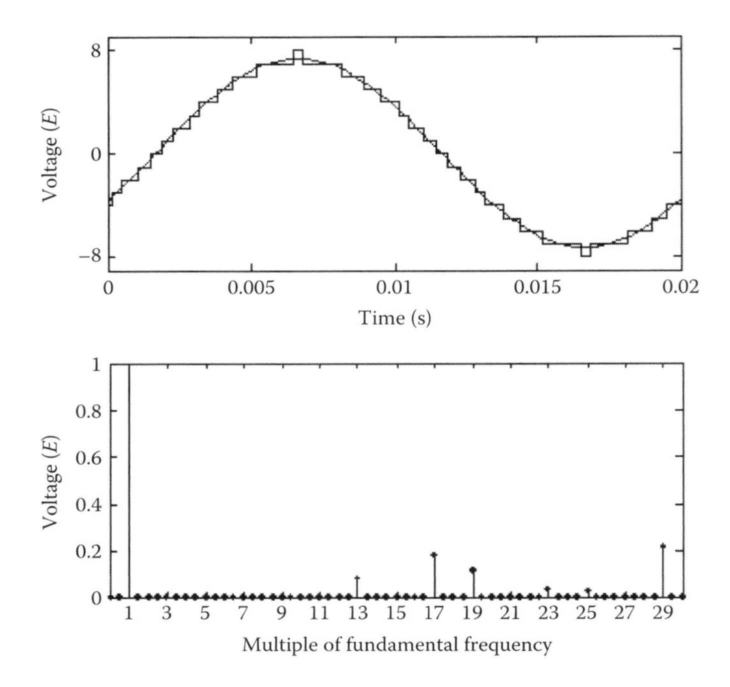

FIGURE 9.18 Synthesized line-to-line voltage waveform and frequency spectrum of a two-HB THMI with step modulation technique.

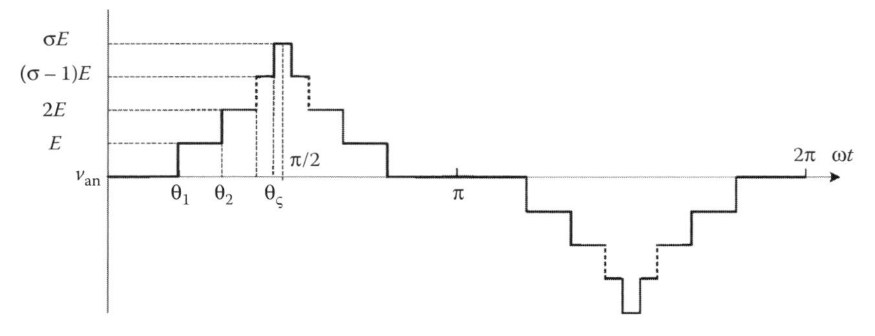

FIGURE 9.19 Limitation to the minimum *MR* in the step modulation.

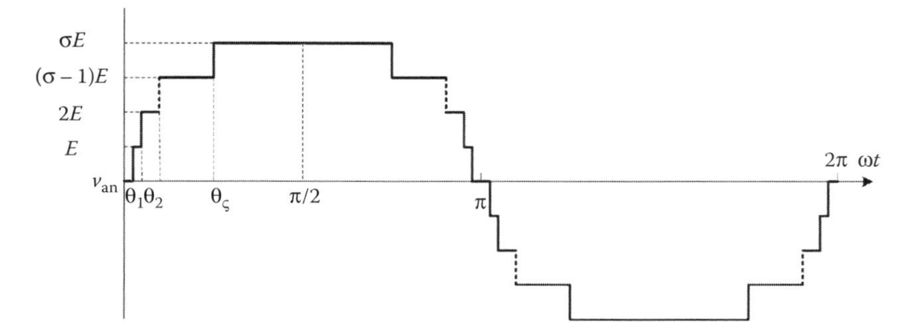

FIGURE 9.20 Limitation to the maximum *MR* in the step modulation.

TABLE 9.8

Range of Modulation Index under Different Output Voltage Levels with the Step Modulation in a Two-HB THMI

σ	*MR* (min)	*MR* (max)	*M* (min)	*M* (max)	Range of *M*
1	0	1	0	0.25	0–0.15
2	0.3	0.9	0.15	0.45	0.15–0.34
3	0.46	0.83	0.34	0.63	0.34–0.55
4	0.55	0.86	0.55	0.86	0.55–0.86
4*	0.3	0.94	0.3	0.94	0.86–0.94

4* = 2.5.

index *MR* has limitations. To extend to the smaller ranges of the modulation index, the inverter will output fewer voltage levels. Consequently, the number of positive/negative voltage levels that the inverter outputs, σ, is smaller than the maximum number of the positive/negative levels, σ_{max}. In the step modulation strategy, the number of switching angles in the quarter wave of v_{an}, ς, is equal to σ. The definition of the relative modulation index, *MR*, is based on σ as shown in Equation 9.27. This definition is easily included in Equation 9.26 to express the nonlinear transcendental equation sets that are used to calculate the switching angles. In practice, the modulation index, *M*, is used. *M* is based on the σ_{max} and can be expressed as

$$M = \frac{\pi\,|v_{an}|_1}{4\sigma_{max}E}. \tag{9.31}$$

The relationship between *MR* and *M* can be expressed as

$$\frac{M}{MR} = \frac{\sigma}{\sigma_{max}}. \tag{9.32}$$

In the two-HB THMI, according to Equation 9.26, the maximum *MR* is calculated as 0.86 and the minimum *MR* is 0.55 as the levels of output voltage are nine. The range of *M* is also from 0.55 to 0.86 with the nine-level output voltage. To extend to the lower modulation index, fewer output voltage levels are synthesized. The range of *MR* is 0.46–0.83 when the output voltage levels are seven. According to Equation 9.32, the range of *M* is 0.34–0.62 when the output voltage levels are seven. Thus, the modulation range is extended to 0.34 by decreasing the levels of output voltage.

Table 9.8 shows the relative modulation index and the modulation index with different output voltage levels in the two-HB THMI. First, the minimum and maximum *MR* are calculated by the optimization method. Second, the minimum and maximum *M* is calculated by using Equation 9.32. It is preferable to use more output voltage levels. The last column of Table 9.8 shows the arrangement of *M* with different output voltage levels. In addition, the maximum limitation of *M* can reach 0.94 irrespective of the elimination of the 11th harmonic, as shown in the last row of Table 9.8.

The scheme of switching angles of the two-HB THMI is shown in Figure 9.21. When the modulation index reaches the lower limitation, such as 0.34, the third switching angle is close to $\pi/2$, which verifies Figure 9.19. When the modulation index reaches the maximum value 0.86 or 0.94, the first angle is close to zero, which verifies Figure 9.20.

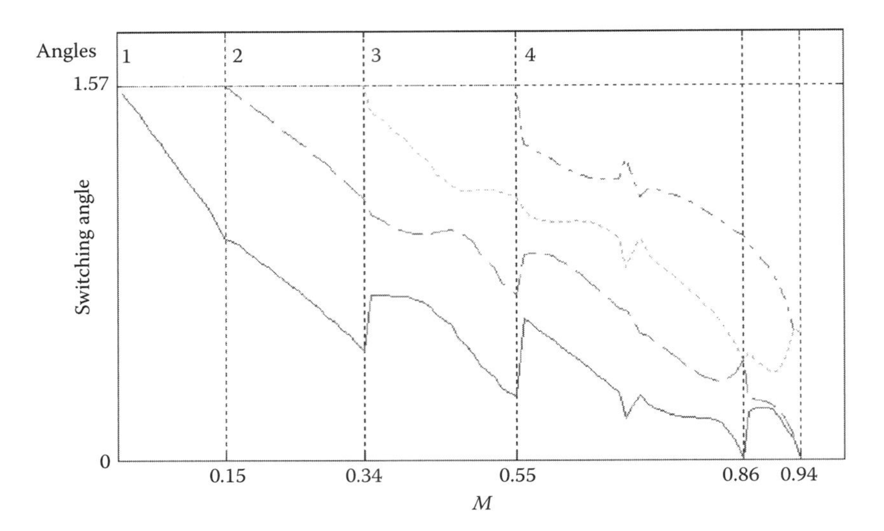

FIGURE 9.21 Scheme of switching angles with the step modulation as a function of modulation index in a two-HB THMI.

9.5.2.3.2 *Virtual Stage Modulation Strategy*

In the step modulation strategy, the output voltage levels of the multilevel inverter limit the amount of eliminated lower-order harmonics. Only three lower-order harmonics can be eliminated by the step modulation in a two-HB THMI. It is not very satisfactory in the applications that required a high-quality sinusoid voltage output. The virtual stage modulation strategy is a new modulation strategy that increases the amount of eliminated lower-order harmonics without increasing the number of output voltage levels. The switching angles can be derived as

$$\begin{cases} \sum_{i=1}^{\alpha} \cos(\eta_1 \theta_{pi}) - \sum_{i=1}^{\beta} \cos(\eta_1 \theta_{ni}) = \sigma \cdot MR \\ \sum_{i=1}^{\alpha} \cos(\eta_2 \theta_{pi}) - \sum_{i=1}^{\beta} \cos(\eta_2 \theta_{ni}) = 0 \\ \vdots \\ \sum_{i=1}^{\alpha} \cos(\eta_\varsigma \theta_{pi}) - \sum_{i=1}^{\beta} \cos(\eta_\varsigma \theta_{ni}) = 0 \end{cases}, \tag{9.33}$$

where σ is the number of positive/negative levels of v_{an} and can be expressed as

$$\sigma = \alpha - \beta, \tag{9.34}$$

where ς is the number of switching angles in the quarter waveform of v_{an} and can be expressed as

$$\varsigma = \alpha + \beta. \tag{9.35}$$

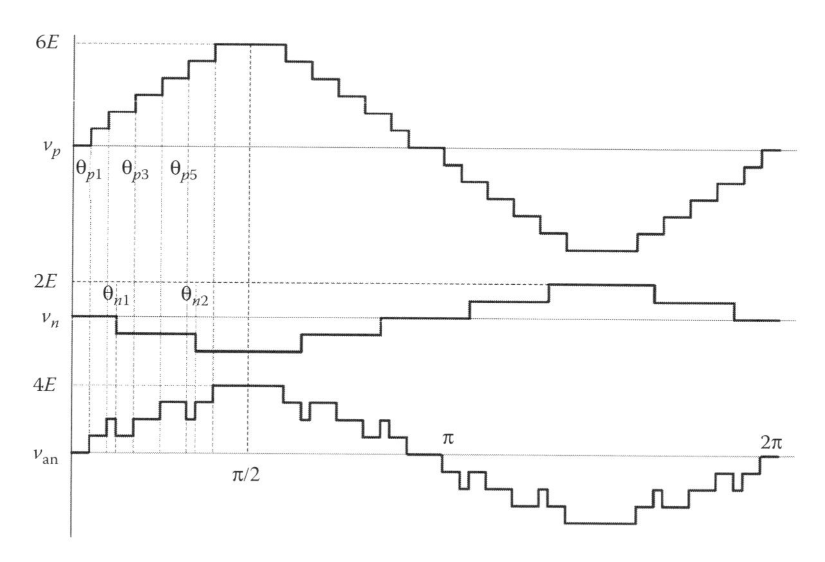

FIGURE 9.22 Waveform using the virtual stage modulation two-HB, nine-level, $\alpha = 6$, $\beta = 2$.

The expression for *MR* is given in Equation 9.27. Equation 9.33 is subject to

$$\begin{aligned} &0 < \theta_{p1} < \theta_{p2} < \cdots < \theta_{p\alpha} < \pi/2, \\ &0 < \theta_{n1} < \theta_{n2} < \cdots < \theta_{n\beta} < \pi/2, \\ &\theta_{nj} < \theta_{p(j+\sigma)}, \quad j = 1, 2, \ldots, \beta. \end{aligned} \tag{9.36}$$

In the two-HB THMI, when the output voltage changes between E and $2E$ or $-E$ and $-2E$, the switching components of the higher-voltage HB will switch-on/off as shown in Figure 9.15. To make high-voltage switching components switch at lower frequency in the THMI, the limitation (Equation 9.37) is added into Equation 9.33 to assure that higher-voltage switching components switch at the fundamental frequency.

$$\theta_{p2} < \theta_{n1}. \tag{9.37}$$

Figure 9.22 illustrates the waveform using the virtual stage modulation for the two-HB THMI whose output voltage levels are nine. The number of virtual stages, β, is two.

Figures 9.23 and 9.24 show the typical synthesized waveform of phase leg voltage, line-to-line voltage waveform, and their frequency spectrum in the virtual stage modulation strategy. The *MR* is 0.83 and the number of virtual stages is two. The values of θ_{p1} to θ_{p6} are 0.1321, 0.3320, 0.5307, 0.6226, 0.9133, and 1.0419. θ_{n1} is 0.5750 and θ_{n2} is 0.9652. Because of two additional virtual stages, four more degrees of freedom of switching angles are created such that the 13th, 17th, 19th, and 23rd harmonics can be eliminated from the phase leg voltage as shown in Figure 9.23. The THD of the phase leg voltage is 10.67%. The triple-order harmonic components of the line-to-line voltage do not exist and the harmonics are pushed to 1250 Hz as shown in Figure 9.24. The THD of the line-to-line voltage is 7.3%.

In the virtual stage modulation strategy, the relative modulation index also has upper and lower limitations. Compared with the step modulation strategy, the optimal computation of the virtual stage modulation strategy endures more unequal restrictions as shown in

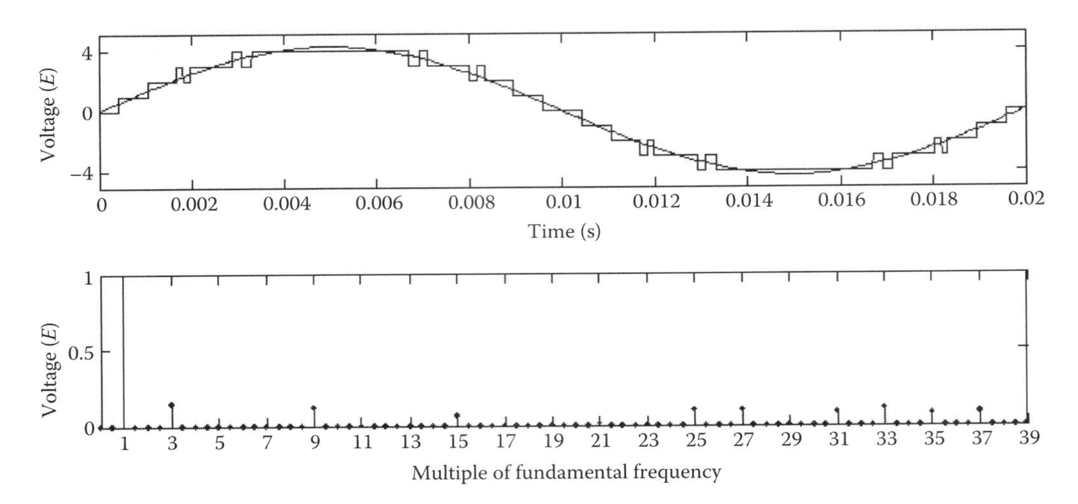

FIGURE 9.23 Synthesized phase leg voltage waveform and frequency spectrum of a two-HB THMI with the virtual stage modulation.

Equations 9.36 and 9.37. When the switching angles do not satisfy these restrictions, the themes of switching angles no longer exist.

The concept of the relative modulation index can be used in the step modulation strategy by a similar method. Table 9.9 shows two cases. One is the nine-level output voltage with two virtual stages and the other is the seven-level output voltage with one virtual stage. With the nine-level output voltage and two virtual stages, the 5th, 7th, 11th, 13th, 17th, 19th, and 23rd harmonics can be eliminated. With the seven-level output voltage and one virtual stage, the 5th, 7th, 11th, and 13th can be eliminated. When the output voltage levels are five or three, the virtual stage modulation strategy is not applicable in the two-HB THMI since the restriction (Equation 9.37) must be violated. Therefore, when M is less than 0.38 in this

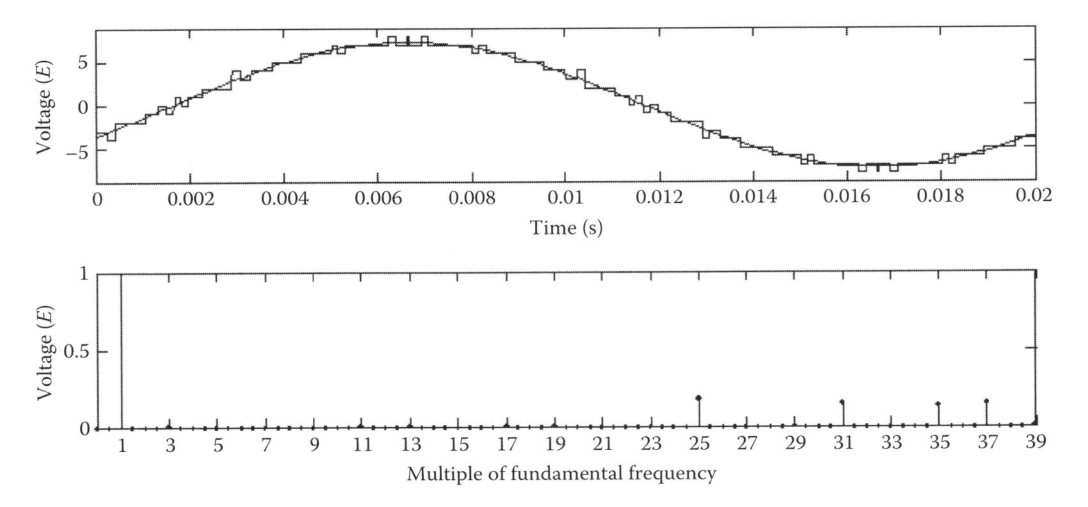

FIGURE 9.24 Synthesized line-to-line voltage waveform and frequency spectrum of a two-HB THMI with the virtual stage modulation.

TABLE 9.9

Range of Modulation Index with the Virtual Stage Modulation in a Two-HB THMI

σ	β	*MR* (min)	*MR* (max)	*M* (min)	*M* (max)	Range of *M*
3	1	0.51	0.92	0.38	0.69	0.38–0.459
4	2	0.459	0.92	0.459	0.92	0.459–0.92

Index in a two-HB THMI p_1 to p_6 mean θ_{p1} to θ_{p6}, and n_1 to n_2 mean θ_{n1} to θ_{n2}.

case, the step modulation strategy will be used. With the virtual stage modulation strategy, the scheme of switching angles is shown in Figure 9.25.

9.5.2.3.3 *Hybrid Modulation Strategy*

The hybrid modulation strategy for the hybrid multilevel inverters has been presented, which incorporates stepped voltage waveform synthesis in higher-power HB cells in conjunction with high-frequency variable PWM in the lowest-power HB cell. Figure 9.26 presents a block diagram of the command circuit utilized to determine the command signals of the power devices of all HBs. As shown in Figure 9.26, the reference signal of the hybrid multilevel inverter, v_{ref}, is the command signal of the HB with the highest DC voltage source ($V_{DC,h}$). This signal is compared with a voltage level corresponding to the sum of all smaller DC voltage sources of the hybrid multilevel inverter, $\sigma_{max,h-1}$. If the command signal is greater than this level, the output of the inverter with the highest DC voltage source must be equal to $V_{DC,h}$. In addition, if the command signal is less than the negative value of $\sigma_{max,h-1}$, the output of this cell must be equal to $-V_{DC,h}$, otherwise the output of this cell must be zero.

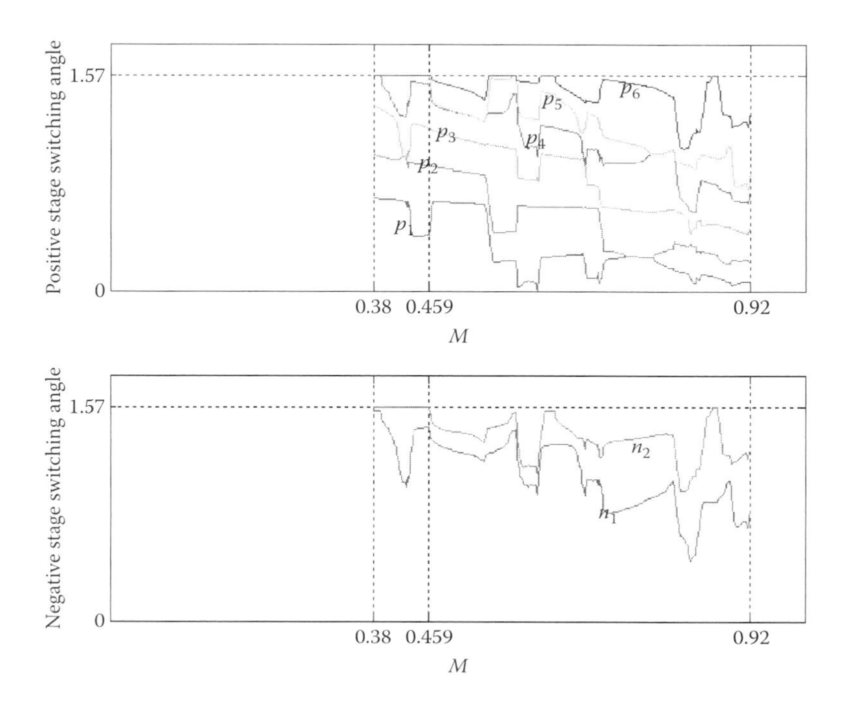

FIGURE 9.25 Scheme of switching angles for the virtual stage modulation as a function of modulation.

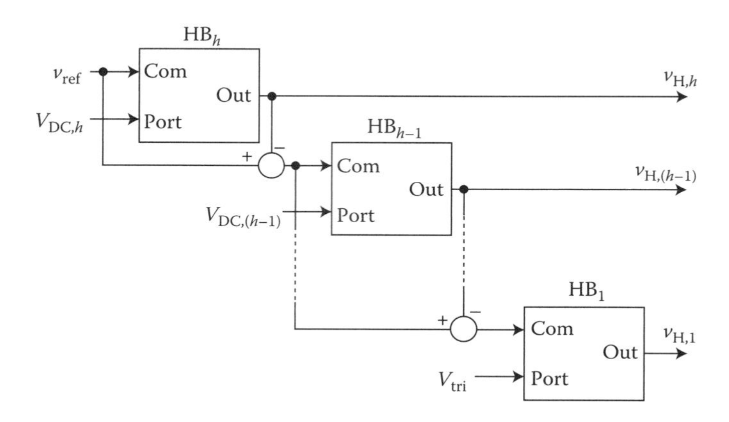

FIGURE 9.26 Hybrid modulation for hybrid multilevel inverters.

The command signal of the ith HB cell is the difference between the command signal of the HB_{i+1} and the output voltage of the HB_{i+1}. In this way, the command signal of the ith cell contains information about the harmonic content of the output voltage of all higher voltage cells. This command signal is compared with a voltage level corresponding to the sum of all voltage sources until the HB_{i-1} ($\sigma_{max,i-1}$). In the same way as that presented for the HB_h, the output voltage of this cell is synthesized from the comparison of these two signals.

Finally, the command signal of HB_1 (the lowest power inverter) is compared with a high-frequency triangle carrier signal, resulting in a high-frequency output voltage. Therefore, the output voltage harmonics will be concentrated around the frequency multiples of the switching frequency of the inverter with the lowest DC voltage source. Consequently, the spectral response of the output voltage depends on the switching frequency of the lowest power inverter, while the power processing depends on the inverter with the highest DC voltage source.

However, with the hybrid modulation strategy, a voltage waveform must be synthesized to modulate at high frequency among all adjacent voltage steps. Only the lower voltage HB can switch at high frequency, so the DC voltages must satisfy the following equation:

$$V_{DCi^*} \leq 2\sum_{k=1}^{i-1} V_{DCk^*}, \quad j = 2, 3, \ldots, h, \tag{9.38}$$

where * means the normalized value. Therefore, the hybrid modulation strategy can be applied in BHMIs and QLMIs. The relationship between the DC voltages of the THMI is shown in Equation 9.21, so the THMI cannot use the hybrid modulation strategy.

9.5.2.3.4 *Subharmonic PWM Strategies*

Subharmonic PWM strategies for multilevel inverters employ extensions of carrier-based techniques used for conventional inverters. It has been reported that the spectral performance of a five-level waveform can be significantly improved by employing alternative dispositions and phase shifts in the carrier signals. This concept can be extended to a nine-level case with the available options for polarity and phase variation. A representative subharmonic PWM waveform with the nine-level phase leg voltage is shown in Figure 9.27.

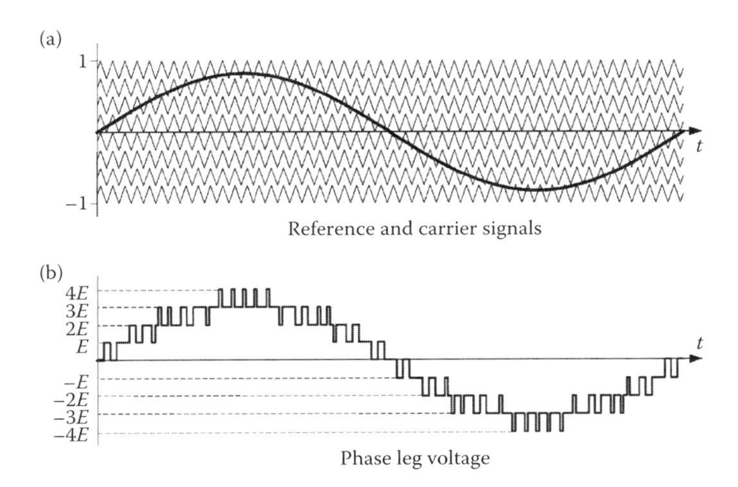

FIGURE 9.27 Representative waveforms for subharmonic PWM waveform with carrier polarity variation. (a) Reference and carrier signals and (b) phase leg voltage.

If a two-HB THMI is used to synthesize the nine-level phase leg voltage as shown in Figure 9.27b, the higher voltage HBs will switch at high frequencies, since the output voltage varies between E and $2E$ or $-E$ and $-2E$ continually in a certain interval. In THMI, it is not appropriate that the higher-voltage HBs switch at a high frequency. Therefore, the subharmonic PWM is not applicable in THMI.

9.5.2.3.5 *Simple Modulation Strategy*

The simple modulation strategy is the simplest modulation strategy with which the switching pattern is determined by comparing a reference signal with stages and then choosing the stages most close to the reference signal. Figure 9.28 shows an illustration of the simple modulation strategy with the nine-level output voltage.

The advantages of this strategy are simple a control algorithm, high flexibility, and dynamic response. The disadvantage of the strategy is that the amplitudes of lower-order harmonic components are relatively higher. The THMI can generate the greatest voltage levels among all multilevel inverters using the same number of components. If the number of voltage levels is high enough, the lower-order harmonic components of output voltages will be very small with the simple modulation strategy. For example, in the case of a four-HB THMI that can generate 81-level voltage, with the simple modulation strategy, the amplitude of each lower-order harmonic component of the output voltage is less than 0.9% of the amplitude of the fundamental component and the THD of the output voltage is less than 2%.

9.5.2.3.6 *Several Modulation Strategies*

Several modulation strategies have been investigated. With the hybrid modulation strategy and modulation strategies working with high switching frequencies, such as the subharmonic PWM strategy, a voltage waveform must be synthesized to modulate at high frequency among all adjacent voltage steps. However, in THMI, it cannot be achieved when only the lowest-voltage HB switches at a high frequency, which can be derived from Equations 9.21 and 9.38. In other words, if a voltage can be synthesized to modulate at high frequency in THMI, the higher-voltage HBs must switch at high frequency. One of the most

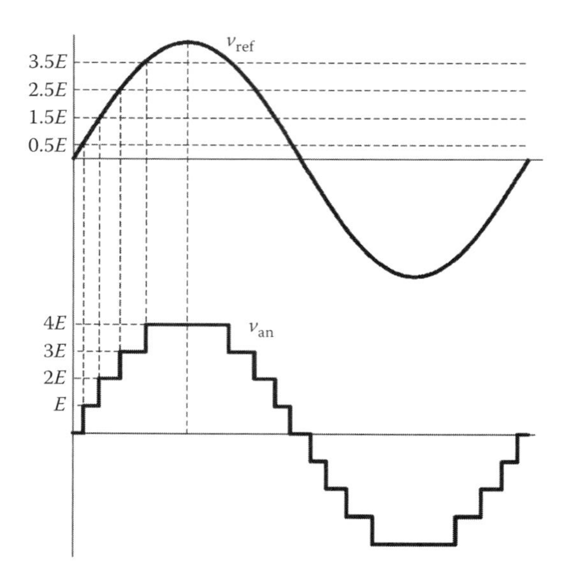

FIGURE 9.28 Illustration of the simple modulation strategy.

important advantages of THMI is that higher-voltage HBs can switch at lower frequency. Therefore, higher-frequency switching of higher-voltage HBs is not only unacceptable in high-power applications but also violates the main advantage of THMI. Therefore, the hybrid modulation strategy and other modulation strategies working with high switching frequencies are not applicable in THMI. The low-frequency modulation strategies such as the step modulation strategy and the virtual stage modulation strategy are suitable for THMI. In the virtual stage modulation, additional constraints, such as Equation 9.37 for two-HB THMI, must be added to ensure the higher HB switch at lower frequency. Additionally, the simple modulation strategy can be used in those THMIs that can generate many voltage levels. At the same time, for those THMIs that can generate many voltage levels, the space vector modulation can achieve a very good linearity between the modulation index and the fundamental component of load voltage and eliminate common-mode voltages.

9.5.2.4 Regenerative Power

The DC sources of the THMI can be batteries or bridge rectifiers. Batteries cannot endure a large reverse current for a long time, which will damage them. Diode bridge rectifiers cannot permit reverse power. Controlled bridge rectifiers can transmit energy to supplies. However, compared with simple diode bridge rectifiers, the controlled bridge rectifiers are much more complex and costly because of complex control circuits and the higher price of controlled semiconductors.

9.5.2.4.1 Analysis of DC-Bus Power Injection

The switching function is involved in the analysis of DC-bus power injection. The switching function, F, is shown in Table 9.9. The relationship between the output voltage of an HB, v_H, and the DC-link voltage of the HB, V_{DC}, can be written as Equation 9.39. The relationship between i_{DC} (current flowing through the DC bus) and i_{an} (output current of the THMI)

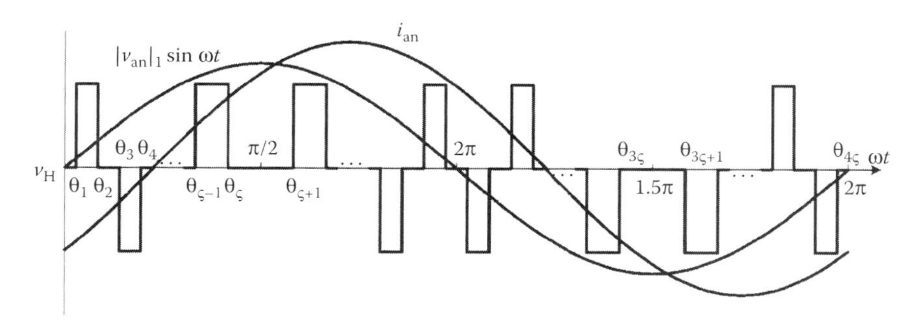

FIGURE 9.29 General waveform of DC-bus voltage and THMI output current.

can be also derived as Equation 9.40.

$$v_H = F \cdot V_{DC}. \tag{9.39}$$

$$i_{DC} = F \cdot i_{an}. \tag{9.40}$$

Only the fundamental component of the output current of the THMI is considered since high-frequency harmonic components do not generate average power. So i_{an} can be expressed as

$$i_{an} = I_{an} \cdot \sin(\omega t + \varphi), \tag{9.41}$$

where I_{an} is the amplitude of i_{an} and φ is the angle of the PF. General waveforms of v_H and i_{an} are shown in Figure 9.29. The average DC power that supplies the HB over a period can be calculated as

$$P_{DC} = \frac{1}{T}\int_0^T V_{DC} \cdot i_{DC} \mathrm{d}t, \tag{9.42}$$

where T is the period of i_{an}. From Equations 9.40 and 9.42, we obtain

$$P_{DC} = \frac{1}{T}\int_0^T V_{DC} \cdot F \cdot i_{an} \mathrm{d}t = \frac{1}{T}\int_0^T v_H \cdot i_{an} \mathrm{d}t. \tag{9.43}$$

The relationship between switching angles in Figure 9.29 can be expressed as

$$\theta_i = \begin{cases} \pi + \theta_{i-2\varsigma}, & i = (2\varsigma + 1) \cdots 4\varsigma, \\ \pi - \theta_{2\varsigma+1-i}, & i = (\varsigma + 1) \cdots 2\varsigma, \end{cases} \tag{9.44}$$

Derived from Equation 9.44, v_H has the following characteristics:

$$v_H(\pi - \omega t) = v_H(\omega t), \tag{9.45}$$

$$v_H(\omega t + \pi) = -v_H(\omega t). \tag{9.46}$$

From Equation 9.41, we obtain

$$i_{an}(\omega t + \pi) = -i_{an}(\omega t). \tag{9.47}$$

Derived from Equations 9.46 and 9.47, the average DC power can be calculated over the half period as

$$P_{DC} = \frac{2}{T}\int_{0}^{T/2} v_H \cdot i_{an} dt. \quad (9.48)$$

Suppose that P_i is the power generated by the voltage pulse from θ_i/ω to θ_{i+1}/ω and the corresponding voltage pulse from $\theta_{2\varsigma-i}/\omega$ to $\theta_{2\varsigma+1-i}/\omega$. Then, P_i can be expressed as

$$P_i = \frac{(-1)^n\omega}{\pi}\left(\int_{\theta_i/\omega}^{\theta_{i+1}/\omega} V_{DC}\cdot I_{an}\cdot \sin(\omega t+\varphi)dt + \int_{\theta_{2\varsigma-i}/\omega}^{\theta_{2\varsigma+1-i}/\omega} V_{DC}\cdot I_{an}\cdot \sin(\omega t+\varphi)dt\right), \quad (9.49)$$

where $i = 2n - 1$, with n being a natural number. Derived from Equations 9.44 and 9.49, P_i is expressed as

$$P_i = \frac{(-1)^n}{\pi} V_{DC}\cdot I_{an}\cdot 2\cdot \cos(\varphi)\cdot(\cos(\theta_i) - \cos(\theta_{i+1})). \quad (9.50)$$

Thus, the average DC power of the HB can be expressed as

$$P_{DC} = \frac{2}{\pi} V_{DC}\cdot I_{an}\cdot \cos(\varphi)\cdot \sum(\cos(\theta_{4n-3}) - \cos(\theta_{4n-2}) - \cos(\theta_{4n-1}) + \cos(\theta_{4n})). \quad (9.51)$$

In Equation 9.51, if θ_j is greater than $\pi/2$, θ_j will be set as $\pi/2$.

In general, the PF angle φ ranges from $-\pi/2$ to $\pi/2$, so $\cos(\phi)$ is greater than zero. V_{DC} and I_{an} are positive. Thus, we can conclude from Equation 9.51 that the power of the DC bus is negative if

$$\sum(\cos(\theta_{4n+1}) - \cos(\theta_{4n+2}) - \cos(\theta_{4n+3}) + \cos(\theta_{4n+4})) < 0. \quad (9.52)$$

Negative power of the DC bus means regenerative power.

9.5.2.4.2 *Regenerative Power in THMI*

Regenerative power may occur in lower-voltage HBs of THMI. Take the example of a two-HB THMI. If the step modulation strategy is applied, the restrictions that are added to Equation 9.36 to ensure that the power of DC buses is always positive are shown in

TABLE 9.10

Additional Restriction to Avoid Regenerative Power of DC Buses in the Step Modulation

σ	Restriction
1	$\cos(\theta_1) > 0$
2	$\cos(\theta_1) - 2\cos(\theta_2) > 0$
3	$\cos(\theta_1) - 2\cos(\theta_2) + \cos(\theta_3) > 0$
4	$\cos(\theta_1) - 2\cos(\theta_2) + \cos(\theta_3) + \cos(\theta_4) > 0$

TABLE 9.11

Range of Modulation Index with the Step Modulation in a Two-HB THMI (Avoid Regenerative Power of DC Buses)

σ	*MR* (min)	*MR* (max)	*M* (min)	*M* (max)	Range of *M*
1	0	1	0	0.25	0–0.15
2	0.3	0.66	0.15	0.33	0.15–0.33
3	0.59	0.68	0.44	0.51	0.44–0.51
4	0.55	0.86	0.55	0.86	0.55–0.86
4*	0.56	0.94	0.56	0.94	0.86–0.94

Table 9.10. With these restrictions, ranges of the relative modulation index are calculated as shown in Table 9.11. The range of the relative modulation index decreases greatly when σ is two or three compared with Table 9.11. The range of the modulation index is not continuous as shown in the last column of Table 9.11. The regenerative power will occur in the lower-voltage HB when M ranges from 0.51 to 0.55 or from 0.33 to 0.44.

Consider that the virtual stage modulation strategy is used in a two-HB THMI. In Table 9.11, two cases are analyzed. One is four-level positive/negative output voltage with two virtual stages and the other is five-level positive/negative output voltage with one virtual stage. It is possible only for the DC bus of the lower-voltage HB to have regenerative power. For the first case, the restriction that ensures positive power can be written as

$$\cos(\theta_{p1}) - 2\cos(\theta_{p2}) + \cos(\theta_{p3}) + \cos(\theta_{p4}) + \cos(\theta_{p5}) + \cos(\theta_{p6}) - \cos(\theta_{n1}) - \cos(\theta_{n2}) > 0. \tag{9.53}$$

For the second case, the restriction can be expressed as

$$\cos(\theta_{p1}) - 2\cos(\theta_{p2}) + \cos(\theta_{p3}) + \cos(\theta_{p4}) + \cos(\theta_{p5}) - \cos(\theta_{n1}) > 0. \tag{9.54}$$

With these restrictions, the range of the relative modulation index decreases as shown in Table 9.12. The regenerative power will occur in the lower-voltage HB when M ranges from 0.53 to 0.62.

9.5.2.4.3 *Method for Avoiding Regenerative Power*

In the previous subsection, the regenerative power that lower-voltage HBs may generate was discussed. In this section, the methods that are usually used to solve this problem will be introduced. A method is proposed that the DC links of lower-voltage HBs are supplied with low-power, isolated power sources fed by a common power supply from the highest-voltage HB. These power sources are bidirectional and a bidirectional DC–DC power supply is used for this purpose. It is also possible to use independent output transformers with a common

TABLE 9.12

Range of Modulation Index Range with the Virtual Stage Modulation Strategy in a Two-HB THMI (Avoid Regenerative Power of DC Bus)

σ	β	*MR* (min)	*MR* (max)	*M* (min)	*M* (max)	Range of *M*
3	1	0.62	0.71	0.46	0.53	0.46–0.53
4	2	0.62	0.92	0.62	0.92	0.62–0.92

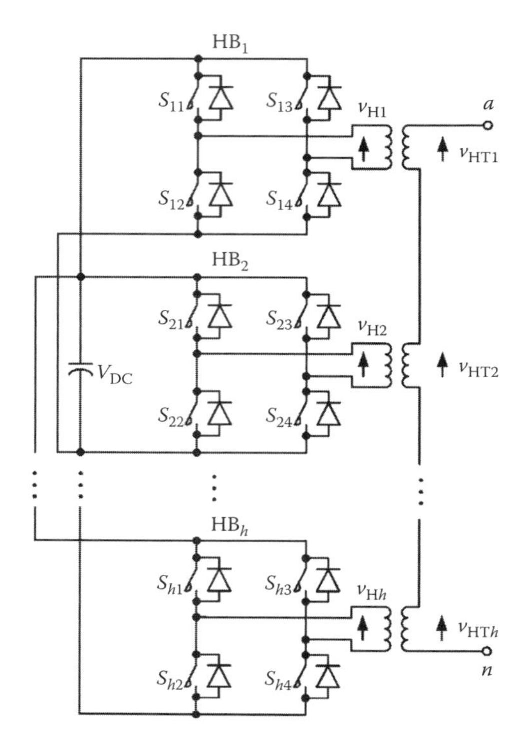

FIGURE 9.30 THMI with output transformers.

DC supply as shown in Figure 9.30. A variation of this configuration was used by ABB in his 16(2/3) Hz substation for railroads in Bremen (Germany). In the system described here, the transformers are smaller for lower-voltage HBs because the voltages are scaled to the power of three. Besides, the switching frequency of the transformers connected with lower-voltage HBs is lower. Then the transformers connected with lower-voltage HBs become smaller for two reasons: voltage and switching frequency.

The above two methods to solve the problem of regenerative power use additional equipment, such as the bidirectional DC/DC converter or output transformers, which increase the cost of the inverter system and power losses. A new method is presented to avoid regenerative power. This method does not use additional devices. The regenerative power is eliminated by avoiding outputting several null-voltage levels, which will be explained with the example of a four-HB THMI in the following.

The average power of the DC bus of an HB can be expressed as Equation 9.51. In general, the PF angle φ ranges from $-\pi/2$ to $\pi/2$; so $\cos(\phi)$ is greater than zero. V_{DC} and I_{an} are positive. Therefore, from Equation 9.51 and Figure 9.28, we can conclude that the reason for the regenerative power is the negative segments of v_H when the fundamental components of v_{an} are positive or the positive segments of v_H when the fundamental components of v_{an} are negative. The segments of v_H resulting in the regenerative power of the HB are called regenerative segments. The basic idea of eliminating regenerative power is to avoid outputting several levels of v_{an}, which will cause regenerative segments in HBs. Table 9.13 shows the voltage levels of v_{an} that cause regenerative segments of HBs in the case of a four-HB THMI. The voltage levels of v_{an} that are not selected to output are called null voltage levels. Table 9.13 also shows the priory of null voltage levels. For example, if the

TABLE 9.13

Voltage Levels of v_{an} which Cause Regenerative Segments of HBs

HB_1	±14, ±17, ±32, ±23, ±5, ±20, ±38, ±29, ±26, ±11, ±2, ±8, ±35
HB_2	±14, ±32, ±23, ±5, ±15, ±34, ±25, ±7, ±16, ±6, ±24, ±33
HB_3	±14, ±17, ±15, ±20, ±19, ±16, ±21, ±18, ±22

regenerative power occurs in the DC link of the HB_1, the voltage level (14) and (−14) are selected as null voltage levels first. If the regenerative power still occurs, the voltage levels (17) and (−17) are also selected as null voltage levels. With the priory shown in Table 9.13, the null voltage levels distribute as evenly as possible, which results in better power quality.

Figure 9.31 shows the flow chart of the algorithm that stabilizes the DC-link voltage of an HB. V_{DC} is the DC-link voltage of an HB. $V_{DC,normal}$ is the normal DC-link voltage. $V_{DC,last}$ is the DC-link voltage in the previous sampling. N_{null} is the number of null voltage levels. In the switch table, the voltage levels are set as null or not based on N_{null} in Table 9.13.

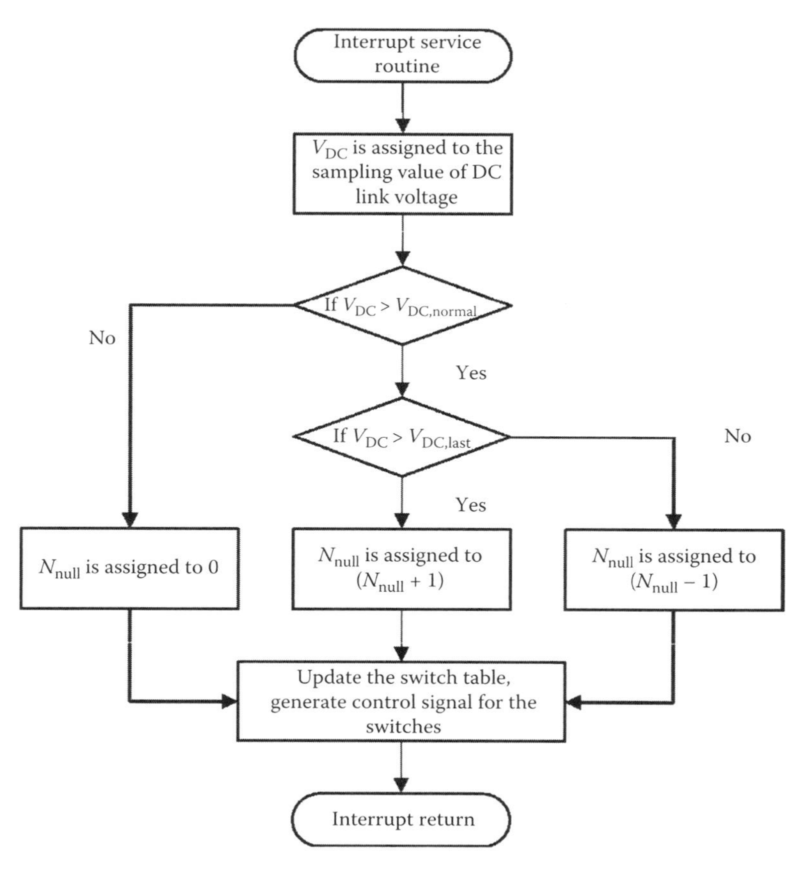

FIGURE 9.31 Flow chart of the algorithm to stabilize DC-link voltages.

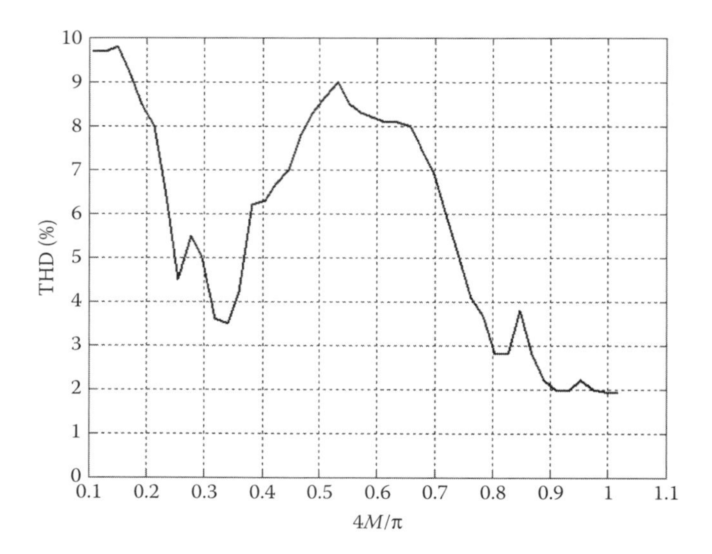

FIGURE 9.32 Relationship between the modulation index and THD.

With a lower modulation index, the power quality that the THMI outputs is a little bit poorer with the proposed control scheme, because more null voltage levels do not devote themselves to the output voltage of the THMI. In the case of the four-HB THMI, with up to 81-level output voltage of the THMI, the simple modulation strategy as shown in Section 9.5.2 is suitable for the THMI. If the simple modulation strategy is used and the new method is applied to eliminate the effect of regenerative power, the relationship between the modulation index and the THD is shown in Figure 9.32.

9.5.2.4.4 *Summary of the Regenerative Power in THMI*

The topology of the THMI has the distinct advantage that less components are used compared with the topologies of other multilevel inverters, but the THMI also has the notable disadvantage that the power of the lower-voltage HBs may be regenerative with a lower modulation index. If the THMI feeds an *RL* or *RC* load and simple diode bridge rectifiers are used as DC sources, then the regenerative power will cause the DC capacitor voltages to increase, which will damage the devices.

Therefore, basically, the THMI is suitable for two applications. The first is the application of reactive power compensation. The average power of the DC link of an HB is zero when the PF angle is zero as shown in Equation 9.51, so the problem of regenerative power is avoided. The second is the application in which the inverter always runs with a higher modulation index. From Table 9.11, we can find that the two-HB THMI runs with step modulation without the problem of regenerative power when M ranges from 0.56 to 0.94. From Table 9.12, we can find that the two-HB THMI runs with virtual stage modulation without the problem of regenerative power when M is from 0.62 to 0.92.

However, the inverter is required to work at any modulation index for active load in most cases. Two methods have been presented to solve the regenerative power problem. The first uses bidirectional DC/DC converters and the second uses additional output transformers. A new method to solve the regenerative power is presented as a cost-effective solution because it does not use additional equipment. The DC capacitor voltages of lower voltage

HBs are kept stable by the new method. The trade-off is that power quality will decrease a little with lower modulation index.

9.5.3 Experimental Results

Some experimental results are shown here to help readers understand the content better.

9.5.3.1 Experiment to Verify the Step Modulation and the Virtual Stage Modulation

The performances of the step modulation strategy and the virtual stage modulation strategy have been verified by the experiment of a single two-HB THMI. In the control circuit, a TMS320F240 DSP is used as the main processor, which provides the gate logic signals. In an HB, four MOSFETs, IRF540, are used as the main switches, which are connected in a full-bridge configuration. The load is a 23.2 Ω resistor. The total ratio of voltage measure is 1:2. The frequency spectrums are analyzed by the FFT function of an oscilloscope. The scale of the Y-axis of the frequency spectrum is 5 dB V/div and that of the reference level is 5 dB V.

The switching pattern of the step modulation technique is programmed and is loaded to the DSP. In the step modulation strategy, when the number of output voltage levels is nine and M is equal to 0.83, the switching angles will be 0.14778, 0.32325, 0.57376, and 0.99696. The THMI output voltage is shown in Figure 9.33. The frequency is 50 Hz and the step voltage is about 5 V. The frequency spectrum is shown in Figure 9.34. The 5th, 7th, and 11th harmonics are less than 0.028 V (−37 dB × 2 V), which means that they are nearly eliminated. This verifies the simulation result as shown in Figure 9.17.

When the number of output voltage levels is seven and M is equal to 0.49, the switching angles are 0.44717, 0.9097, and 1.1215. The output voltage of the THMI is shown in Figure 9.35 and the frequency spectrum is shown in Figure 9.36. The 5th and 7th harmonics are less than 0.02 V (−40 dB × 2 V), which means that they are nearly eliminated.

When the number of output voltage levels is five and M is equal to 0.32, the switching angles are 0.51847 and 1.1468. The output voltage of the THMI is shown in Figure 9.37 and

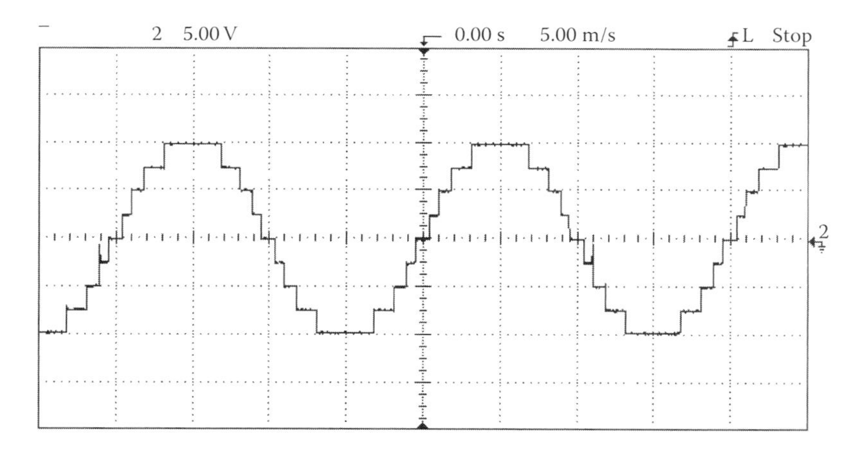

FIGURE 9.33 Output voltage of the THMI with the step modulation $M = 0.83$ (10 V/div).

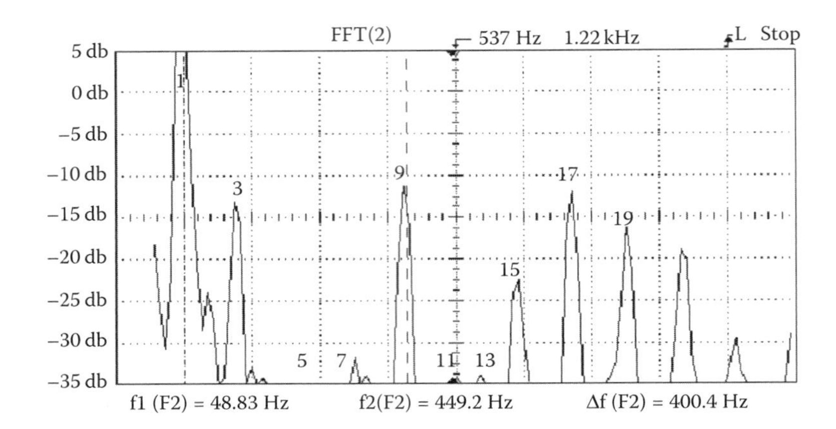

FIGURE 9.34 Frequency spectrum with the step modulation $M = 0.83$.

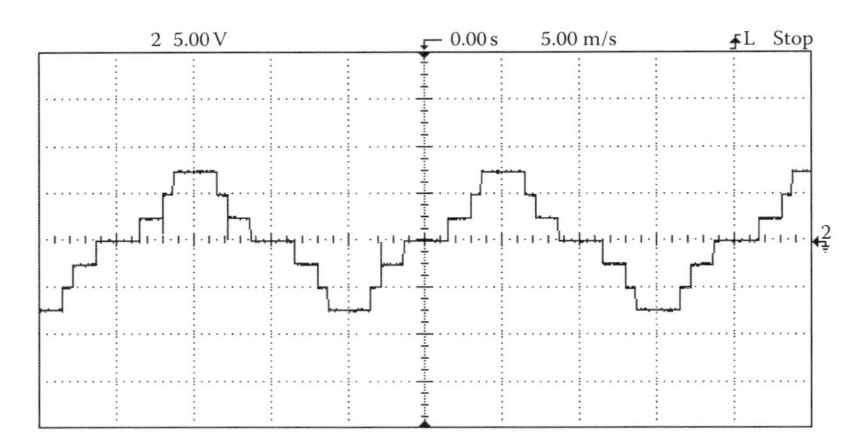

FIGURE 9.35 Output voltage of the THMI with the step modulation $M = 0.49$ (10 V/div).

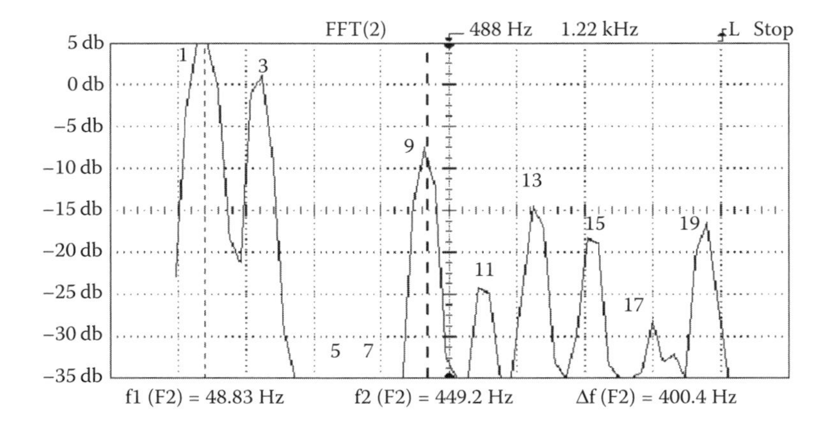

FIGURE 9.36 Frequency spectrum with the step modulation $M = 0.49$.

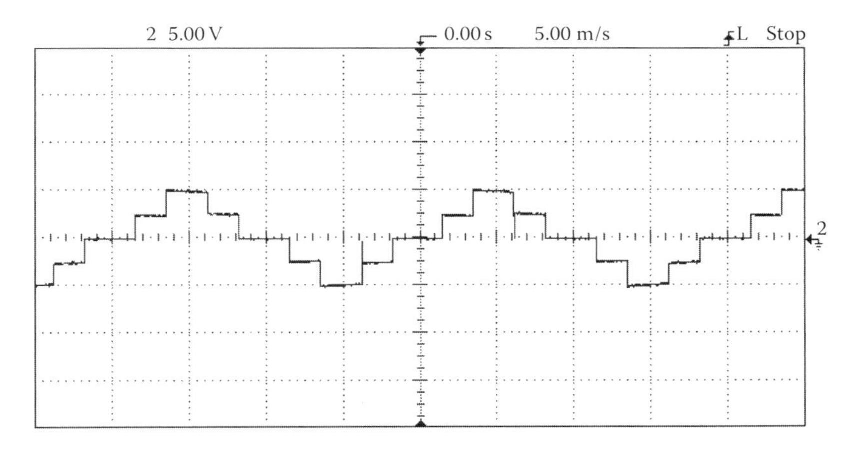

FIGURE 9.37 Output voltage of the THMI with the step modulation $M = 0.32$ (10 V/div).

the frequency spectrum is shown in Figure 9.38. The 5th harmonics are less than 0.02 V (−40 dB×2 V), which means that they are nearly eliminated.

The switching pattern of the modified virtual stage modulation technique is programmed and is loaded to the DSP. In virtual stage modulation, when the number of output voltage levels is nine and M is 0.83, the switching angles are 0.13177, 0.33186, 0.52855, 0.6202, 0.91294, 1.0423, 0.57124, and 0.96573. The output voltage of the THMI is shown in Figure 9.39 and the frequency spectrum is shown in Figure 9.40. The 5th, 7th, 11th, 13th, 17th, 19th, and 23rd harmonics are less than 0.035 V (−35 dB × 2 V), which means that they are nearly eliminated. It verifies the simulation result as shown in Figure 9.23.

When the number of output voltage levels is seven and M is equal to 0.49, the switching angles are 0.40549, 0.88038, 1.1497, 1.5318, and 1.5082. The output voltage of the THMI is shown in Figure 9.41 and the frequency spectrum is shown in Figure 9.42. The 5th, 7th, 11th, and 13th harmonics are less than 0.02 V (−40 dB × 2 V), which means that they are nearly eliminated.

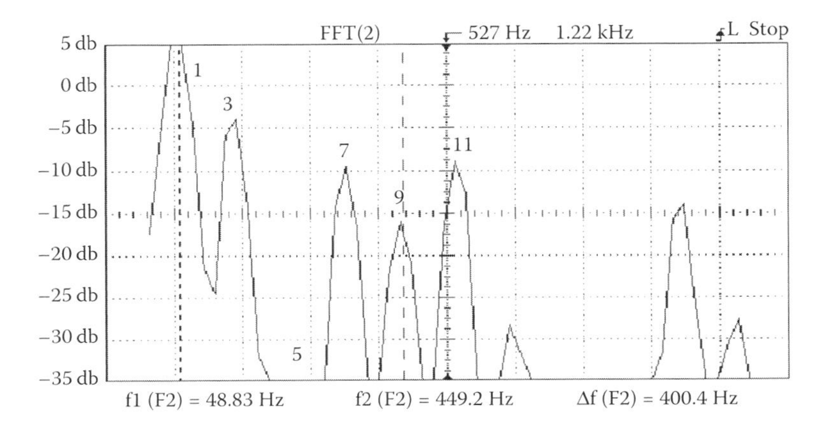

FIGURE 9.38 Frequency spectrum with the step modulation $M = 0.32$.

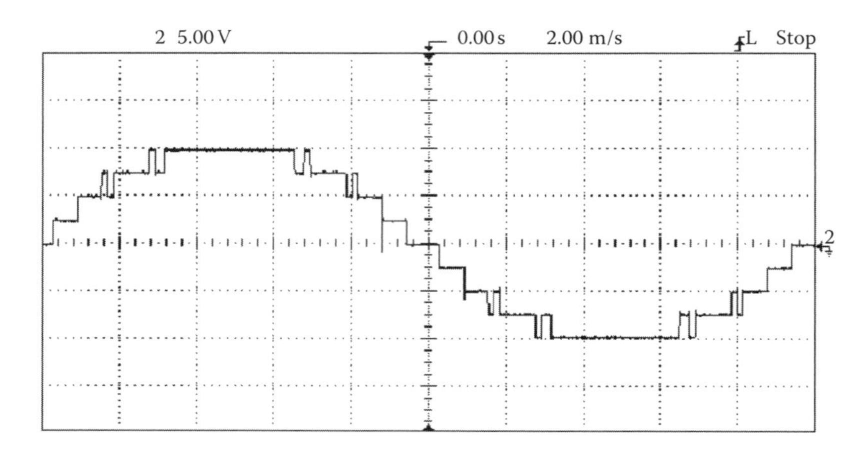

FIGURE 9.39 Output voltage of the THMI with the virtual stage modulation $M = 0.83$ (10 V/div).

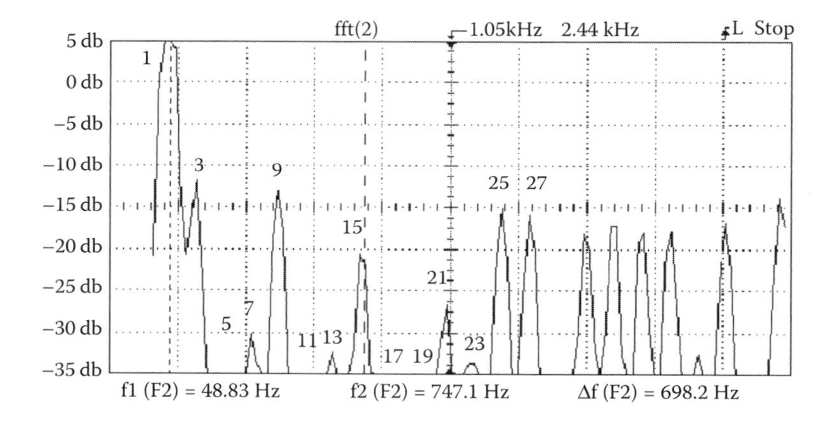

FIGURE 9.40 Frequency spectrum with the virtual stage modulation $M = 0.83$.

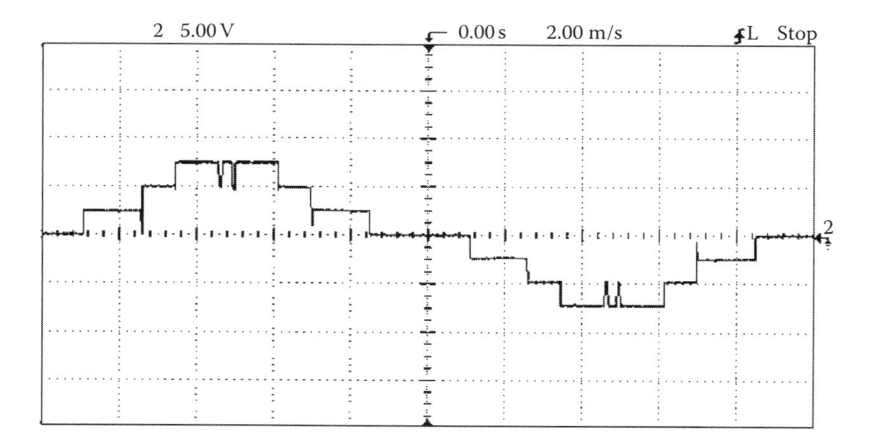

FIGURE 9.41 Output voltage of the THMI with the virtual stage modulation $M = 0.49$ (10 V/div).

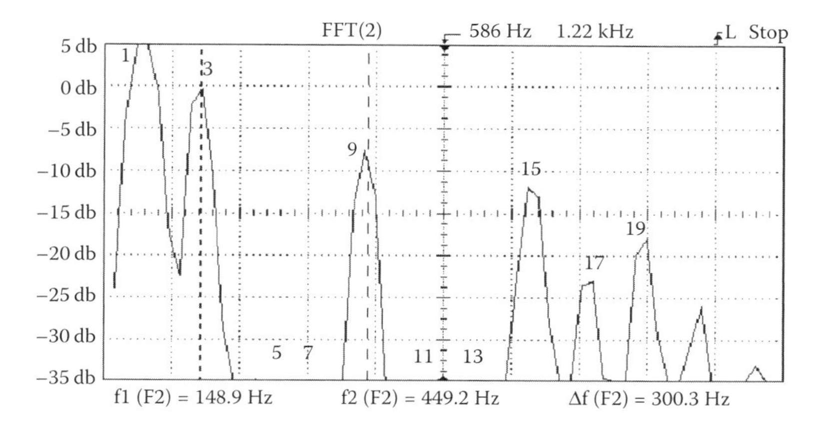

FIGURE 9.42 Frequency spectrum with the virtual stage modulation $M = 0.49$.

9.5.3.2 Experiment to Verify the New Method of Eliminating the Regenerative Power

The performance of the methods of eliminating the effect of regenerative power by avoiding outputting the null voltage levels is verified by the experiment of a four-HB THMI, in which diode bridge rectifiers are used as the DC sources of HBs. The step voltage is 5.9 V. The frequency of the output voltage is set at 50 Hz and the sampling frequency is set at 10 kHz. The inverter has up to 81 output voltage levels, so the simple modulation strategy is used as shown in Section 9.5.2. The control algorithm to stabilize the DC-link voltages is shown in Figure 9.26. A TMS320F240 DSP-controlled card is used to control the inverter. The configuration of the experimental system is shown in Figure 9.43.

Figure 9.44 shows the waveform of the output voltage of the four-HB THMI with the simple modulation strategy when the modulation index is 0.79. The power quality is good due to a large number of voltage levels.

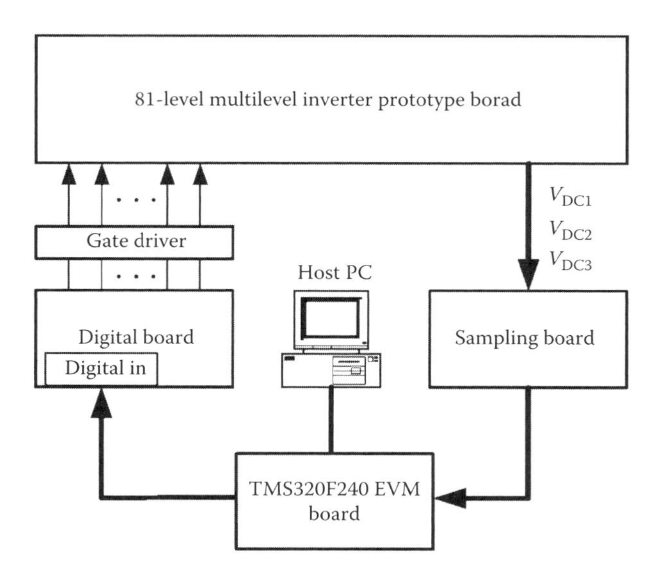

FIGURE 9.43 General representation of experimental test system.

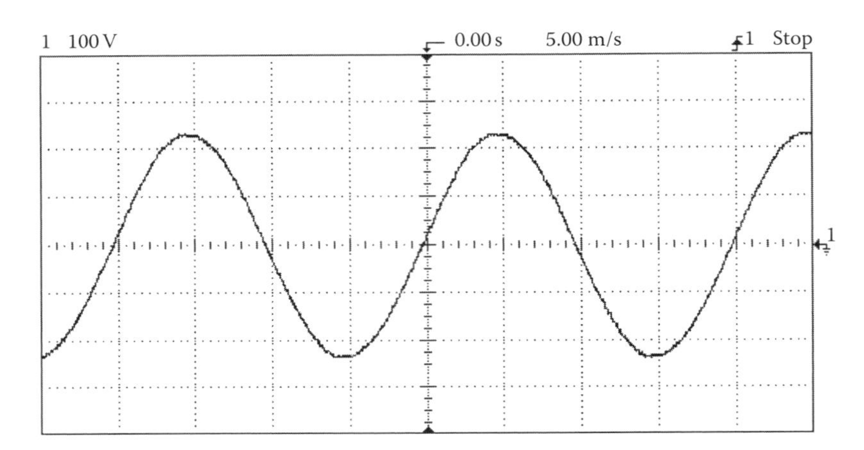

FIGURE 9.44 Waveform of the output voltage of the inverter with the simple modulation strategy $M = 0.79$ (100 V/div), frequency = 50 Hz, and THD = 1.94%.

Figure 9.45 shows the output voltage waveform with some null voltage levels when the modulation index is 0.7. From the enlarged figure, we can observe that some voltage levels are not generated. Moreover, the step voltages are kept nearly the same, which means that the voltages of DC capacitors are kept stable. Figure 9.46 shows the worst case when the modulation index is 0.53. In this case, the null voltage levels include ±5, ±14, ±15, ±16, ±17, ±19, ±20, ±21, ±23, ±32, and ±34.

9.5.4 Trinary Hybrid 81-Level Multilevel Inverters

A trinary hybrid 81-level multilevel inverter is designed for a motor drive with zero common-mode voltage. Figure 9.47 shows the power circuit topology of the THMI for motor drive. Bidirectional DC/DC converters connected to DC link of HBs feed HBs. To get the maximum output voltage levels of the inverter, the ratio of the DC-link voltages is arranged as 1:3:9:27, so the inverter can output 81 voltage levels at each phase. With four HBs per phase, however, a CMI [7–9] can output only nine voltage levels at each phase [6] and a BHMI [10] can output only 31 voltage levels at each phase. The more output voltage levels a multilevel inverter has, the more nearly a sinusoidal waveform can be synthesized. Therefore, with the trinary hybrid topology, THD can be reduced to a great extent.

The three phases of the inverter are controlled separately and the operating principle of each phase is identical. In the following, the A-phase of the inverter is analyzed. HB_{ak} represents the kth HB in the A-phase leg of the inverter as shown in Figure 9.48. $v_{H,ak}$ and $v_{C,ak}$ represent the output voltage and the DC-link voltage of the HB_{ak}, respectively. A switching function, F_{ak}, is used to relate $v_{H,ak}$ and $v_{C,ak}$ as

$$v_{H,ak} = F_{ak} \cdot v_{C,ak}, \quad k = 1, 2, 3, 4. \tag{9.55}$$

The value of F_{ak} can be either 1 or −1 or 0. For the value 1, the upper switch of the left leg and the lower switch of the right leg in an HB need to be turned on. For the value −1, the lower switch of the left leg and the upper switch of the right leg in an HB need to be turned on. For the value 0, the upper switches of both legs or the lower switches of both legs need

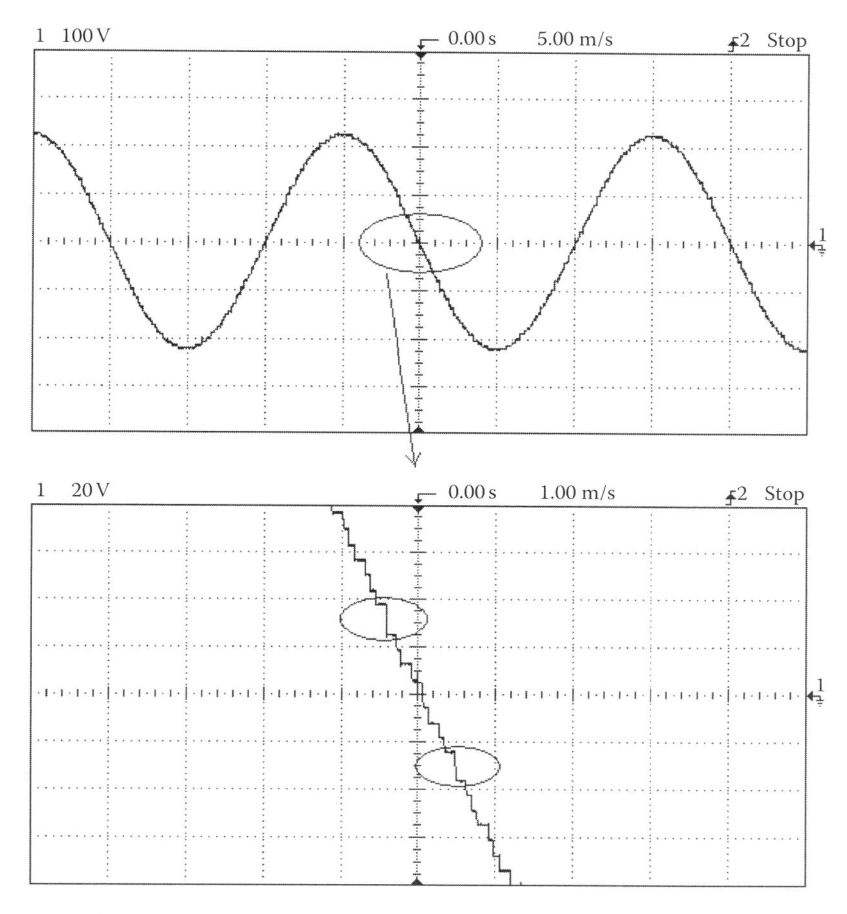

FIGURE 9.45 Waveform of the output voltage of the inverter $M = 0.7$ (100 V/div).

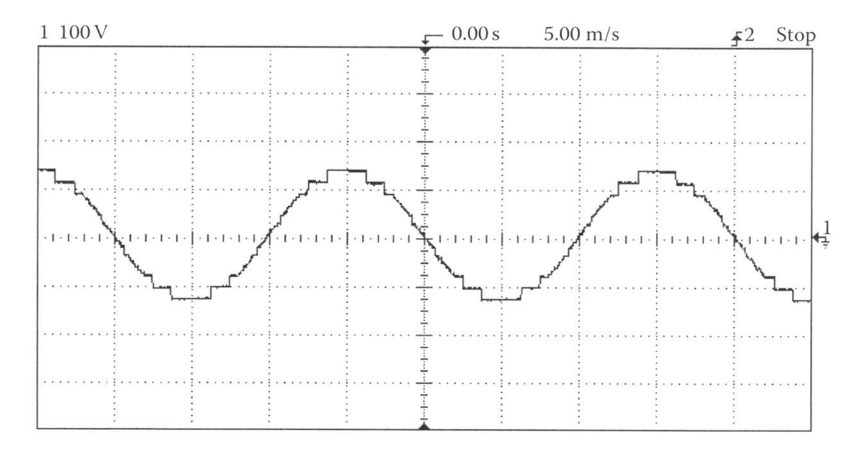

FIGURE 9.46 Waveform of the output voltage of the inverter $M = 0.42$ (100 V/div).

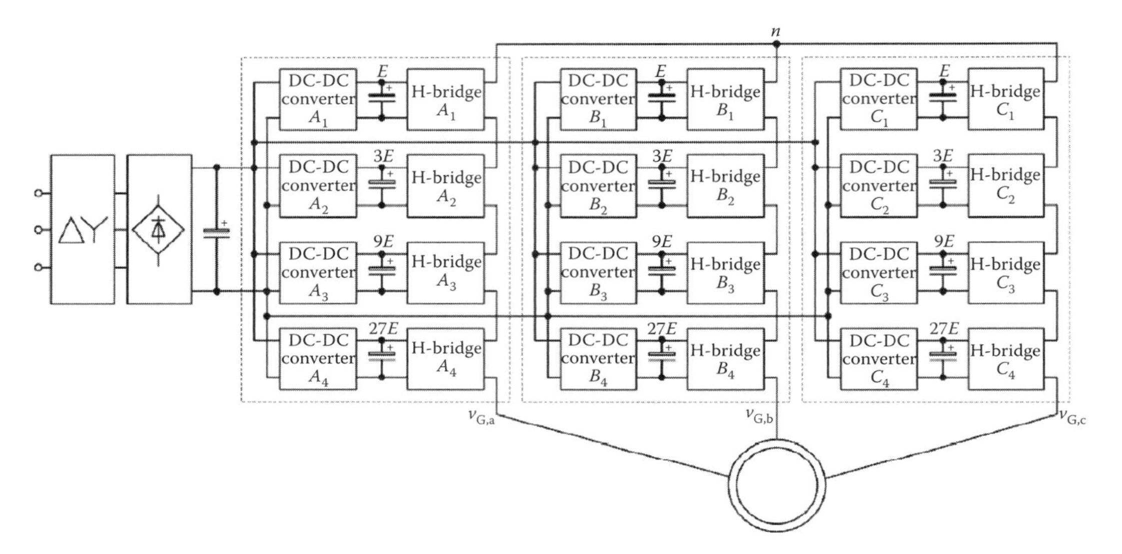

FIGURE 9.47 Power circuit topology of the trinary hybrid 81-level inverter for motor drive.

to be turned on. The A-phase voltage of the inverter, $v_{G,a}$, is represented as

$$v_{G,a} = \sum_{1}^{4} (F_{ak} \cdot v_{C,ak}). \tag{9.56}$$

Suppose that l is the ordinal of the expected voltage level that the inverter outputs. If l is positive or zero, the inverter outputs the positive lth voltage level. If l is negative, the inverter outputs the negative $(-l)$th voltage level. The value of F_{ak} can be determined by

$$\begin{aligned}
F_{a4} &= B_u(\text{ABS}(l) - 13) \cdot \text{ABS}(l)/l, \quad l_3 = l - F_{a4} \cdot 27, \\
F_{a3} &= B_u(\text{ABS}(l_3) - 4) \cdot \text{ABS}(l_3)/l_3, \quad l_2 = l_3 - F_{a3} \cdot 9, \\
F_{a2} &= B_u(\text{ABS}(l_2) - 1) \cdot \text{ABS}(l_2)/l_2, \quad l_1 = l_2 - F_{a2} \cdot 3, \\
F_{a1} &= B_u(\text{ABS}(l_1) \cdot \text{ABS}(l_1)/l_1,
\end{aligned} \tag{9.57}$$

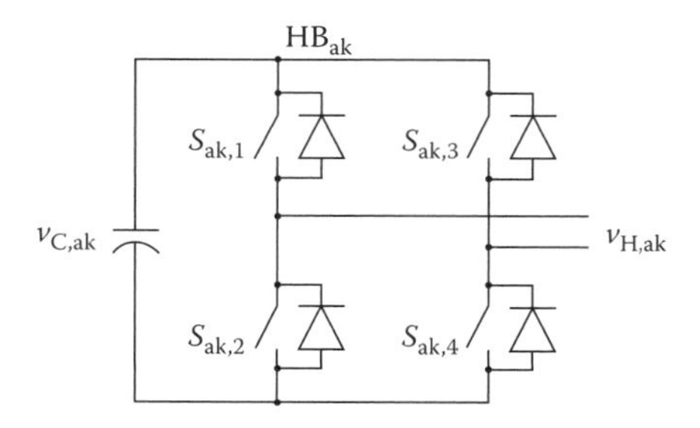

FIGURE 9.48 H-bridge.

where ABS is the function of the absolute value and B_u is defined as

$$B_u(\tau) = \begin{cases} 1, & \tau > 0, \\ 0, & \tau \leq 0. \end{cases} \quad (9.58)$$

From Equation 9.57, we obtain the relationship between the output voltage of a phase leg and the values of the switching functions of HBs in a phase leg.

9.5.4.1 Space Vector Modulation

$v_{G,a}$, $v_{G,b}$, and $v_{G,c}$ are the voltages of terminals a, b, and c of the inverter with respect to the neutral n. Three-phase inverter output voltages can be represented by a space vector in an x–y plane using the following transformation:

$$v = v_x + jv_y = \frac{2}{3}(v_{G,a} + \alpha v_{G,b} + \alpha^2 v_{G,c}), \quad (9.59)$$

where

$$\alpha = -\frac{1}{2} + j\frac{\sqrt{3}}{2}. \quad (9.60)$$

Equation 9.59 can be expressed as a function of their real and imaginary components:

$$v_x = \frac{1}{3}(2v_{G,a} - v_{G,b} - v_{G,c}), \quad (9.61)$$

$$v_y = \frac{1}{\sqrt{3}}(v_{G,b} - v_{G,c}). \quad (9.62)$$

The number of different voltage vectors is represented as

$$N_v = 2N_l - 1 + \sum_{i=1}^{2(N_l-1)} 2i, \quad (9.63)$$

where N_l is the number of voltage levels. Each phase can generate 81 different voltages, so totally 19,411 different voltage vectors can be generated as shown in Figure 9.49.

The common-mode voltage is defined as

$$v_{cm} = \frac{1}{3}(v_{G,a} + v_{G,b} + v_{G,c}). \quad (9.64)$$

Considering this definition, we can find vectors generated by three-phase voltages, which produce zero common-mode voltage as shown in Figure 9.50. The use of only vectors that generate zero common-mode voltages to the load reduces the density of vectors available to be applied. The number of different authorized licensed voltage vectors with zero common-mode voltage is represented as

$$N_{vz} = \frac{3N_l^2 + 1}{4}. \quad (9.65)$$

Therefore, there are still 4921 different voltage vectors available.

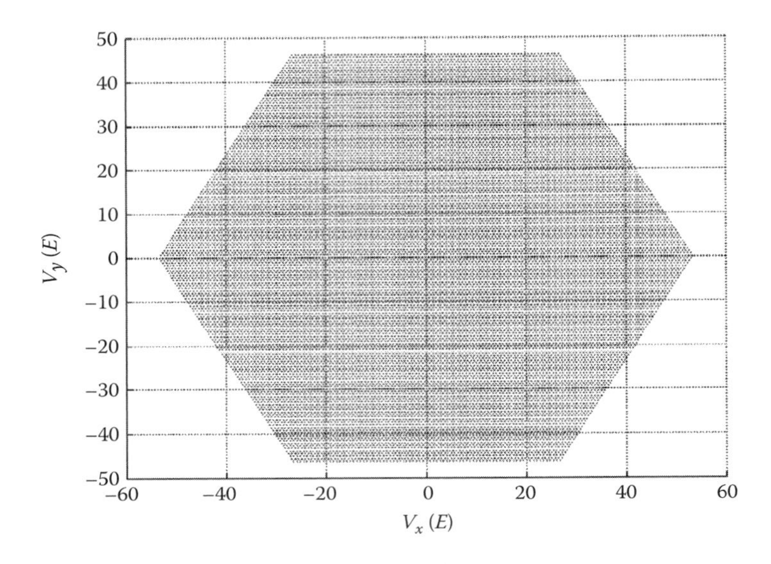

FIGURE 9.49 Voltage vectors of a three-phase 81-level inverter.

In Figure 9.51, the nearest voltage vector with respect to the reference vector v_{ref} is delivered. The following algorithm is used to select the appropriate vector based on the information about the reference vector.

Step 1. Normalize the reference vector $v_{ref} = v_{xref} + jv_{yref}$;

$$v'_{ref} = \frac{1}{E}v_{xref} + j\frac{\sqrt{3}}{E}v_{yref} = x + jy. \tag{9.66}$$

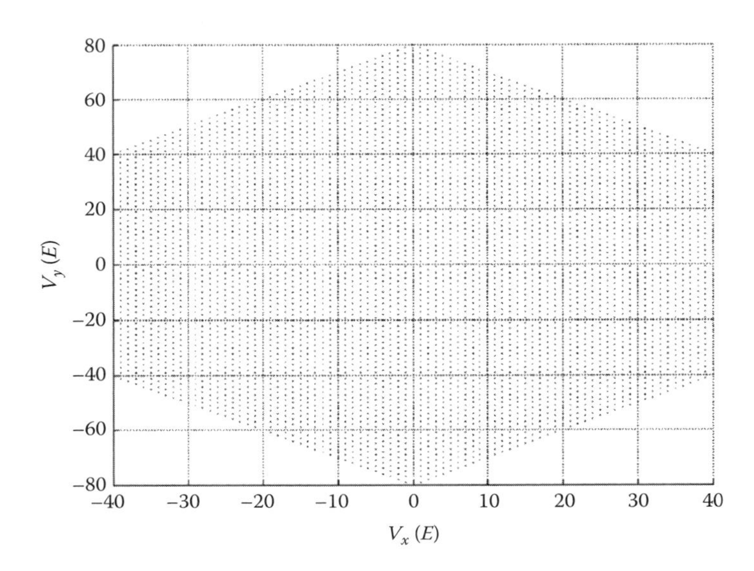

FIGURE 9.50 Voltage vectors of a three-phase 81-level inverter with zero common-mode voltage.

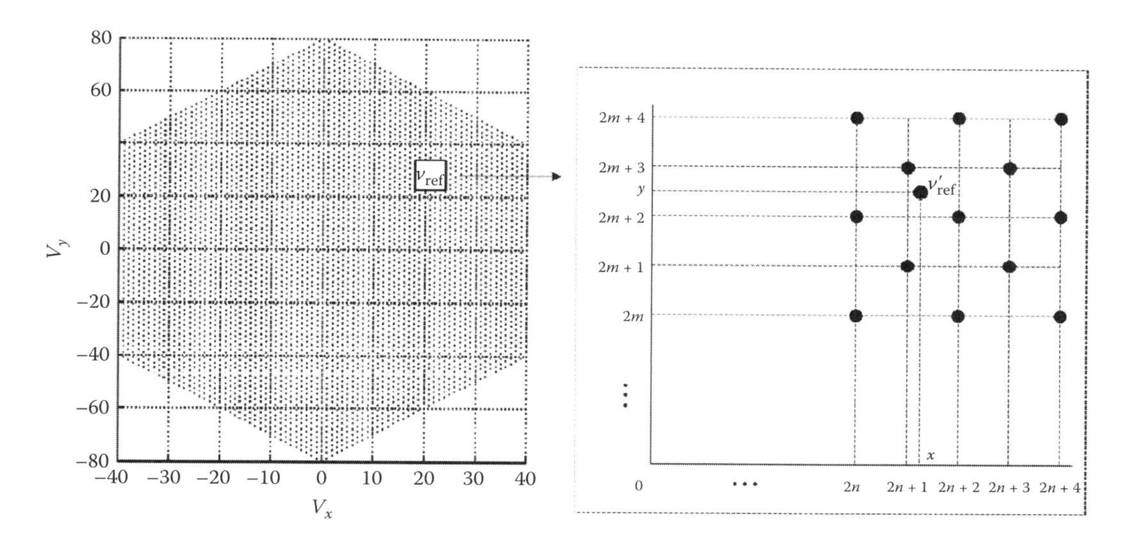

FIGURE 9.51 Normalized voltage vectors of a three-phase 81-level inverter with zero common-mode voltage.

Step 2. Normalize the candidate space vector with the transformation (Equation 9.66), converting them into integer values. After conversion, the space vectors with zero common-mode voltage are shown in Figure 9.51. The addition of the x-axis value and the y-axis value of each space vector with zero common-mode voltage is even.

Step 3. v'_{ref} will lie in one of the rectangles defined by two normalized candidate space vectors, as shown in the right part of Figure 9.51. The rectangle is identified by the values of the left bottom point of the rectangle. v'_{ref} (x, y) lies in the rectangle (floor(x), floor(y)), where floor (α) is the function that rounds the elements of α to the nearest integer that is less than or equal to α. In the rectangle (floor(x), floor(y)), there are two normalized voltage vectors, (floor(x), floor(y)) and (floor(x)+1, floor(y)+1), if the addition of floor(x) and floor(y) is even. The two vectors are (floor(x)+1, y) and (x, floor(y)+1), if the addition of floor(x) and floor(y) is odd. Suppose that the reference vector, v'_{ref} (x, y), lies in the rectangle with two normalized voltage vectors, v_1 and v_2. The nearest vector is selected by comparing the distances of each candidate vector, v_1 and v_2, with respect to v'_{ref}, using the following equations:

$$d_1 = \sqrt{(3(x - \mathrm{Re}(v_1))^2 - (y - \mathrm{Im}((v_1))^2}, \tag{9.67}$$

$$d_2 = \sqrt{(3(x - \mathrm{Re}(v_2))^2 - (y - \mathrm{Im}(v_2))^2}. \tag{9.68}$$

The selection is done by using the following equation:

$$\text{if} \quad d_1 < d_2 \quad \text{then} \quad v_{sel} = v_1; \qquad \text{otherwise} \quad v_{sel} = v_2. \tag{9.69}$$

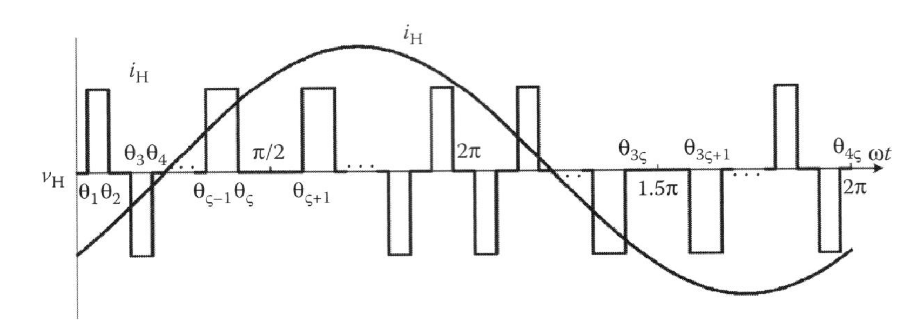

FIGURE 9.52 General waveform of the output voltage and current of an HB.

Step 4. Three-phase output voltages with zero common-mode voltage are generated by an inverse transformation for v_{sel} as

$$\begin{aligned} v_{G,a} &= \text{round}(\text{Re}(v_{sel})), \\ v_{G,b} &= v_{G,a} + \frac{\text{Im}(v_{sel}) - 3\text{Re}(v_{sel})}{2}, \\ v_{G,c} &= v_{G,a} - \frac{\text{Im}(v_{sel}) + 3\text{Re}(v_{sel})}{2}. \end{aligned} \tag{9.70}$$

9.5.4.2 DC Sources of HBs

There are three reasons to set DC sources of HBs as bidirectional DC/DC converters in the proposed topology. The first reason is that the bidirectional DC/DC converter can transfer the regenerative power from the HB to the rectifier. In an HB, the output voltage is v_H and the current flowing through the HB is i_H, as shown in Figure 9.52.

Only the fundamental component of the output current of the inverter is considered since high-frequency harmonic components do not generate average power. Thus, the average power flowing through the DC link of the HB, $P_{H,dc}$, can be expressed as

$$P_{H,dc} = \frac{2}{\pi} v_C I_H \cos\varphi \sum (\cos(\theta_{4n-3}) - \cos(\theta_{4n-2}) - \cos(\theta_{4n-1}) - \cos(\theta_{4n})), \quad n = 1, 2, \ldots, \tag{9.71}$$

where v_C is the DC-link voltage of the HB and I_H is the amplitude of i_H. Here, φ is the angle of PF for the fundamental components of v_H and i_H. In Equation 9.71, if θ_j is greater than $\pi/2$, θ_j will be set as $\pi/2$. In general, the PF angle φ ranges from $-\pi/2$ to $\pi/2$, so $\cos(\phi)$ is greater than zero. v_C and I_H are positive. Thus, we can conclude from Equation 9.71 that the power of the DC link is negative if

$$\sum (\cos(\theta_{4n-3}) - \cos(\theta_{4n-2}) - \cos(\theta_{4n-1}) - \cos(\theta_{4n})) < 0, \quad n = 1, 2, \ldots. \tag{9.72}$$

When the inverter feeds the motor, the power of the DC link of the HB with the highest DC voltages is always positive. However, the power of the DC link of other HBs may be negative with a lower modulation index. Therefore, the bidirectional DC/DC converter is necessary here to transfer the regenerative power of the DC link back to the rectifier to avoid the increase of the DC-link voltage.

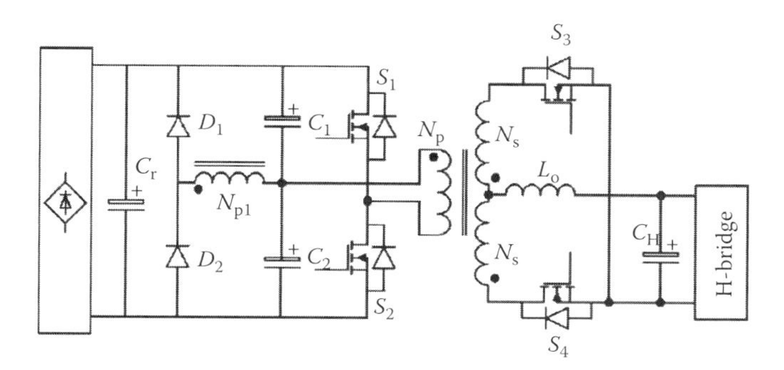

FIGURE 9.53 Bidirectional DC/DC converter.

The second reason is that the variation of the DC-link voltage of an HB is required to be very small. For example, the variation of the DC-link voltage of the HB with the DC-link voltage of $27E$ must be less than $0.5/27 = 0.019$. Otherwise, the contribution of the HB with a DC-link voltage of E for the power quality will be almost nothing. The DC/DC converters with high bandwidth closed-loop control can stabilize the DC-link voltages of the HBs.

The third reason is that the transformers used in the bidirectional converters are small, cost effective, and highly efficient. In other topologies of hybrid multilevel inverters for motor drives, the output ports of HBs are connected together by transformers. However, these low-frequency transformers are bulky and have low efficiency. Compared with the configurations with low-frequency transformers, the efficiency of the DC/DC converter is higher. The efficiency of the DC/DC converter measured in the low-power experiments is around 90%. In practical high-power applications, it can reach 97% [11], which is much higher than that of the traditional configuration of low-frequency transformers and rectifiers.

Several topologies of bidirectional DC/DC converters were proposed [12,13]. The topology of a bidirectional DC/DC converter [13] is used in the proposed system shown in Figure 9.53. The transformer provides galvanic isolation between the input and the output. The primary side of the converter is a half-bridge and is connected to the DC link of a rectifier. All DC/DC converters share a diode rectifier as shown in Figure 9.53. The secondary side, connected to the DC link of the HB, forms a current-fed push–pull. The converter has two modes of operation. In the forward mode, the DC link of an HB is powered by the DC link of the rectifier. In the backward mode, the DC link of an HB provides energy to the DC link of a rectifier.

The left part of Figure 9.54 shows the idealized waveforms in the forward mode: *Interval* $t_0 - t_1$: Switch S_2 is off and S_1 is on at time t_0. A voltage across the primary winding is $v_{Cr}/2$. The body diode of switch S_4, D_{S4}, is forward biased. The current flow through S_1, i_{S1} contributes to the linearly increasing inductor current and the transformer primary magnetizing current. *Interval* $t_1 - t_2$: Switch S_1 is turned off at time t_1 and S_2 remains on. No power is transferred to the secondary side during this dead-time interval since there is zero voltage across the primary side. The energy stored in L_o results in the freewheeling of the current i_{Lo}, equally through the body diodes D_{S3} and D_{S4}. *Interval* $t_2 - t_3$: Switch S_2 is turned on at time t_2 and S_1 remains off. The operation is similar to that during interval $t_0 - t_1$, but now D_{S3} conducts and provides secondary side rectification. Inductor current rises linearly again. *Interval* $t_3 - t_4$: Switch S_2 is turned off at time t_2 and S_1 remains off. The operation is similar to that in the interval $t_1 - t_2$. Figure 9.54 shows a balancing winding N_{p1} and two diodes D_1 and D_2 on the primary side of the half-bridge. They maintain the center point voltage at the junction of C_1 and C_2 to one-half of the input voltage and prevent a

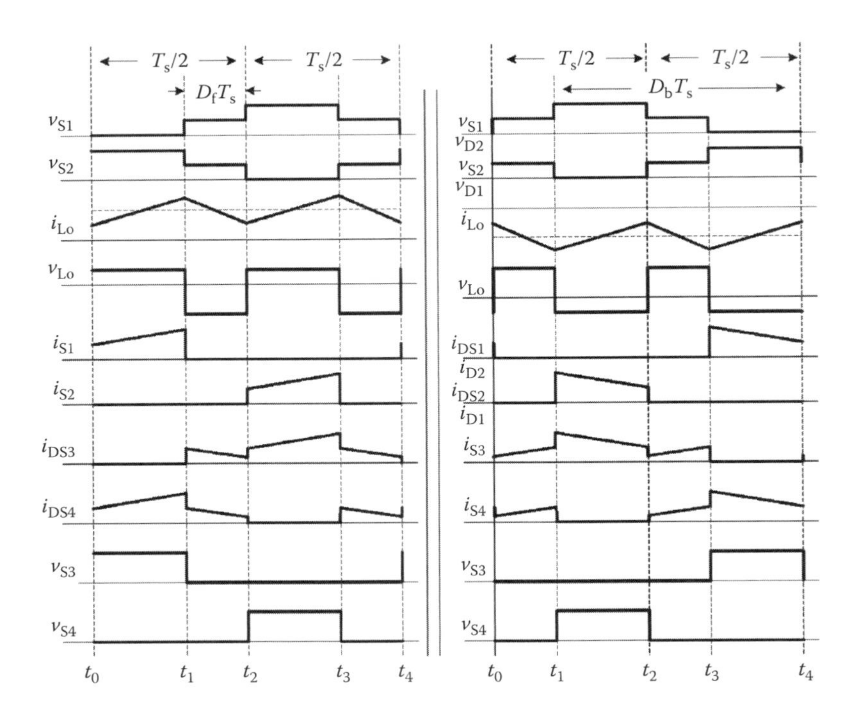

FIGURE 9.54 Waveforms of bidirectional DC/DC converter during the forward/backward mode.

runaway condition of a staircase situation of the transformer core. N_{p1} has the same number of turns as the winding N_p and is phrased in series with it through the on time of S_1 and S_2.

In the backward mode, the switches S_3 and S_4 of the current-fed push–pull topology are driven at duty ratios greater than 0.5. The converter operation during this mode is shown in the right part of Figure 9.54. *Interval* $t_0 - t_1$: Switch S_3 is turned on and S_4 remains on at time t_0. N_S is subject to a short circuit, which causes the inductor L_o to store energy as the DC-link voltage of the HB appears across it. i_{Lo} ramps up linearly and is shared equally by both S_3 and S_4. During this interval, C_1 and C_2 provide the output power. *Interval* $t_1 - t_2$: Switch S_4 is turned off and S_3 remains on at time t_1. The energy stored in the inductor during the previous interval is now transferred to the load through D_{S2} and D_1. Voltages across N_{p1} and N_p are identical due to their series phasing and equal number of turns. This allows simultaneous and equal charging of both C_1 and C_2 through D_1 and D_{S2}, respectively. *Interval* $t_2 - t_3$: Switch S_4 is turned on and S_3 remains on at time t_2. This interval is similar to the interval $t_0 - t_1$. The duty ratio for S_3 is therefore greater than 0.5. *Interval* $t_3 - t_4$: Switch S_3 is turned off and S_4 remains on at time t_3. The stored energy of L_o is transferred to the primary side of the converter through S_4, D_{S1}, and D_2. The conduction of D_{S1} and D_2 results in equal charging of C_1 and C_2, respectively. Current-mode control is used for both modes of converter operations. Small signal analyses for both modes under mode control is performed to generate the transfer functions to design and evaluate the control loop [13].

9.5.4.3 Motor Controller

The proposed multilevel inverter is used to feed an induction motor. The vector control technique is applied to the motor controller. Vector control implies independent control of flux-current and torque/current components of the stator current through a coordinated

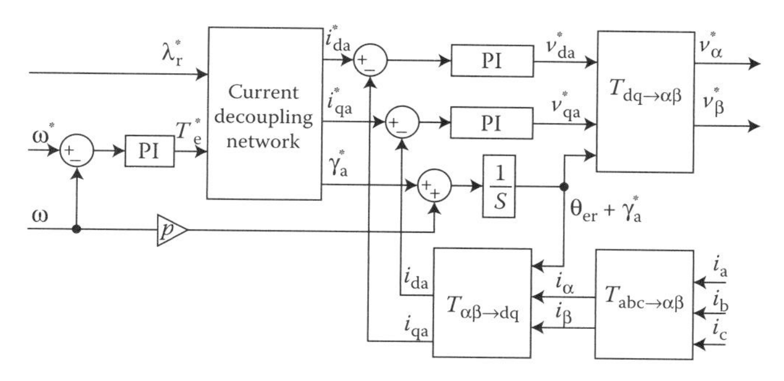

FIGURE 9.55 Motor controller.

change in the supply voltage amplitude, phase, and frequency. As the flux variation tends to be slow, constancy of flux should produce a fast torque/current response and finally a fast speed (position) response.

The controller is shown in Figure 9.55 and the current decoupling network in the controller is shown in Figure 9.56. To simplify the current decoupling network, the rotor flux orientation is used in the current decoupling network. Once the reference $d-q$ current i_{da}^*, i_{qa}^* and flux orientation angle $\theta_{er}+\gamma_a^*$ are known, the DC current controllers are used to translate these commands to v_{da}^* and v_{qa}^*, and use Park transformation to translate v_{da}^* and v_{qa}^* to v_α^* and v_β^*. The output signal of the motor controller, v_α^* and v_β^*, will be sent to the inverter controller to control the multilevel inverter to provide the appropriate voltages to feed the motor.

9.5.4.4 Simulation and Experimental Results

The performance of the 81-level THMI for the motor drive presented above has been verified by simulation. The simulation investigations were performed with MATLAB®/Simulink®. The unit voltage of the multilevel inverter, E, is set as 10 V. The modulation index is defined as

$$m = \frac{\pi|v_{an}|_1}{4 \times 40E}, \tag{9.73}$$

where $|v_{an}|_1$ is the fundamental amplitude of the output voltage. Based on the simulation results, the relationship between $|v_{an}|_1$ and the modulation index is shown in Figure 9.57.

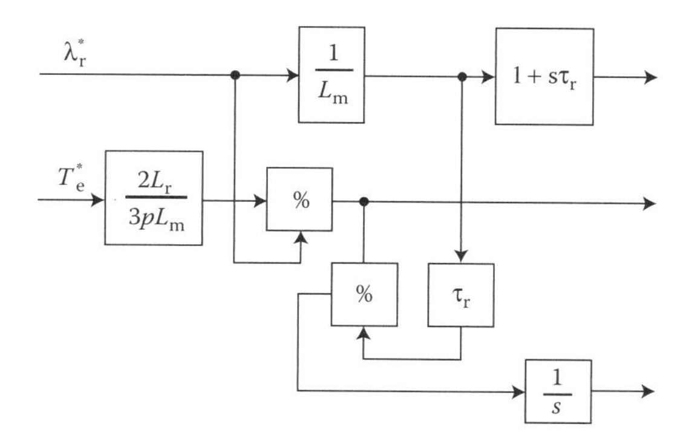

FIGURE 9.56 Current decoupling network.

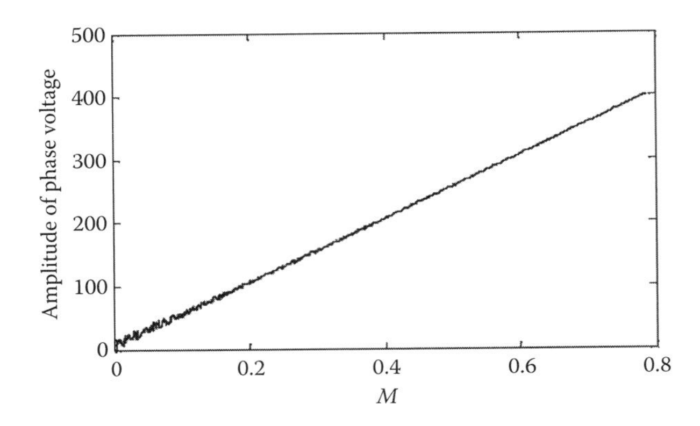

FIGURE 9.57 Amplitude of phase voltage versus modulation index.

In the range of very low modulation index, it does not have a very good linear relationship. However, due to a great number of voltage steps, the relationship becomes satisfied linearly with higher modulation index.

When the inverter drives an induction motor, a command of speed step changes from 715 to 1430 rmp in 1 ms. Figure 9.58 shows the simulation results of speed, output voltage of the inverter, output current of the inverter, DC-link voltages of HBs in the A-phase, and common-mode voltages. The speed has a rapid response. The common-mode voltage is always zero except during the short transition time. The THD of the output voltage is as low as 1%. Figure 9.59 shows the detailed waveforms of the output voltage of inverters. Figure 9.60 shows the simulation results of torque, output voltages, and output currents of

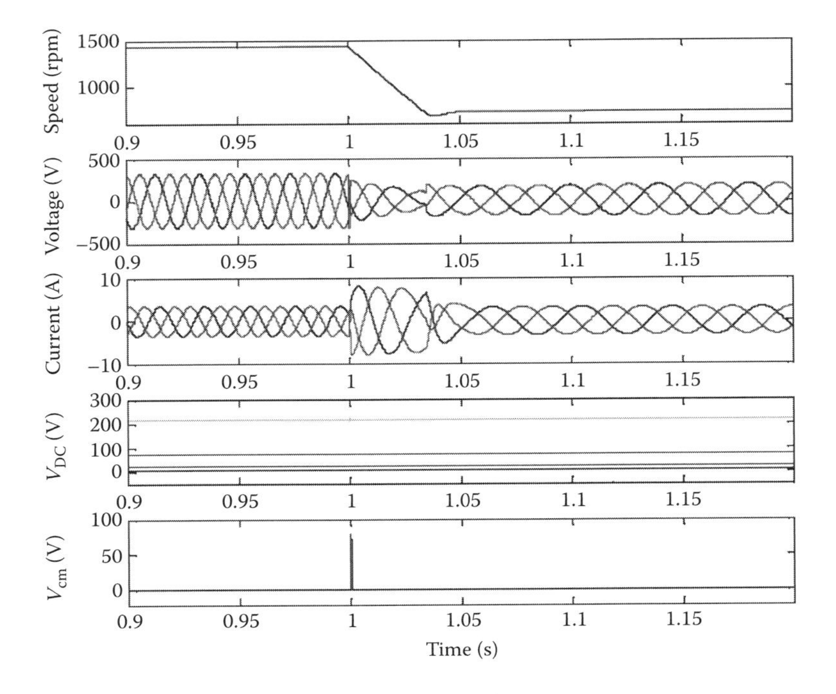

FIGURE 9.58 Simulation waveforms for a step change of speed.

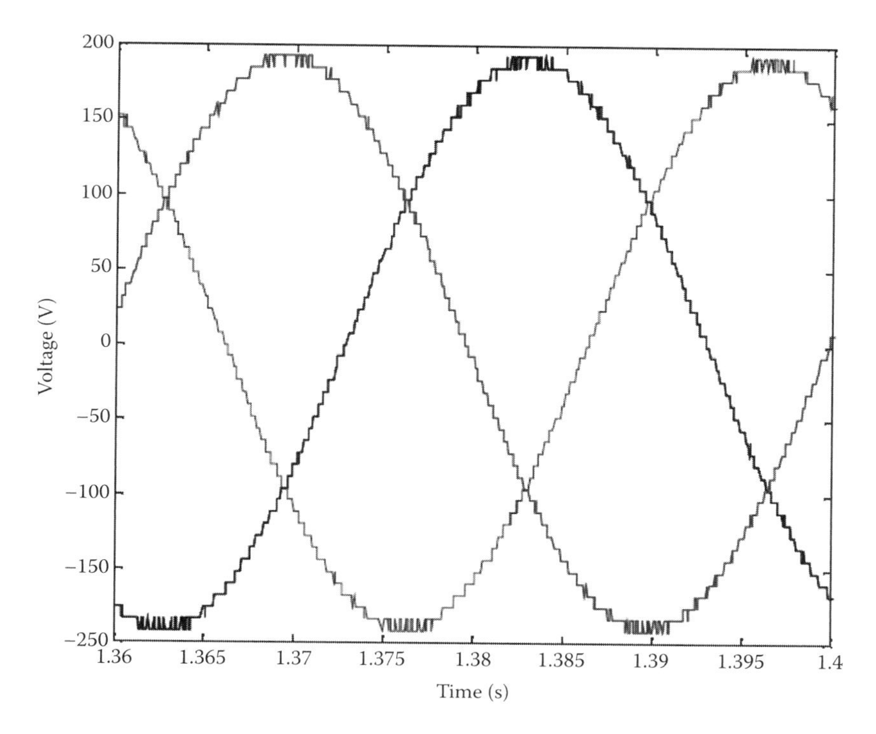

FIGURE 9.59 Simulation waveforms of the output voltages of the inverter.

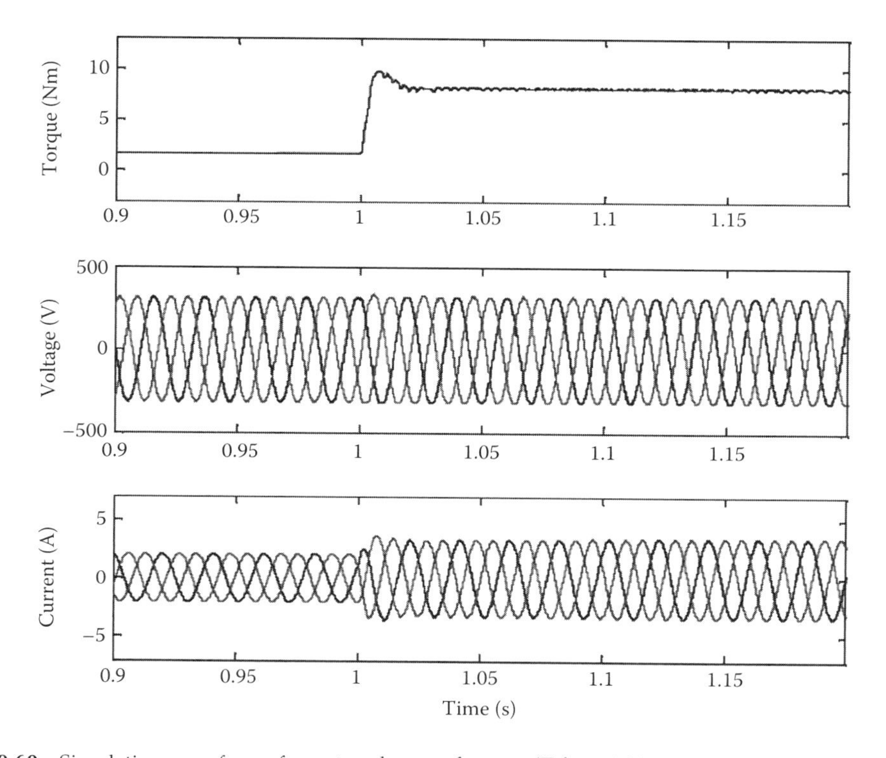

FIGURE 9.60 Simulation waveforms for a step change of torque (T from 1.29 to 7.74 Nm).

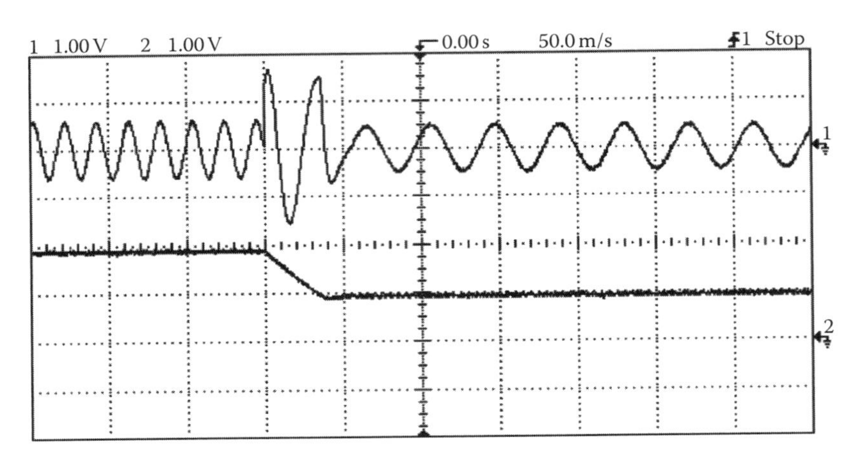

FIGURE 9.61 Experiment waveforms for a step change of speed. CH1: speed (750 rad/s/div); CH2: phase current (2 A/div).

the inverter, when the reference torque has a step change from 1.29 to 7.74 Nm. The motor drive system also has a good dynamic response for the step change of torque.

To verify the performance of the proposed inverter experimentally, a hardware prototype has been built in the laboratory. The experimental setup of the proposed control system consists of a three-phase, 380 V, 50 Hz, 4 pole, 3-kW induction motor and a power circuit using a trinary hybrid multilevel inverter. The inverter and the motor are controlled using TMS320F240 controller cards. Current-mode controller of the DC/DC converters is implemented by UC 3846 and UCC 3804, for the forward mode and backward mode, respectively. Figures 9.61 and 9.62 show the waveforms of speed, phase current, phase voltage, and line-to-line voltage when the reference speed of the motor has a step change, which verify the simulation results as shown in Figure 9.58. Figure 9.63 shows the detailed waveforms of phase voltage and common-mode voltage. As shown in Figure 9.63, the phase voltage is synthesized by many stable step voltages and the common-mode voltage is almost zero.

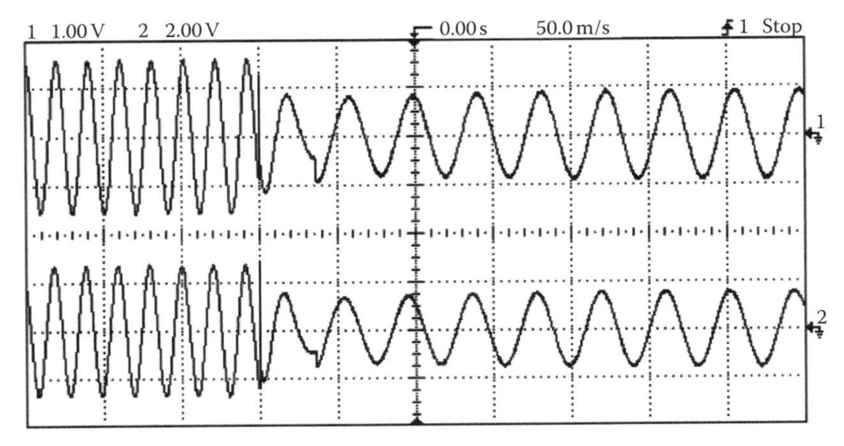

FIGURE 9.62 Experiment waveforms for a step change of speed. CH1: phase voltage (200 V/div); CH2: line-to-line voltage (400 V/div).

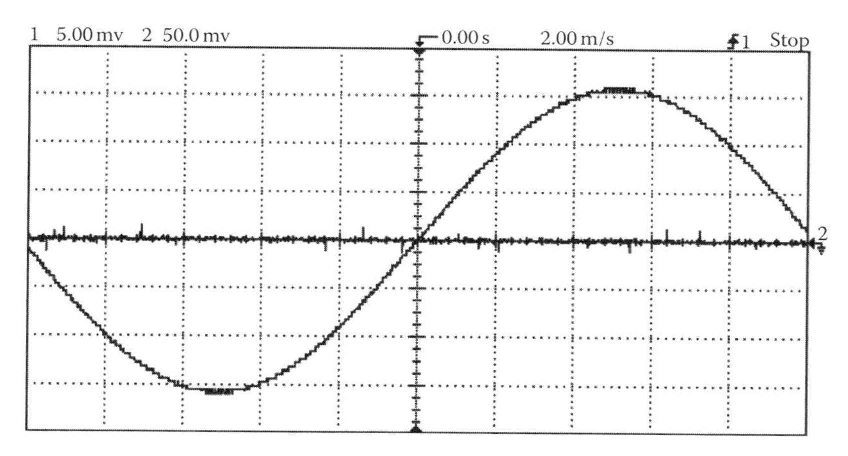

FIGURE 9.63 Experiment detailed waveforms. CH1: phase voltage (100 V/div); CH2: common-mode voltage (20 V/div).

9.6 Other Kinds of Multilevel Inverters

Several other kinds of multilevel inverters are introduced in this subsection [14].

9.6.1 Generalized Multilevel Inverters

A GMI topology has been presented previously. The existing multilevel inverters, such as DCMIs and CCMIs, can be derived from this GMI topology. Moreover, the GMI topology can balance each voltage level by itself, regardless of load characteristics. Therefore, the GMI topology provides a true multilevel structure that can balance each DC voltage level automatically at any number of levels, regardless of active or reactive power conversion, and without any assistance from other circuits. Thus, in principle, it provides a complete multilevel topology that embraces the existing multilevel inverters.

Figure 9.64 shows the GMI structure per phase leg. Each switching device, diode, or capacitor's voltage is E, that is, $1/(m-1)$ of the DC-link voltage. Any inverter with any number of levels, including the conventional two-level inverter, can be obtained using this generalized topology.

As an application example, a four-level bidirectional DC/DC converter, shown in Figure 9.65, is suitable for the dual-voltage system to be adopted in future automobiles. The four-level DC/DC converter has a unique feature, which is that no magnetic components are needed. From this GMI inverter topology, several new multilevel inverter structures can be derived.

9.6.2 Mixed-Level Multilevel Inverter Topologies

For high-voltage high-power applications, it is possible to adopt multilevel diode-clamped or capacitor-clamped inverters to replace the full-bridge cell in a CMI. The reason for doing so is to reduce the amount of separate DC sources. The nine-level cascaded inverter requires four separate DC sources for one phase leg and 12 for a three-phase inverter. If a three-level

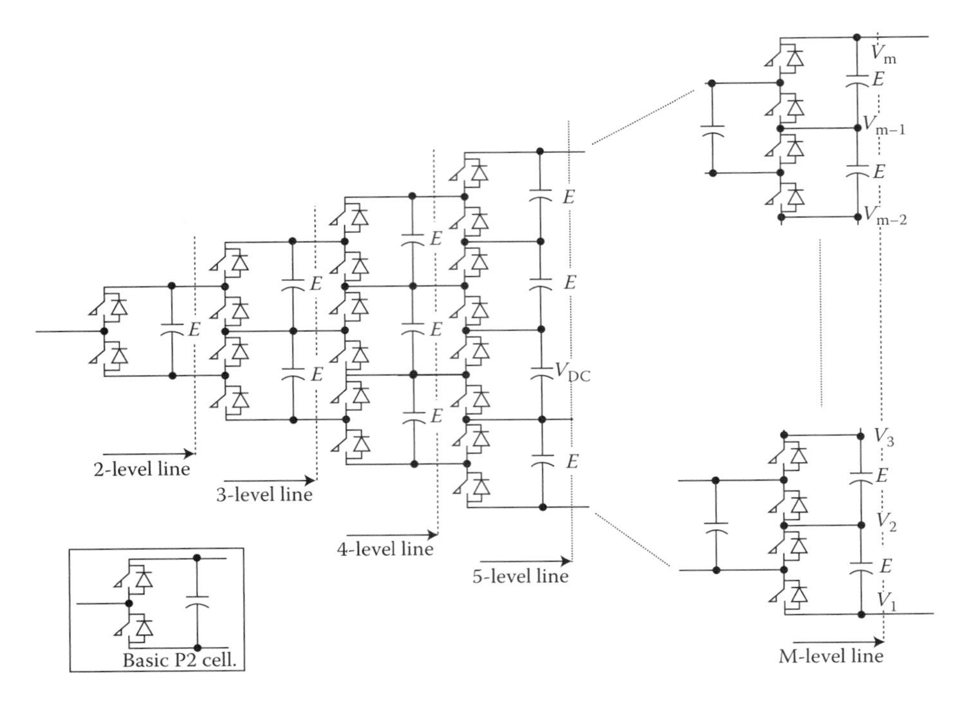

FIGURE 9.64 GMI structure.

inverter replaces the full-bridge cell, the voltage level is effectively doubled for each cell. Thus, to achieve the same nine voltage levels for each phase, only two separate DC sources are needed for one phase leg and six for a three-phase inverter. The configuration can be considered as having mixed-level multilevel cells because it embeds multilevel cells as the building block of the CMI.

9.6.3 Multilevel Inverters by Connection of Three-Phase Two-Level Inverters

Standard three-phase two-level inverters are connected by transformers as shown in Figure 9.66. In order for the inverter output voltages to be added up, the inverter outputs of

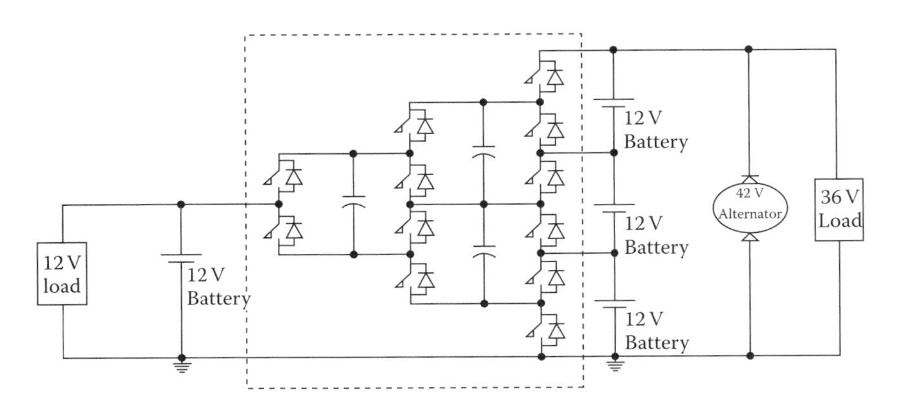

FIGURE 9.65 Application example: a four-level inverter for the dual-voltage system in automobiles.

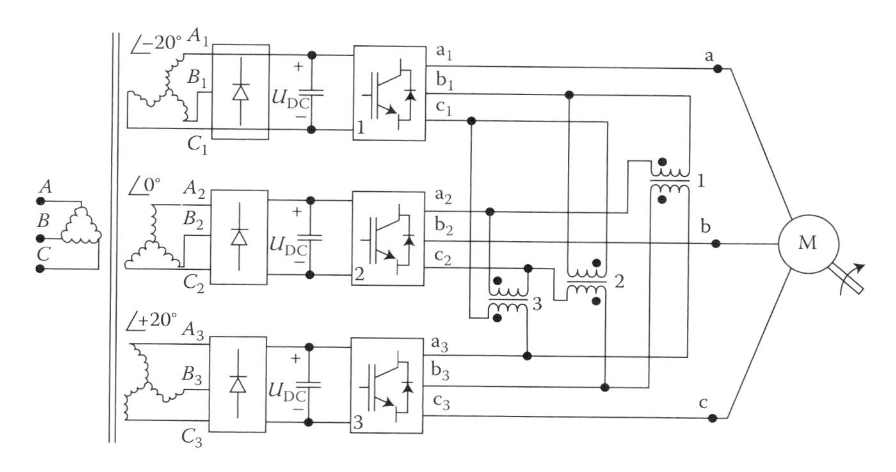

FIGURE 9.66 Cascaded inverter with three-phase cells.

the three modules need to be synchronized with a separation of 120° between each phase. For example, obtaining a three-level voltage between outputs a and b, the voltage is synthesized by $V_{ab} = V_{a1-b1} + V_{a1-b1} + V_{a1-b1}$. The phase between b_1 and a_2 is provided by a_3 and b_3 through an isolated transformer. With three inverters synchronized, the voltages V_{a1-b1}, V_{a1-b1}, and V_{a1-b1} are all in phase; thus, the output level is simply tripled.

9.7 Soft-Switching Multilevel Inverters

There are numerous ways of implementing soft-switching methods, such as ZVS and ZCS, to reduce the switching losses and to increase efficiency for different multilevel inverters. For the CMI, because each inverter cell is a two-level circuit, the implementation of soft switching is not at all different from that of conventional two-level inverters. For capacitor- or diode-clamped inverters, however, the choices of soft-switching circuits can be found with different circuit combinations. Although ZCS is possible, most literature works proposed ZVS types including the auxiliary resonant commutated pole (ARCP), the coupled inductor with zero-voltage transition (ZVT), and their combinations.

9.7.1 Notched DC-Link Inverters for Brushless DC Motor Drive

The brushless DC motor (BDCM) has been widely used in industrial applications because of its low inertia, fast response, high power density, and high reliability and because it is maintenance free. It exhibits the operating characteristics of a conventional commutated DC permanent magnet motor, but eliminates the mechanical commutator and brushes. Hence many problems associated with brushes are eliminated such as radio-frequency interference and sparking, which is the potential source of ignition in the inflammable atmosphere. It is usually supplied by a hard-switching PWM inverter, which normally has low efficiency since the power losses across the switching devices are high. In order to reduce the losses, many soft-switching inverters have been designed [15].

The soft-switching operation of the power inverter has attracted much attention in the recent decade. In electric motor drive applications, soft-switching inverters are usually classified into three categories, namely resonant pole inverters, resonant DC-link inverters, and resonant AC-link inverters [16]. The resonant pole inverter has the disadvantage of containing a considerably large number of additional components, in comparison with other hard- and soft-switching inverter topologies. The resonant AC-link inverter is not suitable for BDCM drivers.

In medium-power applications, the resonant DC-link concept [17] offered the first practical and reliable way to reduce commutation losses and to eliminate individual snubbers. Thus, it allows high operating frequencies and improved efficiency. The inverter is quite simple enough to get the ZVS condition of the six main switches only by adding one auxiliary switch. However, the inverter has the drawbacks of high voltage stress of the switches, high voltage ripple of the DC link, and the frequency of the inverter relating to the resonant frequency. Furthermore, the inductor power losses of the inverter are also considerable as current flows in the inductor always. In order to overcome the drawbacks of high-voltage stress of the switches, an actively clamped resonant DC-link inverter was introduced [18–21]. The control scheme of the inverter is too complex and the output contains subharmonics that, in some cases, cannot be accepted. These inverters still do not overcome the drawbacks of high inductor power losses.

In order to generate voltage notches of the DC link at controllable instants and reduce the power losses of the inductor, several quasi-parallel resonant schemes were proposed [22–24]. As a dwell time is generally required after every notch, severe interferences occur, mainly in multiphase inverters, appreciably worsening the modulation quality. A novel DC-rail parallel resonant ZVT voltage source inverter [25] is introduced; it overcomes the many drawbacks mentioned above. However, it requires two ZVTs per PWM cycle; it would worsen the output and limit the switch frequency of the inverter.

On the other hand, the majority of soft-switching inverters proposed in recent years have been aimed at the induction motor drive applications. So it is necessary to study the novel topology of the soft-switching inverter and the special control circuit for BDCM drive systems. This chapter proposed a novel resonant DC-link inverter for the BDCM drive system that can generate voltage notches of the DC link at controllable instant and width. And the inverter possesses the advantage of low switching power loss, low inductor power loss, low voltage ripple of the DC link, low device voltage stress, and a simple control scheme.

The construction of the soft-switching inverter is shown in Figure 9.67. There is a front uncontrolled rectifier to obtain DC supply. The input AC supply can be single phase for low/medium power or three phases for medium/high power. It contains a resonant circuit, a conventional circuit and a control circuit. The resonant circuit contains three auxiliary switches (one IGBT and two fast switching thyristors), a resonant inductor, and a resonant capacitor. All auxiliary switches work under the ZVS or ZCS condition. This generates voltage notches of the DC link to guarantee that the main switches S_1–S_6 of the inverter operate in the ZVS condition. The fast switching thyristor is the proper device for use as an auxiliary switch. We need not control the turn off of a thyristor and it has higher surge current capability than any other power semiconductor switche.

9.7.1.1 Resonant Circuit

The resonant circuit consists of three auxiliary switches, one resonant inductor, and one resonant capacitor. The auxiliary switches are controlled at a certain instant to obtain the

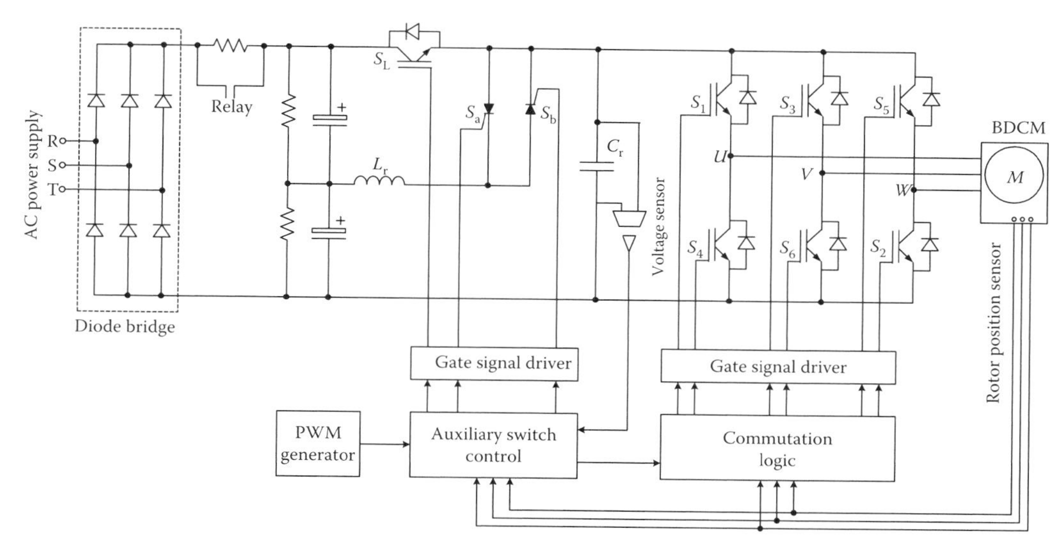

FIGURE 9.67 Construction of the soft-switching for BDCM drive system.

resonance between the inductor and the capacitor. Thus, the voltage of the DC link reaches zero temporarily (voltage notch) and the main switches of the inverter get ZVS condition for commutation.

Since the resonant process is very short, the load current can be supposed to be constant. The equivalent circuit is shown in Figure 9.68. The corresponding waveforms of the auxiliary switches gate signal, resonant capacitor voltage (u_{Cr}), inductor current (i_{Lr}), and current of switch S_L (i_{SL}) are illustrated in Figure 9.69. The operation of the ZVT process can be divided into six modes.

Mode 0 (as shown in Figure 9.70a) $0 < t < t_0$. Its operation is the same as that of the conventional inverter. Current flows from the DC source through S_L to the load. The voltage across C_r (u_{Cr}) is equal to the voltage of the supply (V_s). The auxiliary switches S_a and S_b are in the off state.

Mode 1 (as shown in Figure 9.70b) $t_0 < t < t_1$. When it is the instant for phase current commutation or the PWM signal is flopped from "1" to "0," the thyristor S_a is fired (ZCS turn on due to L_r) and IGBT S_L is turned off (ZVS turn off due to C_r) at the same time. The capacitor C_r resonates with inductor L_r and the voltage across capacitor C_r is decreased.

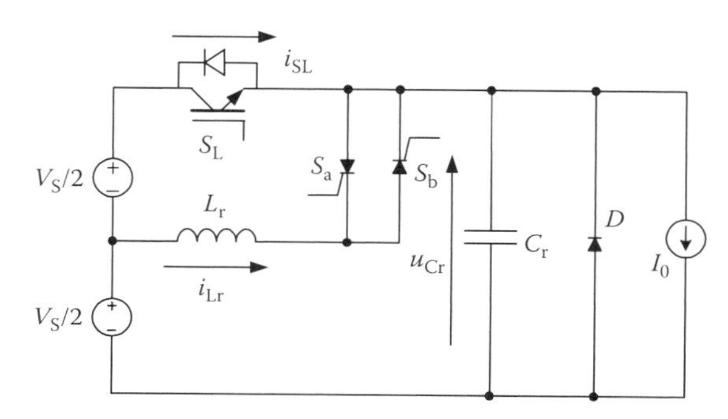

FIGURE 9.68 Equivalent circuit.

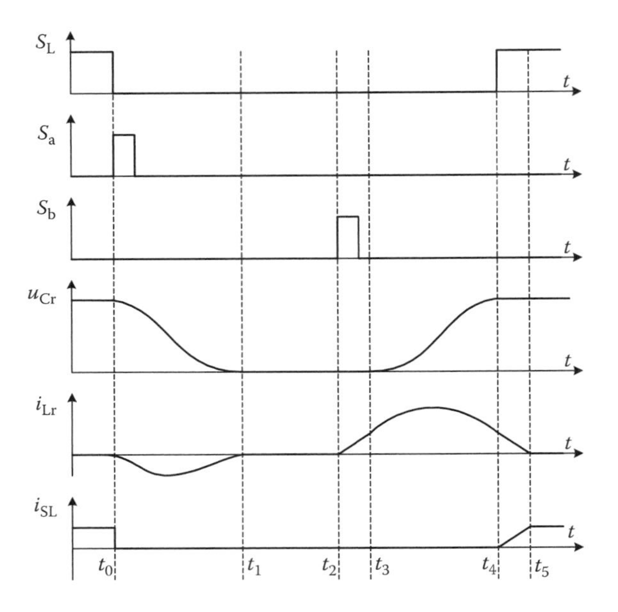

FIGURE 9.69 Some waveforms of the equivalent circuit.

Redefining the initial time, we have

$$\begin{aligned} u_{Cr}(t) + R_{Lr}i_{Lr}(t) + L_r\frac{di_{Lr}(t)}{dt} &= \frac{V_S}{2}, \\ I_O - i_{Lr}(t) + C_r\frac{du_{Cr}(t)}{dt} &= 0, \end{aligned} \tag{9.74}$$

where R_{Lr} is the resistance of the inductor L_r, I_O is the load current, V_S is the DC power supply voltage, with the initial conditions $u_{Cr}(0) = V_S$ and $i_{Lr}(0) = 0$. Solving Equation 9.74, we obtain

$$\begin{aligned} u_{Cr}(t) &= \left(\frac{V_S}{2} - R_{Lr}I_O\right) + \left(\frac{V_S}{2} - R_{Lr}I_O\right)e^{-t/\tau}\cos(\omega t) \\ &\quad + \frac{1}{L_rC_r\omega}e^{-t/\tau}\left(\frac{1}{4}R_{Lr}C_rV_S - L_rI_O + \frac{1}{2}R_{Lr}^2C_rI_O\right)\sin(\omega t), \\ i_{Lr}(t) &= I_O - I_Oe^{-t/\tau}\cos(\omega t) - \frac{V_S + R_{Lr}I_O}{2L_r\omega}e^{-t/\tau}\sin(\omega t), \end{aligned} \tag{9.75}$$

where

$$\tau = \frac{2L_r}{R_{Lr}}, \qquad \omega = \sqrt{\frac{1}{L_rC_r} - \frac{1}{\tau^2}}.$$

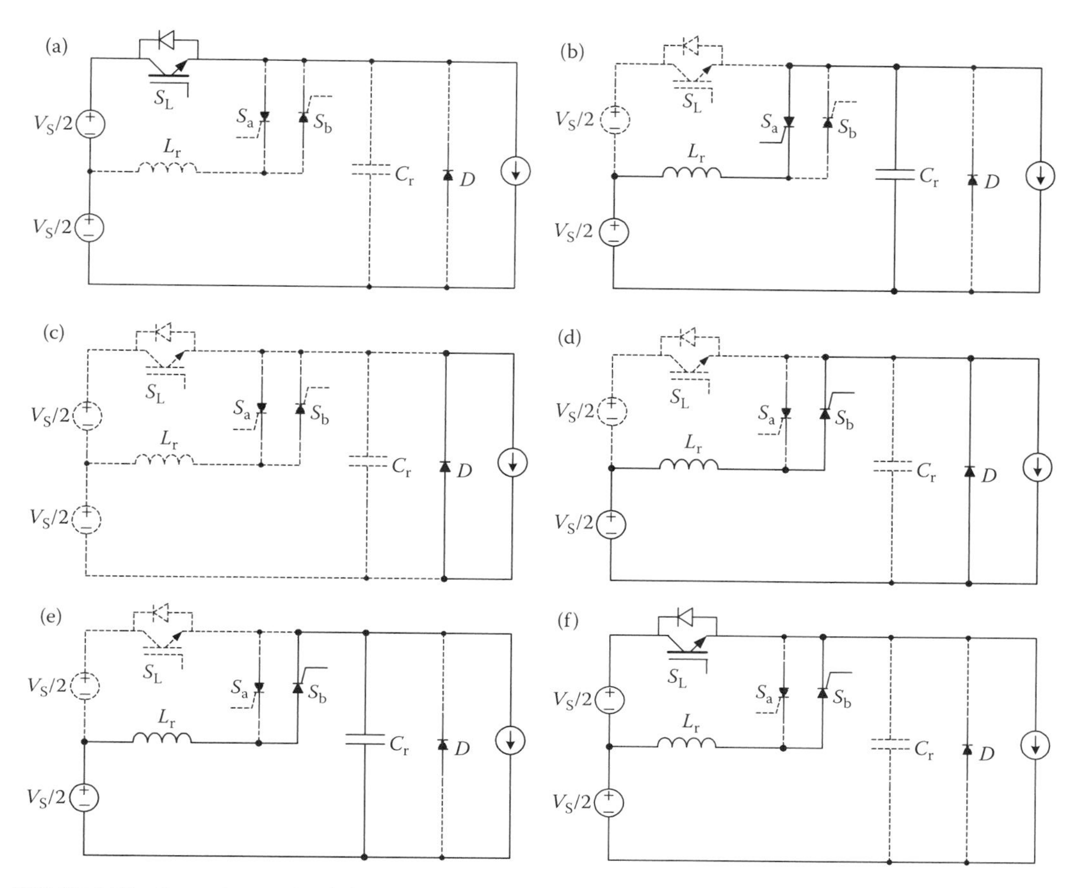

FIGURE 9.70 Operation mode of the ZVS process: (a) mode 0, (b) mode 1, (c) mode 2, (d) mode 3, (e) mode 4, and (f) mode 5.

As the resonant frequency is very high (several hundreds of kHz), $\omega L_r \gg R_{Lr}$, resonant inductor resistance R_{Lr} can be neglected. Then Equation 9.75 can be simplified as

$$
\begin{aligned}
u_{Cr}(t) &= \frac{V_S}{2} - I_O\sqrt{\frac{L_r}{C_r}}\sin\left(\frac{1}{\sqrt{L_rC_r}}t\right) + \frac{V_S}{2}\cos\left(\frac{1}{\sqrt{L_rC_r}}t\right), \\
i_{Lr}(t) &= I_O - I_O\cos\left(\frac{1}{\sqrt{L_rC_r}}t\right) - \frac{V_S}{2}\sqrt{\frac{C_r}{L_r}}\sin\left(\frac{1}{\sqrt{L_rC_r}}t\right),
\end{aligned}
\tag{9.76}
$$

that is,

$$
\begin{aligned}
u_{Cr}(t) &= \frac{V_S}{2} + K\cos(\omega_r t + \alpha), \\
i_{Lr}(t) &= I_O - K\sqrt{\frac{C_r}{L_r}}\sin(\omega_r t + \alpha),
\end{aligned}
\tag{9.77}
$$

where

$$K = \sqrt{\frac{V_S^2}{4} + \frac{I_O^2 L_r}{C_r}}, \quad \omega_r = \sqrt{\frac{1}{L_r C_r}}, \quad \alpha = tg^{-1}\left(\frac{2I_O}{V_S}\sqrt{\frac{L_r}{C_r}}\right).$$

Let $u_{Cr}(t) = 0$; then we obtain

$$\Delta T_1 = t_1 - t_0 = \frac{\pi - 2\alpha}{\omega_r}. \tag{9.78}$$

$i_{Lr}(t)$ is zero at $t = t_1$. Then the thyristor S_a is self-turned-off.

Mode 2 (as shown in Figure 9.70c) $t_1 < t < t_2$. None of the auxiliary switches is fired and the voltage of the DC link (u_{Cr}) is zero. The main switches of the inverter can now be either turned on or turned off under ZVS condition during the interval. The load current flows through the freewheeling diode D.

Mode 3 (as shown in Figure 9.70d) $t_2 < t < t_3$. As the main switches have turned on or turned off, the thyristor S_b is fired (ZCS turn on due to L_r) and i_{Lr} starts to build up linearly in the auxiliary branch. The current in the freewheeling diode D begins to fall linearly. The load current is slowly diverted from the freewheeling diodes to the resonant branch. But u_{Cr} is still equal to zero. We have

$$\Delta T_2 = t_3 - t_2 = \frac{2I_O L_r}{V_S}. \tag{9.79}$$

At t_3, i_{Lr} equals the load current I_O and the current through the diode becomes zero. Thus the freewheeling diode turns off under zero-current condition.

Mode 4 (as shown in Figure 9.70e) $t_3 < t < t_4$. i_{Lr} is increased continuously from I_O and u_{Cr} is increased from zero when the freewheeling diode D is turned off. Redefining the initial time, we obtain the same equation as Equation 9.74. But the initial conditions are $u_{Cr}(0) = 0$ and $i_{Lr}(0) = I_O$; neglecting the inductor resistance and solving the equation; we obtain

$$\begin{aligned} u_{Cr}(t) &= \frac{V_S}{2} - \frac{V_S}{2}\cos\left(\frac{1}{\sqrt{L_r C_r}}t\right), \\ i_{Lr}(t) &= I_O + \frac{V_S}{2}\sqrt{\frac{C_r}{L_r}}\sin\left(\frac{1}{\sqrt{L_r C_r}}t\right), \end{aligned} \tag{9.80}$$

that is,

$$\begin{aligned} u_{Cr}(t) &= \frac{V_S}{2}[1 - \cos(\omega_r t)], \\ i_{Lr}(t) &= I_O + \frac{V_S}{2}\sqrt{\frac{C_r}{L_r}}\sin(\omega_r t). \end{aligned} \tag{9.81}$$

When

$$\Delta T = t_4 - t_3 = \frac{\pi}{\omega_r}, \tag{9.82}$$

$u_{Cr} = E$, IGBT S_L is fired (ZVS turn on), and $i_{Lr} = I_O$ again. The peak inductor current can be derived from Equation 9.81, that is,

$$i_{Lr-m} = I_O + \frac{V_S}{2}\sqrt{\frac{C_r}{L_r}}. \quad (9.83)$$

Mode 5 (as shown in Figure 9.70f) $t_4 < t < t_5$. When the DC-link voltage is equal to the supply voltage, the auxiliary switch S_L is turned on (ZVS turned on due to C_r). i_{Lr} is decreased linearly from I_O to zero at t_5 and the thyristor S_b is self-turned-off.

Then go back to mode 0 again. The operation principle of the other procedure is the same as that of a conventional inverter.

9.7.1.2 Design Consideration

The design of the resonant circuit is to determine the resonant capacitor C_r, the resonant inductor L_r, and the switching instants of the auxiliary switches S_a, S_b, and S_L. It is assumed that the inductance of BDCM is much higher than resonant inductance L_r. From the analysis presented previously, the design considerations can be summarized as follows:

The auxiliary switch S_L works under ZVS condition, the voltage stress is DC power supply voltage V_S. The current flow through it is load current. The auxiliary switches S_a and S_b work under the ZCS condition, the voltage stress is $V_S/2$ and the peak current flow through them is i_{Lr-m}. As the resonant auxiliary switches S_a and S_b carry the peak current only during switch transitions, they can be rated as lower continuous currents.

The resonant period is expressed as $T_r = 1/f_r = 2\pi\sqrt{L_rC_r}$; for high switching frequency inverters, T_r should be as short as possible. For getting the expected T_r, the resonant inductor and capacitor values have to be selected. The first component to be designed is the resonant inductor. Small inductance values can yield small T_r, but the rising slope of the inductor current $\mathrm{d}i_{Lr}/\mathrm{d}t = V_S/2L_r$ should be small to guarantee that the freewheeling diode turns off. For the 600–1200 V power diode, the reverse recovery time is about 50–200 ns, and the rule to select an inductor is [11]

$$\frac{\mathrm{d}i_{Lr}}{\mathrm{d}t} = \frac{V_S}{2L_r} = 75\text{–}150\,\mathrm{A}/\mu\mathrm{s}. \quad (9.84)$$

Certainly inductance is as high as possible. This implies that a high inductance value is necessary. Thus an optimum value of the inductance has to be chosen that would reduce the inductor current rise slope, while T_r would be small enough.

The capacitance value is inversely proportional to the ascending or descending slope of the DC-link voltage. It means that capacitance is as high as possible for the switch S_L to get the ZVS condition, but as the capacitance increases, more and more energy gets stored in it. This energy should be charged or discharged via the resonant inductor; with high capacitance, the peak value of the inductor current will be high. The peak value of i_{Lr} should be limited to twice the peak load current. From Equations 9.76 through 9.83, we obtain

$$\sqrt{\frac{C_r}{L_r}} \le \frac{2I_{O\max}}{V_S}. \quad (9.85)$$

Thus an optimum value of the capacitance has to be chosen that would limit the peak inductor current, while the ascending or descending slope of the DC-link voltage is low enough.

9.7.1.3 Control Scheme

When the duty of PWM is 100%, that is, when there is no PWM, the main switches of the inverter work under commutation frequency. When it is the instant to commutate the phase current of the BDCM, we control the auxiliary switches S_a, S_b, and S_L and resonance occurs between L_r and C_r. The voltage of the DC link reaches zero temporarily; thus the ZVS condition of the main switches is obtained. When the duty of PWM is less than 100%, the auxiliary switch S_L works as a chop. The main switches of the inverter do not switch within a PWM cycle when the phase current does not need to commutate. It has the benefit of reducing the phase current drop when the PWM is off. The phase current is commutated when the DC-link voltage becomes zero. So there is only one DC-link voltage notch per PWM cycle. It is very important especially for very low or very high duty of PWM where the interval between two voltage notches is very short, even overlapping, which will limit the tuning range.

The commutation logical circuit of the system is shown in Figure 9.71. It is similar to the conventional BDCM commutation logical circuit except for adding six D flip-flops to the output. Thus the gate signal of the main switches is controlled by the synchronous pulse CK that will be mentioned later, and the commutation can be synchronized with the auxiliary switches control circuit. The operation of the inverter can be divided into the PWM operation and non-PWM operation.

1. *Non-PWM operation*

When the duty of PWM is 100%, that is, when there is no PWM, the whole ZVT process (modes 1 through 5) occurs when the phase current commutation is ongoing. The control scheme for the auxiliary switches in this operation is illustrated in Figure 9.72a. When mode 1 begins, the pulse signal for the thyristor S_a is generated by a monostable flip-flop and the gate signal for IGBT S_L is decreased to a low level (i.e., turn off the S_L) at the

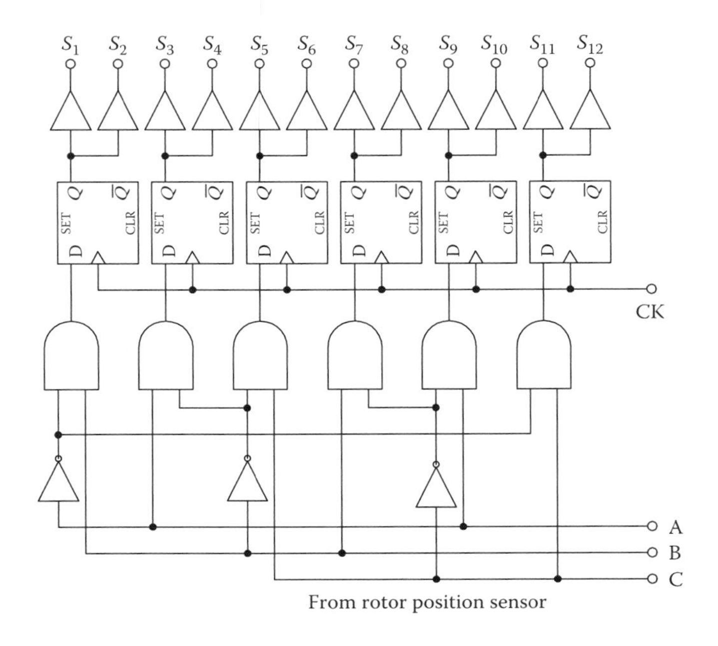

FIGURE 9.71 Commutation logical circuit for main switches.

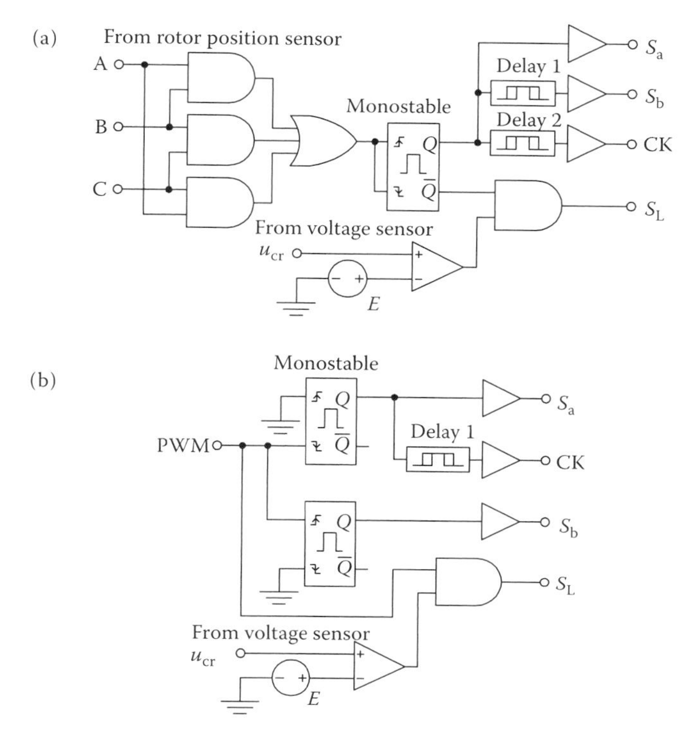

FIGURE 9.72 Control scheme for the auxiliary switches in (a) non-PWM operation and (b) PWM operation.

same time. Then, the pulse signal for the thyristor S_b and the synchronous pulse CK can be obtained after two short delays (delay1 and delay2, respectively). Obviously delay1 is longer than delay 2. Pulse CK is generated during mode 2 when the voltage of the DC link is zero and the main switches of the inverter get the ZVS condition. Then modes 3 through 5 occur and the voltage of the DC link is increased to that of the supply again.

2. *PWM operation*

In this operation, the auxiliary switch S_L works as a chop, but the main switches of the inverter do not turn on or turn off within a single PWM cycle when the phase current does not need to commutate. The load current is commutated when the DC-link voltage becomes zero, that is, when the PWM signal is "0" (as the PWM cycle is very short, it does not affect the operation of the motor). The control scheme for the auxiliary switches in PWM operation is illustrated in Figure 9.72b.

- When the PWM signal is flopped from "1" to "0," mode 1 begins, the pulse signal for the thyristor S_a is generated, and the gate signal for IGBT S_L is decreased to a low level. But the voltage of the DC link does not increase until the PWM signal is flipped from "0" to "1." Pulse CK is generated during mode 2.
- When the PWM signal is flipped from "0" to "1," mode 3 begins, and the pulse signal for the thyristor S_b is generated at the moment (mode 3). Then, when the voltage of the DC link is increased to E (the voltage of the supply), the gate signal for IGBT S_L is flipped to a high level (modes 4 and 5).

Thus, only one ZVT occurs per PWM cycle: modes 1 and 2 for PWM turned off and modes 3 through 5 for PWM turned on. And the switching frequency would not be greater than the PWM frequency.

Normally, a drive system requires a speed or position feedback signal to get high speed or position precision and to be less susceptible to disturbances of load and power supply. The speed feedback signal can be derived from a tachometer-generator, an optical encoder, a resolver or a rotor position sensor. Quadrature encoder pulse (QEP) is a standard digital speed or position signal and can be inputted to many devices (e.g., the special DSP for the drive system TMS320C24x has a QEP receive circuit). The QEP can be easily derived from the rotor position sensor of a BDCM. The converter digital circuit and interesting waveforms are shown in Figure 9.73. Some single-chip computers have a digital counter and may require only direction and pulse signals thus the converter circuit can be simpler. The circuit can be implemented by a complex programmer logical device and can only occupy the partial resources of one chip. The circuit can also be implemented by a gate array logic (GAL) IC (e.g., 16V8) and some D flip-flop IC (e.g., 74LS74). With the circuit, a high-precision speed or position signal can be obtained when the motor speed is high or the drive system has a high-ratio speed reduction mechanism. In high-performance systems, the rotor position sensor may be a resolver or optical encoder, with special-purpose decoding circuitry. At this level of control sophistication, it is possible to fine-tune the firing angles and the PWM control as a function of speed and load, to improve various aspects of performance such as efficiency, dynamic performance, or speed range.

9.7.1.4 Simulation and Experimental Results

The proposed topology is verified by Psim simulation software. The schematic circuit of the soft-switching inverter is shown in Figure 9.74. The left bottom of the figure shows the auxiliary switches gate signal generator circuit (see Figure 9.72), which is made up of monostable flip-flop, delay, and logical gate. The gate signals of auxiliary switches S_a and S_b in PWM and non-PWM operation modes are combined by the OR gate. The gate signal of S_L in the two operation modes is combined by the AND gate, and the synchronous signal (CK) is combined by a date selector. The middle bottom of the diagram shows the commutation logical circuit of the BDCM (see Figure 9.71); it is synchronized (by CK) with the auxiliary switches control circuit.

Waveforms of the DC-link voltage u_{Cr}, resonant inductor current i_{Lr}, BDCM phase current, inverter output line–line voltage, and gate signal of the auxiliary switches are shown in Figure 9.75. The value of the resonant inductor L_r is 10 μH and the resonant capacitor C_r is 0.047 μF; so the period of the resonant circuit is about 4 μs. The frequency of the PWM is 20 kHz. From the figure, we can see that the output of the simulation matches the theoretical analysis. The waveforms in Figure 9.75b through h are the same as those in Figure 9.76.

In order to verify the theoretical analysis and simulation results, the proposed soft-switching inverter was tested on an experimental prototype rated as

DC link voltage: 240 V
Power of the BDCM: 3.3 hp
Switching frequency: 10 kHz.

A polyester capacitor of 47 nF and 1500 V was adopted as the DC-link resonant capacitor C_r. The resonant inductor was of 6 μH/20 A with ferrite core. The design of the auxiliary switches control circuit was referenced from Figure 9.74. The monostable flip-flop can be

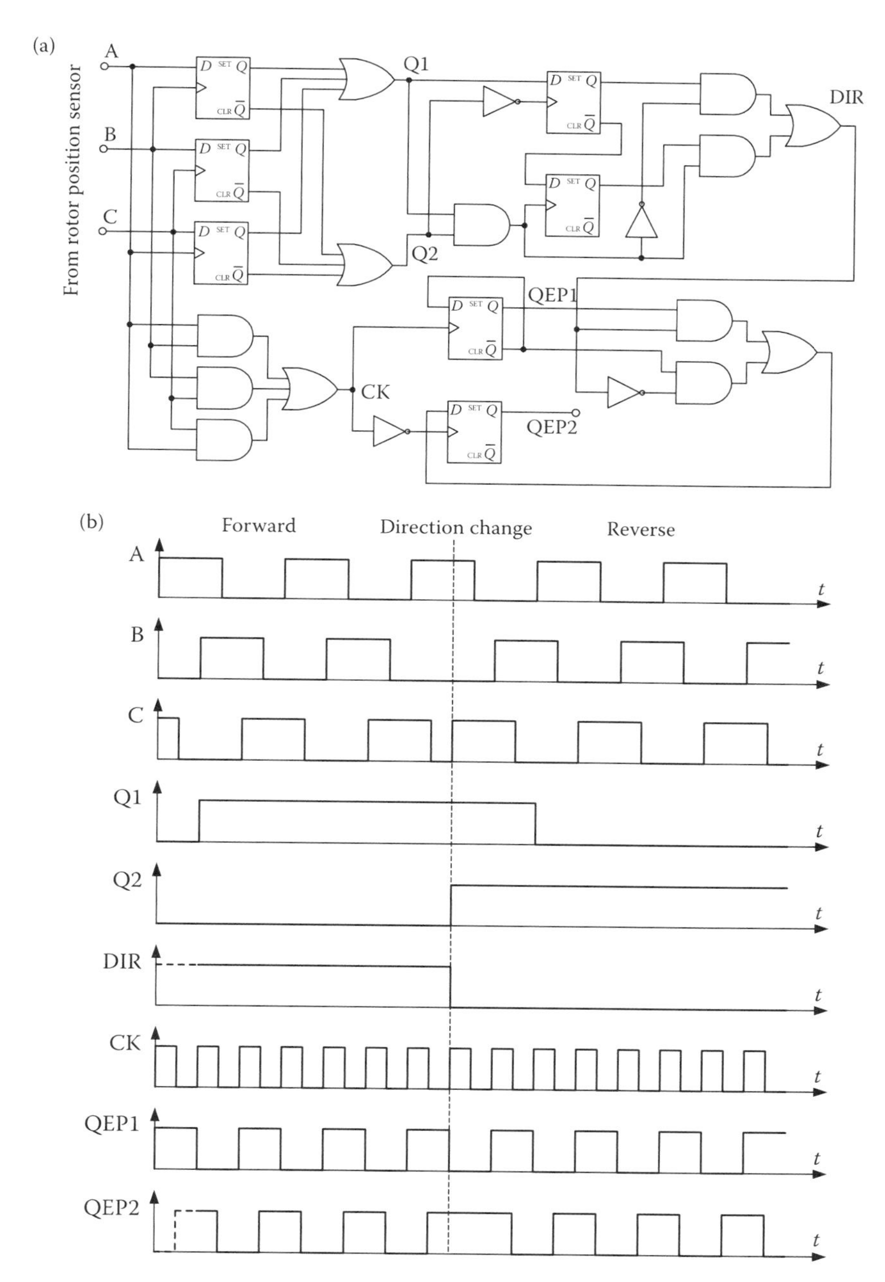

FIGURE 9.73 Circuit of derive QEP from Hall signal and waveforms. (a) The logic diagram and (b) the corresponding waveforms.

implemented by IC 74LS123, the delay can be implemented by Schmitt Trigger and RC circuit, and the logical gate can be replaced by a programmable logical device to reduce the number of ICs.

The waveforms of the voltage across the switch and the current under hard switching and soft switching are shown in Figures 9.76a and 9.76b, respectively. All the voltage signals come from differential probes, and there is a gain of 20. For voltage waveform,

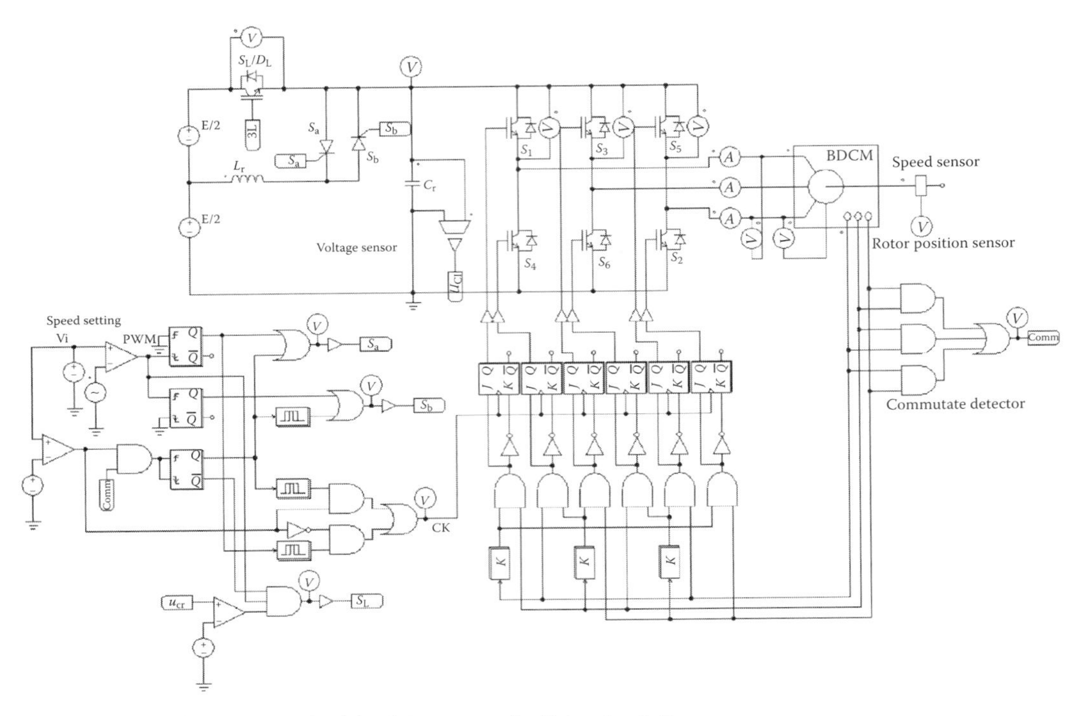

FIGURE 9.74 Schematic circuit of the drive system for Psim simulation.

5.00 V/div = 100 V/div, which is the same for Figure 9.77. It can also be seen that there is a considerable overlap between the voltage and current waveforms during the switching under hard switching. The overlap is much less with soft switching.

A serial of key waveforms with the soft-switching inverter is shown in Figure 9.77. The default scale is DC-link voltage, 100 V/div, and the current is 20 A/div. The default switching frequency is 10 kHz. The DC-link voltage is fixed at 240 V. These experimental waveforms are similar to the simulation waveforms in Figure 9.75.

9.7.2 Resonant Pole Inverter

The resonant pole inverter is a soft-switching DC/AC inverter circuit and is shown in Figure 9.78. Each resonant pole comprises resonant inductor and a pair of resonant capacitors at each phase leg. These capacitors are directly connected in parallel to the main inverter switches in order to achieve (ZVS) condition. In contrast to the resonant DC-link inverter, the DC-link voltage remains unaffected during the resonant transitions. The resonant transitions occur separately at each resonant pole when the corresponding main inverter switch needs switching. Therefore, the main switches in the inverter phase legs can switch independently from each other and choose the commutation instant freely. Moreover, there is no additional main conduction path switch. Thus, the normal operation of the resonant pole inverter is entirely the same as that of the conventional hard-switching inverter [26].

The ARCP inverter [27] and the ordinary resonant snubber inverter [28] provide a ZVS condition without increasing the device voltage and current stress. These inverters can achieve real PWM control. However, they require a stiff DC-link capacitor bank that is center-taped to accomplish commutation. The center voltage of the DC link is susceptible

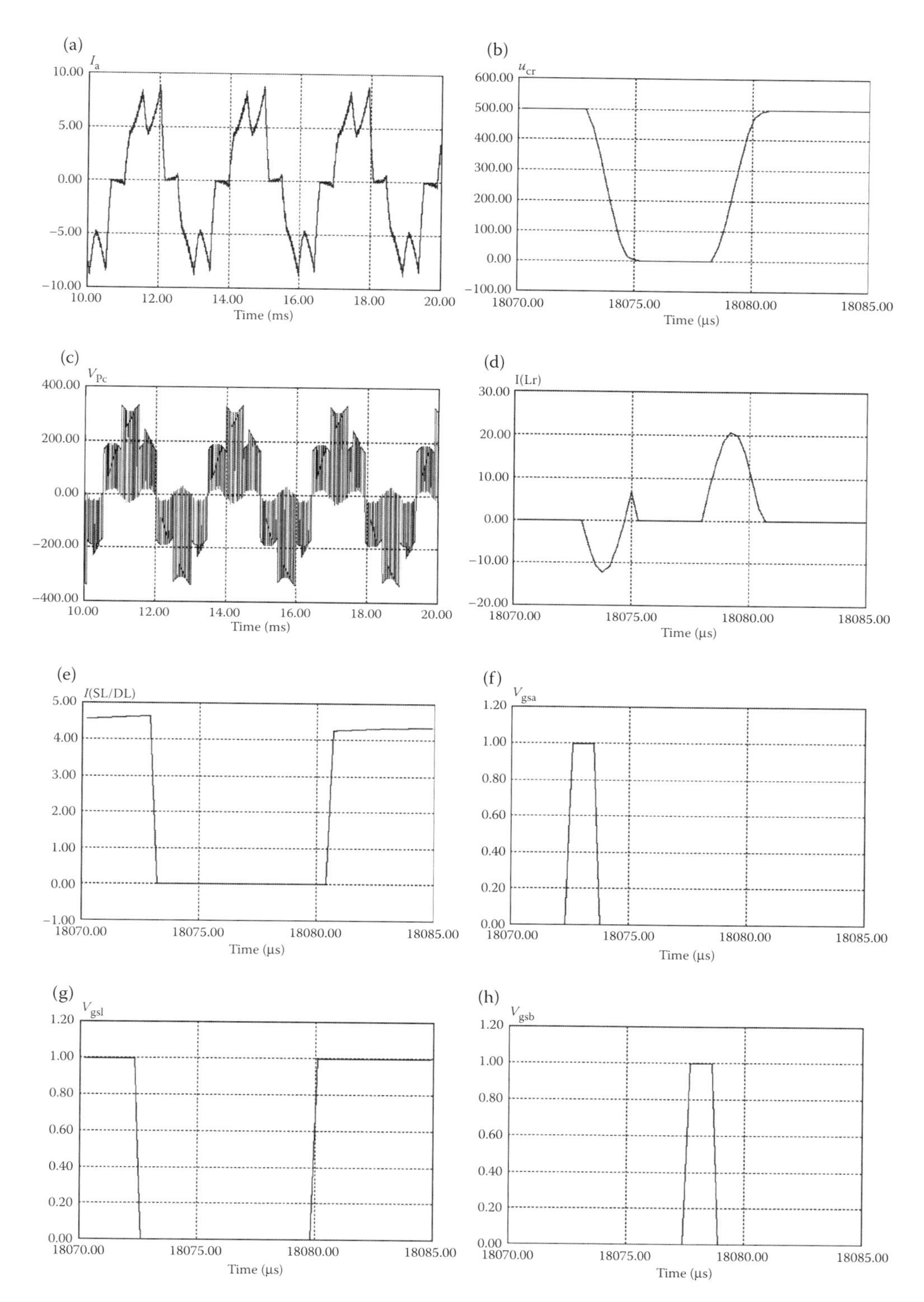

FIGURE 9.75 Simulation results: (a) current of phase a, (b) resonant capacitor voltage μ_{Cr}, (c) voltage of phase a, (d) resonant inductor current (i_{Lr}), (e) current of S_L, (f) S_a gate signal, (g) S_L gate signal, and (h) S_b gate signal.

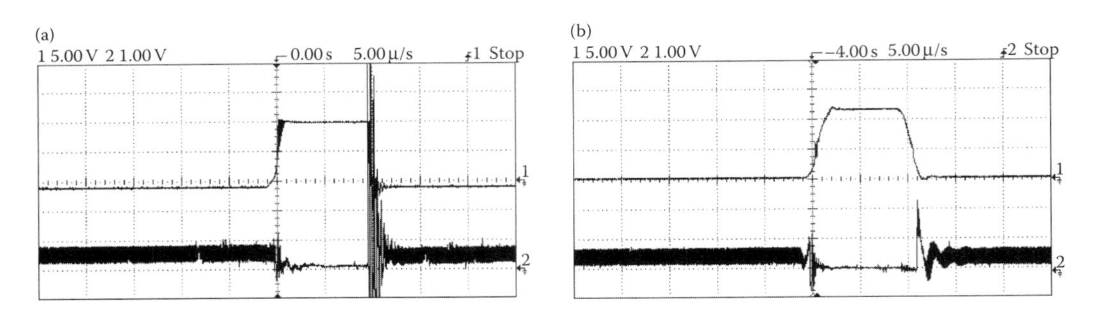

FIGURE 9.76 Voltage and current waveforms of switch S_L in hard switching and soft switching inverter: (a) waveform of switch voltage and current with hard switching (10A/div) and (b) waveform of switch voltage and current with soft switching (10A/div).

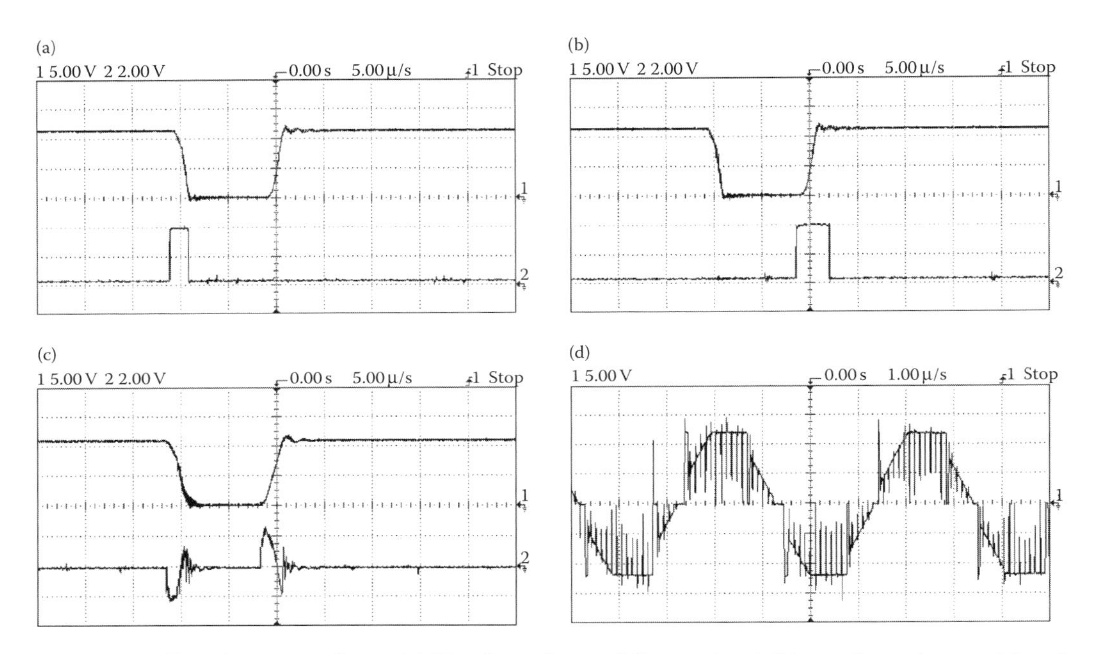

FIGURE 9.77 Experiment waveforms: (a) Waveform of u_{cr} and S_a gate signal, (b) waveform of u_{Cr} and S_b gate signal, (c) waveform of u_{Cr} and i_r gate signal, and (d) waveform of phase voltage (L-L).

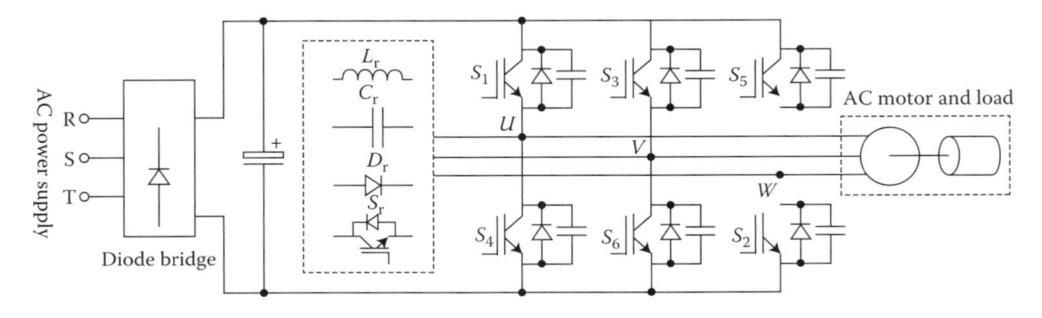

FIGURE 9.78 Resonant pole inverter.

to drift that may affect the operation of the resonant circuit. The resonant transition inverter [29,30] uses only one auxiliary switch, whose switching frequency is much higher than that applied to the main switches. Thus, it will limit the switching frequency of the inverter. Furthermore, the three resonant branches of the inverter work together and will be affected by each other. A Y-configured resonant snubber inverter [31] has a floating neutral voltage that may cause overvoltage failure of the auxiliary switches. A delta (Δ)-configured resonant snubber inverter [32] avoids the floating neutral voltage and is suitable for multiphase operation without circulating currents between the off-state branch and its corresponding output load. However, the inverter requires three inductors and six auxiliary switches.

Moreover, resonant pole inverters have been applied in induction motor drive applications. They are usually required to change two-phase switch states at the same time to obtain a resonant path. It is not suitable for a BDCM drive system as only one switch is needed to change the switching state in a PWM cycle. The switching frequency of three upper switches (S_1, S_3, and S_5) is different from that of three lower switches (S_2, S_4, and S_6) in an inverter for a BDCM drive system. All the switches have the same switching frequency in a conventional inverter for induction motor applications. Therefore, it is necessary to develop a novel topology of the soft-switching inverter and special control circuit for BDCM drive systems. This chapter proposes a special designed resonant pole inverter that is suitable for BDCM drive systems and is easy to apply in industry. In addition, this inverter possesses the following advantages: low switching power losses, low inductor power losses, low switching noise, and a simple control scheme.

9.7.2.1 *Topology of the Resonant Pole Inverter*

A typical controller for the BDCM drive system [33] is shown in Figure 9.79.

The rotor position can be sensed by a Hall-effect sensor or a slotted optical disk, providing three square-waves with phase shift in 120°. These signals are decoded by a combinatorial logic to provide the firing signals for 120° conduction on each of the three phases. The basic forward control loop is the voltage control implemented by PWM (the voltage reference signal compared with a triangular wave or a wave generated by a microprocessor). The PWM is applied only to the lower switches. This not only reduces the current ripple but also avoids the need for a wide bandwidth in the level-shifting circuit that feeds the upper switches. The three upper switches work under commutation frequency (typically several hundreds of Hz) and the three lower switches work under PWM frequency (typically tens of kHz). So it is not important that the three upper switches work under soft-switching condition. The switching power losses can be reduced significantly and the auxiliary circuit would be simpler if only three lower switches work under soft-switching condition. Thus a special design resonant pole inverter for the BDCM drive system is introduced for this purpose. The structure of the proposed inverter is shown in Figure 9.80.

The system contains a diode bridge rectifier, a resonant circuit, a conventional three-phase inverter, and a control circuitry. The resonant circuit consists of three auxiliary switches (S_a, S_b, and S_c), one transformer with turn ratio 1:n, and two diodes D_{fp} and D_r. Diode D_{fp} is connected in parallel to the primary winding of the transformer and diode D_r is serially connected with secondary winding across the DC link. There is one snubber capacitor connected in parallel to each lower switch of the phase leg. The snubber capacitor resonates with the primary winding of the transformer. The emitters of the three auxiliary switches are connected together. Thus, the gate drive of these auxiliary switches can use one common output DC power supply.

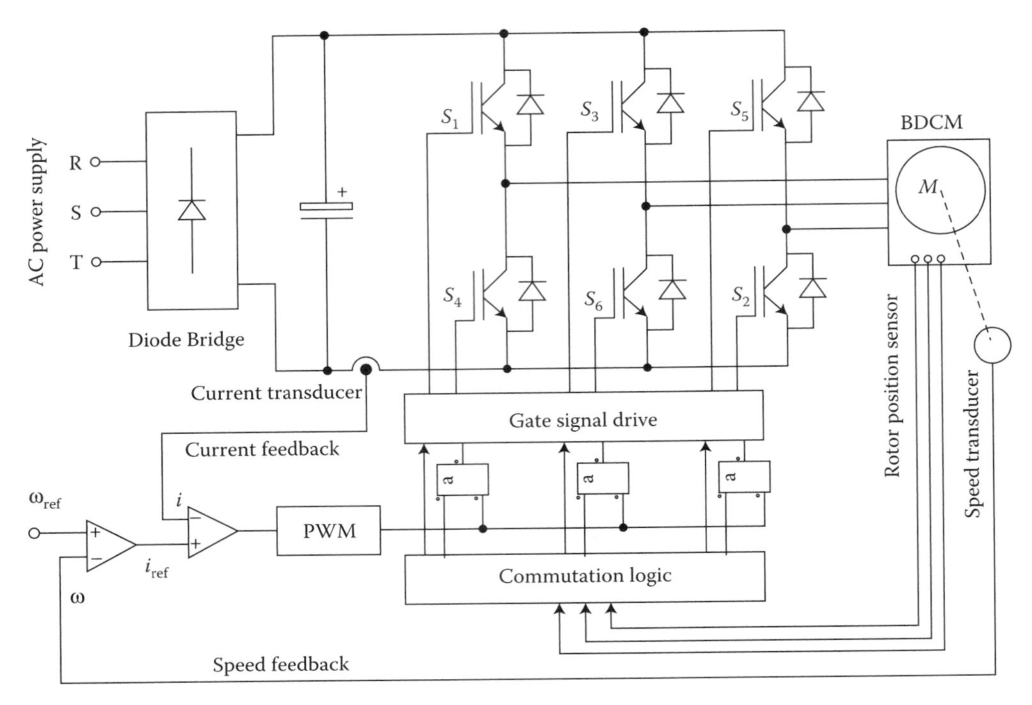

FIGURE 9.79 Typical controller for BDCM drive system.

In the whole PWM cycle, the three lower switches (S_2, S_4, and S_6) can be turned off in the ZVS condition as the snubber capacitors (C_{ra}, C_{rb}, and C_{rc}) can slow down the voltage rise rate. The turn-off power losses can be reduced and the turn-off voltage spike is eliminated. Before turning on the lower switch, the corresponding auxiliary switch (S_a, S_b, or S_c)

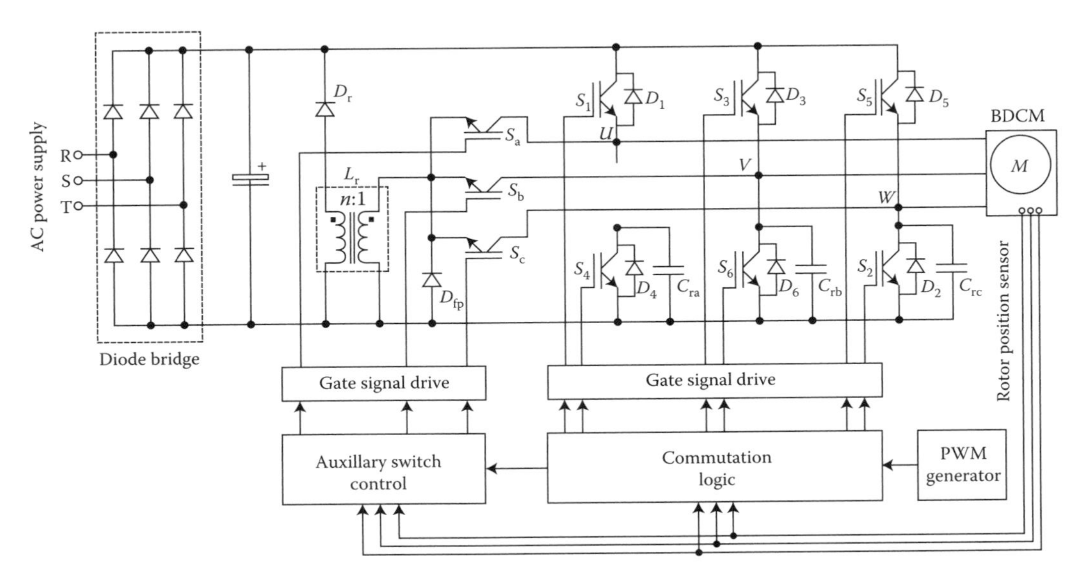

FIGURE 9.80 Structure of the resonant pole inverter for BDCM drive system.

must be turned on ahead of time. The snubber capacitor is then discharged and the lower switches get the ZVS condition. During phase current commutation, the switching state is changed from one lower switch to another (e.g., turn off S_6 and turn on S_2), S_6 can be turned off directly in the ZVS condition, and by turning on the auxiliary switch S_c to discharge the snubber capacitor C_{rc}, the switch S_2 can get the ZVS condition. During phase current commutation, if the switching state is changed from one upper switch to another upper switch, the operation is the same as that of the hard-switching inverter, as the switching power losses of the upper switches are much smaller than that of the lower switches.

9.7.2.2 Operation Principle

For the sake of convenience, to describe the operation principle, we investigate the period of time when the switch S_1 is always turned on, when switch S_6 works under PWM frequency, and when other main inverter switches are turned off. Since the resonant transition is very short, it can be assumed that the load current is constant. The equivalent circuit is shown in Figure 9.81. Where V_S is the DC-link voltage, i_{Lr} is the transformer primary winding current, u_{S6} is the voltage drop across the switch S_6 (i.e., snubber capacitor C_{rb} voltage), and I_O is the load current. The waveforms of the switches (S_6 and S_b) gate signal, PWM signal, the main switch S_6 voltage drop (u_{S6}), and the transformer primary winding current (i_{Lr}) are illustrated in Figure 9.82, and the details will be explained below. Accordingly, at the instant t_0–t_6, the operation of one switching cycle can be divided into seven modes.

Mode 0 (as shown in Figure 9.83a) $0 < t < t_0$: After the lower switch S_6 is turned off, the load current flows through the upper freewheeling diode D_3, and the voltage drop u_{S6} (i.e., snubber capacitor C_{rb} voltage) across the switch S_6 is the same as that of the DC-link voltage. The auxiliary resonant circuit does not operate.

Mode 1 (as shown in Figure 9.83b) $t_0 < t < t_1$: If the switch S_6 is turned on directly, the capacitor discharge surge current will also flow through switch S_6; thus, switch S_6 may face the risk of a second breakdown. The energy stored in the snubber capacitor must be discharged ahead of time. Thus, the auxiliary switch S_b is turned on (ZCS turn on as the current i_{Lr} cannot change suddenly due to the transformer inductance). As the transformer primary winding current i_{Lr} begins to increase, the current flowing through the freewheeling diode decays. The secondary winding current i_{Lrs} also begins to conduct through diode D_r to the DC link. Both of the terminal voltages of the primary and secondary windings are equal to the DC-link voltage V_S. By neglecting the resistances of the windings

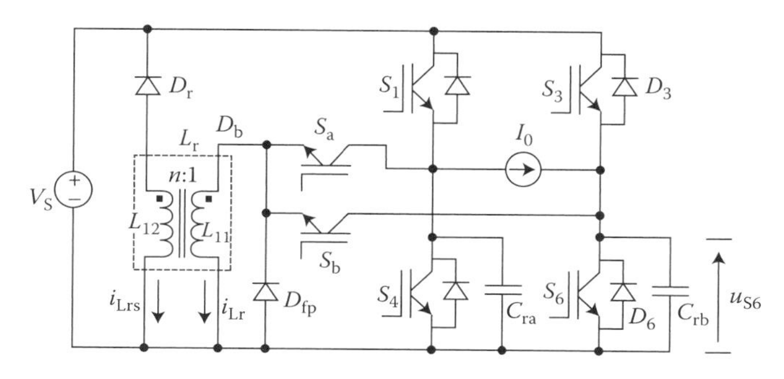

FIGURE 9.81 Equivalent circuit.

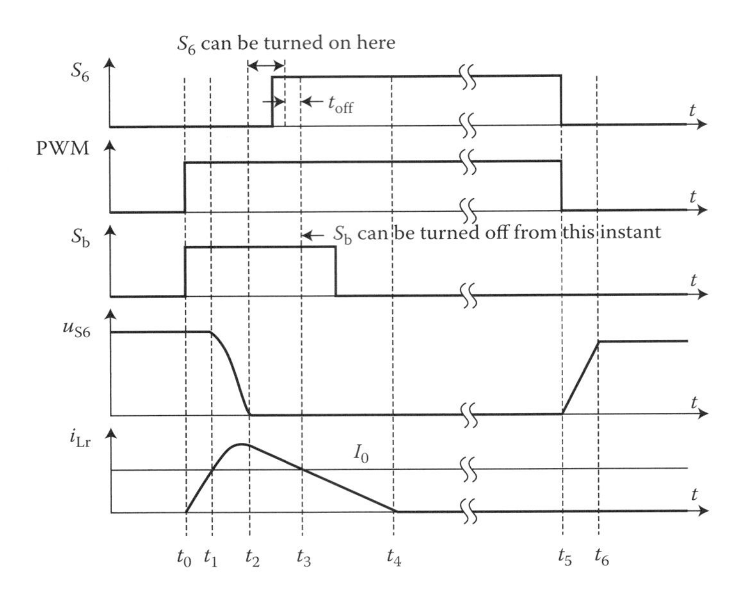

FIGURE 9.82 Key waveforms of the equivalent circuit.

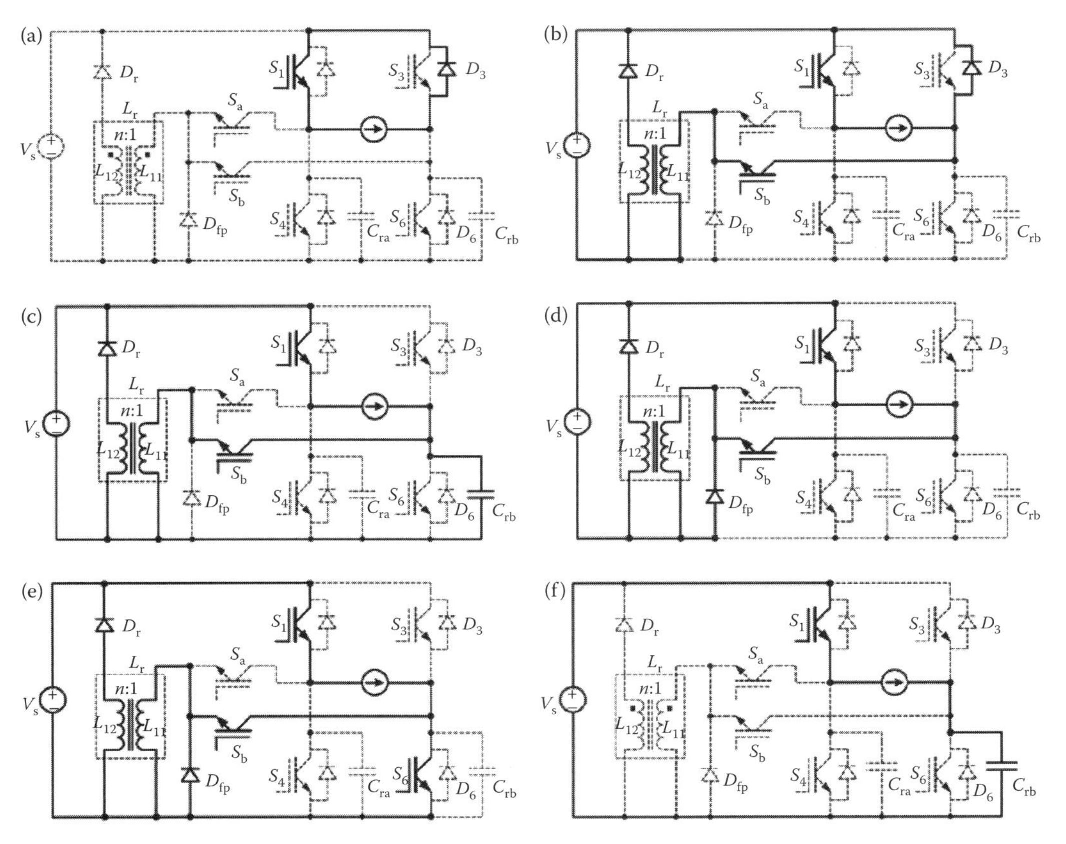

FIGURE 9.83 Operation modes of the resonant pole inverter: (a) mode 0, (b) mode 1, (c) mode 2, (d) mode 3, (e) mode 4, and (f) mode 6.

and using the transformer equivalent circuit (referred to as the primary side) [34], we obtain

$$V_S = L_{l1}\frac{di_{Lr}(t)}{dt} + a^2L_{l2}\frac{d[i_{Lrs}(t)/a]}{dt} + aV_S, \tag{9.86}$$

where L_{l1} and L_{l2} are the primary and secondary winding leakage inductances, respectively, and the transformer turn's ratio is 1:n. The transformer has a high magnetizing inductance. We can assume that $i_{Lrs} = i_{Lr}/n$, and rewrite Equation 9.86 as

$$\frac{di_{Lr}}{dt} = \frac{(n-1)V_S}{n\left(L_{l1} + (1/n^2)L_{l2}\right)} = \frac{(n-1)V_S}{nL_r}, \tag{9.87}$$

where L_r is the equivalent inductance of the transformer $L_{l1} + L_{l2}/n^2$. The transformer primary winding current i_{Lr} increases linearly and the mode is ended when $i_{Lr} = I_O$. The interval of this mode can be determined by

$$\Delta t_1 = t_1 - t_0 = \frac{nL_rI_O}{(n-1)V_S}. \tag{9.88}$$

Mode 2 (as shown in Figure 9.83c) $t_1 < t < t_2$: At $t = t_1$, all the load current flows through the transformer primary winding and the freewheeling diode D_3 is turned off in the ZCS condition. The freewheeling diode reverse recovery problems are reduced greatly. The snubber capacitor C_{rb} resonates with the transformer, and the voltage drop u_{S6} across the switch S_6 decays. By redefining the initial time, the transformer currents i_{Lr} and i_{Lrs} and the capacitor voltage u_{S6} obey the equation

$$\begin{cases} u_{S6}(t) = L_{l1}\dfrac{di_{Lr}(t)}{dt} + a^2L_{l2}\dfrac{d[i_{Lrs}(t)/a]}{dt} + aV_S, \\ -C_r\dfrac{du_{S6}(t)}{dt} = i_{Lr}(t) - I_O, \end{cases} \tag{9.89}$$

where C_r is the capacitance of the snubber capacitor C_{rb}. The transformer current $i_{Lrs} = i_{Lr}/n$, as in mode 1, with initial conditions $u_{S6}(0) = V_S$, $i_{Lr}(0) = I_O$; then the solution of Equation 9.89 is

$$\begin{aligned} u_{S6}(t) &= \frac{(n-1)V_S}{n}\cos(\omega_r t) + \frac{V_S}{n}, \\ i_{Lrs}(t) &= I_O + \frac{(n-1)V_S}{n}\sqrt{\frac{C_r}{L_r}}i_{Lr}\sin(\omega_r t), \end{aligned} \tag{9.90}$$

where $\omega_r = \sqrt{1/(L_rC_r)}$. Let $u_{Cr}(t) = 0$; this yields the duration of the resonance

$$\Delta t_2 = t_2 - t_1 = \frac{1}{\omega_r}\arccos\left(-\frac{1}{n-1}\right). \tag{9.91}$$

The interval is independent of the load current. At $t = t_2$, the corresponding transformer primary current is

$$i_{Lr}(t_2) = I_O + V_S\sqrt{\frac{(n-2)C_r}{nL_r}}. \tag{9.92}$$

The peak value of the transformer primary current can also be determined:

$$i_{Lr-m} = I_O + \frac{n-1}{n} V_S \sqrt{\frac{C_r}{L_r}}. \tag{9.93}$$

Mode 3 (as shown in Figure 9.83d) $t_2 < t < t_3$: When the capacitor voltage u_{S6} reaches zero at $t = t_2$, the freewheeling diode D_{pf} begins to conduct. The current flowing through the auxiliary switch S_b is the load current I_O. The sum current flowing through switch S_b and diode D_{pf} is the transformer primary winding current i_{Lr}. The transformer primary voltage is zero and the secondary voltage is V_S. By redefining the initial time, we obtain

$$0 = L_{l1}\frac{di_{Lr}(t)}{dt} + a^2 L_{l2}\frac{d[i_{Lrs}(t)/a]}{dt} + aV_S. \tag{9.94}$$

Since the transformer current $i_{Lrs} = i_{Lr}/n$ as in mode 1, we can deduce Equation 9.94 to Equation 9.95.

$$\frac{di_{Lr}}{dt} = -\frac{V_S}{nL_r}. \tag{9.95}$$

The transformer primary current decays linearly, and the mode is ended when $i_{Lr} = I_O$.With the initial condition given by Equation 9.92, the interval of this mode can be determined by

$$\Delta t_3 = t_3 - t_2 = \sqrt{n(n-2)L_r C_r}. \tag{9.96}$$

The interval is also independent of the load current. During this mode, the switch is turned on in the ZVS condition.

Mode 4 (as shown in Figure 9.83e) $t_3 < t < t_4$: The transformer primary winding current i_{Lr} decays linearly from the load current I_O to zero. Partial load current flows through the main switch S_6. The sum current flowing through switches S_6 and S_b is equal to the load current I_O. The sum current flowing through switch S_b and diode D_{fp} is the transformer primary winding current i_{Lr}. By redefining the initial time, the transformer winding current obeys Equation 9.95 with the initial condition $i_{Lr}(0) = I_O$. The interval of this mode is

$$\Delta t_4 = t_4 - t_3 = \frac{nL_r I_O}{V_S}. \tag{9.97}$$

The auxiliary switch S_b can be turned off in the ZVS condition. In this case, after switch S_b is turned off, the transformer primary winding current flows through the freewheeling diode D_{fp}. The auxiliary switch S_b can also be turned off in ZVS and ZCS conditions after i_{Lr} decays to zero.

Mode 5 $t_4 < t < t_5$: The transformer primary winding current decays to zero and the resonant circuit idles. This state is probably the same operational state as the conventional hard-switching inverter. The load current flows from the DC link through the two switches S_1 and S_6, and the motor.

Mode 6 (as shown in Figure 9.83f) $t_5 < t < t_6$: The main inverter switch S_6 is turned off directly and the resonant circuit does not work. The snubber capacitor C_{rb} can slow down the rising rate of u_{S6}, while the main switch S_6 operates in the ZVS condition. The duration of the mode is

$$\Delta t_7 = t_7 - t_6 = \frac{C_r V_S}{I_O}. \tag{9.98}$$

The next period starts from mode 0 again, but the load current flows through the free-wheeling diode D_3. During phase current commutation, the switching state is changed from one lower switch to another (e.g., turn off S_6 and turn on S_2), S_6 can be turned off directly in the ZVS condition (similar to mode 6), and by turning on the auxiliary switch S_c to discharge the snubber capacitor C_{rc}, the switch S_2 can get the ZVS condition (similar to modes 1 through 4).

9.7.2.3 Design Considerations

It is assumed that the inductance of BDCM is much higher than the transformer leakage inductance. From the previous analysis, the design considerations can be summarized as follows:

1. Determine the value of the snubber capacitor C_r, and the parameter of the transformer.
2. Select the main and auxiliary switches.
3. Design the control circuitry for the main and auxiliary switches.

The turn ratio (1:n) of the transformer can be determined ahead of time. Equation 9.91 must satisfy

$$n > 2 \tag{9.99}$$

On the other hand, from Equation 9.97, the transformer primary winding current i_{Lr} will take a long time to decay to zero if n is too big. So n must be a moderate number. The equivalent inductance of the transformer $L_r = L_{l1} + L_{l2}/n^2$ is inversely proportional to the rise rate of the switch current when the auxiliary switches are turned on. This means that the equivalent inductance L_r should be big enough to limit the rising rate of the switch current to work in the ZCS condition. The selection of L_r can be referenced from the rule depicted in reference [35].

$$L_r \approx \frac{4t_{on}V_S}{I_{O\max}}, \tag{9.100}$$

where t_{on} is the turn-on time of an IGBT, and $I_{O\max}$ is the maximum load current. The snubber capacitance C_r is inversely proportional to the rise rate of the switch voltage drop when the lower main inverter switches are turned off. This means that the capacitance is as high as possible to limit the rising rate of the voltage to work in the ZVS condition. The selection of the snubber capacitor can be determined as

$$C_r \approx \frac{4t_{on}I_{O\max}}{V_S}, \tag{9.101}$$

where t_{off} is the turn-off time of an IGBT. However, as the capacitance increases, more energy is stored in it. This energy should be discharged when the lower main inverter switches are turned on. With high capacitance, the peak value of the transformer current will also be high. The peak value of i_{Lr} should be restricted to twice that of the maximum load current. From Equation 9.93, we obtain

$$\sqrt{\frac{C_r}{L_r}} \le \frac{nI_{O\max}}{(n-1)V_S}. \tag{9.102}$$

Three lower switches of the inverter (i.e., S_4, S_6, and S_2) are turned on during mode 3 (i.e., lag the rising edge of PWM at the time range $\Delta t_1 + \Delta t_2 \sim \Delta t_1 + \Delta t_2 + \Delta t_3$). In order to turn on these switches at a fixed time (say ΔT_1), lagging the rising edge of PWM under various load currents for control convenience, the following condition should be satisfied.

$$\Delta t_1 + \Delta t_2 + \Delta t_3|_{I_O=0} > (\Delta t_1 + \Delta t_2)|_{I_O=I_{O\max}} + t_{\text{off}}. \tag{9.103}$$

Substitute Equations 9.88, 9.91, and 9.96 into Equation 9.103

$$\sqrt{n(n-2)L_r C_r} > \frac{nL_r I_{O\max}}{(n-1)V_S} + t_{\text{off}}. \tag{9.104}$$

The whole switching transition time is expressed as

$$T_W = \Delta t_1 + \Delta t_2 + \Delta t_3 + \Delta t_4 = \frac{nL_r I_O}{(n-1)V_S} + \sqrt{L_r C_r} \times \left[\arccos\left(-\frac{1}{n-1}\right) + \sqrt{n(n-2)}\right]. \tag{9.105}$$

For high switching frequencies, T_W should be as short as possible. Select the equivalent inductance L_r and the snubber capacitance C_r to satisfy Equations 9.99 through 9.104, and L_r and C_r should be as small as possible.

As the transformer operates at high frequency (20 kHz), the magnetic core material can be ferrite. The design of the transformer needs the parameters of form factor, frequency, input/output voltage, input/output maximum current, and ambient temperature. From Figure 9.61, the transformer current can be simplified as triangle waveforms and then the form factor can be determined as $2/\sqrt{3}$. Ambient temperature is dependent on the application field. Other parameters can be obtained from the previous section. The transformer only carries current during the transition of turning on a switch in one cycle; so the winding can be of a smaller diameter.

The main switches S_{1-6} work under the ZVS condition; therefore the voltage stress is equal to the DC-link voltage V_S. The device current rate can be load current. The auxiliary switches S_{a-c} work under ZCS or ZVS conditions, while the voltage stress is also equal to the DC-link voltage V_S. The peak current flowing through them is limited to double maximum load current. As the auxiliary switches S_{a-c} carry the peak current only during switch transitions, they can be rated with a lower continuous current rating. The additional cost will not be too much.

The gate signal generator circuit is shown in Figure 9.84. The rotor position signal decode module produces the typical gate signal of the main switches. The inputs of the module are rotor position signals, rotating direction of the motor, which "enable" the signal and PWM pulse-train. The rotor position signals are three square-waves with a phase shift in 120°. The "enable" signal is used to disable all outputs in case of emergency (e.g., over current, over voltage, and over heat). The PWM signal is the output of the comparator, comparing the reference voltage signal with the triangular wave. The reference voltage signal is the output of the speed controller. The speed controller is a processor (a single chip computer or a digital signal processor) and the PWM signal can be produced by software. The outputs (G_1–G_6) of the module are the gate signals applied to the main inverter switches. The outputs $G_{1,3,5}$ are the required gate signals for the three upper main inverter switches.

The gate signals of the three lower main inverter switches and the auxiliary switches can be deduced from the outputs $G_{4,6,2}$ as shown in Figure 9.85. The trailing edge of the gate signals for the three lower main inverter switches $G_{S4,6,2}$ is the same as that of $G_{4,6,2}$, and the leading edge of $G_{S4,6,2}$ lags $G_{4,6,2}$ for a short time ΔT_1. The gate signals for the auxiliary

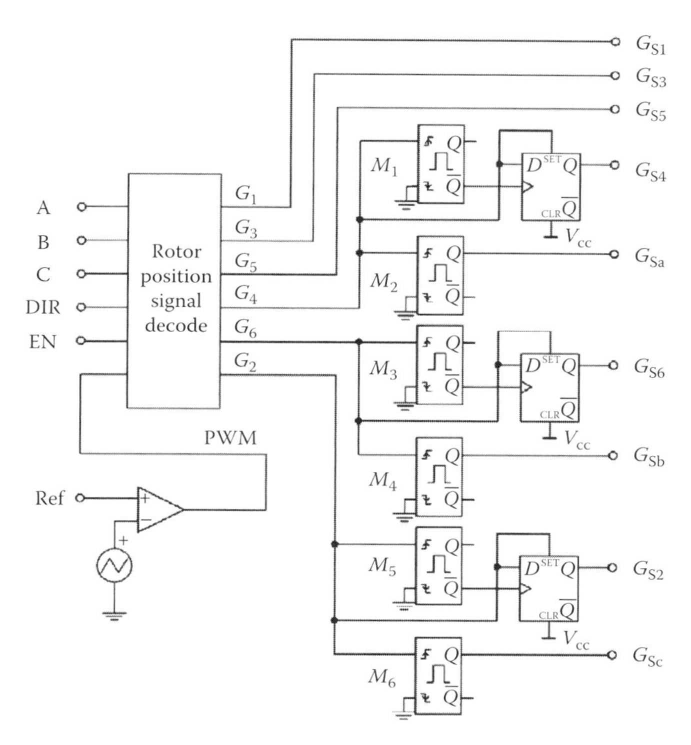

FIGURE 9.84 Gate signal generator circuit.

switches $G_{S4,6,2}$ have a fixed pulse width (ΔT_2) with the same leading edge as that of $G_{4,6,2}$. In Figure 9.84, the gate signals $G_{Sa,b,c}$ are the outputs of monostable flip-flops $M_{4,6,2}$with the inputs $G_{4,6,2}$. The three monostable flip-flops $M_{4,6,2}$have the same pulse width ΔT_2. The gate signals $G_{S4,6,2}$ are combined by the negative outputs of monostable flip-flops $M_{1,3,5}$ and $G_{4,6,2}$. The combining logical controller can be implemented by a D flip-flop with "preset" and "clear" terminals. The three monostable flip-flops $M_{4,6,2}$ have the same pulse width ΔT_1. Determination of the pulse widths of ΔT_1 and ΔT_2 is referenced from the theoretical analysis in Section 9.7.2.2. In order to get the ZVS condition of the main inverter switches under various load currents, the lag time should satisfy

$$(\Delta t_1 + \Delta t_2)|_{I_O=I_{O\max}} < \Delta T_1 < (\Delta t_1 + \Delta t_2 + \Delta t_3)|_{I_O=0} - t_{\text{off}}. \quad (9.106)$$

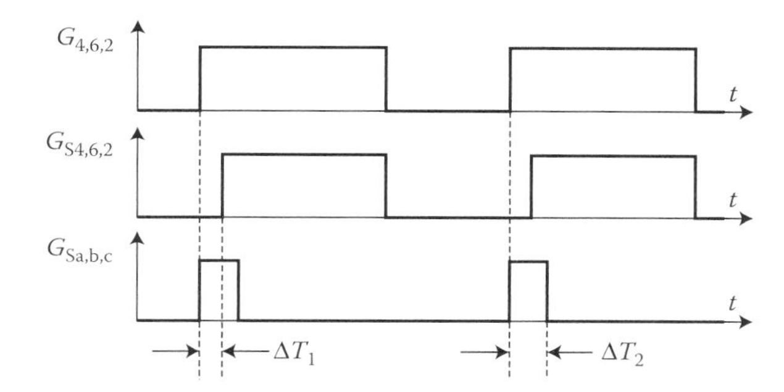

FIGURE 9.85 Gate signals $G_{S4,6,2}$ and $G_{Sa,b,c}$ from $G_{4,6,2}$.

In order to get a soft-switching condition of the auxiliary switches, the pulse width need only satisfy

$$\Delta T_2 > (\Delta t_1 + \Delta t_2 + \Delta t_3)|_{I_O = I_{O\max}}. \tag{9.107}$$

9.7.2.4 Simulation and Experimental Results

The proposed topology is verified by PSim simulation software. The DC-link voltage is 300 V, and the maximum load current is 25 A. The parameters of the resonant circuit were determined from Equations 9.99 through 9.105. The transformer turn ratio is 1:4, and the leakage inductances of the primary and secondary windings are 6 μH and 24 μH, respectively. Therefore, the equivalent transformer inductance L_r is 7.5 μH. The resonant capacitance C_r is 0.047 μF. Then, $\Delta t_1 + \Delta t_2$ and $\Delta t_1 + \Delta t_2 + \Delta t_3$ can be determined under various load currents I_O, as shown in Figure 9.86, considering the turn-off time of a switch lagging time ΔT_1 and the pulse width ΔT_2 are set as 2.1 μs and 5 μs, respectively. The frequency of the PWM is 20 kHz. Waveforms of the transformer primary winding current i_{Lr}, the switch S_6 voltage drop u_{S6}, PWM, the main switch S_6, the auxiliary switch S_b, and the gate signal under low and high load currents are shown in Figure 9.87. The figure shows that the inverter worked well under various load currents. In order to verify the theoretical analysis and simulation results, the inverter was tested by experiment. The test conditions are

1. DC-link voltage: 300 V
2. Power of the BDCM: 3.3 hp
3. Rated phase current: 10.8 A
4. Switching frequency: 20 kHz.

Select 50 A, 1200 V BSM 35 GB 120 DN2 dual IGBT module as the main inverter switches, and 30 A, 600 V IMBH30D-060 IGBT as auxiliary switches. With the datasheets of these switches and Equations 9.99 through 9.105, the values of inductance and capacitance can

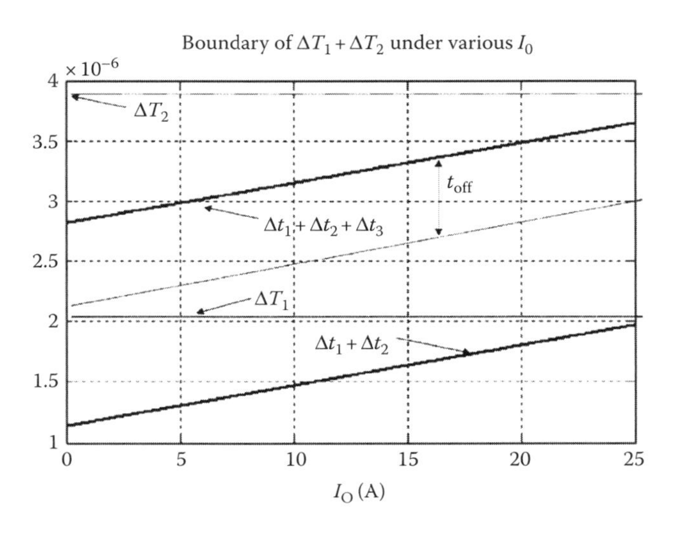

FIGURE 9.86 Boundary of ΔT_1 and ΔT_2 under various load current I_O.

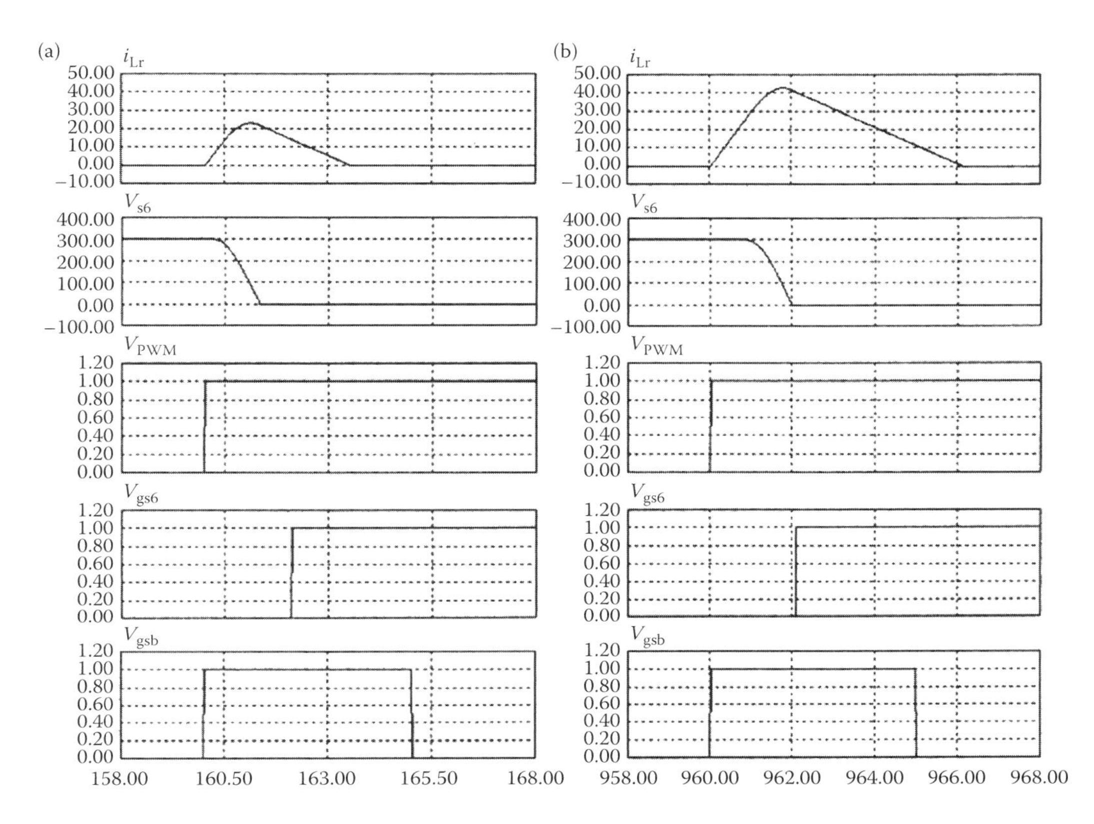

FIGURE 9.87 Simulation waveforms of i_{Lr}, V_{S6}, PWM, S_6, and S_b gate signal under various load current (a) under low load current ($I_O = 5$ A) and (b) under high load current ($I_O = 25$ A).

be determined. Three polyester capacitors of 47 nF/630 V were adopted as the snubber capacitor for the three lower switches of the inverter. A high magnetizing inductance transformer with the turn ratio 1:4 was employed in the experiment. Fifty-two turns wires with size AWG 15 were selected as primary winding, and 208 turns wires with size AWG 20 were selected as secondary winding. The equivalent inductance is about 7 μH. The switching frequency is 20 kHz. The rotor position signal decode module is implemented by a 20 leads GAL IC GAL16V8. The monostable flip-flop was set up by IC 74LS123, a variable resistor, and a capacitor. With (21) and (22), lag time and pulse width are determined to be 2.5 μs and 5 μs, respectively.

The system is tested in light load and full load currents. The voltage waveforms across the main inverter switch u_{S6} and its gate signal in low and high load currents are shown in Figure 9.88a and (b), respectively. All the voltage signals are measured by a differential probe with a gain of 20; for voltage waveform, 5.00 V/div = 100 V/div. The waveforms of u_{S6} and its current i_{S6} are shown in Figure 9.88c, and dv/dt and di/dt are reduced significantly. The waveforms of u_{S6} and the transformer primary winding current i_{Lr} are shown in Figure 9.88d. The phase current is shown in Figure 9.88e. It can be seen that the resonant pole inverter works well under various load currents, and there is little overlap between the voltage and current waveforms during the switching under soft-switching condition; therefore, the switching power losses is low. The efficiency of hard switching and soft switching under rated speed and various load torques (p.u.) is shown in Figure 9.89. The efficiency improves with the soft-switching inverter. Therefore, the design of the system is successful.

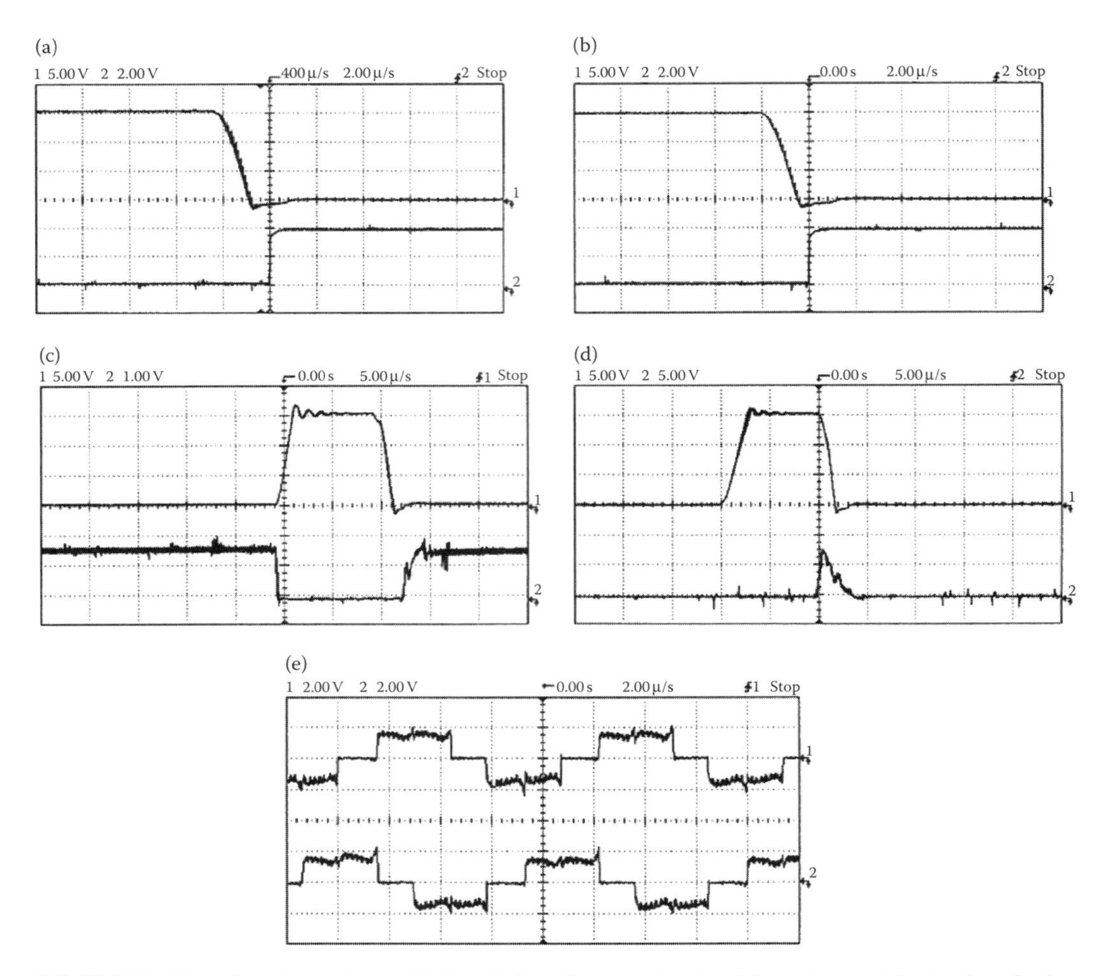

FIGURE 9.88 Experiment waveforms. (a) Switch S_6 voltage u_{S6} (top) and its gate signal (bottom) under low load current (100 V/div). (b) Switch S_6 voltage u_{S6} (top) and its gate signal (bottom) under high load current (100 V/div). (c) Switch S_6 voltage u_{S6} (top) and its current i_{S6} (bottom) (100 V/div, 5 A/div). (d) Switch S_6 voltage u_{S6} (top) and transformer current i_{S6} (bottom) (100 V/div, 25 A/div). (e) Waveforms of phase current (10 A/div).

9.7.3 Transformer-Based Resonant DC-Link Inverter

In order to generate voltage notches of the DC link at controllable instants and reduce the power losses of the inductor, several quasi-parallel resonant schemes were proposed [19,20,36]. As a dwell time is generally required after every notch, severe interferences occur, mainly in multiphase inverters, appreciably worsening the modulation quality. A novel DC-rail parallel resonant ZVT voltage source inverter [37] is introduced; it overcomes the many drawbacks mentioned above. However, it requires a stiff DC-link capacitor bank that is center taped to accomplish commutation. The center voltage of the DC link is susceptible to drift that may affect the operation of the resonant circuit. In addition, it requires two ZVTs per PWM cycle; it would worsen the output voltage and limit the switch frequency of the inverter.

On the other hand, the majority of soft-switching inverters proposed in recent years have been aimed at the induction motor drive applications. So it is necessary to conduct

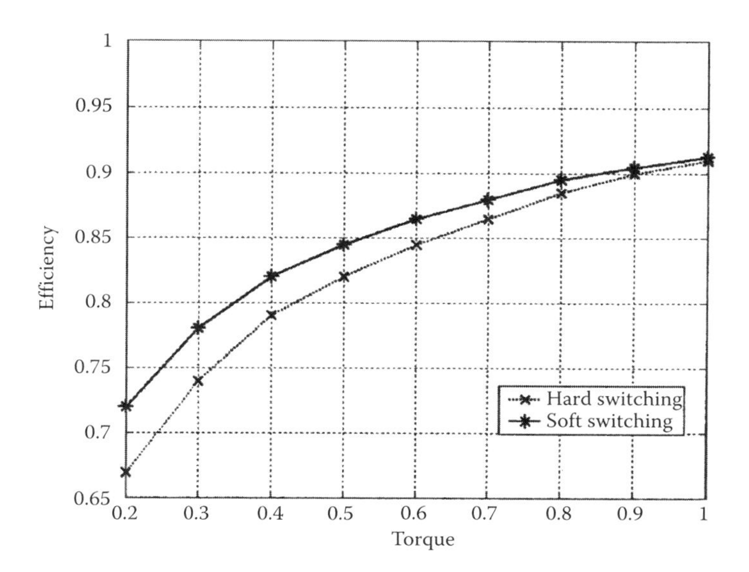

FIGURE 9.89 Efficiency of hard switching and soft switching under various load torques (p.u.).

research on the novel topology of the soft-switching inverter and the special control circuit for BDCM drive systems. This chapter proposed a resonant DC-link inverter based on the transformer for the BDCM drive system to solve the problems mentioned earlier. The inverter possesses the advantages of low switching power loss, low inductor power loss, low DC-link voltage ripple, small device voltage stress, and simple control scheme. The structure of the soft-switching inverter is shown in Figure 9.90 [38]. The system contains a diode

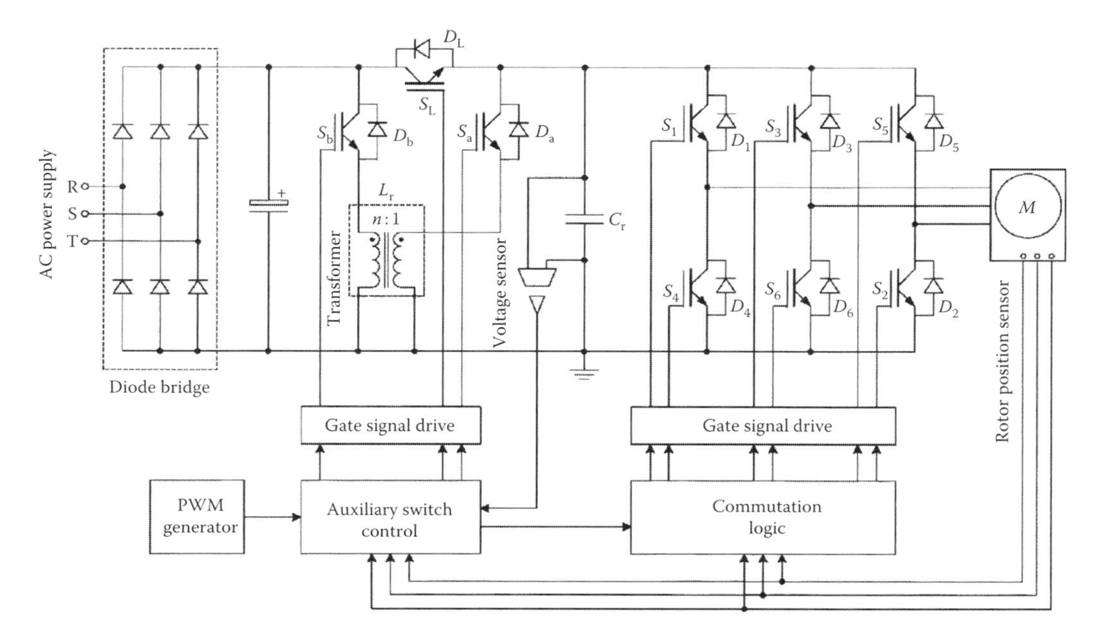

FIGURE 9.90 Structure of the resonant DC = link inverter for BDCM drive system.

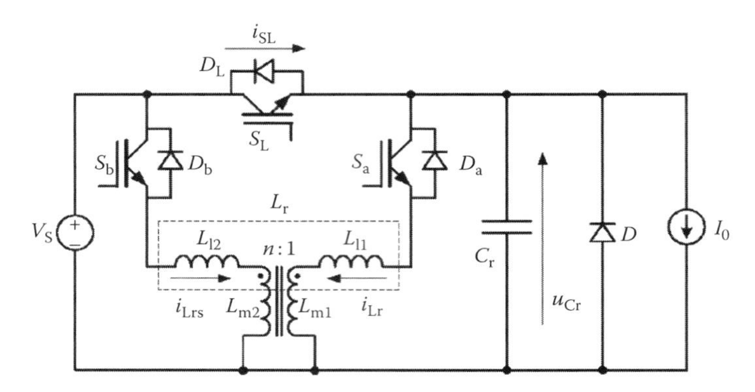

FIGURE 9.91 Equivalent circuit of the inverter.

bridge rectifier, a resonant circuit, a conventional three-phase inverter, and a control circuit. The resonant circuit consists of three auxiliary switches (S_L, S_a, and S_b) and corresponding built-in freewheeling diodes (D_L, D_a, and D_b), one transformer with turn ratio 1:n, and one resonant capacitor. All auxiliary switches work under the ZVS or ZCS condition. It generates voltage notches of the DC link to guarantee that the main switches (S_1–S_6) of the inverter are operating in the ZVS condition.

9.7.3.1 *Resonant Circuit*

The resonant circuit consists of three auxiliary switches, one transformer, and one resonant capacitor. The auxiliary switches are controlled at a certain instant to obtain the resonance between a transformer and a capacitor. Thus, the DC-link voltage reaches zero temporarily (voltage notch) and the main switches of the inverter get the ZVS condition for commutation. Since the resonant process is very short, the load current can be assumed to be constant. The equivalent circuit of the inverter is shown in Figure 9.91. Where V_S is the DC power supply voltage and I_O is the load current. The corresponding waveforms of the auxiliary switches gate signal, PWM signal, resonant capacitor voltage u_{Cr} (i.e., DC-link voltage), and the transformer primary and secondary winding currents i_{Lr} and i_{SL} of a switch (S_L) are illustrated in Figure 9.92. The DC-link voltage is reduced to zero and then rises to the supply voltage again; this process is called a ZVT process or a DC-link voltage notch. The operation of the ZVT process in a PWM cycle can be divided into eight modes.

Mode 0 (as shown in Figure 9.93a) $0 < t < t_0$: Its operation is the same as the conventional inverter. Current flows from the DC power supply through S_L to the load. The voltage u_{Cr} across the resonant capacitor C_r is equal to the supply voltage V_S. The auxiliary switches S_a and S_b are turned off.

Mode 1 (as shown in Figure 9.93b) $t_0 < t < t_1$: When it is the instant for phase current commutation or PWM, the signal is flopped from high to low, the auxiliary switch S_a is turned on with ZCS (as the i_{Lr} cannot suddenly change due to the transformer inductance), and switch S_L is turned off with ZVS (as it cannot change suddenly due to the resonant capacitor C_r) at the same time. The transformer primary winding current i_{Lr} begins to increase and the secondary winding current i_{Lrs} also begins to build up through the diode D_b to the DC link. The terminal voltages of primary and secondary windings of the transformer are the DC-link voltage u_{Cr} and the supply voltage V_S, respectively. Capacitor C_r resonates with the transformer, and the DC-link voltage u_{Cr} is decreased. Neglecting the resistances

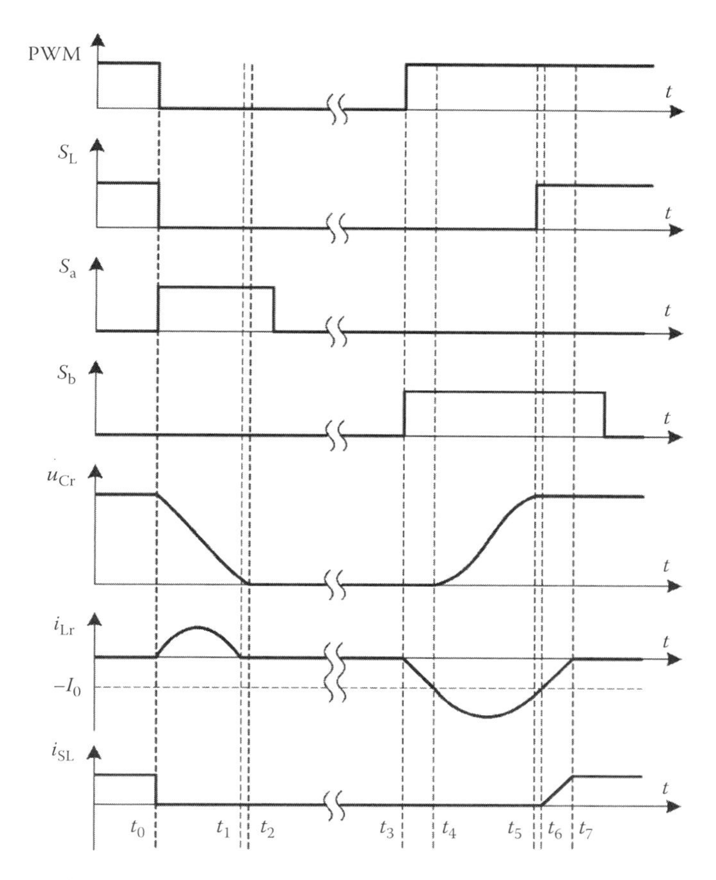

FIGURE 9.92 Key waveforms of the equivalent circuit.

of windings, using the transformer equivalent circuit (referred to as the primary side) [39], the transformer currents i_{Lr} and i_{Lrs} and the DC-link voltage u_{Cr} obey the equation

$$u_{Cr}(t) = L_{l1}\frac{di_{Lr}(t)}{dt} + a^2 L_{l2}\frac{d[i_{Lrs}(t)/a]}{dt} + aV_S,$$
$$i_{Lr}(t) + I_O + C_r\frac{du_{Cr}(t)}{dt} = 0, \tag{9.108}$$

where L_{l1} and L_{l2} are the primary and secondary winding leakage inductances, respectively, the transformer turn's ratio is 1:n. The transformer has a high magnetizing inductance. We can assume that $i_{Lrs} = i_{Lr}/n$, with the initial condition $u_{Cr}(0) = V_S, i_{Lr}(0) = 0$; solving Equation 9.108, we obtain

$$u_{Cr}(t) = \frac{(n-1)V_S}{n}\cos(\omega_r t) - I_O\sqrt{\frac{L_r}{C_r}}\sin(\omega_r t) + \frac{V_S}{n},$$
$$i_{Lr}(t) = I_O\cos(\omega_r t) - I_O + \frac{(n-1)V_S}{n}\sqrt{\frac{L_r}{C_r}}\sin(\omega_r t), \tag{9.109}$$

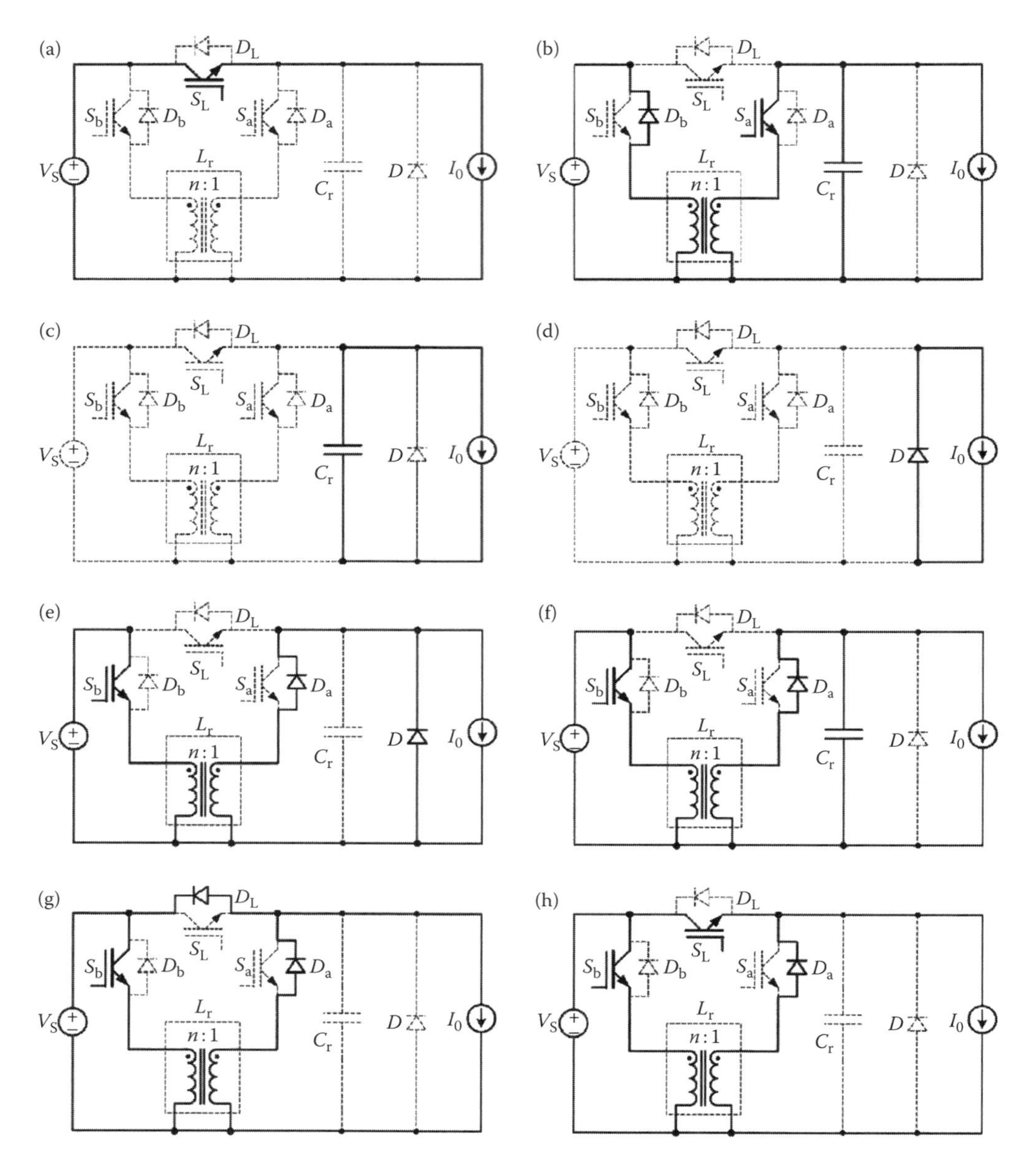

FIGURE 9.93 Operation mode of the resonant DC-link inverter: (a) mode 0, (b) mode 1, (c) mode 2, (d) mode 3, (e) mode 4, (f) mode 5, (g) mode 6, and (h) mode 7.

where $L_r = L_{l1} + L_{l2}/n^2$ is the equivalent inductance of the transformer and $\omega_r = \sqrt{(1/L_r C_r)}$ is the natural angular resonance frequency. Rewriting Equation 9.109, we obtain

$$u_{Cr}(t) = K\cos(\omega_r t + \alpha) + \frac{V_S}{n},$$
$$i_{Lr}(t) = K\sqrt{\frac{C_r}{L_r}}\sin(\omega_r t + \alpha) - I_O, \tag{9.110}$$

where $K = \sqrt{((n-1)^2 V_S^2/n^2 + (I_O^2 L_r/C_r)}$ and $\alpha = \arctan[(nI_O\sqrt{L_r/C_r}/(n-1)V_S]$. Here, n is a number slightly smaller than 2 (the selection of such a number will be explained later), and i_{Lr} will decay to zero faster than u_{Cr}. Let $i_{Lr}(t) = 0$; then the duration of the resonance can be determined by

$$\Delta t_1 = t_1 - t_0 = \frac{\pi - \alpha}{\omega_r}. \quad (9.111)$$

When i_{Lr} is reduced to zero, the auxiliary switch S_a can be turned off with the ZCS condition. At $t = t_1$, the corresponding DC-link voltage u_{Cr} is

$$u_{Cr}(t_1) = \frac{2-n}{n} V_S. \quad (9.112)$$

Mode 2 (as shown in Figure 9.93c) $t_1 < t < t_2$: When the transformer current is reduced to zero, the resonant capacitor is discharged through load from the initial condition as in Equation 9.112. The interval of this mode can be determined by

$$\Delta t_2 = t_2 - t_1 = \frac{C_r V_S (2-n)}{nI_O}. \quad (9.113)$$

As mentioned earlier, n is a number slightly smaller than 2; therefore the interval is normally very short.

Mode 3 (as shown in Figure 9.93d) $t_2 < t < t_3$: The DC-link voltage u_{Cr} is zero. The main switches of the inverter can now be either turned on or turned off under the ZVS condition during this mode. The load current flows through the freewheeling diode D.

Mode 4 (as shown in Figure 9.93e) $t_3 < t < t_4$: As the main switches have turned on or turned off, the auxiliary switch S_b is turned on with ZCS condition (as the i_{Lrs} cannot suddenly change due to the transformer inductance) and the transformer secondary current i_{Lrs} starts to build up linearly. The transformer primary current i_{Lr} also begins to conduct through diode D_a to the load. The current in the freewheeling diode D begins to fall linearly. The load current is slowly diverted from the freewheeling diodes to the resonant circuit. The DC-link voltage u_{Cr} is still equal to zero before the transformer primary current is greater than the load current. The terminal voltages of the transformer primary and secondary windings are equal to zero and the DC power supply voltage V_S, respectively. Redefining the initial time, we obtain

$$0 = L_{11}\frac{di_{Lr}(t)}{dt} + a^2 L_{12}\frac{d[i_{Lrs}(t)/a]}{dt} + aV_S. \quad (9.114)$$

Since the transformer current $i_{Lrs} = i_{Lr}/n$ as in mode 1, rewrite Equation 9.114 as

$$\frac{di_{Lr}}{dt} = -\frac{V_S}{nL_r}. \quad (9.115)$$

The transformer primary current is increased reverse linearly from zero; the mode is ended when $i_{Lr} = -I_O$ and the interval of this mode can be determined by

$$\Delta t_4 = t_4 - t_3 = \frac{nL_r I_O}{V_S}. \quad (9.116)$$

At t_4, i_{Lr} equals the negative load current $-I_O$ and the current through the diode D becomes zero. Thus, the freewheeling diode turns off under ZCS condition, and the diode reverse recovery problems are reduced.

Mode 5 (as shown in Figure 9.93f) $t_4 < t < t_5$: The absolute value of i_{Lr} is continuously increased from I_O, and u_{Cr} is increased from zero when the freewheeling diode D is turned off. Redefining the initial time, we obtain the same equation as Equation 9.108. The initial condition is $u_{Cr}(0) = 0$, $i_{Lr}(0) = -I_O$; neglect the inductor resistance; solving the equation, we obtain

$$u_{Cr}(t) = -\frac{V_S}{n}\cos(\omega_r t) + \frac{V_S}{n},$$
$$i_{Lr}(t) = -I_O - \frac{V_S}{n}\sqrt{\frac{C_r}{L_r}}\sin(\omega_r t). \tag{9.117}$$

When

$$\Delta t_5 = t_5 - t_4 = \frac{1}{\omega_r}\arccos(1-n), \tag{9.118}$$

and $u_{Cr} = V_S$, the auxiliary switch S_L is turned on with ZVS (due to C_r). The interval is independent of the load current. At $t = t_5$, the corresponding transformer primary current i_{Lr} is

$$i_{Lr}(t_5) = -I_O - V_S\sqrt{\frac{(2-n)C_r}{nL_r}}. \tag{9.119}$$

The peak value of the transformer primary current can also be determined:

$$i_{Lr-m} = \left|-I_O - \frac{V_S}{n}\sqrt{\frac{C_r}{L_r}}\right| = I_O + \frac{V_S}{n}\sqrt{\frac{C_r}{L_r}}. \tag{9.120}$$

Mode 6 (as shown in Figure 9.93g) $t_5 < t < t_6$: Both the terminal voltages of primary and secondary windings are equal to the supply voltage V_S after the auxiliary switch S_L is turned on. Redefining the initial time, we obtain

$$V_S = L_{l1}\frac{di_{Lr}(t)}{dt} + a^2 L_{l2}\frac{d[i_{Lrs}(t)/a]}{dt} + aV_S. \tag{9.121}$$

Since the transformer current $i_{Lrs} = i_{Lr}/n$ as in mode 1, rewrite Equation 9.121 as

$$\frac{di_{Lr}}{dt} = \frac{(n-1)V_S}{nL_r}. \tag{9.122}$$

The transformer primary current i_{Lr} decays linearly, and the mode is ended when $i_{Lr} = -I_O$ again. With initial condition (Equation 9.119), the interval of this mode can be determined:

$$\Delta t_6 = t_6 - t_5 = \frac{\sqrt{n(2-n)L_r C_r}}{n-1}. \tag{9.123}$$

The interval is also independent of the load current. As mentioned earlier, n is a number slightly smaller than 2; therefore the interval is also very short.

Mode 7 (as shown in Figure 9.93h) $t_6 < t < t_7$: The transformer primary winding current i_{Lr} decays linearly from the negative load current $-I_O$ to zero. Partial load current flows through the switch S_L. The total current flowing through the switch S_L and transformer is equal to the load current I_O. Redefining the initial time, the transformer winding current obeys Equation 9.122 with the initial condition $i_{Lr}(0) = -I_O$. The interval of this mode is

$$\Delta t_7 = t_7 - t_6 = \frac{nL_r I_0}{(n-1)V_S}. \tag{9.124}$$

Then the auxiliary switch S_b can also be turned off with the ZCS condition after i_{Lr} decays to zero (at any time after t_7).

9.7.3.2 Design Consideration

It is assumed that the inductance of BDCM is much higher than the transformer leakage inductance. From the analysis presented previously, the design considerations can be summarized as follows.

1. Determine the value of the resonant capacitor C_r and the parameters of the transformer.
2. Select the main switches and auxiliary switches.
3. Design the gate signal for the auxiliary switches.

The turn ratio $1:n$ of the transformer can be determined ahead of time. From Equation 9.118, n must satisfy

$$n < 2. \tag{9.125}$$

On the other hand, from Equations 9.112 and 9.113, it is expected that it is as close to 2 as possible so that the duration of mode 2 would not be very long and would be small enough at the end of mode 1.

Normally, n can be selected in the range of 1.7–1.9. The equivalent inductance of the transformer $L_r = L_{l1} + L_{l2}/n^2$ is inversely proportional to the rising rate of switch current when the auxiliary switches are turned on. This means that the equivalent inductance L_r should be big enough to limit the rising rate of the switch current to work in the ZCS condition. The selection of L_r can be referenced from the rule depicted in reference [40].

$$L_r \geq \frac{4t_{on}V_S}{I_{O\max}}, \tag{9.126}$$

where t_{on} is the turn-on time of switch S_a and $I_{O\max}$ is the maximum load current. The resonant capacitance C_r is inversely proportional to the rising rate of switch voltage drop when the switch S_L is turned off. This means that the capacitance is as high as possible to limit the rising rate of the voltage to work in ZVS condition. The selection of the resonant capacitor can be determined by

$$C_r \geq \frac{4t_{off}I_{O\max}}{V_S}, \tag{9.127}$$

where t_{off} is the turn-off time of the switch S_L. However, as the capacitance increases, more energy is stored in it, and the peak value of the transformer current will also be high. The peak value of i_{Lr} should be limited to twice the peak load current. From Equation 9.120, we obtain

$$\sqrt{\frac{C_r}{L_r}} \le \frac{nI_{O\,max}}{V_S}. \tag{9.128}$$

The DC-link voltage rising transition time is expressed as

$$T_w = \Delta t_4 + \Delta t_5 = \frac{nL_r I_{O\,max}}{V_S} + \sqrt{L_r C_r}\arccos(1-n). \tag{9.129}$$

For high switching frequency, T_w should be as short as possible. Select the equivalent inductance L_r and resonant capacitance C_r to satisfy Inequalities 9.125 through 9.128; L_r and C_r should be as small as possible. L_r and C_r selection area is illustrated in Figure 9.94 to determine their values; the valid area is shadowed, where B_1–B_3 is the boundary, which is defined according to Inequalities 9.125 through 9.128.

$$B_1 : L_r = \frac{4t_{on}V_S}{I_{O\,max}}, \tag{9.130}$$

$$B_2 : C_r = \frac{4t_{off}I_{O\,max}}{V_S}, \tag{9.131}$$

$$B_3 : \sqrt{\frac{C_r}{L_r}} = \frac{nI_{O\,max}}{V_S}. \tag{9.132}$$

If boundary B_3 intersects B_1 first as shown in Figure 9.94a, the values of L_r and C_r in the intersection A_1 can be selected. Otherwise, the values of L_r and C_r in the intersection A_2 are selected as shown in Figure 9.94b.

The main switches S_1–S_6 work under the ZVS condition; the voltage stress is equal to the DC power supply voltage V_S. The device current rate can be load current. The auxiliary switch S_L works under the ZVS condition; its voltage and current stress are the same as that of the main switches. The auxiliary switches S_a and S_b work under the ZCS or ZVS condition; the voltage stress is also equal to the DC power supply voltage V_S. The peak current flowing through them is limited to double the maximum load current. As the auxiliary switches S_a and S_b carry the peak current only during switch transitions, they can be rated as lower continuous current rating.

The design of gate signal for the auxiliary switches can be referenced from Figure 9.92. The trailing edge of the gate signal for the auxiliary switch S_L is the same as that of the PWM; the leading edge is determined by the output of the DC-link voltage sensor. The gate signal for the auxiliary switch S_a is a positive pulse with a leading edge the same as that of the PWM trailing edge; its width ΔT_a should be greater than Δt_1. From Equation 9.111, $\Delta t_1$is maximum when the load current is zero. So ΔT_a can be a fixed value determined by

$$\Delta T_a > \Delta t_1|_{max} = \frac{\pi}{\omega_r} = \pi\sqrt{L_r C_r}. \tag{9.133}$$

The gate signal for the auxiliary switch S_b is also a pulse with leading a edge the same as that of the PWM; its width ΔT_b should be longer than $t_7 - t_3$ (i.e., $\Delta t_4 + \Delta t_5 + \Delta t_6 + \Delta t_7$).

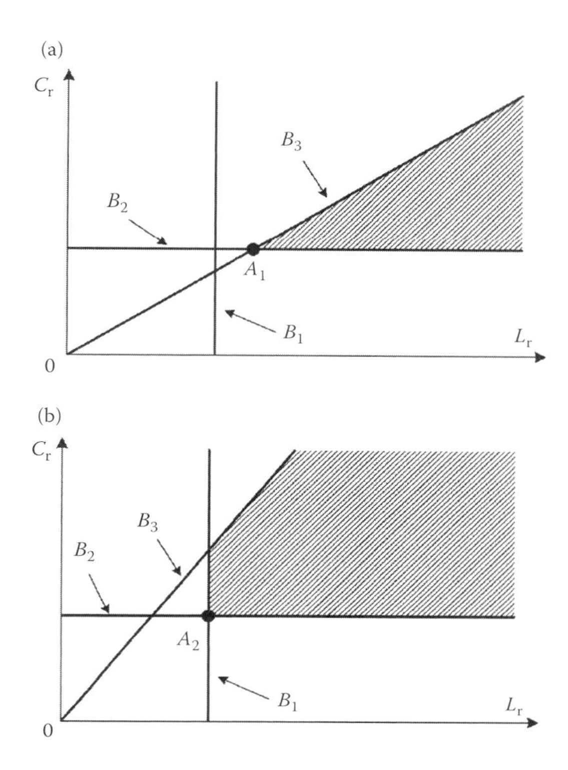

FIGURE 9.94 L and C selection area: (a) Case 1: B2 intersects B3 first and (b) Case 2: B2 intersects B1 first.

ΔT_b can be determined from Equations 9.116,9.118,9.123, and 9.124; that is,

$$\Delta T_b > \sum_{i=4}^{7} \Delta t_i|_{\max} = \frac{n^2 L_r I_{O\max}}{(n-1)V_S} + \sqrt{L_r C_r} \times \left[\arccos(1-n) + \frac{\sqrt{n(2-n)}}{n-1}\right]. \quad (9.134)$$

9.7.3.3 Control Scheme

When the duty of PWM is 100%, that is, a full duty cycle, the main switches of the inverter work under commutation frequency. When it is the instant to commutate the phase current of the BDCM, we control the auxiliary switches S_a, S_b, and S_L, and resonance occurs between the transformer inductor L_r and capacitor C_r. The DC-link voltage reaches zero temporarily; thus ZVS condition of the main switches is obtained. When the duty of PWM is less than 100%, the auxiliary switch S_L works as a chopper. The main switches of the inverter do not switch within a PWM cycle when the phase current does not need to commutate. It has the benefit of reducing phase current drop when the PWM is off. The phase current is commutated when the DC-link voltage becomes zero. There is only one DC-link voltage notch per PWM cycle. It is very important, especially for a very low or very high duty of PWM. Otherwise, the interval between two voltage notches is very short, even overlapping, which will limit the tuning range.

The commutation logical circuit of the system is shown in Figure 9.95. It is similar to the conventional BDCM commutation logical circuit except for adding six D flip-flops to the output. Thus the gate signal of the main switches is controlled by the synchronous pulse CK

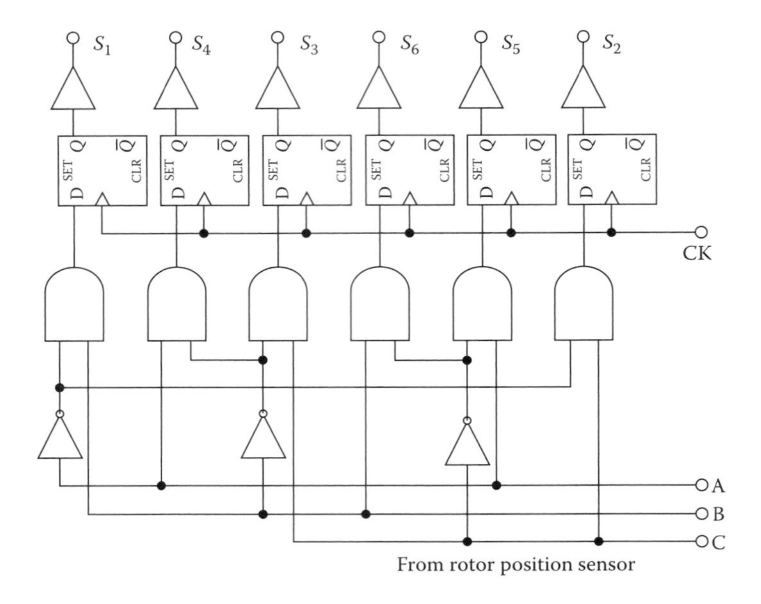

FIGURE 9.95 Commutation logical circuit for the main switches.

that will be mentioned later and the commutation can be synchronized with the auxiliary switches control circuit (shown in Figure 9.96). The operation of the inverter can be divided into PWM operation and full duty cycle operation.

9.7.3.3.1 *Full Duty Cycle Operation*

When the duty of the PWM is 100%, that is, a full duty cycle, the whole ZVT process (modes 1 through 7) occurs when the phase current commutation is ongoing. The monostable flip-flop

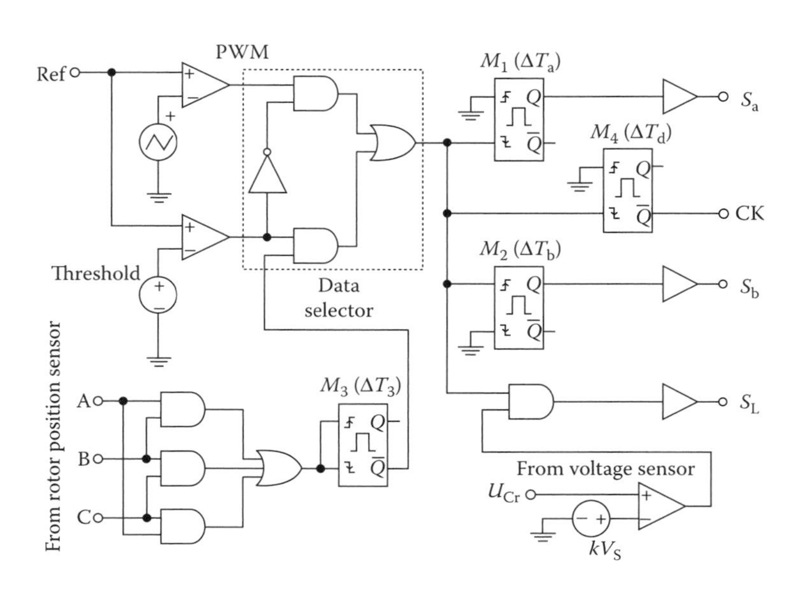

FIGURE 9.96 Control circuit for the auxiliary switches.

M_3 will generate one narrow negative pulse. The width of the pulse ΔT_3 is determined by $(\Delta t_1 + \Delta t_2 + T_c')$, where T_c' is a constant considering the turn-on/off time of the main switches. If n is close to 2, Δt_2 would be very short or u_{Cr} would be small enough at the end of mode 1; ΔT_3 can be determined by

$$\Delta T_3 = \Delta t_1|_{max} + T_c = \pi\sqrt{L_r C_r} + T_c, \tag{9.135}$$

where T_c is a constant that is greater than T_c'. The data selector makes the output of monostable flip-flop M_3 active. The monostable flip-flop M_1 generates a positive pulse when the trailing edge of the M_3 negative pulse is coming. The pulse is the gate signal for the auxiliary switch S_a and its width is ΔT_a, which is determined by Inequality 9.133. The gate signal for switch S_L is flopped to low at the same time. Then mode 1 begins and the DC-link voltage is reduced to zero. Synchronous pulse CK is also generated by a monostable flip-flop M_4, the pulse width ΔT_d should be greater than maximum Δt_1 (i.e., $\pi\sqrt{L_r C_r}$). If the D flip-flops are rising edge active, then CK is connected to the negative output of the M_4, otherwise CK is connected to the positive output. Thus the active edge of pulse CK is within mode 3 when the voltage of the DC link is zero and the main switches of the inverter get ZVS condition. The monostable flip-flop M_2 generates a positive pulse when the leading edge of the negative pulse is coming. The pulse width of M_2 is ΔT_d, which is determined by Inequality 9.134. Then modes 4 through 7 occur and the DC-link voltage is increased to that of the supply again. The leading edge of the gate signal for the switch S_L is determined by the DC-link voltage sensor signal. In other words, in full cycle operation when the phase current commutation is ongoing, the resonant circuit generates a DC-link voltage notch to let the main switches of the inverter switch under the ZVS condition.

9.7.3.3.2 *PWM Operation*

In this operation, the data selector makes the PWM signal active. The auxiliary switch S_L works as a chop, but the main switches of the inverter do not turn on or turn off within a single PWM cycle when the phase current does not need to commutate. The load current is commutated when the DC-link voltage becomes zero. (As the PWM cycle is very short, it does not affect the operation of the motor.)

1. When the PWM signal is flopped down, mode 1 begins, and the pulse signal for the switch S_a is generated by M_1 and the gate signal for the switch S_L is decreased to a low level. However, the voltage of the DC link does not increase until the PWM signal is flipped up. Pulse CK is also generated by M_4 to let the active edge of CK get located in mode 3.
2. When the PWM signal is flipped up, mode 4 begins, and the pulse signal for switch S_b is generated at the moment. Then, when the voltage of the DC link is increased to supply voltage V_S, the gate signal for switch S_L is flipped to a high level.

Thus, only one ZVT occurs per PWM cycle: modes 1 and 2 for PWM turning-off, and modes 4 through 7 for PWM turning-on. And the switching frequency would not be greater than the PWM frequency.

9.7.3.4 Simulation and Experimental Results

The proposed system is verified by PSim simulation software. The DC power supply voltage V_S is 240 V; the maximum load current is 12 A. The transformer turn ratio n is 1:1.8; the leakage inductances of the primary secondary windings are selected as 4 μH and 12.96 μH, respectively. So the equivalent transformer inductance L_r is about 8 μH. The resonant capacitance C_r is 0.1 μF. Switch $S_{a,b}$ gate signal widths ΔT_a and ΔT_b are set as 3 μs and 6 μs, respectively. The narrow negative pulse width ΔT_3 in a full duty cycle is set as 4.5 μs; the delay time for synchronous pulse CK is set as 3.5 μs. The frequency of the PWM is 20 kHz. Waveforms of the DC-link voltage u_{Cr}, the transformer primary winding current i_{Lr}, the switch S_L and diode D_L currents i_{SL} and i_{DL}, PWM, and the auxiliary switch gate signal under low and high load currents are shown in Figure 9.97. The figure shows that the inverter worked well under various load currents.

In order to verify the theoretical analysis and simulation results, the proposed soft-switching inverter was tested on an experimental prototype. The DC-link voltage is 240 V, the rated phase current is 10.8 A, and the switching frequency is 20 kHz. Select 50 A/1200 V BSM 35 GB 120 DN2 dual IGBT module as the main inverter switches S_1–S_6 and the auxiliary switch S_L; another switch in the same module of S_L can be adopted as the auxiliary switch S_a, and 30 A/600 V IMBH30D-060 IGBT can be adopted as the auxiliary switch S_b. With the datasheets of these switches and Equations 9.125 through 9.128, the value of the capacitance and the parameter of the transformer can be determined. A polyester capacitor of 0.1 μF, 1000 V was adopted as the DC-link resonant capacitor C_r. A high magnetizing inductance transformer with turn ratio 1:1.8 was employed in the experiment. The equivalent inductance is about 8 μH under a short-circuit test [39]. The switching frequency is 20 kHz. The monostable flip-flop is set up using IC 74LS123, a variable resistor, and a capacitor. The logical gate can be replaced by a programmable logical device to reduce the number of ICs. ΔT_a, ΔT_b, ΔT_c, and ΔT_d are set as 3 μs, 6 μs, 4.5 μs, and 3.5 μs, respectively.

The system is tested in light and heavy loads. The waveforms of DC-link voltage u_{Cr} and the transformer primary winding current i_{Lr} in low and high load currents are shown in Figure 9.98a and b, respectively. The transformer-based resonant DC-link inverter works well under various load currents. The waveforms of auxiliary switch S_L voltage u_{SL} and its current i_{SL} are shown in Figure 9.98c. There is little overlap between the switch S_L voltage and its current during the switching under the soft-switching condition; so the switching power losses are low. The waveforms of resonant DC-link voltage u_{Cr} and synchronous signal CK are shown in Figure 9.98d, in which the main switches can switch under the ZVS condition during commutation. The phase current of BDCM is shown in Figure 9.98e. The design of the system is successful.

Homework

9.1. A diode-clamped five-level inverter shown in Figure 9.3b operates in the state with best THD. Determine the corresponding switching angles, switch status, and THD.

9.2. A capacitor-clamped three-level inverter is shown in Figure 9.8b. It operates in the equal-angle state, that is, the operation time in each level is 45°. Determine the status of the switches and the corresponding THD.

9.3. A three-HB multilevel inverter is shown in Figure 9.9. The output voltage is v_{an}. It is implemented as a THMI. Explain the inverter working operation, draw the corresponding waveforms, and indicate the source voltages arrangement and how many levels can be implemented.

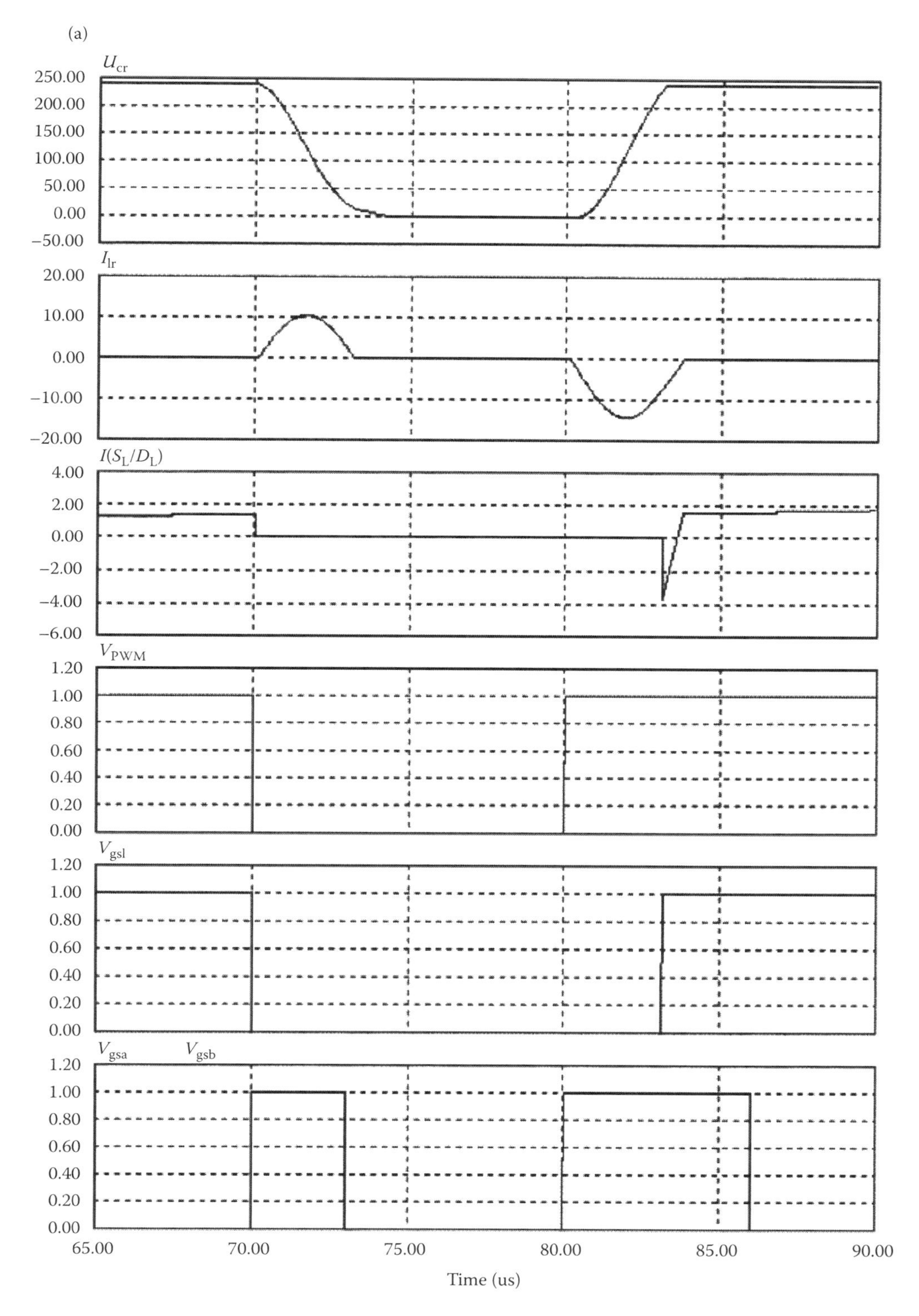

FIGURE 9.97 Waveforms of u_{Cr}, i_{Lr}, i_{SL}/i_{DL}, PWM, and auxiliary switches gate signal under various load current: (a) under low load current ($I_O = 2\,A$) and (b) under high load current ($I_O = 8\,A$).

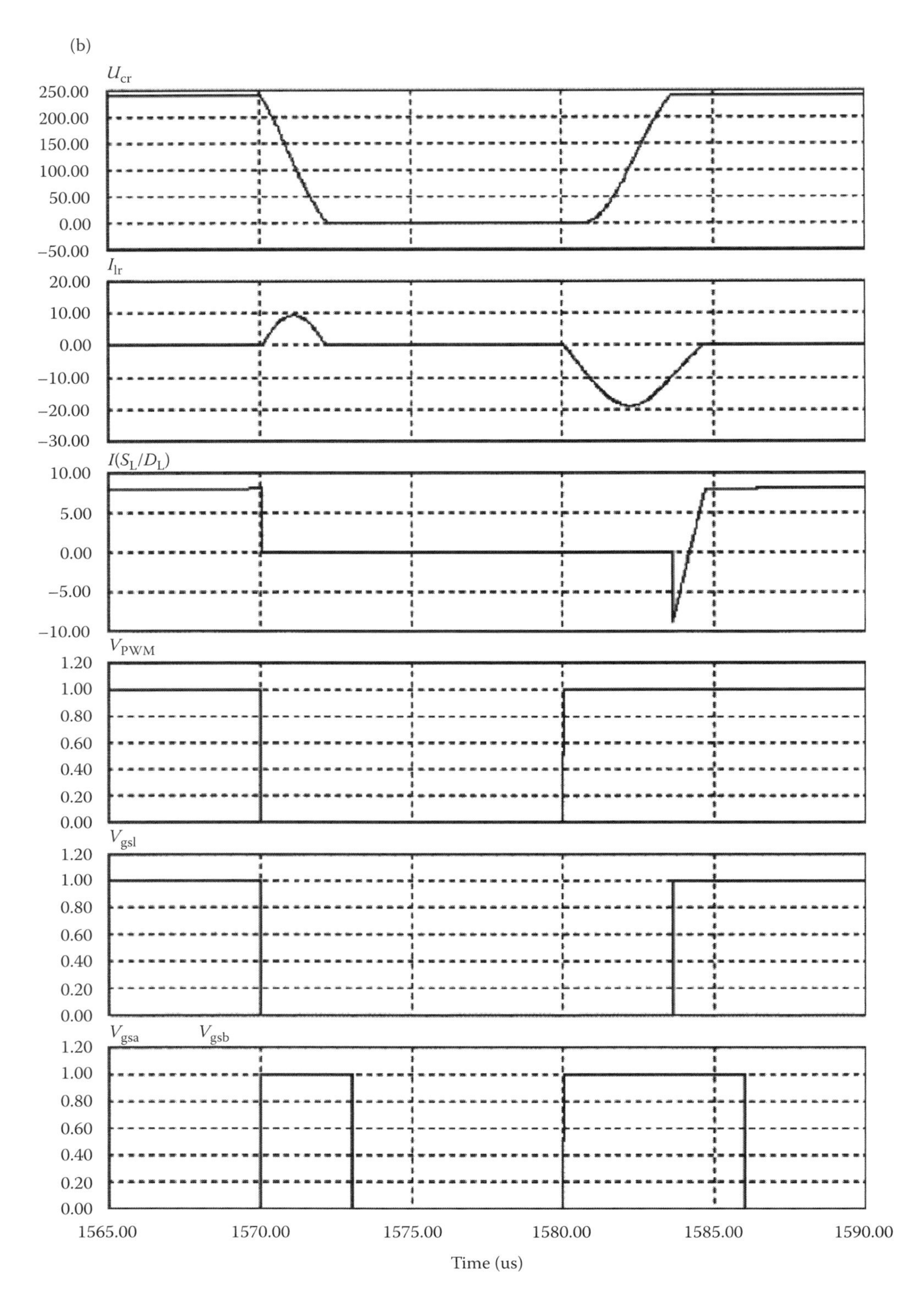

FIGURE 9.97 Continued.

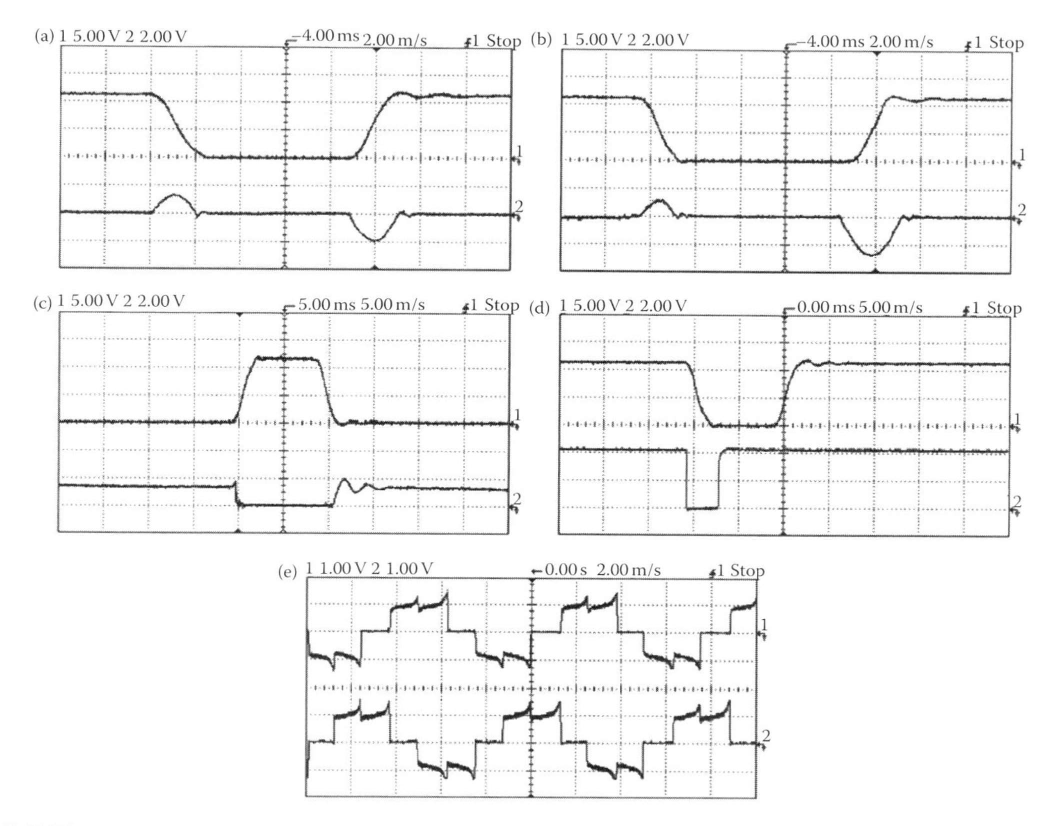

FIGURE 9.98 Experiment waveforms: (a) the DC-link voltage u_{Cr} (top) and transformer current i_{Lr} (bottom) under low load current (100 V/div, 10 A/div), (b) the DC-link voltage u_{Cr} (top) and transformer current i_{Lr} (bottom) under high load current (100 V/div, 10 A/div), (c) switch S_L voltage (top) and current (bottom) (100 V/div, 10 A/div), (d) the DC-link voltage u_{Cr} (top) and the synchronous signal CK (bottom) (100 V/div), and (e) the phase current of BDCM (5 A/div).

References

1. Nabae, A., Takahashi, I., and Akagi, H. 1980. A neutral-point clamped PWM inverter. *Proceedings of IEEE APEC'80 Conference*, pp. 761–766.
2. Nabae, A., Takahashi, I., and Akagi, H. 1981. A neutral-point clamped PWM inverter. *IEEE Transactions on Industry Applications*, 17, 518–523.
3. Mohan, N., Undeland, T. M., and Robbins, W. P. 2003. *Power Electronics: Converters, Applications and Design*. New York: Wiley.
4. Trzynadlowski, A. M. 1998. *Introduction to Modern Power Electronics*. New York: Wiley.
5. Peng, F. Z. 2001. A generalized multilevel inverter topology with self voltage balancing. *IEEE Transactions on Industry Applications*, 37, 611–618.
6. Liu, Y. and Luo, F. L. 2008. Trinary hybrid 81-level multilevel inverter for motor drive with zero common-mode voltage. *IEEE-Transactions on Industrial Electronics*, 55, 1014–1021.
7. Hammond, P. W. 1997. New approach to enhance power quality for medium voltage AC drives. *IEEE Transactions on Industry Applications*, 33, 202–208.
8. Baker, R. H. and Bannister, L. H. 1975. Electric power converter, *U.S. Patent 3 867 643*.

9. Cengelci, E., Sulistijo, S. U., Woo, B. O., Enjeti, P., Teoderescu, R., and Blaabjerg, F. 1999. A new medium-voltage PWM inverter topology for adjustable-speed drives. *IEEE Transactions on Industry Applications*, 35, 628–637.
10. Manjrekar, M. D., Steimer, P. K., and Lipo, T. A. 2000. Hybrid multilevel power conversion system: A competitive solution for high-power applications. *IEEE Transactions on Industry Applications*, 36, 834–841.
11. Akagi, H. 2006. Medium-voltage power conversion systems in the next generation. *Proceedings of IEEE-IPEMC* 2006, pp. 23–30.
12. Inoue, S. and Akagi, H. 2007. A bidirectional isolated DC–DC converter as a core circuit of the next-generation medium-voltage power conversion system. *IEEE Transactions on Power Electronics*, 22, 535–542.
13. Jain, M., Daniele, M., and Jain, P. K. 2000. A bidirectional DC–DC converter topology for low power application. *IEEE Transactions on Power Electronics*, 15, 595–606.
14. Liu, Y. and Luo, F. L. 2006. Multilevel inverter with the ability of self voltage balancing. *IEE-Proceedings on Electric Power Applications*, 153, pp. 105–115.
15. Pan, Z. Y. and Luo, F. L. 2004. Novel soft-switching inverter for brushless DC motor variable speed drive system. *IEEE Transactions on Power Electronics*, 19, 280–288.
16. Divan, D. M. 1989. The resonant DC link converter—a new concept in static power conversion. *IEEE Transactions on Industry Applications*, 25, 317–325.
17. Divan, D. M. and Skibinski, G. 1989. Zero-switching-loss inverters for highpower applications. *IEEE Transactions on Industry Applications*, 25, 634–643.
18. Yi, W., Liu, H. L., Jung, Y. C., Cho, J. G., and Cho, G. H. 1992. Program-controlled soft switching PRDCL inverter with new space vector PWM algorithm. *Proceedings of IEEE PESC'92*, pp. 313–319.
19. Malesani, L., Tenti, P., Tomasin, P., and Toigo, V. 1995. High efficiency quasiresonant DC link three-phase power inverter for full-range PWM. *IEEE Transactions on Industry Applications*, 31, 141–148.
20. Jung, Y. C., Liu, H. L., Cho, G. C., and Cho, G. H. 1995. Soft switching space vector PWM inverter using a new quasiparallel resonant DC link. *Proceedings of IEEE PESC*, pp. 936–942.
21. Zhengfeng, M. and Yanru, Z. 2001. A novel DC-rail parallel resonant ZVT VSI for three-phases AC motor drive, *Proceedings of International Conference on Electric Machines Systems* (*ICEMS* 2001), pp. 492–495.
22. Murai, Y., Kawase, Y., Ohashi, K., Nagatake, K., and Okuyama, K. 1989. Torque ripple improvement for brushless DC miniature motors. *IEEE Transactions on Industry Applications*, 25, 441–450.
23. Chang-heeWon, C., Joong-ho Song, J., and Choy, I. 2002. Commutation torque ripple reduction in brushless DC motor drives using a single DC current sensor. *Proceedings of IEEE PESC*, pp. 985–990.
24. Sebastian, T. and Gangla, V. 1996. Analysis of induced EMF waveforms and torque ripple in a brushless permanent magnet machine. *IEEE Transactions on Industry Applications*, 32, 195–200.
25. Pillay, P. P. and Krishnan, R. 1988. Modeling of permanent magnet motor drives. *IEEE Transactions on Industrial Electronics*, 35, 537–541.
26. Pan, Z. Y. and Luo, F. L. 2005. Novel resonant pole inverter for brushless DC motor drive system. *IEEE Transactions on Power Electronics*, 20, 173–181.
27. De Doncker, R.W. and Lyons, J. P. 1990. The auxiliary resonant commutated pole converter. *Proceedings of IEEE Industry Applications Society Annual Meeting*, pp. 1228–1235.
28. McMurray, W. 1989. Resonant snubbers with auxiliary switches. *Proceedings of IEEE Industry Applications Society Annual Meeting*, pp. 289–834.
29. Vlatkovic, V., Borojevic, D., Lee, F., Cuadros, C., and Gataric, S. 1993. A new zero-voltage transition, three-phase PWM rectifier/inverter circuit. *Proceedings of IEEE PESC*, pp. 868–873.
30. Cuadros, C., Borojevic, D., Gataric, S., and Vlatkovic, V. 1994. Space vector modulated, zero-voltage transition three-phase to DC bidirectional converter. *Proceedings of IEEE PESC*, pp. 16–23.

31. Lai, J. S., Young, Sr., R.W., Ott, Jr., G.W., White, C. P., McKeever, J. W., and Chen, D. 1995. A novel resonant snubber based soft-switching inverter. *Proceedings of Applied Power Electronics Conference* pp. 797–803.
32. Lai, J. S., Young, Sr., R. W., Ott, Jr., G.W., McKeever, J. W., and Peng, F. Z. 1996. A delta-configured auxiliary resonant snubber inverter. *IEEE Transactions on Industry Applications*, 32, 518–525.
33. Miller, T. J. E. 1989. *Brushless Permanent-Magnet and Reluctance Motor Drives*. Oxford, UK: Clarendon.
34. Sen, P. C. 1997. *Principles of Electric Machines and Power Electronics*. New York: Wiley.
35. Divan, D. M., Venkataramanan, G., and De Doncker, R. W. 1987. Design methodologies for soft switched inverters. *Proceedings of IEEE Industry Applications Society. Annual Meeting*, pp. 626–639.
36. Yi, W., Liu, H. L., Jung, Y. C., Cho, J. G., and Cho, G. H. 1992. Program-controlled soft switching PRDCL inverter with new space vector PWM algorithm. *Proceedings of IEEE PESC*, pp. 313–319.
37. Ming, Z. Z. and Zhong, Y. R. 2001. A novel DC-rail parallel resonant ZVT VSI for three-phases AC motor drive. *Proceedings of International Conference Electronic Machines Systems*, pp. 492–495.
38. Pan, Z. Y. and Luo, F. L. 2005. Transformer based resonant DC link inverter for brushless DC motor drive system. *IEEE Transactions on Power Electronics*, 20, 939–947.
39. Sen, P. C. 1997. *Principles of Electric Machines and Power Electronics*. New York: Wiley.
40. Wang, K. R., Jiang, Y. M., Dubovsky, S., Hua, G. C., Boroyevich, D., and Lee, F. C. 1997. Novel DC-rail soft-switched three-phase voltage-source inverters. *IEEE Transactions on Industry Applications*, 33, 509–517.

10

Traditional AC/AC Converters

AC/AC conversion technology is an important subject area in research and industrial applications. In recent decades, the AC/AC conversion technique has been developed to a great extent. We can sort them into two parts. The converters developed in the last century can be called the traditional AC/AC converters that are introduced in this chapter. The new technologies of AC/AC conversion technology will be introduced in the next chapter [1–6].

10.1 Introduction

A power electronic AC/AC converter accepts electric power from one system and converts it for delivering it to another AC system with a different *amplitude, frequency*, and *phase*. They may be of single-phase or three-phase type depending on their power ratings. The AC/AC converters employed to vary the rms voltage across the load at constant frequency are known as *AC voltage controllers* or *AC regulators*. The voltage control is accomplished either by (i) *phase control* under natural commutation using pairs of *triacs, SCRs*, or thyristors; or by (ii) *on/off control* under forced commutation using fully controlled self-commutated switches such as gate turn-off thyristors (GTOs), power bipolar transistors (BTs), insulated gate bipolar transistors (IGBTs), MOS-controlled thyristors (MCTs), and so on [7–8].

The AC/AC power converters in which the AC power at one frequency is directly converted to an AC power at another frequency *without any intermediate DC conversion link* are known as *cycloconverters*, the majority of which use naturally commutated SCRs for their operation when the maximum output frequency is limited to a fraction of the input frequency. With the rapid advancement of fast-acting fully controlled switches, the force-commutated cycloconverters (FCCs) or the recently developed *matrix converters* (MCs) with bidirectional on/off control switches provide independent control of the magnitude and frequency of the generated output voltage as well as sinusoidal modulation of the output voltage and current.

While typical applications of AC voltage controllers include lighting and heating control, online transformer tap changing, soft-starting, and speed control of pump and fan drives, the cycloconverters are used mainly for high-power low-speed large AC motor drives for application in cement kilns, rolling mills, and ship propellers. The power circuits, control methods, and the operation of the AC voltage controllers, cycloconverters, and MCs are introduced in this section. A brief review regarding their applications is also given.

The input voltage of a diode rectifier is an AC voltage, which can be single-phase or three-phase voltages. They are usually a pure sinusoidal wave. For a single-phase input voltage, the input voltage can be expressed as

$$v_s = \sqrt{2}V_{rms} \sin \omega t = V_m \sin \omega t,$$

where v_s is the instantaneous input voltage, V_m is its amplitude, and V_{rms} is its rms value.

Traditional AC/AC converters are sorted in three groups:

- Voltage-regulation converters
- Cycloconverters
- MCs.

Each group has single-phase and three-phase converters.

10.2 Single-Phase AC/AC Voltage-Regulation Converters

The basic power circuit of a single-phase AC/AC voltage converter, as shown in Figure 10.1a, is composed of a pair of SCRs connected back-to-back (also known as inverse-parallel or antiparallel) between the AC supply and the load. This connection provides a *bidirectional full-wave symmetrical* control and the SCR pair can be replaced by a Triac in Figure 10.1b for low-power applications. Alternate arrangements are as shown in Figure 10.1c with two diodes and two SCRs to provide a common cathode connection for simplifying the gating circuit without needing isolation, and in Figure 10.1d with one SCR and four diodes to reduce the device cost but with increased device conduction loss. An SCR and diode combination, known as a *thyrode controller*, as shown in Figure 10.1e, provides a *unidirectional half-wave asymmetrical voltage* control with device economy but introduces a DC component and more harmonics, and thus is not very practical to use except for a very low power heating load [1–5].

With *phase control*, the switches conduct the load current for a chosen period of each input cycle of voltage, and with *on/off* control, the switches connect the load either for a few cycles of input voltage and disconnect it for the next few cycles (*integral cycle control*) or the switches are turned on and off several times within alternate half-cycles of input voltage (*AC chopper or PWM AC voltage controller*).

10.2.1 Phase-Controlled Single-Phase AC/AC Voltage Controller

For a full-wave, symmetrical phase control, the SCRs T_1 and T_2 shown in Figure 10.1a are gated at α and $\pi + \alpha$, respectively, from the zero crossing of the input voltage, and by varying α, the power flow to the load is controlled through voltage control in alternate half-cycles. As long as one SCR is carrying current, the other SCR remains reverse biased by the voltage drop across the conducting SCR. The principle of operation in each half-cycle is similar to that of the controlled half-wave rectifier and one can use the same approach for the analysis of the circuit.

10.2.1.1 Operation with R load

Figure 10.2 shows the typical voltage and current waveforms for the single-phase bi-directional phase-controlled AC voltage controller of Figure 10.1a with resistive load.

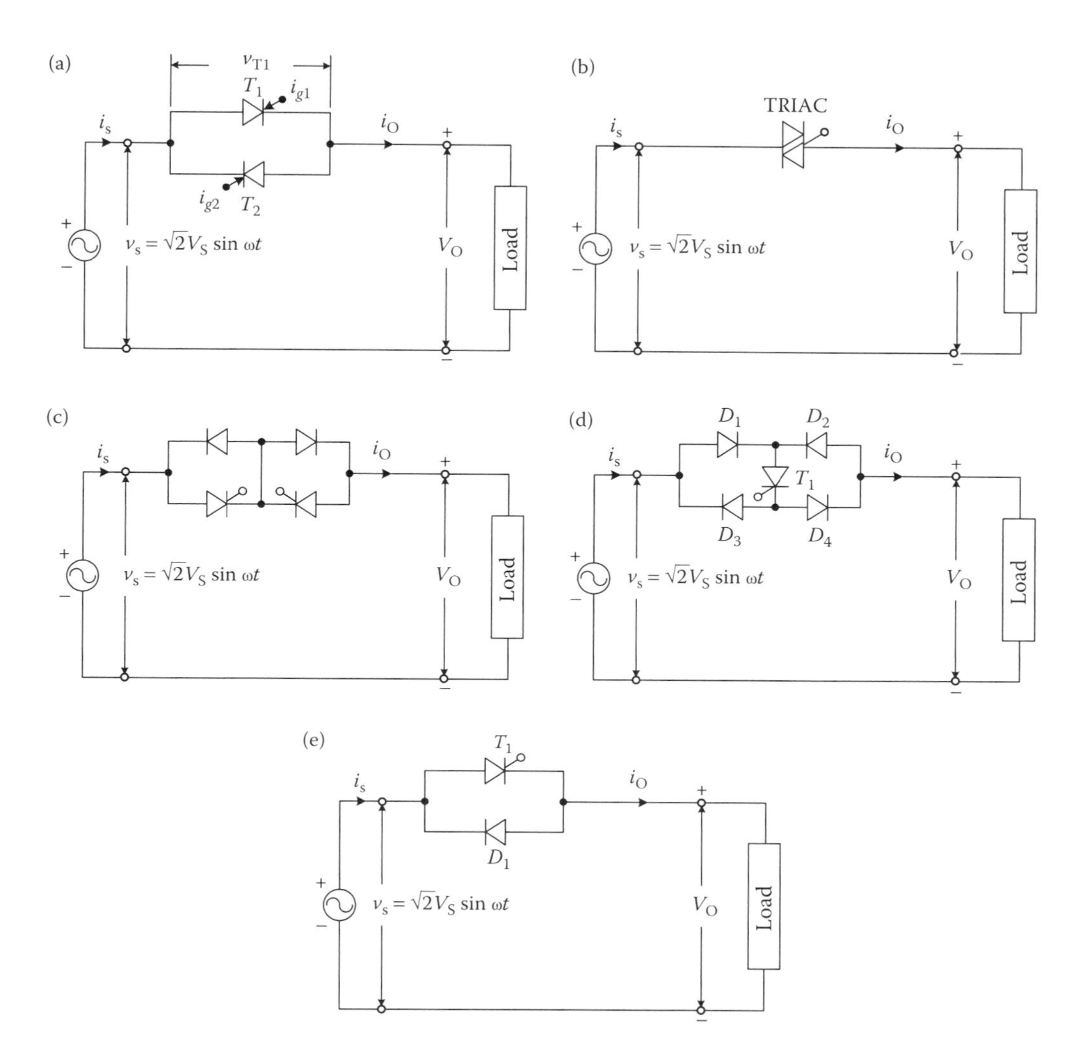

FIGURE 10.1 Single-phase AC voltage controllers: (a) full-wave with two SCRs in inverse parallel, (b) full-wave with Triac, (c) full-wave with two SCRs and two diodes, (d) full-wave with four diodes and one SCR, and (e) half-wave with one SCR and one diode in antiparallel. (Reprinted from Luo, F. L., Ye, H., and Rashid, M. H. 2005. *Digital Power Electronics and Applications*. Boston: Academic Press, Elsevier. With permission.)

The output voltage and current waveforms have half-wave symmetry and thus no DC component.

If $v_S = \sqrt{2}V_S \sin \omega t$ is the source voltage, then the rms output voltage with T_1 triggered at α can be found from the half-wave symmetry as

$$V_O = \left[\frac{1}{\pi}\int_{\alpha}^{\pi} 2V_S^2 \sin^2 \omega t \, d(\omega t)\right]^{1/2} = V_S\left[1 - \frac{\alpha}{\pi} + \frac{\sin 2\alpha}{2\pi}\right]^{1/2}. \tag{10.1}$$

Note that V_O can be varied from V_S to 0 by varying α from 0 to π.

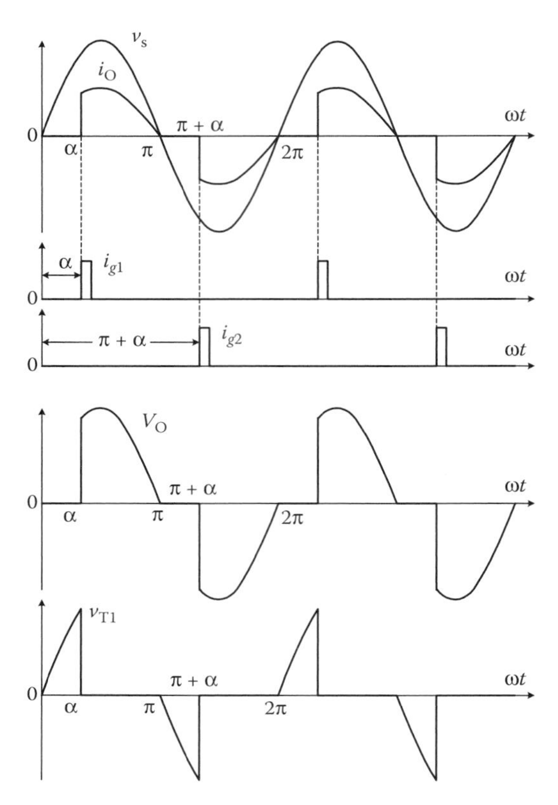

FIGURE 10.2 Waveforms of the single-phase AC full-wave voltage controller with *R* load. (Reprinted from Luo, F. L., Ye, H., and Rashid, M. H. 2005. *Digital Power Electronics and Applications*. Boston: Academic Press, Elsevier. With permission.)

The rms value of load current is

$$I_O = \frac{V_O}{R}. \tag{10.2}$$

The input power factor is

$$\frac{P_O}{VA} = \frac{V_O}{V_S} = \left[1 - \frac{\alpha}{\pi} + \frac{\sin 2\alpha}{2\pi}\right]^{1/2}. \tag{10.3}$$

The average SCR current is

$$I_{A,SCR} = \frac{1}{2\pi R}\int_{\alpha}^{\pi} \sqrt{2}V_S \sin \omega t \, d(\omega t). \tag{10.4}$$

As each SCR carries half the line current, the rms current in each SCR is

$$I_{O,SCR} = \frac{I_O}{\sqrt{2}}. \tag{10.5}$$

Example 10.1

A single-phase full-wave AC/AC voltage controller shown in Figure 10.1a has input rms voltage $v_S = 220\,\text{V}/50\,\text{Hz}$, load $R = 100\,\Omega$, and the firing angle $\alpha = 60°$ for the thyristors T_1 and T_2. Determine the output rms voltage V_O and current I_O, and the DPF.

Solution

From Equation 10.1, the output rms voltage is

$$V_O = V_S\left(1 - \frac{\alpha}{\pi} + \frac{\sin 2\alpha}{2\pi}\right)^{1/2} = 220\left(1 - \frac{1}{3} + \frac{\sqrt{3}}{4\pi}\right)^{1/2}$$

$$= 220(1 - 0.33333 + 0.13783)^{1/2} = 197.33\,\text{V}.$$

The output rms current is

$$I_O = \frac{V_O}{R} = \frac{197.33}{100} = 1.9733\,\text{A}.$$

The fundamental harmonic wave is delayed to the supply voltage by the firing angle $\alpha = 60°$. Therefore, DPF $= \cos\alpha = 0.5$.

From this example, we can recognize the fact that if the firing angle is greater than 90°, it is possible to obtain leading PF.

10.2.1.2 Operation with RL Load

Figure 10.3 shows the voltage and current waveforms for the controller in Figure 10.1a with RL load. Due to the inductance, the current carried by the SCR T_1 may not fall to zero at $\omega t = \pi$ when the input voltage goes negative, and may continue until $\omega t = \beta$, the extinction angle, as shown in Figure 10.3.

The conduction angle

$$\theta = \beta - \alpha \tag{10.6}$$

of the SCR depends on the firing delay angle α and the load impedance angle ϕ. The expression for the load current $I_O(\omega t)$ when conducting from α to β can be derived in the same way as that used for a phase-controlled rectifier in a DCM by solving the relevant Kirchhoff voltage equation:

$$i_O(\omega t) = \frac{\sqrt{2}V}{Z}\left[\sin(\omega t - \phi) - \sin(\alpha - \phi)e^{(\alpha - \omega t)/\tan\phi}\right], \quad \alpha < \omega t < \beta, \tag{10.7}$$

where Z (load impedance) $= (R^2 + \omega^2 L^2)^{1/2}$, and ϕ (load impedance angle) $= \tan^{-1}(\omega L/R)$. The angle β, when the current I_O falls to zero, can be determined from the following transcendental equation obtained by inserting $i_O(\omega t = \beta) = 0$ in Equation 10.7:

$$\sin(\beta - \phi) = \sin(\alpha - \phi)e^{(\alpha - \beta)/\tan\phi}. \tag{10.8}$$

From Equations 10.6 and 10.8, one can obtain a relationship between θ and α for a given value of ϕ, as shown in Figure 10.4, which shows that as α is increased, the conduction

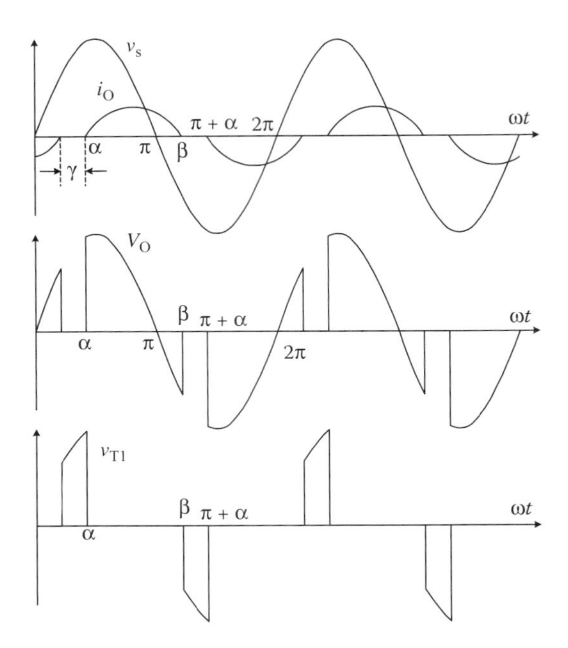

FIGURE 10.3 Typical waveforms of the single-phase AC voltage controller with *RL* load. (Reprinted from Luo, F. L., Ye, H., and Rashid, M. H. 2005. *Digital Power Electronics and Applications*. Boston: Academic Press, Elsevier. With permission.)

angle θ decreases and the rms value of the current decreases. The rms output voltage is

$$V_O = \left[\frac{1}{\pi} \int_{\alpha}^{\beta} 2V_S^2 \sin^2 \omega t \, d(\omega t) \right]^{1/2} = \frac{V_S}{\sqrt{\pi}} \left[\beta - \alpha + \frac{\sin 2\alpha}{2} + \frac{\sin 2\beta}{2} \right]^{1/2}. \tag{10.9}$$

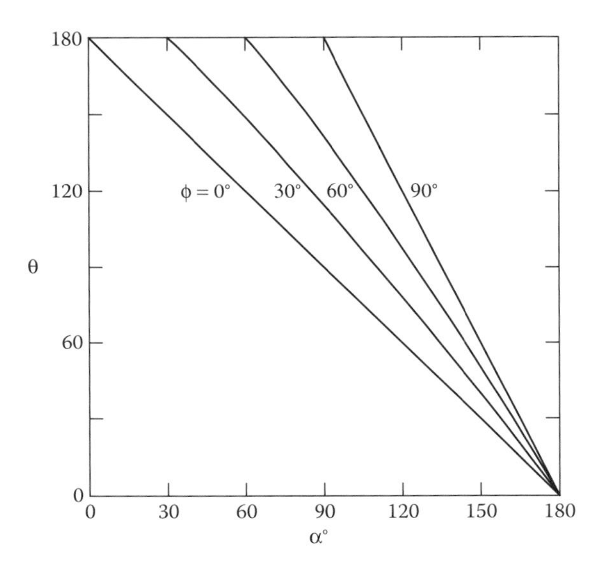

FIGURE 10.4 θ versus α curves for the single-phase AC voltage controller with *RL* load. (Reprinted from Rashid, M. H. 2001. *Power Electronics Handbook*, New York: Academic Press, pp. 307–333. With permission.)

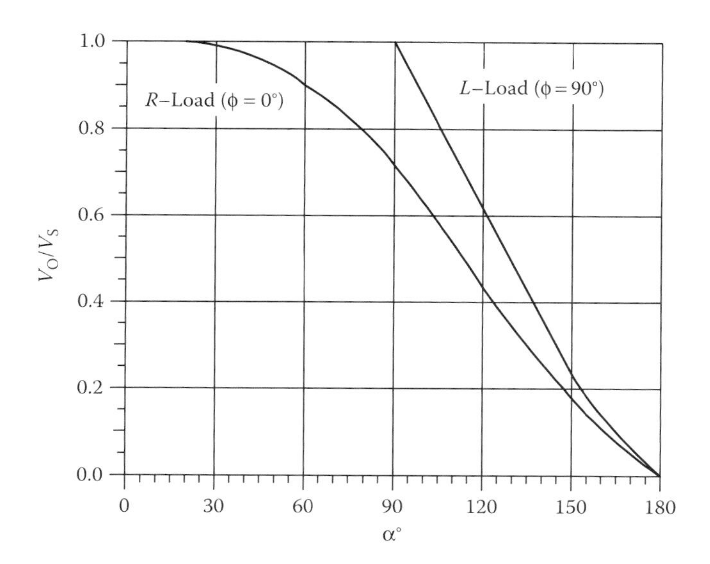

FIGURE 10.5 Envelope of control characteristics of a single-phase AC voltage controller with *RL* load. (Reprinted from Rashid, M. H. 2001. *Power Electronics Handbook*, New York: Academic Press, pp. 307–333. With permission.)

V_O can be evaluated for the two possible extreme values of $\phi = 0$ when $\beta = \pi$, and $\phi = \pi/2$ when $\beta = 2\pi - \alpha$, and the envelope of the voltage-control characteristics for this controller is shown in Figure 10.5.

The rms SCR current can be obtained from Equation 10.7 as

$$I_{O,SCR} = \left[\frac{1}{2\pi} \int_{\alpha}^{\beta} i_O^2 \, d(\omega t) \right]. \tag{10.10}$$

The rms load current is

$$I_O = \sqrt{2} I_{O,SCR}. \tag{10.11}$$

The average value of the SCR current is

$$I_{A,SCR} = \frac{1}{2\pi} \int_{\alpha}^{\beta} i_O \, d(\omega t). \tag{10.12}$$

Example 10.2

A single-phase full-wave AC/AC voltage controller shown in Figure 10.1a has input rms voltage $V_S = 220\text{V}/50\,\text{Hz}$, load $R = 100\,\Omega$, and $L = 183.78\,\text{mH}$, and the firing angle $\alpha = 60°$ for the thyristors T_1 and T_2. Determine the extinction angle β, the output rms voltage V_O and current I_O, and the DPF.

Solution

Since the load is an *RL* load, the output voltage is shown in Figure 10.3. The load impedance angle ϕ is

$$\phi = \tan^{-1}\frac{\omega L}{R} = \tan^{-1}\frac{100\pi \times 183.78\text{m}}{100} = \tan^{-1} 0.57735 = 30^\circ.$$

The conduction angle θ is determined by Equation 10.6, or check the value from Figure 10.4. The conduction angle θ is about 150° (or $5\pi/6$). Therefore, the extinction angle β is

$$\beta = \theta + \alpha = \frac{5\pi}{6} + \frac{\pi}{3} = \frac{7}{6}\pi\ \text{rad}.$$

From Equation 10.1, the output rms voltage is

$$V_O = V_s\left(1 - \frac{\alpha}{\pi} + \frac{\sin 2\alpha}{2\pi}\right)^{1/2} = 220\left(1 - \frac{1}{3} + \frac{\sqrt{3}}{4\pi}\right)^{1/2}$$
$$= 220(1 - 0.33333 + 0.13783)^{1/2} = 197.33\,\text{V}.$$

The output rms current is

$$I_O = \frac{V_O}{R} = \frac{197.33}{100} = 1.9733\ \text{A}.$$

The fundamental harmonic wave is delayed to the supply voltage by the firing angle $\alpha = 60^\circ$. Therefore, the DPF is given by the equation DPF $= \cos\alpha = 0.5$.

10.2.1.3 Gating Signal Requirements

For the inverse-parallel SCRs as shown in Figure 10.1a, the gating signals of SCRs must be isolated from one another as there is no common cathode. For *R* load, each SCR stops conducting at the end of each half-cycle, and under this condition, single short pulses may be used for gating as shown in Figure 10.2. With *RL* load, however, this single short pulse gating is not suitable as shown in Figure 10.6. When SCR T_2 is triggered at $\omega t = \pi + \alpha$, SCR T_1 is still conducting due to the load inductance. By the time the SCR T_1 stops conducting at β, the gate pulse for SCR T_2 has already ceased and T_2 will fail to turn on, causing the converter to operate as a single-phase rectifier with conduction of only T_1. This necessitates the application of a sustained gate pulse either in the form of a continuous signal for the half-cycle period, which increases the dissipation in the SCR gate circuit and a large isolating pulse transformer, or better, a *train of pulses* (*carrier frequency gating*) to overcome these difficulties.

10.2.1.4 Operation with $\alpha < \phi$

If $\alpha = \phi$, then from Equation 10.8,

$$\sin(\beta - \phi) = \sin(\beta - \alpha) = 0 \quad (10.13)$$

and

$$\beta - \alpha = \theta = \pi. \quad (10.14)$$

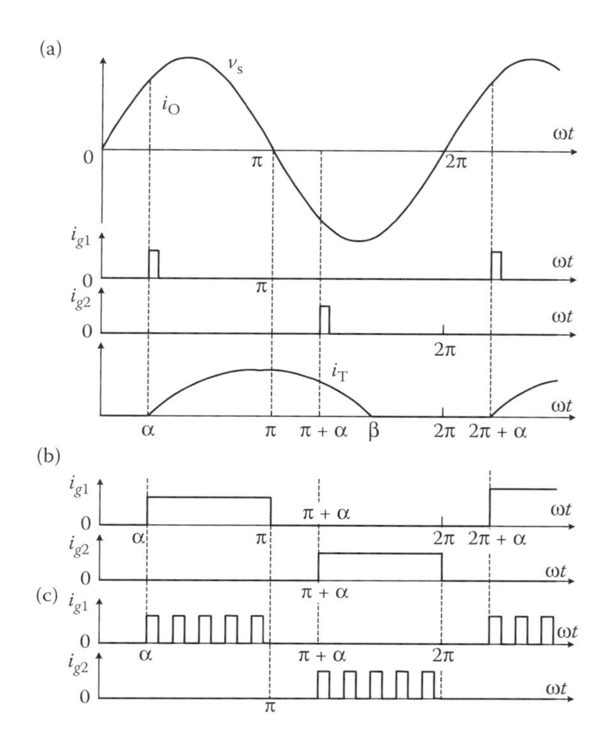

FIGURE 10.6 Single-phase full-wave controller with *RL* load: gate pulse requirements. (a) Source voltage and firing pulse and thyrister current, (b) thyristers' gate currents, and (c) other thyristers' gate currents. (Reprinted from Rashid, M. H. 2001. *Power Electronics Handbook*, New York: Academic Press, pp. 307–333. With permission.)

As the conduction angle θ cannot exceed π and the load current must pass through zero, the control range of the firing angle is $\phi \leq \alpha \leq \pi$. With narrow gating pulses and $\alpha < \phi$, only one SCR will conduct, resulting in a rectifier action as shown. Even with a train of pulses, if $\alpha < \phi$, the changes in the firing angle will not change the output voltage and current, but both SCRs will conduct for the period π with T_1 turning on at $\omega t = \pi$ and T_2 at $\omega t + \pi$. This *dead zone* ($\alpha = 0$ to ϕ), whose duration varies with the load impedance angle ϕ, is not a desirable feature in closed-loop control schemes. An alternative approach to the phase control with respect to the input voltage zero crossing has been reported in which the firing angle is defined with respect to the instant when it is the load current (not the input voltage) that reaches zero, this angle being called *the hold-off angle* (γ) *or the control angle* (as marked in Figure 10.3). This method requires sensing the load current, which may otherwise be required anyway in a closed-loop controller for monitoring or control purposes.

10.2.1.5 Power Factor and Harmonics

As in the case of phase-controlled rectifiers, the important limitations of the phase-controlled AC voltage controllers are the poor PF and the introduction of harmonics in the source currents. As seen from Equation 10.3, the input PF depends on α, and as α increases, the PF decreases.

The harmonic distortion increases and the quality of the input current decreases with an increase of the firing angle. The variations of low-order harmonics with the firing angle as computed by Fourier analysis of the voltage waveform of Figure 10.2 (with *R* load) are

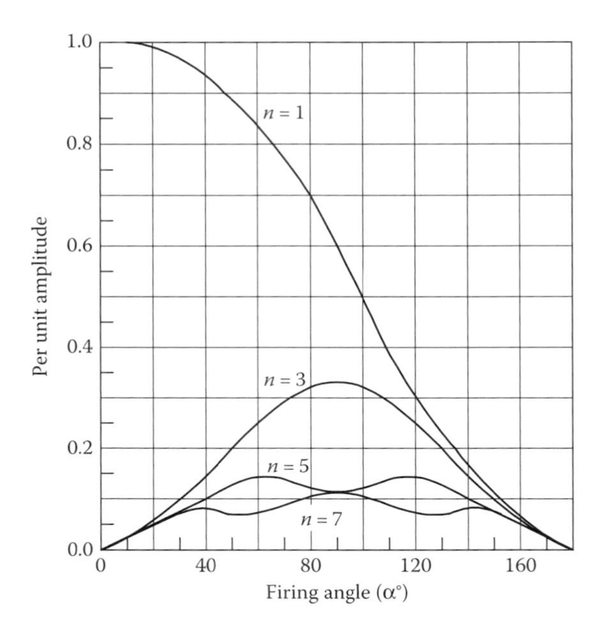

FIGURE 10.7 Harmonic content as a function of the firing angle for a single-phase voltage controller with *RL* load. (Reprinted from Rashid, M. H. 2001. *Power Electronics Handbook*, New York: Academic Press, pp. 307–333. With permission.)

shown in Figure 10.7. Only odd harmonics exist in the input current because of half-wave symmetry.

10.2.2 Single-Phase AC/AC Voltage Controller with On/Off Control

Figure 10.8 shows an on/off AC/AC voltage-regulation controller. In a period T, n cycles are on and m cycles are off. The conduction duty cycle k is

$$k = \frac{n}{n+m}. \tag{10.15}$$

10.2.2.1 Integral Cycle Control

As an alternative to the phase control, the method of integral cycle, control or burst firing is used for heating loads. Here, the switch is turned on for a time t_n with n integral cycles and turned off for a time t_m with m integral cycles (Figure 10.8). As the SCRs or Triacs used here are turned on at the zero crossing of the input voltage and turn off occurs at zero current, supply harmonics and radio-frequency interference are very low.

However, subharmonic frequency components may be generated that are undesirable as they may set up subharmonic resonance in the power supply system, cause lamp flicker, and may interfere with the natural frequencies of motor loads causing shaft oscillations.

For sinusoidal input voltage $v = \sqrt{2}V_S \sin \omega t$, the rms output voltage is

$$V_O = V_S\sqrt{k}, \tag{10.16}$$

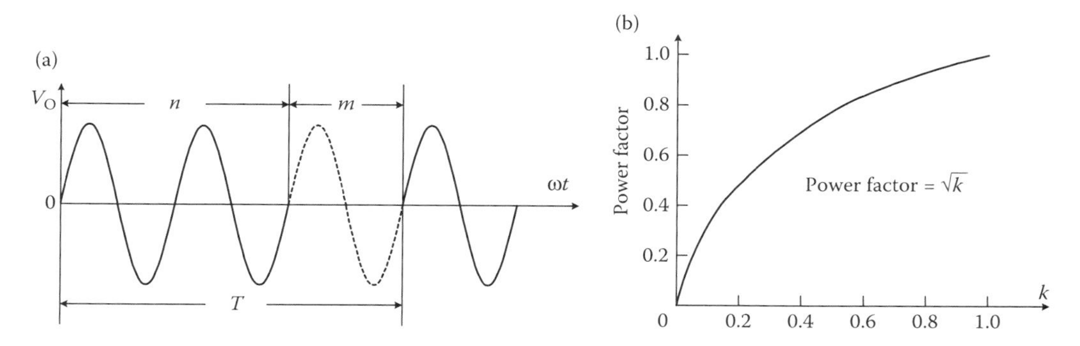

FIGURE 10.8 Integral cycle control: (a) typical load-voltage waveforms and (b) power factor with the duty cycle k. (Reprinted from Luo, F. L., Ye, H., and Rashid, M. H. 2005. *Digital Power Electronics and Applications*. Boston: Academic Press, Elsevier. With permission.)

where $k = n/(n+m)$ = duty cycle and V_S = rms phase voltage. The PF is given by

$$\mathrm{PF} = \sqrt{k}, \tag{10.17}$$

which is poorer for lower values of the duty cycle k.

Example 10.3

A single-phase integral cycle-controlled AC/AC controller has input rms voltage $V_S = 240\mathrm{V}$. It is turned on and off with a duty cycle $k = 0.4$ at five cycles (see Figure 10.8). Determine the output rms voltage V_O and the input-side PF.

SOLUTION

Since the input rms voltage is 240V and the duty cycle $k = 0.4$, the output rms voltage is

$$V_O = V_S\sqrt{k} = 240 \times \sqrt{0.4} = 151.79\mathrm{V}.$$

The power factor is

$$\mathrm{PF} = \sqrt{k} = \sqrt{0.4} = 0.632.$$

10.2.2.2 PWM AC Chopper

As in the case of the controlled rectifier, the performance of AC voltage controllers can be improved in terms of harmonics, quality of output current, and input PF by PWM control in PWM AC choppers. The circuit configuration of one such a single-phase unit is shown in Figure 10.9.

Here, fully controlled switches S_1 and S_2 connected in antiparallel are turned on and off many times during the positive and negative half-cycles of the input voltage, respectively; S_1 and S_2 provide the freewheeling paths for the load current when S_1 and S_2 are off. An input capacitor filter may be provided to attenuate the high switching frequency current drawn from the supply and also to improve the input PF. Figure 10.10 shows the typical output voltage and load-current waveform for a single-phase PWM AC chopper. It can be

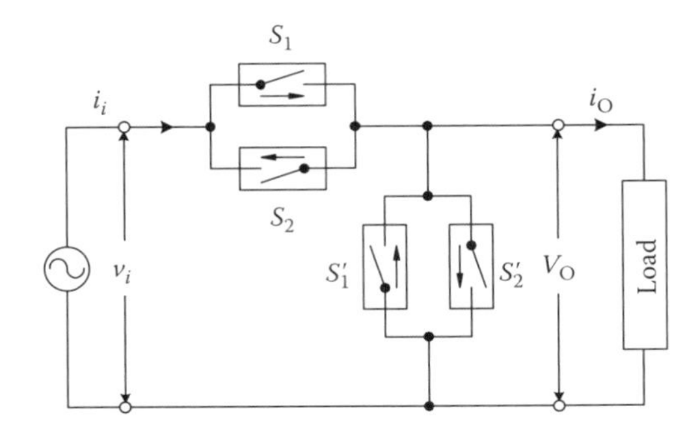

FIGURE 10.9 Single-phase PWM as chopper circuit. (Reprinted from Luo, F. L., Ye, H., and Rashid, M. H. 2005. *Digital Power Electronics and Applications*. Boston: Academic Press, Elsevier. With permission.)

shown that the control characteristics of an AC chopper depend on the modulation index k, which theoretically varies from zero to unity. The relation between input and output voltages is expressed in Equation 10.16 and the PF is calculated by using Equation 10.17. Applying a low-pass filter in the output side of a PWM AC chopper, a good sine wave can be obtained.

Example 10.4

A single-phase PWM AC chopper has input rms voltage $V_S = 240$V. Its modulation index $k = 0.4$ (see Figure 10.10). Determine the output rms voltage V_O and the input-side PF.

SOLUTION

Since the input rms voltage is 240V and the modulation index $k = 0.4$, the output rms voltage is

$$V_O = V_S\sqrt{k} = 240 \times \sqrt{0.4} = 151.79\text{V}.$$

The power factor is

$$\text{PF} = \sqrt{k} = \sqrt{0.4} = 0.632.$$

Analogously, a three-phase PWM chopper consists of three single-phase choppers that are either delta connected or four-wire star connected.

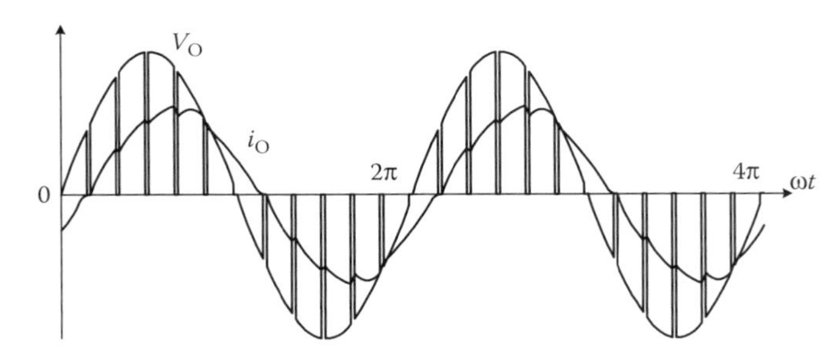

FIGURE 10.10 Typical output voltage and current waveforms of a single-phase PWM AC chopper. (Reprinted from Luo, F. L., Ye, H., and Rashid, M. H. 2005. *Digital Power Electronics and Applications*. Boston: Academic Press, Elsevier. With permission.)

10.3 Three-Phase AC/AC Voltage-Regulation Converters

Three-phase AC/AC voltage controllers have various circuits and configurations.

10.3.1 Phase-Controlled Three-Phase AC Voltage Controllers

Several possible circuit configurations for three-phase phase-controlled AC regulators with star- or delta-connected loads are shown in Figure 10.11a–h. The configurations in Figures 10.11a and 10.11b can be realized by three single-phase AC regulators operating independently of each other and they are easy to analyze. In Figure 10.11a, the SCRs should be rated to carry line currents and withstand phase voltages, whereas in Figure 10.11b they should be capable of carrying phase currents and withstanding the line voltages. Also, in Figure 10.11b the line currents are free from triplen harmonics, while these are present in the closed delta. The PF in Figure 10.11b is slightly higher. The firing angle control range for both these circuits is 0–180° for R load.

The circuits in Figures 10.11c and 10.11d are three-phase three-wire circuits and are difficult to analyze. In both these circuits, at least two SCRs (one in each phase) must be gated simultaneously to cause the controller to start by establishing a current path between the supply lines. This necessitates two firing pulses spaced at 60° apart per cycle for firing each SCR. The operation modes are defined by the number of SCRs conducting in these modes. The firing control range is 0–150°. The triplen harmonics are absent in both these configurations.

Another configuration is shown in Figure 10.11e when the controllers are delta connected and the load is connected between the supply and the converter. Here, current can flow between two lines even if one SCR is conducting, so each SCR requires one firing pulse per cycle. The voltage and current ratings of SCRs are nearly the same as those of the circuit in Figure 10.11b. It is also possible to reduce the number of devices to three SCRs in delta, as shown in Figure 10.11f, by connecting one source terminal directly to one load circuit terminal. Each SCR is provided with gate pulses in each cycle spaced 120° apart. In Figure 10.11e and 10.11f, each end of each phase must be accessible. The number of devices in Figure 10.11f is lower, but their current ratings must be higher.

As in the case of the single-phase phase-controlled voltage regulator, the total regulator cost can be reduced by replacing six SCRs by three SCRs and three diodes, resulting in three-phase *half-wave controlled* unidirectional AC regulators, as shown in Figure 10.11g and 10.11h for star- and delta-connected loads. The main drawback of these circuits is the large harmonic content in the output voltage, particularly the second harmonic, because of the asymmetry. However, the DC components are absent in the line. The maximum firing angle in the half-wave controlled regulator is 210°.

10.3.2 Fully Controlled Three-Phase Three-Wire AC Voltage Controller

10.3.2.1 Star-Connected Load with Isolated Neutral

The analysis of the operation of the full-wave controller with isolated neutral as shown in Figure 10.11c is, as mentioned, quite complicated in comparison with that of a single-phase controller, particularly for an RL or motor load. As a simple example, the operation of this controller is considered here with a simple star-connected R load. The six SCRs are turned

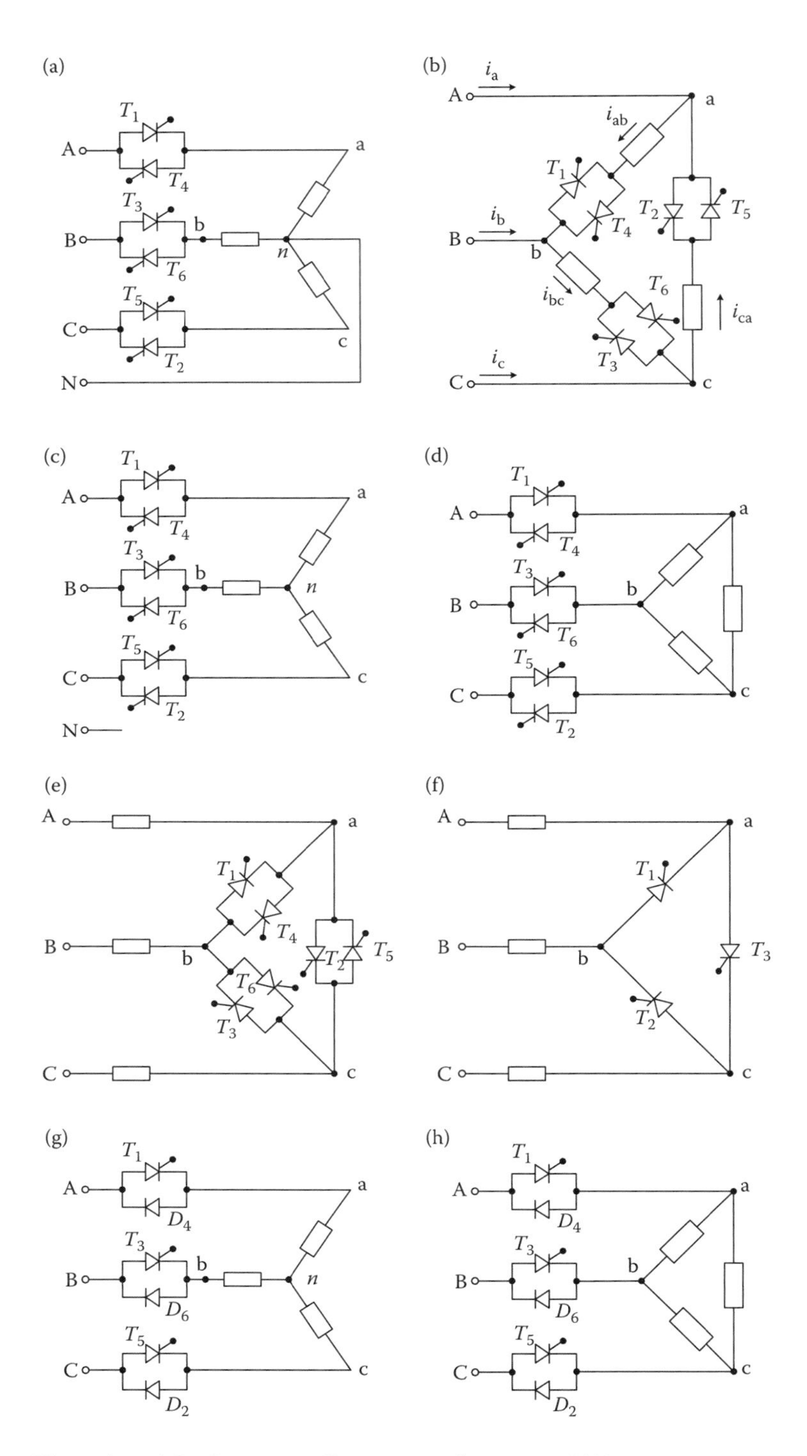

FIGURE 10.11 Three-phase AC voltage-controller circuit configurations. (a) Y-connection circuit with neutral, (b) delta-connection circuit with phase-control, (c) Y-connection circuit without neutral, (d) delta-connection circuit with line-control, (e) delta-connection circuit with line-load plus full-control, (f) delta-connection circuit with line-load plus half-control, (g) Y-connection circuit with half-control, and (h) delta-connection circuit with half-control. (Reprinted from Luo, F. L., Ye, H., and Rashid, M. H. 2005. *Digital Power Electronics and Applications*. Boston: Academic Press, Elsevier. With permission.)

on in the sequence 1-2-3-4-5-6 at 60° intervals and the gate signals are sustained throughout the possible conduction angle.

The output phase voltage waveforms for $\alpha = 30°, 75°$, and 120° for a balanced three-phase *R* load are shown in Figure 10.12. At any interval, either three SCRs or two SCRs or no SCRs may be on, and the instantaneous output voltages to the load are either line-to-neutral voltages (three SCRs on) or one-half of the line-to-line voltage (two SCRs on) or zero (no SCR on).

Depending on the firing angle α, there may be *three* operating modes.

Mode I (also known as Mode 2/3): $0 < \alpha < 60°$. There are periods when *three* SCRs are conducting, one in each phase for either direction and periods when just *two* SCRs are conducting.

For example, with $\alpha = 30°$ in Figure 10.12a, assume that at $\omega t = 0$, SCRs T_5 and T_6 are conducting, and the current through the *R* load in a-phase is zero, making $v_{an} = 0$. At $\omega t =$

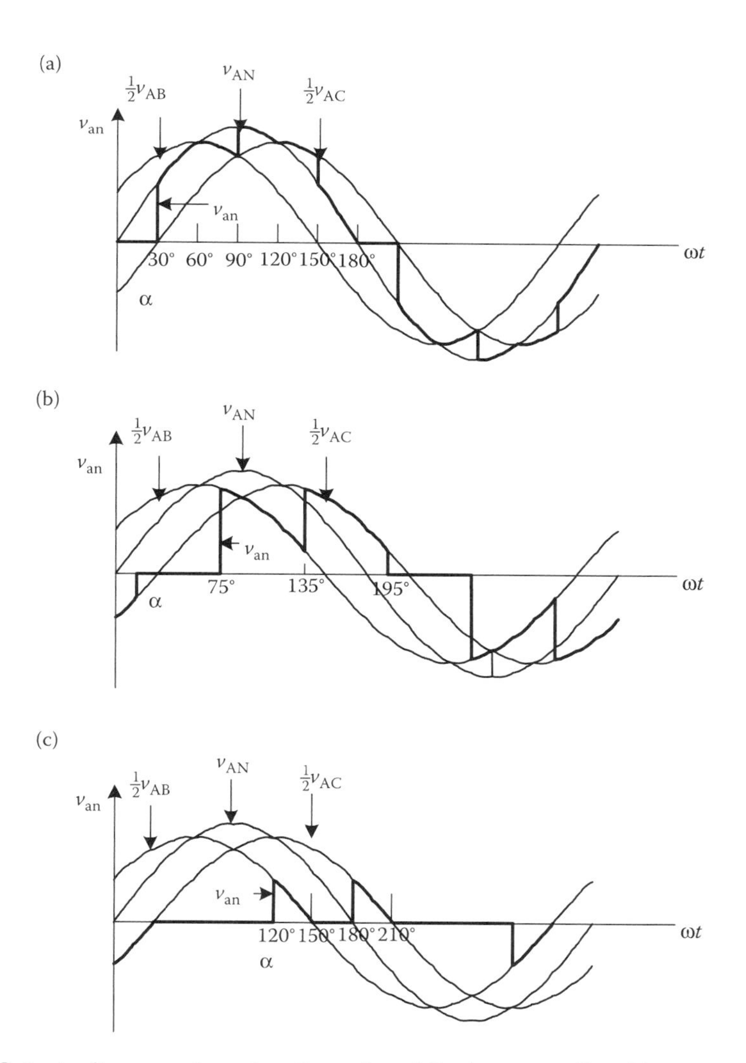

FIGURE 10.12 Output voltage waveforms for a three-phase AC voltage controller with star-connected *R* load: (a) v_{an} for $\alpha = 30°$, (b) v_{an} for $\alpha = 75°$, and (c) v_{an} for $\alpha = 120°$. (Reprinted from Rashid, M. H. 2001. *Power Electronics Handbook*, New York: Academic Press, pp. 307–333. With permission.)

30°, T_1 receives a gate pulse and starts conducting; T_5 and T_6 remain on and $v_{an} = v_{AN}$. The current in T_5 reaches zero at 60°, turning T_5 off. With T_1 and T_6 staying on, $v_{an} = (1/2)v_{AB}$. At 90°, T_2 is turned on, the three SCRs T_1, T_2, and T_6 are then conducting and $v_{an} = v_{AN}$. At 120°, T_6 turns off, leaving $T_1$and T_2 on, so $v_{an} = (l/2)v_{AC}$. Thus with the progress of firing in sequence until $\alpha = 60°$, the number of SCRs conducting at a particular instant alternates between two and three.

Mode II (also known as Mode 2/2): $60° < \alpha < 90°$. *Two* SCRs, one in each phase, always conduct.

For $\alpha = 75°$ as shown in Figure 10.12b, just prior to $\alpha = 75°$, SCRs T_5 and T_6 were conducting and $v_{an} = 0$. At 75°, T_1 is turned on; T_6 continues to conduct while T_5 turns off as v_{CN} is negative; $v_{an} = (1/2)v_{AB}$. When T_2 is turned on at 135°, T_6 is turned off and $v_{an} = (1/2)v_{AC}$. The next SCR to turn on is T_3, which turns off T_1 and $v_{an} = 0$. One SCR is always turned off when another is turned on in this range of α and the output is either one-half line-to-line voltage or zero.

Mode III (also known as Mode 0/2): $90° < \alpha < 150°$. When *none* or *two* SCRs conduct.

For $\alpha = 120°$ (Figure 10.12c), earlier no SCRs were on and $v_{an} = 0$. At $\alpha = 120°$, SCR T_1 is given a gate signal while T_6 has a gate signal already applied. As v_{AB} is positive, T_1 and T_6 are forward biased, and they begin to conduct and $v_{an} = (1/2)v_{AB}$. Both T_1 and T_6 turn off when v_{AB} becomes negative. When a gate signal is given to T_2, it turns on, and T_1 turns on again.

For $\alpha > 150°$, there is no period when two SCRs are conducting and the output voltage is zero at $\alpha = 150°$. Thus, the range of the firing angle control is $0 \le \alpha \le 150°$.

For *star-connected R load*, assuming the instantaneous phase voltages as

$$\begin{aligned} v_{AN} &= \sqrt{2}V_S \sin \omega t, \\ v_{BN} &= \sqrt{2}V_S \sin(\omega t - 120°), \\ v_{CN} &= \sqrt{2}V_S \sin(\omega t - 240°), \end{aligned} \tag{10.18}$$

the expressions for the rms output phase voltage V_O can be derived for the three modes as

$$0 \le \alpha \le 60°, \quad V_O = V_S\left[1 - \frac{3\alpha}{2\pi} + \frac{3}{4\pi}\sin 2\alpha\right]^{1/2}, \tag{10.19}$$

$$60° \le \alpha \le 90°, \quad V_O = V_S\left[\frac{1}{2} + \frac{3}{4\pi}\sin 2\alpha + \sin(2\alpha + 60°)\right]^{1/2}, \tag{10.20}$$

$$90° \le \alpha \le 150°, \quad V_O = V_S\left[\frac{5}{4} - \frac{3\alpha}{2\pi} + \frac{3}{4\pi}\sin(2\alpha + 60°)\right]^{1/2}. \tag{10.21}$$

For *star-connected pure L load*, the effective control starts at $\alpha > 90°$ and the expressions for two ranges of α are

$$90° \le \alpha \le 120°, \quad V_O = V_S\left[\frac{5}{2} - \frac{3\alpha}{\pi} + \frac{3}{2\pi}\sin 2\alpha\right]^{1/2} \tag{10.22}$$

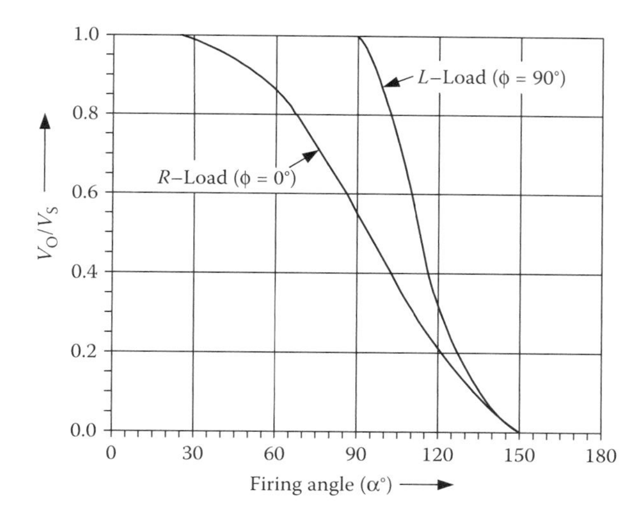

FIGURE 10.13 Envelope of control characteristics for a three-phase full-wave AC voltage controller. (Reprinted from Rashid, M. H. 2001. *Power Electronics Handbook*, New York: Academic Press, pp. 307–333. With permission.)

and

$$120^\circ \leq \alpha \leq 150^\circ, \quad V_O = V_S \left[\frac{5}{2} - \frac{3\alpha}{\pi} + \frac{3}{2\pi} \sin(2\alpha + 60^\circ) \right]^{1/2}. \tag{10.23}$$

The control characteristics for these two limiting cases ($\phi = 0$ for R load and $\phi = 90^\circ$ for L load) are shown in Figure 10.13. Here also, as in the single-phase case, the dead zone may be avoided by controlling the voltage with respect to the control angle or hold-off angle (γ) from the zero crossing of the current in place of the firing angle α.

10.3.2.2 RL Load

The analysis of the three-phase voltage controller with star-connected RL load with isolated neutral is quite complicated as the SCRs do not cease to conduct at voltage zero and the extinction angle β is to be found out by solving the transcendental equation for the case. The Mode-II operation, in this case, disappears and the operation shift from Mode I to Mode III depends on the so-called critical angle α_{crit}, which can be evaluated from a numerical solution of the relevant transcendental equations. Computer simulation either by PSpice program or a switching-variable approach coupled with an iterative procedure is a practical means of obtaining the output voltage waveform in this case. Figure 10.14 shows typical simulation results, using the latter approach for a three-phase voltage-controller-fed RL load for $\alpha = 60^\circ, 90^\circ$, and 105°, which agree with the corresponding practical oscillograms given.

10.3.2.3 Delta-Connected R load

The configuration is shown in Figure 10.11b. The voltage across an R load is the corresponding line-to-line voltage when one SCR in that phase is on. Figure 10.15 shows the line and phase currents for $\alpha = 120^\circ$ and 90° with an R load. The firing angle α is measured from the zero crossing of the line-to-line voltage and the SCRs are turned on in the sequence as they are numbered. As in the single-phase case, the range of the firing angle is $0 \leq \alpha \leq 180^\circ$.

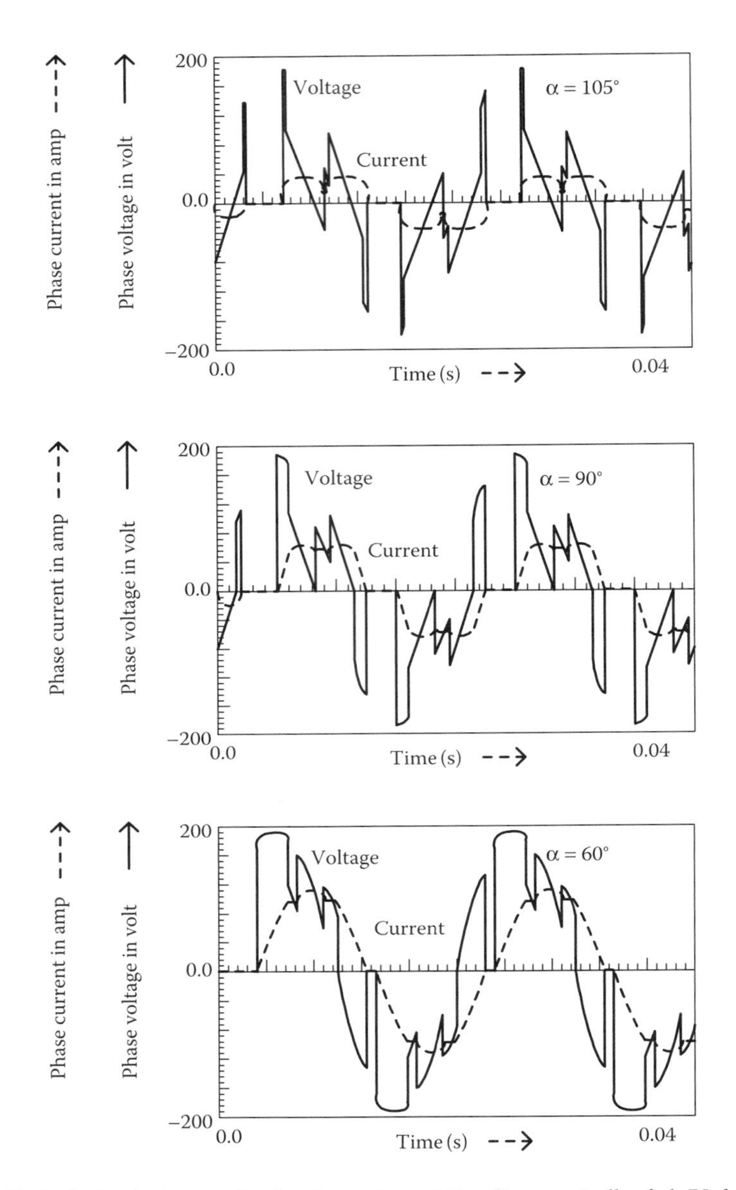

FIGURE 10.14 Typical simulation results for three-phase AC voltage-controller-fed *RL* load ($R = 1\,\Omega, L = 3.2\,\text{mH}$) for $\alpha = 60°$, $90°$, and $105°$. (Reprinted from Rashid, M. H. 2001. *Power Electronics Handbook*, New York: Academic Press, pp. 307–333. With permission.)

The line currents can be obtained from the phase currents as

$$\begin{aligned} i_a &= i_{ab} - i_{ca}, \\ i_b &= i_{bc} - i_{ab}, \\ i_c &= i_{ca} - i_{bc}. \end{aligned} \tag{10.24}$$

The line currents depend on the firing angle and may be discontinuous as shown. Due to the delta connection, the triplen harmonic currents flow around the closed delta and do

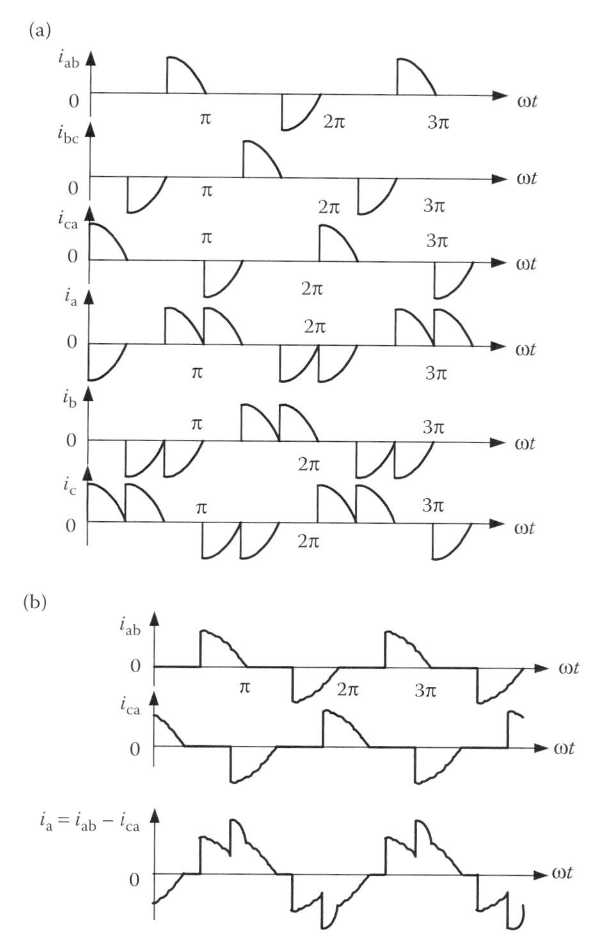

FIGURE 10.15 Waveforms of a three-phase AC voltage controller with a delta-connected R load: (a) $\alpha = 120°$; (b) $\alpha = 90°$. (Reprinted from Rashid, M. H. 2001. *Power Electronics Handbook*, New York: Academic Press, pp. 307–333. With permission.)

not appear in the line. The rms value of the line current varies in the range

$$\sqrt{2}I_{\Delta} \leq I_{L,rms} \leq \sqrt{3}I_{\Delta,rms} \tag{10.25}$$

as the conduction angle varies from a very small value (large α) to 180° ($\alpha = 0$).

10.4 Cycloconverters

In contrast to the AC voltage controllers operating at constant frequency discussed so far, a cycloconverter operates as a direct AC/AC frequency changer with an inherent voltage control feature. The basic principle of this converter to construct an alternating voltage wave of lower frequency from successive segments of voltage waves of higher frequency AC supply by a switching arrangement was conceived and patented in the 1920s. Grid-controlled mercury-arc rectifiers were used in these converters installed in Germany in the 1930s to obtain $16\frac{2}{3}$ Hz single-phase supply for AC series traction motors from a three-phase 50 Hz system, while at the same time a cycloconverter using 18 thyratrons supplying a 400-hp synchronous motor was in operation for some years as a power station auxiliary drive in the United States. However, the practical and commercial utilization of these schemes did not take place until the SCRs became available in the 1960s. With the development of large power SCRs and microprocessor-based control, the cycloconverter today is a mature practical converter for application in large-power low-speed variable-voltage variable-frequency (VVVF) AC drives in cement and steel rolling mills as well as in variable-speed constant-frequency (VSCF) systems in aircraft and naval ships [9–11].

A cycloconverter is a naturally commuted converter with the inherent capability of bidirectional power flow and there is no real limitation on its size unlike an SCR inverter with commutation elements. Here, the switching losses are considerably low, the regenerative operation at full power over the complete speed range is inherent, and it delivers a nearly sinusoidal waveform resulting in minimum torque pulsation and harmonic heating effects. It is capable of operating even with the blowing out of an individual SCR fuse (unlike the inverter), and the requirements regarding turn-off time, current rise time, and dv/dt sensitivity of SCRs are low. The main limitations of a naturally commutated cycloconverter (NCC) are (i) limited frequency range for sub-harmonic-free and efficient operation and (ii) poor input displacement/power factor, particularly at low output voltages.

10.4.1 Single-Phase/Single-Phase (SISO) Cycloconverters

Although rarely used, the operation of a single-phase input to single-phase output (SISO) cycloconverter is useful to demonstrate the basic principle involved. Figure 10.16a shows the power circuit of a single-phase bridge-type cycloconverter, which has the same arrangement as that of the dual converter.

The firing angles of the individual two-pulse two-quadrant bridge converters are continuously modulated here so that each ideally produces the same fundamental AC voltage at its output terminals as marked in the simplified equivalent circuit in Figure 10.16b. Because of the unidirectional current-carrying property of the individual converters, it is inherent that the positive half-cycle of the current is carried by the P-converter and the negative half-cycle of the current by the N-converter, regardless of the phase of the current with respect to the voltage. This means that for a reactive load, each converter operates in both the rectifying and inverting region during the period of the associated half-cycle of the low-frequency output current.

10.4.1.1 Operation with R Load

Figure 10.17 shows the input and output voltage waveforms with a pure R load for a 50 to $16\frac{2}{3}$ Hz cycloconverter. The P- and N-converters operate for all alternate $T_O/2$ periods. The output frequency $(1/T_O)$ can be varied by varying T_O, and the voltage magnitude by varying the firing angle α of the SCRs. As shown in the figure, three cycles of the AC input wave are combined to produce one cycle of the output frequency to reduce the supply frequency to one-third across the load.

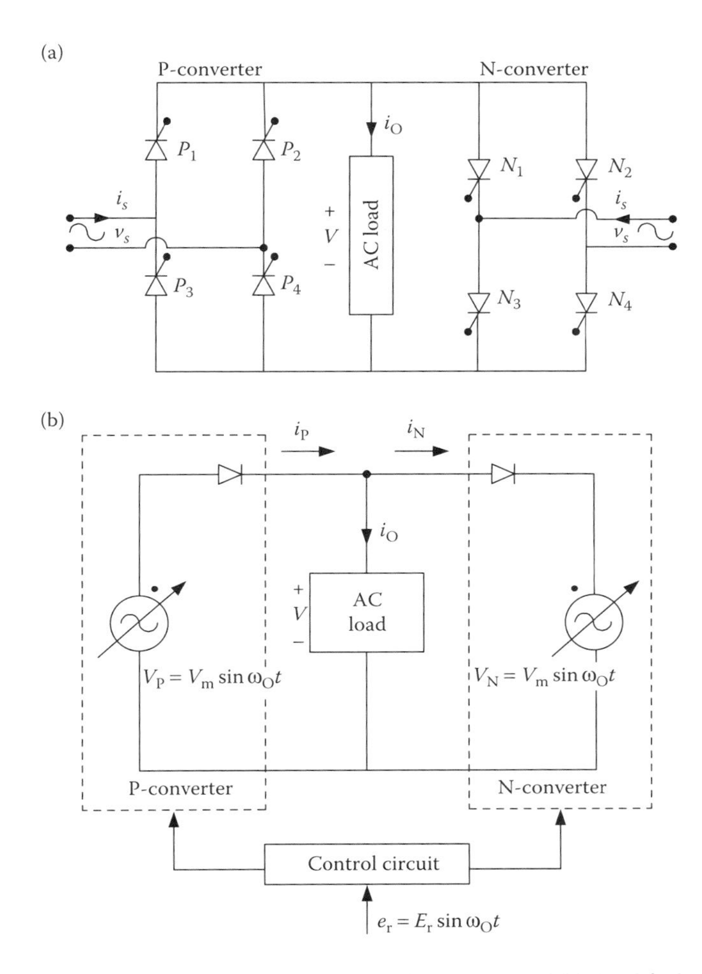

FIGURE 10.16 (a) Power circuit for a single-phase bridge cycloconverter and (b) simplified equivalent circuit of a cycloconverter. (Reprinted from Luo, F. L., Ye, H., and Rashid, M. H. 2005. *Digital Power Electronics and Applications*. Boston: Academic Press, Elsevier. With permission.)

For example, the waveforms of a SISO AC/AC cycloconverter with $T_O = 3T_S$ are shown in Figure 10.17. The firing angle α is listed in Tables 10.1 and 10.2 (the blank means no-firing pulse applied).

Assuming the input voltage amplitude $\sqrt{2}V_S$ and the output voltage amplitude $\sqrt{2}V_O$ keep the relation given below for *full regulation*:

$$\frac{\sqrt{2}V_O}{\pi/3}\int_{\pi/3}^{2\pi/3} \sin\alpha \, d\alpha \leq \sqrt{2}V_S\frac{1}{\pi}\int_0^{\pi} \sin\alpha \, d\alpha, \tag{10.26}$$

that is,

$$3V_O \leq 2V_S. \tag{10.27}$$

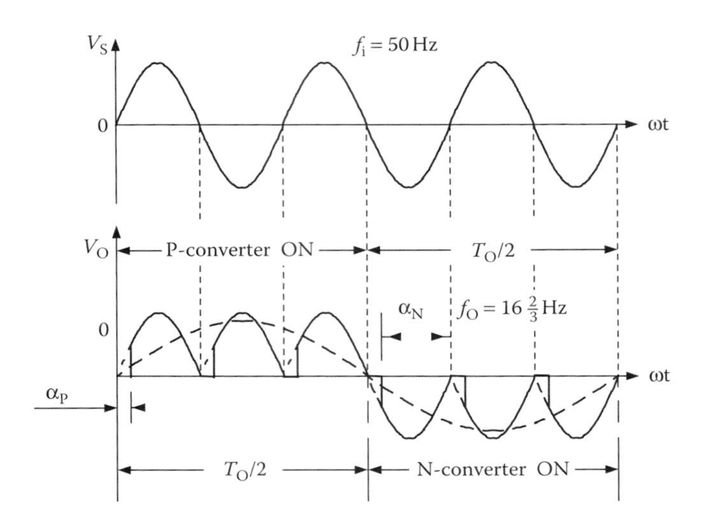

FIGURE 10.17 Input and output waveforms of a 50–16$\frac{2}{3}$ Hz cycloconverter with R load. (Reprinted from Luo, F. L., Ye, H., and Rashid, M. H. 2005. *Digital Power Electronics and Applications*. Boston: Academic Press, Elsevier. With permission.)

We then obtain the following firing angles calculation formulae:

$$\sqrt{2}V_O \frac{3}{\pi} \int_0^{\pi/3} \sin\theta \, d\theta = \sqrt{2}V_S \frac{1}{\pi} \int_{\alpha 1}^{\pi} \sin\theta \, d\theta, \tag{10.28}$$

$$\alpha_1 = \cos^{-1}\left(\frac{3V_O}{2V_S} - 1\right) \tag{10.29}$$

and

$$\sqrt{2}V_O \frac{3}{\pi} \int_{\pi/3}^{2\pi} \sin\theta \, d\theta = \sqrt{2}V_S \frac{1}{\pi} \int_{\alpha 2}^{\pi} \sin\theta \, d\theta, \tag{10.30}$$

$$\alpha_2 = \cos^{-1}\left(\frac{3V_O}{V_S} - 1\right). \tag{10.31}$$

We also obtain

$$\alpha_3 = \alpha_1 = \cos^{-1}\left(\frac{3V_O}{2V_S} - 1\right). \tag{10.32}$$

TABLE 10.1

The Firing Angle Set of the Positive Rectifier

Half-Cycle No. in f_O	1	2	3	4	5	6
SCR	P_1P_4	P_2P_3	P_1P_4	P_2P_3	P_1P_4	P_2P_3
α_P	α_1	α_2	α_1			

TABLE 10.2

The Firing Angle Set of the Negative Rectifier

Half-Cycle No. in f_O	1	2	3	4	5	6
SCR	N_1N_4	N_2N_3	N_1N_4	N_2N_3	N_1N_4	N_2N_3
α_N				α_1	α_2	α_1

The phase-angle shift (delay) in the frequency f_S is

$$\sigma = \frac{\alpha_1}{2} = \frac{1}{2}\cos^{-1}\left(\frac{3V_O}{2V_S} - 1\right) \tag{10.33}$$

and in the frequency f_O, it is

$$\sigma' = \frac{1}{3}\frac{\alpha_1}{2} = \frac{1}{6}\cos^{-1}\left(\frac{3V_O}{2V_S} - 1\right). \tag{10.34}$$

If the full regulation condition (Equation 10.27) is not satisfied, the modulation can still be done by other ways; the limitation condition is usually

$$V_O \leq 1.2V_S. \tag{10.35}$$

If α_P is the firing angle of the P-converter, then the firing angle of the N-converter α_N is $\pi - \alpha_P$, and the average voltage of the P-converter is equal to and opposite of that of the N-converter. The inspection of the waveform with α remaining fixed in each half-cycle generates a square wave having a large low-order harmonic content. A near approximation to sine wave can be synthesized by a phase modulation of the firing angles as shown in Figure 10.18 for a 50–10 Hz cycloconverter. The harmonics in the load-voltage waveform are fewer when compared with the earlier waveform. The supply current, however, contains a subharmonic at the output frequency for this case as shown.

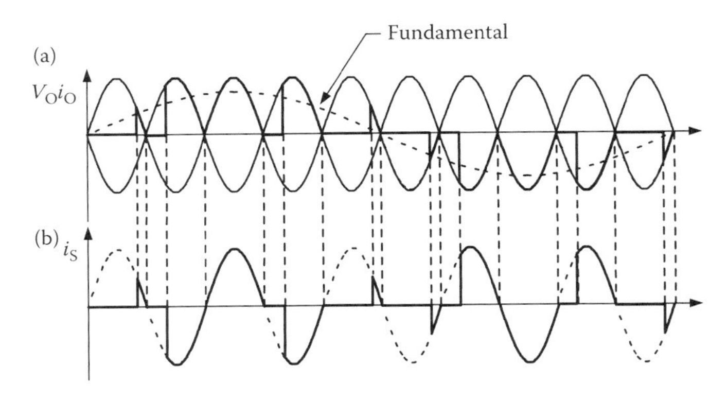

FIGURE 10.18 Waveforms of a single-phase/single-phase cycloconverter (50–10 Hz) with R load: (a) load voltage and load current and (b) input supply current. (Reprinted from Luo, F. L., Ye, H., and Rashid, M. H. 2005. *Digital Power Electronics and Applications*. Boston: Academic Press, Elsevier. With permission.)

Example 10.5

Consider a full-wave SISO AC/AC cycloconverter. The input rms voltage $V_S = 140$V/50 Hz and the output voltage $V_O = 90$V/$16\frac{2}{3}$ Hz, and the load is a resistance R with a low-pass filter. Assuming that the filter is appropriately designed, only the fundamental component ($f_O = 16\frac{2}{3}$ Hz) remains in the output voltage. Tabulate the firing angle (α in the period $T_S = 1/f_S = 20$ ms) of both rectifiers' SCRs in a full period $T_O = 1/f_O = 60$ ms, and calculate the phase-angle shift σ in the input voltage over the period $T_S = 1/f_S$.

SOLUTION

The table is shown below (the blank means no-firing pulse applied)

Half-Cycle No. in f_O	1	2	3	4	5	6
			Positive Rectifier			
SCR	P_1P_4	P_2P_3	P_1P_4	P_2P_3	P_1P_4	P_2P_3
α_P	α_1	α_2	α_1			
			Negative Rectifier			
SCR	N_1N_4	N_2N_3	N_1N_4	N_2N_3	N_1N_4	N_2N_3
α_N				α_1	α_2	α_1

The full regulation condition is

$$\frac{\sqrt{2}V_O}{\pi/3}\int_{\pi/3}^{2\pi/3}\sin\alpha\, d\alpha \le \sqrt{2}V_S\frac{1}{\pi}\int_0^{\pi}\sin\alpha\, d\alpha,$$

$$V_S \ge 3V_O\cos\frac{\pi}{3} = 1.5V_O,$$

that is,

$$V_S = 140 \ge 3V_O\cos\frac{\pi}{3} = 1.5V_O = 135\,\text{V}.$$

$$\sqrt{2}V_O\frac{3}{\pi}\int_0^{\pi/3}\sin\theta\, d\theta = \sqrt{2}V_S\frac{1}{\pi}\int_{\alpha 1}^{\pi}\sin\theta\, d\theta,$$

$$3\left(1-\cos\frac{\pi}{3}\right)V_O = (1+\cos\alpha_1)V_S,$$

$$\alpha_1 = \cos^{-1}\left(\frac{1.5V_O}{V_S}-1\right) = \cos^{-1}(-0.0357) = 92.05°.$$

$$\sqrt{2}V_O\frac{3}{\pi}\int_{\pi/3}^{2\pi/3}\sin\theta\, d\theta = \sqrt{2}V_S\frac{1}{\pi}\int_{\alpha 2}^{\pi}\sin\theta\, d\theta,$$

$$3\left(\cos\frac{\pi}{3}-\cos\frac{2\pi}{3}\right)V_O = (1+\cos\alpha_2)V_S,$$

$$\alpha_2 = \cos^{-1}\left(\frac{3V_O}{V_S}-1\right) = \cos^{-1}(0.9286) = 21.79°.$$

The phase-angle shift σ in the input voltage over the period $T_S = 1/f_S$ is

$$\sigma = \frac{\alpha_1}{2} = \frac{1}{2} \times 92.05 = 46.02°.$$

10.4.1.2 Operation with RL Load

The cycloconverter is capable of supplying the loads of any PF. Figure 10.19 shows the idealized output voltage and current waveforms for a lagging-power-factor load where both the converters are operating as rectifiers and inverters at the intervals marked. The load current lags the output voltage and the load-current direction determines which converter is conducting. Each converter continues to conduct after its output voltage changes polarity, and during this period, the converter acts as an inverter and the power is returned to the AC source. The inverter operation continues until the other converter starts to conduct. By controlling the frequency of oscillation and the *depth of modulation* of the firing angles of the converters (as will be shown later), it is possible to control the frequency and the amplitude of the output voltage.

The load current with *RL* load may be continuous or discontinuous depending on the load phase angle ϕ. At light load inductance or for $\phi \leq \alpha \leq \pi$, there may be discontinuous load current with short zero-voltage periods. The current wave may contain even harmonics as well as subharmonic components. Further, as in the case of a dual converter, although the mean output voltages of the two converters are equal and opposite, the instantaneous values may be unequal, and a circulating current can flow within the converters. This circulating current can be limited by having a center-tapped reactor connected between the converters or can be completely eliminated by logical control similar to the dual converter case in which the gate pulses to the idle converter are suppressed when the other converter is active. A zero current interval of short duration is needed between the P- and N-converters to ensure that the supply lines of the two converters are not short-circuited.

For the circulating-current scheme, the converters are kept in virtually continuous conduction over the whole range and the control circuit is simple. To obtain a reasonably good sinusoidal voltage waveform using the line-commutated two-quadrant converters, and to eliminate the possibility of the short circuit of the supply voltages, the output frequency of the cycloconverter is limited to a much lower value of the supply frequency.

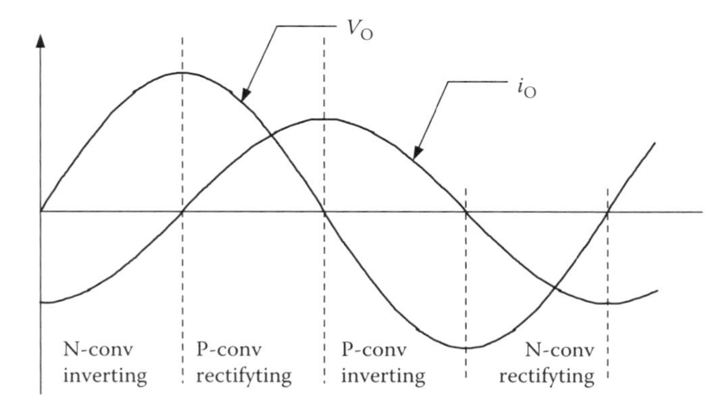

FIGURE 10.19 Load voltage and current waveform for a cycloconverter with *RL* load. (Reprinted from Rashid, M. H. 2001. *Power Electronics Handbook*, New York: Academic Press, pp. 307–333. With permission.)

The output voltage waveform and the output frequency range can be improved further by using converters of higher pulse numbers.

10.4.2 Three-Phase Cycloconverters

Three-phase cycloconverters have several circuits. For example, there are the three-pulse cycloconverters, 6-pulse cycloconverters, and 12-pulse cycloconverters.

10.4.2.1 Three-Phase Three-Pulse Cycloconverter

Figure 10.20a shows a schematic diagram of a three-phase half-wave (three-pulse) cycloconverter feeding a single-phase load, and Figure 10.20b shows the configuration of a three-phase half-wave (three-pulse) cycloconverter feeding a three-phase load. The basic process of a three-phase cycloconversion is illustrated in Figure 10.20c at 15 Hz, 0.6 PF lagging load from a 50-Hz supply. As the firing angle α is cycled from 0° at "a" to 180° at "j," half a cycle of output frequency is produced (the gating circuit is to be suitably designed to introduce this oscillation of the firing angle). For this load, it can be seen that although the mean output voltage reverses at X, the mean output current (assumed sinusoidal) remains positive until Y. During XY, the SCRs A, B, and C in the P-converter are "inverting." A similar period exists at the end of the negative half-cycle of the output voltage when D, E, and F SCRs in the N-converter are "inverting." Thus, the operation of the converter follows in the order of "rectification" and "inversion" in a cyclic manner, with the relative durations being dependent on the load power factor. The output frequency is that of the firing angle oscillation, about a quiescent point of 90° (the condition when the mean output voltage, given by $V_O = V_{dO} \cos \alpha$, is zero). For obtaining the positive half-cycle of the voltage, firing angle α is varied from 90° to 0° and then to 90°, and for the negative half-cycle, from 90° to 180° and back to 90°. Variation of α within the limits of 180° automatically provides for "natural" line commutation of the SCRs. It is shown that a complete cycle of low-frequency output voltage is fabricated from the segments of the three-phase input voltage by using the phase-controlled converters. The P- or N-converter SCRs receive firing pulses that are timed such that each converter delivers the same mean output voltage. This is achieved, as in the case of the single-phase cycloconverter or the dual converter, by maintaining the firing angle constraints of the two groups as $\alpha_P = (180° - \alpha_N)$. However, the instantaneous voltages of two converters are not identical, and a large circulating current may result unless limited by an intergroup reactor as shown (*circulating-current cycloconverter*) or completely suppressed by removing the gate pulses from the nonconducting converter by an intergroup blanking logic (*circulating-current-free cycloconverter*).

10.4.2.1.1 *Circulating-Current-Mode Operation*

Figure 10.21 shows typical waveforms of a three-pulse cycloconverter operating with circulating current. Each converter conducts continuously with rectifying and inverting modes as shown and the load is supplied with an average voltage of two converters reducing some of the ripple in the process, with the intergroup reactor behaving as a potential divider. The reactor limits the circulating current, with the value of its inductance to the flow of load current being one-fourth of its value to the flow of circulating current as the inductance is proportional to the square of the number of turns. The fundamental waves produced by both the converters are the same. The reactor voltage is the instantaneous difference between the converter voltages, and the time integral of this voltage divided by the inductance

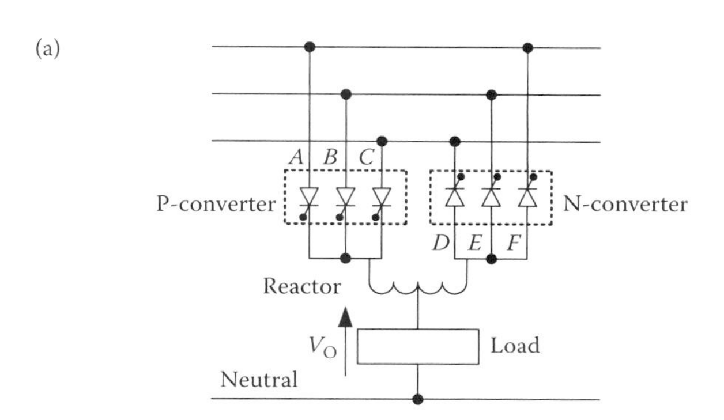

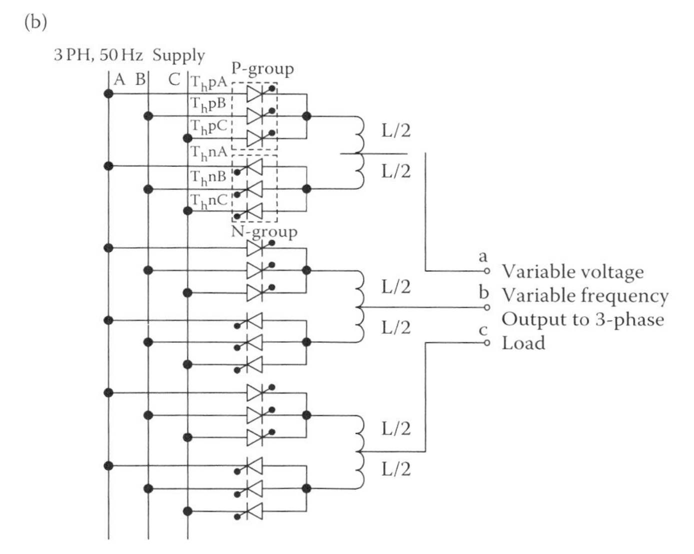

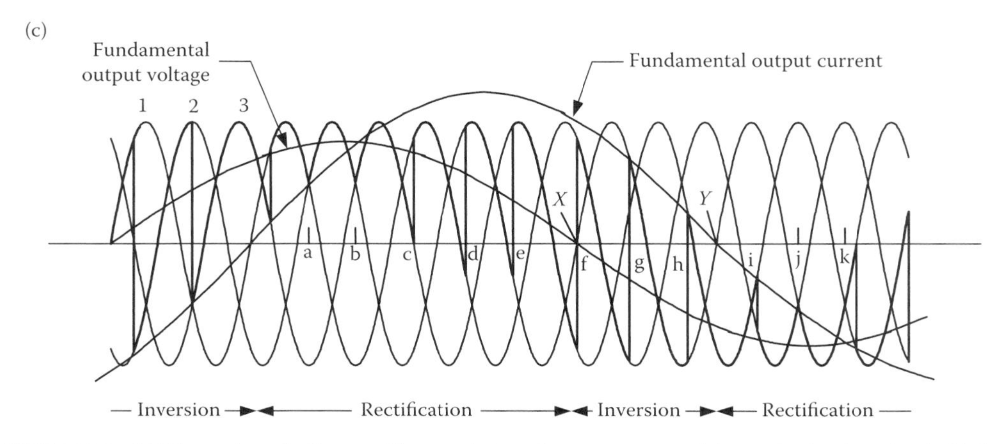

FIGURE 10.20 (a) Three-phase half-wave (three-pulse) cycloconverter supplying a single-phase load, (b) three-pulse cycloconverter supplying a three-phase load, and (c) output voltage waveform for one phase of a three-pulse cycloconverter operating at 15 Hz from a 50-Hz supply and 0.6 power factor lagging load. (Reprinted from Rashid, M. H. 2001. *Power Electronics Handbook*, New York: Academic Press, pp. 307–333. With permission.)

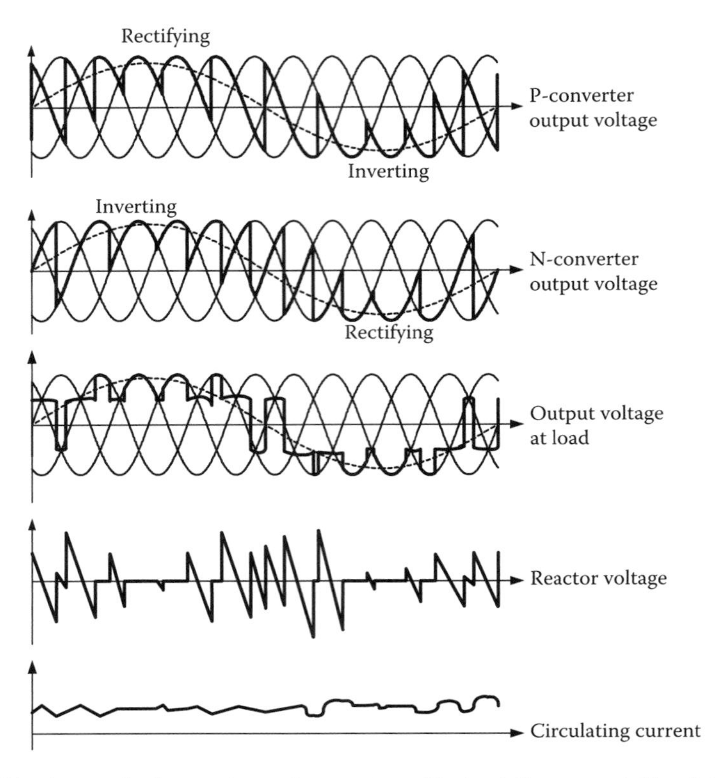

FIGURE 10.21 Waveforms of a three-pulse cycloconverter with circulating current. (Reprinted from Luo, F. L., Ye, H., and Rashid, M. H. 2005. *Digital Power Electronics and Applications*. Boston: Academic Press, Elsevier. With permission.)

(assuming negligible circuit resistance) is the circulating current. For a three-pulse cycloconverter, it can be observed that this current reaches its peak value when $\alpha_P = 60°$ and $\alpha_N = 120°$.

10.4.2.1.2 Output Voltage Equation

A simple expression for the fundamental rms output voltage of the cycloconverter and the required variation of the firing angle α can be derived with the assumptions that (i) the firing angle α in successive half-cycles is varied slowly resulting in a low-frequency output; (ii) the source impedance and the commutation overlap are neglected; (iii) the SCRs are ideal switches; and (iv) the current is continuous and ripple-free. The average DC output voltage of a *p*-pulse dual converter with fixed α is

$$V_{dO} = V_{dOmax} \cos \alpha, \quad (10.36)$$

where $V_{dOmax} = \sqrt{2} V_{ph} \frac{p}{\pi} \sin \frac{p}{\pi}$.

For the *p*-pulse dual converter operating as a cycloconverter, the average phase voltage output at any point of the low frequency should vary according to the equation

$$V_{O,av} = V_{O1,max} \sin \omega_O t, \quad (10.37)$$

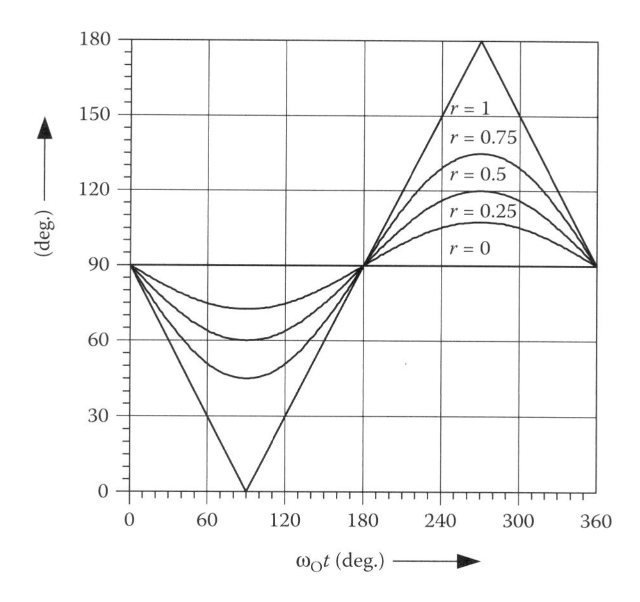

FIGURE 10.22 Variations of the firing angle (α) with r in a cycloconverter. (Reprinted from Rashid, M. H. 2001. *Power Electronics Handbook*, New York: Academic Press, pp. 307–333. With permission.)

where $V_{O1,max}$ is the desired maximum value of the fundamental output voltage of the cycloconverter. Comparing Equation 10.36 with Equation 10.37, the required variation of α to obtain a sinusoidal output is given by

$$\alpha = \cos^{-1}\left[\left(\frac{V_{O1,max}}{V_{dO\,max}}\right)\sin\omega_O t\right] = \cos^{-1}[r\sin\omega_O t], \tag{10.38}$$

where r is the ratio $(V_{O1,max}/V_{dO\,max})$, called the *voltage magnitude control ratio.* Equation 10.38 shows α as a nonlinear function of r (≤ 1) as shown in Figure 10.22.

However, the firing angle α_P of the P-converter cannot be reduced to 0° as this corresponds to $\alpha_N = 180°$ for the N-converter, which, in practice, cannot be achieved because of allowance for commutation overlap and finite turnoff time of the SCRs. Thus the firing angle α_P can be reduced to a certain finite value α_{min} and the maximum output voltage is reduced by a factor $\cos\alpha_{min}$.

The fundamental rms voltage per phase of either converter is

$$V_{Or} = V_{ON} = V_{OP} = rV_{ph}\frac{p}{\pi}\sin\frac{\pi}{p}. \tag{10.39}$$

Although the rms value of the low-frequency output voltage of the P-converter and that of the N-converter are equal, the actual waveforms differ, and the output voltage at the midpoint of the circulating-current limiting reactor (Figure 10.21), which is the same as the load voltage, is obtained as the mean of the instantaneous output voltages of the two converters.

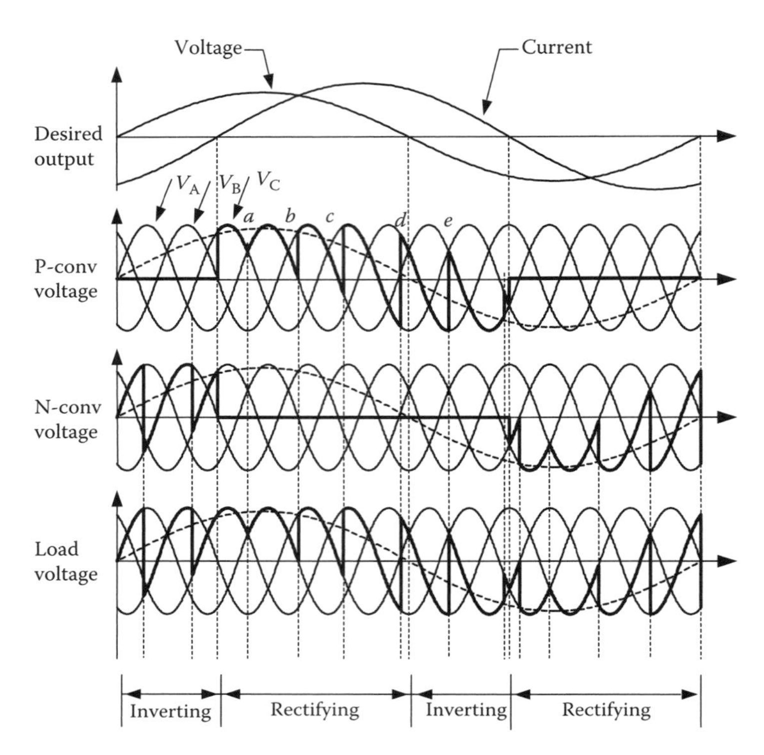

FIGURE 10.23 Waveforms for a three-pulse circulating current-free cycloconverter with *RL* load. (Reprinted from Rashid, M. H. 2001. *Power Electronics Handbook*, New York: Academic Press, pp. 307–333. With permission.)

10.4.2.1.3 Circulating-Current-Free-Mode Operation

Figure 10.23 shows the typical waveforms for a three-pulse cycloconverter operating in this mode with *RL* load assuming continuous current operation. Depending on the load-current direction, only one converter operates at a time and the load voltage is the same as the output voltage of the conducting converter. As explained earlier, in the case of the single-phase cycloconverter, there is a possibility of a short circuit of the supply voltages at the crossover points of the converter unless care is taken in the control circuit. The waveforms drawn also neglect the effect of overlap due to the AC supply inductance. A reduction in the output voltage is possible by retarding the firing angle gradually at the points *a*, *b*, *c*, *d*, and *e* in Figure 10.23 (this can easily be implemented by reducing the magnitude of the reference voltage in the control circuit). The circulating current is completely suppressed by blocking all the SCRs in the converter that is not delivering the load current. A current sensor is incorporated into each output phase of the cycloconverter that detects the direction of the output current and feeds an appropriate signal to the control circuit to inhibit or blank the gating pulses to the nonconducting converter in the same way as in the case of a dual converter for DC drives. The circulating-current-free operation improves the efficiency and the displacement factor of the cycloconverter and also increases the maximum usable output frequency. The load voltage transfers smoothly from one converter to the other.

10.4.2.2 Three-Phase 6-Pulse and 12-Pulse Cycloconverters

A six-pulse cycloconverter circuit configuration is shown in Figure 10.24. Typical load-voltage waveforms for 6-pulse (with 36 SCRs) and 12-pulse (with 72 SCRs) cycloconverters

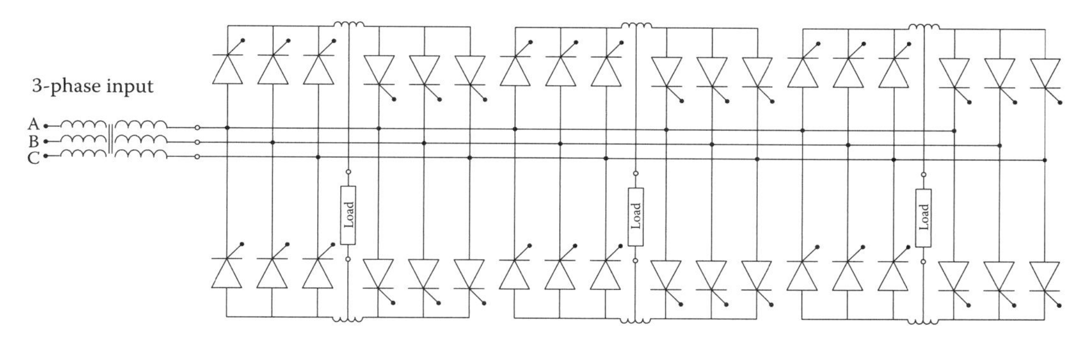

FIGURE 10.24 Three-phase six-pulse cycloconverter with isolated loads. (Reprinted from Luo, F. L., Ye, H., and Rashid, M. H. 2005. *Digital Power Electronics and Applications*. Boston: Academic Press, Elsevier. With permission.)

are shown in Figure 10.25. The 12-pulse converter is obtained by connecting two 6-pulse configurations in series and appropriate transformer connections for the required phase–shift. It may be seen that the higher pulse numbers will generate waveforms closer to the desired sinusoidal form and thus permit higher-frequency output. The phase loads may be isolated from each other as shown or interconnected with suitable secondary winding connections.

10.4.3 Cycloconverter Control Scheme

Various possible control schemes (analog as well as digital) for deriving trigger signals for controlling the basic cycloconverter have been developed over the years.

Output of the several possible signal combinations: It has been shown that a sinusoidal reference signal ($e_r = E_r \sin \omega_O t$) at desired output frequency f_O and a cosine modulating signal ($e_m = E_m \cos \omega_i t$) at input frequency f_i is the best combination possible for comparison to derive the trigger signals for the SCRs (Figure 10.26), which produces the output waveform with the lowest total harmonic distortion. The modulating voltages can be obtained as the phase-shifted voltages (B-phase voltage for A-phase SCRs, C-phase voltage for B-phase SCRs, and so on) as explained in Figure 10.27, where at the intersection point "a"

$$E_m \sin(\omega_i t - 120°) = -E_r \sin(\omega_O t - \phi)$$

or

$$\cos(\omega_i t - 30°) = \left(\frac{E_r}{E_m}\right) \sin(\omega_O t - \phi).$$

From Figure 10.27, the firing delay for A-phase SCR $\alpha = (\omega_i t - 30°)$. Thus,

$$\cos \alpha = \left(\frac{E_r}{E_m}\right) \sin(\omega_O t - \phi).$$

The cycloconverter output voltage for continuous current operation is

$$V_O = V_{dO} \cos \alpha = V_{dO} \left(\frac{E_r}{E_m}\right) \sin(\omega_O t - \phi), \tag{10.40}$$

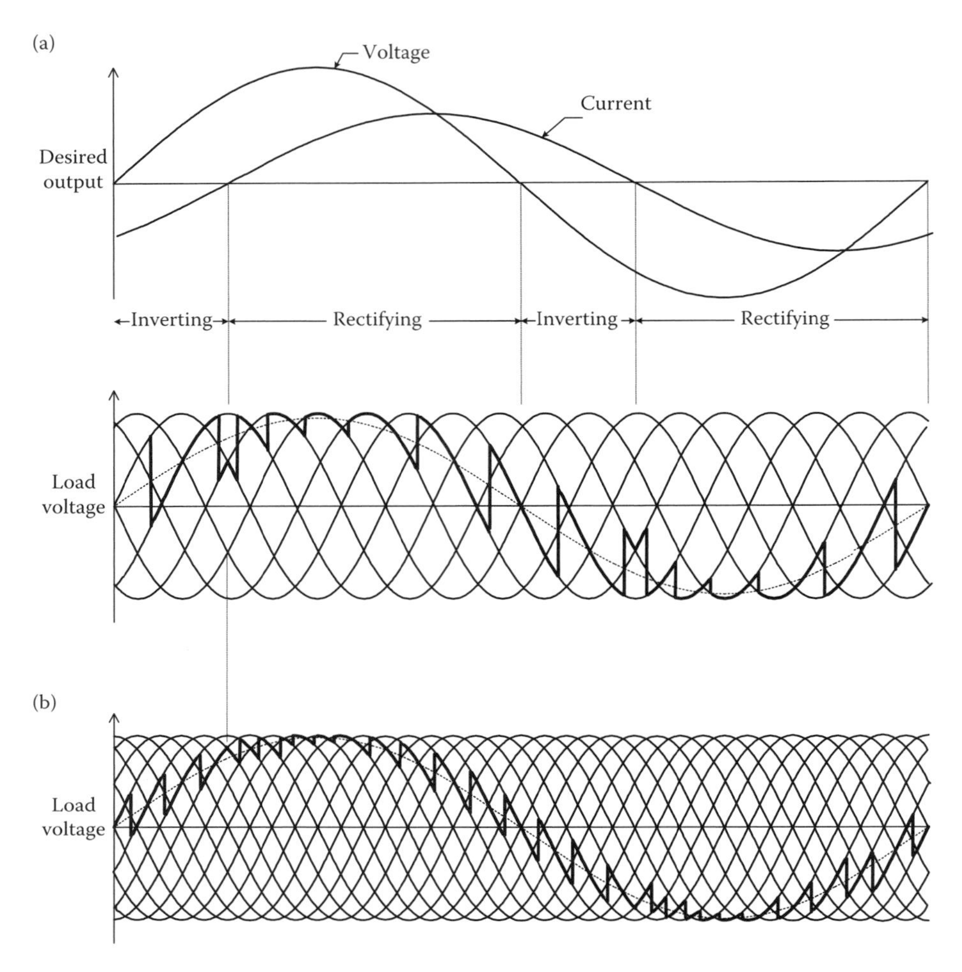

FIGURE 10.25 Cycloconverter load-voltage waveforms with lagging power factor load: (a) 6-pulse connection and (b) 12-pulse connection (Reprinted from Luo, F. L., Ye, H., and Rashid, M. H. 2005. *Digital Power Electronics and Applications*. Boston: Academic Press, Elsevier. With permission.)

which shows that the amplitude, frequency, and phase of the output voltage can be controlled by controlling corresponding parameters of the reference voltage, thus making the transfer characteristic of the cycloconverter linear. The derivation of the two complementary voltage waveforms for the P-group or N-group converter "blanks" in this way is illustrated in Figure 10.28. The final cycloconverter output waveshape is composed of alternate half-cycle segments of the complementary P-converter and N-converter output voltage waveforms that coincide with the positive and negative current half-cycles, respectively.

10.4.3.1 Control Circuit Block Diagram

Figure 10.29 shows a simplified block diagram of the control circuit for a circulating-current-free cycloconverter. The same circuit is also applicable to a circulating-current cycloconverter with the omission of the *converter group selection and blanking circuit*.

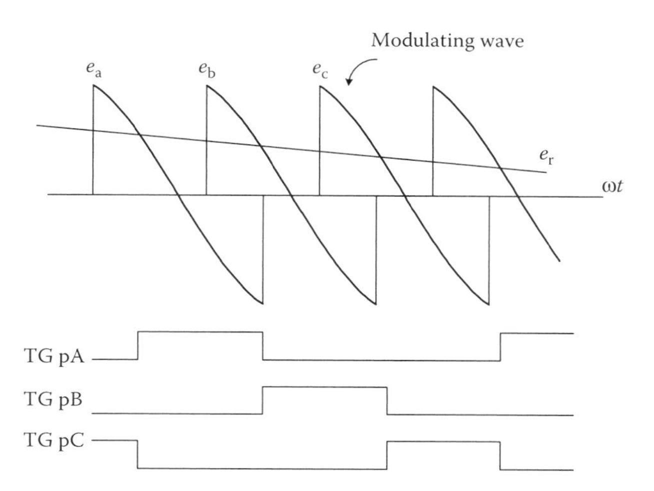

FIGURE 10.26 Deriving firing signals for a converter group of a three-pulse cycloconverter. (Reprinted from Rashid, M. H. 2001. *Power Electronics Handbook*, New York: Academic Press, pp. 307–333. With permission.)

The *synchronizing circuit* produces the modulating voltages ($e_a = -Kv_b, e_b = -Kv_c, e_c = -Kv_a$), synchronized with the mains through step-down transformers and proper filter circuits.

The *reference source* produces a VVVF reference signal (e_{ra}, e_{rb}, e_{rc}) (three-phase for a three-phase cycloconverter) for comparison with the modulation voltages. Various ways (analog or digital) have been developed to implement this reference source as in the case of the PWM inverter. In one of the early analog schemes (Figure 10.30) for a three-phase cycloconverter, a variable-frequency unijunction transistor (UJT) relaxation oscillator of the frequency $6f_d$ triggers a ring counter to produce a three-phase square-wave output of frequency (f_d), which is used to modulate a single-phase fixed frequency (f_c) variable amplitude sinusoidal voltage in a three-phase full-wave transistor chopper. The three-phase output contains ($f_c - f_d$), ($f_c + f_d$), ($3f_d + f_c$), and so forth, frequency components from where the "wanted" frequency component ($f_c - f_d$) is filtered out for each phase using a low-pass filter. For example, with $f_c = 500$ Hz and the frequency of the relaxation oscillator

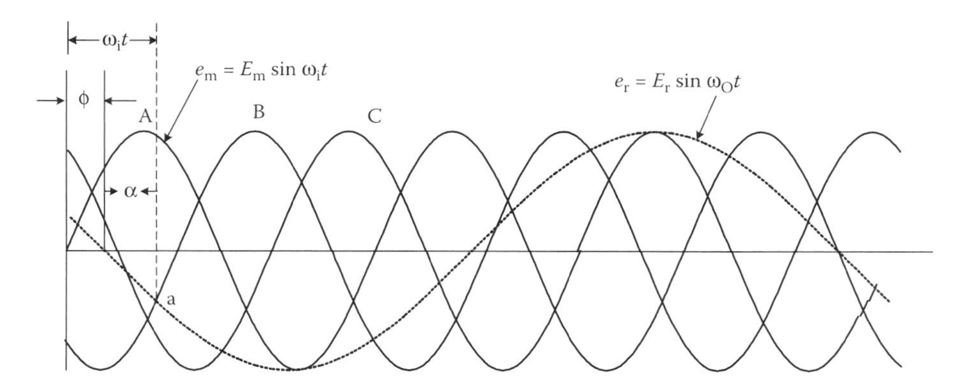

FIGURE 10.27 Derivation of the cosine modulating voltages. (Reprinted from Rashid, M. H. 2001. *Power Electronics Handbook*, New York: Academic Press, pp. 307–333. With permission.)

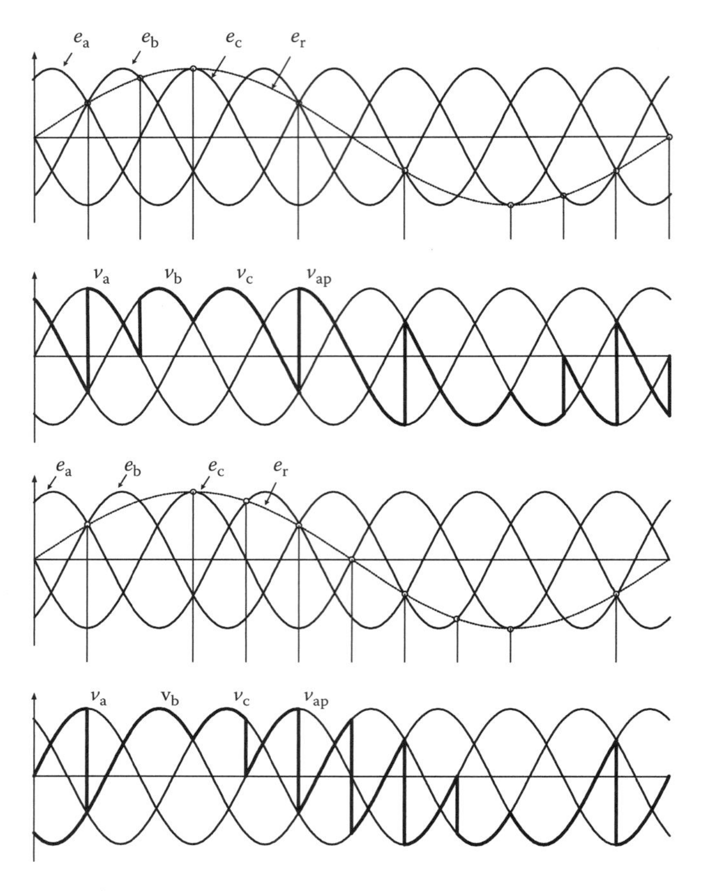

FIGURE 10.28 Derivation of P-converter and N-converter output voltages. (Reprinted from Rashid, M. H. 2001. *Power Electronics Handbook*, New York: Academic Press, pp. 307–333. With permission.)

varying between 2820 and 3180 Hz, a three-phase 0–30 Hz reference output can be obtained with the facility for phase sequence reversal.

The *logic* and *trigger* circuit for each phase involves comparators for comparison of the reference and modulating voltages and inverters acting as buffer stages. The outputs of the comparators are used to clock the flip-flops or latches whose outputs in turn feed the SCR gates through AND gates and pulse amplifying and isolation circuit. The second input to the AND gates is from the *converter group selection and blanking circuit*.

In the *converter group selection and blanking circuit*, the zero crossing of the current at the end of each half-cycle is detected and is used to regulate the control signals to either P-group or N-group converters depending on whether the current goes to zero from negative to positive or positive to negative, respectively. However, in practice, the current that is discontinuous passes through multiple zero crossings while changing direction, which may lead to undesirable switching of the converters. Therefore, in addition to the current signal, the reference voltage signal is also used for the group selection and a *threshold* band is introduced in the current signal detection to avoid inadvertent switching of the converters. Further, a delay circuit provides a blanking period of appropriate duration between the converter group switchings to avoid line-to-line short circuits. In some schemes, the delays are not introduced when a small circulating current is allowed during crossover instants

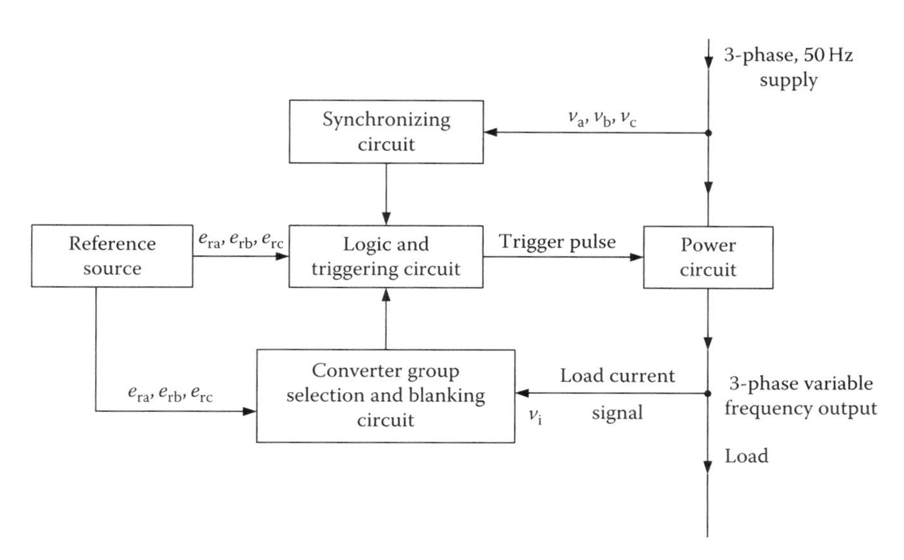

FIGURE 10.29 Block diagram of a circulating current-free cycloconverter control circuit. (Reprinted from Rashid, M. H. 2001. *Power Electronics Handbook*, New York: Academic Press, pp. 307–333. With permission.)

limited by reactors of limited size, and this scheme operates in the so-called dual mode—circulating current as well as circulating-current-free mode for minor and major portions of the output cycle, respectively. A different approach to the converter group selection, based on the closed-loop control of the output voltage where a bias voltage is introduced between the voltage transfer characteristics of the converters to reduce the circulating current, is discussed.

10.4.3.2 Improved Control Schemes

With the development of microprocessors and PC-based systems, digital software control has taken over many tasks in modern cycloconverters, particularly in replacing the low-level reference waveform generation and analog signal comparison units. The reference waveforms can easily be generated in the computer, stored in the EPROMs and accessed under the control of a stored program and microprocessor clock oscillator. The analog signal voltages can be converted to digital signals by using analog-to-digital converters (ADCs). The

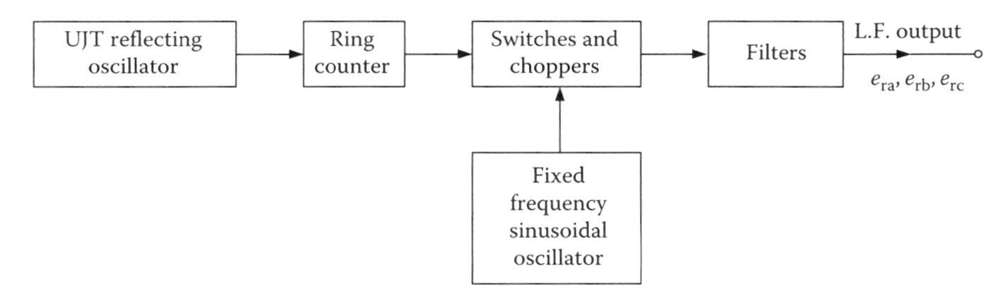

FIGURE 10.30 Block diagram of a VVVF three-phase reference source. (Reprinted from Rashid, M. H. 2001. *Power Electronics Handbook*, New York: Academic Press, pp. 307–333. With permission.)

waveform comparison can then be made with the comparison features of the microprocessor system. The addition of time delays and intergroup blanking can also be achieved with digital techniques and computer software. A modification of the cosine firing control, using communication principles such as *regular sampling* in preference to the *natural sampling* of the reference waveform yielding a stepped sine wave before comparison with the cosine wave has been shown to reduce the presence of *subharmonics* (to be discussed later) in the circulating-current cycloconverter and to facilitate microprocessor-based implementation, as in the case of PWM inverters.

10.4.4 Cycloconverter Harmonics and Input Current Waveform

The exact waveshape of the output voltage of the cycloconverter depends on (i) the pulse number of the converter; (ii) the ratio of the output frequency to the input frequency (f_O/f_i); (iii) the relative level of the output voltage; (iv) load displacement angle; (v) circulating current or circulating-current-free operation; and (vi) the method of control of the firing instants. The harmonic spectrum of a cycloconverter output voltage is different from and more complex than that of a phase-controlled converter. It has been revealed that because of the continuous "to-and-fro" phase modulation of the converter firing angles, the harmonic distortion components (known as *necessary distortion terms*) have frequencies that are sums of, and differences between, multiples of output and input supply frequencies.

10.4.4.1 Circulating-Current-Free Operations

A derived general expression for the output voltage of a cycloconverter with circulating-current-free operation shows the following spectrum of harmonic frequencies for the 3-pulse, 6-pulse, and 12-pulse cycloconverters employing the cosine modulation technique:

$$\begin{aligned} \text{3-pulse: } f_{OH} &= \left|3(2k-1)f_i \pm 2nf_O\right| \quad \text{and} \quad \left|6kf_i \pm (2n+1)f_O\right|, \\ \text{6-pulse: } f_{OH} &= \left|6kf_i \pm (2n+1)f_O\right|, \\ \text{6-pulse: } f_{OH} &= \left|6kf_i \pm (2n+1)f_O\right|, \end{aligned} \tag{10.41}$$

where k is any integer from unity to infinity and n is any integer from zero to infinity. It may be observed that for certain ratios of f_O/f_i, the order of harmonics may be less than or equal to the desired output frequency. All such harmonics are known as *subharmonics* as they are not higher multiples of the input frequency. These subharmonics may have considerable amplitudes (e.g., with a 50-Hz input frequency and a 35-Hz output frequency, a subharmonic of frequency $3 \times 50 - 4 \times 35 = 10$ Hz is produced whose magnitude is 12.5% of that of the 35-Hz component) and are difficult to filter and thus are objectionable. Their spectrum increases with an increase of the ratio f_O/f_i and thus limits its value at which a tolerable waveform can be generated.

10.4.4.2 Circulating-Current Operation

For circulating-current operation with continuous current, the harmonic spectrum in the output voltage is the same as that of the circulating-current-free operation except that each harmonic family now terminates at a definite term, rather than having an infinite number of components. They are

$$\begin{aligned}
&\text{3-pulse: } f_{\mathrm{OH}} = \begin{cases} |3(2k-1)f_{\mathrm{i}} \pm 2nf_{\mathrm{O}}|, & n \le 3(2k-1)+1, \\ |6kf_{\mathrm{i}} \pm (2n+1)f_{\mathrm{O}}|, & (2n+1) \le (6k+1), \end{cases} \\
&\text{6-pulse: } f_{\mathrm{OH}} = \left|6kf_{\mathrm{i}} \pm (2n+1)f_{\mathrm{O}}\right|, \quad (2n+1) \le (6k+1), \\
&\text{12-pulse: } f_{\mathrm{OH}} = \left|6kf_{\mathrm{i}} \pm (2n+1)f_{\mathrm{O}}\right|, \quad (2n+1) \le (12k+1).
\end{aligned} \tag{10.42}$$

The amplitude of each harmonic component is a function of the output voltage ratio for the circulating-current cycloconverter and the output voltage ratio as well as the load displacement angle for the circulating-current-free mode.

From the point of view of maximum useful attainable output-to-input frequency ratio ($f_{\mathrm{O}}/f_{\mathrm{i}}$) with the minimum amplitude of objectionable harmonic components, a guideline is available for it as 0.33, 0.5, and 0.75 for the 3-, 6-, and 12-pulse cycloconverters, respectively. However, with a modification of the cosine wave modulation timings such as *regular sampling* in the case of only circulating-current cycloconverters and using a *subharmonic detection and feedback control concept* for both the circulating-current and circulating-current-free cases, the subharmonics can be suppressed and the useful frequency range for the NCCs can be increased.

10.4.4.3 Other Harmonic Distortion Terms

Besides the harmonics mentioned, other harmonic distortion terms consisting of frequencies of integral multiples of desired output frequency appear if the transfer characteristic between the output and reference voltages is not linear. These are called *unnecessary distortion terms*, which are absent when the output frequencies are much less than the input frequency. Further, some *practical distortion terms* may appear due to practical nonlinearities and imperfections in the control circuits of the cycloconverter, particularly at relatively lower levels of output voltages.

10.4.4.4 Input Current Waveform

Although the load current, particularly for higher-pulse cycloconverters, can be assumed to be sinusoidal, the input current is more complex as it is made up of pulses. Assuming the cycloconverter to be an ideal switching circuit without losses, it can be shown from the instantaneous power balance equation that in a cycloconverter supplying a single-phase load, the input current has harmonic components of frequencies ($f_1 \pm 2f_{\mathrm{O}}$), called *characteristic harmonic frequencies*, which are independent of the pulse number, and they result in an oscillatory power transmittal to the AC supply system. In the case of a cycloconverter feeding a balanced three-phase load, the net instantaneous power is the sum of the three oscillating instantaneous powers when the resultant power is constant and the net harmonic component is greatly reduced when compared with that of the single-phase load case. In general, the total rms value of the input current waveform consists of three components: in-phase, quadrature, and harmonic. The in-phase component depends on the active power output, while the quadrature component depends on the net average of the oscillatory firing angle and is always lagging.

10.4.5 Cycloconverter Input Displacement/Power Factor

The input supply performance of a cycloconverter such as displacement factor or fundamental power factor, input power factor, and the input current distortion factor are defined similarly to those of the phase-controlled converter. The harmonic factor for the case of a cycloconverter is relatively complex as the harmonic frequencies are not simple multiples of the input frequency but are sums of, and differences between, multiples of output and input frequencies.

Irrespective of the nature of the load, leading, lagging, or unity power factor, the cycloconverter requires reactive power decided by the average firing angle. At low output voltage, the average phase displacement between the input current and the voltage is large and the cycloconverter has a low input displacement and power factor. Besides the load displacement factor and output voltage ratio, another component of the reactive current arises due to the modulation of the firing angle in the fabrication process of the output voltage. In a phase-controlled converter supplying DC load, the maximum displacement factor is unity for maximum DC output voltage. However, in the case of the cycloconverter, the maximum input displacement factor (IDF) is 0.843 with unity power factor load. The displacement factor decreases with reduction in the output voltage ratio. The distortion factor of the input current is given by (I_1/I), which is always less than unity, and the resultant power factor, (=distortion factor × displacement factor) is thus much lower (around 0.76 at the maximum) than the displacement factor, which is a serious disadvantage of the NCC.

10.4.6 Effect of Source Impedance

The source inductance introduces commutation overlap and affects the external characteristics of a cycloconverter similar to the case of a phase-controlled converter with DC output. It introduces delay in the transfer of current from one SCR to another, and results in a voltage loss at the output and a modified harmonic distortion. At the input, the source impedance causes "rounding off" of the steep edges of the input current waveforms, resulting in a reduction of the amplitudes of higher-order harmonic terms as well as a decrease in the IDF.

10.4.7 Simulation Analysis of Cycloconverter Performance

The nonlinearity and discrete time nature of practical cycloconverter systems, particularly for discontinuous current conditions, make an exact analysis quite complex, and a valuable design and analytical tool is digital computer simulation of the system. Two general methods of computer simulation of the cycloconverter waveforms for *RL* and induction motor loads with circulating-current and circulating-current-free operation have been suggested; one of the methods is the *crossover point method*, which is very fast and convenient. This method gives the crossover points (intersections of the modulating and reference waveforms) and the conducting phase numbers for both P- and N-converters from which the output waveforms for a particular load can be digitally computed at any interval of time for a practical cycloconverter.

10.4.8 Forced-Commutated Cycloconverter

The NCC with SCRs as devices, discussed so far, is sometimes referred to as a *restricted frequency changer* as, in view of the allowance for the output voltage quality ratings, the maximum output voltage frequency is restricted ($f_O \ll f_i$), as mentioned earlier. With devices replaced by fully controlled switches such as forced-commutated SCRs, power transistors, IGBTs, GTOs, and so on, an FCC can be built where the desired output frequency is given by $f_O = |f_S - f_i|$, where f_S is the switching frequency, which may be larger or smaller than f_i. In the case when $f_O \geq f_i$, the converter is called an *unrestricted frequency changer* (UFC), and when $f_O \leq f_i$, the converter is called a *slow switching frequency changer* (SSFC). The early FCC structures have been treated comprehensively. It has been shown that in contrast to the NCC, where the IDF is always lagging, in the UFC, the IDF is leading when the load displacement factor is lagging and vice versa, and in SSFC, the IDF is identical to that of the load. Further, with proper control in an FCC, the IDF can be made either unity (UDFFC) with a concurrent composite voltage waveform, or controllable (CDFFC) where P-converter and N-converter voltage segments can be shifted relative to the output current wave for controlling the IDF continuously from lagging via unity to leading.

In addition to allowing bilateral power flow, UFCs offer an unlimited output frequency range and good input voltage utilization, do not generate input current and output voltage subharmonics, and require only nine bidirectional switches (Figure 10.31) for a three-phase to three-phase conversion. The main disadvantage of the structures treated is that they generate large unwanted low-order input current and output voltage harmonics that are difficult to filter out, particularly for low-output voltage conditions. This problem has largely been solved with the introduction of an imaginative PWM voltage control scheme, which is the basis of a newly designated converter called the *MC (also known as the PWM cycloconverter)*, which operates as a *generalized solid-state transformer* with significant improvement in voltage and input current waveforms, resulting in sine wave input and sine wave, as will be discussed in the next subsection.

10.5 Matrix Converters

The MC is a development of the FCC based on bidirectional fully controlled switches, incorporating PWM voltage control, as mentioned earlier. This technique was developed by Venturine in 1980 [12]. With the initial progress reported, it has received considerable attention as it provides a good alterative to the double-sided PWM voltage source rectifier–inverter having the advantages of being a single-stage converter with only nine switches for three-phase to three-phase conversion and inherent bidirectional power flow, sinusoidal input/output waveforms with moderate switching frequency, the possibility of a compact design due to the absence of DC-link reactive components and controllable input power factor independent of the output load current [12–21]. The main disadvantages of the MCs developed so far are the inherent restriction of the *voltage transfer ratio* (0.866), a more complex control and protection strategy, and above all, the nonavailability of a fully controlled bidirectional high-frequency switch integrated in a silicon chip (Triac, although bilateral, cannot be fully controlled).

The power circuit diagram of the most practical three-phase to three-phase (3ϕ–3ϕ) MC is shown in Figure 10.31a, which uses nine bidirectional switches so arranged that any of three input phases can be connected to any output phase as shown in the switching matrix symbol in Figure 10.31b. Thus, the voltage at any input terminal may be made to appear at any output terminal or terminals while the current in any phase of the load may be drawn from any phase or phases of the input supply. For the switches, the inverse parallel

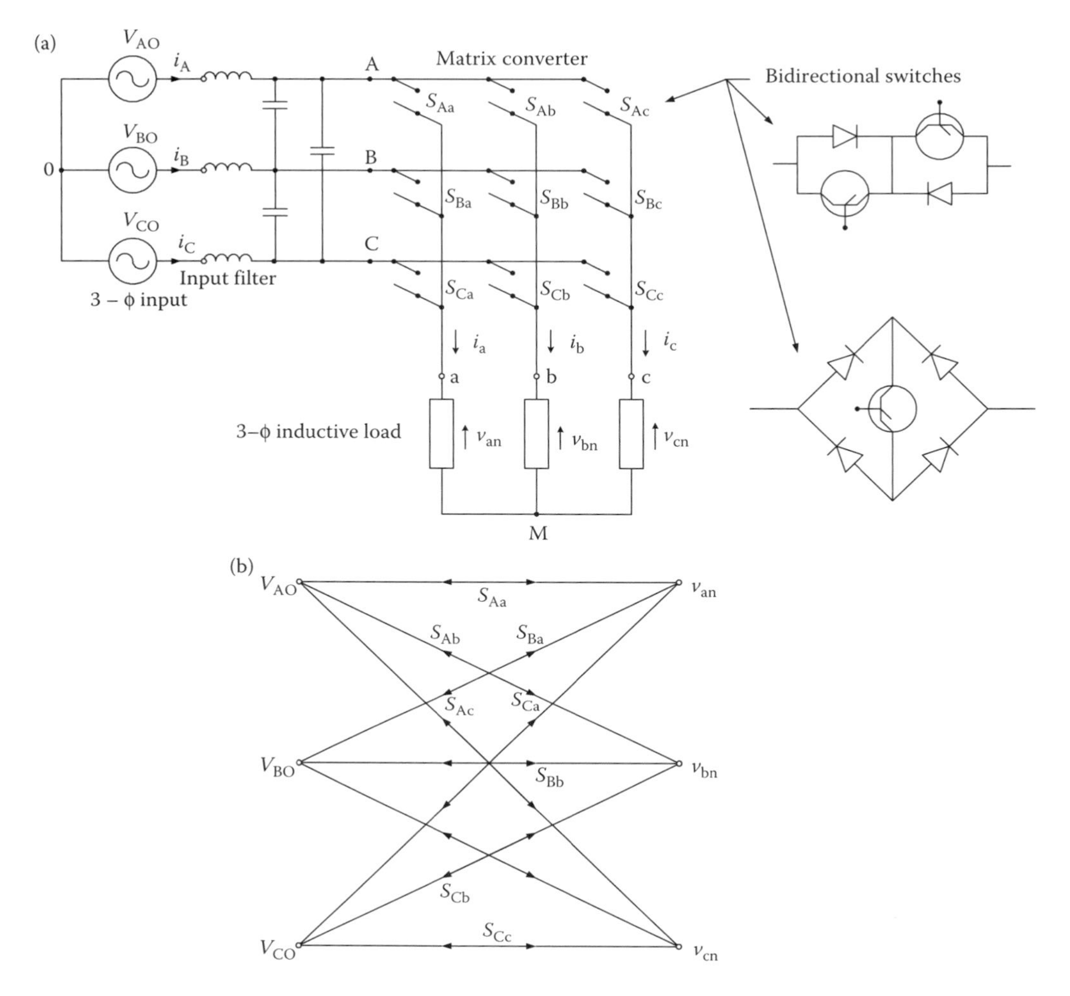

FIGURE 10.31 (a) The 3ϕ–3ϕ MC (FCC) circuit with input filter and (b) switching matrix symbol for the converter. (Reprinted from Luo, F. L., Ye, H., and Rashid, M. H. 2005. *Digital Power Electronics and Applications*. Boston: Academic Press, Elsevier. With permission.)

combination of reverse-blocking self-controlled devices such as Power MOSFETs or IGBTs or transistor-embedded diode bridge as shown has been used so far. The circuit is called an MC as it provides exactly one switch for each of the possible connections between the input and the output. The switches should be controlled in such a way that at any point of time, one and only one of the three switches connected to an output phase must be closed to prevent "short-circuiting" of the supply lines or interrupting the load-current flow in an inductive load. With these constraints, it can be visualized that from the possible 512 ($=2^9$) states of the converter, only 27 switch combinations are allowed, as given in Table 10.3, which includes the resulting output line voltages and input phase currents. These combinations are divided into three groups. Group I consists of six combinations where each output phase is connected to a different input phase. In Group II, there are three subgroups, each having six combinations with two output phases short-circuited (connected to the same input phase). Group III includes three combinations with all output phases short-circuited.

TABLE 10.3

Three-Phase/Three-Phase Matrix Converter Switching Combinations

Group	A	B	C	v_{ab}	v_{bc}	v_{ca}	i_A	i_B	i_C	S_{Aa}	S_{Ab}	S_{Ac}	S_{Ba}	S_{Bb}	S_{Bc}	S_{Ca}	S_{Cb}	S_{Cc}
	A	B	C	v_{AB}	v_{BC}	v_{CA}	i_a	i_b	i_c	1	0	0	0	1	0	0	0	1
	A	C	B	$-v_{CA}$	$-v_{BC}$	$-v_{AB}$	i_a	i_c	i_b	1	0	0	0	0	1	0	1	0
	B	A	C	$-v_{AB}$	$-v_{CA}$	$-v_{BC}$	i_b	i_a	i_c	0	1	0	1	0	0	0	0	1
I	B	C	A	v_{BC}	v_{CA}	v_{AB}	i_c	i_a	i_b	0	1	0	0	0	1	0	1	0
	C	A	B	v_{CA}	v_{AB}	v_{BC}	i_b	i_c	i_a	0	0	1	1	0	0	0	1	0
	C	B	A	$-v_{BC}$	$-v_{AB}$	$-v_{CA}$	i_c	i_b	i_a	0	0	1	0	1	0	1	0	0
	A	C	C	$-v_{CA}$	0	v_{CA}	i_a	0	$-i_a$	1	0	0	0	0	1	0	0	1
	B	C	C	v_{BC}	0	$-v_{BC}$	0	i_a	$-i_a$	0	1	0	0	0	1	0	0	1
	B	A	A	$-v_{AB}$	0	$-v_{AB}$	$-i_a$	i_a	0	0	1	0	1	0	0	1	0	0
II-A	C	A	A	v_{CA}	0	$-v_{CA}$	$-i_a$	0	i_a	0	0	1	1	0	0	1	0	0
	C	B	B	$-v_{BC}$	0	v_{BC}	0	$-i_a$	i_a	0	0	1	0	1	0	0	1	0
	A	B	B	v_{AB}	0	$-v_{AB}$	i_a	$-i_a$	0	1	0	0	0	1	0	0	1	0
	C	A	C	$-v_{CA}$	$-v_{CA}$	0	i_b	0	$-i_b$	0	0	1	1	0	0	0	0	1
	C	B	C	$-v_{BC}$	v_{BC}	0	0	i_b	$-i_b$	0	0	1	0	1	0	0	0	1
	A	B	A	v_{AB}	$-v_{AB}$	0	$-i_b$	i_b	0	1	0	0	0	1	0	1	0	0
II-B	A	C	A	$-v_{CA}$	v_{CA}	0	$-i_b$	0	i_b	1	0	0	0	0	1	1	0	0
	B	C	B	v_{BC}	$-v_{BC}$	0	0	$-i_b$	i_b	0	1	0	0	0	1	0	1	0
	B	A	B	$-v_{AB}$	v_{AB}	0	i_b	$-i_b$	0	0	1	0	1	0	0	0	1	0
	C	C	A	0	v_{CA}	$-v_{CA}$	i_c	0	$-i_c$	0	0	1	0	0	1	1	0	0
	C	C	B	0	$-v_{BC}$	v_{BC}	0	i_c	$-i_c$	0	0	1	0	0	1	0	1	0
	A	A	B	0	v_{AB}	$-v_{AB}$	$-i_c$	i_c	0	1	0	1	0	1	0	0	1	0
II-C	A	A	C	0	$-v_{CA}$	v_{CA}	$-i_c$	0	i_c	1	0	1	0	0	0	0	0	1
	B	B	C	0	v_{BC}	$-v_{BC}$	0	$-i_c$	i_c	0	1	0	0	1	0	0	0	1
	B	B	A	0	$-v_{AB}$	v_{AB}	i_c	$-i_c$	0	0	1	0	0	1	0	1	0	0
	A	A	A	0	0	0	0	0	0	0	0	0	1	0	0	1	0	0
III	B	B	B	0	0	0	0	0	0	0	0	0	0	1	0	0	1	0
	C	C	C	0	0	0	0	0	0	0	0	0	0	0	1	0	0	1

Source: Data from Luo, F. L., Ye, H., and Rashid, M. H. 2005. *Digital Power Electronics and Applications*, New York: Academic Press, p. 238.

With a given set of input three-phase voltages, any desired set of three-phase output voltages can be synthesized by adopting a suitable switching strategy. However, it has been shown that regardless of the switching strategy, there are physical limits on the achievable output voltage with these converters as the maximum peak-to-peak output voltage cannot be greater than the minimum voltage difference between two phases of the input.

To have complete control of the synthesized output voltage, the envelope of the three-phase reference or target voltages must be fully contained within the continuous envelope of the three-phase input voltages. The initial strategy with the output frequency voltages as references reported the limit as 0.5 of the input, as shown in Figure 10.32a. This value can be increased to 0.866 by adding a third harmonic voltage of input frequency $(V_i/4)\cos 3\omega_i t$ to all target output voltages and subtracting from them a third harmonic voltage of output frequency $(V_O/6)\cos 3\omega_O t$, as shown in Figure 10.32b. However, this process involves a considerable amount of additional computations in synthesizing the output voltages. The other alternative is to use the space vector modulation (SVM) strategy as used in PWM

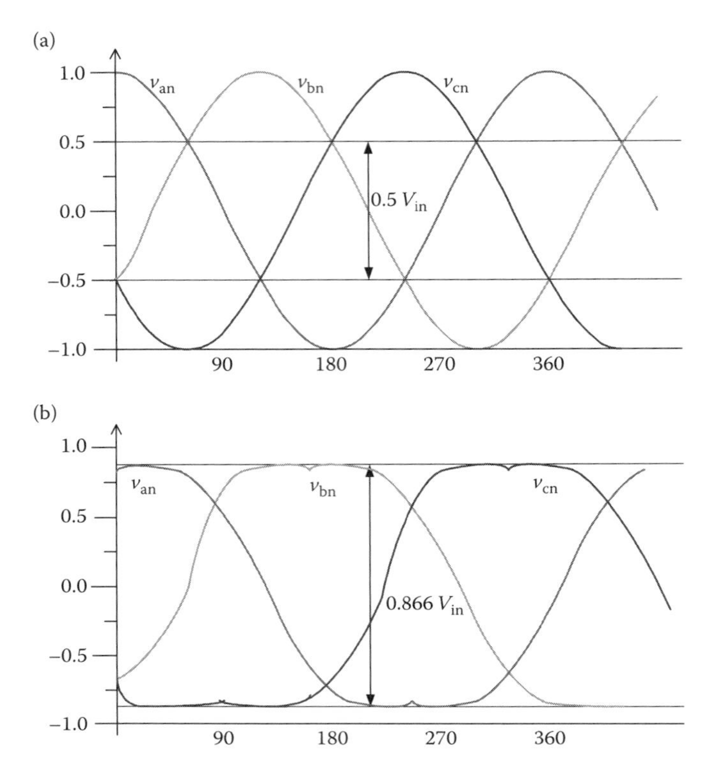

FIGURE 10.32 Output voltage limits for a three-phase AC/AC MC: (a) basic converter input voltages and (b) maximum attainable with inclusion of third harmonic voltages of input and output frequency to the target voltages. (Reprinted from Luo, F. L., Ye, H., and Rashid, M. H. 2005. *Digital Power Electronics and Applications*. Boston: Academic Press, Elsevier. With permission.)

inverters without adding third harmonic components, but it also yields the maximum voltage transfer ratio as 0.866.

An AC input LC filter is used to eliminate the switching ripples generated in the converter and the load is assumed to be sufficiently inductive to maintain the continuity of the output currents.

10.5.1 Operation and Control Methods of the MC

The converter in Figure 10.31 connects any input phase (A, B, and C) to any output phase (a, b, and c) at any instant. When connected, the voltages v_{an}, v_{bn}, and v_{cn} at the output terminals are related to the input voltages V_{AO}, V_{BO}, and V_{CO} as

$$\begin{bmatrix} v_{an} \\ v_{bn} \\ v_{cn} \end{bmatrix} = \begin{bmatrix} S_{Aa} & S_{Ba} & S_{Ca} \\ S_{Ab} & S_{Bb} & S_{Cb} \\ S_{Ac} & S_{Bc} & S_{Cc} \end{bmatrix} \begin{bmatrix} v_{AO} \\ v_{BO} \\ v_{CO} \end{bmatrix}, \tag{10.43}$$

where S_{Aa} through S_{Cc} are the switching variables of the corresponding switches shown in Figure 10.31. For a balanced linear star-connected load at the output terminals, the input phase currents are related to the output phase currents by

$$\begin{bmatrix} i_A \\ i_B \\ i_C \end{bmatrix} = \begin{bmatrix} S_{Aa} & S_{Ab} & S_{Ac} \\ S_{Ba} & S_{Bb} & S_{Bc} \\ S_{Ca} & S_{Cb} & S_{Cc} \end{bmatrix} \begin{bmatrix} i_a \\ i_b \\ i_c \end{bmatrix}. \tag{10.44}$$

Note that the matrix of the switching variables in Equation 10.44 is a transpose of the respective matrix in Equation 10.43. The MC should be controlled using a specific and appropriately timed sequence of the values of the switching variables, which will result in the balanced output voltages having the desired frequency and amplitude, while the input currents are balanced and in phase (for unity IDF) or at an arbitrary angle (for controllable IDF) with respect to the input voltages. As the MC, in theory, can operate at any frequency, at the output or input, including zero, it can be employed as a three-phase AC/DC converter, DC/three-phase AC converter, or even a buck–boost DC chopper, and thus as a *universal power converter.*

The control methods adopted so far for the MC are quite complex and are the subject of continuing research. Of the methods proposed for independent control of the output voltages and input currents, two are in widespread use and will be reviewed briefly here. They are (i) the *Venturini* method based on a mathematical approach of transfer function analysis; and (ii) the SVM approach (as has been standardized now in the case of PWM control of the DC-link inverter).

10.5.1.1 Venturini Method

Given a set of three-phase input voltages with constant amplitude V_i and frequency $f_i = \omega_i/2\pi$, this method calculates a switching function involving the duty cycles of each of the nine bidirectional switches and generates the three-phase output voltages by sequential piecewise sampling of the input waveforms. These output voltages follow a predetermined set of reference or target voltage waveforms and with a three-phase load connected, a set of input currents I_i, and angular frequency ω_i, should be in phase for unity IDF or at a specific angle for controlled IDF.

A transfer function approach is employed to achieve the previously mentioned features by relating the input and output voltages and the output and input currents as

$$\begin{bmatrix} V_{O1}(t) \\ V_{O2}(t) \\ V_{O3}(t) \end{bmatrix} = \begin{bmatrix} m_{11}(t) & m_{12}(t) & m_{13}(t) \\ m_{21}(t) & m_{22}(t) & m_{23}(t) \\ m_{31}(t) & m_{32}(t) & m_{33}(t) \end{bmatrix} \begin{bmatrix} V_{i1}(t) \\ V_{i2}(t) \\ V_{i3}(t) \end{bmatrix}, \tag{10.45}$$

$$\begin{bmatrix} I_{i1}(t) \\ I_{i2}(t) \\ I_{i3}(t) \end{bmatrix} = \begin{bmatrix} m_{11}(t) & m_{21}(t) & m_{31}(t) \\ m_{12}(t) & m_{22}(t) & m_{32}(t) \\ m_{13}(t) & m_{23}(t) & m_{33}(t) \end{bmatrix} \begin{bmatrix} I_{O1}(t) \\ I_{O2}(t) \\ I_{O3}(t) \end{bmatrix}, \tag{10.46}$$

where the elements of the modulation matrix $m_{ij}(t)$ $(i, j = 1, 2, 3)$ represent the duty cycles of a switch connecting output phase i to input phase j within a sample switching interval. The elements of $m_{ij}(t)$ are limited by the constraints

$$0 \le m_{ij}(t) \le 1 \quad \text{and} \quad \sum_{j=1}^{3} m_{ij}(t) = 1, \quad (i = 1, 2, 3).$$

The set of three-phase target or reference voltages to achieve the maximum voltage transfer ratio for unity IDF is

$$\begin{bmatrix} V_{O1}(t) \\ V_{O2}(t) \\ V_{O3}(t) \end{bmatrix} = V_{Om} = \begin{bmatrix} \cos \omega_O t \\ \cos(\omega_O t - 120°) \\ \cos(\omega_O t - 240°) \end{bmatrix} + \frac{V_{im}}{4} \begin{bmatrix} \cos 3\omega_i t \\ \cos 3\omega_i t \\ \cos 3\omega_i t \end{bmatrix} - \frac{V_{Om}}{6} \begin{bmatrix} \cos 3\omega_O t \\ \cos 3\omega_O t \\ \cos 3\omega_O t \end{bmatrix} \quad (10.47)$$

where V_{Om} and V_{im} are the magnitudes of output and input fundamental voltages of angular frequencies ω_O and ω_i, respectively. With $V_{Om} \leq 0.866 V_{im}$, a general formula for the duty cycles $m_{ij}(t)$ is derived. For unity IDF condition, a simplified formula is

$$m_{ij} = \frac{1}{3} \left\{ 1 + 2q \cos(\omega_i t - 2(j-1)60°) \left[\cos(\omega_O t - 2(i-1)60°) + \frac{1}{2\sqrt{3}} \cos(3\omega_i t) - \frac{1}{6} \cos(3\omega_O t) \right] - \frac{2q}{3\sqrt{3}} [\cos(4\omega_i t - 2(j-1)60°) - \cos(2\omega_i t - 2(1-j)60°)] \right\} \quad (10.48)$$

where $i, j = 1, 2$, and 3 and $q = V_{Om}/V_{im}$.

The method developed as in the preceding is based on a *direct transfer function* (DTF) approach using a single modulation matrix for the MC, employing the switching combinations of all three groups in Table 10.3. Another approach called the *indirect transfer function* (ITF) approach considers the MC as a combination of a PWM voltage source rectifier and a PWM voltage source inverter (VSR-VSI) and employs the already well-established VSR and VSI PWM techniques for MC control utilizing the switching combinations of only Group II and Group III of Table 10.3. The drawback of this approach is that the IDF is limited to unity and the method also generates higher and fractional harmonic components in the input and output waveforms.

10.5.1.2 The SVM Method

The SVM is now a well-documented inverter PWM control technique that yields high voltage gain and less harmonic distortion compared with the other modulation techniques discussed earlier. Here, the three-phase input currents and output voltages are represented as space vectors, and the SVM is applied simultaneously to the output voltage and input current space vectors. Applications of the SVM algorithm to control MCs have appeared in the literature and have been shown to have the inherent capability to achieve full control of the instantaneous output voltage vector and the instantaneous current displacement angle even under supply voltage disturbances. The algorithm is based on the concept that the MC output line voltages for each switching combination can be represented as a voltage space vector denoted by

$$V_O = \frac{2}{3}[v_{ab} + v_{bc} \exp(j120°) + v_{ca} \exp(-j120°)]. \quad (10.49)$$

Of the three groups in Table 10.3, only the switching combinations of Group II and Group III are employed for the SVM method. Group II consists of switching state voltage vectors having constant angular positions and are called *active* or *stationary* vectors. Each subgroup of Group II determines the position of the resulting output voltage space vector, and the six state space voltage vectors form a six-sextant hexagon used to synthesize the desired output voltage vector. Group III comprises the *zero* vectors positioned at the center of the output voltage hexagon, and these are suitably combined with the active vectors for the output voltage synthesis.

The modulation method involves selection of the vectors and their on-time computation. At each sampling period T_S, the algorithm selects four active vectors related to any possible combination of output voltage and input current sectors in addition to the zero vector to construct a desired reference voltage. The amplitude and the phase angle of the reference voltage vector are calculated and the desired phase angle of the input current vector is determined in advance. For the computation of the on-time periods of the chosen vectors, these are combined into two sets leading to two new vectors adjacent to the reference voltage vector in the sextant and having the same direction as the reference voltage vector. Applying the standard SVM theory, the general formulae derived for the vector on-times, which satisfy, at the same time, the reference output voltage and input current displacement angle, are

$$\begin{aligned}
t_1 &= \frac{2qT_S}{\sqrt{3}\cos\phi_i}\sin(60^\circ - \theta_O)\sin(60^\circ - \theta_i),\\
t_2 &= \frac{2qT_S}{\sqrt{3}\cos\phi_i}\sin(60^\circ - \theta_O)\sin\theta_i,\\
t_3 &= \frac{2qT_S}{\sqrt{3}\cos\phi_i}\sin\theta_O\sin(60^\circ - \theta_i),\\
t_4 &= \frac{2qT_S}{\sqrt{3}\cos\phi_i}\sin\theta_O\sin\theta_i,
\end{aligned} \tag{10.50}$$

where q is the voltage transfer ratio, ϕ_i is the input displacement angle chosen to achieve the desired input power factor (when $\phi_i = 0$, the maximum value of $q = 0.866$ is obtained), and θ_O and θ_i are the phase displacement angles of the output voltage and input current vectors, respectively, whose values are limited to the range 0–60°. The on-time of the zero vector is

$$t_O = T_S - \sum_{i=1}^{4} t_i. \tag{10.51}$$

The integral value of the reference vector is calculated over one sample time interval as the sum of the products of the two adjacent vectors and their on-time ratios. The process is repeated at every sample instant.

10.5.1.3 Control Implementation and Comparison of the Two Methods

Both methods need a digital signal processor (DSP)-based system for their implementation. In one scheme for the Venturini method, the programmable timers, as available, are used to time out the PWM gating signals. The processor calculates the six-switch duty cycles in each sampling interval, converts them to integer counts, and stores them in the memory for the next sampling period. In the SVM method, an EPROM is used to store the selected sets of active and zero vectors, and the DSP calculates the on-times of the vectors. Then with an identical procedure as in the other method, the timers are loaded with the vector on-times to generate PWM waveforms through suitable output ports. The total computation time of the DSP for the SVM method has been found to be much less than that of the Venturini method.

Comparison of the two schemes shows that while in the SVM method the switching losses are lower, the Venturini method shows better performance in terms of input current and output voltage harmonics.

10.5.2 Commutation and Protection Issues in an MC

As the MC has no DC-link energy storage, any disturbance of the input supply voltage will affect the output voltage immediately, and a proper protection mechanism has to be incorporated, particularly against over voltage from the supply and over current in the load side. As mentioned, two types of bidirectional switch configurations have hitherto been used—one, the transistor (now IGBT) embedded in a diode bridge, and the other, the two IGBTs in antiparallel with reverse voltage blocking diodes (shown in Figure 10.31). In the latter configuration, each diode and IGBT combination operates in only two quadrants, which eliminates the circulating currents otherwise built up in the diode-bridge configuration that can be limited by only bulky commutation inductors in the lines.

The MC does not contain freewheeling diodes that usually achieve safe commutation in the case of other converters. To maintain the continuity of the output current as each switch turns off, the next switch in sequence must be immediately turned on. In practice, with bidirectional switches, a momentary short circuit may develop between the input phases when the switches crossover and one solution is to use a *semisoft current commutation* using a multistepped switching procedure to ensure safe commutation. This method requires independent control of each two-quadrant switches, sensing the direction of the load current and introducing a delay during the change of switching states.

A clamp capacitor connected through two three-phase full-bridge diode rectifiers involving an additional 12 diodes (a new configuration with the number of additional diodes reduced to six using the antiparallel switch diodes has been reported) at the input and

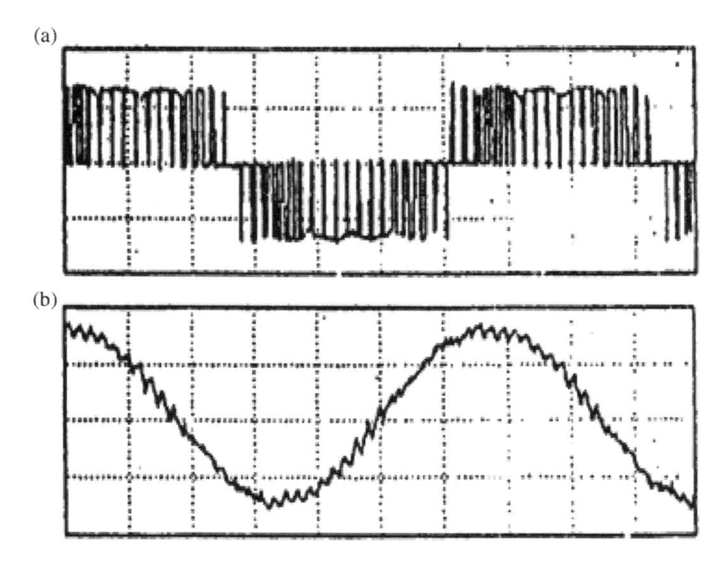

FIGURE 10.33 Experimental waveforms for an MC at 30-Hz frequency from 50-Hz input: (a) output line voltage and (b) output line current. (Reprinted from Rashid, M. H. 2001. *Power Electronics Handbook*, New York: Academic Press, pp. 307–333. With permission.)

output lines of the MC serves as a voltage clamp for possible voltage spikes under normal and fault conditions.

A three-phase single-stage LC filter consisting of three capacitors in star and three inductors in the line is used to adequately attenuate the higher-order harmonics and render sinusoidal input current. Typical values of L and C based on a 415-V converter with a maximum line current of 6.5 A and a switching frequency of 20 kHz are 3 mH and 1.5 μF only. The filter may cause a minor phase shift in the input displacement angle that needs correction. Figure 10.33 shows typical experimental waveforms of the output line voltage and line current of an MC. The output line current is mostly sinusoidal except for a small ripple when the switching frequency is around only 1 kHz.

Homework

10.1. A single-phase full-wave AC/AC voltage controller shown in Figure 10.1a has input rms voltage $V_S = 220\,\text{V}/50\,\text{Hz}$, load $R = 200\,\Omega$, and the firing angle $\alpha = 75°$ for the thyristors T_1 and T_2. Determine the output rms voltage V_O and current I_O, and the DPF.

10.2. A single-phase full-wave AC/AC voltage controller shown in Figure 10.1a has input rms voltage $V_S = 220\,\text{V}/50\,\text{Hz}$, load $R = 100\,\Omega$, and the output rms voltage $V_O = 155.56\,\text{V}$. Determine the firing angle α for the thyristors T_1 and T_2 and the DPF.

10.3. A single-phase integral cycle controlled AC/AC controller has input rms voltage $V_S = 120\,\text{V}$. It turned on and off with a duty cycle $k = 0.6$ at five cycles (see Figure 10.8). Determine the output rms voltage V_O and the input-side PF.

10.4. A single-phase PWM AC chopper has input rms voltage $V_S = 120\,\text{V}$. Its modulation index $k = 0.6$ (see Figure 10.10). Determine the output rms voltage V_O and the input-side PF.

10.5. Consider a full-wave SISO AC/AC cycloconverter. The input rms voltage $V_S = 140\,\text{V}/50\,\text{Hz}$ and the output voltage $V_O = 90\,\text{V}/10\,\text{Hz}$, and the load is a resistance $R = 1\,\Omega$ with a low-pass filter. Assuming that the filter is appropriately designed, only the fundamental component ($f_O = 10\,\text{Hz}$) remains in the output voltage. Tabulate the firing angle (α in the period $T_S = 1/f_S = 20\,\text{ms}$) of the SCRs of both rectifiers in a full period $T_O = 1/f_O = 100\,\text{ms}$, and calculate the phase-angle shift σ in the input voltage over the period $T_S = 1/f_S$.

References

1. Luo, F. L., Ye, H., and Rashid, M. H. 2005. *Digital Power Electronics and Applications*. Boston: Academic Press, Elsevier.
2. Rashid, M. H. 2001. *Power Electronics Handbook*, New York: Academic Press, pp. 307–333.
3. Agrawal, J. P. 2001. *Power Electronics Systems*, Englewood Cliffs, NJ: Prentice-Hall, pp. 355–389.
4. Rombaut, C., Seguier, G., and Bausiere R. 1987. *Power Electronics Converters—AC/AC Converters*. New York: McGraw-Hill.

5. Lander, C. W. 1993. *Power Electronics*. London: McGraw-Hill.
6. Dewan, S. B. and Straughen, A. 1975. *Power Semiconductor Circuits*. New York: Wiley.
7. Hart, D. W. 1997. *Introduction to Power Electronics*. Englewood Cliffs, NJ: Prentice-Hall.
8. Williams, B. W. 1987. *Power Electric Devices, Drivers and Applications*. London: Macmillan.
9. Pelly, B. R. 1971. *Thyristor Phase-Controlled Converters and Cycloconverters*. New York: Wiley.
10. McMurray, W. 1972. *The Theory and Design of Cycloconverters*. Cambridge: MIT Press.
11. Syam, P., Nandi, P. K., and Chattopadhyay, A. K. 1998. An improvement feedback technique to suppress sub-harmonics in a naturally commutated cycloconverter. *IEEE Transactions on Industrial Electronics*, 45, 950–962.
12. Venturine, M. 1980. A new sine-wave in sine-wave out converter technique eliminated reactor elements. *Proceedings of Powercon'80*, pp. E3-1–E3-15.
13. Alesina, A. and Venturine, M. 1980. The generalized transformer: A new bidirectional waveform frequency converter with continuously adjustable input power factor. *Proceedings of IEEE-PESC'80*, pp. 242–252.
14. Alesina, A. and Venturine, M. 1989. Analysis and design of optimum amplitude nine-switch direct AC–AC converters. *IEEE Transactions on Power Electronics*, 4, 101–112.
15. Ziogas, P. D., Khan, S. I., and Rashid, M. 1985. Some improved forced commutated cycloconverter structures. *IEEE Transactions on Industry Applications*, 21, 1242–1253.
16. Ziogas, P. D., Khan, S. I., and Rashid, M. 1986. Analysis and design of forced commutated cycloconverter structures and improved transfer characteristics. *IEEE Transactions on Industrial Electronics*, 33, 271–280.
17. Ishiguru, A., Furuhashi, T., and Okuma, S. 1991. A novel control method of forced commutated cycloconverter using instantaneous values of input line voltages. *IEEE Transactions on Industrial Electronics*, 38, 166–172.
18. Huber, L., Borojevic, D., and Burani, N. 1992. Analysis, design and implementation of the space-vector modulator for commutated cycloconverters. *IEE-Proceedings Part B*, 139, pp. 103–113.
19. Huber, L. and Borojevic, D. 1995. Space-vector modulated three-phase to three-phase matrix converter with input power factor correction. *IEEE Transactions on Industry Applications*, 31, 1234–1246.
20. Zhang, L., Watthanasarn, C., and Shepherd, W. 1998. Analysis and comparison of control strategies for AC–AC matrix converters. *IEE-Proceedings on Electric Power Applications*, 144, pp. 284–294.
21. Das, S. P. and Chattopadhyay, A. K. 1997. Observer based stator flux oriented vector control of cycloconverter-fed synchronous motor drive. *IEEE Transactions on Industry Applications*, 33, 943–955.

11

Improved AC/AC Converters

Traditional methods of AC/AC converters have been introduced in Chapter 10. All those methods have some general drawbacks:

1. The output voltage is lower than the input voltage
2. The input-side THD is high
3. The output voltage frequency is lower than the input voltage frequency when voltage regulation and cycloconversion methods are used.

Some new methods to construct AC/AC converters can overcome the above-mentioned drawbacks. The following converters are introduced in this chapter:

- DC-modulated single-phase single-stage AC/AC converters
- DC-modulated single-phase multistage AC/AC converters
- DC-modulated multiphase AC/AC converters
- Subenvelope modulation method to reduce the THD for matrix AC/AC converters.

11.1 DC-Modulated Single-Phase Single-Stage AC/AC Converters

Single-stage AC/AC converters are the most popular structure widely used in various industrial applications [1–4]. These AC/AC converters are traditionally implemented by the voltage regulation technique, cycloconverters, and matrix converters. However, they have high THD, low PF, and poor power transfer efficiency (η). A typical single-stage AC/AC converter implemented with voltage regulation technique and the corresponding waveforms are shown in Figure 11.1. The devices can be thyristors, IGBTs, and MOSFETs. For a clear example, MOSFETs are applied in the circuit with a pure resistive load R [5]. The input voltage is

$$v_S(t) = \sqrt{2}V_S \sin \omega t,$$

where V_S is the rms value and ω is the input voltage frequency $\omega = 2\pi f = 100\pi$. The PF is calculated by using the formula [6]

$$\mathrm{PF} = \mathrm{DPF} \Big/ \sqrt{1 + \mathrm{THD}^2},$$

where DPF $= \cos \Phi_1$ is the displacement power factor, THD is the total harmonic distortion, and the delay angle Φ_1 is the phase delay angle of the fundamental harmonic component.

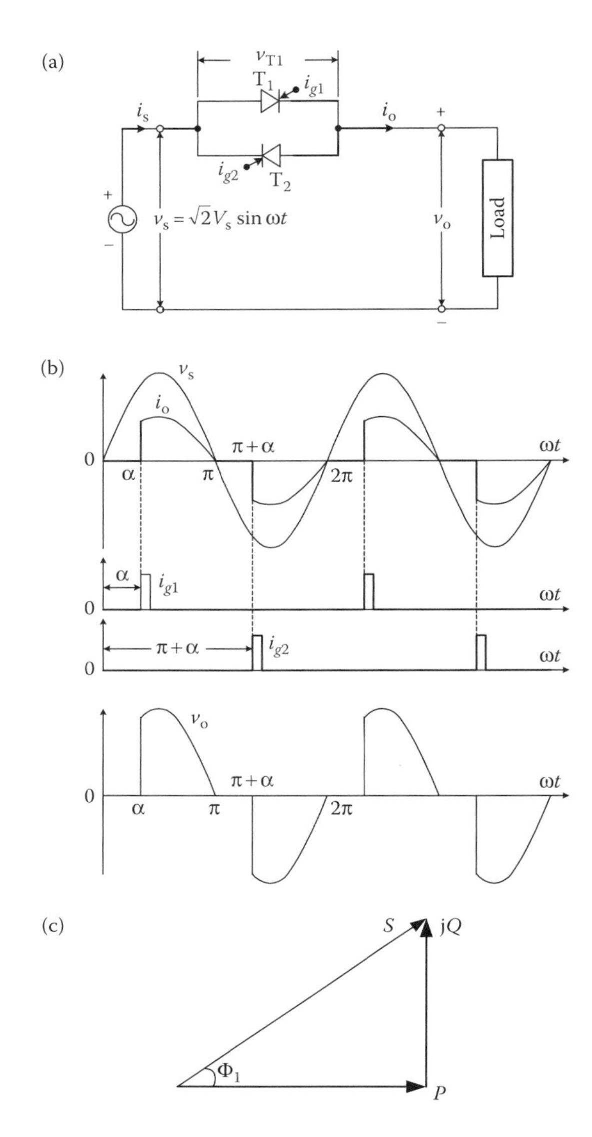

FIGURE 11.1 Typical single-stage AC/AC converter with voltage regulation technique: (a) circuit diagram, (b) waveforms, and (c) power vectors.

For example, if the firing angle α is 30° (i.e., the fundamental harmonic phase angle Φ_1 is 30°), the typical values are DPF = cos 30° = 0.866 and THD = 0.15 (or 15%). Therefore, PF = 0.856. The PF has a low value. The power vector diagram is shown in Figure 11.1c, where **P** is the real power, j**Q** is the reactive power, and **S** is the apparent power.

$$\mathbf{S} = \mathbf{P} + j\mathbf{Q}.$$

DC/DC conversion technology [7–10] can implement fast response and high efficiency. Our novel approach to PFC is called *DC-modulation power factor correction AC/AC conversion* [11–14]. Using this technique, a high PF can be achieved. A DC-modulated single-stage buck-type AC/AC converter is shown in Figure 11.2.

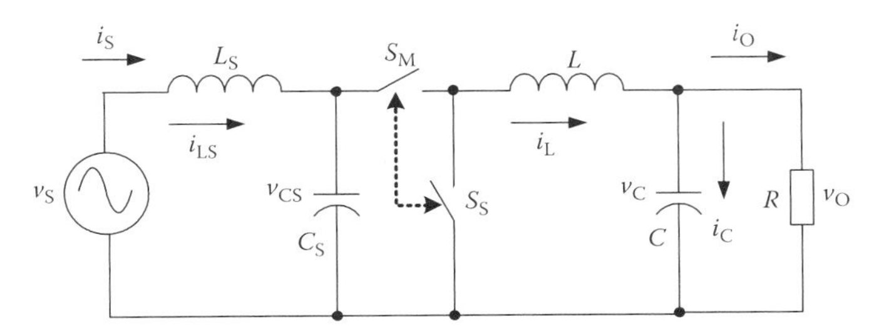

FIGURE 11.2 DC-modulated single-stage buck-type AC/AC converter.

We assume that the input power supply rms voltage is 240 V with the frequency f=50 Hz. The master switch S_M and the slave switch S_S are bidirectional switches. They are working in the exclusive states. The DC-modulation switching frequency f_m is usually high, say f_m = 20 kHz. Therefore, the input power supply voltage is a quasi-static DC voltage (positive or negative value) in a DC-modulation period $T_m = 1/f_m$ (50 μs). In a DC-modulation period T_m, the input voltage is a quasi-constant DC value. Therefore, the converter performs a DC/DC conversion. We can use all the conclusions on DC/DC conversion technology in this operation. The key device is the bidirectional exclusive switches S_M–S_S (even more multi bidirectional switches) for the DC-modulation operation. Since the buck converter input current is a pulse train with the repeating frequency f_m, a low-pass input filter L_S–C_S is required.

11.1.1 Bidirectional Exclusive Switches S_M–S_S

The switching devices for bidirectional exclusive switches can be MOSFETs or IGBTs. MOSFETs were selected for our design. The bidirectional exclusive switches S_M–S_S for the DC-modulation operation were designed, and they have the following technical features:

1. The master switch S_M is controlled by a PWM pulse train and it conducts the input current to flow in the forward direction in the positive input voltage. On the other hand, the S_M conducts the input current to flow in the reverse direction in the negative input voltage.
2. The slave switch S_S is conducted when the master switch S_M is switched off exclusively. It is the freewheeling device to conduct the current to flow.

Figure 11.3 shows the circuit of the bidirectional exclusive switches S_M–S_S for the DC-modulation operation. The switching control signal is a PWM pulse train that has an adjustable frequency f_m and pulse width. The repeating period $T_m = 1/f_m$ and the conduction duty cycle k = (pulse width)/T_m.

If some converters require more than one bidirectional exclusive slave switches, the construction of the further S_S needs only to copy/repeat the existing one. If some converters require more than one bidirectional slave switches and one synchronously bidirectional slave switch, the construction of the synchronously bidirectional slave switch S_{S-S} needs only to copy/repeat the master switch S_M. A group comprising a master switch S_M with a synchronously bidirectional slave switch S_{S-S} and two bidirectional exclusive slave switches S_{S1} and S_{S2} is shown in Figure 11.4.

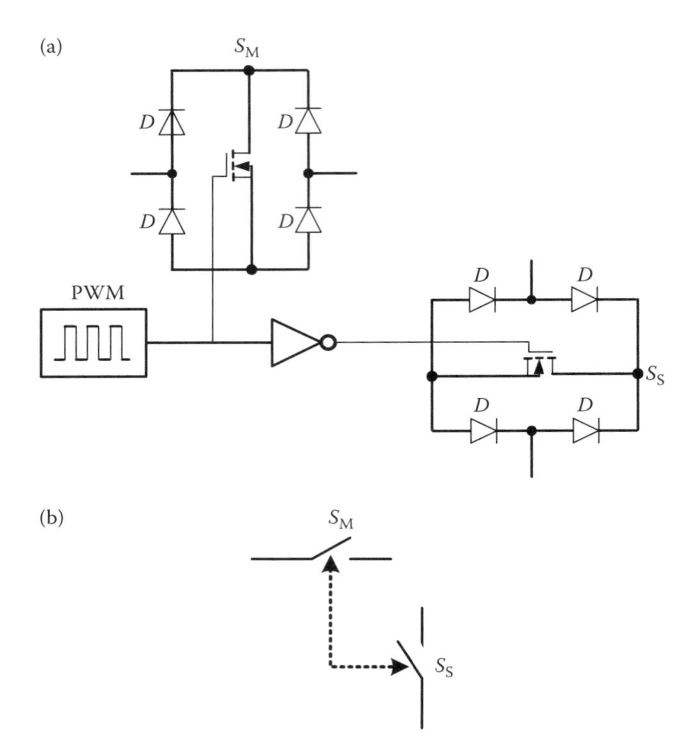

FIGURE 11.3 Bidirectional exclusive switches S_M–S_S for the DC-modulation operation: (a) circuit of a bidirectional exclusive switches S_M–S_S and (b) symbol of a bidirectional exclusive switches S_M–S_S

11.1.2 Mathematical Modeling of DC/DC Converters

The mathematical modeling of DC/DC converters is an important topic of the DC/DC converter development. This topic has been well discussed by Luo and Ye [8–10]. The main points are as follows:

1. The input pumping energy is

$$PE = \int_0^{T_m} V_S i_S(t)\, dt = V_S \int_0^{T_m} i_S(t)\, dt = V_S I_S T_m, \tag{11.1}$$

   where the average current I_S is

$$I_S = \frac{1}{T_m} \int_0^{T_m} i_S(t)\, dt. \tag{11.2}$$

2. The SE in an inductor is

$$W_L = \frac{1}{2} L I_L^2. \tag{11.3}$$

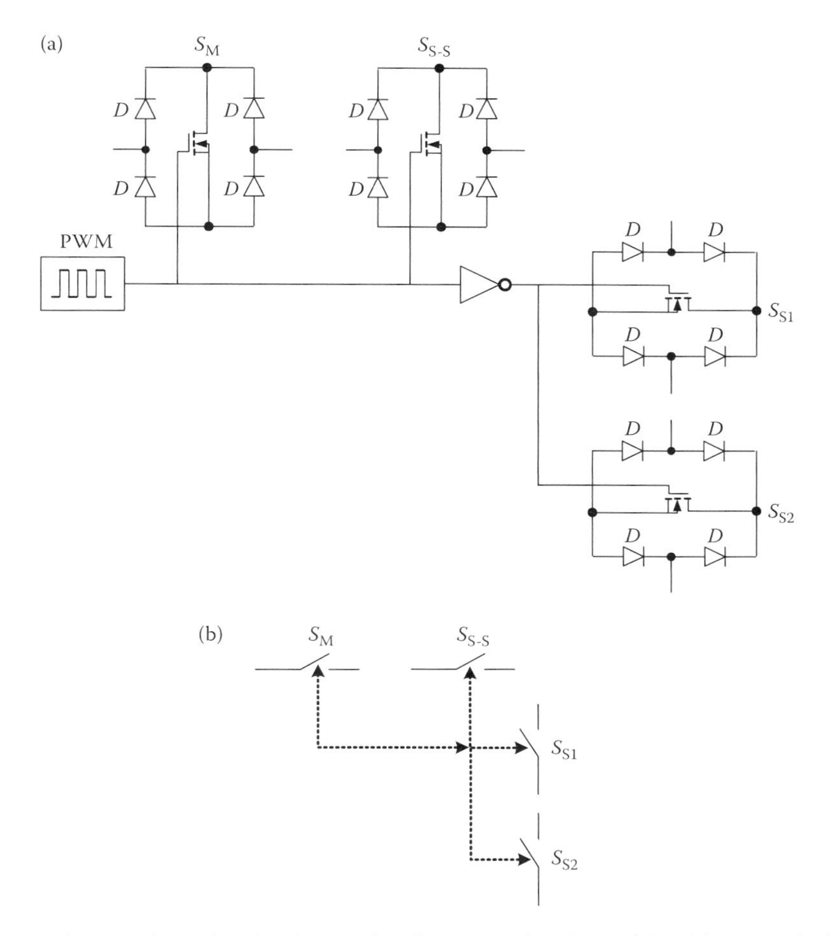

FIGURE 11.4 Bidirectional switches S_M–S_{S-S} and exclusive switches S_{S1} and S_{S2}: (a) circuit of a bidirectional switches S_M–S_{S-S} plus exclusive switches S_{S1} and S_{S2} and (b) symbol of a bidirectional switches S_M–S_{S-S} plus exclusive switches S_{S1} and S_{S2}.

The SE in a capacitor is

$$W_C = \frac{1}{2} C V_C^2. \tag{11.4}$$

Therefore, if there are n_L inductors and n_C capacitors, the total *SE* in a DC/DC converter is

$$SE = \sum_{j=1}^{n_L} W_{Lj} + \sum_{j=1}^{n_C} W_{Cj}. \tag{11.5}$$

3. The energy factor is

$$EF = \frac{SE}{PE} = \frac{SE}{V_1 I_1 T_m} = \frac{\sum_{j=1}^{m} W_{Lj} + \sum_{j=1}^{n} W_{Cj}}{V_1 I_1 T_m}. \tag{11.6}$$

4. The capacitor/inductor stored energy ratio (CIR) is defined as [7]

$$\mathrm{CIR} = \frac{\sum_{j=1}^{n_C} W_{Cj}}{\sum_{j=1}^{n_L} W_{Lj}}. \tag{11.7}$$

5. The time constant τ is defined as

$$\tau = \frac{2T_m \times \mathrm{EF}}{1+\mathrm{CIR}}\left(1+\mathrm{CIR}\frac{1-\eta}{\eta}\right), \tag{11.8}$$

where η is the power transfer efficiency. If there are no power losses, $\eta = 1$.

$$\tau = \frac{2T_m \times \mathrm{EF}}{1+\mathrm{CIR}}. \tag{11.9}$$

6. The damping time constant τ_d is defined as

$$\tau_d = \frac{2T_m \times \mathrm{EF}}{1+\mathrm{CIR}} \frac{\mathrm{CIR}}{\eta+\mathrm{CIR}(1-\eta)}. \tag{11.10}$$

If there are no power losses,

$$\tau_d = \frac{2T_m \times \mathrm{EF}}{1+1/\mathrm{CIR}}. \tag{11.11}$$

7. The time constant ratio ξ is defined as

$$\xi = \frac{\tau_d}{\tau} = \frac{\mathrm{CIR}}{\eta(1+\mathrm{CIR}(1-\eta/\eta))^2}. \tag{11.12}$$

If there are no power losses,

$$\xi = \frac{\tau_d}{\tau} = \mathrm{CIR}. \tag{11.13}$$

8. A DC/DC converter has the transfer function

$$G(s) = \frac{M}{1+s\tau+s^2\tau\tau_d} = \frac{M}{1+s\tau+\xi s^2\tau^2}, \tag{11.14}$$

where M is the voltage transfer gain in a steady state, for example, $M = k$ for a buck converter.

Example 11.1

A buck converter is shown in Figure 11.2 having the following components; $L = 1$ mH; $C = 0.4\,\mu$F; the load $R = 100\,\Omega$; the input voltage and current are v_S and i_S, respectively; the output voltage and current are v_O and i_O, respectively; there are no power losses, that is, $\eta = 1$; the switching frequency is f_m (the switching period $T_m = 1/f_m$); the conduction duty cycle is k. Calculate the transfer function and its step-response.

Solution

We obtain the following data:

$$v_O = kv_S, \quad i_S = ki_O;$$

$$v_O = Ri_O, \qquad P_{in} = v_S i_S = v_O i_O = P_O \text{ with } \eta = 1,$$

$$PE = \int_0^{T_m} V_S i_S(t)\,dt = V_S \int_0^{T_m} i_S(t)\,dt = V_S I_S T_m, \qquad W_L = \frac{1}{2}LI_L^2 = \frac{1}{2}Li_O^2,$$

$$W_C = \frac{1}{2}CV_C^2 = \frac{1}{2}Cv_O^2, \qquad SE = \frac{1}{2}(Li_O^2 + Cv_O^2) = \frac{1}{2}(L + CR^2)i_O^2,$$

$$EF = \frac{SE}{PE} = \frac{(L + CR^2)i_O^2}{2v_O i_O T_m} = \frac{L/R + CR}{2T_m}, \qquad CIR = \frac{0.5Li_O^2}{0.5Cv_O^2} = \frac{L}{CR^2} = \frac{1}{4},$$

$$\tau = \frac{2T_m \times EF}{1 + CIR} = \frac{L/R + RC}{1 + CIR} = 40\,\mu s, \qquad \tau_d = \frac{2T_m \times EF}{1 + 1/CIR} = \frac{L/R + RC}{1 + 1/CIR} = 10\,\mu s.$$

Therefore, the transfer function is

$$G(s) = \frac{k}{1 + s\tau + 0.25s^2\tau^2} = \frac{M}{(1 + 0.00002s)^2}. \tag{11.15}$$

This transfer function is in the critical condition with two folded poles. The corresponding step-response in the time-domain has fast response without overshot and oscillation.

$$g(t) = k\left[1 - \left(1 + \frac{2t}{\tau}\right)e^{-2t/\tau}\right] = k\left[1 - \left(1 + \frac{2t}{0.00004}\right)e^{-2t/0.00004}\right]. \tag{11.16}$$

The settling time from one steady state to another is about $2.4\tau = 0.000096$ s $= 0.096$ ms. This time period is much smaller than the power supply period $T = 1/f = 20$ ms. The corresponding radian distance is only 1.73°. The statistic average delay angle Φ_1 is its 1/e time, that is, $\Phi_1 = 0.24\tau/e = 0.0353$ ms or 0.636°, and DPF $= \cos\Phi_1 = 0.999938$. Consequently, we can assume that the output voltage can follow the input voltage waveform.

11.1.3 DC-Modulated Single-Stage Buck-Type AC/AC Converter

The DC-modulated single-stage buck-type AC/AC converter is shown in Figure 11.2. We have to investigate the operation during both positive and negative half-cycles of the input voltage.

11.1.3.1 Positive Input Voltage Half-Cycle

When the input voltage is positive, the buck converter operates in the usual manner. The equivalent circuits during the switch-on and switch-off periods are shown in Figure 11.5.

The output voltage is calculated by

$$v_O = kv_S = k\sqrt{2}V_S \sin\omega t, \quad 0 \le \omega t < \pi, \tag{11.17}$$

where k is the conduction duty cycle of the buck converter, V_S is the rms value of the input voltage, and ω is the power supply radian frequency.

11.1.3.2 Negative Input Voltage Half-Cycle

When the input voltage is negative, the buck converter operates in the reverse manner. The equivalent circuits during the switch-on and switch-off periods are shown in Figure 11.6.

The output voltage is calculated by

$$v_O = -k|v_S| = k\sqrt{2}V_S \sin \omega t, \quad \pi \leq \omega t < 2\pi, \tag{11.18}$$

where k is the conduction duty cycle, V_S is the rms value, and ω is the power supply radian frequency.

11.1.3.3 Whole-Cycle Operation

Combining the above two state operations, we can summarize the whole-cycle operation. The output voltage is calculated by

$$v_O = kv_S = k\sqrt{2}V_S \sin \omega t, \tag{11.19}$$

where k is the conduction duty cycle, V_S is the rms value, and ω is the power supply radian frequency. The whole-cycle input and output voltage waveforms are shown in Figure 11.7a with conduction duty cycle $k = 0.75$. The voltage transfer gain $M = V_O/V_S$ versus the

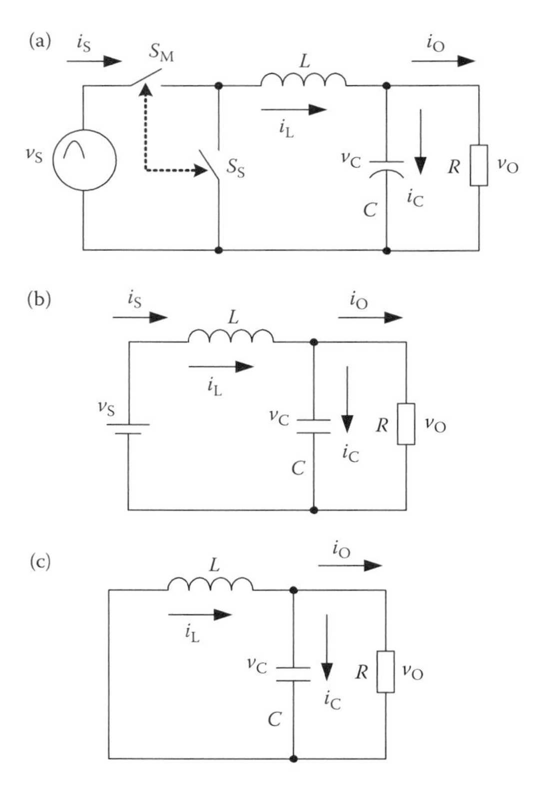

FIGURE 11.5 DC-modulated buck-type AC/AC converter working in a positive half-cycle: (a) circuit diagram, (b) equivalent circuit during switch-on and (c) equivalent circuit during switch-off.

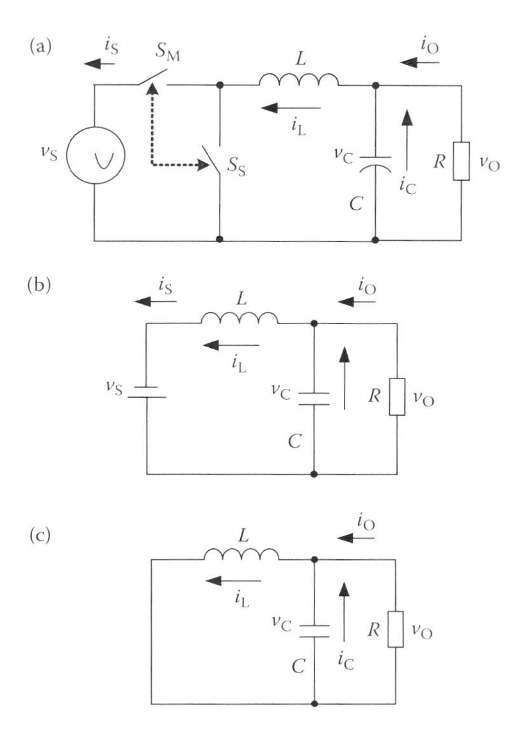

FIGURE 11.6 DC-modulated buck-type AC/AC converter working in a negative half-cycle: (a) circuit diagram, (b) equivalent circuit during switch-on and (c) equivalent circuit during switch-off.

DC/DC converter conduction duty cycle k is shown in Figure 11.7b. It is easy to obtain the variable output voltage with very high PF and high efficiency.

The whole-cycle input voltage and current waveforms are shown in Figure 11.8a. The spectrum of the input current is shown in Figure 11.8b. The spectrum is very clean, and the little distortion from the harmonic component I_M at 20 kHz is far from the fundamental frequency component I_S at 50 Hz. Its value is only 0.5%, that is, $I_M/I_S = 0.005$. Therefore, THD = 1.0000125. Considering that DPF = 0.9998, we obtain the final PF as 0.99979.

11.1.3.4 Simulation and Experimental Results

The simulation and experimental results are shown below in order to verify the design.

11.1.3.4.1 Simulation Results

The DC-modulated buck-type AC/AC converter has the following components: $L_S = 1\,\text{mH}$, $C_S = 10\,\mu\text{F}$, $L = 10\,\text{mH}$, and $C = 3\,\mu\text{F}$. The conduction duty cycle is selected as $k = 0.75$. The simulation results are shown in Figure 11.9. The output voltage $V_O = 0.75 \times V_S = 150V_{rms}$ (the peak value is approximately 212 V) with the frequency $f = 50$ Hz. The waveforms of the input and output voltages $v_S(t)$ and $v_O(t)$ are shown as Channels 1 and 2 in Figure 11.9a. It can be seen that there is no phase delay, although there may be about 3.374° phase angle delay from our analysis. The output current I_O should be 1 A and the output power $P_O = V_O^2/R = 150\,\text{W}$.

The input current is measured as $I_S = 0.95\,\text{A}$ (the peak value is approximately 1.34 A) with the frequency $f = 50$ Hz. The waveforms of the input voltage $v_S(t)$ and current $i_S(t)$

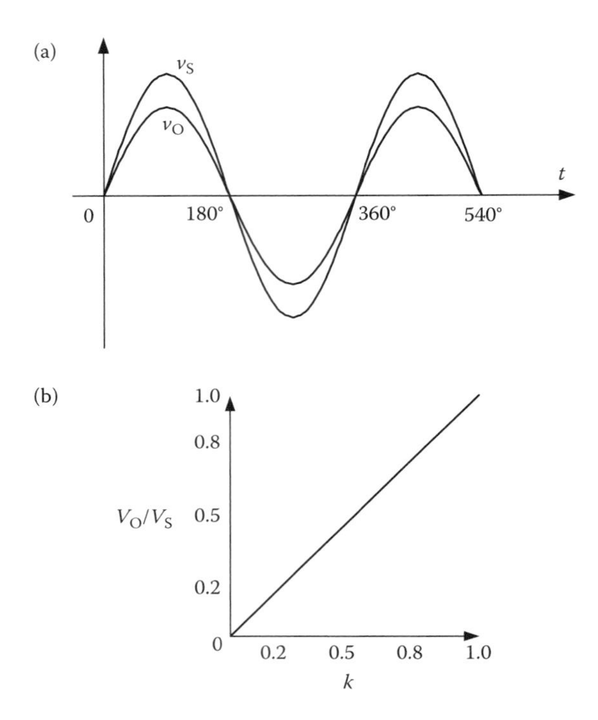

FIGURE 11.7 Input/output voltage waveforms of the DC-modulated buck-type AC/AC converter: (a) input/output voltage waveforms and (b) voltage transfer gain versus conduction duty cycle k.

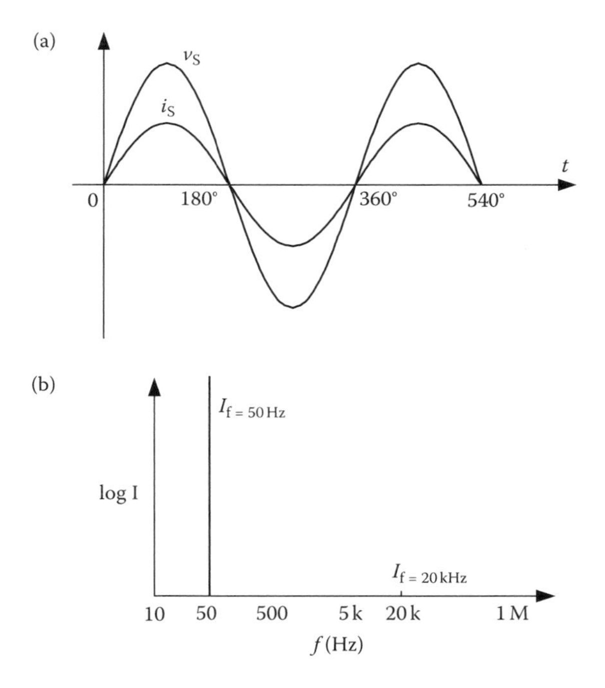

FIGURE 11.8 Input voltage/current waveforms of the DC-modulated buck-type AC/AC converter: (a) input voltage and current waveforms and (b) spectrum of input current.

are shown as Channels 1 and 2 in Figure 11.9b. It can be seen that there is almost no phase delay, although there may be about 3.374° phase angle delay from our analysis. The Fast Fourier Transform (FFT) spectrum of the input current is shown in Figure 11.9c, and THD = 0.015. The input power $P_{in} = V_S \times I_S = 190$ W. Although the theoretical analysis has no power losses for the ideal condition ($\eta = 1$), the particular test shows that there are power losses, which are mainly caused by the power losses of the switches. From the test results, we obtain the final PF as 0.9979 and the power transfer efficiency $\eta = P_o/P_{in} 190/200 = 0.95$ or 95%.

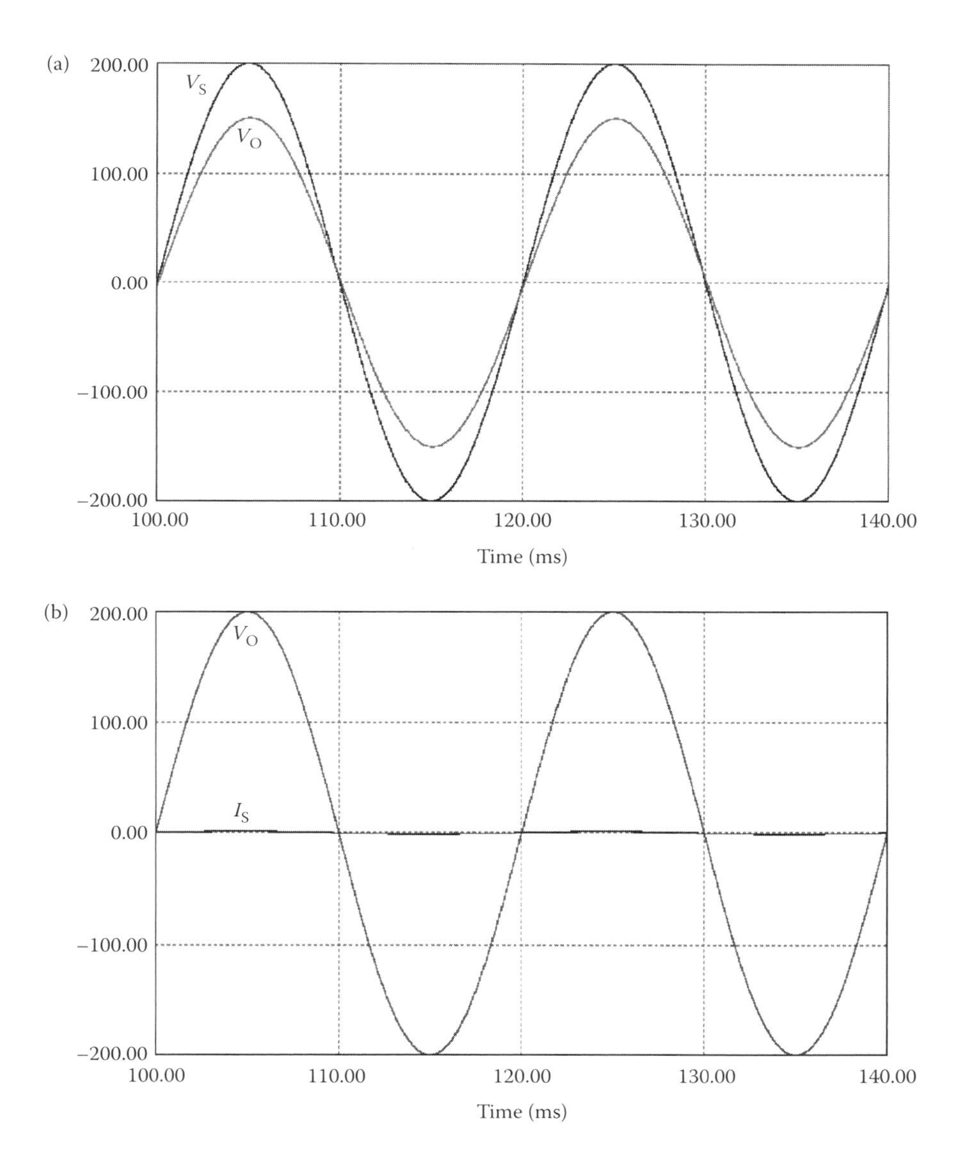

FIGURE 11.9 Test results of the DC-modulated buck-type AC/AC converter: (a) input voltage waveforms of the DC-modulated buck-type AC/AC converter, (b) input voltage and current waveforms of the DC-modulated buck-type AC/AC converter, and (c) spectrum of input current of the DC-modulated buck-type AC/AC converter.

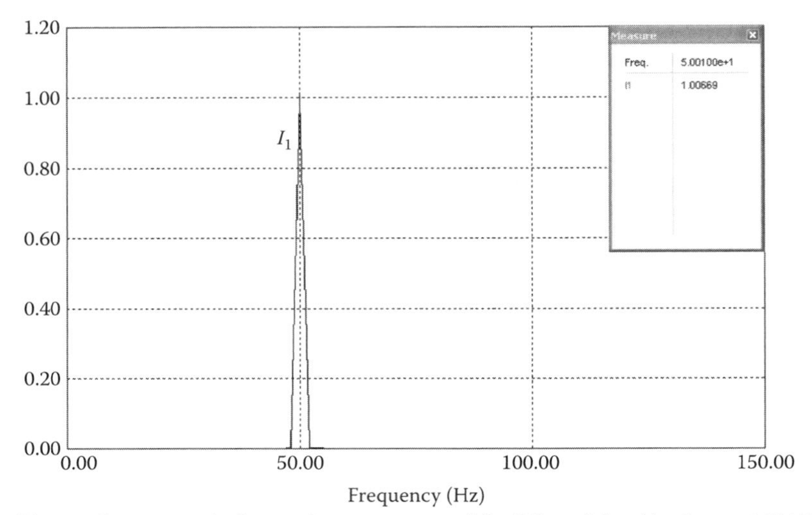

Test result spectrum 1st harmonic measurement of the DC-modulated buck-type AC/AC converter f = 50 kHz, k = 0.75 (Zoom 1)

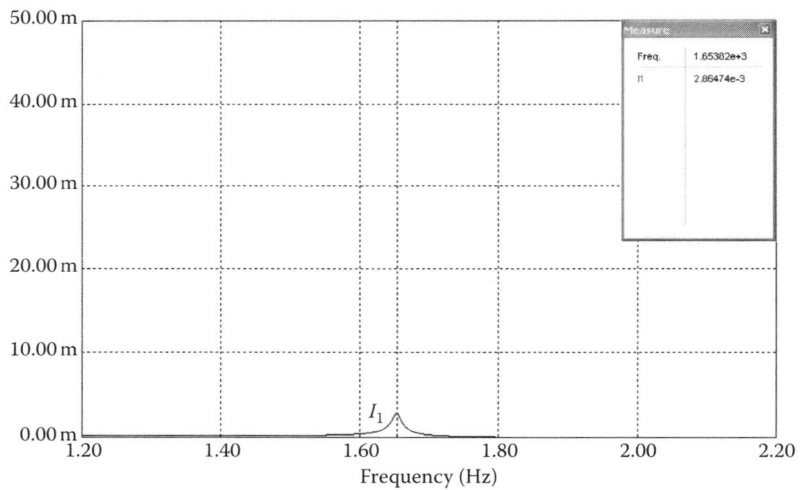

Test result spectrum 2nd harmonic measurement of the DC-modulated buck-type AC/AC converter f = 50 kHz, k = 0.75 (Zoom 2)

FIGURE 11.9 Continued.

11.1.3.4.2 Experimental Results

The experimental results of the DC-modulated buck-type AC/AC converter are shown in Figure 11.10.

Example 11.2

A buck-type DC-modulated AC/AC converter in Figure 11.2 has input rms voltage $v_S = 240$V and a dimmer load with $R = 100\,\Omega$. In order to adjust the light, the output rms voltage v_O varies in the range of 100–200V. Calculate the range of the conduction duty cycle k and the output current and power.

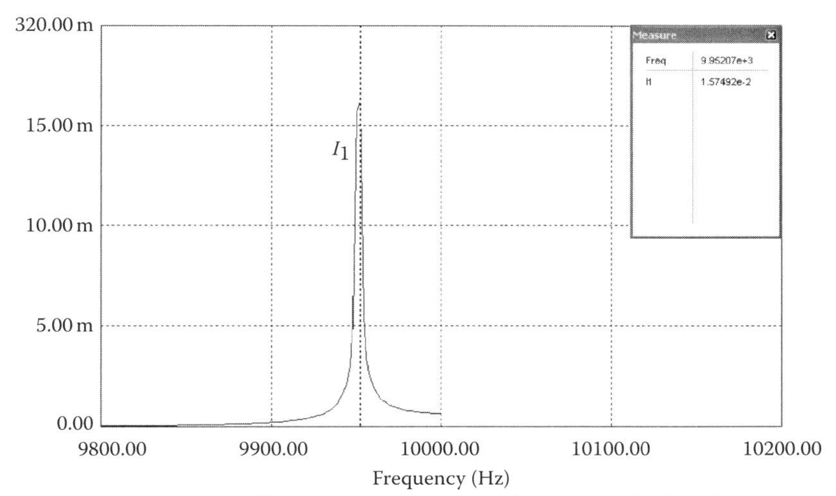

Test result spectrum 3rd harmonic measurement of the DC-modulated buck-type AC/AC converter $f = 50\,\text{kHz}$, $k = 0.75$ (Zoom 3)

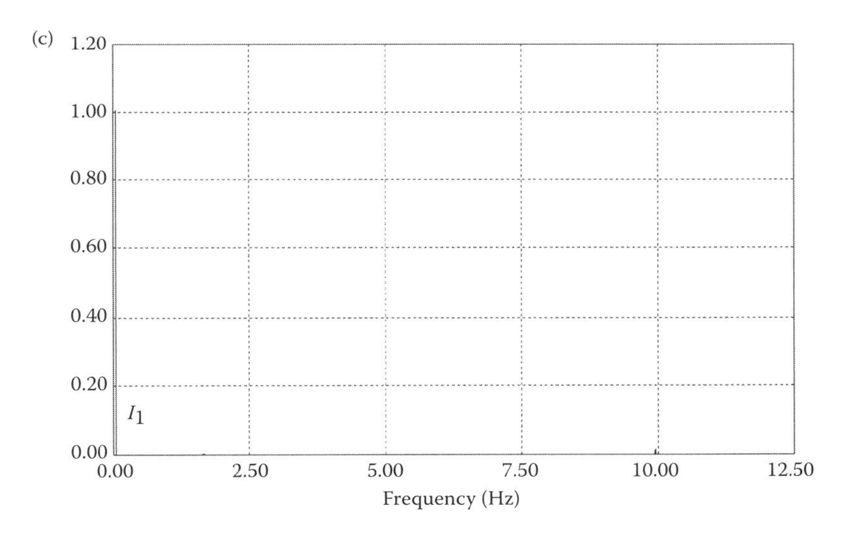

FIGURE 11.9 Continued.

Solution

Since the output voltage is calculated as

$$v_O = kv_S = 240\,k,$$

the conduction duty cycle k is calculated as

$$k = \frac{v_O}{v_S} = \begin{cases} \dfrac{100}{240} = 0.42 \\ \dfrac{200}{240} = 0.83 \end{cases}.$$

The range of k is 0.42–0.83.

The output rms current is 1–2 A, and the output power is 100–400 W.

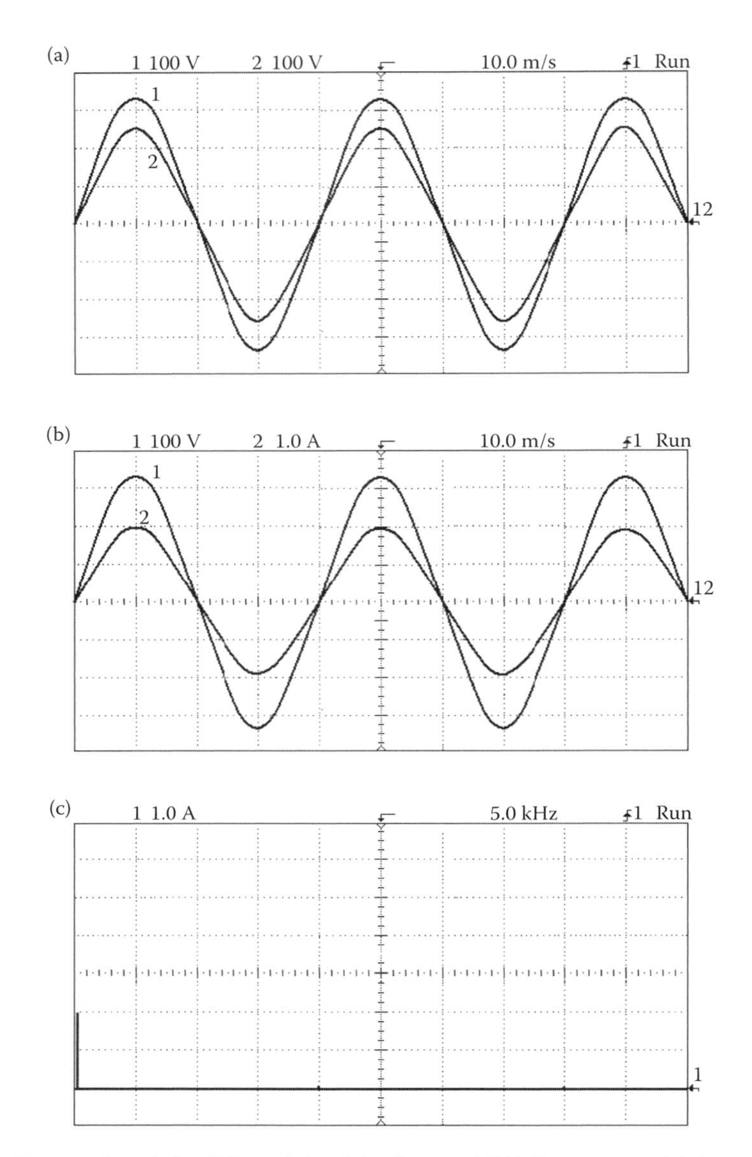

FIGURE 11.10 Test results of the DC-modulated buck-type AC/AC converter: (a) input/output voltage waveforms, (b) input voltage and current waveforms, and (c) spectrum of the input current.

11.1.4 DC-Modulated Single-Stage Boost-Type AC/AC Converter

The DC-modulated single-stage buck-type AC/AC converter can only convert an input voltage to a lower output voltage. For certain applications, the output voltage is required to be higher than the input voltage. For this purpose, the DC-modulated single-stage boost-type AC/AC converter has been designed, and is shown in Figure 11.11. Since the input current is a continuous current, there is no need to set a low-pass filter. The operation should be investigated during both positive and negative half-cycles of the input voltage.

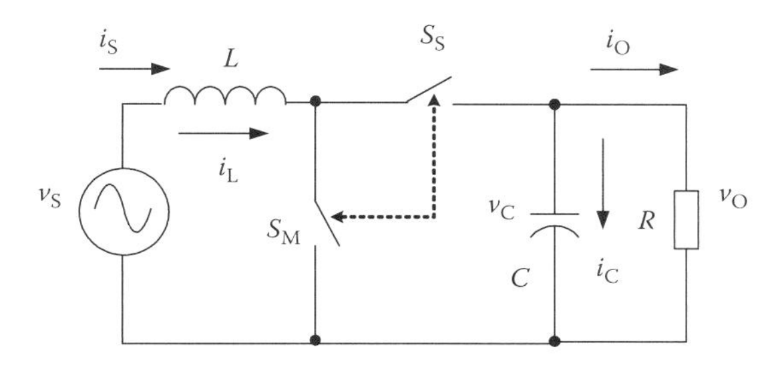

FIGURE 11.11 DC-modulated single-stage boost-type AC/AC converter.

11.1.4.1 Positive Input Voltage Half-Cycle

When the input voltage is positive, the boost converter operates in the usual manner. The equivalent circuits during the switch-on and switch-off periods are shown in Figure 11.12.

The output voltage is calculated by

$$v_O = \frac{v_S}{1-k} = \frac{\sqrt{2}}{1-k} V_S \sin \omega t, \quad 0 \le \omega t < \pi, \tag{11.20}$$

where k is the conduction duty cycle of the boost converter, V_S is the rms value of the input voltage, and ω is the power supply radian frequency.

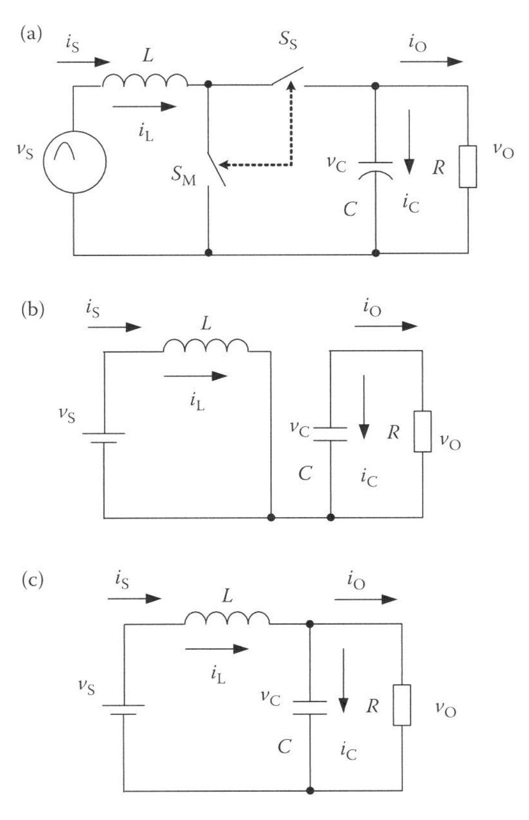

FIGURE 11.12 DC-modulated boost-type AC/AC converter working in a positive half-cycle: (a) circuit diagram, (b) equivalent circuit during switch-on, and (c) equivalent circuit during switch-off.

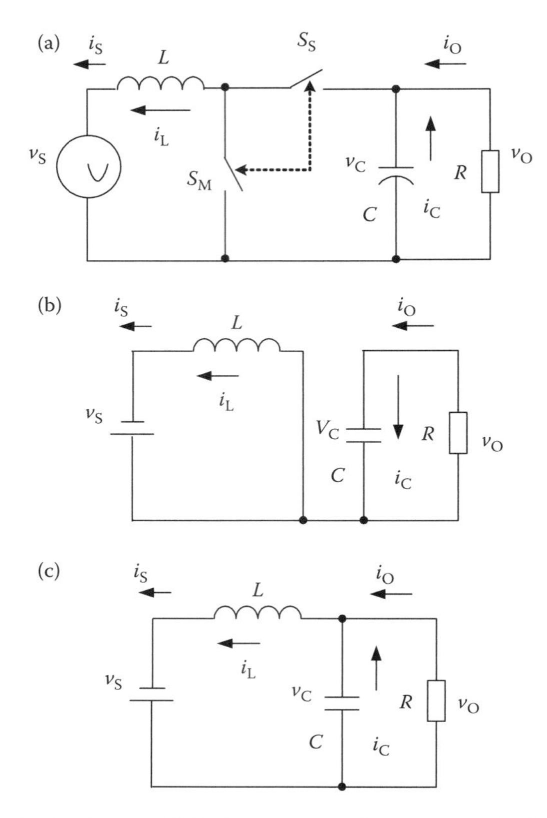

FIGURE 11.13 DC-modulated buck-type AC/AC converter working in a negative half-cycle: (a) circuit diagram, (b) equivalent circuit during switch-on, and (c) equivalent circuit during switch-off.

11.1.4.2 Negative Input Voltage Half-Cycle

When the input voltage is negative, the boost converter operates in the reverse manner. The equivalent circuits during the switch-on and switch-off periods are shown in Figure 11.13.

The output voltage is calculated by

$$v_O = -\frac{|v_S|}{1-k} = \frac{\sqrt{2}}{1-k}V_S \sin \omega t, \quad \pi \le \omega t < 2\pi, \tag{11.21}$$

where k is the conduction duty cycle, V_S is the rms value, and ω is the power supply radian frequency.

11.1.4.3 Whole-Cycle Operation

Combining the above two cycles of operations, we can summarize the whole-cycle operation. The output voltage is calculated by

$$v_O = \frac{v_S}{1-k} = \frac{\sqrt{2}}{1-k}V_S \sin \omega t, \tag{11.22}$$

where k is the conduction duty cycle, V_S is the rms value, and ω is the power supply radian frequency. The whole-cycle input and output voltage waveforms are shown in Figure 11.14a with the duty cycle $k = 0.25$. The voltage transfer gain $M = V_O/V_S$ versus the DC/DC converter conduction duty cycle k is shown in Figure 11.14b. It is easy to obtain the variable output voltage higher than the input voltage with very high PF and high efficiency.

The whole-cycle input voltage and current waveforms are shown in Figure 11.15a. The spectrum of the input current is shown in Figure 11.15b. The spectrum is very clean and the little distortion from the harmonic component I_M at 20 kHz is far from the fundamental frequency component I_S at 50 Hz. Its value is only 0.5%, that is, $I_M/I_S = 0.005$. Therefore, THD = 1.0000125. Considering that DPF = 0.9998, we obtain the final PF as 0.99979.

11.1.4.4 Simulation and Experimental Results

The simulation and experimental results are shown in order to verify the design.

11.1.4.4.1 Simulation Results

The DC-modulated boost-type AC/AC converter shown in Figure 11.11 has the following components: $L = 10$ mH and $C = 3\,\mu$F. The conduction duty cycle is selected as $k = 0.25$. The experimental results are shown in Figure 11.16. The output voltage $V_O = V_S/(1-k) = 267V_{rms}$ (the peak value is approximately 377 V) with the frequency $f = 50$ Hz. The

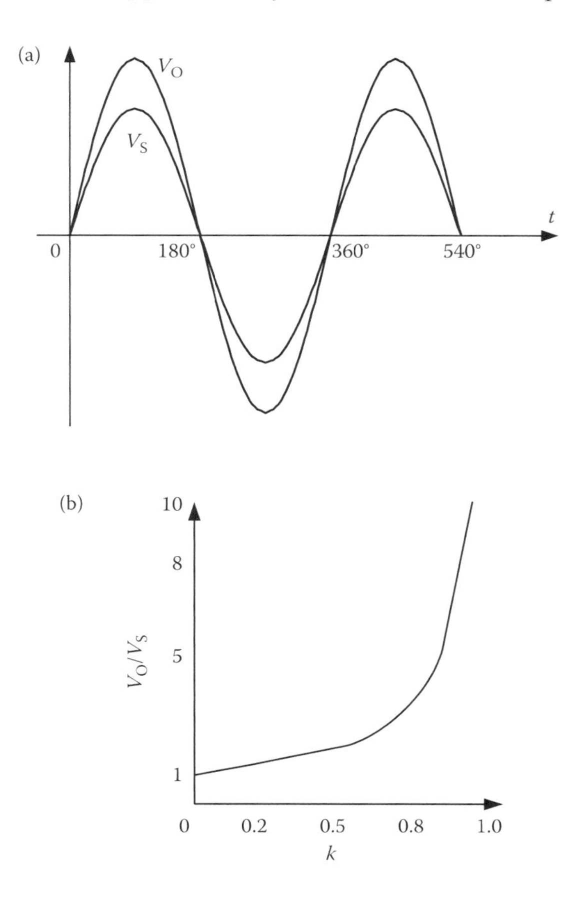

FIGURE 11.14 Input/output voltage waveforms of the DC-modulated boost-type AC/AC converter: (a) input/output voltage waveforms and (b) voltage transfer gain versus conduction duty cycle k.

waveforms of the input and output voltages $v_S(t)$ and $v_O(t)$ are shown as Channels 1 and 2 in Figure 11.16a. From the figure, it can be seen that there is no phase delay, although there may be about 3.374° phase angle delay from our analysis. The output current I_O should be 1.8 A, and the output power $P_O = V_O^2/R = 475$ W.

The input current is measured as $I_S = 2.4$ A (the peak value is approximately 3.39 A) with the frequency $f = 50$ Hz. The waveforms of the input voltage $v_S(t)$ and current $i_S(t)$ are shown as Channels 1 and 2 in Figure 11.16b. From the figure, it can be seen that there is no phase delay, although there may be about 3.374° phase angle phase delay from our analysis in Section 11.1.3.4. The FFT spectrum of the input current is shown in Figure 11.16c, and THD = 0.015. The input power $P_{in} = V_S \times I_S = 480$ W. Although the theoretical analysis has no power losses for the ideal condition ($\eta = 1$), the particular test shows the power losses. From the results, we obtain the final PF as 0.9979 and the power transfer efficiency $\eta = P_o/P_{in} = 475/480 = 0.989$ or 98.9%.

11.1.4.4.2 Experimental Results

The experimental results are shown in Figure 11.17.

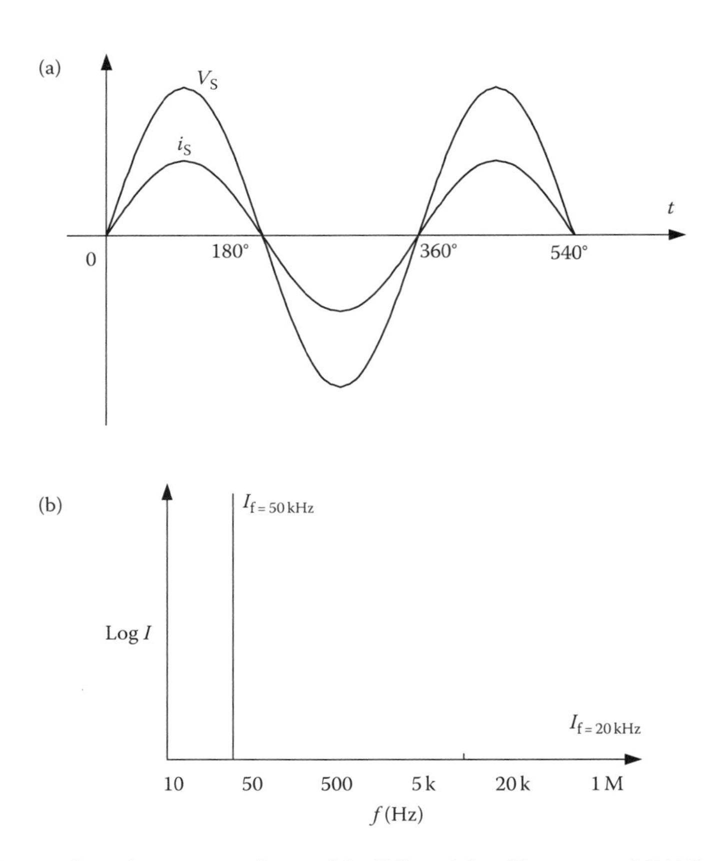

FIGURE 11.15 Input voltage/current waveforms of the DC-modulated boost-type AC/AC converter: (a) input voltage and current waveforms and (b) spectrum of input current.

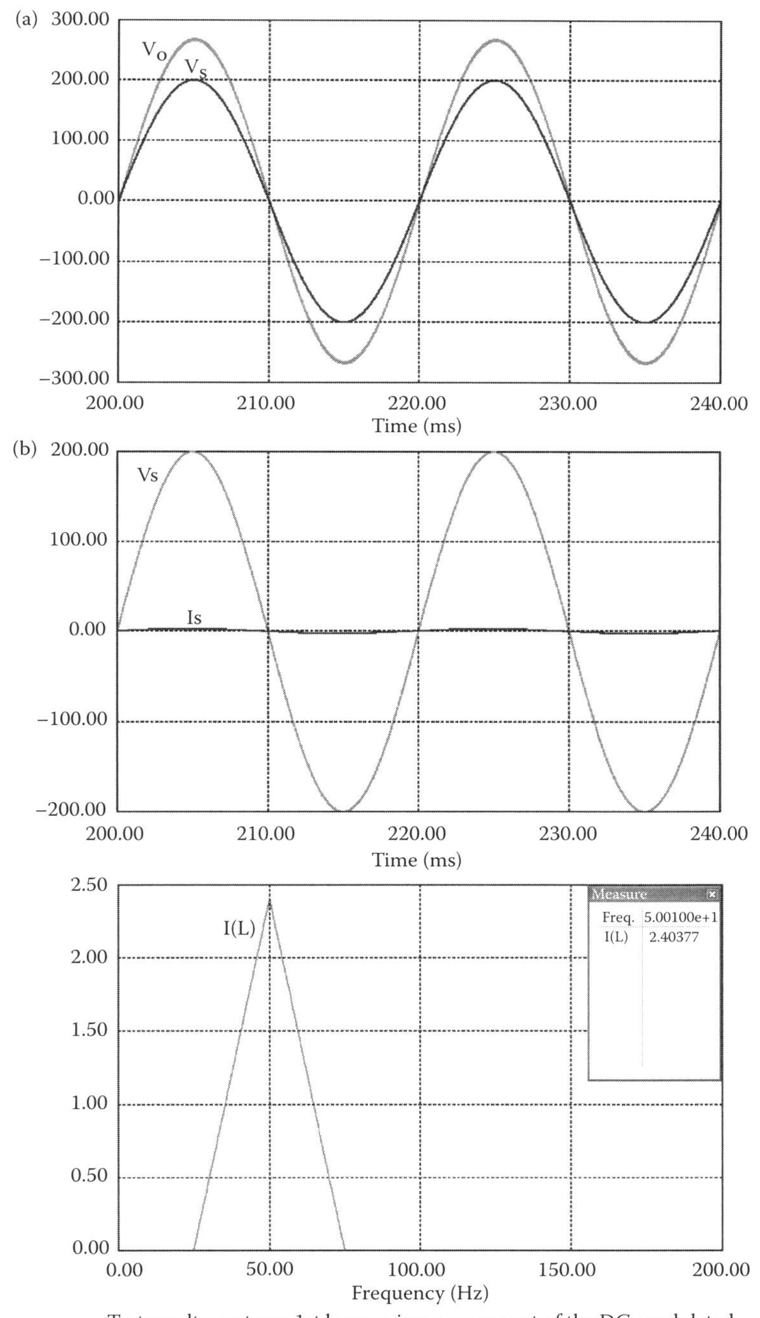

Test result spectrum 1st harmonic measurement of the DC-modulated boost-type AC/AC converter $f = 50$ kHz, $k = 0.25$ (Zoom 1)

FIGURE 11.16 Test results of the DC-modulated boost-type AC/AC converter: (a) Input/output voltage waveforms of the DC-modulated boost-type AC/AC converter, (b) input voltage and current waveforms of the DC-modulated boost-type AC/AC converter, and (c) spectrum of input current of the DC-modulated boost-type AC/AC converter.

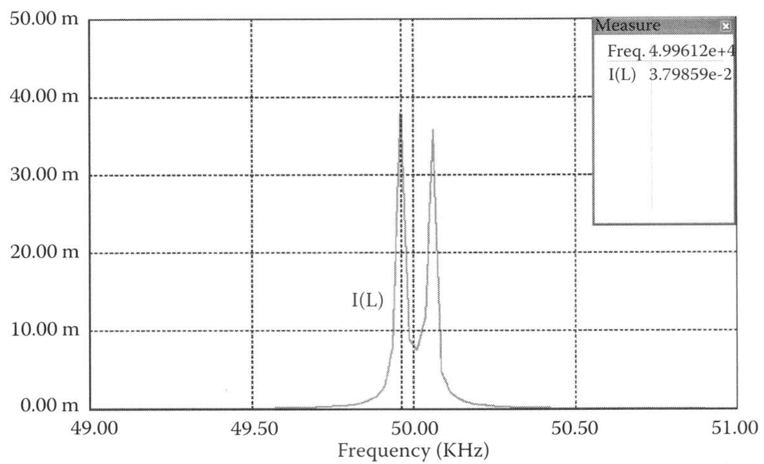

Test result spectrum 2nd harmonic measurement of the DC-modulated boost-type AC/AC converter $f = 50$ kHz, $k = 0.25$ (Zoom 2)

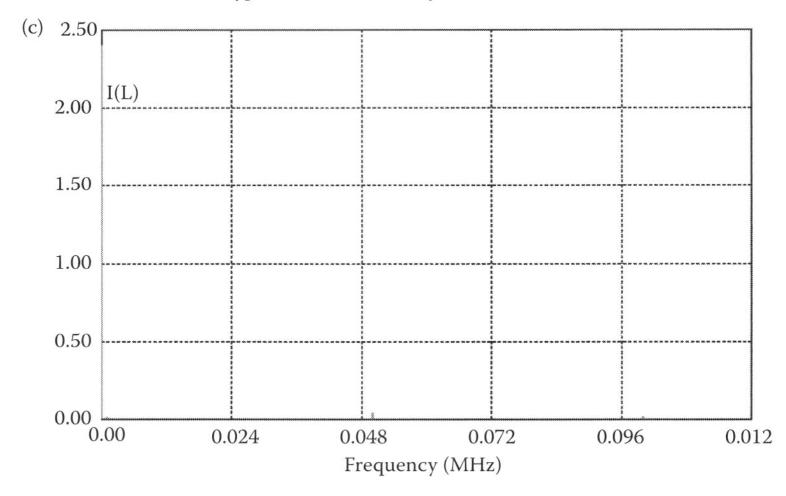

FIGURE 11.16 Continued.

11.1.5 DC-Modulated Single-Stage Buck–Boost-Type AC/AC Converter

For certain applications, the output voltage is required to be lower and higher than the input voltage. For this purpose, the DC-modulated single-stage buck–boost-type AC/AC converter has been designed, and is shown in Figure 11.18. Since the input current is a pulse train with the repeating frequency f_m, a low-pass filter L_S–C_S is required. We have to investigate the operation during both positive and negative half-cycles of the input voltage.

11.1.5.1 Positive Input Voltage Half-Cycle

When the input voltage is positive, the buck–boost converter operates in the usual manner. The equivalent circuits during the switch-on and switch-off periods are shown in Figure 11.19.

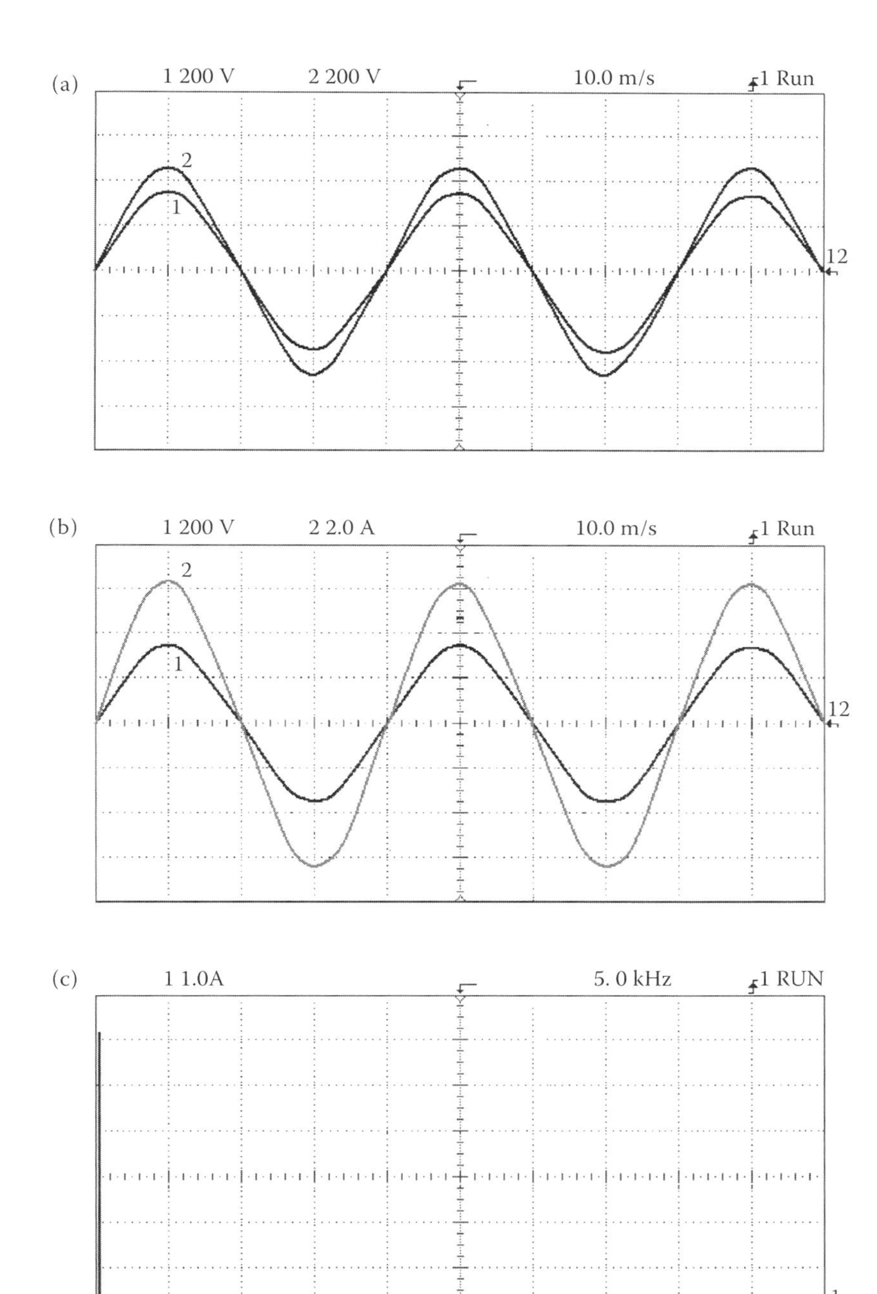

FIGURE 11.17 Test results of the DC-modulated boost-type AC/AC converter: (a) Input/output voltage waveforms, (b) Input voltage and current waveforms, and (c) spectrum of the input current.

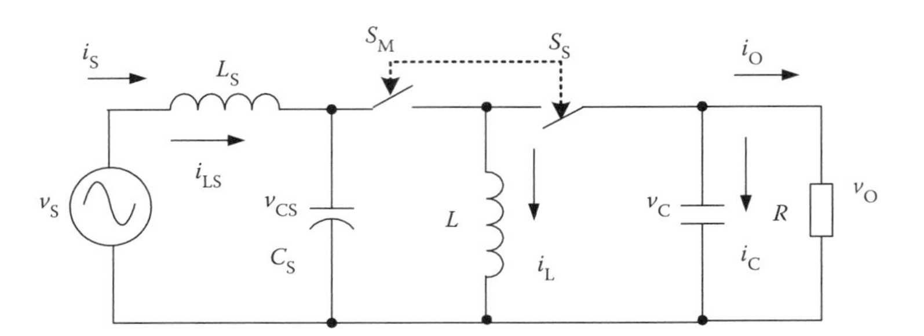

FIGURE 11.18 DC-modulated single-stage buck–boost-type AC/AC converter.

The output voltage is calculated by

$$v_O = \frac{kv_S}{1-k} = \frac{k\sqrt{2}}{1-k} V_S \sin \omega t, \quad 0 \le \omega t < \pi, \tag{11.23}$$

where k is the conduction duty cycle of the buck converter, V_S is the rms value of the input voltage, and ω is the power supply radian frequency.

11.1.5.2 Negative Input Voltage Half-Cycle

When the input voltage is negative, the buck–boost converter operates in the reverse manner. The equivalent circuits during the switch-on and switch-off periods are shown in Figure 11.20.

The output voltage is calculated by

$$v_O = -\frac{k}{1-k}|v_S| = \frac{k\sqrt{2}}{1-k} V_S \sin \omega t, \quad \pi \le \omega t < 2\pi, \tag{11.24}$$

where k is the conduction duty cycle, V_S is the rms value, and ω is the power supply radian frequency.

11.1.5.3 Whole-Cycle Operation

Combining the above two state operations, we can summarize the whole-cycle operation. The output voltage is calculated by

$$v_O = kv_S = \frac{k\sqrt{2}}{1-k} V_S \sin \omega t, \tag{11.25}$$

where k is the conduction duty cycle, V_S is the rms value, and ω is the power supply radian frequency.

The whole-cycle input and output voltage waveforms are shown in Figure 11.21a with the conduction duty cycle $k = 0.43$. The voltage transfer gain $M = V_O/V_S$ versus the DC/DC converter conduction duty cycle k is shown in Figure 11.21b. It is easy to obtain the variable output voltage higher or lower than the input voltage with very high PF and high efficiency. There is polarity reversal between input and output voltages, or a phase angle shift of 180°. Usually, this phase angle shift does not affect most industrial applications.

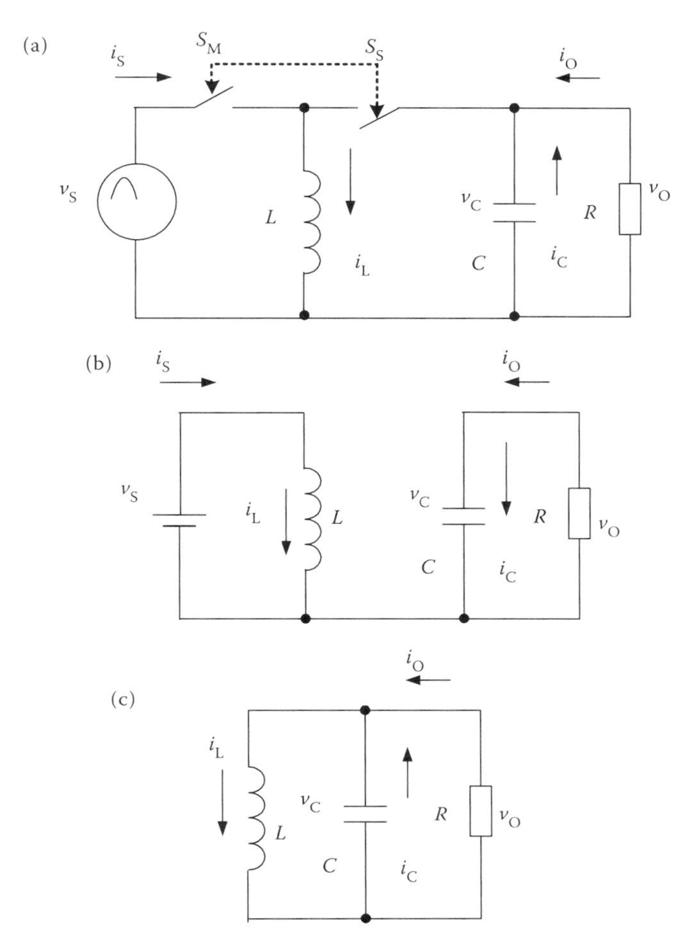

FIGURE 11.19 DC-modulated buck–boost-type AC/AC converter working in a positive half-cycle: (a) circuit diagram, (b) equivalent circuit during switch-on, and (c) equivalent circuit during switch-off.

The whole-cycle input voltage and current waveforms are shown in Figure 11.22a. The spectrum of the input current is shown in Figure 11.22b. The spectrum is very clean and the little distortion from the harmonic component I_M at 20 kHz is far from the fundamental frequency component I_S at 50 Hz. Its value is only 0.5%, that is, $I_M/I_S = 0.005$. Therefore, THD = 1.0000125. Considering that DPF = 0.9998, we obtain the final PF as 0.99979.

11.1.5.4 Simulation and Experimental Results

The simulation and experimental results are shown in order to verify the design.

11.1.5.4.1 Simulation Results

The DC-modulated buck–boost-type AC/AC converter shown in Figure 11.18 has the following components: $L_S = 1\,\text{mH}$, $C_S = 10\,\mu\text{F}$, $L = 10\,\text{mH}$, and $C = 3000\,\text{nF}$. The conduction duty cycle is $k = 0.45$. The simulation results are shown in Figure 11.23. The output voltage $V_O = k/(1-\text{k}) \times V_S = 0.818 \times V_S = 163.6V_{\text{rms}}$ (the peak value is approximately 231.4 V)

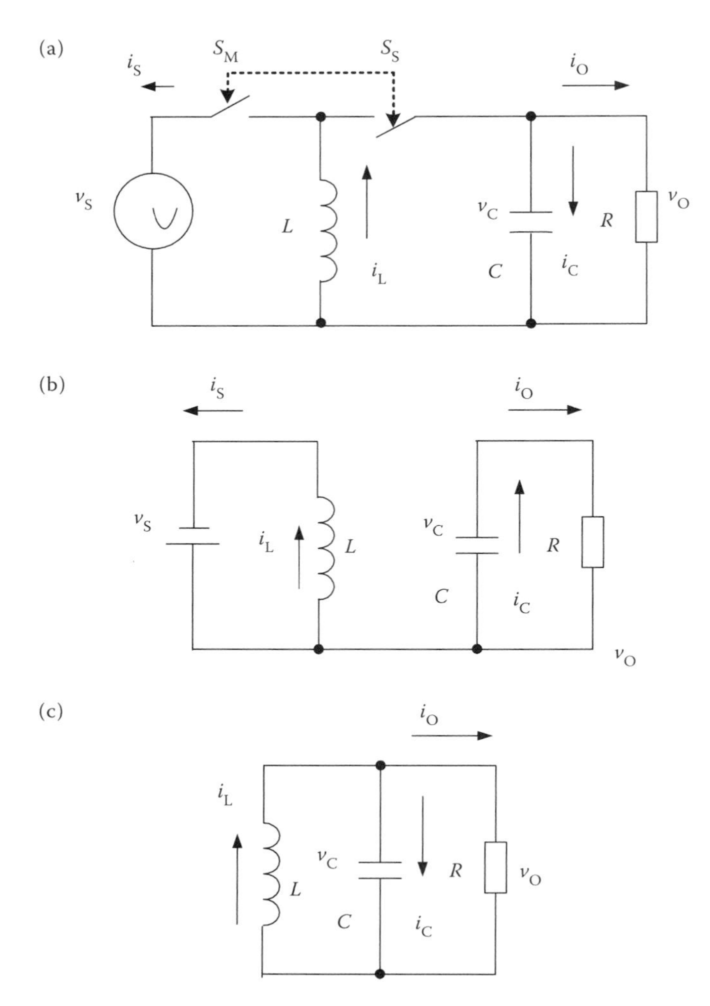

FIGURE 11.20 DC-modulated buck–boost-type AC/AC converter working in a negative half-cycle: (a) circuit diagram, (b) equivalent circuit during switch-on, and (c) equivalent circuit during switch-off.

with the frequency $f = 50\,\text{Hz}$. The waveforms of the input and output voltages $v_S(t)$ and $v_O(t)$ are shown as Channels 1 and 2 in Figure 11.23a. It can be seen that there is no phase delay, although there may be about 3.374° phase angle delay from our analysis in Section 11.3. The output current I_O should be 1.1 A, and the output power $P_O = V_O^2/R = 178.5\,\text{W}$.

The input current is measured as $I_S = 0.9\,\text{A}$ (the peak value is approximately 1.27 A) with the frequency $f = 50\,\text{Hz}$. The waveforms of the input voltage $v_S(t)$ and current $i_S(t)$ are shown as Channels 1 and 2 in Figure 11.23b. It can be seen that there is nearly no phase delay, although there may be about 3.374° phase angle delay from our analysis. The FFT spectrum of the input current is shown in Figure 11.23c, and THD $= 0.1375$. The input power $P_{in} = V_S \times I_S = 180\,\text{W}$. Although the theoretical analysis shows no power losses for the ideal condition ($\eta = 1$), the particular test shows there are power losses, which are mainly caused by the power losses of the switches. From the test results, we obtain the final PF as 0.9939 and the power transfer efficiency $\eta = P_O/P_{in} = 178.5/180 = 0.9917$ or 99.17%.

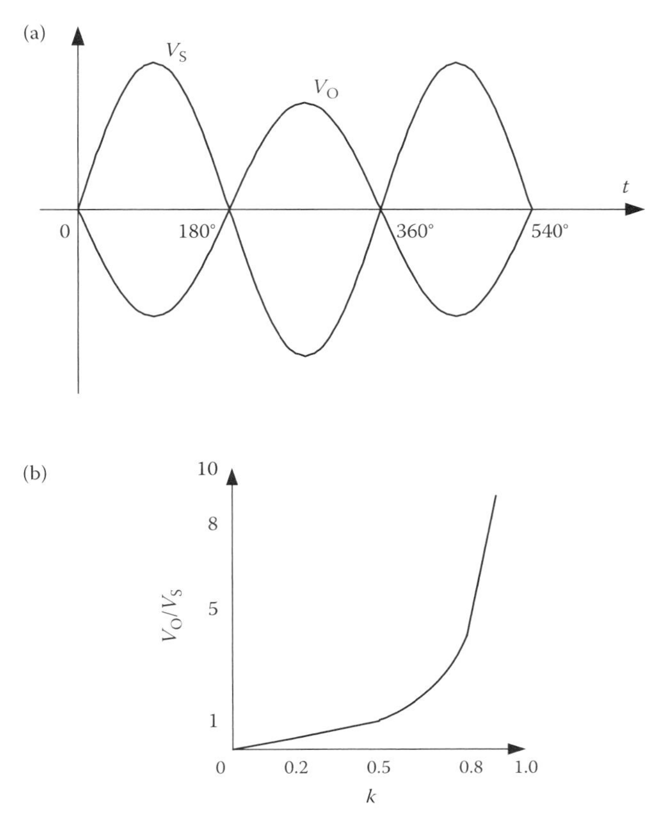

FIGURE 11.21 Input/output voltage waveforms of the DC-modulated boost-type AC/AC converter: (a) input/output voltage waveforms and (b) voltage transfer gain versus conduction duty cycle k.

11.1.5.4.2 Experimental Results

The experimental results are shown in Figure 11.24.

11.2 Other Types of DC-Modulated AC/AC Converters

Understanding the clue to the design and construction of the DC-modulated single-stage AC/AC converter, we can easily design and construct two-stage AC/AC converters. Some converters have a more complex structure such as Luo-converters, super-lift Luo-converters, and multistage cascaded boost converters [8,15–19]. In order to offer more information to readers, a DC-modulated positive output Luo-converter and a two-stage boost-type AC/AC converter have been designed.

11.2.1 DC-Modulated P/O Luo-Converter-Type AC/AC Converter

The DC-modulated P/O Luo-converter-type AC/AC converter is shown in Figure 11.25. Its output voltage has the same polarity as the input voltage.

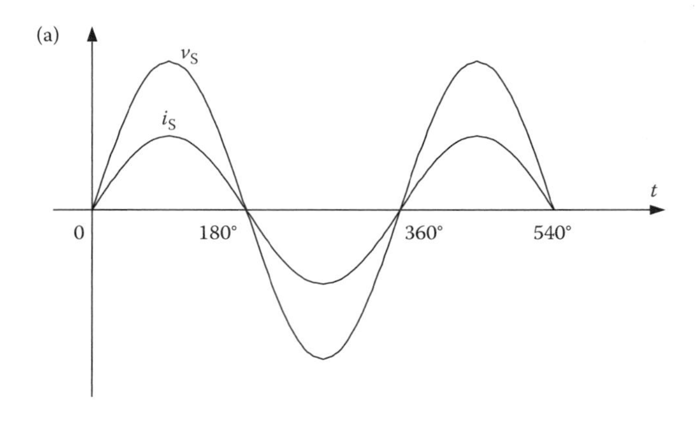

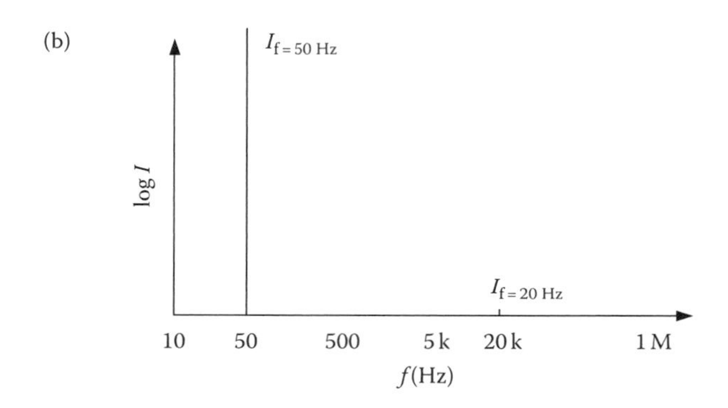

FIGURE 11.22 Input voltage/current waveforms of the DC-modulated boost-type AC/AC converter: (a) input voltage and current waveforms and (b) spectrum of input current.

The P/O Luo-converter has a more complex circuit than the buck–boost converter. An input low-pass filter is required. The values of all components are shown in the circuit diagram in Figure 11.25. The simulation circuit is shown in Figure 11.26.

The simulation waveforms with $k = 0.7$ and $f = 80\,\text{kHz}$ are shown in Figure 11.27.

Example 11.3

The DC-modulated positive output Luo-converter-type AC/AC converter shown in Figure 11.25 has an input rms voltage $v_S = 240\text{V}$ and a dimmer load with $R = 100\,\Omega$. In order to obtain the output rms voltage, v_O varies in the range of 100–400V (higher and lower than the input voltage). Calculate the range of the conduction duty cycle k and the output current and power.

Solution

Since the output voltage is calculated as

$$v_O = \frac{k}{1-k} v_S = \frac{k}{1-k} 240,$$

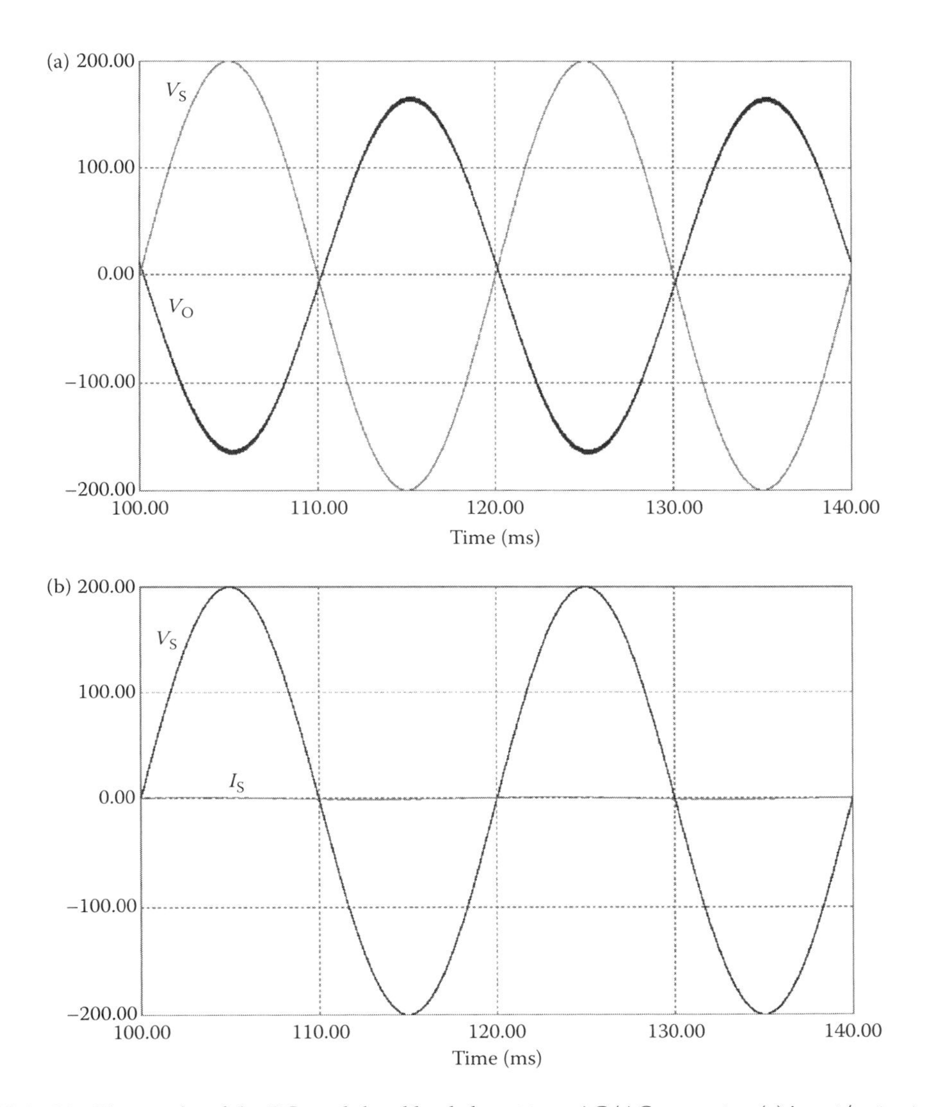

FIGURE 11.23 Test results of the DC-modulated buck–boost-type AC/AC converter: (a) input/output voltage waveforms of the DC-modulated buck-boost-type AC/AC converter, (b) input voltage and current waveforms of the DC-modulated buck-boost-type AC/AC converter, and (c) spectrum of input current of the DC-modulated buck-boost-type AC/AC converter.

the conduction duty cycle k is calculated as

$$k = \frac{v_O}{v_O + v_S} = \begin{cases} \dfrac{100}{240 + 100} = 0.294 \\ \dfrac{400}{240 + 400} = 0.625 \end{cases}$$

The range of k is 0.294–0.625.

The output rms current is 1–4 A, and the output power is 100–1600 W.

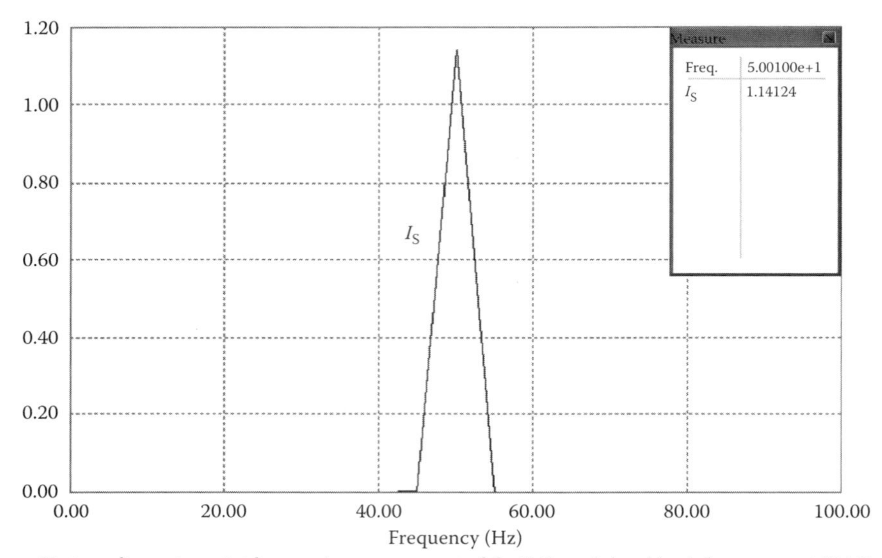

Test result spectrum 1st harmonic measurement of the DC-modulated buck-boost-type AC/AC converter $f = 50\,\text{kHz}$, $k = 0.45$ (Zoom 1)

(c)

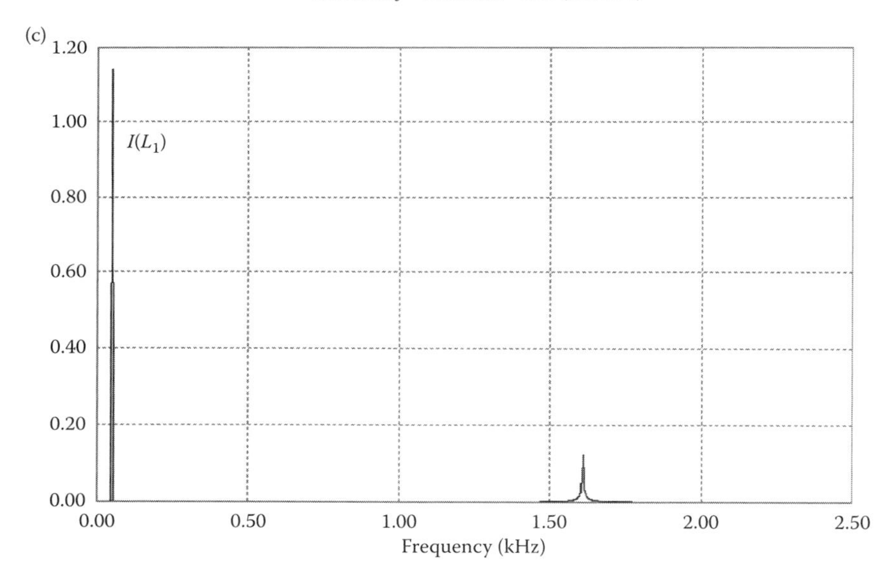

FIGURE 11.23 Continued

11.2.2 DC-Modulated Two-Stage Boost-Type AC/AC Converter

A DC-modulated two-stage boost-type AC/AC converter is shown in Figure 11.28.

The four bidirectional switches (S_M–S_{S-S} plus S_{S1}–S_{S2}) in Figure 11.4 are applied. The output voltage is

$$v_O(t) = \left(\frac{1}{1-k}\right)^2 v_S = \left(\frac{1}{1-k}\right)^2 \sqrt{2}V_S \sin \omega t. \tag{11.26}$$

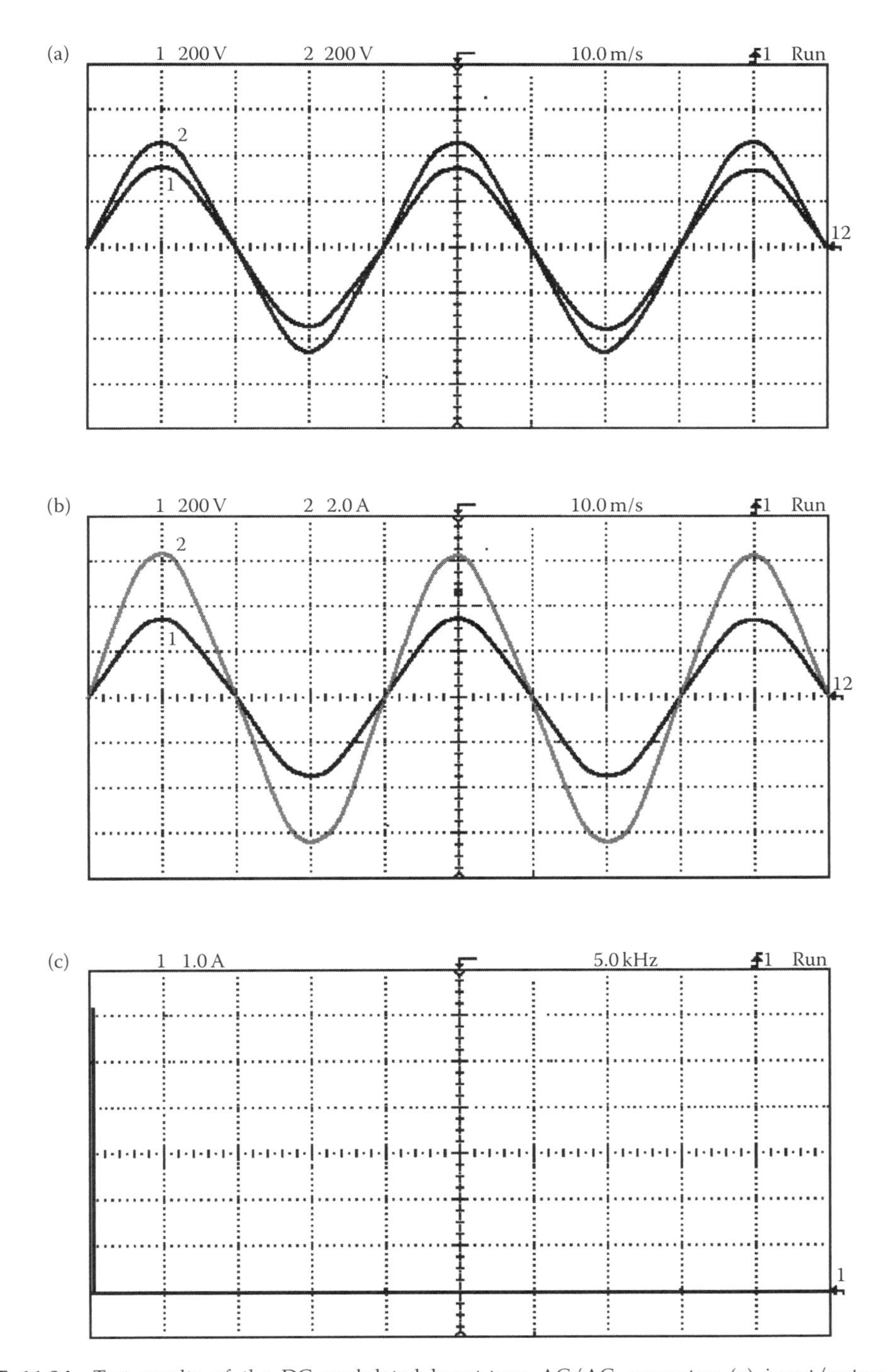

FIGURE 11.24 Test results of the DC-modulated boost-type AC/AC converter: (a) input/output voltage waveforms, (b) input voltage and current waveforms, and (c) spectrum of the input current.

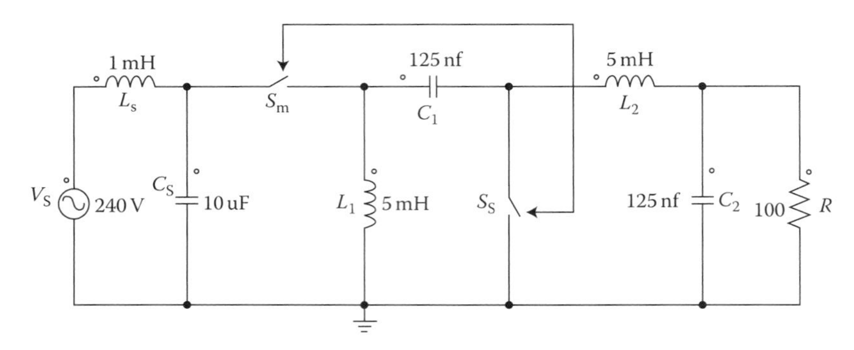

FIGURE 11.25 DC-modulated positive output Luo-converter-type AC/AC converter.

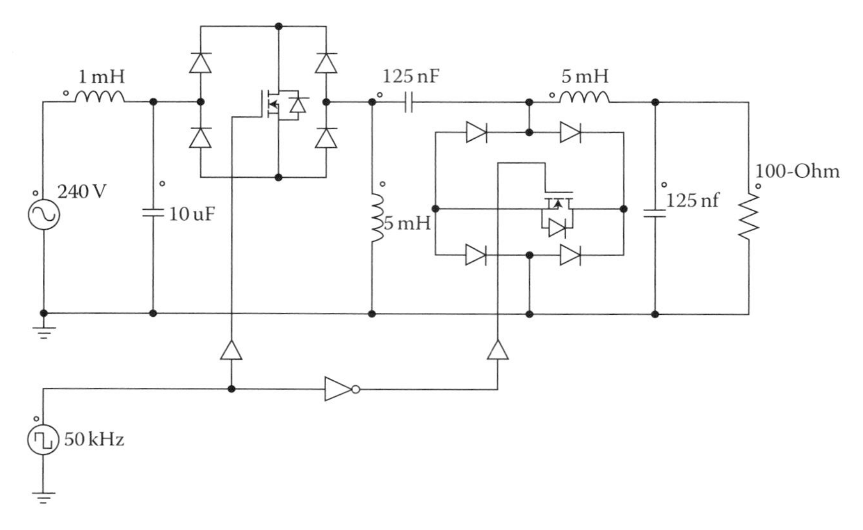

FIGURE 11.26 Simulation circuit of the DC-modulated P/O Luo-converter-type AC/AC converter.

The voltage transfer gain is

$$M = \frac{v_O(t)}{v_S(t)} = \left(\frac{1}{1-k}\right)^2. \tag{11.27}$$

From this calculation formula, the output voltage can be easily increased to a high voltage. For example, $k = 0.7$ results in the voltage transfer gain $M = 11.11$.

11.3 DC-Modulated Multiphase AC/AC Converters

Using the same technique, we can construct DC-modulated multiphase AC/AC converters.

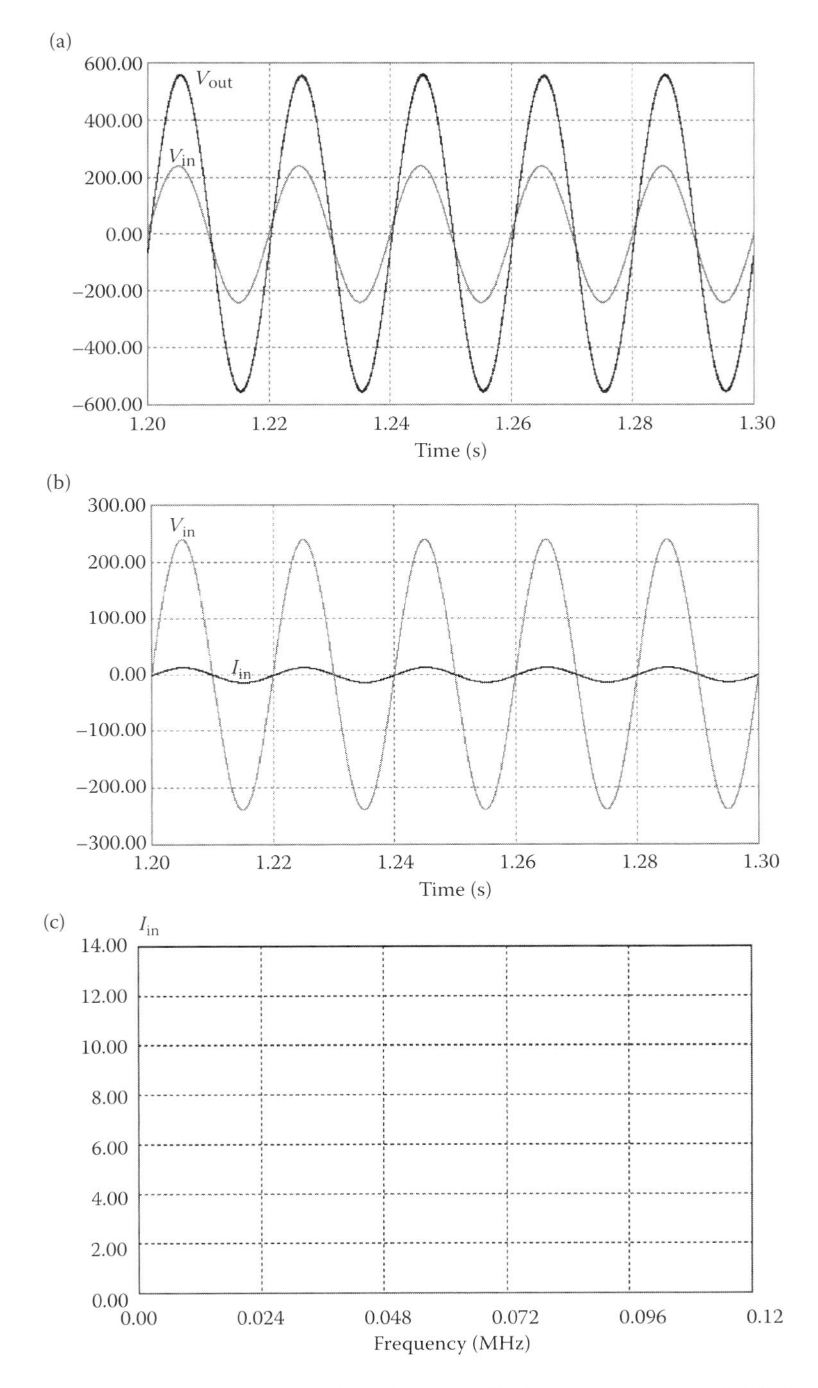

FIGURE 11.27 Simulation results of the DC-modulated P/O Luo-converter-type AC/AC converter: (a) input/output voltage waveforms, (b) input voltage and current waveforms, and (c) spectrum of input current.

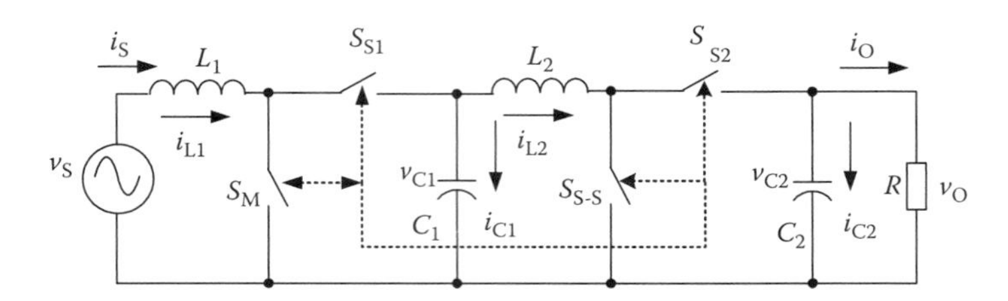

FIGURE 11.28 Circuit diagram of the DC-modulated two-stage boost-type AC/AC Converter.

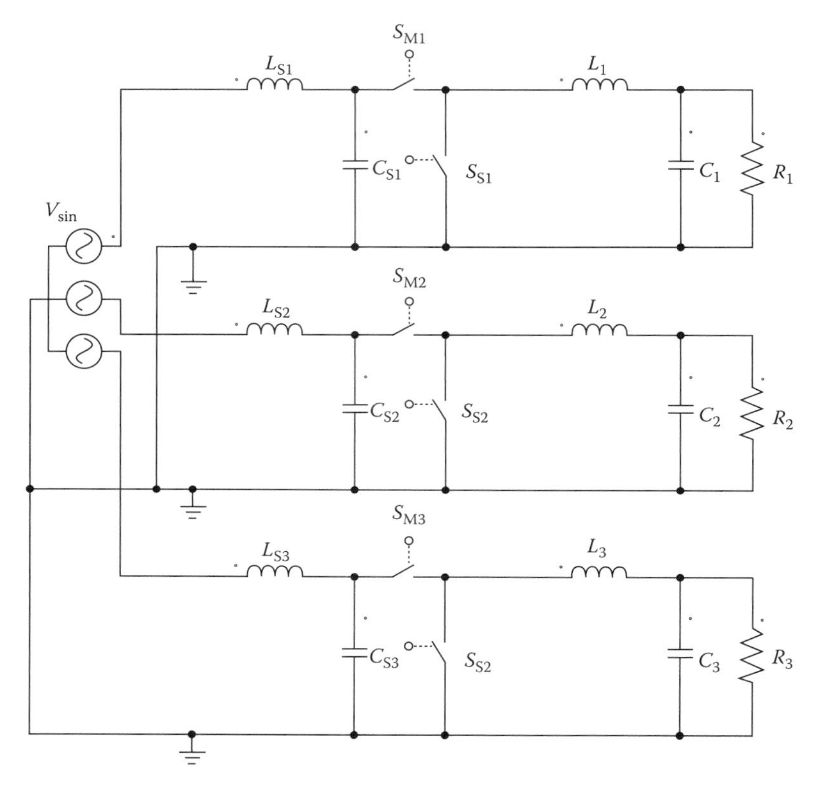

FIGURE 11.29 DC-modulated three-phase buck-type AC/AC converter.

11.3.1 DC-Modulated Three-Phase Buck-Type AC/AC Converter

A DC-modulated three-phase buck-type AC/AC converter is shown in Figure 11.29. The simulation results are shown in Figure 11.30.

11.3.2 DC-Modulated Three-Phase Boost-Type AC/AC Converter

A DC-modulated three-phase boost-type AC/AC converter is shown in Figure 11.31. The simulation results are shown in Figure 11.32.

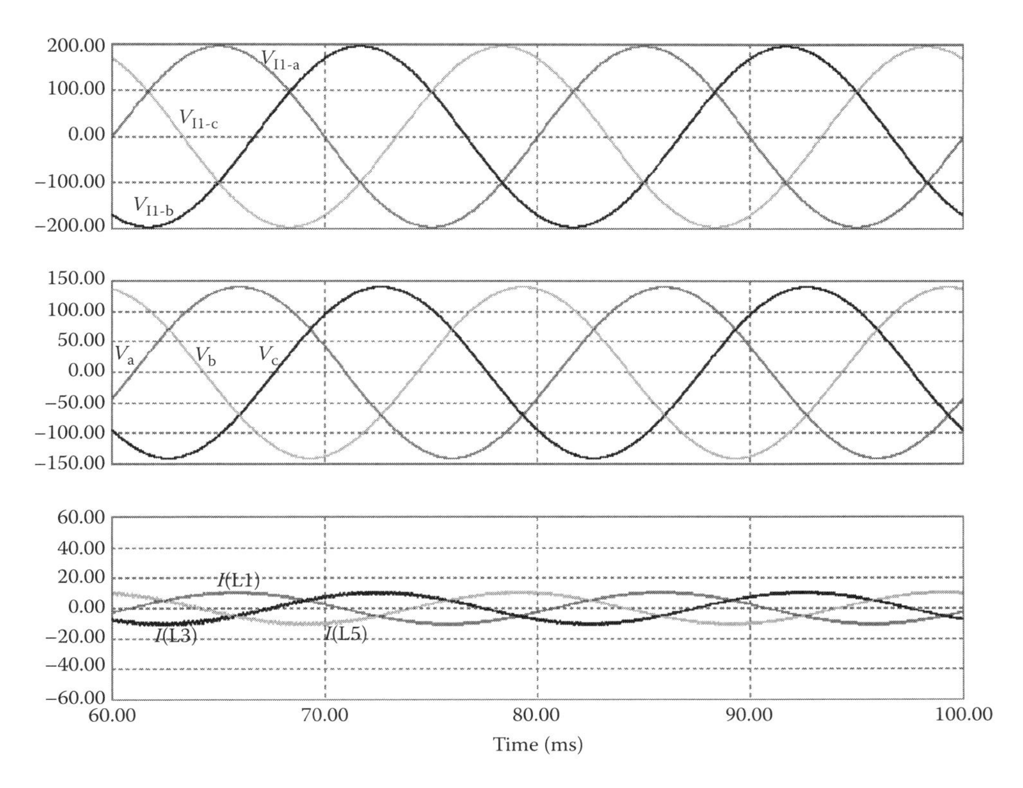

FIGURE 11.30 Simulation results of the DC-modulated three-phase buck-type AC/AC converter.

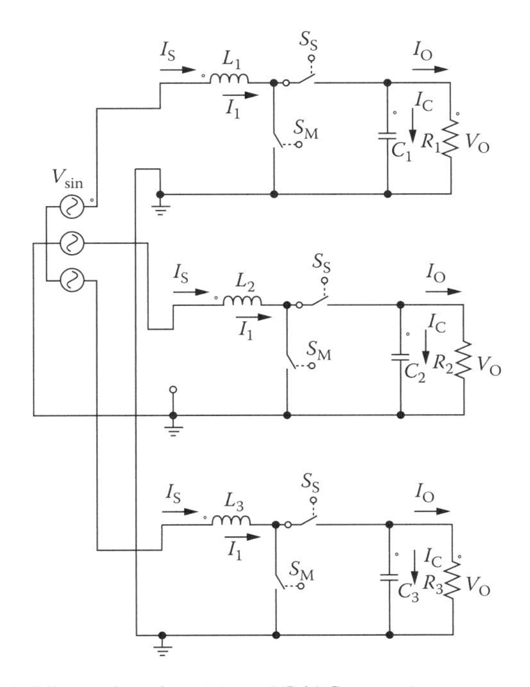

FIGURE 11.31 DC-modulated three-phase boost-type AC/AC converter.

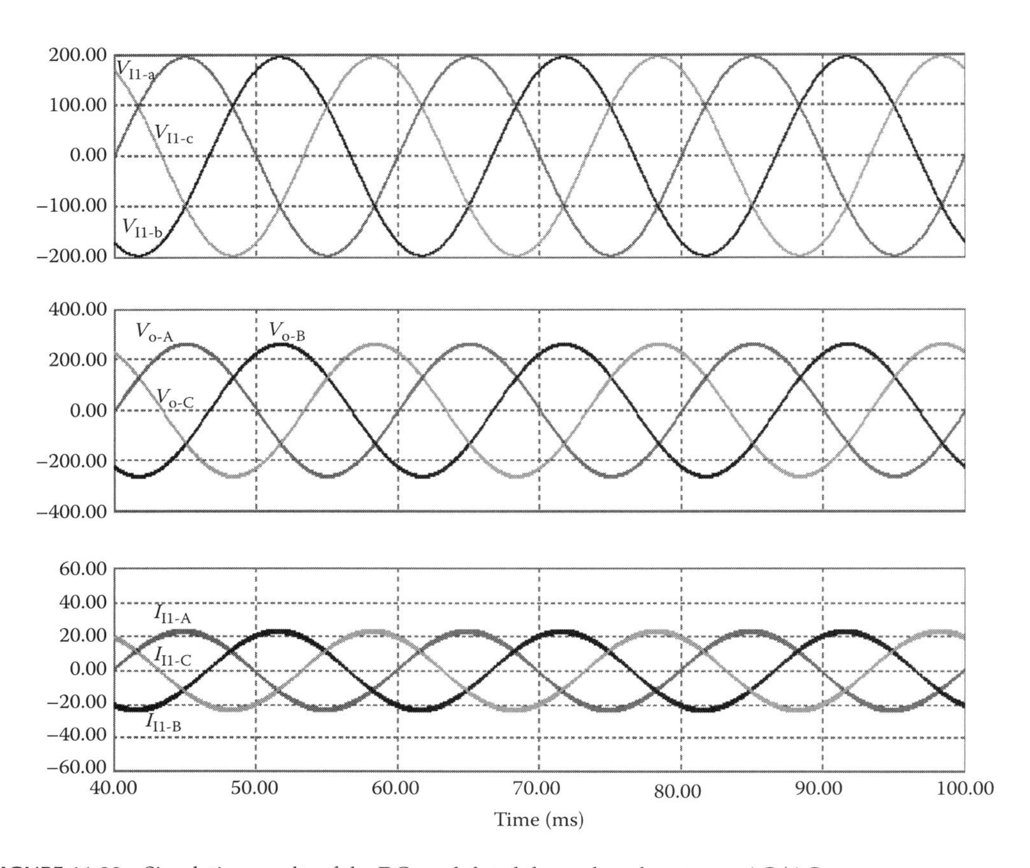

FIGURE 11.32 Simulation results of the DC-modulated three-phase boost-type AC/AC converter.

11.3.3 DC-Modulated Three-Phase Buck–Boost-Type AC/AC Converter

A DC-modulated three-phase buck–boost-type AC/AC converter is shown in Figure 11.33. The simulation results are shown in Figure 11.34.

11.4 Subenvelope Modulation Method to Reduce the THD of AC/AC Matrix Converters

An AC/AC matrix converter is an array of power semiconductor switches that directly connects a three-phase AC source to another three-phase load. It can convert an AC power source with a certain voltage and frequency to another AC load with variable voltage and variable frequency directly without the DC link and the bulk energy storage component. Classical modulation methods, such as the Venturini method and the SVM method using AC-network maximum-envelope modulation, implement matrix conversion successfully. However, in the meantime, they cause very high THD. This chapter presents a novel approach, the *subenvelope modulation* (SEM) method, to reduce the THD of matrix converters effectively [20,21]. The approach is extended to an improved version of matrix converters and the THD can be reduced further. The algorithm of the SEM method is described in

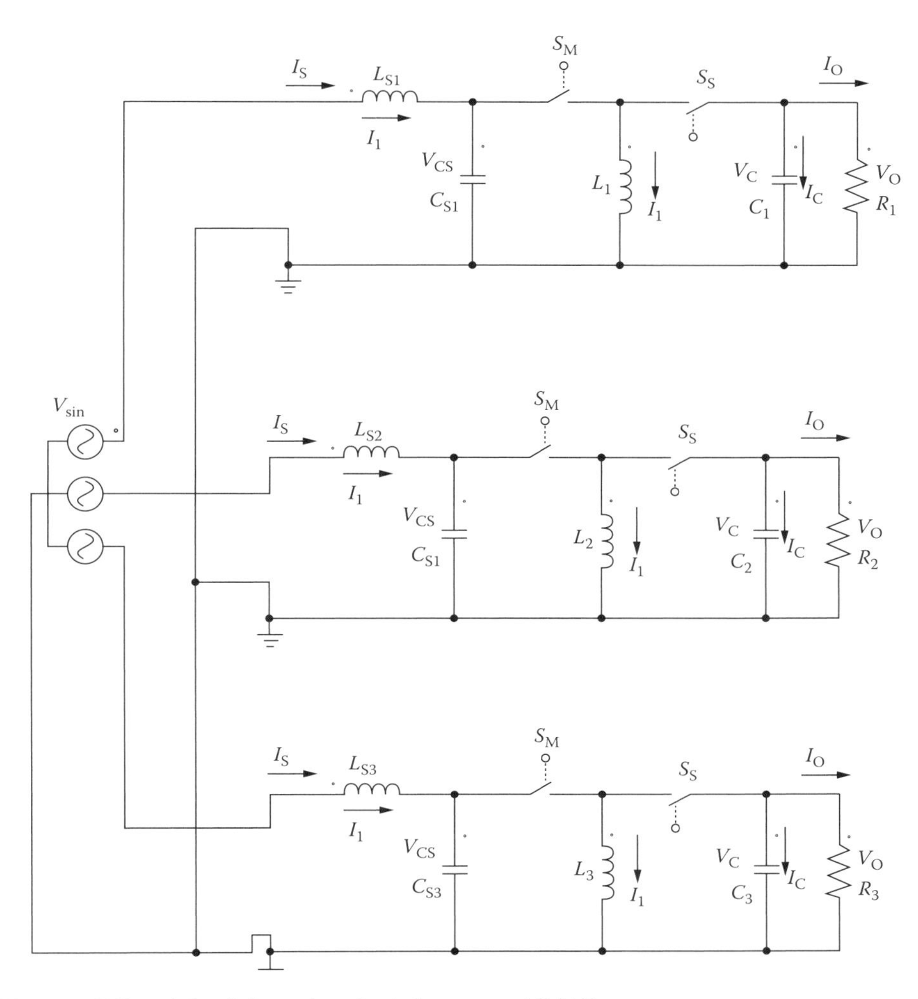

FIGURE 11.33 DC-modulated three-phase buck–boost-type AC/AC converter.

detail. Simulation and experimental results are also presented to verify the feasibility of the SEM approach. The results will be very helpful for industrial applications.

An AC/AC matrix converter [22–24] is an array of power semiconductor switches that directly connects a three-phase AC source to another three-phase load. This converter has several attractive features that have been investigated in recent decades. It can convert an AC power source with a certain voltage and frequency to another AC load with variable voltage and variable frequency directly without a DC link and a bulk energy storage component. It eliminates a large energy storage component, that is, a bulk inductor or an electrolytic capacitor. The structure of the classical 3×3 matrix converter is shown in Figure 11.35. The semiconductor switches are marked with S_{Jk}, which means that the switch is connected between input phase J and output phase k, where J = {A,B,C}, k = {a,b,c}.

All the switches S_{Jk} in matrix converters require a bidirectional-switch capability of blocking voltage and conducting current in both directions. Unfortunately, there are no such devices available now; so discrete devices need to be used to construct suitable switch cells. One option is the diode bridge bidirectional switch cell arrangement, which consists of an

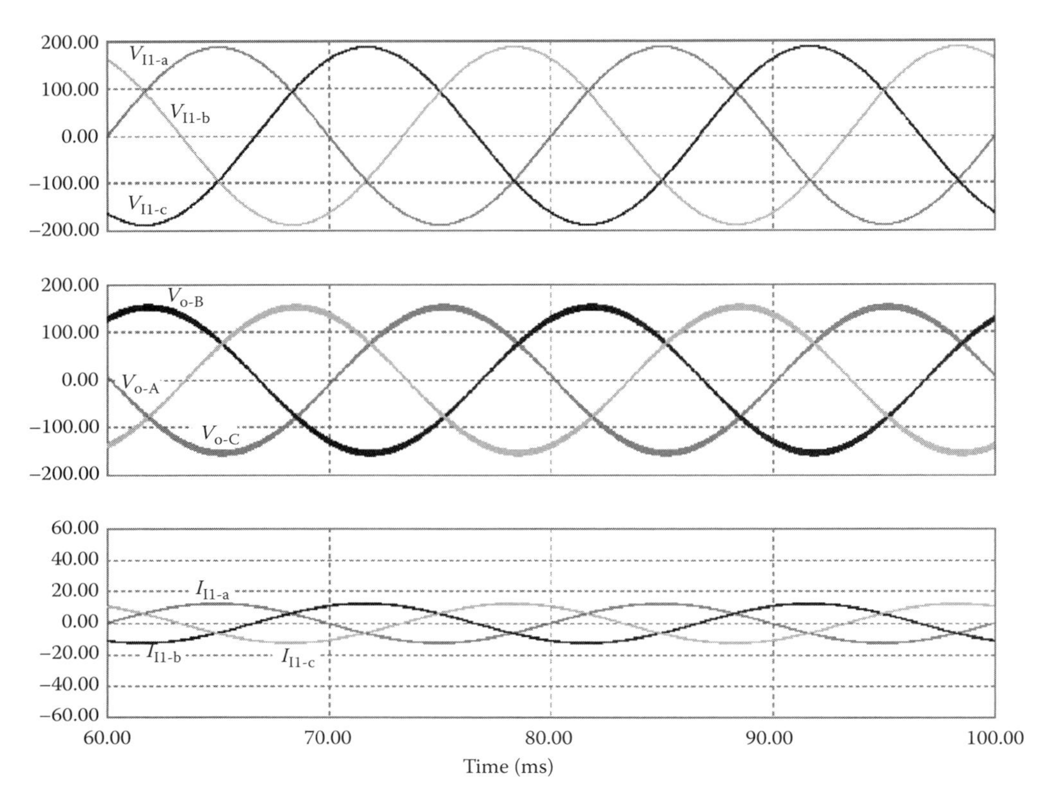

FIGURE 11.34 Simulation results of the DC-modulated three-phase buck–boost-type AC/AC converter.

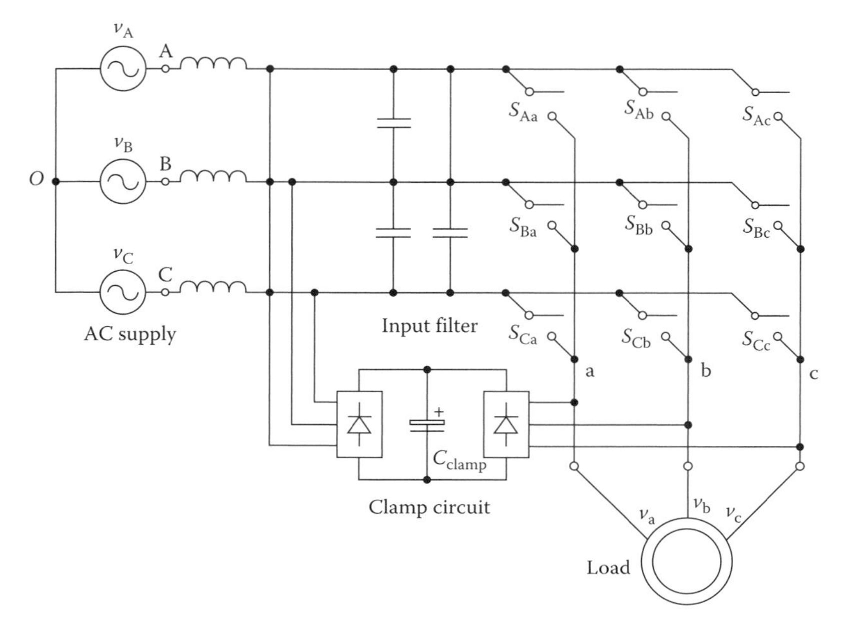

FIGURE 11.35 Structure of conventional matrix converter.

IGBT (or other full control power semiconductor switches) at the center of a single-phase diode bridge [25]. The main advantage is that both current directions are carried by the same switching device; therefore, only one gate driver is required per switch cell. Device power losses are relatively high since there are three devices in each conduction path. The current direction through the switch cell cannot be controlled. This is a disadvantage since many advanced commutation methods require the current direction of the switch cell to be controllable.

The common-emitter bidirectional switch cell arrangement consists of two IGBTs and two diodes in another scheme. The diodes provide the reverse blocking capability. There are several advantages to using this arrangement when compared with the diode bridge bidirectional switch cell. First, it is possible to independently control the direction of the current. Second, the conduction power losses are also reduced since only two devices carry the current. Third, each bidirectional switch cell requires an isolated power supply for the gate drive. The switch cell can be connected to a common collector. The conduction power losses are the same as that of the common-emitter configuration. An often-quoted advantage of this method is that only six isolated power supplies are needed to supply the gate driver [26]. Therefore, the common-emitter configuration is generally preferred for creating the matrix converter bidirectional switch cells.

Normally, the matrix converter is fed by a three-phase voltage source and, for this reason, the input terminals should not be in short circuit (rule 1). On the other hand, the load typically has an inductive nature and, for this reason, an output phase must never be in an open circuit (rule 2). Reliable current commutation between switches in matrix converters is more difficult to achieve than in the conventional VSI since there are no natural freewheeling paths [22]. The commutation has to be actively controlled at all times with respect to the two basic rules. These rules can be visualized by considering just two switch cells and one output phase of a matrix converter. It is important that no two bidirectional switches are switched on at any point of time, as shown in Figure 11.36a. This would result in line-to-line short-circuiting and the destruction of the converter due to rush current. Also, the bidirectional switches for each output phase should not all be turned off at any instant, as shown in Figure 11.36b. This would result in the absence of a path for the inductive load current, causing rush voltage. These two considerations cause a conflict since semiconductor devices cannot be switched instantaneously due to propagation delays and finite switching times. There are some successful approaches to avoid these two cases: basic current commutation [22], current direction-based commutation [26–28], relative voltage magnitude-based commutation [29–31], and soft-switching techniques [32,33].

Classical modulation methods, such as the Venturini method [23,24] and the SVM method [30] using AC-network maximum-envelope modulation, implement matrix conversion successfully. However, in the meantime, they cause very high THD. This chapter presents a

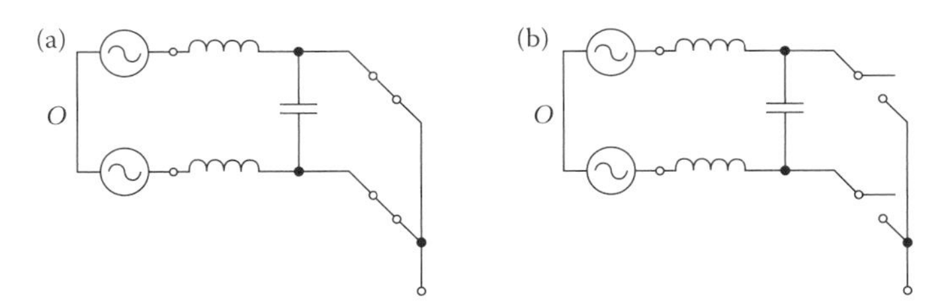

FIGURE 11.36 Two cases that the matrix converter should avoid: (a) short circuits on the matrix converter input lines, and (b) open circuits on the matrix converter output lines.

novel approach, the SEM method, to reduce the THD for matrix converters effectively. The approach is extended to an improved version of matrix converters and the THD can be reduced further. The algorithm of the SEM method is described in detail. Simulation and experimental results are also presented to verify the feasibility of the SEM approach. The results will be very helpful in industrial applications.

In the following description, we assume that

- The three phases of the input AC supply are balanced; the input phase voltages are v_A, v_B, and v_C.
- The input AC supply frequency is f_i; the corresponding angular speed is $\omega_i = 2\pi f_i$.
- The output phase voltages are v_a, v_b, and v_c.
- The output frequency is f_O; the corresponding angular speed is $\omega_O = 2\pi f_O$.
- The switching frequency is f; the period is T. Usually, $f \gg f_i$ and f_O.
- V_{DC} is the *imaginary* DC-link voltage, corresponding to the maximum-envelope rectifying average voltage.

11.4.1 SEM Method

One commonly used modulation method for matrix converters is maximum-envelope modulation, which is shown in Figure 11.37a, that is, the output phase voltage is pulse width modulated between the maximum input phase and the minimum input phase. The disadvantages are obvious: the magnitude of the output pulse is the difference between the maximum input phase and the minimum input phase; so the output pulse has a high magnitude and a narrow width. Therefore, there are many high-frequency components in the output voltage, and these will result in very high THD. Moreover, there is a high dv/dt, which will induce severe EMI.

Actually, in matrix converters, the three output phases can be connected to any input phases. So the output phase can be modulated between any two input phases. If the output phase is modulated between two *adjacent* input phases as shown in Figure 11.37b, the pulse magnitude of the output voltage can be low. Correspondingly, the high-frequency components of the output voltage can be reduced. Thus the THD and dv/dt are also reduced. The input line current pulses are smaller and wider, and the THD of the input line current is also reduced. The approach is called the SEM method.

(a)

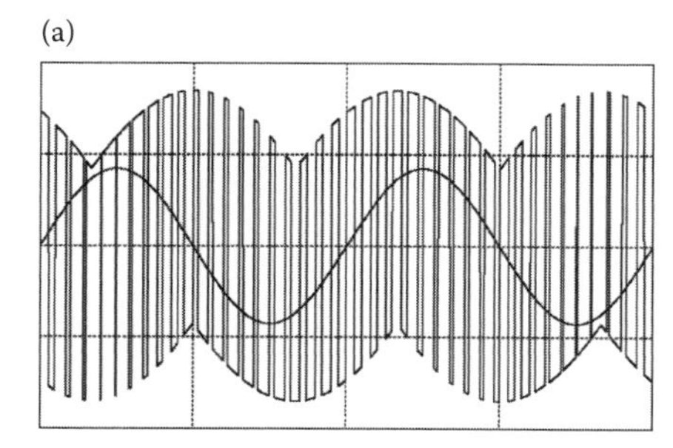

(b)

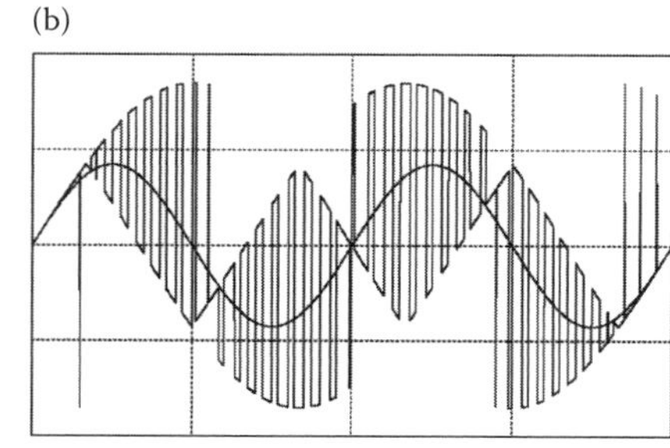

FIGURE 11.37 Modulation method for the conventional matrix converter: (a) Maximum envelope modulation method and (b) SEM method.

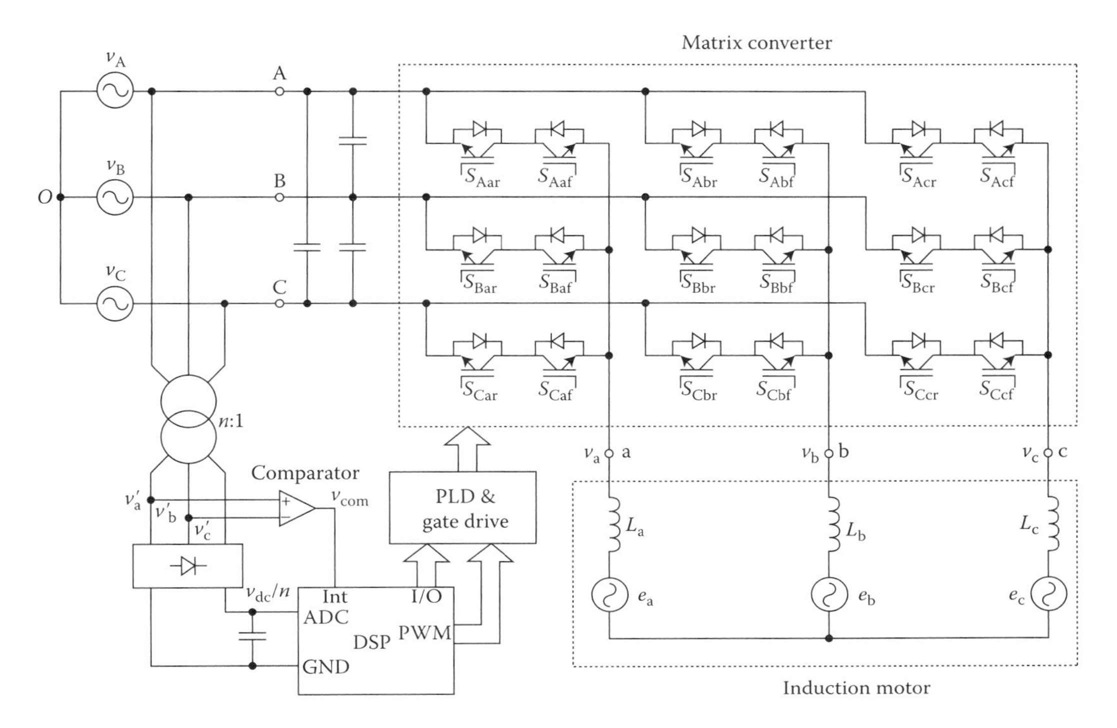

FIGURE 11.38 Structure of a matrix converter drive system.

The structure of the matrix converter implementing the SEM method is shown in Figure 11.38. The matrix converter comprises 18 power semiconductor switches (nine switch cells) so that all the output phases can be connected to any input phases with bidirectional current capability. The switches are marked with S_{Kjr} or S_{Kjf}, where K = {A,B,C} is the input phase, j = {a,b,c} is the output phase, r denotes the switch that carries the current from the output to the input (reverse), and f denotes the switch that carries the current from the input to the output (forward). It was mentioned that the common collector switch cell has the advantage of requiring fewer isolated DC power supplies for the gate drives. The nine switch cells (18 switches) constitute the common collector configuration. The 18 switches can be divided into six switch groups, S_{Ajr}, S_{Bjr}, S_{Cjr}, S_{Kaf}, S_{Kbf}, and S_{Kcf}. Each group comprises three common-emitter switches and uses a common gate drive output DC floating power supply.

11.4.1.1 Measure the Input Instantaneous Voltage

It is necessary to know the instantaneous phase voltage of AC supply. One approach is to measure the input voltage with three voltage sensors. If the AC supply is a balanced sinusoidal supply, one simple approach to get the instantaneous phase voltage is applicable, that is, calculating the input voltage in real time. If the magnitude and time base of the three-phase supply are known, the instantaneous phase voltage can be determined. Thus a three-phase transformer and a rectifier are adopted. The turn ratio of the transformer is n:1. The adoption of the transformer is to insulate the control circuit from the power stage. The scaled DC-link voltage V_{DC}/n can be obtained by a small rate rectifier and an electrolytic capacitor. In order to get the time base, a comparator is introduced. The input of the comparator can be either of two input phases (such as phase A and phase B). The output waveform v_{com} of the comparator is shown in Figure 11.39.

At the falling edge of v_{com} (such as t_1), the instantaneous phase voltage of the AC supply can be obtained:

$$v_A(t_1) = V_m \sin\left(\frac{5\pi}{6}\right),$$
$$v_B(t_1) = V_m \sin\left(\frac{\pi}{6}\right), \quad (11.28)$$
$$v_C(t_1) = V_m \sin\left(\frac{-\pi}{2}\right),$$

where $V_m = \pi V_{DC}/3\sqrt{3}$ and V_{DC} is the imaginary DC-link voltage. The frequency of the AC power supply is known as f_i and the angular frequency $\omega_i = 2\pi f_i$. Redefining the initial time, the instantaneous input voltage during one cycle $(1/f_i = T_i)$ can be expressed as

$$v_A(t) = \frac{\pi V_{DC}}{3\sqrt{3}} \sin\left(\omega_i t + \frac{5\pi}{6}\right),$$
$$v_B(t) = \frac{\pi V_{DC}}{3\sqrt{3}} \sin\left(\omega_i t + \frac{\pi}{6}\right), \quad (11.29)$$
$$v_C(t) = \frac{\pi V_{DC}}{3\sqrt{3}} \sin\left(\omega_i t - \frac{\pi}{2}\right).$$

In a discrete system with sampling frequency f (sampling time T), the voltage sequence of the AC power supply can be obtained from Equation 11.29:

$$v_A(kT) = \frac{\pi V_{DC}}{3\sqrt{3}} \sin\left(\omega_i kT + \frac{5\pi}{6}\right),$$
$$v_B(kT) = \frac{\pi V_{DC}}{3\sqrt{3}} \sin\left(\omega_i kT + \frac{\pi}{6}\right), \quad (11.30)$$
$$v_C(kT) = \frac{\pi V_{DC}}{3\sqrt{3}} \sin\left(\omega_i kT - \frac{\pi}{2}\right).$$

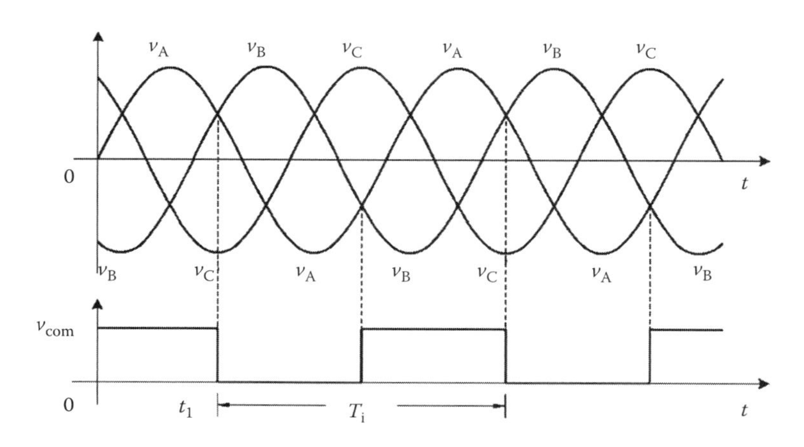

FIGURE 11.39 Output of the comparator.

With the help of signal v_{com}, the voltage sequence of the AC power supply can be calculated rigorously without error accumulation.

11.4.1.2 Modulation Algorithm

A SEM example (only v_a is illustrated) is shown in Figure 11.40. The output is modulated between two adjacent input phases.

The modulation rule of the example is shown in Table 11.1. v_{high} is the smallest one that is greater than v_{ra} (v_{ra} is the reference voltage of the output phase a), v_{low} is the biggest one that is less than v_{ra}, that is, the output phase a is connected to two adjacent phases; alternately, the duty cycle of PWM, δ, can be determined:

$$\delta = \frac{v_a - v_{low}}{v_{high} - v_{low}}. \tag{11.31}$$

Assume that the output frequency is f_O, the magnitude of the reference output phase voltage is V_O, the angular frequency is $\omega_O = 2\pi f_O$, and the initial phase angle is ϕ_0. The

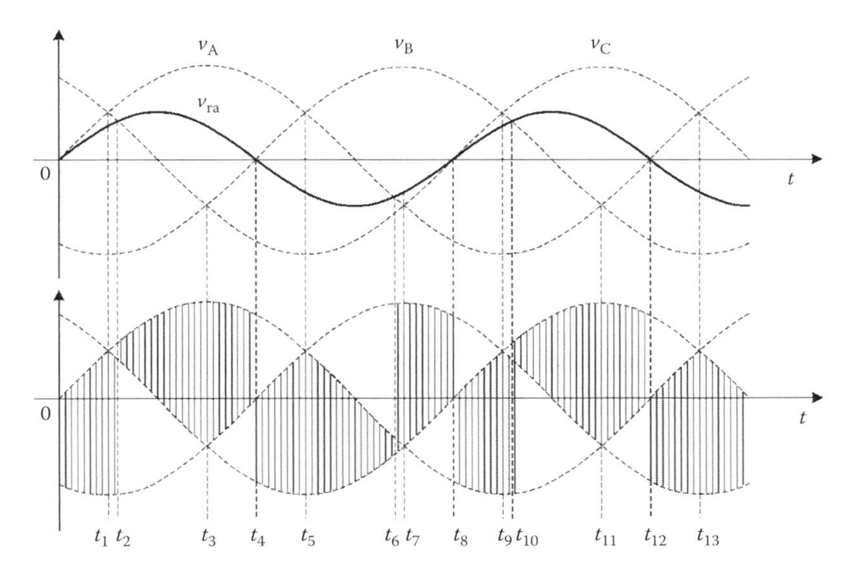

FIGURE 11.40 One SEM example.

TABLE 11.1
SEM Rule of the Example in Figure 11.40

Time		$\sim t_1$	$t_1 \sim t_2$	$t_2 \sim t_3$	$t_3 \sim t_4$	$t_4 \sim t_5$	$t_5 \sim t_6$	$t_6 \sim t_7$
Modulation	v_{high}	v_A	v_C	v_A	v_A	v_B	v_A	v_B
Phase	v_{low}	v_B	v_B	v_C	v_B	v_C	v_C	v_A
		$t_7 \sim t_8$	$t_8 \sim t_9$	$t_9 \sim t_{10}$	$t_{10} \sim t_{11}$	$t_{11} \sim t_{12}$	$t_{12} \sim t_{13}$	$t_{13} \sim$
		v_B	v_C	v_B	v_C	v_C	v_A	v_C
		v_C	v_A	v_A	v_B	v_A	v_B	v_B

PWM switching frequency is f and the period is T. The modulation algorithm of the system can be accomplished by

ALGORITHM 1

Define the variables $\boldsymbol{\theta}_{\mathrm{i}}$, and $\boldsymbol{\theta}_{\mathrm{O}}$, and their initial values are set to zero, that is,

$$\boldsymbol{\theta}_{\mathrm{i}} = 0, \quad \boldsymbol{\theta}_{\mathrm{O}} = 0.$$

Define array $v_{\mathrm{i}}[3]$ to store the input phase voltage.

Define array ***order***[3] to map v_{i} to the input phase, and initialize the array with {1,2,3}, representing {A,B,C}.

Do nothing until the first falling edge of the signal v_{com} appears. When the first falling edge appears, for every PWM cycle T, do the following loop:

a. Calculate the AC power supply voltage $\boldsymbol{v}_{\mathrm{i}}[1] = v_{\mathrm{A}}(kT)$, $\boldsymbol{v}_{\mathrm{i}}[2] = v_B(kT)$, $\boldsymbol{v}_i[3] = v_C(kT)$ according to Equation 11.30

$$\begin{aligned} v_{\mathrm{i}}[1] &= v_{\mathrm{A}}(kT) = V_{\mathrm{m}} \sin\left(\frac{\theta_{\mathrm{i}} + 5\pi}{6}\right), \\ v_{\mathrm{i}}[2] &= v_{\mathrm{B}}(kT) = V_{\mathrm{m}} \sin\left(\frac{\theta_{\mathrm{i}} + \pi}{6}\right), \\ v_{\mathrm{i}}[3] &= v_{\mathrm{C}}(kT) = V_{\mathrm{m}} \sin\left(\frac{\theta_{\mathrm{i}} - \pi}{2}\right), \end{aligned} \tag{11.32}$$

where $V_{\mathrm{m}} = \pi V_{\mathrm{DC}}/3\sqrt{3}$ is the magnitude of the input phase voltage, which is measured from the transformer. If the AC power supply is measured by three voltage sensors, this step can be ignored.

b. Sort the voltages $\boldsymbol{v}_{\mathrm{i}}[1]$, $\boldsymbol{v}_{\mathrm{i}}[2]$, and $\boldsymbol{v}_{\mathrm{i}}[3]$ in descending order:
   i. If $\boldsymbol{v}_{\mathrm{i}}[1] < \boldsymbol{v}_{\mathrm{i}}[2]$, then exchange $\boldsymbol{v}_{\mathrm{i}}[1]$ and $\boldsymbol{v}_{\mathrm{i}}[2]$ and also exchange ***order***[1] and ***order***[2]; else do nothing.
   ii. If $\boldsymbol{v}_{\mathrm{i}}[1] < \boldsymbol{v}_{\mathrm{i}}[3]$, then exchange $\boldsymbol{v}_{\mathrm{i}}[1]$ and $\boldsymbol{v}_{\mathrm{i}}[3]$ and also exchange ***order***[1] and ***order***[3]; else do nothing.
   iii. If $\boldsymbol{v}_{\mathrm{i}}[2] < \boldsymbol{v}_{\mathrm{i}}[3]$, then exchange $\boldsymbol{v}_{\mathrm{i}}[2]$ and $\boldsymbol{v}_{\mathrm{i}}[3]$ and also exchange ***order***[2] and ***order***[3]; else do nothing.

   Thus if condition $\boldsymbol{v}_{\mathrm{i}}[1] \geq \boldsymbol{v}_{\mathrm{i}}[2] \geq \boldsymbol{v}_{\mathrm{i}}[3]$ is satisfied, then the variable ***order*** will also map the input phases.

c. Calculate the three output voltages v_{a}, v_{b}, and v_{c} with the following equation:

$$\begin{aligned} v_{\mathrm{a}} &= V_{\mathrm{O}} \sin(\theta_{\mathrm{O}} + \phi_0), \\ v_{\mathrm{b}} &= V_{\mathrm{O}} \sin\left(\frac{\theta_{\mathrm{O}} - 2\pi}{3 + \phi_0}\right), \\ v_{\mathrm{c}} &= V_{\mathrm{O}} \sin\left(\frac{\theta_{\mathrm{O}} - 4\pi}{3 + \phi_0}\right), \end{aligned} \tag{11.32}$$

where V_{O} is the magnitude of the reference output phase voltage and ϕ_0 is the initial phase angle.

d. For the value of v_{a}, do the following:

i. If $v_a \geq \boldsymbol{v}_\mathbf{i}[2]$, it means that v_a is between $\boldsymbol{v}_\mathbf{i}[1]$ and $\boldsymbol{v}_\mathbf{i}[2]$; then output phase a is modulated between input phase ***order***[1] and ***order***[2], and the PWM duty cycle δ is

$$\delta = \frac{v_a - v_i[2]}{v_i[1] - v_i[2]}. \tag{11.33}$$

ii. Else it means that v_a is between $\boldsymbol{v}_i[2]$ and $\boldsymbol{v}_i[3]$; then output phase a is modulated between input phase ***order***[2] and ***order***[3], and the PWM duty cycle δ is

$$\delta = \frac{v_a - v_i[3]}{v_i[2] - v_i[3]}. \tag{11.34}$$

e. Do the same procedure as (d) for v_b and v_c.
f. Increase $\boldsymbol{\theta}_\mathbf{i}$ by $\boldsymbol{\omega}_i t$.
g. Add $\boldsymbol{\theta}_O$ with $\boldsymbol{\omega}_O T$; if $\boldsymbol{\theta}_O$ is greater than 2π, then subtract 2π from $\boldsymbol{\theta}_O$.
h. Wait for the next PWM cycle, and do (a)–(g) again.

In the meantime, if the falling edge of the signal v_{com} appears, set variable $\boldsymbol{\theta}_\mathbf{i}$ to zero.

The algorithm can be implemented easily by a microprocessor.

11.4.1.3 Improve Voltage Ratio

Assume that the input AC supply phase voltage is

$$\begin{aligned} v_A(t) &= V_m \sin(\omega_i t), \\ v_B(t) &= V_m \sin\left(\frac{\omega_i t - 2\pi}{3}\right), \\ v_C(t) &= V_m \sin\left(\frac{\omega_i t - 4\pi}{3}\right). \end{aligned} \tag{11.35}$$

The output phase voltage is

$$\begin{aligned} v_a(t) &= qV_m \sin(\omega_O t), \\ v_b(t) &= qV_m \sin\left(\frac{\omega_O t - 2\pi}{3}\right), \\ v_c(t) &= qV_m \sin\left(\frac{\omega_O t - 4\pi}{3}\right), \end{aligned} \tag{11.36}$$

where q is the voltage ratio of the output voltage (voltage transfer gain, usually $q < 1$). The direct phase voltage modulation with Equation 11.36 has a maximum voltage ratio of 50%, as illustrated in Figure 11.41.

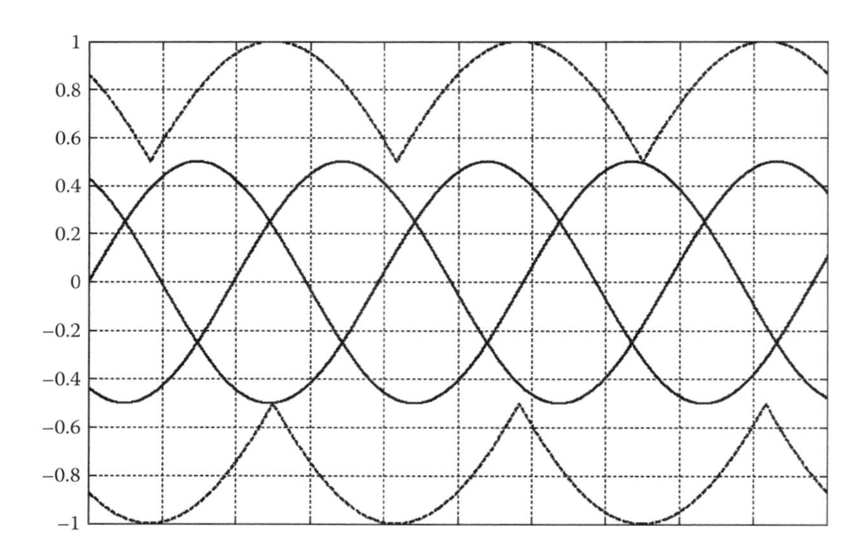

FIGURE 11.41 Illustration of the maximum voltage ratio of 50%.

An improvement of the voltage ratio to $\sqrt{3}/2$ (or 87%) is possible [34,35] by adding common-mode voltages to the target outputs as

$$\begin{cases} v_a(t) = qV_m\left[\sin(\omega_O t) + \dfrac{\sin(3\omega_O t)}{6} - \dfrac{\sin(3\omega_i t)}{2\sqrt{3}}\right], \\ v_b(t) = qV_m\left[\sin\left(\dfrac{\omega_O t - 2\pi}{3}\right) + \dfrac{\sin(3\omega_O t)}{6} - \dfrac{\sin(3\omega_i t)}{2\sqrt{3}}\right], \\ v_c(t) = qV_m\left[\sin\left(\dfrac{\omega_O t - 4\pi}{3}\right) + \dfrac{\sin(3\omega_O t)}{6} - \dfrac{\sin(3\omega_i t)}{2\sqrt{3}}\right]. \end{cases} \tag{11.37}$$

The common-mode voltages have no effect on the output line-to-line voltages, but allow the target outputs to fit within the input voltage envelope with a value of up to 87%, as illustrated in Figure 11.42.

The improvement of the voltage ratio is achieved by redistributing the null output states of the converter (all output lines connected to the same input line) and is analogous to the similar well-established technique in conventional DC-link PWM converters. It should be noted that a voltage ratio of 87% is the intrinsic maximum for any modulation method. Venturini provides a rigorous proof of this fact in references [34,35].

To increase the voltage ratio, Equation11.32 in algorithm 1 should be changed to

$$\begin{aligned} v_{CO} &= \frac{\sin(3\theta_O)/6 - \sin(3\theta_i)}{2\sqrt{3}}, \\ v_a &= V_O[\sin(\theta_O + \phi_0) + v_{CO}], \\ v_b &= V_O\left[\sin\left(\frac{\theta_O - 2\pi}{3 + \phi_0}\right) + v_{CO}\right], \\ v_c &= V_O\left[\sin\left(\frac{\theta_O - 4\pi}{3 + \phi_0}\right) + v_{CO}\right]. \end{aligned} \tag{11.38}$$

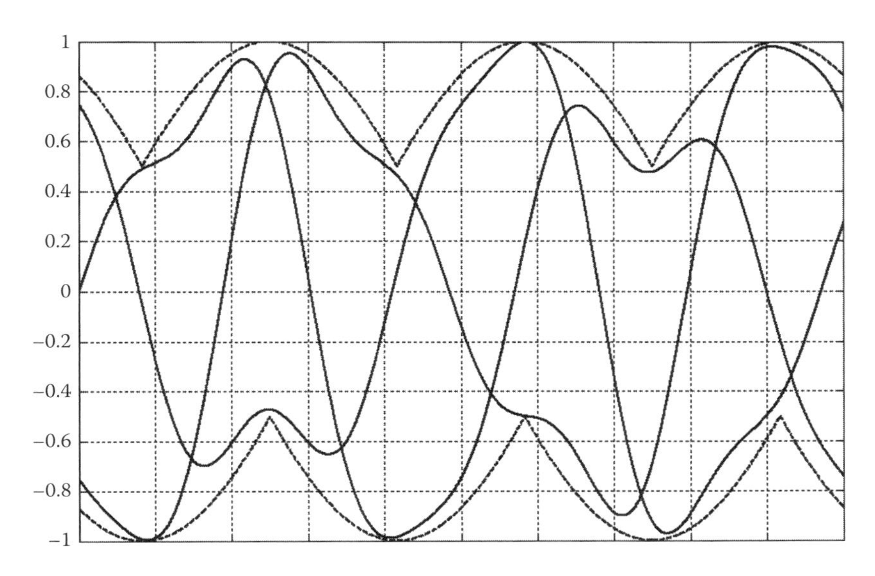

FIGURE 11.42 Illustration of the maximum voltage ratio of 87%.

11.4.2 24-Switch Matrix Converter

From Figure 11.40, it can be seen that if the reference voltage is greater than zero, the output is modulated between two positive phases or a neutral point and one positive phase, and vice versa; then the THD can be further reduced. One example of the modulation is shown in Figure 11.43.

The modulation can be accomplished by a 12-switch-cell (24 switches) matrix converter, as shown in Figure 11.44. The structure is similar to Figure 11.38 except for adding phase

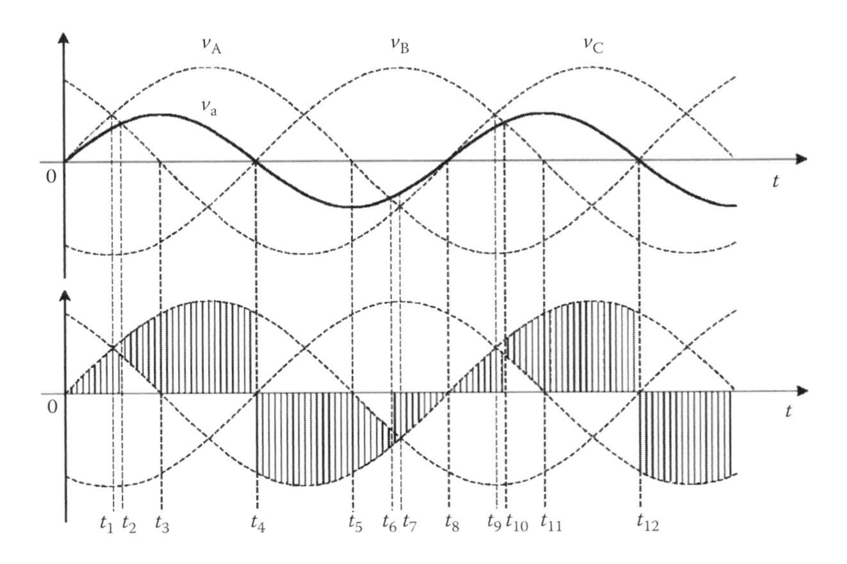

FIGURE 11.43 One SEM example for a 12 switch cells matrix converter.

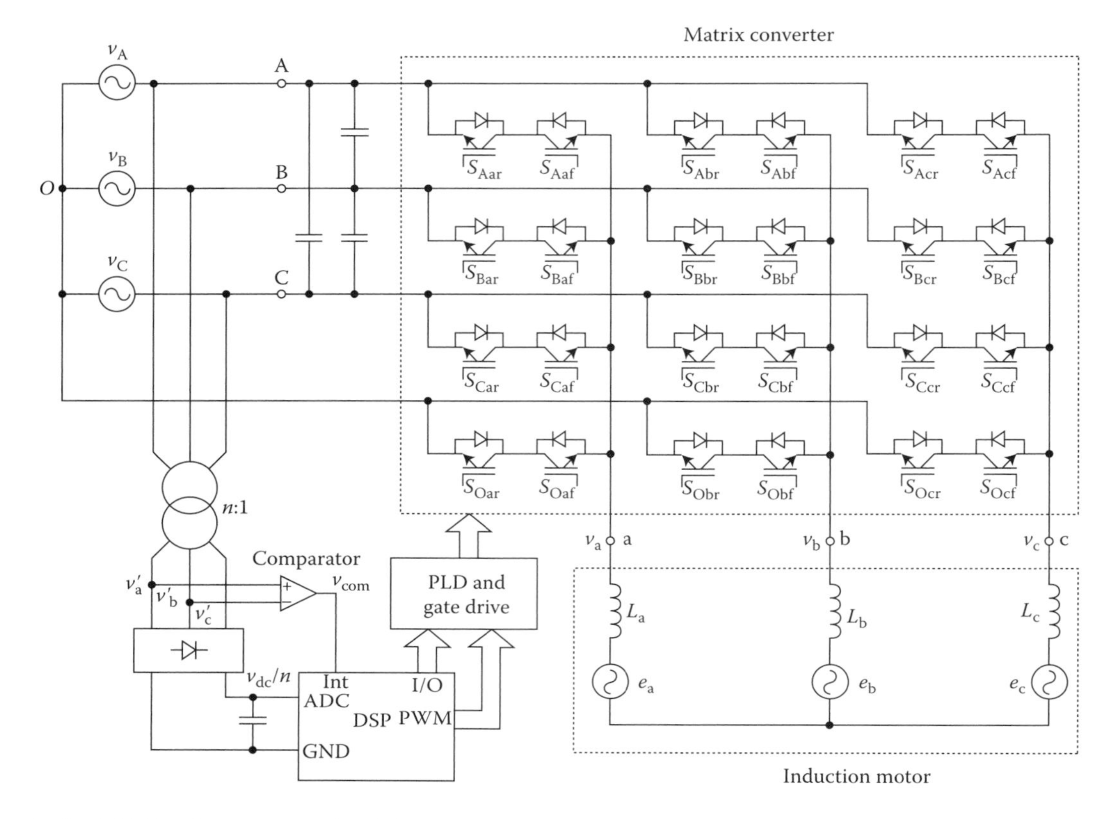

FIGURE 11.44 Structure of a 24-switches matrix converter.

lag from all the outputs to the neutral point. The added switches are S_{Ojf} and S_{Ojr}, where j = {a,b,c} is the output phase. The modulation rule of the example is shown in Table 11.2.

The modulation algorithm is similar to algorithm 1 except for procedure (d).

ALGORITHM 2

Define the variables $\boldsymbol{\theta}_i$ and $\boldsymbol{\theta}_O$, and their initial values are set to zero, that is,

$\boldsymbol{\theta}_i = 0, \boldsymbol{\theta}_O = 0$.

Define array $v_i[3]$ to store the input phase voltage.

TABLE 11.2

Modulation Rule of a 24-Switches Matrix Converter

Time		$\sim t_1$	$t_1 \sim t_2$	$t_2 \sim t_3$	$t_3 \sim t_4$	$t_4 \sim t_5$	$t_5 \sim t_6$	$t_6 \sim t_7$
Modulation	v_{high}	v_A	v_C	v_A	v_A	0	v_A	0
Phase	v_{low}	0	0	v_C	0	v_C	v_C	v_A
			$t_7 \sim t_8$	$t_8 \sim t_9$	$t_9 \sim t_{10}$	$t_{10} \sim t_{11}$	$t_{11} \sim t_{12}$	$t_{12} \sim$
			0	v_C	v_B	v_C	v_C	0
			v_C	0	0	v_B	0	v_B

Define array ***order***[3] to map v_i to the input phase, and initialize the array with {1,2,3}, representing {A,B,C}.

Define variables $\mathbf{v_{high}}$ and $\mathbf{v_{low}}$ to store the voltages that have been modulated.

Do nothing until the first falling edge of the signal v_{com} appears. When the first falling edge appears, for every PWM cycle T, do the following loop:

(a), (b), and (c) are the same as that of Algorithm 1.

(d) For the value of v_a, do the following:

i. If $v_a \geq v_i[2]$, it means that v_a is between $\mathbf{v}_i[1]$ and $\mathbf{v}_i[2]$; then store the $\mathbf{v}_i[1]$ to $\mathbf{v}_{high}$, and store $\mathbf{v}_i[2]$ to $\mathbf{v}_{low}$.

ii. Else it means that v_a is between $\mathbf{v}_i[2]$ and $\mathbf{v}_i[3]$; then store the $\mathbf{v}_i[2]$ to $\mathbf{v}_{high}$, and store $\mathbf{v}_i[3]$ to $\mathbf{v}_{low}$.

iii. If $v_a > 0$ and $\mathbf{v_{low}} < 0$, then store $\mathbf{v_{low}}$ with zero. It means that v_a is modulated between the neutral point and the lower positive phase that is higher than v_a.

vii. Else do nothing. It means that v_a is modulated between two positive phases.

iv. If $v_a < 0$ and $\mathbf{v}_{high} > 0$, then store $\mathbf{v}_{high}$ with zero. It means that v_a is modulated between the neutral point and one higher negative phase that is lower than v_a.

ix. Else do nothing. It means that v_a is modulated between two negative phases.

v. Calculate the PWM conduction duty cycle δ:

$$\delta = \frac{v_a - v_{low}}{v_{high} - v_{low}}. \tag{11.39}$$

(e), (f), (g), and (h) are also the same as that of Algorithm 1.

11.4.3 Current Commutation

We need to investigate current commutation between input phases.

11.4.3.1 Current Commutation between Two Input Phases

The current commutation must obey two rules: avoid two input phases being in short circuit and avoid any output phase being in open circuit. Relative voltage magnitude-based commutation [30–32] will be introduced to this system. For one output phase, it is always modulated between two input phases (the neutral point is also considered as one phase, so total four phases). When the PWM signal is high, it is connected to the smallest input phase (v_{high}), which is higher than the reference voltage. When the PWM signal is low, it is connected to the largest input phase (v_{low}), which is lower than the reference voltage. For convenience, assume that the output phase is a, v_{high} is phase A, and v_{low} is phase B; then the gate signal for one PWM cycle is shown in Figure 11.45a. All switches related to the commutation are also shown in Figure 11.45b. Switching state transition is also illustrated in Figure 11.45c.

Commutation details are given below:

- When the PWM signal is high, switches S_{Aaf}, S_{Aar}, and S_{Baf} are all turned on. As $v_A > v_B$ and switch S_{Bar} is turned off, there is no short circuit for two input phases. The output phase current i_a always flows through phase A: when $i_a > 0$, i_a flows from phase A because of $v_A > v_B$; when $i_a < 0$, i_a flows to phase A because switch S_{Bar} is turned off.

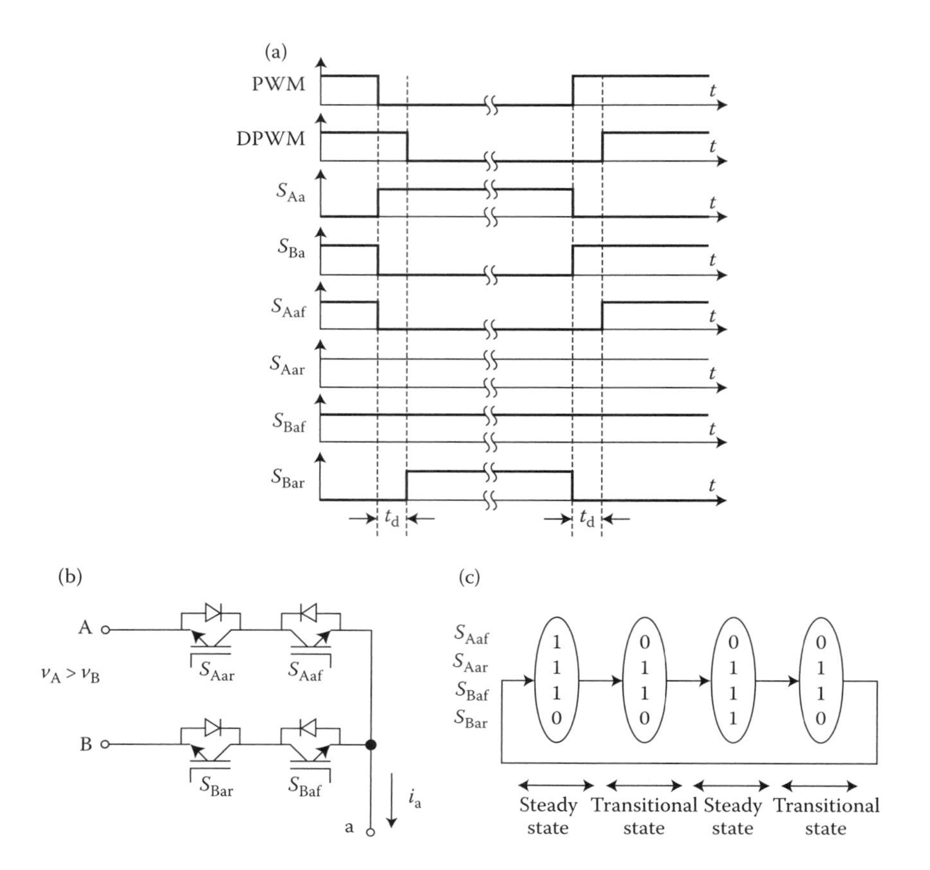

FIGURE 11.45 Gate signal of current commutation: (a) gate signal, (b) switches related to the commutation, and (c) switching state transition

- When the PWM signal is flopped down, switch S_{Aaf} is turned off immediately; the output current i_a transfers from phase A to phase B, and there is no open circuit in phase a. After a short delay t_d, switch S_{Bar} is turned on, and there is no short circuit in the input phases. The delay t_d between each switching event is determined by the device characteristics, such as turn-off time of a device.
- When the PWM signal is low, switches S_{Aar}, S_{Baf}, and S_{Bar} are all turned on. As $v_A > v_B$ and switch S_{Aaf} is turned off, there is no short circuit in the two input phases. Similarly, the output current i_a always flows through phase B.
- When the PWM signal is flipped up, switch S_{Bar} is turned off immediately; the output phase current i_a transfers from phase B to phase A, and there is no open circuit in phase a. After a short delay t_d, switch S_{Aaf} is turned on, and there is no short circuit in the input phases.

11.4.3.2 Current Commutation-Related Three Input Phases

Within a PWM cycle, that is, when the PWM signal is flipped up to a high value, current commutation takes place only between two adjacent input phases. However, when the PWM signal is flopped down to a low value, current commutation may take place among

TABLE 11.3

All Possible Switching State Transition

Condition	M > P > N				P > M > N				M > N > P				M > P > NP			
Mode	**Mode 1**				**Mode 2**				**Mode 3**				**Mode 4**			
State	0	1	2	3	0	1	2	3	0	1	2	3	0	1	2	3
S_{Mjf}	1	0	0		1	1			1	0	0	0	1	0	0	0
S_{Mjr}	1	1	1		1	1			1	1	1	0	1	1	1	0
S_{Njf}	1	1	0		1	0			1	1	0	0	1	1	1	1
S_{Njr}	0	0	0		0	0			0	0	1	1	0	0	0	1
S_{Pjf}	0	1	1		0	0			0	1	1	1	0	0	0	0
S_{Pjr}	0	0	1		0	1			0	0	0	1	0	0	1	1
Phase	MN → MP				MN → PM				MN → NP				MN → PN			

the other two of the three input phases and neutral O. For example, before t_4 in Figure 11.43, modulation takes place between phases A and O; after t_4, modulation takes place between phases O and C. We assumed that during the last PWM cycle, the modulation takes place between M and N (voltage of M > voltage of N). In the next PWM cycle, phase P will be related to the modulation. Then all the possible switching states among M, N, and P are shown in Table 11.3, where State 0 is the initial state of the switches.

Rewriting the switching state transition in Table 11.4, the functions of modes 1–4 are the same as those presented in Table 11.3; mode 5 is the current commutation that takes place only between two input phases, where State 0 is the initial state. State 4 is the destination state. From Table 11.4, it can be found that

- *For switch* S_{Mjf}: If the destination state is "1," then its state is kept unchanged; if the destination state is "0," then its state is always flopped to "0" immediately.
- *For switch* S_{Mjr}: If the destination state is "1," then its state is kept unchanged; if the destination state is "0," then its state is always flopped to "0" in step 4.
- *For switch* S_{Njf}: If the destination state is "1," then its state is kept unchanged; if the destination state is "0," then its state is always flopped to "0" in step 2.

TABLE 11.4

All Possible Switching State Transition

Condition	M > P > N					P > M > N					M > N > P					M > P > N					M > N				
Mode	**Mode 1**					**Mode 2**					**Mode 3**					**Mode 4**					**Mode 5**				
State	0	1	2	3	4	0	1	2	3	4	0	1	2	3	4	0	1	2	3	4	0	1	2	3	4
S_{Mjf}	1	0	0	0	0	1	1	1	1	1	1	0	0	0	0	1	0	0	0	0	1	0	0	0	0
S_{Mjr}	1	1	1	1	1	1	1	1	1	1	1	1	1	1	0	1	1	1	1	0	1	1	1	1	1
S_{Njf}	1	1	0	0	0	1	1	0	0	0	1	1	0	0	0	1	1	1	1	1	1	1	1	1	1
S_{Njr}	0	0	0	0	0	0	0	0	0	0	0	0	0	1	1	0	0	0	1	1	0	0	0	1	1
S_{Pjf}	0	1	1	1	1	0	0	0	0	0	0	1	1	1	1	0	0	0	0	0	0	0	0	0	0
S_{Pjr}	0	0	0	1	1	0	0	0	1	1	0	0	0	1	1	0	0	0	1	1	0	0	0	0	0
Phase	MN → MP					MN → PM					MN → NP					MN → PN					MN → MN				

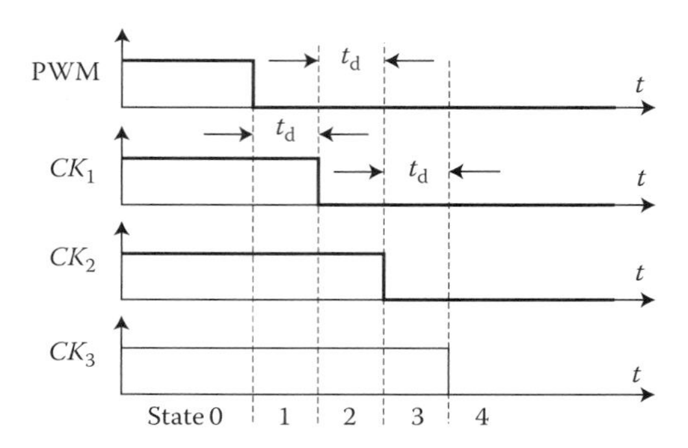

FIGURE 11.46 Signal of delayed PWM.

- *For switch* S_{Njr}: If the destination state is "0," then its state is kept unchanged; if the destination state is "1," then its state is always flipped to "1" in step 3.
- *For switch* S_{Pjf}: If the destination state is "0," then its state is kept unchanged, if the destination state is "1," then its state is always flipped to "1" immediately.
- *For switch* S_{Pjr}: If the destination state is "0," then its state is kept unchanged, if the destination state is "1," then its state is always flipped to "1" in step 3.

For all switches S_{Kjf} and S_{Kjr}, the general switching state transition true table can be summarized as Table 11.4. Q_{fn} is the initial state of the switches S_{Kjf}, Q_{rn} is the initial state of the switches S_{Kjr}, Q_{fn+1} is the destination state of the switches S_{Kjf}, and Q_{rn+1} is the destination state of the switches S_{Kjr}.

The state transition can be implemented by a combinational logic circuit by adding three delayed PWM signals CK_1–CK_3, as shown in Figure 11.46. Then the logical equation of the switches S_{Kjf} and S_{Kjr} can be obtained as

$$S_{Kjf} = Q_{fn} \cdot Q_{fn+1} + Q_{fn} \cdot Q_{rn} \cdot \overline{Q}_{fn+1} \cdot \text{PWM} + Q_{fn} \cdot \overline{Q}_{rn} \cdot \overline{Q}_{fn+1} \cdot CK_1 + \overline{Q}_{fn} \cdot \overline{Q}_{rn} \cdot Q_{fn+1} \cdot \overline{PWM}, \tag{11.40}$$

$$S_{Kjr} = Q_{fn} \cdot Q_{rn} \cdot Q_{rn+1} + Q_{fn} \cdot Q_{rn} \cdot \overline{Q}_{rn+1} \cdot CK_3 + \overline{Q}_{rn} \cdot Q_{rn+1} \cdot CK_2. \tag{11.41}$$

11.4.4 Simulation and Experimental Results

The direct phase voltage modulation has a maximum voltage ratio of 50%. The simulation and experimental results are based on three-phase voltage modulation methods:

- *Modulation I*: Maximum-envelope modulation for a conventional 9-switch-cell matrix converter.
- *Modulation II*: The SEM method for a conventional 9-switch-cell matrix converter.
- *Modulation III*: The SEM method for a 12-switch-cell matrix converter.

11.4.4.1 Simulation Results

The simulation results of output phase voltage (only phase a is illustrated), output line–line voltage, input line current, and their FFT under three modulation methods are shown in Figures 11.47 through 11.49, respectively. The input phase voltage is 240 V (rms value, the same for the following), the frequency is 50 Hz; the output phase voltage is 120 V, the frequency is 100 Hz; and the switching frequency is 2 kHz. From the FFT figure, we can see that the harmonics with a 24-switch converter are reduced significantly.

With the improved ratio modulation, the voltage ratio can reach about 87%. The simulation results of the phase voltage, the line-line voltage, and the FFT of the line-line voltage with modulations I and III are shown in Figure 11.50e and f. The corresponding THD of these two waveforms are 63.4% and 20.8%, respectively.

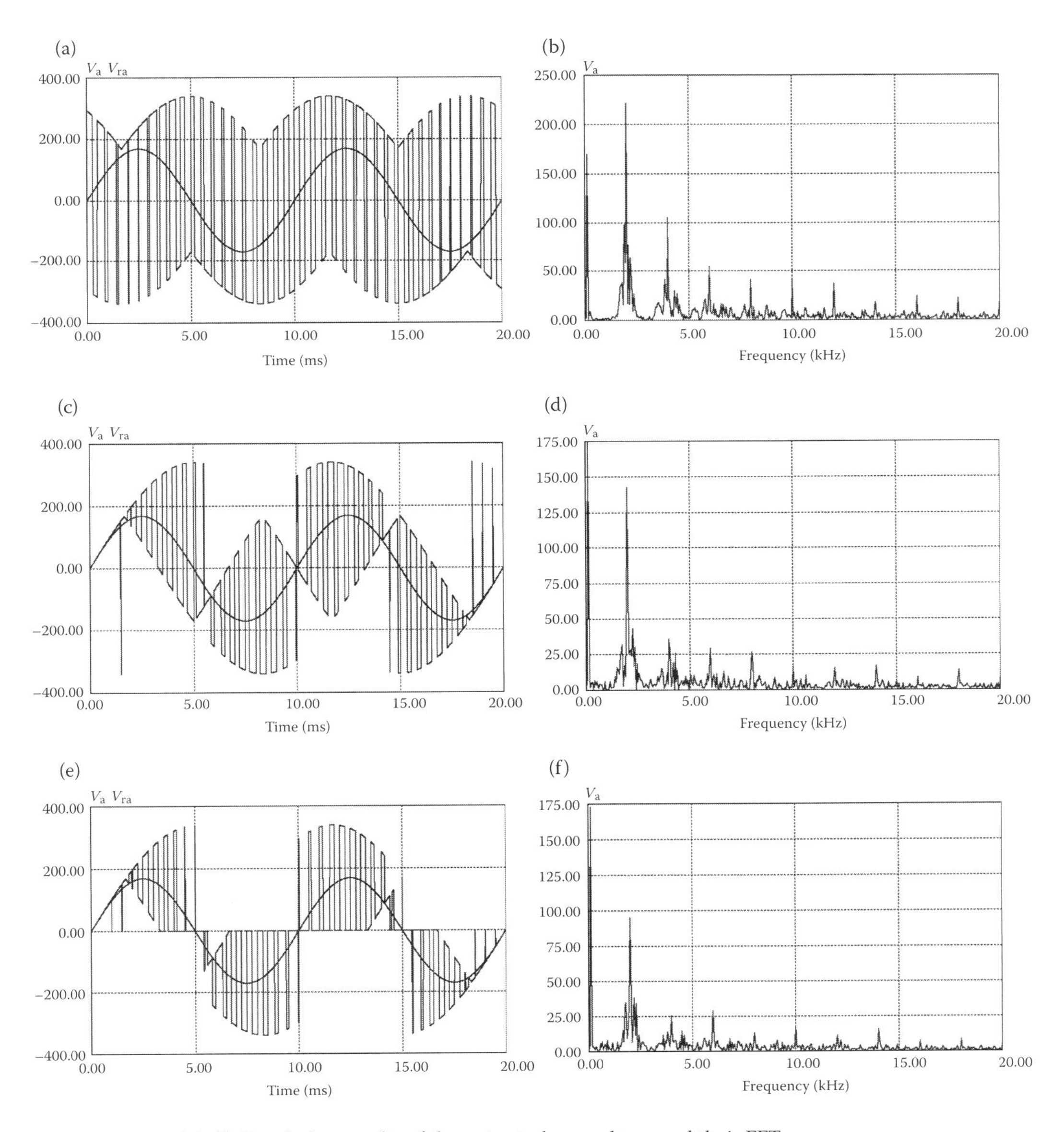

FIGURE 11.47 (a)–(f) Simulation results of the output phase voltage and their FFT.

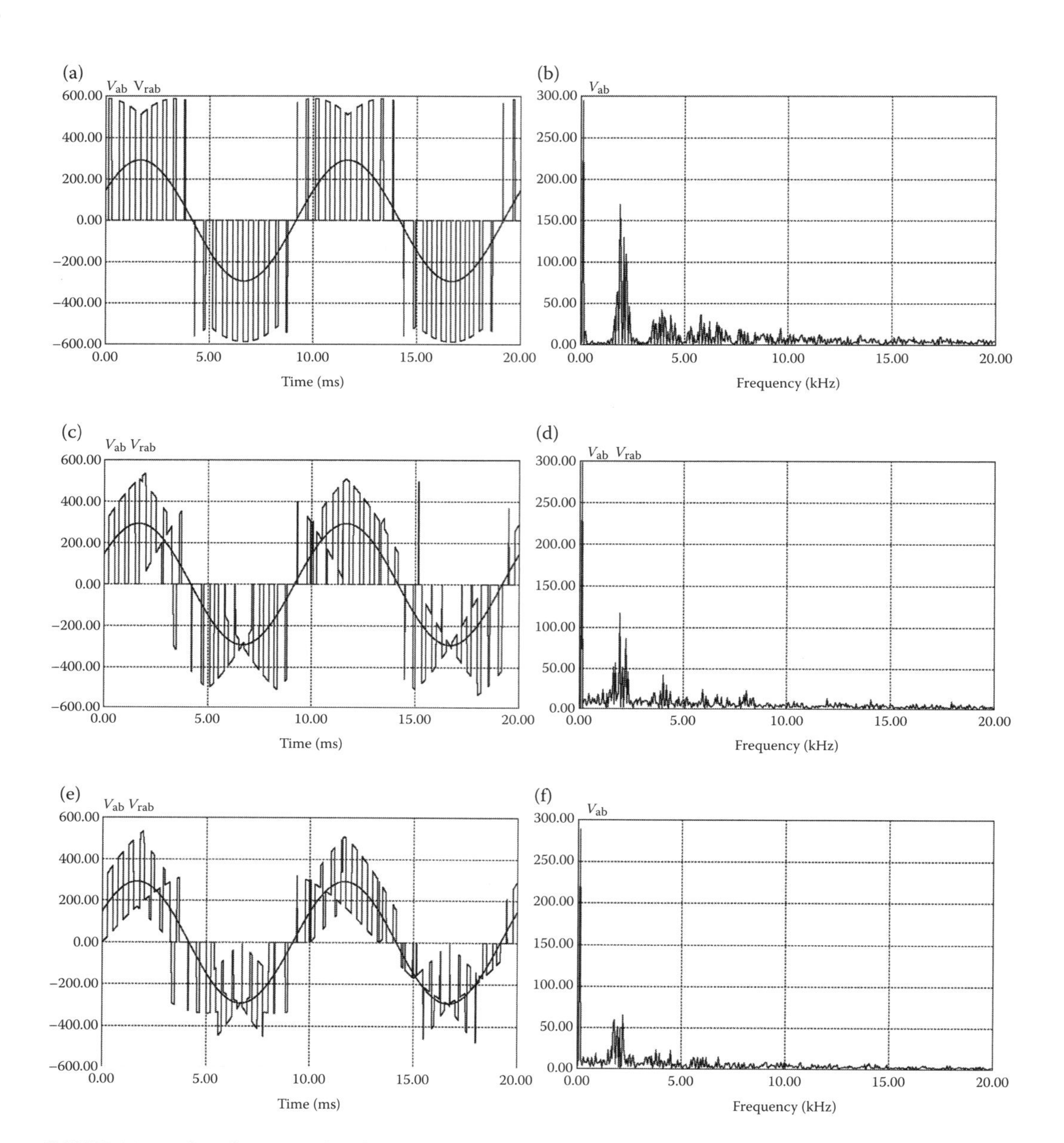

FIGURE 11.48 Simulation results of the output line–line voltage and their FFT: (a) output line-line voltage with modulation I, (b) FFT of figure a, (c) output line-line voltage with modulation II, (d) FFT of figure c (e) output line-line voltage modulation II, and (f) FFT of figure e.

11.4.4.2 Experimental Results

In order to verify the feasibility of the proposed scheme, a 12-switch-cell matrix converter is built up. The modulation algorithm is implemented by a DSP TMS320F2407, which is specially designed for power electronics and electric drive. The digital-signal-processor (DSP) comprises a dual 10-bit 16-channel analog-to-digital convertor (ADC), a PWM generator, a digital I/O, and other modules. So the DSP can measure input voltages, generate

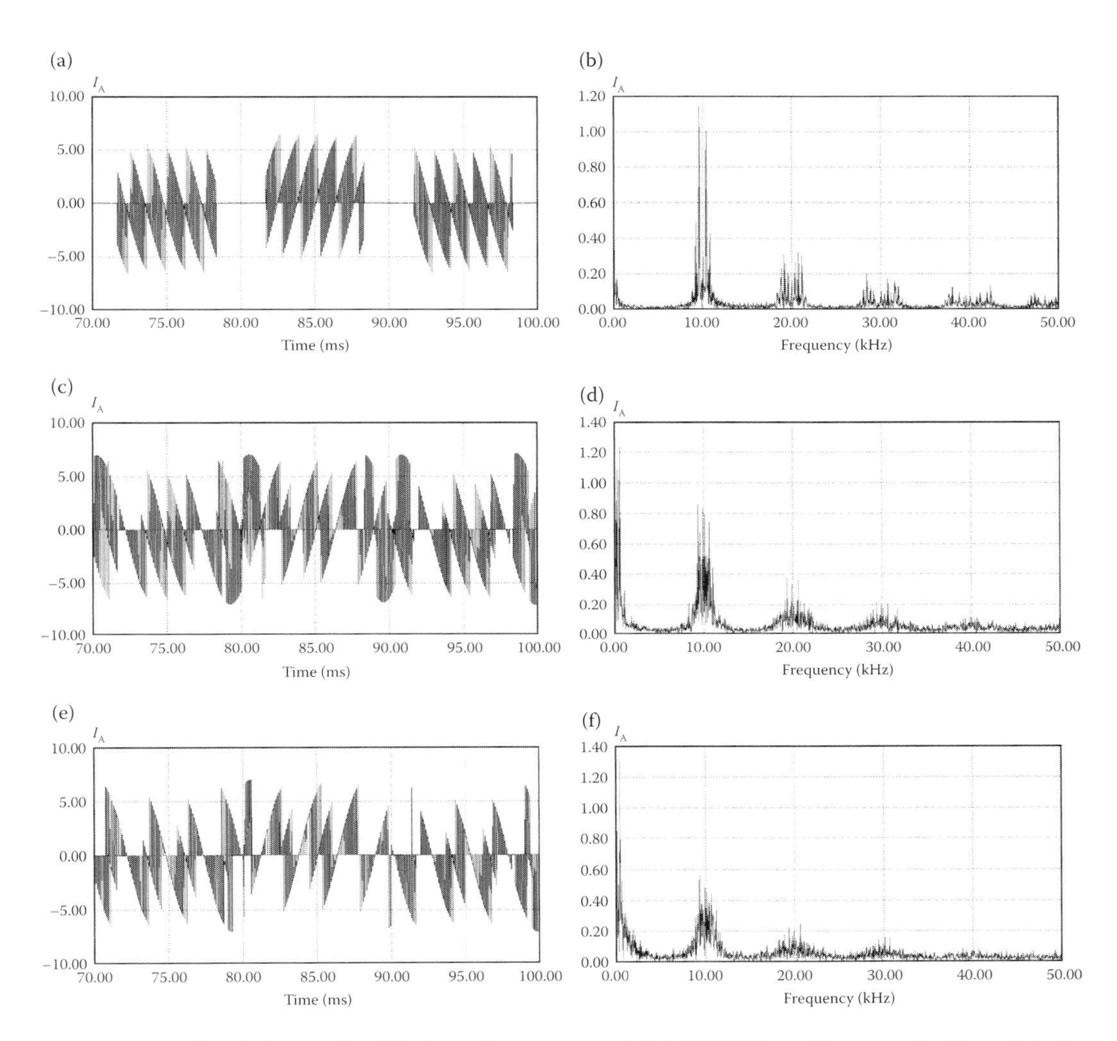

FIGURE 11.49 Simulation results of the input line current and their FFT: (a) input line current with modulation I, (b) FFT of figure a, (c) input line current with modulation II, (d) FFT of figure c, (e) input line current with modulation III, and (f) FFT of figure e.

three required PWM signals for three output phases, and indicate the peripheral circuit whose phases are to be modulated. The peripheral circuit is to generate a gate signal for the matrix converter (including current commutation), which is built up by GAL PLD, logic gate, monostable flip-flops, and so on. 15A/1200 V 1MBH15-120 IGBT is adopted as main switches. A photo-coupled gate driver, TLP250, is used to implement the gate driver circuit. This gate driver provides a peak output current of 1.5 A. It also isolates the input signal from the output, and thus common-mode noise is reduced. An IGBT needs +15 to +20 V voltage to turn on and −5 to −10 V to turn off. To get the required voltage with single DC power supply, a +7 V stabilivolte tube (Zener diode) is used to get the required negative turnoff voltage. Thus the driver circuit can provide +17 V for turnon and −7 V for turnoff with single +24 V output DC power supply. The gate driver circuit for one IGBT is shown in Figure 11.51. The switches with emitter connected together can use the same gate driver

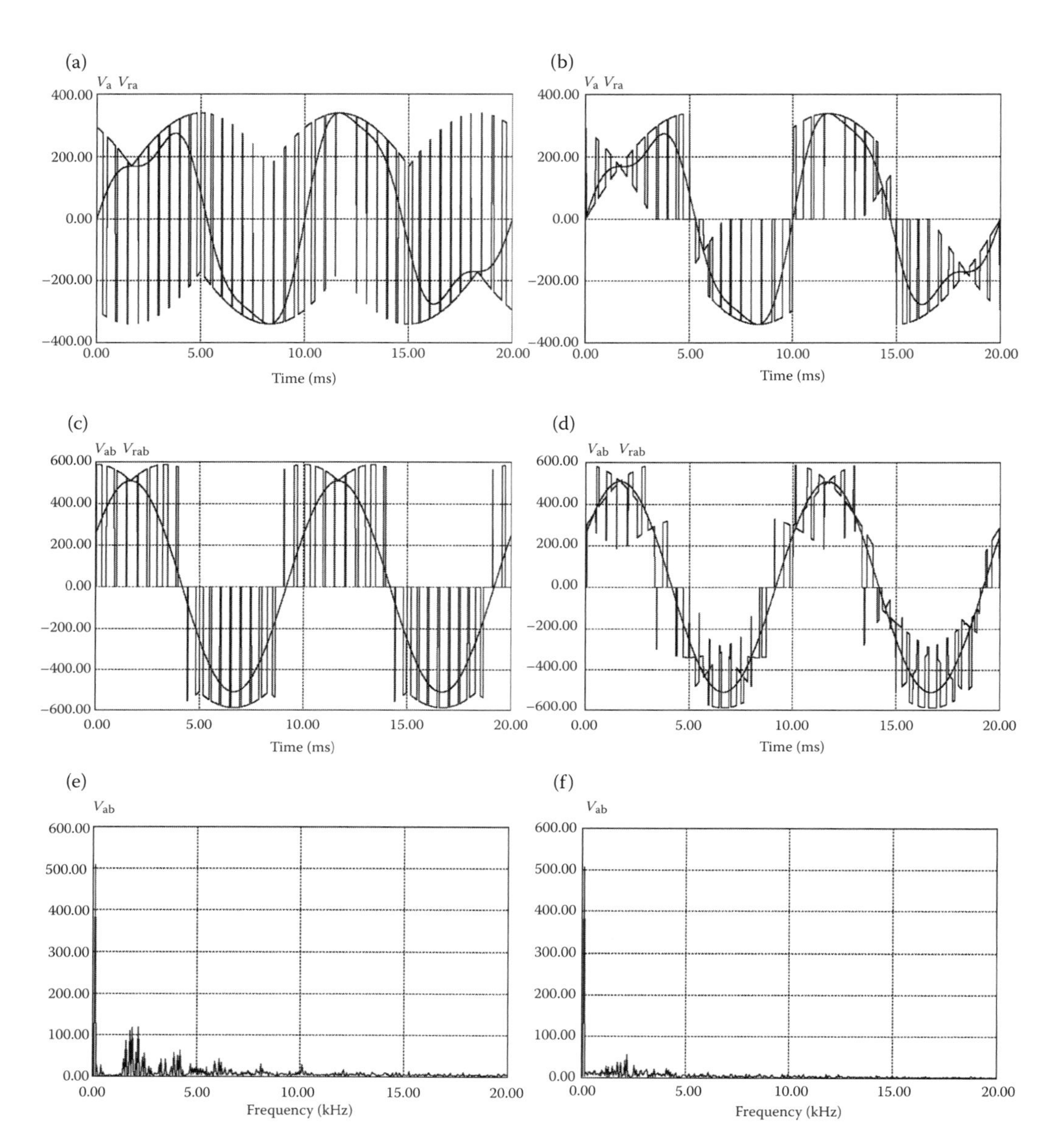

FIGURE 11.50 Simulations results with enhanced ratio modulation: (a) phase voltage of with modulation I, (b) phase voltage of with modulation III, (c) line-line voltage of with modulation I, (d) line-line voltage of with modulation III, (e) FFT of figure c (THD: 63.4%), and (f) FFT of figure d (THD: 20.8%).

output DC power supply. For a 12-switch-cell matrix converter, seven insulated DC power supplies are required.

The phase voltage of the AC power supply is 240 V, the line–line voltage is around 415 V, and the frequency f_i is 50 Hz. The switching frequency f is 10 kHz. Modulation methods I–III can be implemented by the same hardware only by changing the software of the DSP. A 2.2 kW three-phase induction motor is connected to the output of the matrix converter as load. Experimental waveforms of the output line–line voltage, the phase current

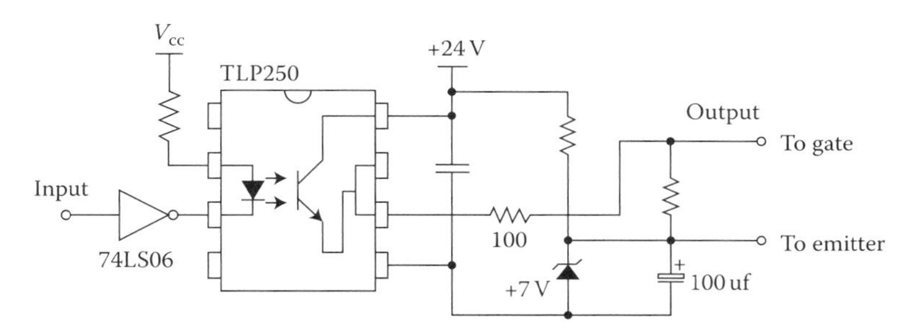

FIGURE 11.51 Gate driver circuit for one IGBT.

of the induction motor, and The FFT of line–line voltage under modulation methods I–III are shown in Figure 11.52. All the voltage signals are measured by a differential probe with a gain of 20, the voltage scale is 400 V/div, the current scale is 8 A/div, the magnitude scale of the FFT waveform is 20 dB/div, and the frequency scale is 5 kHz/div. Figure 11.53a is with modulation I, the output frequency f_O is 130 Hz, the voltage ratio δ is $\sqrt{3}/2$, and the magnitude of the output line–line voltage is 360 V with THD = 62.7%. The experimental results implementing modulations II and III with the same frequency and magnitude are shown in Figures 11.52b and 11.52c, respectively. From the figure, we can see that the output under the SEM method has lower THD (28.3% and 17.2%) when compared with maximum-envelope modulation. If the method is applied to the 12-switch-cell

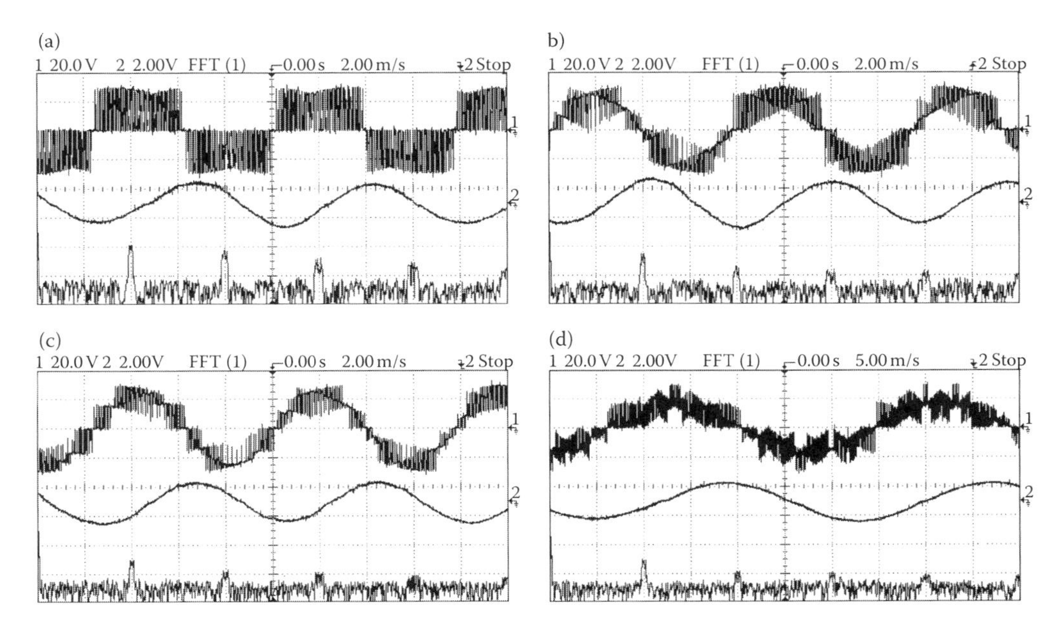

FIGURE 11.52 Experimental waveforms of the output line–line voltage (top, 400 V/div), the phase current (middle, 8 A/div), and the FFT of line–line voltage (bottom, 20 dB/div, 5 kHz/div) under various modulation methods: (a) Modulation method I (output frequency: 130 Hz, line-line voltage 360 V with THD = 62.7%), (b) Modulation method II (output frequency: 130 Hz, line-line voltage 360 V with THD = 28.3%), (c) Modulation method III (output frequency: 130 Hz, line-line voltage 360 V with THD = 17.2%), and (d) Modulation method III (output frequency: 35 Hz, line-line voltage 252 V with THD = 15.3%).

matrix converter, the THD can be further reduced. Figure 11.52d is also with modulation III, the output frequency is 35 Hz, and the magnitude of the line–line voltage is 252 V with THD = 15.3%.

Homework

11.1. A boost-type DC-modulated AC/AC converter is shown in Figure 11.11 having $L = 0.25$ mH and $C = 0.4\,\mu$F; the load $R = 100\,\Omega$; the input voltage and current are v_S and i_S, respectively; the output voltage and current are v_O and i_O, respectively; there are no power losses, that is, $\eta = 1$; the switching frequency is f_m (the switching period $T_m = 1/f_m$); and the conduction duty cycle is k. Calculate the transfer function and its step-response.

11.2. A boost-type DC-modulated AC/AC converter is shown in Figure 11.11 having the following components: the input rms voltage $v_S = 240$ V and a load with $R = 100\,\Omega$. In order to adjust the light, the output rms voltage v_O varies in the range of 500–1000 V. Calculate the range of the conduction duty cycle k, and the output current and power.

11.3. A DC-modulated two-stage boost-type AC/AC converter is shown in Figure 11.25 having the following components: the input rms voltage $v_S = 240$ V and a dimmer load with $R = 100\,\Omega$. In order to obtain the high output rms voltage, v_O varies in the range of 1000–3600 V. Calculate the range of the conduction duty cycle k and the output current and power.

References

1. Luo, F. L. 2009. Switched-capacitorized DC/DC converters. *Proceedings of the IEEE-ICIEA 2009*, pp. 377–382.
2. Shen, M. and Qian, Z. 2002. A novel high-efficiency single-stage PFC converter with reduced voltage stress. *IEEE Transactions on Industry Application*, 38, 507–513.
3. Qiu, M., Moschopoulos, G., Pinheiro, H., and Jain, P. 1999. Analysis and design of a single stage power factor corrected full-bridge converter. *Proceedings of the IEEE APEC*, pp. 119–125.
4. Qiao, C. and Smedley, K. M. 2001. A topology survey of single-stage power factor correction with a boost type input-current-shaper. *IEEE Transactions on Power Electronics*, 16, 360–368.
5. Rashid, M. H. 2001. *Power Electronics Handbook*. San Diego: Academic Press.
6. Mohan, N., Undeland, T. M., and Robbins, W. P. 2003. *Power Electronics* (3rd edition). New York: Wiley.
7. Cheng, K. W. E. 2003. Storage energy for classical switched mode power converters. *Proceedings of IEE-EPA*, pp. 439–446.
8. Luo, F. L. and Ye, H. 2004. *Advanced DC/DC Converters*. Boca Raton: CRC Press LLC.
9. Luo, F. L. and Ye, H. 2005. Energy factor and mathematical modeling for power DC/DC converters. *IEE-Proceedings on EPA*, vol. 152, pp. 191–198.
10. Luo, F. L., Ye, H., and Rashid, M. H. 2005. *Digital Power Electronics and Applications*. Boston: Academic Press, Elsevier.
11. Luo, F. L. and Ye, H. 2007. Research on DC-modulated power factor correction AC/AC converters. *Proceedings of IEEE IECON 2007*, pp. 1478–1484.

12. Luo, F. L. and Ye, H. 2007. DC-modulated single-stage power factor correction AC/AC converters. *Proceedings of IEEE ICIEA 2007*, pp. 1477–1483.
13. Luo, F. L. and Ye, H. 2006. DC-modulated power factor correction on AC/AC Luo-converter. *Proceedings of ICARCV 2006*, pp. 1791–1796.
14. Luo, F. L. and Ye, H. 2006. DC-modulated single-stage power factor correction AC/AC converters (key notes). *Proceeding of the 10th CPESAM*, pp. 21–32.
15. Luo, F. L. and Ye, H. 2002. Positive output super-lift Luo-converters. *Proceedings of the IEEE International Conference PESC 2002*, pp. 425–430.
16. Luo, F. L. and Ye, H. 2003. Positive output super-lift converters. *IEEE Transactions on Power Electronics*, 18, 105–113.
17. Luo, F. L. and Ye, H. 2003. Negative output super-lift Luo-converters. *Proceedings of the IEEE International Conference PESC 2003*, pp. 1361–1366.
18. Luo, F. L. and Ye, H. 2003. Negative output super-lift converters. *IEEE Transactions on Power Electronics*, 1113–1121.
19. Luo, F. L. and Ye, H. 2004. Positive output cascade boost converters. *IEE-EPA Proceedings*, vol. 151, pp. 590–606.
20. Luo, F. L. and Pan, Z. Y. 2006. Sub-envelope modulation method to reduce total harmonic distortion of AC/AC matrix converters. *IEE-Proceedings on Electric Power Applications*, vol. 153, pp. 856–863.
21. Luo, F. L. and Pan, Z. Y. 2006. Sub-envelope modulation method to reduce total harmonic distortion of AC/AC matrix converters. *Proceeding of the IEEE Conference PESC 2006*, pp. 2260–2265.
22. Wheeler, P. W., Rodríguez, J., Clare, J. C., Empringham, L., and Weinstein, A. 2002. Matrix converters: A technology review. *IEEE Transactions on Industrial Electronics*, 38, 276–288.
23. Venturini, M. 1980. A new sine wave in sine wave out, conversion technique which eliminates reactive elements. *Proceedings of POWERCON 7*, pp. E3_1–E3_15.
24. Venturini, M. and Alesina, A. 1980. The generalized transformer: A new bidirectional sinusoidal waveform frequency converter with continuously adjustable input power factor. *Proceedings of IEEE PESC'80*, pp. 242–252.
25. Neft, C. L. and Schauder, C. D. 1992. Theory and design of a 30-HP matrix converter. *IEEE Transactions on Industrial Application*, 28, 546–551.
26. Wheeler, P. and Grant, D. 1997. Optimized input filter design and low loss switching techniques for a practical matrix converter. *Proceedings by Institution of Electrical Engineers*, pt. B, vol. 44, pp. 53–60.
27. Svensson, T. and Alakula, M. 1991. The modulation and control of a matrix converter synchronous machine drive. *Proceedings of EPE'91*, pp. 469–476.
28. Empringham, L., Wheeler, P., and Clare, J. 1998. Intelligent commutation of matrix converter bi-directional switch cells using novel gate drive techniques. *Proceedings of IEEE PESC'98*, pp. 707–713.
29. Ziegler, M. and Hofmann, W. 1998. Semi natural two steps commutation strategy for matrix converters. *Proceedings of IEEE PESC'98*, pp. 727–731.
30. Casadei, D., Serra, G., Tani, A., and Zarri, L. 2002. Matrix converter modulation strategies: A new general approach based on space-vector representation of the switch state. *IEEE Transactions on Industrial Electronics*, 49, 370–381.
31. Kwon, B. H., Min, B. H., and Kim, J. H. 1998. Novel commutation technique of AC–AC converters. *Proceedings of Inst. Elect. Eng.*, pt. B, pp. 295–300.
32. Pan, C. T., Chen, T. C., and Shieh, J. J. 1993. A zero switching loss matrix converter. *Proceedings of IEEE PESC'93*, pp. 545–550.
33. Villaça, M. V. M. and Perin, A. J. 1995. A soft switched direct frequency changer. *Conf. Rec. IEEE-IAS Annual Meeting*, pp. 2321–2326.
34. Alesina, A. and Venturini, M. 1988. Intrinsic amplitude limits and optimum design of 9-switches direct PWM AC–AC converters. *Proceedings of IEEE PESC'88*, pp. 1284–1291.
35. Alesina, A. and Venturini, M. G. B. 1989. Analysis and design of optimum amplitude nine-switch direct AC–AC converters. *IEEE Transactions on Power Electronics*, 4, 101–112.

12

AC/DC/AC and DC/AC/DC Converters

AC/DC/AC and DC/AC/DC conversion technologies are a special subject area in research and industrial applications. AC/DC/AC converters are usually applied in synchronous and asynchronous AC motor ASDs. In recent years, they have also been widely used in renewable energy systems, especially in wind turbine energy systems. DC/AC/DC converters are usually applied in high-voltage equipment to isolate the source side and the load side. They are also adopted in medium and small power systems such as solar panels, photovoltaic cells, and fuel cell energy systems.

12.1 Introduction

Renewable energy sources have been a hot topic in recent years. Most renewable energy sources are DGs. These sources, such as fuel cells, solar panels, photovoltaic cells, and wind turbines, are not standard general sources with stable output voltage and frequency. Some renewable energy sources are AC voltage sources like wind turbines. The output AC voltage (single-phase or three-phase) of a wind turbine depends on wind speed and other factors. Definitely, its output AC voltage amplitude is unstable, and the output frequency and phase are also unstable. Consequently, direct use of this output AC voltage of a wind turbine is inconvenient.

The necessity of using an AC/DC/AC converter is given below:

- The AC source voltage is unstable
- The AC source frequency is unstable
- The AC source phase number does not match the load requirement.

Some renewable energy sources are DC voltage sources, for example, fuel cells and solar panels. The output DC voltage of a solar panel depends on the weather, temperature, and sunlight. Definitely, its output DC voltage and power are unstable. Consequently, direct use of this output DC voltage is inconvenient. Normalized DC/DC converters are restricted by the power limitation. DC/AC/DC converters can transfer large power and they are better than normal DC/DC converters.

The necessity of using a DC/AC/DC converter is given below:

- The DC source voltage is unstable
- The DC source power is unstable
- The DC source impedance does not match the requirement.

12.2 AC/DC/AC Converters Used in Wind Turbine Systems

Wind turbines are one of the most promising energy sources, which have gained attention in recent decades and have penetrated utility systems deeply compared with other renewable sources [1–3]. Unfortunately, the output voltage and frequency of wind turbines are unstable since the wind speed is variable. These turbines are installed onshore or offshore, or sometimes as a wind farm where a large number of turbines are installed and connected together. A single wind-tower structure is shown in Figure 12.1.

German scientist Albert Betz proved that the wind turbine is most efficient when the wind slows down to $2v/3$ of its speed just before the rotor and decreases to $v/3$ after the rotor, before regaining its original speed v due to surrounding winds. Therefore, Betz's law states that the maximum power that can be extracted from wind is 59% of the total power available in the wind, ignoring mechanical and aerodynamic losses. Wind turbines transfer the linear moving wind energy into rotational energy by the function

$$P = 0.5\rho\pi R^2 v^3 Cp, \tag{12.1}$$

where P is power, ρ is air density, R is turbine radius, v is wind speed, and Cp is turbine power coefficient. Cp is a function only of the tip speed ratio λ where the variation of Cp with λ is given. The tip speed has to be maintained at an optimal value in order to extract maximum power.

Typically, wind turbines consist of three aerodynamically designed blades that are positioned in the horizontal axis, and the whole system is mounted on a tower. The rotational mechanical energy is converted to electrical energy by using a generator. In some cases, energy is transmitted through a gearbox to change the speed. Basically, wind turbines are controlled mechanically, by either pitch controlling or stall controlling. Pitch controlling is more complex when the wind speed is continuously measured and blades are adjusted accordingly in order to capture energy efficiently. Also, it would protect the turbine from high wind speeds. This control method is more efficient compared with stall control. The stall-controlled blades are fixed at a constant pitch angle, which is not changed during the operation. Stall is a simple aerodynamic effect that separates airflow from the aerofoil when

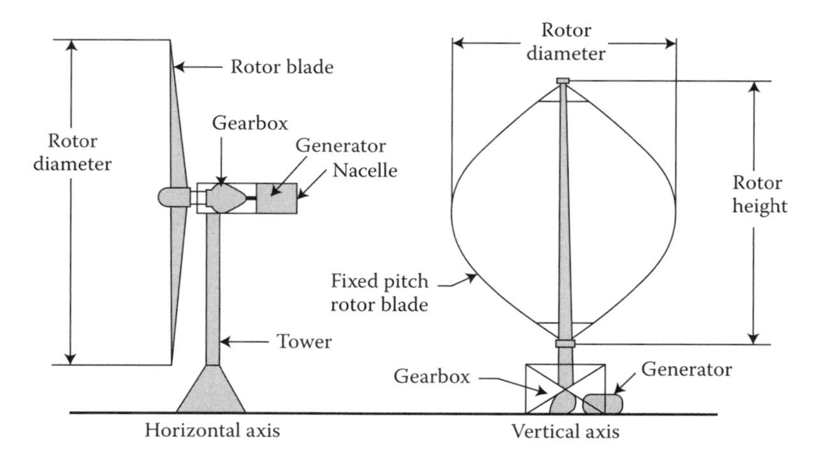

FIGURE 12.1 Wind turbine configuration.

the turbine runs at a constant speed and when the wind speed increases. This changes the angle of attack and limits the wind power captured, thereby protecting the turbine from high wind speeds. However, due to the randomness of wind availability and also, when these wind turbines are operated to capture maximum power, the operating voltage and frequency tend to vary, making the output unsuitable for grid connection demanding power conditioning before being consumed.

There are many generator topologies that are commonly used in wind turbines such as induction generators, synchronous generators, and permanent magnet synchronous generators. Some generators are connected directly to the grid while others use power electronic interfaces. Power electronic interfaces have to be selected depending on the generator used and the adopted controlling method. Generally, induction generators are used with fixed-speed wind turbines and power is limited mechanically with pitch or stall controlling. The other type is the variable-speed wind turbine, which controls the pitch and uses a power electronic interface at the output of the generator (which can be a synchronous generator, a permanent magnet synchronous generator, or a doubly fed induction generator). There are different power electronic converter topologies that are employed in interfacing these wind generators to overcome problems of variation in frequency and voltage. In the case of synchronous generators, full-rated power electronic converters are used. Usually, they can be AC to DC converters followed by inverters or simple rectifiers followed by DC to DC converters and then inverters. For induction generators, there are two possibilities; they can have AC to DC converters followed by DC to AC inverters in both the stator and rotor or only AC to DC inverters followed by DC to AC inverters connected in the rotor of induction generators. In summary, all these topologies use a combination of two or more power electronic converters, making the overall process inefficient and difficult to control, as identified in the first section of this chapter. This leaves space for the development of single-stage topologies in integrating wind power-generating systems.

The output AC voltage of a wind turbine can be single phase, three phase, or multiphase. Its output voltage and frequency are usually not stable. Some industrial applications require AC/AC converters to transfer an unstable AC energy source to a stable AC load. Most AC/AC converters are not suitable for these applications. We need to use AC/DC/AC converters to implement the work.

12.2.1 Review of Traditional AC/AC Converters

Traditional AC/AC converters have been introduced in Chapter 10. There are three methods to implement AC/AC conversion:

- VR converters
- Cycloconverters
- MCs

VR converters are usually used in applications with a stable input AC source, unchanged output frequency (fundamental harmonic frequency), and adjustable output voltage. These converters have the following advantages: simple structure, lower cost, and easy control. The drawbacks are poor PF, heavy distorted waveform (poor THD), low power transfer efficiency, and low voltage transfer gain.

Cycloconverters are usually used in applications with a stable input AC source and adjustable output voltage and output frequency (lower than half of the input frequency).

These converters have the following advantages: good PF, slight distorted waveform (good THD), and adjustable output voltage and frequency. The drawbacks are complex structure, complex control circuitry, higher cost, and low efficiency. The output voltage and frequency are lower than the input voltage and frequency.

MCs are usually used in applications with a stable input AC source. These converters have the following advantages: adjustable output voltage and frequency, simple structure, and lower cost. Although the maximum output voltage is lower than 0.866 times the input voltage, the output voltage can be easily adjustable. The output frequency can be either higher or lower than the input frequency. The drawbacks are poor PF, heavy distorted waveform (poor THD), bidirectional switches required, heavy network pollution, and complex control circuitry.

12.2.2 New AC/DC/AC Converters

AC/DC/AC converters can absorb the energy from a random input AC voltage source with unstable voltage and frequency to a fixed DC-link voltage (AC/DC conversion), and then convert the energy to a required AC output voltage with adjustable frequency and voltage (DC/AC conversion). The uncertainty of the input voltage from a random input AC voltage source has been dispelled by a controlled AC/DC converter, whose output DC voltage is DC-link voltage. The DC-link voltage is stable and can be used for DC/AC inverters. The typical application is ASD. In this subsection we introduce four circuits:

- AC/DC/AC boost-type converters
- Three-level AC/DC/AC converters
- Two-level AC/DC/AC ZSI
- Three-level diode-clamped AC/DC/AC ZSI.

12.2.2.1 AC/DC/AC Boost-Type Converters

An AC/DC/AC boost-type converter is shown in Figure 12.2. The input AC source may be a single-phase or three-phase energy source with unstable voltage and frequency. If it is a wind turbine, its voltage can vary in ±25% and its frequency can change in ±15%. The AC/DC converter has two blocks. The first block is an AC/DC rectifier, which can be a diode rectifier. Its output DC voltage is unstable with high efficiency, but independent of frequency. It may rarely be a controllable rectifier, and the output DC voltage can be slightly stable with lower efficiency, but independent of frequency as well. The second block is a

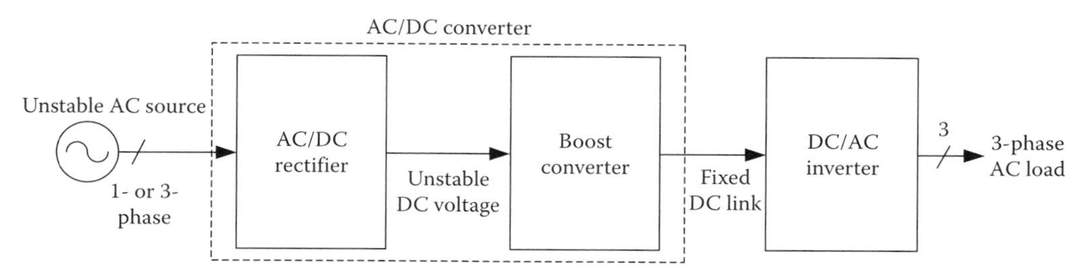

FIGURE 12.2 AC/DC/AC boost-type converter.

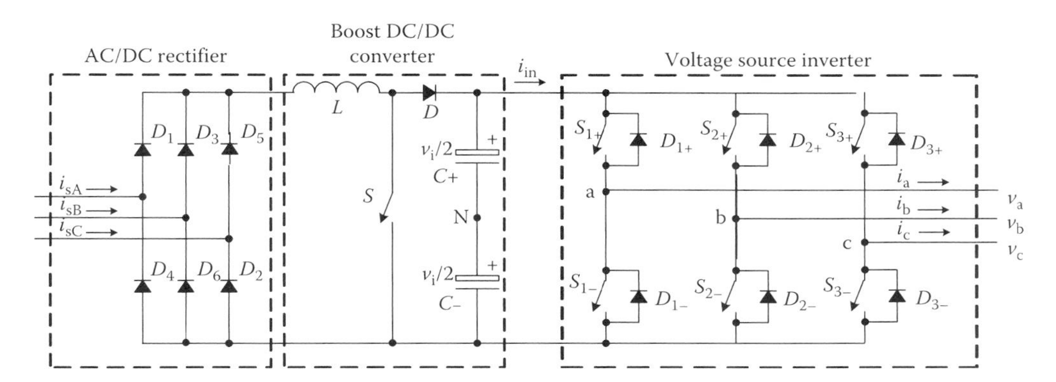

FIGURE 12.3 Circuit of an AC/DC/AC boost-type converter.

boost DC/DC converter. It can convert an unstable DC voltage to a fixed DC-link voltage, for example, 660 V DC.

The real-end block is a DC/AC inverter. Usually, it is a VSI with three-phase output AC voltage with 400 V/60 Hz.

A particular circuit diagram is shown in Figure 12.3.

Example 12.1

A wind turbine has three-phase output voltage 230 V ± 25% and frequency 60 Hz ± 15%; the power rate is 5 kW. The end user is a three-phase load with voltage 400V. Design an AC/DC/AC boost-type converter for this application.

Solution

Use a diode rectifier to rectify the input AC voltage to an unstable output DC voltage; the efficiency η can be 92–97%. The wind turbine three-phase output voltage is 230 V ± 25% independent of the frequency 60 Hz ± 15%. The rectified output DC voltage can be 311 V ± 25%, that is, 233–389V DC.

Use a boost DC/DC converter to convert the unstable 233–389V DC to a fixed 660V DC, that is, V_{in} is 233–389V and V_O is 660V DC. V_O is the fixed DC-link voltage. The corresponding duty cycle k can be set in the range of

$$k = \frac{V_O - V_{in}}{V_O} = \begin{cases} \dfrac{660 - 233}{660} = 0.647, \\ \dfrac{660 - 389}{660} = 0.410. \end{cases} \quad (12.2)$$

A VSI is selected for a DC/AC inverter. In the linear-operation region, the maximum output line-to-line peak voltage is 0.866 × DC-link voltage. Therefore, the maximum output line-to-line rms voltage is

$$V_{AC} = \frac{0.866 V_{link}}{\sqrt{2}} = \frac{0.866 \times 660}{\sqrt{2}} = 404\text{V}. \quad (12.3)$$

This output three-phase voltage is satisfactory.

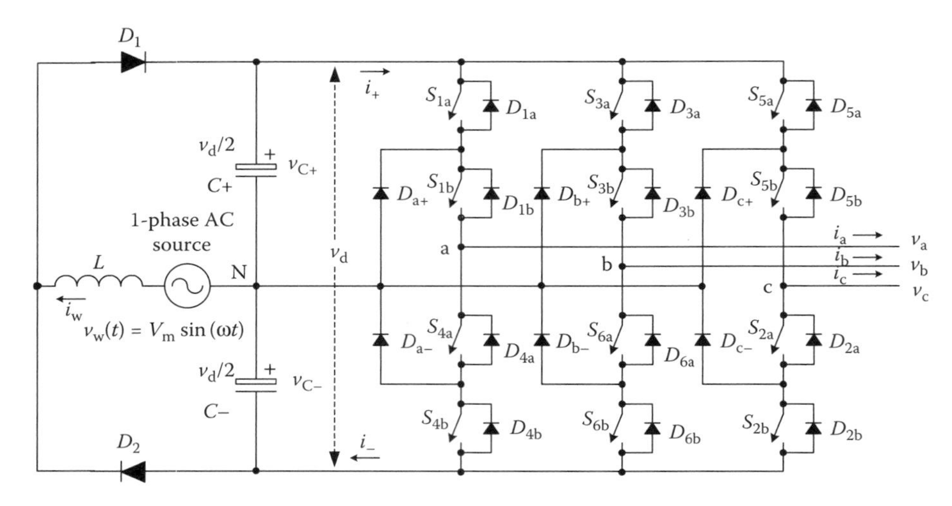

FIGURE 12.4 Three-level diode-clamped AC/DC/AC converter.

12.2.2.2 Three-Level Diode-Clamped AC/DC/AC Converter

A three-level AC/DC/AC converter is shown in Figure 12.4 [4]. The AC source can be a single-phase wind turbine generator. A single-phase half-wave AC/DC diode rectifier is used to obtain the DC-link voltage v_d. Two balanced capacitors $C+$ and $C-$ ($C+ = C- = C$) are charged to voltage $v_d/2$. A three-phase three-level diode-clamped voltage source DC/AC inverter converts the DC-link voltage to the load.

Usually, if the single-phase wind turbine output voltage has smaller voltage variation, for example, ±5–10%, and the applications are not so serious, we can directly link the AC/DC rectifier to the DC/AC diode-clamped inverter. Therefore, this is the simplest AC/DC/AC converter, but it works well and is easily controlled.

12.2.2.2.1 AC/DC Half-Wave Rectifier

The diode AC/DC rectifier has a source inductor L and two identical half-wave diode rectifiers. It converts the wind turbine voltage

$$v_W(t) = V_m \sin(\omega t) \tag{12.4}$$

to the DC-link voltage v_d. Since there is a source inductor and two capacitors, the DC-link voltage is

$$v_d \approx 0.9\, V_m. \tag{12.5}$$

The average voltage across both capacitors C_+ and C_- is half the DC-link voltage $v_d/2$. The differential coefficients of the source current and the capacitor voltages are

$$\begin{aligned}
\frac{di_W}{dt} &= \frac{v_W - sv_{C+} + (1-s)v_{C-}}{L},\\
\frac{dv_{C+}}{dt} &= \frac{si_W - i_+}{C},\\
\frac{dv_{C-}}{dt} &= \frac{-(1-s)i_W - i_-}{C},
\end{aligned} \tag{12.6}$$

where $s = 1$ for the positive half-cycle of wind turbine voltage and $s = 0$ for the negative half-cycle of wind turbine voltage.

12.2.2.2.2 Three-Level Diode-Clamped DC/AC Inverter

The three-level diode-clamped DC/AC inverter is shown in the right-hand part of Figure 12.4. There are two fast recovery diodes, four power switch devices, and four freewheeling diodes in each leg of the three-level inverter. There are eight (2^3) switching states in the traditional two-level inverter. However, there are 27 (3^3) switching states in the three-level inverter. The switching states of each phase of the three-level inverter are expressed as

$$v_{xN} = \begin{cases} \dfrac{v_d}{2}, & v_{ref,x} > v_{tri,1}, \\ 0, & v_{tri,1} > v_{ref,x} > v_{tri,2}, \\ \dfrac{v_d}{2}, & v_{tri,2} > v_{ref,x}, \end{cases} \tag{12.7}$$

where $x =$ a, b, or c, $v_{ref,x}$ is the phase x reference, $v_{tri,1}$ is the upper triangle pulse, and $v_{tri,2}$ is the lower triangle pulse. A space voltage vector $\overline{V}$ can be used to represent the output voltages of the three-phase inverter:

$$\overline{V} = \sqrt{\frac{2}{3}}\left(v_{aN} + v_{bN}e^{j120^\circ} + v_{cN}e^{j240^\circ}\right). \tag{12.8}$$

The space vector representation of the output voltages of the inverter in the two-axis coordinate system is shown in Figure 12.5. According to the magnitude of the voltage vectors, the possible switching states can be classified into four groups: large voltage vector $\left[|\overline{V}| = \sqrt{(2/3)v_d}\right]$ such as $(+--)$ $(++-)$ $(-+-)$ $(-++)$ $(--+)$ $(+-+)$, middle voltage vector $\left(|\overline{V}| = v_d/\sqrt{2}\right)$ such as $(+0-)$ $(0+-)$ $(-+0)$ $(-0+)$ $(0-+)$ $(+-0)$, small voltage vector $\left(|\overline{V}| = v_d/\sqrt{6}\right)$ such as $(+00)$ $(++0)$ $(0+0)$ $(00+)$ $(0++)$ $(+0+)$, and zero voltage vector $\left(|\overline{V}| = 0\right)$ such as $(+++)/(---)/(000)$. The DC-link capacitor

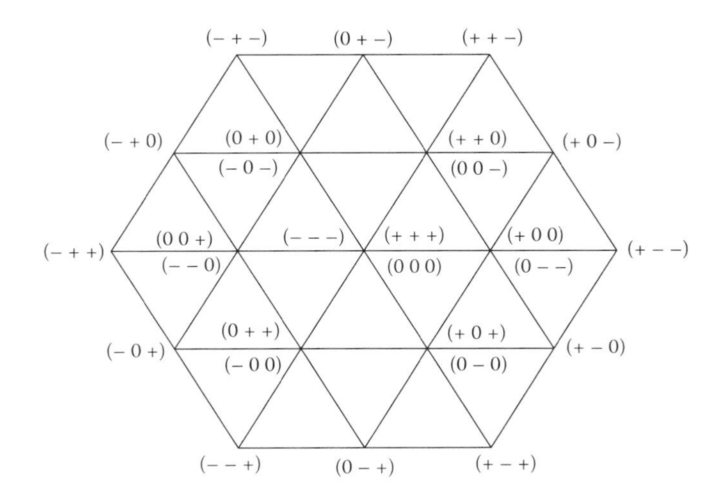

FIGURE 12.5 Space vector representation of the three-phase three-level inverter.

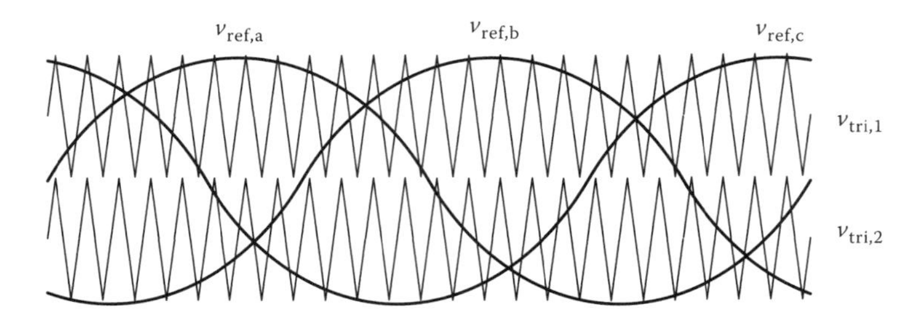

FIGURE 12.6 Waveforms.

voltages are usually regulated and maintained in a balanced condition in the three-level inverter. To reduce voltage unbalance on the capacitors, the redundant switching states can be used to provide some degrees of freedom.

12.2.2.2.3 Waveforms

The three-level pulse-width-modulated waveforms are generated by comparing three reference control signals with two triangular carrier waves, shown in Figure 12.6.

12.2.3 Two-Level AC/DC/AC ZSI

A two-level AC/DC/AC ZSI is shown in Figure 12.7 [5]. A front-end diode rectifier is competitively appealing because of its low cost, albeit at the expense of a loss in controllability at its AC input regardless of the DC/AC inverter topology connected after the front-end rectifier. In addition, using a passive diode rectifier is known to constrain the converter to operate in only the forward power transfer mode and generally support only voltage buck or boost operation, which to some extent constrains its disturbance (e.g., voltage sag and surge) ride-through ability.

The real end is formed by cascading a two-level Z-source DC/AC inverter as depicted in Figure 12.7. The ZSI allows the system output voltage to be stepped down or up as desired by inserting a unique X-shaped LC impedance network, comprising two inductors and two capacitors, between the rectifier and inverter circuitries. On the diode rectifier AC input side, three small delta-connected filter capacitors are added for filtering the three-phase AC input currents so as to eliminate frequent high current peaks that are drawn by

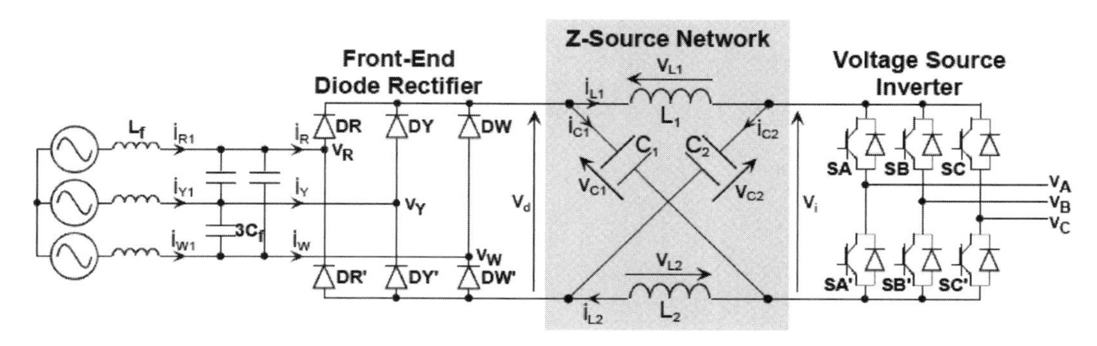

FIGURE 12.7 Topology of the two-level AC/DC/AC Z-source converter. [Reprinted from Loh, P. C., et al. 2007. *Proceedings of the IEEE PESC* 2007, pp. 2691–2697. (©2007 IEEE). With permission.]

traditional diode rectifiers with large DC capacitive storages. The proposed ZSI therefore appears as a compromising solution between the traditional AC/DC/AC diode-front-end converter and the back-to-back controlled converter, since it supports voltage buck–boost operation at a low cost, while constrained by its unidirectional energy conversion and a less than perfect input current waveform, which is still harmonically less distorted as compared with that drawn by a traditional diode rectifier.

12.2.4 Three-Level Diode-Clamped AC/DC/AC ZSI

A three-level diode-clamped AC/DC/AC ZSI is shown in Figure 12.8 [5]. This three-level AC/DC/AC ZSI can produce a harmonically less distorted three-level output waveform with the use of only minimal passive and semiconductor components for implementation. Specifically, the proposed converter uses only six diodes for its AC/DC rectification and only a single X-shaped LC impedance network for performing its buck–boost operation with no increase in its front-end element count as compared with that needed by the two-level converter documented earlier in Figure 12.7. On its AC output end, the proposed converter uses a three-level DC/AC inverter, which can be of either NPC or DC-link cascaded topology (DLC). Although not intuitively obvious, the neutral potential needed by the three-level inverter circuitry can uniquely be tapped from the wye-connected filter capacitors placed before the front-end diode rectifier for input current filtering.

The real end is a three-level diode-clamped Z-source DC/AC inverter as depicted in Figure 12.8. The resulting AC/DC/AC converter therefore offers a low-cost alternative that can ride through deep voltage sags, while producing an improved three-level voltage waveform for powering the externally connected AC load.

12.2.5 Linking a Wind Turbine System to a Utility Network

DG sources do not usually have standard and stable output voltage and frequency. A wind turbine is a typical example, although it is an AC voltage source. In order to link its output

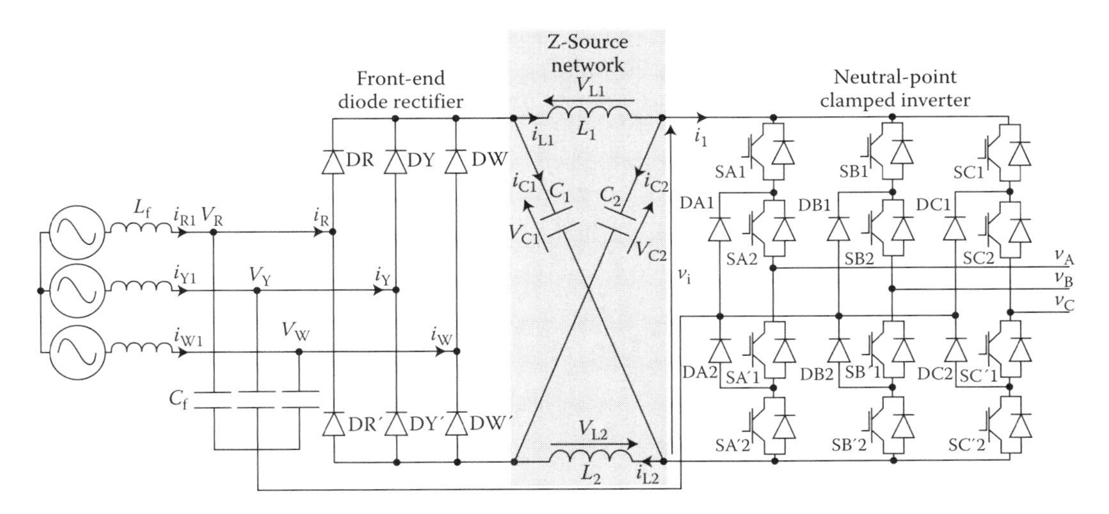

FIGURE 12.8 Topology of the proposed AC/DC/AC Z-source NPC converter. [Reprinted from Loh, P. C., et al. 2007. *Proceedings of the IEEE PESC* 2007, pp. 2691–2697. (©2007 IEEE). With permission.]

power to a utility network, an AC/DC/AC converter is necessarily required to implement the synchronization. The following are the synchronization conditions for an AC generator to link to a utility network:

- The output voltage amplitude of the AC generator is the same as the voltage amplitude of the utility network.
- The frequency of the AC generator is the same as the frequency of the utility network.
- The voltage phase of the AC generator is the same as the voltage phase of the utility network.

To link a wind turbine system to a utility network, the AC/DC/AC inverter has to adjust its output voltage, frequency, and phase angle. If one carefully controls the DC/AC inverter, the synchronization condition is not difficult to achieve.

12.3 DC/AC/DC Converters

There are more than 600 topologies of DC/DC converters existing for DC voltage conversion [6]. Usually, this is enough for research and industrial applications. A DC/AC/DC converter is required for some special applications such as high power transformation.

12.3.1 Review of Traditional DC/DC Converters

Three traditional converters are the buck converters, boost converters, and buck–boost converters. They have the simplest structure and a clear operation process. One inductor plays the role of a pumping circuit. The maximum power transferred from the source to the load is restricted by the PE.

For example, a buck converter as shown in Figure 12.9a converts the energy from source V_1 to load R (the voltage is V_2). The inductor current increases when switch S is on and decreases when switch S is off. In the steady state, the inductor current changes from I_{min} to I_{max} when switch S is on and from I_{max} to I_{min} when switch S is off. In a switching cycle T in the steady state, the energy that inductor L absorbs from the source is

$$\mathrm{PE} = \frac{1}{2}L\left(I_{max}^2 - I_{min}^2\right). \tag{12.9}$$

The total power transferred from source to load is

$$P = f \times \mathrm{PE} = \frac{1}{2}fL\left(I_{max}^2 - I_{min}^2\right). \tag{12.10}$$

The maximum power corresponds to $I_{min} = 0$. This means that the converter works in the DCM.

$$P_{max} = \frac{1}{2}fLI_{max}^2. \tag{12.11}$$

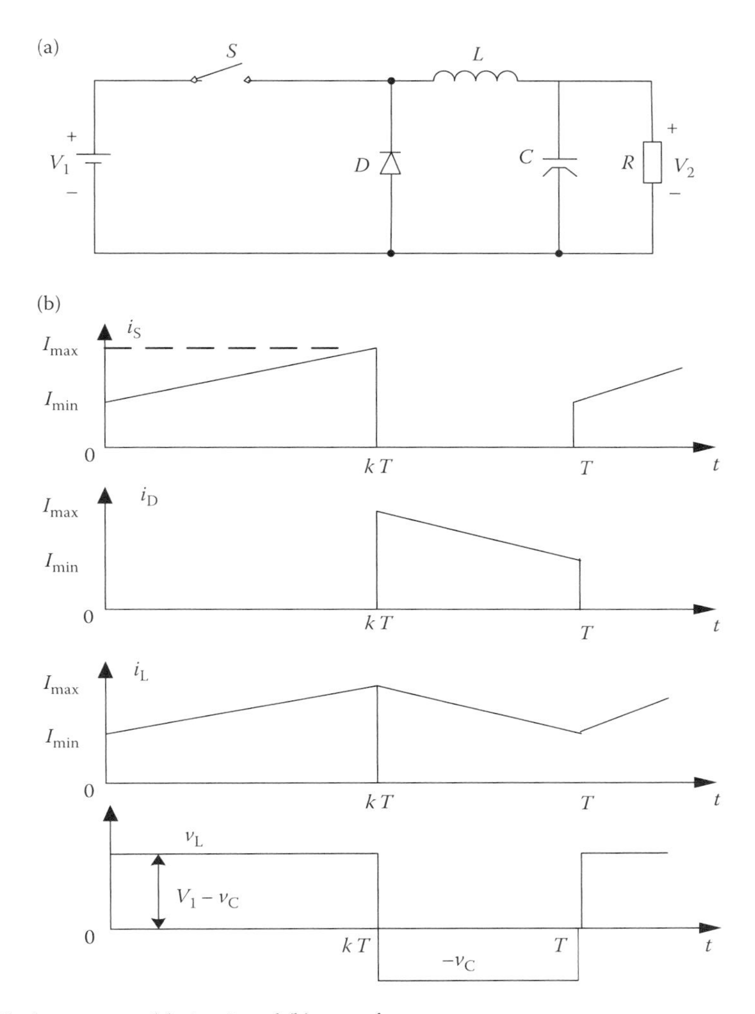

FIGURE 12.9 Buck converter: (a) circuit and (b) wave forms.

Example 12.2

A buck converter has $V_1 = 40\text{V}$, $L = 10\,\text{mH}$, $C = 20\,\mu\text{F}$, $R = 10\,\Omega$, switching frequency $f = 20\,\text{kHz}$, and duty cycle $k = 0.5$. Calculate the power transferred to the load.

SOLUTION

The output voltage V_2 is

$$V_2 = kV_1 = 0.5 \times 40 = 20\text{V}.$$

Therefore, the power is

$$P = \frac{V_2^2}{R} = \frac{20^2}{10} = 40\,\text{W}.$$

From the known data $T = 1/f = 50\,\mu s$ and using Equations 5.13 and 5.14, we obtain

$$I_{max} = kV_1\left(\frac{1}{R} + \frac{1-k}{2L}T\right) = 20\frac{41}{400} = 2.05\,\text{A},$$

$$I_{min} = kV_1\left(\frac{1}{R} - \frac{1-k}{2L}T\right) = 20\frac{39}{400} = 1.95\,\text{A}.$$

Substituting the values into Equation 12.10, we obtain the output power as

$$P = \frac{1}{2}fL\left(I_{max}^2 - I_{min}^2\right) = \frac{20\,\text{k} \times 10\,\text{m}}{2}(2.05^2 - 1.95^2) = 40\,\text{W}.$$

It is verified.

The same operation is available for boost and buck–boost converters. From this example, we know that the power delivered from source to load is restricted by the pumping circuit.

12.3.2 Chopper-Type DC/AC/DC Converters

In order to increase the power delivered from source to load, we need to avoid using an inductor-pumping circuit. A good way is to apply the choppers to chop the DC source voltage to the AC pulse train, and then rectify the AC waveform back to DC voltage. The rectifier can be diode rectifiers or a transformer plus diode rectifiers. Figure 12.10 shows a DC/AC/DC converter with a dual-polarity chopper plus a diode rectifier circuit [7]. The chopper has two pairs of switches (S_{1+}, S_{2-}) and (S_{2+}, S_{1-}). Each pair of switches switch-on and switch-off simultaneously. The output AC voltage v_{AC} is an AC voltage with positive and negative peak values, $+v_i$ and $-v_i$. A diode rectifier (D_1–D_4) is applied to rectify the AC voltage v_{AC} to the DC output voltage v_O.

Using this DC/AC/DC converter, the power delivered from source v_i to load has no restriction since there is no pumping circuit. The output voltage v_O is lower than the input voltage v_i. The switching duty cycle of the pair (S_{1+}, S_{2-}) is k_1 and the pairs switching duty cycle of the pair (S_{2+}, S_{1-}) is k_2. Usually, $k_1 + k_2 \leq 1$. The output voltage v_O is

$$v_O = (k_1 + k_2)v_i. \tag{12.12}$$

We can add a transformer in the circuit, and then obtain random output voltage depending on the transformer turn-ratio. Figure 12.11 shows the DC/AC/DC converter with a dual-polarity chopper plus a transformer and diode rectifier circuit. Using this DC/AC/DC converter the power delivered from source v_i to load has no restriction since there is no pumping circuit. The output voltage v_O can be higher or lower than the input voltage v_i.

The switching duty cycle of the pair (S_{1+}, S_{2-}) is k_1 and the pairs switching duty cycle of the pair (S_{2+}, S_{1-}) is k_2. Usually, $k_1 + k_2 \leq 1$. The transformer winding turn's ratio is n; n can be greater or smaller than unity. If the turn's ratio is greater than unity, it is very easy to obtain the output voltage v_O, which is higher than the input voltage v_i. The output voltage v_O is

$$v_O = n(k_1 + k_2)v_i. \tag{12.13}$$

12.3.3 Switched-Capacitor DC/AC/DC Converters

Switched capacitors can be used to build DC/AC/DC converters [8]. Since switched capacitors can be integrated into a power IC chip, these converters have small sizes and high

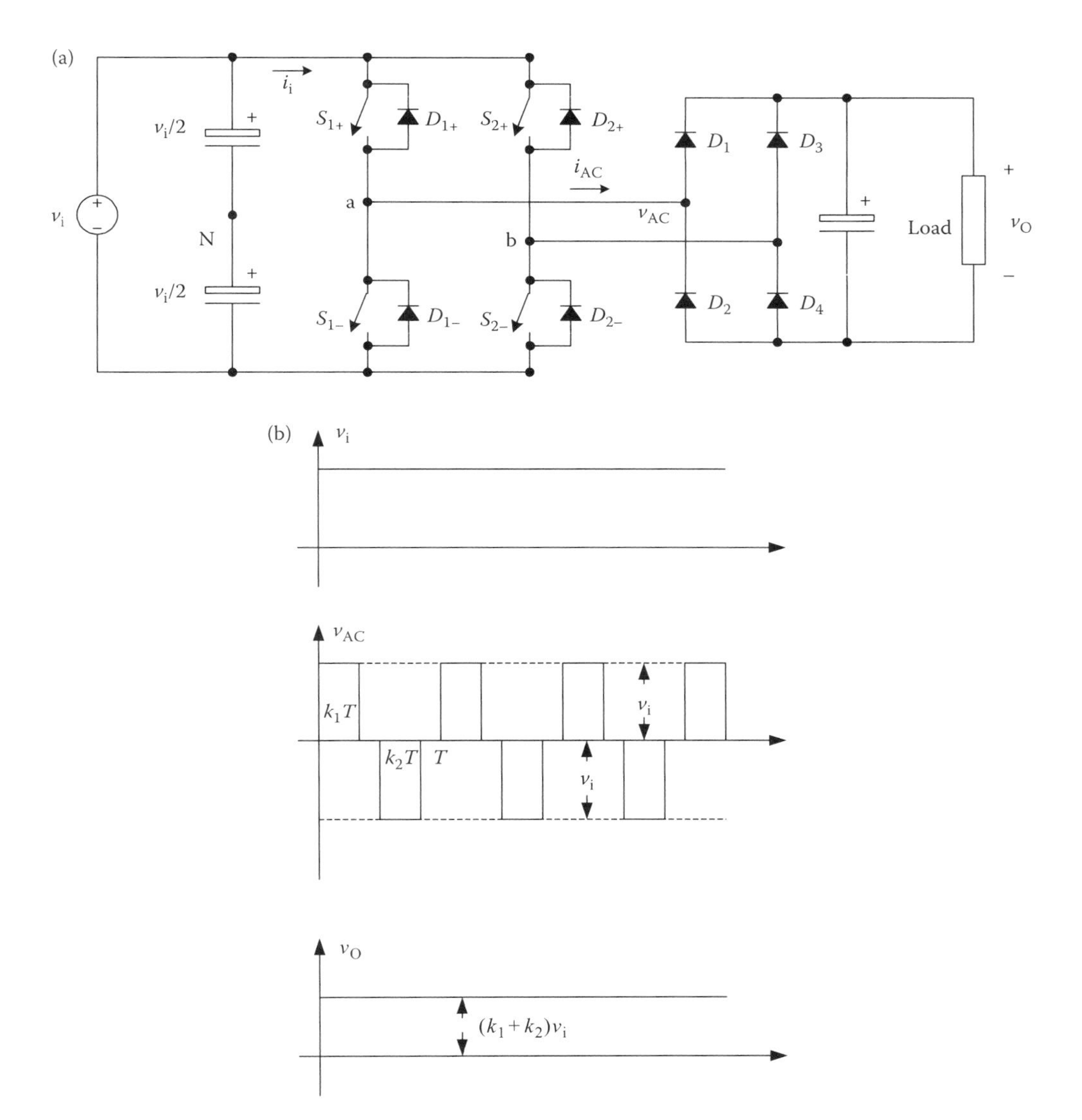

FIGURE 12.10 DC/AC/DC converter with a dual-polarity chopper plus a diode rectifier: (a) circuit and (b) waveforms.

power densities. In this subsection, we introduce several switched-capacitor DC/AC/DC converters:

- Single-stage switched-capacitor DC/AC/DC converters
- Three-stage switched-capacitor DC/AC/DC converters
- Four-stage switched-capacitor DC/AC/DC converters.

12.3.3.1 Single-Stage Switched-Capacitor DC/AC/DC Converter

A single-stage switched-capacitor DC/AC/DC converter is shown in Figure 12.12.

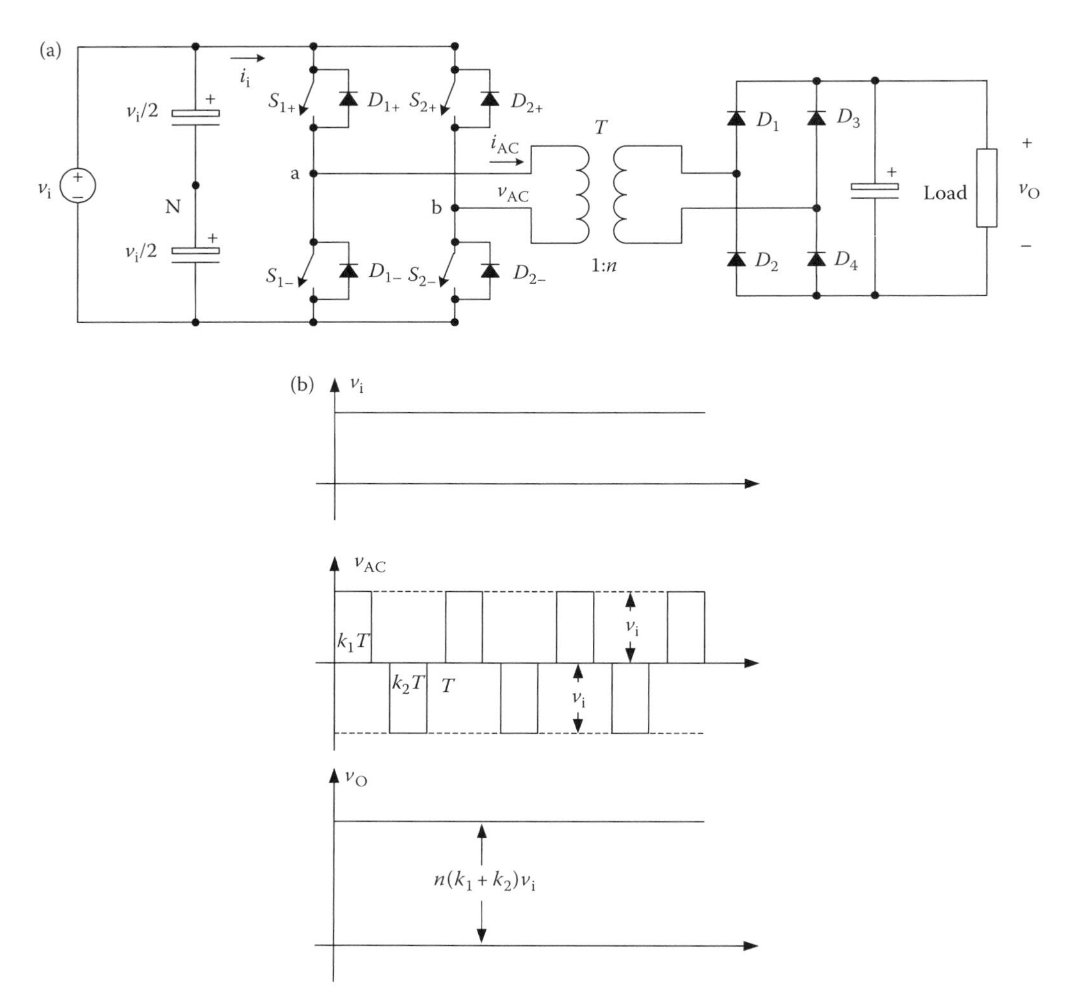

FIGURE 12.11 DC/AC/DC converter with a dual-polarity chopper plus a transformer and diode rectifier circuit: (a) circuit and (b) waveforms.

This single-stage switched-capacitor DC/AC/DC converter has input voltage v_i, middle AC voltage v_{AC}, and output DC voltage v_O. The switching frequency is f, and the corresponding period is $T = 1/f$. The conduction duty cycle of the main switches S_1 and S_2 is k_1. Therefore, the main switches S_1 and S_2 switch-on during the period k_1T. In the meantime, the auxiliary switches S_3 and S_4 switch-on, to change the switched capacitor voltage v_C to the source voltage v_i. From $t = k_1T$ to T, the main switches S_1 and S_2 switch-off. The auxiliary switches S_5 and S_6 switch-on in the period k_2T. In the meantime, the auxiliary switches S_3 and S_4 must switch-off. We can arrange the main switches S_1 and S_2 to be switched on at $t = 0$ to k_1T, and the auxiliary switches S_5 and S_6 switch-on at $t = 0.5T$ to $(0.5 + k_2)T$. The auxiliary switches S_3 and S_4 can switch-on from $t = 0$ to $0.5T$. In other words, the auxiliary switches S_3 and S_4 can switch-on simultaneously with the main switches S_1 and S_2, but may not necessarily switch-off simultaneously with the main switches S_1 and S_2. The auxiliary switches S_3 and S_4 can switch-off at any moment from $t > 0$ to $0.5T$. Usually, $(k_1 + k_2) \leq 1$.

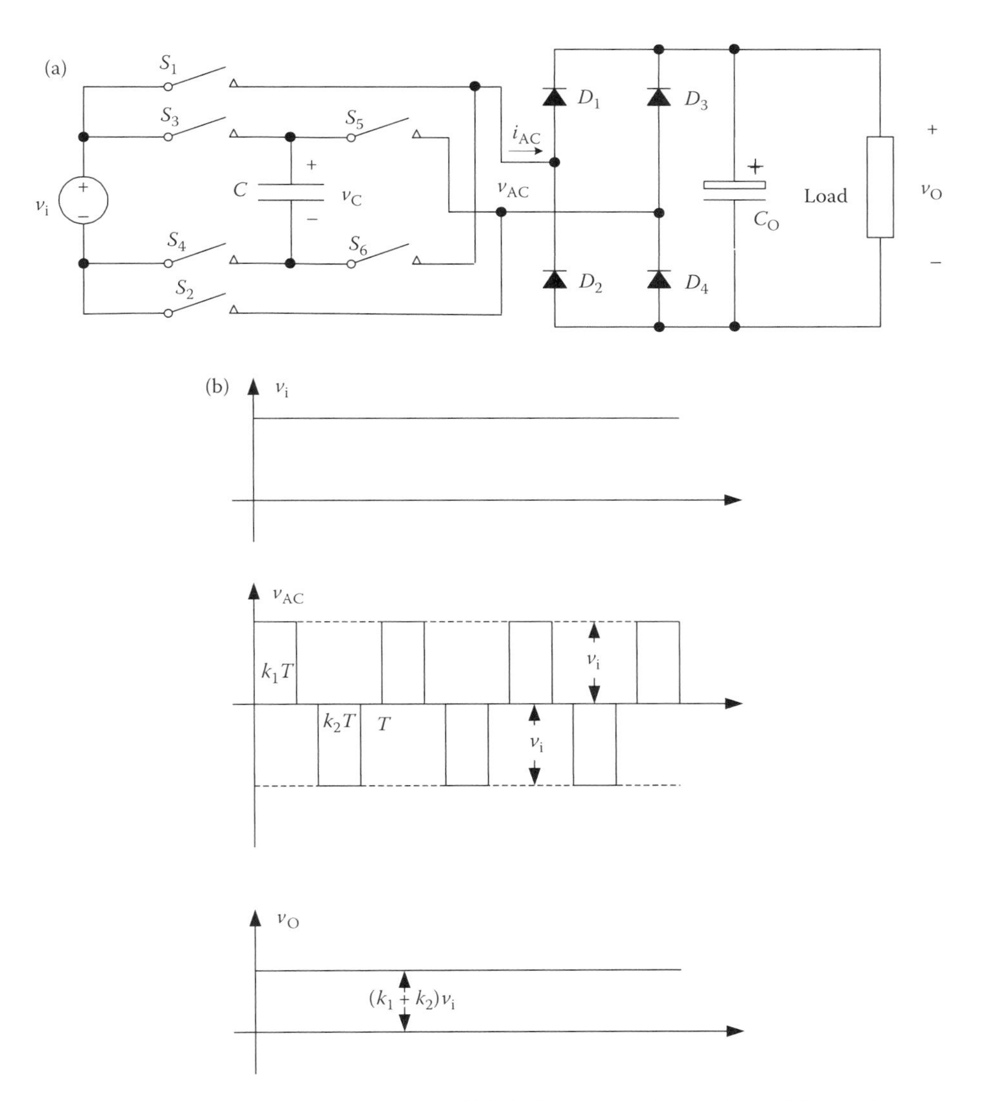

FIGURE 12.12 Single-stage switched-capacitor DC/AC/DC converter: (a) circuit and (b) waveforms.

The output voltage v_O is

$$v_O = (k_1 + k_2)v_i. \tag{12.14}$$

The corresponding waveforms are shown in Figure 12.12b.

We can add a transformer in the circuit, and then obtain random output voltage depending on the transformer turn-ratio. Figure 12.13 shows the DC/AC/DC converter with a switched capacitor plus a transformer and diode rectifier circuit. Using this DC/AC/DC converter, the power delivered from source v_i to load has no restriction since there is no inductor-pumping circuit. The output voltage v_O can be higher or lower than the input voltage v_i. The transformer winding turn's ratio is n; n can be greater or smaller than unity. If the turn's ratio n is greater than unity, it is very easy to obtain the output voltage v_O,

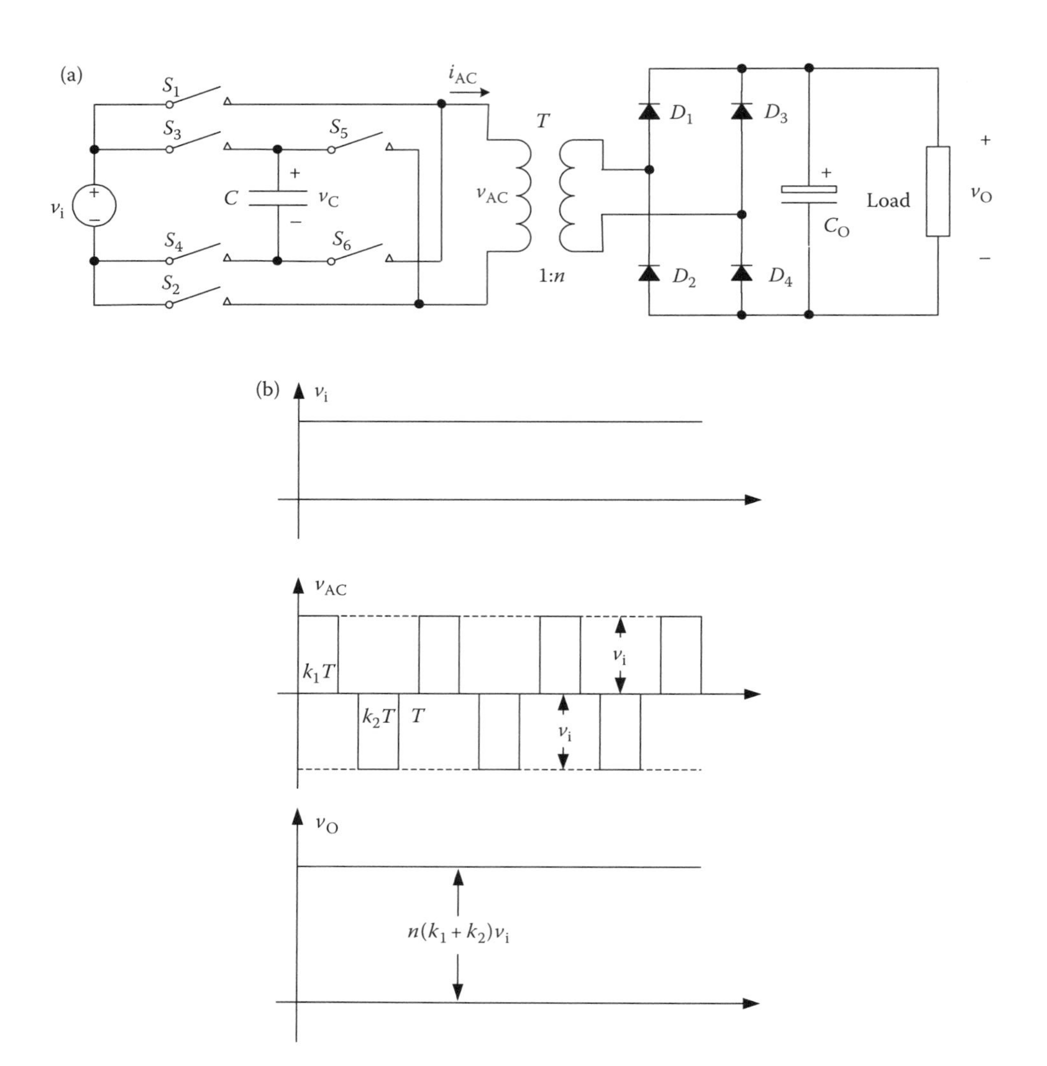

FIGURE 12.13 DC/AC/DC converter with a switched capacitor plus a transformer and diode rectifier circuit: (a) circuit and (b) waveforms.

which is higher than the input voltage v_i. The output voltage v_O is

$$v_O = n(k_1 + k_2)v_i. \tag{12.15}$$

The corresponding waveforms are shown in Figure 12.13b.

12.3.3.2 Three-Stage Switched-Capacitor DC/AC/DC Converter

Using more switched capacitors, we can design other switched-capacitor DC/AC/DC converters. A three-stage switched-capacitor DC/AC/DC converter is shown in Figure 12.14.

This three-stage switched-capacitor DC/AC/DC converter has three switched capacitors C_1, C_2, and C_3, and input voltage v_i, middle AC voltage v_{AC}, and output DC voltage v_O. The switching frequency is f, and the corresponding period is $T = 1/f$. The conduction duty

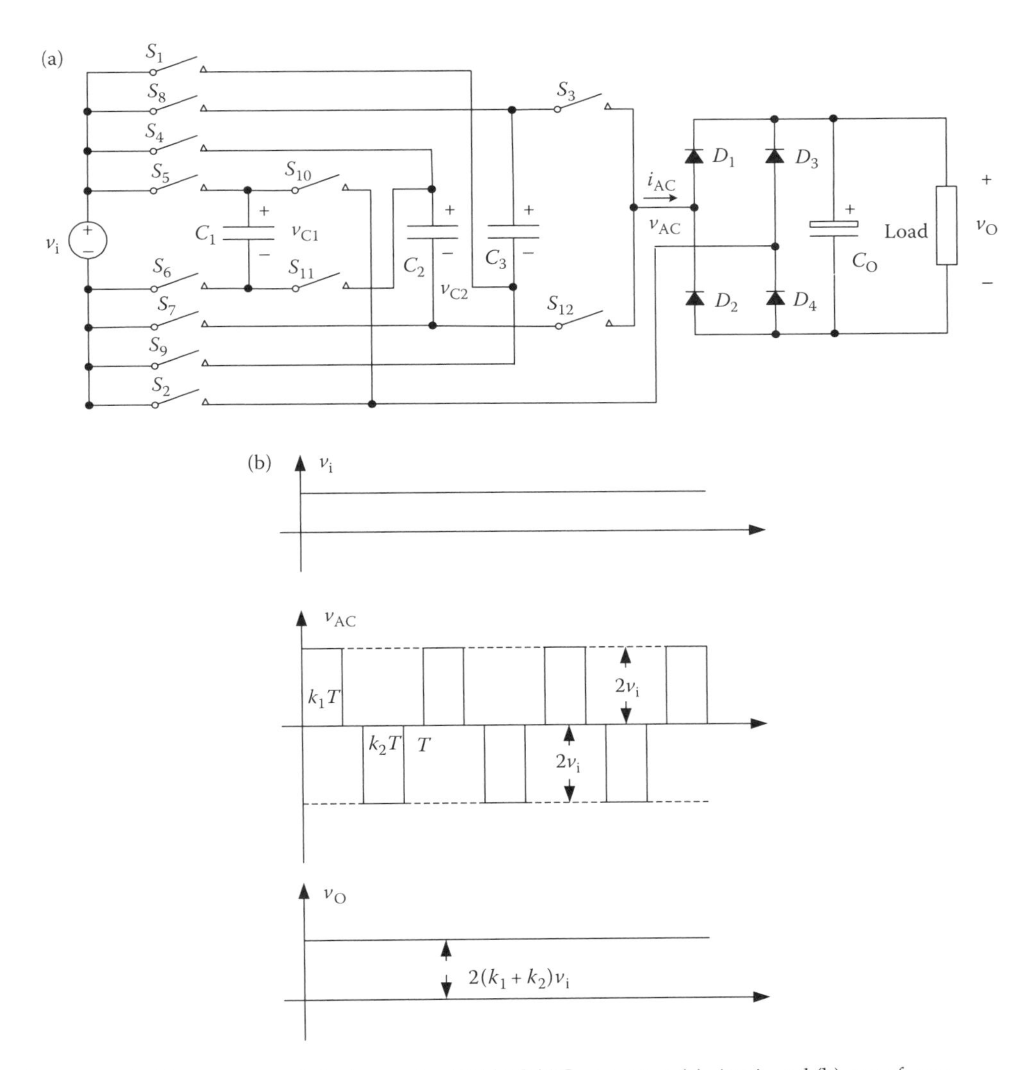

FIGURE 12.14 Three-stage switched-capacitor DC/AC/DC converter: (a) circuit and (b) waveforms.

cycle of the main switches S_1, S_2, and S_3 is k_1. Therefore, the main switches S_1, S_2, and S_3 switch-on during the period k_1T to provide AC voltage $v_{AC} = 2v_i$. In the meantime, the auxiliary switches S_4–S_7 switch-on, to change the switched capacitors C_1 and C_2 (voltages v_{C1} and v_{C2}) to source voltage v_i. From $t = k_1T$ to T, the main switches S_1–S_3 switch-off. The auxiliary switches S_{10}–S_{12} switch-on from $t = 0.5T$ (in the period k_2T) to provide AC voltage $v_{AC} = -2v_i$. In the meantime, the auxiliary switches S_8 and S_9 switch-on to charge the switched capacitor C_3 to source voltage v_i. The auxiliary switches S_4–S_7 must switch-off in this period. We can arrange the main switches S_1–S_3 to be switched on at $t = 0$ to k_1T, and the auxiliary switches S_{10}–S_{12} switch-on at $t = 0.5T$ to $(0.5 + k_2)T$. The auxiliary switches S_4–S_7 can switch-on from $t = 0$ to $0.5T$. In other words, the auxiliary switches S_4–S_7 can switch-on simultaneously with the main switches S_1–S_3, but may not necessarily switch-off simultaneously with the main switches S_1–S_3. The auxiliary switches S_4–S_7 can switch-off at any moment from $t > 0$ to $0.5T$. Similarly, the auxiliary switches S_8 and S_9 can switch-off at any moment from $t > 0.5T$ to T. Usually, $(k_1 + k_2) \leq 1$. The output voltage v_O is

$$v_O = 2(k_1 + k_2)v_i. \tag{12.16}$$

The corresponding waveforms are shown in Figure 12.14b.

We can add a transformer in the circuit, and then obtain random output voltage depending on the transformer turn-ratio. Figure 12.15 shows the DC/AC/DC converter with a three-stage switched capacitor plus a transformer and diode rectifier circuit. Using this DC/AC/DC converter, the power delivered from source v_i to load has no restriction since there is no inductor-pumping circuit. The output voltage v_O can be higher or lower than the input voltage v_i. The transformer winding turn's ratio is n; n can be greater or smaller than unity. If the turn's ratio n is greater than unity, it is very easy to obtain the output voltage

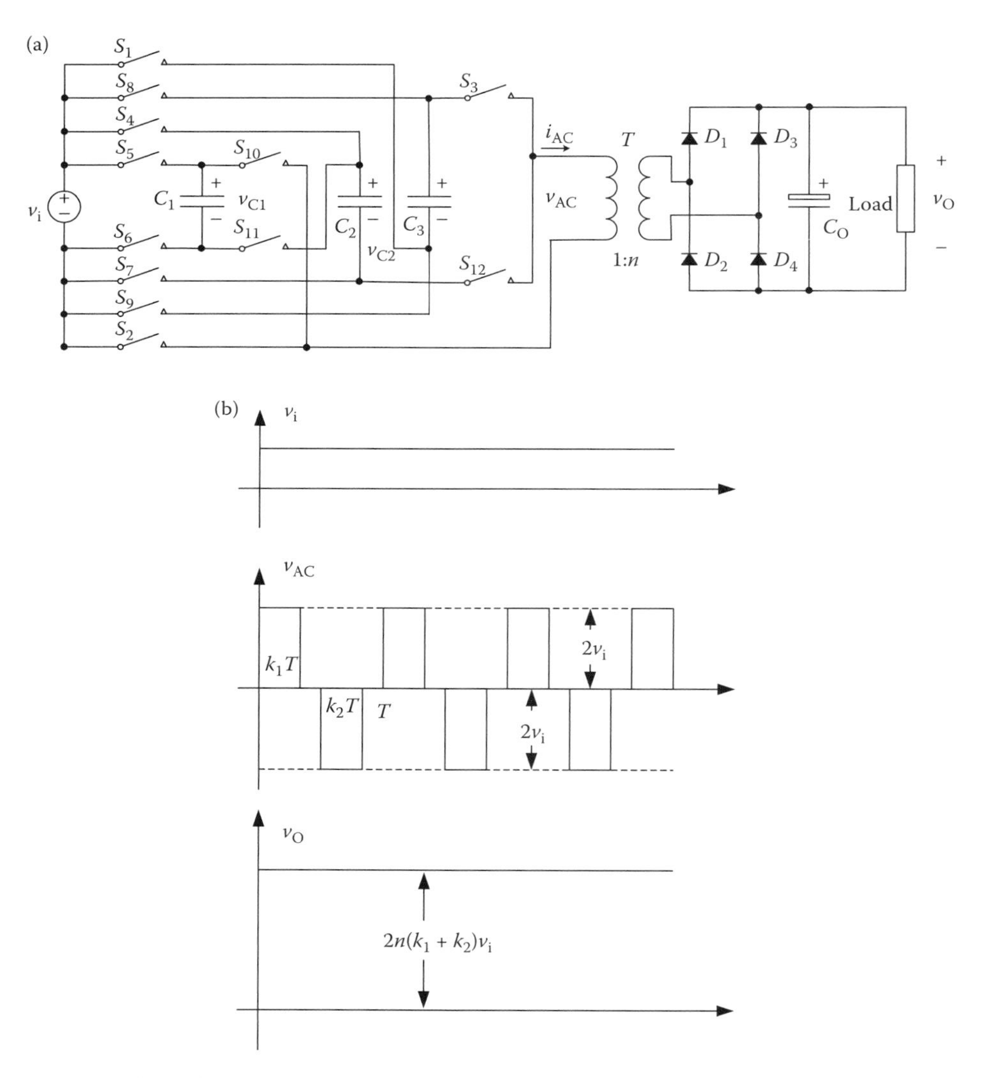

FIGURE 12.15 DC/AC/DC converter with a three-stage switched capacitor plus a transformer and diode rectifier circuit: (a) circuit and (b) waveforms.

v_O, which is higher than the input voltage v_i. The output voltage v_O is

$$v_O = n(k_1 + k_2)v_i. \quad (12.17)$$

The corresponding waveforms are shown in Figure 12.15b.

It is possible to design other odd number (n is the odd number, $m > 3$) stage switched-capacitor DC/AC/DC converters. $(m - 1)/2$ switched capacitors plus the source voltage supply the positive half-cycle of the intermediate AC voltage; other $(m + 1)/2$ switched capacitors supply the negative half-cycle of the intermediate AC voltage.

12.3.3.3 Four-Stage Switched-Capacitor DC/AC/DC Converter

It is possible to design even number (n is the even number, $m \geq 2$) stage switched-capacitor DC/AC/DC converters. Considering the symmetry of the intermediate AC voltage, half

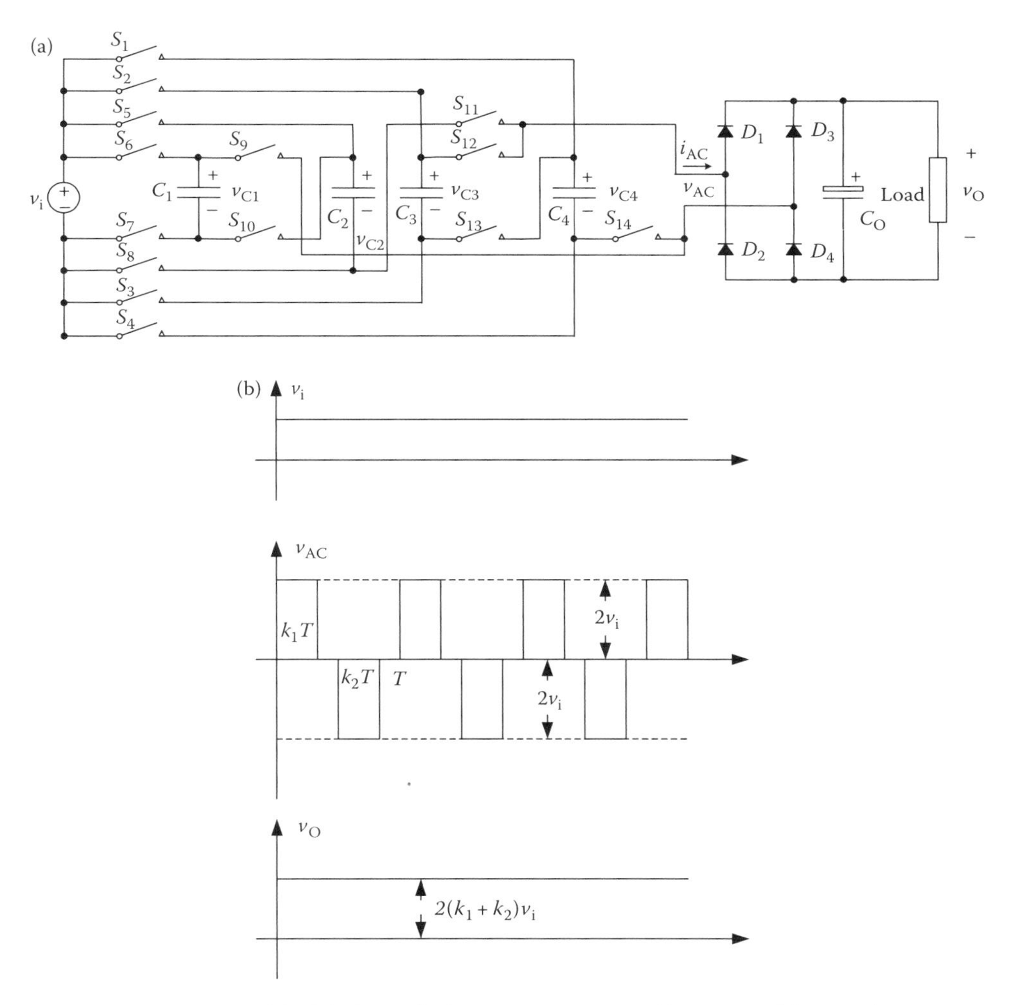

FIGURE 12.16 Four-stage DC/AC/DC switched-capacitor converter: (a) circuit and (b) waveforms.

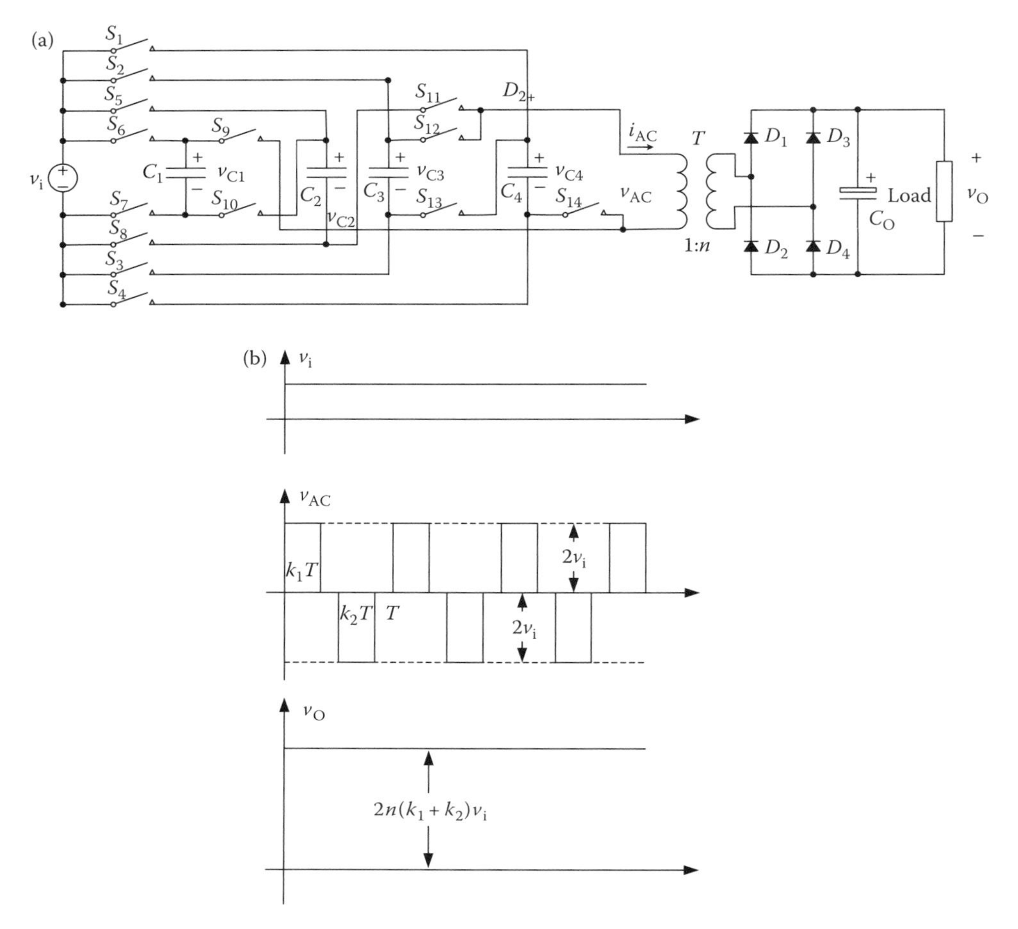

FIGURE 12.17 Four-stage switched-capacitor with a transformer DC/AC/DC converter: (a) circuit and (b) waveforms.

of the switched capacitors ($m/2$ switched capacitors) supply the positive half-cycle of the intermediate AC voltage; the other half of the switched capacitors supply the negative half-cycle of the intermediate AC voltage. The source voltage can only be used to change the two groups of switched capacitors alternatively. Figure 12.16 shows the four-stage switched-capacitor DC/AC/DC converter.

When $t = 0$, the switches S_5–S_8 switch-on to charge capacitors C_1 and C_2 to the source voltage v_i. The switches S_{12}–S_{14} switch-on, and the capacitors C_3 and C_4 supply $+2v_i$ to v_{AC} in the period of k_1T. When $t = 0.5T$, the switches S_1–S_4 switch-on to charge the capacitors C_3 and C_4 to the source voltage v_i. The switches S_9–S_{11} switch-on, and the capacitors C_1 and C_2 supply $-2v_i$ to v_{AC} in the period of k_2T. The waveforms are shown in Figure 12.16b. After the diode rectifier, we obtain the output voltage v_O as

$$v_O = (k_1 + k_2)v_i. \tag{12.18}$$

We can still add a transformer in the circuit to enlarge the output voltage. A four-stage switched-capacitor with a transformer DC/AC/DC converter is shown in Figure 12.17. The waveforms are shown in Figure 12.17b.

The transformer turn's ratio is n. The output voltage is

$$v_O = n(k_1 + k_2)v_i. \qquad (12.19)$$

Homework

12.1. A wind turbine has single-phase output voltage $300\,V \pm 25\%$ and frequency $50\,Hz \pm 15\%$; the power rate is 1 kW. The end user is a three-phase load with voltage 110 V/50 Hz. Design a three-level diode-clamped AC/DC/AC converter for this application.

12.2. A boost converter has $V_1 = 40\,V$, $L = 10\,mH$, $C = 20\,\mu F$, $R = 10\,\Omega$, switching frequency $f = 20\,kHz$, and duty cycle $k = 0.5$. Calculate the power transferred to the load.

12.3. A single-stage switched-capacitor DC/AC/DC converter plus a transformer (with turn-ratio $n = 5$) is shown in Figure 12.11. The input voltage $v_i = 40\,V$, load $R = 10\,\Omega$, and duty cycles $k_1 = k_2 = 0.4$. Calculate the output voltage and power transferred to the load.

12.4. A three-stage switched-capacitor DC/AC/DC converter is shown in Figure 12.12. The input voltage $V_i = 40\,V$, load $R = 10\,\Omega$, and duty cycles $k_1 = k_2 = 0.4$. Calculate the output voltage and power transferred to the load.

References

1. Masters, G. M. 2005. *Renewable and Efficient Electric Power Systems*. New York: Wiley.
2. Ackermann, T. 2005. *Wind Power in Power Systems*. New York: Wiley.
3. Johnson, G. L. 1985. *Wind Energy Systems*. Englewood Cliffs, NJ: Prentice-Hall.
4. Lin, B. R., Lu, H. H., and Chen, Y. M. 1998. Implementation of three-level AC/DC/AC converter with power factor correction and harmonic reduction. *Proceedings of IEEE PEDES*, pp. 768–773.
5. Loh, P. C., Gao, F., Tan, P. C., and Blaabjerg, F. 2007. Three-level AC-DC-AC Z-source converter using reduced passive component count. *Proceedings of the IEEE PESC* 2007, pp. 2691–2697.
6. Luo, F. L. and Ye, H. 2004. *Advanced DC/DC Converters*. Boca Raton: CRC Press.
7. Luo, F. L. and Ye, H. 2009. Chopper-type DC/AC/DC converters. *Technical Talk of ICIEA* 2009, pp. 1356–1368.
8. Luo, F. L. and Ye, H. 2009. Switched-capacitor DC/AC/DC converters. *Technical Talk of ICIEA* 2009, pp. 163–168.

Index

E

F

Q

R

S

W

Z